Advances in Organic Chemistry
Methods and Results
VOLUME 9

Iminium Salts in Organic Chemistry
Part 2

Advisory Board

D. H. R. BARTON
Imperial College of Science and Technology, London, England

K. W. BENTLEY
Reckitt and Sons Ltd., Hull, England

A. J. BIRCH
The Australian National University, Canberra, Australia

CARL DJERASSI
Stanford University, California

A. ESCHENMOSER
Eidg. Technische Hochschule, Zürich, Switzerland

R. HUISGEN
Ludwig-Maximilians-Universität, Munich, Germany

E. C. KOOYMAN
Rijksuniversiteit, Leiden, The Netherlands

E. LEDERER
Institut de Chimie des Substances Naturelles, Gif-sur-Yvette, France

G. OURISSON
Université de Strasbourg, France

A. QUILICO
Instituto de Chimica Generale del Politecnico, Milan, Italy

R. A. RAPHAEL
University Chemical Laboratory, Cambridge, England

S. STALLBERG-STENHAGEN
University of Gothenberg, Sweden

G. STORK
Columbia University, New York

LORD TODD
University Chemical Laboratory, Cambridge, England

M. VISCONTINI
University of Zürich, Switzerland

A. WEISSBERGER
Eastman Kodak Company, Rochester, New York

K. WIESNER
University of New Brunswick, Fredericton, Canada

R. B. WOODWARD
Harvard University, Cambridge, Massachusetts

H. WYNBERG
University of Groningen, Holland

ADVANCES IN ORGANIC CHEMISTRY: Methods and Results
E. C. TAYLOR, *Editor*

Iminium Salts in Organic Chemistry
Part 2

Edited by

H. BÖHME

Universität Marburg/Lahn, Germany

H. G. VIEHE

Université de Louvain, Belgium

An Interscience ® Publication

JOHN WILEY & SONS   New York   •   London   •   Sydney   •   Toronto

An Interscience® Publication

Copyright © 1979 by John Wiley & Sons, Inc.

All rights reserved. Published simultaneously in Canada.

Reproduction or translation of any part of this work
beyond that permitted by Sections 107 or 108 of the 1976
United States Copyright Act without the permission of the
copyright owner is unlawful. Requests for permission or
further information should be addressed to the Permissions
Department, John Wiley & Sons, Inc.

Library of Congress Cataloging in Publication Data:

Main entry under title:

Iminium salts in organic chemistry.

   (Advances in organic chemistry; v. 9)
   "An Interscience publication."
   Includes bibliographical references and indexes.
   1. Imines. I. Böhme, Horst, 1908– II. Viehe,
Heinz Günter. III. Series.

QD251.A36 vol. 9 [QD401] 547'.008s [547'.04]
ISBN 0-471-90692-1 (v. 1)   76-16155
ISBN 0-471-90693-X (v. 2)

Printed in the United States of America

10 9 8 7 6 5 4 3 2 1

Foreword

The exceptional role of the iminium grouping in many reactions that occur both in the laboratory and in nature has been recognized for a long time. Since then, however, organic chemistry has become such an extremely broad and diversified science that the enormous progress attained meanwhile in iminium chemistry, including new methods, reagents, ideas, new ways, and fields of application, may have escaped general attention. People engaged in this area have become aware that an urgent need exists for a book which not only gathers the vast amount of new material but reintegrates all of the recent achievements into a more general framework in terms of modern concepts of organic chemistry.

I strongly believe that the present work fulfills these requirements. Although I was engaged in the early discussions during the conception of this book, I am now very impressed at seeing the final result. Both the editors and the authors have succeeded in creating a book from which, I am sure, the chemical community will profit for a long time.

Z. ARNOLD

Prague, Czechoslovakia
February 1976

Series Editor's Note

Although most volumes in the *Advances in Organic Chemistry* series will continue to be multiauthored works presenting authoritative, critical, and timely discussions of new developments in synthetic and instrumental methodology, in line with the general objectives of the series as set forth in the Preface to previous volumes, the present volume, which will appear in two parts, marks a further expansion of the concept of *Advances*. The first departure from the normal format, as outlined above, will be found in Volume 7, which was a single-authored research monograph. The present volume is likewise devoted to a single topic, but is multiauthored and prepared under the general editorship of outside experts in the field. We hope that the rapidity of publication of the two types of research monographs in the *Advances* series will be attractive both to readers and to authors, and that the series as a whole will continue to present in a challenging, provocative, and stimulating manner new ideas, new techniques, and new methods that will become part of the classical repertoire of the practicing organic chemist.

EDWARD C. TAYLOR
Series Editor
Advances in Organic Chemistry

Preface

Research workers in nitrogen chemistry have felt the need for an adequate coverage of modern iminium salt chemistry. This book, we think, will satisfy this need.

Many discussions preceded the 1972 meeting in Marburg at which it was decided to "launch" this book. The project started with an encounter of H. G. Viehe with Z. Arnold in Prague, 1972, followed by others with L. Ghosez in Louvain, with H. Eilingsfeld, H. Pommer, and M. Pape in BASF-Ludwigshafen, with H. Bredereck in Stuttgart, with E. Kühle and E. Grigat in Bayer-Leverkusen, and with C. Jutz in Munich. We feel honored and thank the authors for their extensive work and for their trust and confidence. To Prof. E. C. Taylor, the series editor, we address our repeated thanks for his masterly streamlining of this book.

May all the work serve well now!

H. BÖHME
H. G. VIEHE

Marburg, Germany
Louvain-la-Neuve, Belgium
August 1976

Contents

The Background of Carboxamide Salt Chemistry 1
    H. Bredereck

Adducts from Acid Amides and Acylation Reagents (Vilsmeier–Haack Adducts) 5
    W. Kantlehner

Chloromethyleniminium Salts (Amide Halides) 65
    W. Kantlehner

Chlorodialkylaminomethyleniminium Salts—Chloroformamidinium Salts 143
    W. Kantlehner

Chloro(alkylmercapto)-methyleniminium Salts (Mercaptochloroformamidinium Salts) 173
    W. Kantlehner

Alkoxymethyleniminium Salts 181
    W. Kantlehner

Alkylmercaptomethyleniminium Salts 279
    W. Kantlehner

Amidinium Salts 321
    W. Kantlehner

Orthoamides: Properties and Reactions 393
    G. SIMCHEN

Iminium and Nitrilium Salts in the Dimerization of Cyano Compounds in Acids 527
    S. YANAGIDA, M. OKAHARA, and S. KOMORI

N-Acylpyridinium Salts 573
    A. N. KOST, S. I. SUMINOV, and A. K. SHEINKMAN

3,3-Rearrangements of Iminium Salts 655
    H. HEIMGARTNER, H.-J. HANSEN, and H. SCHMID

Iminium Salts in Nature 733
    J. KNABE

Epilogue: Complements and Perspectives for Iminium Salt Chemistry 759
    H. BÖHME and H. G. VIEHE

Author Index 775

Subject Index 827

Advances in Organic Chemistry

Methods and Results

VOLUME 9

Iminium Salts in Organic Chemistry
Part 2

THE BACKGROUND OF CARBOXAMIDE SALT CHEMISTRY

By H. BREDERECK, *Institut für Organische Chemie Biochemie und Isotopenforschung der Universität, D-7-Stuttgart 80, Pfaffenwaldring 55, Germany*

As an introduction to the reviews by W. Kantlehner and G. Simchen on acid amide–acid halide and acid amide–dialkyl sulfate adducts and their subsequent reactions, we wish to describe briefly how we became involved in this research area. Such an historical review describing our initial experiments seems to be well justified, since this area has turned out to be a very rich and profitable one.

We had developed, some years ago, new syntheses of imidazoles and oxazoles through an extensive research program (1). Among other features of this program, it was found that α-haloketones reacted with formamide to give either imidazoles or oxazoles, depending on the reaction conditions. In order to obtain more detailed information about the control of halogens in these reactions, we treated numerous halogen compounds with formamide at 150°C (Table I). The alkyl halides yielded either formylamino compounds (N-substituted formamides) (**1a**) or formates (**1b**). Formation of a relatively stable

$$R-Hal + 2HCONH_2 \xrightarrow{a} RNHCHO + CO + NH_4Hal$$

(1a)

$$R-Hal + 2HCOHN_2 \xrightarrow{b} ROCHO + HCN + NH_4Hal$$

(1b)

TABLE I

Reactions of Formamide with Halogen Compounds

Halide	Products
Octyl bromide	Octyl formate
Benzyl chloride, benzyl bromide, benzyl iodide	Benzyl formate + N-benzylformamide
Benzhydryl chloride	N-Benzhydrylformamide
Trityl chloride	N-(Triphenylmethyl)formamide

Translated from German by Dr. W.-J. Richter, MPI für Kohleforschung, D 4330 Mühlheim.

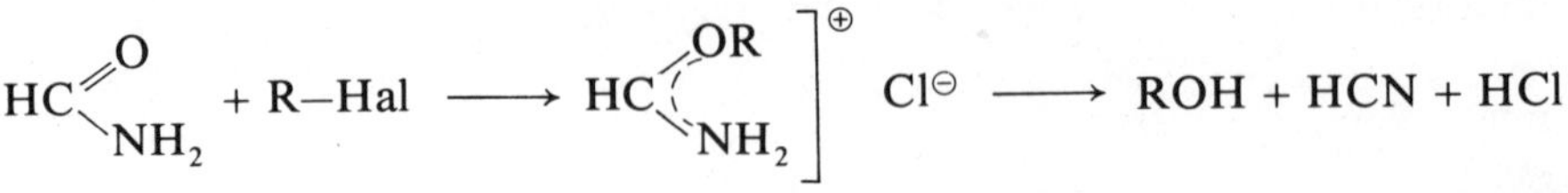

$$HC\begin{smallmatrix}O\\NH_2\end{smallmatrix} + R-Hal \longrightarrow \left[HC\begin{smallmatrix}OR\\NH_2\end{smallmatrix}\right]^{\oplus} Cl^{\ominus} \longrightarrow ROH + HCN + HCl$$

(2)

carbonium ion from the alkyl halide is essential to the formation of a formylamine (path *a*). Formates are formed from intermediate imino esters (path *b*), very likely by dissociation of the alkyl halide; the imino ester (2) decomposes to ROH and HCN at the high reaction temperatures used during the reactions. Formates then form by formylation of the hydroxy compound with formamide/HCl:

$$ROH + HCONH_2 + HCl \longrightarrow ROCHO + NH_4Cl$$

At the beginning of this research program, we restricted ourselves to an experimental test of this proposed mechanism for the formation of formates. We failed to trap the formimino ester, our proposed intermediate, by quenching the reaction, but we did succeed in isolating the thioformimino ester hydrohalides (3) from alkyl halides and thioformamide. Although formamides require a reaction temperature of about 150°C, thioformamides reacted satisfactorily at about 60°C. At 150°C thioformimino ester hydrohalides undergo decomposition to give mercapto compounds and HCN.

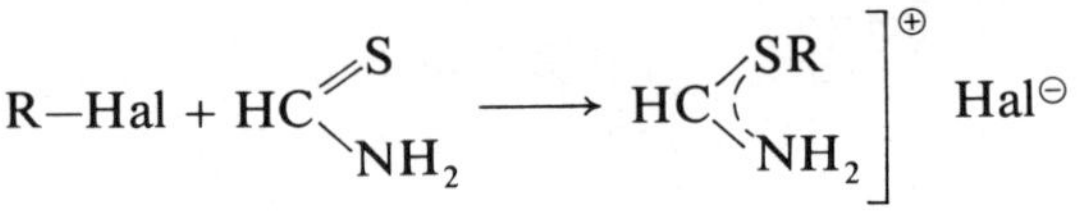

$$R-Hal + HC\begin{smallmatrix}S\\NH_2\end{smallmatrix} \longrightarrow \left[HC\begin{smallmatrix}SR\\NH_2\end{smallmatrix}\right]^{\oplus} Hal^{\ominus}$$

(3)

Obviously the next step was to attempt the synthesis of formimino esters from formamides and alkyl halides at temperatures lower than 150°C. For the first time, we were able to isolate tris(formamido)methane using an excess of formamide; this compound crystallizes from the reaction mixture. The formation of tris(formamido)methane (6) follows Scheme 1. Using only 1 or 2

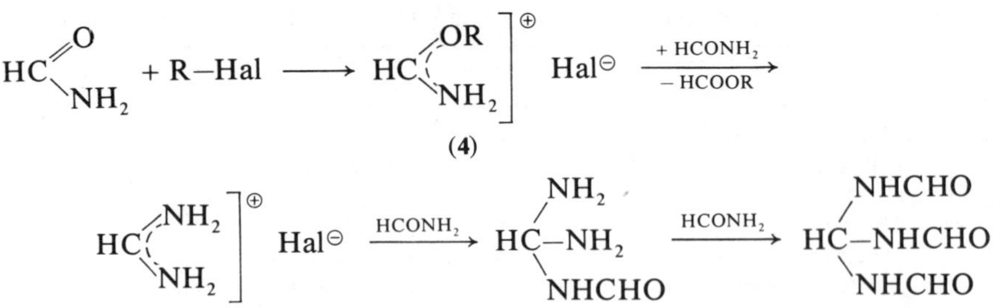

$$HC\begin{smallmatrix}O\\NH_2\end{smallmatrix} + R-Hal \longrightarrow \left[HC\begin{smallmatrix}OR\\NH_2\end{smallmatrix}\right]^{\oplus} Hal^{\ominus} \xrightarrow[-HCOOR]{+HCONH_2}$$

(4)

$$\left[HC\begin{smallmatrix}NH_2\\NH_2\end{smallmatrix}\right]^{\oplus} Hal^{\ominus} \xrightarrow{HCONH_2} HC\begin{smallmatrix}NH_2\\-NH_2\\NHCHO\end{smallmatrix} \xrightarrow{HCONH_2} HC\begin{smallmatrix}NHCHO\\-NHCHO\\NHCHO\end{smallmatrix}$$

(5)   (6)

Scheme 1

moles of formamide, one obtains formimino ester (**4**) and formamidine salts (**5**) as well as tris(formamido)methane. According to the above scheme, formimino esters as well as formamidine salts are converted to tris(formamido)methane (**6**) by an excess of formamide.

Subsequently we reacted the now readily accessible tris(formamido)methane in numerous ways. Since the intermediate formation of formimino esters is essentially an *O*-alkylation, we expected that formation of tris(formamido)methane from formamide and alkylating agents would be a general reaction. Thus we obtained this compound not only from alkyl halides but also in a very smooth reaction from dialkyl sulfates (dimethyl, diethyl, and diisopropyl sulfate) and from alkyl sulfonates (benzene- and *p*-toluenesulfonic acid esters). Triethyloxonium tetrafluoroborate may also be used as a strong alkylating agent; in this case, the reaction proceeds at room temperature. In addition to alkylating agents, acylating agents (acid chlorides) also yield tris(formamido)methane. We employed the following acid chlorides: acetyl chloride, benzoyl chloride, ethyl chloroformate, phosphorus oxychloride, phosphorus trichloride, and sulfuryl chloride. In all these cases, the reaction proceeds via compounds analogous to formimino esters. We isolated such compounds as the hygroscopic crystalline product **7**, using formamide and benzoyl chloride, and the formamide–phosphorus oxychloride adduct **8** analogously. The structures of these compounds were determined by IR analysis of the DMF–phosphorus oxychloride adduct **9**, which was isolated crystalline. This adduct (the same one on which the Vilsmeier–Haack–Arnold aldehyde synthesis is based) was used for many subsequent reactions.

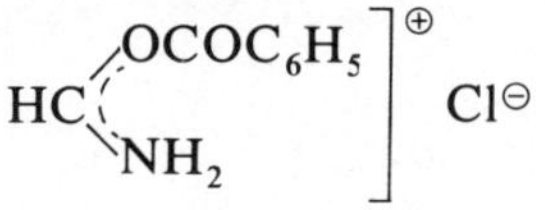

$$\left[HC\begin{matrix}OCOC_6H_5\\NH_2\end{matrix} \right]^{\oplus} Cl^{\ominus} \quad \left[HC\begin{matrix}OPOCl_2\\NH_2\end{matrix} \right]^{\oplus} Cl^{\ominus} \quad \left[HC\begin{matrix}OPOCl_2\\N(CH_3)_2\end{matrix} \right]^{\oplus} Cl^{\ominus}$$

(**7**) 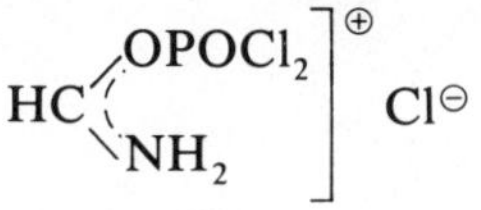 (**8**) 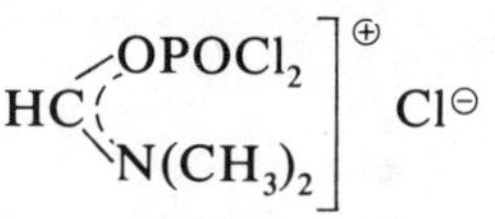 (**9**)

In the course of the formation of tris(formamido)methane, the formamide–alkylating agent (the formimino ester salt) appeared to be the key intermediate. Consequently, we became interested in the behaviour of *N*-alkylated formamides in the presence of alkylating agents, especially dialkyl sulfates. Despite some initial negative results, we were subsequently able to isolate the *N*,*N*-dimethylformiminomethylester salt (**10**) from dimethylformamide and dimethyl sulfate.

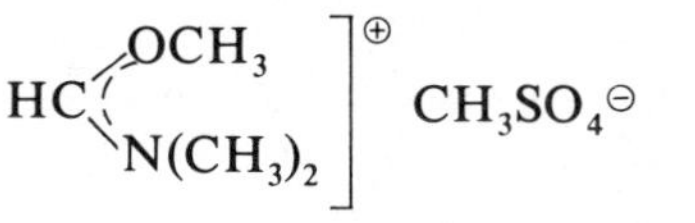

$$\left[HC\begin{matrix}OCH_3\\N(CH_3)_2\end{matrix} \right]^{\oplus} CH_3SO_4^{\ominus}$$

(**10**)

This marked the beginning of what has turned out to be an extensive research program on the synthesis and properties of acid amide–dialkyl sulfate and acid amide–acid chloride adducts, which are the subjects of the chapters by W. Kantlehner and G. Simchen.

REFERENCE

1. H. Bredereck, R. Gompper, H. G. v. Schuh, and G. Theilig, *Angew. Chem.* **71**, 753–774 (1959).

ADDUCTS FROM ACID AMIDES AND ACYLATION REAGENTS (VILSMEIER–HAACK ADDUCTS)

By W. KANTLEHNER, *D-7080-AALEN-Dewangen, Laachweg 14, Germany*

CONTENTS

I. Acid Amide–Phosphorus Oxychloride Adducts 6
   A. Structure of Acid Amide–Phosphorus Oxychloride Adducts 6
      1. Dimethylformamide–Phosphorus Oxychloride Adduct 6
      2. Adducts of Higher Amides and Urea with Phosphorus Oxychloride . . . 8
      3. Thioamide–Phosphorus Oxychloride and Amide–Phosphorus Sulfochloride Adducts 9
   B. Reactions of Acid Amide–Phosphorus Oxychloride Adducts 9
      1. Reactions with Amines 9
      2. Reactions with Hydrazine Derivatives and Hydroxylamine 17
      3. Reactions with Azomethines 19
      4. Reactions with Bases: Isonitrile Syntheses 19
      5. Reactions with Epoxides, Cyclic Ethers, Acyloins, and Polyvalent Alcohols . 20
      6. Reactions with Heterocycles Containing Hydroxyl Groups 21
      7. Reactions with Alkali Metal Alkoxides and Mercaptans 22
      8. Reactions with Acid Amides 22
II. Acid Amide–Triphenylphosphine Dihalide Adducts 23
   A. Reactions of Acid Amide–Triphenylphosphine Dihalide Adducts 24
III. Adducts of Acid Amide–Chlorophosphoric Acid Ester Adducts 27
   A. Reactions of Acid Amide–Chlorophosphoric Acid Ester Adducts 28
IV. Acid Amides–Phosphorus Trihalide Adducts 30
V. Acid Amide–Phosphorus Nitrile Chloride Adducts 33
VI. Acid Amide–Cyanuric and Cyanogen Chloride Adducts 33
   A. Reactions of Acid Amide–Cyanuric and Cyanogen Chloride Adducts . . . 35
VII. Acid Amide–Sulfuryl Chloride Adducts 37
VIII. Acid Amide–Sulfonyl Chloride Adducts 38
IX. Acid Amide–Chlorosulfonamide Adducts 42
X. Acid Amide–Carboxylic Acid Chloride Adducts 42
   A. Reactions of Acid Amide–Acid Chloride Adducts 46
XI. Acid Amide–Acid Anhydride Adducts 49
   References 50
   Addendum 54

Translated from German by Dr. W.-J. Richter, MPI für Kohleforschung, D 4330 Mühlheim.

I. Acid Amide–Phosphorus Oxychloride Adducts

A. STRUCTURE OF ACID AMIDE–PHOSPHORUS OXYCHLORIDE ADDUCTS

1. Dimethylformamide–Phosphorus Oxychloride Adduct

From the studies by Dimroth and Zoeppritz (1) and Vilsmeier and Haack (2, 3), it has long been known that aromatic and heteroaromatic compounds are formylated when treated with N,N-disubstituted formamides and $POCl_3$ (Vilsmeier–Haack reaction). Reviews have been published in various languages (4–10), but only rather recently has the nature of the formylating agents been studied.

Vilsmeier and Haack (2) did synthesize the N-methylformanilide–phosphorus oxychloride adduct in a pure state. Later H. Bredereck et al. (11) successfully obtained the N,N-dimethylformamide–phosphorus oxychloride adduct in a crystalline form; it was characterized by elementary analysis. Thus the suggestion of several authors (2, 12–15) was experimentally verified, namely, formation of the acid amide–phosphorus oxychloride adduct in a 1:1 molar ratio.

Lorenz and Winzinger (16) postulated that the N-methylformanilide–$POCl_3$ adduct was a chloromethyleneiminium dichlorophosphate (1), whereas Smith (13), Silverstein et al. (14), and Jutz (15) proposed a dichlorophosphato-iminium chloride structure (2) for the $DMF/POCl_3$ adduct.

$$
\text{H–C}\begin{array}{c}\text{Cl}\\\oplus\\\text{N}\end{array}\begin{array}{c}\text{CH}_3\\\\\text{C}_6\text{H}_5\end{array} \quad PO_2Cl_2^{\ominus} \qquad \text{H–C}\begin{array}{c}\text{OPOCl}_2\\\oplus\\\text{N(CH}_3)_2\end{array} \quad Cl^{\ominus} \qquad \text{H–C}\begin{array}{c}\text{Cl}\\\oplus\\\text{N(CH}_3)_2\end{array} \quad Cl^{\ominus}
$$

(1) (2) (3)

To establish the true structure the IR spectrum of the $DMF/POCl_3$ adduct was studied (11). The experiments apparently indicated that the phosphorus of the $POCl_3$ is bound to the amide oxygen, thus indicating structure 2 for the amide–$POCl_3$ adducts. Bosshard and Zollinger (17) reached the same conclusion by comparing the formylating activity of chloromethyleniminium chlorides like 3 with that of the amide–$POCl_3$ adducts. Both formylating agents should display roughly the same reactivity toward activated aromatic aromatic compounds provided that the gegenion does not influence the reactivity. Actually the DMF–$POCl_3$ adduct was found to be much more reactive. These qualitative results were confirmed by kinetic measurements of the formylation of thiophene derivatives (18, 19). The difference in reactivity between 3 and DMF–$POCl_3$ may be attributed to the different solubilities of the iminium salts or to different degrees of dissociation of the ion pairs; thus these observations are not unequivocal proof for either structure 1 or 2. Additionally it was shown that the reaction of the DMF–$POCl_3$ adduct and aluminum chloride yields a salt-like compound, whose IR spectrum displays a band attributable to the C—O—P valence vibration; the same band is observed in the IR spectrum of the DMF–POCl$_3$ adduct (17). Ziegenbein and Francke (20) proposed the same structure **2** for the DMF–POCl$_3$ adduct. They were also able to isolate the primary addition product (**4**) from phenylacetylene and the DMF–POCl$_3$ adduct, and elucidated its structure. The nature of the reaction product (**4**) permitted a conclusion as to the nature of the formylating agent and the assignment of structure **2** to it, as already proposed by Bredereck (11).

$$C_6H_5-C\equiv CH \xrightarrow{DMF/POCl_3} C_6H_5-\underset{Cl}{\overset{Cl}{C}}=CH-\underset{N(CH_3)_2}{\overset{OPOCl_2}{CH}}$$

(**4**)

Arnold and Holy (21) studied extensively halomethyleniminium salts and the DMF–POCl$_3$ adduct by IR measurements. They pointed out that the IR bands at 1040 and 1160 cm^{-1}—attributed to the P—O—C vibrations by Bredereck et al. (11) and by Bosshard and Zollinger (17)—also appear in the spectra of the iminium salts **5–8**, and result from the C=N vibrations of the iminium structure —C=N$^\oplus$(CH$_3$)$_2$.

$$H-C\overset{X}{\underset{N(CH_3)_2}{\overset{\oplus}{\diagup}}} \quad Y^\ominus$$

(**5**) X = Y = Cl
(**6**) X = Y = Br
(**7**) X = Y = I
(**8**) X = Cl, Y = SbCl$_6$

Arnold and Holy (21) further demonstrated that both chloride and dichlorophosphate anions exist in the DMF–POCl$_3$ adduct. These results are interpreted most satisfactorily by assuming an equilibrium of forms **2** and **9**. Such an equilibrium had been suspected by Bosshard and Zollingen (17), and the NMR studies by Martin et al. (22–24) do not rule out such an equilibrium. However, in solution the equilibrium is almost completely on the side of the N,N-dimethylchloromethyleniminium dichlorophosphate (**9**), as the ^{31}P NMR spectrum shows (24). No conclusions can be made about the equilibrium in the solid state, since solvation effects are lacking here. The ^{1}H NMR spectrum of the DMF–POCl$_3$ adduct in chloroform shows signals at $\tau_{HC} = -0.17$ and

$$H-C\overset{OPOCl_2}{\underset{N(CH_3)_2}{\overset{\oplus}{\diagup}}} \quad Cl^\ominus \rightleftarrows H-C\overset{Cl}{\underset{N(CH_3)_2}{\overset{\oplus}{\diagup}}} \quad PO_2Cl_2^\ominus$$

(**2**) \qquad\qquad (**9**)

$\tau_{N^{\oplus}(CH_3)_2}$ = 6.07 at room temperature (23); contrarily, Smith (25) reports a τ value of 0.25 for the formyl hydrogen. More recently several authors have attributed solely the chloromethyleniminium structure **9** to the DMF–POCl$_3$ adduct, based on spectroscopic results (23) and on studies of vinylogous amides (26).

Martin et al. (22, 24) were the first to report kinetic studies on the DMF–POCl$_3$ adduct formation. The reaction is first order with respect to DMF and POCl$_3$. The reaction velocity is solvent dependent and decreases in the order CH$_2$Cl$_2$ > CH$_2$ClCH$_2$Cl > CHCl$_3$. According to these studies the activation energy in chloroform is ~8 kcal/mole and in methylene chloride ~5 kcal/mole (22, 24). In the course of these studies it was also shown, by NMR measurements, that the POCl$_3$ is not irreversibly bound in the adduct, but can be further transferred to other DMF molecules (22, 24, 27).

In a more recent kinetic study of the adduct formation of DMF and POCl$_3$ in 1,2-dichloroethane, Alumi et al. (18) determined the activation energy to be 15.8 kcal/mole and the activation entropy to be −20.7 kcal/mole. Their values differ from those reported by Martin (22) by nearly 100%. Fritz and Oehl (26) demonstrated halide–oxygen exchange between DMF and the DMF–POCl$_3$ adduct:

$$D-C\overset{Cl}{\underset{N(CD_3)_2}{\overset{\oplus}{\diagdown}}} \;\; PO_2Cl_2^{\ominus} + H-C\overset{O}{\underset{N(CH_3)_2}{\diagdown}} \rightleftharpoons$$

$$D-C\overset{O}{\underset{N(CD_3)_2}{\diagdown}} + H-C\overset{Cl}{\underset{N(CH_3)_2}{\overset{\oplus}{\diagdown}}} \;\; PO_2Cl_2^{\ominus}$$

The kinetics of this reaction have been studied by Martin et al. (27).

2. Adducts of Higher Amides and Urea with Phosphorus Oxychloride

The complexes of higher acid amides and urea with POCl$_3$ have not been isolated, but have commonly been characterized by their subsequent reaction, for example, with primary amines (11), amides (28–31), hydrazine derivatives (11), or activated aromatic compounds (15, 17) (see also the chapter on "Amidinium Salts" in this volume). Thus it was shown that aliphatic and aromatic monocarboxylic acid dialkylamides do form adducts with POCl$_3$ (11).

With dicarboxylic acid bis(dialkylamides) the type of adduct formed depends on the distance between the two carboxamide functions in the molecule (29). Complex formation does not occur with oxalic acid bis(dimethylamide), but malonic acid bis(dimethylamide) does form a 1:1 adduct with POCl$_3$ (29). Succinic acid bis(dimethylamide) gives, in low yield a 1:2 adduct with POCl$_3$. There are several other unidentified reactions as well.

On the other hand, adipic acid bis(dimethylamide) and terephthalic acid bis(dimethylamide) form 1:2 adducts without remarkable side reactions. Five-, six-, and seven-membered lactams as well as their N-alkylated and arylated derivatives (α-pyrrolidones, α-piperidones, ε-caprolactams, and 1-methylquinolone) are capable of adduct formation with $POCl_3$ (29).

Bis adduct formation with $POCl_3$ was demonstrated for 2,5-diketopiperazine and sarcosine anhydride by subsequent reactions (29).

3. Thioamide–Phosphorus Oxychloride and Amide–Phosphorus Sulfochloride Adducts

N,N-Dimethylthioformamide also forms an adduct with $POCl_3$, which can be used for formylating heterocycles, and is said to be more reactive than the DMF–$POCl_3$ adduct (32).

DMF undergoes self-condensation in the presence of phosphorus sulfochloride; the following mechanism has been proposed for this reaction (33):

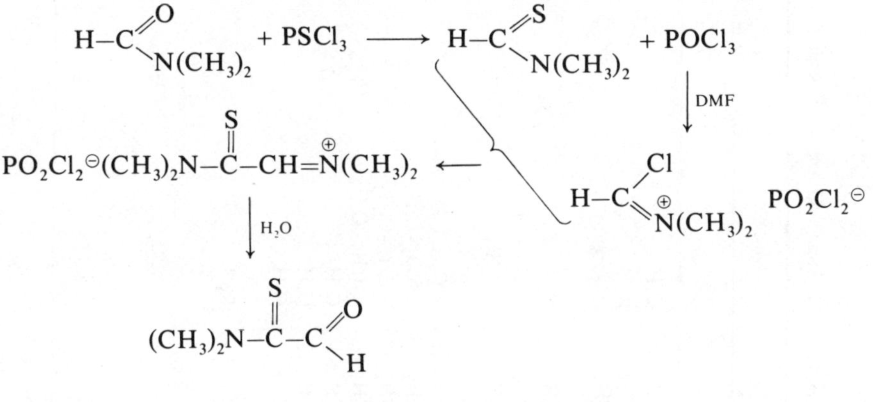

$$H-C(=O)N(CH_3)_2 + PSCl_3 \longrightarrow H-C(=S)N(CH_3)_2 + POCl_3$$

$$\downarrow DMF$$

$$PO_2Cl_2^{\ominus} (CH_3)_2N-\overset{S}{\underset{\|}{C}}-CH=\overset{\oplus}{N}(CH_3)_2 \longleftarrow H-C(Cl)=\overset{\oplus}{N}(CH_3)_2 \quad PO_2Cl_2^{\ominus}$$

$$\downarrow H_2O$$

$$(CH_3)_2N-\overset{S}{\underset{\|}{C}}-C(=O)H$$

B. REACTIONS OF ACID AMIDE–PHOSPHORUS OXYCHLORIDE ADDUCTS

1. Reactions with Amines

Bureš and Kundera (34) and Davis and Yelland (35) were the first to report experiments investigating the formylation of primary amines with amides and $POCl_3$. Davis and Yelland succeeded in transforming n-butylamine into N,N'-di-n-butylformamidine—isolated as the picrate—with the aid of N-n-butylformamide–$POCl_3$ complex:

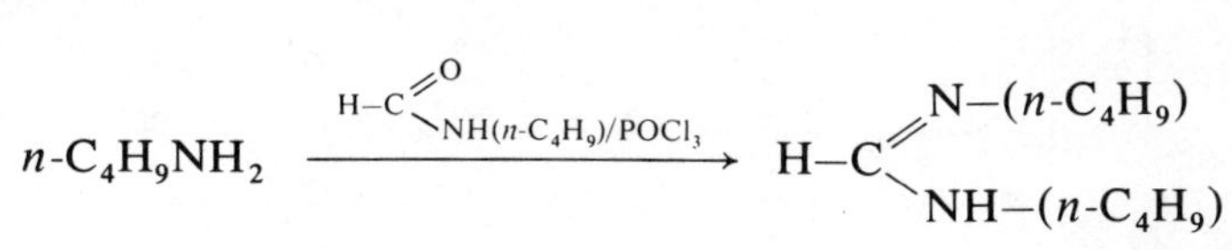

$$n\text{-}C_4H_9NH_2 \xrightarrow{H-C(=O)NH(n\text{-}C_4H_9)/POCl_3} H-C(=N-(n\text{-}C_4H_9))(NH-(n\text{-}C_4H_9))$$

TABLE I

N,N,N'-Trisubstituted Amidines (12) from Acid Amide–Phosphorus Oxychloride Adducts and Primary Amines

R	Amidine			Yield, %	b.p./pressure, °C/torr m.p., °C	n_D^{20}	Ref.
	R^1	R^2	R^3				
H	CH_3	CH_3	C_6H_5	85	119–122/10	1.5955	11
H	CH_3	CH_3	p-CH_3–C_6H_4	56	94–0.15	1.5855	11
H	CH_3	CH_3	α-Naphthyl	73	159–160/0.3 m.p. 40	1.6690	11
H	CH_3	CH_3	C_6H_5–CH_2	70	82/0.5	1.5452	11
H	CH_3	CH_3	C_6H_5–CH_2–CH_2	82	74–76/0.01	1.5349	11
H	CH_3	CH_3	c-C_6H_{11}	63	84–86/12 m.p. 40–43	—	11
H	CH_3	CH_3	n-C_4H_9	45	55–57/12	1.4502	11
H	CH_3	CH_3	p-NO_2–C_6H_4	70	m.p. 80–82	—	53
H	CH_3	CH_3	p-CH_3–C_6H_4	85	161–162/12	1.5900	53
H	CH_3	CH_3	m-CH_3–C_6H_4	78	164/12	1.5918	53
CH_3	CH_3	CH_3	C_6H_5	58	75–76/0.005	1.5775	11
C_5H_{11}	CH_3	CH_3	C_6H_5	56	158/0.006	1.5221	11
C_6H_5	CH_3	CH_3	C_6H_5	55	128–130/0.2 67–70	—	11
H	CH_3	C_6H_5	C_6H_5	56	146–147/0.3	1.6555	11
H	CH_3	CH_3	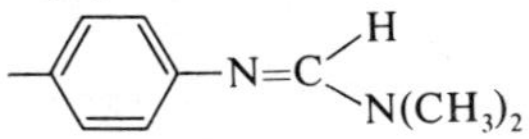 ![p-tolyl-N=C(H)-N(CH3)2]	70	m.p. 121	—	29
$(CH_3)_2N\overset{\underset{\parallel}{O}}{-}C-CH_2$	CH_3	CH_3	C_6H_5	34	m.p. 75	—	29

I. ACID AMIDE–PHOSPHORUS/OXYCHLORIDE ADDUCTS

Detailed studies of the acylation of primary and secondary amines originate from Raison (36), Bredereck et al. (11, 29), and Jutz et al. (31). According to their work the adducts from N,N-disubstituted acid amides and $POCl_3$ (**10**) react with primary aliphatic and aromatic amines to give amidine hydrochlorides (**11**). The free amidines (**12**) may be obtained from them by treatment with alkaline hydroxides (11, 29–31, 36) (see Table I).

$$R-C{\overset{\oplus}{\underset{NR^1R^2}{\diagup Cl}}}\ \ PO_2Cl_2^{\ominus} + R^3-NH_2 \longrightarrow$$

(**10**)

$$R-C{\overset{\oplus}{\underset{NR^1R^2}{\diagup NHR^3}}}\ Cl^{\ominus} \xrightarrow{NaOH} R-C{\underset{NR^1R^2}{\diagup NR^3}}$$

(**11**)   (**12**)

Diamines like p-phenylenediamine are formylated as follows (29):

$$H_2N-\!\!\!\left\langle\!\!\!\bigcirc\!\!\!\right\rangle\!\!\!-NH_2 \xrightarrow[POCl_3]{DMF} {\underset{(CH_3)_2N}{H\diagdown}}\!\!C{=}N-\!\!\!\left\langle\!\!\!\bigcirc\!\!\!\right\rangle\!\!\!-N{=}C{\underset{N(CH_3)_2}{\diagup H}}$$

Sulfonamides, as well, can be transformed into N-sulfonylformamidines (37, 38):

$$R-SO_2NH_2 \xrightarrow{DMF/POCl_3} H-C{\underset{N(CH_3)_2}{\diagup N-SO_2R}}$$

Iodinated anilines (39), aminopyrimidines (40, 41), amino-1,2,3-triazole-carboxaldehydes (42), and naphthylamines (43) are also formylated at their amino groups. These condensation reactions do not require the isolation of the acid amide–$POCl_3$ adducts, which may instead be generated *in situ* and reacted directly with amines. The highest yields of amidines are obtained using 1:1:2.3 ratio of amine : $POCl_3$: acid amide, respectively (11). Besides N-formylation one finds occasionally also C-formylation (44) when using DMF–$POCl_3$:

$$\underset{\underset{C_6H_5}{H^{N-N}}}{CH_3\diagdown\!\!\!\diagup NH_2} \xrightarrow{DMF/POCl_3} \underset{\underset{C_6H_5}{N-N}}{CH_3\diagdown\!\!\!\diagup\overset{CHN(CH_3)_2}{\diagup\!\!\!\diagdown N-CHO}}$$

The bisamidines of dicarboxylic acids (e.g., **13**) (29) are obtained from treatment with primary amines of the bis-$POCl_3$–acid adduct, generated *in situ*

TABLE II

Bisamidines of Dicarboxylic Acids (**13**) from Dicarboxylic Acid Dimethylamides, Phosphorus Oxychloride, and Aniline (29)

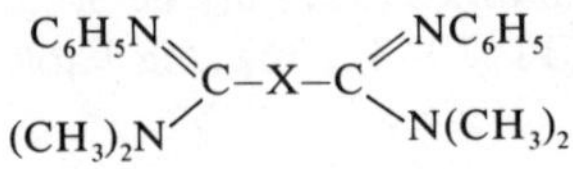

$$\underset{(CH_3)_2N}{\overset{C_6H_5N}{>}}C-X-C\underset{N(CH_3)_2}{\overset{NC_6H_5}{<}}$$

Bisamidine X	Yield, %	m.p., °C
–(CH$_2$)$_2$–	6	188
–(CH$_2$)$_4$–	63	112
 –C$_6$H$_4$– (p-phenylene)	42	211

(see Table II). A number of heterocycle syntheses were performed via acid amide–POCl$_3$ adducts. The following survey shows several examples:

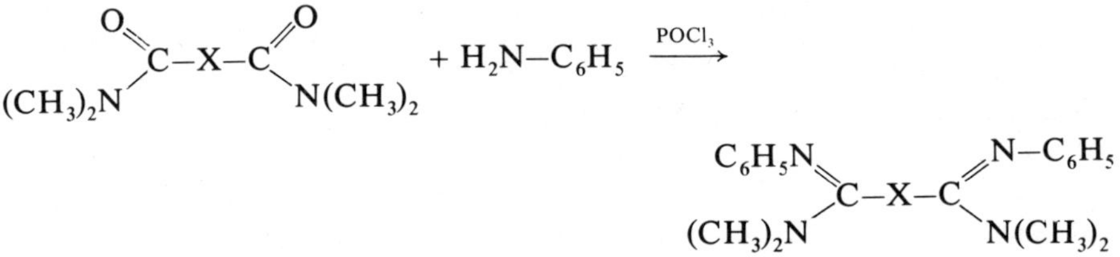

$$\underset{(CH_3)_2N}{\overset{O}{>}}C-X-C\underset{N(CH_3)_2}{\overset{O}{<}} + H_2N-C_6H_5 \xrightarrow{POCl_3}$$

$$\underset{(CH_3)_2N}{\overset{C_6H_5N}{>}}C-X-C\underset{N(CH_3)_2}{\overset{N-C_6H_5}{<}}$$
(**13**)

[Scheme: 2,4,6-triaminopyrimidine-2-thiol + DMF/POCl$_3$ → thiazolo-pyrimidine products with N=CH–N(CH$_3$)$_2$ and N–CH=N$^⊕$(CH$_3$)$_2$ Cl$^⊖$ substituents]

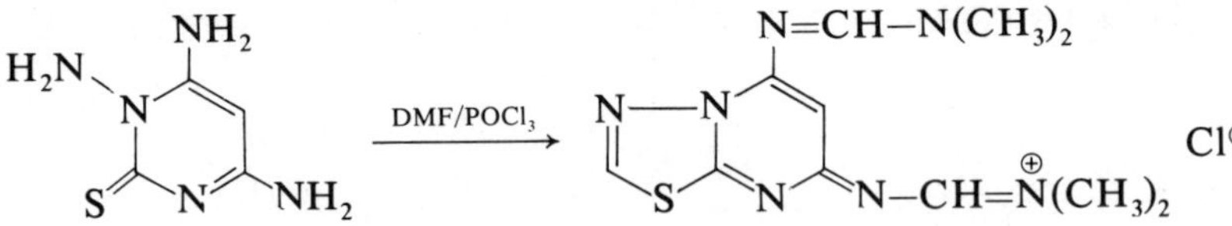

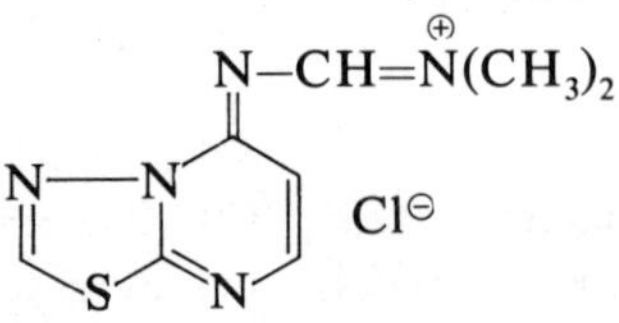

(ref. 45)

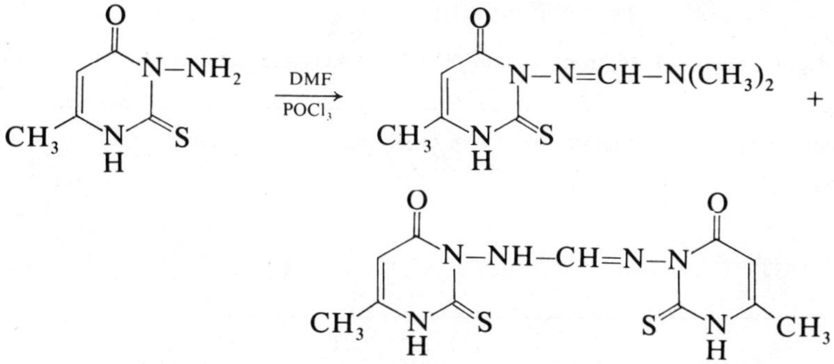

I. ACID AMIDE–PHOSPHORUS/OXYCHLORIDE ADDUCTS

(ref. 46)

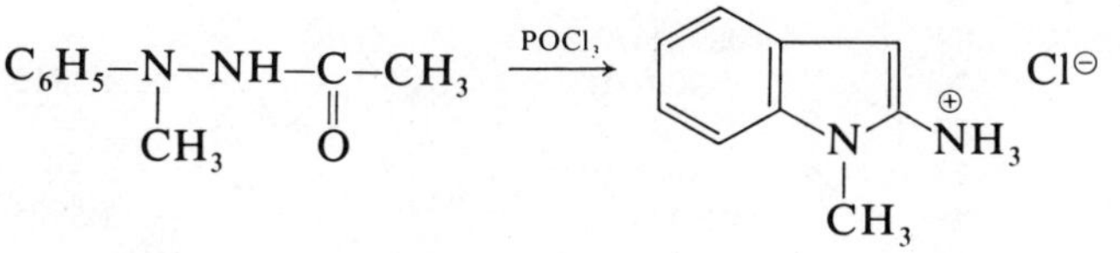

(ref. 47)

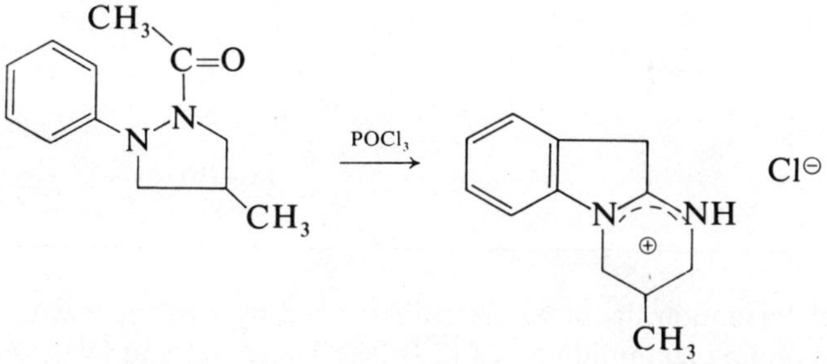

(ref. 47)

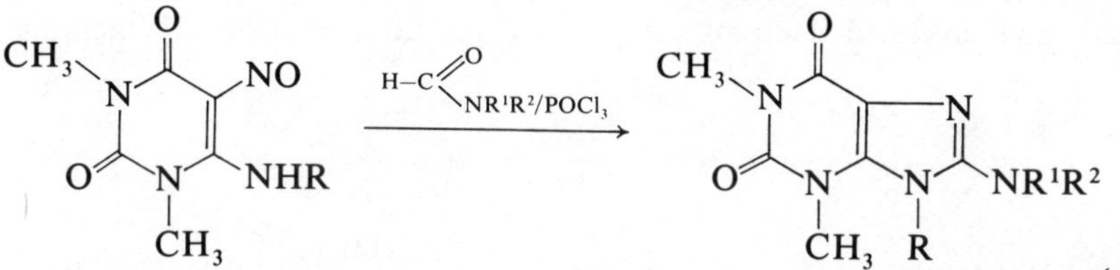

(ref. 48)

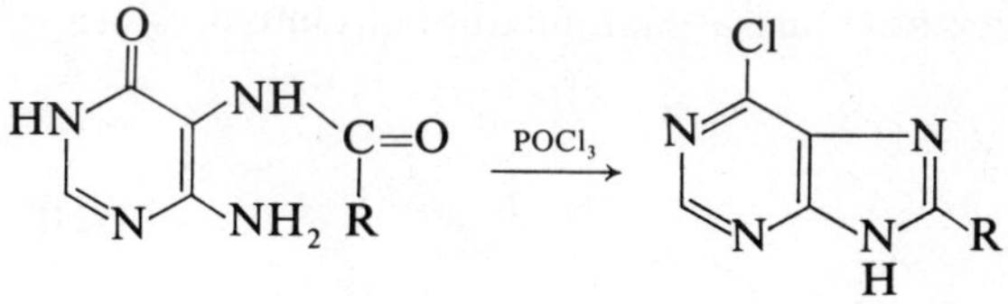

(ref. 49)

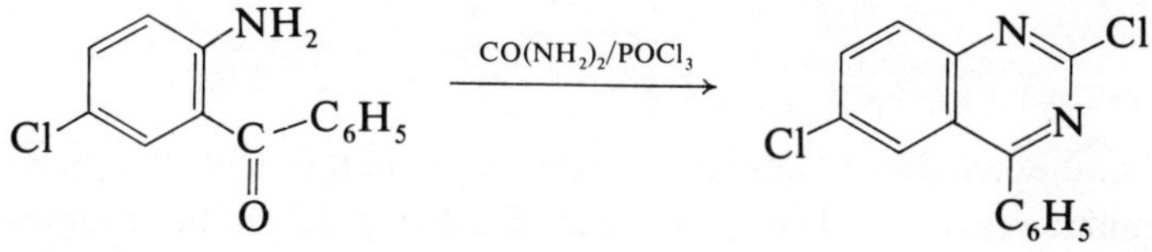

(ref. 50)

TABLE III

Lactam Imides (**14–16**) from Lactam–POCl$_3$ Adducts and Aniline (**29**)

Lactam imide 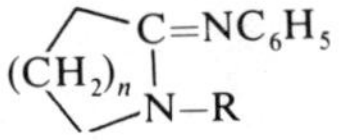 $(CH_2)_n \genfrac{}{}{0pt}{}{C=NC_6H_5}{N-R}$		Yield, %	b.p./pressure, °C/torr	m.p., °C	n_D^{20}
n	R				
3	H	65	117–121/0.001	114–115	—
4	H	70	—	95–96	—
5	H	68	108–111/0.2	107	—
3	CH$_3$	76	90/0.01	—	1.5915
3	o-CH$_3$–C$_6$H$_4$	30	—	106	—
4	CH$_3$	77	109/0.05	—	—
5	CH$_3$	78	140–141/0.9	—	1.5805

[Structure: tetrahydroquinoline with =C(H)–C$_6$H$_5$ and N–R]

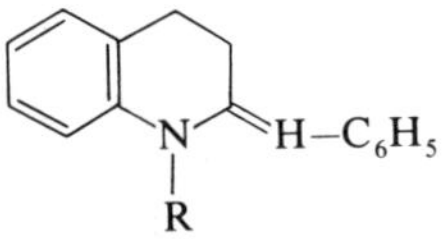					
R = H		59	—	159–161	—
R = CH$_3$		64	—	75	—

After POCl$_3$ adduct formation, lactams are transformed by treatment with amines into the corresponding lactamidines **14** (29) (see Tables III and IV). N-Alkylated and arylated lactams may undergo this reaction. If lactams

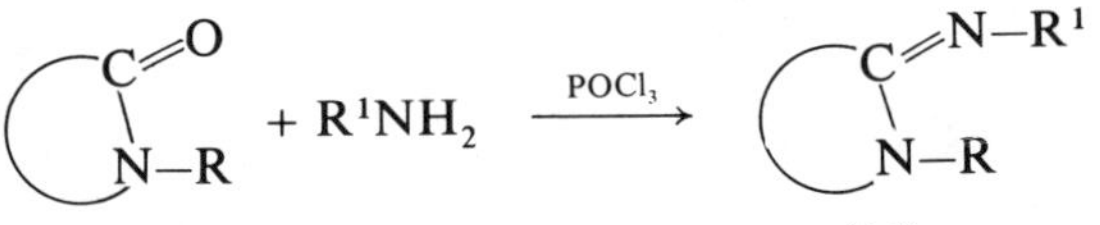

[Scheme: lactam + R^{1}NH$_2$ → (via POCl$_3$) lactamidine] (**14**)

unsubstituted at the nitrogen atom are used, a mixture of tautomers (**15** ⇌ **16**) forms (29).

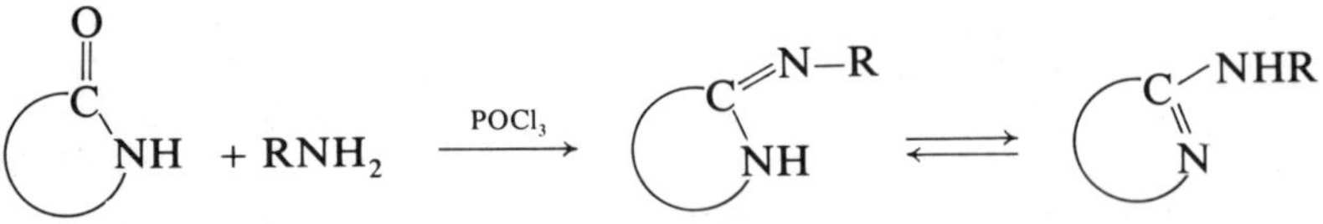

[Scheme: lactam-NH + RNH$_2$ → (via POCl$_3$) **15** ⇌ **16**]

The case of vinylpyrrolidone is unusual. Adduct formation with POCl$_3$ and reaction with aniline results in cleavage of the vinyl group during the course of

I. ACID AMIDE–PHOSPHORUS/OXYCHLORIDE ADDUCTS

TABLE IV
Lactam Imides (**14**, **17**) from Lactam–POCl$_3$ Adducts and Primary and Secondary Amines (29)

Lactam imide	Yield, %	b.p./pressure, °C/torr	m.p., °C	n_D^{20}
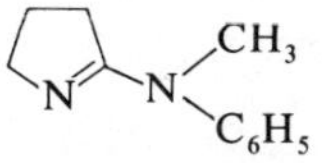 pyrrolidine-N(CH$_3$)(C$_6$H$_5$) derivative	58	65/0.01	—	1.5547
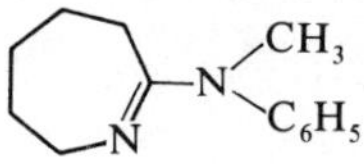 azepane-N(CH$_3$)(C$_6$H$_5$) derivative	46	82/0.05	—	—
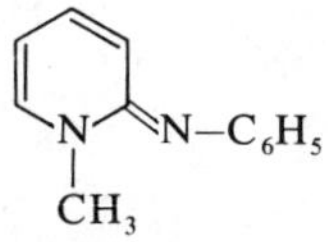 1-methyl-2-(phenylimino)pyridine	27	—	71–72	—
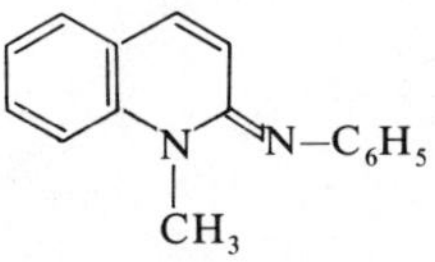 1-methyl-2-(phenylimino)quinoline	64	—	75	—
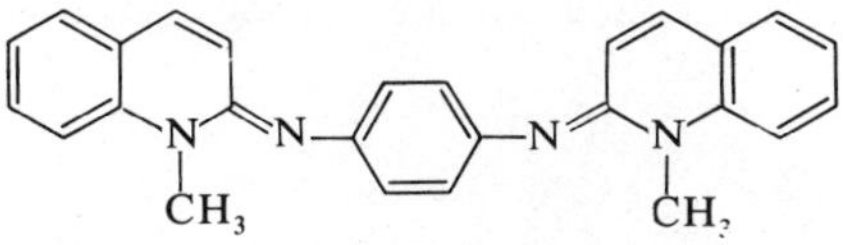 bis-quinoline bridged by phenylene (N-CH$_3$)	56	—	316–317	—
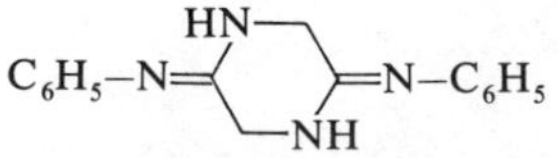 C$_6$H$_5$–N= piperazine =N–C$_6$H$_5$ (HN, NH)	91	—	245–250	—
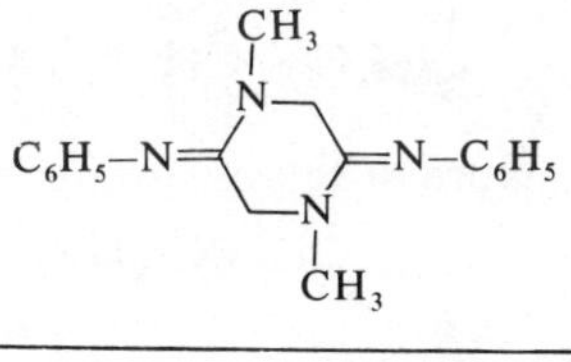 C$_6$H$_5$–N= piperazine =N–C$_6$H$_5$ (N-CH$_3$, N-CH$_3$)	83	—	170–180 (dec)	—

the reaction; the same product as from the reaction of pyrrolidone–POCl$_3$ and aniline is obtained (29):

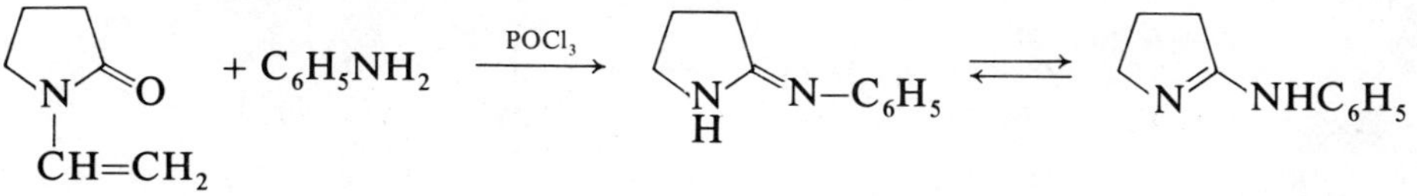

pyrrolidone (N-CH=CH$_2$, C=O) + C$_6$H$_5$NH$_2$ $\xrightarrow{POCl_3}$ pyrrolidine=N–C$_6$H$_5$ (NH) $\rightleftarrows$ pyrrolidine–NHC$_6$H$_5$

Lactams unsubstituted at the nitrogen atom react with secondary amines in the presence of $POCl_3$ to yield lactam imides **17** (29) (see also Table IV). In the course of the reaction of tertiary amines like dimethylaniline with pyrrolidone

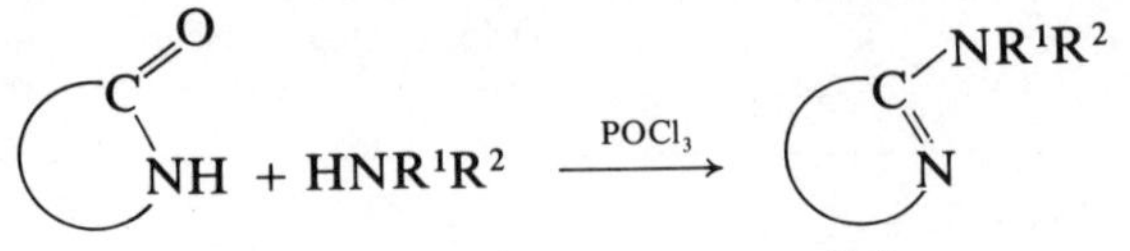

$$\text{[lactam C(=O)NH]} + HNR^1R^2 \xrightarrow{POCl_3} \text{[cyclic C(NR}^1R^2\text{)=N]}$$

(**17**)

and $POCl_3$ a cleavage analogous to the von Braun degradation occurs: the same product as from the reaction between pyrrolidone–$POCl_3$ and N-methylaniline is obtained (29):

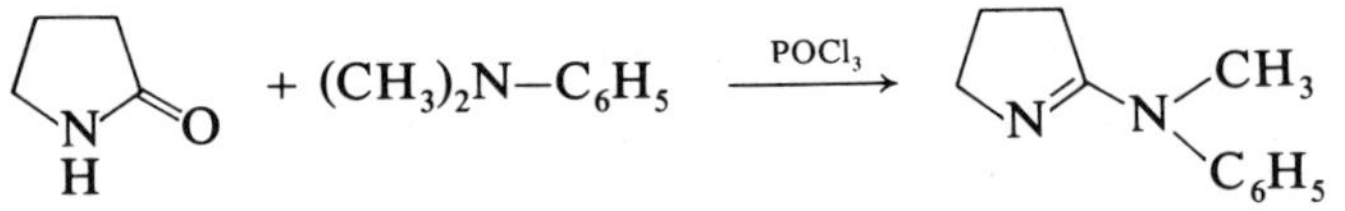

$$\text{pyrrolidone} + (CH_3)_2N-C_6H_5 \xrightarrow{POCl_3} \text{product with } N(CH_3)(C_6H_5)$$

Benzoxazole is synthesized from *o*-aminophenol and DMF–$POCl_3$ (51). Here also the amidine is the proposed intermediate:

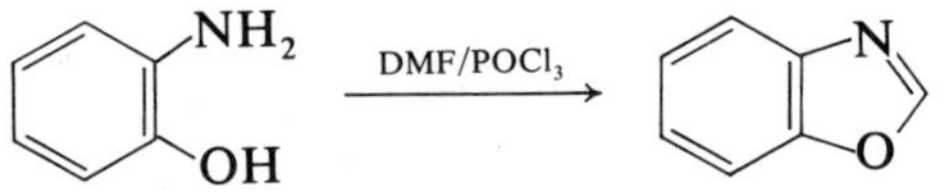

$$\text{o-aminophenol} \xrightarrow{DMF/POCl_3} \text{benzoxazole}$$

Amino acids are formylated by DMF–$POCl_3$ at both the carbon and the nitrogen atoms (52):

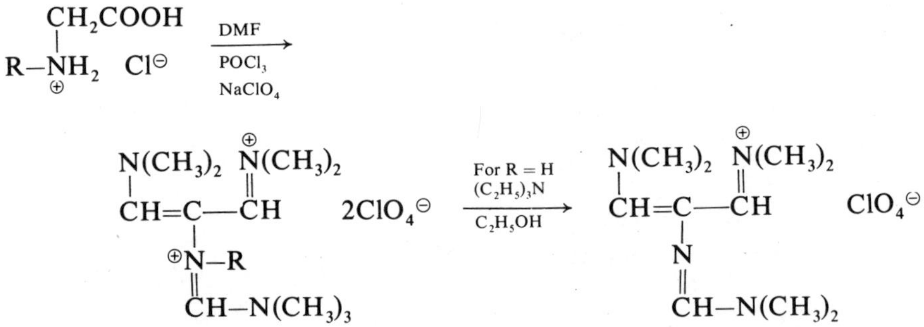

$$R-\overset{\oplus}{N}H_2\; Cl^{\ominus} \;\; \overset{CH_2COOH}{|} \xrightarrow[NaClO_4]{DMF, POCl_3}$$

$$\begin{array}{c} N(CH_3)_2 \;\; \overset{\oplus}{N}(CH_3)_2 \\ | \quad\quad\quad \| \\ CH=C\!-\!CH \\ | \\ \overset{\oplus}{N}\!-\!R \\ \| \\ CH\!-\!N(CH_3)_3 \end{array} \; 2ClO_4^{\ominus} \xrightarrow[C_2H_5OH]{\text{For } R=H \\ (C_2H_5)_3N} \begin{array}{c} N(CH_3)_2 \;\; \overset{\oplus}{N}(CH_3)_2 \\ | \quad\quad\quad \| \\ CH=C\!-\!CH \\ | \\ N \\ \| \\ CH\!-\!N(CH_3)_2 \end{array} ClO_4^{\ominus}$$

If N,N,N'-trisubstituted, N,N,N',N'-tetrasubstituted, and N,N'-symmetrically disubstituted ureas are reacted with primary amines in the presence of $POCl_3$, guanidines (**18**) are obtained (29). Although complexes

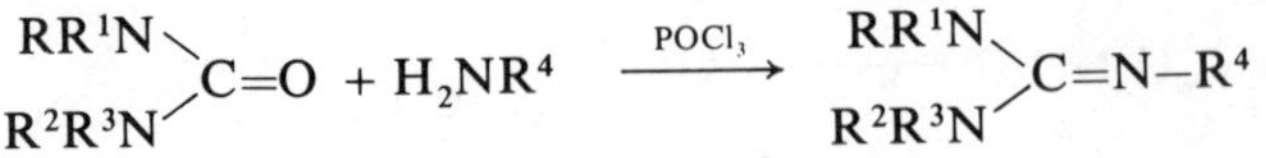

$$\begin{array}{c} RR^1N \\ \diagdown \\ C=O \\ \diagup \\ R^2R^3N \end{array} + H_2NR^4 \xrightarrow{POCl_3} \begin{array}{c} RR^1N \\ \diagdown \\ C=N-R^4 \\ \diagup \\ R^2R^3N \end{array}$$

(**18**)

from ureas and $POCl_3$ of type **19** have not yet been isolated, it is very likely that they function as the carboxylating agent in these condensations. The guanidines (**18**) synthesized in this way are summarized in Table V.

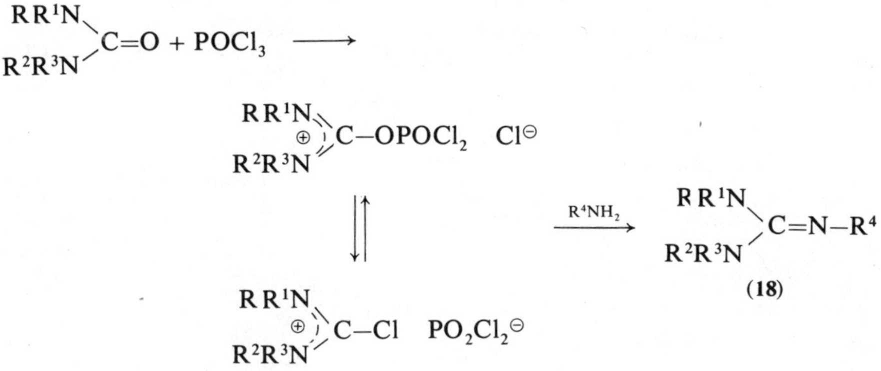

$$\begin{array}{c} RR^1N \\ R^2R^3N \end{array}\!\!\!C=O + POCl_3 \longrightarrow$$

$$\begin{array}{c} RR^1N \\ R^2R^3N \end{array}\!\!\!\overset{\oplus}{C}-OPOCl_2 \quad Cl^{\ominus}$$

$$\updownarrow \qquad \xrightarrow{R^4NH_2} \quad \begin{array}{c} RR^1N \\ R^2R^3N \end{array}\!\!\!C=N-R^4$$

$$\begin{array}{c} RR^1N \\ R^2R^3N \end{array}\!\!\!\overset{\oplus}{C}-Cl \quad PO_2Cl_2^{\ominus} \qquad (\mathbf{18})$$

(**19**)

TABLE V

Substituted Guanidines (**18**) from Urea–$POCl_3$ Adducts and Primary Aromatic Amines (29)

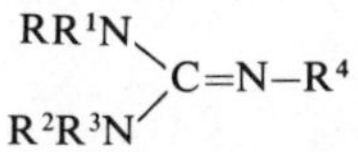

$$\begin{array}{c} RR^1N \\ R^2R^3N \end{array}\!\!\!C=N-R^4$$

		Guanidine			Yield, %	b.p./pressure, °C/torr	m.p., °C	n_D^{20}
R	R^1	R^2	R^3	R^4				
H	CH_3	CH_3	H	C_6H_5	60	—	115–116	—
CH_3	CH_3	CH_3	CH_3	C_6H_5	52	130/10	—	1.5696
C_2H_5	C_2H_5	C_2H_5	C_2H_5	C_6H_5	49	88–90/0.01	—	1.5382
CH_3	CH_3	CH_3	CH_3	$p\text{-}CH_3\text{-}C_6H_4$	68	144/11	—	1.5621
C_2H_5	C_2H_5	H	C_6H_5	C_6H_5	50	—	96–97	—
CH_3	CH_3	CH_3	CH_3	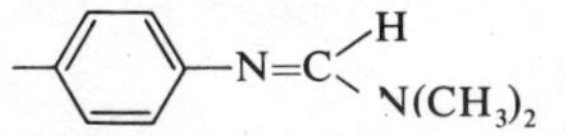 –C$_6$H$_4$–N=CH–N(CH$_3$)$_2$	56	—	154	—

2. Reactions with Hydrazine Derivatives and Hydroxylamine

If the DMF–$POCl_3$ adduct is reacted with N-monosubstituted hydrazines, N-formylformamidrazones (**20**) are formed (11). In the first reaction step formamidrazones (**21**) are generated—analogously to the reaction of amide–$POCl_3$ adducts with primary amines—which are again formylated by additional DMF–$POCl_3$ adduct yielding the salts **22**. If **22** is hydrolyzed, N-formylformamidrazones (**20**) are liberated. If N,N-disubstituted hydrazines like

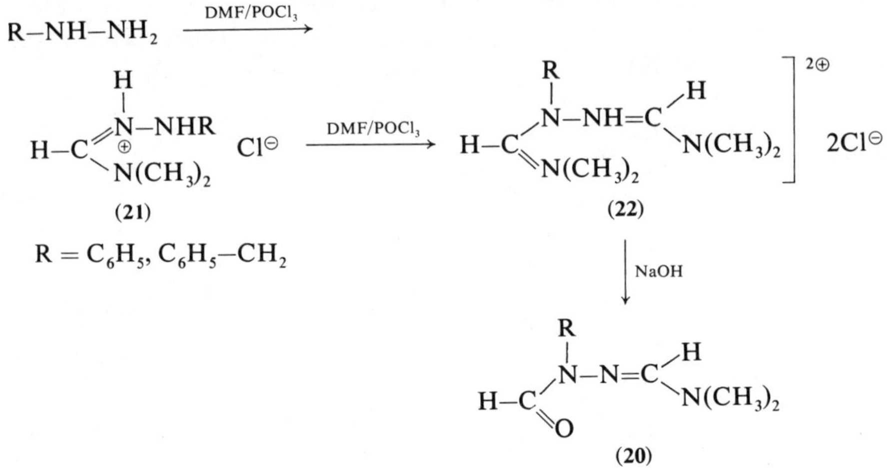

ADDUCTS FROM ACID AMIDES AND ACYLATION REAGENTS

$$R-NH-NH_2 \xrightarrow{DMF/POCl_3}$$

[structure (21): H-C(=N-NHR)(N(CH$_3$)$_2$)·H, Cl$^\ominus$, ⊕] $\xrightarrow{DMF/POCl_3}$ [structure (22): H-C(=N-NH=C(H)(N(CH$_3$)$_2$))(N(CH$_3$)$_2$) with R on N, 2⊕, 2Cl$^\ominus$]

$R = C_6H_5, C_6H_5-CH_2$

$\downarrow$ NaOH

[structure (20): H-C(=O)-N(R)-N=C(H)(N(CH$_3$)$_2$)]

N-methyl-N-phenylhydrazine are used, reaction occurs only at the unsubstituted nitrogen yielding the formamidrazone 23 (11).

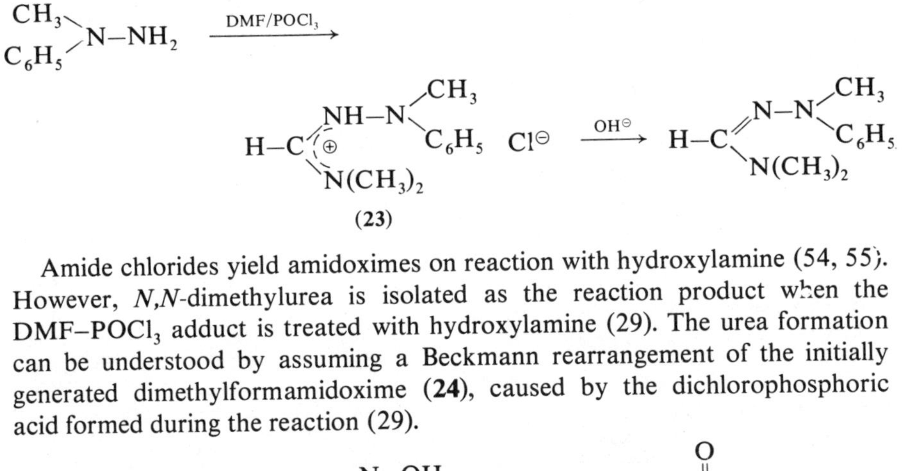

$$\begin{matrix}CH_3\\C_6H_5\end{matrix}N-NH_2 \xrightarrow{DMF/POCl_3}$$

[structure: H-C(⊕)(=NH-N(CH$_3$)(C$_6$H$_5$))(N(CH$_3$)$_2$), Cl$^\ominus$] $\xrightarrow{OH^\ominus}$ [structure (23): H-C(=N-N(CH$_3$)(C$_6$H$_5$))(N(CH$_3$)$_2$)]

Amide chlorides yield amidoximes on reaction with hydroxylamine (54, 55). However, N,N-dimethylurea is isolated as the reaction product when the DMF–POCl$_3$ adduct is treated with hydroxylamine (29). The urea formation can be understood by assuming a Beckmann rearrangement of the initially generated dimethylformamidoxime (24), caused by the dichlorophosphoric acid formed during the reaction (29).

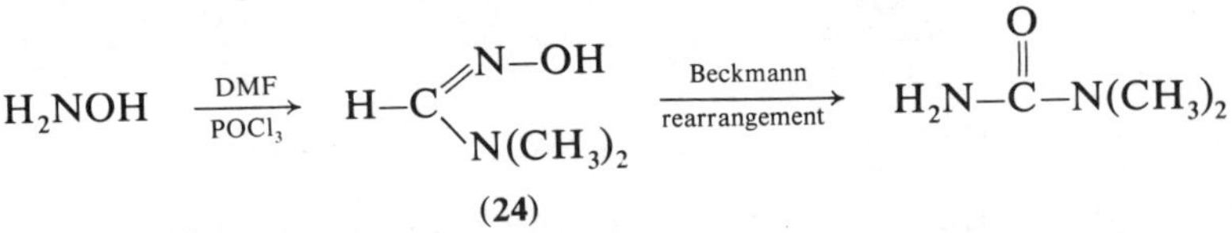

$$H_2NOH \xrightarrow[POCl_3]{DMF} \underset{(24)}{H-C(=N-OH)(N(CH_3)_2)} \xrightarrow{\text{Beckmann rearrangement}} H_2N-\overset{O}{\underset{\|}{C}}-N(CH_3)_2$$

Amide oximes are cyclized with DMF–POCl$_3$ to give 1,2,4-oxadiazoles (56):

$$R-C(=N-OH)(NH_2) \xrightarrow{DMF/POCl_3} \text{[1,2,4-oxadiazole with R]}$$

3. Reactions with Azomethines

Schiff bases are formylated at the nitrogen atom by the DMF–POCl$_3$ adduct, yielding amidinium salts **25**; these may be isolated as the perchlorates (57).

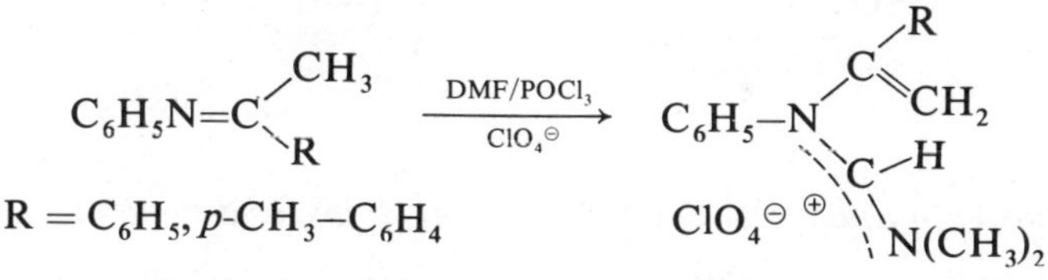

$$C_6H_5N=C\begin{smallmatrix}CH_3\\R\end{smallmatrix} \xrightarrow[ClO_4^\ominus]{DMF/POCl_3} C_6H_5-N\begin{smallmatrix}R\\C\\\diagdown CH_2\\C-H\end{smallmatrix}$$

$$R = C_6H_5, p\text{-}CH_3\text{-}C_6H_4 \qquad ClO_4^\ominus \quad \oplus \quad N(CH_3)_2$$

(**25**)

4. Reactions with Bases: Isonitrile Syntheses

Isonitriles (**26**) are obtained from *N*-monosubstituted formamides, pyridine, and POCl$_3$ (58–60). The reaction proceeds via two steps: adduct formation with phosphorus oxychloride, followed by an α-elimination. The yields of

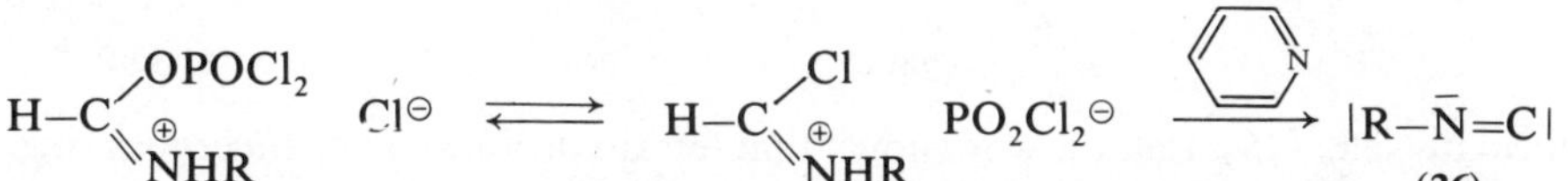

$$H-C\begin{smallmatrix}OPOCl_2\\\oplus\\NHR\end{smallmatrix} \; Cl^\ominus \rightleftarrows H-C\begin{smallmatrix}Cl\\\oplus\\NHR\end{smallmatrix} \; PO_2Cl_2^\ominus \xrightarrow{\text{pyridine}} |R-\bar{N}=C|$$
(**26**)

aliphatic isonitriles range from 58 to 95%, whereas aromatic compounds yield only 7–54% via this method. However, if potassium *t*-butoxide is used as a base, aromatic isonitriles are isolated in much higher yields (56–88%). Diisonitriles are also accessible in this way.

The increase in yield of this aromatic isonitrile synthesis if potassium *t*-butoxide is used is explained by the intermediate formation of the amide anion, which is more readily attacked at the oxygen atom by POCl$_3$ than the amide itself (59, 60):

$$2\,R-\overset{\ominus}{N}-CH-O \;\; K^\oplus + POCl_3 + 2(CH_3)_3COK \longrightarrow$$

$$2\,RNC + KPO_3 + 3KCl + 2(CH_3)_3COH$$

Ugi et al. (59) have published a review of aliphatic and aromatic isonitriles synthesized in this way. Formylhydrazones can be dehydrated to isocyanamide derivatives (**27**) by the phosphorus oxychloride method (61–65).

$$H-C\begin{smallmatrix}O\\NHNR_2\end{smallmatrix} \xrightarrow{POCl_3/NR_3} |C=\bar{N}-NR_2$$
(**27**)

N-Arylsulfonylformamides and POCl$_3$/pyridine do not yield sulfonylisonitriles, but rather *N,N'*-bis(arylsulfonyl)formamidines (**28**) (66).

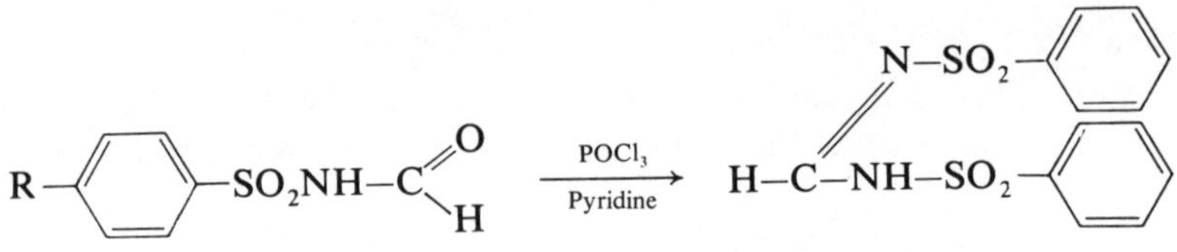

5. Reactions with Epoxides, Cyclic Ethers, Acyloins, and Polyvalent Alcohols

Ziegenbein and Franke (20) isolated 1-acyloxy-2-chloroalkanes and -cycloalkanes (**30**) resp. from the reaction of dimethylformamide–or dimethylacetamide–$POCl_3$ adducts and epoxides followed by hydrolysis of the intermediate salts (**29**). Later it was shown that tetrahydrofuran also undergoes ring cleavage, but 1,4-dichlorobutane is the isolated product (67):

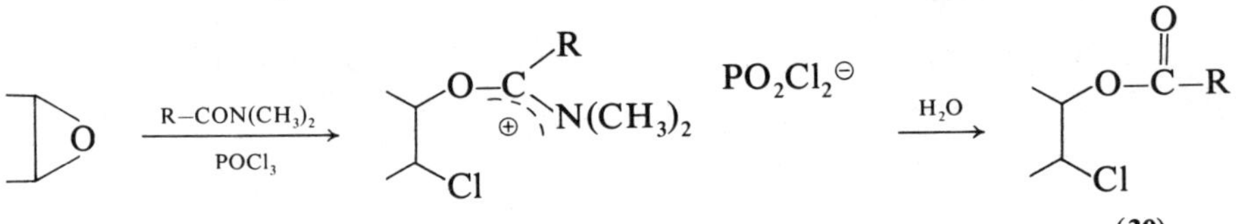

More recent studies demonstrate that DMF–$POCl_3$ cleaves cyclic ethers like dioxane as well as open-chain ethers like **31** (68).

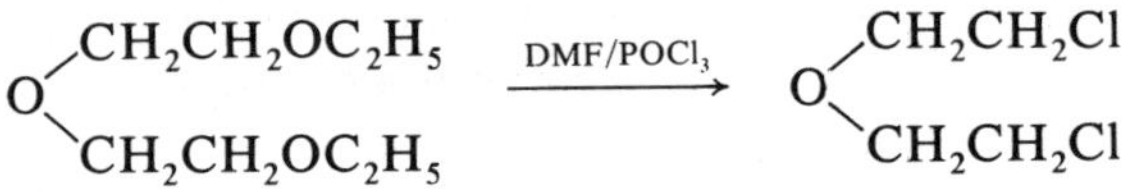

On reaction with DMF–$POCl_3$, acyloin (**32**) yields a mixture of **34** and **35**, and **33** yields **36** (69).

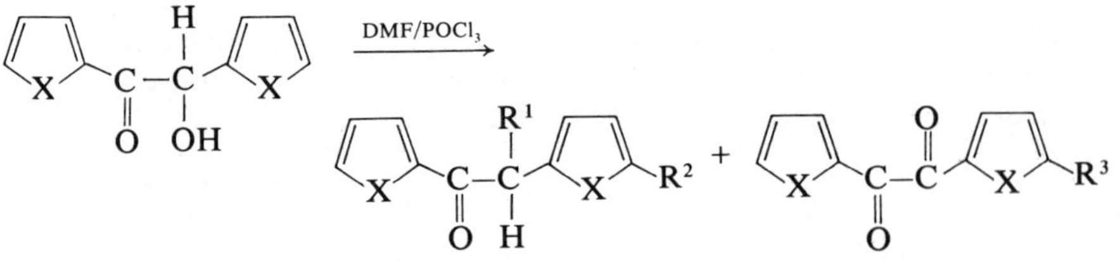

(32) X = O ⟶ (34) R^1 = H, R^2 = Cl + (35) R^3 = CHO
(33) X = S ⟶ (36) R^1 = Cl, R^2 = H

DMF–POCl$_3$ only substitutes the hydroxy group with chlorine in benzoin (69):

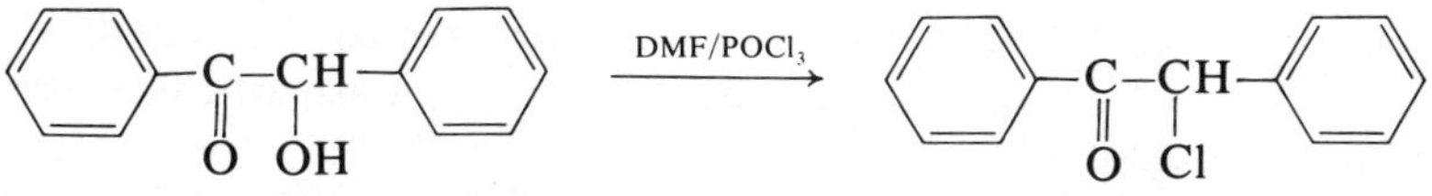

C$_6$H$_5$–CO–CH(OH)–C$_6$H$_5$ $\xrightarrow{\text{DMF/POCl}_3}$ C$_6$H$_5$–CO–CH(Cl)–C$_6$H$_5$

Vicinal diols **37** and **38** give **39** and **40**, respectively (69):

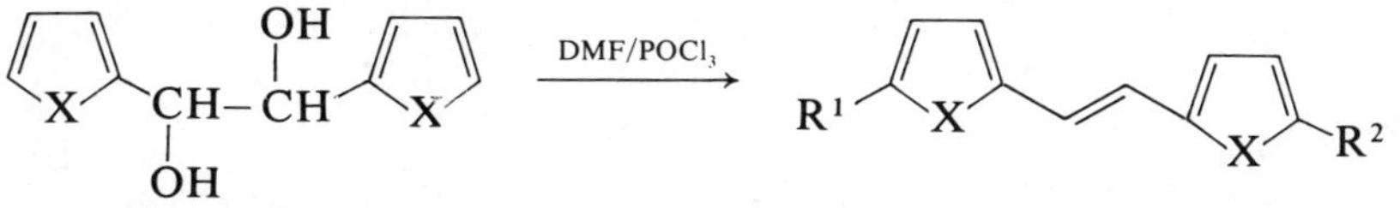

furyl-CH(OH)–CH(OH)-furyl $\xrightarrow{\text{DMF/POCl}_3}$ R^1–(X)–CH=CH–(X)–R^2

(37) X = O ⟶ (39) X = O, R^1 = Cl, R^2 = CHO
(38) X = S ⟶ (40) X = S, R^1 = Cl, R^2 = H

The hydroxy groups of diphenylglycol are formylated (69):

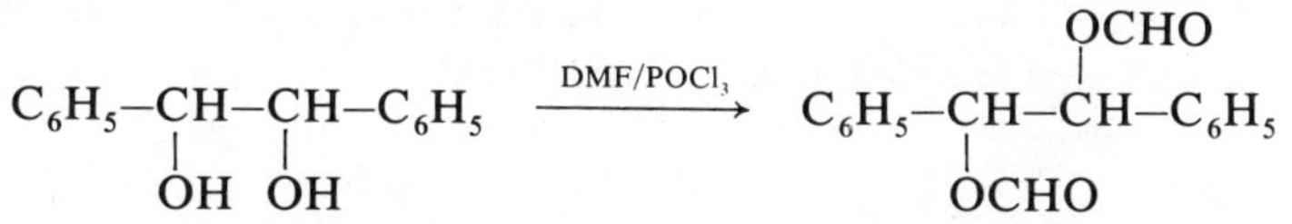

C$_6$H$_5$–CH(OH)–CH(OH)–C$_6$H$_5$ $\xrightarrow{\text{DMF/POCl}_3}$ C$_6$H$_5$–CH(OCHO)–CH(OCHO)–C$_6$H$_5$

6. Reactions with Heterocycles Containing Hydroxyl Groups

Heterocyclic OH groups may be substituted by chlorine with the aid of the DMF–POCl$_3$ adduct. A number of these reactions are described in various patents. The following survey gives some examples:

2-phenyl-4,6-dihydroxy-1,3,5-triazine $\xrightarrow{\text{DMF/POCl}_3}$ 2-phenyl-4,6-dichloro-1,3,5-triazine (ref. 70)

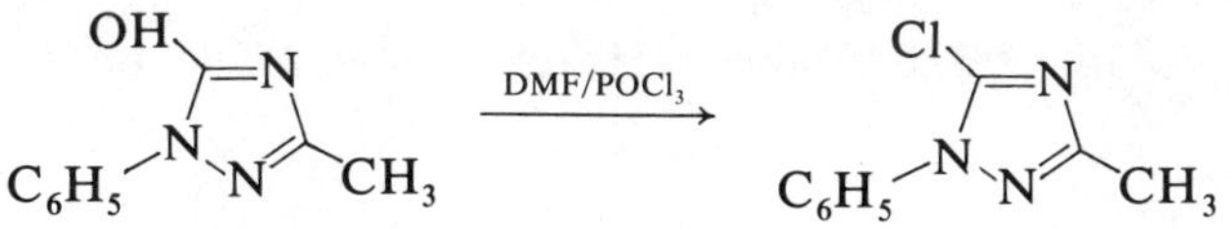

1-phenyl-3-methyl-5-hydroxy-1,2,4-triazole $\xrightarrow{\text{DMF/POCl}_3}$ 1-phenyl-3-methyl-5-chloro-1,2,4-triazole (ref. 71)

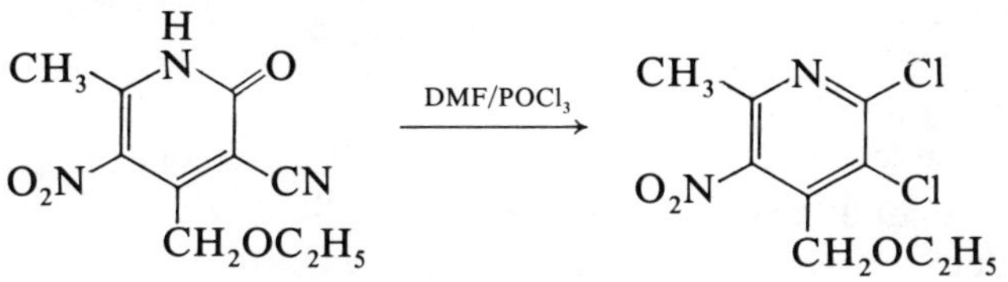

6-methyl-3-cyano-4-(ethoxymethyl)-5-nitro-2-pyridone $\xrightarrow{\text{DMF/POCl}_3}$ 6-methyl-2,3-dichloro-4-(ethoxymethyl)-5-nitropyridine (ref. 72)

7. Reactions with Alkali Metal Alkoxides and Mercaptans

The DMF–POCl$_3$ adduct and sodium methoxide give dimethylformamide acetal in low yields (7%); phosphoric acid esters are by-products (29, 73). Since several better routes to amide acetals are known, this particular one is of little synthetic interest.

On the other hand, the reaction of the DMF–POCl$_3$ adduct with mercaptans and thioglycols provides a useful method for the synthesis of amide mercaptals (**41, 42**) (74).

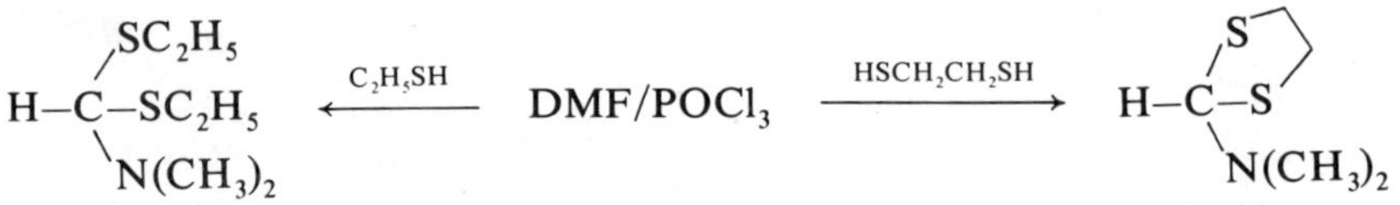

$$\underset{(41)}{H-C(SC_2H_5)_2 N(CH_3)_2} \xleftarrow{C_2H_5SH} DMF/POCl_3 \xrightarrow{HSCH_2CH_2SH} \underset{(42)}{H-C\underset{N(CH_3)_2}{\overset{S}{\diagup}}\!\!\!-S}$$

Thiophenol reacts with N,N-dimethylbenzamide–phosphorus oxychloride via the intermediate **43** to yield phenyl thiobenzoate (36).

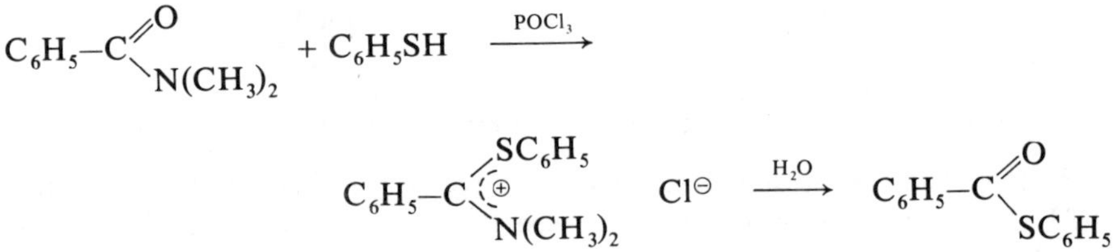

$$C_6H_5-C(=O)N(CH_3)_2 + C_6H_5SH \xrightarrow{POCl_3}$$

$$\underset{(43)}{C_6H_5-\overset{+}{C}(SC_6H_5)N(CH_3)_2}\ Cl^{\ominus} \xrightarrow{H_2O} C_6H_5-C(=O)SC_6H_5$$

8. Reactions with Acide Amides

N,N-Disubstituted acid amides and N-monosubstituted lactams undergo self-condensation in the presence of POCl$_3$ to yield β-ketocarboxylic acid amides (**44**) (29, 54, 75–77). This reaction is discussed elsewhere (see the chapter in Part 1 by C. Jutz, "The Vilsmeier–Haack–Arnold Acylations."

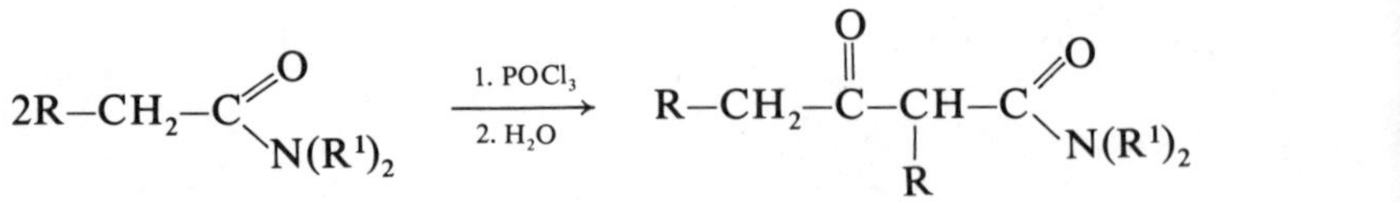

$$2R-CH_2-C(=O)N(R^1)_2 \xrightarrow[2.\ H_2O]{1.\ POCl_3} \underset{(44)}{R-CH_2-\overset{O}{\overset{\|}{C}}-\overset{R}{\underset{|}{C}H}-C(=O)N(R^1)_2}$$

Kobayashi and Morita et al. (78–81) studied the condensation of unsubstituted amides and lactams in the presence of POCl$_3$; pyrimidines and condensed pyrimidines are generated in a one-step reaction:

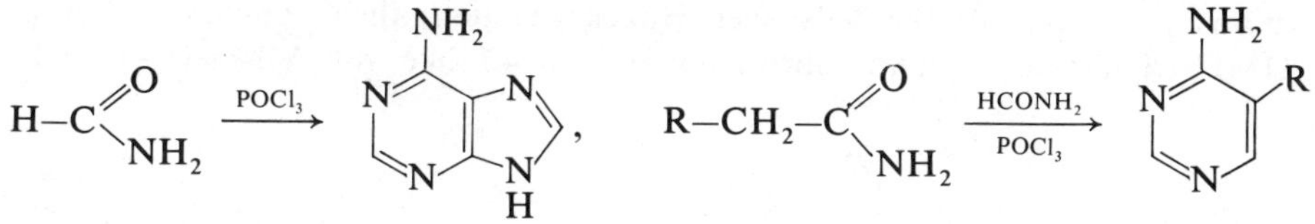

II. ACID AMIDE–TRIPHENYLPHOSPHINE DIHALIDE ADDUCTS

[Scheme: formamide + POCl₃ → adenine-type bicyclic; R–CH₂–CONH₂ + HCONH₂/POCl₃ → 5-substituted 4-aminopyrimidine]

$R = H, C_nH_{2n+1}: n = 1\text{–}4, 6, 8, 14, 16$

[Scheme: diamide (CH₂)₄ with two CONH₂ groups + POCl₃ → cyclopentapyrimidine; lactam (CH₂)ₙ + HCONH₂/POCl₃ → fused pyrimidine]

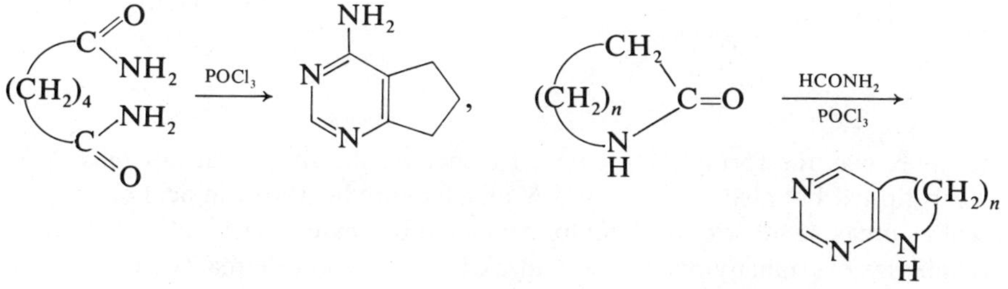

N-Monosubstituted formamides condense with *N,N'*-disubstituted ureas in the presence of POCl₃ to yield amidinium salts like **45** (82).

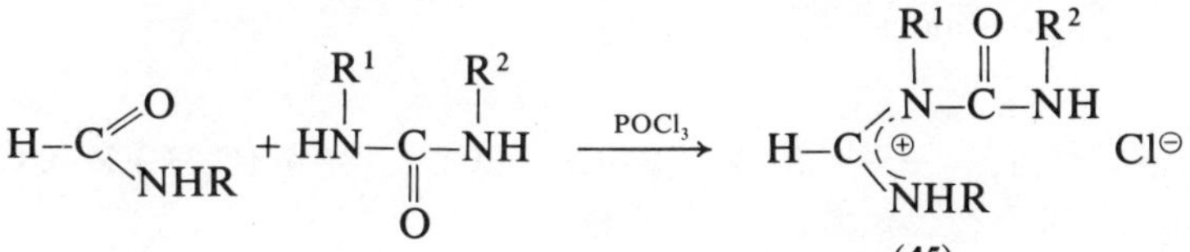

$$H-C(=O)NHR + HN(R^1)-C(=O)-NH(R^2) \xrightarrow{POCl_3} H-C^{\oplus}(NHR)-N(R^1)-C(=O)-NH(R^2) \quad Cl^{\ominus}$$

(**45**)

N-Arylsulfonyl formamides are converted into *N,N'*-bis-(arylsulfonyl)formamidines (**46**) using POCl₃/pyridine (66). Similarly, the reaction of POCl₃ with formanilide yields *N,N'*-diphenylformamidine (83):

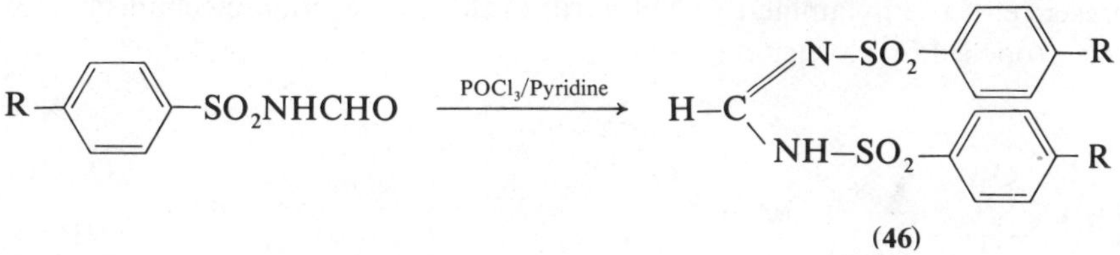

$$R\text{–}C_6H_4\text{–}SO_2NHCHO \xrightarrow{POCl_3/\text{Pyridine}} H-C(=N-SO_2-C_6H_4-R)(NH-SO_2-C_6H_4-R)$$

(**46**)

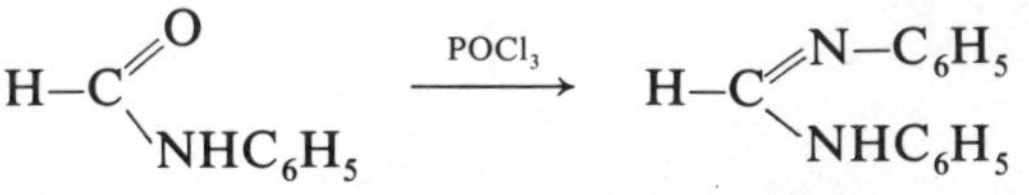

$$H-C(=O)NHC_6H_5 \xrightarrow{POCl_3} H-C(=N-C_6H_5)(NHC_6H_5)$$

II. Acid Amide–Triphenylphosphine Dihalides Adducts

Horner et al. (84), as well as Bestmann et al. (85), reported reactions with acide amide–triphenylphosphine dibromide adducts as intermediates. According to these authors DMF and triphenylphosphine dibromide allow formylations analogous to the Vilsmeier–Haack reaction; they concluded that a DMF–triphenylphosphine dibromide adduct **47** like the Vilsmeier–Haack

$$\text{H–C}\underset{\text{N(CH}_3)_2}{\overset{\text{O}}{\diagup\!\!\!\diagdown}} + (C_6H_5)_3\overset{\oplus}{P}\text{–Br} \quad Br^{\ominus} \longrightarrow \text{H–C}\underset{\overset{\oplus}{N}(CH_3)_2}{\overset{\text{O–P(C}_6H_5)_3}{\diagup\!\!\!\diagdown}} \quad Br^{\ominus}$$

(47)

complex was the formylating agent. Further results show that not only *N,N*-disubstituted but also *N*-mono and *N*-unsubstituted carboxylic acid amides, as well as ureas, react with triphenylphosphine dibromide to yield adducts like **47**. A mixture of triphenylphosphine and carbon tetrachloride reacts as triphenylphosphine dihalides do (86–92). The two components initially react to give the adduct **48**, which rearranges into the more stable (trichloromethyl)triphenylphosphonium chloride (**49**). **48** and **49** are capable of adduct formation with carboxylic acid amides.

$$(C_6H_5)_3P + CCl_4 \longrightarrow (C_6H_5)_3\overset{\oplus}{P}\text{–Cl} \quad CCl_3^{\ominus} \longrightarrow (C_6H_5)_3\overset{\oplus}{P}\text{–CCl}_3 \quad Cl^{\ominus}$$
(48)   (49)

A. REACTIONS OF ACID AMIDE–TRIPHENYLPHOSPHINE DIHALIDE ADDUCTS

Primary amides are dehydrated by triphenylphosphine dihalides in the presence of triethylamine to yield nitriles (**50**); the reaction mechanism shown was proposed by Horner et al. (84).

$$R^1\text{–C}\underset{NH_2}{\overset{O}{\diagup\!\!\!\diagdown}} \xrightarrow{R_3PX_2} R^1\text{–C}\underset{\overset{\oplus}{NH_2}}{\overset{O–PR_3}{\diagup\!\!\!\diagdown}} \; X^{\ominus} \xrightarrow[-H\overset{\oplus}{N}(C_2H_5)_3X^{\ominus}]{+N(C_2H_5)_3} R^1\text{–C}\underset{NH}{\overset{O–PR_3}{\diagup\!\!\!\diagdown}}$$

$$\downarrow \begin{array}{l}+N(C_2H_5)_3 \\ -H\overset{\oplus}{N}(C_2H_5)_3X^{\ominus} \\ -R_3PO\end{array}$$

$$R^1\text{–}C\equiv N$$
(50)

Benzhydrazides and thiobenzhydrazides (**51**) are converted into diaryltetrazines (**52**), when heated with phenyltetrachlorophosphorane (93).

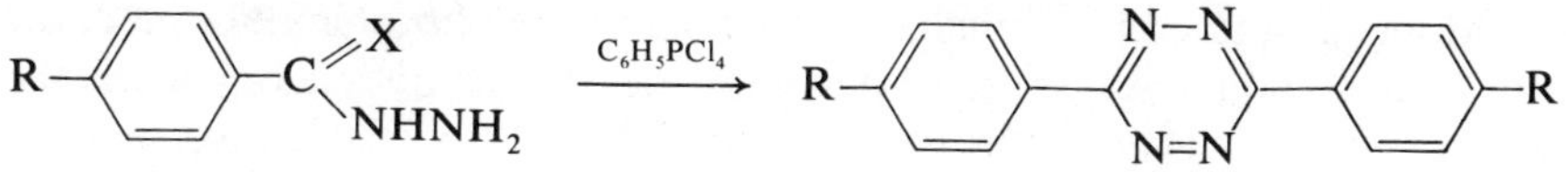

II. ACID AMIDE–TRIPHENYLPHOSPHINE DIHALIDE ADDUCTS

$$R-C_6H_4-C(=X)-NHNH_2 \xrightarrow{C_6H_5PCl_4} R-C_6H_4-\text{[1,2,4,5-tetrazine]}-C_6H_4-R$$

X = O, S; R = H, CH$_3$, Cl

(51) → (52)

Similarly, N,N-disubstituted cyanamides (**53**) are synthesized from N,N-disubstituted ureas (85).

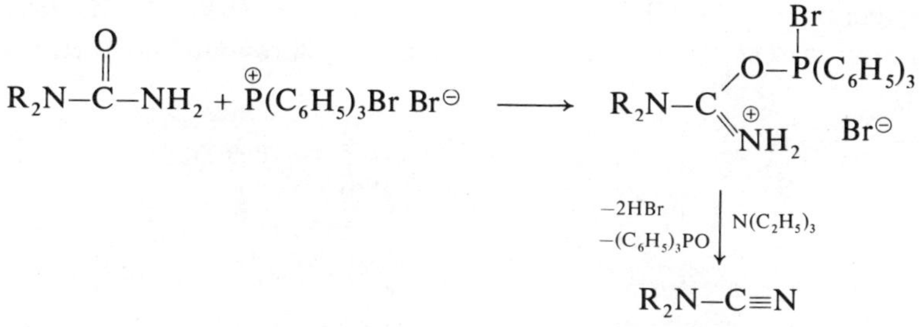

$$R_2N-\overset{O}{\underset{\|}{C}}-NH_2 + \overset{\oplus}{P}(C_6H_5)_3Br\ Br^{\ominus} \longrightarrow R_2N-\underset{\overset{\oplus}{N}H_2}{\overset{O-P(C_6H_5)_3}{\underset{|}{C}}} Br^{\ominus}$$

$$\xrightarrow[-(C_6H_5)_3PO]{-2HBr\quad N(C_2H_5)_3} R_2N-C\equiv N$$

(53)

N,N'-Disubstituted ureas are transformed into carbodiimides (**54**) using triphenylphosphine dibromide and triethylamine (85).

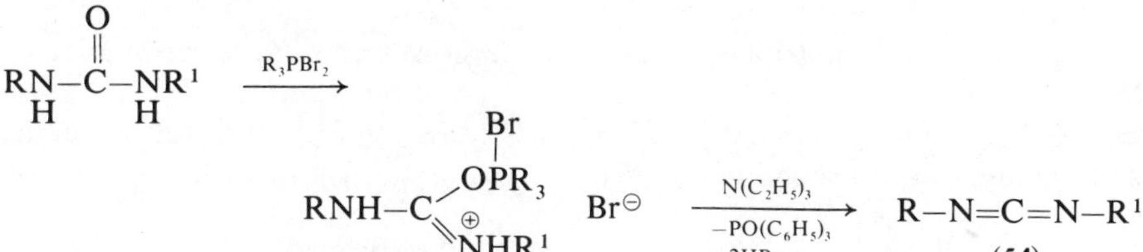

$$RN-\overset{O}{\underset{\|}{C}}-NR^1 \xrightarrow{R_3PBr_2}$$
H H

$$RNH-\underset{\overset{\oplus}{N}HR^1}{\overset{OPR_3}{\underset{|}{C}}}\ Br^{\ominus} \xrightarrow[-PO(C_6H_5)_3]{N(C_2H_5)_3} R-N=C=N-R^1$$
$$-2HBr \qquad (54)$$

Chloroformamidines (**55**) are obtained by treatment of N,N,N'-trisubstituted ureas or thioureas with triphenyl- or trimorpholinophosphine/CCl$_4$ or triphenylphosphine dichloride/triethylamine (90, 91).

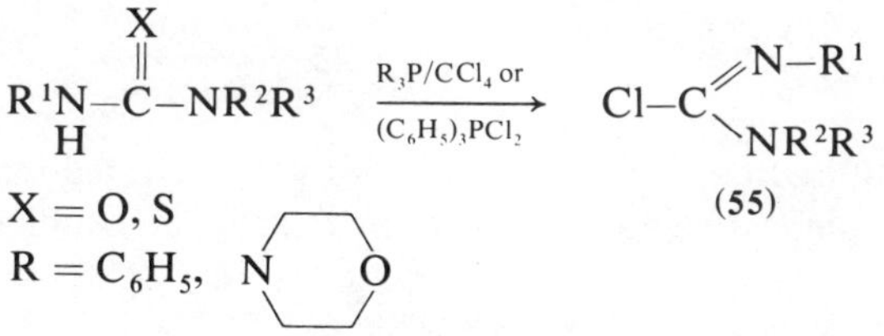

$$R^1N-\overset{X}{\underset{\|}{C}}-NR^2R^3 \xrightarrow[(C_6H_5)_3PCl_2]{R_3P/CCl_4\ or} Cl-C\underset{NR^2R^3}{\overset{N-R^1}{\diagdown}}$$
H

X = O, S
R = C$_6$H$_5$, N(morpholino)

(55)

Isocyanates are accessible from N-monosubstituted carbamoyl halides and triphenylphosphine/CCl$_4$ (92):

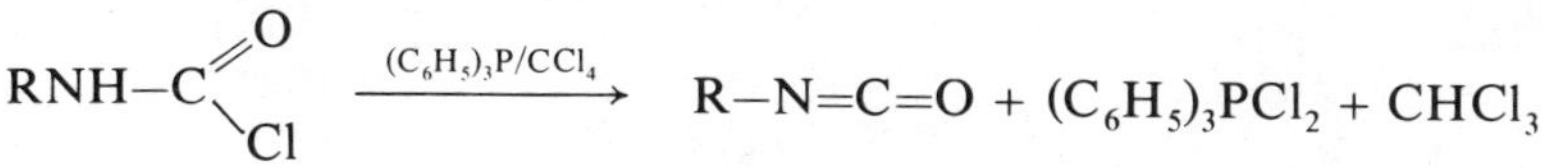

$$RNH-C\overset{O}{\underset{Cl}{\diagdown}} \xrightarrow{(C_6H_5)_3P/CCl_4} R-N=C=O + (C_6H_5)_3PCl_2 + CHCl_3$$

N-Monosubstituted formamides react with triphenylphosphine dibromide and triethylamine to give isonitriles (**56**) in good yields (85). By analogy with other isonitrile syntheses (59) adduct formation in the amide oxygen is assumed.

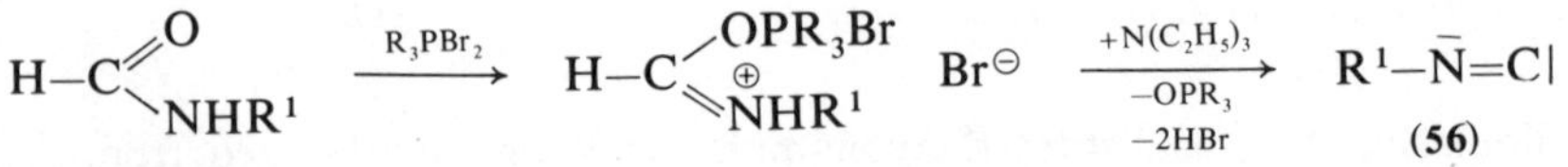

$$H-C\overset{O}{\underset{NHR^1}{\diagdown}} \xrightarrow{R_3PBr_2} H-C\overset{OPR_3Br}{\underset{NHR^1}{\diagdown}}{}^{\oplus}\ Br^{\ominus}\ \xrightarrow[-OPR_3\atop -2HBr]{+N(C_2H_5)_3} R^1-\bar{N}{=}C|\quad (\mathbf{56})$$

Ketenimines (**57**) are obtained via dehydration of *N*-monosubstituted acid amides containing α CH bonds adjacent to the carbonyl function, the dehydration was effected by triphenylphosphine dibromide and triethylamine (85, 94).

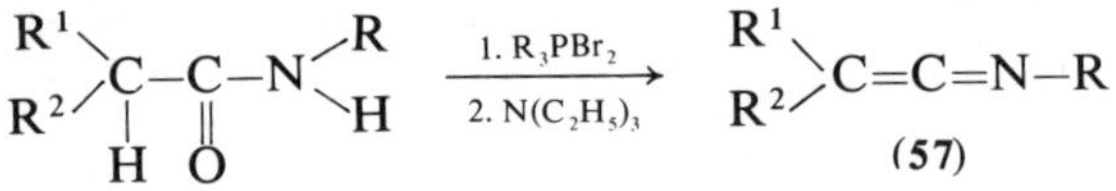

$$\underset{R^2}{\overset{R^1}{\diagdown}}\underset{\underset{H}{|}}{C}-\underset{\underset{O}{\|}}{C}-N\overset{R}{\underset{H}{\diagdown}} \xrightarrow[2.\ N(C_2H_5)_3]{1.\ R_3PBr_2} \underset{R^2}{\overset{R^1}{\diagdown}}C{=}C{=}N{-}R \quad (\mathbf{57})$$

Alkyl bromides are formed by alcoholysis of triphenylphosphine dibromide in DMF (95). Cycloaliphatic alcohols and phenols also undergo this reaction (95):

$$ROH \xrightarrow{DMF/(C_6H_5)_3PBr_2}$$

When treated with triphenylphosphine/carbon tetrachloride, primary carboxylic acid amides and thioamides are converted into nitriles (86), *N*-monosubstituted formamides into isonitriles (88), and *N*,*N*-disubstituted ureas and thioureas into carbodiimides (87):

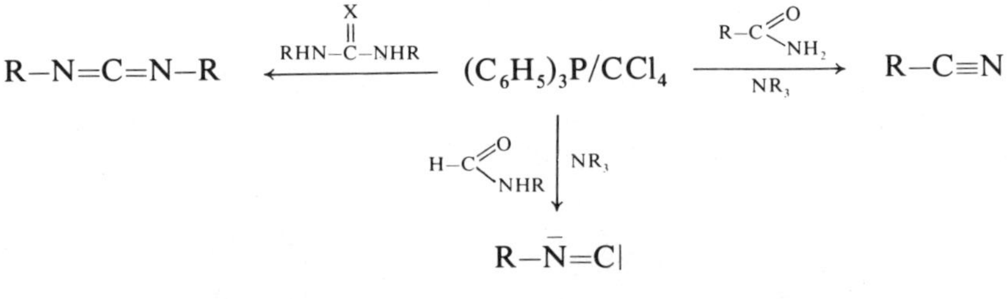

$$R-N{=}C{=}N-R \xleftarrow{RHN-\overset{\overset{X}{\|}}{C}-NHR} (C_6H_5)_3P/CCl_4 \xrightarrow{R-C\overset{O}{\underset{NH_2}{\diagdown}} \atop NR_3} R-C{\equiv}N$$

$$\Bigg\downarrow{H-C\overset{O}{\underset{NHR}{\diagdown}}}\ NR_3$$

$$R-\bar{N}{=}C|$$

N-Monosubstituted carboxylic acid amides are converted into imide halides by triphenylphosphine dihalides as well as by triphenylphosphine/CCl$_4$ or CBr$_4$ (89):

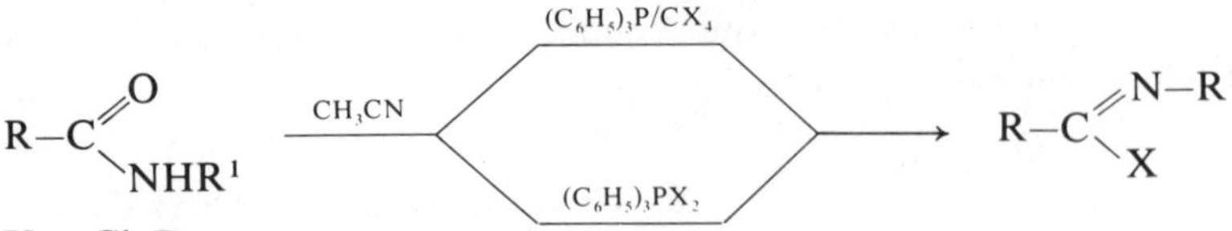

$$R-C\overset{O}{\underset{NHR^1}{\diagdown}} \xrightarrow[\underset{NR_3}{(C_6H_5)_3PX_2}]{\overset{(C_6H_5)_3P/CX_4}{CH_3CN}} R-C\overset{N-R^1}{\underset{X}{\diagdown}}$$

X = Cl, Br

III. Acid Amide–Chlorophosphoric Acid Ester Adducts

It was shown by conductivity experiments that DMF and phenyl dichlorophosphate form adducts in a 1 : 1 ratio. The adduct formation from N-methylpyrrolidone and dichlorophosphoric acid ester was demonstrated by the same method (96).

Cramer and Winter (97) studied the reactivity of the adducts from acid amides and chlorophosphoric acid diaryl esters and dichlorophosphoric acid aryl esters. They proposed structures **58** and **59**, respectively, for the

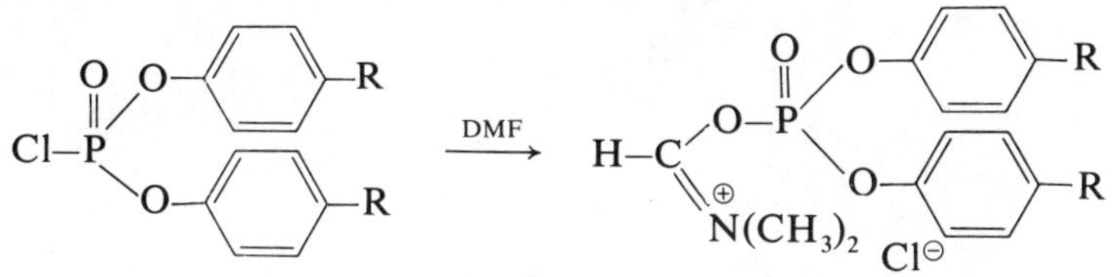

(58)

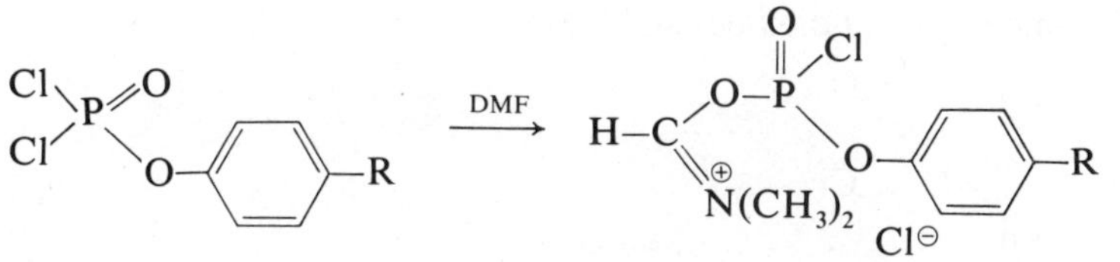

(59)

dimethylformamide adducts. The adduct of formamide and chlorophosphoric acid ester is unstable; it decomposes into hydrocyanic acid, hydrogen chloride, and partially esterified phosphoric acid (97):

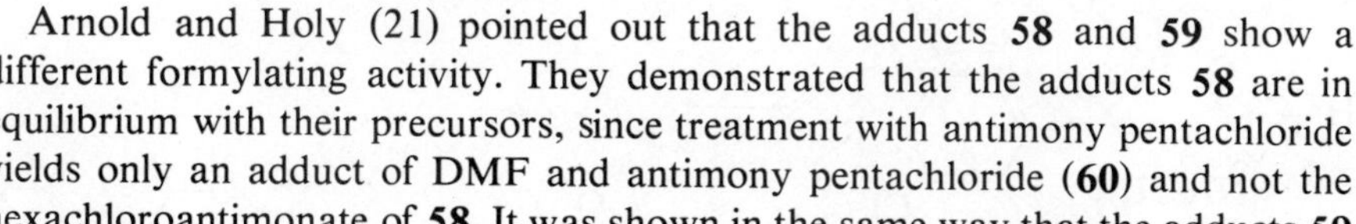

$$\longrightarrow HCN + HCl + O=P-O-\text{...}$$

Arnold and Holy (21) pointed out that the adducts **58** and **59** show a different formylating activity. They demonstrated that the adducts **58** are in equilibrium with their precursors, since treatment with antimony pentachloride yields only an adduct of DMF and antimony pentachloride (**60**) and not the hexachloroantimonate of **58**. It was shown in the same way that the adducts **59** are in equilibrium with dimethyl chloromethyleniminium arylchlorophosphates (**61**), since the adducts react with SbCl$_5$ to yield N,N-dimethylchloromethyleniminium hexachloroantimonate (**62**).

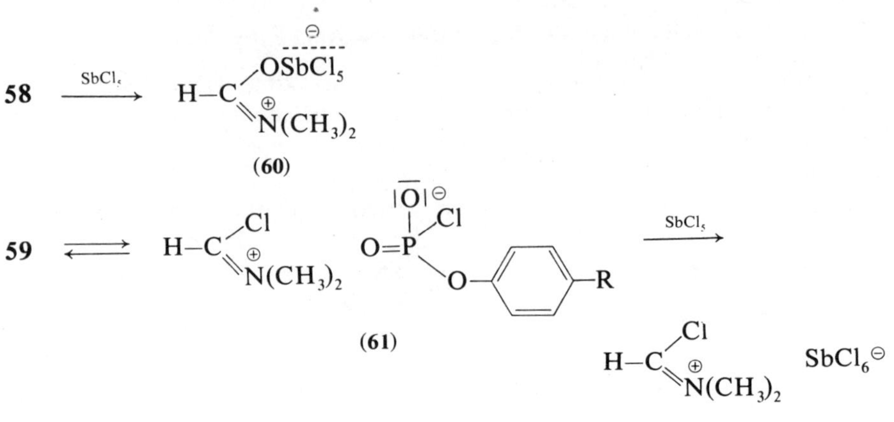

58 $\xrightarrow{SbCl_5}$ H—C$\begin{smallmatrix}OSbCl_5^\ominus\\\oplus\\N(CH_3)_2\end{smallmatrix}$

(60)

59 $\rightleftarrows$ H—C$\begin{smallmatrix}Cl\\\oplus\\N(CH_3)_2\end{smallmatrix}$ · O=P(O⁻)(Cl)—O—C₆H₄—R $\xrightarrow{SbCl_5}$

(61)

H—C$\begin{smallmatrix}Cl\\\oplus\\N(CH_3)_2\end{smallmatrix}$ SbCl$_6^\ominus$

(62)

Raetz and Sweeting (98) described the adduct **63** from the spiro chlorophosphoric acid ester **64** (3,9-dichloro-2,4,8-10-tetraoxa-3,9-diphosphaspiro[5,5]undecane 3,9-dioxide) and DMF.

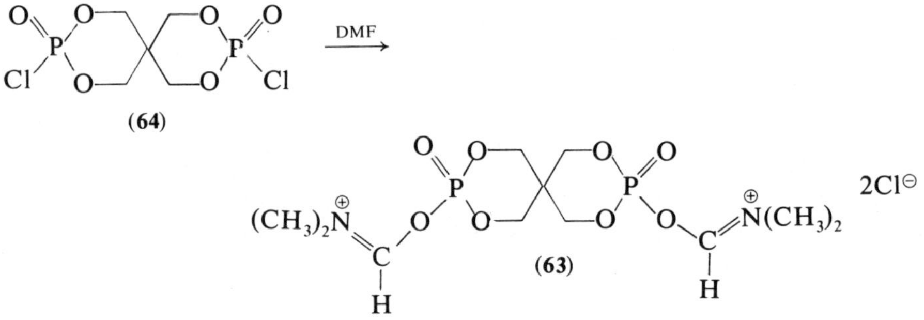

(64) $\xrightarrow{DMF}$

(63) 2Cl$^\ominus$

A. REACTIONS OF ACID AMIDE–CHLOROPHOSPHORIC ACID ESTER ADDUCTS

The adducts **58** and **59** are extremely efficient phosphorylating agents. Hydrolysis of **58** yields DMF and diarylphosphates (97); the hydrolysis occurs

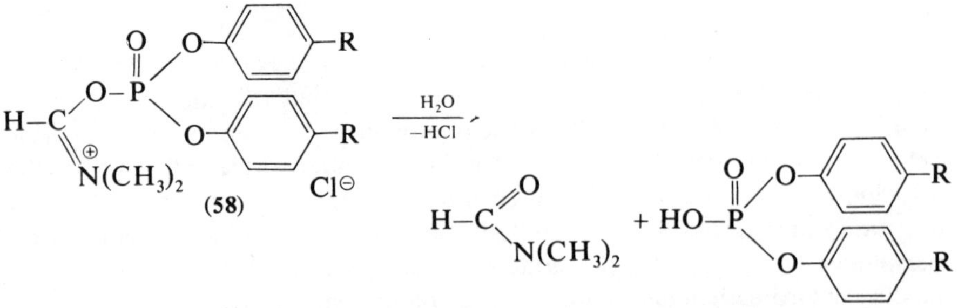

(58) $\xrightarrow[-HCl]{H_2O}$ H—C(=O)N(CH₃)₂ + HO—P(=O)(O—C₆H₄—R)₂

III. ACID AMIDES–CHLOROPHOSPHORIC ACID ESTER ADDUCTS

much more rapidly than that of chlorophosphoric acid esters, the starting material for the adduct (97). The adducts **58** and **59** are transformed into phosphoric acid esters **65** and **66**, respectively, by phenols or alcohols (96).

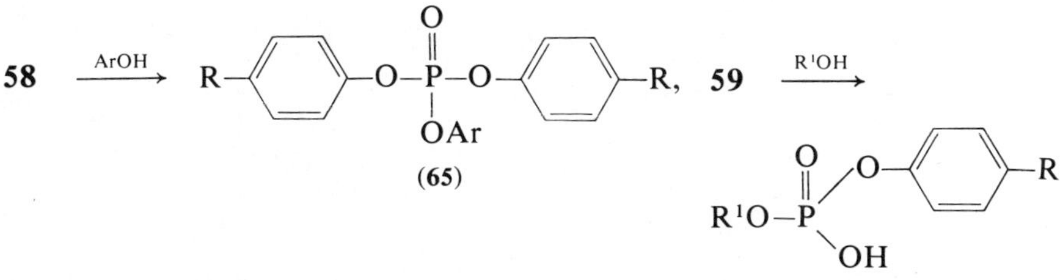

$$58 \xrightarrow{ArOH} \text{R–C}_6\text{H}_4\text{–O–P(=O)(OAr)–O–C}_6\text{H}_4\text{–R, }59 \xrightarrow{R'OH}$$

(65)

$$R'O–P(=O)(OH)–O–C_6H_4–R$$

(66)

The adduct **63** reacts with butanediol-(1,4) to give the spiro phosphate **67** (98).

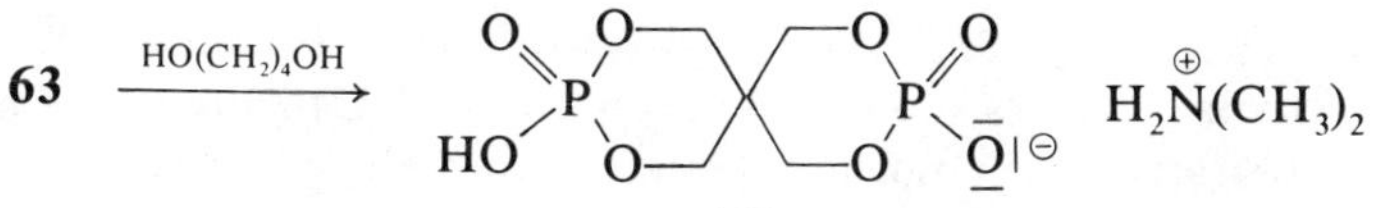

$$63 \xrightarrow{HO(CH_2)_4OH} (67) \quad H_2\overset{\oplus}{N}(CH_3)_2$$

Mixed anhydrides from phosphoric acids and carboxylic acids (**68**) can be obtained from the adducts **58** and carboxylic acids (**69**) in the presence of triethylamine (97). In place of carboxylic acids partially esterified phosphoric

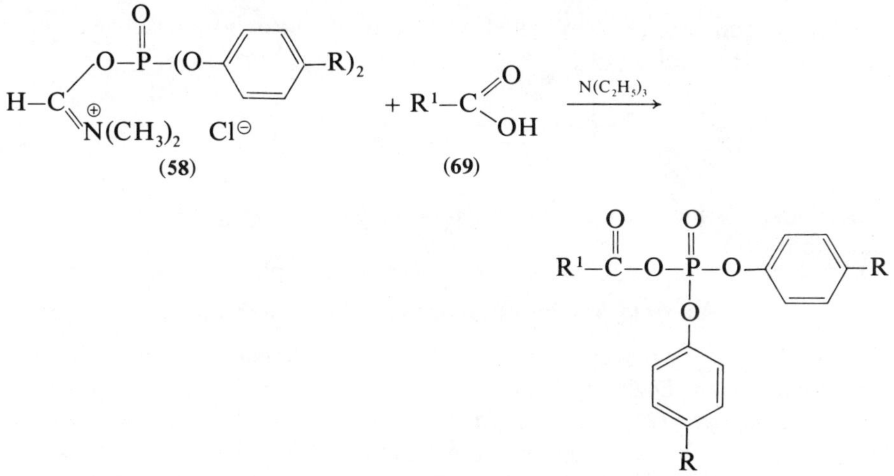

$$H–C\overset{O–P(=O)(O–C_6H_4–R)_2}{\underset{\overset{\oplus}{N}(CH_3)_2 \; Cl^{\ominus}}{}} + R^1–C\overset{O}{\underset{OH}{}} \xrightarrow{N(C_2H_5)_3}$$

(58) (69)

$$R^1–C(=O)–O–P(=O)(O–C_6H_4–R)–O–C_6H_4–R$$

(68)

acids may also be used; the diesters of the pyrophosphoric acid **70** are thus obtained (97). To synthesize mixed diesters of pyrophosphoric acid in pure state the partially esterified phosphoric acid has to be used as starting material

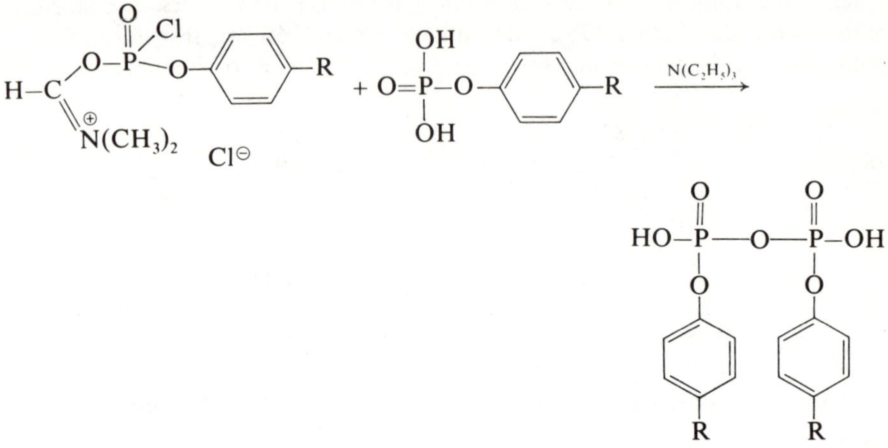

30 ADDUCTS FROM ACID AMIDES AND ACYLATION REAGENTS

(97). In acetonitrile esters of chlorophosphoric acid, dimethylformamide, and sodium perchlorate yield symmetrical esters of pyrophosphoric acid (**71**) (21).

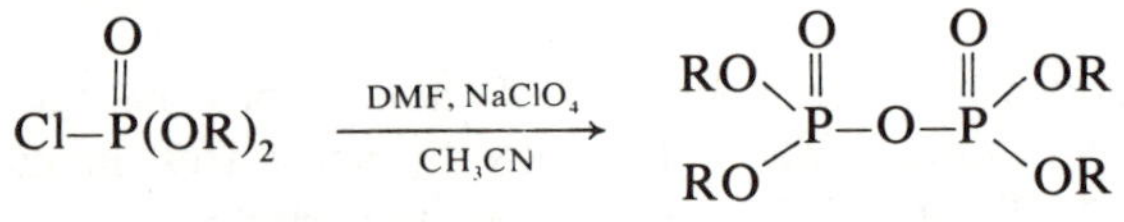

Primary amides are dehydrated with diethylchlorophosphoric acid and triethylamine to nitriles (99):

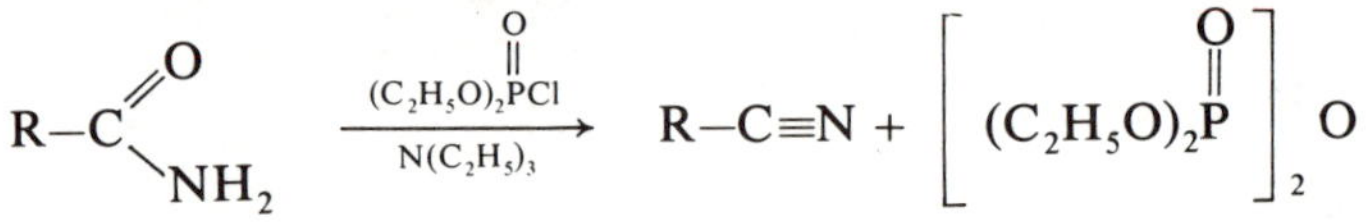

Secondary formamides are converted into isonitriles using the same reaction conditions (99).

IV. Acid Amide–Phosphorus Trihalide Adducts

Amides have long been suspected of forming adducts with phosphorustrichloride. From DMF and PCl_3, Smith (25) was able to synthesize a crystalline adduct (**72**) consisting of two molecules of DMF and one molecule of PCl_3; it may be recrystallized from CCl_4, and is soluble in chloroform, ether, and benzene. The IR spectrum shows a band at 1660 cm^{-1}; the NMR spectrum proves that the two DMF molecules are equally bound (τ_{CH} = 1.9, $\tau_{N(CH_3)_2}$ = 7.00 and 7.09 in $CHCl_3$).

Smith chose **72** from the various possible structures. However, the solubility and the low conductivity of chloroform solutions of the adduct do not support

IV. ACID AMIDE–PHOSPHORUS TRIHALIDE ADDUCTS

this conclusion. Further studies will be required to clarify whether or not PCl_3 is bound to DMF as other Lewis acids are, such as BCl_3, BF_3, and $SbCl_5$, for example, as in **73**.

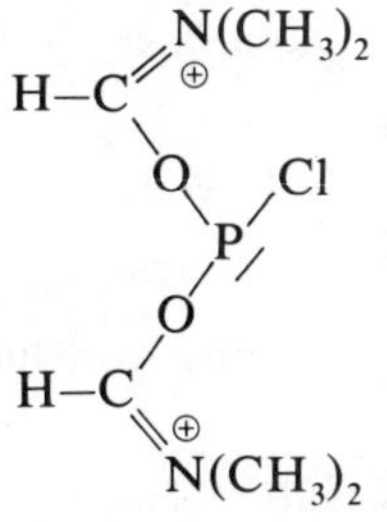

(72) 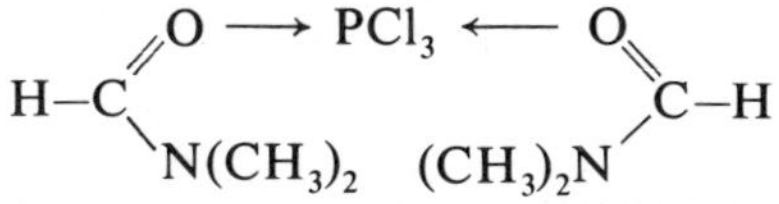 (73)

A large number of amidines were synthesized from carboxylic acid amides and primary and secondary amines in the presence of phosphorus trichloride (27, 28, 35, 36, 77, 78, 80, 81, 100–104). It is still uncertain whether this reaction proceeds via acid amide–PCl_3 adducts or via phosphonic acid trialkyl- or dialkylamides, which may be formed as intermediates from PCl_3 and amines. Formamidines have been successfully obtained from iminophosphoranes (**74**) and DMF (105).

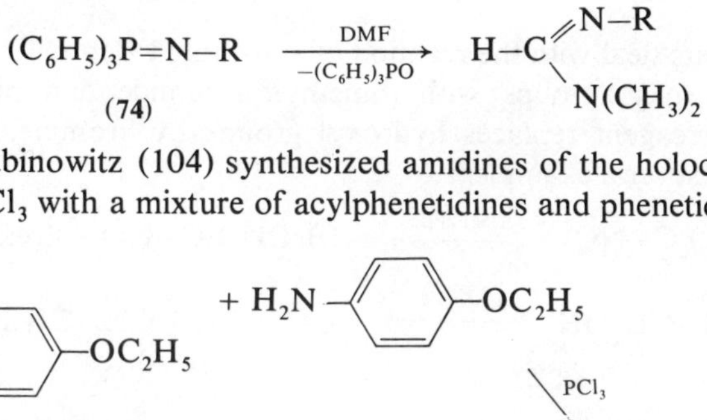

$$(C_6H_5)_3P=N-R \xrightarrow[-(C_6H_5)_3PO]{DMF} H-C\begin{smallmatrix}N-R\\N(CH_3)_2\end{smallmatrix}$$

(74)

Hill and Rabinowitz (104) synthesized amidines of the holocaine type (**75**) by reacting PCl_3 with a mixture of acylphenetidines and phenetidine; the yields

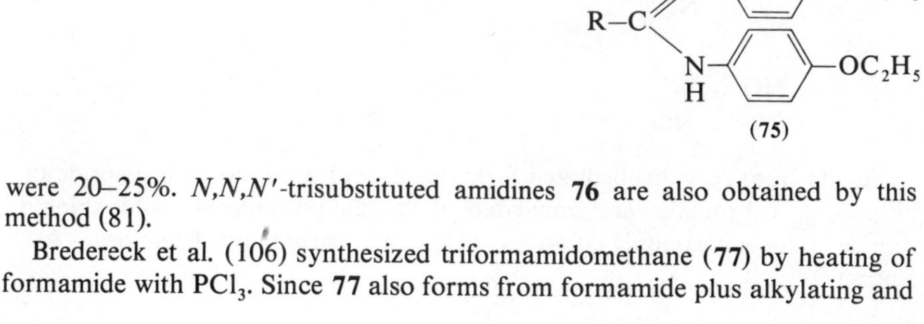

$R = C_2H_5, n\text{-}C_3H_7, i\text{-}C_4H_9$

(75)

were 20–25%. N,N,N'-trisubstituted amidines **76** are also obtained by this method (81).

Bredereck et al. (106) synthesized triformamidomethane (**77**) by heating of formamide with PCl_3. Since **77** also forms from formamide plus alkylating and

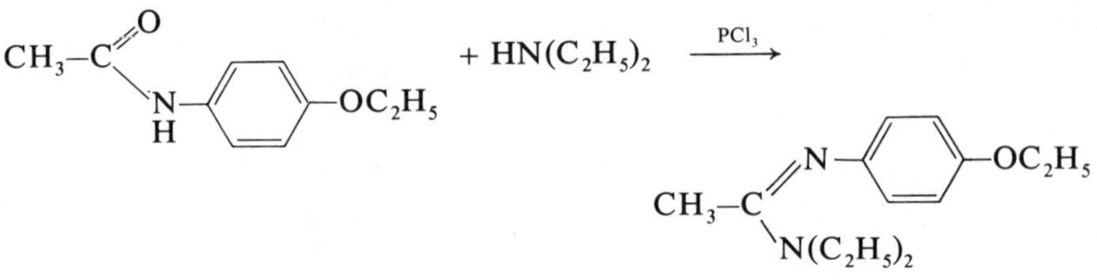

$$CH_3-C\underset{\underset{H}{N}}{\overset{O}{\diagup}}-\!\!\!\!\!\!\!\!\!\!\!\!\!\!\!\!\bigcirc\!\!\!\!\!-OC_2H_5 \quad + HN(C_2H_5)_2 \quad \xrightarrow{PCl_3}$$

$$CH_3-C\underset{N(C_2H_5)_2}{\overset{N-\!\!\!\!\bigcirc\!\!\!\!-OC_2H_5}{\diagup}}$$

(76)

acylating agents—attacking at the amide oxygen—one may conclude that PCl_3 initially forms an adduct at the formamide oxygen.

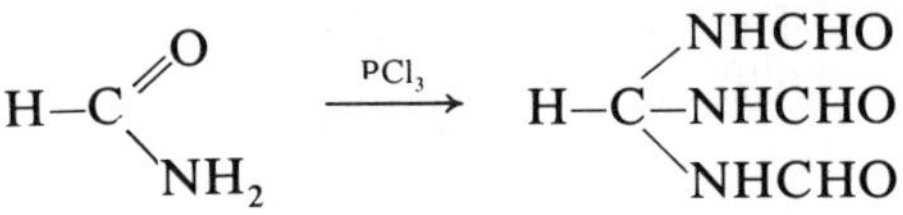

$$H-C\underset{NH_2}{\overset{O}{\diagup}} \xrightarrow{PCl_3} H-C\underset{NHCHO}{\overset{NHCHO}{-}}NHCHO$$

(77)

As with $POCl_3$, N-monosubstituted formamides are dehydrated by PCl_3, PBr_3, and BCl_3 in the presence of tertiary bases to yield isonitriles **78** (48, 49).

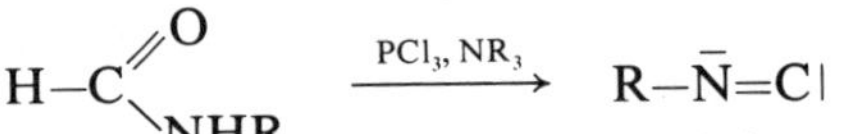

$$H-C\underset{NHR}{\overset{O}{\diagup}} \xrightarrow{PCl_3, NR_3} R-\bar{N}=C|$$

(78)

Several patents deal with the reaction of substituted alcohols or heterocycles containing hydroxyl groups with dimethylformamide and phosphorus tribromide. This reagent replaces hydroxyl groups by bromine. The following summary gives several examples:

$$(HOCH_2)_3C-NO_2 \xrightarrow{DMF/PBr_3} (BrCH_2)_3C-NO_2 \quad \text{(ref. 107)}$$

$$CH_3-CH_2-\underset{NO_2}{\overset{|}{CH}}-CH_2OH \xrightarrow{DMF/PBr_3} CH_3-CH_2-\underset{NO_2}{\overset{|}{CH}}-CH_2Br \quad \text{(ref. 108)}$$

[pyridazinone with C_6H_5, HO, NO_2 substituents] $\xrightarrow{DMF/PBr_3}$ [pyridazinone with C_6H_5, Br, NO_2 substituents] (ref. 109)

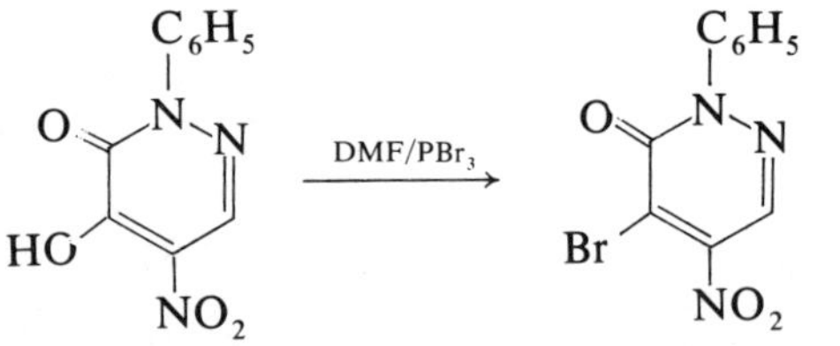

Adducts from N,N-disubstituted formamides and phosphorus trichloride or phosphorus tribromide are converted with thionyl chloride and thionyl bromide into carbamoyl chlorides (**79**) and carbamoyl bromides (**80**), respectively (110, 111).

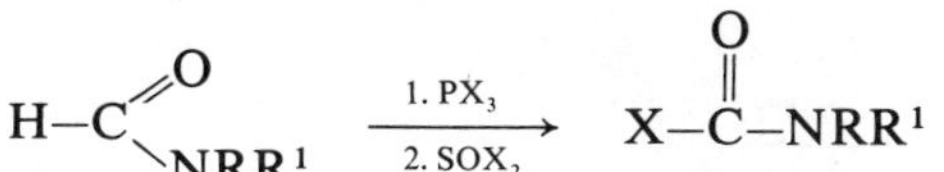

$$H-C\overset{O}{\underset{NRR^1}{\diagdown}} \xrightarrow[\text{2. SOX}_2]{\text{1. PX}_3} X-\overset{O}{\underset{}{\overset{\|}{C}}}-NRR^1$$

(79) X = Cl
(80) X = Br

V. Acid Amide–Phosphorus Nitrile Chloride Adducts

Phosphorus nitrile chloride (PNCl$_2$)$_3$ gives an adduct (81) with DMF corresponding to that from cyanuric chloride and DMF (112); it is useful for the formylation of activated aromatic compounds as well as for the synthesis of amidines (113–115); 81 reacts with potassium iodide to give the aza vinylogue of tetramethylformamidinium iodide (82) (116).

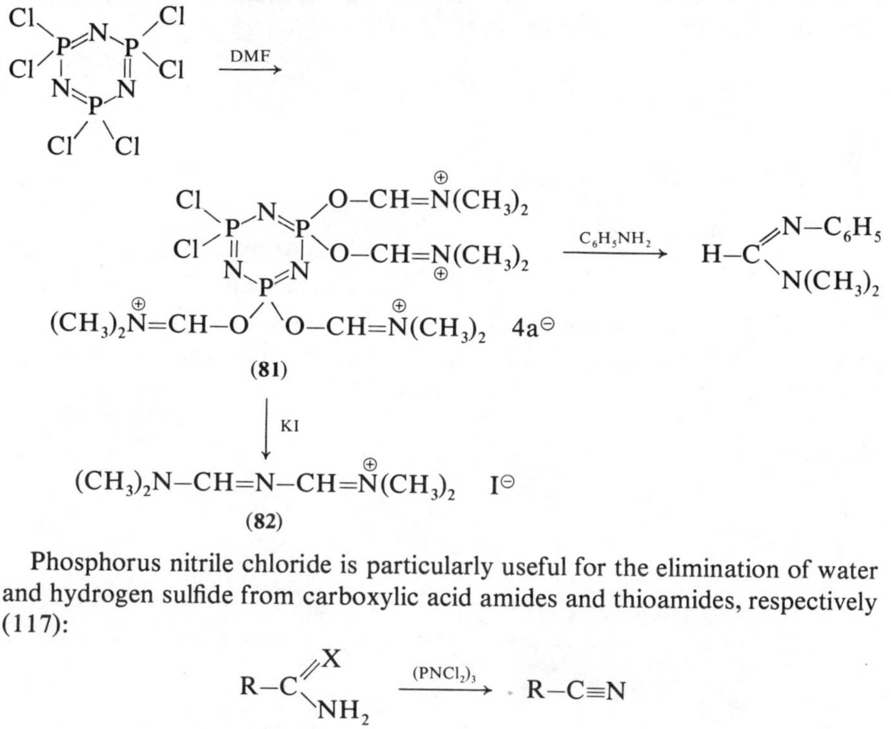

[Structure of (PNCl$_2$)$_3$ with Cl substituents] $\xrightarrow{\text{DMF}}$

[Structure of adduct 81 with O–CH=N(CH$_3$)$_2$ groups, 4a$^\ominus$] $\xrightarrow{C_6H_5NH_2}$ $H-C\overset{N-C_6H_5}{\underset{N(CH_3)_2}{\diagdown}}$

(81)

$\downarrow$ KI

$(CH_3)_2N-CH=N-CH=\overset{\oplus}{N}(CH_3)_2 \quad I^\ominus$
(82)

Phosphorus nitrile chloride is particularly useful for the elimination of water and hydrogen sulfide from carboxylic acid amides and thioamides, respectively (117):

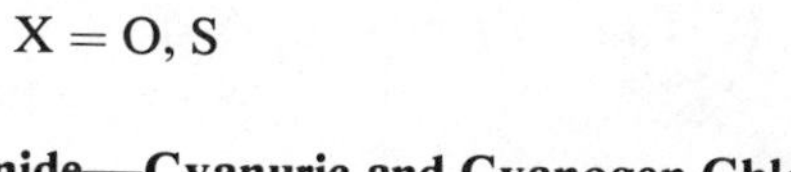

$$R-C\overset{X}{\underset{NH_2}{\diagdown}} \xrightarrow{(PNCl_2)_3} R-C\equiv N$$

X = O, S

VI. Acid Amide—Cyanuric and Cyanogen Chloride Adducts

In a weakly exothermic reaction with 2 moles of DMF, cyanuric chloride forms an adduct (83) of the Vilsmeier–Haack type (118). It is formed via nucleophilic substitution of the chlorine in the triazine ring by the carbonyl

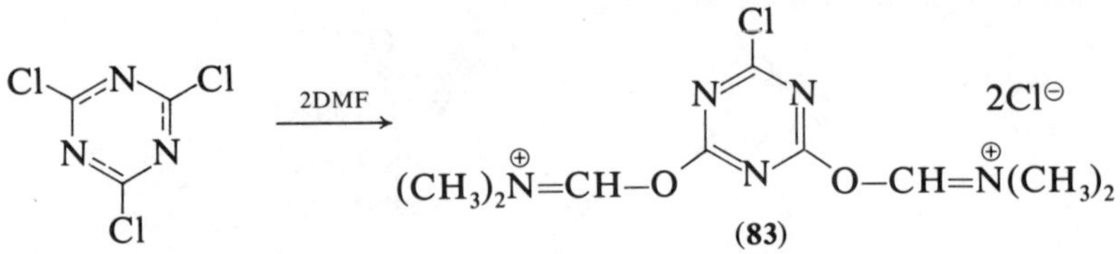

Cyanuric chloride structure + 2DMF → (83) with $(CH_3)_2\overset{\oplus}{N}=CH-O$ and $O-CH=\overset{\oplus}{N}(CH_3)_2$ groups on triazine, $2Cl^{\ominus}$

oxygen of the acid amide. The structure of the adduct was established by hydrolysis, alcoholysis, and aminolysis. The finding that the adduct **83** formylates aromatic compounds and CH-acidic compounds further supports this assumption (119–121).

An adduct from cyanuric chloride and DMF in a molar ratio of 1:3 (**84**) is also known; it is unstable, and it decomposes in the presence of dimethylformamide with elimination of CO_2 to yield (3-dimethylamino-2-azaprop-2-en-1-ylidene)dimethylammonium chloride (**85**) (118).

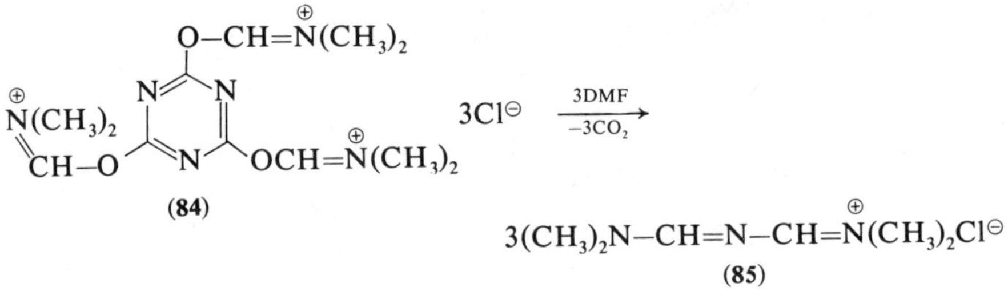

Structure (84): triazine with three $O-CH=\overset{\oplus}{N}(CH_3)_2$ substituents, $3Cl^{\ominus}$

$\xrightarrow[-3CO_2]{3DMF}$

$3(CH_3)_2\overset{}{N}-CH=N-CH=\overset{\oplus}{N}(CH_3)_2 Cl^{\ominus}$
(**85**)

With the exception of formylpyrrolidine, 1:2 adducts of cyanuric chloride and other *N,N*-disubstituted formamides cannot be isolated, but rather decompose into the aza salts analogous to **85** (118).

Cyanuric bromide and cyanuric fluoride do not react with *N,N*-disubstituted formamides in this way (118).

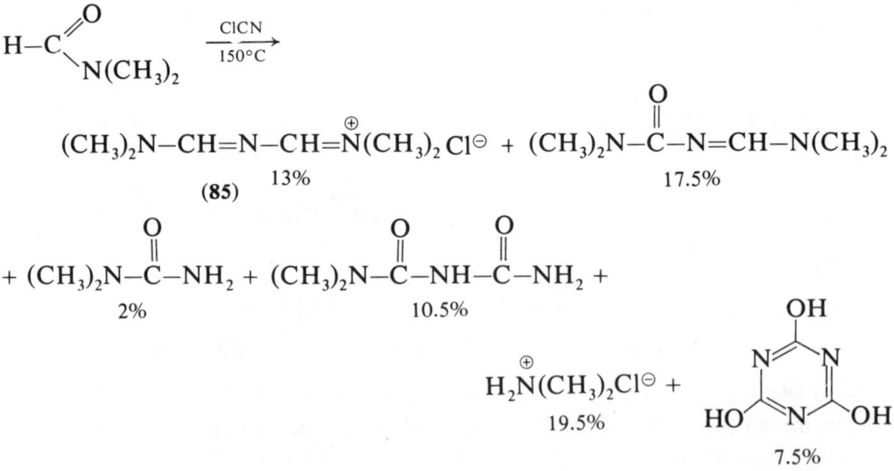

$$H-\overset{O}{\underset{N(CH_3)_2}{C}} \xrightarrow{ClCN,\ 150°C}$$

$(CH_3)_2N-CH=N-CH=\overset{\oplus}{N}(CH_3)_2\ Cl^{\ominus}$ + $(CH_3)_2N-\overset{O}{\underset{}{C}}-N=CH-N(CH_3)_2$
(**85**) 13%   17.5%

+ $(CH_3)_2N-\overset{O}{\underset{}{C}}-NH_2$ + $(CH_3)_2N-\overset{O}{\underset{}{C}}-NH-\overset{O}{\underset{}{C}}-NH_2$ +
   2%   10.5%

$H_2\overset{\oplus}{N}(CH_3)_2Cl^{\ominus}$ + (cyanuric acid trihydroxy triazine)
19.5%   7.5%

VI. ACID AMIDE—CYANURIC AND CYANOGEN CHLORIDE ADDUCTS

However, cyanogen chloride appears to form a labile adduct (122). Thus the formation of the iminium salt **85** was established among many other products (122) deriving from cyanogen chloride and DMF on heating.

A. REACTIONS OF ACID AMIDE–CYANURIC AND CYANOGEN CHLORIDE ADDUCTS

Water decomposes the adduct **83** into DMF, cyanuric acid, and hydrogen chloride (118); alcoholysis yields the same products as well as alkylhalides (118). As with amide chlorides and amide–$POCl_3$ adducts, the adduct **83** may be used to transform primary amines into amidinium salts like **86** (118).

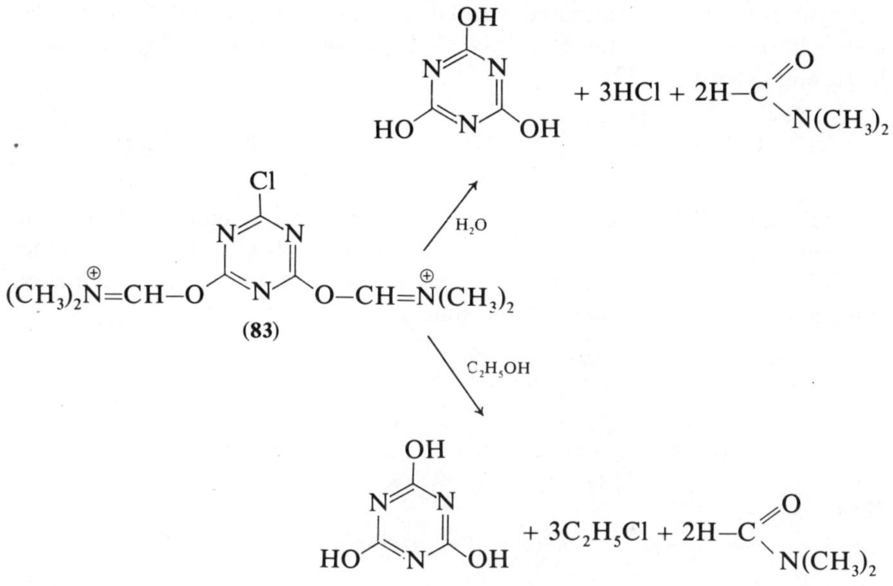

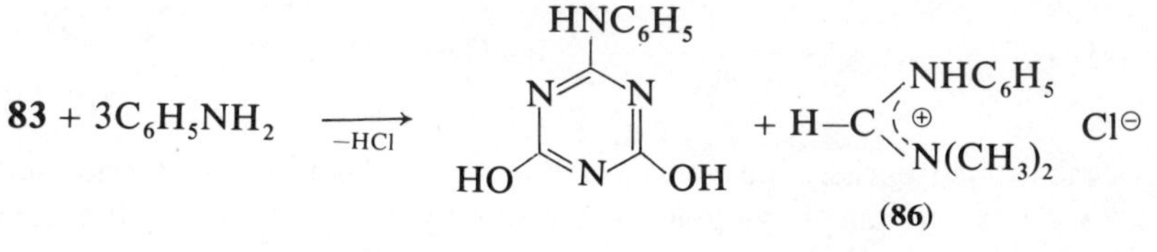

Secondary amines may also be used. In that case aqueous alkaline solutions are used for the workup, yielding not the amidinium salt **87**, but rather its hydrolysis product, diphenylformamide (122).

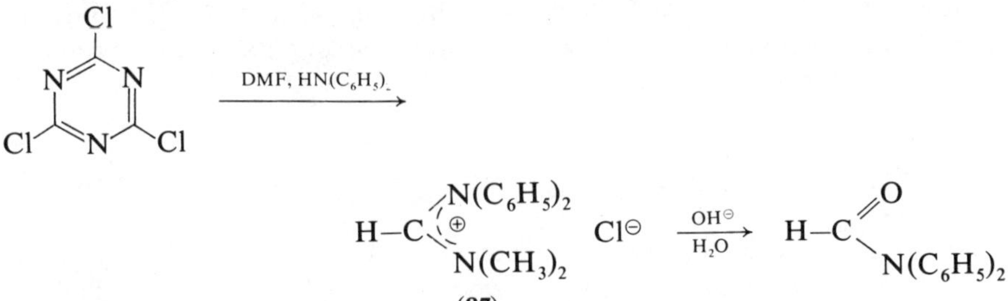

$$\text{(cyanuric chloride)} \xrightarrow{\text{DMF, HN(C}_6\text{H}_5)} \quad H-C\begin{smallmatrix}N(C_6H_5)_2 \\ \oplus \\ N(CH_3)_2\end{smallmatrix} \ Cl^{\ominus} \xrightarrow{\text{OH}^{\ominus}}_{\text{H}_2\text{O}} H-C\begin{smallmatrix}O \\ N(C_6H_5)_2\end{smallmatrix}$$

(87)

If cyanogen chloride is treated with DMF in the presence of diphenylamine, diphenylformamide is obtained as well as diphenylcyanamide on hydrolysis of the reaction mixture (122):

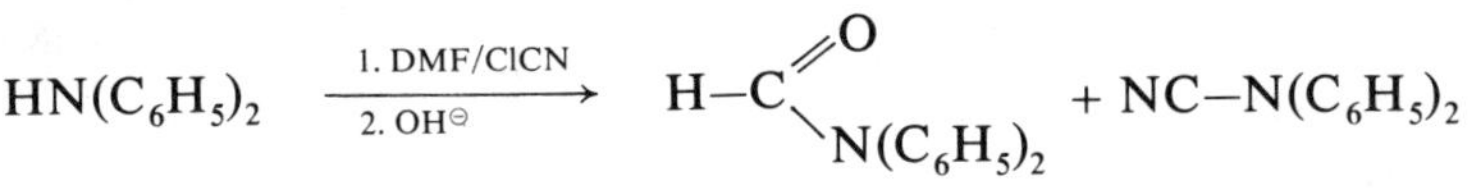

$$HN(C_6H_5)_2 \xrightarrow[\text{2. OH}^{\ominus}]{\text{1. DMF/ClCN}} H-C\begin{smallmatrix}O \\ N(C_6H_5)_2\end{smallmatrix} + NC-N(C_6H_5)_2$$

N-Monosubstituted formamides are dehydrated with cyanuric chloride and potassium carbonate in acetone to yield isonitriles (123). By analogy to other isonitrile synthesis this reaction clearly proceeds via primary adduct formation at the amide oxygen followed by α-elimination:

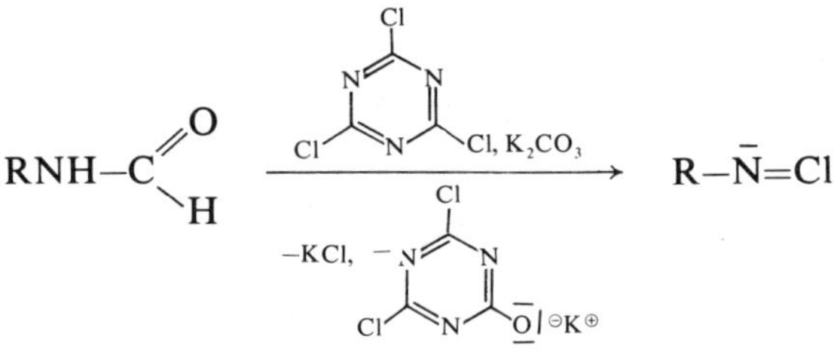

$$RNH-C\begin{smallmatrix}O \\ H\end{smallmatrix} \xrightarrow[-\text{KCl},\ -\text{(triazinol-K)}]{\text{cyanuric chloride, K}_2\text{CO}_3} R-\bar{N}=Cl$$

The conversion of *N,N'*-disubstituted thiourea with cyanuric chloride to carbodiimides is similar (124):

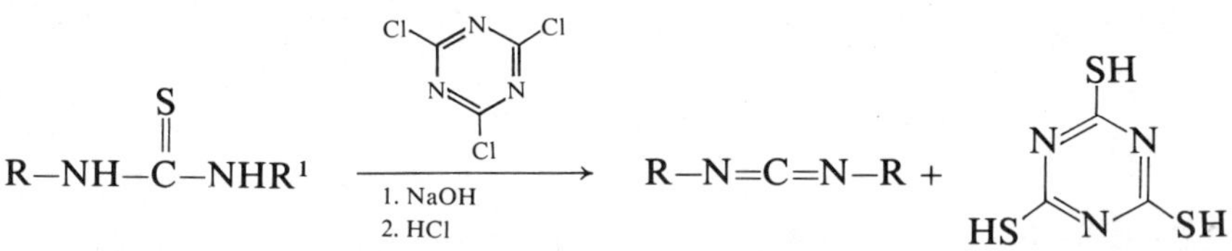

$$R-NH-\overset{S}{\underset{\|}{C}}-NHR^1 \xrightarrow[\text{2. HCl}]{\text{1. NaOH}} R-N=C=N-R + \text{(trimercapto-triazine)}$$

As mentioned earlier, the action of dimethylformamide on the adducts **84** and **83** yields the aza vinylogues of tetramethylformamidinium chloride (**85**) (118). The adduct **83** does not require isolation. The same salt **85** forms by reaction of cyanuric chloride with excess dimethylformamide (118). By this route a series of aza-vinylogous *N,N,N',N'*-tetrasubstituted formamidines has been synthesized from *N,N*-disubstituted formamides and cyanuric chloride (see Table VI).

TABLE VI
Aza Vinylogues of Formamidinium Salts from Cyanuric Chloride and
N,N-Disubstituted Formamides (118)

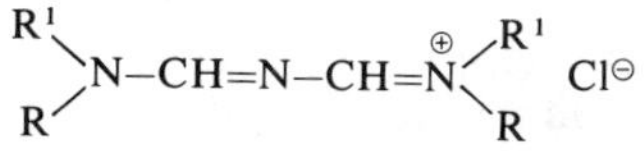

$$\begin{array}{c}R^1\\ \\ R\end{array}\!\!N-CH=N-CH=\overset{\oplus}{N}\!\!\begin{array}{c}R^1\\ \\ R\end{array}\quad Cl^\ominus$$

R	R¹	m.p., °C	Yield, %
CH_3	CH_3	103	86
C_2H_5	C_2H_5	94	76
CH_3	$c\text{-}C_6H_{11}$	155	40
CH_3	C_6H_5	104	50
$-(CH_2)_4-$		145	68
$-(CH_2)_5-$		124	72
$-(CH_2)_2-O-(CH_2)_2-$		110	67

VII. Acid Amide–Sulfuryl Chloride Adducts

Kühle (125) was the first to synthesize an adduct from sulfuryl chloride and DMF; he described the crystalline compound as hygroscopic, melting at 40–41°C. Based on the IR spectrum (bands at 808, 920, 1045, 1155, 1612, and 1725 cm⁻¹) the structure **88** was attributed to it; later studies by Kojtscheff et al. (126) confirmed these assumptions. **88** decomposes into dimethylcarbamoyl chloride on heating to 170–180°C (125). Based on the observation

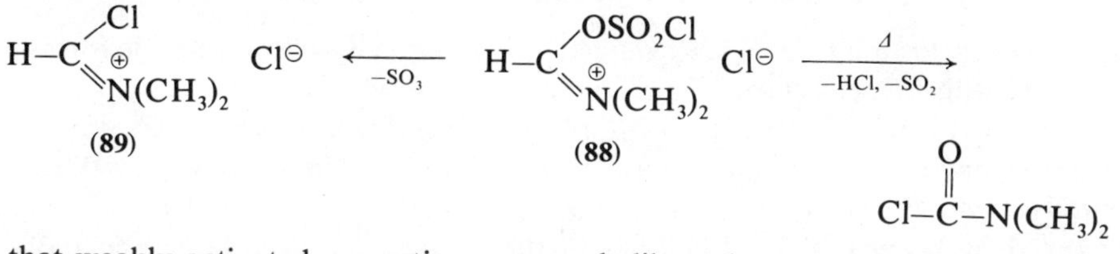

$$H-\overset{\oplus}{C}\!\!\begin{array}{c}Cl\\ \\ N(CH_3)_2\end{array}\ Cl^\ominus\ \xleftarrow{-SO_3}\ H-\overset{\oplus}{C}\!\!\begin{array}{c}OSO_2Cl\\ \\ N(CH_3)_2\end{array}\ Cl^\ominus\ \xrightarrow[-HCl,-SO_2]{\Delta}\ Cl-\overset{O}{\overset{\|}{C}}-N(CH_3)_2$$

(89)   (88)

that weakly activated aromatic compounds like anisole are not formylated by **88** but rather chlorosulfonated, Kojtscheff et al. (126) proposed a thermal degradation of **88** into chloromethyleniminium chloride (**89**) and SO_3. The intermediate sulfonic acid is transformed by **89** into the sulfonyl chloride:

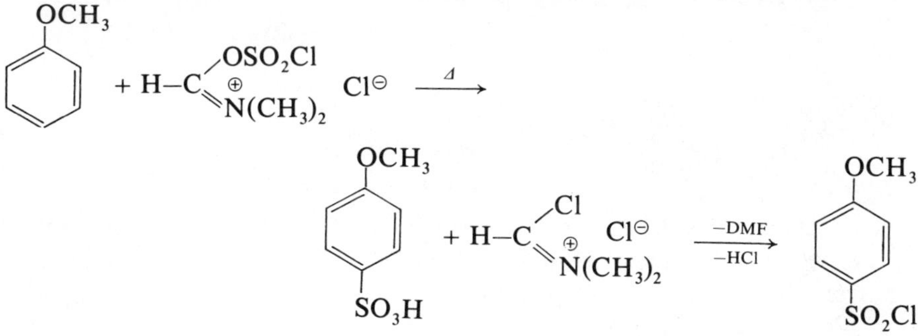

$$\text{anisole} + H-\overset{\oplus}{C}\!\!\begin{array}{c}OSO_2Cl\\ \\ N(CH_3)_2\end{array}\ Cl^\ominus\ \xrightarrow{\Delta}$$

$$\text{4-MeO-C}_6H_4\text{-SO}_3H + H-\overset{\oplus}{C}\!\!\begin{array}{c}Cl\\ \\ N(CH_3)_2\end{array}\ Cl^\ominus\ \xrightarrow[-HCl]{-DMF}\ \text{4-MeO-C}_6H_4\text{-SO}_2Cl$$

This decomposition in the presence of nucleophilic reagents is further supported by the finding that a mixture of the DMF–sulfur trioxide adduct and *N,N*-dimethylchloromethyleniminium chloride undergoes the same reaction with anisole (126).

N-Monoalkylated formamides react faster with sulfuryl chloride than with thionyl chloride (125). Very likely SO_2Cl_2 forms adducts which transform to carbamoyl chlorides (**90**); on heating, these are converted into isocyanates (**91**) by elimination of HCl (125, 127, 128).

$$H-C\overset{O}{\underset{NHR}{\diagdown}} \xrightarrow{SO_2Cl_2/SOCl_2} R-NH-\underset{(90)}{\overset{O}{\overset{\|}{C}}}-Cl \xrightarrow[-HCl]{\Delta} \underset{(91)}{R-N=C=O}$$

With aliphatic substituents such as methyl and ethyl, the carbamoyl chlorides **90** were isolated (125).

Triformamidomethane (**77**) is formed by the reaction of formamide and sulfurylchloride (106). Presumably an adduct analogous to **88** is the intermediate:

$$H-C\overset{O}{\underset{NH_2}{\diagdown}} \xrightarrow{SO_2Cl_2} H-\underset{\underset{(77)}{}}{C}\begin{smallmatrix}\diagup NHCHO \\ -NHCHO \\ \diagdown NHCHO\end{smallmatrix}$$

VIII. Acid Amide–Sulfonyl Chloride Adducts

Studies in 1956 by Hall (129) indicated that benzenesulfonyl chloride forms adducts with dimethylformamide, dimethylacetamide, and *N*-methylpyrrolidone. In 1965 the structure of those adducts from DMF and sulfonyl chlorides had been formulated as **92** (130), despite the demonstration by Witte and Huisgen (131) in 1958, that secondary amides (**93**) and lactams are attacked by benzenesulfonyl chloride at the amide oxygen in the presence of pyridine yielding structures of the type **94**. The adducts **92** have not yet been

$$H-C\overset{O}{\underset{N(CH_3)_2}{\diagdown}} + Cl-SO_2-Ar \rightleftarrows H-C\overset{O-SO_2Ar}{\underset{N(CH_3)_2}{\underset{\oplus}{\diagdown}}}\quad Cl^{\ominus}$$
$$(92)$$

$$R-C\overset{O}{\underset{NHR^1}{\diagdown}} \xrightarrow[pyridine]{C_6H_5SO_2Cl} R-C\overset{O-SO_2-C_6H_5}{\underset{N-R^1}{\diagdown}}$$
$$(93) \qquad\qquad (94)$$

isolated. There is evidence that the adducts are formed in an equilibrium reaction. Many reactions are readily explained by assuming intermediates such as **92** from the reaction of sulfonyl chlorides and acid amides.

Hertler and Corey (132) as well as Hagedorn and Tönjes (133) treated N-monosubstituted formamides with p-toluenesulfonyl chloride or benzenesulfonyl chloride in the presence of pyridine to give isonitriles in good yields. This reaction was used later by several authors (134, 135) to prepare isonitriles. To synthesize low-boiling isonitriles quinoline is preferred as a base, since its high boiling point facilitates the separation of the reaction products (59):

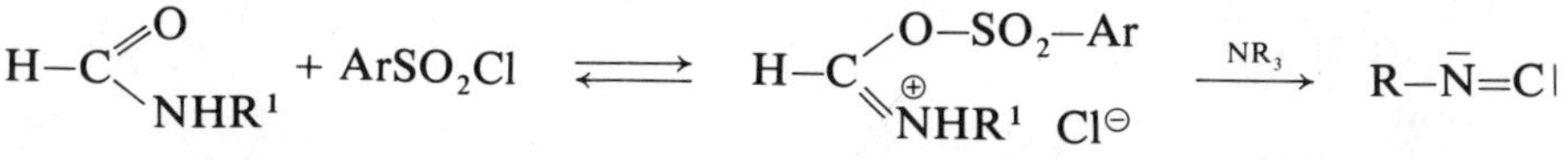

$$H-C\begin{matrix}O\\NHR^1\end{matrix} + ArSO_2Cl \rightleftharpoons H-C\begin{matrix}O-SO_2-Ar\\\overset{\oplus}{N}HR^1\ Cl^{\ominus}\end{matrix} \xrightarrow{NR_3} R-\bar{N}=C|$$

On heating with p-toluenesulfonyl chloride or methanesulfonyl chloride, formamide gives adenine (**95**) in 9 or 1.4% yield, respectively (79). Other

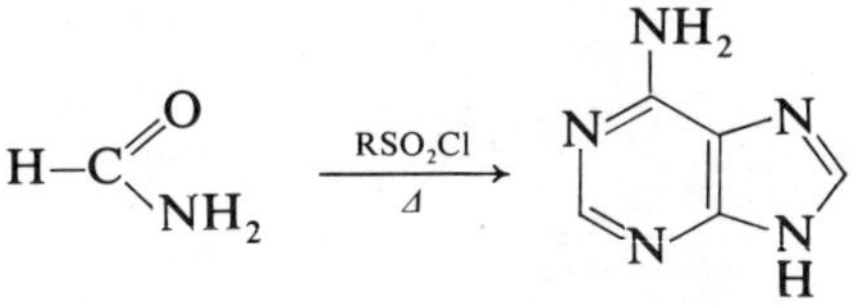

$$H-C\begin{matrix}O\\NH_2\end{matrix} \xrightarrow[\Delta]{RSO_2Cl} \text{[adenine structure]}$$

(**95**)

primary amides are dehydrated to nitriles in excellent yields by toluenesulfonyl chloride and pyridine. The following mechanism has been proposed for this reaction (136):

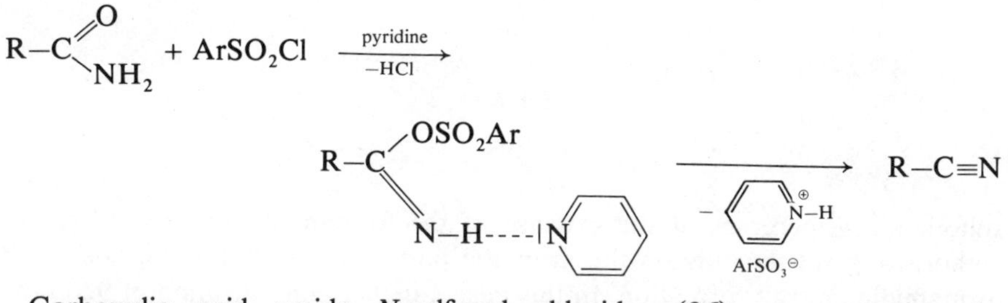

$$R-C\begin{matrix}O\\NH_2\end{matrix} + ArSO_2Cl \xrightarrow[-HCl]{pyridine}$$

$$R-C\begin{matrix}OSO_2Ar\\N-H----|N\bigcirc\end{matrix} \longrightarrow R-C\equiv N$$

$$-\bigcirc\overset{\oplus}{N}-H$$
$$ArSO_3^{\ominus}$$

Carboxylic acid amide N-sulfonyl chlorides (**96**) are spontaneously decomposed into nitriles by dimethylformamide or other tertiary carboxylic acid amides (137, 138). The following mechanism, also presuming adduct formation at the amide oxygen, was proposed (137):

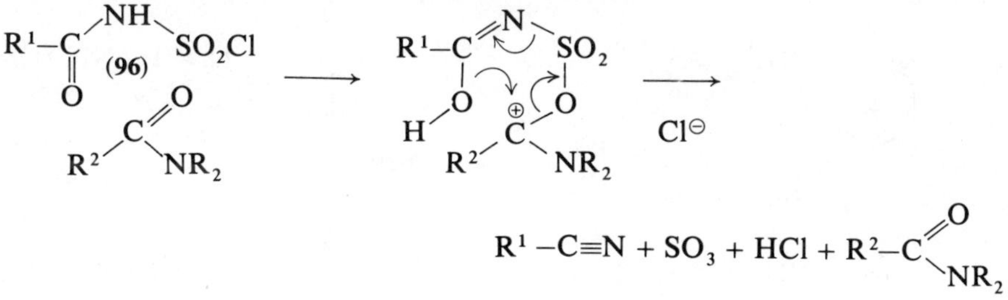

$$\begin{matrix}R^1-C\begin{matrix}NH\\(96)\end{matrix}SO_2Cl\\\parallel\\O\\C\begin{matrix}\\NR_2\end{matrix}\\R^2\end{matrix} \longrightarrow \begin{matrix}R^1-C\begin{matrix}N\\\\O\end{matrix}SO_2\\H\overset{\oplus}{C}\\R^2NR_2\end{matrix} Cl^{\ominus} \longrightarrow$$

$$R^1-C\equiv N + SO_3 + HCl + R^2-C\begin{matrix}O\\NR_2\end{matrix}$$

On heating DMF and arylsulfonyl chlorides form tetramethylformamidinium salts (139–141). Here also the reaction is initiated by acylation of the amide oxygen, followed by attack on the adduct **92** by another mole of DMF. With weakly basic amines ($p_k < 3.5$) **92** gives formamidinium salts (**97**) (130). Sulfonamides (**98**) are by-products. With more basic amines the yield of

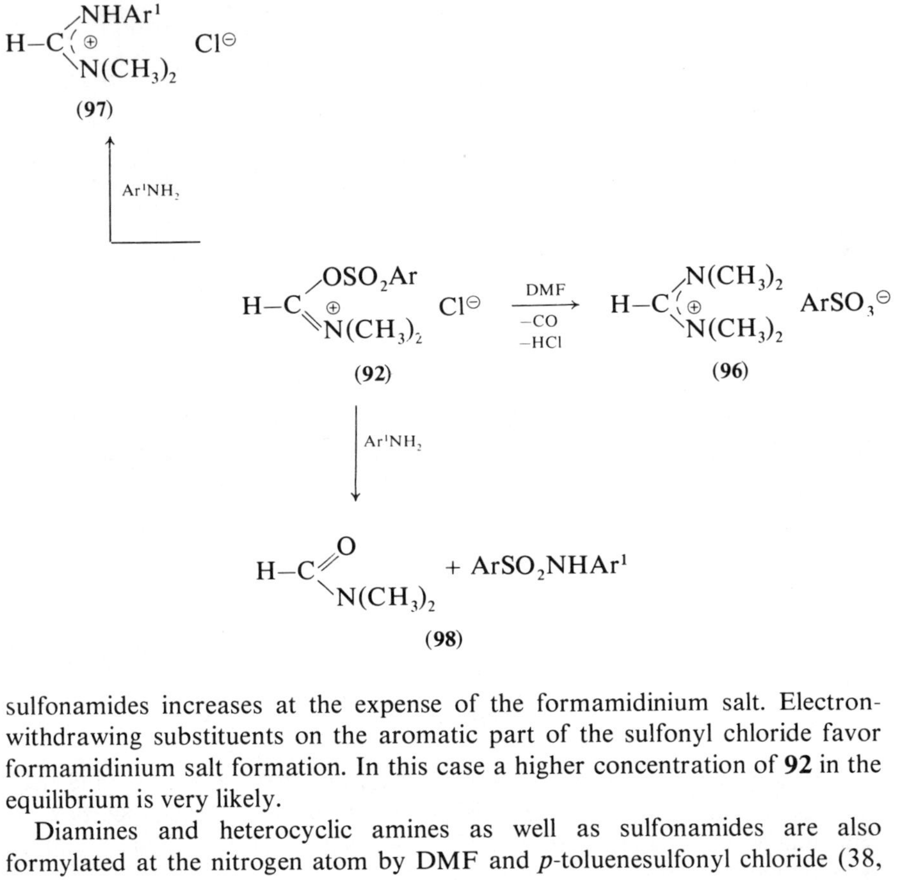

sulfonamides increases at the expense of the formamidinium salt. Electron-withdrawing substituents on the aromatic part of the sulfonyl chloride favor formamidinium salt formation. In this case a higher concentration of **92** in the equilibrium is very likely.

Diamines and heterocyclic amines as well as sulfonamides are also formylated at the nitrogen atom by DMF and p-toluenesulfonyl chloride (38, 142, 143):

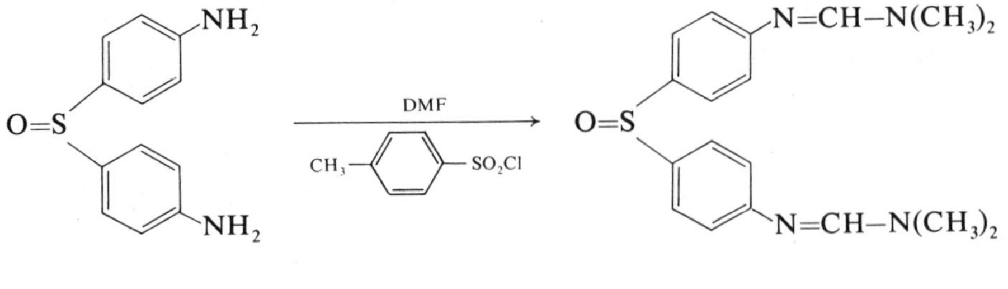

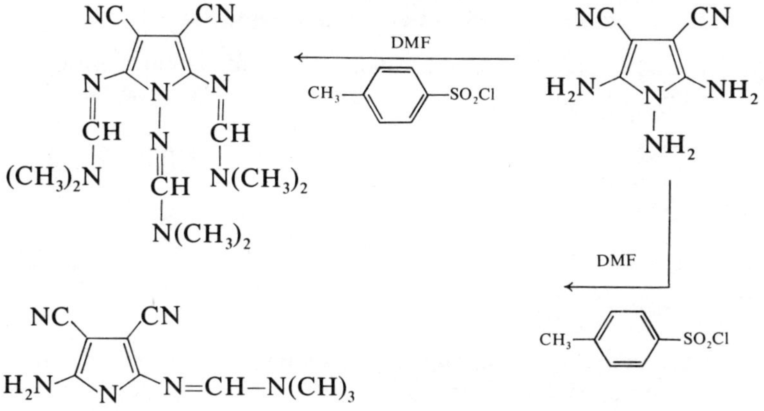

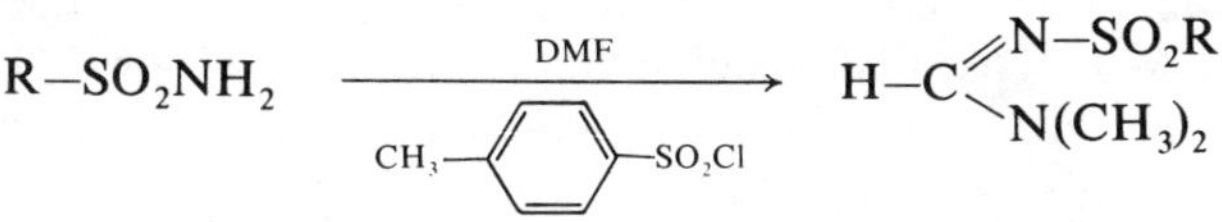

VIII. ACID AMIDE–SULFONYL CHLORIDE ADDUCTS

Acid amide–sulfonyl chloride adducts are labile, according to observations by Ohme and Schmitz (144), since only the formamide–benzoyl chloride adduct but not the formamide–*p*-toluenesulfonyl chloride adduct yields formimino ester hydrochlorides upon alcoholysis. However, DMF–sulfonyl chloride adducts **92** and alcohols give formates **99** after hydrolysis (130). This is explained by assuming the primary formation of *N,N*-dimethylalkoxymethyleniminium chloride (**100**) from **92** and the alcohol. If the alcohol used is weakly nucleophilic (e.g., *p*-nitrobenzyl alcohol, picric acid), *N,N*-dimethylalkoxymethyleniminium chloride (**100**) decomposes into DMF and alkyl chloride in a competitive reaction (130).

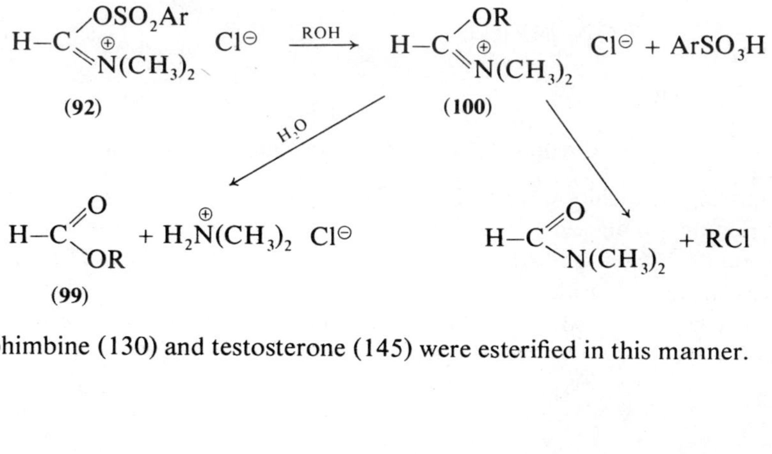

Yohimbine (130) and testosterone (145) were esterified in this manner.

IX. Acid Amide–Chlorosulfonamide Adducts

A mixture of DMF and N,N-disubstituted chlorosulfonamides form an equilibrium with the corresponding adducts (101) (146). These adducts are

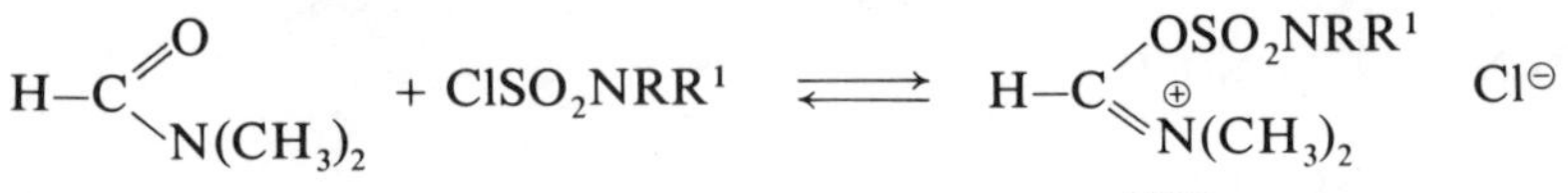

$$\text{H–C}\begin{smallmatrix}\nearrow O \\ \searrow N(CH_3)_2\end{smallmatrix} + ClSO_2NRR^1 \rightleftharpoons \text{H–C}\begin{smallmatrix}\nearrow OSO_2NRR^1 \\ \overset{\oplus}{\searrow} N(CH_3)_2\end{smallmatrix} \quad Cl^\ominus$$

(101)

$R = R^1 = CH_3$
$R-R^1 = -(CH_2)_2-O-(CH_2)_2-$

useful for the formylation of the nitrogen atom in primary aromatic amines and hydrazines. The amidines 102 and amidrazones 103 form in markedly better yields (~70–85%) than those from the corresponding DMF–toluenesulfonyl chloride adducts (146). Dimethylaniline is not formylated by the adduct, but rather sulfonated. The dialkylsulfamate anion of 104 or SO_3 had been proposed as the sulfonating agent, which possibly forms from the adduct in the following way (146):

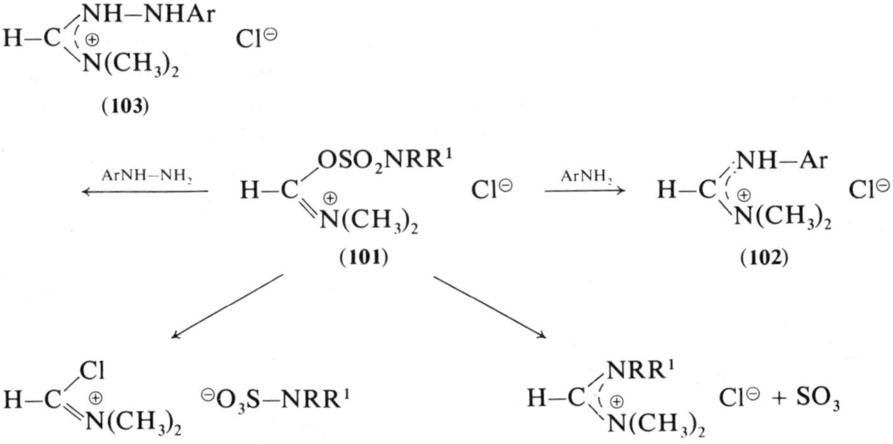

$$\text{H–C}\begin{smallmatrix}\nearrow NH-NHAr \\ \overset{\oplus}{\searrow} N(CH_3)_2\end{smallmatrix} \quad Cl^\ominus$$

(103)

$$\xleftarrow{ArNH-NH_2} \text{H–C}\begin{smallmatrix}\nearrow OSO_2NRR^1 \\ \overset{\oplus}{\searrow} N(CH_3)_2\end{smallmatrix} Cl^\ominus \xrightarrow{ArNH_2} \text{H–C}\begin{smallmatrix}\nearrow NH-Ar \\ \overset{\oplus}{\searrow} N(CH_3)_2\end{smallmatrix} Cl^\ominus$$

(101) \hspace{3cm} (102)

$$\text{H–C}\begin{smallmatrix}\nearrow Cl \\ \overset{\oplus}{\searrow} N(CH_3)_2\end{smallmatrix} \quad {}^\ominus O_3S-NRR^1 \hspace{2cm} \text{H–C}\begin{smallmatrix}\nearrow NRR^1 \\ \overset{\oplus}{\searrow} N(CH_3)_2\end{smallmatrix} \quad Cl^\ominus + SO_3$$

(104)

X. Acid Amide–Carboxylic Acid Chloride Adducts

On mixing acid chlorides and DMF, Hall (129) observed a marked increase of the electric conductivity of the solution, which he attributed to an acylation of the amide oxygen and formation of the ionized adducts 105. The reaction products of acid chlorides and DMF could not be isolated, but the crystalline adducts 105 from DMF and acetyl or benzoyl bromide were successfully synthesized (11, 129).

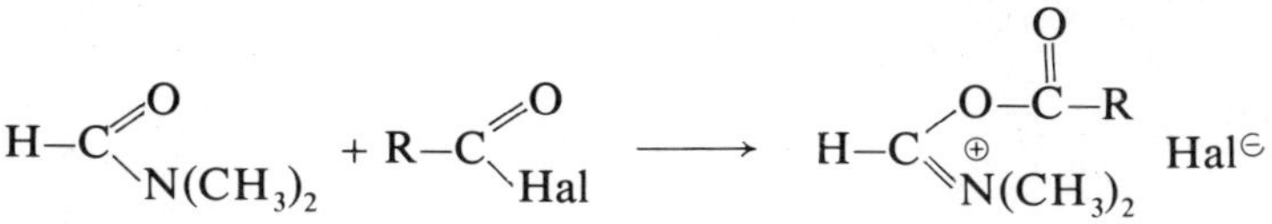

X. ACID AMIDE–CARBOXYLIC ACID CHLORIDE ADDUCTS

$$\text{H–C}\underset{N(CH_3)_2}{\overset{O}{\diagup}} + \text{R–C}\underset{Hal}{\overset{O}{\diagup}} \longrightarrow \text{H–C}\underset{N(CH_3)_2}{\overset{O–\overset{O}{\overset{\|}{C}}–R}{\diagup}} \; Hal^{\ominus}$$

$R = CH_3, C_6H_5$
$Hal = Cl, Br$

(105)

Smith (25) studied the adduct from DMF and acetyl chloride by NMR techniques. In chloroform the signals of those protons bound to the formyl carbon are said to be remarkably shifted upfield compared with DMF:

	τ_{CH}	τ_{NCH_3}	τ_{COCH_3}
DMF	1.90	6.73; 6.84	—
DMF–acetyl chloride adduct	2.04	7.19; 7.29	7.03

As with DMF, the dimethylamino groups in the adduct give a doublet owing to restricted rotation about the partial C=N double bond. The reaction of benzoyl chloride with formamide yields a crystalline, hygroscopic product; the proposed structure is O-benzoylisoformamide hydrochloride (**106**) (106).

Thioformamide forms an analogous crystalline adduct (**107**) with benzoyl chloride (147). N-Benzylthiobenzamide is also acylated by p-nitrobenzoyl chloride at the sulfur atom (148).

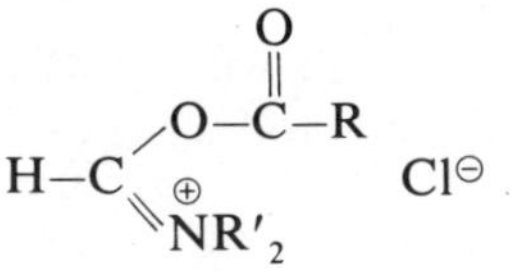

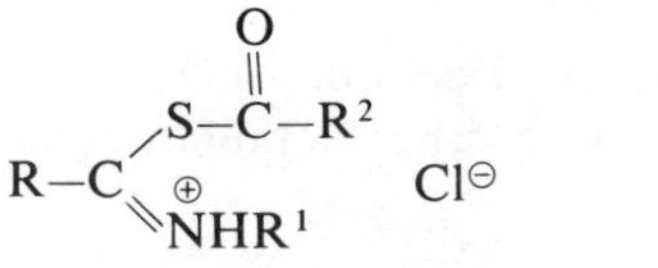

$$\text{H–C}\underset{NR'_2}{\overset{O–\overset{O}{\overset{\|}{C}}–R}{\diagup}} \; Cl^{\ominus} \qquad \text{R–C}\underset{NHR^1}{\overset{S–\overset{O}{\overset{\|}{C}}–R^2}{\diagup}} \; Cl^{\ominus}$$

$R = C_6H_5, R' = H$ $R = H, C_6H_5, R^1 = H, C_6H_5$
(**106**)   (**107**)

Ethyl chloroformate and DMF initially form the adduct **108**, which loses CO_2 to form adduct **109**; this in turn decomposes, yielding DMF and ethyl chloride (149). The intermediate formation of the salt **109** was established by the reaction of ethyl chloroformate with DMF in the presence of $NaBF_4$; the salt **110** can be isolated in this case (150).

Formylations of CH acids like malononitrile and ethyl cyanoacetate can be performed with the DMF–chloroformate adduct (150). If N-unsubstituted and

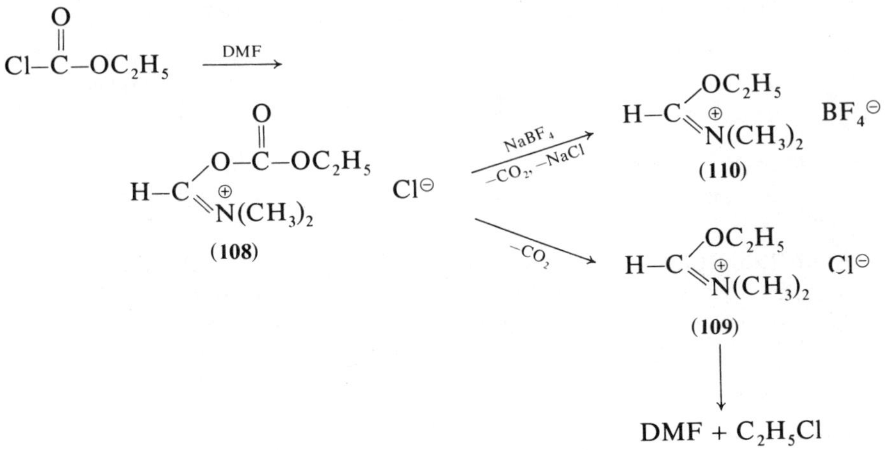

$$Cl-\overset{O}{\underset{\|}{C}}-OC_2H_5 \xrightarrow{DMF}$$

[structure **108**: $H-C(O-CO-OC_2H_5)=\overset{+}{N}(CH_3)_2 \quad Cl^-$]

$\xrightarrow{NaBF_4, -CO_2, -NaCl}$ [structure **110**: $H-C(OC_2H_5)=\overset{+}{N}(CH_3)_2 \quad BF_4^-$]

$\xrightarrow{-CO_2}$ [structure **109**: $H-C(OC_2H_5)=\overset{+}{N}(CH_3)_2 \quad Cl^-$]

$\downarrow$

$DMF + C_2H_5Cl$

N-monosubstituted carboxamides are used, the adducts **111**, analogous to **109**, are stable and isolable (151). Thioamides may also be acylated in this way

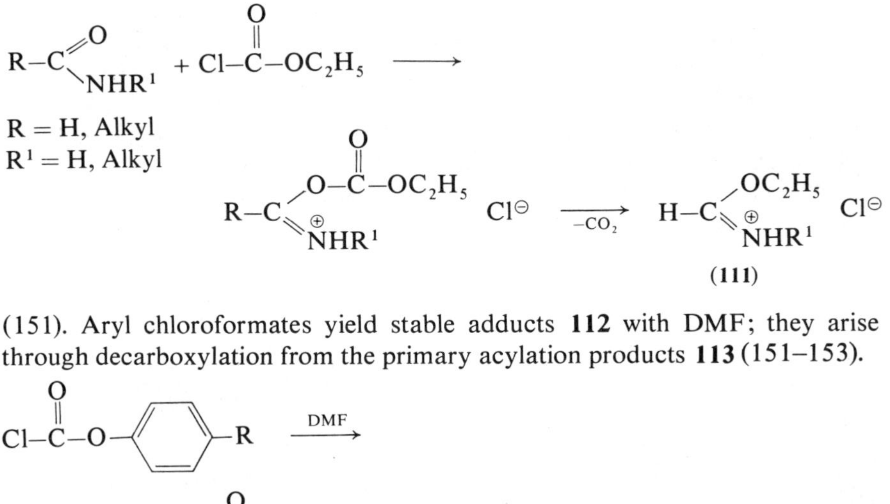

$$R-\overset{O}{\underset{NHR^1}{C}} + Cl-\overset{O}{\underset{\|}{C}}-OC_2H_5 \longrightarrow$$

R = H, Alkyl
R¹ = H, Alkyl

[structure: $R-C(O-CO-OC_2H_5)=\overset{+}{N}HR^1 \quad Cl^-$] $\xrightarrow{-CO_2}$ [structure **111**: $H-C(OC_2H_5)=\overset{+}{N}HR^1 \quad Cl^-$]

(151). Aryl chloroformates yield stable adducts **112** with DMF; they arise through decarboxylation from the primary acylation products **113** (151–153).

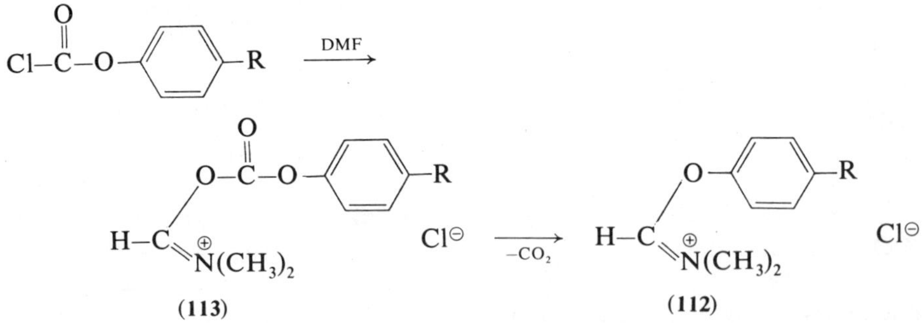

$$Cl-\overset{O}{\underset{\|}{C}}-O-\text{C}_6\text{H}_4-R \xrightarrow{DMF}$$

[structure **113**: $H-C(O-CO-O-C_6H_4-R)=\overset{+}{N}(CH_3)_2 \quad Cl^-$] $\xrightarrow{-CO_2}$ [structure **112**: $H-C(O-C_6H_4-R)=\overset{+}{N}(CH_3)_2 \quad Cl^-$]

The primary acylation products from *N*,*N*-disubstituted formamides and *N*,*N*-dialkylated carbamoyl chlorides (**114**) also decompose with CO_2 evolution, yielding the formamidinium salts **115** (154–157).

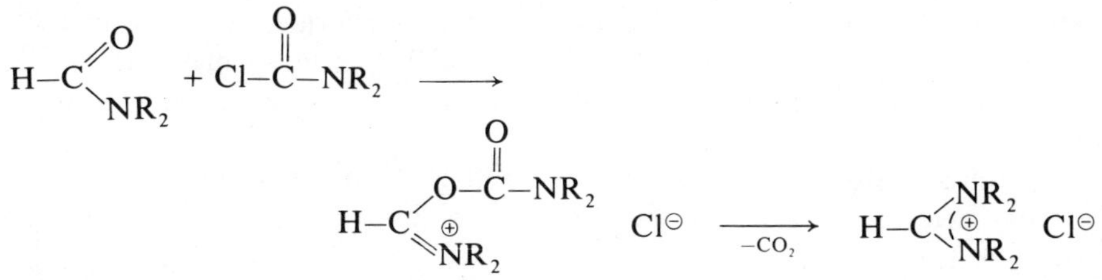

$$H-C\overset{O}{\underset{NR_2}{\diagup}} + Cl-\overset{O}{\underset{\|}{C}}-NR_2 \longrightarrow$$

$$H-C\overset{O-\overset{O}{\underset{\|}{C}}-NR_2}{\underset{NR_2}{\diagup}} \quad Cl^{\ominus} \xrightarrow{-CO_2} H-C\overset{NR_2}{\underset{NR_2}{\diagup}} \quad Cl^{\ominus}$$

$$(114) \qquad\qquad (115)$$

Perchlorobutyryl chloride reacts with DMF via an initial *O*-acylation product **116** to yield dimethylchloromethyleniminium chloride (**117**) (158).

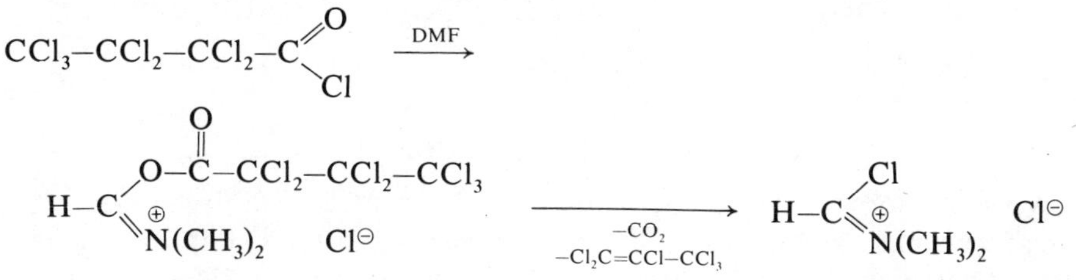

$$CCl_3-CCl_2-CCl_2-C\overset{O}{\underset{Cl}{\diagup}} \xrightarrow{DMF}$$

$$H-C\overset{O-\overset{O}{\underset{\|}{C}}-CCl_2-CCl_2-CCl_3}{\underset{N(CH_3)_2}{\diagup}} \quad Cl^{\ominus} \xrightarrow[-Cl_2C=CCl-CCl_3]{-CO_2} H-C\overset{Cl}{\underset{N(CH_3)_2}{\diagup}} \quad Cl^{\ominus}$$

$$(116) \qquad\qquad\qquad (117)$$

Adduct **118** is proposed as an intermediate in the acylation of *N,N*-diethylphenylacetamide by trichloroacetyl chloride; elimination of HCl yields **119**, which is then further acylated to yield **120** (159).

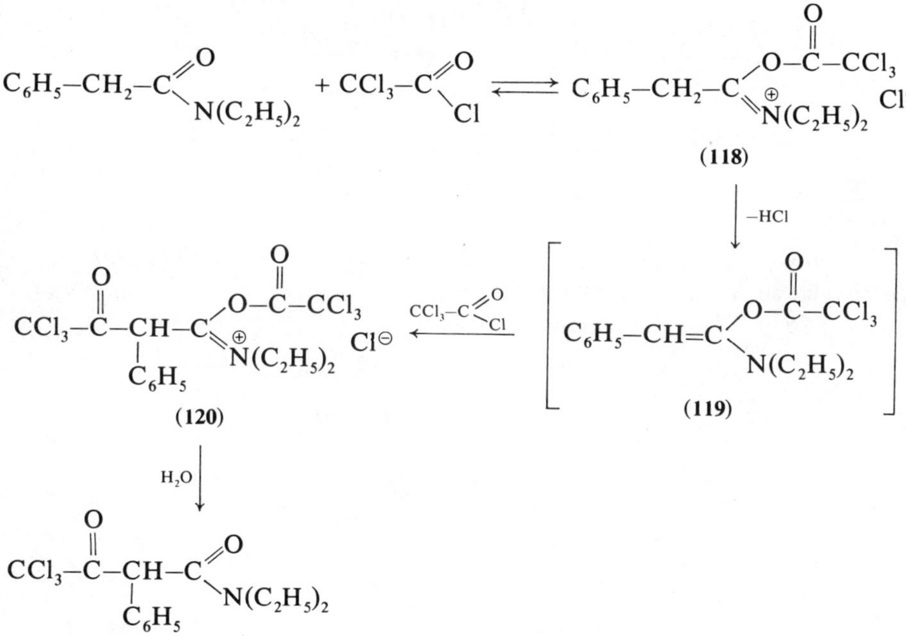

$$C_6H_5-CH_2-C\overset{O}{\underset{N(C_2H_5)_2}{\diagup}} + CCl_3-C\overset{O}{\underset{Cl}{\diagup}} \rightleftharpoons C_6H_5-CH_2-C\overset{O-\overset{O}{\underset{\|}{C}}-CCl_3}{\underset{N(C_2H_5)_2}{\diagup}} \quad Cl^{\ominus}$$

$$(118)$$

$$\Big\downarrow -HCl$$

$$CCl_3-\overset{O}{\underset{\|}{C}}-\underset{C_6H_5}{\underset{|}{CH}}-C\overset{O-\overset{O}{\underset{\|}{C}}-CCl_3}{\underset{N(C_2H_5)_2}{\diagup}} \quad Cl^{\ominus} \xleftarrow{CCl_3-C\overset{O}{\underset{Cl}{\diagup}}} \left[C_6H_5-CH=C\overset{O-\overset{O}{\underset{\|}{C}}-CCl_3}{\underset{N(C_2H_5)_2}{\diagup}}\right]$$

$$(120) \qquad\qquad\qquad (119)$$

$$\Big\downarrow H_2O$$

$$CCl_3-\overset{O}{\underset{\|}{C}}-\underset{C_6H_5}{\underset{|}{CH}}-C\overset{O}{\underset{N(C_2H_5)_2}{\diagup}}$$

Other acylating agents such as phosgene, thionyl chloride, PCl$_5$, and oxalyl chloride are discussed in Part 1 in the chapter called "The Vilsmeier–Haack–Arnold Acylations" as well as in Part II, in the chapter called "Amide Halides."

According to Horning and Muchowski (160) adducts **121** with any group R can be generated *in situ* from **117** and the sodium or ammonium salt of a carboxylic acid.

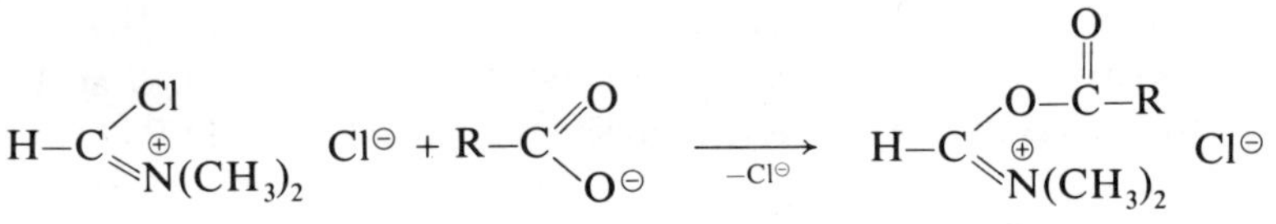

$$\text{H-C}\overset{\text{Cl}}{\underset{\overset{\oplus}{\text{N(CH}_3)_2}}{\diagdown}} \text{Cl}^\ominus + \text{R-C}\overset{\text{O}}{\underset{\text{O}^\ominus}{\diagdown}} \xrightarrow{-\text{Cl}^\ominus} \text{H-C}\overset{\text{O-C-R}}{\underset{\overset{\oplus}{\text{N(CH}_3)_2}}{\diagdown}} \text{Cl}^\ominus$$

(117)   (121)

A. REACTIONS OF ACID AMIDE–ACID CHLORIDE ADDUCTS

When the formamide benzoyl chloride adduct **106** is hydrolyzed, benzoic acid, formamide, and hydrogen chloride are formed (106). On alcoholysis of **106** formimino ester hydrochlorides (**122**) are obtained in 70–99% yield. The adduct from formamide and acetyl chloride does not undergo this reaction (144).

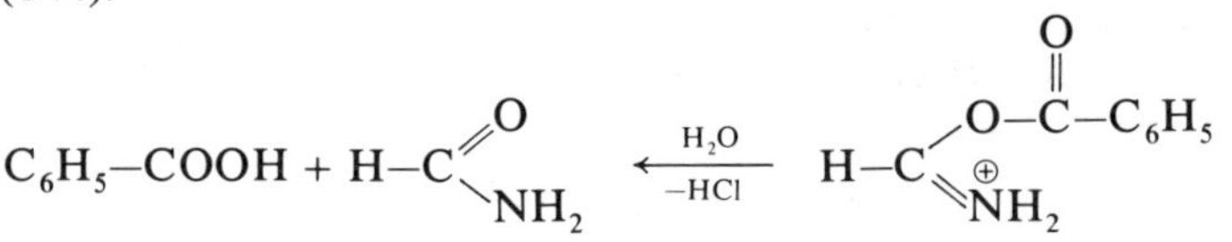

$$C_6H_5-\text{COOH} + \text{H-C}\overset{\text{O}}{\underset{\text{NH}_2}{\diagdown}} \xleftarrow[-\text{HCl}]{H_2O} \text{H-C}\overset{\text{O-C-C}_6\text{H}_5}{\underset{\overset{\oplus}{\text{NH}_2}}{\diagdown}} \text{Cl}^\ominus$$

(106)

$$\textbf{106} \xrightarrow{\text{ROH}} \text{H-C}\overset{\text{OR}}{\underset{\overset{\oplus}{\text{NH}_2}}{\diagdown}} \text{Cl}^\ominus + C_6H_5\text{COOH}$$

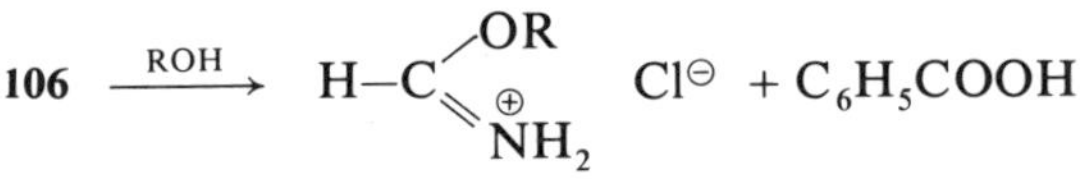

(122)

The adducts **113** react with alcohols to give alkyl aryl ethers **123** (152); with carboxylic acids esters (**124**) are obtained. On the other hand, alcoholysis of adduct **125** yields alkyl bromides (**152**).

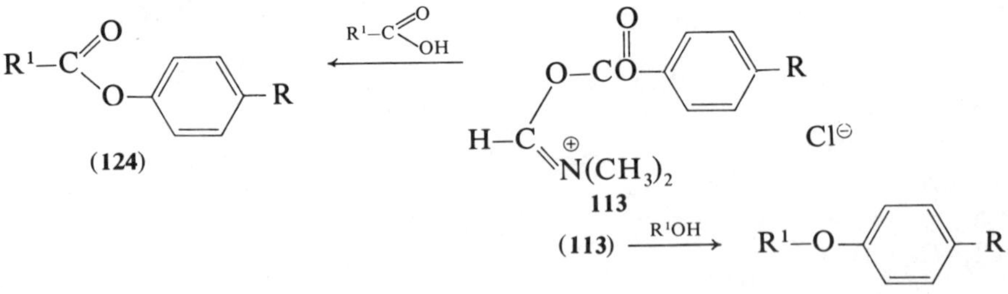

(124) ← R^1–C(=O)OH — (113) — R^{1}OH → (123)

X. ACID AMIDE–CARBOXYLIC ACID CHLORIDE ADDUCTS

$$\text{H–C}\underset{\overset{\oplus}{N(CH_3)_2}}{\overset{O-\overset{\overset{O}{\|}}{C}-R}{\diagup}} \quad Br^{\ominus} \quad \xrightarrow{R^1OH} \quad R^1\text{–Br}$$

(125)   (152)

Amidines as well may be obtained from amide acid halide adducts; thus DMF, benzoyl bromide, and aniline yield N,N-dimethyl-N'-phenylformamidine (126) (11); sulfonamides, DMF, and benzoyl chloride or phenyl chloroformate give N-sulfonylformamidines (127) (38).

$$\text{H–C}\underset{N(CH_3)_2}{\overset{N-C_6H_5}{\diagup}} \qquad \text{H–C}\underset{N(CH_3)_2}{\overset{N-SO_2R}{\diagup}} \qquad \text{R–C}\underset{NHC_6H_5}{\overset{O}{\diagup}}$$

(126)   (127)   (128)

Adduct formation of acetyl chloride and benzoyl chloride with DMF was studied by treating the adducts generated *in situ* with aniline, the product being N,N-dimethyl-N'-phenylformamidine (126) (160). The corresponding carboxylic acid anilides 128 are always observed as by-products. The acid chloride which is in equilibrium with the acid amide–DMF adduct is assumed to yield the anilide. The yield of amidine is increased by high concentrations of DMF; low reaction temperatures have the same effect.

Aromatic amines react with the adducts 113 to yield carbamates (129). Whereas chloromethyleniminium chlorides dehydrate primary amides to nitriles, the DMF–benzoyl chloride adduct 130 and the DMF–acetyl chloride adduct 131 react with benzamide and urea to give amidinium salts 132. Other carboxylic acid amides gave no reaction under these conditions (161).

$$\text{H–C}\underset{\overset{\oplus}{N(CH_3)_2}}{\overset{O-\overset{\overset{O}{\|}}{C}-O-\!\!\bigcirc\!\!-R}{\diagup}} \quad Cl^{\ominus} \quad \xrightarrow{R^1NH_2} \quad R^1NH-\overset{\overset{O}{\|}}{C}-O-\!\!\bigcirc\!\!-R$$

113   129

$$\text{H–C}\underset{\overset{\oplus}{N(CH_3)_2}}{\overset{O-\overset{\overset{O}{\|}}{C}-R}{\diagup}} Cl^{\ominus} + H_2N-\overset{\overset{O}{\|}}{C}-R^1 \xrightarrow{-R-C\overset{O}{\underset{OH}{\diagup}}} \text{H–C}\underset{\overset{\oplus}{N(CH_3)_2}}{\overset{NH-\overset{\overset{O}{\|}}{C}-R^1}{\diagup}} Cl^{\ominus}$$

(130) R = C_6H_5 (132)
(131) R = CH_3

If benzoyl chloride is treated with a tenfold excess of formamide, benzoic acid, ammonium chloride, formamidinium chloride (133), and triformamidomethane (**77**) are isolated. To explain this variety of compounds, the following scheme has been proposed (106):

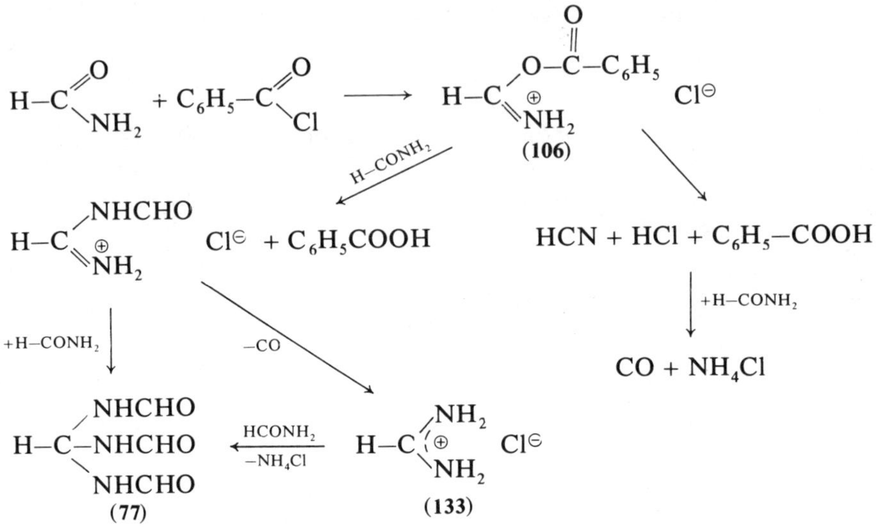

$$H-C\underset{NH_2}{\overset{O}{\diagup}} + C_6H_5-C\underset{Cl}{\overset{O}{\diagup}} \longrightarrow H-C\underset{NH_2}{\overset{O-C(=O)-C_6H_5}{\diagup}}\ Cl^{\ominus}$$
(**106**)

From (106), three pathways:

- (via H–CONH$_2$) → $H-C\underset{NH_2}{\overset{NHCHO}{\diagup}}\ Cl^{\ominus}\ +\ C_6H_5COOH$
- → HCN + HCl + C$_6$H$_5$–COOH → (+H–CONH$_2$) → CO + NH$_4$Cl
- (−CO) → $H-C\underset{NH_2}{\overset{NH_2}{\diagup}}\ Cl^{\ominus}$ (**133**)

$$H-C\underset{NH_2}{\overset{NHCHO}{\diagup}}\ Cl^{\ominus}\ \xrightarrow{+H-CONH_2}\ H-C\underset{NHCHO}{\overset{NHCHO}{-NHCHO}}$$
(**77**)

(**133**) $\xrightarrow[-NH_4Cl]{HCONH_2}$ (**77**)

The treatment of formamide with acetyl chloride or ethyl chloroformate yielding tri(formamido)methane (**77**) may be formulated in a similar way (106).

p-Nitrobenzoic anhydride (**134**) was synthesized from *p*-nitrobenzoyl chloride, DMF, and sodium perchlorate in acetonitrile (21). This reaction

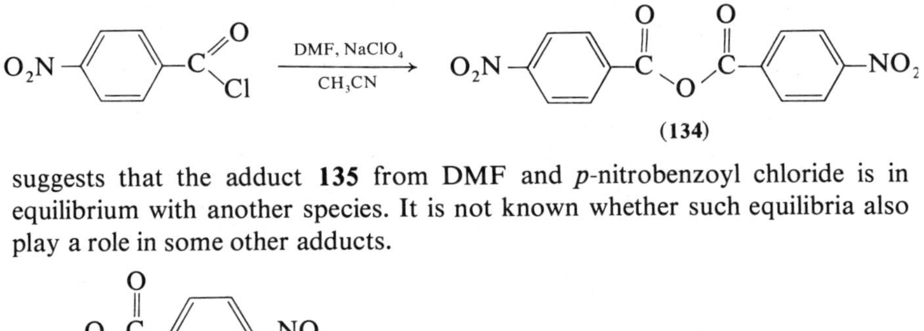

$$O_2N-C_6H_4-C(=O)Cl \xrightarrow[CH_3CN]{DMF,\ NaClO_4} O_2N-C_6H_4-C(=O)-O-C(=O)-C_6H_4-NO_2$$
(**134**)

suggests that the adduct **135** from DMF and *p*-nitrobenzoyl chloride is in equilibrium with another species. It is not known whether such equilibria also play a role in some other adducts.

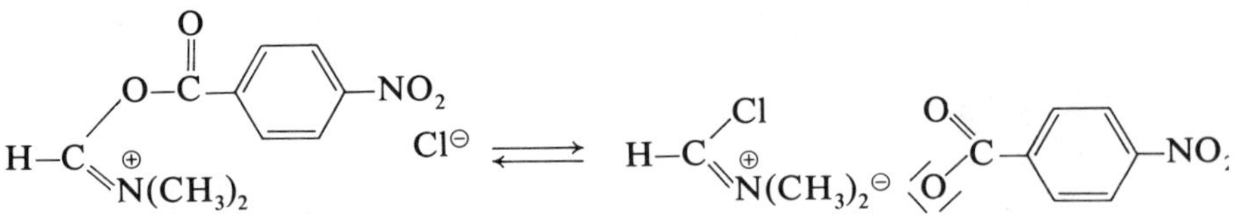

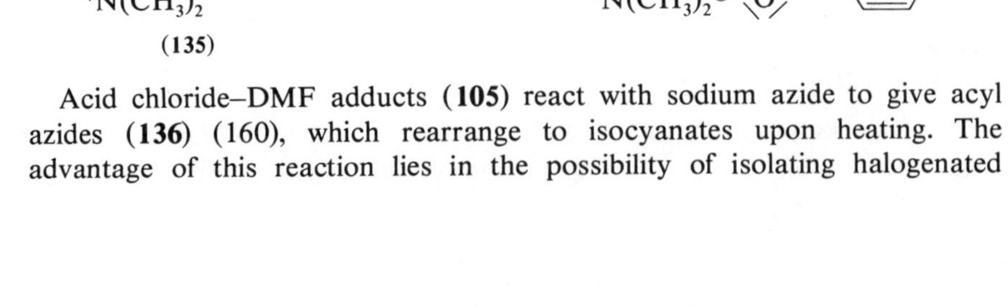

$$H-C\underset{N(CH_3)_2}{\overset{O-C(=O)-C_6H_4-NO_2}{\diagup}}\ Cl^{\ominus}\ \rightleftharpoons\ H-C\underset{N(CH_3)_2^{\ominus}}{\overset{Cl}{\diagup}}\ \cdot\ O_2\!\!<\!\!C-C_6H_4-NO_2$$
(**135**)

Acid chloride–DMF adducts (**105**) react with sodium azide to give acyl azides (**136**) (160), which rearrange to isocyanates upon heating. The advantage of this reaction lies in the possibility of isolating halogenated

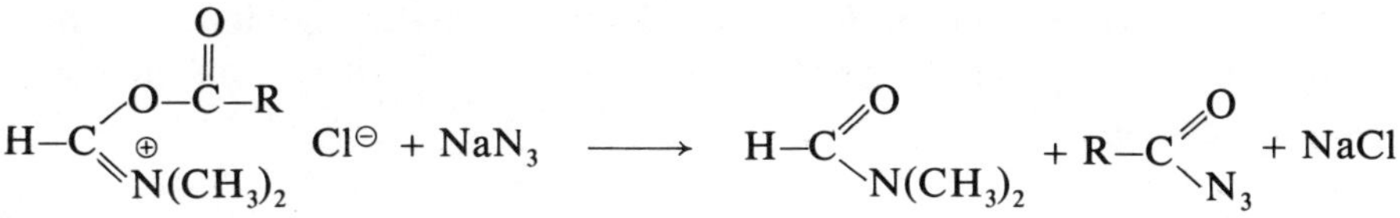

XI. ACID AMIDE–ACID ANHYDRIDE ADDUCTS

$$\text{H-C}\begin{smallmatrix}\text{O-C(=O)-R}\\\oplus\\\text{N(CH}_3)_2\end{smallmatrix}\;\text{Cl}^\ominus + \text{NaN}_3 \longrightarrow \text{H-C}\begin{smallmatrix}\text{O}\\\text{N(CH}_3)_2\end{smallmatrix} + \text{R-C}\begin{smallmatrix}\text{O}\\\text{N}_3\end{smallmatrix} + \text{NaCl}$$

(105)   (136)

acylazides or isocyanates without any substitution of the halide by azide, for example:

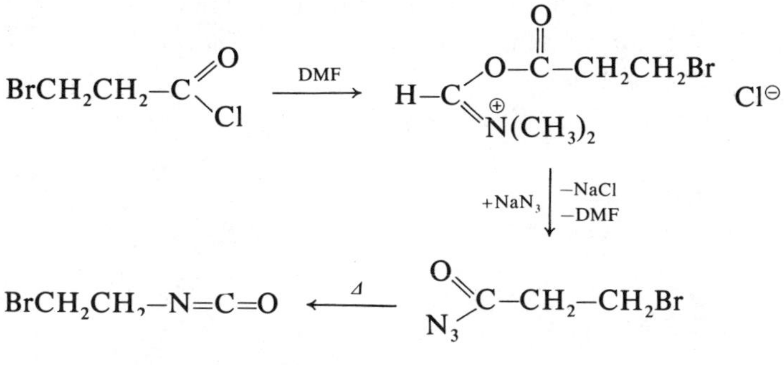

$$\text{BrCH}_2\text{CH}_2\text{-C}\begin{smallmatrix}\text{O}\\\text{Cl}\end{smallmatrix} \xrightarrow{\text{DMF}} \text{H-C}\begin{smallmatrix}\text{O-C(=O)-CH}_2\text{CH}_2\text{Br}\\\oplus\\\text{N(CH}_3)_2\end{smallmatrix}\;\text{Cl}^\ominus$$

$$+\text{NaN}_3 \Bigg|\begin{smallmatrix}-\text{NaCl}\\-\text{DMF}\end{smallmatrix}$$

$$\text{BrCH}_2\text{CH}_2\text{-N=C=O} \xleftarrow{\Delta} \text{N}_3\text{-C(=O)-CH}_2\text{-CH}_2\text{Br}$$

XI. Acid Amide–Acid Anhydride Adducts

Kröhnke et al. (162–164) were able to demonstrate that vinylogous acid amides are acylated by acetic anhydride at the oxygen atom, and the adducts act as vinylogous Vilsmeier reagents. Acylation of amides with carboxylic acid anhydrides yields imides and nitriles. To explain these reaction products, it is presumed that O-acylated acid amides (137) are the primary intermediates (165).

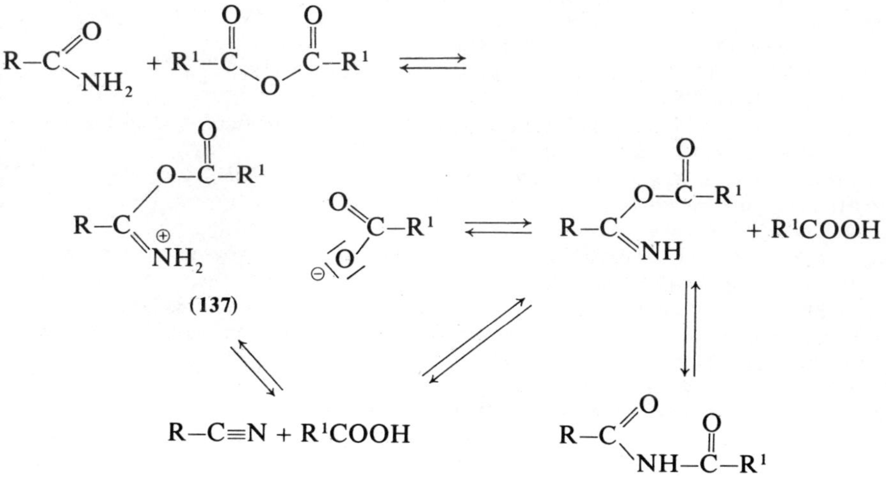

$$\text{R-C}\begin{smallmatrix}\text{O}\\\text{NH}_2\end{smallmatrix} + \text{R}^1\text{-C(=O)-O-C(=O)-R}^1 \rightleftarrows$$

$$\text{R-C}\begin{smallmatrix}\text{O-C(=O)-R}^1\\\oplus\\\text{NH}_2\end{smallmatrix} \quad \begin{smallmatrix}\text{O}\\\text{C-R}^1\\\text{O}^\ominus\end{smallmatrix} \rightleftarrows \text{R-C}\begin{smallmatrix}\text{O-C(=O)-R}^1\\\text{NH}\end{smallmatrix} + \text{R}^1\text{COOH}$$

(137)

$$\text{R-C}\equiv\text{N} + \text{R}^1\text{COOH} \quad\quad \text{R-C}\begin{smallmatrix}\text{O}\\\text{NH-C(=O)-R}^1\end{smallmatrix}$$

With DMF–acetic anhydride, dimethylamino-methylene groups can be introduced into pyrylium salts (166, 167) and other CH-active compounds (168), for example:

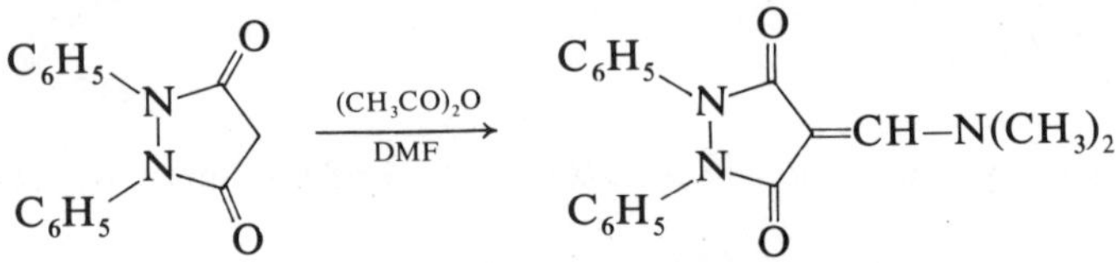

It is, therefore, a reasonable assumption that acid amides and carboxylic acid anhydrides also form adducts of the Vilsmeier type in an equilibrium reaction:

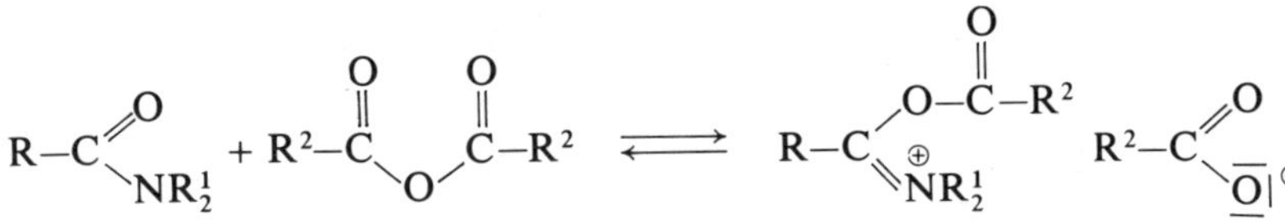

ACKNOWLEDGMENT

To Dr. Heinz Eilingsfeld BASF AG, Ludwigshafen, and to Dr. Bernd Funke as well as Dr. Erwin Haug, both of Fachhochschule Aalen, Aalen, I am very indebted for fruitful discussions and critical revisions of the text.

REFERENCES

1. O. Dimroth and E. Zoeppritz *Ber. Deut. Chem. Ges.*, **55**, 995 (1902).
2. A. Vilsmeier and A. Haack, *Ber. Deut. Chem. Ges.*, **60**, 119 (1927).
3. A. Vilsmeier, *Chemiker-Ztg.*, **75**, 133 (1951).
4. O. Bayer, *Houben-Weyl, Methoden der Organischen Chemie*, Vol. VII/1, *Sauerstoff-Verbindungen II*, Georg Thieme Verlag, Stuttgart, 1954, p. 29.
5. V. I. Minkin and G. N. Dorofeenko, *Russ. Chem. Rev.*, **29**, 599 (1960).
6. M. R. de Meheas, *Bull. Soc. Chim. France* **1962**, 1989.
7. G. A. Olah and St. J. Kuhn, in *Friedel Crafts and Related Reactions*, G. A. Olah, Ed., Vol. III, Part 2, John Wiley & Sons, New York, 1964, p. 1211.
8. Michihiko Ochiai, *Kagaku No Ryoiki*, **19**, 900 (1965); *C.A.*, **67**, 32076d.
9. H. Ulrich, in *The Chemistry of Imidoyl-Halides*, Plenum Press, New York, 1968, pp. 87–97.
10. R. Bonnett, "Imidoyl Halides," *The Chemistry of the Carbon–Nitrogen Double Bond*, S. Patai, Ed., John Wiley & Sons, New York, 1970.
11. H. Bredereck, R. Gompper, K. Klemm, and H. Rempfer, *Chem. Ber.*, **92**, 837 (1959).
12. E. Compaigne and W. L. Archer, *Organic Syntheses*, Vol. 33, John Wiley & Sons, New York, 1953, p. 27.
13. G. F. Smith, *J. Chem. Soc.*, **1954**, 3842.
14. R. N. Silverstein, E. E. Ryskiewicz, C. Willert, and R. C. Kocher, *J. Org. Chem.*, **20**, 668 (1955).
15. C. Jutz, *Chem. Ber.*, **91**, 850 (1958).

REFERENCES

16. A. Lorenz and R. Winzinger, *Helv. Chim. Acta,* **28,** 600 (1945).
17. H. H. Bosshard and H. Zollinger, *Helv. Chim. Acta,* **42,** 1659 (1959).
18. S. Alumi, P. Linda, G. Marino, S. Santini, and G. Savelli, *J. Chem. Soc. Perkin II,* **1972,** 2070.
19. P. Linda, A. Luccarelli, G. Marino, and G. Savelli, *J. Chem. Soc. Perkin II,* **1974,** 1610.
20. W. Ziegenbein and W. Franke, *Chem. Ber.* **93,** 1681 (1960).
21. Z. Arnold and A. Holy, *Collection Czech. Chem. Commun.,* **27,** 2886 (1962).
22. G. J. Martin, S. Poignant, M. L. Filleux, and M. T. Quemeneuer, *Tetrahedron Letters,* **58,** 5061 (1970).
23. G. Martin and M. Martin, *Bull. Soc. Chim. France,* **1963,** 1637.
24. G. J. Martin and S. Poignant, *J. Chem. Soc. Perkin II,* **1972,** 1964.
25. T. D. Smith, *J. Chem. Soc. A,* **1966,** 841.
26. H. Fritz and R. Oehl, *Ann. Chem.,* **749,** 159 (1971).
27. G. J. Martin and S. Poignant, *J. Chem. Soc. Perkin II,* **1974,** 642.
28. H. Bredereck, R. Gompper, and K. Klemm, *Chem. Ber.* **92,** 1456 (1959).
29. H. Bredereck and K. Bredereck, *Chem. Ber.,* **94,** 2278 (1961).
30. H. Bredereck, F. Effenberger, H. Botsch, and H. Rehn, *Chem. Ber.,* **98,** 1981 (1965).
31. C. Jutz and H. Amschler, *Chem. Ber.,* **96,** 2100 (1963).
32. J. G. Dingwall, D. H. Reid, and K. Wade, *J. Chem. Soc. C,* **1969,** 913.
33. E. Günther, F. Wolf, and G. Wolter, *Z. Chem.,* **8,** 63 (1968).
34. E. E. Bureš and M. Kundera, *Cas. Ceks. Lek.,* **14,** 272 (1934); *C.A.,* **29,** 4750. (1935).
35. T. L. Davis and W. E. Yelland, *J. Am. Chem. Soc.,* **59,** 1998 (1937).
36. C. G. Raison, *J. Chem. Soc.,* **1949,** 3319.
37. E. Enders, Ger. Pat. 949,285 (Sept. 20, 1956); *C.A.,* **53,** 95250e (1959).
38. G. R. Pettit and R. E. Kadunce, *J. Org. Chem.* **27,** 4566 (1962).
39. K. Dierbach, East. Ger. Pat. 26,918 (Feb. 5, 1964), *C.A.,* **61,** 13236h.
40. W. Klocker and M. Herberz, *Monatsh. Chem.,* **96,** 1567 (1965).
41. E. C. Taylor and R. W. Morrison, Jr., *Angew. Chem. Intern. Ed. Engl.,* **4,** 868 (1965).
42. A. Albert and H. Taguchi, *J. Chem. Soc. Perkin Trans. I,* **1973,** 2037.
43. E. M. Roberts, J. M. Grisar, R. D. McKenzie, G. P. Claxton, and T. R. Blohm, *J. Med. Chem.,* **15,** 1270 (1972).
44. Ya. Kvitko and T. M. Loginova, *Zh. Org. Khim.,* **10,** 1088 (1974); *Cheminform,* **1974** 34–264, p. 96.
45. T. Tsuji and Y. Kamo, *Chem. Lett.,* **1972,** 641; *Cheminform,* **1973,** 48–270, p. 130.
46. T. Tsuji, *Chem. Pharm. Bull.,* **22,** 471 (1974).
47. Yu. N. Portnov, G. A. Golubeva, A. N. Kost, and V. S. Volkov, *Khim. Geterotsikl. Soedin,* **1973,** 647; *Cheminform,* **1973,** 35–304, p. 152.
48. Fumio Yoneda, Masatsuga Higuchi, Takafumi Matsumura, and Keitaro Senga, *Bull. Chem. Soc. Japan,* **46,** 1837 (1973).
49. C. V. Z. Smith, R. K. Robins, and R. L. Tolman, *J. Chem. Soc. Perkin Trans. I,* **1973,** 1855.
50. S. Kwon, F. Ikeda, and K. Isegawa, *Nippon Kagaku Kaishi,* **1973,** 1944; *Cheminform,* **1974,** 14–342, p. 127.
51. C. S. Davis, A. D. Kneval, and G. L. Jenkins, *J. Org. Chem.,* **27,** 1919 (1962).
52. Z. Arnold, J. Sauliova, and V. Krehuak, *Collection Czech. Chem. Commun.,* **38,** 2633 (1973).
53. H. Bredereck, F. Effenberger, and H. Botsch, *Chem. Ber.,* **97,** 3397 (1964).
54. H. Eilingsfeld, M. Seefelder, and H. Weidinger, *Angew. Chem.,* **72,** 840 (1960).
55. H. Eilingsfeld, M. Seefelder, and H. Weidinger, *Chem. Ber.,* **96,** 2672 (1963).
56. C. A. Ainsworth, W. E. Butin, J. Davenport, M. E. Callender, and M. C. McCowen, *J. Med. Chem.,* **10,** 208 (1967).

57. M. A. Kira, Z. N. Nofal, and K. Z. Gadalla, *Tetrahedron Letters*, **1970**, 4215.
58. I. Ugi and R. Meyr, *Angew. Chem.*, **70**, 702 (1958).
59. I. Ugi, U. Fetzer, U. Eholzer, H. Knupfer, and K. Offermann, *Angew. Chem.*, **77**, 492 (1965).
60. I. Ugi and R. Meyr, *Chem. Ber.*, **93**, 239 (1960).
61. H. Bredereck, B. Föhlisch, and K. Walz, *Angew. Chem.*, **74**, 388 (1962); *Angew. Chem. Intern. Ed. Engl.*, **1**, 224 (1962).
62. B. Föhlisch, H. Bredereck, and K. Walz, *Angew. Chem*, **76**, 580 (1964); *Angew. Chem. Intern. Ed. Engl.*, **3**, 647 (1964).
63. H. Bredereck, B. Föhlisch, and K. Walz, *Ann. Chem.*, **686**, 92 (1965).
64. H. Bredereck, B. Föhlisch, and K. Walz, *Ann. Chem.*, **688**, 93 (1965).
65. I. Hagedorn, *Angew. Chem.*, **75**, 305 (1963).
66. I. Hagedorn, H. Etling, and K. E. Lichtel, *Chem. Ber.*, **99**, 520 (1966).
67. W. Ziegenbein and K. H. Hornung, *Chem. Ber.*, **95**, 2976 (1962).
68. T. I. Lonshehakova, B. I. Buzykin, and V. S. Tsivunin, *Zh. Org. Khim.*, **10**, 2459 (1974); *Cheminform*, **1975**, 7–126, p. 32.
69. S. Pennanen, *Acta Chem. Scand.*, **27**, 3133 (1973).
70. H. Weidinger and G. Wellenreuther, Brit. Pat. 927,974 (1963); *C.A.*, **60**, 2987 (1964).
71. H. Balli and F. Kersting, *Ann. Chem.*, **647**, 1 (1961).
72. Y. Murakami, A. Kurita, and O. Yoneya, Jap. Pat. 22,886 (1963); *C.A.* **60**, 5466e (1964).
73. K. Bredereck, Dissertation, University of Stuttgart, 1961.
74. F. M. Stoyanovich, B. P. Federov, G. M. Adrianova, *Dokl. Akad. Nauk SSSR*, **145**, 584 (1962); *C.A.*, **58**, 4448 (1963).
75. H. Bredereck, R. Gompper, and K. Klemm, *Chem. Ber.*, **92**, 1456 (1959).
76. H. Eilingsfeld, M. Seefelder, and H. Weidinger, *Chem. Ber.*, **96**, 2899 (1963).
77. M. Seefelder (to BASF AG), Ger. Pat. 1,089,760 (1958); *C.A.* 5621 (1961).
78. K. Morita et al., *Chem. Ind. London*, **1968**, 1117.
79. M. Ochiai, R. Marumoto, S. Kobayashi, H. Shimadzu, and K. Morita, *Tetrahedron*, **24**, 5731 (1968).
80. K. Morita, S. Kobayashi, H. Shimadzu, and M. Ochiai, *Tetrahedron Letters*, **1970**, 861.
81. S. Kobayashi, *Bull. Chem. Soc. Japan* **46**, 2835 (1973).
82. W. Jentsch and M. Seefelder, *Chem. Ber.*, **98**, 274 (1965).
83. S. Seshadri, M. S. Sardessai, and M. A. Betrabet, *J. Indian Chem. Soc.*, **7**, 667 (1969); *C.A.*, **71**, 112732n.
84. L. Horner, H. Oediger, and H. Hoffmann, *Ann. Chem.*, **626**, 26 (1959).
85. H. J. Bestmann, J. Lienert, and L. Mott, *Ann. Chem.*, **718**, 24 (1968).
86. R. Appel, R. Kleinstück, and K. D. Ziehn, *Chem. Ber.*, **104**, 1030 (1971).
87. R. Appel, R. Kleinstück, and K. D. Ziehn, *Chem. Ber.*, **104**, 1335 (1971).
88. R. Appel, R. Kleinstück, and K. D. Ziehn, *Angew. Chem.*, **83**, 143 (1971).
89. R. Appel, K. Warning, and K. D. Ziehn, *Chem. Ber.*, **106**, 3450 (1973).
90. R. Appel, K. D. Ziehn, and K. Warning, *Chem. Ber.*, **106**, 2093 (1973).
91. R. Appel, K. Warning, and K. D. Ziehn, *Chem. Ber.*, **107**, 698 (1974).
92. R. Appel, K. Warning, K. D. Ziehn, and A. Gilak, *Chem. Ber.*, **107**, 2671 (1974).
93. G. I. Matyushecheva, V. S. Mikhailov, and L. M. Yagupol'skii, *Zh. Org. Khim.*, **10**, 124 (1974); *Cheminform* **1974**, 19–344, p. 145.
94. G. Buono, *Tetrahedron Letters*, **1973**, 5167.
95. G. A. Wiley, R. L. Herschkowitz, B. M. Rein, and B. C. Chung, *J. Am. Chem. Soc.*, **86**, 964 (1964).
96. F. Cramer, S. Rittner, W. Reinhard, and P. Desai, *Chem. Ber.*, **99**, 2252 (1966).
97. F. Cramer and M. Winter, *Chem. Ber.*, **94**, 989 (1961).
98. R. Raetz and O. J. Sweeting, *J. Org. Chem.*, **28**, 1608 (1963).

REFERENCES

99. P. I. Alimov, L. N. Antokhina, and I. V. Cheplanova, *Izv. Akad. Nauk SSSR Ser. Khim.*, **1972**, 147; *C.A.*, **77**, 4834n.
100. E. Bamberger and J. Lorenzen, *Ann. Chem.*, **273**, 269 (1893).
101. A. J. Hill and M. V. Cox, *J. Am. Chem. Soc.*, **48**, 3214 (1926).
102. T. L. Davis and W. E. Yelland, *J. Am. Chem. Soc.*, **59**, 1998 (1937).
103. M. Sen and J. N. Ray, *J. Chem. Soc.*, **1926**, 646.
104. A. J. Hill and I. Rabinowitz, *J. Am. Chem. Soc.*, **48**, 732 (1926).
105. J. M. Muchowski, *Can. J. Chem.*, **52**, 2255 (1974).
106. H. Bredereck, R. Gompper, H. Rempfer, K. Klemm, and H. Keck, *Chem. Ber.*, **92**, 329 (1959).
107. G. B. Bachmann and R. W. Dowens, U.S. Pat. 3,054,829 (1962); *C.A.*, **58**, 3730g (1963).
108. G. B. Bachmann, U.S. Pat. 3,169,150 (1965); *C.A.*, **62**, 14500d.
109. F. Reicheneder, K. Dury, and P. Dimroth, Fr. Pat. 1,413,606 (1965); *C.A.* **64**, 5107h.
110. N. Schindler and W. Plöger, *Chem. Ber.*, **104**, 969 (1971).
111. N. Schindler, Ger. Pat. 2,053,840 (1972); *C.A.*, **77**, 34006m.
112. B. I. Stepanov and G. I. Migachev, *Zh. Vses. Khim. Obshchestva im. D. I. Mendeleeva*, **11**, 472 (1966); *C.A.*, **65**, 18493b (1966).
113. B. I. Stepanov and G. I. Migachev, *Z. Obsch. Chim.*, **38**, 194 (1968); *C.A.*, **69**, 35624h (1968).
114. B. I. Stepanov and G. I. Migachev, Russ. Pat. 313,828 (1968); *C.A.*, **70**, 11372u (1969).
115. Z. B. Kiro and B. I. Stepanov, *Zh. Org. Khim.*, **7**, 2196 (1971).
116. H. Hartel, H. Pritzkow, and J. Jander, *Chem. Ber.*, **103**, 652 (1970).
117. J. C. Graham and D. H. Morr, *Can. J. Chem.*, **50**, 3857 (1972).
118. H. Gold, *Angew. Chem.*, **72**, 956 (1960).
119. R. R. Schmidt, Dissertation, University of Stuttgart, 1962.
120. R. R. Schmidt, *Chem. Ber.*, **103**, 3791 (1970).
121. Ryohei Oda and Keiji Yamamoto, *Nippon Kagaku Zasshi*, **83**, 1292 (1962); *C.A.*, **59**, 11399.
122. M. D. Scott and H. Spedding, *J. Chem. Soc. C*, **1968**, 1603.
123. R. Wittmann, *Angew. Chem.*, **73**, 219 (1961).
124. S. Furumoto, *Nippon Kagaku Zasshi*, **92**, 1055 (1971); *C.A.*, **76**, 153720d (1972).
125. E. Kühle, *Angew. Chem.*, **74**, 861 (1962).
126. T. Kojtscheff, F. Wolf, and G. Wolter, *Z. Chem.*, **6**, 148 (1966).
127. K. Harsanyl, D. Korbonits, and P. Kiss, *Acta Chim. Acad. Sci. Hung.*, **77**, 333 (1973).
128. P. A. S. Smith and N. W. Kalenda, *J. Org. Chem.*, **23**, 1599 (1958).
129. H. K. Hall, *J. Am. Chem. Soc.*, **78**, 2717 (1956).
130. J. D. Albright, E. Benz, A. E. Lanzilotti, and L. Goldman, *Chem. Commun.*, **17**, 413 (1965).
131. J. Witte and R. Huisgen, *Chem. Ber.*, **91**, 1129 (1958).
132. W. Hertler and E. J. Corey, *J. Org. Chem.*, **23**, 1221 (1958).
133. J. Hagedorn and H. Tönjes, *Pharmazie*, **12**, 567 (1957).
134. J. Casanova, R. E. Schuster, N. D. Werner, *J. Chem. Soc.*, **1963**, 4280.
135. M. Lipp, F. Dallacker, and I. Meier zu Kocker, *Monatsh. Chem.*, **90**, 41 (1959).
136. C. R. Stephens, E. J. Bianoc, and F. J. Pilgrim, *J. Am. Chem. Soc.*, **77**, 1701 (1955).
137. G. Lohaus, *Chem. Ber.*, **100**, 2719 (1967).
138. J. K. Rasmussen and A. Hassner, *Synthesis*, **1973**, 682.
139. H. E. Ulery, *J. Org. Chem.*, **30**, 2464 (1965).
140. A. E. Kulikova and F. A. Ekstrin, *Z. Org. Chim.*, **4**, 1945 (1968); *C.A.*, **69**, 86541 (1969).
141. H. Schindlbauer, *Monatsh. Chem.*, **100**, 1590 (1969).
142. N. Steiger, U.S. Pats. 3,073,851 (Jan. 15, 1963), 3,133,078 (May 12, 1964), 3,182,053 (May 4, 1965).

143. C. L. Dickinson, W. J. Middleton, and V.A. Engelhardt, *J. Org. Chem.*, **27**, 2470 (1962).
144. R. Ohme and E. Schmitz, *Angew. Chem.*, **79**, 531 (1967).
145. K. Morita, S. Magouchi, and M. Nichikawa, *Chem. Pharm. Bull. Japan*, **7**, 896 (1959).
146. P. L. Scott and J. A. Barry, *Tetrahedron Letters*, **1968**, 2457.
147. H. Bredereck, R. Gompper, and H. Seiz, *Chem. Ber.*, **90**, 1837 (1957).
148. R. Boudet, *Compt. Rend.*, **239**, 1803 (1954).
149. H. Bredereck, F. Effenberger, and G. Simchen *Chem. Ber.*, **96**, 1350 (1963).
150. K. Ikawa, F. Takami, Y. Fukui, and K. Tokuyama, *Tetrahedron Letters*, **1969**, 3279.
151. F. H. Suydam, W. E. Greth, and N. R. Langermann, *J. Org. Chem.*, **34**, 292 (1969); M. Itoh, *Chem. Pharm. Bull (Tokyo)*, **18**, 784 (1970).
152. R. R. Koganty, M. B. Shambhue, and G. A. Digenis, *Tetrahedron Letters*, **1973**, 4511.
153. V. A. Pattison, J. G. Solson, and R. L. K. Carr, *J. Org. Chem.*, **33**, 1084 (1968).
154. Z. Arnold, *Collection Czech. Chem. Commun.*, **24**, 760 (1959).
155. H. Gold and O. Bayer, *Chem. Ber.*, **94**, 2594 (1961).
156. H. Gold, Ger. Pat. 1,146,892 (1963); *C.A.*, **59**, 10070h (1963).
157. K. H. König and H. Pommer, Ger. Pat. 1,173,087 (July 2, 1964); *C.A.*, **61**, 11899b (1964).
158. A. Roedig and W. Wenzel, *Angew. Chem.*, **81**, 36 (1969).
159. A. J. Speziale, L. R. Smith, and J. E. Fedder, *J. Org. Chem.*, **30**, 4303 (1965).
160. D. E. Horning and J. M. Muchowski, *Can. J. Chem.*, **45**, 1247 (1967).
161. H. Finkbeiner, *J. Org. Chem.*, **30**, 2861 (1965).
162. K. Dickore and F. Kröhnke, *Chem. Ber.*, **93**, 1068, 2479 (1960).
163. H. G. Nordmann and F. Kröhnke, *Angew. Chem.*, **81**, 747 (1969).
164. H. G. Nordmann and F. Kröhnke, *Ann. Chem.*, **731**, 80 (1970).
165. D. Davidson and H. Skovronek, *J. Am. Chem. Soc.*, **80**, 376 (1958).
166. G. A. Reynolds and J. A. V. Allan, *J. Org. Chem.*, **34**, 2736 (1969).
167. H. Khedija, H. Strzelecka, and M. Simalty, *Bull. Soc. Chim. France*, **1973**, 218.
168. F. Eiden, *Arch. Pharm. (Weinheim)*, **295**, 533 (1962).

ADDENDUM*

To Section I-A

Further ^{1}H, ^{13}C, ^{35}Cl-NMR-spectroscopic investigations on Vilsmeier-Haack adducts (dealing with kinetics of formation, restriction of rotation around the partial carbon-nitrogen double bond) have been performed (169, 170).

To Section I-B-1

A lot of amidines and heterocycles were prepared with the aid of the DMF/POCl$_3$ adduct or of POCl$_3$. Some more recent results are listed below.

$$\underset{NH_2}{\overset{NHCO_2C_2H_5}{\bigcirc}} \xrightarrow[\text{2. NaOH,H}_2\text{O}]{\text{1. DMF/POCl}_3} \underset{N=CH-N(CH_3)_2}{\overset{NHCO_2C_2H_5}{\bigcirc}} \quad \text{(ref. 171)}$$

$$C_6H_5-N\underset{O}{\overset{N}{\diagup}}\!\!\!\diagdown NH_2 \xrightarrow{\text{DMF/POCl}_3} C_6H_5-N\underset{Cl}{\overset{N}{\diagup}}\!\!\!\diagdown\underset{CHO}{N=CH-N(CH_3)_2} \quad \text{(ref. 172)}$$

*Translated from German by W. Kantlehner, and by J. H. Benirschke, Fachhochschule Aalen, 708-Aalen, Hohenstaufenstrasse 1, Germany.

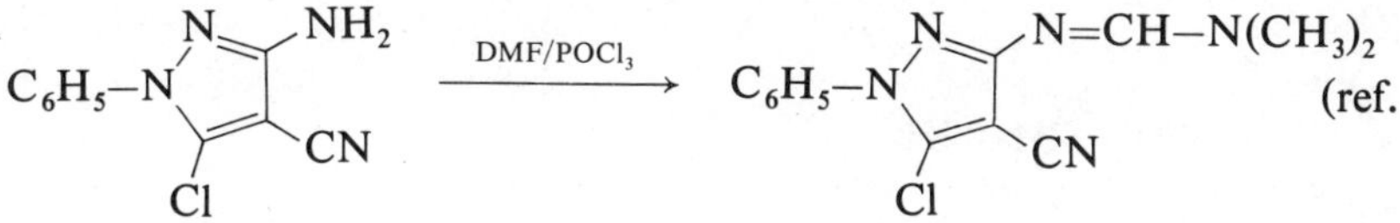

C_6H_5-N-pyrazole(NH_2, Cl, CN) $\xrightarrow{DMF/POCl_3}$ C_6H_5-N-pyrazole($N=CH-N(CH_3)_2$, Cl, CN) (ref. 172)

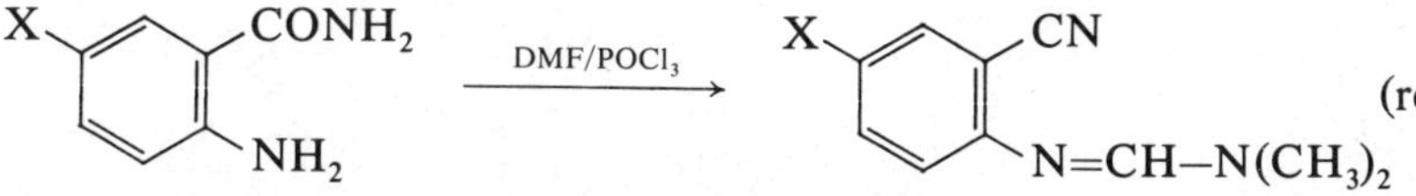

X-C$_6$H$_3$(CONH$_2$)(NH$_2$) $\xrightarrow{DMF/POCl_3}$ X-C$_6$H$_3$(CN)(N=CH-N(CH$_3$)$_2$) (ref. 173)

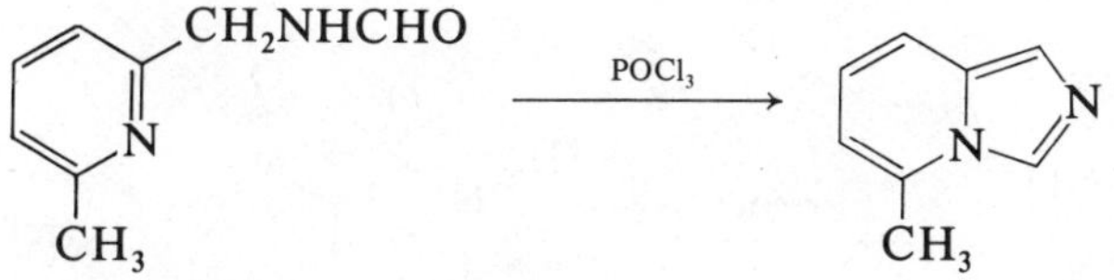

2-(CH$_2$NHCHO)-6-methylpyridine $\xrightarrow{POCl_3}$ imidazo[1,5-a]pyridine with CH$_3$ (Ref. 174)

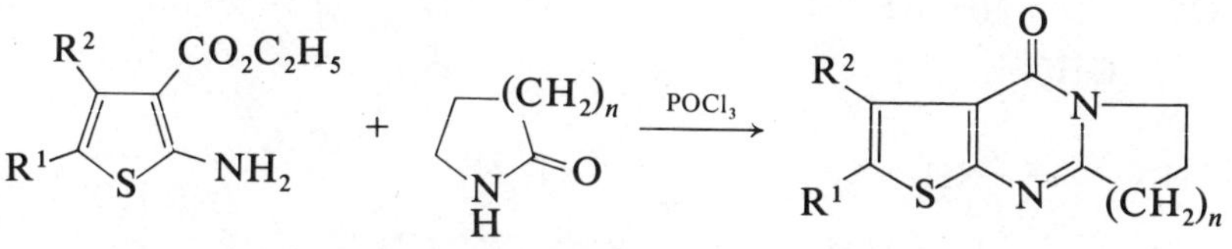

thiophene(R^2, $CO_2C_2H_5$, R^1, NH_2) + pyrrolidinone-(CH$_2$)$_n$ $\xrightarrow{POCl_3}$ fused thieno-pyrimidinone (R^1, R^2, (CH$_2$)$_n$)

$n = 1, 2, 3$ (Ref. 175)

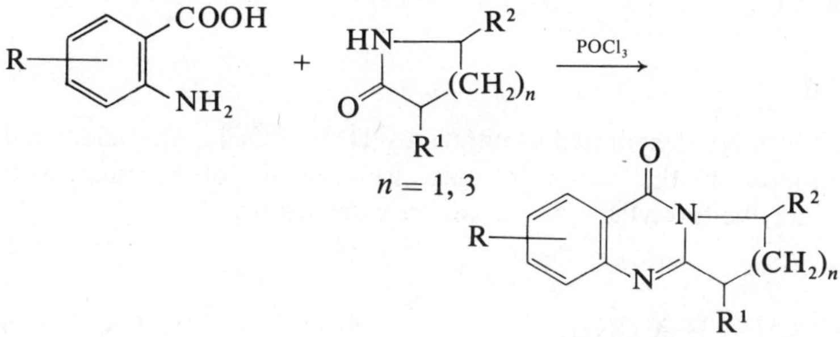

R-C$_6$H$_3$(COOH)(NH$_2$) + HN-CH(R^2)-(CH$_2$)$_n$-CH(R^1)-C(=O) $\xrightarrow{POCl_3}$

$n = 1, 3$

fused quinazolinone (R, R^1, R^2, (CH$_2$)$_n$)   (ref. 176)

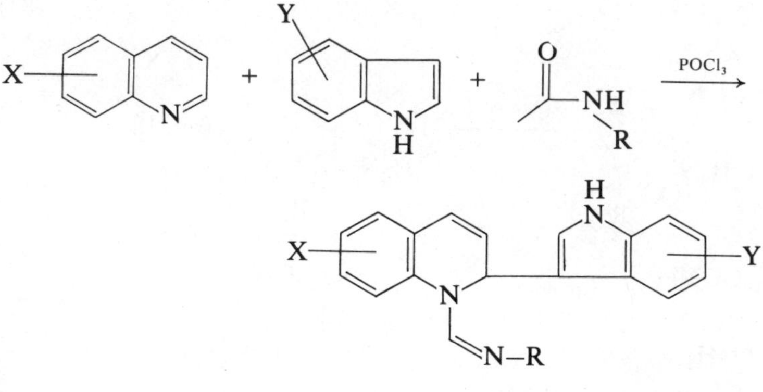

X-quinoline + Y-indole + CH$_3$C(=O)-NH-R $\xrightarrow{POCl_3}$

product: X-quinoline-2-yl-3-indolyl-Y with N-CH=N-R   (ref. 177)

(ref. 178)

(ref. 179)

(ref. 180)

X = Cl, I

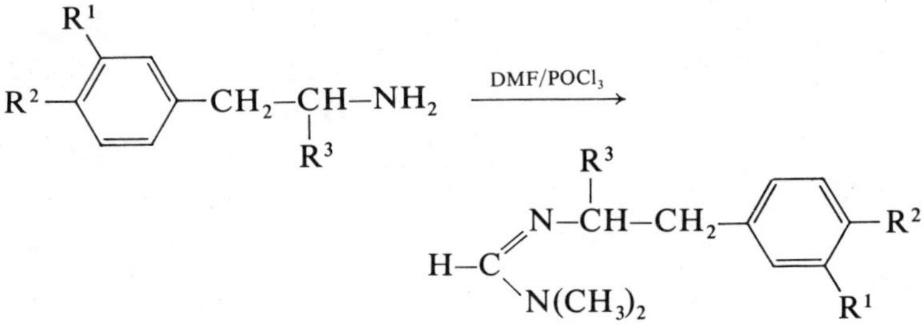

To Section I-B-3

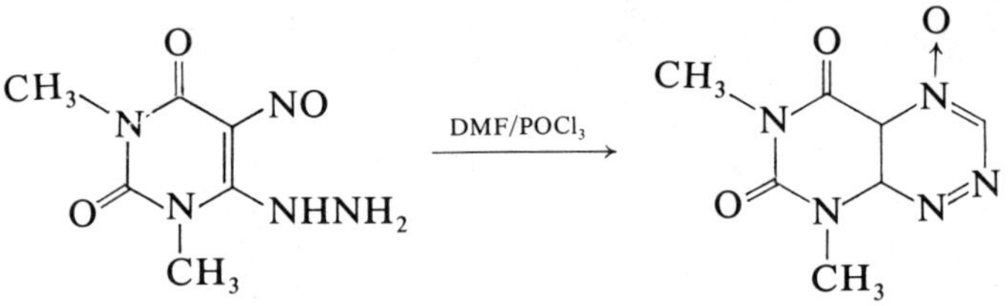

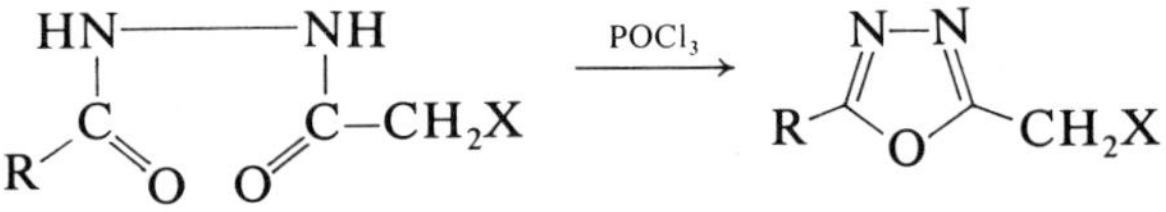

Primary amides are dehydrated to nitriles by DMF/$POCl_3$. Any additional functional groups in the molecule that are capable of reacting with DMF/$POCl_3$ may be formylated. Examples are given below.

(ref. 181)

(ref. 182)

(ref. 173)

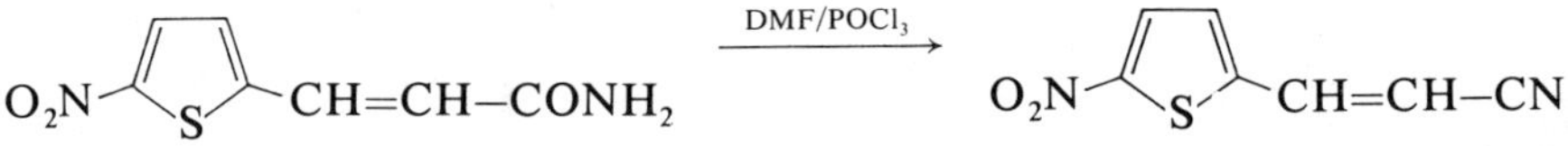

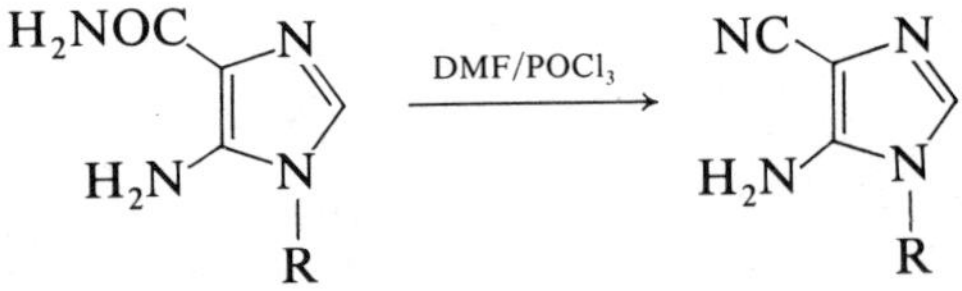

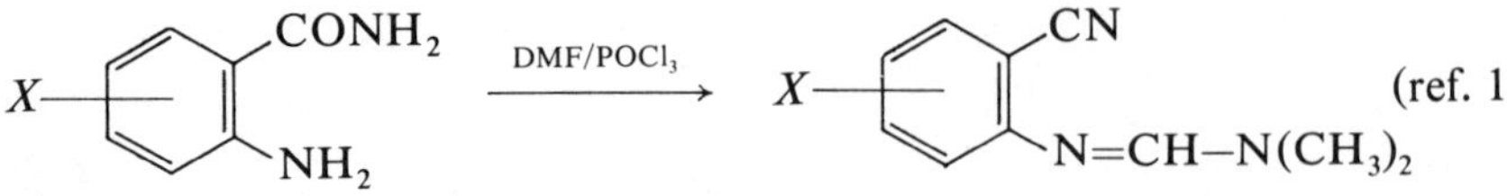

Dehydration of carboxylic acid amides by means of $POCl_3$/tertary amine is a standard procedure for the preparation of nitriles. This reaction is very likely to run via amide–$POCl_3$ adducts. Reviews, showing the scope of the reaction, have been published (183, 184). More recent examples are given by the following formulas:

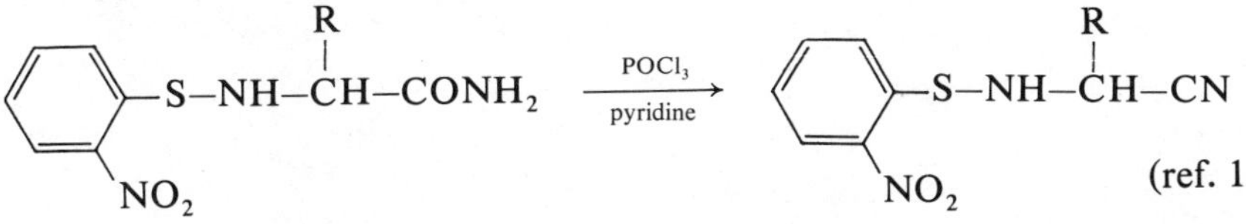

$$\underset{NO_2}{\underset{|}{C_6H_4}}-S-NH-\underset{\underset{R}{|}}{CH}-CONH_2 \xrightarrow[\text{pyridine}]{POCl_3} \underset{NO_2}{\underset{|}{C_6H_4}}-S-NH-\underset{\underset{R}{|}}{CH}-CN \quad \text{(ref. 185)}$$

benzothiazole-CONH$_2$ $\xrightarrow[\text{pyridine}]{POCl_3}$ benzothiazole-CN (ref. 186)

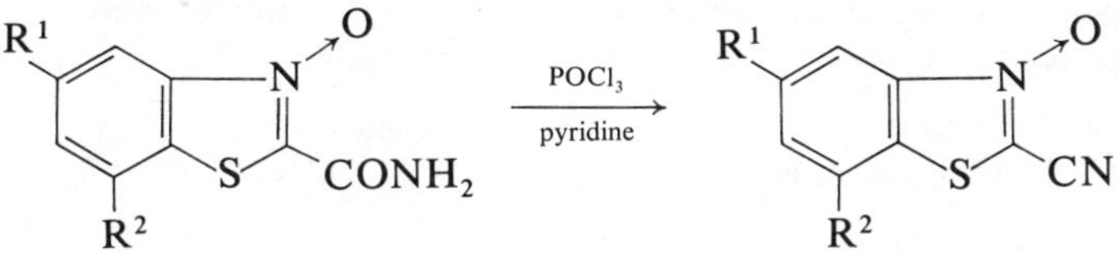

To Section I-B-4

Further isonitriles have been prepared from N-formylamines and $POCl_3$/triethylamine (187, 188), whereby special interest should be extended to α-isocyanonitriles, which have proved to be useful synthetic tools (189).

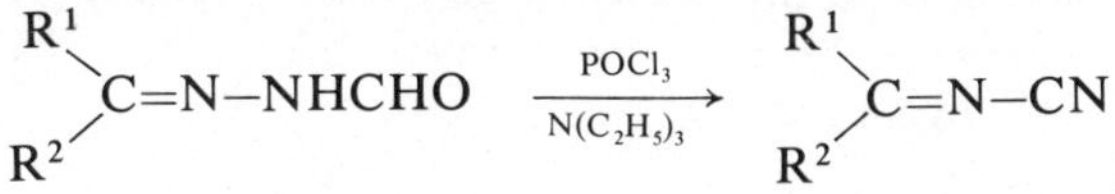

$$\underset{R^2}{\overset{R^1}{>}}C=N-NHCHO \xrightarrow[N(C_2H_5)_3]{POCl_3} \underset{R^2}{\overset{R^1}{>}}C=N-CN \quad \text{(ref. 188)}$$

$$R^1-\underset{\underset{NHCHO}{|}}{CH}-CN \xrightarrow[N(C_2H_5)_3]{POCl_3} R^1-\underset{\underset{NC}{|}}{CH}-CN \quad \text{(ref. 187)}$$

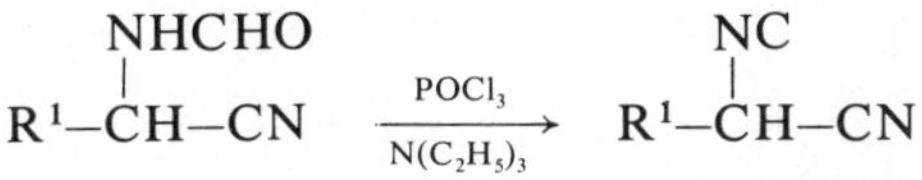

To Section I-B-7

Further examples for the preparation of carboxylic acid amide thioacetals from amide–$POCL_3$ adducts are given in ref. 190.

To Section I-B-8

Action of DMF–$POCl_3$ on (N-formyl)-3-amino-thiophene results in the formulation of the heteroaromatic nucleus; additionally, the N-formyl group is transformed to an amidine function to some extent (191).

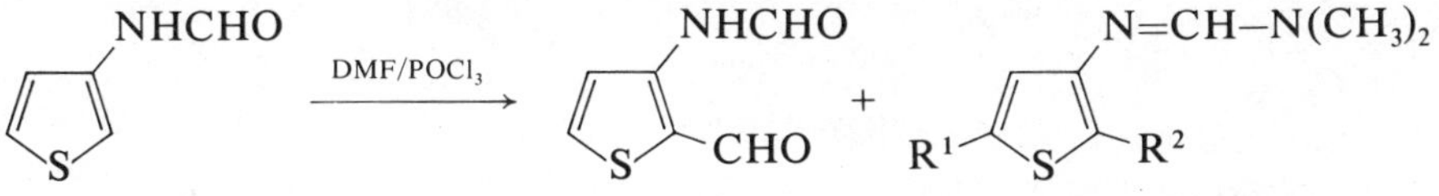

thiophene-NHCHO $\xrightarrow{DMF/POCl_3}$ thiophene(NHCHO)(CHO) + thiophene(N=CH–N(CH$_3$)$_2$)(R^1)(R^2)

(a) $R^1 = R^3 = H$ (b) $R^2 = CHO, R^4 = H$ (c) $R^2 = H, R^4 = CHO$

1,4-Benzodiazepines (**137**) react with formamide/$POCl_3$ to form pyrimidine-annealed-1,4-benzodiazepines (**138**) (192).

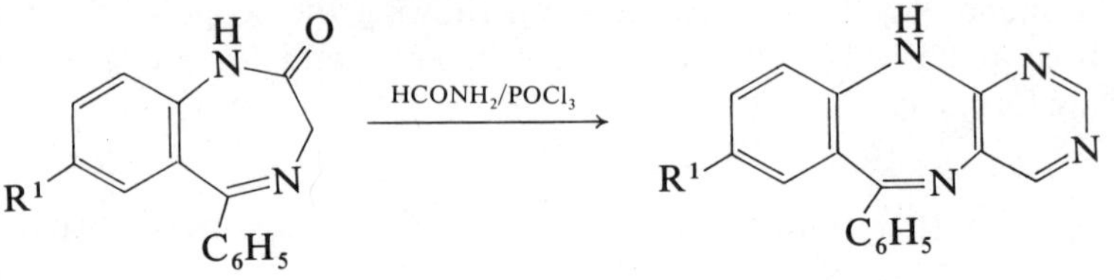

$POCl_3$ cyclizes the *N*-acylated aminoacid amide (**139**) to the corresponding oxazole (**140**) (193).

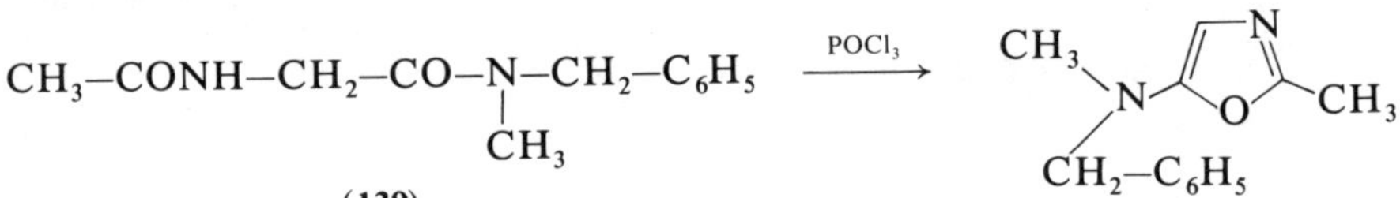

$CH_3-CONH-CH_2-CO-N-CH_2-C_6H_5$
$\quad\quad\quad\quad\quad\quad\quad\quad\quad\;\; |$
$\quad\quad\quad\quad\quad\quad\quad\quad\quad\; CH_3$
(**139**)

$\xrightarrow{POCl_3}$

(**140**)

Section I-B-9

Reaction with naphtholes. In the presence of $POCl_3$, malonic acid–ester–dialkylamides react with naphtholes to yield keten-*O,N*-acetals (**141**), which may be cyclized to form (**142**) (194, 195).

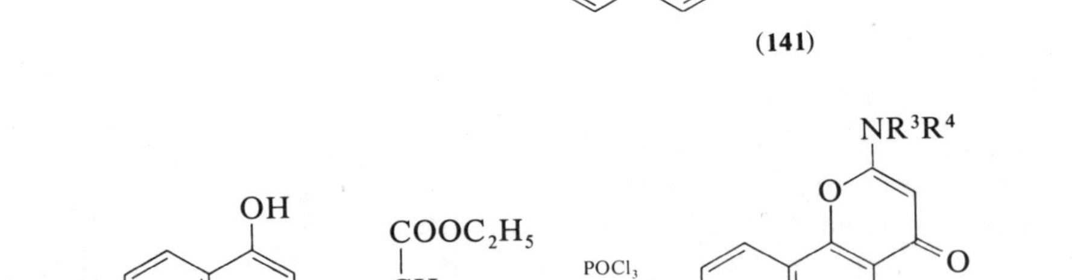

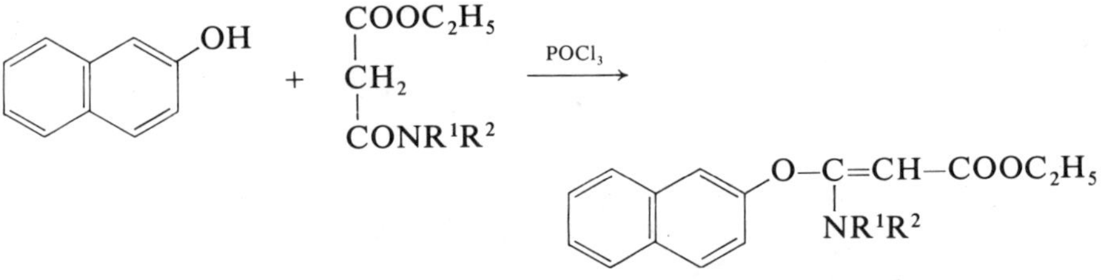

Section I-B-10

Reaction with nitriles. Iminium salts (**143, 144**) are formed by the reaction of nitriles and *NN*-dialkyl-formamide–POCl$_3$ adducts in the presence of HCl (196).

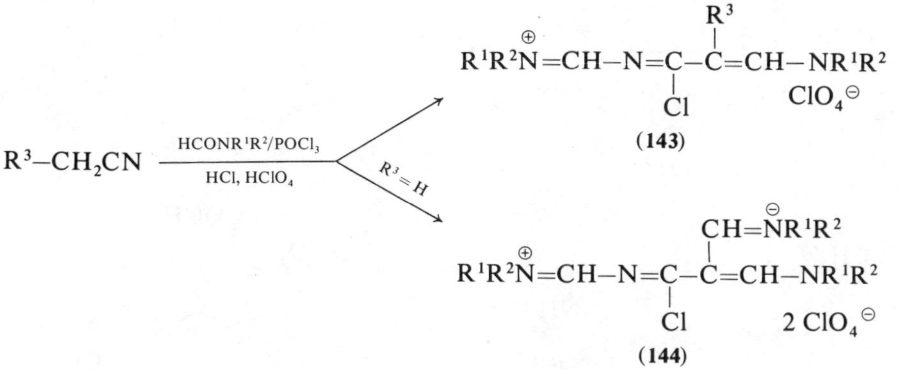

$$R^3-CH_2CN \xrightarrow[\text{HCl, HClO}_4]{\text{HCONR}^1R^2/\text{POCl}_3}$$

$$R^1R^2\overset{\oplus}{N}=CH-N=C-\underset{\underset{Cl}{|}}{\overset{\overset{R^3}{|}}{C}}=CH-NR^1R^2 \quad ClO_4^{\ominus}$$

(**143**)

$R^3 = H$

$$R^1R^2\overset{\oplus}{N}=CH-N=C-\underset{\underset{Cl}{|}}{\overset{\overset{\overset{\ominus}{CH}=NR^1R^2}{|}}{C}}=CH-NR^1R^2 \quad 2\,ClO_4^{\ominus}$$

(**144**)

The nitrile (**145**) is reported to react with DMF/POCl$_3$ to give the adduct (**146**), which is highly reactive toward nucleophilic reagents (197).

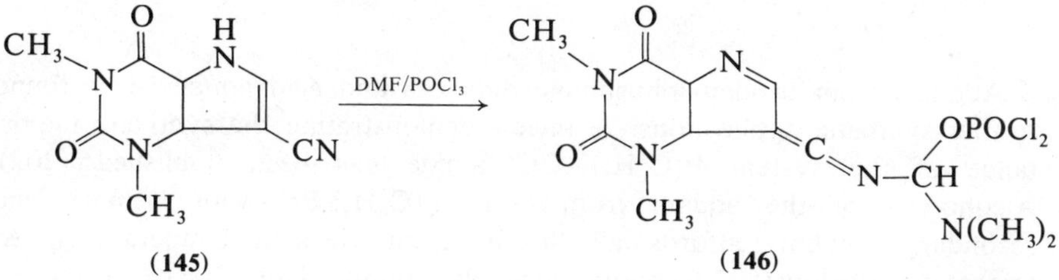

[Structure of **145**] →(DMF/POCl$_3$)→ [Structure of **146**]

(**145**)   (**146**)

Section I-B-11

Reduction of carboxylic acid amide–POCl$_3$ adducts. Aducts from POCl$_3$ and aromatic carboxylic acid amides may be reduced to amines or aldehydes, depending on the reducing reagents used. Amines are obtained by NaBH$_4$ (198) or Zn/ethanol (199), respectively. Aldehydes are formed by reduction with Zn/H$_2$O (200).

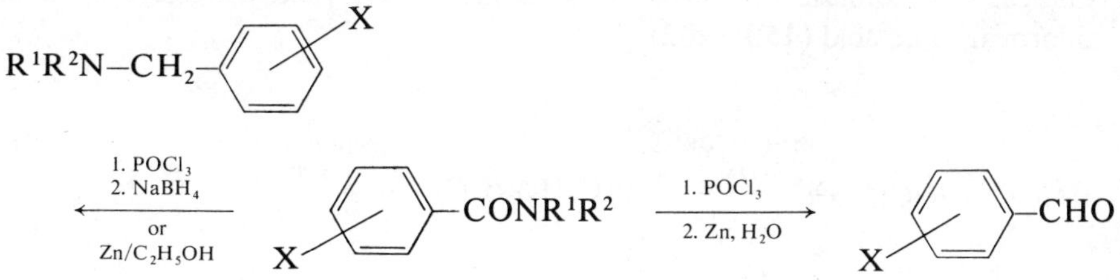

$$R^1R^2N-CH_2-C_6H_4-X \xleftarrow[\text{Zn/C}_2\text{H}_5\text{OH}]{\substack{\text{1. POCl}_3 \\ \text{2. NaBH}_4 \\ \text{or}}} X-C_6H_4-CONR^1R^2 \xrightarrow[\text{2. Zn, H}_2\text{O}]{\text{1. POCl}_3} X-C_6H_4-CHO$$

Section 1-B-12

Miscellaneous reactions. Adducts from heterocyclic annealed lactams, such as **147**, undergo acylation at the aromatic nucleus on reaction with acetic acid-anhydride (201).

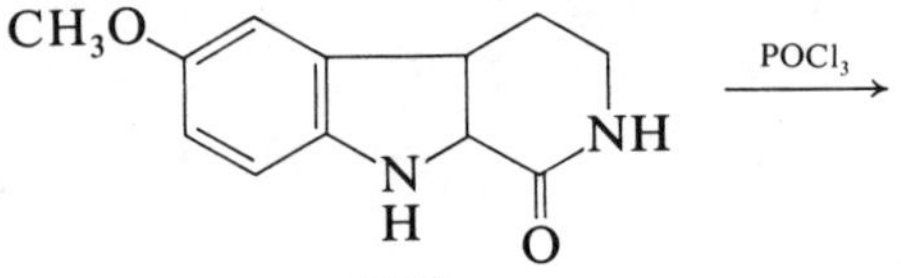

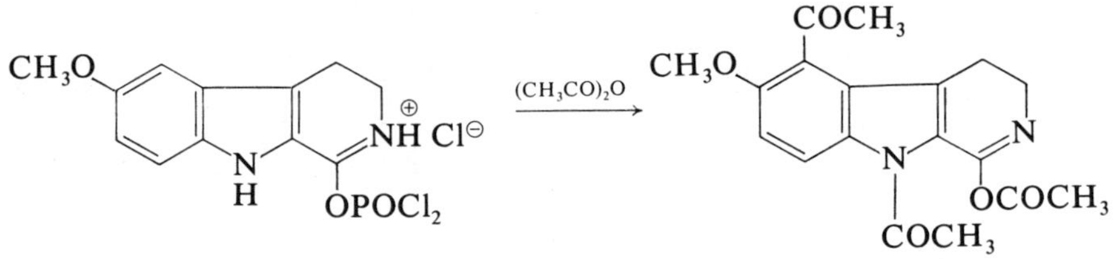

To Section II

Adducts from triphenylphosphane dihalides and acid amides have found further synthetic applications. A review demonstrating the synthetic importance of the system $P(C_6H_5)_3$–CCl_4-amide has been published (202). Alcoholysis of the adduct from DMF–$P(C_6H_5)_3Br_2$ with primary and secondary alcohols affords alkylbromides at elevated temperatures. At temperatures below 0°C secondary alcohols yield formiates (203).

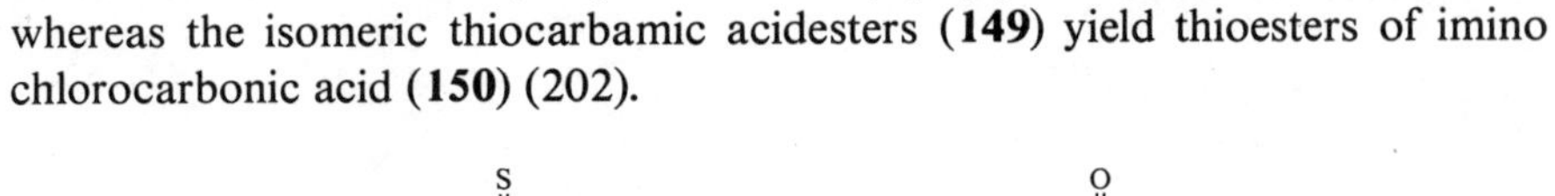

The system DMF–$P[N(CH_3)_2]_3$–CCl_4 has proved to be useful for esterification of hydroxy groups of partially protected carbohydrates (204). Thiocarbamidic acid-*O*-esters (**148**) react with $/(C_6H_5)_3$–CCl_4 to form isocyanates, whereas the isomeric thiocarbamic acidesters (**149**) yield thioesters of imino chlorocarbonic acid (**150**) (202).

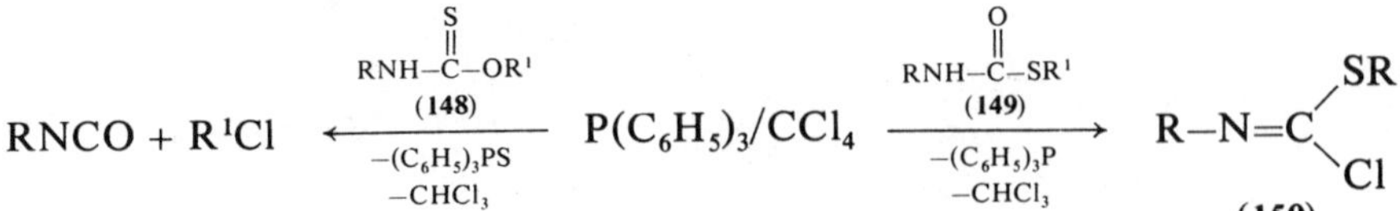

Ketene imines were synthesized from acid amides and $P(C_6H_5)_3$–CCl_4 (205, 206).

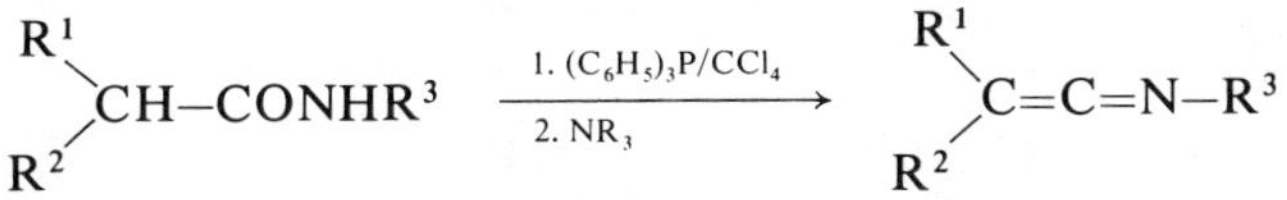

$$R^1R^2CH-CONHR^3 \xrightarrow[\text{2. NR}_3]{\text{1. }(C_6H_5)_3P/CCl_4} R^1R^2C=C=N-R^3$$

Further examples for preparations of imidoyl-halides using the system $P(C_6H_5)_3/CX_4$ ($x =$ Cl, Br) have been reported (207, 208, 209).

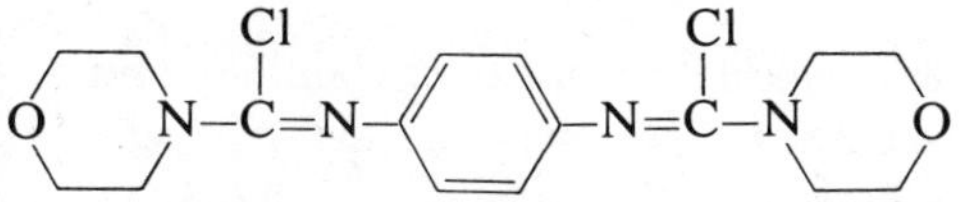

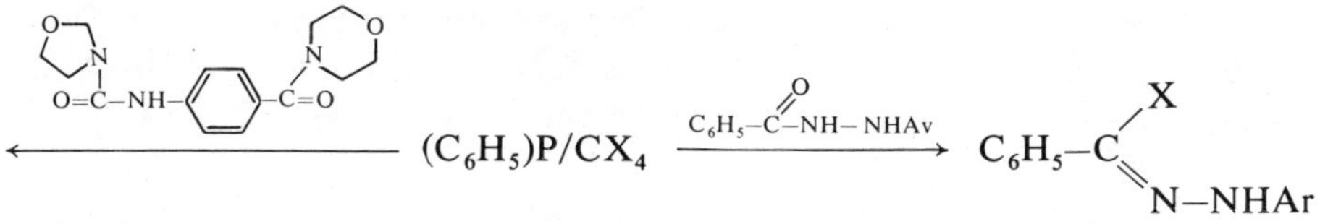

$$\xleftarrow{} (C_6H_5)P/CX_4 \xrightarrow{C_6H_5-\overset{O}{\overset{\|}{C}}-NH-NHAv} C_6H_5-C\underset{N-NHAr}{\overset{X}{\diagup}}$$

Succinimides (151) react with $P(C_6H_5)_4/CCl_4$ to form a mixture of compounds 152 and 153 (210).

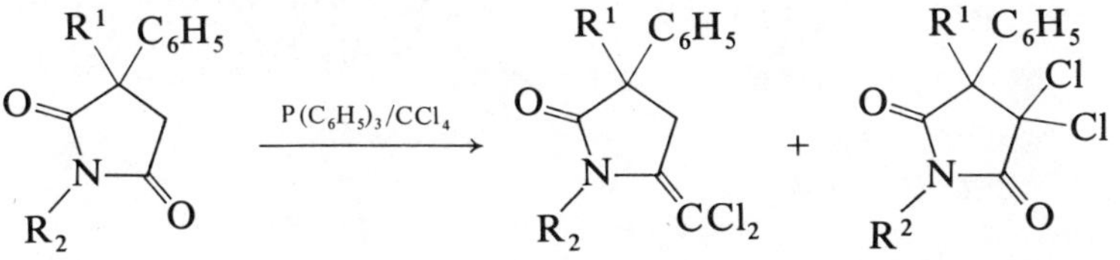

(151) $\xrightarrow{P(C_6H_5)_3/CCl_4}$ (152) + (153)

To Section IV

Alcoholysis of the DMF–PCl_3-adduct yields alkylchlorides; remarkably, only two chlorine atoms of the adduct (assuming a 1:1 composition) can be utilized in this reaction (211).

$$ROH \xrightarrow{DMF/POCl_3} RCl$$

Based on this finding two new structures (154, 155) for the adduct have been proposed (211).

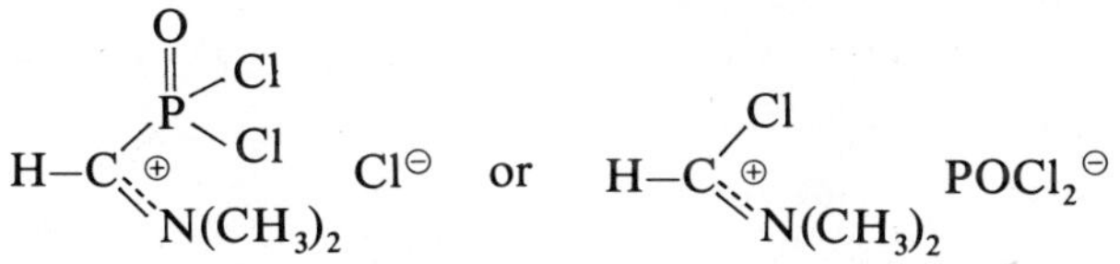

(154)   or   (155)

Dialkylformamide/PX_3 adducts ($x = Cl$, Br) react with acetic acid, followed by hydrolysis, to afford dialkylaminomethylene diphosphonic acids (**156**) (212).

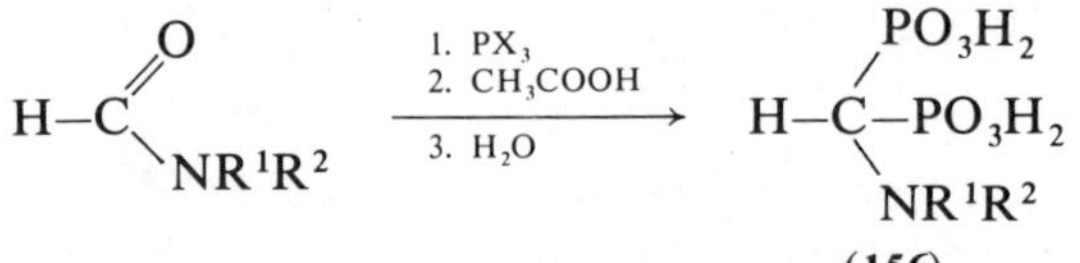

$$H-\underset{NR^1R^2}{\overset{O}{C}} \xrightarrow[3.\ H_2O]{1.\ PX_3 \quad 2.\ CH_3COOH} H-\underset{NR^1R^2}{\overset{PO_3H_2}{\underset{|}{C}}-PO_3H_2}$$

(**156**)

This reaction seems to point out that in adducts from dialkylamides and phosphorous trihalides there are two molecules PX_3 to one molecule of amide. With this assumption for amide–PCl_3 adducts two other structures (**157, 158**), which form perhaps an equilibrium should be taken into consideration. These structures would explain also why there are only two halogen atoms (based on PCl_3) available for alkylhalogenide formation.

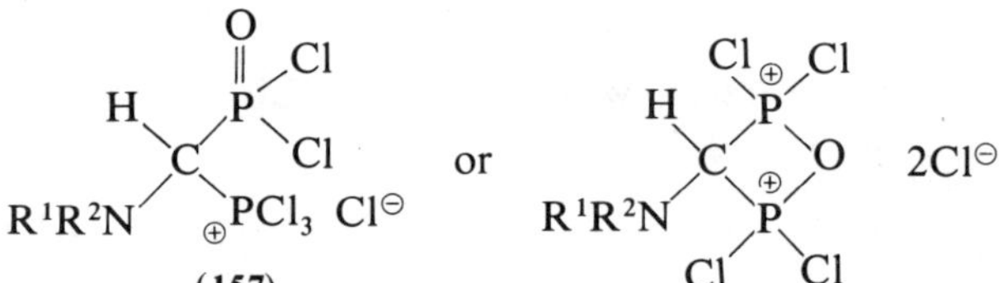

(**157**)   or   (**158**)

To Section VIII

Toluenesulfonylchloride/pyridine was reacted with the cyclic carbamate **159** to give **160** (213).

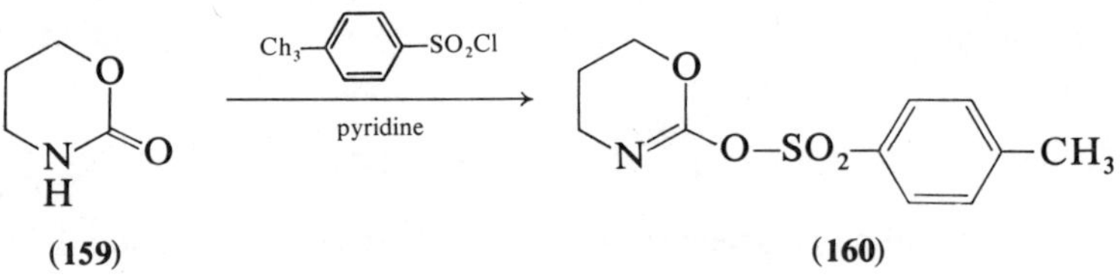

(**159**) → (**160**)

NN'-Disubstituted ureas can be dehydrated by toluenesulfonylchloride to form carbodiimides (214).

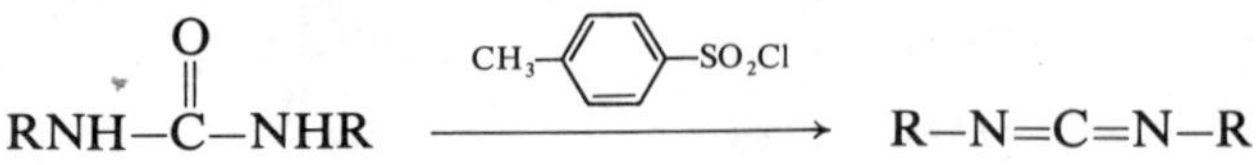

$$RNH-\overset{O}{\underset{\|}{C}}-NHR \xrightarrow{CH_3-\bigcirc-SO_2Cl} R-N=C=N-R$$

To Section X

Nitriles can be prepared from carboxylic acid amides using chloroformic acid ester as dehydrating reagent (215).

$$\underset{H}{\overset{H}{>}}C\overset{CONH_2}{\underset{\parallel}{=}}\overset{}{\underset{C}{\underset{\diagdown}{\diagup}}}\overset{}{\underset{COOH}{\diagdown}} \quad \xrightarrow{ClCOOC_2H_5} \quad \underset{H}{\overset{H}{>}}C\overset{CN}{\underset{\parallel}{=}}\overset{}{\underset{C}{\underset{\diagdown}{\diagup}}}\overset{}{\underset{COOC_2H_5}{\diagdown}}$$

REFERENCES

169. G. Jugie, J. A. S. Smith, and G. J. Gèrard, *J. Chem. Soc. Perkin. II*, **1975**, 925.
170. M. Martin, G. Ricolleau, S. Poignant, and G. Martin, *J. Chem. Soc., Perkin. II*, **1976**, 182.
171. W. Schulze, P. Held, and A. Junov, *Z. Chem.*, **15**, 184 (1975).
172. S. B. Barnela, R. S. Pandit, and S. Seshadri, *Indian J. Chem., Sect.*, **B14**, 668 (1976); *C. I.*, 10-255 (1977).
173. C. H. Foster and E. U. Elam, *J. Org. Chem.*, **41**, 2646 (1976).
174. O. Fuentes and W. W. Paudler, *J. Org. Chem.*, **40**, 1210 (1975).
175. V. I. Shvedov, I. A. Kharizomenova, and A. N. Grinev, *Khim. Geterosikl. Soedin*, **1975**, 765; *C. I.*, 39-235 (1975).
176. Kh. M. Shakhidoyatov, A. Irisbaev, L. M. Yun, E. Oripov, and Ch. Sh. Kadyrov, *Khim. Geterotsikl. Soedin*, **1976**, 1564; *C. I.*, 13-278 (1977).
177. R. P. Ryan, R. A. Hamby, and Y. H. Wu, *J. Org. Chem.*, **40**, 724 (1975).
178. N. B. Marchenko, V. G. Granik, R. G. Glushkov, L. I. Budarova, V. A. Kuzovkin, and R. A. Al'tshuler, *Khim. Farm. Zh.*, **10**, 46 (1976); *C. I.*, 2-141 (1977).
179. M. Ichiba, K. Senga, S. Nishigaki, and F. Yoneda, *J. Heterocycl. Chem.*, **14**, 175 (1977).
180. I. Cojocuriu, Z. Cojokaru, and C. Nistor, *Rev. Chim. (Bucaresti)*, **28**, 15 (1977); *C. I.*, 29-224 (1977).
181. M. El-Kerdawy, M. N. Tolba, A.-G. El-Agamey, *Acta Pharm. Jugosl.*, **26**, 141 (1976); *C. I.*, 33-214 (1976).
182. A. K. Sen and S. Ray, *Indian J. Chem. Sect.*, **B14**, 346 (1976); *C. I.*, 43-209 (1976).
183. P. Kurtz, in Houben–Weyl–Müller, *Methoden der Org. Chemie*, Vol. 8, Georg Thieme Verlag, Stuttgart, 1952, p. 330.
184. C. A. Buehler and D. E. Pearson, *Survey of Organic Synthesis*, Wiley-Interscience, New York, 1970, p. 951.
185. K. Kawashiro, H. Yoshida, and S. Morimoto, *Chem. Lett.*, **1976**, 417.
186. K. Wagner and L. Oehlmann, *Chem. Ber.*, **109**, 611 (1976).
187. K. Hantke, U. Schöllkopf, and H. H. Hausberg, *Liebigs Ann. Chem.*, **1975**, 1531.
188. P. Jacobsen, *Acta Chem. Scand.*, **B30**, 995 (1976).
189. U. Schöllkopf, *Angew. Chem.*, **89**, 351 (1977).
190. B. P. Federov and F. M. Stoyanovich, *Izv. Akad. SSSR*, **1960**, 1828; C. A., **55**, 14298 (1961).
191. G. Ah-Kow, C. Paulmier, and P. Pastour, *Bull. Soc. Chim. Fr.*, **1976**, 151.
192. S. Kobayashi, *Bull. Chem. Soc. Jap.*, **48**, 302 (1975); *C. I.*, 15-286 (1975).
193. R. G. Harrison, M. R. J. Jolley, and J. C. Saunders, *Tetrahedron Lett.*, **1976**, 293.
194. G. Roma, A. Ermilii, and M. Mazzei, *J. Heterocycl. Chem.*, **12**, 31 (1975).
195. A. Ermilii, A. Balbi, G. Roma, A. Ambrosini, and N. Passerini, *Farm. Ed. Sci.*, **31**, 627 (1976); *C. I.*, 2-190 (1977).

196. J. Liebscher and H. Hartmann, *Collect. Czech. Chem. Commun.*, **41,** 1565 (1976).
197. Y. Okamoto and T. Ueda, *Tetrahedron Lett.*, **1976,** 2317.
198. A. Rahman, A. Basha, N. Waheed, and S. Ahmed, *Tetrahedron Lett.*, **1976,** 219.
199. A. Basha and A. Rahman, *Experientia*, **33,** 101 (1977).
200. Atta-Ur-Rahman and A. Basha, *J. Chem. Soc., Chem. Commun.*, **1976,** 594.
201. L. N. Yakhontov, M. Ya. Uritskaya, O. S. Anisimova, T. Ya. Filipenko, K. F. Turchin, E. M. Peresleni, and Yu. N. Sheinker, *Khim. Geterotsikl. Soedin,* **1975,** 1270; *C. I.,* 51–241 (1975).
202. R. Appel, *Angew. Chem.*, **87,** 863 (1975).
203. R. K. Boeckmann, Jr. and B. Ganem, *Tetrahedron Lett.*, **1974,** 913.
204. S. Czernecki and C. Georgoulis, *C. R. Acad. Sci.*, **C280,** 305 (1975).
205. J. Goerdeler and C. Lindner, *Tetrahedron Lett.*, **1972,** 1519.
206. H. Teichmann and Am Thai, *Z. Chem.*, **17,** 93 (1977).
207. P. Wolkoff, *Can. J. Chem.*, **53,** 1333 (1975).
208. W. Ried, N. Kothe, R. Schweitzer, and A. Höhle, *Chem. Ber.*, **109,** 2921 (1976).
209. C. R. Harrison, P. Hodge, and W. J. Rogers, *Synthesis*, **1977,** 41.
210. C. Gadreau and A. Foucaud, *Bull. Chem. Soc. Chim. Fr.*, **1976,** 2068.
211. A. G. Anderson, Jr., N. E. T. Owen, F. J. Freenor, and D. Erickson, *Synthesis*, **1976,** 398.
212. M. Fukuda, Y. Okamoto, and H. Sakurai, *Bull. Chem. Soc. Japan,* **48,** 1030 (1975); *C. I.,* 23–361 (1975).
213. A. Krantz and B. Hoppe, *Tetrahedron Lett.*, **1975,** 695.
214. J. C. Sheehan, P. A. Cruickshank, and G. L. Boshart, *J. Org. Chem.*, **26,** 2525 (1961).
215. C. K. Sauers and R. J. Cotter, *J. Org. Chem.*, **26,** 6 (1961).

CHLOROMETHYLENIMINIUM SALTS (AMIDE HALIDES)

By W. KANTLEHNER, *D-7080-AALEN-Dewangen, Laachweg 14, Germany*

CONTENTS

I. Synthesis	66
A. From Acid Amides and Halide-Transferring Compounds	66
B. Side Reactions in the Synthesis of Amide Chlorides from Amides	69
1. Addition of Hydrogen Halides to Nitriles and Imidoyl Halides	72
2. Addition of Hydrogen Halides and Halogens to α-Chloroenamines and Ynamines	74
C. Addition of Amines to α,α-Dihalogenated Olefins	75
D. Addition of Dihalocarbenes to Compounds Containing C=N Groups	75
E. Chloromethyleniminium Halides from Phosgeniminium Salts (Dichloromethyleniminium Chlorides)	76
F. Amide Halides via Amide Chlorides	76
1. Chlorination of Amide Chlorides	76
2. Via Coupling of Diazonium Salts with Amide Chlorides	77
3. Via Reactions of Chloromethyleniminium Chlorides with Hydrogen Halides	77
4. Via Self-Condensation of Chloromethyleniminium Chlorides	78
II. Structure and Properties of Halomethyleniminium Salts	78
A. Physical Properties of Amide Halides	80
III. Reactions of Halomethyleniminium Salts	82
A. Reactions with Hydroxy Compounds	82
1. Hydrolysis	82
2. Alcoholysis	83
3. Reactions of Amide Chlorides with Phenols or Other Heterocyclic Compounds Containing Hydroxyl Groups	86
4. Reactions of Amide Chlorides with Carboxylic and Sulfonic Acids	88
B. Reaction of Amide Chlorides with Cyclic Ethers and Acetals	92
C. Reaction with Compounds Containing Thiol Groups	94
1. Thiolysis	94
2. Mercaptolysis	95
D. Reactions with Compounds Containing NH_2 Groups	96
1. Reactions with Primary and Secondary Amines, Hydroxylamines, and Hydrazine Derivatives	96
2. Reactions with Amides and Nitriles	99
3. Reactions with Ureas, Thioureas, and Urethanes	103

Translated from German by Dr. W.-J. Richter, MPI für Kohleforschung, D 4330 Mülheim.

E. Reactions of Amide Chlorides with Tertiary Amines 105
   1. N-Monosubstituted Chloromethyleniminium Salts Without α-CH Bonds . . 105
   2. Reaction of Tertiary Amines with Chloromethyleniminium Salts Containing α-CH Bonds 106
F. Reactions of Amide Chlorides with Alkoxides 107
G. Reactions of Amide Chlorides with Alkali Metal Amides and Imides 111
H. Reactions of Amide Chlorides with Lithium Phosphides 113
I. Anion Exchange with Halomethyleniminium Halides 113
J. Reactions of Halides and Diazonium Salts with Amide Chlorides 115
   1. Reaction Mechanism 115
   2. Halogenation of Amide Chlorides 115
   3. Coupling of Amide Chlorides with Diazonium Salts 116
K. Reactions Proceeding via Oxidation of the Amide Chloride Function 117
L. Thermal Degradation of Halomethyleniminium Halides 118
M. Reduction of Chloromethyleniminium Salts 119
References 120
Addendum 125

I. Synthesis

A. FROM ACID AMIDES AND HALIDE-TRANSFERRING COMPOUNDS

Chloromethyleniminium salts, formerly called amide halides, are generated from amides and halogenating agents. The following compounds are used for halide transfer: phosphorus pentachloride (1–8), phosgene (6–18), thionyl chloride (6, 7, 17, 18, 19, 20), oxalyl chloride (6, 19), perchlorobutyryl chloride (21), pyrocatechol dichloromethylene ether (2,2-dichlorobenzodioxole) (22), carbonyl dibromide (23), and phosphorus pentabromide (23). These reagents attack the oxygen of the amide (**1**) giving polar adducts, for example, **2–6**, which at higher temperatures decompose to chloromethyleniminium salts (**7**) via loss of neutral molecules (6, 7). These reactions are performed in solvents such as ether, chloroform, carbon tetrachloride, toluene, methylene chloride, and benzene.

Chloromethyleniminium chlorides have also been formed from pyrocatechol phosphorus trichloride (**8**) and amides. These compounds were not isolated, but were decomposed in a von Braun type of reaction (24). Perchlorobutyryl

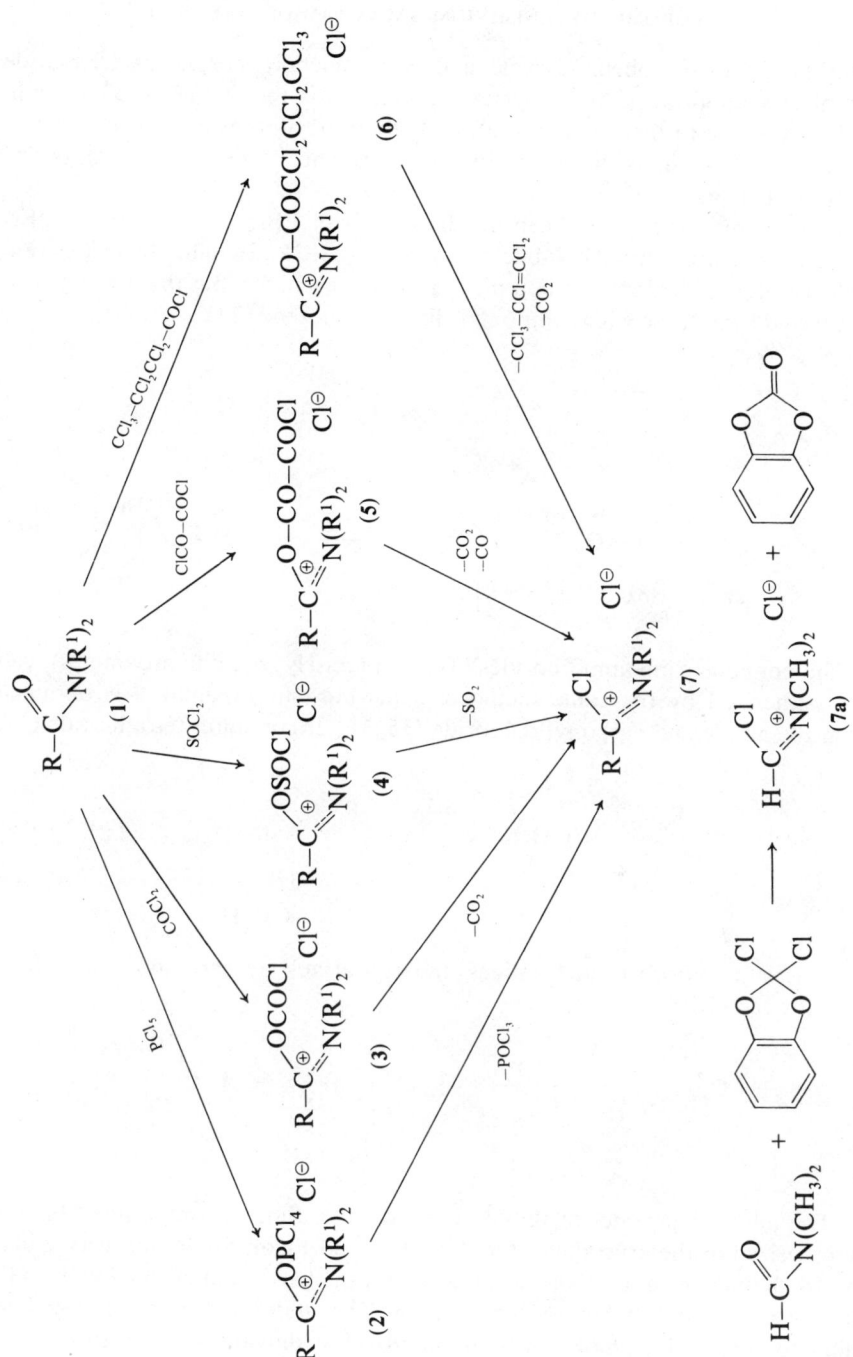

chloride, 2,2-dichlorobenzodioxole and pyrocatechol phosphorus trichloride are of less importance as halogenating agents for the synthesis of iminium chlorides owing to their inaccessibility. Among the aforementioned halogenating agents phosgene is the most advantageous since it gives the purest amide chlorides (7).

Iminium bromides have been synthesized with PBr_5 (23, 25) and $COBr_2$ (23), but they are extremely labile compounds (23, 25). Iminium bromides (**8a**) are very often obtained with complex bromide gegenions, but the bromine can be cleaved from these with compounds like cyclohexene (23).

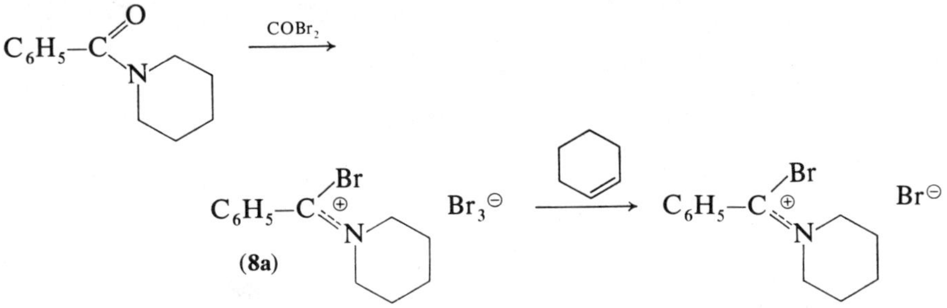

$$C_6H_5-C(=O)-N(piperidine) \xrightarrow{COBr_2} [C_6H_5-C(Br)=N^+(piperidine)]\ Br_3^{\ominus} \xrightarrow{cyclohexene} [C_6H_5-C(Br)=N^+(piperidine)]\ Br^{\ominus}$$

(**8a**)

The nonpolar "iminium fluorides" (more precisely, α,α-difluoroamines), can be synthesized by the same methods. Thus α,α-difluoroamine **9** is obtained from the amide and carbonyl difluoride (15, 16, 26) or sulfur tetrafluoride (15,

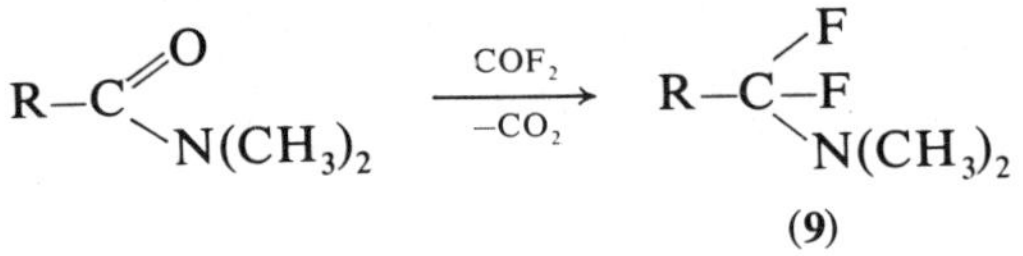

$$R-C(=O)-N(CH_3)_2 \xrightarrow[-CO_2]{COF_2} R-CF_2-N(CH_3)_2$$

(**9**)

$R = H, C_6H_5$

27, 28). N-monosubstituted amides undergo attack by carbonyl difluoride at the nitrogen (26):

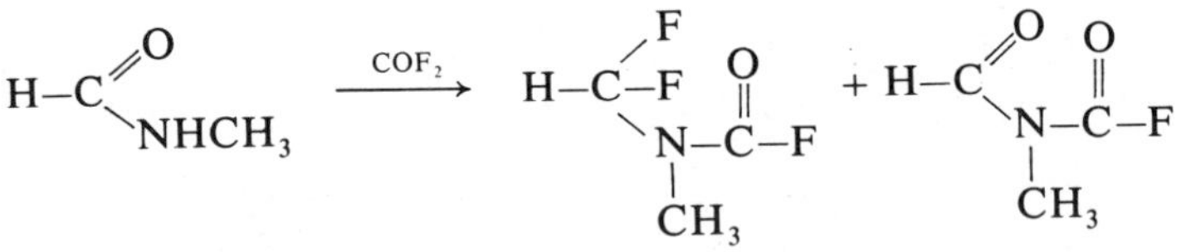

$$H-C(=O)-NHCH_3 \xrightarrow{COF_2} H-CF_2-N(CH_3)-C(=O)-F\ +\ H-C(=O)-N(CH_3)-C(=O)-F$$

The addition product of thionyl chloride and dimethylformamide (**10**) was synthesized in the crystalline state by Ferre and Palomo (29) and was studied by IR. Kikugawa et al. (30, 31) showed that the loss of sulfur dioxide from the primary product **10** proceeds reversibly. The adduct **10** is re-formed from dimethylformamide chloride (**7a**) in liquid sulfur dioxide. Adduct **10** was also studied by NMR (32, 33). In chloroform, the formyl hydrogen gives a signal at

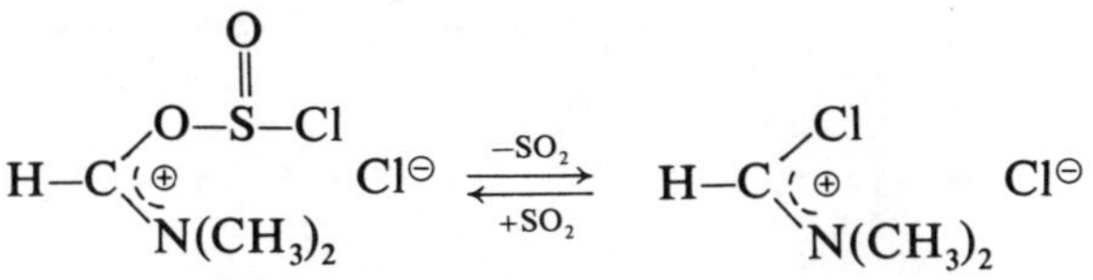

$$\underset{(10)}{H-C\overset{O-S(=O)-Cl}{\underset{N(CH_3)_2}{\oplus}}}\;Cl^{\ominus}\;\underset{+SO_2}{\overset{-SO_2}{\rightleftarrows}}\;\underset{(7a)}{H-C\overset{Cl}{\underset{N(CH_3)_2}{\oplus}}}\;Cl^{\ominus}$$

$\tau = 2.20$ ppm, whereas the dimethylamino groups give a doublet at $\tau = 7.54$ and 7.75 ppm owing to restricted rotation around the $C=N^{\oplus}$ double bond.

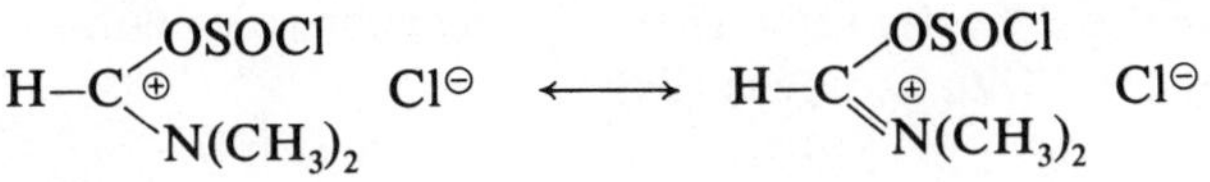

$$H-C\overset{OSOCl}{\underset{N(CH_3)_2}{\oplus}}\;Cl^{\ominus}\;\longleftrightarrow\;H-C\overset{OSOCl}{\underset{N(CH_3)_2}{\oplus}}\;Cl^{\ominus}$$

B. SIDE REACTIONS IN THE SYNTHESIS OF AMIDE CHLORIDES FROM AMIDES

When phosphorus pentachloride reacts with α-halo-N,N-dialkylacetamides, α-chloroenamines (**12a, 12b**) are formed as well as amide chlorides (**11a, 11b**); additional α-chlorination, giving **13**, may also occur (4, 34). This reaction has already been applied to the synthesis of α,α-dichlorolactams such as **14** (35).

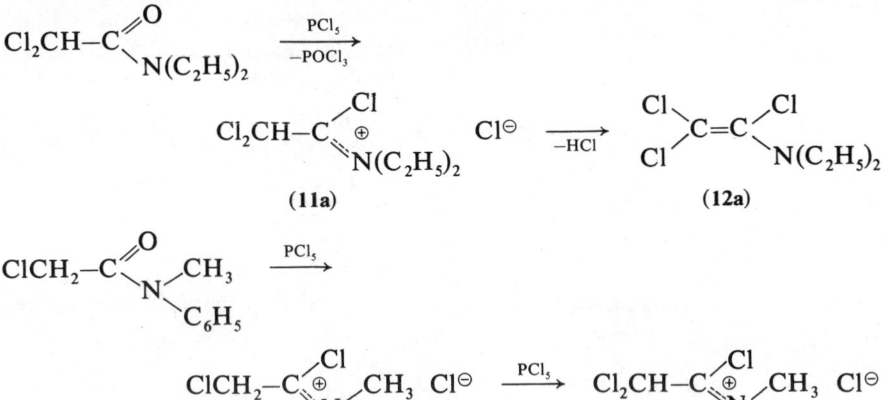

$$Cl_2CH-C\overset{O}{\underset{N(C_2H_5)_2}{\diagdown}}\;\xrightarrow[-POCl_3]{PCl_5}$$

$$\underset{(11a)}{Cl_2CH-C\overset{Cl}{\underset{N(C_2H_5)_2}{\oplus}}}\;Cl^{\ominus}\;\xrightarrow{-HCl}\;\underset{(12a)}{\overset{Cl}{\underset{Cl}{\diagup}}C=C\overset{Cl}{\underset{N(C_2H_5)_2}{\diagdown}}}$$

$$ClCH_2-C\overset{O}{\underset{N\diagdown C_6H_5}{\diagup CH_3}}\;\xrightarrow{PCl_5}$$

$$\underset{(11b)}{ClCH_2-C\overset{Cl}{\underset{N\diagdown C_6H_5}{\oplus\;\diagup CH_3}}}\;Cl^{\ominus}\;\xrightarrow{PCl_5}\;\underset{(13)}{Cl_2CH-C\overset{Cl}{\underset{N\diagdown C_6H_5}{\oplus\;\diagup CH_3}}}\;Cl^{\ominus}$$

$$\downarrow$$

$$\underset{(12b)}{\overset{Cl}{\underset{Cl}{\diagup}}C=C\overset{Cl}{\underset{N\diagdown C_6H_5}{\diagup CH_3}}}$$

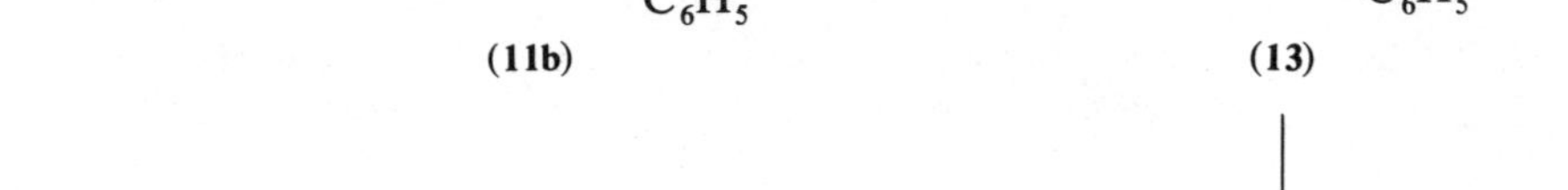

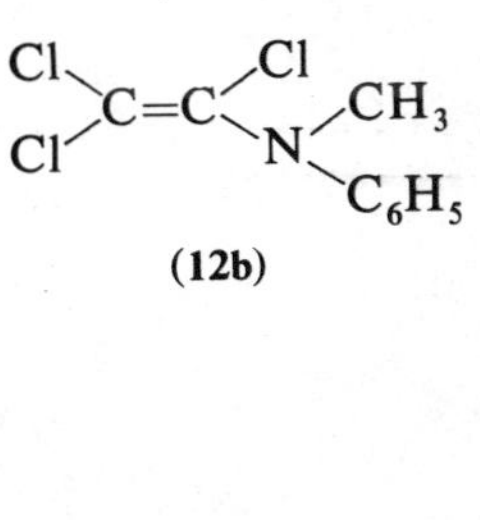

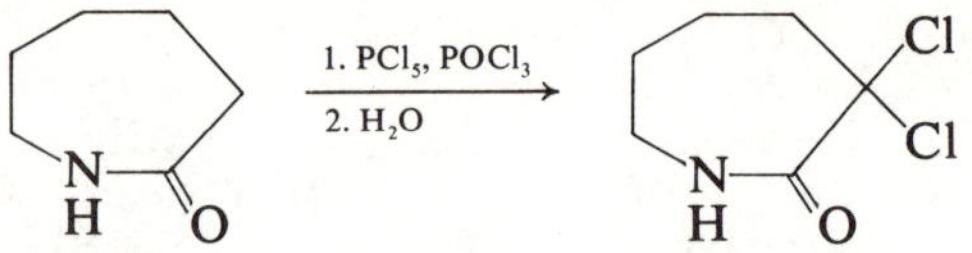

Buyle and Viehe (36) made similar observations. If N,N-dialkylated amides containing an α-CH_2 group are treated with phosgene, α-chloro-β-(chlorocarbonyl)enamines (16) are formed as by-products in yields of 25–35% in addition to amide chlorides (15). These side reactions also occur if N-alkylated lactams are reacted with phosgene (36).

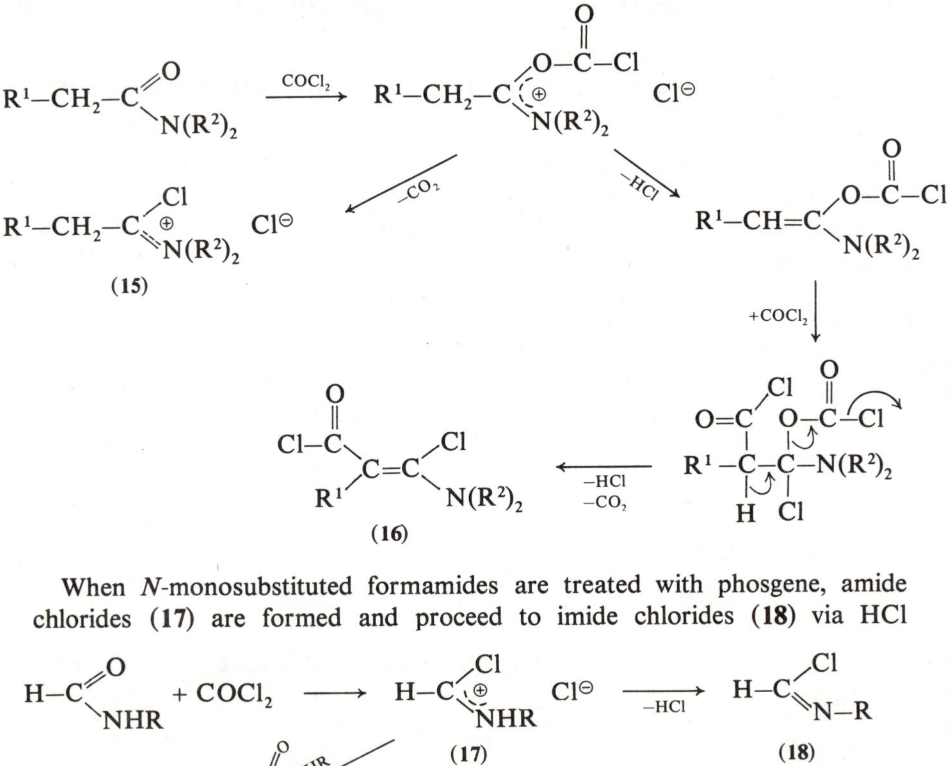

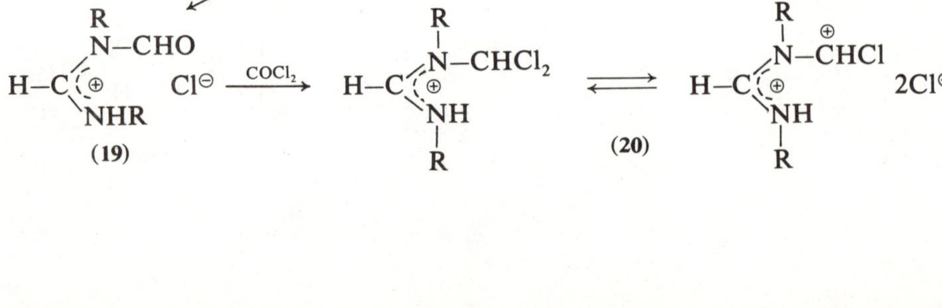

When N-monosubstituted formamides are treated with phosgene, amide chlorides (17) are formed and proceed to imide chlorides (18) via HCl elimination. Excess amide also yields *N*-formylformamidinium salts (**19**), which undergo repeated attack of phosgene and finally give *N,N'*-dialkyl-*N'*-dichloromethylformamidinium chlorides (**20**); these compounds exist partly as amide chlorides (37).

Thionyl chloride is less effective than phosphorus pentachloride or phosgene in transforming amides into amide chlorides. Thus *N,N*-diethylchloroacetamide is inert toward SOCl$_2$ (34). Additionally the Vilsmeier formylation of activated aromatics with dimethylformamide and thionyl chloride sometimes yields diaryl sulfides (38). This disulfide formation also occurs with the adducts of thionyl chloride and other amides (except formamide and *N*-methylformanilide) (38).

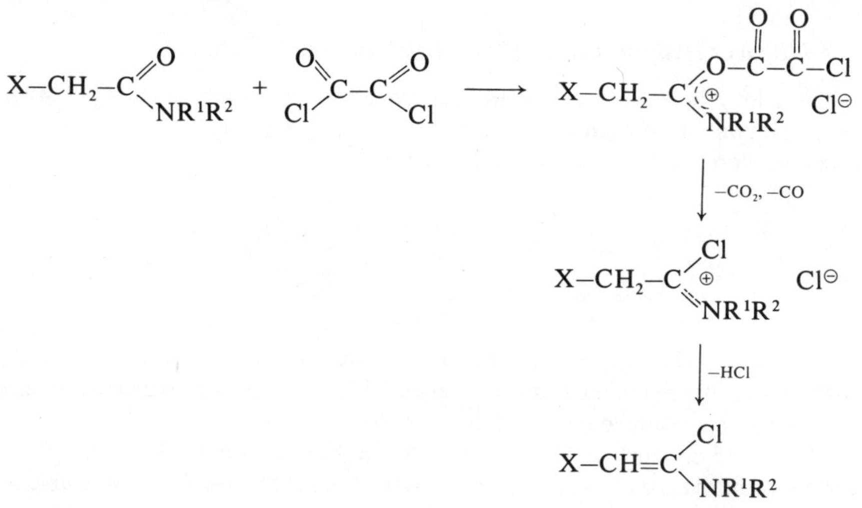

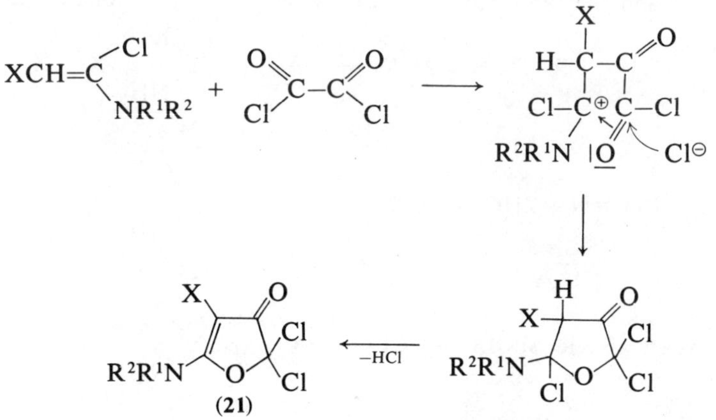

If oxalyl chloride is used to synthesize α-chlorinated amide chlorides, attack at the amide oxygen is immediately followed by cyclization to give dichlorofuranones (**21**) (34, 39). Thioamides react the same way (39):

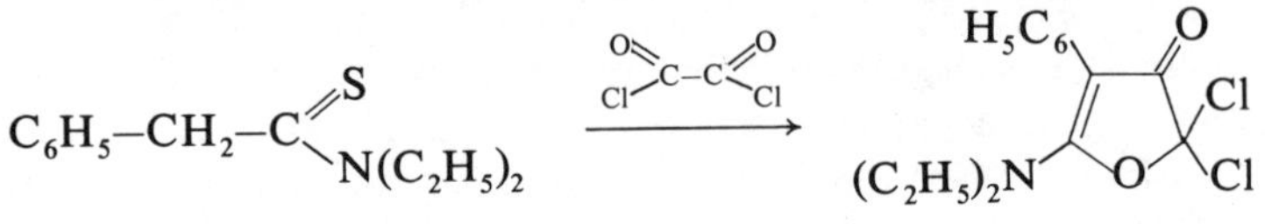

$$C_6H_5-CH_2-C\overset{S}{\underset{N(C_2H_5)_2}{\diagdown}} \quad \overset{O}{\underset{Cl}{\diagdown}}C-C\overset{O}{\underset{Cl}{\diagdown}} \longrightarrow \quad \begin{array}{c} H_5C_6 \quad O \\ \diagup \quad \diagdown \\ (C_2H_5)_2N \quad O \quad Cl \\ \diagdown \quad \diagup \\ Cl \end{array}$$

Primary amides yield acylisocyanates (**22**) with oxalyl chloride (40–42).

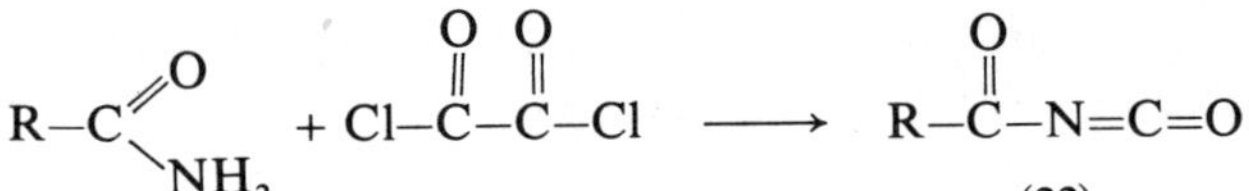

$$R-C\overset{O}{\underset{NH_2}{\diagdown}} + Cl-\overset{O}{\underset{\|}{C}}-\overset{O}{\underset{\|}{C}}-Cl \longrightarrow R-\overset{O}{\underset{\|}{C}}-N=C=O$$
$$(22)$$

1. Addition of Hydrogen Halides to Nitriles and Imidoyl Halides

It has long been known that nitriles react with hydrogen halides (43–46); these reactions have been reviewed repeatedly (47, 48). The reaction products were considered to be α,α-dihaloamines (**23**).

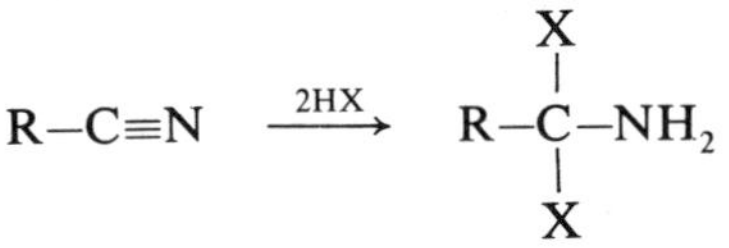

$$R-C\equiv N \xrightarrow{2HX} R-\underset{X}{\overset{X}{\underset{|}{C}}}-NH_2$$
$$(23)$$

Hantzsch (49) erroneously described the adducts as nitrilium salts; furthermore he postulated the existence of a 1:1 nitrile–hydrogen halide adduct with an assumed imide halide structure.

Klages and co-workers (50, 51) also studied the addition of HCl and HBr to nitriles, to nitrile–Lewis acid adducts, and to imidoyl halides. They were able to demonstrate that stable 1:1 adducts of nitriles and hydrogen halides were not formed and that amide halides (**24**) or hexachloroantimonates (**25, 26**) were

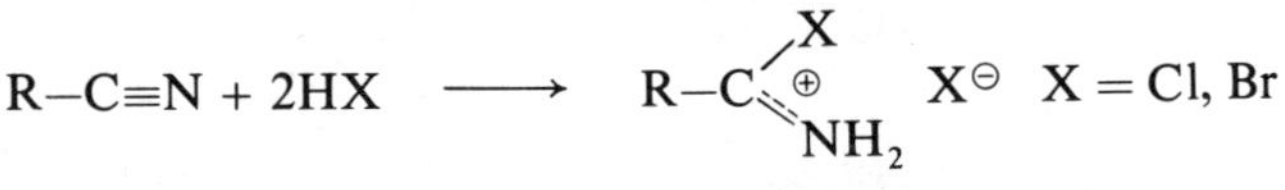

$$R-C\equiv N + 2HX \longrightarrow R-\overset{X}{\underset{NH_2}{C^{\oplus}}} \quad X^{\ominus} \quad X = Cl, Br$$
$$(24)$$

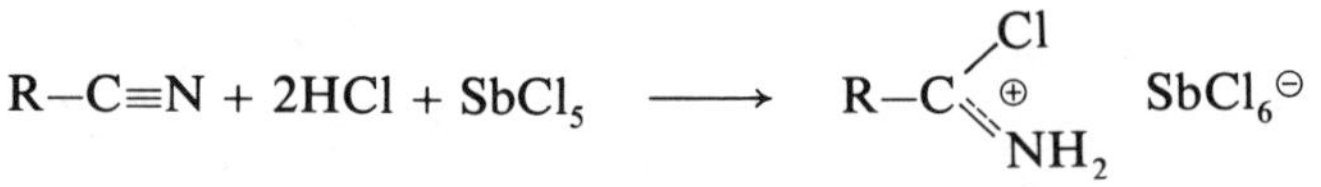

$$R-C\equiv N + 2HCl + SbCl_5 \longrightarrow R-\overset{Cl}{\underset{NH_2}{C^{\oplus}}} \quad SbCl_6^{\ominus}$$
$$(25)$$

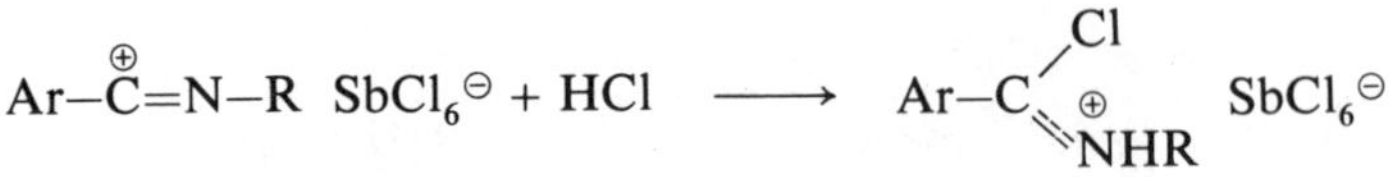

$$Ar-\overset{\oplus}{C}=N-R \; SbCl_6^{\ominus} + HCl \longrightarrow Ar-\overset{Cl}{\underset{NHR}{C^{\oplus}}} \quad SbCl_6^{\ominus}$$
$$(26)$$

I. SYNTHESIS

formed [the anions may be complex, such as $HCl_2^{\ominus}$ (6)]. Several subsequent publications (52–55) proposed alternative structures for the adducts of nitriles and hydrogen halides. Allenstein (56–65) later performed very detailed IR studies on various nitrile–hydrogen halide adducts which confirmed Klages' opinion that these adducts have an amide halide structure. The following survey gives several examples.

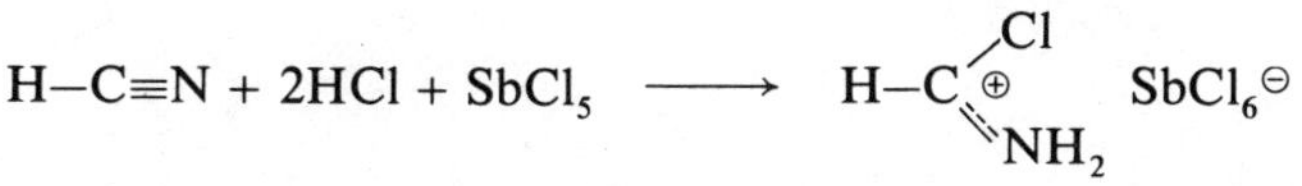

$$H-C\equiv N + 2HCl + SbCl_5 \longrightarrow H-C\overset{\oplus}{\underset{NH_2}{\diagup\hspace{-0.3em}Cl}} \quad SbCl_6^{\ominus} \qquad \text{(ref. 65)}$$

$$CH_3-CN + 2HX \longrightarrow CH_3-C\overset{\oplus}{\underset{NH_2}{\diagup\hspace{-0.3em}X}} \quad X^{\ominus} \quad X = Br, I \qquad \text{(ref. 61)}$$

$$CH_3-CN + 2HCl + SbCl_5 \longrightarrow CH_3-C\overset{\oplus}{\underset{NH_2}{\diagup\hspace{-0.3em}Cl}} \quad SbCl_6^{\ominus} \qquad \text{(ref. 61)}$$

$$Cl-CN + 2HCl + SbCl_5 \longrightarrow Cl-\overset{Cl}{\underset{\oplus}{C}}-NH_2 \ SbCl_6^{\ominus} \qquad \text{(ref. 62)}$$

$$NC-CH_2-CN + 3HBr \longrightarrow \overset{H_2\overset{\oplus}{N}}{\underset{Br}{\diagup}}C-CH=C\overset{NH_2}{\underset{Br}{\diagdown}} \ Br^{\ominus} \qquad \text{(ref. 63)}$$

$$R-S-CN + 2HBr \longrightarrow R-S-C\overset{\overset{\oplus}{NH_2}}{\underset{Br}{\diagdown}} \ Br^{\ominus} \qquad \text{(ref. 64)}$$

$$\overset{H_2N}{\underset{H_2N}{\diagdown}}C=N-CN + 2HCl \longrightarrow \overset{H_2N}{\underset{H_2N}{\diagdown}}C=N-C\overset{\overset{\oplus}{NH_2}}{\underset{Cl}{\diagdown}} \ Cl^{\ominus} \qquad \text{(ref. 57)}$$

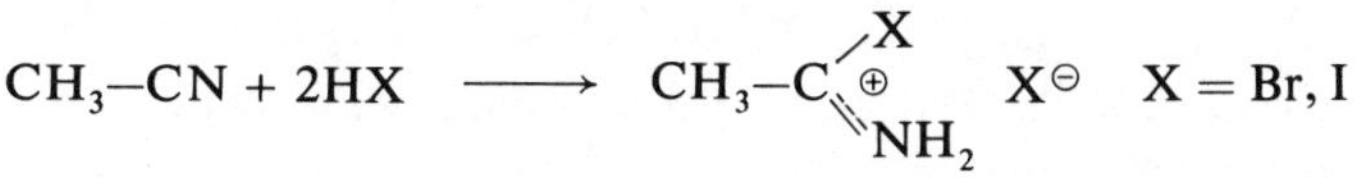

$$\overset{R_2N}{\underset{R_2N}{\diagdown}}C=C\overset{CN}{\underset{CN}{\diagdown}} + 2HX \longrightarrow \overset{R_2N}{\underset{R_2N}{\diagdown}}C=C\overset{C\overset{NH_2}{\underset{X}{\diagdown}}\oplus}{\underset{CN}{\diagdown}} \ X^{\ominus} \qquad \text{(ref. 58)}$$

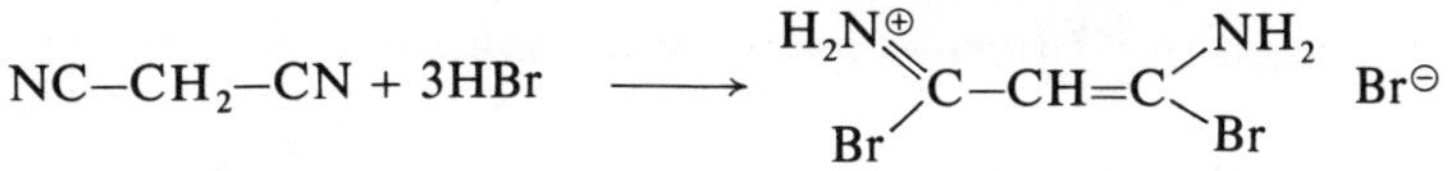

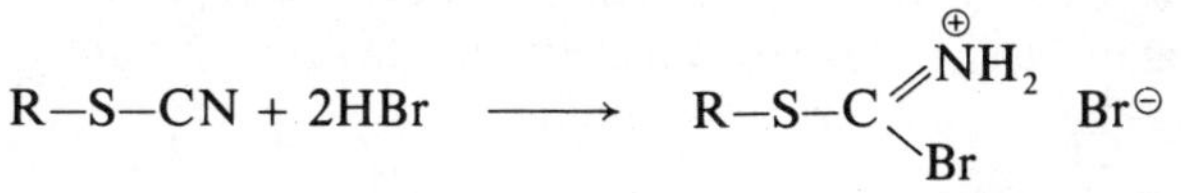

$$H-\overset{CN}{\underset{CN}{C}}-CN + 3HX \longrightarrow \overset{H_2\overset{\oplus}{N}}{\underset{X}{\diagdown}}C-\overset{}{\underset{CN}{C}}=C\overset{NH_2}{\underset{X}{\diagdown}} \ X^{\ominus} \qquad \text{(ref. 60)}$$

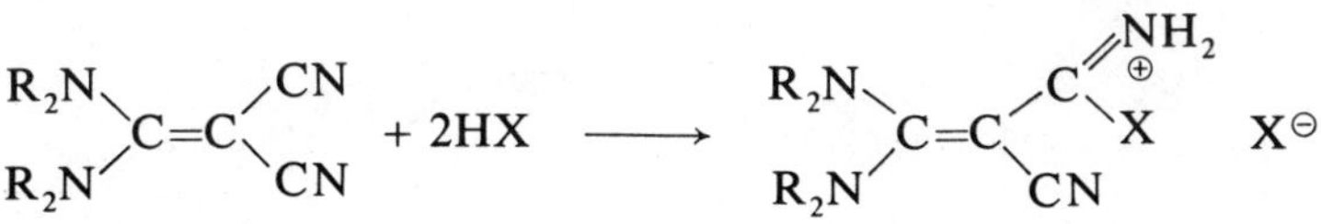

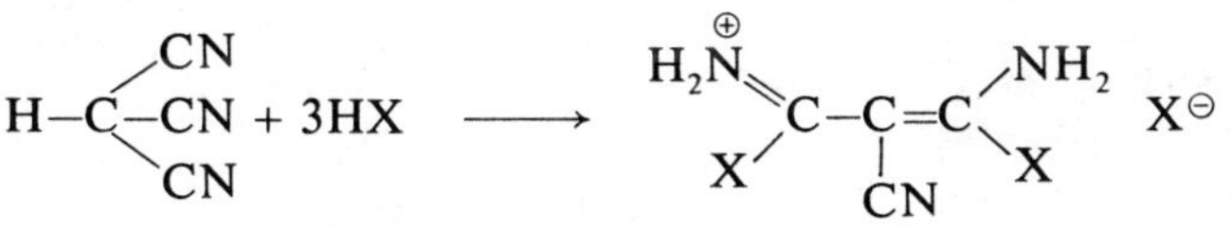

The results of these experiments demonstrated that the stability of the adducts strongly depends on the acidity of the hydrogen halide. Thus dichloromethyleniminium chloride is unstable at temperatures above 6°C (55), whereas the analogous bromide and iodide are stable at room temperature (61). The iminium chloride may be stabilized by complexation with $SbCl_5$ (61). Amide fluorides (**27**) are also stabilized by complexation (66).

$$R-C\overset{\oplus}{\underset{NH_2}{\diagup^F}}\quad (HF)_n F^{\ominus}$$
(**27**)

During the course of this investigation, the much-discussed (67–74, 76, 77) sesquihalide of hydrogen cyanide, $2HCN \cdot 3HCl$, was found to be N-dichloromethyleneformamidinium chloride (**28**) (75).

$$\underset{Cl}{\overset{Cl}{\diagdown}}CH-NH-CH=\overset{\oplus}{N}H_2 \quad Cl^{\ominus}$$
(**28**)

Grdinic and Hahn (8) investigated the addition of HCl, HBr, DCl, and DBr to N-alkylformimidoyl halides. Based on IR studies, they also reached the conclusion that amide halides are formed.

2. Addition of Hydrogen Halides and Halogens to α-Chloroenamines and Ynamines

The addition of HCl or Cl_2 to trichlorovinylamines (**29**) yields chloromethyleniminium salts **30** and **31** (34, 78–80). Interestingly, the reaction may

$$CCl_3-C\overset{\oplus}{\underset{NR^1R^2}{\diagup^{Cl}}} Cl^{\ominus} \xleftarrow{Cl_2} \underset{Cl}{\overset{Cl}{\diagdown}}C=C\underset{NR^1R^2}{\diagup^{Cl}} \xrightleftharpoons{HCl} CHCl_2-C\overset{\oplus}{\underset{NR^1R^2}{\diagup^{Cl}}} Cl^{\ominus}$$

(**31**)   (**29**)   (**30**)

be reversible. The addition of 1 mole of HCl to α-chloroenamines (**33**) (79) or 2 moles of HCl to ynamines (**34**) (81) also yields amide chlorides, **32** and **35**.

$$\underset{CH_3}{\overset{CH_3}{\diagdown}}C=C\underset{N\diagdown}{\diagup^{Cl}} \xrightarrow{HCl} \underset{CH_3}{\overset{CH_3}{\diagdown}}CH-C\overset{\oplus}{\underset{N\diagdown}{\diagup^{Cl}}} Cl^{\ominus}$$

(**33**)   (**32**)

$$R^1-C\equiv C-N(R^2)_2 \xrightarrow{2HCl} R^1-CH_2-C\overset{\oplus}{\underset{N(R^2)_2}{\diagup^{Cl}}} Cl^{\ominus}$$

(**34**)   (**35**)

C. ADDITIONS OF AMINES TO α,α-DIHALOGENATED OLEFINS

Until now this type of reaction has been applied only to tetrafluoroethylene (36). ($\alpha,\alpha,\beta,\beta$-Tetrafluoroethyl)dimethylamine (37) has been synthesized by addition of dimethylamine to 36 (82).

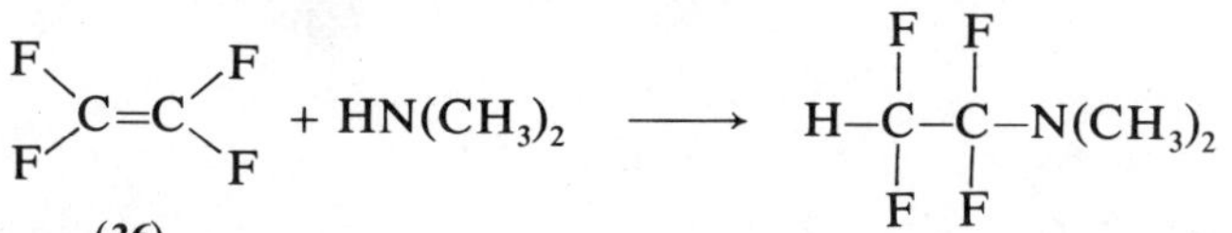

$$\underset{F}{\overset{F}{>}}C=C\underset{F}{\overset{F}{<}} + HN(CH_3)_2 \longrightarrow H-\underset{\underset{F}{|}}{\overset{\overset{F}{|}}{C}}-\underset{\underset{F}{|}}{\overset{\overset{F}{|}}{C}}-N(CH_3)_2$$

(36) (37)

D. ADDITION OF DIHALOCARBENES TO COMPOUNDS CONTAINING C=N GROUPS

The reaction of C=N double bonds with dichlorocarbene (83–94) and dibromocarbene (95) yields 2,2-dihaloaziridines such as 38 and 39. Azomethines (83, 84, 92, 95), ketimines (90), and benzophenone anils (83, 84) have been used.

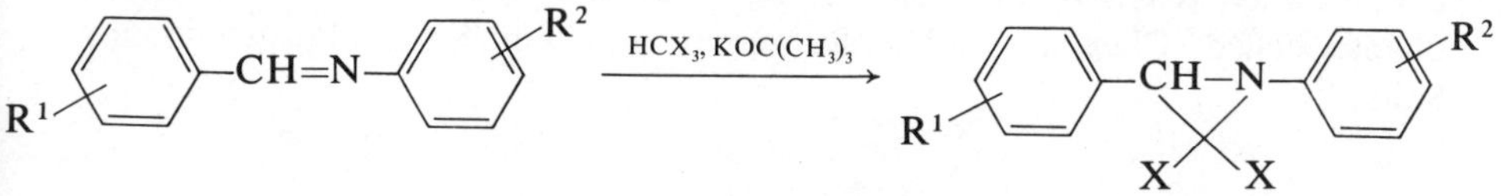

[Reaction scheme showing ArCH=NAr' + HCX₃, KOC(CH₃)₃ → aziridine (38), X = Cl, Br]

[Reaction scheme showing (C₆H₅)₂C=N-C₆H₅ + HCCl₃, KOC(CH₃)₃ → aziridine (39)]

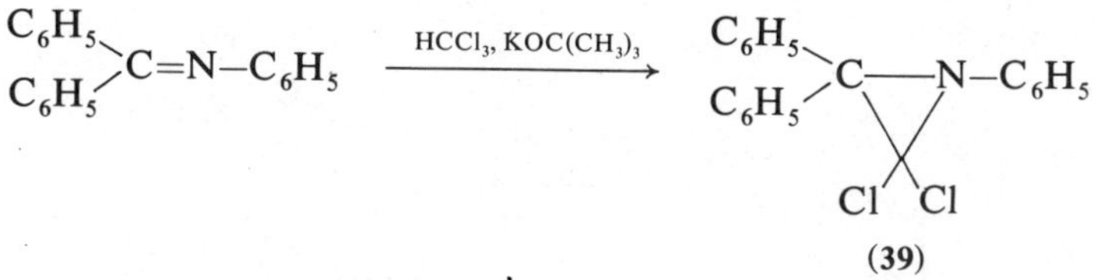

If isocyanide dichlorides (40) are reacted with phenylmercury trihalomethanes 41 and 42, 2,2,3,3-tetrachloroaziridines (43) are formed (94). N-phenylbenzimide chloride reacts with 41 in the same way, giving aziridine 44 (94). Azobenzene, as well as azoxybenzene, also yields an aziridine (45), among other products, when reacting with 41 (94).

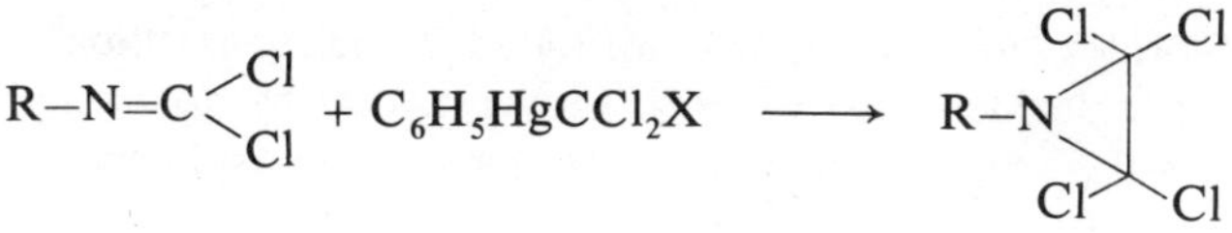

$$R-N=C\underset{Cl}{\overset{Cl}{<}} + C_6H_5HgCCl_2X \longrightarrow R-N\underset{Cl}{\overset{Cl}{<}}\underset{Cl}{\overset{Cl}{>}}$$

(40)   (41) X = Br   (43)
       (42) X = F

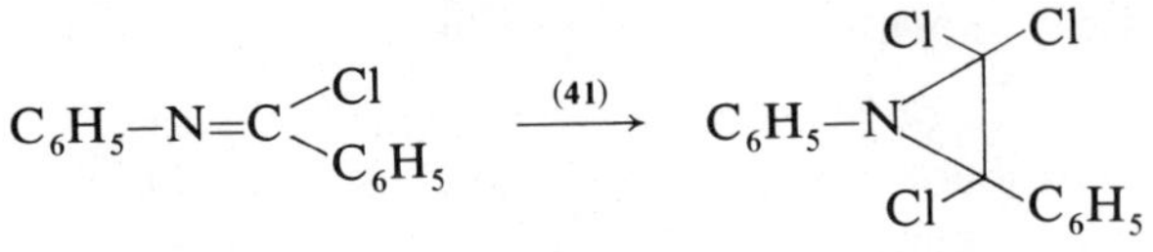

$$C_6H_5-N=C\begin{smallmatrix}Cl\\C_6H_5\end{smallmatrix} \xrightarrow{(41)} C_6H_5-N\begin{smallmatrix}Cl&Cl\\&\\Cl&C_6H_5\end{smallmatrix}$$

(44)

$$C_6H_5-\overset{O\uparrow}{N}=N-C_6H_5 \xrightarrow{(41)} C_6H_5-N\begin{smallmatrix}Cl&Cl\\&\\Cl&Cl\end{smallmatrix} \xleftarrow{(41)} C_6H_5-N=N-C_6H_5$$

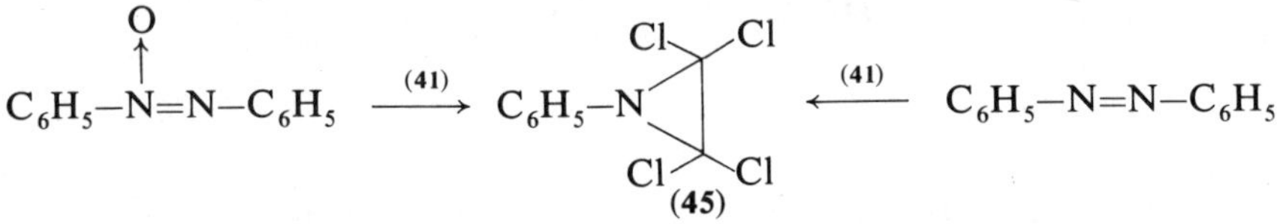

(45)

E. CHLOROMETHYLENIMINIUM HALIDES FROM PHOSGENIMINIUM SALTS (DICHLOROMETHYLENIMINIUM CHLORIDES)

When amides, lactams, enamines, and CH_2-acidic compounds are reacted with dichloromethyleniminium chlorides, **46** and **47** are formed. These condensation reactions have already been reviewed (96) [see also Part 1, the chapter called "Chemistry of Dichloromethyleniminium Salts (Phosgeniminium Salts)"].

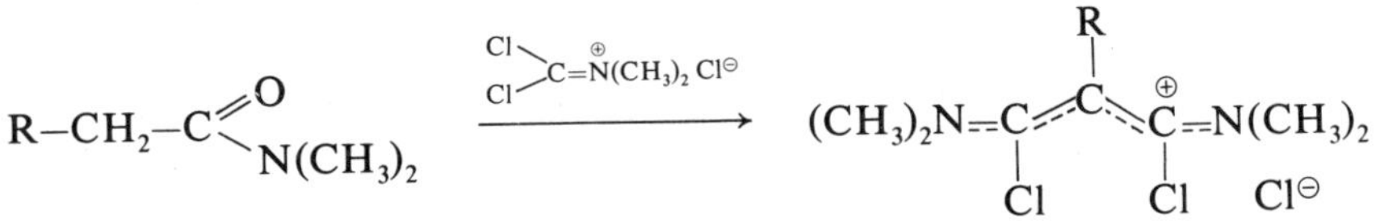

$$R-CH_2-C\begin{smallmatrix}O\\N(CH_3)_2\end{smallmatrix} \xrightarrow{\begin{smallmatrix}Cl\\Cl\end{smallmatrix}C=\overset{\oplus}{N}(CH_3)_2, Cl^{\ominus}} (CH_3)_2N=C\overset{R}{\underset{Cl}{-}}C\overset{}{\underset{Cl}{-}}\overset{\oplus}{C}=N(CH_3)_2 \quad Cl^{\ominus}$$

(46)

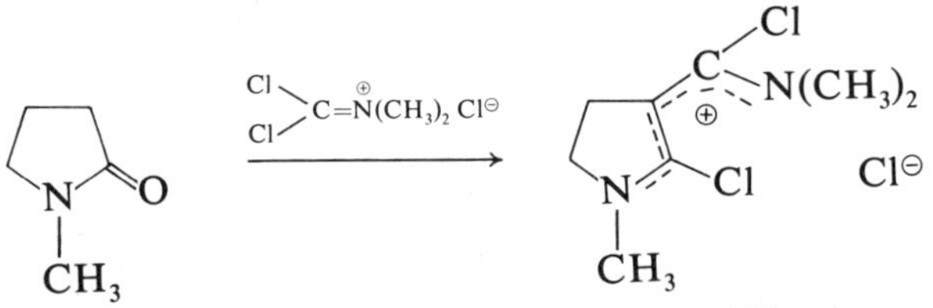

(47)

F. AMIDE HALIDES VIA AMIDE CHLORIDES

1. Chlorination of Amide Chlorides

Amide chlorides (**48**) are easily chlorinated in their α-positions, giving α,α-dichloroamide chlorides (**49**) (7, 97). However, up to now only hydrolysis products, α,α-dichlorinated amides (**50**), have been isolated, not the α,α-dichloroamide chlorides (**49**) themselves.

α-Monochlorinated amide chlorides can also be synthesized if proper reaction conditions are applied (80, 81).

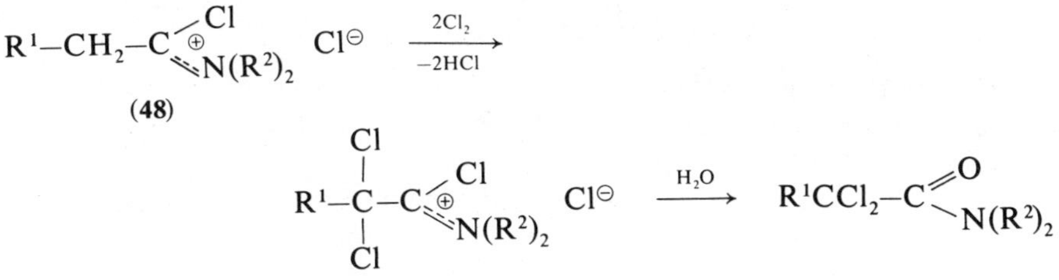

I. SYNTHESIS

$$R^1-CH_2-C\overset{\oplus}{\underset{N(R^2)_2}{\diagup Cl}} \; Cl^\ominus \xrightarrow[-2HCl]{2Cl_2}$$
(48)

$$R^1-\underset{Cl}{\overset{Cl}{|}}C-C\overset{\oplus}{\underset{N(R^2)_2}{\diagup Cl}} \; Cl^\ominus \xrightarrow{H_2O} R^1CCl_2-C\overset{\diagup O}{\underset{N(R^2)_2}{}}$$
(49)   (50)

2. Via Coupling of Diazonium Salts with Amide Chlorides

Diazonium salts couple with the α-methylene group of amide chlorides (51) and give arylhydrazones derived from α-ketoamide chlorides (52) (97).

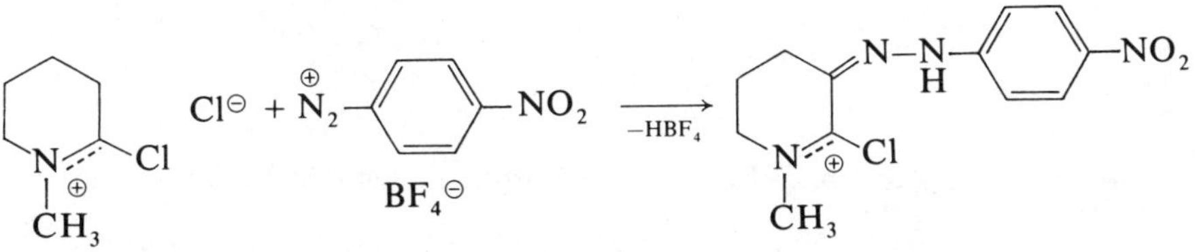

3. Via Reactions of Chloromethyleniminium Chlorides with Hydrogen Halides

If dry hydrogen halides, for example, HF (98), HBr (99), and HI (13), are passed through a chloroform solution of N,N-dimethylchloromethyleniminium chloride (53), halide exchange occurs (13). N,N-Dimethylbromomethyleniminium bromide (54) and N,N-dimethyliodomethyleniminium iodide (55) were first synthesized this way.

If hydrogen fluoride is used, a distillable adduct 56 from (difluoromethyl)dimethylamine with 3 moles of hydrogen fluoride is generated initially. This is transformed to pure (difluoromethyl)dimethylamine (57) by distillation over potassium fluoride. BF$_3$ converts compound 57 into N,N-dimethylfluoromethyleniminium tetrafluoroborate (58).

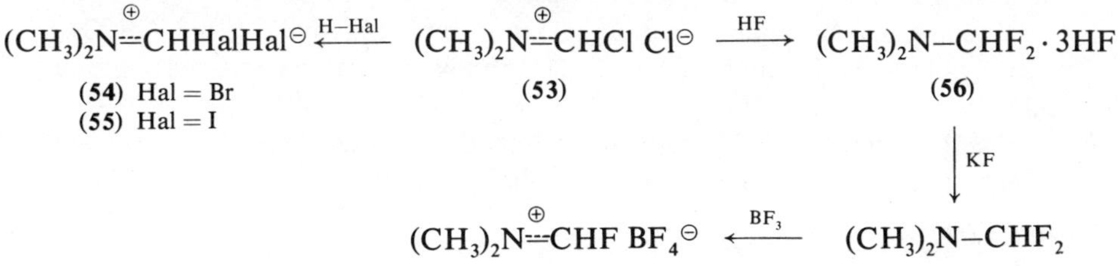

$$(CH_3)_2\overset{\oplus}{N}=CHHalHal^\ominus \xleftarrow{H-Hal} (CH_3)_2\overset{\oplus}{N}=CHCl\;Cl^\ominus \xrightarrow{HF} (CH_3)_2N-CHF_2\cdot 3HF$$
(54) Hal = Br   (53)   (56)
(55) Hal = I

$$\downarrow KF$$

$$(CH_3)_2\overset{\oplus}{N}=CHF\;BF_4^\ominus \xleftarrow{BF_3} (CH_3)_2N-CHF_2$$
(58)   (57)

4. Via Self-Condensation of Chloromethyleniminium Chlorides

Chloromethyleniminium chlorides containing an α-CH$_2$ group, such as **59**, give chloromethyleniminium salts (**60**) via self-condensation when heated in inert solvents (7, 97) (see also Part 1, the chapter by C. Jutz, "The Vilsmeier-Haack-Arnold Acylations"). However, most iminium salts **60** formed that way have not been isolated, but hydrolyzed to give β-ketoamides (**61**).

$$2R-CH_2-C\overset{Cl}{\underset{N(R^1)_2}{\overset{\oplus}{\diagdown}}} \; Cl^{\ominus} \quad \xrightarrow{\Delta}$$

(**59**)

$$R-CH_2-\underset{N(R^1)_2}{\overset{R}{C}}=C\overset{R}{\underset{C\overset{\oplus}{\diagdown}N(R^1)_2}{\diagdown Cl}} \; Cl^{\ominus} \quad \xrightarrow{H_2O} \quad R-CH_2-\overset{O}{\overset{\|}{C}}-\overset{R}{\overset{|}{CH}}-C\overset{O}{\diagdown N(R^1)_2}$$

(**60**)   (**61**)

II. Structure and Properties of Halomethyleniminium Salts

Contrary to older references, chloromethyleniminium chlorides, bromomethyleniminium bromides, and iodomethyleniminium iodides are polar compounds. In the case of N,N-unsubstituted iminium salts **24–26**, the existence of ions was proved by conductivity measurements (50).

Additionally, iminium chlorides are insoluble in nonpolar solvents such as petroleum ether, ether, and benzene, but soluble in polar ones such as chloroform and methylene chloride; this points to the ionic structure of the iminium chlorides (6, 7, 19). The IR spectra of iminium chlorides, iminium bromides, and iminium iodides also support this concept. The spectra of the iminium chlorides show a band at $\sim 6\mu$, which is attributed to the C=N$^{\oplus}$ vibration (6–8, 13, 19, 100).

The location of the band is practically independent of the anion, as expected from the polar structure of the iminium salts. Thus N,N-dimethylchloromethyleniminium chloride, as well as N,N-dimethylchloromethyleniminium hexachloroantimonate, show an absorption band at 6.01 μ (13).

With iminium bromides, we find the bands in the range of 6.1–6.35 μ (23).

In Table I, the location of the C=N$^{\oplus}$ bands of some halomethyleniminium salts are listed.

The IR spectra of N,N-unsubstituted amide chlorides indicate the existence of strong hydrogen bonds between the halide anions and the NH$_2$ hydrogens (57–61).

The NMR spectra also confirm the proposed structure of the bromomethyleniminium bromides (23).

II. STRUCTURE AND PROPERTIES OF HALOMETHYLENIMINIUM SALTS

TABLE I
Location of the C=N$^\oplus$ bands of Halomethyleniminium Salts

$$R-C\underset{NR^1R^2}{\overset{X}{\underset{\oplus}{\diagup}}}\ Y^\ominus$$

R	R^1	R^2	X	Y	$\lambda_{C=N}^{\oplus}, \mu$	Ref.
H	CH$_3$	CH$_3$	Cl	Cl	5.95	100
					6.01	13
H	CH$_3$	CH$_3$	Cl	AlCl$_4$	6.0	100
H	CH$_3$	CH$_3$	Cl	SbCl$_6$	6.01	13
CH$_3$	CH$_3$	CH$_3$	Cl	Cl	6.05	100
CH$_3$	CH$_3$	CH$_3$	Cl	AlCl$_4$	6.07	100
C$_6$H$_5$	CH$_3$	H	Cl	Br	6.18	8
C$_6$H$_5$	CH$_3$	CH$_3$	Cl	Cl	6.12	100
					6.09	8
C$_6$H$_5$	CH$_3$	CH$_3$	Cl	AlCl$_4$	6.13	100
2-C$_4$H$_3$O	CH$_3$	CH$_3$	Cl	Cl	6.17	8
H	CH$_3$	CH$_3$	Br	Br	6.05	13
H	CH$_3$	CH$_3$	I	I	6.14	13
H	CH$_3$	CH$_3$	F	BF$_4$	5.68	98
					5.76	
C$_6$H$_5$	–(CH$_2$)$_4$–		Br	Br$_3$	6.17	23
C$_6$H$_5$	–(CH$_2$)$_5$–		Br	Br$_3$	6.27	23
C$_6$H$_5$	–CH–(CH$_2$)$_4$– | CH$_3$		Br	Br$_3$	6.31	23
C$_6$H$_5$	–CH–(CH$_2$)$_3$–CH– | | CH$_3$ CH$_3$		Br	Br$_3$	6.34	23
C$_6$H$_5$	CH$_3$	CH$_3$	Br	Br	6.10	23
C$_6$H$_5$	–(CH$_2$)$_4$–		Br	Br	6.04	23
					6.25	
C$_6$H$_5$	–(CH$_2$)$_5$–		Br	Br	6.19	23
C$_6$H$_5$	–CH–(CH$_2$)$_4$– | CH$_2$–CH$_2$–CH$_3$		Br	SbF$_6$	6.27	23

The nonionic nature of "amide fluorides" (13) is indicated by comparison with amide chlorides. So (difluoromethyl)dimethylamine (**57**), for example, is a liquid boiling at 47–51°C (98), whereas N,N-dimethylchloromethyleniminium chloride (**7a**) melts at 140–145°C (7). Only if BF$_3$ is added to (difluoromethyl)-dimethylamine (**57**) is ionization of the compound achieved (98):

$$H-C\underset{N(CH_3)_2}{\overset{F}{\diagup}}F\ +\ BF_3\ \longrightarrow\ H-C\underset{N(CH_3)_2}{\overset{F}{\underset{\oplus}{\diagup}}}\ BF_4^\ominus$$

(**57**)   (**58**)

The iminium chloride of the squaric acid derivative **62** shows similar features. It is accessible from squaric acid diamide (**63**) and phosgene or thionyl chloride. It is completely ionized only if $SbCl_5$ is added (18).

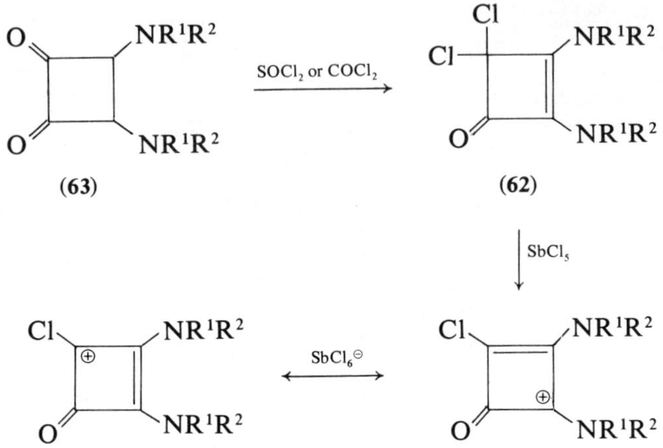

Zollinger et al. (6) pointed out that amide chlorides may exist partly as free ions and partly as ion pairs. A quantitative proof, for example, conductivity measurement, is still lacking.

Mass spectroscopic investigations on iminium bromides have also been reported (23, 101).

The structure of nitrile–hydrogen halide adducts, which had been determined as halomethyleniminium halides by extensive spectroscopic investigations (56–65), was confirmed by X-ray analysis (102) and neutron diffraction studies (103).

A. PHYSICAL PROPERTIES OF AMIDE HALIDES

The amide halides

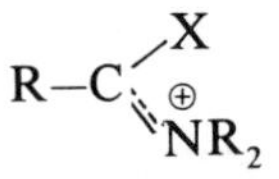

$$R-C\overset{X}{\underset{NR_2}{\oplus}} \quad X^\ominus \ (X = Cl, Br, I)$$

are mostly colorless, extremely hygroscopic solids. They are insoluble in nonpolar organic solvents such as saturated and unsaturated hydrocarbons and ether, but soluble in chlorinated hydrocarbons such as methylene chloride and chloroform. Polar protic solvents such as alcohols (7, 14) or dipolar aprotic solvents such as dimethylformamide (104) and homologous amides (7, 97) are of limited use since they are capable of reacting with amide chlorides at higher temperatures.

The melting points of the amide chlorides decrease with the increasing number of alkyl groups in the molecule; melting points of amide halides are listed in Table II.

II. STRUCTURE AND PROPERTIES OF HALOMETHYLENIMINIUM SALTS

TABLE II
Melting Points of Amide Halides

$$R-C{\overset{X}{\underset{NR^1R^2}{\oplus}}}\ Y^{\ominus}$$

R	R^1	R^2	X	Y	m.p., °C	Ref.
H	CH$_3$	CH$_3$	Cl	Cl	140–145	7, 14, 17
H	CH$_3$	CH$_3$	Br	Br	156–158	23, 99
H	CH$_3$	CH$_3$	I	I	110	13
H	—(CH$_2$)$_5$—		Cl	Cl	58–66	7
H	CH$_3$	C$_6$H$_5$	Cl	Cl	Oil	17
CH$_3$	CH$_3$	CH$_3$	Cl	Cl	150–120	6, 7
C$_2$H$_5$	CH$_3$	CH$_3$	Cl	Cl	68–70	7
C$_2$H$_5$	—(CH$_2$)$_5$—		Cl	Cl	82–85	14
n-C$_3$H$_7$	CH$_3$	CH$_3$	Cl	Cl	82–84	7, 14
n-C$_3$H$_7$	C$_2$H$_5$	C$_2$H$_5$	Cl	Cl	~20	7
i-C$_3$H$_7$	C$_2$H$_5$	C$_2$H$_5$	Cl	Cl	~20	7
n-C$_4$H$_9$	CH$_3$	CH$_3$	Cl	Cl	50–55	7
—(CH$_2$)$_3$—		CH$_3$	Cl	Cl	75–79	14
C$_6$H$_5$	CH$_3$	H	Cl	Cl	93–95	8
C$_6$H$_5$	CH$_3$	H	Cl	Br	115–117	8
C$_6$H$_5$	CH$_3$	D	Cl	Br	115–118	8
C$_6$H$_5$	CH$_3$	CH$_3$	Cl	Cl	95–96	7, 14
C$_6$H$_5$	—(CH$_2$)$_5$—		Cl	Cl	136–140	14
C$_6$H$_5$	—(CH$_2$)$_4$—		Br	Br$_3$	64	23
C$_6$H$_5$	—(CH$_2$)$_5$—		Br	Br$_3$	97	23
C$_6$H$_5$	—CH—(CH$_2$)$_4$—CH— $\mid$ CH$_3$		Br	Br$_3$	125	23
C$_6$H$_5$	—CH—(CH$_2$)$_3$—CH— $\mid$ $\quad\quad\quad\quad$ $\mid$ CH$_3$ $\quad\quad\quad\quad$ CH$_3$		Br	Br$_3$	124	23
C$_6$H$_5$	CH$_3$	CH$_3$	Br	Br	120 (dec)	23
C$_6$H$_5$	—(CH$_2$)$_4$—		Br	Br	178 (dec)	23
C$_6$H$_5$	—(CH$_2$)$_5$—		Br	Br	195 (dec)	23
C$_6$H$_5$	—CH—(CH$_2$)$_4$— $\mid$ CH$_3$		Br	Br	160	23
C$_6$H$_5$	—CH—(CH$_2$)$_3$—CH— $\mid$ $\quad\quad\quad\quad$ $\mid$ CH$_3$ $\quad\quad\quad\quad$ CH$_3$		Br	Br	—	23
C$_6$H$_5$	—CH—(CH$_2$)$_4$— $\mid$ n-C$_3$H$_7$		Br	Br	85–95	23
p-OCH$_3$—C$_6$H$_4$	—(CH$_2$)$_5$—		Cl	Cl	85	14
p-NO$_2$—C$_6$H$_4$—	CH$_3$	CH$_3$	Cl	Cl	90 (dec)	14

TABLE II (cont.)

R	R¹	R²	X	Y	m.p., °C	Ref.
p-NO$_2$-C$_6$H$_4$-	-(CH$_2$)$_5$-		Cl	Cl	117–119	14
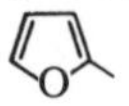 (2-furyl)	CH$_3$	H	Cl	Cl	103–104	8
	CH$_3$	H	Cl	Br	132–134 (dec)	8
	CH$_3$	D	Cl	Cl	103–105	8
	CH$_3$	D	Cl	Br	133–134 (dec)	8
	C$_2$H$_5$	H	Cl	Cl	93–94	8
	C$_2$H$_5$	H	Cl	Br	151–152	8
	CH$_3$	CH$_3$	Cl	Cl	181–182	8
	C$_2$H$_5$	C$_2$H$_5$	Cl	Cl	99–103	8

III. Reactions of Halomethyleniminium Salts

Reactions of N,N-unsubstituted halomethyleniminium halides (nitrile–hydrogen halide adducts) with nucleophiles (47, 48) as well as their use for heterocyclic synthesis (105, 106) have already been reviewed (see also the chapter "Iminium and Nitrilium Salts in the Dimerization of Cyano Compounds in Acids," by S. Yanagida, M. Okahara, and S. Komori in this volume).

A. REACTIONS WITH HYDROXY COMPOUNDS

1. Hydrolysis

Chloromethyleniminium salts **64** violently react with water to re-form the amide (5, 7, 9). This reaction is of little synthetic importance (14).

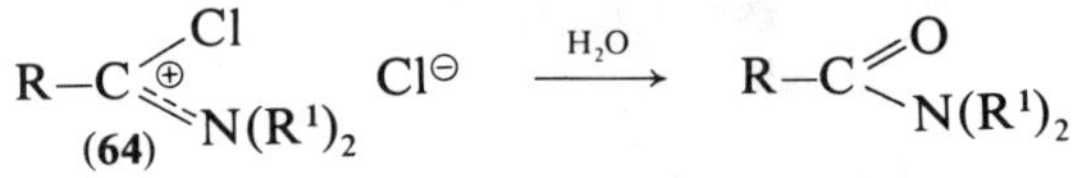

$$R-\overset{\oplus}{C}\underset{N(R^1)_2}{\overset{Cl}{\diagdown}} \quad Cl^{\ominus} \quad \xrightarrow{H_2O} \quad R-C\underset{N(R^1)_2}{\overset{O}{\diagdown}}$$

(64)

The iminium salt **65** reacts anomalously with water giving squaric acid chloride amide **66** (18), whereas the squaric acid derivative **67** hydrolyzes to the amide **68**, as do other iminium chlorides (17).

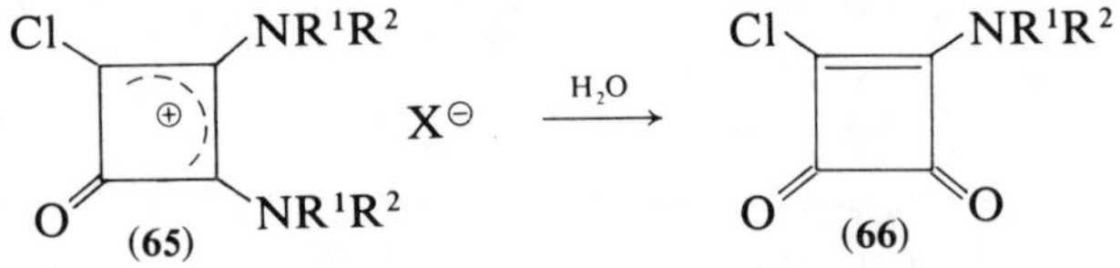

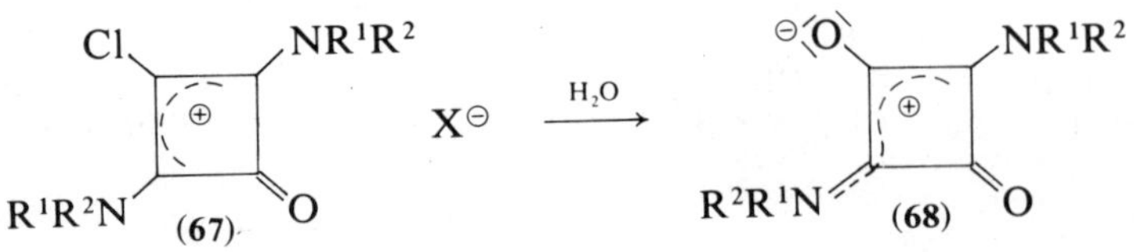

III. REACTIONS OF HALOMETHYLENIMINIUM SALTS

The hydrolysis of 2,2-dihaloaziridines (**69, 70**) occurs via ring opening to yield α-haloamides (**71**) (83, 92, 93, 107) and α-hydroxyamides (**72**) (87, 92, 93, 95).

The iminium salt **73** is converted to the amidinium salt **74** by the addition of water (75, 108).

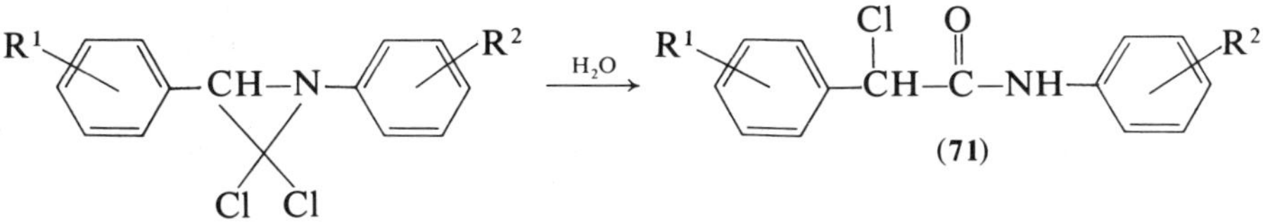

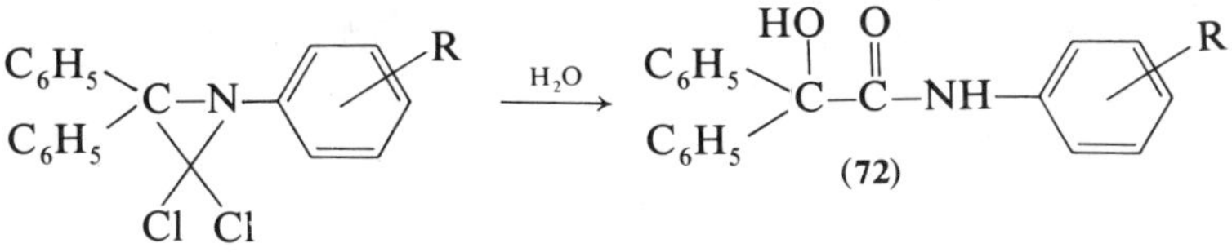

[Structures **69** → **71**]

[Structures **70** → **72**]

[Structures **73** → **74**]

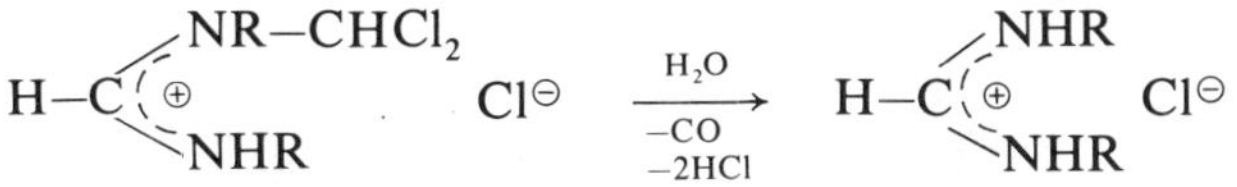

The amide bromide function of the iminium salt **75** is transferred to an amide structure with DMSO (75).

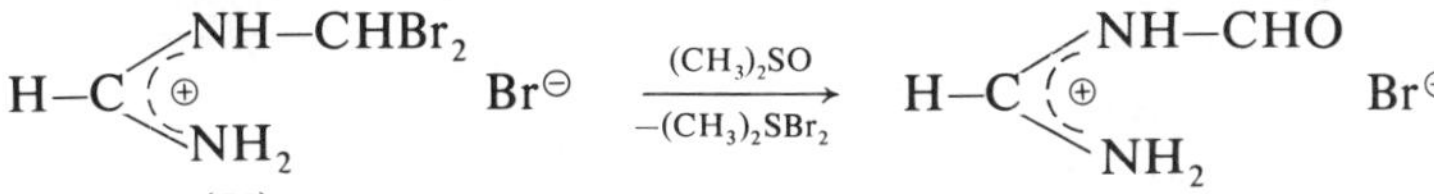

[Structure **75** with (CH$_3$)$_2$SO]

2. Alcoholysis

Chloromethyleniminium ions with complex anions such as **76** are transformed into stable N,N-dimethylalkoxymethyleniminium salts (**77**) by primary alcohols (13, 109).

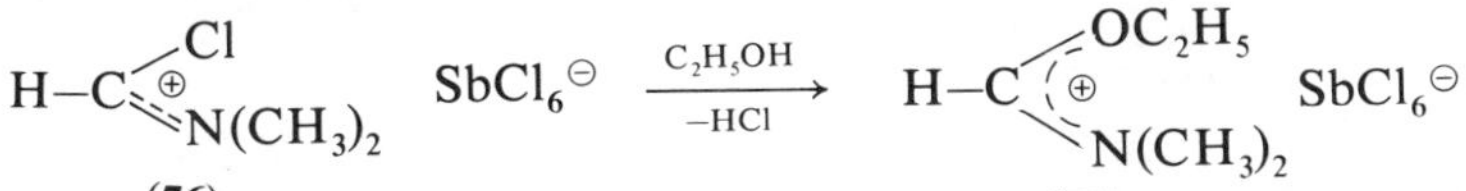

[Structures **76** → **77**]

The use of tertiary alcohols leads to relatively stable N,N-dimethylalkoxymethyleniminium chlorides (**78**), which may be easily isolated as perchlorate salts (110).

CHLOROMETHYLENIMINIUM SALTS (AMIDE HALIDES)

$$H-C\!\!\begin{array}{c}\diagup Cl \\ \oplus \\ \diagdown N(CH_3)_2 \end{array} Cl^\ominus \quad \xrightarrow{(CH_3)_2\overset{R}{C}-OH}$$

$R = CH_3, C_6H_5$

$$H-C\!\!\begin{array}{c}\diagup O-\overset{CH_3}{\underset{R}{C}}-CH_3 \\ \oplus \\ \diagdown N(CH_3)_2 \end{array} Cl^\ominus \quad \xrightarrow{NaClO_4} \quad H-C\!\!\begin{array}{c}\diagup O-\overset{CH_3}{\underset{R}{C}}-CH_3 \\ \oplus \\ \diagdown N(CH_3)_2 \end{array} ClO_4^\ominus$$

(78)   (79)

Alcohols and amide chlorides similarly form NN dialkyl(alkoxyalkyl)-iminium chlorides, primarily via hydrogen chloride elimination. These compounds easily decompose on heating into the amide and alkyl chloride (14, 111). This reaction shows strong S_N2 behavior, since the use of 2-octanol leads to 2-chlorooctane with 100% inversion (111).

$$R-C\!\!\begin{array}{c}\diagup Cl \\ \oplus \\ \diagdown N(R^1)_2 \end{array} Cl^\ominus + R^2OH \quad \xrightarrow{-HCl}$$

$$R-C\!\!\begin{array}{c}\diagup O-R^2 \\ \oplus \\ \diagdown N(R^1)_2 \end{array} Cl^\ominus \quad \xrightarrow{\Delta} \quad R-C\!\!\begin{array}{c}\diagup O \\ \diagdown N(R^1)_2 \end{array} + R^2Cl$$

(80)

This cleavage is also known to occur for imido ester hydrochlorides (112). Owing to the good yields of alkyl halides, the reaction finds wide synthetic application (7, 14, 111).

If primary alcohols are used in the alcoholysis, one always obtains the corresponding alkyl halide without isomerization of the carbon skeleton (111). Alcohols cleave the amide chlorides **81** into alkyl chlorides and amidinium salts (**82**) (108). The alcoholysis of the salt **83** proceeds similarly (75).

$$H-C\!\!\begin{array}{c}\diagup NRCHCl_2 \\ \oplus \\ \diagdown NHR \end{array} Cl^\ominus \quad \xrightarrow[-CO, -HCl]{+R'OH} \quad H-C\!\!\begin{array}{c}\diagup NHR \\ \oplus \\ \diagdown NHR \end{array} Cl^\ominus + RCl$$

(81)   (82)

$$H-C\!\!\begin{array}{c}\diagup NH-CHBr_2 \\ \oplus \\ \diagdown NH_2 \end{array} Br^\ominus \quad \xrightarrow{C_2H_5OH} \quad H-C\!\!\begin{array}{c}\diagup NH_2 \\ \oplus \\ \diagdown NH_2 \end{array} Br_3^\ominus + H-C\!\!\begin{array}{c}\diagup O \\ \diagdown OC_2H_5 \end{array}$$

(83)

III. REACTIONS OF HALOMETHYLENIMINIUM SALTS

The chloromethyleniminium chlorides not only are useful for synthesizing *n*- and *iso*-alkyl halides, but also allow the conversion of polyfunctional, unsaturated, and heterocyclic alcohols into the corresponding alkyl chlorides such as **84–87** (7, 14, 113, 114). Very recently, alkyl bromides were also synthesized from bromomethyleniminium bromides (111).

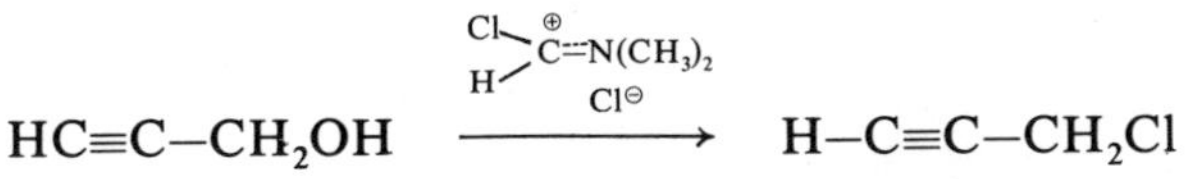

$$HC\equiv C-CH_2OH \xrightarrow{\underset{H}{\overset{Cl}{>}}C=\overset{\oplus}{N}(CH_3)_2 \; Cl^{\ominus}} H-C\equiv C-CH_2Cl$$
(84)

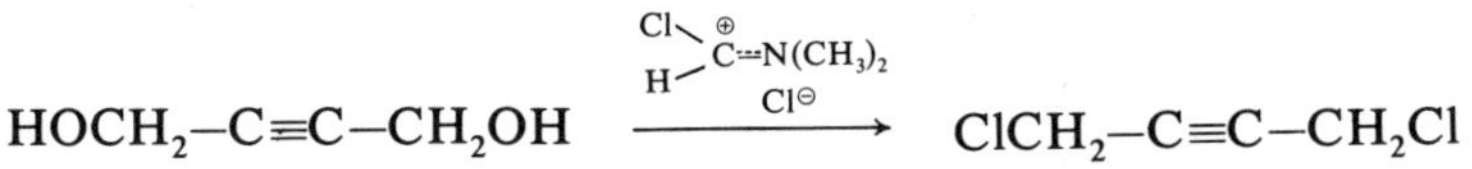

$$HOCH_2-C\equiv C-CH_2OH \xrightarrow{\underset{H}{\overset{Cl}{>}}C=\overset{\oplus}{N}(CH_3)_2 \; Cl^{\ominus}} ClCH_2-C\equiv C-CH_2Cl$$
(85)

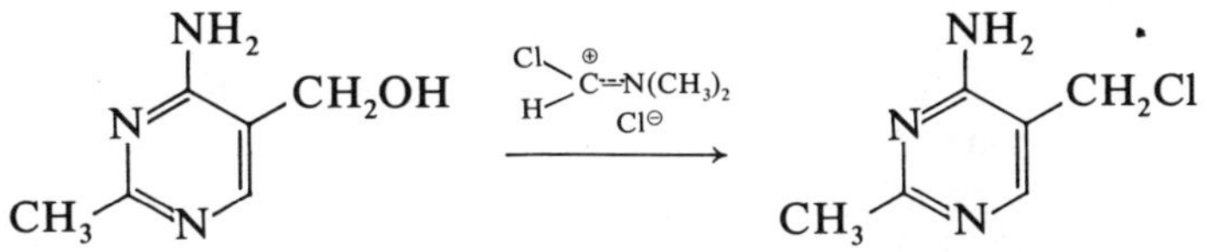

(86)

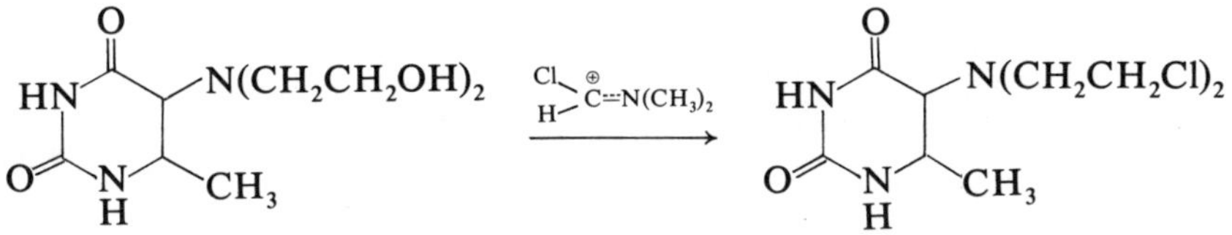

(87)

None of these reactions requires the isolation of **7a**. Even catalytic amounts of DMF are sufficient (14), since the halogenating agent (phosgene or thionyl chloride) regenerates the chloride from the dimethylformamide recovered during the reaction. The amide chloride in turn reacts with the alcohol again.

When chloromethyleniminium chlorides (**7**) are alcoholyzed, the alkoxymethyleniminium chloride intermediate (**88**) may react via dialkylamine elimination with water, giving esters (**89**), or with hydrogen sulfide, giving thio esters (**90**) (14). Polyfunctional alcohols like erythritol or mannitol and dimethylformamide chloride also form the corresponding formates (110).

Bifunctional compounds (NH$_2$, OH) are cyclized with *N,N*-dialkylalkoxymethyleniminium chlorides (**88**) generated *in situ*. Several 2-substituted imidazoles, oxazoles, and 1,3,4-oxadiazoles were synthesized this way (14).

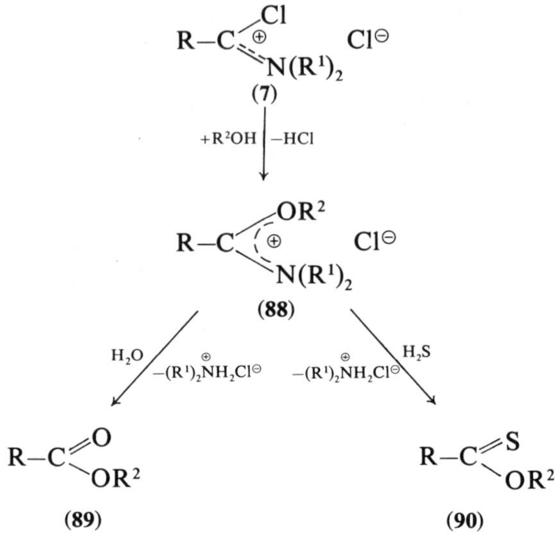

If furoine (**91**) is treated with DMF/SOCl$_2$, **92** is isolated in 54% yield, whereas the reaction of **91** with DMF/COCl$_2$ forms **93** in 47% yield (115).

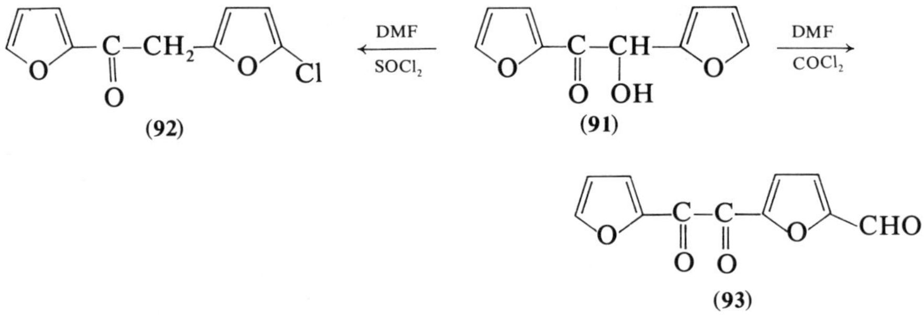

2,2-Dichloroaziridines (**94**) react with alcohols via ring opening to give α-chloro esters (**95**) (87).

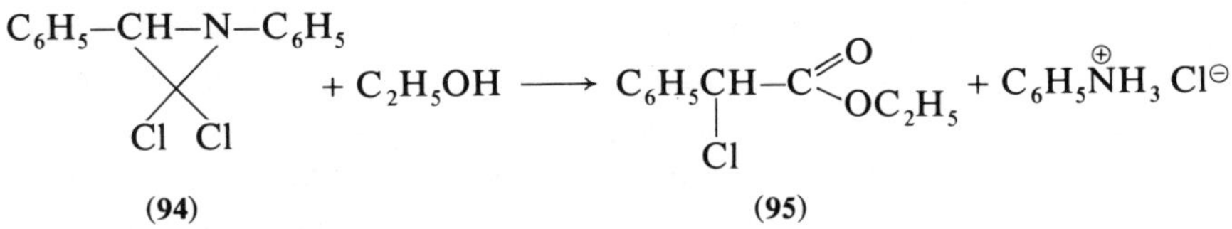

3. Reactions of Amide Chlorides with Phenols and other Heterocyclic Compounds Containing Hydroxyl Groups

Strongly acidic phenols, like picric acid, exchange their OH group with chlorine when reacted with dimethylformamide chloride (**7a**) (77, 116, 117).

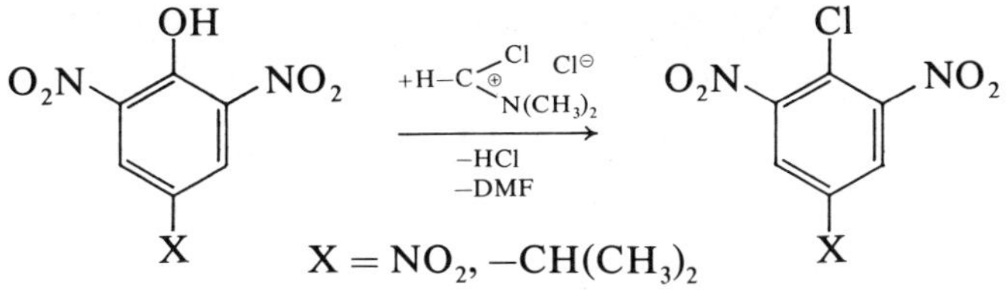

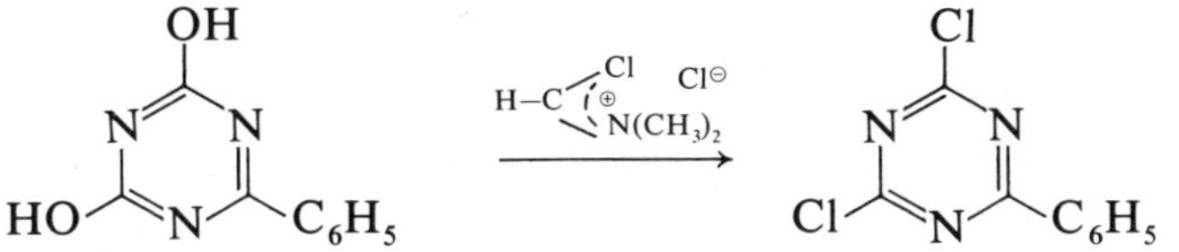

III. REACTIONS OF HALOMETHYLENIMINIUM SALTS

[Scheme: 2,6-dinitrophenol derivative + H–C(Cl)=N(CH$_3$)$_2$ Cl$^⊖$ →(−HCl, −DMF) 2-chloro-1,3-dinitrobenzene derivative; X = NO$_2$, −CH(CH$_3$)$_2$]

[Scheme: 2,4-dihydroxy-6-phenyl-1,3,5-triazine + H–C(Cl)=N(CH$_3$)$_2$ Cl$^⊖$ → 2,4-dichloro-6-phenyl-1,3,5-triazine]

Analogous reactions have been studied with several other heterocycles and have been patented (7, 118, 119):

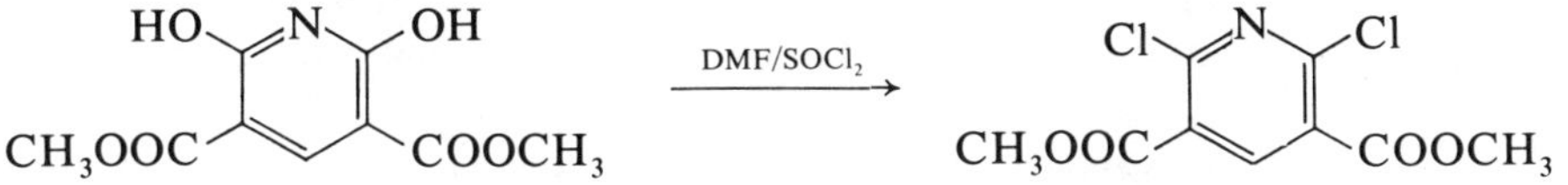

[Scheme: 2,6-dihydroxy-3,5-bis(methoxycarbonyl)pyridine →(DMF/SOCl$_2$) 2,6-dichloro-3,5-bis(methoxycarbonyl)pyridine]

[Scheme: 1-phenyl-4-hydroxy-5-nitro-pyridazin-6(1H)-one →(DMF/SOCl$_2$) 1-phenyl-4-chloro-5-nitro-pyridazin-6(1H)-one]

Generally, catalytic amounts of dimethylformamide are sufficient for these reactions.

Electron-rich phenols behave very much like aliphatic alcohols. Ege and Frey (120) studied the reactions of pyrocatechols (**96**) with dimethyl-

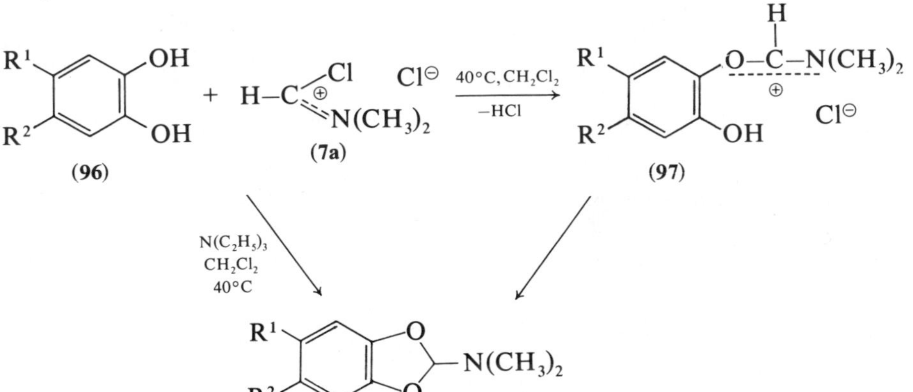

[Scheme: (**96**) pyrocatechol with R^1, R^2 + H–C(Cl)=N(CH$_3$)$_2$ Cl$^⊖$ (**7a**) →(40°C, CH$_2$Cl$_2$, −HCl) (**97**) aryl-O–CH(N(CH$_3$)$_2$)... OH, Cl$^⊖$ →(N(C$_2$H$_5$)$_3$, CH$_2$Cl$_2$, 40°C) (**98**) benzo[1,3]dioxole-2-yl-N(CH$_3$)$_2$; R^1 = R^2 = H, Cl]

formamide chloride (**7a**) in methylene chloride. In the absence of base, *N,N*-dimethyl(2-hydroxyphenoxy)methyleniminium chloride **97** is obtained, which is transformed into 2-(dimethylamino)benzodioxoles **98** by base. Contrarily, direct cyclization occurs in the presence of triethylamine.

From phenol and *N,N*-dimethylchloromethyleniminium chloride, the salt **99** could be isolated (111).

$$H-C\underset{N(CH_3)_2}{\overset{OC_6H_5}{\lessgtr}}\quad Cl^{\ominus}$$
(**99**)

4. Reactions of Amide Chlorides with Carboxylic and Sulfonic Acids

The reaction of carboxylic and sulfonic acids with thionyl chloride or phosgene to yield carboxylic acid chlorides or sulfonic acid chlorides is catalyzed by *N,N*-disubstituted formamides (121–138).

Bosshard et al. examined the reaction and were able to show that amide chlorides were the intermediates (19, 124). The acylating agents (SOCl$_2$, COCl$_2$) form the amide chloride with dimethylformamide; the amide chloride reacts with the carboxylic acid or sulfonic acid via HCl loss. The addition product (**100**) finally decomposes into acid chloride (**101**) and DMF. The

$$H-C\underset{N(CH_3)_2}{\overset{O}{\lessgtr}} \xrightarrow[-SO_2]{SOCl_2} H-C\underset{N(CH_3)_2}{\overset{Cl}{\lessgtr}}\quad Cl^{\ominus}$$

$$H-C\underset{N(CH_3)_2}{\overset{Cl}{\lessgtr}}\quad Cl^{\ominus} + R-C\underset{OH}{\overset{O}{\lessgtr}} \xrightarrow{-HCl}$$

$$\underset{\underset{N(CH_3)_2}{|}}{H-C-O-C-R}\overset{Cl\ \ \ O}{} \longrightarrow H-C\underset{N(CH_3)_2}{\overset{O}{\lessgtr}} + R-C\underset{Cl}{\overset{O}{\lessgtr}}$$
(**100**) \hspace{4cm} (**101**)

kinetics of the reaction were studied by reacting isomeric naphthalene mono- and dicarboxylic acids with thionyl chloride in the presence of DMF (125). The reaction proceeds with sufficient speed at room temperature or even lower temperatures. This elegant acid chloride synthesis—the yields are generally better than 90%— has been patented many times.

By this method, mono as well as dicarboxylic acids (9, 124, 126, 127), α,α-dihalo acids (128), and heterocyclic acids (129, 130) are converted into the corresponding acid chlorides. Even the salts of sulfonic acids give satisfactory yields (19).

III. REACTIONS OF HALOMETHYLENIMINIUM SALTS

Acid bromides can also be synthesized this way if thionyl bromide is used (123).

In peptide syntheses this reaction is also used for the activation of amino acids and peptides with free carboxylic groups (131, 132).

In Table III several examples indicating the range of applicability are summarized. Eilingsfeld, Seefelder, and Weidinger describe more reactions in ref. 7.

TABLE III
A Survey of Acid Chlorides from Acids or Their Salts and Amide Chlorides

Reaction	Ref.
R–C(=O)OH $\xrightarrow{SOCl_2 \text{ or } COCl_2/DMF}$ R–C(=O)Cl R = C_nH_{2n+1} (n = 3, 4, 5, ...), C_6H_5, CCl_3	7
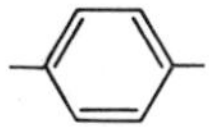 HOOC–R–COOH ⟶ Cl(O=)C–R–C(=O)Cl R = –C$_6$H$_4$–, –(CH$_2$)$_4$–, etc.	7, 122, 124, 127
[phthalimide-N-CH(COOH)CH$_2$CH$_2$COOH] $\xrightarrow{SOCl_2/DMF}$ [phthalimide-N-CH(COCl)CH$_2$CH$_2$COCl]	133
C$_6$H$_5$–SO$_3$H ⟶ C$_6$H$_5$–SO$_2$Cl	124
[4-methyl-imidazole-2-COOH] $\xrightarrow{SOCl_2/DMF}$ [4-methyl-imidazole-2-COCl]	129
HOOC(CH$_2$)$_2$–[5-methoxy-isoxazole] $\xrightarrow{SOCl_2/DMF}$ ClC(=O)(CH$_2$)$_2$–[5-methoxy-isoxazole]	134

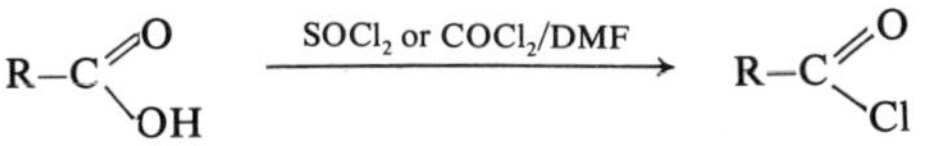

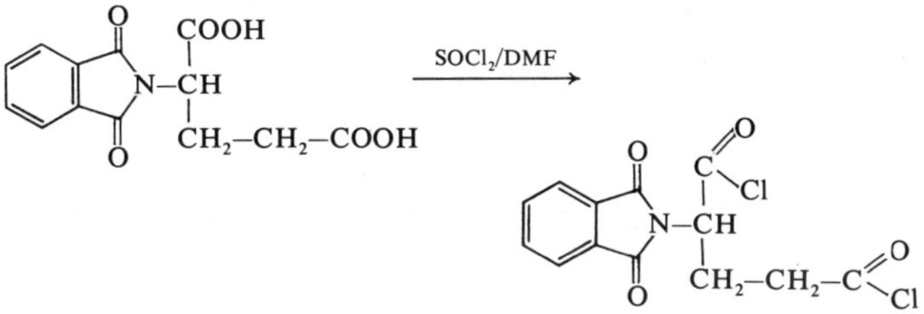

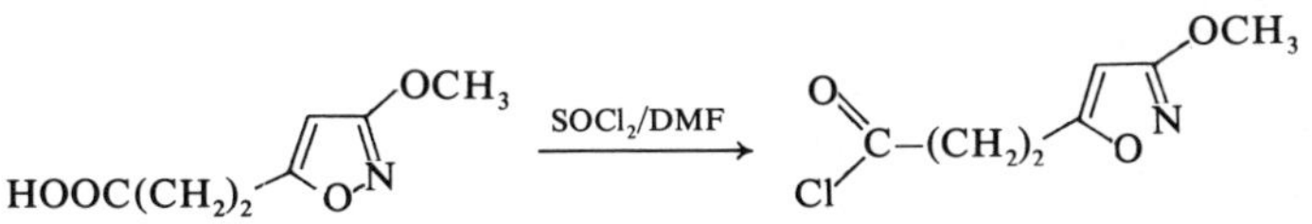

TABLE III (cont.)

Reaction	Ref.
$(C_2H_5)_2C(SO_3Na)(COONa) \xrightarrow{SOCl_2/DMF} (C_2H_5)_2C(SO_2Cl)(COCl)$ with C=O	135
4-HOOC, 2,6-(HOOC)$_2$-pyridine $\xrightarrow{SOCl_2/DMF}$ 4-(ClC=O), 2,6-(ClC=O)$_2$-pyridine	130
Phthalimide-N-CH$_2$COOH $\xrightarrow[\text{2. H}_2\text{N-CH}_2\text{-COOC}_2\text{H}_5]{\text{1. COCl}_2/\text{DMF}}$ Phthalimide-N-CH$_2$-C(O)-NH-CH$_2$-C(O)-OC$_2$H$_5$	131, 132
C$_4$H$_9$-CH(ONH$_3^{\oplus}$ Cl$^{\ominus}$)-COOH $\xrightarrow{SOCl_2/DMF}$ C$_4$H$_9$-CH(ONH$_3^{\oplus}$ Cl$^{\ominus}$)-C(O)Cl	136
$(O_2N{-}C_6H_4{-})_2P(O)OH \xrightarrow{SOCl_2/DMF} (O_2N{-}C_6H_4{-})_2P(O)Cl$	13
benzoxazine-2,4-dione ⇌ 2-(NCO)-C$_6$H$_4$-COOH $\xrightarrow{COCl_2/DMF}$ 2-(NCO)-C$_6$H$_4$-C(O)Cl	137
2-CH$_3$-3-CH(CH$_3$)$_2$-6-(COOR)-C$_6$H$_2$-OCH$_2$COOH $\xrightarrow{SOCl_2/DMF}$ 2-CH$_3$-3-CH(CH$_3$)$_2$-6-(COOR)-C$_6$H$_2$-OCH$_2$-C(O)Cl	138

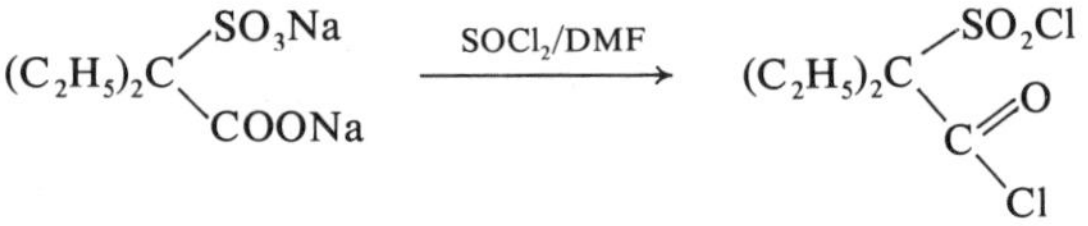

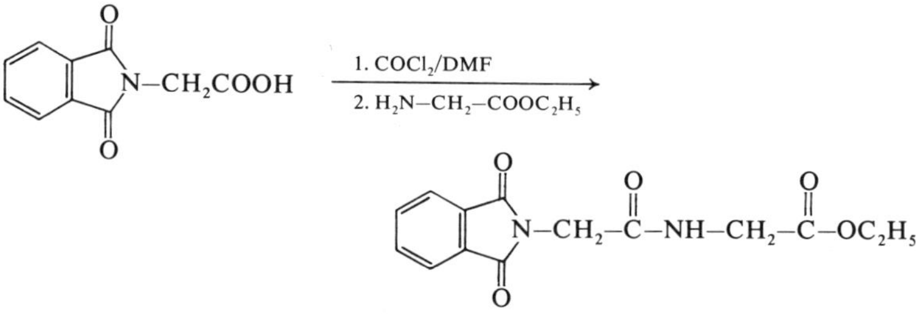

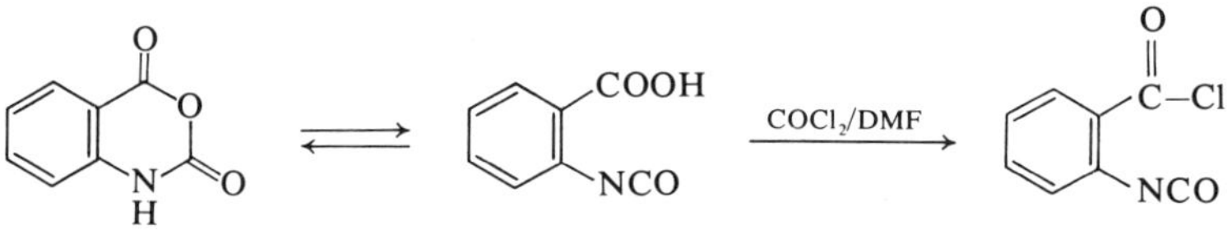

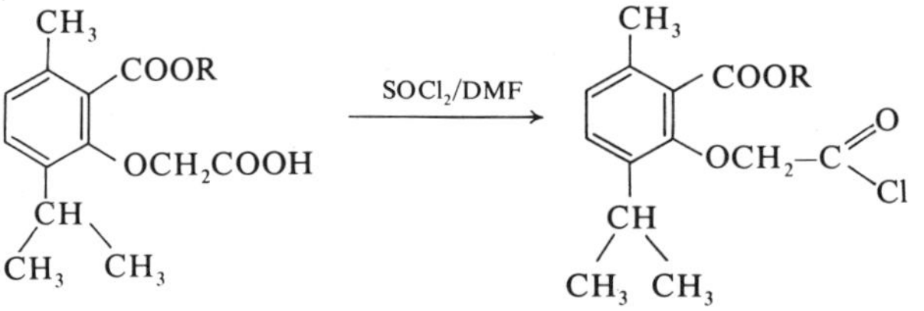

III. REACTIONS OF HALOMETHYLENIMINIUM SALTS

During this acid chloride synthesis some side reactions are possible, if the carboxylic acid contains other functional groups that are capable of reacting with the amide chloride or the hydrogen chloride formed. If, for example, α,β-unsaturated carboxylic acids (**102**) are reacted, β-chloro saturated acid chlorides (**103**) are formed by addition of the HCl generated *in situ* to the α,β-unsaturated chloride or to the free acid (7).

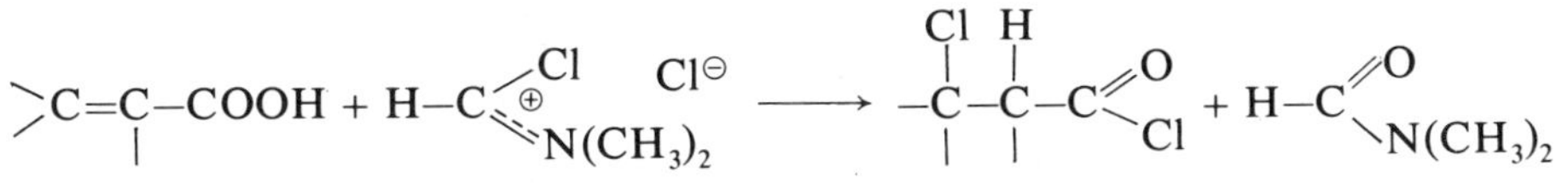

$$>\!\!C\!=\!\!C\!-\!COOH + H\!-\!C\overset{\oplus}{\underset{N(CH_3)_2}{\diagup Cl}}\quad Cl^{\ominus} \longrightarrow -\overset{Cl}{\underset{|}{C}}\!-\!\overset{H}{\underset{|}{C}}\!-\!C\overset{\diagup O}{\diagdown Cl} + H\!-\!C\overset{\diagup O}{\diagdown N(CH_3)_2}$$

(**102**)   (**103**)

Alkyl or aryl substituents in the β position, however, suppress HCl addition (7).

With unsaturated dicarboxylic acids, cyclization and chlorination are possible as well (139):

$$\begin{array}{c} C-COOH \\ ||| \\ C-COOH \end{array} \xrightarrow{DMF/SOCl_2} \text{(dichloromaleic anhydride)}$$

If the carboxylic acid contains strongly acidic phenolic or heterocyclic OH (140, 141) or SH groups (142), they are substituted by chlorine. Furthermore, compounds containing thiol groups may undergo dimerization giving disulfides, depending on the amount of DMF used (142).

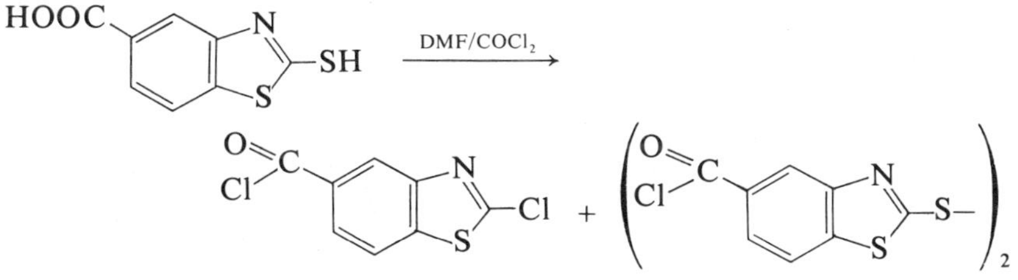

[Reaction schemes showing conversion of HOOC-quinoxaline-diol to Cl-acyl-dichloroquinoxaline with DMF/COCl$_2$; quinoxaline-COOH-OH to quinoxaline-COCl-Cl with DMF/SOCl$_2$; and HOOC-benzothiazole-SH to chloroacyl-benzothiazole-Cl plus disulfide dimer with DMF/COCl$_2$.]

The bis(p-nitrophenyl)chlorophosphoric acid ester was synthesized from bis-(p-nitrophenyl)phosphoric acid using thionyl chloride and DMF (13). Generally, the diester of orthophosphoric acid (**104**) is formed if the monoester of phosphoric acid (**105**) reacts with dimethylformamide chloride in the presence of pyridine, and the adduct formed (**106**) is alcoholyzed (143).

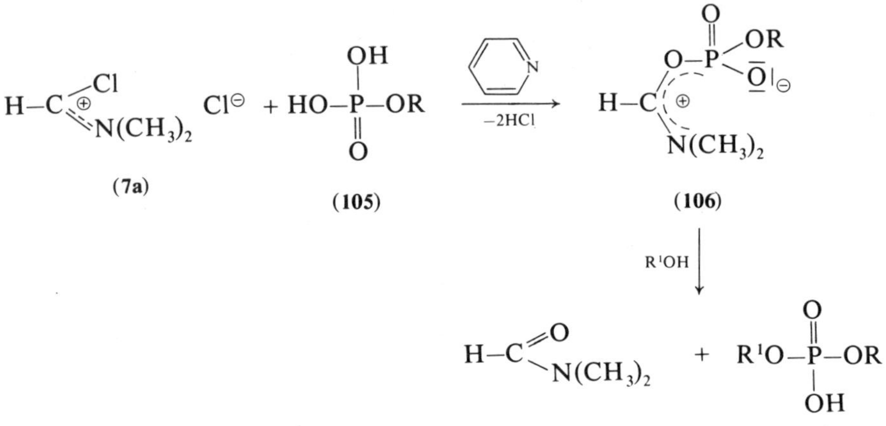

Oligo- and polynucleotides are obtained using dimethylformamide chloride as the condensing reagent (143–145).

2,2-Dichloroaziridines (**107**) also generate acid chlorides from carboxylic acids (87).

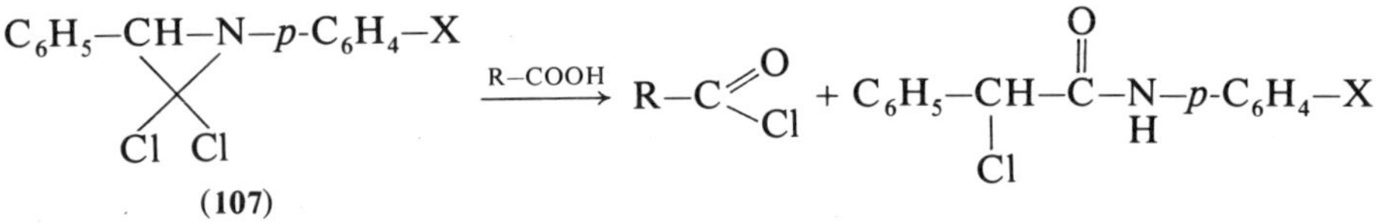

Finally, carboxylic acid fluorides are formed from carboxylic acids and (difluoromethyl)dimethylamine, even under the mildest reaction conditions (0°C) (98):

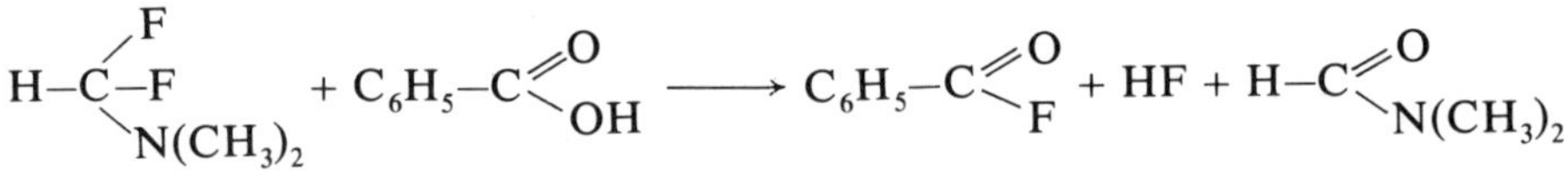

B. REACTION OF AMIDE CHLORIDES WITH CYCLIC ETHERS AND ACETALS

Ziegenbein and Hornung (146–148) studied the reaction of dimethylformamide chloride with cyclic ethers. These are cleaved by amide chlorides, giving dihalo compounds. Catalytic amounts of DMF are sufficient in these reactions.

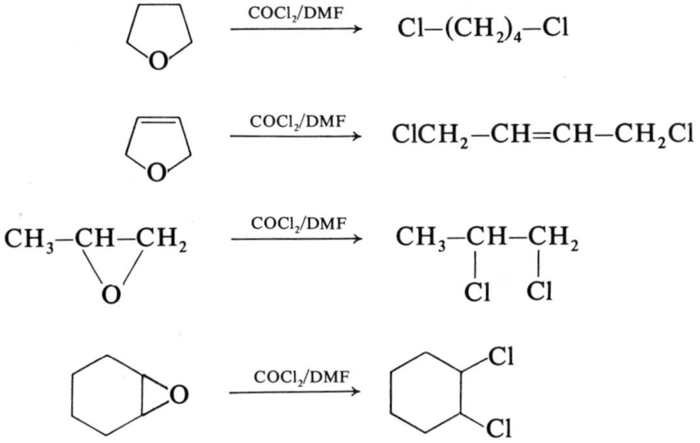

III. REACTIONS OF HALOMETHYLENIMINIUM SALTS

$$\text{tetrahydrofuran} \xrightarrow{\text{COCl}_2/\text{DMF}} \text{Cl-(CH}_2)_4\text{-Cl}$$

$$\text{2,5-dihydrofuran} \xrightarrow{\text{COCl}_2/\text{DMF}} \text{ClCH}_2\text{-CH=CH-CH}_2\text{Cl}$$

$$\text{CH}_3\text{-CH-CH}_2\text{(epoxide)} \xrightarrow{\text{COCl}_2/\text{DMF}} \text{CH}_3\text{-CH-CH}_2 \;\; (\text{Cl, Cl})$$

$$\text{cyclohexene oxide} \xrightarrow{\text{COCl}_2/\text{DMF}} \text{1,2-dichlorocyclohexane}$$

When epoxides are reacted with amide chlorides, 1-acyloxy-2-chloro compounds (**108**) are formed as intermediates after hydrolysis.

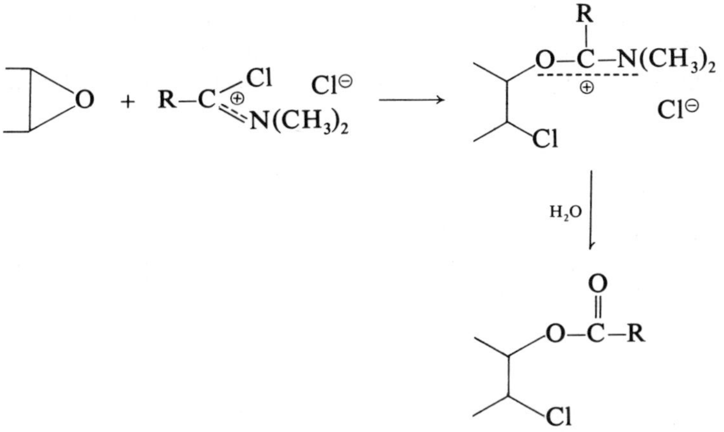

$$\triangleright\!\!\!\text{O} + R-C \overset{\text{Cl}}{\underset{N(CH_3)_2}{\oplus}} \;\; Cl^{\ominus} \longrightarrow \underset{Cl}{\overset{O-\overset{R}{\underset{\oplus}{C}}-N(CH_3)_2}{\diagup\!\!\!\diagdown}} \;\; Cl^{\ominus}$$

$$\downarrow H_2O$$

$$\underset{Cl}{\overset{O-\overset{O}{\overset{\|}{C}}-R}{\diagup\!\!\!\diagdown}}$$

(**108**)

Tetrahydrofuran is transformed into 1,4-dichlorobutane by the amidinium chloride **109** (108).

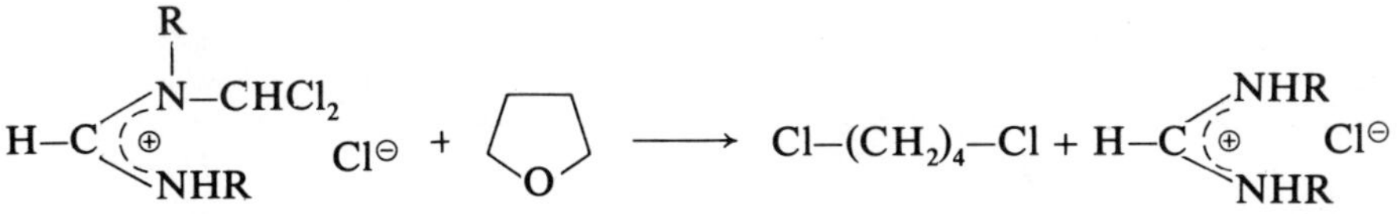

$$H-C\overset{N-CHCl_2}{\underset{NHR}{\diagup\!\!\!\!\oplus}}\;\; Cl^{\ominus} \;+\; \square\text{O} \longrightarrow Cl-(CH_2)_4-Cl + H-C\overset{NHR}{\underset{NHR}{\diagup\!\!\!\!\oplus}}\;\; Cl^{\ominus}$$

(**109**)

Acetals of aliphatic aldehydes are formylated by amide chlorides (11, 12) (see also Part 1, the chapter by C. Jutz, "the Vilsmeier-Haack-Arnold Acylations").

$$\text{CH}_3\text{-CH}_2\text{-CH} \begin{array}{c} \text{OC}_2\text{H}_5 \\ \text{OC}_2\text{H}_5 \end{array} \xrightarrow{\text{COCl}_2/\text{DMF}} (\text{CH}_3)_2\text{N-CH=C-CHO} \\ \phantom{(\text{CH}_3)_2\text{N-CH=C-CH}} | \\ \phantom{(\text{CH}_3)_2\text{N-CH=C-C}} \text{CH}_3$$

Some patents mention the transformation by amide chlorides of acetals into geminal dihalo compounds, where the two alkoxy groups are substituted (149).

$$\text{CH}_3\text{O-CH}_2\text{-CH} \begin{array}{c} \text{OCH}_3 \\ \text{OCH}_3 \end{array} \xrightarrow{\text{COCl}_2/\text{DMF}} \text{CH}_3\text{O-CH}_2\text{-CH} \begin{array}{c} \text{Cl} \\ \text{Cl} \end{array}$$

C. REACTIONS WITH COMPOUNDS CONTAINING THIOL GROUPS

1. Thiolysis

When chloroform or methylene chloride solutions of amide chlorides (**7**) are exposed to hydrogen sulfide, thioamides (**110**) (7, 14, 78, 150–153) are obtained in good yields. This sulfurization method allows the synthesis not only of aliphatic and aromatic *N,N*-disubstituted thioamides, but also *N*-alkylated thiolactams. Generally, the amide chlorides are not isolated. The thioamides of

$$\underset{(\mathbf{7})}{\text{R-C}^{\oplus} \begin{array}{c} \text{Cl} \\ \text{N(R}^1)_2 \end{array}} \text{Cl}^{\ominus} + \text{H}_2\text{S} \longrightarrow \underset{(\mathbf{110})}{\text{R-C} \begin{array}{c} \text{S} \\ \text{N(R}^1)_2 \end{array}} + 2\text{HCl}$$

the squaric acid **111** were synthesized in the same way (17). Thioamides and thiolactams generated by this method are listed in Table IV.

TABLE IV

Thioamides (**110**) from Amide Chlorides (**7**) and Hydrogen Sulfide (14)

Thioamide

$$\text{R-C} \begin{array}{c} \text{S} \\ \text{NR}^1\text{R}^2 \end{array}$$

R	R^1	R^2	Yield, %	m.p., °C	b.p./pressure, °C
H	CH_3	CH_3	85	—	111–113/10
n-C_3H_7	C_2H_5	C_2H_5	63	—	128–132/11
C_6H_5	CH_3	CH_3	83	67	—
p-NO_2-C_6H_4	CH_3	CH_3	83	146	—
–$(CH_2)_3$–		CH_3	83	—	144–145/15
–$(CH_2)_5$–		CH_3	70	—	159–163/13

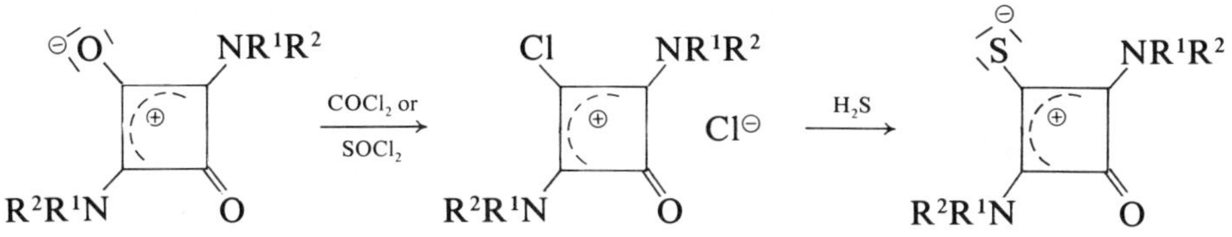

III. REACTIONS OF HALOMETHYLENIMINIUM SALTS 95

$R^1 = R^2 = CH_3$
$R^1-R^2 = (CH_2)_2-O-(CH_2)_2$

Chloromethyleniminium chlorides, generated *in situ* from chloroform and amines, are successfully reacted with H_2S to yield thioformamides (**112**) (154).

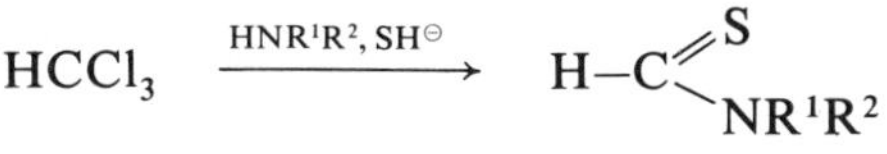

$$HCCl_3 \xrightarrow{HNR^1R^2, SH^\ominus} H-C\underset{NR^1R^2}{\overset{S}{\lessgtr}}$$
(**112**)

2. Mercaptolysis

Amide chlorides (**7**) react with mercaptans, via HCl elimination, to give N,N-dialkylmercaptoalkylmethyleniminium chlorides (**113**). They are not isolated, but are directly converted into thioesters (**114**) and dithioesters (**115**) by hydrolysis or thiolysis (14), respectively.

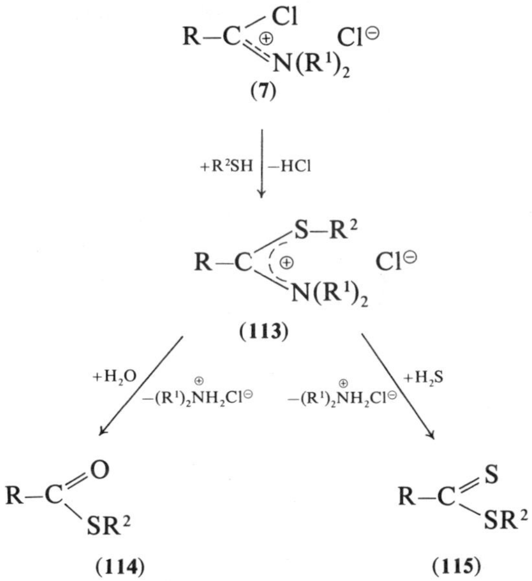

D. REACTIONS WITH COMPOUNDS CONTAINING NH$_2$ GROUPS

1. Reactions with Primary and Secondary Amines, Hydroxylamines, and Hydrazine Derivatives

Wallach (150), Beckmann and Fellrath (155), as well as J. v. Braun et al. (20, 156–158) discovered the reaction of aromatic amide chlorides with aromatic amines to yield N,N,N'-trisubstituted amidines (**116**).

$$R-\overset{\oplus}{C}\!\!\begin{array}{c}Cl\\N(R^1)_2\end{array}\ Cl^{\ominus}\ +\ H_2NR^2\ \xrightarrow{-HCl}$$

$$R-\overset{\oplus}{C}\!\!\begin{array}{c}NHR^2\\N(R^1)_2\end{array}\ Cl^{\ominus}\ \xrightarrow{H_2O,\ OH^{\ominus}}\ R-C\!\!\begin{array}{c}N-R^2\\N(R^1)_2\end{array}$$
$$(116)$$

Previously reported reactions, which give N-mono- and N,N'-disubstituted amidines and proceed via amide chlorides as intermediates, have been reviewed by Shriner and Neumann (159).

Eilingsfeld et al. (7), as well as other authors (160), converted a series of aminoanthraquinones into the corresponding amidines using dimethylformamide chloride.

All these reactions generate hydrogen chloride, which partly converts the starting amine into the hydrochloride. Since aromatic amine hydrochlorides smoothly react with amide chlorides, the synthesis is not hampered (7).

The amine hydrochlorides of strongly basic amines react between 150 and 180°C. Using these reaction conditions, Eilingsfeld et al. (7) synthesized several N,N-disubstituted amidinium chlorides (**117**) from ammonium chloride and amide chlorides.

$$R-\overset{\oplus}{C}\!\!\begin{array}{c}Cl\\N(R^1)_2\end{array}\ Cl^{\ominus}\ +\ NH_4Cl\ \xrightarrow[-2HCl]{150-170°C}\ R-\overset{\oplus}{C}\!\!\begin{array}{c}NH_2\\N(R^1)_2\end{array}\ Cl^{\ominus}$$
$$(117)$$

The reaction of amide chlorides with aliphatic amines is less appropriate for the synthesis of N,N,N',N'-tetraalkylformamidinium salts, since the amine hydrochloride is formed along with the amidinium salt (**118**). The former compound is not separable by crystallization, as Böhme and Soldan (161) have shown for the reaction of DMF chloride (**7a**) with dimethylamine.

$$H-\overset{\oplus}{C}\!\!\begin{array}{c}Cl\\N(CH_3)_2\end{array}\ Cl^{\ominus}\ +\ HN(CH_3)_2\ \longrightarrow$$

$$H-\overset{\oplus}{C}\!\!\begin{array}{c}N(CH_3)_2\\N(CH_3)_2\end{array}\ Cl^{\ominus}\ +\ H_2\overset{\oplus}{N}(CH_3)_2\ Cl^{\ominus}$$
$$(118)$$

III. REACTIONS OF HALOMETHYLENIMINIUM SALTS

The cyclobutenylium salt **120** was isolated in low yields from **119** and morpholine (**118**).

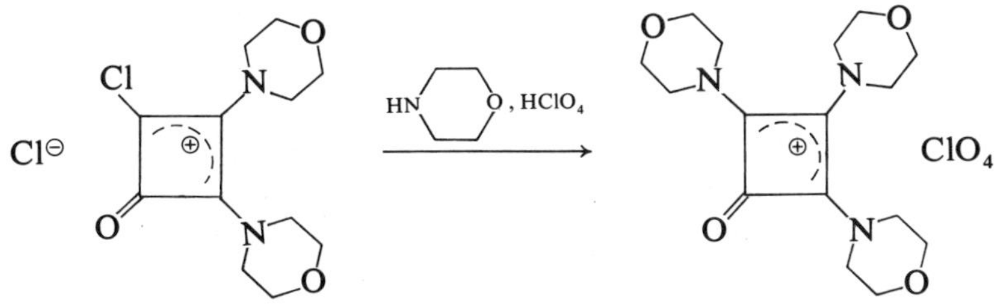

(119) (120)

Grdinič and Hahn (8) studied the reaction of aniline with a series of N-mono- and N,N-disubstituted amide chlorides (**121**). (In general, 2-furan-carboxylic acid amide chlorides were used.) In these cases, the corresponding amidines (**122**) were also obtained, partly characterized as picrates.

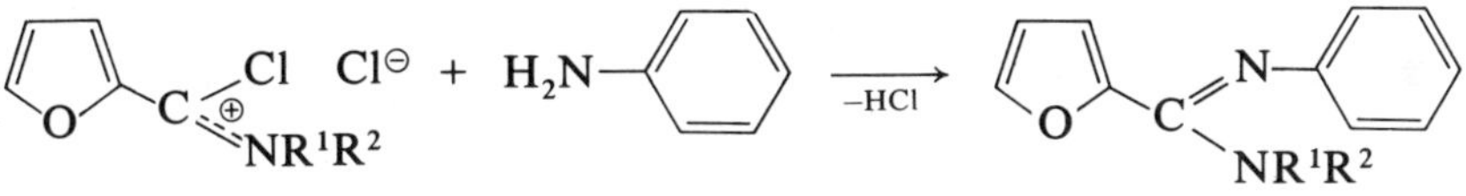

(121) (122)

R^1	H	CH_3
R^2	CH_3	C_2H_5

When DMF and $SOCl_2$ are reacted with aniline, N,N-dimethyl-N'-phenyl-formamidine (**123**) forms, along with N,N'-diphenylformamidine (**124**) (162). The analogous reaction with diphenylamine gives N,N-diphenylformamide (**125**).

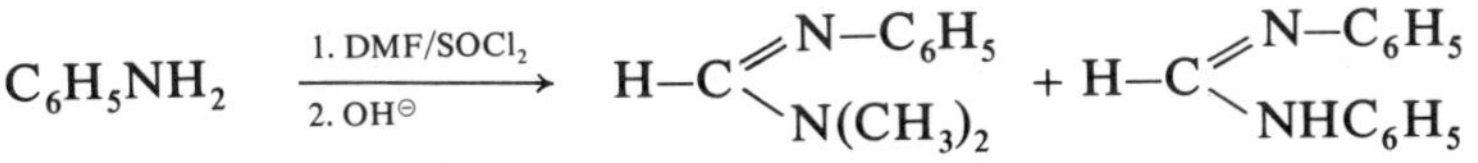

$C_6H_5NH_2$ $\xrightarrow{\text{1. DMF/SOCl}_2}{\text{2. OH}^\ominus}$ $H-C{\displaystyle{\nwarrow N-C_6H_5 \atop \swarrow N(CH_3)_2}}$ + $H-C{\displaystyle{\nwarrow N-C_6H_5 \atop \swarrow NHC_6H_5}}$

(123) (124)

$HN(C_6H_5)_2$ $\xrightarrow{\text{1. DMF/SOCl}_2}{\text{2. OH}^\ominus}$ $H-C{\displaystyle{\nwarrow O \atop \swarrow N(C_6H_5)_2}}$

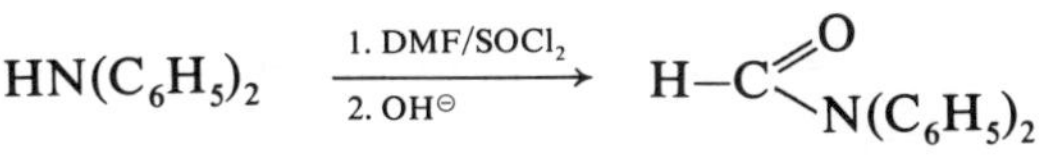

(125)

Amidinium salt **127** was synthesized from bromomethyleniminium bromide **126** (23).

Csürös et al. (163–165) reacted a large number of o-substituted anilines with amide chlorides to yield amidinium salts (**128**) which were cyclized to form

(126) (127)

pyrimidine derivatives (**129**); further examples of these reactions are given in the chapter on amidinium salts.

(**128**)

(**129**)

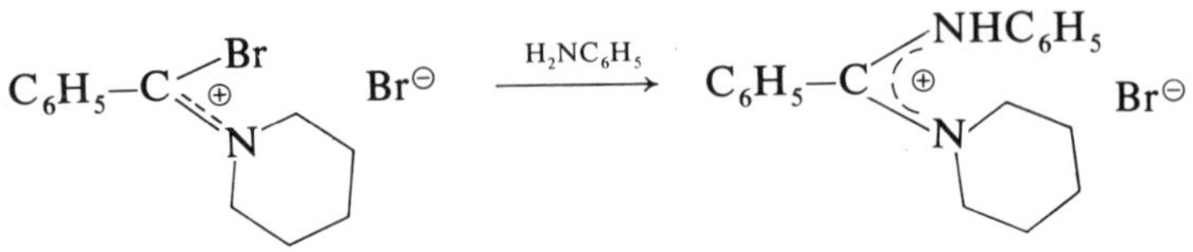

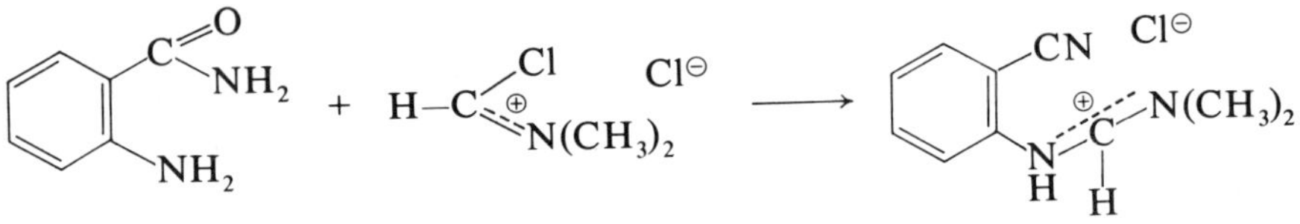

α-Ketoamide chloride arylhydrazones (**130**) are also converted into amidines (**131**) by aromatic amines (97).

(**130**) (**131**)

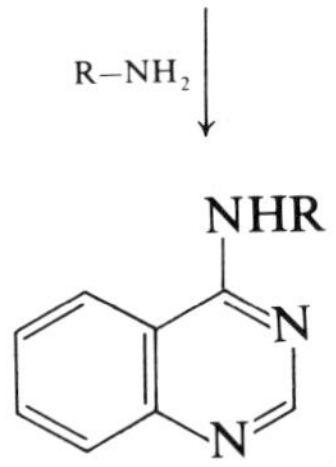

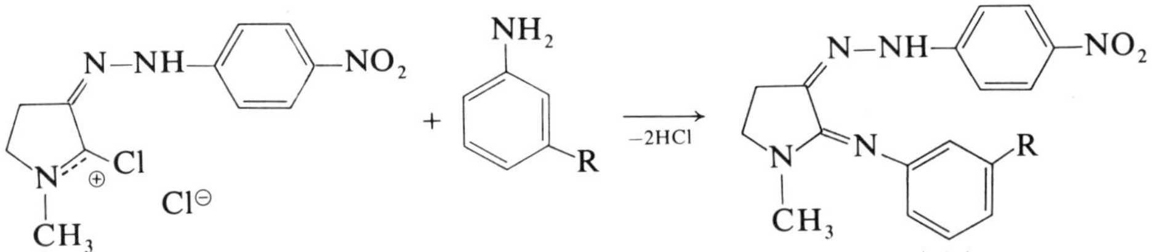

Primary amines give amidinium salts with amide chlorides like **132** as well (108).

(**132**)

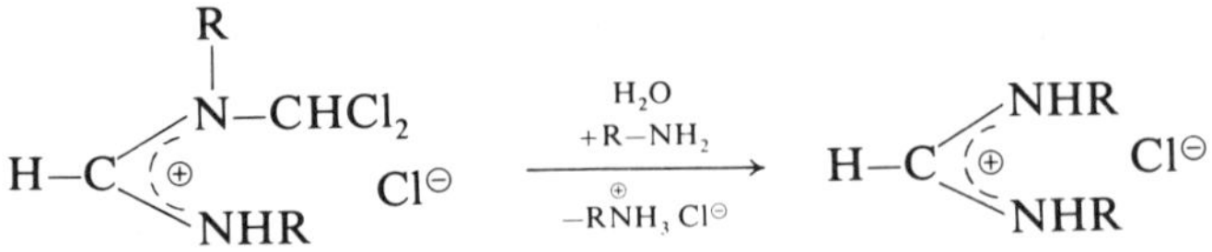

The aminolysis of the iminium bromide **133** proceeds similarly (75).

$$\text{H-C}\underset{\text{NH}_2}{\overset{\text{NHCHBr}_2}{\diagup}}\oplus \xrightarrow[\text{H}_2\text{O}]{\text{HNR}_2} \text{H-C}\underset{\text{NR}^2}{\overset{\text{O}}{\diagup}} + \text{H-C}\underset{\text{NH}_2}{\overset{\text{NH}_2}{\diagup}}\oplus \quad \text{Br}^{\ominus}$$
(133)

Hydroxylamine and hydrazine or their hydrochlorides show an analogous reactivity toward amide chlorides; amide oximes (**134**) and amide azines (**135**) (7, 14) are obtained.

$$\text{R-C}\underset{\text{N(R}^1)_2}{\overset{\text{Cl}}{\diagup}}\oplus \quad \text{Cl}^{\ominus} + \text{H}_2\text{NOH} \xrightarrow[-2\text{HCl}]{} \text{R-C}\underset{\text{N(R}^1)_2}{\overset{\text{N-OH}}{\diagup}}$$
(134)

$$2\text{R-C}\underset{\text{N(R}^1)_2}{\overset{\text{Cl}}{\diagup}}\oplus \quad \text{Cl}^{\ominus} + \text{H}_2\text{N-NH}_2 \xrightarrow[-4\text{HCl}]{} \underset{\underset{\text{N(R}^1)_2}{|}}{\text{R-C}}=\text{N-N}=\underset{\underset{\text{N(R}^1)_2}{|}}{\text{C-R}}$$
(135)

2,2-Dichloroaziridines undergo ring cleavage to form α-chloroamidines (**136**) when reacted with aliphatic primary amines (87).

$$\underset{\underset{\text{C}_6\text{H}_5}{|}}{\overset{\text{C}_6\text{H}_5-\text{CH-CCl}_2}{\diagdown\diagup}}\text{N} + \text{R-NH}_2 \longrightarrow \text{C}_6\text{H}_5-\underset{\underset{\text{Cl}}{|}}{\text{CH}}-\text{C}\underset{\text{NHR}}{\overset{\text{N-C}_6\text{H}_5}{\diagup}}$$
(136)

2. Reactions with Amides and Nitriles

Primary amides (**137**) are dehydrated to nitriles (**138**) by amide chlorides (7). Dimethylformamide chloride (**7a**) is especially effective; it gives rise to dehydration within a few minutes at room temperature (7). Normally the yields are better than 80%. Since the dimethylformamide is regenerated during

$$\text{R-C}\underset{\text{NH}_2}{\overset{\text{O}}{\diagup}} + \text{H-C}\underset{\text{N(CH}_3)_2}{\overset{\text{Cl}}{\diagup}}\oplus \quad \text{Cl}^{\ominus} \longrightarrow$$
(137) (7a)
$$\text{R-C}\equiv\text{N} + 2\text{HCl} + \text{H-C}\underset{\text{N(CH}_3)_2}{\overset{\text{O}}{\diagup}}$$
(138)

the course of the reaction, catalytic amounts (5–10%) are sufficient to dehydrate amides with thionyl chloride or phosgene. Mono-, di-, and tetra-amides have been converted into the corresponding nitriles by this route.*

---
* Table V summarizes this kind of reaction.

TABLE V
Survey of Nitriles via Dehydration of Carboxylic Acid Amides with Dimethylformamide Chloride

Reaction	Ref.
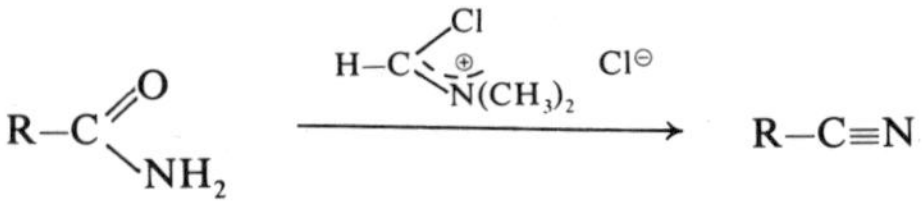 $R-CONH_2 \xrightarrow{[H-CHCl-N(CH_3)_2]^+ Cl^-} R-C\equiv N$; $R = C_6H_5, C_6H_5-CH_2-$	7
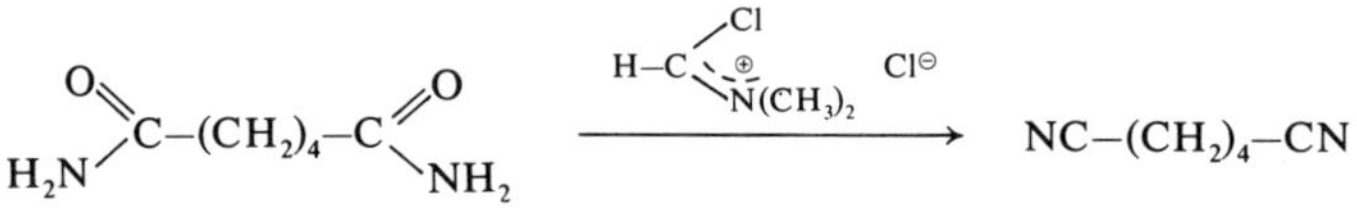 $H_2N-CO-(CH_2)_4-CO-NH_2 \xrightarrow{[H-CHCl-N(CH_3)_2]^+ Cl^-} NC-(CH_2)_4-CN$	7
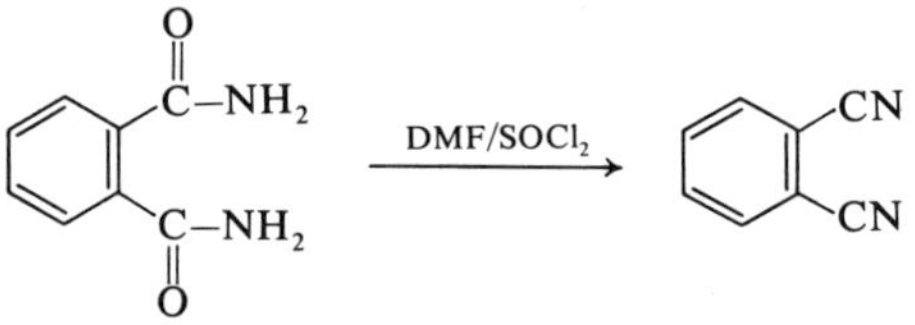 o-C₆H₄(CONH₂)₂ →(DMF/SOCl₂)→ o-C₆H₄(CN)₂	166
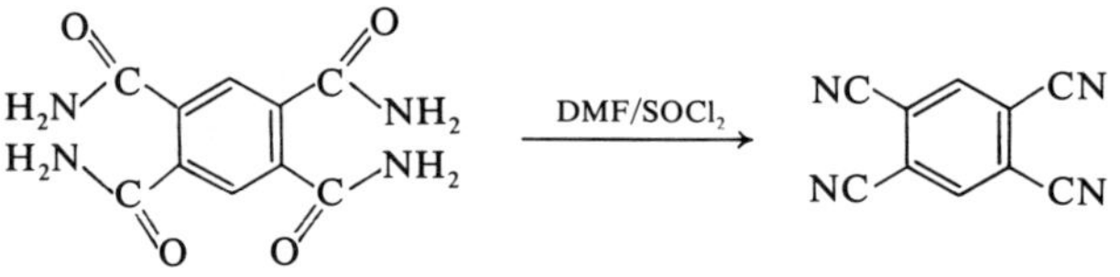 1,2,4,5-C₆H₂(CONH₂)₄ →(DMF/SOCl₂)→ 1,2,4,5-C₆H₂(CN)₄	166, 167
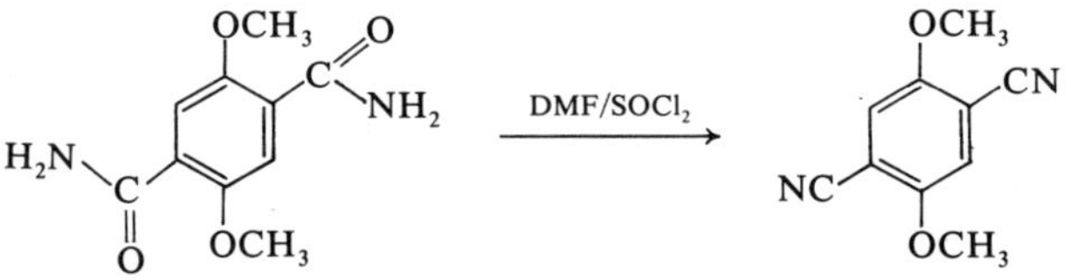 4,6-dimethoxy-1,3-benzenedicarboxamide →(DMF/SOCl₂)→ 4,6-dimethoxy-1,3-dicyanobenzene	168
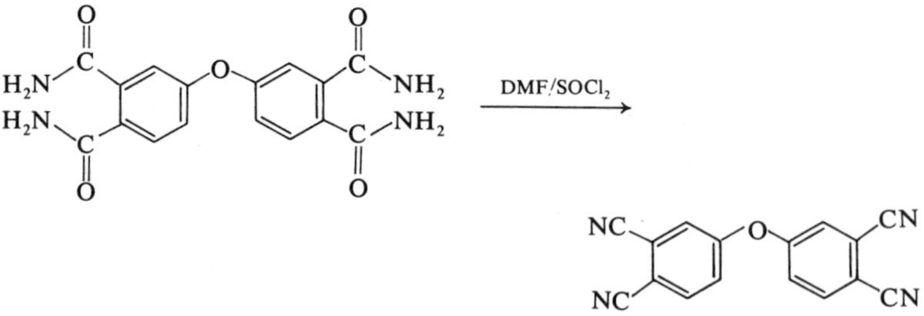 bis(3,4-dicarboxamidophenyl) ether →(DMF/SOCl₂)→ bis(3,4-dicyanophenyl) ether	169

III. REACTIONS OF HALOMETHYLENIMINIUM SALTS

The dehydration of oximes by amide chlorides to give nitriles also seems to be possible, as the following example shows (170):

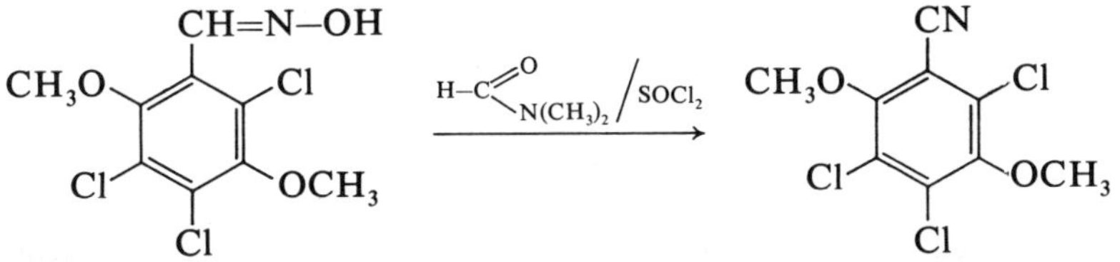

[Reaction: 2,4-dichloro-3,5-dimethoxybenzaldoxime + H-C(=O)N(CH₃)₂ / SOCl₂ → corresponding nitrile]

Amide chlorides of primary amides (**139**) dimerize to amidinium salts (**140**), which react further with phosgene to yield pyrimidine derivatives (**141**) (171).

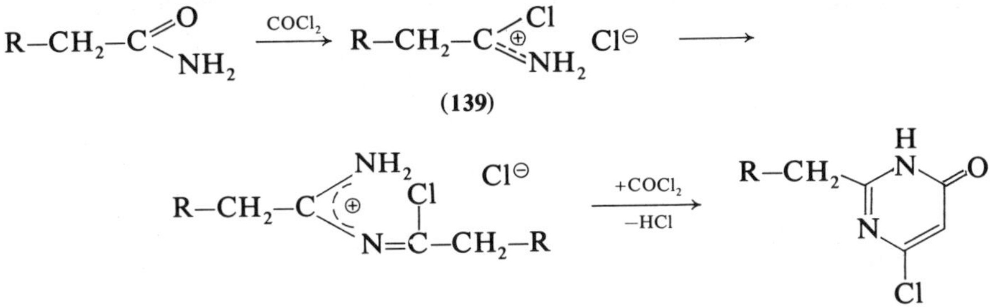

$$R-CH_2-C(=O)NH_2 \xrightarrow{COCl_2} R-CH_2-C(Cl)(\overset{\oplus}{=}NH_2)\ Cl^{\ominus} \longrightarrow$$

(139)

$$R-CH_2-C(\overset{\oplus}{NH_2})(Cl)-N=C(Cl)-CH_2-R\ Cl^{\ominus} \xrightarrow[-HCl]{+COCl_2}$$

(140)

[pyrimidinone **141**: R-CH₂ substituted, with N-H, C=O, and Cl]

(141)

The dimerization of nitriles under pressure in the presence of HCl should be interpreted the same way (172). In the presence of HCl, nitriles react with *N,N*-

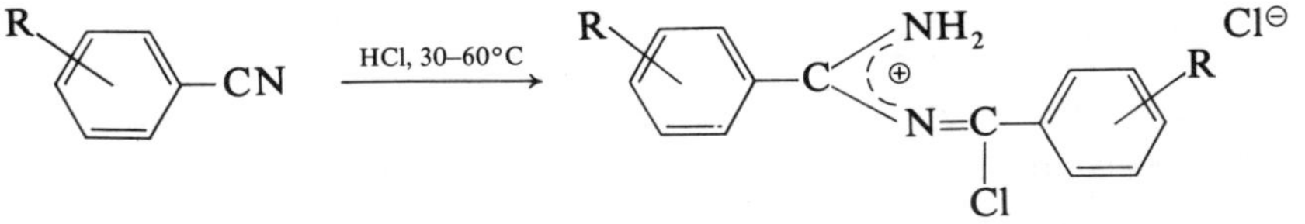

[Ar-CN $\xrightarrow{HCl,\ 30-60°C}$ Ar-C(NH₂)(⊕)-N=C(Cl)-Ar   Cl⁻]

dimethylchloromethyleniminium chloride to give the salts **142** (173). In a similar way, five-, six-, and seven-membered heterocycles (**144**) (pyridones,

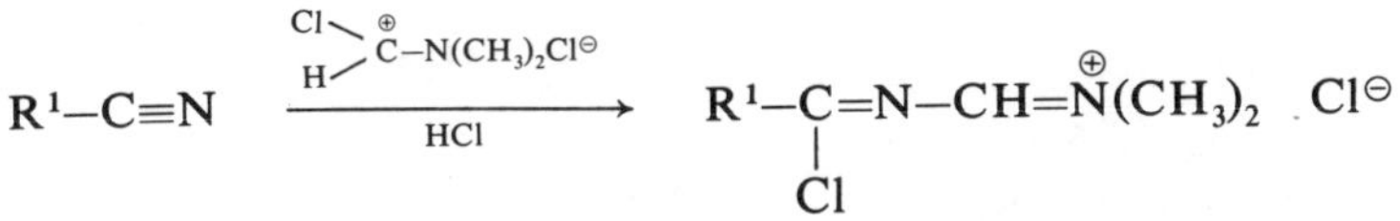

$$R^1-C\equiv N \xrightarrow[HCl]{\underset{H}{\overset{Cl\diagdown\ \oplus}{C-N(CH_3)_2\ Cl^{\ominus}}}} R^1-C(Cl)=N-CH=\overset{\oplus}{N}(CH_3)_2\ \ Cl^{\ominus}$$

(142)

isoquinolines, quinazolones, 1,3-benzthiazinones) were synthesized from ω-cyano acid halides (**143**) and hydrogen halides. These reactions have already been reviewed by Simchen and Entenmann (105) and Yanagida and Komori (106) (see also the chapter in this volume by S. Yanagida, M. Okahara, and S. Komori "Iminium and Nitrilium Salts in the Dimerization of Cyano Compounds in Acids").

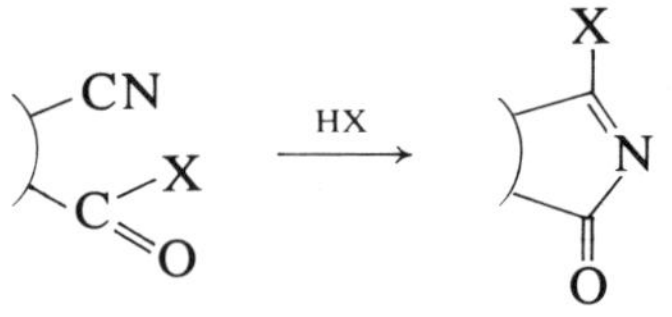

(143) → (144)

When heated with thionyl chloride (174) or phosphorus pentachloride (1), formanilides yield N,N'-diarylformamidinium salts (145); benzanilide reacts successfully, too, but acetanilide (174) does not. The course of the reaction was studied by Kühle (175), who found that amide chlorides are intermediates.

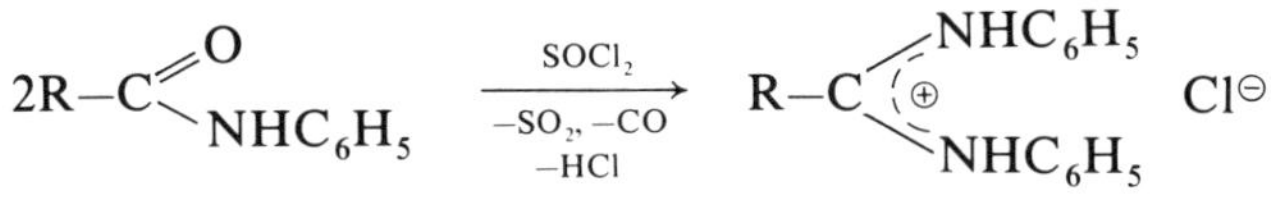

$$2R-C(=O)NHC_6H_5 \xrightarrow[-HCl]{SOCl_2, -SO_2, -CO} R-C(NHC_6H_5)_2^{\oplus} \; Cl^{\ominus}$$

R = H, C_6H_5 (145)

The amide chloride 146, made from thionyl chloride and formanilide, was also reacted with DMF to yield N,N-dimethyl-N'-phenylformamidinium chloride (147) (175).

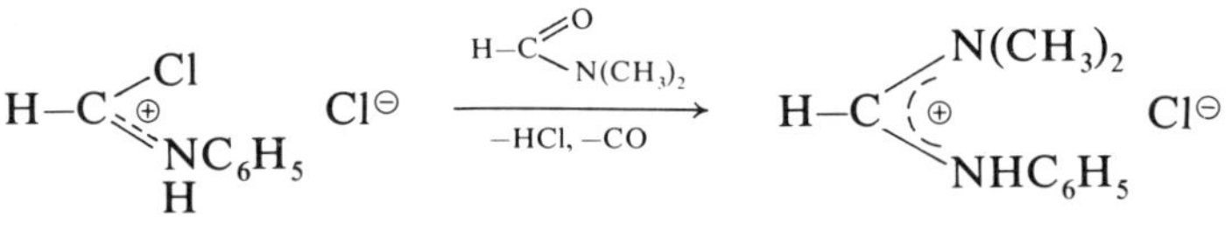

$$H-C(Cl)(NC_6H_5)H^{\oplus} \; Cl^{\ominus} \xrightarrow[-HCl, -CO]{H-C(=O)N(CH_3)_2} H-C(N(CH_3)_2)(NHC_6H_5)^{\oplus} \; Cl^{\ominus}$$

(146)   (147)

N,N,N',N'-tetramethylformamidinium chloride (148) has been successfully synthesized in an analogous reaction from DMF and thionyl chloride (104).

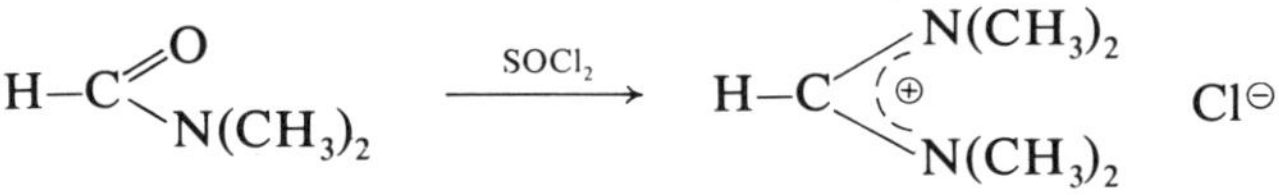

$$H-C(=O)N(CH_3)_2 \xrightarrow{SOCl_2} H-C(N(CH_3)_2)_2^{\oplus} \; Cl^{\ominus}$$

(148)

The reactions of N-monosubstituted formamides with dimethylformamide chloride (7a) show a different pattern. Here the amide chloride dehydrates the formamide, and isonitriles (149) are formed (176). However, the reaction of N-

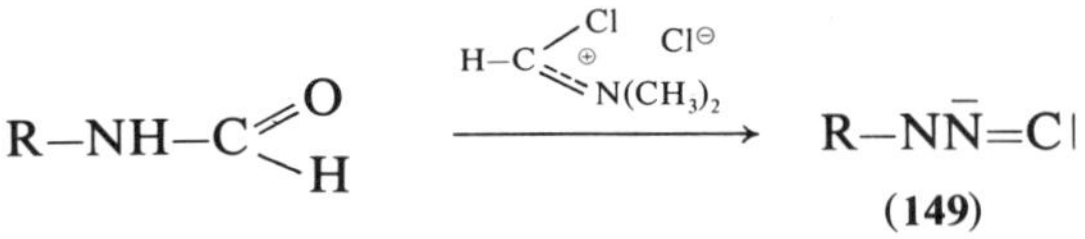

$$R-NH-C(=O)H \xrightarrow{H-C(Cl)N(CH_3)_2 \, Cl^{\ominus}} R-N\bar{N}=Cl$$

(149)

monosubstituted formamide with phosgene to yield amidinium salts was successful (37, 177).

III. REACTIONS OF HALOMETHYLENIMINIUM SALTS

At room temperature dimethylformamide chloride and DMF show an halide–oxygen exchange (178):

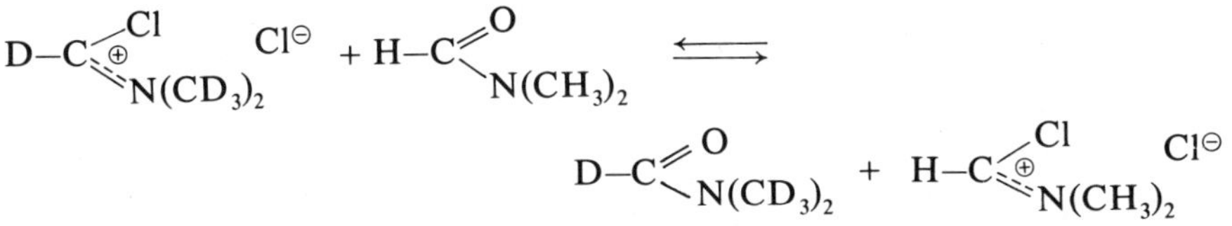

$$D-C{\overset{\oplus}{\underset{N(CD_3)_2}{<}}}^{Cl} \; Cl^{\ominus} + H-C{\overset{O}{\underset{N(CH_3)_2}{<}}} \rightleftharpoons$$

$$D-C{\overset{O}{\underset{N(CD_3)_2}{<}}} + H-C{\overset{\oplus}{\underset{N(CH_3)_2}{<}}}^{Cl} \; Cl^{\ominus}$$

The following mechanism has been proposed for this reaction (178):

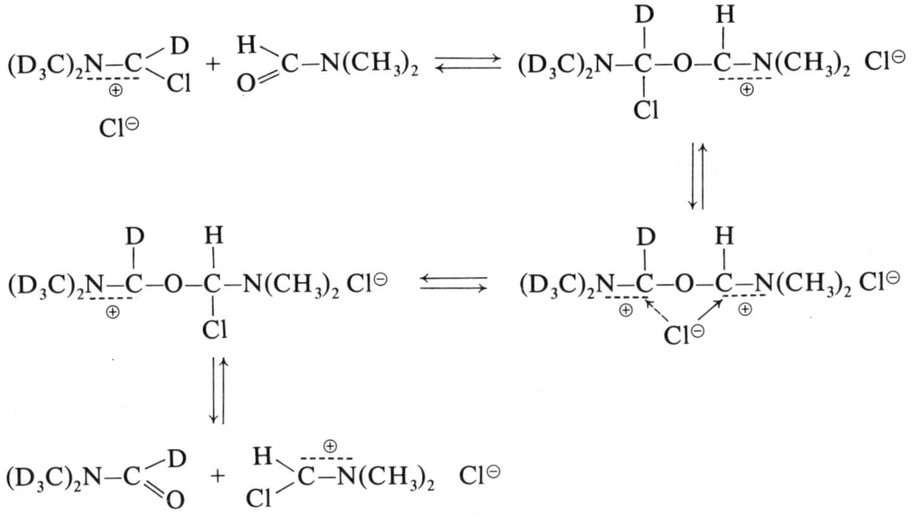

$$(D_3C)_2N\underset{\oplus}{-}C{\overset{D}{<}}_{Cl} + {\overset{H}{\underset{O}{>}}}C-N(CH_3)_2 \rightleftharpoons (D_3C)_2N-\underset{Cl}{\overset{D}{\underset{|}{C}}}-O-\underset{\oplus}{\overset{H}{\underset{|}{C}}}-N(CH_3)_2 \; Cl^{\ominus}$$

$$Cl^{\ominus}$$

$$(D_3C)_2N\underset{\oplus}{-}\underset{Cl}{\overset{D}{\underset{|}{C}}}-O-\overset{H}{\underset{|}{C}}-N(CH_3)_2 \; Cl^{\ominus} \rightleftharpoons (D_3C)_2N\underset{\oplus}{-}\overset{D}{\underset{|}{C}}-O-\underset{\oplus}{\overset{H}{\underset{|}{C}}}-N(CH_3)_2 \; Cl^{\ominus}$$

$$Cl^{\ominus}$$

$$(D_3C)_2N-C{\overset{D}{<}}_O + {\overset{H}{\underset{Cl}{>}}}C\overset{\oplus}{-}N(CH_3)_2 \; Cl^{\ominus}$$

3. Reactions with Ureas, Thioureas, and Urethanes

Amide chlorides are converted by urea and its derivatives into amidinium salts (**150, 151**) (7, 179–181). In this reaction, the amide chlorides are

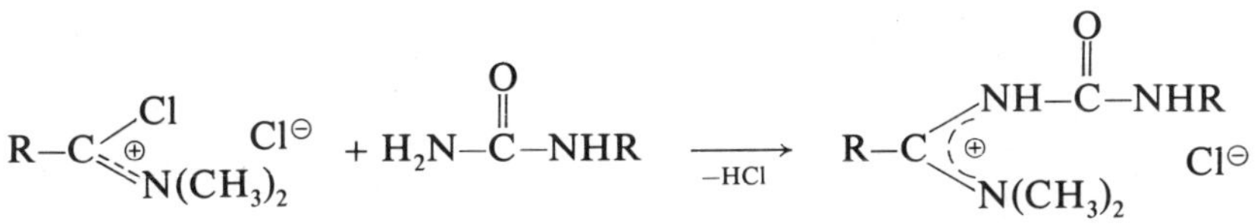

$$R-C{\overset{\oplus}{\underset{N(CH_3)_2}{<}}}^{Cl} \; Cl^{\ominus} + H_2N-\overset{O}{\underset{\|}{C}}-NHR \xrightarrow{-HCl} R-C{\overset{NH-\overset{O}{\underset{\|}{C}}-NHR}{\underset{N(CH_3)_2}{<}}}^{\oplus} \; Cl^{\ominus}$$

(**150**)

$$H-C{\overset{\oplus}{\underset{NHR}{<}}}^{Cl} + HN{\overset{R}{\underset{|}{-}}}\overset{O}{\underset{\|}{C}}-NR^2R^3 \xrightarrow{-HCl} H-C{\overset{N-\overset{R}{\underset{|}{}}\overset{O}{\underset{\|}{C}}-NR^2R^3}{\underset{NHR}{<}}}^{\oplus} \; Cl^{\ominus}$$

(**151**)

TABLE VI
Amidinium Salts from Ureas, Urea Derivatives, and Thioureas (182)

Amidinium salt

$$R-C{\overset{\oplus}{\underset{NR^3R^4}{\diagup NR^1R^2}}} \quad Cl^{\ominus}$$

R	R¹	R²	R³	R⁴	m.p., °C
H	H	−C(=O)−NH₂	CH₃	CH₃	190–192
H	H	−C(=O)−NHC₂H₅	CH₃	CH₃	176–178 (dec)
H	C₆H₅	−C(=O)−NHC₆H₅	CH₃	CH₃	124–129 (dec)
H	H	−C(=O)−NH−NH₂	CH₃	CH₃	115 (dec)
H	−CH₂−CH₂−NH−C(=O)−		CH₃	CH₃	129–130 (dec)
H	H	−C(=O)−NH−(c−C₈H₁₅)	CH₃	CH₃	—
H	H	−C(=O)−NH−C₆H₅	CH₃	CH₃	203–207 (dec)
H	H	−C(=S)−NH−C₆H₅	CH₃	CH₃	158–159 (dec)
H	H	−C(=NH)−NHC₆H₅	CH₃	CH₃	164–166 (dec)
H	H	−C(=O)−NH−C(=O)−NH₂	CH₃	CH₃	158–160
−(CH₂)₃−	CH₃		H	−C(=O)−NH₂	167–169 (dec)

generated *in situ* by reacting the amide–urea mixture with phosgene or thionyl chloride.

Thioureas (7), urethanes (7), biuret (182), and guanidine carbonate (182) are capable of reacting analogously. Tetrahydrofuran, acetonitrile, 1,1,2,2-tetrachloroethane, and carbon tetrachloride are suitable solvents for these reactions (182).

Table VI summarizes the amidinium salts that have been prepared from urea and its analogues.

Ulrich et al. (183) discovered the conversion of *N*-monoalkylated urethanes into isocyanates (152) by phosgene when DMF is present as a catalyst. To explain the reaction, they proposed that the iminol structure of the urethane (153) reacted with the intermediate dimethylformamide chloride to yield imido chloroformate (154), which decomposed to isocyanate (152).

$$R-\underset{H}{N}-\underset{\|}{C}-OC_2H_5$$
$$O$$

$$\updownarrow \quad + H-C\underset{N(CH_3)_2}{\overset{Cl}{\oplus}} \quad Cl^{\ominus} \quad \xrightarrow{-HCl} \quad R-N=C\underset{Cl}{\overset{OC_2H_5}{\diagup}} \quad + H-C\underset{N(CH_3)_2}{\overset{O}{\diagup}}$$

$$R-N=C-OC_2H_5 \qquad\qquad (154)$$
$$\underset{OH}{|}$$
$$(153) \qquad\qquad\qquad\qquad \downarrow -C_2H_5Cl$$

$$R-N=C=O$$
$$(152)$$

E. REACTIONS OF AMIDE CHLORIDES WITH TERTIARY AMINES

1. *N-Monosubstituted Chloromethyleniminium Salts Without α-CH Bonds*

When phosphorus pentachloride, thionyl chloride, or phosgene reacts with *N*-monosubstituted formamides in the presence of tertiary amines, isonitriles (155) are formed. This effective and widely applicable reaction was discovered by Ugi et al. (184–186). The reaction proceeds in two steps: the amide chloride (156) is formed, and the isonitrile (155) is formed via α-elimination of amine hydrochloride.

$$H-C\underset{NHR}{\overset{O}{\diagup}} \quad \xrightarrow{COCl_2, SOCl_2, \text{ or } PCl_5} \quad H-C\underset{NHR}{\overset{Cl}{\oplus}} \quad Cl^{\ominus} \quad \xrightarrow[-2[HN(R^1)_3Cl^{\ominus}]]{+2N(R^1)_3} \quad R-\overset{-}{N}\overset{+}{N}=C|$$
$$(156) \qquad\qquad\qquad (155)$$

The following amines are used for the dehydrohalogenation: trimethylamine, triethylamine, tri-*n*-butylamine, *N,N*-dimethylcyclohexylamine, *N,N*-diethylaniline, pyridine, and quinoline. As long as the starting *N*-substituted formamide does not contain any functional group that may decompose the isonitrile formed or react in any other way with the halogenating agent (PCl_5, $COCl_2$, $SOCl_2$), any isonitrile may be synthesized this way. Thus N-formyl compounds of primary, aliphatic, and aromatic amines were converted into isonitriles; diisonitriles were generated analogously. Even β- and γ-hydroxypropyl(and butyl)isonitriles have recently been synthesized in this way (187). Ugi et al. (186) have published an extensive review of isonitriles obtained by this method.

Arylthiocyanates (**158**) are obtained from N-formylsulfenamides (**157**) and phosgene/triethylamine (188).

$$R\text{-}C_6H_4\text{-SNHCHO} \xrightarrow[N(C_2H_5)_3]{COCl_2} R\text{-}C_6H_4\text{-}S\text{-}\overset{\oplus}{N}\!\!\equiv\!\!\overset{\ominus}{C}l \longrightarrow R\text{-}C_6H_4\text{-}S\text{-}C\!\!\equiv\!\!N$$
(**157**)   (**158**)

2. Reaction of Tertiary Amines with Chloromethyleniminium Salts Containing α-CH bonds

Ghosez et al. (80) were able to synthesize alkyl- or aryl-substituted α-chloroenamines (**160**) from amide chlorides (**159**) containing at least one α-CH bond when HCl was removed by triethylamine or pyridine. The hydrogen

$$\underset{R^2}{\overset{R^1}{>}}CH-\overset{\oplus}{C}\underset{NR^3R^4}{\overset{Cl}{<}} \; Cl^{\ominus} \xrightarrow[-HCl]{base} \underset{R^2}{\overset{R^1}{>}}C=C\underset{NR^3R^4}{\overset{Cl}{<}}$$
(**159**)   (**160**)

chloride loss occurs at temperatures between −20 and +20°C. The α-chloroenamines (**160**) are obtained in yields of 60–70%; Table VII summarizes some of the α-chloroenamines generated this way.

TABLE VII

α-Chloroenamines (**160**) from Amide Chlorides and Tertiary Amines (80)

α-Chloroenamines
$$\underset{R^2}{\overset{R^1}{>}}C=C\underset{NR^3R^4}{\overset{Cl}{<}}$$

R^1	R^2	R^3	R^4	Yield, %	b.p./pressure, °C
CH_3	CH_3	\multicolumn{2}{c}{$-(CH_2)_5-$}	70	95–100/15
CH_3	H	\multicolumn{2}{c}{$-(CH_2)_5-$}	57	85–90/14
C_2H_5	H	\multicolumn{2}{c}{$-(CH_2)_5-$}	57	98/8
C_6H_5	H	C_2H_5	C_2H_5	60–70	—
C_6H_5	H	\multicolumn{2}{c}{$-(CH_2)_5-$}	60–70	—
C_6H_5	H	\multicolumn{2}{c}{$-(CH_2)_2-O-(CH_2)_2-$}	60–70	—

III. REACTIONS OF HALOMETHYLENIMINIUM SALTS

Further reactions are discussed in Part 1 in the chapter "α-Haloenamines and Keteniminium Salts," by L. Ghosez and J. Marchand-Brynaert. The ethynyl amidinium salt **162** is obtained from the iminium salt **161** via HCl loss (189).

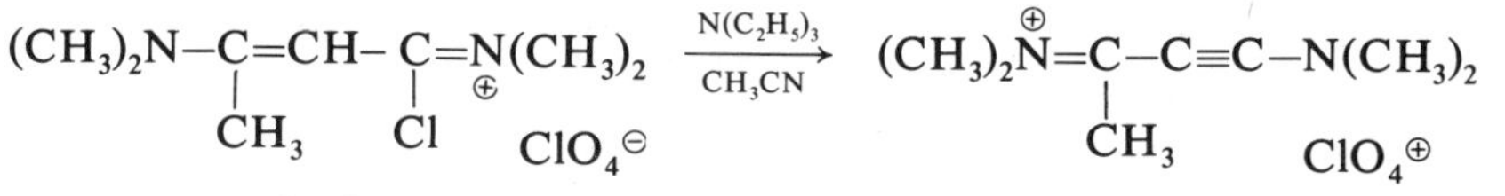

$$(CH_3)_2N-\underset{\underset{CH_3}{|}}{C}=CH-\underset{\underset{Cl}{|}}{\overset{\oplus}{C}}=N(CH_3)_2 \quad ClO_4^{\ominus} \xrightarrow[CH_3CN]{N(C_2H_5)_3} (CH_3)_2\overset{\oplus}{N}=\underset{\underset{CH_3}{|}}{C}-C\equiv C-N(CH_3)_2 \quad ClO_4^{\oplus}$$

(161) (162)

N-Alkenylamides **163** and **164** react with phosgene to yield amide chlorides, for example, **165**, which give ketenimines, for example, **166**, on treatment with triethylamine. Furthermore, they may be converted into nitriles **167** and **168** via Claisen rearrangement (190).

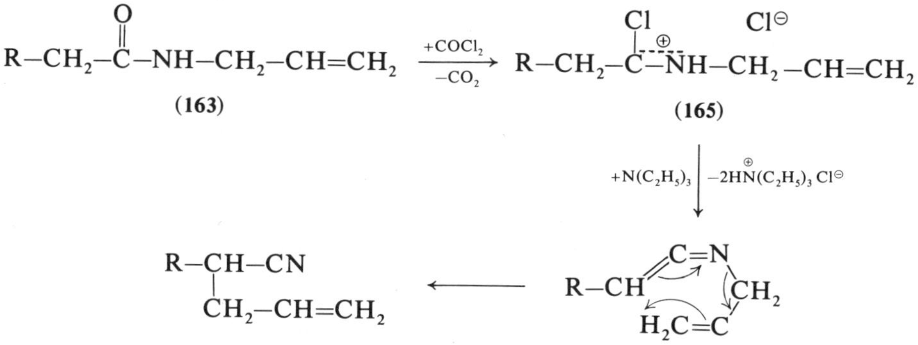

$$R-CH_2-\overset{O}{\overset{\|}{C}}-NH-CH_2-CH=CH_2 \xrightarrow[-CO_2]{+COCl_2} R-CH_2-\underset{\underset{}{|}}{\overset{Cl}{\overset{\oplus}{C}}}-\overset{}{N}H-CH_2-CH=CH_2 \quad Cl^{\ominus}$$

(163) (165)

$$\Bigg\downarrow {+N(C_2H_5)_3}\ {-2H\overset{\oplus}{N}(C_2H_5)_3\ Cl^{\ominus}}$$

$$\underset{\underset{CH_2-CH=CH_2}{|}}{R-CH-CN} \quad \longleftarrow \quad R-CH\underset{H_2C=C}{\overset{C=N}{\diagup}\diagdown}CH_2$$

(167) (166)

$$CH_3-CH_2-\overset{O}{\overset{\|}{C}}-NH-CH_2-C\equiv CH \xrightarrow[N(C_2H_5)_3]{COCl_2} CH_2=C=CH-\underset{\underset{CH_3}{|}}{CH}-CN$$

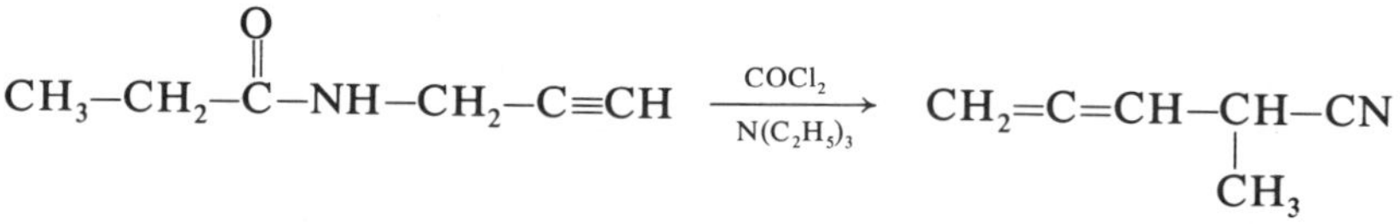

(164) (168)

F. REACTIONS OF AMIDE CHLORIDES WITH ALKOXIDES

Amide chlorides react with alkoxides to yield the highly reactive amide acetals (**169**) (7, 14, 100, 108, 191–193). *N,N*-Dialkylalkoxymethyleniminium ions (**170**) are the intermediates. Although these compounds with halide gegenions are much more labile than the corresponding tetrafluoroborates or alkylsulfates (generated during the amide acetal syntheses of Meerwein et al. (194) and Bredereck et al. (195)) and easily decompose to alkyl chlorides (see

Section III-A-2), they give amide acetals in good yield owing to the high nucleophilicity of the alkoxide anion.

$$R-C{\overset{Cl}{\underset{NR^1R^2}{\oplus}}} \;Cl^{\ominus} \xrightarrow[-NaCl]{+NaOR^3} R-C{\overset{OR^3}{\underset{NR^1R^2}{\oplus}}} \;Cl^{\ominus} \xrightarrow[-NaCl]{+NaOR^3} R-C{\overset{OR^3}{\underset{NR^1R^2}{-OR^3}}}$$
$$\qquad\qquad\qquad\qquad\qquad\qquad (170) \qquad\qquad\qquad\qquad (169)$$

The amide acetal synthesis is performed by adding the amide chloride to an alcoholic solution of the alkoxide (7, 14) or to a suspension of the alkoxide in methylene chloride (15, 196, 197). Instead of an alcoholic alkoxide, an alcoholic solution of tertiary amine may also be used (192).

In order to avoid dealkylation of the N,N-dialkylalkoxymethyleniminium chlorides (170) by the alkoxide (ether formation!) (7, 14), maintenance of reaction temperatures of 0°C or below is essential.

TABLE VIII

Amide Acetals from Amide Chlorides and Alkoxides

Amide acetal $R-C{\overset{OR^3}{\underset{NR^1R^2}{-OR^3}}}$							
R	R^1	R^2	R^3	Yeild, %	b.p./pressure, °C/torr	n_D^{20}	Ref.
H	CH_3	CH_3	CH_3	55	104	—	7, 14, 193, 195
H	CH_3	CH_3	C_2H_5	60	132–137	1.4075	7, 14, 100, 196
H	CH_3	CH_3	n-C_3H_7	27	96/50	—	7, 14, 194
H	C_2H_5	C_2H_5	C_2H_5	35	160–162	—	7, 193
H	—$(CH_2)_4$—		CH_3	—	67–69/26	1.4350/25	15
H	—$(CH_2)_5$—		C_2H_5	73	65–67/13	—	7
H	CH_3	CH_3	C_6H_5	—	80/85/0.5	—	196
H	CH_3	CH_3	—CH_2–C_6H_{11}	35	75–80/11	—	100, 196
H	CH_3	CH_3	CH_2–C_6H_5	30	95–100/11	—	100, 196, 197
CH_3	CH_3	CH_3	CH_3	—	118–20	1.4047/25	15
CH_3	CH_3	CH_3	C_2H_5	—	56–58/13	—	100, 196
C_2H_5	CH_3	CH_3	C_2H_5	40	60/0.5	—	14, 100, 193, 196
—$(CH_2)_3$—		CH_3	CH_3	70	51–52/18	—	7, 14, 192
—$(CH_2)_3$—		CH_3	C_2H_5	50	65–67/18	1.4365	7, 193

III. REACTIONS OF HALOMETHYLENIMINIUM SALTS

Table VIII summarizes the amide acetals (**169**) generated from amide chlorides.

The amide acetals **172** were obtained from the amide chlorides **171** and alkoxides (108).

$$\underset{(\mathbf{171})}{\text{H-C}\underset{\text{NHR}}{\overset{\oplus}{\overset{\text{N-CHCl}_2}{\diagup}}} \text{Cl}^{\ominus}} \quad \xrightarrow{\text{NaOR}^1} \quad \underset{(\mathbf{172})}{\text{H-C}\underset{\text{N-R}}{\overset{\text{N-CH(OR}^1)_2}{\diagup}}}$$

If amide chlorides act on alkoxide solutions which have been neutralized by acetic acid, ortho esters (**174**) are generated (7, 14). The initial formation of amide acetals has been demonstrated.

We would like to mention that amide chlorides react with carboxylate anions via the formation of amide–acid chloride adducts (**173**) (198). These may then react with alcohols to give ortho esters, as has been shown in the case of the formamide–benzoyl chloride adduct (199).

$$\text{H-C}\underset{\text{N(CH}_3)_2}{\overset{\oplus}{\overset{\text{Cl}}{\diagup}}} \text{Cl}^{\ominus} + \text{R-C}\underset{\text{O}^{\ominus}}{\overset{\text{O}}{\diagdown\!\!/}} \xrightarrow{-\text{Cl}^{\ominus}} \text{H-C}\underset{\text{N(CH}_3)_2}{\overset{\oplus}{\overset{\text{O-C(O)-R}}{\diagup}}} \text{Cl}^{\ominus}$$

(**173**)

If amide chlorides are added to alcoholic solutions of tertiary amines, ortho esters (**174**) are also isolated from the reaction mixture after it stands for some time (7, 14).

$$\text{R-C}\underset{\text{N(R}^1)_2}{\overset{\oplus}{\overset{\text{Cl}}{\diagup}}} \text{Cl}^{\ominus} \xrightarrow[-2\text{HNR}_3\text{Cl}^{\ominus}]{+2\text{R}^2\text{OH} + 2\text{NR}_3} \text{R-C}\underset{\text{N(R}^1)_2}{\overset{\text{OR}^2}{\diagup}} \xrightarrow[-\text{H}_2\text{N(R}^1)_2\text{Cl}^{\ominus}]{\text{R}^2\text{OH, HNR}_3\text{Cl}^{\ominus}} \text{R-C}\underset{\text{OR}^2}{\overset{\text{OR}^2}{\diagup}}$$

(**174**)

Up to now, only ethyl orthopropionate and ethyl orthobenzoate have been synthesized from amide chlorides by this route (14). A formamide acetal synthesis, discovered by Stilz, presumably proceeds via amide chlorides (7). Here dichlorocarbene, generated by trichloroacetic acid ester and sodium alkoxide, is reacted with a mixture of a secondary amine and sodium alkoxide:

$$(\text{R}^1)_2\text{NH} + |\text{CCl}_2 \longrightarrow \left[(\text{R}^1)_2\overset{\oplus}{\underset{\text{H}}{\text{N}}}-\text{C}\underset{\text{Cl}}{\overset{\text{Cl}}{\diagup}} \right]^{\ominus} \longrightarrow$$

$$(\text{R}^1)_2\overset{\oplus}{\text{N}}-\text{C}\underset{\text{H}}{\overset{\text{Cl}}{\diagup}} \text{Cl}^{\ominus} \xrightarrow[-2\text{NaCl}]{+2\text{NaOR}^2} \text{H-C}\underset{\text{N(R}^1)_2}{\overset{\text{OR}^2}{\diagup}}$$

Several authors have considered the formation of amide chloride intermediates when chloroform reacts with mixtures of amines and alkoxides (200–202). Appropriate reaction conditions allow the isolation of amide acetals (**175**), aminal esters (**176**), and triaminomethanes (**177**). The formation of **176** and **177** is explained by the synchronous transamination of the amide acetals (**175**) (202).

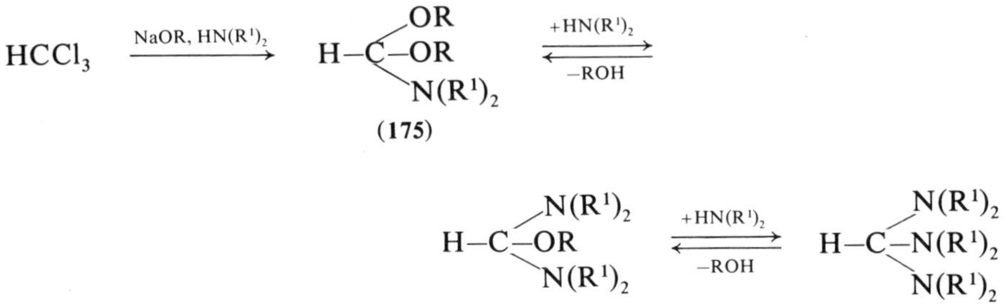

$$HCCl_3 \xrightarrow{NaOR, HN(R^1)_2} H-\underset{N(R^1)_2}{\overset{OR}{\underset{|}{C}}}-OR \xrightleftharpoons[-ROH]{+HN(R^1)_2}$$

(**175**)

$$H-\underset{N(R^1)_2}{\overset{N(R^1)_2}{\underset{|}{C}}}-OR \xrightleftharpoons[-ROH]{+HN(R^1)_2} H-\underset{N(R^1)_2}{\overset{N(R^1)_2}{\underset{|}{C}}}-N(R^1)_2$$

(**176**)   (**177**)

A mixture of triaziridinomethane (**179**), diaziridinomethoxymethane (**180**), and aziridinodimethoxymethane (**181**) is formed from chloroform and aziridine (**178**) in the same way in the presence of sodium methoxide. The reaction of

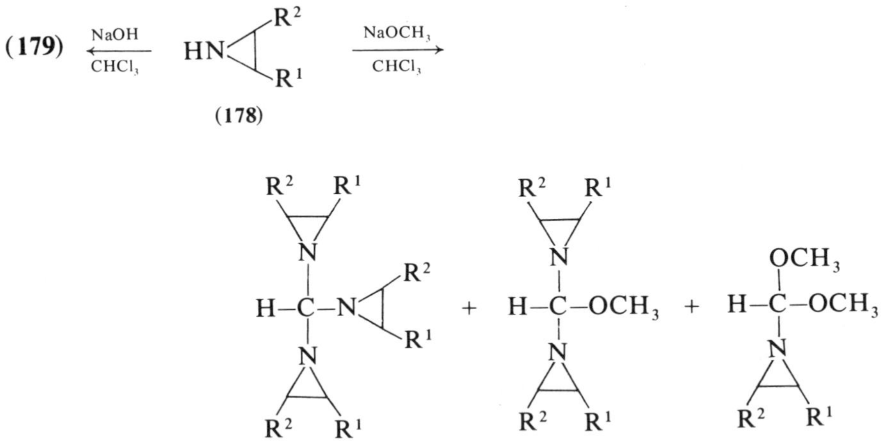

$$(\mathbf{179}) \xleftarrow[CHCl_3]{NaOH} HN\!\!\begin{array}{c}R^2\\R^1\end{array} \xrightarrow[CHCl_3]{NaOCH_3}$$

(**178**)

(**179**)   +   (**180**)   +   (**181**)

aziridine, or *C*-substituted aziridines, with chloroform in the presence of NaOH yields only triaziridinomethanes (**179**) (203). Dichloromethyl methyl ether (**182**) also reacts with aziridines and NaOH to give the corresponding diaziridinomethoxymethanes (**180**) (203).

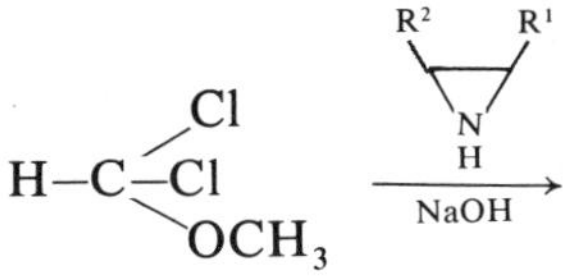

III. REACTIONS OF HALOMETHYLENIMINIUM SALTS

$$\underset{(182)}{H-C\underset{OCH_3}{\overset{Cl}{\overset{|}{-}}}Cl} \xrightarrow{NaOH} \underset{(180)}{\overset{R^2\diagup\underset{H}{N}\diagdown R^1}{\triangle}}$$

Tris(dialkylamino)methanes (**183**) were synthesized, among other compounds such as **184** and **185**, from chloroform and secondary amines using phase-transfer catalysis (204).

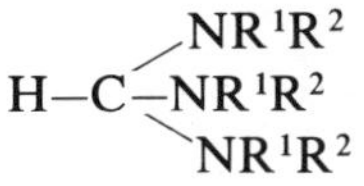

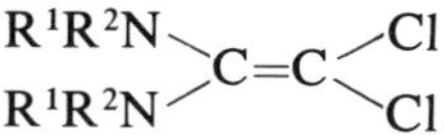

$$\underset{(183)}{H-C\underset{NR^1R^2}{\overset{NR^1R^2}{\overset{|}{-}}}NR^1R^2} \qquad \underset{(184)}{\overset{R^1R^2N}{\underset{Cl}{>}}C=C\underset{Cl}{\overset{Cl}{<}}} \qquad \underset{(185)}{\overset{R^1R^2N}{\underset{R^1R^2N}{>}}C=C\underset{Cl}{\overset{Cl}{<}}}$$

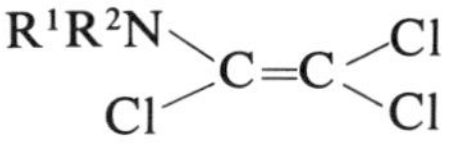

At low temperatures, amide acetals **188** and **189** are accessible from (difluoromethyl)dialkylamines **186** and **187** and sodium alkoxide suspended in ether (15, 16).

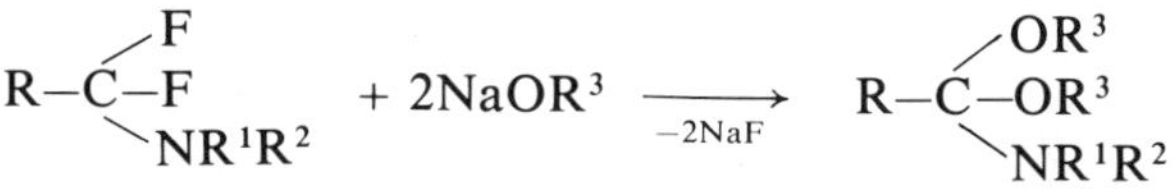

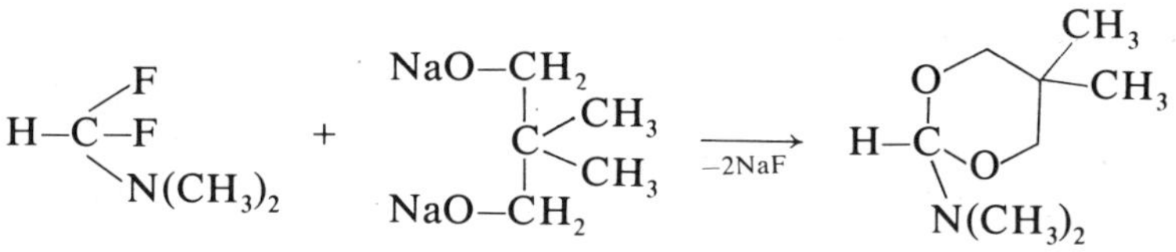

$$\underset{(186)}{R-C\underset{NR^1R^2}{\overset{F}{\overset{|}{-}}}F} + 2NaOR^3 \xrightarrow{-2NaF} \underset{(188)}{R-C\underset{NR^1R^2}{\overset{OR^3}{\overset{|}{-}}}OR^3}$$

$$\underset{(187)}{H-C\underset{N(CH_3)_2}{\overset{F}{\overset{|}{-}}}F} + \underset{NaO-CH_2}{\overset{NaO-CH_2}{\underset{|}{C}}\underset{CH_3}{\overset{CH_3}{<}}} \xrightarrow{-2NaF} \underset{(189)}{H-C\underset{N(CH_3)_2}{\overset{O\diagdown\diagup CH_3}{\underset{O\diagup}{\overset{|}{<}}CH_3}}}$$

G. REACTIONS OF AMIDE CHLORIDES WITH ALKALI METAL AMIDES AND IMIDES

Böhme and Soldan (161) were the first to report the reactions of dimethylformamide chloride (**7a**) and metal imides and amides. Thus (dimethylamino)diphthalimidomethane (**190**) and (dimethylamino)-disaccharinomethane (**191**) were synthesized from dimethylformamide chloride (**7a**) and potassium phthalimide and sodium saccharin. The authors failed to carry out the analogous reaction of dimethylformamide chloride and lithium dimethylamide where tris(dimethylamino)methane was the desired product. Tris(dimethylamino)methane (**192**) is obtained in 40% yield, if dimethylformamide chloride is first reacted with dimethylamine and then with lithium dimethylamide (205). During this reaction, tetramethylformamidinium chloride (**193**) is initially formed from the amide chloride and dimethylamine. This compound reacts with lithium dimethylamide to yield tris(dimethylamino)-methane (**192**), in a well-known reaction (206, 207).

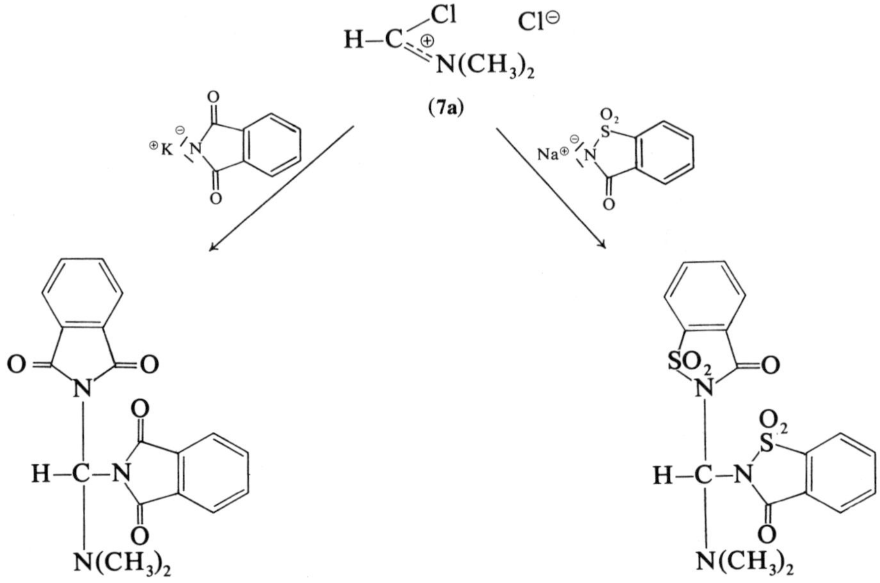

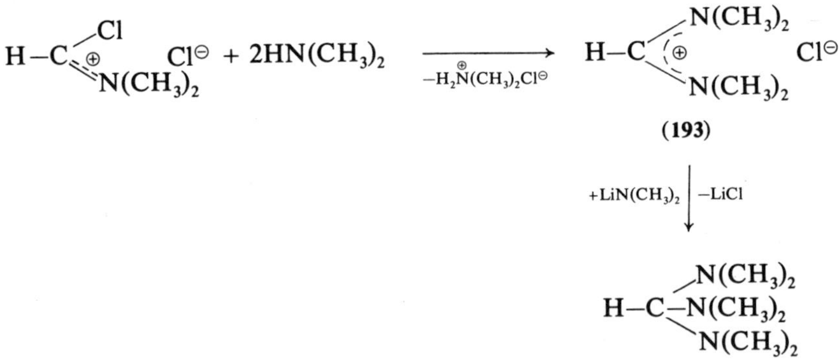

Buijle and Viehe (81) demonstrated that 2 moles of HCl are cleaved from amide chlorides (**194**) by lithium dialkylamides (**195**), to form ynamines (**196**). This reaction provides an easy approach to the highly reactive ynamines. The yields are strongly amide dependent. Bulky amides, like lithium dicyclohexyl-amide, give the highest yields (see Table IX).

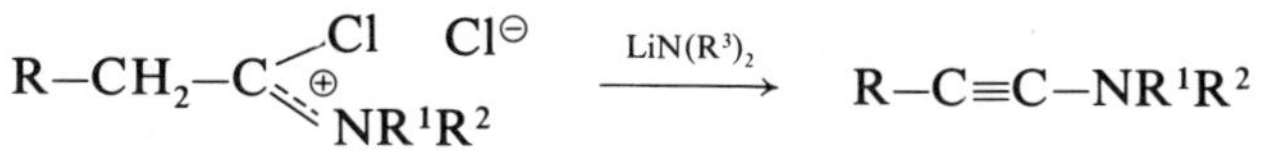

III. REACTIONS OF HALOMETHYLENIMINIUM SALTS

$$R-CH_2-C\underset{NR^1R^2}{\overset{Cl}{\diagup}}{}^{\oplus}\ Cl^{\ominus} \xrightarrow{LiN(R^3)_2} R-C\equiv C-NR^1R^2$$

(194) (195) (196)

Table IX summarizes some of the ynamines synthesized this way.

TABLE IX

Ynamines (196) from Amide Chlorides (194) and Lithium Dialkyl Amides (81)

	Ynamine $R-C\equiv C-NR^1R^2$		Base $LiNR^3$		
R	R^1	R^2	R^3	Yield, %	b.p./pressure, °C/torr
CH_3	C_2H_5	C_2H_5	C_2H_5	39	60–62/90
CH_3	C_2H_5	C_2H_5	C_6H_{11}	58	60–62/90
CH_3	$-(CH_2)_5-$		CH_3	12	72–73/15
CH_3	$-(CH_2)_5-$		C_2H_5	39	72–73/15
CH_3	$-(CH_2)_5-$		C_3H_7	0	—
CH_3	$-(CH_2)_5-$		C_6H_{11}	46	72–73/15
C_2H_5	C_2H_5	C_2H_5	C_2H_5	38	50–51/30
			C_6H_{11}	77	
$n\text{-}C_4H_9$	C_2H_5	C_2H_5	C_2H_5	29	77–78/15

H. REACTIONS OF AMIDE CHLORIDES WITH LITHIUM PHOSPHIDES

If dimethylformamide chloride (7a) is added to an ether solution, or suspension, of lithium diorganophosphide (197), bis(diorganophosphino)-(dimethylamino)methanes (198) are formed in yields of 41–49% (208). However, if the lithium phosphide is added to the amide chloride, only unidentified products are obtained.

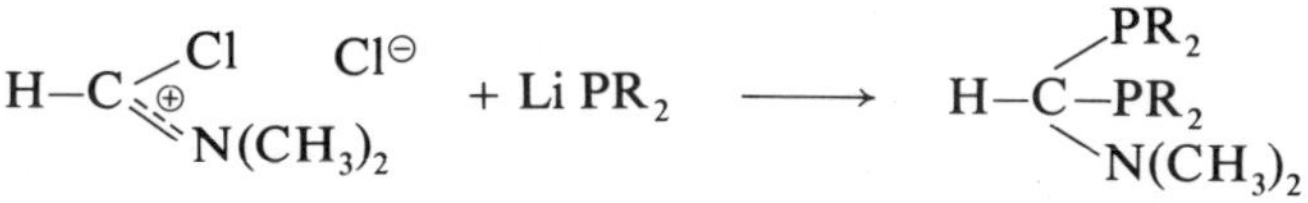

$$H-C\underset{N(CH_3)_2}{\overset{Cl}{\diagup}}{}^{\oplus}\ Cl^{\ominus} + Li\ PR_2 \longrightarrow H-\underset{N(CH_3)_2}{\overset{PR_2}{\diagup}}C-PR_2$$

(7a) (197) (198)

$R = CH_3, C_2H_5, n\text{-}C_4H_9, C_6H_5$

I. ANION EXCHANGE WITH HALOMETHYLENIMINIUM HALIDES

The anions of chloromethyleniminium chlorides react with Lewis acids like $AlCl_3$ (6, 100, 209), $SbCl_5$ (13, 17, 18), and BF_3 (13) to yield iminium salts with complex anions like 199–201. The anions of bromomethyleniminium

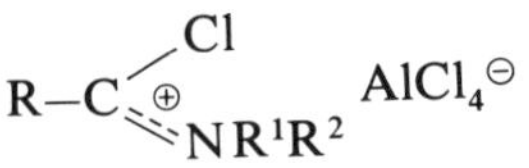

(199)

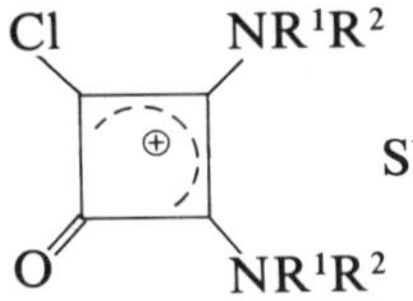

(200)

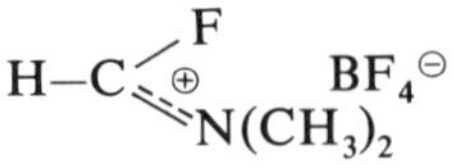

(201)

bromides (**202, 203**) were exchanged with silver hexafluoroantimonate and triethyloxonium fluoroborate to give the salts **204** and **205** (23), respectively.

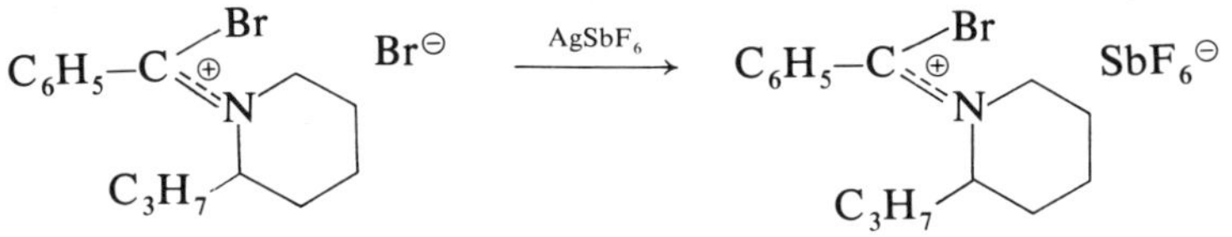

(202) (204)

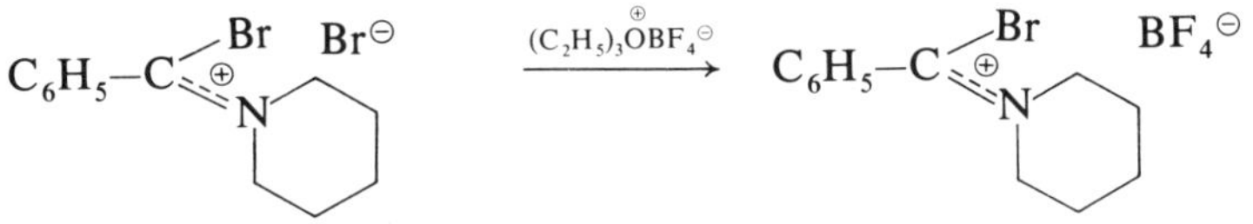

(203) (205)

The less polarizable anions yield less hygroscopic salts, which can be more easily handled and characterized than the amide chlorides themselves.

Table X summarizes the properties of several such iminium salts.

TABLE X

Iminium Salts from Amide Chlorides and Lewis Acids

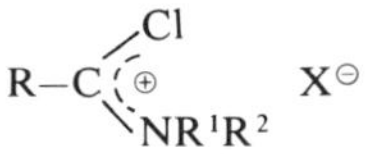

Iminium salt $R{-}C(Cl){=}\overset{\oplus}{N}R^1R^2\ X^{\ominus}$					
R	R^1	R^2	X	m.p., °C	Ref.
H	CH_3	CH_3	$AlCl_4$	225–227	6
H	CH_3	CH_3	$SbCl_6$	162–164	13
H	C_2H_5	CH_3	$AlCl_4$	95–98	6
C_2H_5	H	C_2H_5	$AlCl_4$	142–145	6
				251	210
C_2H_5	CH_3	CH_3	$AlCl_4$	oil	6

J. REACTIONS OF HALIDES AND DIAZONIUM SALTS WITH AMIDE CHLORIDES

1. Reaction Mechanism

Owing to the neighboring positive charge, the α-hydrogens of amide chlorides are strongly activated towards HCl loss in an equilibrium reaction. Thus α-chloroenamines (**206**) are formed (34, 35, 79) which are capable of reacting with electrophiles. This explains why amide chlorides on the one hand

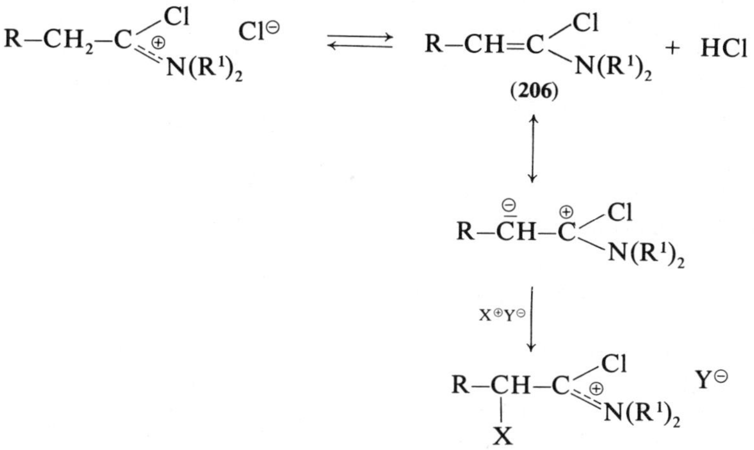

$$R-CH_2-C\underset{N(R^1)_2}{\overset{Cl}{\oplus}} \; Cl^{\ominus} \rightleftarrows R-CH=C\underset{N(R^1)_2}{\overset{Cl}{\diagup}} + HCl$$
$$(206)$$

$$\updownarrow$$

$$R-\overset{\ominus}{CH}-\overset{\oplus}{C}\underset{N(R^1)_2}{\overset{Cl}{\diagup}}$$

$$\downarrow X^{\oplus}Y^{\ominus}$$

$$R-\underset{X}{\overset{\;}{CH}}-C\underset{N(R^1)_2}{\overset{Cl}{\oplus}} \; Y^{\ominus}$$

undergo self-condensation (7, 97) and on the other are easily halogenated in their α-positions (7, 97), as are imide chlorides (211). The coupling reaction with diazonium salts falls into the same pattern (7, 97).

2. Halogenation of Amide Chlorides

Chlorine reacts even at 0°C with amide chlorides (**207**) in chloroform. Here α,α-dichloroamide chlorides (**208**) are formed which may be hydrolyzed to the

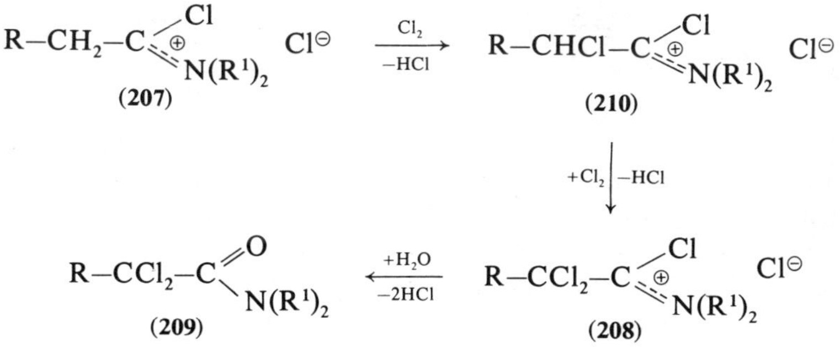

$$R-CH_2-C\underset{N(R^1)_2}{\overset{Cl}{\oplus}} \; Cl^{\ominus} \xrightarrow[-HCl]{Cl_2} R-CHCl-C\underset{N(R^1)_2}{\overset{Cl}{\oplus}} \; Cl^{\ominus}$$
$$(207) \hspace{4cm} (210)$$

$$+Cl_2 \Big| -HCl$$

$$R-CCl_2-C\underset{N(R^1)_2}{\overset{O}{\diagdown}} \xleftarrow[-2HCl]{+H_2O} R-CCl_2-C\underset{N(R^1)_2}{\overset{Cl}{\oplus}} \; Cl^{\ominus}$$
$$(209) \hspace{4cm} (208)$$

corresponding α,α-dihaloamides (209) (7, 97). Until recently, this was the simplest approach to α-haloamides. α-Monohaloamide chlorides (210) are also obtained by this method (80, 81).

N-Alkylated lactams (211) are also chlorinated in their α positions, relative to the carbonyl group, after prior conversion to amide chlorides (7).

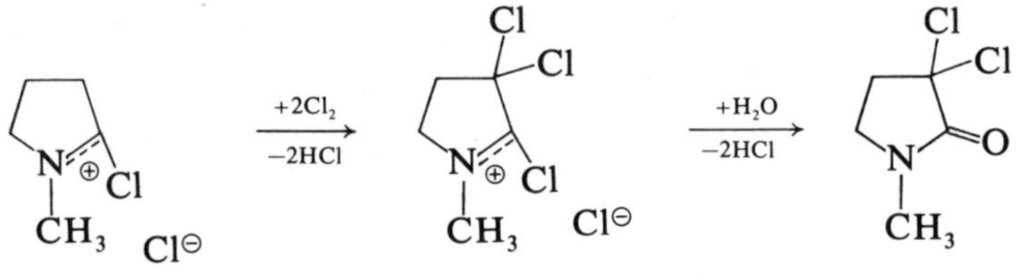

In contrast with chlorination, bromination gives rise to side reactions, yielding mostly monobromo derivatives contaminated with chlorine (97).

3. Coupling of Amide Chlorides with Diazonium Salts

In dry organic solvents, amide chlorides containing α-CH$_2$ groups couple with diazonium salts; arylhydrazones of α-ketoamide chlorides (212) are isolated (97). Subsequent hydrolysis yields arylhydrazones of α-ketoamides (213). In weakly alkaline solvents like DMF or *N*-methylpyrrolidone, the coupling reaction proceeds smoothly with strongly electrophilic diazonium salts like *p*-nitrophenyldiazonium fluoroborate.

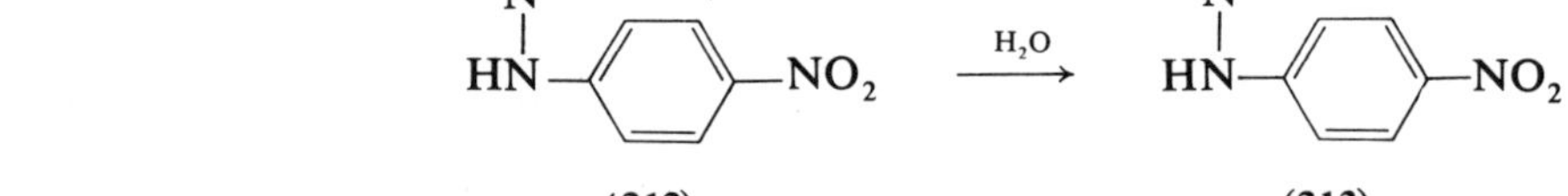

More weakly electrophilic diazonium salts, like phenyldiazonium fluoroborate, demand stronger bases as catalysts for the coupling reaction (97). Amide chlorides obtained from lactams also couple very easily with weakly electrophilic diazonium salts (97). Arylhydrazones (213) from α-ketoamides obtained by this route are summarized in Table XI.

TABLE XI
Arylhydrazones (213) of α-Ketoamides from Amide Chlorides and Diazonium Salts (97)

Arylhydrazone

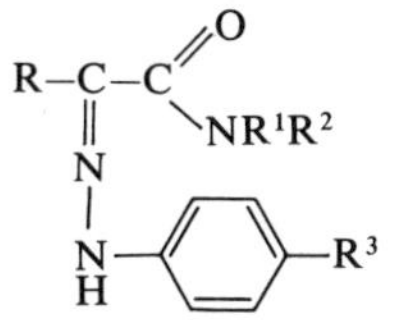

$$R-\underset{\underset{\underset{H}{N}}{\overset{\|}{N}}}{C}-C\overset{O}{\underset{NR^1R^2}{\diagdown}}$$

with N–H–C₆H₄–R³

R	R¹	R²	R³	Yield, %	m.p., °C
CH₃		–(CH₂)₅–	NO₂	94	197–100 (dec)
C₂H₅	CH₃	CH₃	NO₂	93	159–161 (dec)
–(CH₂)₂–	—	CH₃	H	49	237–239 (dec)
–(CH₂)₂–	—	CH₃	NO₂	88	262–264 (dec)
–(CH₂)₃–	—	CH₃	NO₂	87	237–239 (dec)

K. REACTIONS PROCEEDING VIA OXIDATION OF THE AMIDE CHLORIDE FUNCTION

If formanilide is treated with sufuryl chloride or thionyl chloride and chlorine, amide chlorides are initially formed, isocyanide dichlorides (214) being the final product.

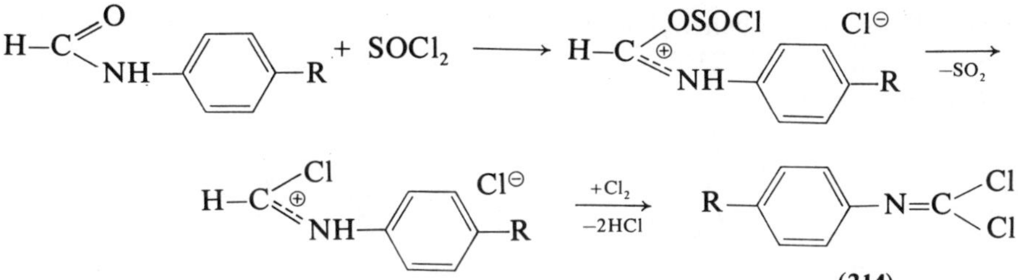

$$H-C\overset{O}{\underset{NH-C_6H_4-R}{\diagdown}} + SOCl_2 \longrightarrow H-\overset{\oplus}{C}\overset{OSOCl}{\underset{NH-C_6H_4-R}{\diagdown}} Cl^{\ominus} \xrightarrow{-SO_2}$$

$$H-\overset{\oplus}{C}\overset{Cl}{\underset{NH-C_6H_4-R}{\diagdown}} Cl^{\ominus} \xrightarrow[-2HCl]{+Cl_2} R-C_6H_4-N=C\overset{Cl}{\underset{Cl}{\diagdown}}$$

(214)

Formanilides containing inductive (-I) or mesomeric (-M) substituents or their aromatic groups react especially easily. If phosgene or oxalyl chloride are used instead of thionyl chloride, the isocyanide dichlorides (214) are isolated in only somewhat lower yields.

Oxygen and thionyl chlorides, reacting with formanilide, also yield isocyanide dichlorides (214) with small amounts of isocyanates as well (175). Presumably, thionyl chloride is partially oxidized by oxygen to give sulfuryl chloride, which transforms the amide chloride into the isocyanide dichloride.

Sulfenyl chlorides react with N-arylformamide chlorides to form iminocarbonic acid thioester chlorides (**215**) (175).

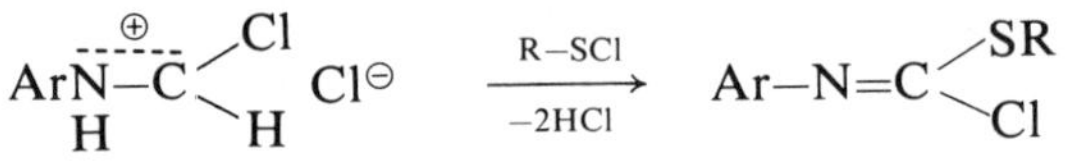

$$\text{ArN}^{\oplus}\text{-CHCl}\cdot\text{H} \cdot \text{Cl}^{\ominus} \xrightarrow[-2\text{HCl}]{\text{R-SCl}} \text{Ar-N=C} \begin{smallmatrix} \text{SR} \\ \text{Cl} \end{smallmatrix}$$

(**215**)

If formamide is treated with thionyl chloride and sulfenyl chlorides, rhodanides (**216**) are the resulting compounds (175).

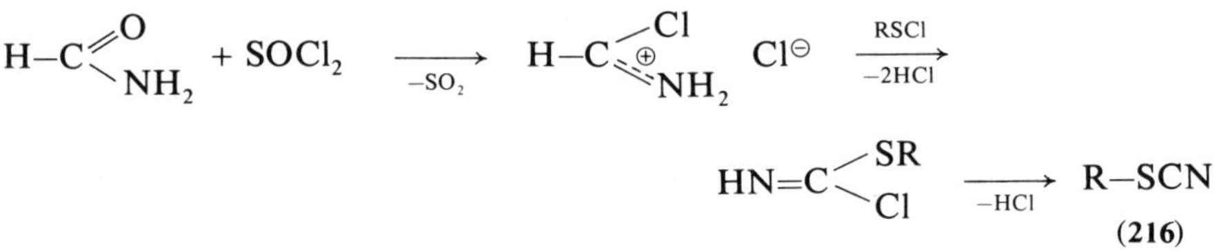

$$\text{H-C}\begin{smallmatrix}\text{O}\\\text{NH}_2\end{smallmatrix} + \text{SOCl}_2 \xrightarrow{-\text{SO}_2} \text{H-C}^{\oplus}\begin{smallmatrix}\text{Cl}\\\text{NH}_2\end{smallmatrix} \cdot \text{Cl}^{\ominus} \xrightarrow[-2\text{HCl}]{\text{RSCl}}$$

$$\text{HN=C}\begin{smallmatrix}\text{SR}\\\text{Cl}\end{smallmatrix} \xrightarrow{-\text{HCl}} \text{R-SCN}$$

(**216**)

L. THERMAL DEGRADATION OF HALOMETHYLENIMINIUM HALIDES

von Braun et al. (4, 210–212) and other authors (213, 214) extensively studied the thermal decomposition of iminium halides (**217**). The decomposition to nitriles (**219**) occurs stepwise via imide halides (**218**). Benzoylated,

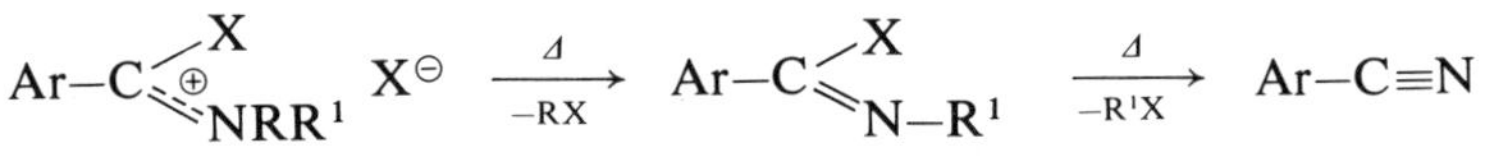

$$\text{Ar-C}^{\oplus}\begin{smallmatrix}\text{X}\\\text{NRR}^1\end{smallmatrix} \text{X}^{\ominus} \xrightarrow[-\text{RX}]{\Delta} \text{Ar-C}\begin{smallmatrix}\text{X}\\\text{N-R}^1\end{smallmatrix} \xrightarrow[-\text{R}^1\text{X}]{\Delta} \text{Ar-C}{\equiv}\text{N}$$

(**217**)   (**218**)   (**219**)

cyclic bases, like N-benzoylpiperidine (**220**), also decompose according to this scheme (23, 24, 215, 216). Here the pure iminium bromides react much better than the chlorides.

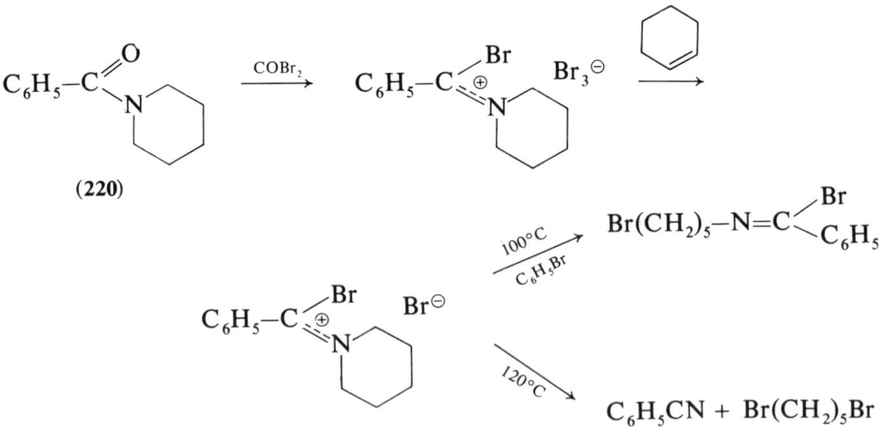

$$C_6H_5\text{-CO-N(piperidine)} \xrightarrow{\text{COBr}_2} C_6H_5\text{-C}^{\oplus}(\text{Br})\text{=N(piperidine)} \cdot \text{Br}_3^{\ominus} \longrightarrow$$

(**220**)

$$\xrightarrow[C_6H_5\text{Br}]{100°C} \text{Br(CH}_2)_5\text{-N=C}\begin{smallmatrix}\text{Br}\\C_6H_5\end{smallmatrix}$$

$$C_6H_5\text{-C}^{\oplus}(\text{Br})\text{=N(piperidine)} \cdot \text{Br}^{\ominus}$$

$$\xrightarrow{120°C} C_6H_5\text{CN} + \text{Br(CH}_2)_5\text{Br}$$

If the alicyclic base is bound to an aromatic ring, as in tetrahydroquinoline (**221**), the reaction stops at the imide chloride (**222**) stage.

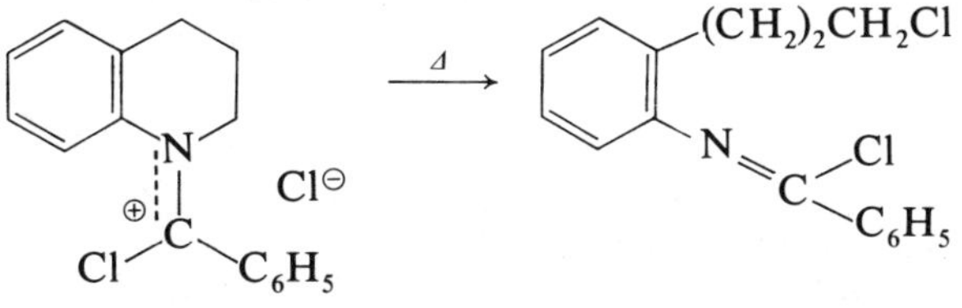

(**221**)   (**222**)

More recently the range of application of the thermal iminium chloride degradation was studied with regard to the synthesis of alkyl halides (217, 218). The thermal degradation of 2,2-dichloroaziridinones (**223**) also yields imide chlorides (**224**) (94).

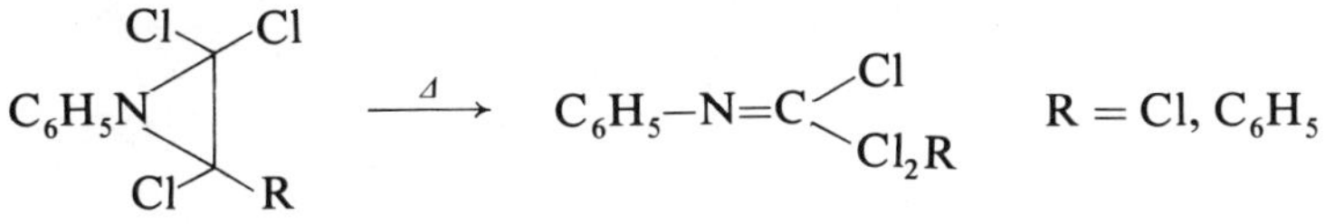

C_6H_5N ... $C_6H_5-N=C$ ... $R = Cl, C_6H_5$

(**223**)   (**224**)

Amide chlorides (**194**) containing α-CH bonds lose HCl upon thermolysis to form α-chloroenamines (**206**) (4).

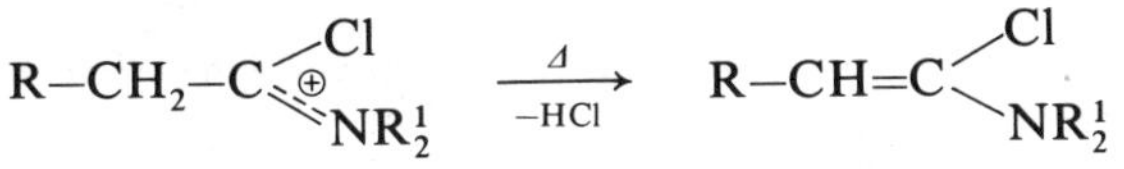

$R-CH_2-C$ ... $\xrightarrow{\Delta, -HCl}$ ... $R-CH=C$

(**194**)   (**206**)

M. REDUCTION OF CHLOROMETHYLENIMINIUM SALTS

2,2-Dichloroaziridines (**225**) are reduced by lithium aluminum hydride to yield the amines (**226**) (93).

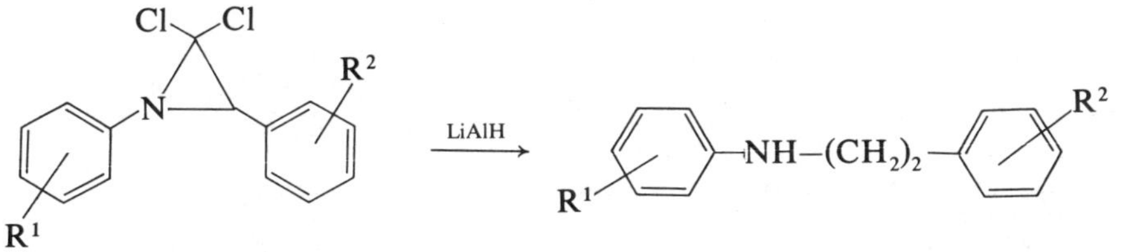

(**225**)   (**226**)

ACKNOWLEDGMENT

To Dr. Heinz Eilingsfeld BASF AG, Ludwigshafen, and to Dr. Bernd Funke as well as Dr. Erwin Haug, both of Fachhochschule Aalen, Aalen, I am very indebted for fruitful discussions and critical revisions of the text.

REFERENCES

1. O. Wallach and O. Kamensky, *Ann. Chem.*, **214**, 233 (1882).
2. O. Wallach, *Ann. Chem.*, **184**, 1 (1877).
2a. R. Gnehm, *Ber. Deut. Chem. Ges.*, **9**, 844 (1876).
2b. H. Klinger, *Ann. Chem.*, **184**, 261 (1877).
3. O. Wallach, *Ann. Chem.*, **214**, 193, 226 (1882).
4. J. v. Braun and A. Heymons, *Ber. Deut. Chem. Ges.*, **62**, 409 (1929).
5. J. v. Braun, *Ber. Deut. Chem. Ges.*, **37**, 2678 (1904).
6. H. H. Bosshard and H. Zollinger, *Helv. Chim. Acta*, **42**, 1653 (1959).
7. H. Eilingsfeld, M. Seefelder, and H. Weidinger, *Angew. Chem.*, **72**, 836 (1960).
8. M. Grdinic and V. Hahn, *J. Org. Chem.* **30**, 2381 (1965).
9. F. Hallmann, *Ber. Deut. Chem. Ges.*, **9**, 846 (1876).
10. N. Roh and G. Kochendörfer (to I. G. Farbenindustrie), DRP 677,207 (1937).
11. Z. Arnold and F. Sorm, *Collection Czech. Chem. Commun.*, **23**, 452 (1958); *Chem. Listy*, **51**, 1082 (1957).
12. Z. Arnold, *Collection Czech. Chem. Commun.*, **24**, 4048 (1959).
13. Z. Arnold and A. Holy, *Collection Czech. Chem. Commun.*, **27**, 2886 (1962).
14. H. Eilingsfeld, M. Seefeler, and H. Weidinger, *Chem. Ber.*, **96**, 2671 (1963).
16. M. H. Brown (to E. I. du Pont de Nemours and Co.), U.S. Pat. 3,214,428 (1965); *C.A.*, **64**, 3501h, 3542 (1966).
15. M. H. Brown (to E. I. du Pont de Nemours and Co.), U.S. Pat. 3,092,637 (1963); *C.A.*, **59**, 12764g.
17. G. Seitz, H. Morck, K. Mann, and R. Schmiedel, *Chem.-Ztg.*, **98**, 459 (1974).
18. G. Seitz and H. Morck, *Chimia*, **26**, 386 (1972).
19. H. H. Bosshard, R. Mory, M. Schmid, and H. Zollinger, *Helv. Chim. Acta*, **42**, 1653 (1959).
20. J. v. Braun and W. Pinkernelle, *Ber. Deut. Chem. Ges.*, **67**, 1218 (1934).
21. A. Roedig and W. Wenzel, *Angew. Chem.*, **81**, 36 (1969).
22. H. Gross, J. Rusche, and H. Bornowski, *Ann. Chem.*, **675**, 142 (1964).
23. B. A. Philipps, G. Fodor, J. Gal, F. Letourneau, and J. J. Ryan, *Tetrahedron*, **29**, 3309 (1973).
24. H. Gross and J. Gloede, *Chem. Ber.*, **96**, 1387 (1963).
25. J. v. Braun and C. Müller, *Ber. Deut. Chem. Ges.*, **39**, 2018 (1906).
26. F. S. Fawcett, C. W. Tullock, and D. D. Coffman, *J. Am. Chem. Soc.*, **84**, 4275 (1962).
27. W. R. Hasek, W. C. Smith, and V. A. Engelhardt, *J. Am. Chem. Soc.*, **82**, 543 (1960).
28. W. C. Smith, *Angew. Chem.*, **74**, 742 (1962).
29. G. Ferre and A. L. Palomo, *Tetrahedron Letters*, **1969**, 2161.
30. K. Kikugawa, M. Ichino, and T. Kawashima, *Chem. Pharm. Bull.*, **19**, 2466 (1971).
31. K. Kikukawa and T. Kawashima, *Chem. Pharm. Bull.*, **19**, 2629 (1971).
32. T. D. Smith, *J. Chem. Soc. A*, **1966**, 841.
33. G. J. Martin, S. Poignant, M. L. Fillieux, and M. T. Quemeneur, *Tetrahedron Letters*, **1970**, 5061.
34. A. J. Speziale and L. R. Smith, *J. Org. Chem.*, **27**, 4361 (1962).
35. R. G. Glushkov, V. G. Smirnova, K. A. Zaitseva, N. A. Novitskaya, M. D. Mashkovskii, and G. N. Pershin, *Khim. Farm. Zh.*, **8**, 14 (1974); *Cheminform*, **1974**, 25–341, p. 133.
36. R. Buyle and H. G. Viehe, *Tetrahedron*, **24**, 4217 (1968).
37. W. Jentsch, *Chem. Ber.*, **67**, 1361 (1964).
38. G. Wolter and W. Kosler, *Z. Chem.*, **10**, 401 (1970).
39. A. J. Speziale, L. R. Smith, J. E. Fedder, *J. Org. Chem.*, **30**, 4303 (1965).
40. A. J. Speziale and L. R. Smith, *J. Org. Chem.*, **27**, 3742 (1962).

REFERENCES

41. A. J. Speziale and L. R. Smith, *J. Org. Chem.*, **28**, 1805 (1963).
42. A. J. Speziale, L. R. Smith, and J. E. Fedder, *J. Am. Chem. Soc.*, **87**, 4306 (1965).
43. A. Gautier, *Ann. Chem.*, **142**, 289 (1867).
44. C. Engler, *Ann. Chem.*, **149**, 297 (1869).
45. L. Henry, *Bull. Soc. Chim. France*, 7 [2], 85 (1867).
46. H. Biltz, *Ber. Deut. Chem. Ges.*, **25**, 2533 (1892).
47. E. N. Zilberman, *Russ. Chem. Rev. Engl. Transl*, **31**, 615 (1962).
48. F. C. Schaefer, *The Chemistry of the Cyano Group*, Z. Rapport, Ed., John Wiley & Sons, New York, 1970.
49. A. Hantzsch, *Ber. Deut. Chem. Ges.*, **64**, 667 (1931).
50. F. Klages and W. Grill, *Ann. Chem.*, **594**, 21 (1955).
51. F. Klages, R. Ruhnau, and W. Hauser, *Ann. Chem.*, **626**, 3492 (1963).
52. W. J. Middleton and V. A. Engelhardt, *J. Am. Chem. Soc.*, **80**, 2788 (1958).
53. W. J. Middleton, E. L. Little, D. D. Coffman, and V. A. Engelhardt, *J. Am. Chem. Soc.*, **80**, 2795 (1958).
54. G. J. Janz and S. S. Danyluk, *J. Am. Chem. Soc.*, **81**, 3850 (1959).
55. L. E. Hinkel and G. N. Treharne, *J. Chem. Soc.*, **1945**, 866.
56. E. Allenstein, *Z. Anorg. Allgem. Chem.*, **322**, 265 (1963).
57. E. Allenstein, *Z. Anorg. Allgem. Chem.*, **322**, 276 (1963).
58. E. Allenstein and P. Quis, *Chem. Ber.*, **96**, 1035 (1963).
59. E. Allenstein and P. Quis, *Chem. Ber.*, **96**, 2918 (1963).
60. E. Allenstein, *Chem. Ber.*, **96**, 3230 (1963).
61. E. Allenstein and A. Schmidt, *Spectrochim. Acta*, **20**, 1451 (1964).
62. E. Allenstein and A. Schmidt, *Chem. Ber.*, **97**, 1286 (1964).
63. E. Allenstein and P. Quis, *Chem. Ber.*, **97**, 1857 (1964).
64. E. Allenstein and P. Quis, *Chem. Ber*, **97**, 3162 (1964).
65. E. Allenstein and A. Schmidt, *Chem. Ber.*, **97**, 1863 (1964).
66. K. Wiechert, H. H. Heilmann, P. Mohr, *Z. Chem.*, **3**, 308 (1963).
67. A. Gautier and H. Gal, *Ann. Chem.*, **138**, 36 (1866).
68. L. Claisen and F. Matthews, *J. Chem. Soc.*, **41**, 264 (1882).
69. L. Claisen and F. Matthews, *Ber. Deut. Chem. Ges.*, **16**, 308 (1883).
70. J. U. Nef, *Ann. Chem.*, **270**, 303 (1892).
71. L. E. Hinkel and R. P. Hullin, *J. Chem. Soc.*, **1949**, 1033.
72. L. E. Hinkel and G. H. Summers, *J. Chem. Soc.*, **1952**, 2813.
73. C. Grundmann and A. Kreutzberger, *J. Am. Chem. Soc.*, **76**, 5646 (1954).
74. L. E. Hinkel, E. E. Ayling, and J. H. Beynon, *J. Chem. Soc.*, **1935**, 674.
75. E. Allenstein, A. Schmidt, and V. Beyl, *Chem. Ber.*, **99**, 431 (1966).
76. L. Gattermann and K. Schnitzspahn, *Ber. Deut. Chem. Ges.*, **31**, 1770 (1898).
77. F. B. Dains, *Ber. Deut. Chem. Ges.*, **35**, 2496 (1902).
78. A. J. Speziale and L. R. Smith, *J. Org. Chem.*, **28**, 3492 (1963).
79. A. J. Speziale and R. C. Freeman, *J. Am. Chem. Soc.*, **82**, 903, 909 (1960).
80. L. Ghosez, B. Haveaux, and H. G. Viehe, *Angew. Chem.*, **81**, 468 (1969).
81. R. Buijle, A. Halleux, and H. G. Viehe, *Angew. Chem.*, **78**, 593 (1966).
82. N. N. Jarovenko and M. A. Raksa, *Z. obsc. chim.*, **29**, 2159 (1969).
83. A. G. Look and E. K. Fields, *J. Org. Chem.*, **27**, 3686 (1962).
84. J. A. Deyrup and R. B. Greenwald, *Tetrahedron Letters*, **1965**, 321.
85. J. A. Deyrup and R. B. Greenwald, *J. Am. Chem. Soc.*, **87**, 4538 (1965).
86. E. K. Fields and J. M. Sandri, *Chem. Ind. (London)*, **1959**, 1216.
87. K. Ichimura and M. Ohta, *Tetrahedron Letters*, **1966**, 807.
88. P. K. Kadaba and J. O. Edwards, *J. Org. Chem.*, **25**, 1431 (1960).
89. D. Klamann, H. Wache, K. Ulm, and F. Nerdel, *Chem. Ber.*, **100**, 1870 (1967).

90. A. Lukasiewicz, *Tetrahedron*, **20**, 1113 (1964).
91. A. Lukasiewicz and J. Lesinka, *Tetrahedron*, **21**, 3247 (1965).
92. B. S. Drach, I. Yu. Dolgushina, and A. D. Sinitsa, *Zh. Org. Khim.*, **9**, 2368 (1973).
93. R. R. Kostikov, A. F. Khlebnikov, and A. Ogloblin, *Zh. Org. Khim.*, **9**, 2346 (1973); *Cheminform*, **1974**, 8–143, p. 37.
94. D. Seyferth, W. Tronich, and Huong-min Shih, *J. Org. Chem.*, **39**, 158 (1974).
95. N. S. Kozlov, V. D. Pak, and V. V. Mashevskii, *Khim. Geterotsikl. Soedin*, **1974**, 84; *Cheminform*, **1975**, 5–225, p. 95.
96. H. G. Viehe and Z. Janousek, *Angew. Chem.*, **85**, 837 (1973); *Angew. Chem. Intern. Ed. Engl.*, **12**, 806 (1973).
97. H. Eilingsfeld, M. Seefelder, and H. Weidinger, *Chem. Ber.*, **96**, 2899 (1963).
98. Z. Arnold, *Collection Czech. Chem. Commun.*, **28**, 2047 (1963).
99. Z. Arnold and A. Holy, *Collection Czech. Chem. Commun.*, **26**, 3059 (1961).
100. H. H. Bosshard, E. J. Jenny, and H. Zollinger, *Helv. Chim. Acta*, **44**, 1203 (1961).
101. J. Gal, B. A. Phillips, and R. Smith, *Can. J. Chem.*, **51**, 132 (1973).
102. B. Matkovic, S. W. Peterson, and J. M. Williams, *Croat. Chem. Acta*, **39**, 139 (1967).
103. S. W. Peterson and J. M. Williams, *J. Am. Chem. Soc.*, **88**, 2866 (1966).
104. W. Kantlehner and P. Speh, *Chem. Ber.*, **104**, 3714 (1971).
105. G. Simchen and G. Entenmann, *Angew. Chem.*, **85**, 155 (1973).
106. S. Yanagida and S. Komori, *Synthesis*, **1973**, 189.
107. H. W. Heine and A. B. Smith, *Angew. Chem. Intern. Ed. Engl.*, **2**, 400 (1963).
108. W. Jentsch, *Chem. Ber.*, **97**, 2755 (1964).
109. E. Allenstein and A. Schmidt, *Z. Anorg. Allgem. Chem.*, **344**, 113 (1966).
110. Z. Arnold, *Collection Czech. Chem. Commun.*, **26**, 1723 (1961).
111. D. R. Hepburne and H. R. Hudson, *Chem. Ind. London*, **1974**, 664.
112. A. Pinner, *Die Iminoäther und ihre Derivate*, Robert Oppenheimer, Berlin, 1892.
113. L. Suranyi and H. Wilk, Ger. Pat. 1,150,987 (July 4, 1963).
114. L. Kh. Fel'dman, G. S. Semicheva, and N. G. Zhokhovets, *Mechenye Biol. Aktion Veshchestv. Sb. Statei*, **1962**, 26.
115. S. Pennanen, *Acta Chem. Scand.*, **27**, 3133 (1973).
116. I. Matsumoto, *Yakugaku Zasshi*, **85**, 544 (1965).
117. E. V. P. Tao and C. F. Christie, *Org. Prep. Proceed. Intern.*, **4**, 73 (1972); *Cheminform*, **1973**, 49–256, p. 99.
118. J. Kuthan, Czech. Pat. 109,895 (Feb. 15, 1964).
119. F. Reicheneder, K. Dury, and P. Dimroth, Fr. Pat. 1,413,603 (Oct. 8, 1965).
120. G. Ege and H. O. Frey, *Tetrahedron Letters*, **1972**, 4217.
121. R. Mory, E. Stöchlin, and M. Schmidt (to Ciba), Ger. Pat. 1,026,7500 (1958); *C.A.*, **55**, 5428 (1961).
122. Hoechst, Belg. Pat. 553,871 (Jan. 15, 1957).
123. W. A. Gregory, U.S. Pat. 2,88,486 (May 26, 1959).
124. H. H. Bosshard and H. Zollinger, *Angew. Chem.*, **71**, 375 (1959).
125. E. G. Freidlin, E. L. Bulakh, S. S. Gitis, and V. S. Buryak, *Kinet. Katal.*, **15**, 911 (1974); *Cheminform*, **1974**, 46–246, p. 85.
126. S. H. Parker and D. D. Rudy, U.S. Pat. 3,184,506 (May 18, 1965).
127. E. R. Talaty, Belg. Pat. 625,809 (March 29, 1963).
128. P. E. Bratachanskii and G. Kommissarova, *Zh. Prikl. Khim.*, **47**, 239 (1974).
129. E. F. Godefroy, C. A. M. van der Eycken, and C. van de Westeringh, *J. Org. Chem.*, **29**, 3707 (1964).
130. W. Ried and G. Neidhardt, *Ann. Chem.*, **666**, 148 (1963).
131. M. Zaoral and Z. Arnold, Czech. Pat. 99,291 (Appl. March 7, 1960).
132. M. Zaoral and Z. Arnold, *Tetrahedron Letters*, **1960**, 9.

133. H. v. Euler, H. Hasselquist, and I. Limuell, *Arkiv Kemi*, **21**, 37 (1963).
134. L. Brehm, P. Kroogsgaard, and H. Hjeds, *Acta Chem. Scand. B*, **28**, 308 (1974); *Cheminform*, **1974**, 32–297, p. 121.
135. B. J. R. Nicolaus, E. Bellasio, and E. Testa, *Helv. Chim. Acta*, **45**, 717 (1962).
136. E. Testa, B. J. R. Nicolaus, L. Mariam, and G. Pagani, *Helv. Chim. Acta*, **46**, 766 (1963).
137. H. Ulrich, B. Tucker, and A. A. R. Sayigh, *J. Org. Chem.*, **32**, 4052 (1967).
138. J. P. Maffraud, M. Amoros, D. Frehel, and F. Eloy, *Chem. Ther.*, **9**, 150 (1974); *Cheminform*, **1974**, 50–206, p. 70.
139. R. N. McDonald and R. A. Krueger, *J. Org. Chem.*, **28**, 2542 (1963).
140. Y.-W. Chang, Fr. Pat. 1,385,595 (Jan. 15, 1965).
141. F. Eiden and G. Bachmann, *Arch. Pharm. Weinheim*, **306**, 401 (1973).
142. Belg. Pat. 618,751 (Dec. 10, 1962).
143. F. Cramer, S. Rittner, W. Reinhard, and P. Desai, *Chem. Ber.*, **99**, 2252 (1966).
144. H. G. Khorana and T. M. Jacob, *J. Am. Chem. Soc.*, **86**, 1630 (1964).
145. M. Ikehara and H. Umo, *Chem. Pharm. Bull.* [*Tokyo*], **12**, 742 (1964).
146. W. Ziegenbein and W. Franke, *Chem. Ber.*, **93**, 1681 (1960).
147. W. Ziegenbein and K. H. Hornung, *Chem. Ber.*, **95**, 2976 (1962).
148. W. Ziegenbein and K. H. Hornung, Ger. Pat. 1,188,570 (March 11, 1965).
149. H. Priewe and K. Gutsche, Ger. Pat. 1,145,622 (March 21, 1963).
150. O. Wallach, *Ber. Deut. Chem. Ges.*, **9**, 1212 (1876).
151. O. Wallach and P. Pirath, *Ber. Deut. Chem. Ges.*, **12**, 1063 (1879).
152. A. Bernthsen, *Ber. Deut. Chem. Ges.*, **10**, 1238 (1877).
153. H. Leo, *Ber. Deut. Chem. Ges.*, **10**, 2133 (1877).
154. W. Walter and G. Maerten, *Ann. Chem.*, **669**, 66 (1963).
155. E. Beckmann and E. Fellrath, *Ann. Chem.*, **273**, 1 (1893).
156. J. v. Braun and K. Weissbach, *Ber. Deut. Chem. Ges.*, **65B**, 1574 (1932).
157. J. v. Braun and G. Lemke, *Ber. Deut. Chem. Ges.*, **55**, 3526 (1922).
158. J. v. Braun, F. Jostes, and A. Heymons, *Ber. Deut. Chem. Ges.*, **60**, 92 (1927).
159. R. L. Shriner and F. W. Neumann, *Chem. Rev.* **35**, 351 (1944).
160. J. Arient and J. Podstata, *Collection Czech. Chem. Commun.*, **39**, 3117 (1974).
161. H. Böhme and F. Soldan, *Chem. Ber.*, **94**, 3112 (1961).
162. M. D. Scott and H. Spedding, *J. Chem. Soc. C*, **1968**, 1603.
163. Z. Csürös, R. Soos, J. Palinkas, and J. Bitter, *Acta Chim. Acad. Sci. Hung.*, **63**, 215 (1970).
164. Z. Csürös, R. Soos, J. Palinkas, and J. Bitter, *Acta Chim. Acad. Sci. Hung.*, **68**, 397 (1971).
165. Z. Csürös, R. Soos, J. Bitter, and J. Palinkas, *Acta Chim. Acad. Sci. Hung.*, **72**, 59 (1972); *C.A.*, **76**, 153706d (1972).
166. J. C. Thurman, *Chem. Ind. London*, **1964**, 752.
167. E. A. Lawton and D. D. McRitchie, *J. Org. Chem.*, **24**, 26 (1958).
168. K. Wallenfels, G. Bachmann, D. Hofmann, and R. Kern, *Tetrahedron*, **21**, 2239 (1965).
169. C. S. Marvel and M. M. Martin, *J. Am. Chem. Soc.*, **80**, 6600 (1958).
170. K. Wallenfels, D. Hofmann, and R. Kern, *Tetrahedron*, **21**, 2231 (1965).
171. S. Yanagida, H. Hayama, and S. Komori, *J. Org. Chem.*, **34**, 4180 (1969).
172. S. Yanagida, T. Fujita, M. Ohoka, J. Katagiri, M. Miyake, and S. Komori, *Bull. Chem. Soc. Japan*, **46**, 292, 303 (1973).
173. J. Liebscher and H. Hartmann, *Z. Chem.*, **14**, 358 (1974).
174. W. H. Warren and F. E. Wilson, *Ber. Deut. Chem. Ges.*, **68**, 957 (1935).
175. E. Kühle, *Angew. Chem.*, **74**, 861 (1962); *Angew. Chem. Intern. Ed. Engl.*, **1**, 647 (1962).
176. H. M. Walborsky and G. E. Niznik, *J. Org. Chem.*, **37**, 187 (1972).
177. A. A. R. Sayigh and H. Ulrich, *J. Chem. Soc.*, **1963**, 3146.

178. H. Fritz and R. Oehl, *Ann. Chem.*, **749**, 159 (1971).
179. Z. Arnold and J. Zemlicka, *Chem. Listy*, **52**, 459 (1958).
180. W. Jentsch and M. Seefelder, *Chem. Ber.*, **98**, 274 (1965).
181. Z. Csürös, R. Soos, J. Bitter, and E. A. Karpati, *Acta Chim. Acad. Sci. Hung.*, **73**, 239 (1972); *C.A.*, **77**, 101082y (1972).
182. K. H. Beyer, H. Eilingsfeld, and H. Weidinger, Ger. Pat. 1,110,625 (1961); *C.A.*, **56**, 3363 (1962).
183. H. Ulrich, B. Tucker, and A. A. Sayigh, *Angew. Chem.*, **79**, 651 (1967); *Angew. Chem. Intern. Ed. Engl.*, **6**, 636 (1967).
184. J. Ugi and R. Meyr, *Angew. Chem.*, **70**, 702 (1958).
185. J. Ugi and R. Meyr, *Chem. Ber.*, **93**, 239 (1960).
186. J. Ugi, U. Fetzer, U. Eholzer, H. Knupfer, and K. Offermann, *Angew. Chem.*, **77**, 492 (1965).
187. K. Bartel and W. P. Fehlhammer, *Angew. Chem.*, **86**, 588 (1974).
188. S. Christophersen and P. Carlsen, *Tetrahedron Letters*, **1973**, 211.
189. J. D. Mee, *J. Am. Chem. Soc.*, **90**, 4712 (1974).
190. K. C. Brannock and R. D. Burpitt, *J. Org. Chem.*, **30**, 2564 (1965).
191. G. D. Lander, *J. Chem. Soc.*, **91**, 968 (1907).
192. H. Bredereck and K. Bredereck, *Chem. Ber.*, **94**, 2278 (1961).
193. H. Eilingsfeld, M. Seefelder, and H. Weidinger, Ger. Pat. 1,119,872 (1961); *C.A.*, **56**, 14083.
194. H. Meerwein, W. Florian, N. Schön, and G. Stopp, *Ann. Chem.*, **641**, 1 (1961).
195. H. Bredereck, F. Effenberger, G. Simchen, *Angew. Chem.*, **73**, 493 (1961).
196. Ciba, Ltd., Brit. Pat. 911,475 (1962); *C.A.*, **58**, 13852x.
197. H. H. Bosshard and H. Zollinger (to Ciba Ltd.), Swiss Pat. 384,564 (1965); *C.A.*, **62**, 16135h (1965).
198. D. E. Horning and J. M. Muchowski, *Can. J. Chem.*, **45**, 1247 (1967).
199. R. Ohme and E. Schmitz, *Angew. Chem.*, **79**, 531 (1967).
200. M. B. Frankel, H. Feuer, and J. Bank, *Tetrahedron Letters*, **1959**, 5.
201. M. Saunders and R. W. Murray, *Tetrahedron*, **11**, 1 (1960).
202. J. W. Scheeren and R. J. Nivard, *Rec. Trav. Chim. Pays-Bas*, **88**, 289 (1969).
203. W. Funke, *Ann. Chem.*, **725**, 15 (1969).
204. J. Graefe, I. Fröhlich, and M. Mühlstädt, *Z. Chem.*, **14**, 434 (1974).
205. H. J. Bredereck, Dissertation, University of Stuttgart, 1969.
206. H. Bredereck, F. Effenberger, and T. Brendle, *Angew. Chem.*, **78**, 147 (1966).
207. H. Bredereck, F. Effenberger, T. Brendle, and H. Muffler, *Chem. Ber.*, **101**, 1885 (1968).
208. K. Issleib and M. Lischewski, *J. Organomet. Chem.*, **46**, 297 (1972).
209. H. Meerwein, P. Laasch, R. Mersch, and J. Spille, *Chem. Ber.*, **89**, 209 (1956).
210. J. v. Braun, F. Jostes, and W. Münch, *Ann. Chem.*, **453**, 113 (1926).
211. J. v. Braun, *Angew. Chem.*, **47**, 611 (1934).
212. J. v. Braun, *Ber. Deut. Chem. Ges.*, **37**, 2812 (1904).
213. O. Wallach and M. Hoffmann, *Ann. Chem.*, **184**, 75 (1877).
214. H. v. Pechmann, *Ber. Deut. Chem. Ges.*, **33**, 611 (1900).
215. J. v. Braun, *Ber. Deut. Chem. Ges.*, **37**, 2915 (1904).
216. G. Fodor, J. J. Ryan, and F. Letourneau, *J. Am. Chem. Soc.*, **91**, 7768 (1969).
217. N. J. Leonard and E. W. Nommensen, *J. Am. Chem. Soc.*, **71**, 2808 (1949).
218. W. R. Vaughan and R. D. Carlson, *J. Am. Chem. Soc.*, **84**, 769 (1962).

ADDENDUM*

To Section I-A

By action of halogenating reagents like PCl_5 (219), $POCl_3/PCl_5$ (220), phosgene (222, 223), chlorine (224), SF_4 (224), onoamides (219, 221, 223), lactames (220, 222), and thioamides (224), further chloromethyleniminium salts have been prepared, which were not always isolated, but were reacted further to form, for example, imid halides (220, 222, 224). A kinetic study concerned with the reaction of $COCl_2$ and $SOCl_2$, respectively with NN-dimethylbenzamide and NN-dimethylacetamide has been performed (225).

To Section I-B

Interesting, preparatively valuable reactions have been found in the reaction of phosgene with N-arylamides. In this case quinolinium salts such as **227** are formed (226).

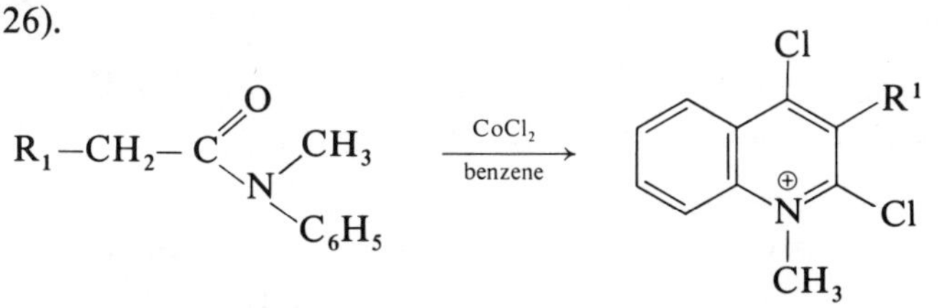

$$R_1-CH_2-C(=O)N(CH_3)(C_6H_5) \xrightarrow[benzene]{COCl_2} \text{(227)}$$

(227)

PCl_5 cyclizes o-methylselenobenzamides (**228**) to give benzisoselenazolium-chlorides (**229**) (227).

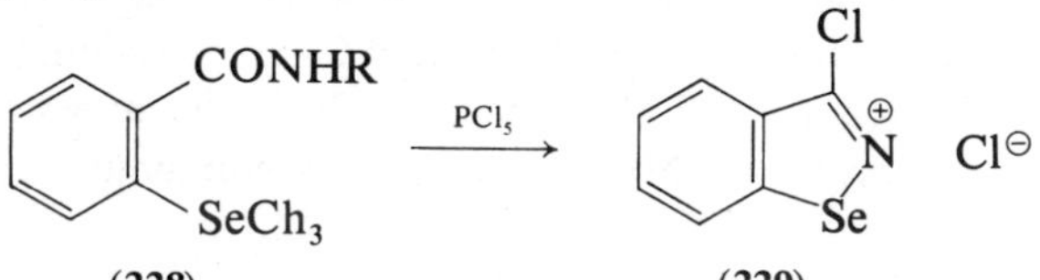

(228) $\xrightarrow{PCl_5}$ (229)

Acylguanidines with phosgene do not form iminiumchlorides; instead, there occurs cyclization forming oxadiazinones (**230**) (228).

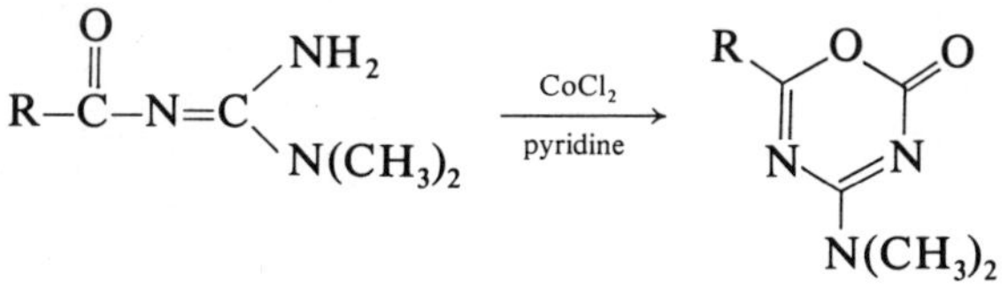

$$R-\overset{O}{\underset{\|}{C}}-N=C(NH_2)(N(CH_3)_2) \xrightarrow[pyridine]{COCl_2} \text{(230)}$$

(230)

---

*Translated from German by W. Kantlehner, and by H. J. Benivschke, Fachhochschule Aalen, 708-Aalen, Hohenstaufenstrasse 1, Germany.

An incorporation of the halogenating compound, instead of iminium salt formation, was observed when several amides were treated with $SOCl_2$ (229, 230, 231) or oxalylchloride (232), respectively.

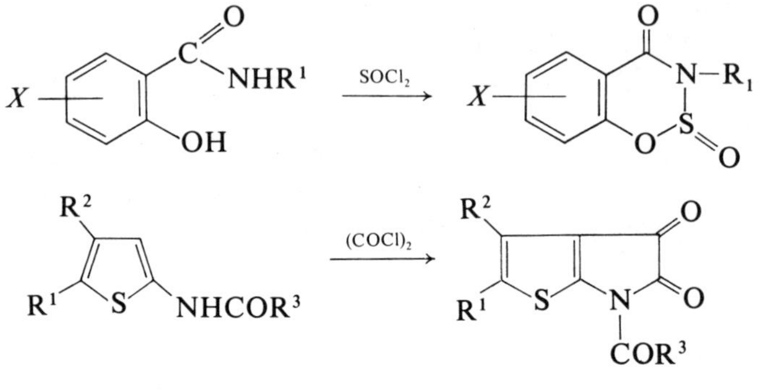

To Section I-B-1

Cyanogene reacts with HCl to yield the iminium salt **231**, if $SbCl_5$ is used in less than stoicheiometric amounts (233).

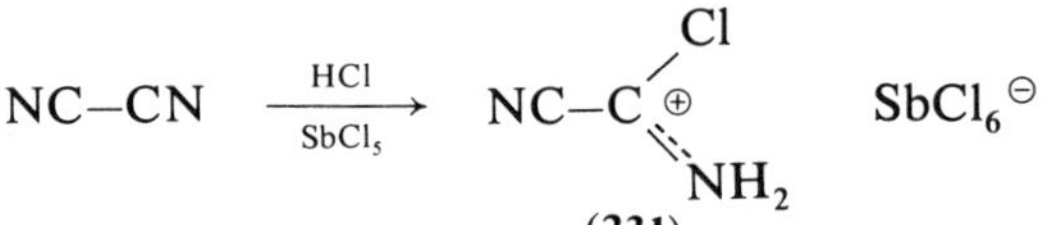

$$NC-CN \xrightarrow[SbCl_5]{HCl} NC-\overset{\oplus}{C}\genfrac{}{}{0pt}{}{Cl}{NH_2} \quad SbCl_6^{\ominus}$$
(231)

To Section I-D

Further 2,2-dihaloaziridines have been synthesized from dihalocarbenes/CXY (x = Cl, Br, F, y = Cl, Br) and C=N-containing compounds (azomethines, ketimines) (234–237). Dihaloaziridines also may be obtained in good yields from chloroform, NaOH and C=N-containing compounds under conditions of phase-transfer catalysis (238–240).

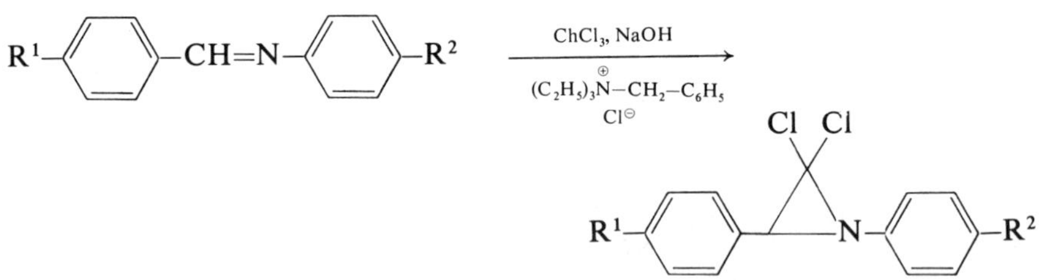

To Section I-E

The reaction between ketones (241), acid amides (242, 243, 244), dicarboxylic acid amides (245), acid chlorides (246), nitriles (247), and phosgene iminium salts is very useful for the preparation of several interesting iminium salts. Some examples are listed below.

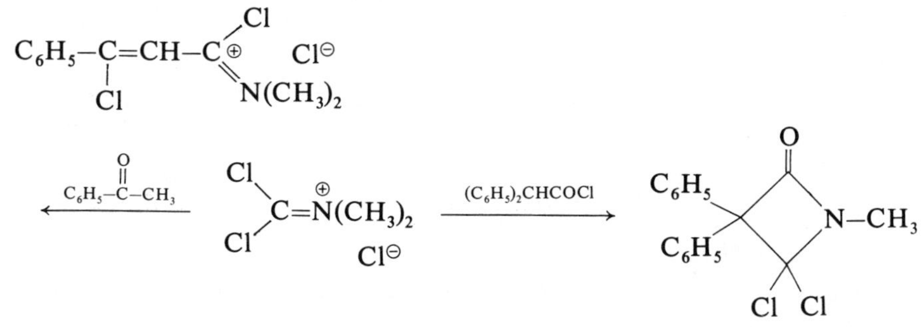

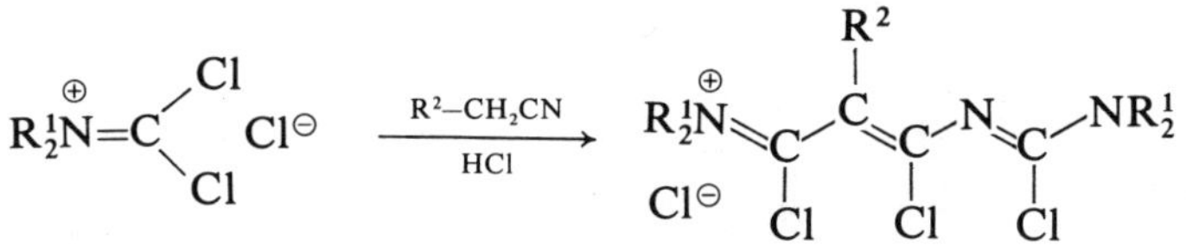

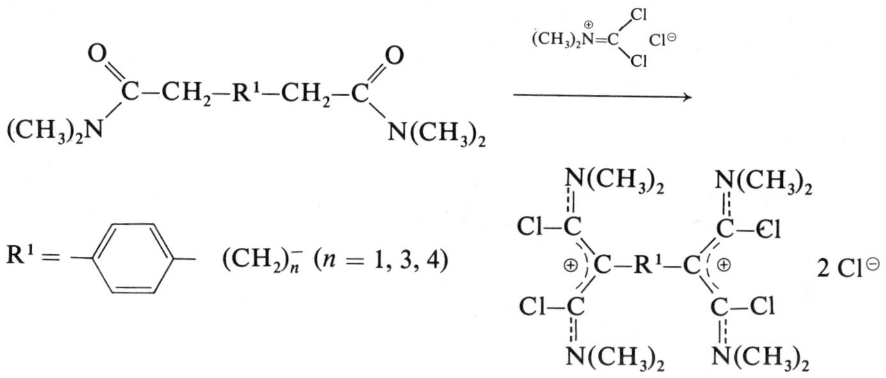

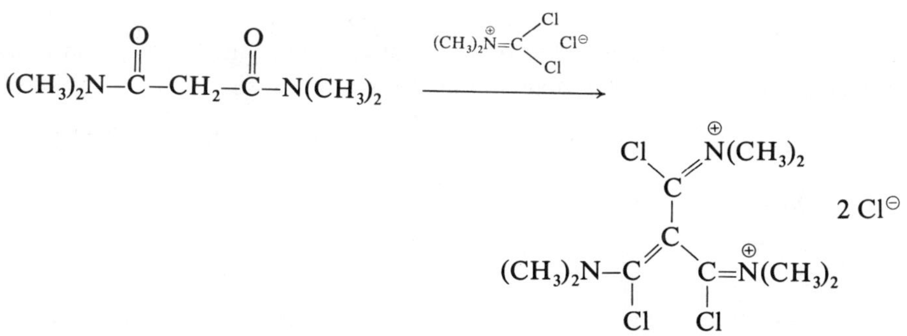

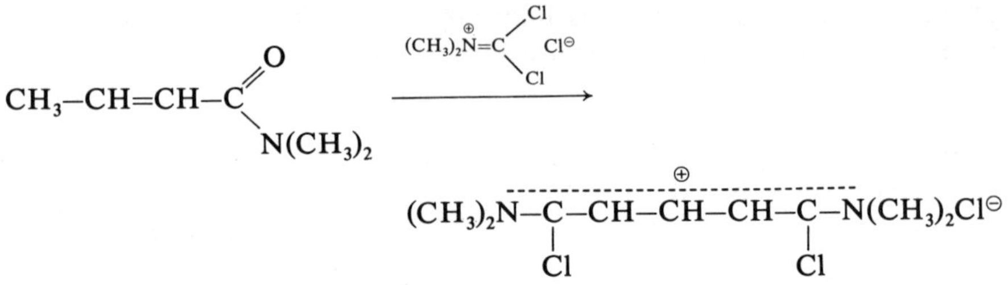

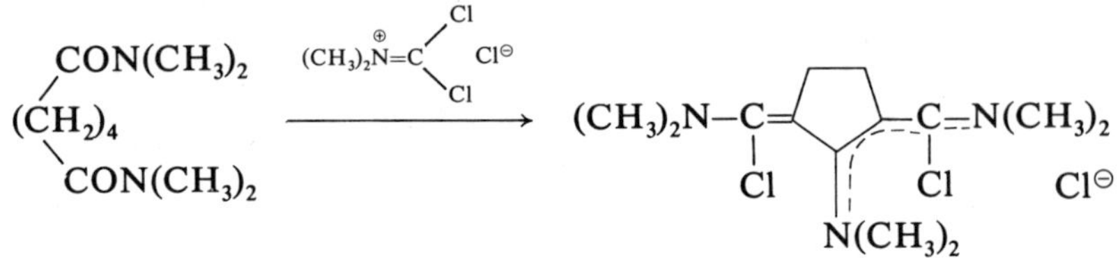

$$CH_3-CH=CH-C\underset{N(CH_3)_2}{\overset{O}{\diagup}} \xrightarrow{(CH_3)_2\overset{\oplus}{N}=C\underset{Cl}{\overset{Cl}{\diagup}} Cl^{\ominus}}$$

$$(CH_3)_2\overset{\oplus}{N}-\underset{Cl}{C}-CH-CH-CH-\underset{Cl}{C}-N(CH_3)_2 Cl^{\ominus}$$

$$\underset{CON(CH_3)_2}{\overset{CON(CH_3)_2}{(CH_2)_4}} \xrightarrow{(CH_3)_2\overset{\oplus}{N}=C\underset{Cl}{\overset{Cl}{\diagup}} Cl^{\ominus}} (CH_3)_2N-\underset{Cl}{C}\!=\!\!\diagup\!\!=\!\!\underset{Cl}{C}\!-\!N(CH_3)_2 \quad Cl^{\ominus}$$
$$N(CH_3)_2$$

To Section II

NMR studies concerned with barriers of rotation around C=N-bond of chloromethyleniminium chlorides (225), as well as ^{35}Cl-NMR data of chloromethyleniminium salts, have been reported (248).

To Section III-A-1

Hydrolysis of iminium salts (**232**) yields amides (**233**) that are suitable for synthesis of propiolic acid amides (241, 243, 244).

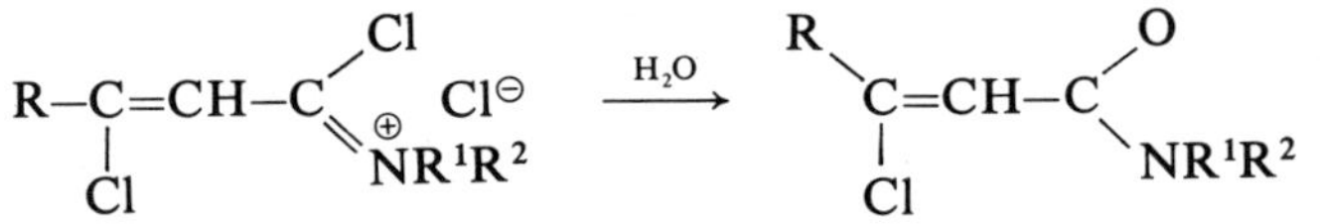

$$R-\underset{Cl}{C}=CH-C\underset{\overset{\oplus}{NR^1R^2}}{\overset{Cl}{\diagup}} Cl^{\ominus} \xrightarrow{H_2O} \underset{Cl}{\overset{R}{C}}=CH-C\underset{NR^1R^2}{\overset{O}{\diagup}}$$
$$(232) \qquad\qquad (233)$$

Other amides that are useful for preparative investigations have also been obtained from hydrolysis of iminium salts (245).

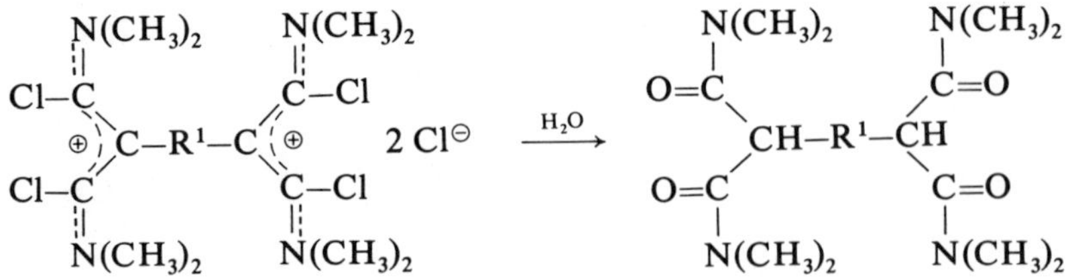

$$\begin{array}{c} N(CH_3)_2 \quad N(CH_3)_2 \\ \| \qquad \| \\ Cl-C \qquad C-Cl \\ \overset{\oplus}{\diagup}C-R^1-C\overset{\oplus}{\diagdown} \quad 2\,Cl^{\ominus} \\ Cl-C \qquad C-Cl \\ \| \qquad \| \\ N(CH_3)_2 \quad N(CH_3)_2 \end{array} \xrightarrow{H_2O} \begin{array}{c} N(CH_3)_2 \quad N(CH_3)_2 \\ | \qquad | \\ O=C \qquad C=O \\ \diagdown CH-R^1-CH\diagup \\ O=C \qquad C=O \\ | \qquad | \\ N(CH_3)_2 \quad N(CH_3)_2 \end{array}$$

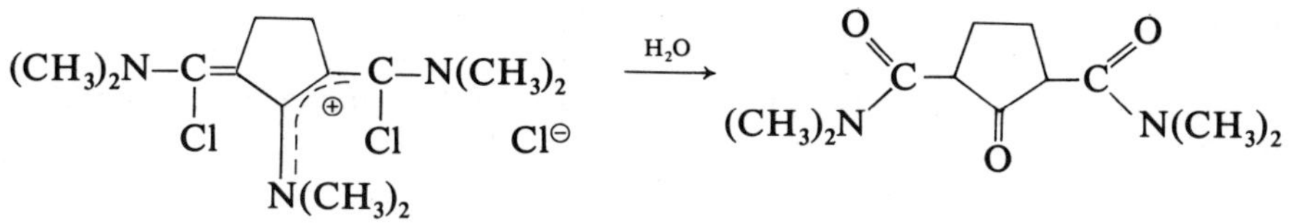

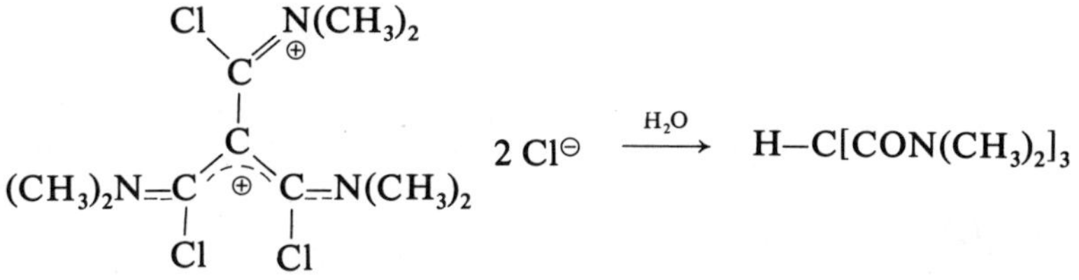

$(CH_3)_2N-C\underset{Cl}{\overset{\oplus}{=}}\underset{N(CH_3)_2}{C}-C-N(CH_3)_2 \quad Cl^\ominus \xrightarrow{H_2O} (CH_3)_2N-\overset{O}{\overset{\|}{C}}-\underset{O}{C}-\overset{O}{\overset{\|}{C}}-N(CH_3)_2$

$\xrightarrow{H_2O} H-C[CON(CH_3)_2]_3$

A number of 2,2-dihaloaziridines were hydrolyzed, leading to α-haloamides or α-hydroxyamides, respectively (234, 237, 239, 249).

To Section III-A-2

Extensive investigations on transformation of alcohols to alkylhalides by means of halomethyleniminium halides demonstrating the preparative scope and the stereochemical course of reaction have been performed (250).

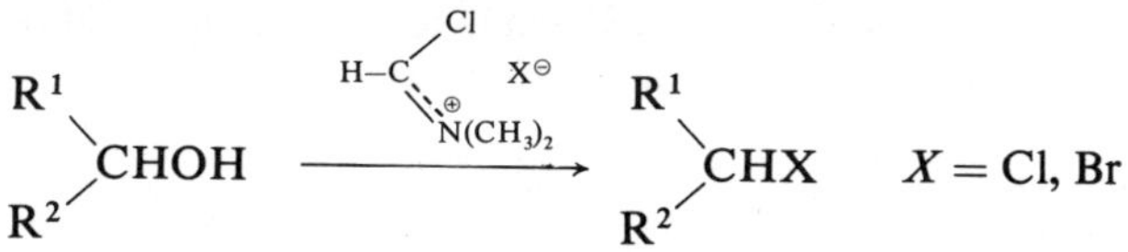

$\underset{R^2}{\overset{R^1}{>}}CHOH \xrightarrow{H-C\overset{Cl}{\underset{N(CH_3)_2}{\overset{\oplus}{<}}} X^\ominus} \underset{R^2}{\overset{R^1}{>}}CHX \quad X = Cl, Br$

Acidic alcoholysis of 2,2-dibrom-1,3-diphenylaziridin (**234**) affords the ester (**235**) (249).

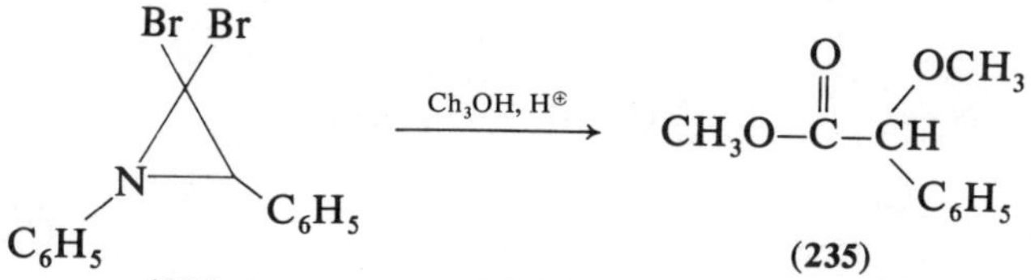

(234) $\xrightarrow{Ch_3OH, H^\oplus}$ $CH_3O-\overset{O}{\overset{\|}{C}}-\underset{C_6H_5}{\overset{OCH_3}{CH}}$ (235)

Practical use of the alcoholysis of amide halides has been made in order to prepare formiates (251).

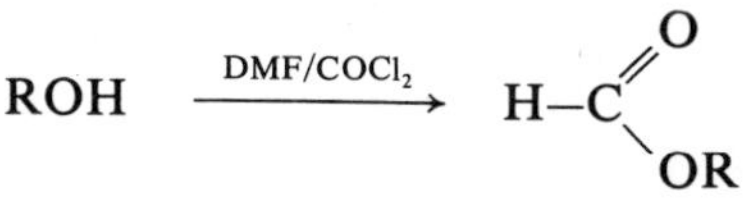

$ROH \xrightarrow{DMF/COCl_2} H-C\overset{O}{\underset{OR}{\diagdown}}$

To Section III-A-4

α-Isocyanoacetic acid chloride—a precursor of isocyanoketene—is obtainable by action of dimethylformamid chloride on salts of α-isocyanaoacetic acid (252).

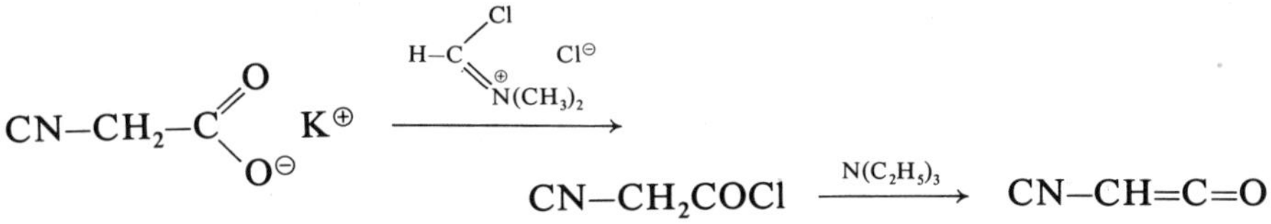

$$CN-CH_2-C(=O)(O^{\ominus}) \; K^{\oplus} \quad \xrightarrow{H-C(Cl)=N^{\oplus}(CH_3)_2 \; Cl^{\ominus}} \quad CN-CH_2COCl \quad \xrightarrow{N(C_2H_5)_3} \quad CN-CH=C=O$$

To Section III-C-1

Iminium salt **236** on thiolysis yields the dithiopyrone **237** (245).

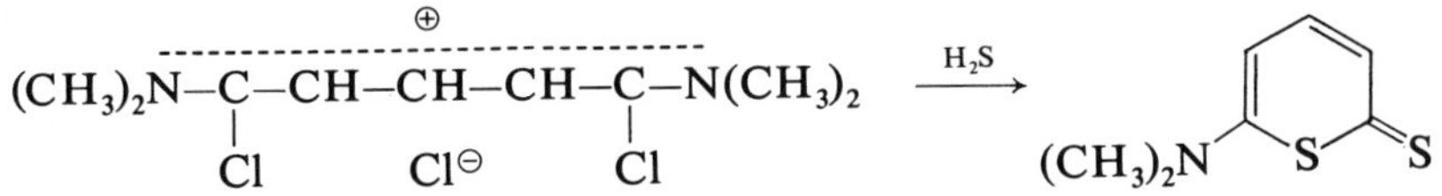

$$(CH_3)_2\overset{\oplus}{N}-C(Cl)-CH-CH-CH-C(Cl)-N(CH_3)_2 \; Cl^{\ominus} \quad \xrightarrow{H_2S} \quad \underset{(237)}{(CH_3)_2N-\text{[dithiopyrone]}-S}$$

(236)   (237)

To Section III-D-1

Aminolysis of the iminium salt **238** leads to interesting amidines (242, 243, 253). Some of these reactions are given by the following formulas.

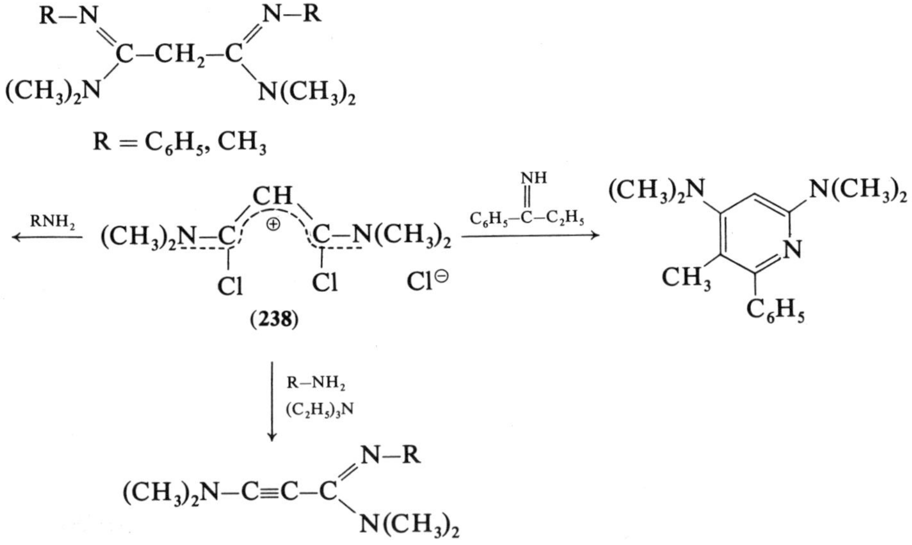

$$(CH_3)_2N-C(=N-R)-CH_2-C(=N-R)-N(CH_3)_2$$

R = C$_6$H$_5$, CH$_3$

$$\xleftarrow{RNH_2} \quad (CH_3)_2N-\overset{\oplus}{C}(Cl)\cdots CH \cdots C(Cl)-N(CH_3)_2 \; Cl^{\ominus} \quad \xrightarrow{C_6H_5-C(=NH)-C_2H_5} \quad \text{pyridine: } (CH_3)_2N, CH_3, C_6H_5, N(CH_3)_2$$

(238)

↓ R–NH$_2$ / (C$_2$H$_5$)$_3$N

$$(CH_3)_2N-C\equiv C-C(=N-R)(N(CH_3)_2)$$

R = alkyl

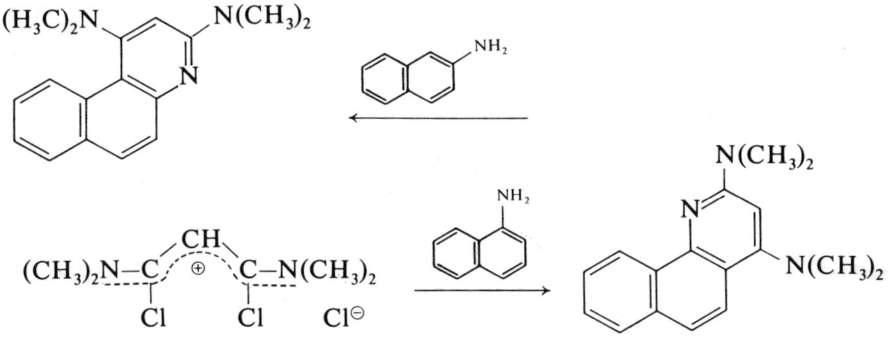

Hydrazinderivatives and aniline are transformed by iminium salts **239** and **236** to heterocycles **240** and **241**, respectively (245).

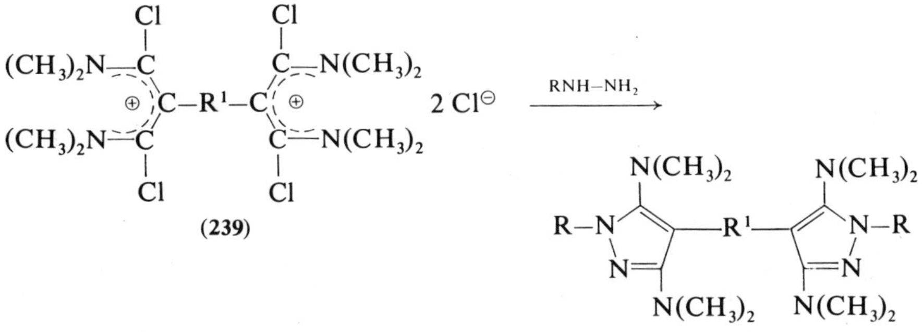

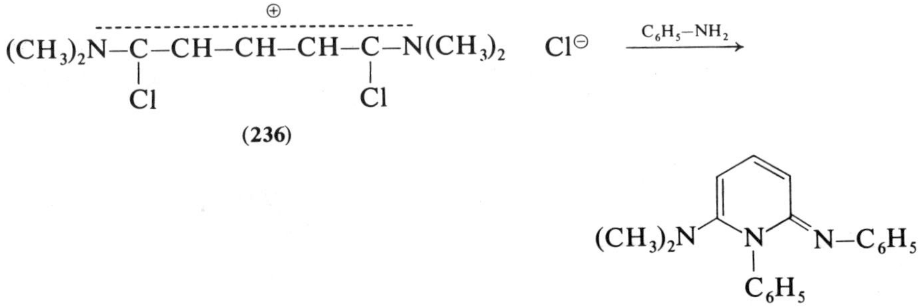

Reaction of **236** with dimethylamine produces the bisamidinium salt **242** (245).

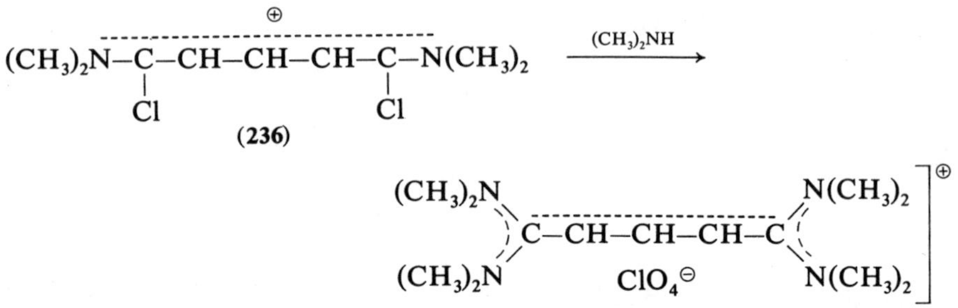

(CH₃)₂N−C−CH−CH−CH−C−N(CH₃)₂ $\xrightarrow{(CH_3)_2NH}$
  | |
  Cl   Cl
        (236)

$$\left[\begin{array}{c} (CH_3)_2N \\ \diagdown \\ C-CH-CH-CH-C \\ (CH_3)_2N ClO_4^{\ominus} \diagup N(CH_3)_2 \\ \diagdown N(CH_3)_2 \end{array} \right]^{\oplus}$$

(242)

The reaction between the uracile **243** and DMF/SOCl₂ yields the isothiazolopyrimidene **244** as the main product and the amidine **245** is formed as a byproduct (254). Analogous compounds show a similar reaction pattern (254).

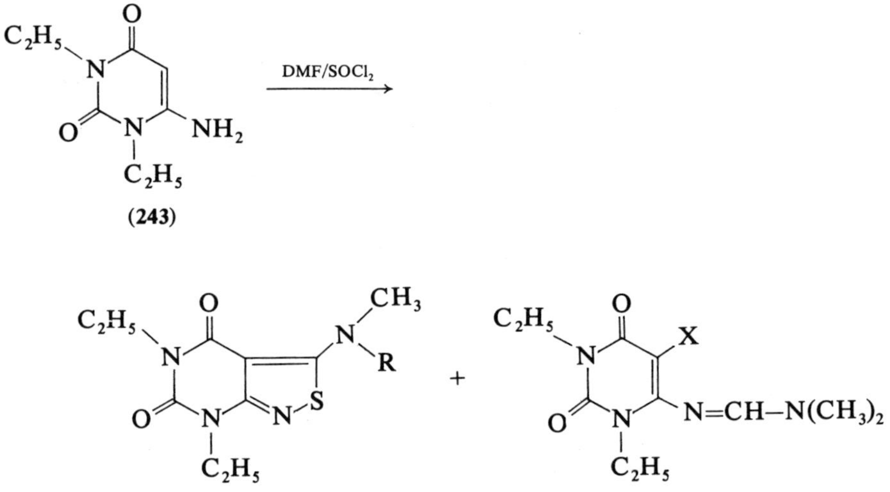

[Structure **243**: N-ethyl uracil derivative with NH₂] $\xrightarrow{DMF/SOCl_2}$

[Structure **244**: isothiazolopyrimidine] + [Structure **245**: uracil with N=CH−N(CH₃)₂]

R	X
CH₃	Cl
H	CHO

By action of DMF/SOCl₂ on the pyrazol derivative **246** under mild reaction conditions the amidine **247** is formed with simultaneous dehydration of the amide function, whereas reaction at elevated temperatures affords **248** (255).

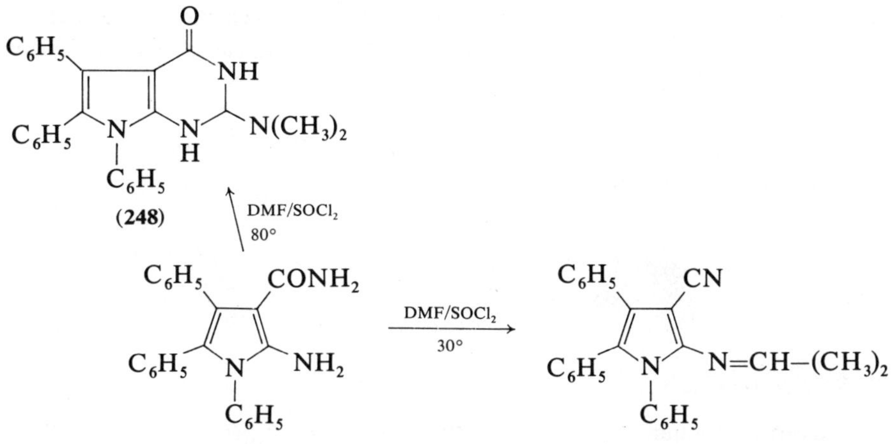

[Structures (248), (246), (247) with reagents DMF/SOCl₂ 80° and DMF/SOCl₂ 30°]

Dihalogenaziridines, on treatment with secondary amines, give rise to α-aminoamidines (**249**), although α, β-unsaturated amidines (**250**) can also be present as the product of a secondary reaction (234, 249).

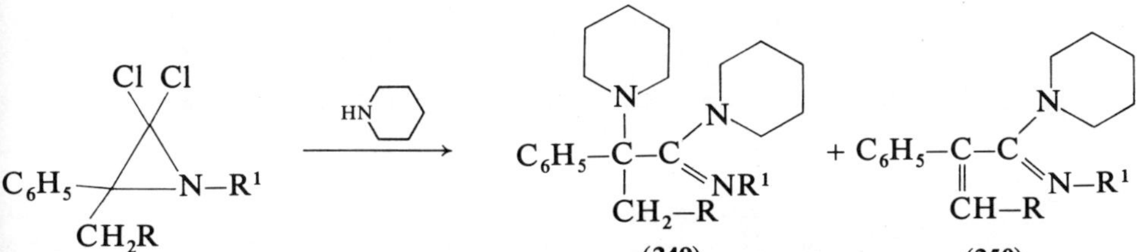

[Reaction scheme showing dichloroaziridine + piperidine (HN) → (249) + (250)]

NN-Dimethyl-*N'*-chlorosulfinylformamidines (**251**) are produced in the reaction of dimethylformamid chloride and **252** (256).

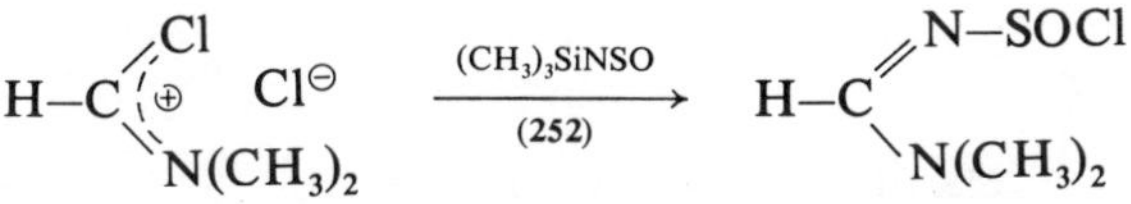

[Reaction scheme: chloro-iminium chloride + (CH₃)₃SiNSO (252) → (251)]

Hydroxylamine transforms iminium salts (**253**) to isooxazoles (**254**) (257).

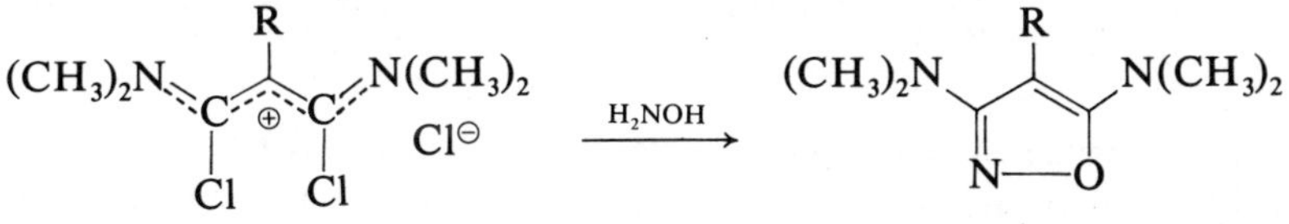

[Reaction scheme: (253) + H₂NOH → (254)]

In the presence of AlCl$_3$ amide chlorides **255** are cyclized by the hydrochlorides of hydroxylamine, trichloracetamidine, and ammonia to form compounds **256** (258).

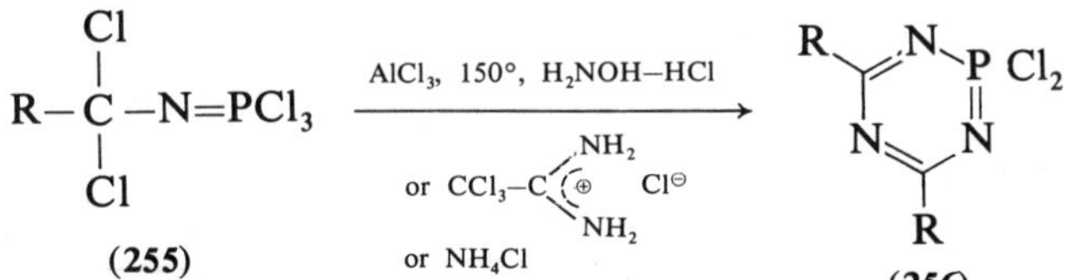

$$R-\underset{\underset{Cl}{|}}{\overset{\overset{Cl}{|}}{C}}-N=PCl_3 \quad \xrightarrow[\substack{\text{or } CCl_3-C\overset{\oplus}{\underset{NH_2}{\diagup^{NH_2}}}\ Cl^{\ominus} \\ \text{or } NH_4Cl}]{AlCl_3,\ 150°,\ H_2NOH-HCl} \quad \underset{(256)}{\underset{R}{\overset{R}{\diagdown}}\underset{N}{\overset{N}{\diagup}}\underset{\|}{\overset{}{}}\underset{N}{\overset{}{}}PCl_2}$$

(255)

To Section III-D-2

A series of aliphatic and aromatic nitriles have been prepared by action of SOCl$_2$ on primary carboxylic acid amides in the presence of catalyctic amounts of DMF (259). *NN'*-Diformylhydrazine reacts with dimethylformamidechloride to yield the iminium salt **257**, which on treatment with alkoxide followed by H$_2$S yields 1,3,4-thiadiazole (260).

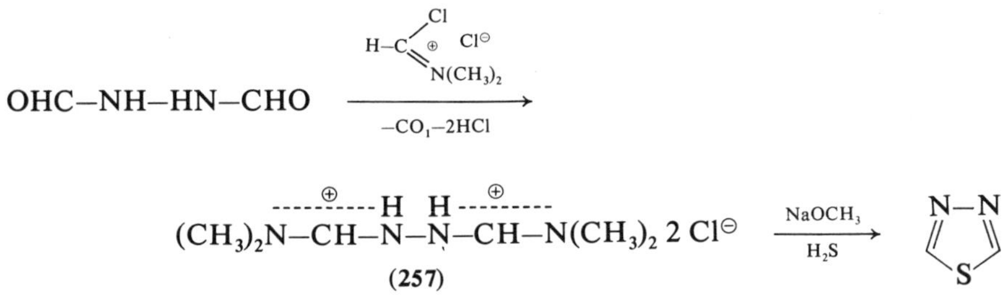

$$OHC-NH-HN-CHO \quad \xrightarrow[-CO,-2HCl]{H-C\overset{Cl}{\diagup}\ Cl^{\ominus}\ \diagdown N(CH_3)_2}$$

$$(CH_3)_2N-\overset{\oplus}{C}H-N-N-\overset{\oplus}{C}H-N(CH_3)_2\ 2\ Cl^{\ominus} \quad \xrightarrow[H_2S]{NaOCH_3} \quad \underset{S}{\overset{N-N}{\diagdown\diagup}}$$

(257)

The amide halogenide **258** is suitable for dehydration of aldoximes, giving nitriles in good yields (258).

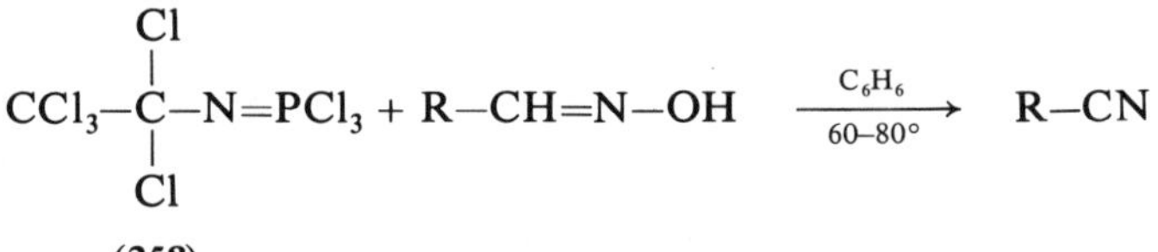

$$CCl_3-\underset{\underset{Cl}{|}}{\overset{\overset{Cl}{|}}{C}}-N=PCl_3 + R-CH=N-OH \quad \xrightarrow[60-80°]{C_6H_6} \quad R-CN$$

(258)

R = CH$_3$, C$_6$H$_5$, *p*-Cl-C$_6$H$_4$-, *p*-NO$_2$-C$_6$H$_4$, *p*-CH$_3$O-C$_6$H$_4$-,

In the presence of hydrogenchloride numerous nitriles have been reacted with chloromethyleneiminium salts (**259**) to form new iminium compounds (**260**), (261–263) which are very reactive and are useful tools for synthetic work (261–269). Alkylrhodanides undergo similar reactions, giving azavinylogous mercaptochlorformamidinium salts **(261)** (270).

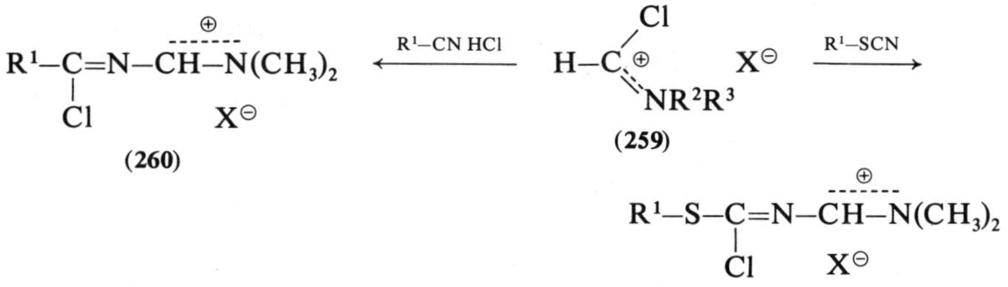

$$R^1-\overset{\oplus}{C}=N-CH-N(CH_3)_2 \quad \xleftarrow{R^1-CN\ HCl} \quad H-\overset{Cl}{\underset{NR^2R^3}{C^{\oplus}}} \quad X^{\ominus} \quad \xrightarrow{R^1-SCN}$$
$$\underset{Cl}{|} \qquad X^{\ominus}$$
$$\textbf{(260)} \qquad\qquad\qquad \textbf{(259)}$$

$$R^1-S-\overset{\oplus}{C}=N-CH-N(CH_3)_2$$
$$\underset{Cl}{|} \qquad X^{\ominus}$$
$$\textbf{(261)}$$

To Section III-E-1

Further transformations of *N*-monosubstituted formamides to isonitriles via amide halides have become known, demonstrating the wide scope of this reaction (271, 272).

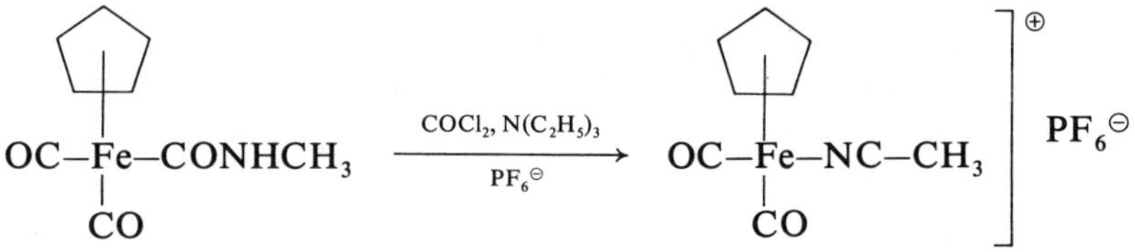

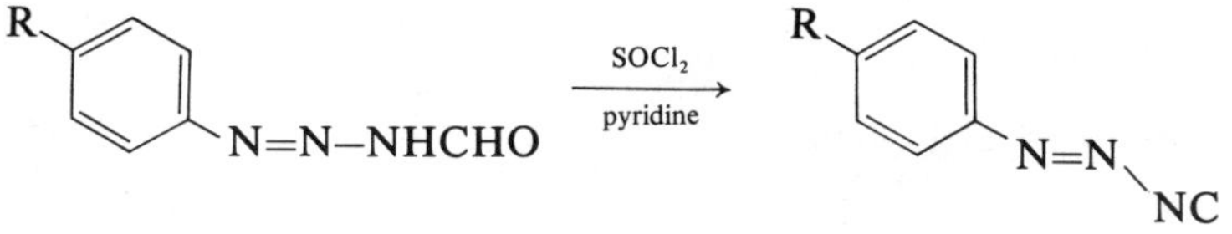

$$\text{OC}-\underset{\text{CO}}{\overset{\text{Cp}}{\text{Fe}}}-\text{CONHCH}_3 \quad \xrightarrow[\text{PF}_6^{\ominus}]{\text{COCl}_2,\ N(C_2H_5)_3} \quad \left[\text{OC}-\underset{\text{CO}}{\overset{\text{Cp}}{\text{Fe}}}-\text{NC}-\text{CH}_3\right]^{\oplus} \text{PF}_6^{\ominus}$$

$$R-\text{C}_6H_4-N=N-NHCHO \quad \xrightarrow[\text{pyridine}]{\text{SOCl}_2} \quad R-\text{C}_6H_4-N=N-NC$$

Investigations concerning the extent of racemization in the course of synthesis of α-isocyanocarboxylic acid derivatives from *N*-formyl-carboxylic acid derivatives by the phosgene method have been performed (273). Recently it was shown that in isonitrile synthesis phosgene can be replaced by diphosgene (chloroformic acid–trichloromethylester), which can be handled more safely and easier than phosgene (274).

$$2\ \text{RNCHO} + \text{Cl}-\overset{\overset{O}{\|}}{C}-\text{OCCl}_3 + 4\ N(C_2H_5)_3 \longrightarrow$$
$$2\text{RNC} + 2\text{CO}_2 + 4\overset{\oplus}{\text{HN}}(C_2H_5)_3\ \text{Cl}^{\ominus}$$

To Section III-E-2

Dehydration of primary carboxylic acid amides by halogenating reagents like PCl_5, $POCl_3$, $COCl_2$, $ClCO_2CCl_3$ (dimeric phosgene), oxalylchloride, $SOCl_2$ in the presence of tertiary amine, proceeds very likely via amide halogenides as reactive intermediates. The reaction is of wide scope and has been reviewed (275, 276).

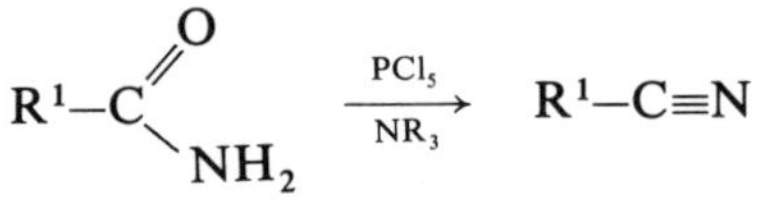

$$R^1-C(=O)NH_2 \xrightarrow{PCl_5, NR_3} R^1-C\equiv N$$

Further examples for preparation of α-chloroenamines **262**—which are precursors for ketene iminium salts—by hydrogen chloride abstraction from chloromethylene iminium salts by means of tertiary amines have been reported (221, 223). Treatment of iminium salts **263** with KCN in acetonitrile or $Zn(CN)_2$/triethylamine in chloroform, respectively, produces α-cyanoenamines (**264**) (221).

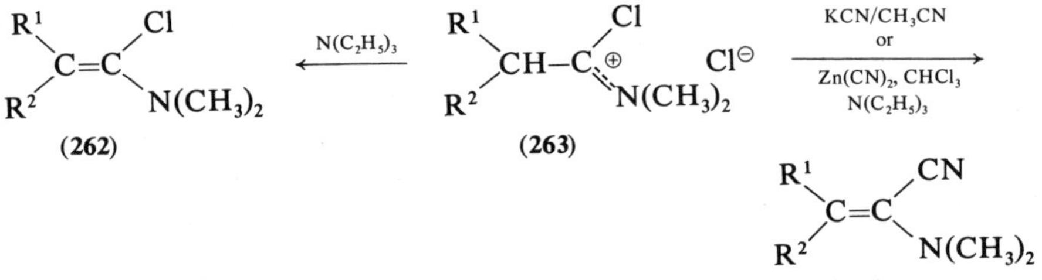

$$\underset{(262)}{R^1R^2C=C(Cl)N(CH_3)_2} \xleftarrow{N(C_2H_5)_3} \underset{(263)}{R^1R^2CH-C^{\oplus}(Cl)N(CH_3)_2 \; Cl^{\ominus}} \xrightarrow[N(C_2H_5)_3]{KCN/CH_3CN \text{ or } Zn(CN)_2, CHCl_3} \underset{(264)}{R^1R^2C=C(CN)N(CH_3)_2}$$

By action of triethylamine on **265** the iminium salt **266** can be obtained (243).

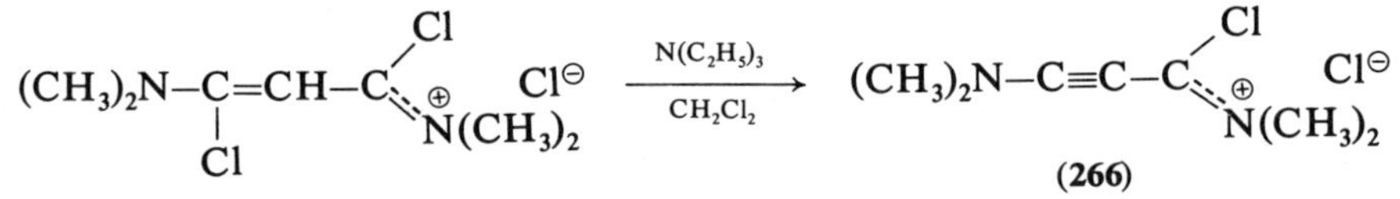

$$\underset{(265)}{(CH_3)_2N-C(Cl)=CH-C^{\oplus}(Cl)N(CH_3)_2 \; Cl^{\ominus}} \xrightarrow[CH_2Cl_2]{N(C_2H_5)_3} \underset{(266)}{(CH_3)_2N-C\equiv C-C^{\oplus}(Cl)N(CH_3)_2 \; Cl^{\ominus}}$$

To Section III-F

Experimental details for the synthesis of amideacetals by action of alkoxides on chloroform may be found in ref. 277. The reaction between the α,α-dibromoaziridine (**234**) and sodium methoxide yields the α-methoxyiminoester (**267**) (249).

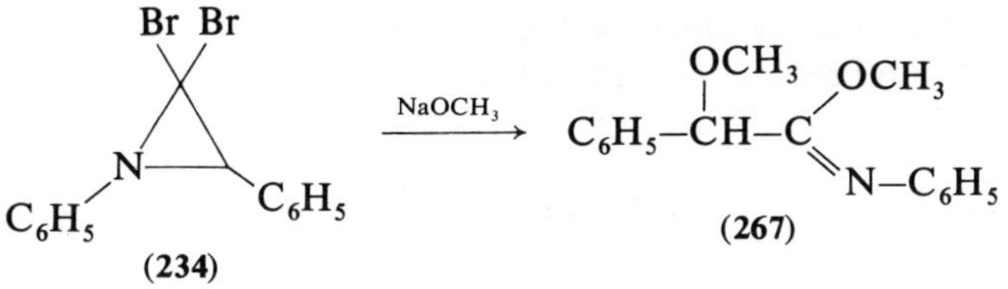

[Structure **(234)** with Br, Br, N-C₆H₅, C₆H₅] →(NaOCH₃)→ $C_6H_5-CH(OCH_3)-C(OCH_3)=N-C_6H_5$ **(267)**

To Section III-J

Treatment of chloromethylene iminium chlorides **(268)** with a mixture of HCN/NaCN (278) or with cuprous I cyanide (279) affords dialkylamino-malondinitriles **(269)**.

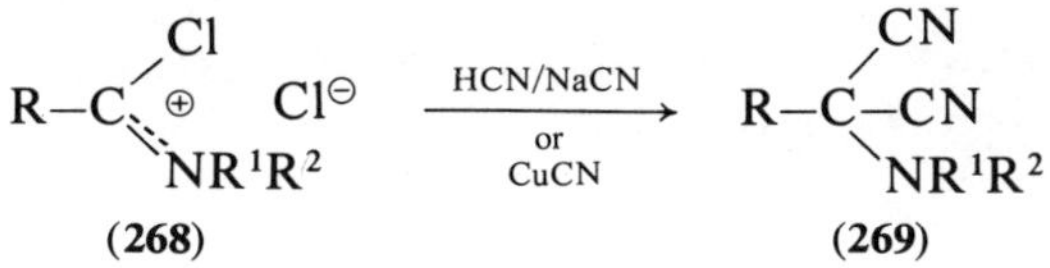

$R-C(Cl)=NR^1R^2 \; Cl^{\ominus}$ **(268)** →(HCN/NaCN or CuCN)→ $R-C(CN)_2-NR^1R^2$ **(269)**

To Section III-J-2

Lactames can be chlorinated in α-position by means of POCl₃/PCl₅, SO₂Cl₂ (280).

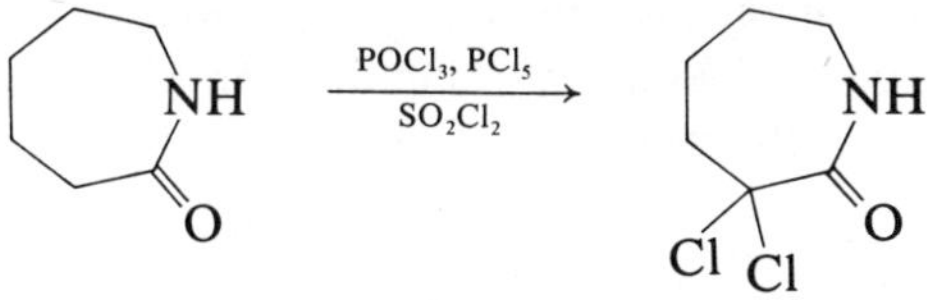

[caprolactam] →(POCl₃, PCl₅, SO₂Cl₂)→ [α,α-dichlorocaprolactam]

Action of ClF/CsF on **270** gives **271** (219).

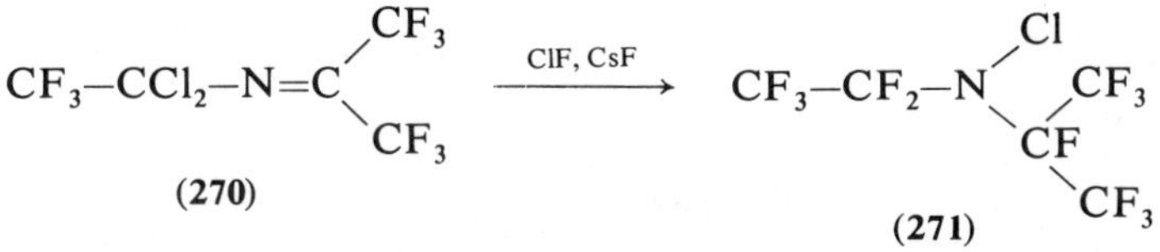

$CF_3-CCl_2-N=C(CF_3)_2$ **(270)** →(ClF, CsF)→ $CF_3-CF_2-N(Cl)-CF(CF_3)_2$ **(271)**

To Section III-L

Thermolysis of dihaloaziridines proceeds via ring opening to yield α-halo-imidhalogenides **(272)** (87, 235, 281–283).

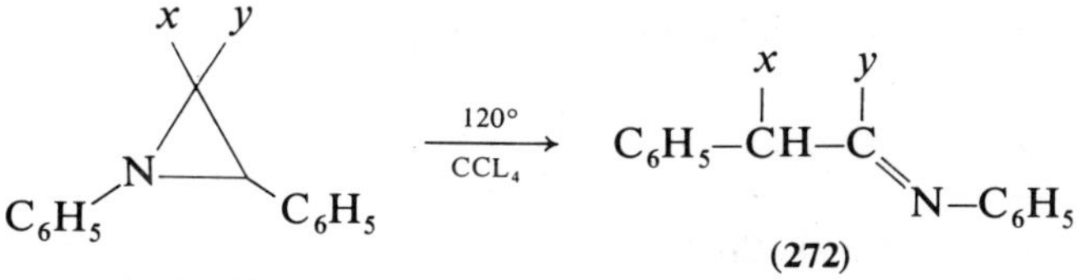

[aziridine with x, y, N-C₆H₅, C₆H₅] →(120°, CCl₄)→ $C_6H_5-CH(x)-C(y)=N-C_6H_5$ **(272)**

x, y = F, Cl, Br

The salts **273** form pyrimidines **274** on heating (247), whereas the salts **275** give rise to chloroformamidines **276** (244).

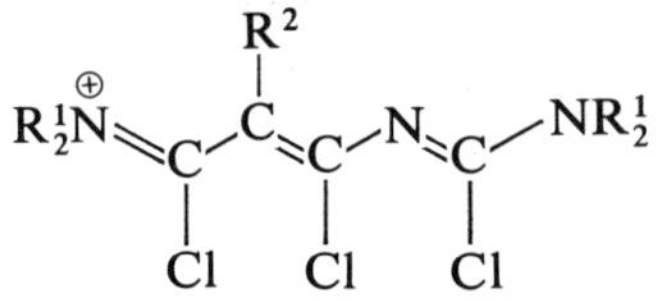

$$R^1_2\overset{\oplus}{N}=\underset{Cl}{\underset{|}{C}}-\underset{Cl}{\underset{|}{\overset{R^2}{\underset{|}{C}}}}=\underset{Cl}{\underset{|}{C}}-N=\underset{}{C}-NR^1_2 \quad \xrightarrow{\Delta} \quad$$

(273)

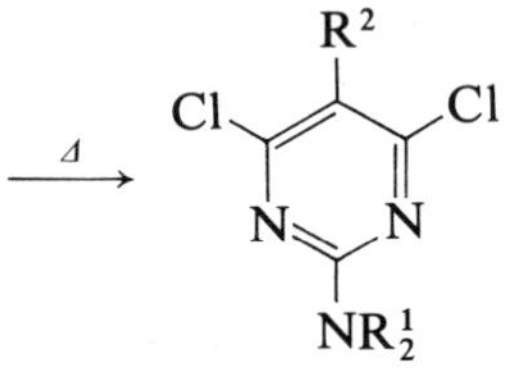

Cl, R², Cl pyrimidine with NR¹₂
(274)

$$R^1-CH=\underset{\underset{CH_3}{\overset{\oplus}{|}}}{\overset{Cl}{\underset{|}{C}}}-N\overset{Cl}{\underset{N(CH_3)_2}{=}}\underset{}{C} \quad Cl^{\ominus} \quad \longrightarrow \quad R^1-CH=\overset{Cl}{\underset{|}{C}}-N=\overset{Cl}{\underset{|}{C}}-N(CH_3)_2$$

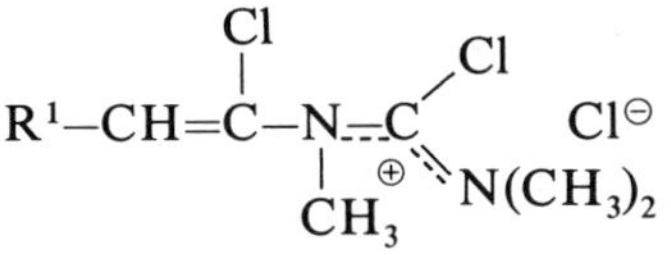

(275) 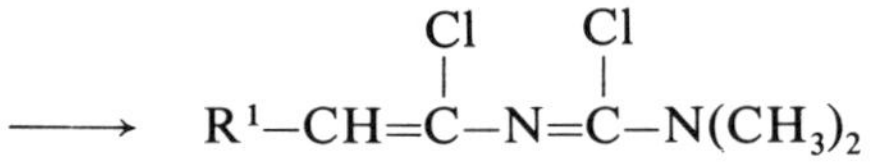   (276)

Section III-M

Reduction of chloromethylen-iminium salts; elimination reactions. It is very likely that in aldehyde synthesis according to Sonn-Müller or Stephen iminium salts **277** and **278** are the reactive intermediates. Both reactions have been reviewed (284–287).

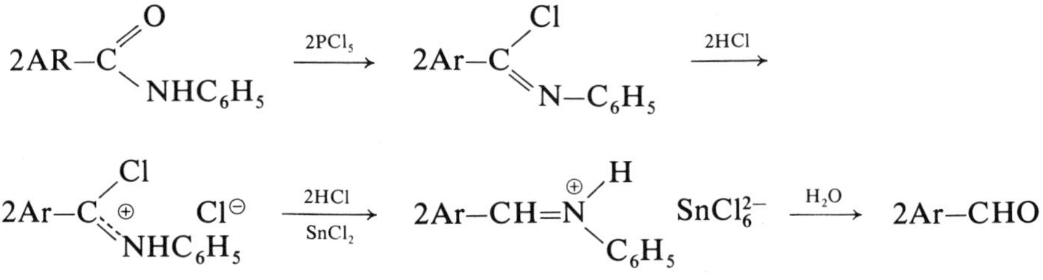

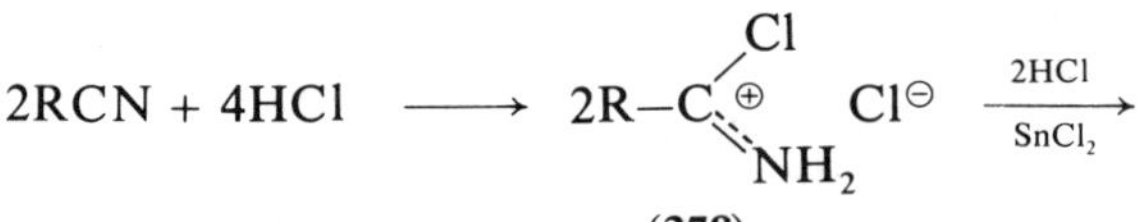

$$2AR-C\overset{O}{\underset{NHC_6H_5}{\diagdown}} \quad \xrightarrow{2PCl_5} \quad 2Ar-C\overset{Cl}{\underset{N-C_6H_5}{\diagdown}} \quad \xrightarrow{2HCl}$$

$$2Ar-\overset{\oplus}{C}\overset{Cl}{\underset{NHC_6H_5}{\diagdown}} \quad Cl^{\ominus} \quad \xrightarrow[SnCl_2]{2HCl} \quad 2Ar-CH=\overset{\oplus}{N}\overset{H}{\underset{C_6H_5}{\diagdown}} \quad SnCl_6^{2-} \quad \xrightarrow{H_2O} \quad 2Ar-CHO$$

$$2RCN + 4HCl \quad \longrightarrow \quad 2R-\overset{\oplus}{C}\overset{Cl}{\underset{NH_2}{\diagdown}} \quad Cl^{\ominus} \quad \xrightarrow[SnCl_2]{2HCl}$$

(278)

$$(R-CH=NH_2)_2S_nCl_6^{2-} \quad \xrightarrow{H_2O} \quad 2R-C\equiv N$$

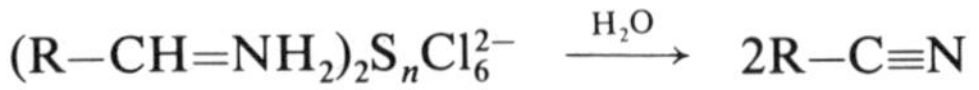

Reduction of **234** with LiAlH₄ yields the amine **279**, whereas reduction of **234** with Zn in acetonitrile affords the amide **280** (249).

$C_6H_5NH(CH_2)_2C_6H_5$
(279)

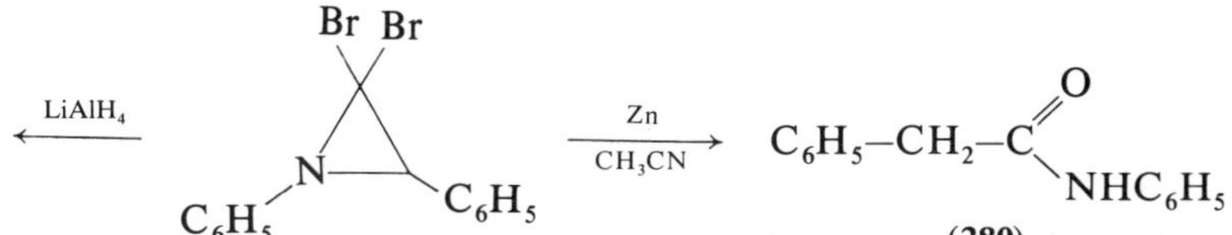

$\xleftarrow{\text{LiAlH}_4}$ [aziridine **(234)**: N-C_6H_5, with Br, Br at top carbon, C_6H_5 at other carbon] $\xrightarrow[\text{CH}_3\text{CN}]{\text{Zn}}$ $C_6H_5-CH_2-\overset{\displaystyle O}{\underset{\displaystyle NHC_6H_5}{C}}$
(280)

By means of tributyltin hydride mixed substituted dihalogenaziridines **281** can be reduced to monohalogenaziridines (236). (The reaction has been shown to be in some cases stereospecific.)

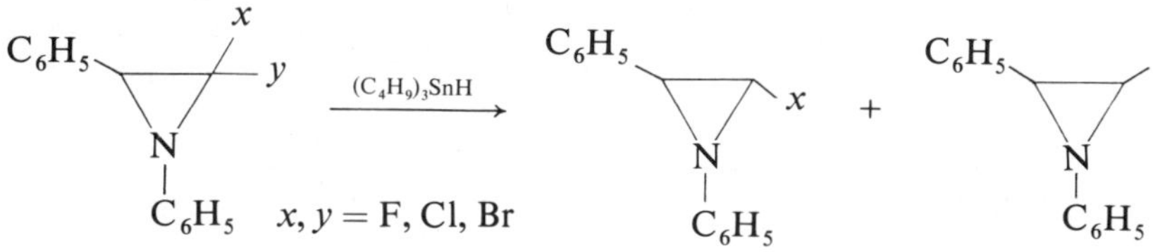

[aziridine **(281)** with C_6H_5 and x,y substituents on carbon, N-C_6H_5; x,y = F, Cl, Br] $\xrightarrow{(C_4H_9)_3\text{SnH}}$ [aziridine with C_6H_5 and x] + [aziridine with C_6H_5 and x]

Treatment of the bis-dichloroaziridine **282** with NaJ in acetone or Zn/HCl in THF, respectively, produces the bis-keteneimine **283** in good yields (240).

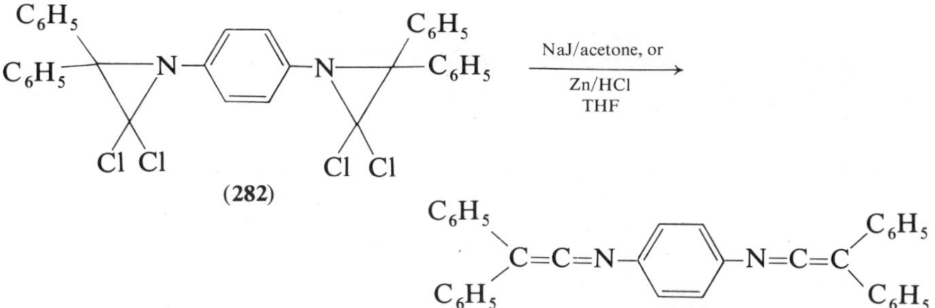

[structure **(282)**: two aziridine rings connected through N–C$_6$H$_4$–N phenylene linker, each aziridine bearing two C_6H_5 groups and two Cl atoms]

$\xrightarrow[\text{THF}]{\text{NaJ/acetone, or Zn/HCl}}$

$\underset{C_6H_5}{\overset{C_6H_5}{>}}C{=}C{=}N{-}\!\!\bigcirc\!\!{-}N{=}C{=}C\underset{C_6H_5}{\overset{C_6H_5}{<}}$
(283)

REFERENCES

219. K. E. Petermann and J. M. Shreeve, *Inorg. Chem.*, **14**, 1223 (1975).
220. F. Alhaique, R. G. E. Cozzani, and F. M. Riccieri, *Farm. Ed. Sci.*, **31**, 845 (1976); *C. I.*, 16–226 (1977).
221. J. Toye and L. Ghosez, *J. Amer. Chem. Soc.*, **97**, 2276 (1975).
222. D. Habeck and W. J. Houlihan, *J. Heterocycl. Chem.*, **13**, 897 (1976).
223. P. R. Ortiz de Montellano and R. Castillo, *Tetrahedron Lett.*, **1976**, 4115.

224. Yu. Yagupol'skii, B. K. Kerzhner, and L. M. Yagupol'skii, *Zh. Org. Khim.*, **12,** 2213 (1976); *C. I.*, 7–119 (1977).
225. M. L. Martin, G. Ricolleau, S. Poignant, and G. Martin, *J. Chem. Soc. Perkin II.*, **1976,** 182.
226. H. Ahlbrecht and C. Vonderheid, *Chem. Ber.*, **108,** 2300 (1975).
227. R. Weber and M. Renson,. *Bull. Soc. Chim. Fr.*, **1976,** 1124.
228. I. T. Kay and I. T. Streeting, *Synthesis*, **1976,** 38.
229. M. Golfier and M. G. Guillerez, *Tetrahedron Lett.*, **1976,** 267.
230. L. R. Morris and L. R. Collins, *J. Heterocycl. Chem.*, **12,** 309 (1975).
231. J. A. Deyrup and H. L. Gingrich, *J. Org. Chem.*, **42,** 1015 (1977).
232. V. I. Shvedov, V. K. Vasil'eva, I. A. Kharizomenova, and A. N. Grinev, *Khim. Geterotsikl. Soedin*, **1975,** 769; *C. I.*, 39–196 (1975).
233. E. Allenstein and K. Löhmar, *Z. Anorg. Allg. Chem.*, **420,** 193 (1976).
234. M. K. Meilahn, L. L. Augstein, and J. L. McManaman, *J. Org. Chem.*, **36,** 3627 (1971).
235. R. R. Kostikov, A. F. Khlebnikov, and K. A. Ogloblin, *Dokl. Akad. Nauk. SSSR, Ser. Khim.*, **223,** 1375 (1975); *C. I.*, 2–99 (1976).
236. H. Yamanaka, J. Kikui, K. Teramura, and T. Ando, *J. Org. Chem.*, **41,** 3794 (1976).
237. M. Furukawa, T. Okawara, and Y. Terawaki, *Chem. Pharm. Bull.*, **25,** 181 (1977).
238. J. Graefe, *Z. Chem.*, **14,** 469 (1974).
239. M. Makosza and A. Kacprowizc, *Rocz. Chem.*, **48,** 2195 (1974); *C. I.*, 12–248 (1975).
240. D. J. Sikkema, E. Molenaar, and D. B. van Guldener, *Recl. Trav. Chim. Pay-Bas*, **95,** 154 (1976).
241. J. S. Baum and H. G. Viehe, *J. Org. Chem.*, **41,** 183 (1976).
242. H. G. Viehe, G. J. de Voghel, and F. Smets, *Chimia*, **30,** 189 (1976).
243. B. Caillaux, P. George, F. Tataruch, Z. Janousek, and H. G. Viehe, *Chimia*, **30,** 387 (1976).
244. E. Goffin, Y. Legrand, and H. G. Viehe, *J. Chem. Res. S*, **1977,** 105.
245. M. Huys-Francotte, Z. Janousek, and H. G. Viehe, *J. Chem. Res. S.*, **1977,** 100.
246. H. G. Viehe, B. Le Clef, and A. Elgavi, *Angew. Chem.*, **89,** 189 (1977).
247. B. Stelander and H. G. Viehe, *Angew. Chem.*, **89,** 182 (1977).
248. G. Jugie, J. A. S. Smith, G. J. Gérard, *J. Chem. Soc. Perkin II*, **1975,** 925.
249. R. R. Kostikov, A. F. Klebnikov, and K. A. Ogloblin, *Zh. Org. Khim.*, **11,** 585 (1975); *C. I.*, 23–281 (1975).
250. D. R. Hepburn and H. R. Hudson, *J. Chem. Soc. Perkin I*, **1976,** 754.
251. K. Morita, S. Noguchi, and M. Nishikawa, *Chem. Pharm. Bull. (Tokyo)*, **7,** 896 (1959).
252. I. Hoppe and U. Schöllkopf, *Chem. Ber.*, **109,** 482 (1976).
253. S. Brenner and H. G. Viehe, *Tetrahedron Lett.*, **1976,** 1617.
254. Y. Furukawa, O. Miyashita, and S. Shima, *Chem. Pharm. Bull.*, **24,** 970 (1976); *C. I.*, 36–236 (1976).
255. M. Kim-Su, K. Eger, and H. Roth, *Arch. Pharm. (Weinheim)*, **309,** 729 (1976).
256. L. N. Markowskii, Yu. G. Shermolovich, V. I. Gorbatenko, and V. I. Shevchenko, *Zh. Org. Khim.*, **11,** 751 (1975).
257. G. J. de Voghel, T. L. Eggerichs, B. Clamot, and H. G. Viehe, *Chimia*, **30,** 191 (1976).
258. V. P. Kukhar and T. N. Kasheva, *Zh. Obshch. Khim.*, **46,** 243 (1976); *C. I.*, 21–284 (1976).
259. R. G. Makitra, R. V. Sendega, and M. N. Seredrnitskaya, *Zh. Org. Khim.*, **11,** 655 (1975); *C.I.*, 24–204 (1975).
260. B. Föhlisch, R. Braun, and K. W. Schultze, *Angew. Chem.*, **79,** 318 (1967).
261. J. Liebscher and H. Hartmann, *Z. Chem.*, **15,** 16 (1975).
262. J. Liebscher and H. Hartmann, *Synthesis*, **1976,** 403.

263. J. Liebscher and H. Hartmann, *Collect. Czech. Chem. Commun.*, **41,** 1565 (1976).
264. J. Liebscher and H. Hartmann, *Z. Chem.*, **15,** 438 (1975).
265. J. Liebscher and H. Hartmann, *Synthesis*, **1976,** 273.
266. J. Liebscher and H. Hartmann, *Synthesis*, **1976,** 521.
267. J. Liebscher and H. Hartmann, *Tetrahedron Lett.*, **1976,** 2005.
268. J. Liebscher and H. Hartmann, *Z. Chem.*, **16,** 268 (1976).
269. J. Liebscher and H. Hartmann, *Tetrahedron*, **33,** 731 (1977).
270. J. Liebscher and H. Hartmann, *Tetrahedron Lett.*, **1975,** 2977.
271. W. P. Fehlhammer and A. Mayr, *Angew. Chem.*, **87,** 776 (1975).
272. T. Ignasiak, J. Suszuko, and B. Ignasiak, *J. Chem. Soc. Perk. I*, **1975,** 2122.
273. R. Urban, D. Marquarding, P. Siedel, I. Ugi, and A. Weinelt, *Chem. Ber.*, **110,** 2012 (1977).
274. G. Skorna and I. Ugi, *Angew. Chem.*, **89,** 267 (1977).
275. P. Kurtz, in Houben-Weyl-Müller, *Methoden der Orgaurschen Chemie*, Vol. 8, Georg Thieme Verlag, Stuttgart, 1952, p. 330.
276. A. Zobáčová, in F. Korte, Ed., Vol. 6, *Methodicum Chimicum*, Georg Thieme Verlag, Stuttgart, 1974, p. 654.
277. W. Stilz, Ger. Pat. 1,161,285 (1964); *C. A.*, **60,** 9156 (1964).
278. Z. Arnold, *Collect. Czech. Chem. Commun.*, **26,** 1113 (1961).
279. Z. Arnold and M. Svoboda, *Collect. Czech. Chem. Commun.*, **42,** 1175 (1977).
280. M. Brenner, H. R. Rickenbacher, *Helv. Chim. Acta*, **41,** 181 (1958).
281. H. W. Heine and A. B. Smith, *Angew. Chem.*, **75,** 669 (1963).
282. R. E. Brooks, J. O. Edwards, G. Levey, and F. Smith, *Tetrahedron*, **22,** 1279 (1966).
283. K. Ichimura and M. Otah, *Bull. Chem. Soc. Japan*, **40,** 1933 (1967).
284. E. Mosettig, in R. Adams, Ed., *Organic Reactions*, Vol. 8, John Wiley and Sons, New York, 1954, pp. 240, 246.
285. O. Bayer, in Houben–Weyl–Müller, *Methoden der organischen Chemie*, Vol. 7/1, Georg Thieme Verlag, Stuttgart, 1954.
286. C. A. Buehler and D. E. Pearson, *Survey of Organic Synthesis*, Wiley–Interscience, New York, 1970, pp. 574, 575.
287. H. Meyer, in J. Falbe, Ed., *Methodicum Chimicum, C–O Verbindungen*, Vol. 5, Georg Thieme Verlag, Stuttgart, 1975, p. 306.

CHLORODIALKYLAMINOMETHYLENIMINIUM SALTS–CHLOROFORMAMIDINIUM SALTS

By W. KANTLEHNER, *D-7080-AALEN-Dewangen, Laachweg 14, Germany*

CONTENTS

I. Synthesis	144
A. Via Halogenation of Ureas	144
1. With Phosgene or Carbonyl Difluoride	144
2. With Phosphorus Pentachloride, Thionyl Chloride, Oxalyl Chloride	145
B. Via Halogenation of Thioureas	146
1. With Phosgene	146
2. With Oxalyl Chloride or Chlorine	147
C. From Carbodiimides and Cyanamides	148
1. Via Addition of Hydrogen Halides to Carbodiimides	148
2. From Carbodiimides and Oxalyl Chloride	148
3. From Cyanamides with Phosgene or Sulfur Dichloride	148
D. From Dichloromethyleniminium Salts	149
II. Structure and Properties of Chloroformamidinium Chlorides	150
III. Reactions of Chloroformamidinium Chlorides	152
A. With Hydroxyl Compounds	152
1. Hydrolysis	152
2. Alcoholysis	153
3. Reactions with Alkoxides	154
B. With SH-Containing Compounds	155
1. With Sodium Hydrogen Sulfide	155
2. With Thiophenols and Thiophenoxides	156
C. With NH-Containing Compounds	157
D. With Organolithium Compounds	160
E. With Organophosphorous Compounds	161
F. Elimination Reactions	162
G. Miscellaneous Reactions	163
References	164
Addendum	165

Translated from German by Dr. W.-J. Richter, MPI für Kohleforschung, D 4330 Mühlheim.

I. Synthesis

A. VIA HALOGENATION OF UREAS

1. With Phosgene or Carbonyl Difluoride

N,N,N',N'-Tetrasubstituted as well as N,N,N'-trisubstituted ureas (**1, 2**) smoothly react with phosgene via the primary O-acylation products (**3**) to yield chloroformamidinium chlorides (**4, 5**) (1–3).

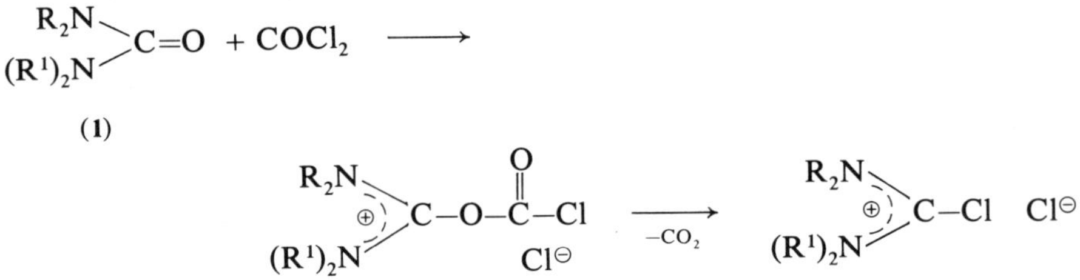

$$\begin{array}{c} R_2N \\ (R^1)_2N \end{array}\!\!\!>\!\!C=O + COCl_2 \longrightarrow$$

(**1**)

$$\begin{array}{c} R_2N \\ (R^1)_2N \end{array}\!\!\!\overset{\oplus}{>}\!\!C-O-\overset{O}{\underset{Cl^{\ominus}}{C}}-Cl \xrightarrow{-CO_2} \begin{array}{c} R_2N \\ (R^1)_2N \end{array}\!\!\!\overset{\oplus}{>}\!\!C-Cl \quad Cl^{\ominus}$$

(**3**)   (**4**)

$$\begin{array}{c} R_2N \\ Ar-N \\ H \end{array}\!\!\!>\!\!C=O \xrightarrow[-CO_2]{+COCl_2} \begin{array}{c} R_2N \\ Ar-N \\ H \end{array}\!\!\!\overset{\oplus}{>}\!\!C-Cl \quad Cl^{\ominus}$$

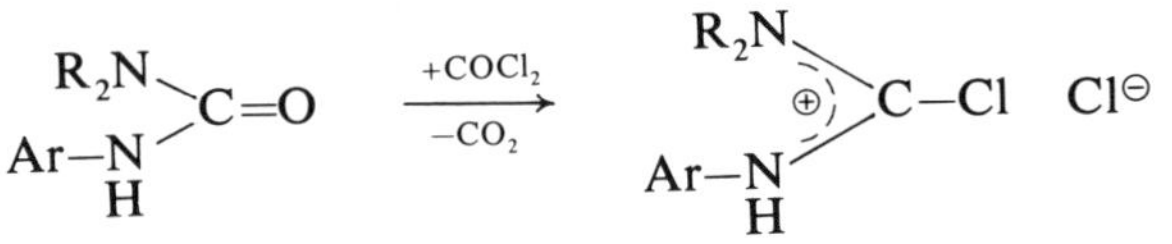

(**2**)   (**5**)

Bis(dimethylamino)difluoromethane (**5a**) was synthesized from N,N,N',N'-tetramethylurea and carbonyl difluoride (4, 5).

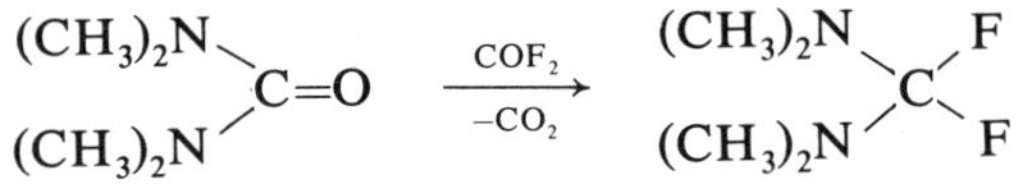

$$\begin{array}{c} (CH_3)_2N \\ (CH_3)_2N \end{array}\!\!\!>\!\!C=O \xrightarrow[-CO_2]{COF_2} \begin{array}{c} (CH_3)_2N \\ (CH_3)_2N \end{array}\!\!\!>\!\!C\!\!\!<\!\!\!\begin{array}{c} F \\ F \end{array}$$

(**5a**)

Bis(difluoroamino)difluoromethane (**6**) is obtained by fluorination of cyanuric chloride (6) or aminoiminomethanesulfinic acid (**7**) (7).

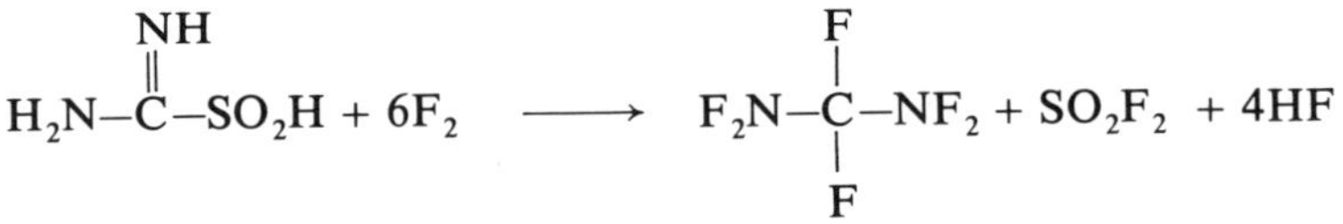

$$\underset{(\mathbf{7})}{H_2N-\overset{NH}{\overset{\|}{C}}-SO_2H} + 6F_2 \longrightarrow \underset{(\mathbf{6})}{F_2N-\overset{F}{\underset{F}{\overset{|}{C}}}-NF_2} + SO_2F_2 + 4HF$$

N,N'-Dialkylated ureas (**8**) react predominantly with phosgene to give chloroformamidinium chlorides (**9**) (~70% yield), if secondary or tertiary alkyl groups are bound to nitrogen (1, 2, 8–11). In addition, acylation at the nitrogen atom takes place to a lesser degree to give **10** from **8**. If the nitrogen atoms of

I. SYNTHESIS

$$\underset{(8)}{\begin{array}{c}R-N^H\\R-N_H\end{array}C=O} \xrightarrow{COCl_2} \underset{(9)}{\begin{array}{c}R-N^H\\R-N_H\end{array}\overset{\oplus}{C}-Cl} \quad Cl^\ominus + \underset{(10)}{R-\underset{H}{N}-\overset{O}{\underset{\|}{C}}-\underset{R}{N}-\overset{O}{\underset{\|}{C}}-Cl}$$

the urea **8** carry only primary alkyl groups, **10** is the principal product (~70% yield) (9). Besides the type of substituents, the solvent also determines the product ratio of **9** and **10** when urea **8** is phosgenated. **10** is formed preferentially in nonpolar solvents, whereas polar solvents favor the formation of chloroformamidinium chlorides (**9**). When N,N'-di-n-butylurea reacts with phosgene in benzene, N,N'-di-n-butylallophanyl chloride forms in nearly 100% yield, as opposed to 48.2% in chloroform plus 47.9% N,N'-dibutylchloroformamidinium chloride (9).

Cyclic ureas (**11**) like ethylene- or propyleneurea exclusively yield allophanyl chlorides (**12**) upon phosgenation (9).

$$\underset{(11)}{\begin{array}{c}R\\ \diagup(CH_2)_n\\HN\diagdown_{\underset{\|}{C}}\diagup NH\\O\end{array}} \xrightarrow[-HCl]{+COCl_2} \underset{(12)}{\begin{array}{c}R\\ \diagup(CH_2)_n\\HN\diagdown_{\underset{\|}{C}}\diagup N-\overset{O}{\underset{\|}{C}}-Cl\\O\end{array}}$$

N,N'-Diarylureas as well as N,N-disubstituted ureas are not transformed into chloroformamidinium chlorides by phosgene (1, 2). With N,N-dimethylurea phosgene attacks at the nitrogen atom (12). Urea itself does not react with phosgene to give chloroformamidinium chloride, but rather carbonyl diurea (13). Phosgene acylates N-hydroxy-N,N',N'-trimethylurea at the hydroxyl group (14).

2. *With Phosphorus Pentachloride, Thionyl Chloride, Oxalyl Chloride*

With phosphorus pentachloride, N,N-disubstituted as well as N,N,N'-trisubstituted ureas are also converted into chloroformamidinium chlorides (8, 10, 11, 15). However, when N,N'-dialkylureas (**8**) are reacted with phosphorus pentachloride, four- and six-membered phosphorus-containing heterocycles like **13** or **14** form as by-products (8, 10, 11, 16), since phosphorus pentachloride also attacks at the nitrogen atom of **8**. Thionyl chloride transforms N,N'-dialkylureas (**8**) into chloroformamidinium chlorides (**9**), if secondary or tertiary alkyl groups are the substituents at the nitrogen atom; otherwise N-sulfinylamines and isocyanates form (11). Oxalyl chloride is said to convert substituted ureas into chloroformamidinium chlorides (1), but

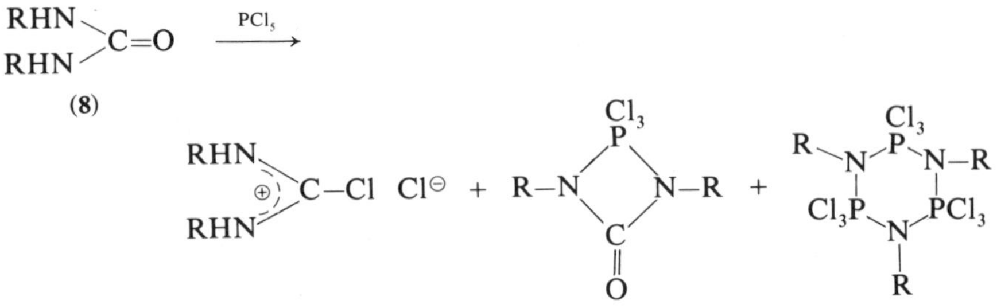

$$\begin{array}{c}\text{RHN}\\\text{RHN}\end{array}\!\!\!\!\!\!\!\!\text{C=O} \quad \xrightarrow{\text{PCl}_5}$$

(8)

$$\begin{array}{c}\text{RHN}\\\text{RHN}\end{array}\!\!\!\!\!\!\!\!\overset{\oplus}{\text{C}}\text{--Cl} \;\; \text{Cl}^{\ominus} \;+\; \text{[R–N–P(Cl}_3\text{)–N–R–C(=O)] ring} \;+\; \text{[R–N–P(Cl}_3\text{)–N–R–P(Cl}_3\text{)–N(R)] ring}$$

(9) (13) (14)

parabanic acids are the sole reaction products, if other than N,N,N',N'-tetraalkylated ureas are reacted with oxalyl chloride (11, 17–19).

B. VIA HALOGENATION OF THIOUREAS

1. With Phosgene

Phosgene primarily acylates thioureas (15) at the sulfur atom; here isolable adducts of the type 16 form (2), which then give chloroformamidinium chlorides (4) via elimination of carbon oxysulfide (1, 2, 8–11).

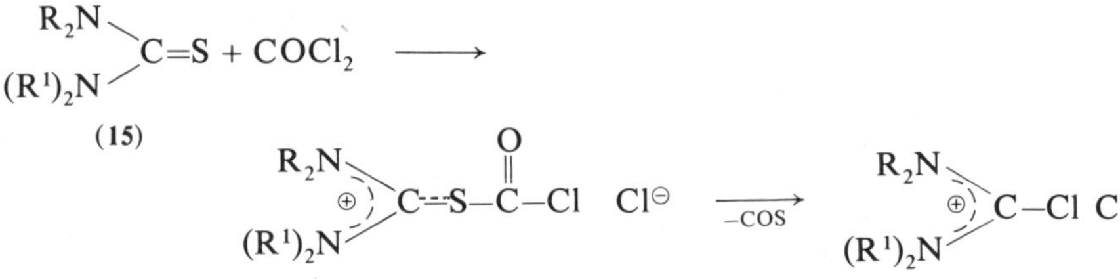

$$\begin{array}{c}\text{R}_2\text{N}\\(\text{R}^1)_2\text{N}\end{array}\!\!\!\!\!\!\!\!\text{C=S} + \text{COCl}_2 \longrightarrow$$

(15)

$$\begin{array}{c}\text{R}_2\text{N}\\(\text{R}^1)_2\text{N}\end{array}\!\!\!\!\!\!\!\!\overset{\oplus}{\text{C}}\text{--S--C(=O)--Cl} \;\; \text{Cl}^{\ominus} \quad \xrightarrow{-\text{COS}} \quad \begin{array}{c}\text{R}_2\text{N}\\(\text{R}^1)_2\text{N}\end{array}\!\!\!\!\!\!\!\!\overset{\oplus}{\text{C}}\text{--Cl} \;\; \text{Cl}^{\ominus}$$

(16) (4)

N,N-Di, N,N,N'-tri-, and N,N,N',N'-tetrasubstituted thioureas also react successfully; N,N'-diarylthioureas behave similarly (1, 2, 11). Even N-t-butylthiourea has been transformed into the corresponding chloroformamidinium chloride (2). Owing to the stronger nucleophilicity of sulfur, the halogenation of thioureas generally proceeds more uniformly than that of ureas (1, 2).

When N,N'-disubstituted thioureas, especially N,N'-diarylureas, are phosgenated, an excess of phosgene must be avoided because it causes the formation of 1,3-thiazetidin-2-ones (17) (1, 2, 18). The compounds 17 have

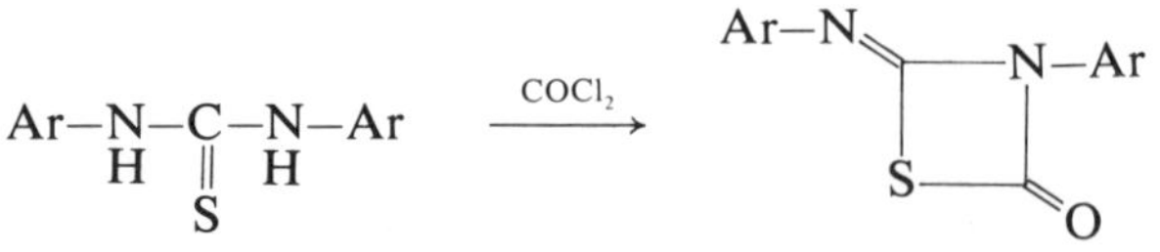

$$\text{Ar--NH--C(=S)--NH--Ar} \quad \xrightarrow{\text{COCl}_2} \quad \text{[Ar–N–C(=N–Ar)–S–C(=O) four-membered ring]}$$

(17)

been synthesized this way (20). Excess phosgene may also attack carbodiimides (**18**) generated from *N,N'*-disubstituted chloroformamidinium chlorides (**9**) via loss of HCl, giving *N*-chlorocarbonylchloroformamidin (**19**) (11, 21).

$$\begin{array}{c}\text{RHN} \\ {}_{\oplus}\!\!\searrow\!\!\text{C--Cl} \ \ \text{Cl}^{\ominus} \\ \text{RHN}\!\!\nearrow \end{array} \underset{-2\text{HCl}}{\rightleftarrows} \begin{array}{c} R \\ | \\ N \\ \| \\ C \\ \| \\ N \\ | \\ R \end{array} \xrightarrow{\text{COCl}_2} \begin{array}{c} O\!=\!C\!\!\diagdown\!\text{Cl} \\ | \\ R\!-\!N\!-\!C\!=\!N\!-\!R \\ | \\ \text{Cl} \end{array}$$

(**9**)   (**18**)   (**19**)

When *N*-alkyl-*N'*-arylsulfonylureas are phosgenated, *N*-alkyl-*N'*-arylsulfonylchloroformamidines (**20**) form, since the chloroformamidinium chlorides initially formed immediately lose HCl (11, 18, 19, 22).

$$\text{ArSO}_2\text{NH--C--NHR} \xrightarrow[-\text{COS},\,-\text{HCl}]{\text{COCl}_2}$$
    ‖
    S

$$\text{ArSO}_2\text{N}\!=\!\text{C}\!\!\diagdown\!\!\begin{array}{c}\text{NHR}\\ \text{Cl}\end{array} \xrightarrow[-\text{HCl}]{\Delta} \text{ArSO}_2\text{N}\!=\!\text{C}\!=\!\text{N--R}$$

     (**20**)   (**21**)

Chloroformamidines (**20**) lose HCl on heating, thus generating *N*-sulfonylcarbodiimides (**21**) (19).

2. *With Oxalyl chloride or Chlorine*

With oxalyl chloride, *N,N'*-dialkylated thioureas are not converted into chloroformamidinium chlorides. Instead, 2-imino-1,3-thiazolidine-4,5-diones (**22**) are formed, and they easily rearrange to thioparabanic acids (**23**) (11, 23).

$$\begin{array}{c}\text{RHN}\\ \searrow\!\!\text{C}\!=\!\text{S}\\ \text{RHN}\!\!\nearrow\end{array} + \begin{array}{c} O\!\!\diagdown\!\!\diagup\!O \\ C\!-\!C \\ \text{Cl}\!\!\diagup\!\!\diagdown\!\text{Cl}\end{array} \longrightarrow \underset{(\mathbf{22})}{\begin{array}{c}N\!-\!R\\ R\!-\!N\!\!\diagup\!\!\diagdown\\ O\!\!=\!\!\diagdown\!\!\diagup\!\!S\\ O\end{array}} \longrightarrow \underset{(\mathbf{23})}{\begin{array}{c}S\\ R\!-\!N\!\!\diagup\!\!\diagdown\\ O\!\!=\!\!\diagdown\!\!\diagup\!N\!-\!R\\ O\end{array}}$$

Elementary chlorine transforms ethylenethiourea (**24**) into the chloroformamidinium chloride (**25**). *N*-(2-Imidazolin-2-yl)ethylenethiourea (**26**) forms as a by-product (24).

$$\underset{(\mathbf{24})}{\begin{array}{c}\text{HN}\!\frown\!\text{NH}\\ \diagdown\!\!\diagup\\ \text{C}\\ \|\\ \text{S}\end{array}} \xrightarrow{\text{Cl}_2} \underset{(\mathbf{25})}{\begin{array}{c}\text{HN}\!\frown_{\!\!\oplus}\!\text{NH}\\ \diagdown\!\!\diagup\\ \text{C}\\ |\\ \text{Cl}\end{array}\text{Cl}^{\ominus}} \ + \ \underset{(\mathbf{26})}{\begin{array}{c}\text{N}\!\frown\\ \diagup\!\!\diagdown\\ \text{C}\text{N}\!\frown\\ \|\diagdown\!\!\diagup\\ \text{N}\text{C}\text{NH}\\ |\|\\ \text{H}\text{S}\end{array}}$$

Treating S-methyl-N,N,N'-trimethylisothiourea (**27**) with chlorine gives the chloroformamidinium chloride **28** (25).

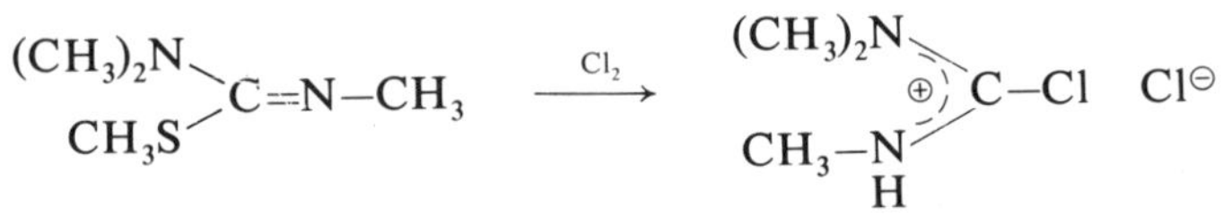

$$(CH_3)_2N\diagdown C=N-CH_3 \quad \xrightarrow{Cl_2} \quad (CH_3)_2N\diagdown \overset{\oplus}{C}-Cl \quad Cl^{\ominus}$$
$$CH_3S\diagup \qquad\qquad\qquad\qquad CH_3-\underset{H}{N}\diagup$$

(**27**)   (**28**)

C. FROM CARBODIIMIDES AND CYANAMIDES

1. Via Addition of Hydrogen Halides to Carbodiimides

Diphenylcarbodiimide (**29**) takes up 2 moles of hydrogen chloride to yield N,N'-diphenylchloroformamidinium chloride (**30**) (26). It has been shown that N,N'-dialkylcarbodiimides react the same way (9, 21).

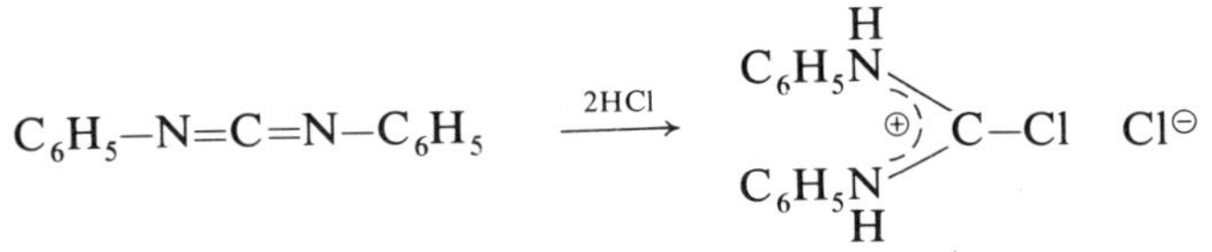

$$C_6H_5-N=C=N-C_6H_5 \quad \xrightarrow{2HCl} \quad \begin{array}{c} H \\ C_6H_5N \diagdown \\ \overset{\oplus}{C}-Cl \quad Cl^{\ominus} \\ C_6H_5N\diagup \\ H \end{array}$$

(**29**)   (**30**)

2. From Carbodiimides and Oxalyl Chloride

Carbodiimides (**18**) react with oxalyl chloride to give 2,2-dichloroimidazoline-4,5-diones (**31**), which are not ionized (11, 18, 23, 27, 28).

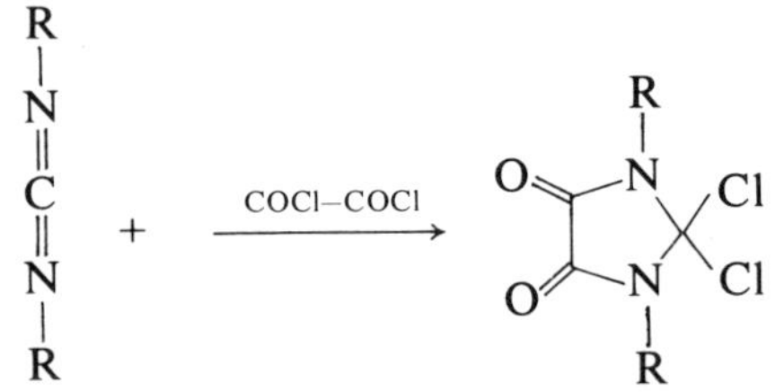

3. From Cyanamides and Phosgene or Sulfur Dichloride

Sulfur dichloride reacts with N,N-disubstituted cyanamides to yield chloroformamidinium chlorides (**32**), which are partly ionized (29).

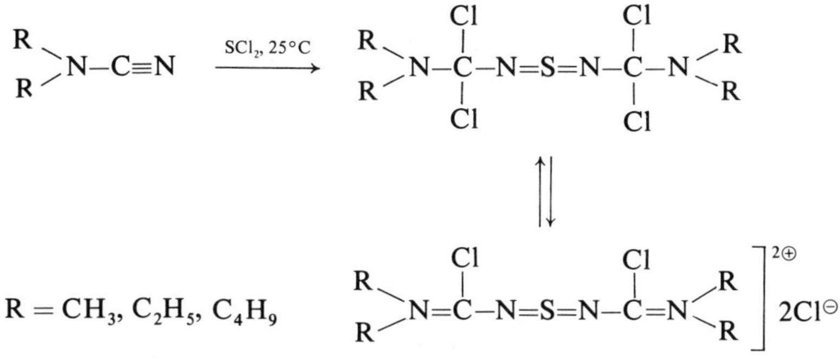

I. SYNTHESIS

$$R_2N-C\equiv N \xrightarrow{SCl_2,\ 25°C} R_2N-\underset{Cl}{\overset{Cl}{C}}-N=S=N-\underset{Cl}{\overset{Cl}{C}}-NR_2$$

$$\updownarrow$$

$$R = CH_3,\ C_2H_5,\ C_4H_9 \qquad \left[R_2N=\underset{Cl}{\overset{Cl}{C}}-N=S=N-\underset{Cl}{\overset{Cl}{C}}=NR_2 \right]^{2\oplus} 2Cl^{\ominus}$$

(32)

Phosgene converts *N,N*-disubstituted cyanamides into chloroformamidinium chlorides (**33**) (30, 31).

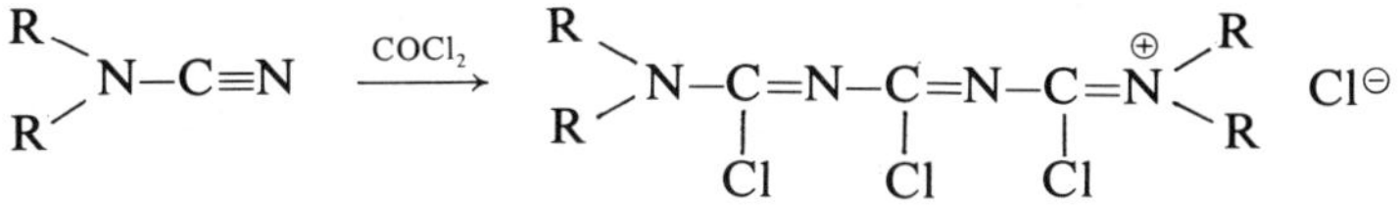

$$R_2N-C\equiv N \xrightarrow{COCl_2} R_2N-\underset{Cl}{\overset{}{C}}=N-\underset{Cl}{\overset{}{C}}=N-\underset{Cl}{\overset{}{C}}=\overset{\oplus}{N}R_2 \quad Cl^{\ominus}$$

(33)

D. FROM DICHLOROMETHYLENIMINIUM SALTS

When phosgeniminium salts like **34** react with hydrazones, the salts **35** are formed, which easily cyclize to pyrazoles (**36**) (32).

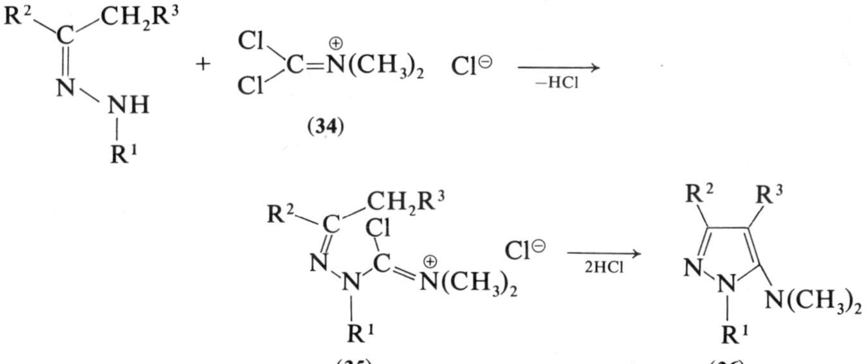

$$\underset{\underset{R^1}{|}}{\overset{R^2}{\underset{N-NH}{C=}}}\hspace{-2pt}CH_2R^3 \quad + \quad \underset{Cl}{\overset{Cl}{C}}=\overset{\oplus}{N}(CH_3)_2\ Cl^{\ominus} \xrightarrow{-HCl}$$

(34)

$$\underset{\underset{R^1}{|}}{\overset{R^2}{\underset{N-N}{C=}}}\hspace{-2pt}\overset{CH_2R^3}{\underset{C=\overset{\oplus}{N}(CH_3)_2}{\overset{|}{Cl}}}\ Cl^{\ominus} \xrightarrow{2HCl} \underset{\underset{R^1}{|}}{\overset{R^2\ \ R^3}{\underset{N-N}{\diagdown\diagup}}}N(CH_3)_2$$

(35) \hspace{80pt} (36)

Ammonia, primary and secondary amines, and secondary amides and *N,N*-disubstituted cyanamides are converted into chloroformamidinium salts (**4, 37, 38**) with the aid of phosgeniminium salts.

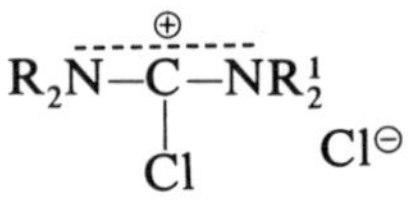

(4) R₂N–C(Cl)–NR'₂ ⊕ ··· Cl⁻

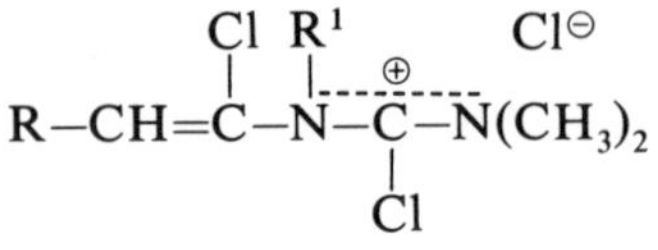

(37) R–CH=C(Cl)–N(R¹)–C(Cl)⊕–N(CH₃)₂ Cl⁻

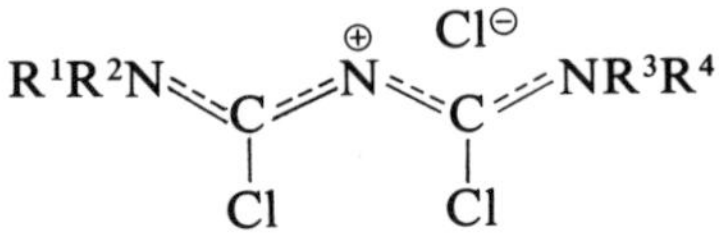

(38) R¹R²N=C(Cl)–N⊕=C(Cl)–NR³R⁴ Cl⁻

These reactions have been reviewed by Viehe (33); see also the chapter in Part 1 called "Chemistry of Dichloromethyleniminium Salts (Phosgeniminium Salts)." Dimethylformamide is converted into chloroformamidinium chloride (**39**) by **34** (34).

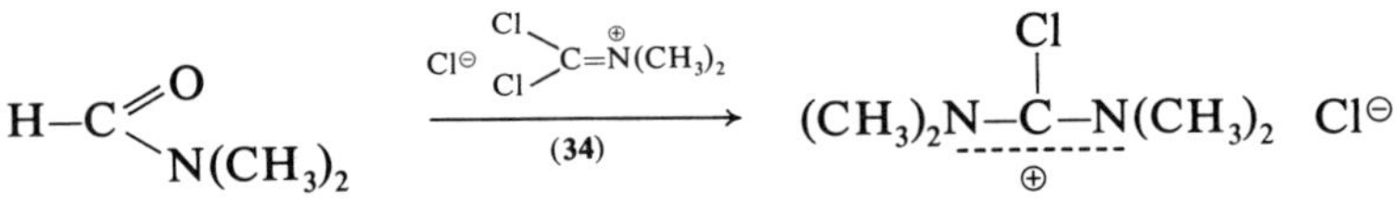

H–C(=O)–N(CH₃)₂ →[(34): Cl₂C=N⊕(CH₃)₂, Cl⁻] (CH₃)₂N–C(Cl)–N(CH₃)₂ ⊕ Cl⁻ **(39)**

Imide halides (**39a**) or iminium salts (**39b**) form from dichloromethyleniminium salts (**34a**) and N,N-dichlorourethane or methyl N-chloroiminobenzoate (34a).

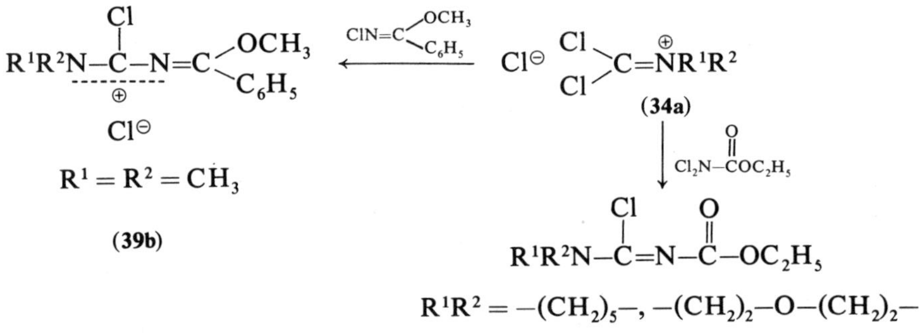

R¹R²N–C(Cl)–N=C(OCH₃)(C₆H₅) ⊕ Cl⁻ ← [ClN=C(OCH₃)(C₆H₅)] ← Cl₂C=N⊕R¹R² Cl⁻ **(34a)**

R¹ = R² = CH₃ **(39b)**

↓ Cl₂N–COC₂H₅

R¹R²N–C(Cl)=N–C(=O)–OC₂H₅

R¹R² = –(CH₂)₅–, –(CH₂)₂–O–(CH₂)₂– **(39a)**

II. Structure and Properties of Chloroformamidinium Chlorides

The relatively high melting points, the insolubility in nonpolar organic solvents, and the good solubility in water—occurring via hydrolysis, however—all point to a polar structure for the chloroformamidinium chlorides **4, 5**, and **9**.

II. STRUCTURE AND PROPERTIES OF CHLOROFORMAMIDINIUM CHLORIDES

All chloroformamidinium chlorides show an IR band between 5.9 and 6.2 μ attributed to the C=N⊕ double bond (1, 2, 9). Thus the chloroformamidinium chlorides may be regarded as completely ionized species. The NMR spectrum of N,N,N',N'-tetramethylchloroformamidinium chloride features a singlet at $\delta = 3.43$ (35). Compared with amide chlorides there are more possible resonance structures, which explains the decrease in electrophilicity of the central carbon atom; the decreased reactivity of the chloroformamidinium chlorides is also understandable.

$$\left[\begin{array}{c} R_2N \\ R_2N \end{array}\!\!\!\overset{\oplus}{>}\!C\!-\!Cl \longleftrightarrow \begin{array}{c} R_2\overset{\oplus}{N}\!\!= \\ R_2N \end{array}\!\!\!>\!C\!-\!Cl \longleftrightarrow \begin{array}{c} R_2N \\ R_2\underset{\oplus}{N}\!\!= \end{array}\!\!\!>\!C\!-\!Cl\right] Cl^{\ominus}$$

The properties of some chloroformamidinium chlorides are summarized in Table I.

TABLE I

Chloroformamidinium Chlorides via Phosgenation of Ureas and Thioureas

$$R^2R^1N\overset{\overset{Cl}{|}}{\underset{\oplus}{-C-}}NR^3R^4 Cl^{\ominus}$$

R¹	R²	R³	R⁴	Yield, %	m.p., °C	Ref.
H	H	H	C(CH₃)₃	—	110–113	1
H	CH₃	H	CH₃	86	162–164	1, 2
H	CH(CH₃)₂	H	CH(CH₃)₂	88	100–105	1, 2
H	CH₂–CH(CH₃)₂	H	CH₂–CH(CH₃)₂	—	60–63	1
H	c-C₆H₁₁	H	c-C₆H₁₁	76	139–141	2
H	CH(CH₃)₂	H	C₆H₅	99	158–160	1, 2
H	c-C₆H₁₁	H	C₆H₅	67	148–150	1, 2
H	C₆H₅	H	C₆H₅	81	126–129	1, 2
H	C₆H₅	H	CH₂C₆H₅	—	141–144	1
H	CH₂–C₆H₅	H	CH₂–C₆H₅	—	108–111	1
H	p-CH₃O–C₆H₄	H	p-CH₃O–C₆H₄–	93	116–118	1, 2
H	m-Cl–C₆H₄	H	m-Cl–C₆H₄–	83	120–122	1, 2
H	β-Naphthyl	H	β-Naphthyl	69	> 300	2
H	C₆H₅	CH₃	CH₃	25	155–158	1, 2
H	p-Cl–C₆H₄	CH₃	CH₃	30	191–197	2
H	c-C₆H₁₁	–(CH₂)₄–		62	146–150	1
H	C₆H₅	–(CH₂)₄–		96	166–171	2
CH₃	CH₃	CH₃	CH₃	96	110–112	1
CH₃	CH₃	–(CH₂)₅–		86	144–145	1, 2
CH₃	CH₃	CH₃	C₆H₅	77	40–50	1, 2
–(CH₂)₅–		–(CH₂)₅–		84	75–80	1, 2

III. Reactions of Chloroformamidinium Chlorides

A. WITH HYDROXYL COMPOUNDS

1. Hydrolysis

N,N,N',N'-Tetrasubstituted chloroformamidinium chlorides (**4**) are hydrolyzed less rapidly than amide chlorides (1, 2). During the course of the hydrolysis the ureas **1** are regenerated (1, 2).

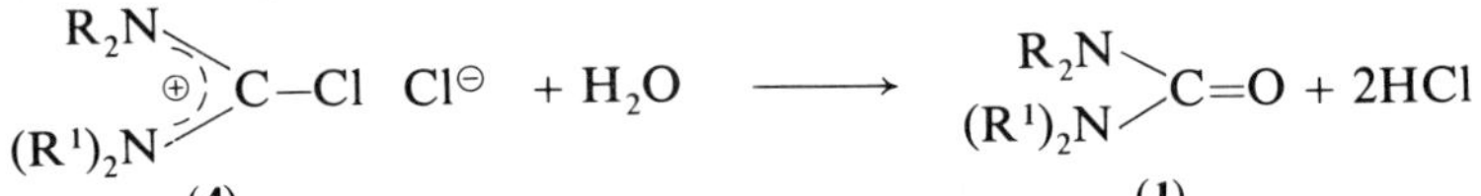

$$\begin{array}{c} R_2N \\ (R^1)_2N \end{array}\!\!\!\overset{\oplus}{C}\!-\!Cl \; Cl^{\ominus} + H_2O \longrightarrow \begin{array}{c} R_2N \\ (R^1)_2N \end{array}\!\!\!C\!=\!O + 2HCl$$
$$(4) \hspace{5cm} (1)$$

On hydrolysis of N,N'-disubstituted chloroformamidinium chlorides (**9**), carbodiimides (**18**) form via hydrogen halide cleavage (1, 2, 9). Dehydrohalogenation of the chloroformamidinium chlorides (**9**) apparently occurs much

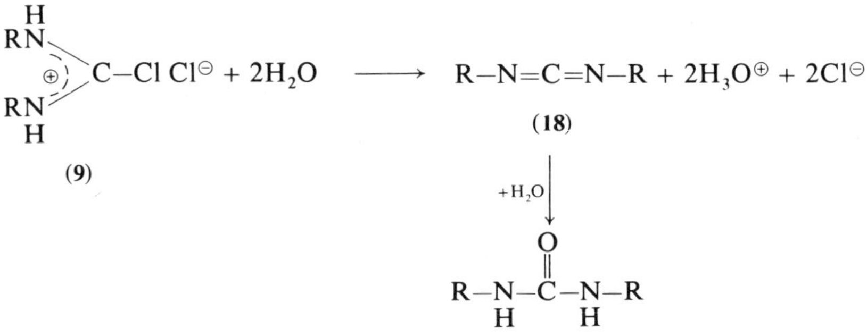

$$\begin{array}{c} H \\ RN \\ RN \\ H \end{array}\!\!\!\overset{\oplus}{C}\!-\!Cl \; Cl^{\ominus} + 2H_2O \longrightarrow R\!-\!N\!=\!C\!=\!N\!-\!R + 2H_3O^{\oplus} + 2Cl^{\ominus}$$
$$(9) \hspace{5cm} (18)$$

$$\downarrow {+H_2O}$$

$$\underset{H \quad\quad H}{R\!-\!N\!-\!\overset{\overset{\displaystyle O}{\|}}{C}\!-\!N\!-\!R}$$
$$(8)$$

faster than hydrolysis of the generated carbodiimides (**18**) to yield ureas (**8**). Since carbodiimides (**18**) add water—both acid- and base-catalyzed—the hydrogen halide acid formed has to be neutralized with base in order to achieve good yields of carbodiimides (1, 2). The carbodiimides are protected against extended action of aqueous base by extracting the mixture with solvents like chloroform (2).

This method allows the isolation even of highly sensitive carbodiimides in good yields. Some carbodiimides synthesized this way are summarized in Table II.

With cyclic chloroformamidinium, chlorides like **25**, formation of carbodiimides is not possible; here hydrolysis exclusively yields ethyleneurea (**40**) (24).

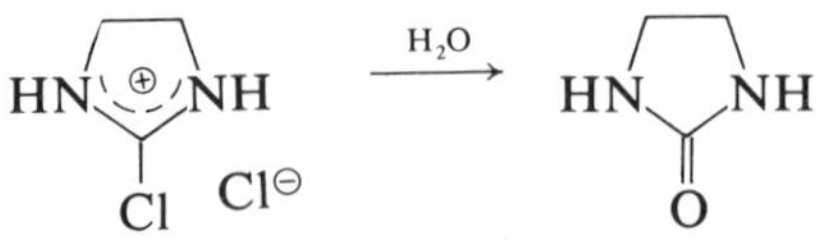

$$\underset{(25)}{\overset{\displaystyle HN\overset{\oplus}{\diagup\!\!\diagdown}NH}{\underset{Cl}{|}\;Cl^{\ominus}}} \xrightarrow{H_2O} \underset{(40)}{\overset{\displaystyle HN\diagup\!\!\diagdown NH}{\underset{O}{\|}}}$$

III. REACTIONS OF CHLOROFORMAMIDINIUM CHLORIDES

TABLE II
Carbodiimides from N,N'-Disubstituted Chloroformamidinium Chlorides

Carbodiimide R–N=C=N–R^1		Yield, %	b.p./pressure, °C/torr	m.p., °C	Ref.
R	R^1				
$CH(CH_3)_2$	$CH(CH_3)_2$	67	35–36/9	—	2
c-C_6H_{11}	c-C_6H_{11}	90	126–128/0.5	—	2
c-C_8H_{15}	c-C_8H_{15}	73	155–157/0.5	—	2
$CH(CH_3)_2$	C_6H_5	89	69–70/0.2	—	2
c-C_6H_{11}	C_6H_5	80	110/0.3	—	2
C_6H_5	C_6H_5	84	131–134/1.0	—	2
p-CH_3O–C_6H_4	p-CH_3O–C_6H_4	67	172–174/0.2	50–52	2
m-Cl–C_6H_4	m-Cl–C_6H_4	94	—	42–44	2
β-Naphthyl	β-Naphthyl	67	—	142–144	2
C_6H_5	$CH_2C_6H_5$	70	—	—	1, 2

Hydrolysis of chloroformamidinium chlorides **33** leads to N,N-bis(dialkylcarbamoyl)ureas (**41**) (12, 35).

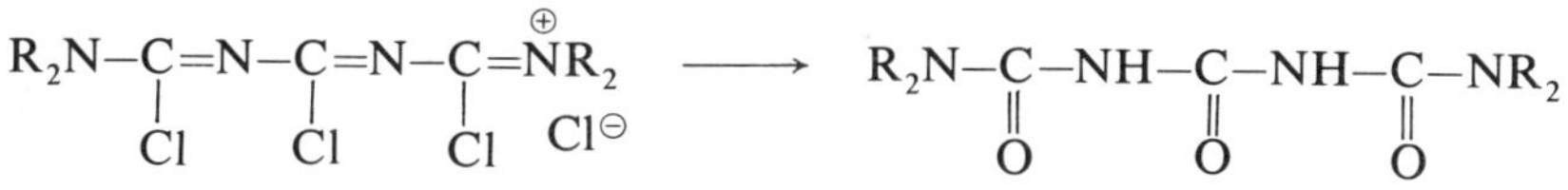

$$R_2N-\underset{Cl}{C}=N-\underset{Cl}{C}=N-\underset{Cl}{\overset{\oplus}{C}}=NR_2 \;\; Cl^{\ominus} \longrightarrow R_2N-\underset{O}{\overset{\|}{C}}-NH-\underset{O}{\overset{\|}{C}}-NH-\underset{O}{\overset{\|}{C}}-NR_2$$

(33)   (41)

2. Alcoholysis

The alcoholysis of chloroformamidinium chlorides has been poorly studied. The reaction of the chloroformamidinium chloride **33** with alcohol yields the isourea **42**, which is cyclized to the triazine **43** by aqueous ammonia (12, 36).

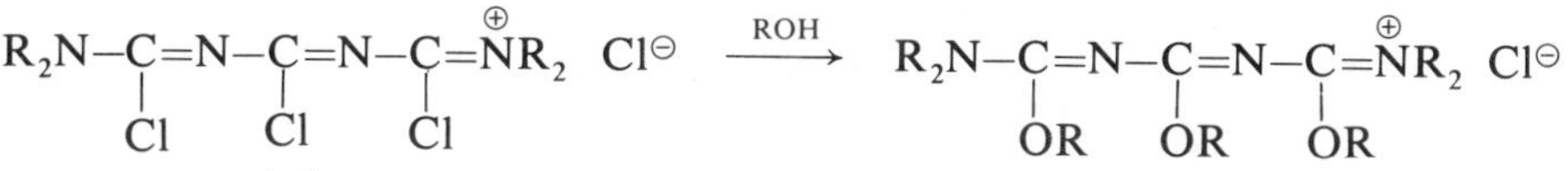

$$R_2N-\underset{Cl}{C}=N-\underset{Cl}{C}=N-\underset{Cl}{\overset{\oplus}{C}}=NR_2 \;\; Cl^{\ominus} \xrightarrow{ROH} R_2N-\underset{OR}{C}=N-\underset{OR}{C}=N-\underset{OR}{\overset{\oplus}{C}}=NR_2 \;\; Cl^{\ominus}$$

(33)   (42)

$\Big\downarrow NH_3, H_2O$

triazine **43** with OR at position 2, R_2N and NR_2 at positions 4 and 6.

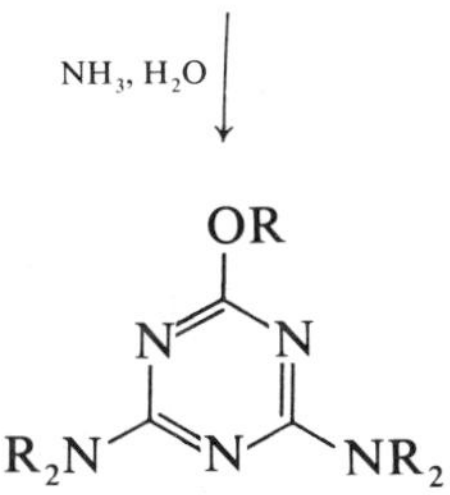

(43)

2,2-Dichloroimidazoline-4,5-diones **44** give 2,2-dimethoxyimidazoline-4,5-diones **45** when methanolyzed (18).

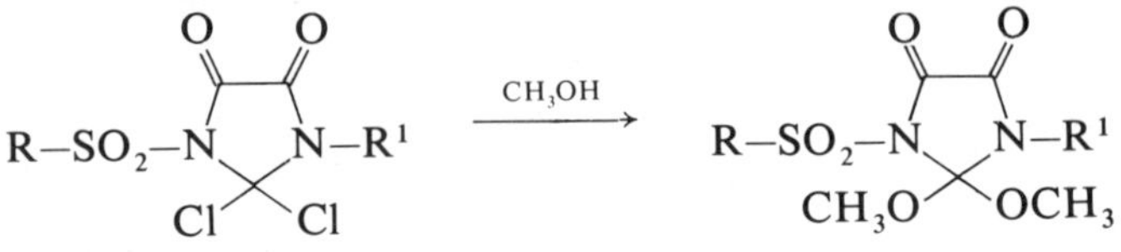

$$R-SO_2-N\underset{Cl\quad Cl}{\overset{O\quad O}{\diagdown N-R^1}} \xrightarrow{CH_3OH} R-SO_2-N\underset{CH_3O\quad OCH_3}{\overset{O\quad O}{\diagdown N-R^1}}$$

(44)   (45)

Phenols convert the chloroformamidinium chloride **25** into 2-aryloxy-1-imidazoline (**46**) (24).

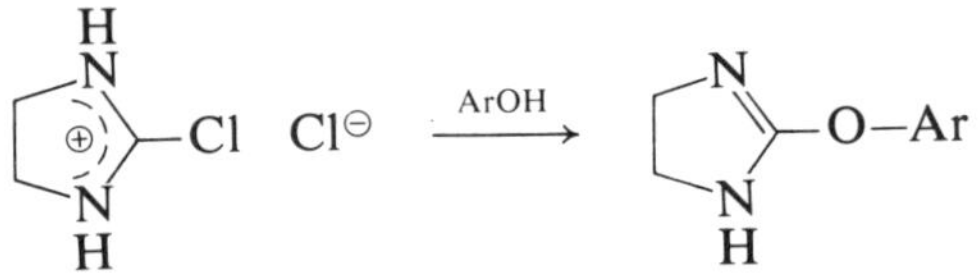

(25) $\xrightarrow{ArOH}$ (46)

3. Reactions with Alkoxides

Alkoxides primarily add to the central carbon atom of the chloroformamidinium chlorides **4** or **5** via formation of the salts **47** or **48**. The further course of the reaction now depends on the degree of substitution at the nitrogen atoms. With salts containing NH groups (**47**) the alkoxide causes elimination of HCl, thus generating isoureas (**49**) (37, 38).

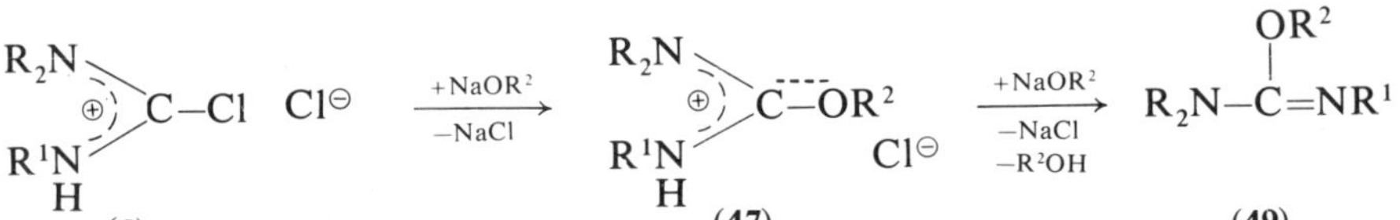

$$\underset{(5)}{R_2N\diagdown\overset{\oplus}{C}-Cl\ Cl^{\ominus}\atop R^1N\diagup\!\!\!\!H}} \xrightarrow[-NaCl]{+NaOR^2} \underset{(47)}{R_2N\diagdown\overset{\oplus}{C}-OR^2\ Cl^{\ominus}\atop R^1N\diagup\!\!\!\!H}} \xrightarrow[-R^2OH]{+NaOR^2\atop -NaCl} \underset{(49)}{R_2N-\overset{OR^2}{\underset{|}{C}}=NR^1}$$

N,N'-Persubstituted salts (**48**) add further alkoxide, leading to the formation of urea acetals (**50**) (1, 2).

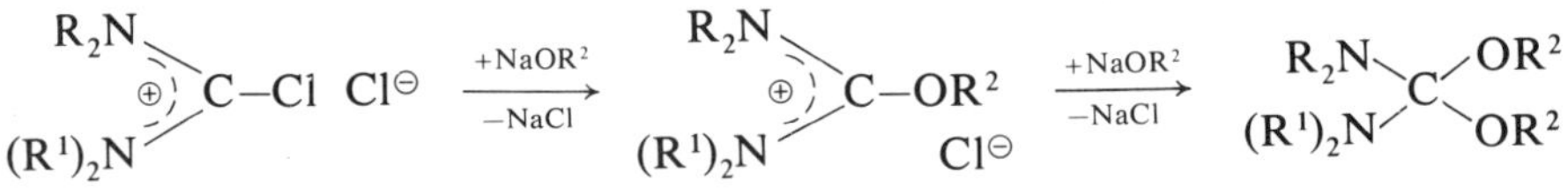

$$\underset{(4)}{R_2N\diagdown\overset{\oplus}{C}-Cl\ Cl^{\ominus}\atop (R^1)_2N\diagup}} \xrightarrow[-NaCl]{+NaOR^2} \underset{(48)}{R_2N\diagdown\overset{\oplus}{C}-OR^2\ Cl^{\ominus}\atop (R^1)_2N\diagup}} \xrightarrow[-NaCl]{+NaOR^2} \underset{(50)}{R_2N\diagdown C\diagup OR^2\atop (R^1)_2N\diagup\diagdown OR^2}}$$

By the same route the tetramethylurea dimethyl acetal **50a** is synthesized from bis(dimethylamino)difluoromethane (**5a**) and sodium methoxide (4).

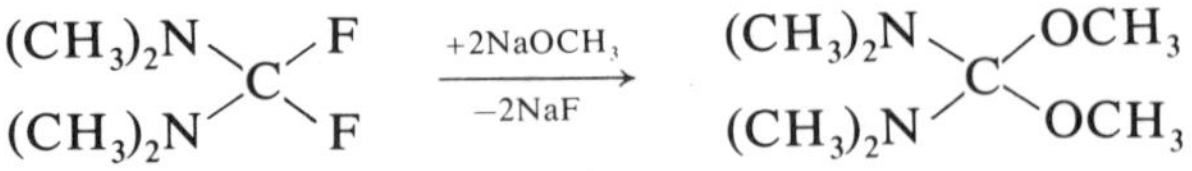

$$\underset{(5a)}{(CH_3)_2N\diagdown C\diagup F\atop (CH_3)_2N\diagup\diagdown F}} \xrightarrow[-2NaF]{+2NaOCH_3} \underset{(50a)}{(CH_3)_2N\diagdown C\diagup OCH_3\atop (CH_3)_2N\diagup\diagdown OCH_3}}$$

III. REACTIONS OF CHLOROFORMAMIDINIUM CHLORIDES

TABLE III

Urea Acetals (50) and Orthocarbamic Acid Esters (51) from Urea Dichlorides and Alkoxides (2)

$$\mathrm{R_2N}\diagdown_C\diagup\mathrm{OR^2}$$
$$\mathrm{X}\diagup{}^{}\diagdown\mathrm{OR^2}$$

X	R	R^2	Yield, %	b.p./pressure, °C/torr
$N(CH_3)_2$	CH_3	C_2H_5	30	61–65/18
OCH_3	CH_3	CH_3	27	130–132
OC_2H_5	CH_3	C_2H_5	73	63/14
OC_2H_5	$-(CH_2)_5-$	C_2H_5	83	100–105/13

The urea acetals are isolated only if alcohol-free alcoholates are used, since the compounds **50** will react with alcohol to yield orthocarbamic acid esters (**51**) (1, 2, 39). Table III summarizes the properties of some orthocarbamic acid derivatives **50** and **51**.

$$\mathrm{R_2N}\diagdown_C\diagup\mathrm{OR^2} \quad \xrightarrow[-HN(R^1)_2]{-R^2OH} \quad \mathrm{R_2N}\diagdown_C\diagup\mathrm{OR^2}$$
$$(R^1)_2N\diagup{}\diagdown\mathrm{OR^2} \quad\quad\quad\quad R^2O\diagup{}\diagdown\mathrm{OR^2}$$

$$\quad\quad (50) \quad\quad\quad\quad\quad\quad\quad\quad\quad (51)$$

B. WITH SH-CONTAINING COMPOUNDS

1. With Sodium Hydrogen Sulfide

Whereas in apolar solvents amide chlorides smoothly react with hydrogen sulfide to give thioamides, the more stable, less reactive chloroformamidinium chlorides exchange their chlorine only when reacted with aqueous sodium hydrogen sulfide; here the substituted thioureas **15** form (1, 2, 12, 40).

$$\mathrm{R_2N}\diagdown\overset{\oplus}{C}-Cl\ \ Cl^\ominus\ +SH^\ominus+OH^\ominus\ \longrightarrow\ R_2N-\overset{\overset{S}{\|}}{C}-NR_2^1+H_2O+2Cl^\ominus$$
$$\mathrm{R_2^1N}\diagup$$

$$\quad (4) \quad\quad\quad\quad\quad\quad\quad\quad\quad\quad\quad (15)$$

In accordance with the different nucleophilicity of $OH^\ominus$ and $SH^\ominus$ thiolysis takes precedence over hydrolysis.

Via hydrogen halide elimination carbodiimides (**18**) are primarily formed from N,N'-disubstituted chloroformamidinium chlorides (**9**); the former are converted into thioureas (**15a**) by addition of hydrogen sulfide (2).

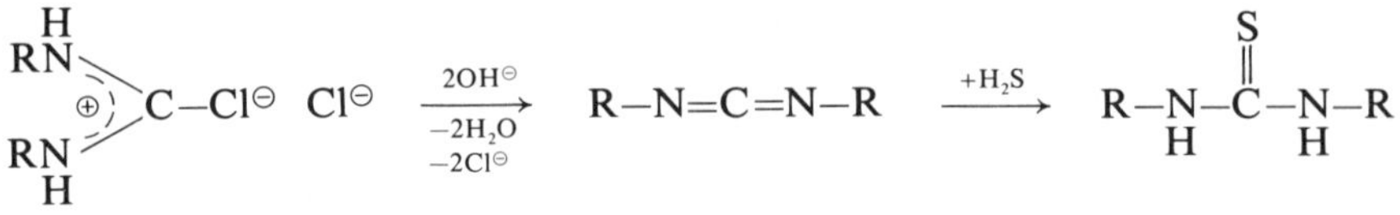

$$\underset{(9)}{\overset{H}{\underset{H}{\overset{RN}{\underset{RN}{>}}}}\text{C-Cl}^{\ominus}\ \text{Cl}^{\ominus}} \xrightarrow[-2\text{H}_2\text{O}\ -2\text{Cl}^{\ominus}]{2\text{OH}^{\ominus}} \underset{(18)}{R-N=C=N-R} \xrightarrow{+\text{H}_2\text{S}} \underset{(15a)}{R-\underset{H}{N}-\overset{S}{\overset{\|}{C}}-\underset{H}{N}-R}$$

Table IV summarizes the properties of some thioureas generated by this method.

TABLE IV

Thioureas from Chloroformamidinium Chlorides and Sodium Hydrogen Sulfide (2)

$$R^1R^2N-\overset{S}{\overset{\|}{C}}-NR^3R^4$$

R^1	R^2	R^3	R^4	Yield, %	m.p., °C	b.p./pressure, °C/torr
CH$_3$	CH$_3$	CH$_3$	CH$_3$	78	78	—
CH$_3$	CH$_3$	—(CH$_2$)$_5$—		75	—	99/0.8
CH$_3$	CH$_3$	CH$_3$	C$_6$H$_5$	82	79	—
—(CH$_2$)$_5$—		—(CH$_2$)$_5$—		86	51	—

2. With Thiophenols and Thiophenoxides

Arylisothiouronium salts (**52**) are synthesized by reacting thiophenols with *N,N'*-persubstituted chloroformamidinium chlorides (**4**) (41, 42). Thallium as well as lead phenoxides may also be used (43).

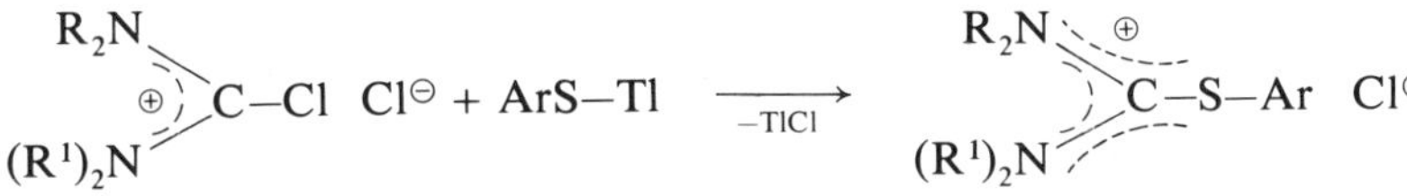

$$\underset{(4)}{\overset{R_2N}{\underset{(R^1)_2N}{>}}\text{C-Cl}\ \text{Cl}^{\ominus} + \text{ArS-Tl}} \xrightarrow{-\text{TlCl}} \underset{(52)}{\overset{R_2N}{\underset{(R^1)_2N}{>}}\overset{\oplus}{\text{C}}-\text{S-Ar}\ \text{Cl}^{\ominus}}$$

From the salt **25** and thiophenols 2-arylthio-1-imidazolines (**53**) are formed (24).

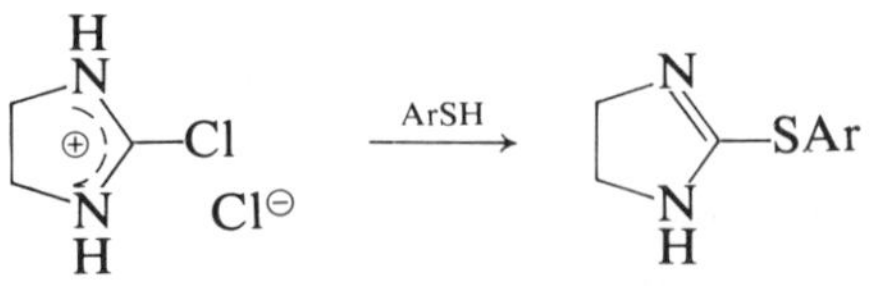

III. REACTIONS OF CHLOROFORMAMIDINIUM CHLORIDES

C. WITH NH-CONTAINING COMPOUNDS

Primary and secondary amines react with chloroformamidinium chlorides to form guanidines and their salts (1, 2, 9, 12, 30, 44, 45). From primary amines and N,N'-disubstituted chloroformamidinium chlorides (9) the carbodiimides 18 are initially generated, which add further amine to give N,N',N''-trisubstituted guanidines 54 (1, 2, 9).

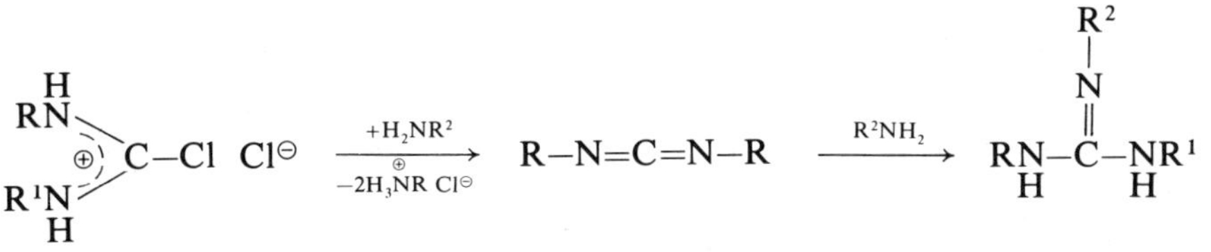

2-Arylamino-1-imidazolines (55) are formed from 25 and aromatic amines (24).

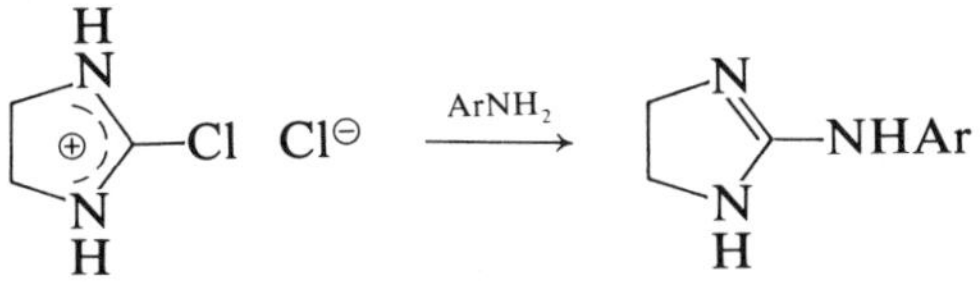

N,N',N''-Pentasubstituted guanidines 56 are obtained from N,N,N',N''-tetrasubstituted chloroformamidinium chlorides 4 and primary amines (1, 2) as well as from cyanamides (46).

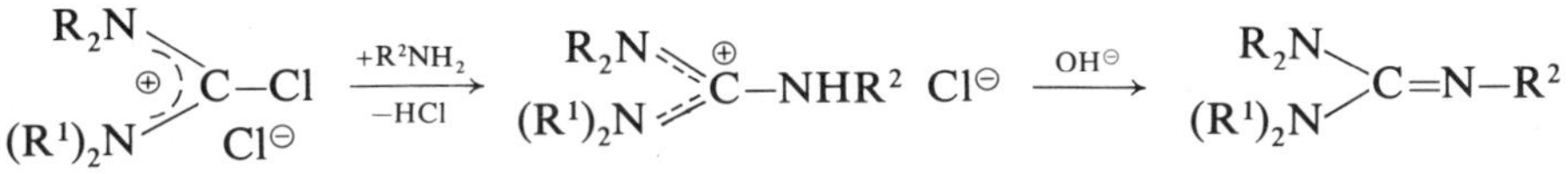

R^2 = Alkyl, Aryl, CN

Hydrazine derivatives like phenylhydrazine or benzhydrazide (1, 2) react as primary amines do with 4 to yield the compounds 57.

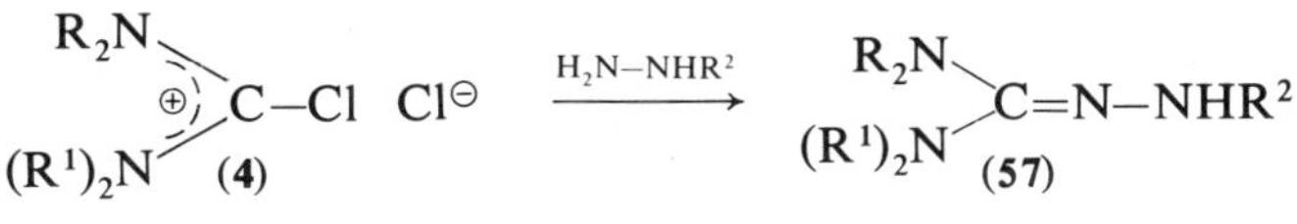

Table V summarizes the guanidine derivatives 54, 56, and 57 from chloroformamidinium chlorides.

TABLE V

Guanidine Derivatives (**54, 56, 57**) from Chloroformamidinium Chlorides (**4**)

Guanidine

$$R-N=C{\overset{NR^1R^2}{\underset{N(R^3)_2}{\diagdown}}}$$

R	R¹	R²	R³	Yield, %	b.p./pressure, °C/torr	m.p., °C	Ref.
CN	CH₃	CH₃	CH₃	54	181–182/2	—	46
C₆H₅	CH₃	CH₃	CH₃	73	87–89/0.3	—	2
m-Cl—C₆H₄	CH₃	CH₃	CH₃	67	110–113/0.5	—	2
p-NO₂—C₆H₄	CH₃	CH₃	CH₃	70	—	83	2
1-Anthraquinoyl	CH₃	CH₃	CH₃	64	—	140	2
1-Anthraquinoyl	—(CH₂)₅—		CH₃	53	—	80	2
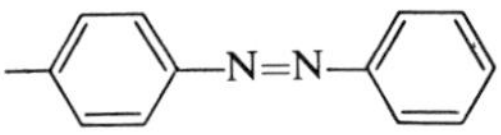 ⟨⟩—N=N—⟨⟩	CH₃	CH₃	CH₃	73	—	90	2
C₆H₅—NH	CH₃	CH₃	CH₃	68	139–142/1.5	—	2
C₆H₅—CONH	CH₃	CH₃	CH₃	59[a]	—	194–196	2

[a] Hydrochloride.

Ammonia causes the chloroformamidinium salt **33** to cyclize to the triazines **58** or **59**, depending on the reaction conditions (12, 31). This kind of ring closure is also possible with primary and secondary aliphatic and aromatic amines (36, 44, 45). The azacyanine **60** is also useful for the synthesis of the heterocycles **61** and **62** (47).

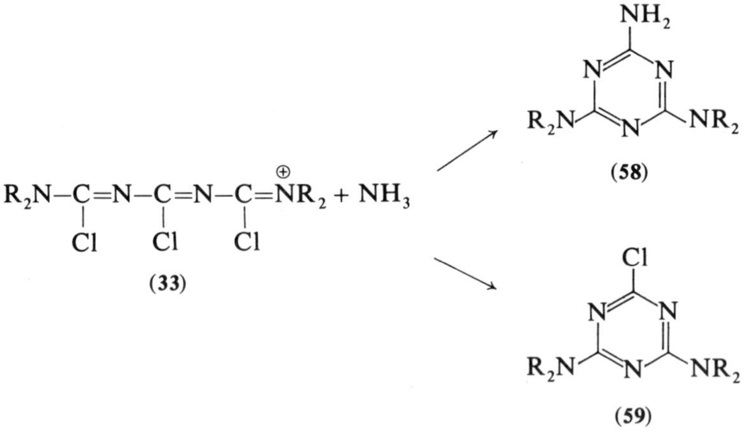

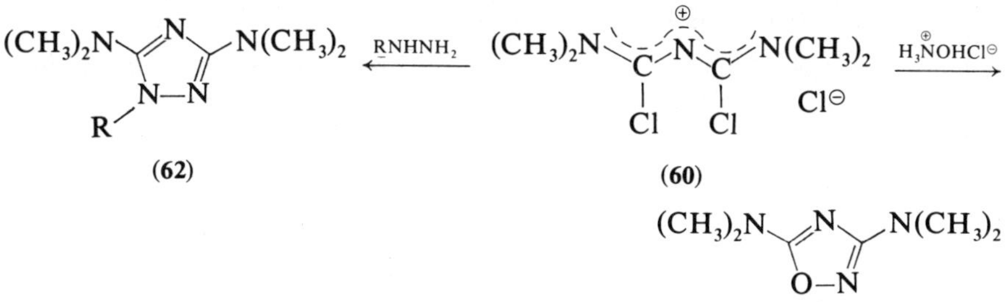

III. REACTIONS OF CHLOROFORMAMIDINIUM CHLORIDES

(62) ⇌ (RNHNH₂) (60) →(H₃NOHCl⁻)

(61)

2,2-Dichloro-1,3-dialkyl-4,5-dioxo-1,3-diazolidines (**31**) and secondary amines like piperidine or morpholine yield the corresponding orthocarbonic acid derivatives **63** (27, 28).

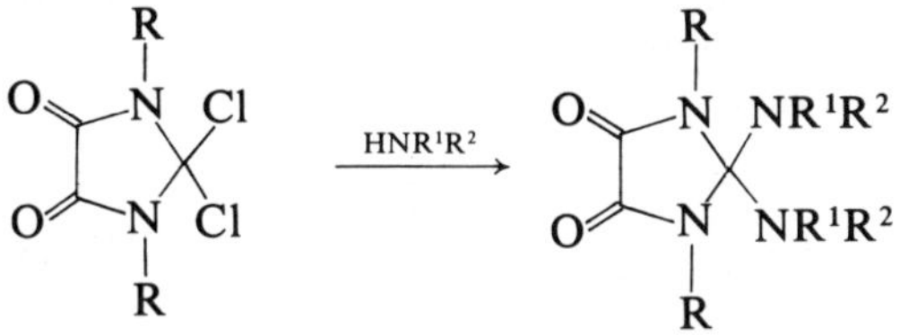

(**31**) →(HNR¹R²) (**63**)

When N,N,N',N'-tetramethylchloroformamidinium chloride (**39**) is reacted with N,N'-diphenylurea, N,N'-diphenylchloroformamidinium chloride (**64**) forms in addition to N,N-dimethyl-N'-phenylformamidinium chloride (**65**) (48).

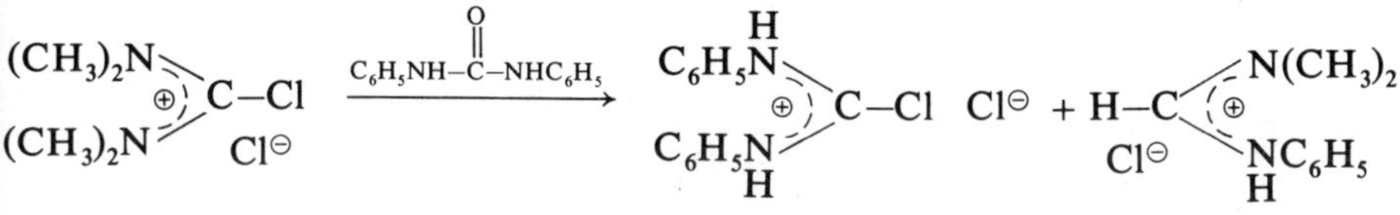

(**39**) →(C₆H₅NH—CO—NHC₆H₅) (**64**) + (**65**)

On the other hand, heating of equimolar amounts of N,N'-dicyclohexylchloroformamidinium chloride (**66**) with N,N'-dicyclohexylurea gives N,N',N''-tricyclohexylguanidinium chloride (**67**) and cyclohexylisocyanate (49).

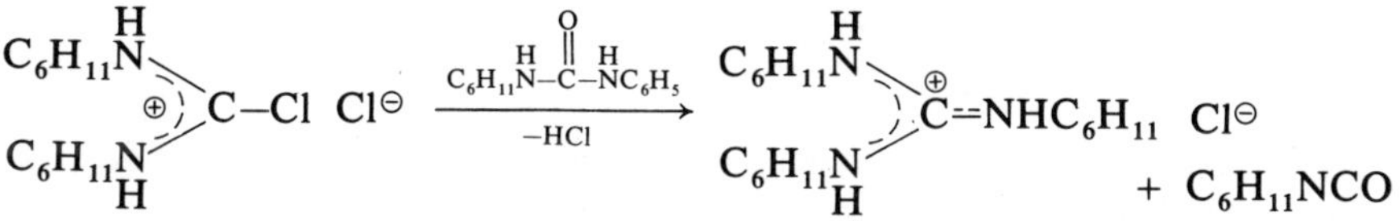

(**66**) →(C₆H₁₁NH—CO—NHC₆H₁₁, −HCl) (**67**) + C₆H₁₁NCO

D. WITH ORGANOLITHIUM COMPOUNDS

Tetrakis(dimethylamino)methane (**68**) was first synthesized from lithium dimethylamide and *N,N,N',N'*-tetramethylchloroformamidinium chloride (**39**) (50).

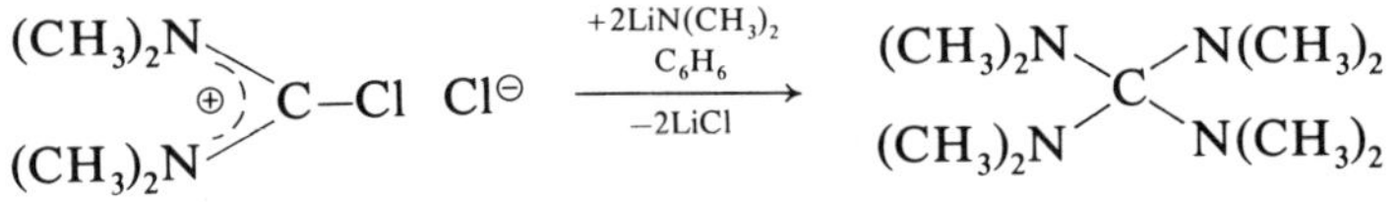

$$(39) \xrightarrow[-2LiCl]{+2LiN(CH_3)_2, C_6H_6} (68)$$

Methyllithium adds to **39** to give *N,N,N',N'*-tetramethylacetamidinium chloride (**69**); this further reacts with excess methyllithium to 1,1-bis-(dimethylamino)ethylene (**70**) (51).

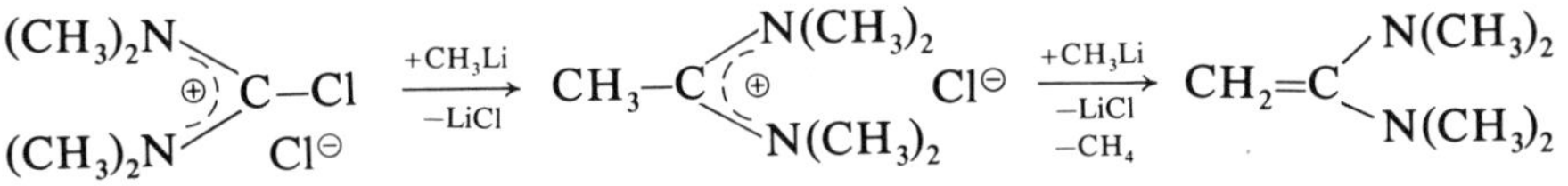

$$(39) \xrightarrow[-LiCl]{+CH_3Li} (69) \xrightarrow[-CH_4]{+CH_3Li, -LiCl} (70)$$

Phenyllithium and **39** primarily generate the benzamidinium salt **71**, which undergoes further reaction with more phenyllithium to yield a mixture of the aminal of benzophenone (**72**) and the ketenaminal **73** (41).

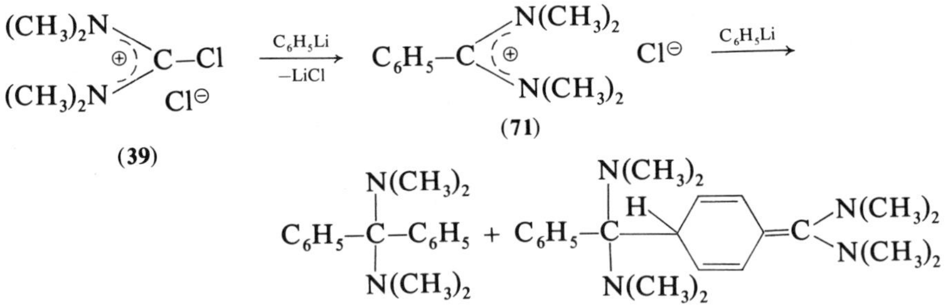

$$(39) \xrightarrow[-LiCl]{C_6H_5Li} (71) \xrightarrow{C_6H_5Li} (72) + (73)$$

The reaction of *N,N,N',N'*-tetramethylchloroformamidinium chloride (**39**) and *t*-butyllithium does not proceed cleanly; tris(dimethylamino)methane (**74**), tetrakis(dimethylamino)methane (**68**), and di-*t*-butyl(dimethylamino)methane

III. REACTIONS OF CHLOROFORMAMIDINIUM CHLORIDES

(75), as well as isobutylenes and isobutanes, were obtained in low yields as the reaction products (51). To rationalize this reaction bis(dimethylamino)carbene has been proposed as the intermediate (51).

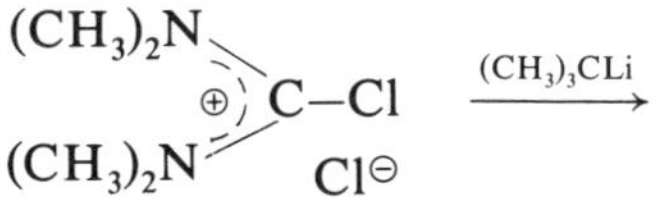

$$(CH_3)_2N\!\!\underset{(CH_3)_2N}{\overset{\oplus}{>}}\!\!C\!-\!Cl \quad Cl^{\ominus} \xrightarrow{(CH_3)_3CLi}$$
(39)

$$H-\underset{N(CH_3)_2}{\overset{N(CH_3)_2}{\underset{|}{C}}}-N(CH_3)_2 \quad + \quad (CH_3)_2N\!\!\underset{(CH_3)_2N}{\overset{}{>}}\!\!C\!\!\underset{N(CH_3)_2}{\overset{N(CH_3)_2}{<}} \quad + \quad [(CH_3)_3C]_2CHN(CH_3)_2$$

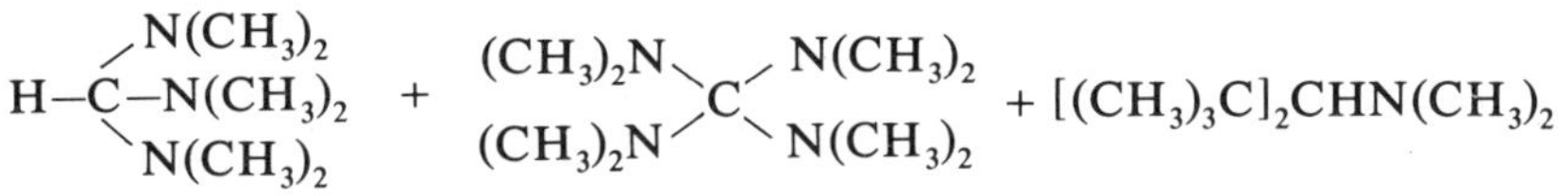

(74)   (68)   (75)

E. WITH ORGANOPHOSPHORUS COMPOUNDS

N,N,N',N'-Tetramethylchloroformamidinium chloride (39) reacts with triethyl phosphite to give the betaine (76), which is converted with water to the salt 77, and with methanol to 78 (35).

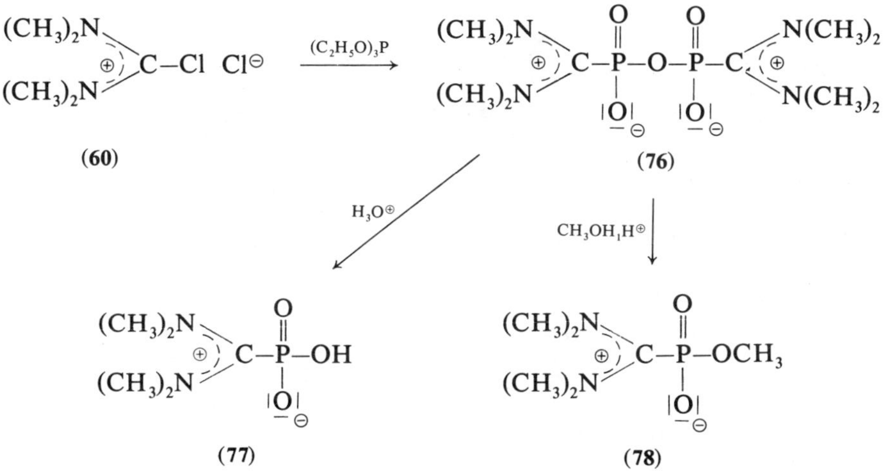

[Reaction scheme showing (60) → (76) via $(C_2H_5O)_3P$; (76) → (77) via $H_3O^\oplus$; (76) → (78) via $CH_3OH, H^\oplus$]

Compound 39 reacts with phenylphosphoric acid diethyl ester (79) to yield N,N,N',N'-tetramethyl(ethoxyphenylphosphinyl)formamidinium chloride (80), which may be further converted to 81 (35).

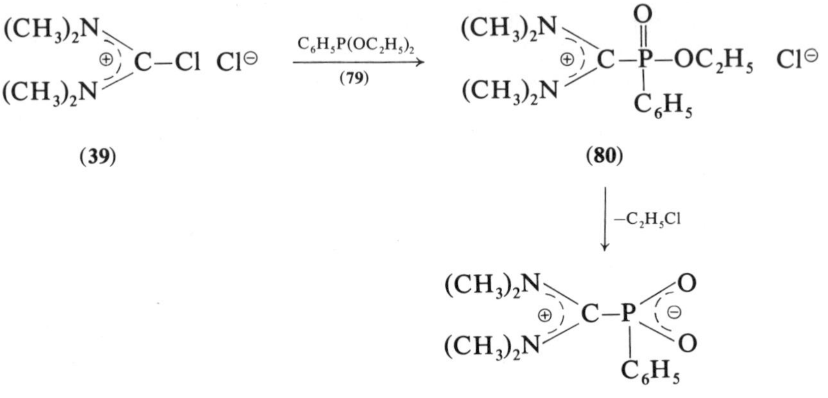

$(CH_3)_2N$
$\overset{\oplus}{C}-Cl\ Cl^{\ominus}$
$(CH_3)_2N$
(39)

$\xrightarrow{C_6H_5P(OC_2H_5)_2}$
(79)

$(CH_3)_2N$
$\overset{\oplus}{C}-\overset{\overset{O}{\|}}{\underset{C_6H_5}{P}}-OC_2H_5\ Cl^{\ominus}$
$(CH_3)_2N$
(80)

$\Big\downarrow -C_2H_5Cl$

$(CH_3)_2N$
$\overset{\oplus}{C}-P\overset{O}{\underset{\underset{C_6H_5}{|}}{\diagdown_O^{\ominus}}}$
$(CH_3)_2N$
(81)

Diphenylphosphinic acid ethyl ester **(82)** gives the amidinium salt **83** when treated with **39** (35).

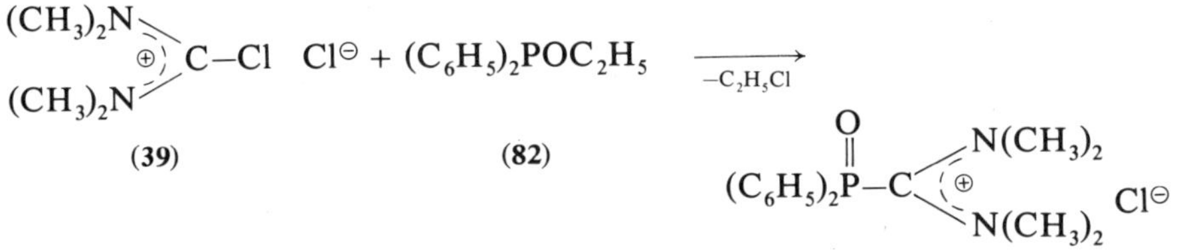

$(CH_3)_2N$
$\overset{\oplus}{C}-Cl\ Cl^{\ominus} + (C_6H_5)_2POC_2H_5 \xrightarrow{-C_2H_5Cl}$
$(CH_3)_2N$
(39) **(82)**

$(C_6H_5)_2\overset{\overset{O}{\|}}{P}-\overset{\oplus}{C}\overset{N(CH_3)_2}{\underset{N(CH_3)_2}{\diagdown}} Cl^{\ominus}$
(83)

F. ELIMINATION REACTIONS

N,N'-Disubstituted chloroformamidinium chlorides **(9)** are converted with base into carbodiimides **(18)** (1, 2, 9). *N,N*-Diarylchloroformamidinium chlorides eliminate HCl on heating in an inert solvent like dichlorobenzene; thus carbodiimides are formed (11).

When the chloroformamidinium salt **33** is heated to 180–190°C, cyclization to the triazine **84** occurs (12).

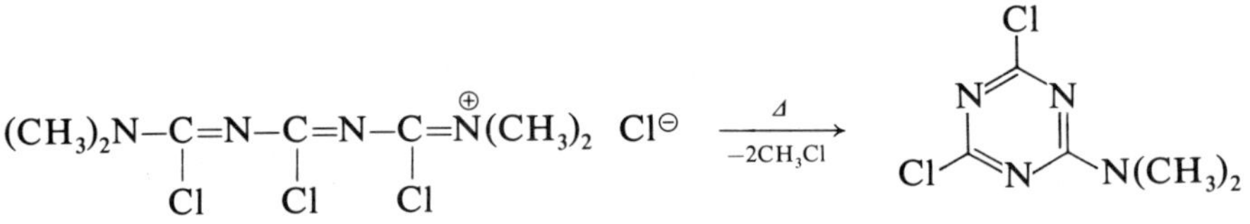

$(CH_3)_2N-\underset{\underset{Cl}{|}}{C}=N-\underset{\underset{Cl}{|}}{C}=N-\underset{\underset{Cl}{|}}{C}=\overset{\oplus}{N}(CH_3)_2\ Cl^{\ominus} \xrightarrow[-2CH_3Cl]{\Delta}$
(33)

[triazine structure with Cl, N(CH_3)_2 substituents]
(84)

Heating of *N,N'*-dialkylchloroformamidinium chlorides yields *N,N',N''*-trisubstituted guanidinium salts like **85**, as demonstrated with the *N,N*-dibutylchloroformamidinium chloride **(86)** (17). The proposed dissociation of **86** into

III. REACTIONS OF CHLOROFORMAMIDINIUM CHLORIDES

butylisocyanide dichloride (**87**) and butylamine may explain the course of the reaction (17).

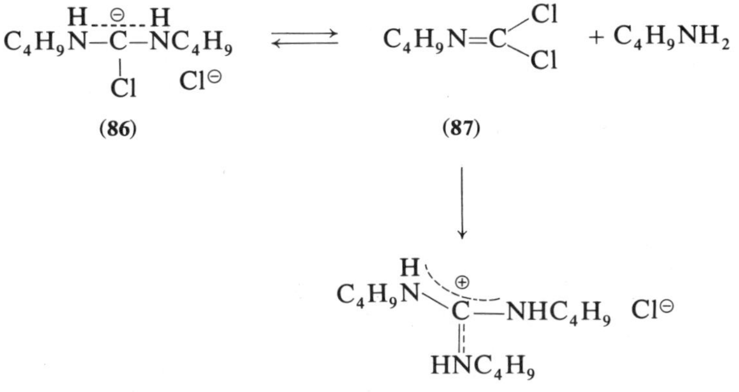

$$C_4H_9\overset{H}{\underset{Cl}{N}}-\overset{\ominus}{C}-\overset{H}{N}C_4H_9 \quad Cl^\ominus \quad (86)$$

$$\rightleftharpoons \quad C_4H_9N=C\overset{Cl}{\underset{Cl}{\diagdown}} + C_4H_9NH_2 \quad (87)$$

$$\downarrow$$

$$C_4H_9\overset{H}{\underset{}{N}}\diagdown \overset{\oplus}{C}-NHC_4H_9 \quad Cl^\ominus$$
$$\overset{\|}{H}NC_4H_9$$
(**85**)

G. MISCELLANEOUS REACTIONS

Trimethylchlorosilane converts **25** into 1-trimethyl-2-chloro-silyl-2-imidazolines (**88**), which trimerize to **89** on heating (24).

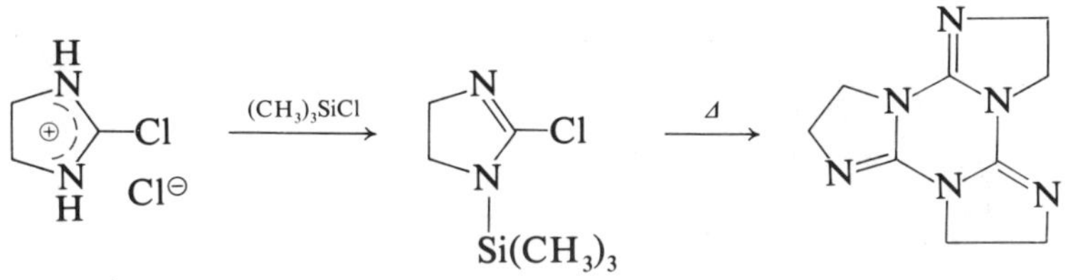

(**25**) →(CH₃)₃SiCl→ (**88**) →Δ→ (**89**)

The iminium salt **39b** eliminates methyl chloride and N,N-dimethylcarbamoyl chloride when heated to give triphenyltriazine (**90**) (34a).

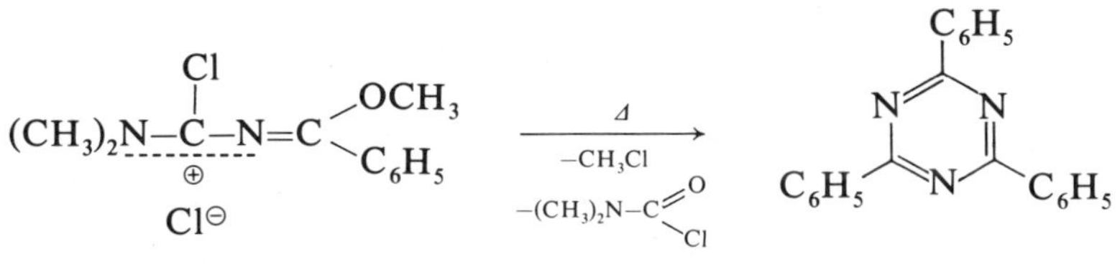

$$(CH_3)_2N-\underset{\oplus}{\overset{Cl}{C}}-N=C\overset{OCH_3}{\underset{C_6H_5}{\diagdown}} \quad Cl^\ominus$$
(**39b**)

$$\xrightarrow[-(CH_3)_2N-C\overset{O}{\underset{Cl}{\diagup}}]{\Delta,\ -CH_3Cl}$$

(**90**) — triphenyltriazine with C₆H₅ groups

ACKNOWLEDGMENT

To Dr. Heinz Eilingsfeld BASF AG, Ludwigshafen, and to Dr. Bernd Funke as well as Dr. Erwin Haug, both of Fachhochschule Aalen, Aalen, I am very indebted for fruitful discussions and critical revisions of the text.

REFERENCES

1. H. Eilingsfeld, M. Seefelder, and H. Weidinger, *Angew. Chem.*, **72**, 836 (1960).
2. H. Eilingsfeld, G. Neubauer, M. Seefelder, and H. Weidinger, *Chem. Ber.*, **97**, 1232 (1964).
3. H. Bredereck and K. Bredereck, *Chem. Ber.*, **94**, 2278 (1961).
4. M. H. Brown, U.S. Pat. 3,092,637 (1963); *C.A.*, **59**, 12764g.
5. F. S. Fawcett, C. W. Tullock, and D. D. Coffman, *J. Am. Chem. Soc.*, **84**, 4275 (1962).
6. M. A. Englin, S. P. Makarov, S. S. Dubov, and A. Ya. Yakubovich, *Zh. Obshch. Khim.*, **35**, 1416 (1965).
7. R. J. Koshar, D. R. Husted, and R. A. Meiklejohn, *J. Org. Chem.*, **31**, 4232 (1966).
8. H. Ulrich and A. A. R. Sayigh, *Angew. Chem. Inter. Ed. Engl.*, **3**, 585 (1964).
9. H. Ulrich, J. N. Tilley, and A. A. R. Sayigh, *J. Org. Chem.*, **29**, 2401 (1964).
10. H. Ulrich and A. A. R. Sayigh, *J. Org. Chem.*, **30**, 2779 (1965).
11. H. Ulrich and A. A. R. Sayigh, *Angew. Chem. Intern. Ed. Engl.*, **5**, 704 (1966).
12. K. Bredereck and R. Richter, *Chem. Ber.*, **99**, 2461 (1966).
13. G. Kränzlein, H. Keller, and H. Schiff (to I. G. Farbenindustrie), DRP 689,421 (1940); *Ann. Chem.* **291**, 374 (1896).
14. P. Gröbner and W. Rudolph, *Eur. J. Med. Chem. Ther.*, **9**, 32 (1974).
15. A. Steindorff, *Ber. Deut. Chem. Ges.*, **37**, 964 (1904).
16. H. P. Latscha, *Z. Anorg. Allgem. Chem.*, **346**, 116 (1966).
17. A. A. R. Sayigh, J. N. Tilley, and H. Ulrich, *J. Org. Chem.*, **29**, 3344 (1964).
18. H. Ulrich, B. Tucker, and A. A. R. Sayigh, *Tetrahedron*, **22**, 1565 (1966).
19. H. Ulrich and A. A. R. Sayigh, *Angew. Chem. Intern. Ed. Engl.*, **3**, 639 (1964).
20. W. Will, *Ber. Deut. Chem. Ges.*, **14**, 1486 (1881).
21. H. Ulrich and A. A. R. Sayigh, *J. Org. Chem.*, **28**, 1427 (1963).
22. Belg. Pat. 628,832 (1963); *C.A.*, **61**, 9438 (1964).
23. H. Ulrich and A. A. R. Sayigh, *J. Org. Chem.*, **30**, 2781 (1965).
24. A. Trani and E. Bellasio, *J. Heterocyl. Chem.*, **11**, 257 (1974).
25. H. Z. Lecher, E. M. Hardy, and C. L. Kosloski, U.S. Pat. 2,845,458 (1958); *C.A.*, **53**, 231 (1959).
26. F. Lengfeld and J. Stieglitz, *Am. Chem. J.*, **17**, 108 (1895).
27. H. D. Stachel, *Angew. Chem.*, **71**, 246 (1958).
28. H. Gross and G. Zinner, *Chem. Ber.*, **106**, 2315 (1973); G. Zinner, R. Volbrath, *Chem. Ber.* **103**, 766 (1970).
29. L. N. Markovskii, Ju. G. Shermolovich, M. I. Povolotskii, and V. I. Shevchenko, *Zh. Org. Khim.*, **9**, 1753 (1973); *Cheminform*, **1973**, 45–283, 102.
30. H. Bredereck and R. Richter, *Chem. Ber.*, **99**, 2454 (1966).
31. Z. Csűrös, R. Soós, A. Antus-Ercsényi, I. Bitter, and J. Tamás, *Acta Chim. Acad. Sci. Hung.*, **77**, 443 (1973).
32. T. v. Vyve and H. G. Viehe, *Angew. Chem.*, **86**, 45 (1974).
33. H. G. Viehe and Z. Janousek, *Angew. Chem.*, **85**, 837 (1973).
34. V. P. Kukhar', V. I. Pasternak, M. I. Povolotskii, and G. N. Pavlenkov, *Zh. Org. Khim.*, **10**, 449 (1974); *Cheminform*, **1974**, 24–202, p. 62.
34a. V. P. Kukhar' and M. Shevchenko, *Zh. Org. Khim.* **11**, 71 (1975); *Cheminform*, **1975**, 14–163.

35. G. H. Birum and J. D. Wilson, *J. Org. Chem.*, **37**, 2730 (1972).
36. Z. Csűrös, R. Soós, A. Ántus-Ercsényi, and I. Bitter, *Acta Chim. Acad. Sci. Hung.*, **77**, 443 (1973).
37. S. O. Winthrop and G. Gavin, *J. Org. Chem.*, **24**, 1936 (1959).
38. E. Kühle and R. Wegler, Ger. Pat. 1,29,020 (1966); *C.A.*, **65**, 13605 (1966).
39. H. Weidinger and M. Seefelder, Ger. Pat. 1,122,936 (1960); *C.A*., **57**, 4552 (1962).
40. H. Weidinger and H. Eilingsfeld, Ger. Pat. 1,119,843 (1961).
41. H. Kessler, H. O. Kalinowski, and C. von Chamier, *Ann. Chem.*, **727**, 228 (1969).
42. G. Seconi, P. Vivarelli, and A. Ricci, *J. Chem. Soc. B*, **1970**, 254.
43. H. Kessler and H. O. Kalinowski, *Angew. Chem.*, **82**, 666 (1970).
44. Z. Csűrös, R. Soós, A. Ántus-Ercsényi, I. Bitter, and J. Tarmás, *Acta Chim. Acad. Sci. Hung.*, **78**, 409 (1973).
46. E. Allenstein and P. Quis, *Chem. Ber.*, **96**, 2918 (1963).
47. Z. Janousek and H. G. Viehe, *Angew. Chem.*, **85**, 90 (1973); *Angew. Chem. Intern. Ed. Engl.*, **12**, 74 (1973).
48. Z. Csűrös, R. Soós, I. Bitter, and E. A. Kărpăti, *Acta Chim. Acad. Sci. Hung.*, **73**, 239 (1972).
49. D. F. Gavin, W. J. Schnabel, E. Kober, and M. A. Robinson, *J. Org. Chem.*, **32**, 2511 (1967).
50. H. Weingarten and W. A. White, *J. Am. Chem. Soc.*, **88**, 2885 (1966).
51. C. F. Hobbs and H. Weingarten, *J. Org. Chem.*, **36**, 2881 (1971).

ADDENDUM

To Sections I-A-1 and I-B-1

A series of chloroformamidinium chlorides have been prepared by action of phosgen on $NNN'N'$-tetrasubstituted and NNN'-trisubstituted ureas and on thioureas, respectively (52).

To Section I-D

Chloroformamidines and chloroformamidinium salts were obtained by treatment of various compounds with phosgeniminium salts (53–57). Some examples are listed below.

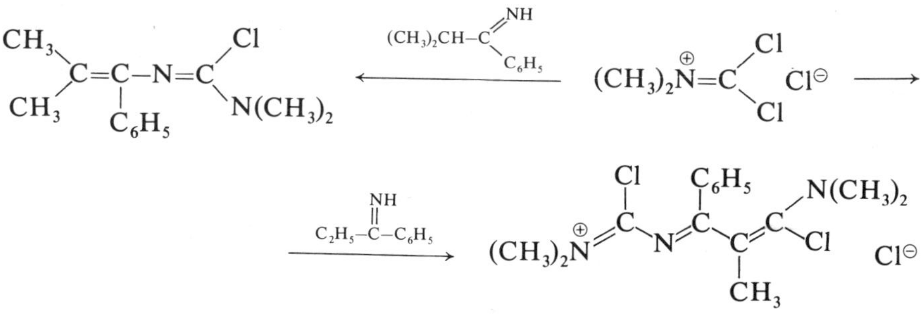

* Translated from German by W. Kantlehner and by H. J. Benivschke, Fachhochschule Aalen, 708-Aalen, Hohenstaufenstrasse 1, Germany.

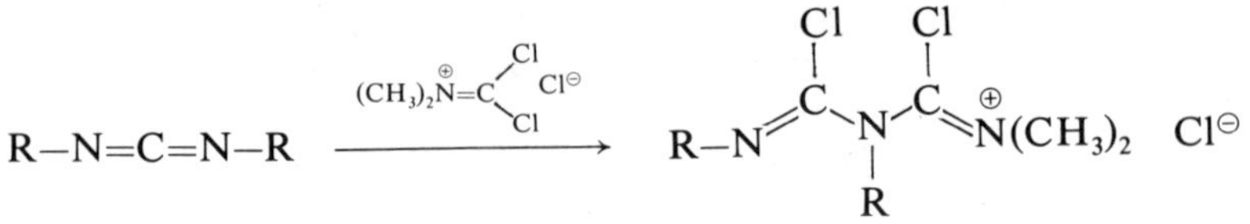

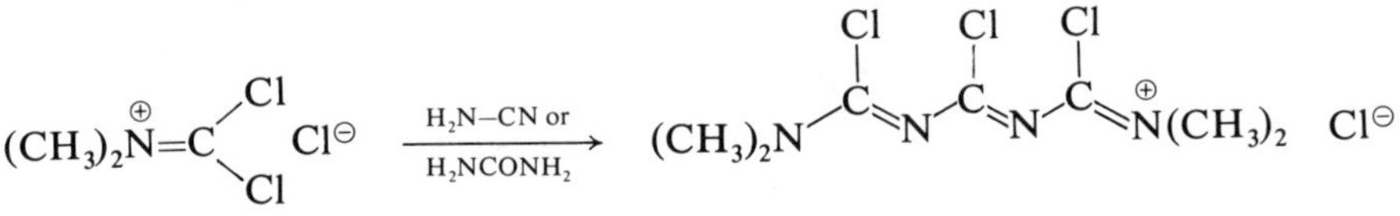

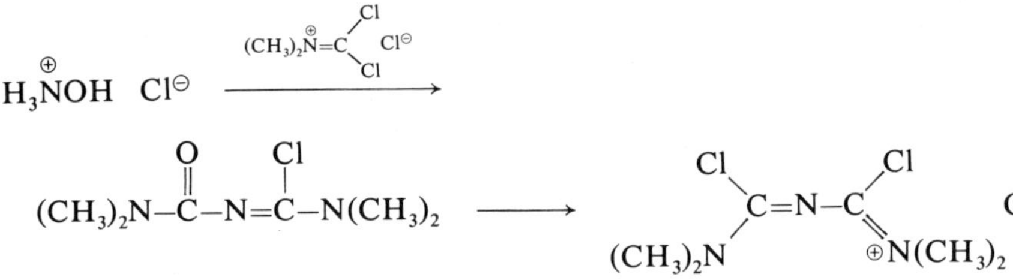

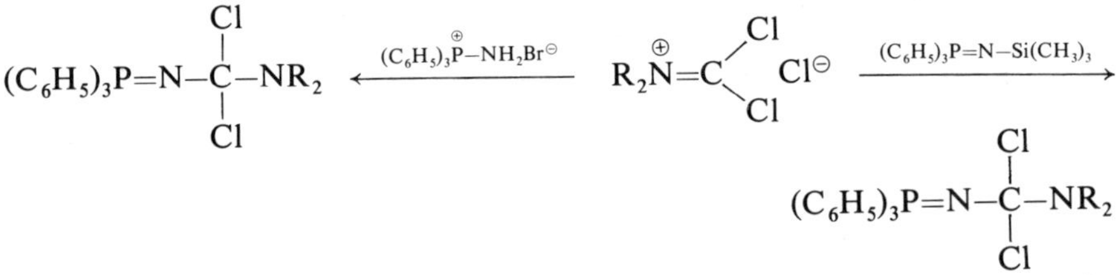

Section I-E

Miscellaneous methods for preparation. Methylisocyanate reacts with $SbCl_5/CCl_4$ to form the chloroformamidinium salt **91**, which affords on methanolysis compounds **92** and **93** (58).

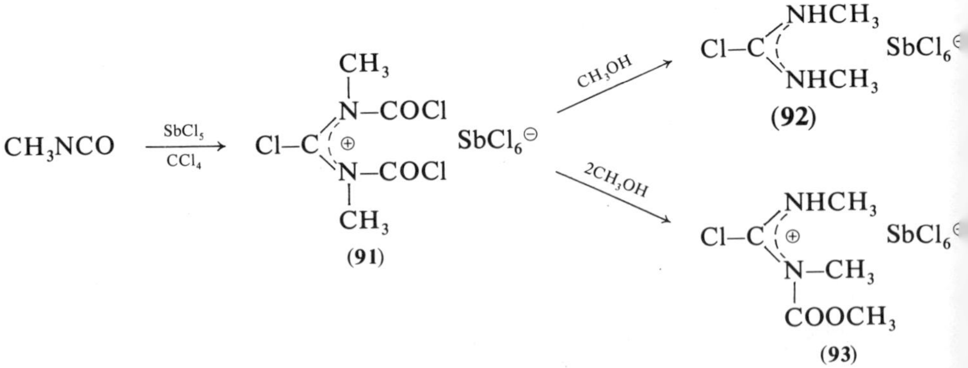

Photolyctic cleavage of elementhalides of the fourth main group of the Periodic Table, in presence of electron-rich olefins **94**, yields chloroformamidinium salts **95** (59).

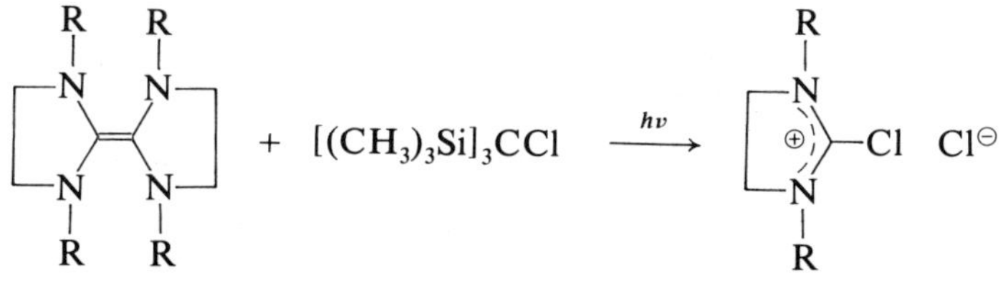

2,2,4,5-Tetrachloro-2-H-imidazol (**96**) is readily accessible from the chlorination of imidazolhydrochloride (60).

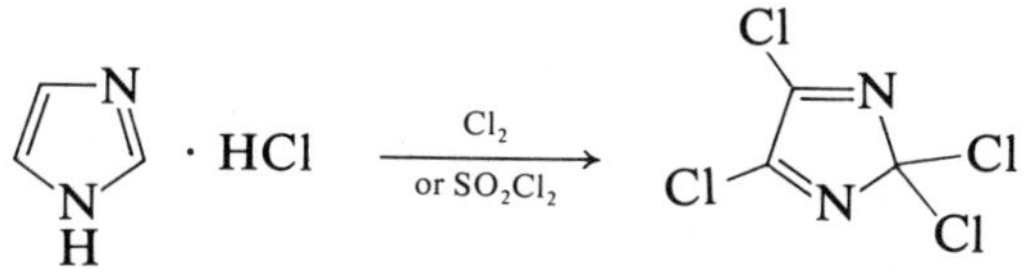

Imidazolium salts **97** can be prepared by alkylation of 2-chlorobenzimidazoles (**98**) with triethyloxoniumfluoroborate (61).

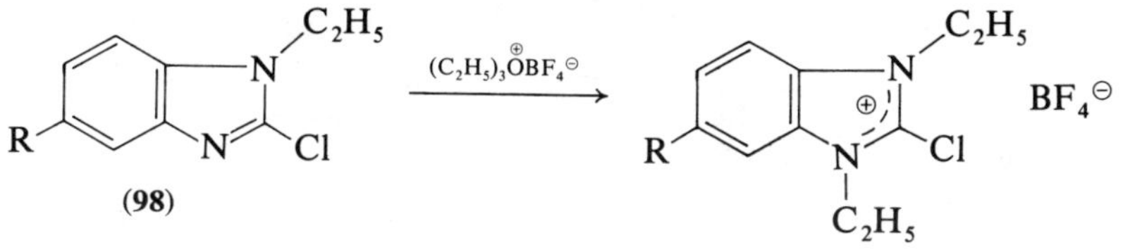

To Section III-A-1

Hydrolysis, alcoholysis, as well as solvolysis with formic and acetic acid of the tetrachloro-2-H-imidazole **96**, produce parabanic acid **99** (60).

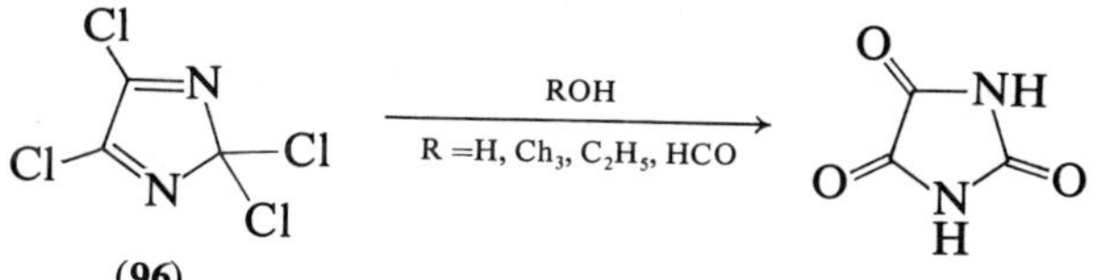

Compound **100** gives the phosphonium salt **101** on hydrolysis at temperatures below 20°C, whereas action of acetic on **100** yields the urea derivative **102** (57).

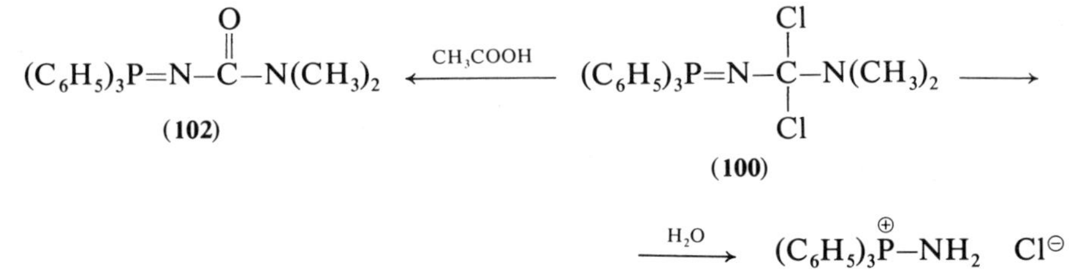

$$(C_6H_5)_3P=N-\underset{(102)}{\overset{O}{\overset{\|}{C}}}-N(CH_3)_2 \xleftarrow{CH_3COOH} (C_6H_5)_3P=N-\underset{\underset{Cl}{|}}{\overset{\overset{Cl}{|}}{C}}-N(CH_3)_2 \xrightarrow{H_2O} (C_6H_5)_3\overset{\oplus}{P}-NH_2 \quad Cl^{\ominus}$$

(100) → (101)

Methanolysis of **103** affords the urea **104** (62).

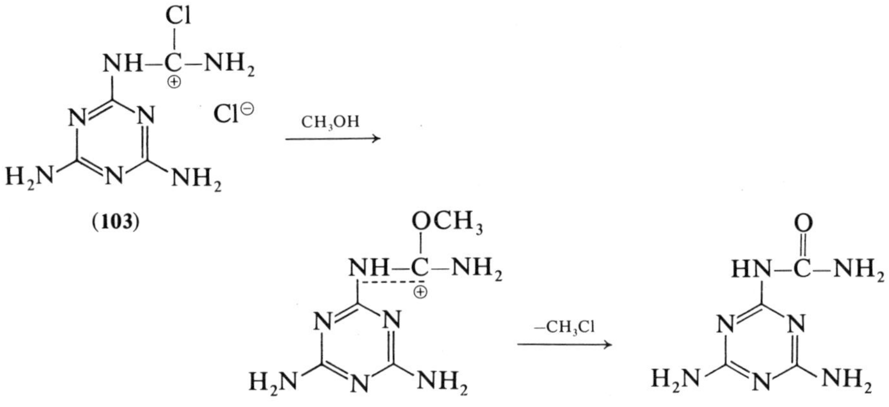

[Structure of 103: triazine with NH–C(Cl)–NH₂ cation, Cl⁻] →(CH₃OH)→ [intermediate with OCH₃] →(−CH₃Cl)→ [Structure of 104: triazine with HN–C(=O)–NH₂]

To Section III-B

Reaction of **33** with Na₂S or NaHS, respectively, affords the 3,4-diazatrithiapentalene **105** (55).

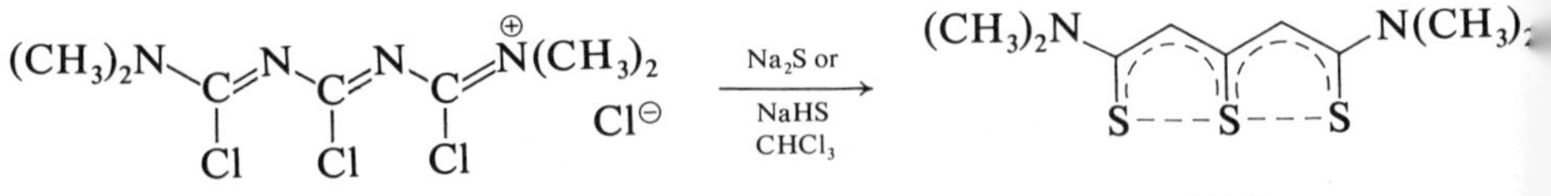

[Structure 33: (CH₃)₂N–C(Cl)=N–C(Cl)=N–C(Cl)=N(CH₃)₂ ⁺ Cl⁻] →(Na₂S or NaHS, CHCl₃)→ [Structure 105: 3,4-diazatrithiapentalene with (CH₃)₂N and N(CH₃)₂ substituents, S---S---S]

From mercaptanes and tetrachlor-2-H-imidazole **96**, imidazolium salts **106** (60) were obtained.

(**96**) (**106**)

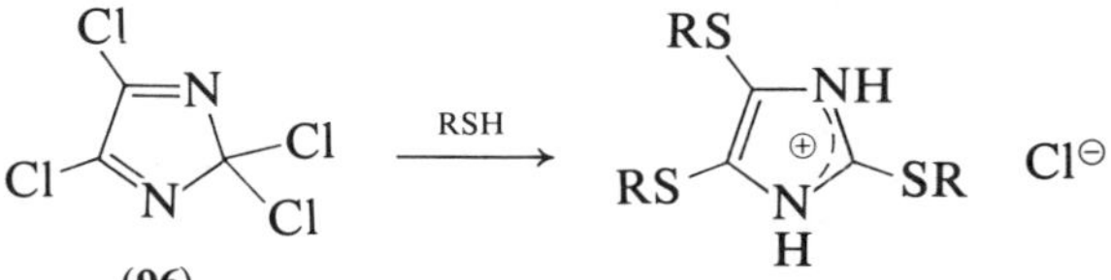

To Section III-C

Nitranilines react with chloroformamidinium chlorides **107** to form guanidines **108**, which may be cyclized, after reduction of the nitro group, to give benzimidazoles **109** (63).

(**108**) (**109**)

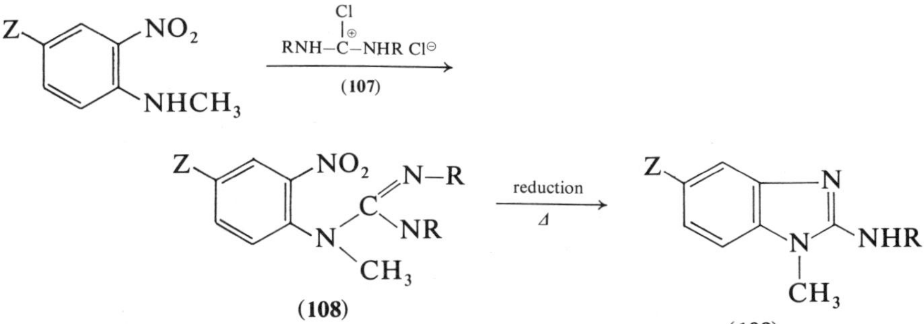

The benzotriazepin **110** was prepared from *o*-phenylenediamine and the iminium compound **111** (54).

(**111**) (**110**)

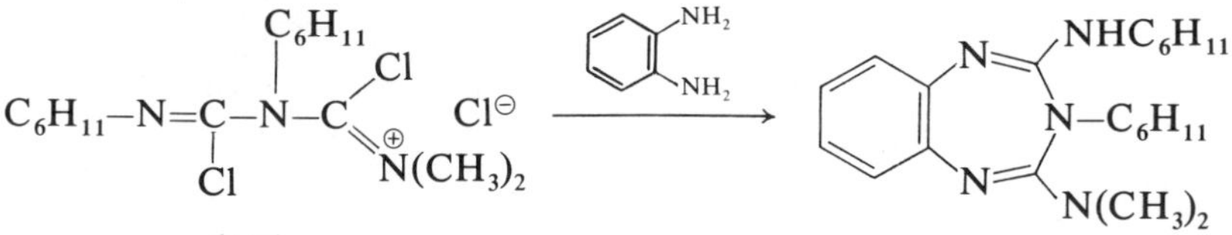

Imidazolium salts **112** are accessible by treatment of **96** with primary amines (60).

(**96**) (**112**)

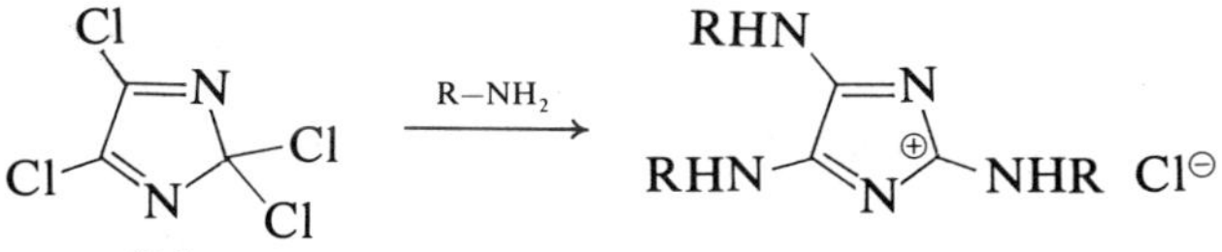

Reaction of chloroformamidinium salts **113** with the hydroxylamin-derivative **114** yields guanidine derivatives **115**, which form *N*-hydroxyguanidines on hydrolysis (52).

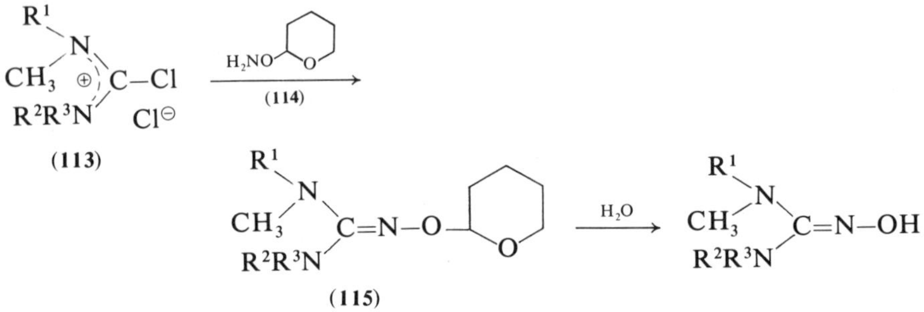

By action of *NNN'N'*-tetrachloroformamidinium chloride on the ketimine **116** the guanidin derivative **117** could be isolated (53).

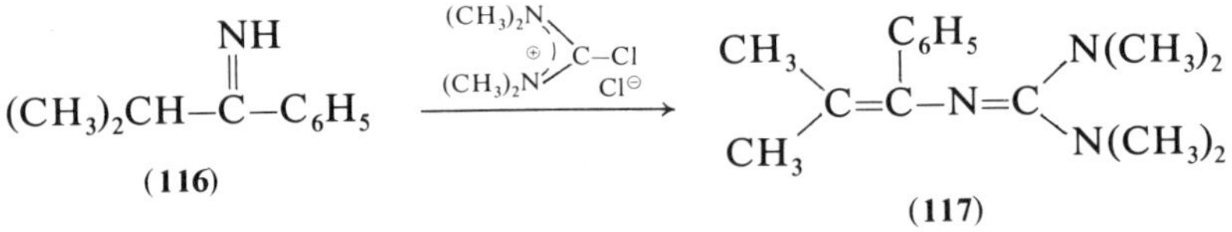

To Section III-G

Tetrachloro-H-imidazol **96** does not undergo Diels-Alder reaction; rather, on heating with maleic acid anhydride, dimerization occurs to **118** (60).

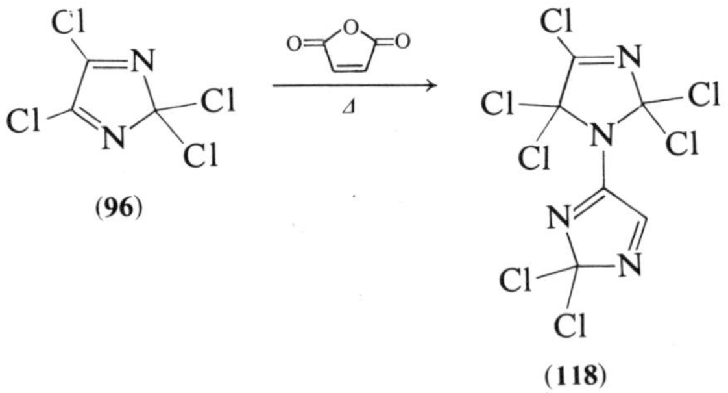

Section III-H

Reactions with inorganic compounds. Bisdimethylaminomalondinitrile **119** forms in the reaction between tetramethylchloroformamidinium chloride and cuprous cyanide (64).

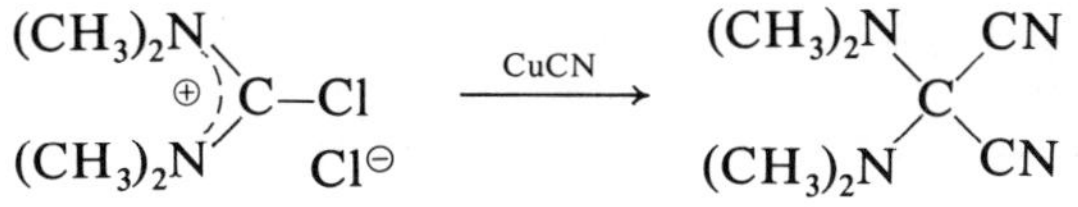

$$(CH_3)_2N\!\!>\!\!\overset{\oplus}{C}\!\!-\!Cl \quad Cl^\ominus \xrightarrow{CuCN} (CH_3)_2N\!\!>\!\!C\!\!<\!\!\overset{CN}{CN}$$
$$(CH_3)_2N \qquad \qquad \qquad (CH_3)_2N$$

(119)

In imidazolium salts **97** the chloratome can be replaced by means of sodium azide (61).

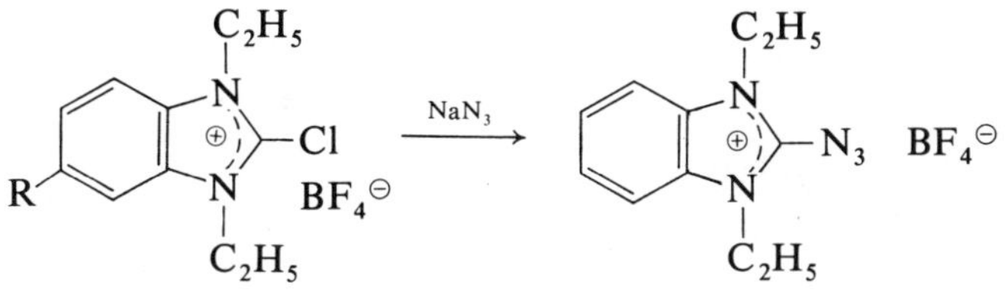

(97)

Silversulfite reacts with the chloroformamidinium salt **39** to yield $S.S.S.$-trioxides of $NNN'N'$-tetramethylthiourea **120** (65).

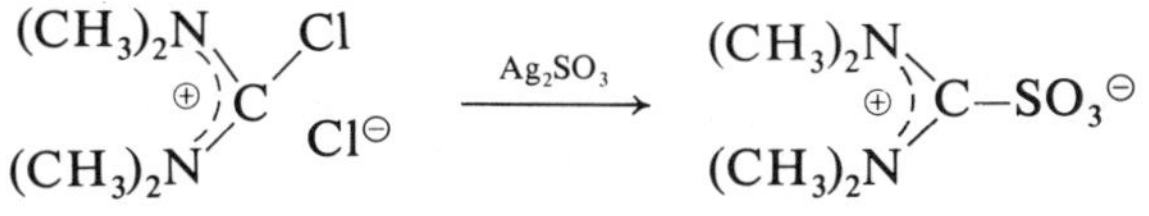

$$(CH_3)_2N\!\!>\!\!\overset{\oplus}{C}\!\!<\!\!\overset{Cl}{Cl^\ominus} \xrightarrow{Ag_2SO_3} (CH_3)_2N\!\!>\!\!\overset{\oplus}{C}\!\!-\!SO_3^\ominus$$
$$(CH_3)_2N \qquad\qquad\qquad (CH_3)_2N$$

(39)   (120)

Bis(dimethylamino)carbenetetracarbonyl iron has been prepared from **39** and iron carbonyl **121** (66).

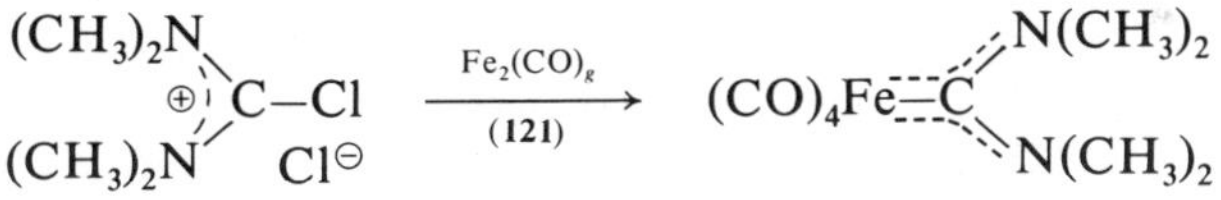

$$(CH_3)_2N\!\!>\!\!\overset{\oplus}{C}\!\!-\!Cl \quad Cl^\ominus \xrightarrow[\text{(121)}]{Fe_2(CO)_8} (CO)_4Fe\!=\!C\!\!<\!\!\overset{N(CH_3)_2}{N(CH_3)_2}$$
$$(CH_3)_2N$$

(39)

REFERENCES

52. S. D. Ziman, *J. Org. Chem.*, **41**, 3253 (1976).
53. J. Baum, D. Scholz, F. Tataruch, and H. G. Viehe, *Chimia*, **29**, 514 (1975).
54. A. Elgavi and H. G. Viehe, *Angew. Chem.*, **89**, 188 (1977).
55. B. Stelander, H. G. Viehe, M. Van Meersche, G. Germain, and J. P. Declercq, *Bull. Soc. Chim. Belg.*, **86**, 291 (1977).
56. V. P. Kukhar' and V. I. Pasternak, *Zh. Org. Khim.*, **11**, 2233 (1975); *C. I.*, 4–115 (1976).
57. V. P. Kukhar', V. I. Pasternak, M. V. Shevchenko, A. S. Shtepanek, and A. V. Krisanov, *Zh. Obshch. Khim.*, **46**, 249 (1976); *C. I.*, 21–278 (1976).
58. G. Birke, H. P. Latscha, and A. Schmidt, *Z. Anorg. Allg. Chem.*, **424**, 287 (1976).
59. M. J. S. Gynane and M. F. Lappert, *J. Organomet. Chem.*, **114**, C4 (1976).

60. K. H. Büchel and H. Erdmann, *Chem. Ber.*, **109**, 1625 (1976).
61. H. Balli and R. Maul, *Helv. Chim. Acta*, **59**, 148 (1976).
62. E. Ichikawa, T. Takizawa, and K. Odo, *Yuki Gosei Kagahu Kyokai Shi*, **32**, 936 (1974); *C. I.*, 20–290 (1975).
63. P. Lugosi, B. Agai, and Gy. Hornyak, *Chem. Eng.*, **19**, 307 (1975).
64. Z. Arnold and M. Svoboda, *Collect. Czech. Chem. Commun.*, **42**, 1175 (1977).
65. W. Walter and C. Rohloff, *Liebigs Ann. Chem.*, **1975**, 295.
66. W. Petz, *Angew. Chem.*, **87**, 288 (1975).

CHLORO(ALKYLMERCAPTO)-METHYLENIMINIUM SALTS (MERCAPTOCHLOROFORMAMIDINIUM SALTS)

By W. KANTLEHNER, *D-7080-AALEN-Dewangen, Laachweg 14, Germany*

CONTENTS

I. Synthesis	173
A. From Dithiocarbamide Acid Esters	173
B. Via Addition of Hydrogen Halides to Thiocyanates	174
II. Structure and Properties of Mercaptochloroformamidinium Chlorides	174
III. Reactions of Mercaptochloroformamidinium Chlorides	175
A. Hydrolysis	175
B. Reactions With Phenols	175
C. Reactions With Primary Amines and Amine Derivatives	175
D. Reactions With Carboxylic Acid Hydrazides and Carboxylic Acid Amide Oximes	176
E. Reactions With Bifunctional Compounds	177
F. Reactions With Heterocycles and CH-Acidic Compounds	178
G. Reactions of Chloro(alkylmercapto)methyleniminium Salts Generated *in situ*	179
References	180

I. Synthesis

A. FROM DITHIOCARBAMIDE ACID ESTERS

Phosgene transforms *N,N*-disubstituted dithiocarbamates (**1**) into crystalline, hygroscopic mercaptochloroformamidinium chlorides (**2**) via elimination of carbon oxysulfide (1, 2).

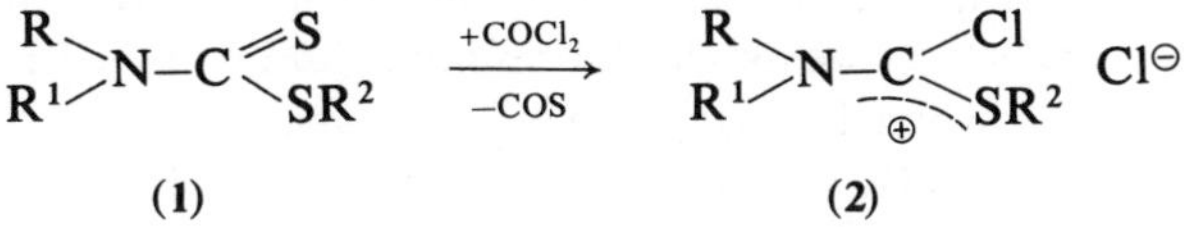

$$\underset{(\mathbf{1})}{\overset{R}{\underset{R^1}{>}}N-C\overset{S}{\underset{SR^2}{\lessgtr}}} \xrightarrow[-COS]{+COCl_2} \underset{(\mathbf{2})}{\overset{R}{\underset{R^1}{>}}N-\overset{+}{C}\overset{Cl}{\underset{SR^2}{\lessgtr}}} Cl^{\ominus}$$

Translated from German by Dr. W.-J. Richter, MPI für Kohleforschung, D 4330 Mühlheim.

The chlorination is performed in apolar solvents like toluene, and it is successful with both aliphatic and aromatic ligands. Cyclic dithiocarbamates like **3** are also transformed into amide chlorides like **4** (1, 2).

$$\underset{(3)}{\underset{S}{\overset{CH_3}{\underset{|}{N}}}\!=\!S} \quad \xrightarrow[-COS]{+COCl_2} \quad \underset{(4)}{\left[\overset{CH_3}{\underset{S}{\underset{|}{N}}}\!\overset{\oplus}{-}Cl\right] Cl^{\ominus}}$$

Other chlorinating agents like PCl_5 and $SOCl_2$ also react successfully with dithiocarbamates (**1**) to yield mercaptochloroformamidinium chlorides (**2**) (1, 2).

Elementary chlorine is also claimed to generate **2** from dithiocarbamates (**1**) (3).

$$\underset{(1)}{\overset{R}{\underset{R^1}{>}}N-C\overset{S}{\underset{SR^2}{<}}} \quad \xrightarrow{Cl_2} \quad \underset{(2)}{\overset{R}{\underset{R^1}{>}}N-\overset{\oplus}{C}\overset{SR^2}{\underset{Cl}{<}} Cl^{\ominus}}$$

B. VIA ADDITION OF HYDROGEN HALIDES TO THIOCYANATES

Hydrogen chloride forms only labile adducts with thiocyanates (4). On the other hand, stable mercaptobromoformamidinium bromides **7** and **8** are obtained from thiocyanates **5** and **6** (4).

$$R-S-C\equiv N + 2HBr \quad \longrightarrow \quad H_2N-\overset{\oplus}{C}\overset{SR}{\underset{Br}{<}} Br^{\ominus}$$

(5) R = CH_3   (7) R = CH_3
(6) R = C_6H_5   (8) R = C_6H_5

Several of these adducts from HCl and thiocyanates are used without isolation for the synthesis of heterocycles (5).

II. Structure and Properties of Mercaptochloroformamidinium Chlorides

Mercaptochloroformamidinium chlorides (**2**) generally are crystalline substances; however, their melting points are not very characteristic (1).

The compounds **2** are insoluble in ethers and hydrocarbons; they are soluble in strongly polar solvents like dimethylformamide and N-methylpyrrolidone.

Some salts also dissolve in methylene chloride and chloroform (1). Owing to its solubility and the IR band at 6 μ from the $C=N^{\oplus}$ vibration of the mercaptochloroformamidinium chlorides, the salt-like structure **2** has been attributed to this class of compounds (1, 4).

III. Reactions of Mercaptochloroformamidinium Chlorides

A. HYDROLYSIS

In the presence of sodium carbonate the hydrolysis of mercaptochloroformamidinium chlorides uniformly proceeds to give thiocarbamic acid S-esters (9); it is considered as further proof for the structure of the class of compounds 2 (1, 2).

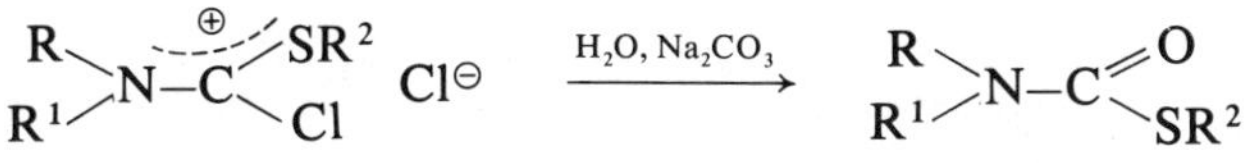

$$\underset{(2)}{\overset{R}{\underset{R^1}{>}}N\overset{\oplus}{-}C\overset{SR^2}{\underset{Cl}{<}}\; Cl^{\ominus}} \xrightarrow{H_2O,\, Na_2CO_3} \underset{(9)}{\overset{R}{\underset{R^1}{>}}N-C\overset{O}{\underset{SR^2}{<}}}$$

B. REACTION WITH PHENOLS

The reaction of 2 with phenols, for example, β-naphthol, yields thiocarbamic acid O-aryl esters, for example, 11, via the adducts 10 (1).

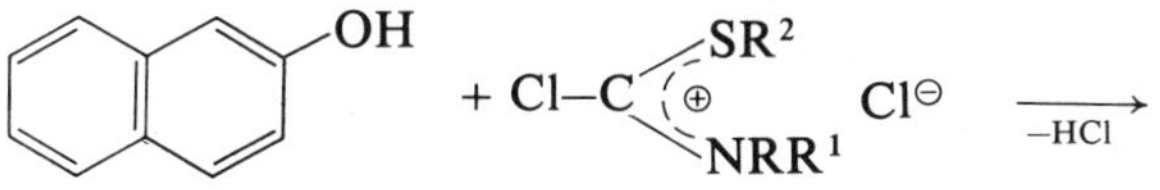

<chemical scheme showing naphthol-OH + Cl-C(SR²)(⊕)(NRR¹) Cl⁻ (2), proceeding with –HCl to adduct 10 (naphthyl-O-C(SR²)(⊕)(NRR¹) Cl⁻), then with –R²Cl to 11 (naphthyl-O-C(=S)-NR¹R²)>

C. REACTIONS WITH PRIMARY AMINES AND AMINE DERIVATIVES

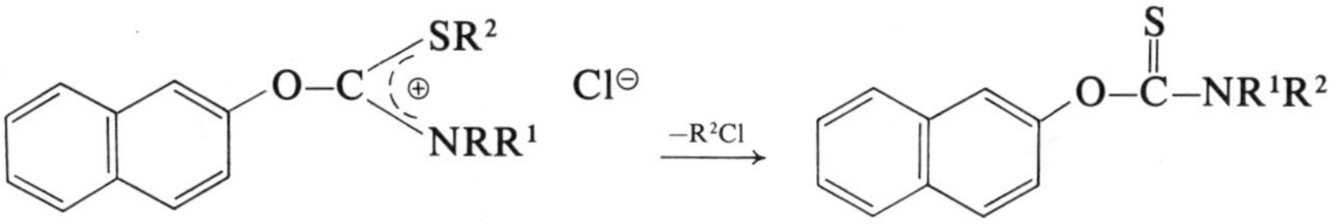

Primary aromatic amines react with 2 to yield isothioureas (12); the reaction is catalyzed by bases, for example, pyridine (1).

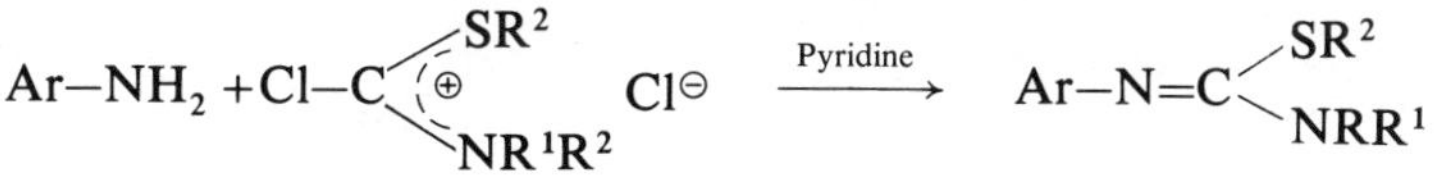

$$\text{Ar—NH}_2 + \underset{(2)}{\text{Cl—C}\overset{SR^2}{\underset{NR^1R^2}{<}}(\oplus)\; Cl^{\ominus}} \xrightarrow{\text{Pyridine}} \underset{(12)}{\text{Ar—N}=C\overset{SR^2}{\underset{NRR^1}{<}}}$$

Tosylamide analogously reacts with mercaptochloroformamidinium chloride 13 to give isothiourea 14 (1).

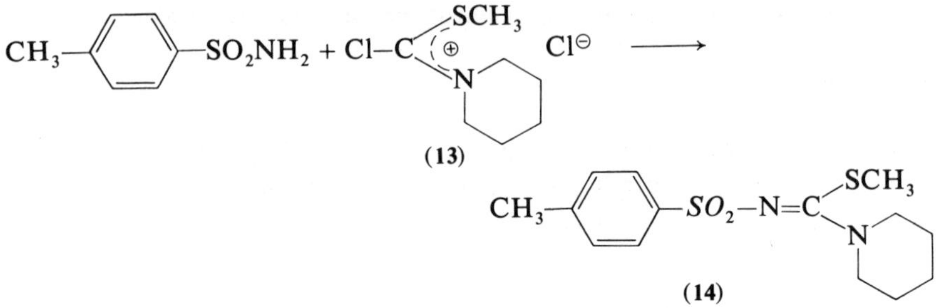

D. REACTIONS WITH CARBOXYLIC ACID HYDRAZIDES AND CARBOXYLIC ACID AMIDE OXIMES

Carboxylic acid hydrazides are cyclized by mercaptochloroformamidinium chlorides (2) to yield 1,3,4-oxadiazoles 16 and 17 via isothiouronium salts 15 (1). The solvent unambiguously influences this kind of reaction.

Strongly basic solvents like pyridine cause cleavage of the alkylmercapto grouping forming 1,3,4-oxadiazoles 16. In weakly basic solvents like N-methylpyrrolidone ring closure via elimination of the amino group occurs to yield 1,3,4-oxadiazoles (17). The route of the cyclization cannot be influenced if the sulfur atom in 2 carries an aryl, a 2-ethynyl, or a 2-cyanoethyl group (1). In these cases the 1,3,4-oxadiazole 16 always forms.

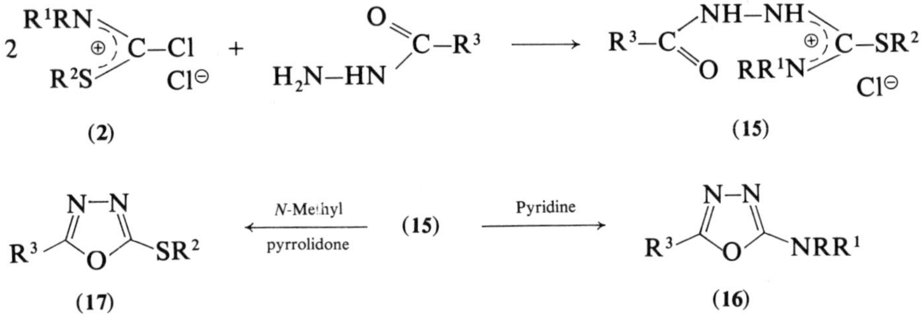

The choice of the solvent allows control of the ring closure of amide oximes with mercaptochloroformamidinium chlorides (13) leading to 1,3,4-oxadiazoles of either type 18 or 19, according to the same principles.

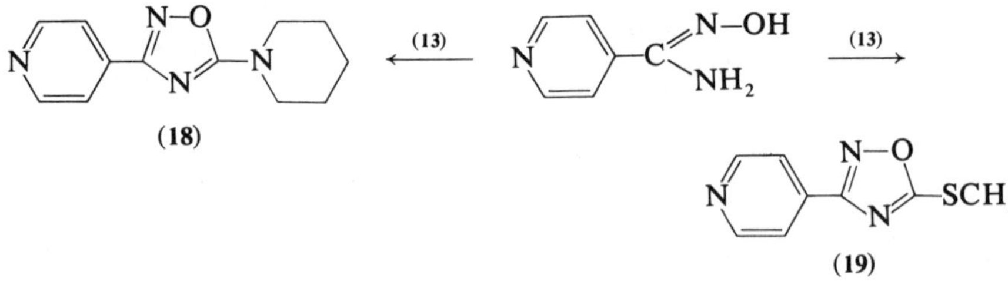

III. REACTIONS OF MERCAPTOCHLOROFORMAMIDINIUM CHLORIDES

The thiosemicarbazone **20** is converted into the 1,3,4-thiadiazole **22** by mercaptochloroformamidinium chloride (**21**) (1).

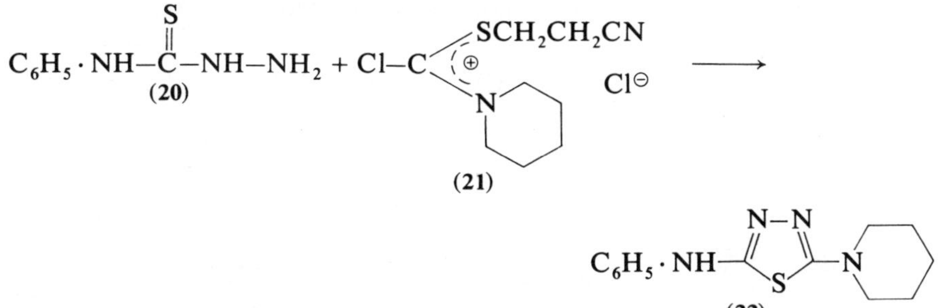

A 1,2,4-triazole (**23**) was synthesized from the imide hydrazide (**24**) and **21** (1).

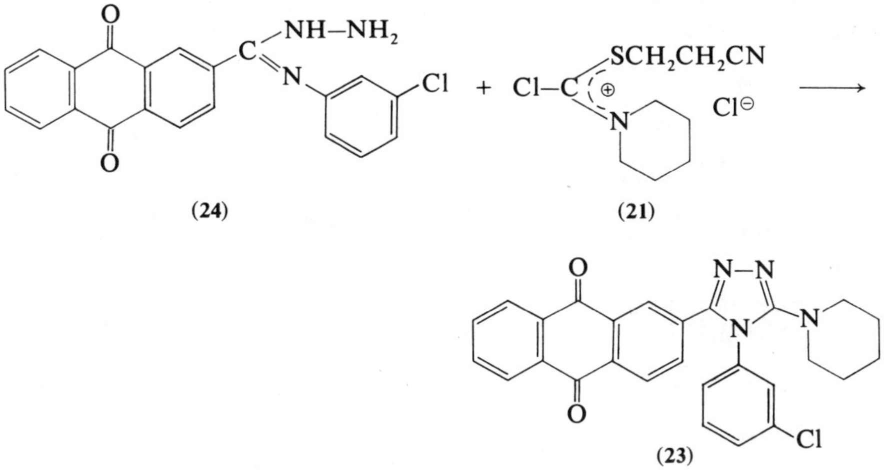

E. REACTIONS WITH BIFUNCTIONAL COMPOUNDS

Bifunctional compounds like o-phenylenediamines, o-aminophenols, and o-aminothiophenols are cyclized by mercaptochloroformamidinium chlorides (**2**) to yield 2-(dialkylamino)benzimidazoles (**25**), benzoxazoles (**26**), and benzthiazoles (**27**) (1).

(**25**) X = NH
(**26**) X = O
(**27**) X = S

If the phenylene group is substituted, the reaction of some bifunctional compounds, for example, **28**, is regulated by the solvent in such a way that the alkylmercapto group remains in the molecule, for example, **29** (1).

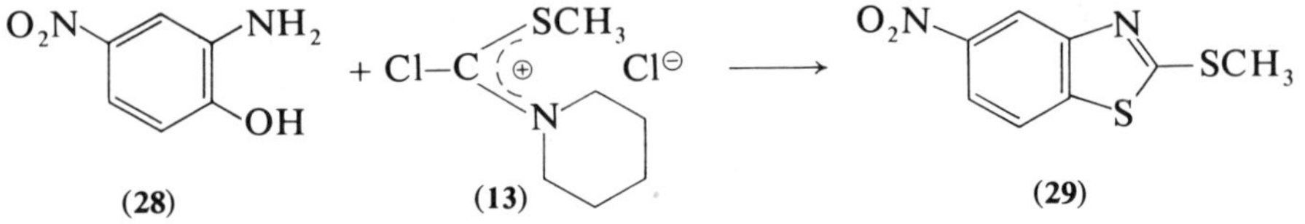

(28) + (13) ⟶ (29)

F. REACTIONS WITH HETEROCYCLES AND CH-ACIDIC COMPOUNDS

Pyrrole and substituted pyrroles, as well as indole, react with *N,N*-dialkyl-chloromethylmercaptomethyleniminium chloride (**30**) to give alkylmethyl-mercaptomethyleniminium chlorides **31** or **32** (6). The reaction seems to be widely applicable, and possibly will become important for the synthesis of salts of the type **31** or **32**.

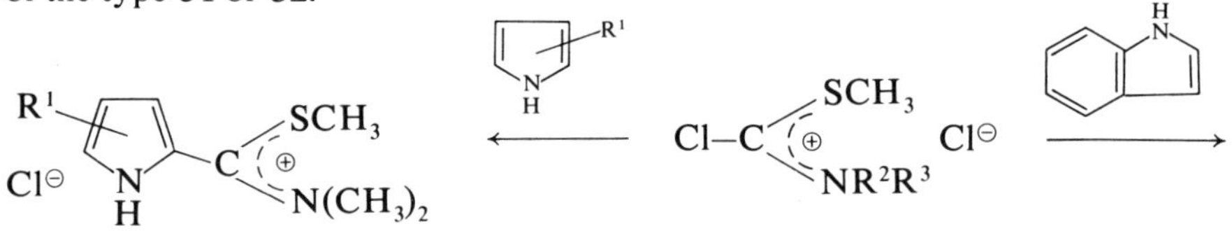

(31)
$R^1 = H$
$R^1 = 3\text{-COOC}_2H_5$
$2,4\text{-}(CH_3)_2$

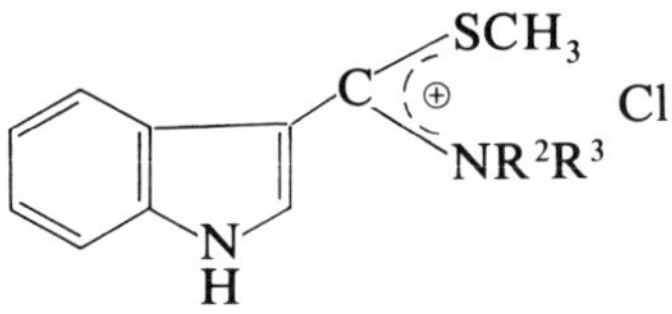

(30) ⟶ (32)

$R^2 = R^3 = CH_3$
$R^2\text{-}R^3 = -(CH_2)_4-$

Carbon acids like malononitrile in the presence of pyridine are converted into ketene *S,N*-acetals like **33** and **34** by chloroalkylmercaptoformamidinium chlorides (1).

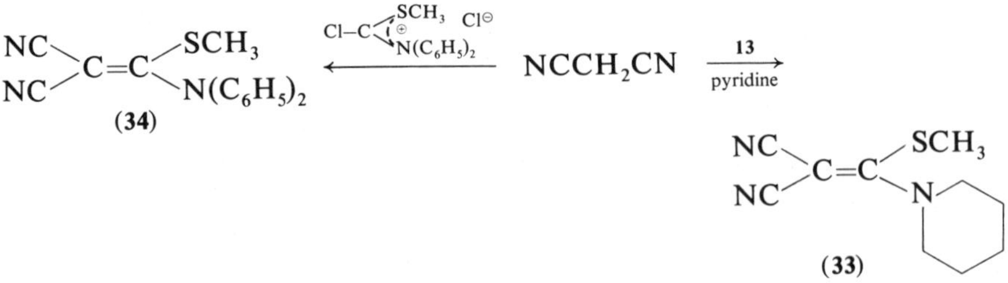

$NCCH_2CN \xrightarrow[\text{pyridine}]{13}$ (33)

⟵ (34)

G. REACTIONS OF CHLOROALKYLMERCAPTOMETHYLENIMINIUM SALTS GENERATED *IN SITU*

a,β-Unsaturated β-thiocyanatocarboxylic acid chlorides may be cyclized with the aid of hydrogen halides. 2-Chloro-1,3-benzthiazin-4-ones (**35**), 2-chloro-1,3-thiazin-4-ones (**36**), 2-chloro-4-thiazolones (**37**), and 2-chloro-5,6-dihydro-1,3-thiazin-4-one (**38**) have been synthesized this way. These reactions presumably proceed via chloroalkylmercaptomethyleniminium salts as reactive intermediates. This type of reaction has been reviewed by Simchen and Entenmann (5).

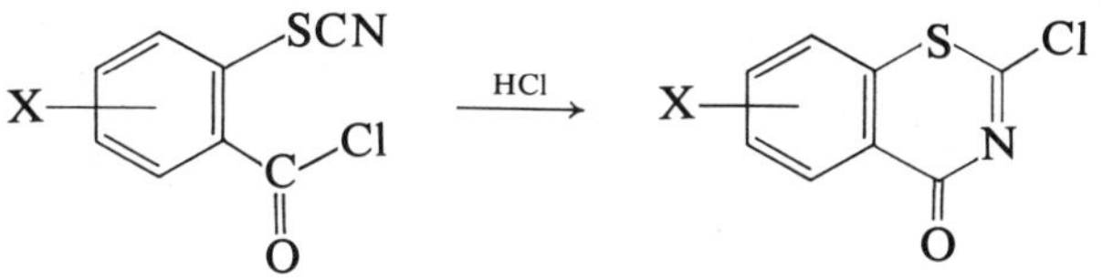

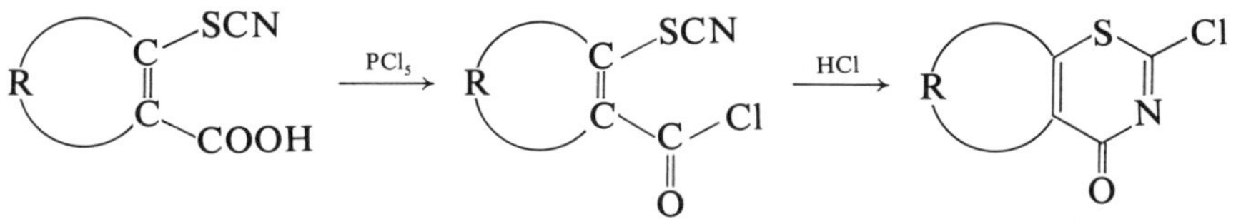

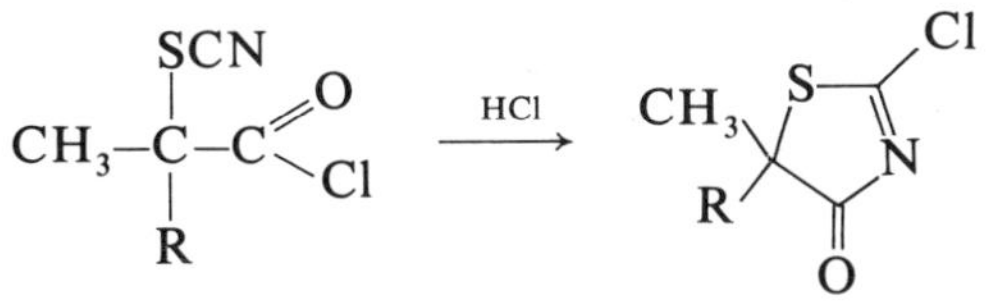

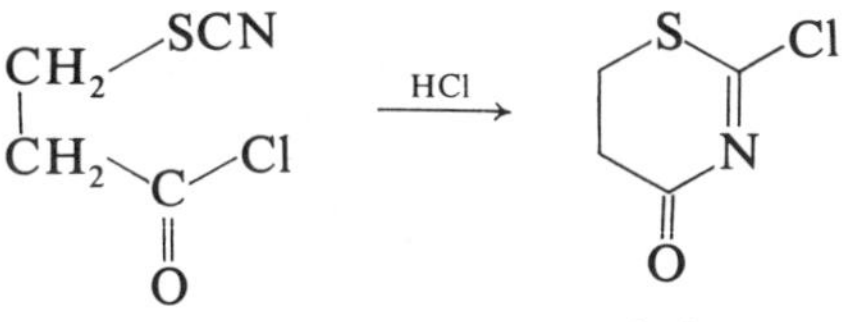

ACKNOWLEDGMENT

To Dr. Heinz Eilingsfeld BASF AG, Ludwigshafen, and to Dr. Bernd Funke as well as Dr. Erwin Haug, both of Fachhochschule Aalen, Aalen, I am very indebted for fruitful discussions and critical revisions of the text.

REFERENCES

1. H. Eilingsfeld and L. Möbius, *Chem. Ber.*, **98**, 1293 (1965).
2. Belg. Pat. 6,60,941 (1965), H. Eilingsfeld, L. Möbius *C.A.*, **64**, 3364 (1966).
3. W. Walter and K. Bode, *Angew. Chem.*, **79**, 292 (1967).
4. E. Allenstein and P. Quis, *Chem. Ber.*, **97**, 3162 (1964).
5. G. Simchen and G. Entenmann, *Angew. Chem.*, **85**, 155 (1973).
6. R. L. N. Harris, *Australian J. Chem.*, **27**, 2635 (1974).

ALKOXYMETHYLENIMINIUM SALTS

By W. KANTLEHNER, *D-7080-AALEN-Dewangen, Laachweg 14, Germany*

CONTENTS

I. Synthesis 182
   A. Via Alkylation of Acid Amides, Lactams, Ureas and Urea Derivatives 182
      1. Alkylation With Oxonium Salts 182
         a. Acid amides 182
         b. Lactams 190
         c. Urea and Urea Derivatives 192
      2. Alkylation with Carboxonium Ions 194
      3. Alkylation with Dialkyl Sulfates 194
         a. Carboxylic Acid Amides 194
         b. Lactams 200
         c. Urea and Urea Derivatives 201
      4. With Alkyl Halides and Sulfonic Acid Esters 203
   B. Via Acylation of Acid Amides 208
   C. Via Alkylation and Acylation of Imino Esters 211
   D. From Chloromethyleniminium Salts (Amide Halides) 213
   E. From Amide Acetals and Lewis Acids 215
   F. Via Reaction of Ketene *O,N*-Acetals with Electrophilic Reagents 216
   G. Via Aminolysis of Carboxonium Ions 217
   H. Via Addition of Hydrogen Halides and Alcohols to Nitriles 218
   I. Addition of Alcohols to Nitrilium Salts 218
II. Structure of Alkoxymethyleniminium Salts 219
III. Reactions of Alkoxymethyleniminium Salts with Nucleophiles 222
   A. Reactions with Hydroxyl Groups 224
      1. Hydrolysis 224
      2. Alcoholysis 226
      3. Reactions with Hemiacetals 228
      4. Acidolysis 229
   B. Reactions with Thiol Groups 230
      1. Thiolysis 230
      2. Reactions with Mercaptans and Thiophenols 231
   C. Reactions with Selenol Groups 232
   D. Reactions with NH-Containing Compounds 232
      1. Ammonia and Primary and Secondary Amines 232
      2. *N*-Monoacylated Amines and Vinylogous Amides 238
      3. Bifunctional Compounds (OH, NH, SH) 239

E. Reactions with Bases	241
F. Reactions with Alkali Metal Alkoxides	248
G. Reactions with Alkali Metal Halides and Cyanides	258
H. Reactions with Complex Hydrides	260
I. Reactions with Alkali Metal Phosphides	261
References	262
Addendum	268

I. Synthesis

A. VIA ALKYLATION OF ACID AMIDES, LACTAMS, UREAS, AND UREA DERIVATIVES

Mesomeric structures as well as molecular orbital calculations (1) of the acid amide function show the highest electron density at the carbonyl oxygen atom.

$$R-C\overset{\bar{O}|}{\underset{NR^1R^2}{}} \longleftrightarrow R-C\overset{O^\ominus}{\underset{\overset{\oplus}{N}R^1R^2}{}} \longleftrightarrow R-C\overset{O^\ominus}{\underset{\overset{\oplus}{N}R^1R^2}{}}$$

Under kinetically controlled reaction conditions, electrophilic reagents attack there, giving rise to iminium ions (1). The polarizability of the O–Y bond as well as the nucleophilicity of X determine the equilibrium.

$$R-C\overset{O}{\underset{NR^1R^2}{}} + XY \rightleftarrows R-C\overset{O-Y}{\underset{NR^1R^2}{\overset{\oplus}{}}} X^\ominus$$

(1)

In accordance with this picture, acid amides are predominantly protonated at the oxygen atom (2–7). In some cases these addition products could be isolated in a crystalline state (8–11), for example, as fluoroborates (2) (11).

$$R-C\overset{O}{\underset{NR^1R^2}{}} \xrightarrow{HF-BF_3} R-C\overset{OH}{\underset{NR^1R^2}{\overset{\oplus}{}}} BF_4^\ominus$$

(2)

If acid chlorides (PCl_5, $POCl_3$, $SOCl_2$, SO_2Cl_2, RSO_2Cl, $RCOCl$) are used as electrophilic reagents, compounds of the Vilsmeier–Haack–Arnold type result; they are dealt with here only as precursors to N,N-dialkylalkoxymethyleniminium salts.

From alkylating agents and acid amides N,N-dialkylalkoxymethyleniminium salts are generated.

1. Alkylation with Oxonium Salts

(a) *Acid Amides.* H. Meerwein (12–14) was the first to alkylate acid amides with oxonium salts. From acetamide and triethyloxonium fluoroborate (3) he obtained a crystalline product with structure 4 attributed to it.

$$CH_3-C\underset{NH_2}{\overset{O}{\diagup}} + (C_2H_5)_3\overset{\oplus}{O}\,BF_4^{\ominus} \xrightarrow[-(C_2H_5)_2O]{} CH_3-C\underset{NH_2}{\overset{OC_2H_5}{\diagup}}{}^{\oplus} \quad BF_4^{\ominus}$$
$$\qquad\qquad\qquad\qquad (3) \qquad\qquad\qquad\qquad\qquad\qquad (4)$$

Later studies by Meerwein and other authors revealed that owing to the extremely high alkylation potential of trialkyloxonium salts, nearly all acid amides (primary as well as secondary and tertiary) are alkylated at the oxygen atom. The reaction normally runs to completion, since ether is cleaved irreversibly, thus forming alkoxymethyleniminium fluoroborates (**4a**).

$$R-C\underset{NR^1R^2}{\overset{O}{\diagup}} + (R^3)_3\overset{\oplus}{O}\,BF_4^{\ominus} \xrightarrow[-R^3-O-R^3]{} R-C\underset{NR^1R^2}{\overset{O-R^3}{\diagup}}{}^{\oplus} \quad BF_4^{\ominus}$$
$$\qquad\qquad\qquad\qquad\qquad\qquad\qquad (4a)$$

For such alkylations the easily accessible trimethyl- or triethyloxonium fluoroborates have been used exclusively. The influence of the ligands R, R^1, and R^2 of the acid amide on the course of the alkylation with triethyloxonium fluoroborate has been studied (15, 16). According to these studies steric effects of the aliphatic ligand R have but little influence on the reaction, since pivalic acid dimethylamide is almost quantitatively converted into **6** by triethyl- as well as by trimethyloxonium fluoroborate (15, 16).

$$(CH_3)_3C-C\underset{N(CH_3)_2}{\overset{O}{\diagup}} + R_3\overset{\oplus}{O}\,BF_4^{\ominus} \xrightarrow[-R_2O]{} (CH_3)_3C-C\underset{N(CH_3)_2}{\overset{O-R}{\diagup}}{}^{\oplus} \quad BF_4^{\ominus}$$
$$\qquad\qquad\qquad (5)\ R = CH_3 \qquad\qquad\qquad\qquad (6)$$
$$\qquad\qquad\qquad (3)\ R = C_2H_5$$

Also electron-withdrawing ligands R like diethoxymethyl, chloromethyl, cyanomethyl, dichloromethyl, dimethylaminocarbonyl, and ethoxycarbonyl, as well as the cyanide group itself, allow alkylation of the acid amide group (see Table I). Only extremely electron-withdrawing groups like the chlorocarbonyl group or steric hindrance in connection with this property, as in the trichloromethyl group, block alkylation of the acid amide function. Thus neither dimethylcarbamoyl chloride nor trichloroacetic acid dimethyl amide is successfully alkylated with triethyloxonium fluoroborate (**3**) (15).

With aromatic groups R no restriction on the alkylation reaction has been reported. Several aromatic acid amides—*o*-, *m*-, and *p*-substituted ones—were methylated with trimethyloxonium fluoroborate to yield **7** (16).

$$X-\underset{}{\langle\text{Ar}\rangle}-C\underset{N(CH_3)_2}{\overset{O}{\diagup}} + (CH_3)_3\overset{\oplus}{O}\,BF_4^{\ominus} \xrightarrow[-CH_3OCH_3]{} X-\underset{}{\langle\text{Ar}\rangle}-C\underset{N(CH_3)_2}{\overset{OCH_3}{\diagup}}{}^{\oplus} \quad BF_4^{\ominus}$$
$$\qquad\qquad\qquad\qquad\qquad\qquad\qquad\qquad (7)$$

X = *o*, *m*, *p*-CH_3, OCH_3, Cl, NO_2

Alkyl or aryl groups at the nitrogen atom display little influence on the alkylation reaction, as shown by the methylation of N,N-diisopropylacetamide (16), N,N-dicyclohexylacetamide (16), or N,N-diphenylformamide (17). Diphenylacetamide was also alkylated with trimethyloxonium fluoroborate (16):

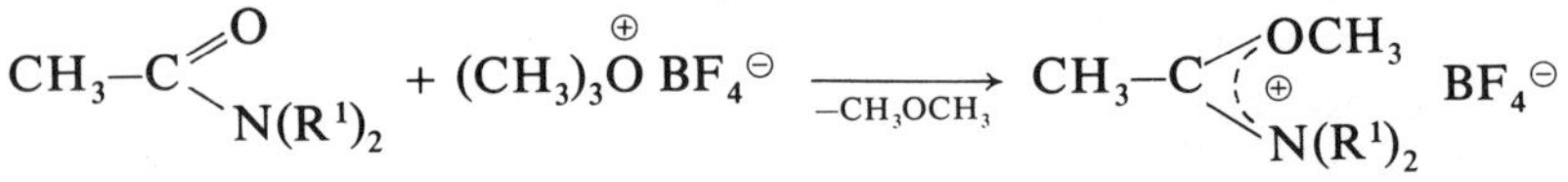

$$CH_3-C\underset{N(R^1)_2}{\overset{O}{\diagdown}} + (CH_3)_3\overset{\oplus}{O}\ BF_4^{\ominus} \xrightarrow{-CH_3OCH_3} CH_3-C\underset{N(R^1)_2}{\overset{OCH_3}{\diagdown}}\ BF_4^{\ominus}$$

$R^1 = i\text{-}C_3H_7, C_6H_{11}, C_6H_5$

With triethyloxonium fluoroborate the N-alkoxy-N-arylcarboxylic acid amides (**8**) were successfully alkylated at the oxygen atom (18).

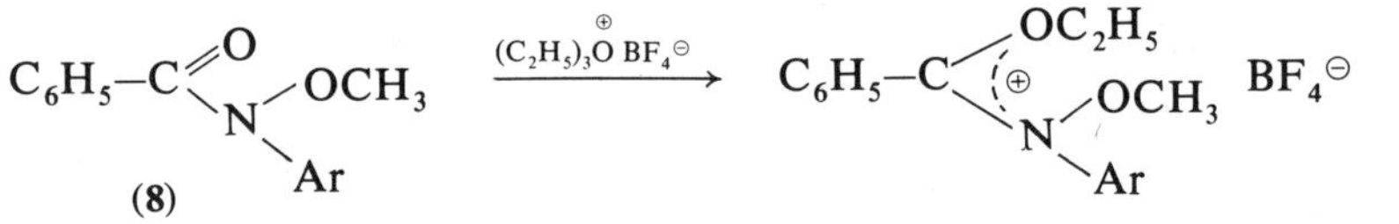

$$C_6H_5-C\underset{\underset{Ar}{N}}{\overset{O}{\diagdown}}-OCH_3 \xrightarrow{(C_2H_5)_3\overset{\oplus}{O}\ BF_4^{\ominus}} C_6H_5-C\underset{\underset{Ar}{N}}{\overset{OC_2H_5}{\diagdown}}-OCH_3\ BF_4^{\ominus}$$
(**8**)

Acylated alicyclic amines like acetylpyrrolidine, acetylpiperidine, benzoylpiperidine, and acetylmorpholine are also alkylated (19–22). With amides carrying an amino function as well, alkylation takes place at the nitrogen atom of the amino group, as shown with N,N-dimethylnicotinamide (**9**) (23).

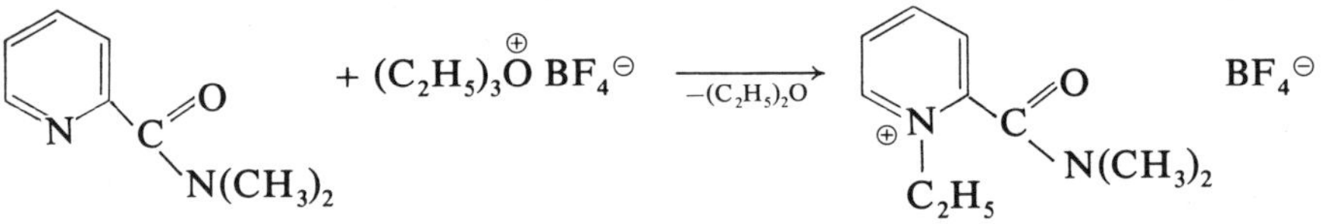

(**9**) + $(C_2H_5)_3\overset{\oplus}{O}\ BF_4^{\ominus}$ $\xrightarrow{-(C_2H_5)_2O}$ ... $BF_4^{\ominus}$

Triethyloxonium fluoroborate predominantly attacks the sodium salt of α-pyridone (**10**) at the nitrogen atom and to a lesser extent at the oxygen atom (24). The O,N-dialkylated product **11** is also obtained by alkylation of both the other products, **12** and **13** although in less amount.

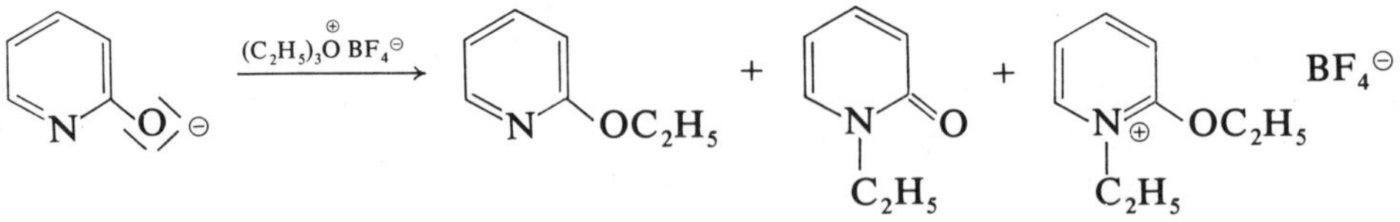

(**10**) $\xrightarrow{(C_2H_5)_3\overset{\oplus}{O}\ BF_4^{\ominus}}$ (**12**) + (**13**) + (**11**) $BF_4^{\ominus}$

If the amide molecule contains an orthoester function, this may be destroyed during the course of the alkylation. When triethoxyacetic acid dimethylamide (**14**), piperidide (**15**), and morpholide (**16**) are reacted with triethyloxonium fluoroborate (3), compound 17 forms via ether cleavage of the orthoester function (21).

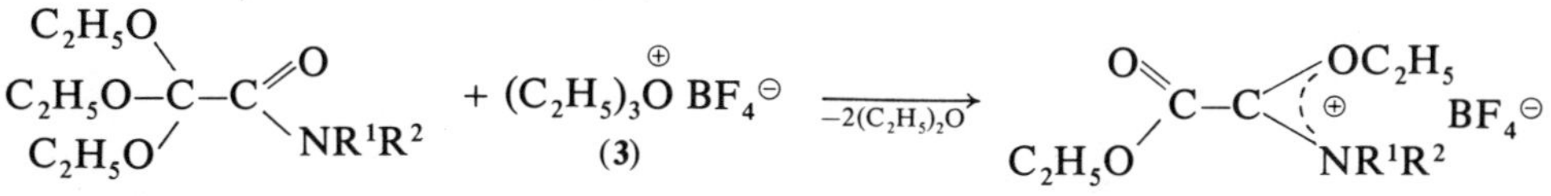

(14) $R^1 = R^2 = CH_3$
(15) $R^1-R^2 = -(CH_2)_5-$
(16) $R^1-R^2 = -(CH_2)_2-O-(CH_2)_2-$

On the other hand, triethoxyacetic acid pyrrolidide (18) is converted by triethyloxonium fluoroborate (3) into 19; here the orthoester function is preserved (21).

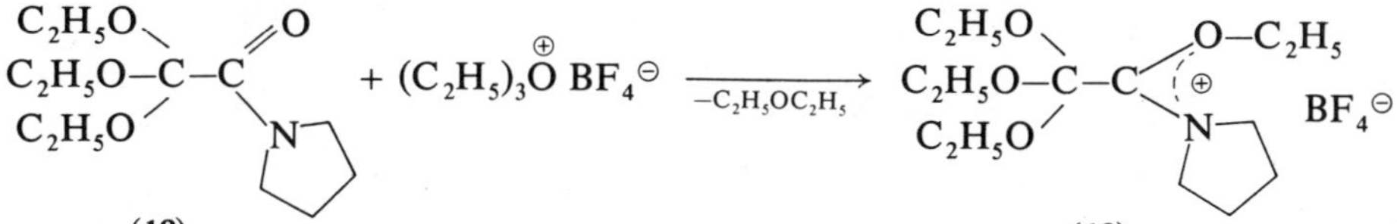

Triethyloxonium fluoroborate does not attack triple bonds if they are present in the amide (25):

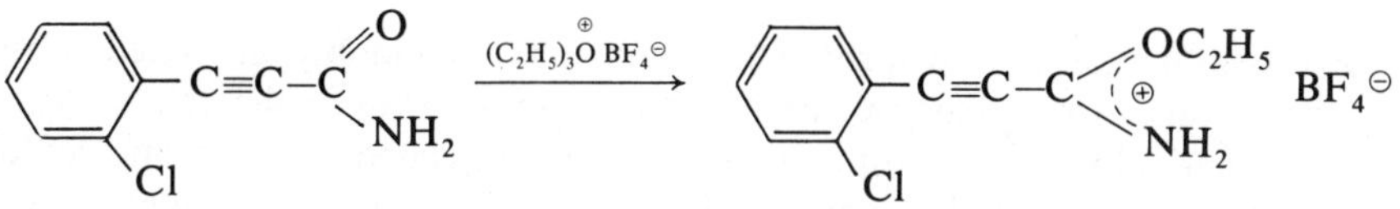

The diazo group as well apparently does not influence the alkylation, as shown by the alkylation of 20 (26).

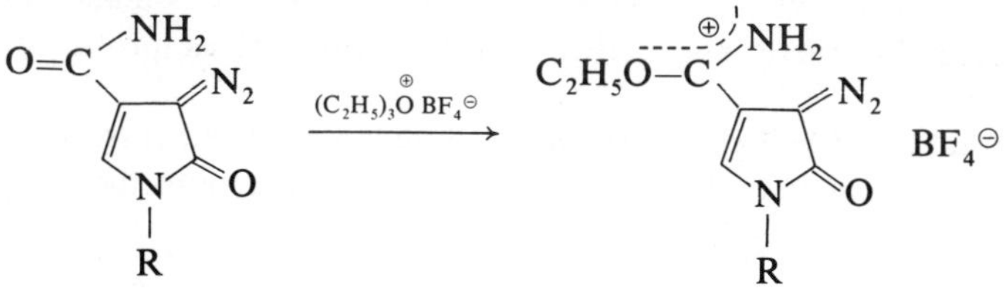

Ferrocenes carrying dialkylaminocarbonyl groups like 21 have also been alkylated with oxonium salts (27).

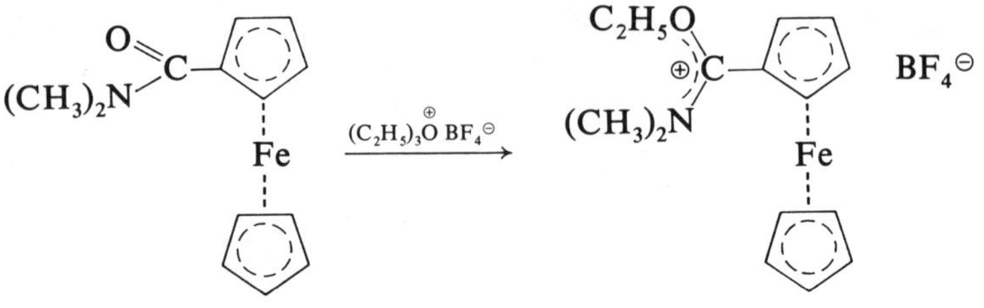

(21)

The cycloheptatriene–norcaradiene equilibrium was studied in the case of the alkylated cycloheptatrienecarboxylic acid dimethylamide (22) (28–30).

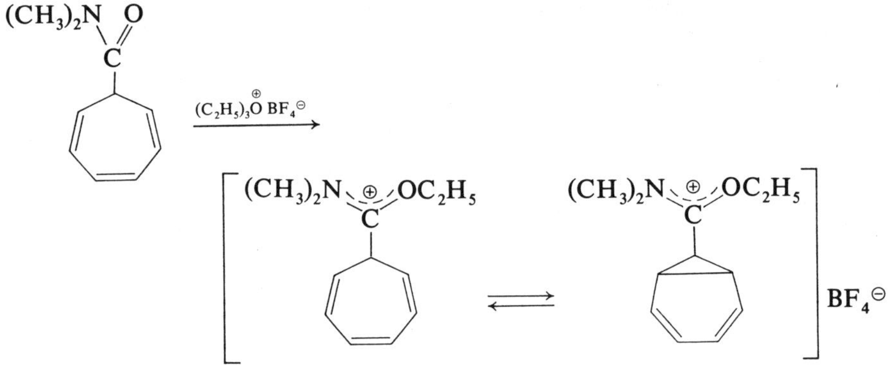

(22)

Bis(dialkylamides) of dicarboxylic acids may be doubly alkylated if the amide functions are at least two carbon atoms apart. Thus oxalic acid (15) and malonic acid (15, 31) bis(dimethylamide) are alkylated only at one amide group. On the other hand, maleic and fumaric acid bis(dimethylamide), as well as terephthalic acid bis(dimethylamide), are successfully doubly alkylated (15). The reaction of succinic acid and adipic acid bis(dimethylamide) with oxonium salts has not been studied so far. However, since adduct formation of dicarboxylic amides with phosphorus oxychloride proceeds analogously (32) and is successful there, these dicarboxylic acid amides probably can be alkylated also.

Squaric acid bis(dialkylamide) 23 is alkylated by triethyloxonium fluoroborate only at a single carbonyl oxygen atom, giving rise to the salts 24 (33, 34).

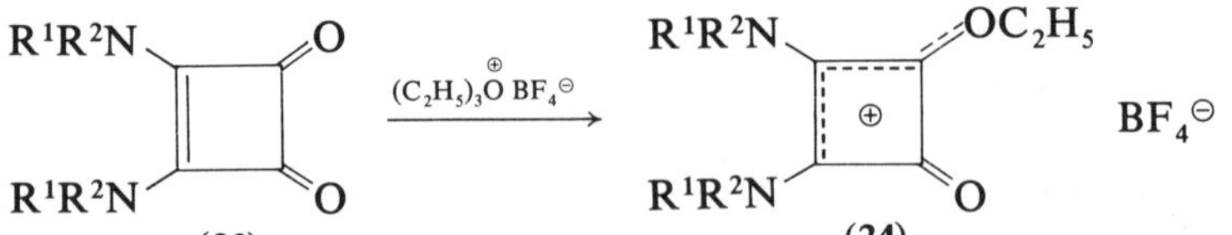

Bis(dialkylcarbamoyl)mercury compounds (**25**) are doubly alkylated by oxonium salts (35).

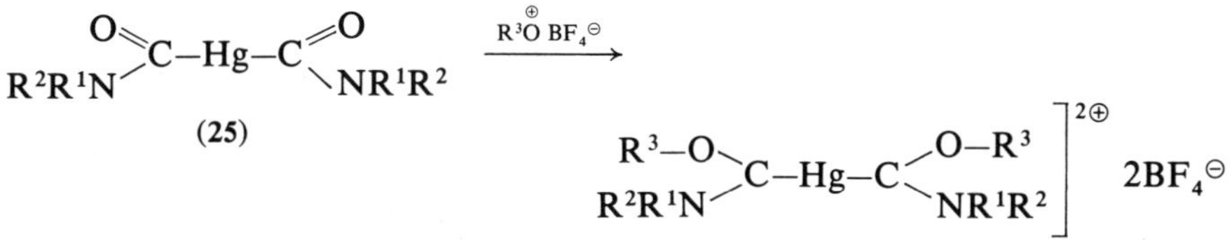

$$\underset{R^2R^1N}{O=}C-Hg-C\underset{NR^1R^2}{=O} \xrightarrow{R^3O^{\oplus} BF_4^{\ominus}} \left[\underset{R^2R^1N}{R^3-O}C-Hg-C\underset{NR^1R^2}{O-R^3}\right]^{2\oplus} 2BF_4^{\ominus}$$

(**25**)

N-Acylated β-aminopropionic acid amides (**26**) are cyclized by triethyloxonium fluoroborate to dihydropyrimidones (**27**) after initial alkylation (36).

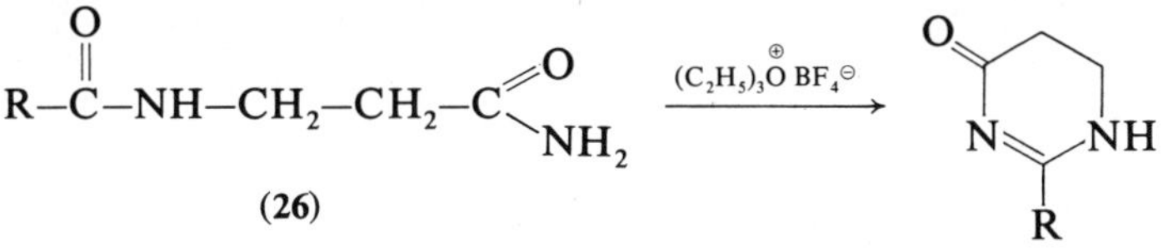

$$R-\overset{O}{\underset{\|}{C}}-NH-CH_2-CH_2-C\underset{NH_2}{=O} \xrightarrow{(C_2H_5)_3O^{\oplus} BF_4^{\ominus}}$$

(**26**) → (**27**) [dihydropyrimidone with R substituent]

Alkylated products such as **28** have been isolated from other dipeptides (36).

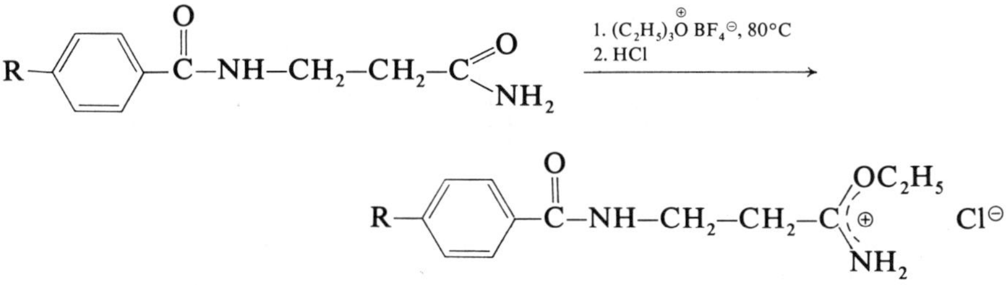

$$R-\text{C}_6H_4-\overset{O}{\underset{\|}{C}}-NH-CH_2-CH_2-C\underset{NH_2}{=O} \xrightarrow[2.\ HCl]{1.\ (C_2H_5)_3O^{\oplus} BF_4^{\ominus},\ 80°C}$$

$$R-\text{C}_6H_4-\overset{O}{\underset{\|}{C}}-NH-CH_2-CH_2-\underset{NH_2}{\overset{OC_2H_5}{C}^{\oplus}} \quad Cl^{\ominus}$$

(**28**)

In a similar fashion *N*-acylated α-amino acid amides react with triethyloxonium fluoroborate to yield **29** (37). If the reaction mixture is allowed to

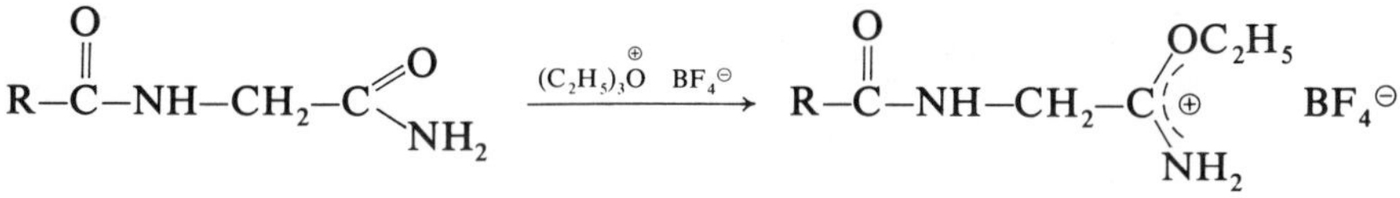

$$R-\overset{O}{\underset{\|}{C}}-NH-CH_2-C\underset{NH_2}{=O} \xrightarrow{(C_2H_5)_3O^{\oplus} BF_4^{\ominus}} R-\overset{O}{\underset{\|}{C}}-NH-CH_2-\underset{NH_2}{\overset{OC_2H_5}{C}^{\oplus}} \quad BF_4^{\ominus}$$

(**29**)

stand, then treated with potassium carbonate, 2-aryl-4-ethoxyimidazoles (**30**) form (37).

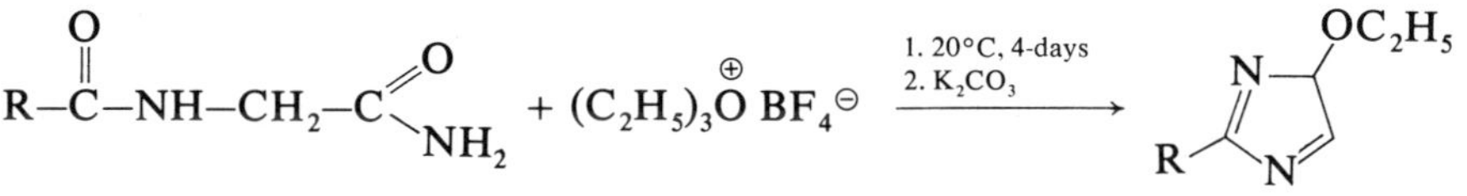

$$R-\overset{O}{\underset{\|}{C}}-NH-CH_2-C\underset{NH_2}{=O} + (C_2H_5)_3O^{\oplus} BF_4^{\ominus} \xrightarrow[2.\ K_2CO_3]{1.\ 20°C,\ 4\text{-days}}$$

(**30**) [2-aryl-4-ethoxyimidazole]

TABLE I

Alkoxymethyleniminium Fluoroborates from Acid Amides and Trialkyloxonium Fluoroborates

$$R-C\underset{NR^1R^2}{\overset{OR^3}{\diagdown}}\quad BF_4^{\ominus}$$

R	R^1	R^2	R^3	Yield, %	m.p., °C	n_D^{20}	Ref.
H	CH$_3$	CH$_3$	CH$_3$	100	—	1.3787	16
H	CH$_3$	CH$_3$	C$_2$H$_5$	89.8	−10	1.3871	13, 14, 16
H	—(CH$_2$)$_2$—O—(CH$_2$)$_2$—		C$_2$H$_5$	97	56–60	—	20
CH$_3$	H	H	C$_2$H$_5$	—	—	—	12, 41
CH$_3$	H	CH$_2$—CH=CH$_2$	CH$_3$	90.5	—	1.4080	16
CH$_3$	H	C$_6$H$_{11}$	CH$_3$	98	121–122	—	16
CH$_3$	H	n-C$_6$H$_{13}$	C$_2$H$_5$	85	Oil	—	42
CH$_3$	H	C$_6$H$_5$	CH$_3$	73	121–122	—	16
CH$_3$	H	C$_6$H$_5$	C$_2$H$_5$	85	157–158	—	42
CH$_3$	CH$_3$	CH$_3$	CH$_3$	97.5	134–138	—	16
CH$_3$	CH$_3$	CH$_3$	C$_2$H$_5$	100	35–37	—	14
CH$_3$	CH$_2$—CH=CH$_2$	CH$_2$—CH=CH$_2$	CH$_3$	92	—	1.4403	16
CH$_3$	C$_2$H$_5$	C$_6$H$_{11}$	CH$_3$	98.5	211–212	—	16
CH$_3$	C$_6$H$_5$	C$_6$H$_5$	CH$_3$	51.5	165–166	—	16
CH$_3$	CH$_3$	C$_6$H$_5$	CH$_3$	65.5	90–94	—	16
C$_2$H$_5$O	H	C$_2$H$_5$	CH$_3$	72	104–105	—	41
C$_2$H$_5$O	H	C$_6$H$_5$	C$_2$H$_5$	87	77–79	—	41
C$_2$H$_5$	CH$_3$	CH$_3$	CH$_3$	93.5	105–113	—	16
C$_2$H$_5$	—(CH$_2$)$_2$—O—(CH$_2$)$_2$—		C$_2$H$_5$	98	72–76	—	20
i-C$_3$H$_7$	CH$_3$	CH$_3$	CH$_3$	90	56–65	—	16
t-C$_4$H$_9$	CH$_3$	CH$_3$	CH$_3$	98	105–110	—	16
t-C$_4$H$_9$	CH$_3$	CH$_3$	C$_2$H$_5$	97	79 (dec)	—	15
C$_6$H$_5$—CH$_2$	CH$_3$	CH$_3$	CH$_3$	87	—	1.4818	16
NC	CH$_3$	CH$_3$	C$_2$H$_5$	68	108	—	15
C$_6$H$_5$OCH$_2$	CH$_3$	CH$_3$	CH$_3$	88.5	40.5–45	—	16
ClCH$_2$	CH$_3$	CH$_3$	CH$_3$	97	58.5–61.5	—	16
Cl$_2$CH	CH$_3$	CH$_3$	CH$_3$	98	72–80	—	16
Cl$_2$CH	CH$_3$	CH$_3$	C$_2$H$_5$	100	—	—	15
NC—CH$_2$	CH$_3$	CH$_3$	C$_2$H$_5$	100	—	1.4127	15
(C$_2$H$_5$O)$_2$CH	CH$_3$	CH$_3$	C$_2$H$_5$	100	—	—	15

				Yield %	mp	n_D	Ref.
(CH$_3$)$_2$N–C(=O)–	CH$_3$	CH$_3$	CH$_3$	99	155–160	—	21
(CH$_3$)$_2$N–C(=O)–	CH$_3$	CH$_3$	C$_2$H$_5$	95	73	—	15
C$_2$H$_5$O–C(=O)–	CH$_3$	CH$_3$	C$_2$H$_5$	98	—	1.4142	15, 21
C$_2$H$_5$O–C(=O)–	–(CH$_2$)$_5$–		C$_2$H$_5$	98	—	—	21
C$_2$H$_5$O–C(=O)–	–(CH$_2$)$_2$–O–(CH$_2$)$_2$–		C$_2$H$_5$	94	—	—	21
(C$_2$H$_5$O)$_3$C	–(CH$_2$)$_4$–		C$_2$H$_5$	91	50–55	—	21
BF$_4^\ominus$ $\overset{C_2H_5O}{\underset{(CH_3)_2N}{}}\!\!\!\overset{\oplus}{C}\!\!-\!\!\overset{H}{\underset{}{C}}\!=\!\overset{H}{\underset{}{C}}$	CH$_3$	CH$_3$	C$_2$H$_5$	100	—	—	15
BF$_4^\ominus$ $\overset{C_2H_5O}{\underset{(CH_3)_2N}{}}\!\!\!\overset{\oplus}{C}\!\!-\!\!\overset{H}{\underset{H}{C}}\!=\!C$	CH$_3$	CH$_3$	C$_2$H$_5$	100	—	—	15
C$_6$H$_5$	H	H	C$_2$H$_5$	80.6	130–131	—	41
C$_6$H$_5$	H	CH$_3$	C$_2$H$_5$	90.1	73	—	41
C$_6$H$_5$	CH$_3$	CH$_3$	CH$_3$	89.5	99–102	—	16
o-CH$_3$–C$_6$H$_4$	H	H	C$_2$H$_5$	80.1	93–94	—	41
o-CH$_3$–C$_6$H$_4$	CH$_3$	CH$_3$	CH$_3$	90	85–88.5	—	16
m-CH$_3$–C$_6$H$_4$	CH$_3$	CH$_3$	CH$_3$	88	38.5–4.2	—	16
p-CH$_3$–C$_6$H$_4$	CH$_3$	CH$_3$	CH$_3$	91	58–64	—	16
o-C$_2$H$_5$O–C$_6$H$_4$	H	H	C$_2$H$_5$	90	139–141	—	41
o-Cl–C$_6$H$_4$	H	H	C$_2$H$_5$	78	101.5–103	—	41
o-Cl–C$_6$H$_4$	CH$_3$	CH$_3$	CH$_3$	94	100–106	—	16
m-Cl–C$_6$H$_4$	CH$_3$	CH$_3$	CH$_3$	91	85–87	—	16
p-Cl–C$_6$H$_4$	CH$_3$	CH$_3$	CH$_3$	90	57–63	—	16
o-NO$_2$–C$_6$H$_4$	CH$_3$	CH$_3$	CH$_3$	94.5	116–120	—	16
m-NO$_2$–C$_6$H$_4$	CH$_3$	CH$_3$	CH$_3$	89	143–145	—	16
p-NO$_2$–C$_6$H$_4$	CH$_3$	CH$_3$	CH$_3$	93.5	115–120	—	16
p-NO$_2$–C$_6$H$_4$	CH$_3$	CH$_3$	C$_2$H$_5$	100	96	—	15
1-Naphthyl	H	H	C$_2$H$_5$	97	88–89	—	41

Very often acid amides are alkylated with oxonium salts, and the salts used directly without isolation (19, 23, 38–40).

Table I summarizes the properties of some of the alkoxymethyleniminium fluoroborates.

(b) Lactams. Lactams are more readily alkylated by oxonium salts at their oxygen atoms than carboxylic acid amides, as shown by Meerwein et al. with the ethylation of N-methylpyrrolidone (**31** → **32**) (13, 14, 43). Since its discovery

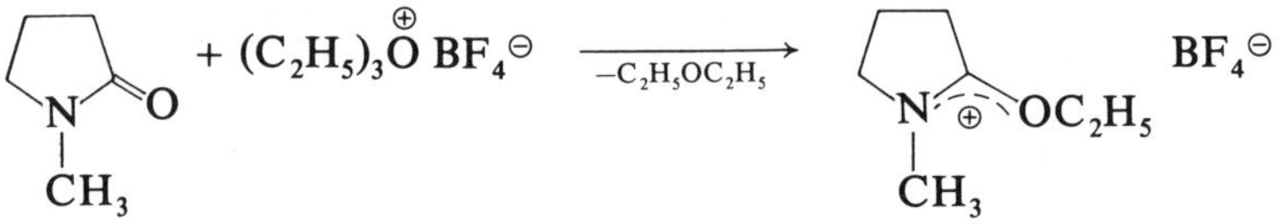

this reaction has been frequently applied; generally the ions are not isolated, but reacted without further purification to give various products. The following conclusions may be drawn from these experiments:

1. N-Substituted as well as N-unsubstituted lactams are alkylated at the oxygen atom (13).

2. The ring size of the lactam does not seem to affect the reaction, since three-membered as well as four- to sixteen-membered ring lactams are alkylated by oxonium salts (**33** → **34**). Also, lactams annellated in various ways undergo alkylation by oxonium salts. Table II gives some examples to

TABLE II

Survey of Alkylation of Lactams by Oxonium Salts

Reaction	Ref.
$n = 1$	44
$n = 2$	44
$n = 3$	13, 14, 36, 40, 42, 45, 46, 47
$n = 4$	36, 40, 45, 46, 48, 49
$n = 5$	36, 40, 42, 47
$n = 6$	23, 40
$n = 7, 9, 10, 11, 14$	40
(β-lactam reaction)	50

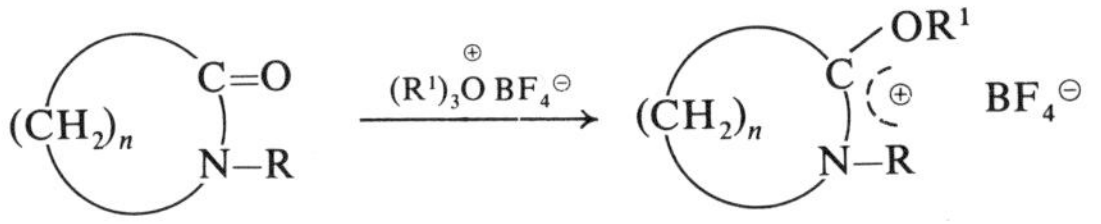

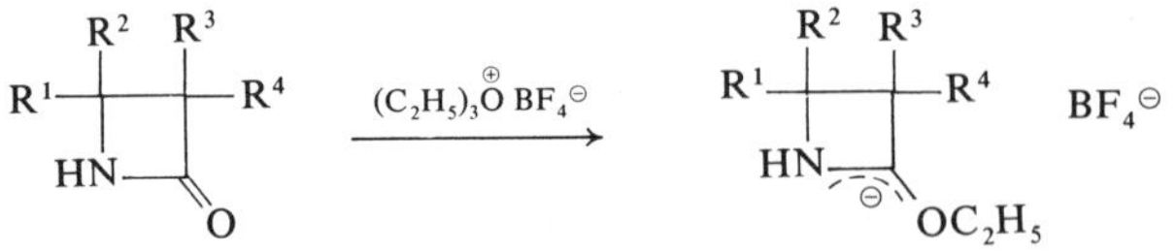

TABLE II (cont.)

Reaction	Ref.
[structure: ethyl pyroglutamate] + $(C_2H_5)_3O^⊕ BF_4^⊖$ → [O-ethyl iminium salt] $BF_4^⊖$	46
[structure: δ-valerolactam with R, R¹] + $(C_2H_5)_3O^⊕ BF_4^⊖$ → [iminium ether] $BF_4^⊖$	48
[structure: 3,4-dihydroquinolin-2(1H)-one with R¹, N-R] + $R_3^⊕O BF_4^⊖$ → [iminium ether, OR²] $BF_4^⊖$	14, 51, 52
[structure: 3,4-dihydroisoquinolin-1(2H)-one, N-R] + $R_3^⊕O BF_4^⊖$ → [iminium ether, OR¹] $BF_4^⊖$	52
[structure: isoindolin-1,3-dione with N-R, R¹] + $R_3^⊕O BF_4^⊖$ → [iminium ether OR²] $BF_4^⊖$	53, 54
[structure: 4-dimethylamino-3-pyrrolin-2-one] + $(C_2H_5)_3O^⊕ BF_4^⊖$ → [N(CH₃)₂ iminium OC₂H₅] $BF_4^⊖$	55
[structure: benzo-fused 8-membered lactam] + $(CH_3)_3O^⊕ BF_4^⊖$ → [OCH₃ iminium] $BF_4^⊖$	56
[structure: benzo-fused azocinone] + $(CH_3)_3O^⊕ BF_4^⊖$ → [OC₂H₅ iminium] $BF_4^⊖$	56
[structure: bromothieno-fused azocinone] + $(C_2H_5)_3O^⊕ BF_4^⊖$ → [OC₂H₅ iminium] $BF_4^⊖$	57

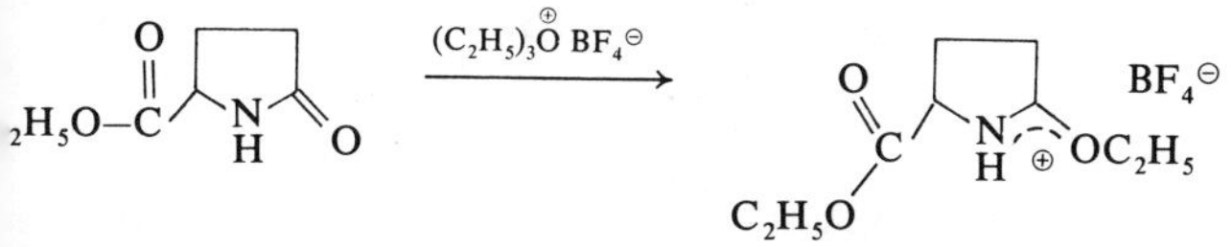

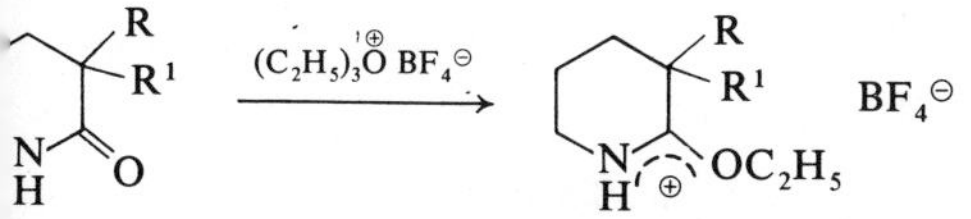

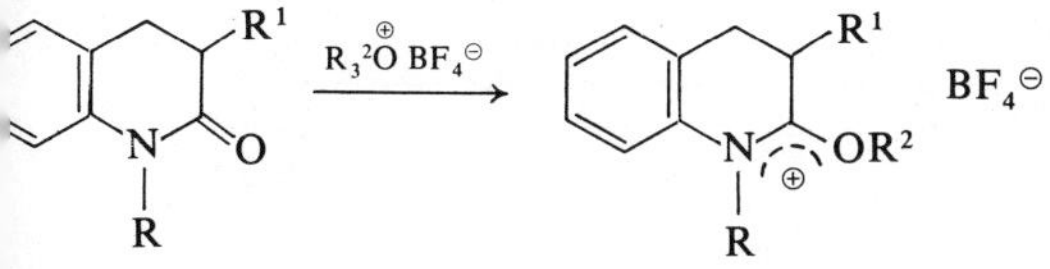

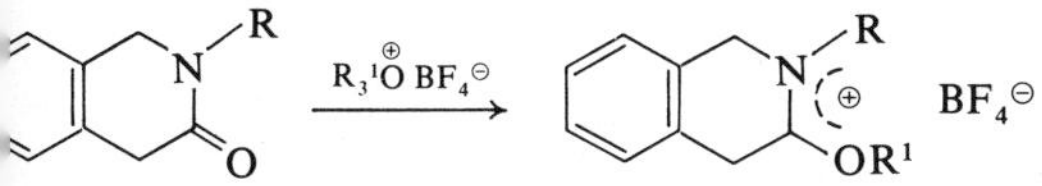

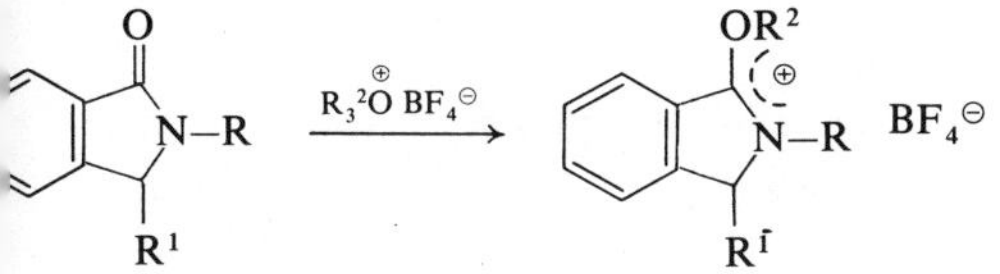

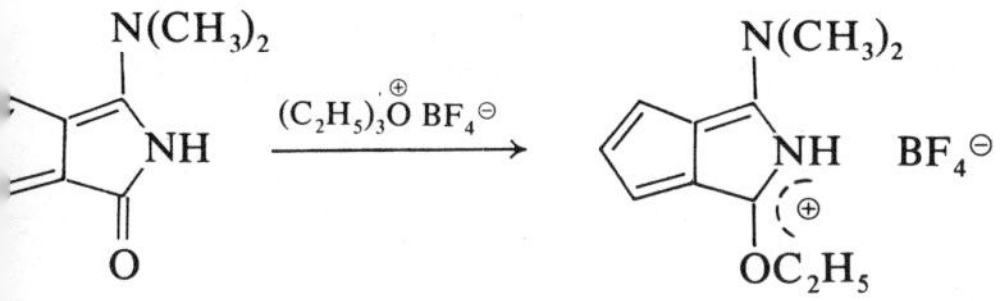

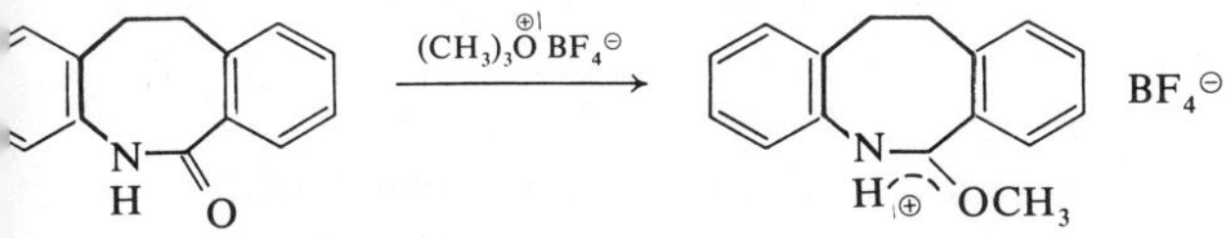

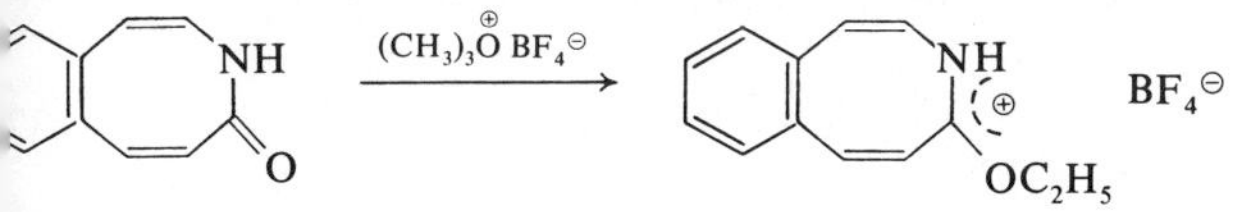

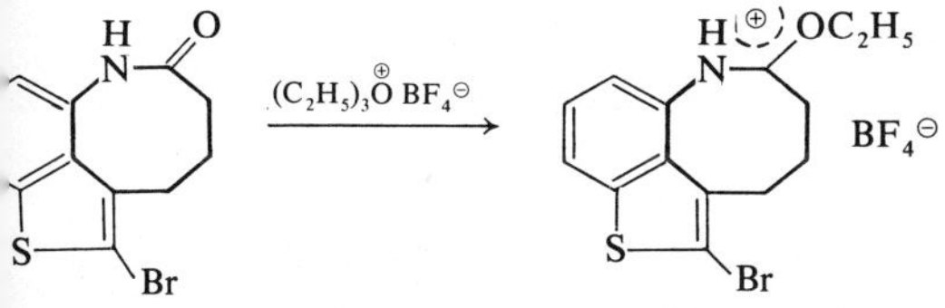

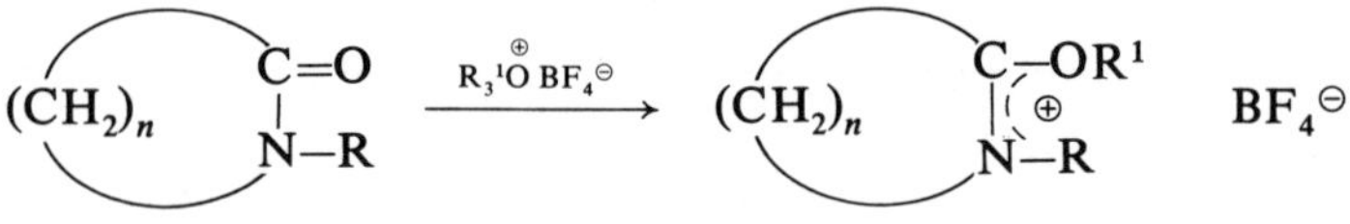

$$\underset{(33)}{(CH_2)_n\begin{array}{c}C=O\\|\\N-R\end{array}} \xrightarrow{R_3{}^1\overset{\oplus}{O}\ BF_4{}^\ominus} \underset{(34)}{(CH_2)_n\begin{array}{c}C-OR^1\\\|\oplus\\N-R\end{array}}\ BF_4{}^\ominus$$

demonstrate the wide range of applications. Table III summarizes the properties of some ions.

TABLE III

Imino Ether Salts from Lactams and Oxonium Salts

Compound	Yield, %	m.p., °C	Ref.
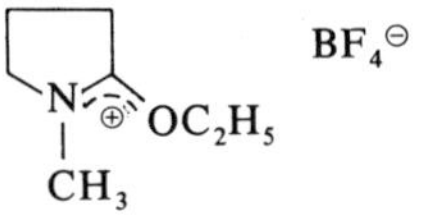 pyrrolidine-N(CH₃)=OC₂H₅ BF₄⁻	83.6	12–14	13
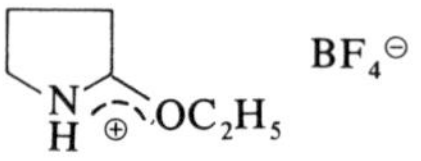 pyrrolidine-NH=OC₂H₅ BF₄⁻	80	48–49.5	42
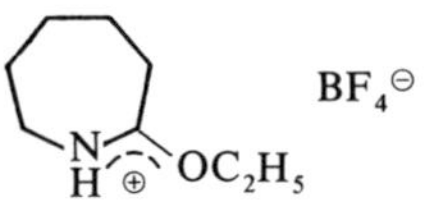 azepane-NH=OC₂H₅ BF₄⁻	88	61–63	42
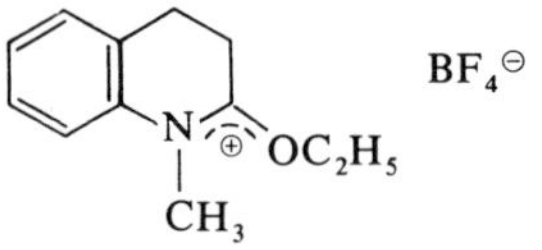 tetrahydroquinoline-N(CH₃)=OC₂H₅ BF₄⁻	96	184–185	14

(c) *Urea and Urea Derivatives.* Meerwein et al. alkylated urea at the oxygen atom with the aid of triethyloxonium fluoroborate (12). The noncrystalline product was isolated as a picrate. Later on, tetramethylurea (14) and other ureas (58) were successfully converted with triethyloxonium fluoroborate into *N,N*-dialkyl(dialkylamino)ethoxymethyleniminium fluoroborates (35).

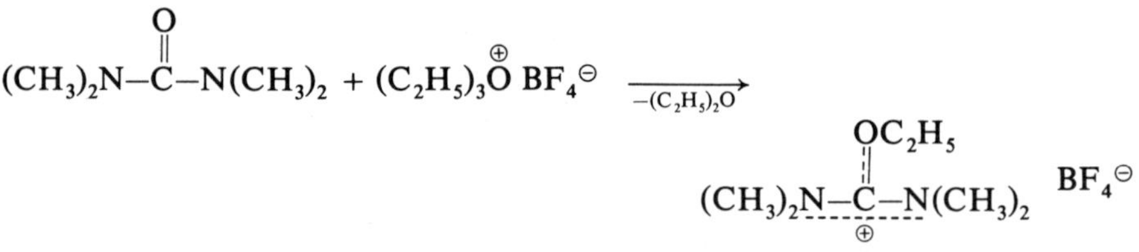

$$(CH_3)_2N-\overset{\overset{O}{\|}}{C}-N(CH_3)_2 + (C_2H_5)_3\overset{\oplus}{O}\ BF_4{}^\ominus \xrightarrow{-(C_2H_5)_2O}$$

$$\underset{(35)}{(CH_3)_2N-\underset{\oplus}{\overset{\overset{OC_2H_5}{\|}}{C}}-N(CH_3)_2}\ BF_4{}^\ominus$$

Cyclic ureas (imidazolones) like **36** are also alkylated with oxonium salts. However, imidazolones (**36**) are not only attacked at the oxygen atom but

I. SYNTHESIS

possibly at the nitrogen atom also. Thus 1,3-dimethyl-2-imidazolone (**36**) is almost evenly alkylated at oxygen (**37**) and at nitrogen (**38**) (14).

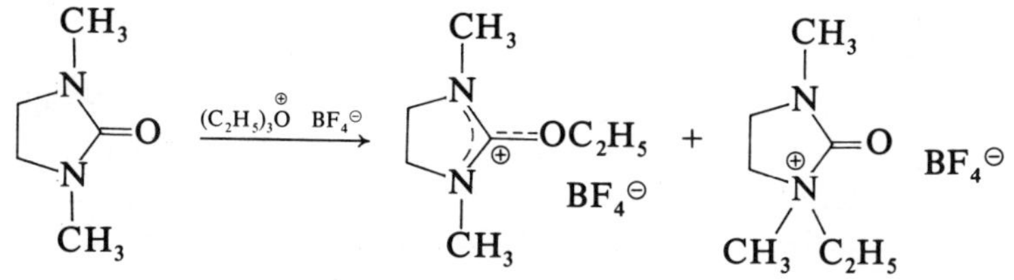

(36)   (37)   (38)

N,N-Dimethylethylurethane is nearly quantitatively ethylated by triethyloxonium fluoroborate at the oxygen atom to yield N,N-dimethyl(diethoxy)methyleniminium fluoroborate (**39**) (14, 15).

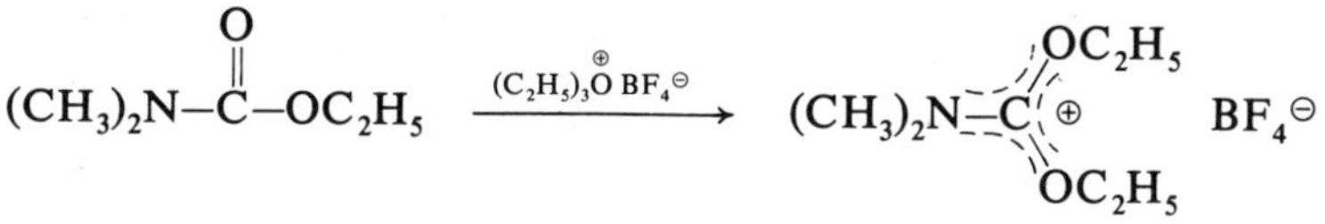

(39)

Triethyloxonium fluoroborate also ethylates N-benzylethylurethane (38).

TABLE IV
Some Products of Alkylation of Ureas

Compound	Yield, %	m.p., °C	Ref.
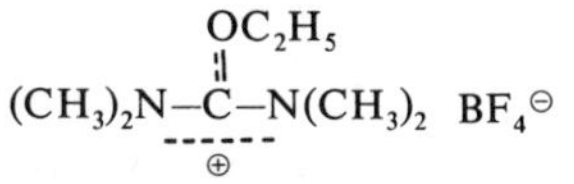 $(CH_3)_2N-\overset{OC_2H_5}{\underset{\oplus}{C}}-N(CH_3)_2$ $BF_4^\ominus$	95	82–83	14
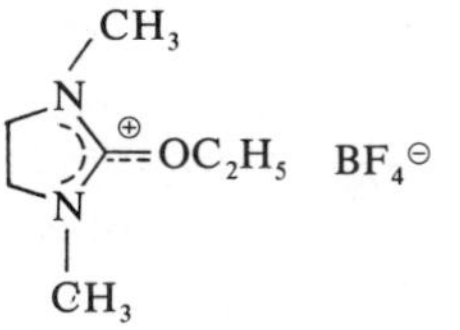 1,3-dimethyl-2-ethoxy-imidazolinium $BF_4^\ominus$	48	49–52	14
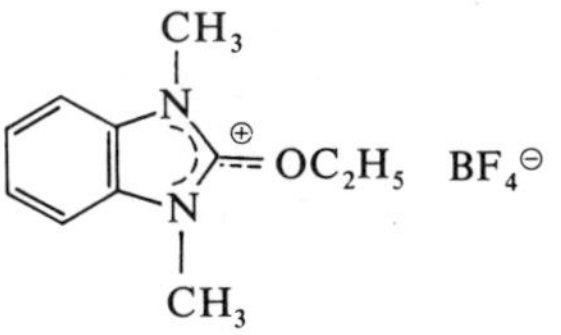 1,3-dimethyl-2-ethoxy-benzimidazolium $BF_4^\ominus$	97	152 (dec)	14
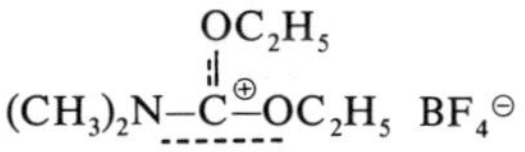 $(CH_3)_2N-\overset{OC_2H_5}{\underset{\oplus}{C}}-OC_2H_5$ $BF_4^\ominus$	98	−52—48	14, 15

Table IV lists some alkoxymethyleniminium salts formed by the alkylation of ureas.

2. Alkylation with Carboxonium Ions

Carboxonium ions overpower trialkyloxonium salts in their alkylating potential (59). Therefore they easily alkylate acid amides at the amide oxygen atom (19, 43, 59). Generally, the salts **40** generated this way are not isolated, but are reacted further. Table V gives some examples.

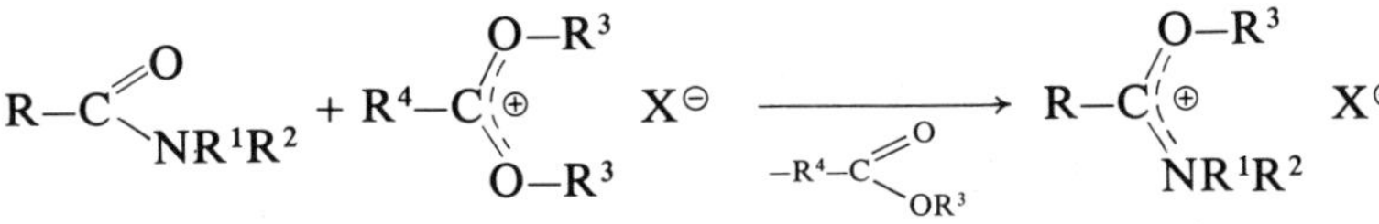

$$R-C\underset{NR^1R^2}{\overset{O}{\diagup}} + R^4-C\underset{O-R^3}{\overset{O-R^3}{\diagup}}\!\!\!{}^{\oplus}\ X^{\ominus} \xrightarrow[-R^4-C\underset{OR^3}{\overset{O}{\diagup}}]{} R-C\underset{NR^1R^2}{\overset{O-R^3}{\diagup}}\!\!\!{}^{\oplus}\ X^{\ominus}$$
$$(40)$$

TABLE V

N,N-Dialkylethoxymethyleniminium Salts (**40**) from Acid Amides and Carboxonium Ions (59)

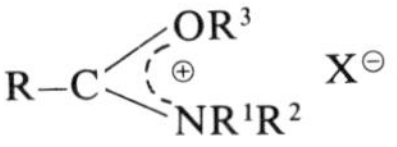

$$R-C\underset{NR^1R^2}{\overset{OR^3}{\diagup}}\!\!\!{}^{\oplus}\ X^{\ominus}$$

R	R^1	R^2	R^3	X	Yield, %	m.p., °C
C_6H_5	C_2H_5	C_2H_5	C_2H_5	$SbCl_6$	95	131–132
CH_3	C_6H_5	C_6H_5	C_2H_5	$SbCl_6$	97	166–168

3. Alkylation with Dialkyl Sulfates

(a) *Carboxylic Acid Amides.* Bühner (60) was the first to study the alkylation of acid amides with dimethyl sulfate. Benzamide as well as acetamide gave crystalline products with dimethyl sulfate, which Bühner called "methylsulfuric acid iminoethers." Oily products were obtained from N-methylbenzamide, dimethylbenzamide, and N-methylacetanilide with dimethyl sulfate; their composition could not be precisely determined.

The fundamental studies by Bredereck et al. (61–68) finally established the facts. According to these workers dialkyl sulfates alkylate primary, secondary, and tertiary carboxylic acid amides at their oxygen atom to form salts **41** in an equilibrium reaction. The equilibrium is determined by the character of the

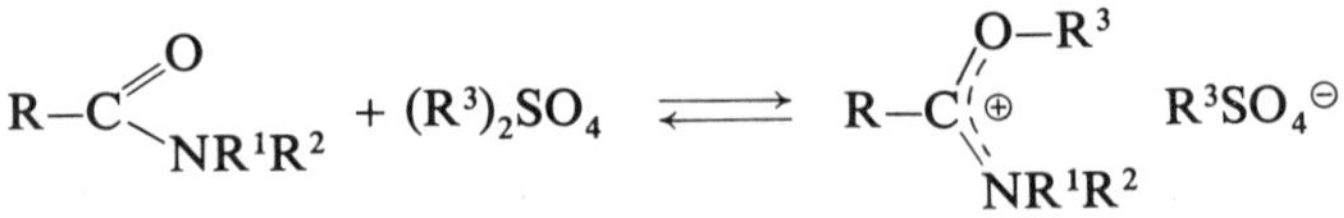

$$R-C\underset{NR^1R^2}{\overset{O}{\diagup}} + (R^3)_2SO_4 \rightleftharpoons R-C\underset{NR^1R^2}{\overset{O-R^3}{\diagup}}\!\!\!{}^{\oplus}\ R^3SO_4^{\ominus}$$
$$(41)$$

substituents R, R^1, and R^2; electron-withdrawing and bulky substituents will displace the equilibrium toward the starting materials (20, 68). Since the reaction is generally slow in both directions, even in unfavorable cases, the salts **41** may be obtained relatively pure. They are further purified by extraction with ether, benzene, etc., or by precipitation from methylene chloride with ether. However, these adducts must be reacted further quickly, since the initial equilibrium is slowly reestablished on storage (16). Fortunately, with the synthetically most important adducts like the dimethylformamide–dimethyl sulfate adduct and the dimethylacetamide–dimethyl sulfate adduct the equilibria are shifted almost completely to the right, so these adducts may be stored in a pure state for prolonged periods of time. Dialkyl sulfates were mainly used to alkylate primary, secondary, and tertiary carboxylic acid amides, the nucleophilicity of which is not diminished by electron-withdrawing ligands at the nitrogen or the acyl carbon atoms.

Thus primary aliphatic and aromatic amides like acetamide (69), propionamide (69), and benzamide have been smoothly converted into the corresponding alkyl or aryl-methoxymethyleniminium ions **42** with dimethyl sulfate.

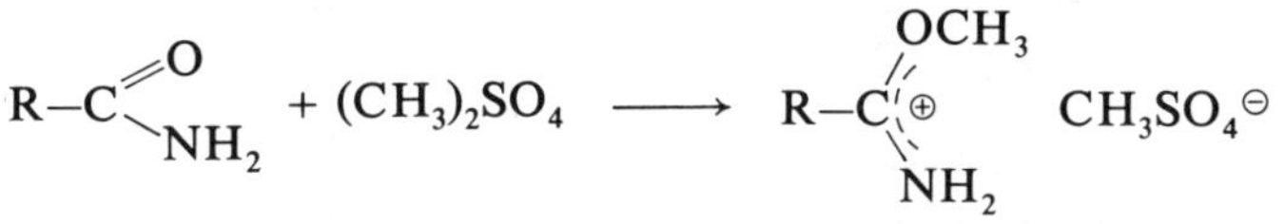

$$R-C\underset{NH_2}{\overset{O}{\diagup\hspace{-0.5em}\diagdown}} + (CH_3)_2SO_4 \longrightarrow R-\overset{OCH_3}{\underset{NH_2}{C^{\oplus}}} \quad CH_3SO_4^{\ominus}$$

(42)

$R = CH_3, C_2H_5, C_6H_5$

$$H-C\underset{NH_2}{\overset{O}{\diagup\hspace{-0.5em}\diagdown}} + (CH_3)_2SO_4 \longrightarrow H-\overset{OCH_3}{\underset{NH_2}{C^{\oplus}}} \quad CH_3SO_4^{\ominus}$$

(43)

$$H-\overset{NH_2}{\underset{NH_2}{C^{\oplus}}} \quad CH_3SO_4^{\ominus}$$

(44)

$$H-\overset{OCH_3}{\underset{NH_3^{\oplus}}{C}}-NHCHO \quad CH_3SO_4^{\ominus}$$

$$H-\underset{NH_3^{\oplus}}{\overset{NH_2}{C}}-NH-CHO \quad CH_3SO_4^{\ominus}$$

$$H-\underset{NHCHO}{\overset{NHCHO}{C}}-NHCHO$$

(45)

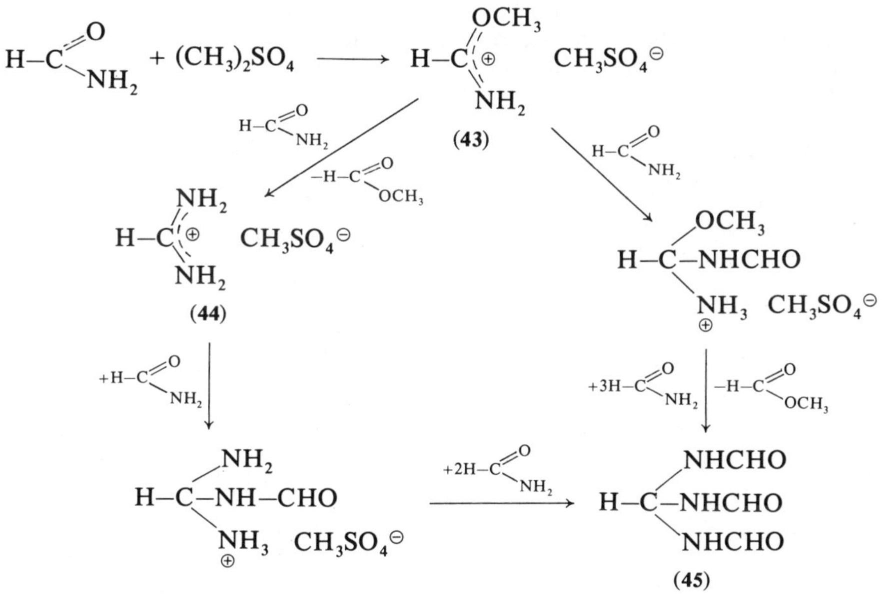

Formamide reacts in a more complex way (61, 62). Initially, the oxygen atom is methylated, but the iminium salt **43** generated reacts further with unreacted formamide to give formamidinium methyl sulfate (**44**); the formation of tris(formamido)methane (**45**) is also possible. Thus if formamide and dimethyl sulfate are reacted in a 2:1 molar ratio, formamidinium methyl sulfate (**44**) forms in 65% yield (61, 62). A five- to tenfold excess of formamide generates tris(formamido)methane (**45**) in good yields (61).

The methylation of *N*-mono- and *N,N*-disubstituted carboxamides, especially formamides and acetamides, has been studied in great detail. As it turns out, the size of the substituent at the nitrogen atom in secondary carboxamides hardly influences the alkylation, since it is possible to convert isopropyl-, isobutyl-, and even *t*-butylacetamide with dimethyl sulfate into *N*-alkyl(alkylmethoxymethylen)iminium methylsulfates (**46**) in yields better than 85% (16).

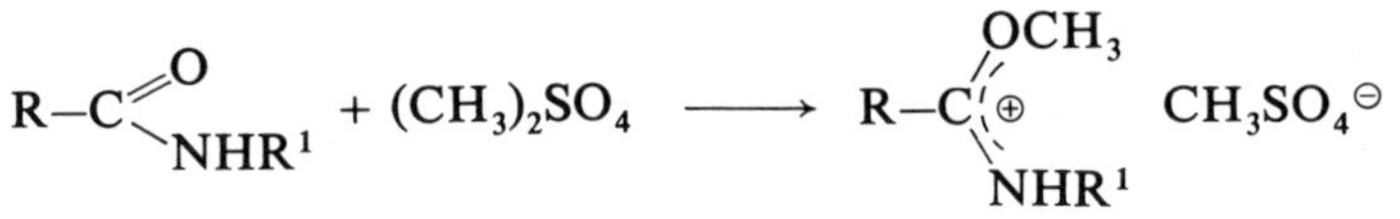

$$R-C\overset{O}{\underset{NHR^1}{\diagup}} + (CH_3)_2SO_4 \longrightarrow R-C\overset{OCH_3}{\underset{NHR^1}{\diagup}}{}^{\oplus} \quad CH_3SO_4^{\ominus}$$

(**46**)

$$R = H, CH_3$$
$$R^1 = CH_3, C_2H_5, n\text{-}C_3H_7, i\text{-}C_3H_7, i\text{-}C_4H_9, t\text{-}C_4H_9$$

When *N*-formylated aromatic amines are methylated, the reaction does not stop at the step of the *N*-arylmethoxymethyleniminium methylsulfates **47**, since they react further with the starting amides to give *N,N'*-diarylformamidinium salts **48** (63).

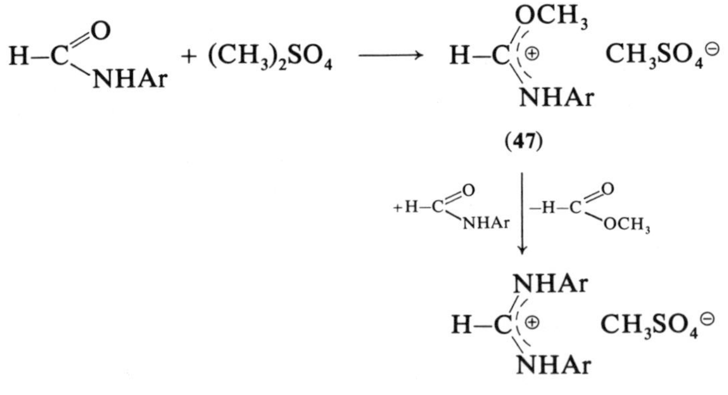

$$H-C\overset{O}{\underset{NHAr}{\diagup}} + (CH_3)_2SO_4 \longrightarrow H-C\overset{OCH_3}{\underset{NHAr}{\diagup}}{}^{\oplus} \quad CH_3SO_4^{\ominus}$$

(**47**)

$$+H-C\overset{O}{\underset{NHAr}{\diagup}} \quad \Big| \quad -H-C\overset{O}{\underset{OCH_3}{\diagup}}$$

$$H-C\overset{NHAr}{\underset{NHAr}{\diagup}}{}^{\oplus} \quad CH_3SO_4^{\ominus}$$

(**48**)

$$Ar = C_6H_5, p\text{-}OCH_3-C_6H_4-, p\text{-}NO_2-C_6H_4-$$

I. SYNTHESIS

Acetanilides do not display this type of reaction (42). Thus acetanilide and p-methoxyacetanilide were successfully methylated with dimethyl sulfate to give **49**, whereas p-nitroacetanilide could not be alkylated in this way (42).

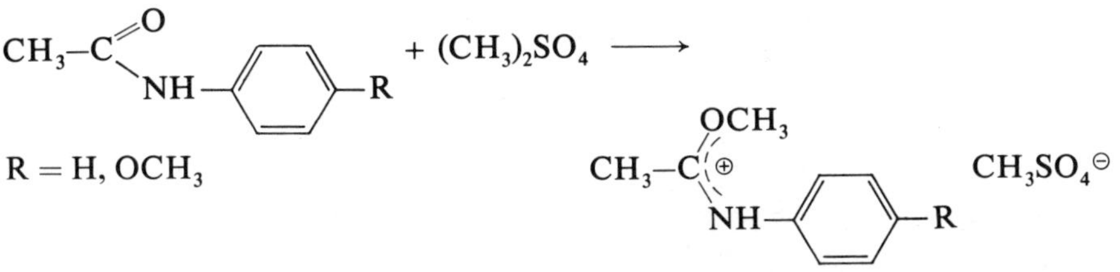

$$CH_3-C(=O)-NH-C_6H_4-R + (CH_3)_2SO_4 \longrightarrow$$

R = H, OCH$_3$

$$CH_3-C(OCH_3)=NH^+-C_6H_4-R \quad CH_3SO_4^{\ominus}$$

(49)

With tertiary amides the steric effect of the substituents at the nitrogen atom is noticeable and the yields decrease; for example, with the methylation of di-n-butylacetamide the yield is only 57% (16).

Electron-withdrawing substituents at the acyl carbon atom further diminish the yields. With cyanoacetic acid dimethylamide the yield falls to 43% on methylation with dimethyl sulfate (68).

Dicarboxylic acid amides with their carboxamide functions less than four carbon atoms apart or not separated by a phenyl group are alkylated at only one of the carboxamide functions (68).

The double methylation with dimethyl sulfate is successful with adipic acid bis(dimethylamide) and terphthalic acid bis(dimethylamide) (70) to give salts **50**. However, the reaction product could not be prepared in an analytically pure form (70).

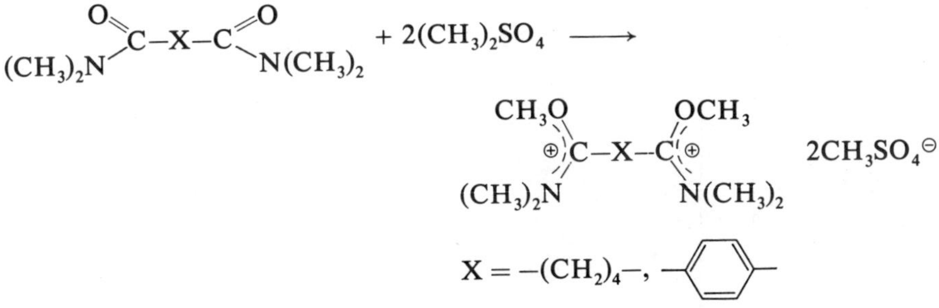

$$(CH_3)_2N-C(=O)-X-C(=O)-N(CH_3)_2 + 2(CH_3)_2SO_4 \longrightarrow$$

$$\overset{\oplus}{C}(OCH_3)(N(CH_3)_2)-X-\overset{\oplus}{C}(OCH_3)(N(CH_3)_2) \quad 2CH_3SO_4^{\ominus}$$

X = –(CH$_2$)$_4$–, –C$_6$H$_4$–

(50)

In most cases, the alkoxymethyleniminium alkylsulfates (**41**) are obtained as yellow to brown oils, which rarely crystallize. Table VI summarizes the properties of some alkyl (or aryl) alkoxymethyleniminium alkyl sulfates.

TABLE VI

Alkoxymethyleniminium Alkylsulfates from Acid Amides and Dialkyl Sulfates

$$R-C\underset{NR^1R^2}{\overset{O-R^3}{\diagup}}\quad R^3SO_4^{\ominus}$$

R	R^1	R^2	R^3	Yield, %	m.p., °C	n_D^{20}	Ref.
H	H	CH_3	CH_3	95	Oil	1.4513	67
H	H	C_2H_5	CH_3	80	Oil	1.4525	67
H	CH_3	CH_3	CH_3	100	Oil	1.4586	65
H	CH_3	CH_3	C_2H_5	90	Oil	1.4562	65
H	C_2H_5	C_2H_5	CH_3	98	Oil	1.4605	16, 71
H	C_3H_7	C_3H_7	CH_3	89	Oil	1.4555	16, 71
H	i-C_3H_7	i-C_3H_7	CH_3	87	Oil	1.4430	71
H	n-C_4H_9	n-C_4H_9	CH_3	92	Oil	1.4570	71
H	—$(CH_2)_4$—		CH_3	100	Oil	1.4780	71
H	—$(CH_2)_5$—		CH_3	98	Oil	1.4845	71
H	—$(CH_2)_2O(CH_2)_2$—		CH_3	100	Oil	1.4811	71
CH_3	CH_3	CH_3	CH_3	100	Oil	1.4636	65
CH_3	C_2H_5	C_2H_5	CH_3	94	Oil	1.4700	16, 65, 68
CH_3	i-C_3H_7	i-C_3H_7	CH_3	84	63–69	—	16
CH_3	n-C_4H_9	n-C_4H_9	CH_3	57	Oil	1.4653	16
CH_3	—$(CH_2)_4$—		CH_3	87	Oil	1.4839	16, 20, 72
CH_3	—$(CH_2)_5$—		CH_3	75	37–44	—	16, 20, 72
CH_3	—$(CH_2)_2$-O-$(CH_2)_2$—		CH_3	87	84–89	—	16, 20, 72
CH_3	H	CH_3	CH_3	100	Oil	1.4564	60, 69, 72
CH_3	H	C_2H_5	CH_3	81	Oil	1.4513	16, 67
CH_3	H	C_2H_5	CH_3	85	Oil	1.4525	16, 67
CH_3	H	n-C_3H_7	CH_3	99	Oil	1.4604	16
CH_3	H	i-C_3H_7	CH_3	86	Oil	1.4589	16
CH_3	H	n-C_4H_9	CH_3	86	Oil	1.4588	16
CH_3	H	sec-C_4H_9	CH_3	94.5	38.5–41	—	16

R¹	R²	R³	Yield (%)	mp/bp (°C)	n_D	Refs.
C_6H_5	H	CH_3	58–66	80–82	—	42, 60
$p\text{-}CH_3O\text{-}C_6H_4$	H	CH_3	82	70–72	—	42
$n\text{-}C_3H_7$	H	CH_3	100	Oil	—	69
C_2H_5	H	CH_3	62	Oil	—	42
C_2H_5	CH_3	CH_3	100	Oil	1.4583	65
$n\text{-}C_3H_7$	CH_3	CH_3	70–93	Oil	1.4675	65, 68
$n\text{-}C_3H_7$	CH_3	CH_3	53	Oil	1.4600	65
$n\text{-}C_5H_{11}$	CH_3	CH_3	79	Oil	1.4640	65
$n\text{-}C_5H_{11}$	C_2H_5	CH_3	37	Oil	1.4585	65
$NC\text{-}CH_2$	CH_3	CH_3	43	Oil	1.5095	68
C_6H_5	H	CH_3	65	Oil	—	42
$C_6H_5\text{-}CH_2$	H	CH_3	60	Oil	—	42
$C_6H_5\text{-}CH_2$	CH_3	CH_3	64	Oil	1.5180	68
$C_6H_5\text{-}CH_2$	H	CH_3	20	Oil	—	42
C_6H_5	H	CH_3	68	108–111	—	60, 69
C_6H_5	H	CH_3	—	Oil	—	60
C_6H_5	CH_3	CH_3	100	Oil	1.5125	60, 65
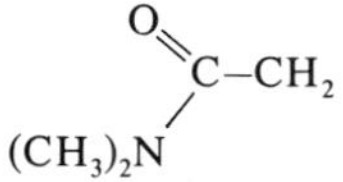 $\underset{(CH_3)_2N}{\underset{\|}{O{=}C}}\text{-}CH_2$	CH_3	CH_3	100	65–67	1.4846	68
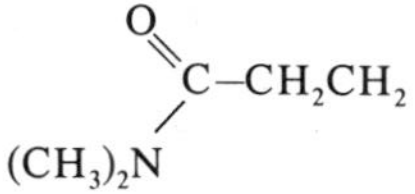 $\underset{(CH_3)_2N}{\underset{\|}{O{=}C}}\text{-}CH_2CH_2$	CH_3	CH_3	88	Oil	1.4826	68
$CH_3SO_4^{\ominus}\ \overset{CH_3O}{\underset{(CH_3)_2N}{C{\oplus}}}\text{-}(CH_2)_4$	CH_3	CH_3	100	108	—	68
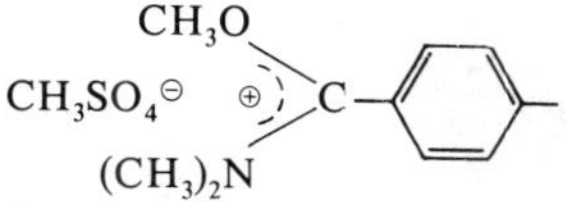 $CH_3SO_4^{\ominus}\ \overset{CH_3O}{\underset{(CH_3)_2N}{C{\oplus}}}\text{-}(p\text{-}C_6H_4\text{-}CH_3)$	CH_3	CH_3	95	130–131	—	70

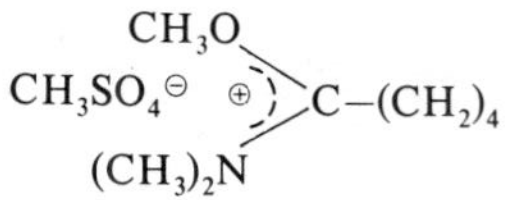

(b) *Lactams.* Benson and Cairns studied the methylation of N-unsubstituted lactams with dimethyl sulfate (73). According to them, lactams are alkylated by dimethyl sulfate at their oxygen atoms. From the methylated products **51** lactim ethers and N-methylated lactams could be synthesized. Later on other

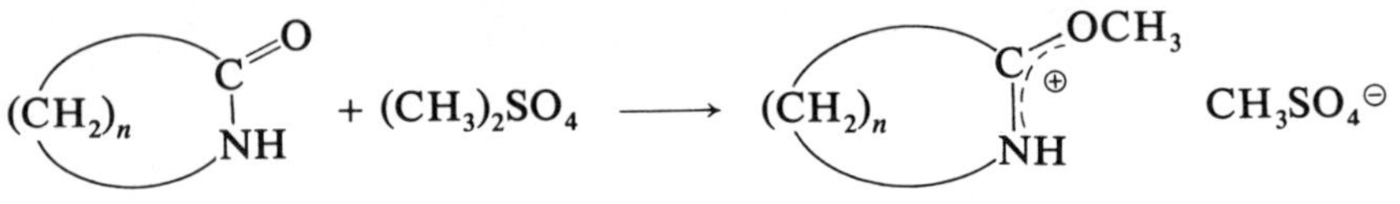

$$(CH_2)_n \overset{C=O}{\underset{NH}{|}} + (CH_3)_2SO_4 \longrightarrow (CH_2)_n \overset{C-OCH_3}{\underset{NH}{\overset{\oplus}{|}}} \quad CH_3SO_4^{\ominus}$$

(51)

authors (74–81) took advantage of this reaction to generate 5-, 7-, 9-, 10-, and 13-membered ring lactim ethers. Lüssi (82) reexamined these reactions while methylating 6-, 7-, 8-, 9-, 11-, and 12-membered ring unsubstituted lactams. In these studies he was able to discredit Cairns' and Benson's assumption that the primary O-alkylated products **51** rearrange into the N-methylated compounds **52** under the influence of dimethyl sulfate. Further, it seemed very likely that

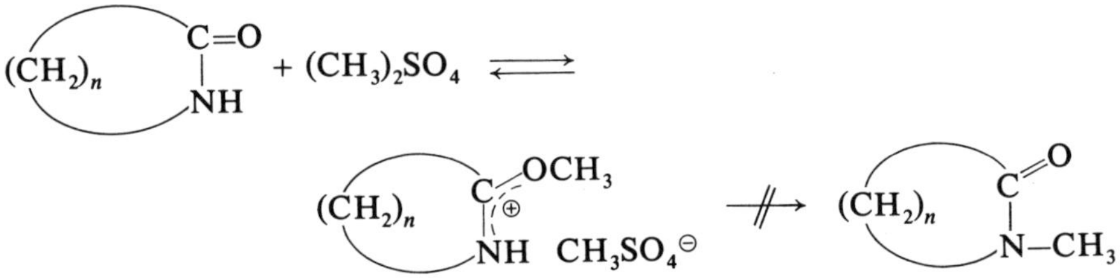

$$(CH_2)_n \overset{C=O}{\underset{NH}{|}} + (CH_3)_2SO_4 \rightleftharpoons$$

$$(CH_2)_n \overset{C-OCH_3}{\underset{NH}{\overset{\oplus}{|}}} CH_3SO_4^{\ominus} \quad \not\longrightarrow \quad (CH_2)_n \overset{C=O}{\underset{N-CH_3}{|}}$$

(52)

the methylation of lactams is indeed an equilibrium reaction. However, the alkylated lactams were neither isolated nor characterized.

Bredereck et al. (68) methylated several N-substituted lactams at their oxygen atoms; the imino lactim ether methyl sulfates **53** and **54** were isolated and characterized.

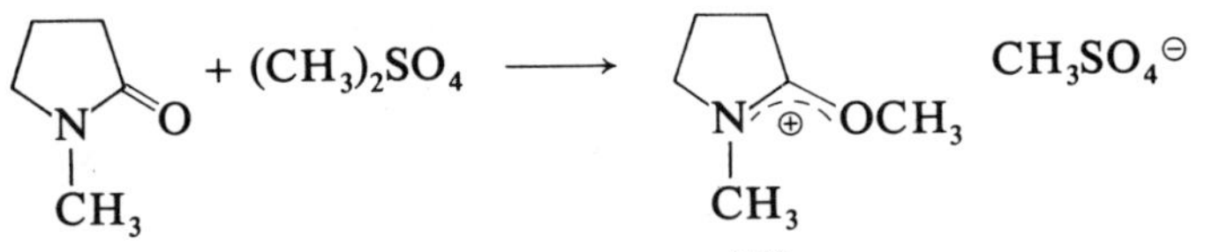

(53)

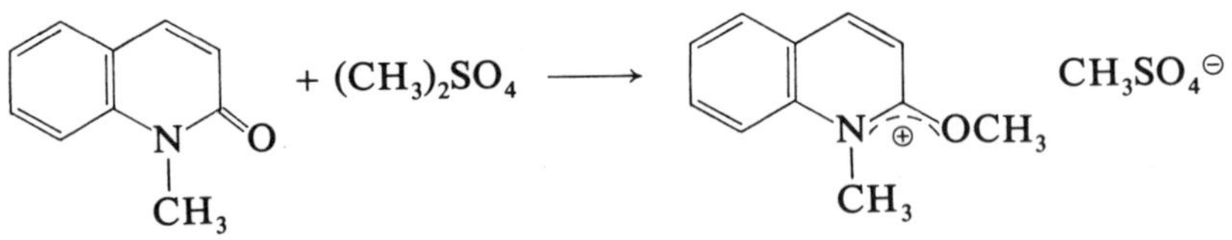

(54)

N-Methylcaprolactam was also methylated with dimethyl sulfate at the oxygen atom, but the methylated product was used without further characterization (83, 84). N-Methylvalerolactam may be methylated with dimethyl sulfate at the oxygen atom, too (85).

Table VII summarizes the properties of several methyl lactimidium acid ester methylsulfates.

TABLE VII

Imino Ether Methylsulfates from Lactams and Dimethyl Sulfate

Compound	Yield, %	m.p., °C	n_D^{20}	Ref.
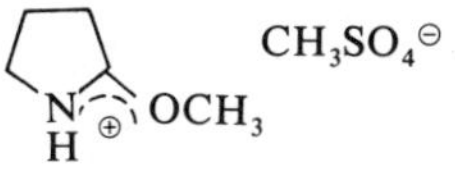 pyrrolidine-OCH₃, N-H, CH₃SO₄⁻	96.7	Oil	1.4713	16, 86
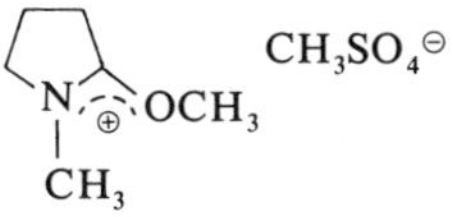 pyrrolidine-OCH₃, N-CH₃, CH₃SO₄⁻	100	Oil	1.4763	16, 68
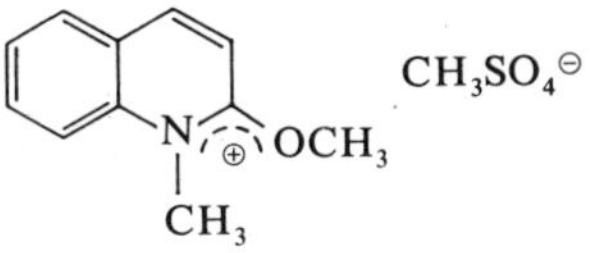 quinoline-OCH₃, N-CH₃, CH₃SO₄⁻	92	110	—	68
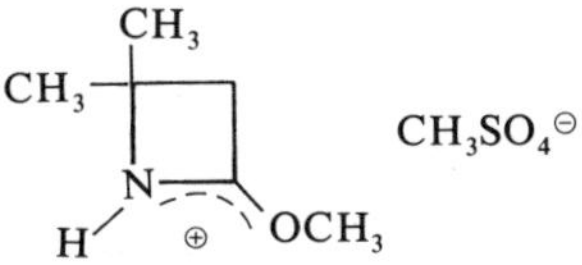 azetidine with CH₃, CH₃, OCH₃, N-H, CH₃SO₄⁻	95	74–75	—	50

(c) *Urea and Urea Derivatives.* Werner (87), Ongly (88), and Janus (89) studied the methylation of urea with dimethyl sulfate; the alkylation occurs at the oxygen atom. The oily reaction product **55** was isolated as a crystalline picrate (89–91).

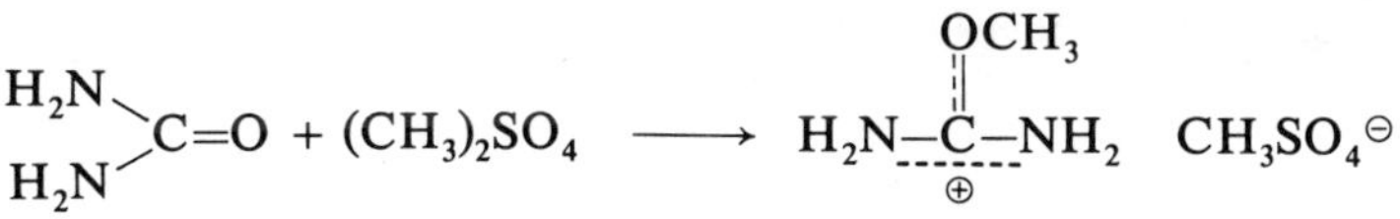

$$\begin{array}{c} H_2N \\ H_2N \end{array}\!\!C\!=\!O + (CH_3)_2SO_4 \longrightarrow H_2N\!-\!\overset{OCH_3}{\underset{\oplus}{C}}\!-\!NH_2 \quad CH_3SO_4^{\ominus}$$

(55)

Werner (87) demonstrated that *N*-monoalkylated ureas and *N,N*-dialkylated ureas are also alkylated at the oxygen atom:

$$\underset{H}{RN}-\overset{O}{\underset{\|}{C}}-NH_2 + (CH_3)_2SO_4 \longrightarrow \underset{H}{R-N}-\overset{OCH_3}{\underset{\oplus}{C}}-NH_2 \quad CH_3SO_4^{\ominus}$$

$$R = CH_3, C_2H_5, C_6H_5$$

$$RR^1N-\overset{O}{\underset{\|}{C}}-NH_2 + (CH_3)_2SO_4 \longrightarrow RR^1N-\overset{OCH_3}{\underset{\oplus}{C}}-NH_2 \quad CH_3SO_4^{\ominus}$$

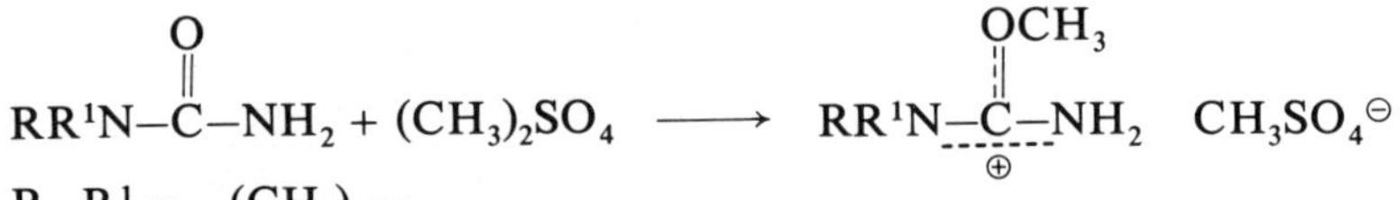

$$R-R^1 = -(CH_2)_5-$$

N,N,N',N'-Tetrasubstituted ureas as well are methylated by dialkyl sulfates at oxygen in an equilibrium reaction (92). Optimum yields of bis(dialkylamino)-

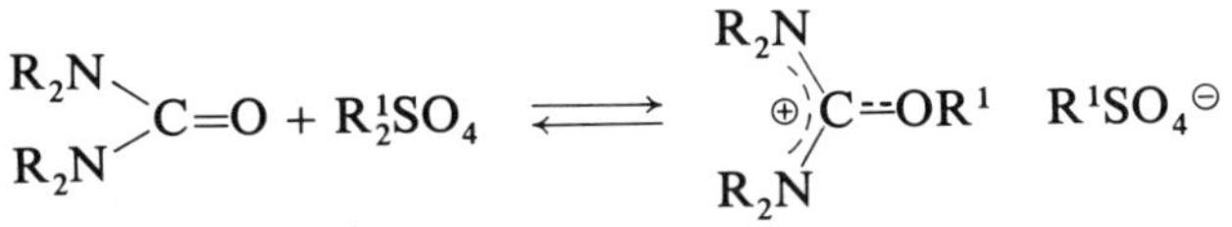

$$\begin{array}{c} R_2N \\ R_2N \end{array}\!\!\!C=O + R_2^1SO_4 \rightleftharpoons \overset{\oplus}{\underset{R_2N}{\overset{R_2N}{\diagdown}}}C-OR^1 \quad R^1SO_4^{\ominus}$$

(56)

alkoxycarbonium alkyl sulfates **56** are obtained in reaction times of about 3 hr at 80°C. The methylation of ethyl *N,N*-dimethylurethane yields tetramethylammonium methylsulfate among other products (68):

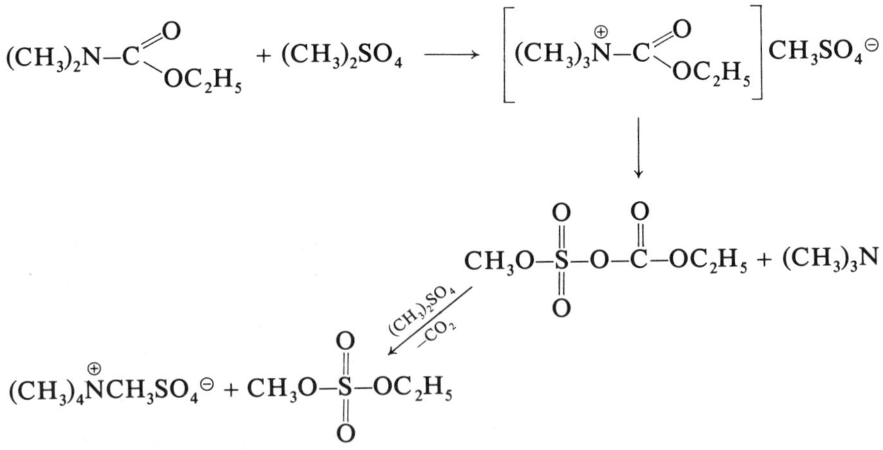

$$(CH_3)_2N-C\!\!\begin{array}{c}\diagup O \\ \diagdown OC_2H_5\end{array} + (CH_3)_2SO_4 \longrightarrow \left[(CH_3)_3\overset{\oplus}{N}-C\!\!\begin{array}{c}\diagup O \\ \diagdown OC_2H_5\end{array}\right] CH_3SO_4^{\ominus}$$

$$\downarrow$$

$$CH_3O-\overset{O}{\underset{\underset{O}{\|}}{S}}-O-\overset{O}{\underset{\|}{C}}-OC_2H_5 + (CH_3)_3N$$

$$(CH_3)_4\overset{\oplus}{N}CH_3SO_4^{\ominus} + CH_3O-\overset{O}{\underset{\underset{O}{\|}}{S}}-OC_2H_5 \xleftarrow{\overset{(CH_3)_2SO_4}{-CO_2}}$$

Table VIII summarizes the properties of several uronium sulfates.

I. SYNTHESIS

TABLE VIII
Adducts (**56**) from Ureas and Dialkyl Sulfates (**92**)

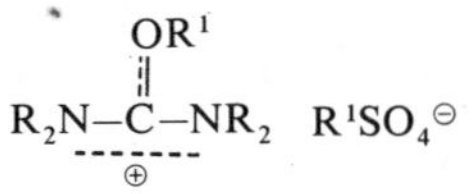

$$\underset{\oplus}{R_2N-\overset{\overset{OR^1}{\|}}{C}-NR_2} \quad R^1SO_4^{\ominus}$$

R	R^1	Yield, %	m.p., °C	n_D^{20}
CH_3	CH_3	87	Oil	1.4806
CH_3	C_2H_5	50	Oil	1.4698
C_2H_5	CH_3	54	Oil	1.4561

4. With Alkyl Halides and Sulfonic Acid Esters

Bredereck et al. (61, 62, 93–95) examined the reaction of alkyl halides with formamide. The reactions of formamide with α-halocarbonyl compounds, giving rise to oxazoles and imidazoles via primary alkylation of the oxygen atom, have been extensively reviewed (96). If alkylating agents predominantly reacting via an S_N1 mechanism are treated with formamide at higher temperatures (150°C), the amide nitrogen atom is preferentially attacked (62, 93–95) to form **57** primarily, but formates (**58**) form as well (62, 93).

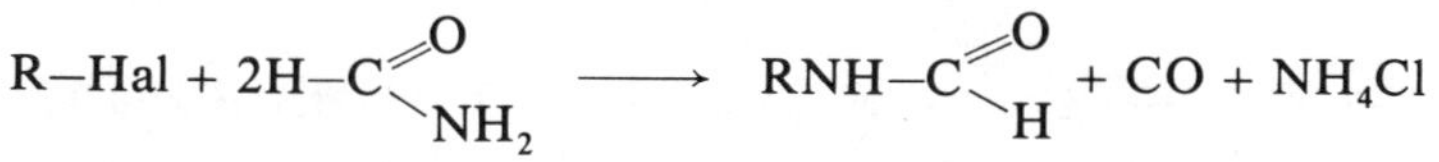

$$R\text{–Hal} + 2H\text{–C}\overset{\displaystyle O}{\underset{\displaystyle NH_2}{\diagup}} \longrightarrow RNH\text{–C}\overset{\displaystyle O}{\underset{\displaystyle H}{\diagup}} + CO + NH_4Cl$$
(**57**)

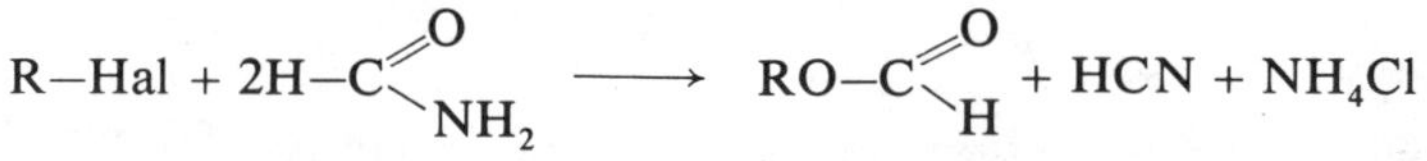

$$R\text{–Hal} + 2H\text{–C}\overset{\displaystyle O}{\underset{\displaystyle NH_2}{\diagup}} \longrightarrow RO\text{–C}\overset{\displaystyle O}{\underset{\displaystyle H}{\diagup}} + HCN + NH_4Cl$$
(**58**)

Those alkylating agents predominantly reacting via an S_N2 mechanism mainly generate formates (**58**) (61, 62) which is explained in the following way: the alkylating agent (alkyl halide, sulfonic acid ester) primarily attacks at the amide oxygen atom via formation of an alkoxymethyleniminium salt (**59**), which decomposes into alcohol, hydrogen cyanide, and hydrogen chloride.

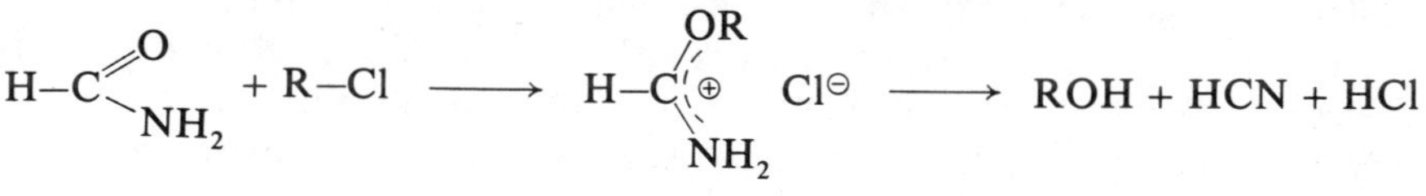

$$H\text{–C}\overset{\displaystyle O}{\underset{\displaystyle NH_2}{\diagup}} + R\text{–Cl} \longrightarrow H\text{–}\overset{\oplus}{C}\overset{\displaystyle OR}{\underset{\displaystyle NH_2}{\diagup}} \; Cl^{\ominus} \longrightarrow ROH + HCN + HCl$$
(**59**)

It was assumed that the formate **58** forms from unreacted formamide, alcohol, and hydrogen chloride in a secondary reaction. However, alcoholysis

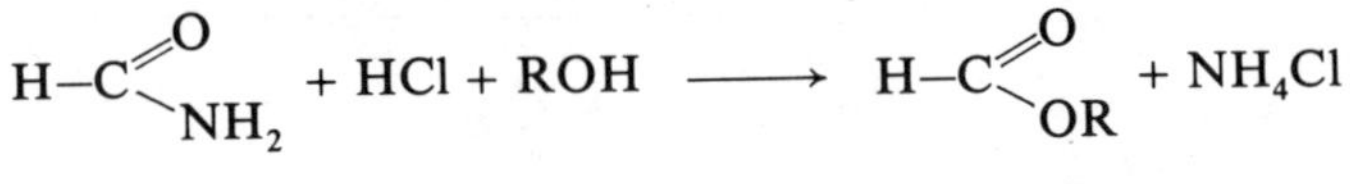

$$\text{H-C(=O)NH}_2 + \text{HCl} + \text{ROH} \longrightarrow \text{H-C(=O)OR} + \text{NH}_4\text{Cl}$$
(58)

of the alkoxymethyleniminium salt **59** is more likely; this yields the carboxonium ion **60**, which in turn alkylates formamide and decomposes into a formate (compare with the alcoholysis of alkoxymethyleniminium salts).

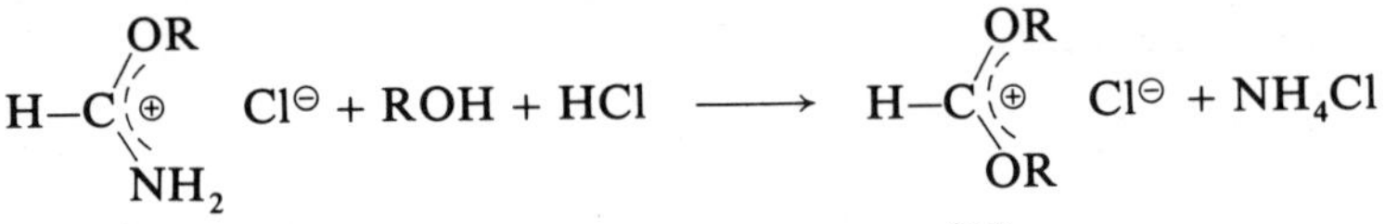

$$\text{H-C}^{\oplus}(\text{OR})(\text{NH}_2)\ \text{Cl}^{\ominus} + \text{ROH} + \text{HCl} \longrightarrow \text{H-C}^{\oplus}(\text{OR})(\text{OR})\ \text{Cl}^{\ominus} + \text{NH}_4\text{Cl}$$
(59) (60)

$$\downarrow \text{H-C(=O)NH}_2$$

$$\text{H-C(=O)OR} + (59)$$

As with dimethyl sulfate, tris(formamido)methane (**45**) is also obtained with other alkylating agents like sulfonic acid esters and excess formamide (62).

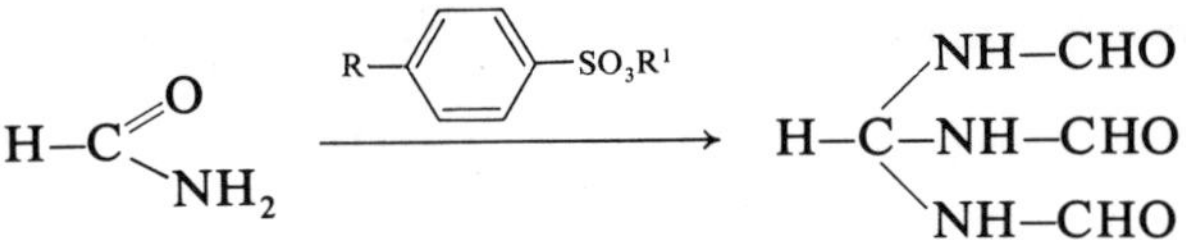

$$\text{H-C(=O)NH}_2 \xrightarrow{\text{R-C}_6\text{H}_4\text{-SO}_3\text{R}^1} \text{H-C(NH-CHO)}_3$$
(45)

Sulfonic acid esters as well as dimethyl sulfate convert formanilides into N,N'-diarylformamidinium sulfonates **61** (63). This reaction is also initiated by

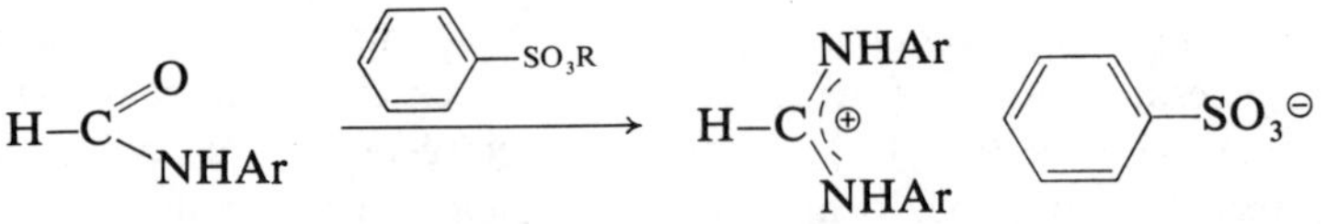

$$\text{H-C(=O)NHAr} \xrightarrow{\text{C}_6\text{H}_5\text{-SO}_3\text{R}} \text{H-C}^{\oplus}(\text{NHAr})(\text{NHAr})\ \text{C}_6\text{H}_5\text{-SO}_3^{\ominus}$$

Ar = C_6H_5, p-CH_3–C_6H_4– (61)

alkylation at the amide oxygen atom. Kornblum and Blackwood (97), as well as Ross and Labes (98), reacted alkyl halides with N,N-disubstituted acid amides, but they did not isolate any intermediates. In dimethylformamide the dehydrohalogenation of the alkyl halides occurs even at room temperature (97).

Neumeyer and Cannon (99) obtained dimethyl-, trimethyl-, and tetramethyl-ammonium bromide from dimethylformamide and methyl bromide:

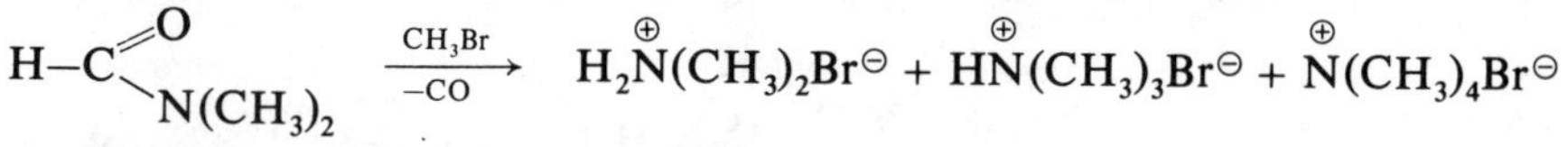

$$H-C\underset{N(CH_3)_2}{\overset{O}{\diagup}} \xrightarrow[-CO]{CH_3Br} H_2\overset{\oplus}{N}(CH_3)_2Br^{\ominus} + H\overset{\oplus}{N}(CH_3)_3Br^{\ominus} + \overset{\oplus}{N}(CH_3)_4Br^{\ominus}$$

Dimethylformamide and dimethylacetamide are spontaneously methylated by methyl fluorosulfonate at 25°C (100). Analysis of the reaction products showed that dimethylformamide is methylated at the oxygen atom (**62**) (95%) and at the nitrogen atom (**63**) (5%), whereas dimethylacetamide is exclusively attacked at the oxygen atom.

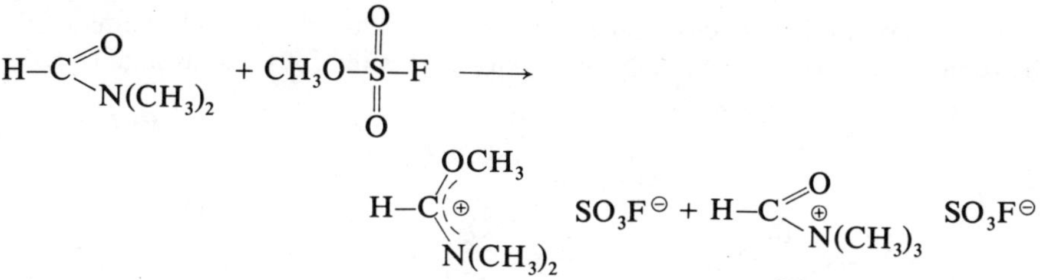

$$H-C\underset{N(CH_3)_2}{\overset{O}{\diagup}} + CH_3O-\underset{O}{\overset{O}{\underset{\|}{\overset{\|}{S}}}}-F \longrightarrow$$

$$H-\overset{\oplus}{C}\underset{N(CH_3)_2}{\overset{OCH_3}{\diagup}} SO_3F^{\ominus} + H-\overset{\oplus}{C}\underset{N(CH_3)_3}{\overset{O}{\diagup}} SO_3F^{\ominus}$$

$$\qquad 95\% \quad (62) \qquad\qquad 5\% \quad (63)$$

It was found that *N*-cyano-*N,N,N*-trialkylammonium salts **64** are also capable of alkylating *N,N*-disubstituted amides at the oxygen atom (101).

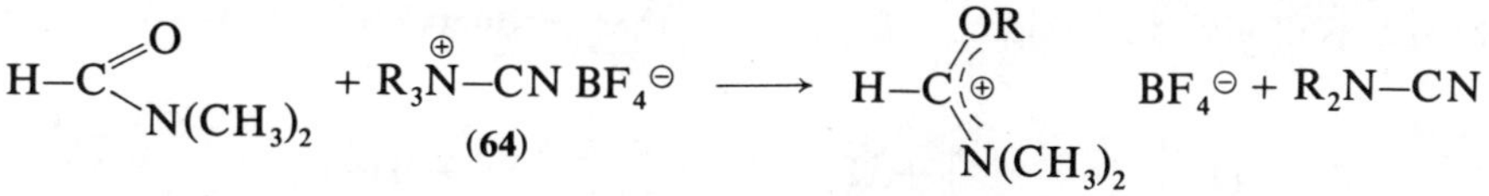

$$H-C\underset{N(CH_3)_2}{\overset{O}{\diagup}} + R_3\overset{\oplus}{N}-CN\ BF_4^{\ominus} \longrightarrow H-\overset{\oplus}{C}\underset{N(CH_3)_2}{\overset{OR}{\diagup}} BF_4^{\ominus} + R_2N-CN$$
$$\qquad\qquad (64)$$

$$R = C_2H_5, C_4H_9$$

Almost synchronously the discovery was made that during the halogenation of alkenes in dimethylformamide (102–104) the latter is alkylated. This reaction was used to synthesize β-chloroalkylformates (**65**) and epoxides (**66**) (102,

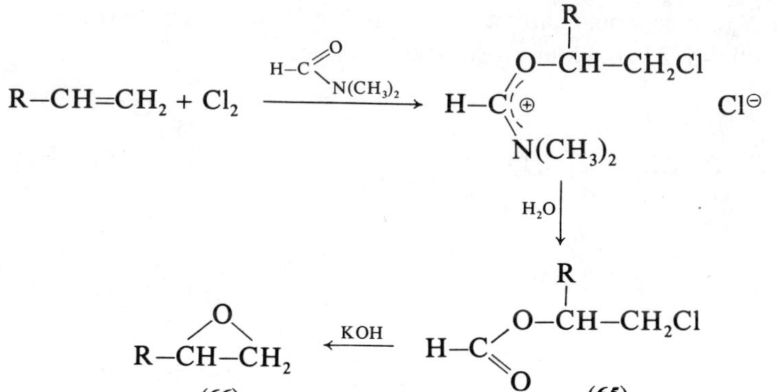

$$R-CH=CH_2 + Cl_2 \xrightarrow{H-C\overset{O}{\diagup}_{N(CH_3)_2}} H-\overset{\oplus}{C}\underset{N(CH_3)_2}{\overset{O-CH-CH_2Cl}{\overset{|}{\diagup}}} Cl^{\ominus}$$

$$\downarrow H_2O$$

$$R-\underset{(66)}{\overset{O}{CH-CH_2}} \xleftarrow{KOH} H-C\underset{O}{\overset{O-\overset{R}{\underset{|}{C}H}-CH_2Cl}{\diagup}}$$
$$\qquad\qquad\qquad\qquad (65)$$

104). This reaction is particularly important, since it makes possible *cis* addition of halides and pseudohalides to olefins (103):

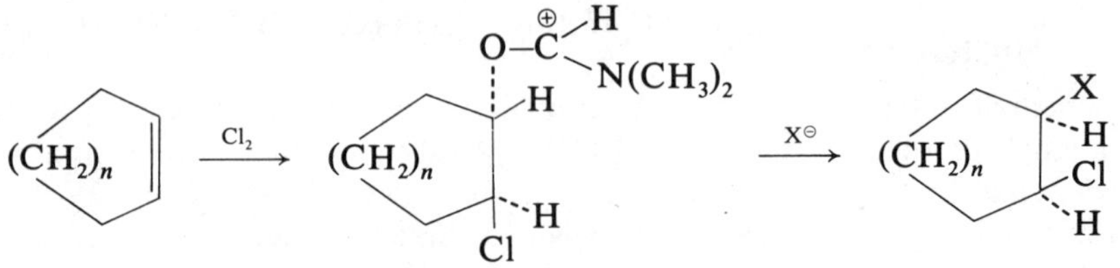

$$(CH_2)_n \text{(cyclohexene)} \xrightarrow{Cl_2} (CH_2)_n \begin{array}{c} O-\overset{\oplus}{C}\diagdown^H_{N(CH_3)_2} \\ \diagdown H \\ \diagdown H \\ Cl \end{array} \xrightarrow{X^\ominus} (CH_2)_n \begin{array}{c} X \\ \diagdown H \\ Cl \\ H \end{array}$$

From formamide, cyclohexene, and *N,N*-dibromobenzenesulfonamide *N*-benzenesulfonylformic acid 2-bromocyclohexyl ester (**67**) was isolated (105)

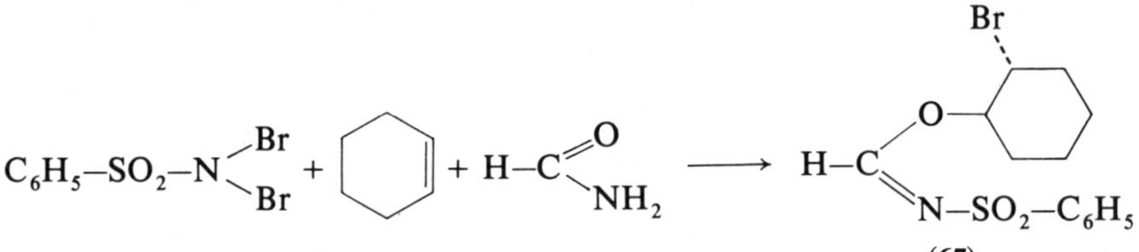

$$C_6H_5-SO_2-N\diagdown^{Br}_{Br} + \bigcirc + H-C\diagdown^O_{NH_2} \longrightarrow H-C\diagdown^{O-\text{[cyclohexyl-Br]}}_{N-SO_2-C_6H_5}$$

(**67**)

among other products. In contrast, dimethylformamide yields *N,N*-dimethyl-*N*-benzenesulfonylformamidine (**68**), among other products (105).

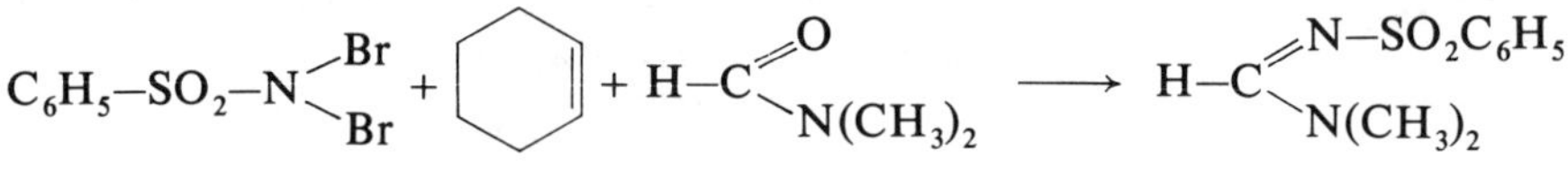

$$C_6H_5-SO_2-N\diagdown^{Br}_{Br} + \bigcirc + H-C\diagdown^O_{N(CH_3)_2} \longrightarrow H-C\diagdown^{N-SO_2C_6H_5}_{N(CH_3)_2}$$

(**68**)

Gompper and Christmann (95) demonstrated the different reactivities of alkyl halides by reacting them with ethyl urethane. Alkylating agents reacting by an S_N1 mechanism attack at the nitrogen atom, whereas those showing preferential S_N2 behaviour mostly act on onygen:

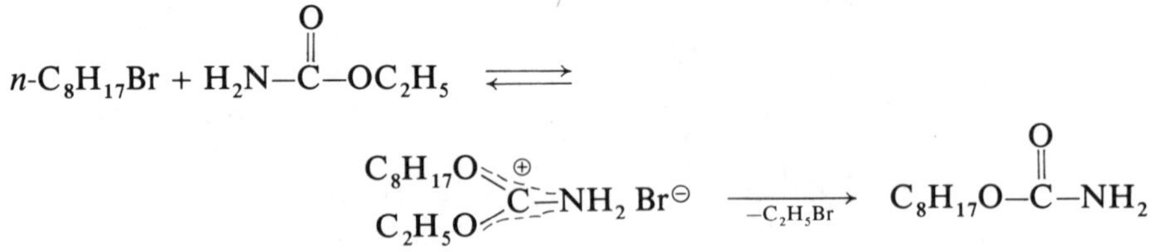

$$H_2N-\overset{O}{\overset{\|}{C}}-OC_2H_5 + (C_6H_5)_3CCl \xrightarrow{-HCl} (C_6H_5)_3C\overset{H}{\overset{|}{N}}-\overset{O}{\overset{\|}{C}}-OC_2H_5$$

$$n\text{-}C_8H_{17}Br + H_2N-\overset{O}{\overset{\|}{C}}-OC_2H_5 \rightleftharpoons$$

$$\begin{array}{c} C_8H_{17}O \\ C_2H_5O \end{array} \overset{\oplus}{C}=NH_2 \text{ Br}^\ominus \xrightarrow{-C_2H_5Br} C_8H_{17}O-\overset{O}{\overset{\|}{C}}-NH_2$$

I. SYNTHESIS

Ahmed and Alder (106) studied the methylation of urethanes by methyl fluorosulfonate. Methylation at oxygen (e.g., **69**) is kinetically controlled, whereas the thermodynamically controlled reaction leads to N-alkylation (e.g., **70**). The ratio of the velocity constants of the O-versus N-methylation ($k_O:k_N$) was determined to be 2:1 for methyl N,N-dimethylcarbamate. With N-methyl-2-oxazolidone $k_O:k_N$ increases to 1000:1! Nevertheless, N-methylated compounds are nearly exclusively isolated after extended reaction times (24 hr) at 100°C.

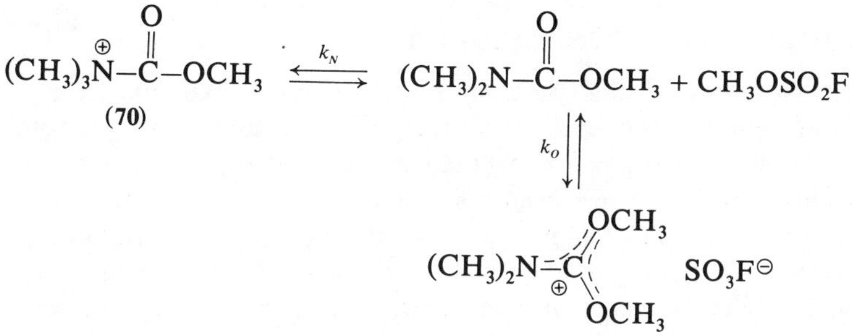

$$(CH_3)_3\overset{\oplus}{N}-\overset{O}{\underset{\|}{C}}-OCH_3 \quad \underset{}{\overset{k_N}{\rightleftarrows}} \quad (CH_3)_2N-\overset{O}{\underset{\|}{C}}-OCH_3 + CH_3OSO_2F$$

$$(\mathbf{70})$$

$$k_O \updownarrow$$

$$(CH_3)_2N\overset{\oplus}{=}C\underset{OCH_3}{\overset{OCH_3}{\diagdown}} \quad SO_3F^{\ominus}$$

$$(\mathbf{69})$$

As the preceding statements have shown, alkyl halides are less suitable for the synthesis of alkoxymethyleniminium ions, since their alkylating potential is very often too low, and the halide anions are sufficiently nucleophilic to dealkylate the already formed cations on heating. N-(2-Haloethenyl)-amides (**71**) represent an exception; they may be cyclized to Δ^2-oxazolines (**72**) via the intermediate formation of oxazolinium ions (**73**). Hellman and his co-workers (107) have published a review of these reactions.

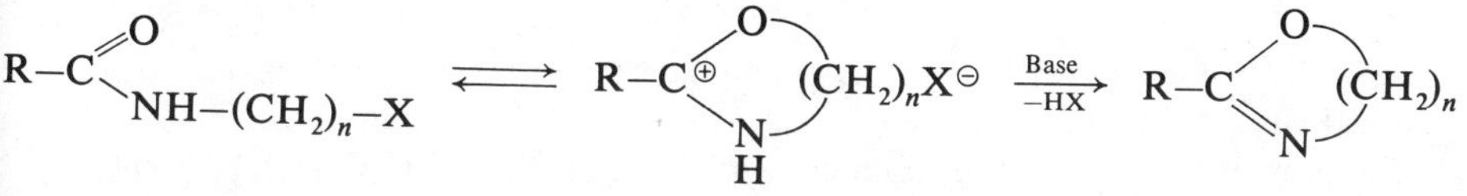

$$R-C\underset{NH-(CH_2)_n-X}{\overset{O}{\diagup}} \rightleftarrows R-\overset{\oplus}{C}\underset{\underset{H}{N}}{\overset{O-}{\diagup}}(CH_2)_n X^{\ominus} \xrightarrow[-HX]{Base} R-C\underset{N}{\overset{O-}{\diagup}}(CH_2)_n$$

(**71**)　　　　　　　　(**73**)　　　　　　　　(**72**)

In combination with Lewis acids, oxazolinium ions (e.g., **74** and **75**) are isolated from N-(2-haloethyl)amides (28, 29, 30, 108).

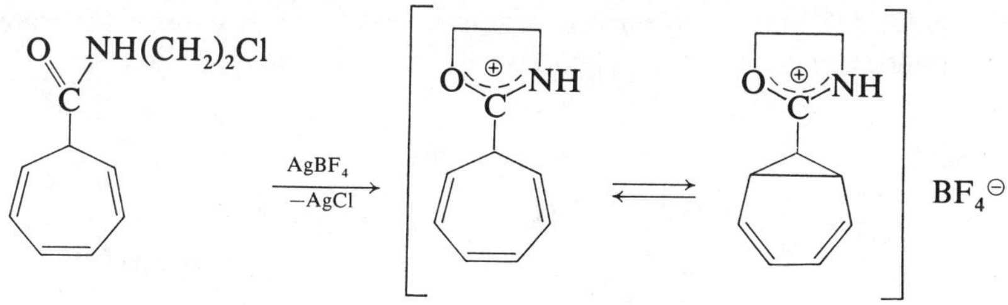

$$\text{cycloheptatrienyl-C(=O)-NH(CH_2)_2Cl} \xrightarrow[-AgCl]{AgBF_4} \left[\text{oxazolinium cation} \rightleftarrows \text{oxazolinium cation} \right] BF_4^{\ominus}$$

(**74**)

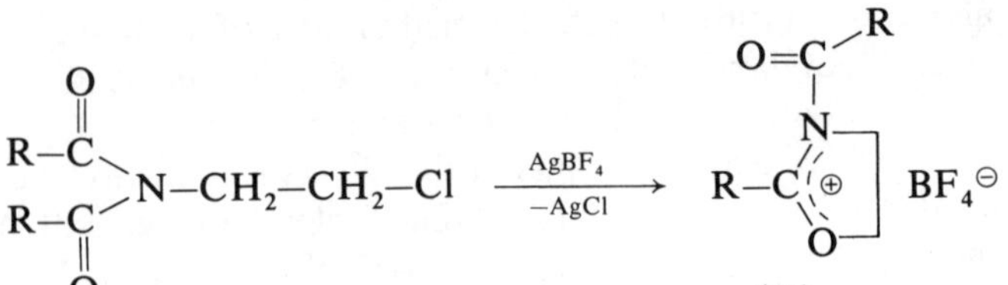

$$\underset{\substack{R-C\\\parallel\\O}}{\overset{\substack{O\\\parallel\\R-C}}{}}N-CH_2-CH_2-Cl \xrightarrow[-AgCl]{AgBF_4} \underset{(75)}{R-C\overset{\overset{\displaystyle O=C-R}{\displaystyle |}}{\underset{\displaystyle O}{\overset{\displaystyle N}{\oplus}}}} \quad BF_4^{\ominus}$$

B. VIA ACYLATION OF ACID AMIDES

Hechelhammer (109) has patented the synthesis of alkoxymethyleniminium salts from primary and secondary carboxamides and ethyl chloroformate. Suydam, Greth, and Langerman (110) studied this reaction extensively. The range of its application has been compared with Pinner's imino ester hydrochloride synthesis (111), and it was shown that the two methods very often are complementary, giving good yields with ethyl chloroformate, where the addition of alcohols and hydrogen chloride to nitriles proceeds poorly (e.g., with long-chain nitriles).

The accessibility of N-alkylalkoxymethyleniminium chlorides (76) is another advantage of this method.

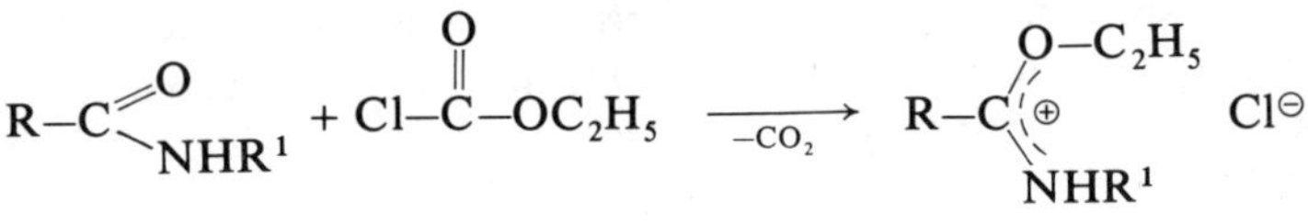

$$R-C\underset{NHR^1}{\overset{O}{\diagdown}} + Cl-\overset{O}{\underset{\parallel}{C}}-OC_2H_5 \xrightarrow{-CO_2} R-\underset{NHR^1}{\overset{O-C_2H_5}{C\oplus}} \quad Cl^{\ominus}$$

$R^1 = H$, alkyl (76)

N-(n-Butyl)butyramide and N-phenylacetamide were not successfully reacted with ethyl chloroformate.

In some cases amide–hydrogen chloride adducts like $(CH_3C(O)NH_2)_2 \cdot HCl$ are found as by-products.

Thioamides also are converted with ethyl chloroformate into the iminium chlorides (e.g., 76) via extrusion of carbon oxysulfide (110). Thioamides react more easily, faster, and more selectively than amides. Some alkoxymethyleniminium chlorides inaccessible from ordinary amides thus can be generated from thioamides, such as N-phenylethoxyethylideniminium chloride (77).

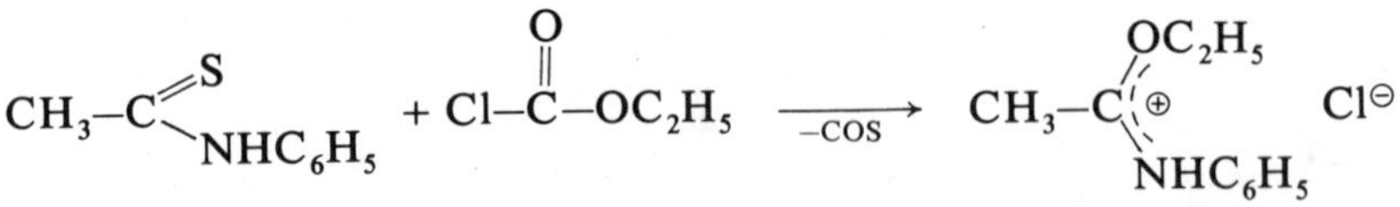

$$CH_3-C\underset{NHC_6H_5}{\overset{S}{\diagdown}} + Cl-\overset{O}{\underset{\parallel}{C}}-OC_2H_5 \xrightarrow{-COS} CH_3-\underset{NHC_6H_5}{\overset{OC_2H_5}{C\oplus}} \quad Cl^{\ominus}$$

(77)

I. SYNTHESIS

Table IX summarizes ethoxymethyleniminium chlorides synthesized by this route.

TABLE IX

Ethoxymethyleniminium Chlorides from Amides and Ethyl Chloroformate (110)

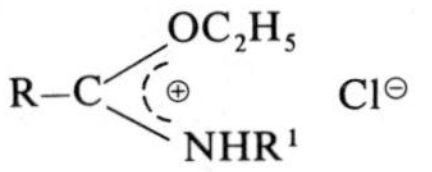

$$R-C\underset{NHR^1}{\overset{OC_2H_5}{\diagup}}{}^{\oplus}\quad Cl^{\ominus}$$

R	R^1	Yield, %	m.p., °C
H	CH_3	88	Oil
H	C_2H_5	92	Oil
CH_3	C_2H_5	78	Oil
CH_3	C_6H_5	60	92–93
C_2H_5	CH_3	67	Oil
C_2H_5	C_2H_5	75	Oil
$n\text{-}C_3H_7$	CH_3	80	Oil
$n\text{-}C_3H_7$	C_2H_5	76	Oil
$n\text{-}C_4H_9$	C_2H_5	69	Oil
$n\text{-}C_5H_{11}$	CH_3	73	Oil
$n\text{-}C_5H_{11}$	C_2H_5	77	Oil
$n\text{-}C_6H_{13}$	C_2H_5	84	Oil
$n\text{-}C_7H_{15}$	CH_3	44	Oil
$n\text{-}C_7H_{15}$	C_2H_5	74	Oil

To explain this reaction a mechanism was proposed by which the oxygen or the sulfur atoms of the carbonyl or thiocarbonyl group are primarily acylated. Then the primary adducts (**78**) give off CO_2 or COS, forming ethoxyalkylideniminium chlorides (**76**).

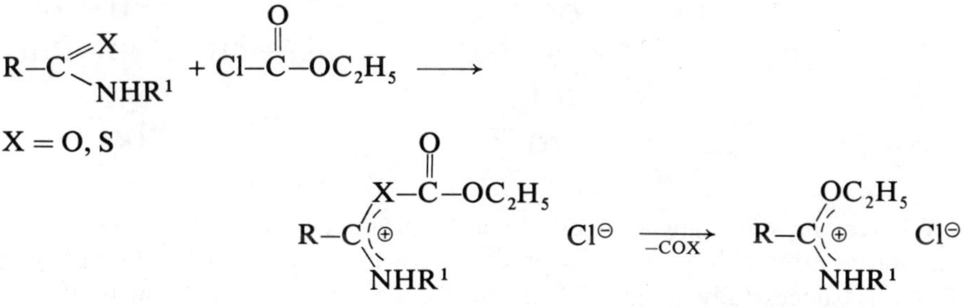

$$R-C\underset{NHR^1}{\overset{X}{\diagup}} + Cl-\overset{O}{\underset{\|}{C}}-OC_2H_5 \longrightarrow$$

X = O, S

$$R-C\underset{NHR^1}{\overset{X-\overset{O}{\underset{\|}{C}}-OC_2H_5}{\diagup}}{}^{\oplus}\quad Cl^{\ominus} \xrightarrow{-COX} R-C\underset{NHR^1}{\overset{OC_2H_5}{\diagup}}{}^{\oplus}\quad Cl^{\ominus}$$

(**78**) (**76**)

Bredereck et al. (65) proposed the same mechanism, when they obtained CO_2 and ethyl chloride as reaction products from dimethylformamide and ethylchloroformate. In the presence of sodium tetrafluoroborate the salt **79** may be isolated from the reaction mixture (112).

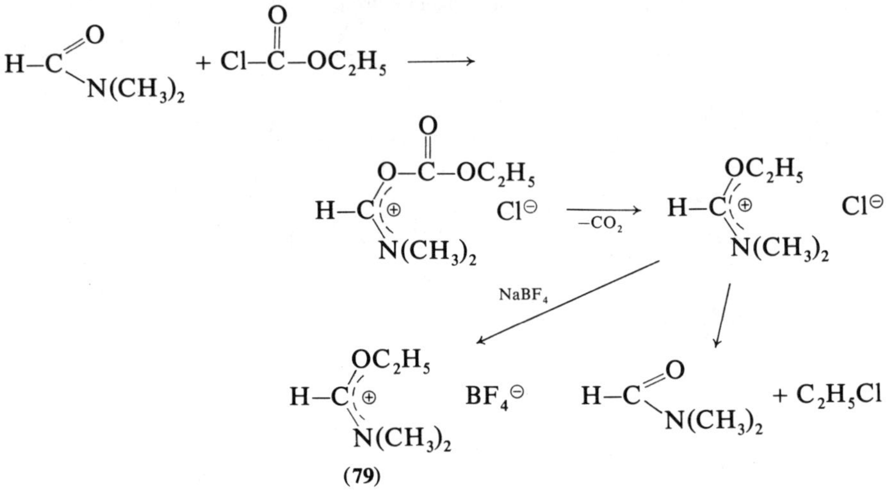

$$H-C\overset{O}{\underset{N(CH_3)_2}{\diagdown}} + Cl-\overset{O}{\underset{\|}{C}}-OC_2H_5 \longrightarrow$$

$$H-\overset{O-\overset{O}{\underset{\|}{C}}-OC_2H_5}{\underset{N(CH_3)_2}{C^{\oplus}}} \quad Cl^{\ominus} \xrightarrow{-CO_2} H-\overset{OC_2H_5}{\underset{N(CH_3)_2}{C^{\oplus}}} \quad Cl^{\ominus}$$

$$\xrightarrow{NaBF_4}$$

$$H-\overset{OC_2H_5}{\underset{N(CH_3)_2}{C^{\oplus}}} \quad BF_4^{\ominus} \qquad H-C\overset{O}{\underset{N(CH_3)_2}{\diagdown}} + C_2H_5Cl$$

(**79**)

The reaction of ethyl chloroformate with formamide gives tris(formamido)-methane (**45**) (62). This may be explained by primary generation of ethoxymethyleniminium chloride (**80**), which reacts with further formamide to tris(formamido)methane.

$$H-C\overset{O}{\underset{NH_2}{\diagdown}} + Cl-\overset{O}{\underset{\|}{C}}-OC_2H_5 \xrightarrow{-CO_2}$$

$$H-\overset{OC_2H_5}{\underset{NH_2}{C^{\oplus}}} \quad Cl^{\ominus} \xrightarrow{H-C\overset{O}{\underset{NH_2}{\diagdown}}} H-\overset{NH-CHO}{\underset{NH-CHO}{C-NH-CHO}}$$

(**80**) (**45**)

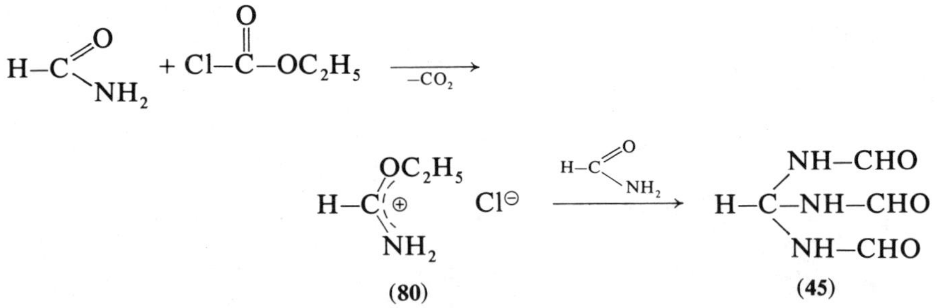

Whereas no stable acylation or alkylation product could be isolated from the reaction of ethyl chloroformate with dimethylformamide (65), product of type (**81**) was successfully obtained from aryl chloroformates and dimethylformamide or dimethylacetamide (113–115).

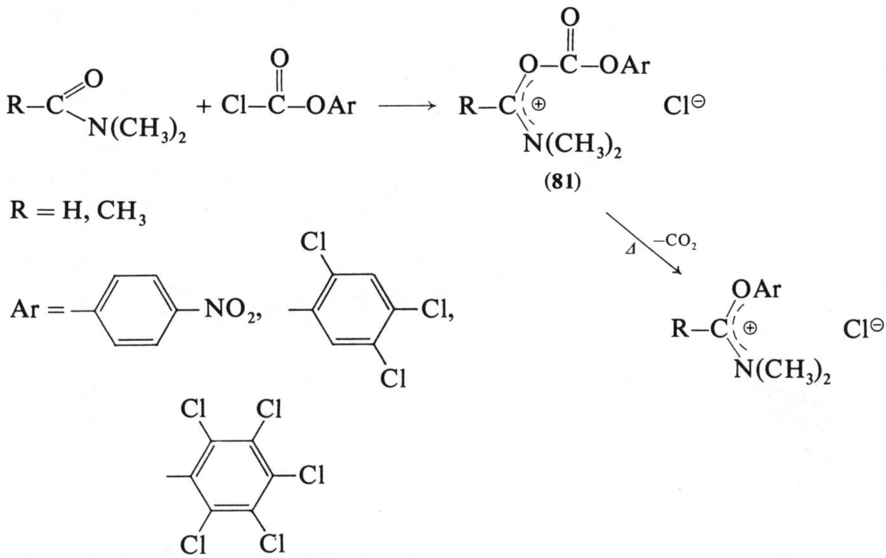

I. SYNTHESIS

$$R-C(=O)-N(CH_3)_2 + Cl-C(=O)-OAr \longrightarrow R-C^{\oplus}(O-C(=O)-OAr)(N(CH_3)_2) \quad Cl^{\ominus}$$
(81)

R = H, CH$_3$

Ar = –C$_6$H$_4$–NO$_2$, 2,5-dichlorophenyl, pentachlorophenyl

$$\xrightarrow{\Delta, -CO_2} R-C^{\oplus}(OAr)(N(CH_3)_2) \quad Cl^{\ominus}$$

Ohme and Schmitz (116) initiated a closely related kind of reaction to synthesize alkoxymethyleniminium chlorides (**82**). They alcoholyzed the formamide/benzoyl chloride adduct **83** already described by Bredereck et al. (62). Here the iminium chlorides are obtained in yields between 60 and 99%. Primary as well as secondary alcohols are suitable for alcoholysis; primary alcohols give better yields of alkoxymethyleniminium chlorides than secondary ones do.

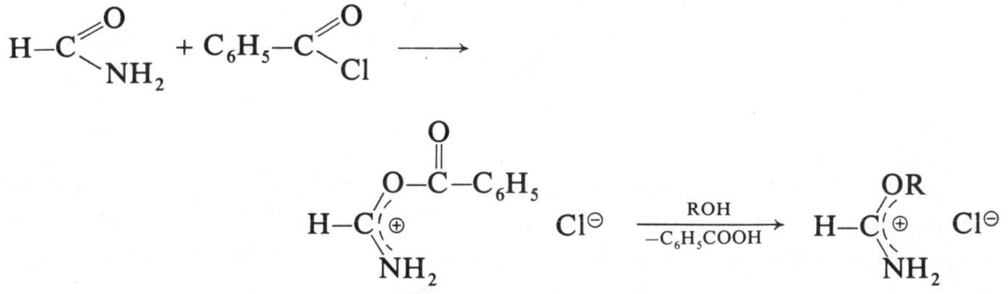

$$H-C(=O)-NH_2 + C_6H_5-C(=O)-Cl \longrightarrow$$

$$H-C^{\oplus}(O-C(=O)-C_6H_5)(NH_2) \quad Cl^{\ominus} \xrightarrow[-C_6H_5COOH]{ROH} H-C^{\oplus}(OR)(NH_2) \quad Cl^{\ominus}$$
(83) (82)

C. VIA ALKYLATION AND ACYLATION OF IMINO ESTERS

Alkylation of imino esters leads to alkoxymethyleniminium salts (**84**), the stabilities of which depend on the corresponding anions. Nucleophilic anions like $Cl^{\ominus}$, $Br^{\ominus}$, and $I^{\ominus}$ attack at the R^1 group, and dealkylate the cation, thus giving rise to *N,N*-disubstituted carboxamides and alkyl halides (117–119):

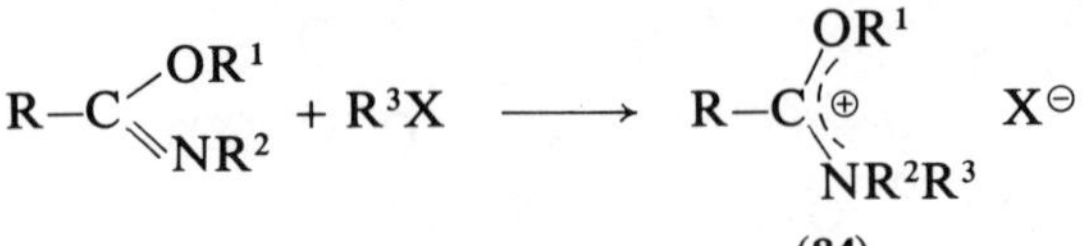

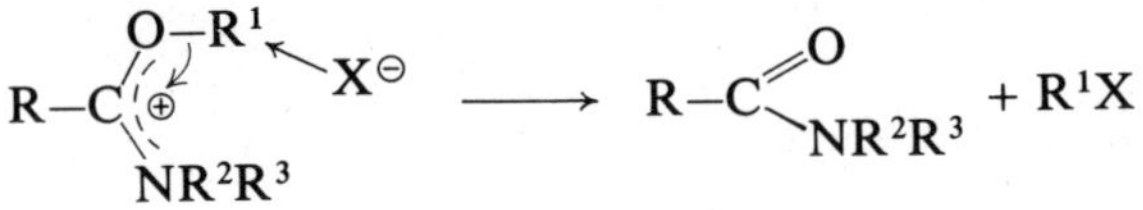

Acylation of imino esters also yields *N*-acyl-*N*-alkylalkoxymethyleniminium salts (**85**), which may react further to imides **86** (120, 121), or to *N*-acylimino esters **87** (121).

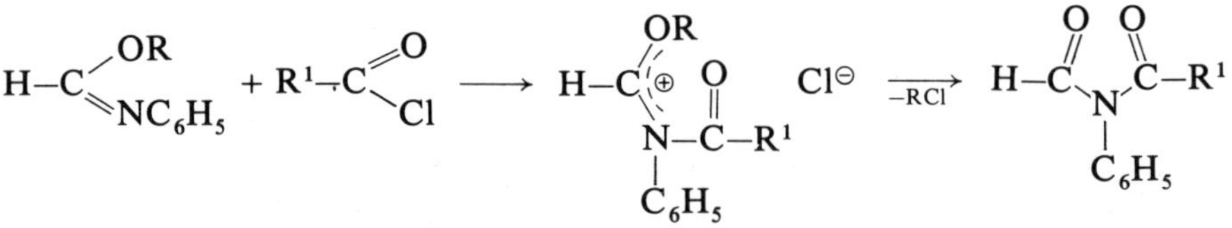

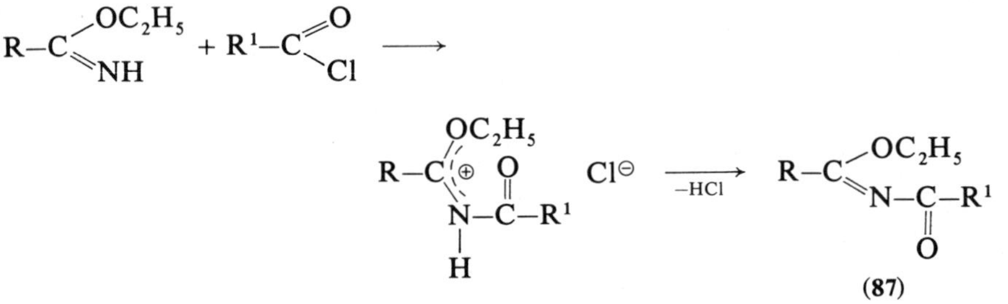

The alkylation and acylation of cyclic imino-ethers like **90** are of greater importance for the synthesis of cyclic ions like the oxazolinium ions **88** and **89**, since these salts are capable of polymerizing Δ^2-oxazolines (**90**). Their synthesis and reactions have been reviewed by Hellmann et al. (107).

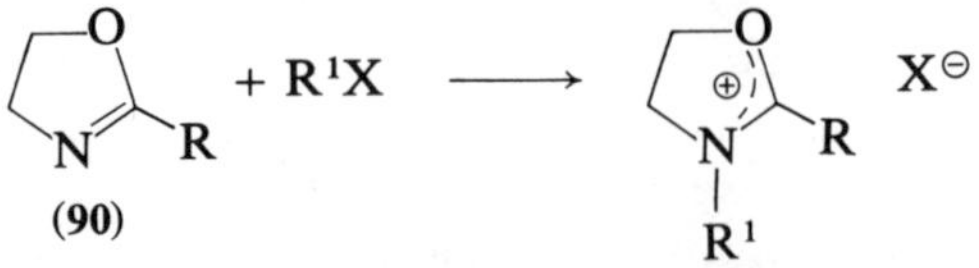

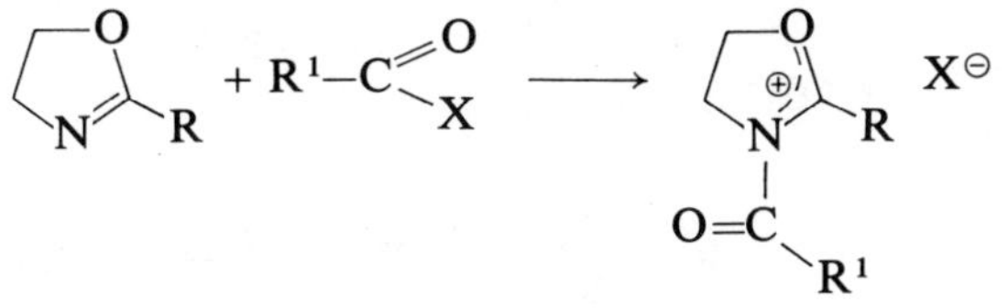

I. SYNTHESIS

$$\text{(89)}$$

Imino ethers have been converted into salts **91–93** by dimethyl sulfate and triethyloxonium fluoroborate (47, 122). Sometimes the methylated products are not isolated, but rather converted into *N*-methylated lactams (47, 123):

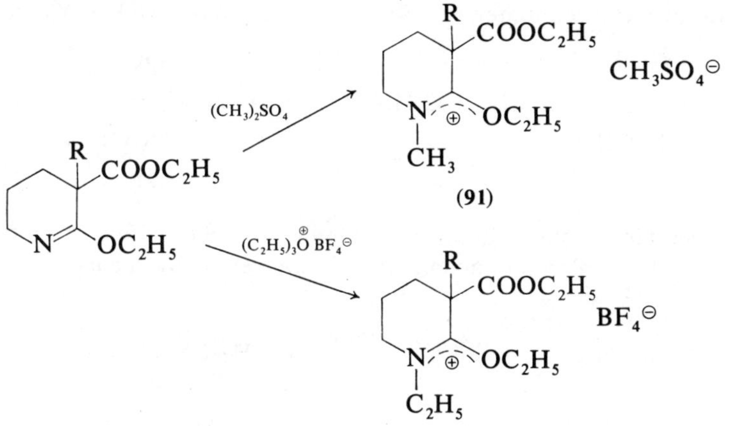

(91)

(92)

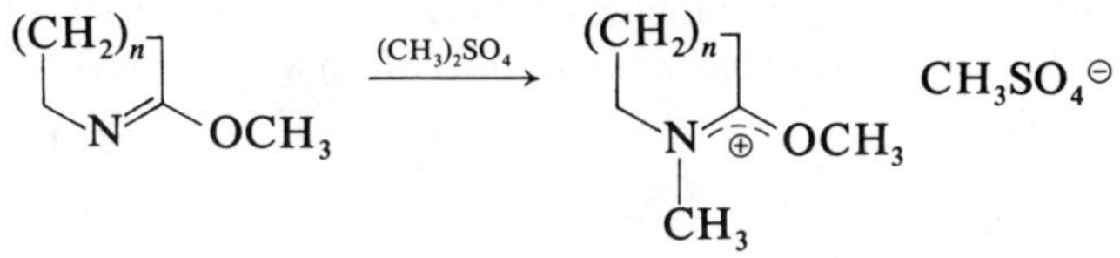

(93)

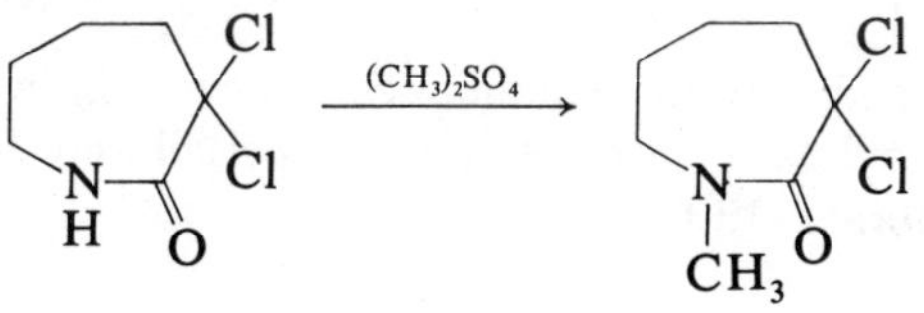

D. FROM CHLOROMETHYLENIMINIUM SALTS (AMIDE HALIDES)

Alcoholysis of amide chlorides (**94**) leads to alkoxymethyleniminium chlorides (**95**), which, however, decompose on heating into amides and alkyl halides (124).

$$\underset{(94)}{\text{R}-\overset{\text{Cl}}{\underset{\text{N}(R^1)_2}{\text{C}\oplus}}} \text{Cl}^\ominus + R^2\text{OH} \xrightarrow{-\text{HCl}} \underset{(95)}{\text{R}-\overset{\text{OR}^2}{\underset{\text{N}(R^1)_2}{\text{C}\oplus}}} \text{Cl}^\ominus \xrightarrow{\Delta} \text{R}-\text{C}\overset{\text{O}}{\underset{\text{N}(R^1)_2}{}} + R^2\text{Cl}$$

In the presence of nucleophilic reagents alcoholysis of the salts leads to a variety of reactions (124, 125). With ethanol, N,N-dimethylchloromethyleniminium hexachloroantimonate (**96**) forms the more stable N,N-dimethyl-

$$\underset{(96)}{\text{H}-\overset{\text{Cl}}{\underset{\text{N}(\text{CH}_3)_2}{\text{C}\oplus}}} \text{SbCl}_6^\ominus + \text{C}_2\text{H}_5\text{OH} \xrightarrow{-\text{HCl}} \underset{(97)}{\text{H}-\overset{\text{OC}_2\text{H}_5}{\underset{\text{N}(\text{CH}_3)_2}{\text{C}\oplus}}} \text{SbCl}_6^\ominus$$

ethoxymethyleniminium hexachloroantimonate (**97**) (126). Chloromethyleniminium hexachloroantimonate (**98**) is converted by alcohol into alkoxy-

$$\underset{(98)}{\text{H}-\overset{\text{Cl}}{\underset{\text{NH}_2}{\text{C}\oplus}}} \text{SbCl}_6^\ominus \xrightarrow[-\text{HCl}]{\text{ROH}} \underset{(99)}{\text{H}-\overset{\text{OR}}{\underset{\text{NH}_2}{\text{C}\oplus}}} \text{Cl}^\ominus$$

methyleniminium hexachloroantimonate (**99**) (127, 128). By the same route the salt **100** is transformed into the bisiminium salt **101** on alcoholysis (129).

$$\underset{(100)}{\text{Br}^\ominus \overset{\text{Br}}{\underset{\text{H}_2\overset{\oplus}{\text{N}}}{}}\text{C}-\text{CH}=\text{C}\overset{\text{Br}}{\underset{\text{NH}_2}{}}} \xrightarrow{\text{C}_2\text{H}_5\text{OH}} \underset{(101)}{\overset{\text{C}_2\text{H}_5\text{O}}{\underset{\text{H}_2\overset{\oplus}{\text{N}}}{}}\text{C}-\text{CH}_2-\overset{\text{OC}_2\text{H}_5}{\underset{\text{NH}_2}{\text{C}\oplus}}} \ 2\text{Br}^\ominus$$

Alcoholysis of dimethylformamide chloride (**102**) with tertiary alcohols yields N,N-dimethylalkoxymethyleniminium perchlorates (**103**) in the presence of sodium perchlorate (130).

$$\underset{(102)}{\text{H}-\overset{\text{Cl}}{\underset{\text{N}(\text{CH}_3)_2}{\text{C}\oplus}}} \text{Cl}^\ominus + \text{HO}-\overset{\text{CH}_3}{\underset{\text{CH}_3}{\text{C}}}-\text{R} \xrightarrow{\text{NaClO}_4} \underset{(103)}{\text{H}-\overset{\text{O}-\overset{\text{CH}_3}{\underset{\text{R}}{\text{C}}}}{\underset{\text{N}(\text{CH}_3)_2}{\text{C}\oplus}} \text{CH}_3} \text{ClO}_4^\ominus$$

I. SYNTHESIS

In contrast to N,N-dialkylalkoxymethyleniminium chlorides (**95**), N,N-dialkyl(aryloxy)methyleniminium chlorides are stable enough to allow isolation. Thus the salt **103a** was synthesized from phenol and dimethylformamide chloride (131).

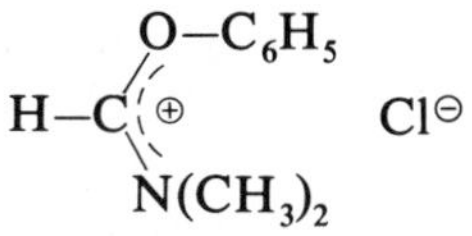

$$H-\overset{O-C_6H_5}{\underset{N(CH_3)_2}{C^\oplus}} \quad Cl^\ominus$$

(**103a**)

Ege and Frey (132) obtained N,N-dimethyl(2-hydroxyphenoxy)methyleniminium chloride (**104**) in boiling methylene chloride from substituted pyrocatechols and dimethylformamide chloride.

$$\underset{R}{\overset{R}{\diagdown}}\text{C}_6\text{H}_2(OH)_2 + H-\overset{Cl}{\underset{N(CH_3)_2}{C^\oplus}} \quad Cl^\ominus \xrightarrow{-HCl} \underset{R}{\overset{R}{\diagdown}}\text{C}_6\text{H}_2(OH)(O-\overset{N(CH_3)_2}{\underset{H}{C^\oplus}}) \quad Cl^\ominus$$

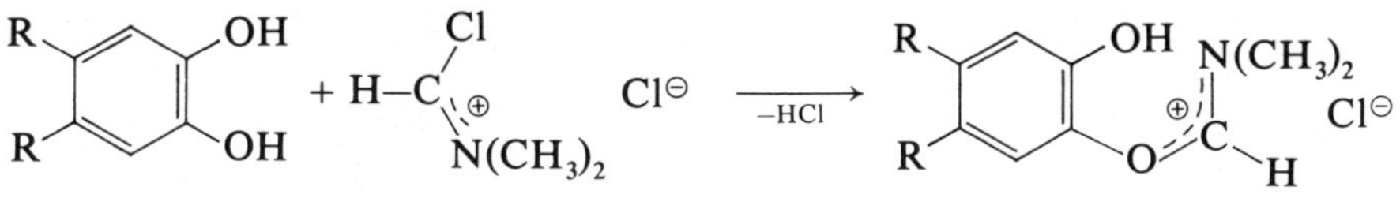

(**104**)

E. FROM AMIDE ACETALS AND LEWIS ACIDS

Amide acetals like **105** and **106** are converted into iminium salts **107** and **108** by Lewis acids, for example, BF$_3$ or SbCl$_5$ (13, 14). From the synthetic

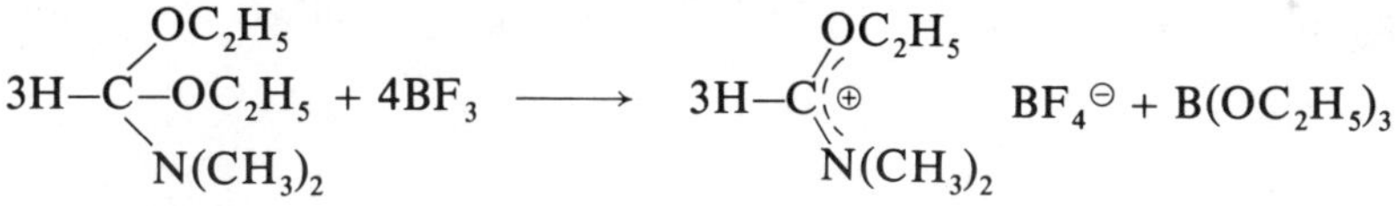

$$3H-\overset{OC_2H_5}{\underset{N(CH_3)_2}{C-OC_2H_5}} + 4BF_3 \longrightarrow 3H-\overset{OC_2H_5}{\underset{N(CH_3)_2}{C^\oplus}} \quad BF_4^\ominus + B(OC_2H_5)_3$$

(**105**)   (**107**)

$$3\underset{\underset{CH_3}{|}}{\overset{OC_2H_5}{\underset{N}{\diagup}}}\hspace{-0.3em}\text{OC}_2\text{H}_5 + 4BF_3 \longrightarrow 3\underset{\underset{CH_3}{|}}{\overset{}{\underset{N^\oplus}{\diagup}}}\hspace{-0.3em}\text{OC}_2\text{H}_5 \quad BF_4^\ominus + B(OC_2H_5)_3$$

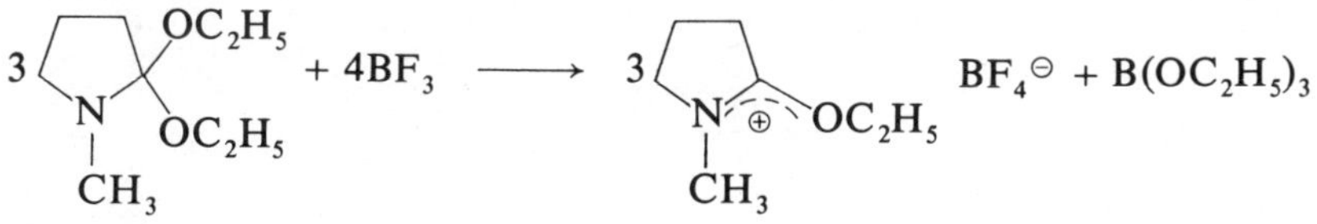

(**106**)   (**108**)

point of view this reaction is uninteresting, since alkoxymethyleniminium salts are the starting materials for the synthesis for amide acetals, but this reaction provided structural proof for amide acetal function. Meerwein et al. (14, 133) demonstrated that carbonium ions may convert amide acetals like **105** into

N,N-dimethylethoxymethyleniminium salts **107** and **108** by removal of the alkoxide ions.

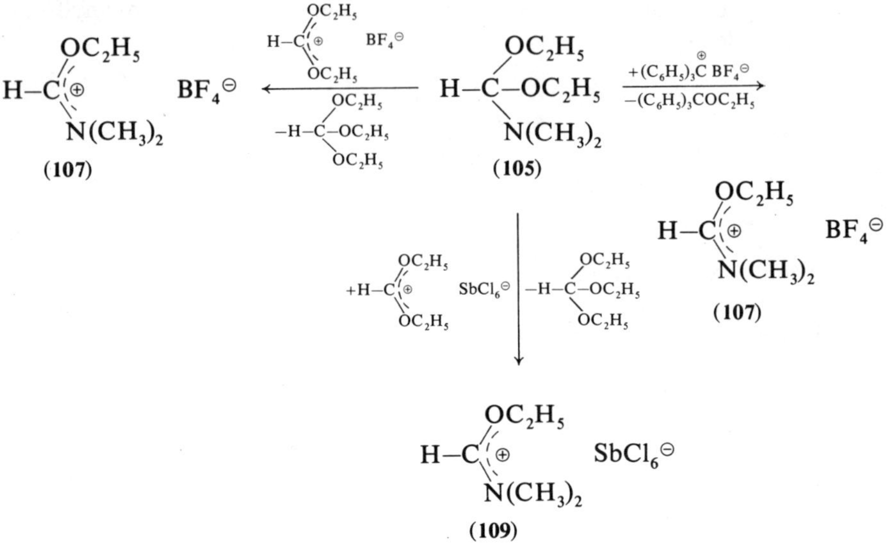

$$H-\overset{OC_2H_5}{\underset{N(CH_3)_2}{C^{\oplus}}} \quad BF_4^{\ominus} \xleftarrow[-H-\overset{OC_2H_5}{\underset{OC_2H_5}{C-OC_2H_5}}]{H-\overset{OC_2H_5}{\underset{OC_2H_5}{C}} \; BF_4^{\ominus}} H-\overset{OC_2H_5}{\underset{N(CH_3)_2}{C-OC_2H_5}} \xrightarrow[-(C_6H_5)_3COC_2H_5]{+(C_6H_5)_3C \; BF_4^{\ominus}}$$

(**107**)   (**105**)

$$\downarrow \; ^{+H-\overset{OC_2H_5}{\underset{OC_2H_5}{C^\oplus}} \; SbCl_6^\ominus \; | \; -H-\overset{OC_2H_5}{\underset{OC_2H_5}{C-OC_2H_5}}} \quad H-\overset{OC_2H_5}{\underset{N(CH_3)_2}{C^\oplus}} \; BF_4^\ominus$$

(**107**)

$$H-\overset{OC_2H_5}{\underset{N(CH_3)_2}{C^\oplus}} \; SbCl_6^\ominus$$

(**109**)

F. VIA REACTION OF KETENE O,N-ACETALS WITH ELECTROPHILIC REAGENTS

The reaction of ketene O,N-acetals like **110** with electrophilic reagents supposedly leads to iminium salts (**111**); here the anion $X^\ominus$ must have a low nucleophilic potential (e.g., $BF_4^\ominus$, $SbCl_6^\ominus$, $ClO_4^\ominus$), since otherwise dealkyl-

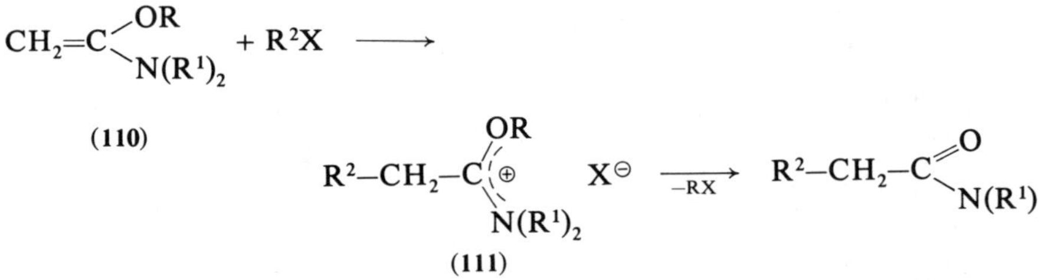

$$CH_2=C\overset{OR}{\underset{N(R^1)_2}{}} + R^2X \longrightarrow$$

(**110**)

$$R^2-CH_2-\overset{OR}{\underset{N(R^1)_2}{C^\oplus}} \; X^\ominus \xrightarrow[-RX]{} R^2-CH_2-C\overset{O}{\underset{N(R^1)}{\diagup\hspace{-0.5em}\diagdown}}$$

(**111**)

ation of **111** will take place. However, this reaction has not been performed up to now, although similar reactions of ketene S,N-acetals and ketene N,N-acetals are known (see the chapters "Alkylmercaptomethyleniminium Salts" and "Amidinium Salts"). If ketene O,N-acetals are reacted with sulfonylisocyanates or sulfonylisothiocyanates, stable 1,4-dipoles **112** or **113** form, which may be converted into the iminium salts **114** or **115** by perchloric acid (134).

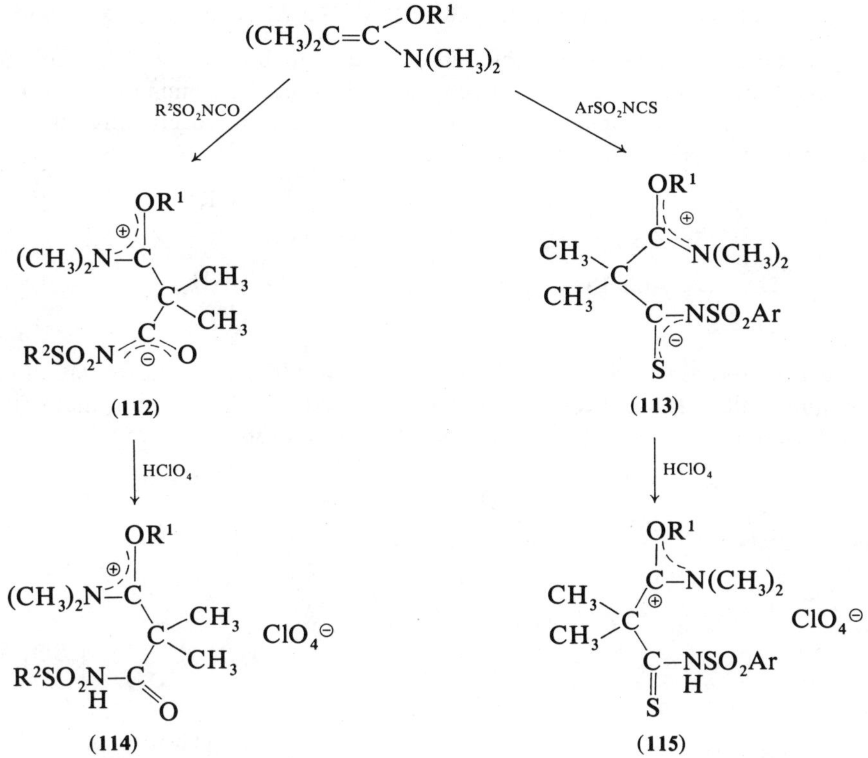

I. SYNTHESIS

[Scheme showing (CH$_3$)$_2$C=C(OR^1)N(CH$_3$)$_2$ reacting with R^{2}SO$_2$NCO to give (112) and with ArSO$_2$NCS to give (113); each then treated with HClO$_4$ to give (114) and (115) respectively with ClO$_4^\ominus$ counterion.]

(112)   (113)

(114)   (115)

G. VIA AMINOLYSIS OF CARBOXONIUM IONS

Aminolysis of carboxonium ions (**116**) would be expected to yield alkoxymethyleniminium ions (**117**). However, this reaction has been described only for salts like **118** by Meerwein (13).

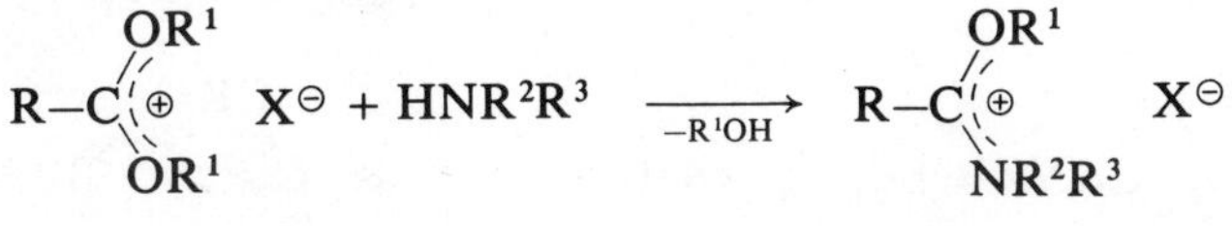

$$R-C(OR^1)_2^{\oplus} \; X^{\ominus} + HNR^2R^3 \xrightarrow{-R^1OH} R-C(OR^1)(NR^2R^3)^{\oplus} \; X^{\ominus}$$

(116)   (117)

When primary aromatic amines act on *O*-ethylbutyrolactonium fluoroborate (**118**), (*N*-arylimino)tetrahydrofuran hydrofluoroborate (**119**) forms.

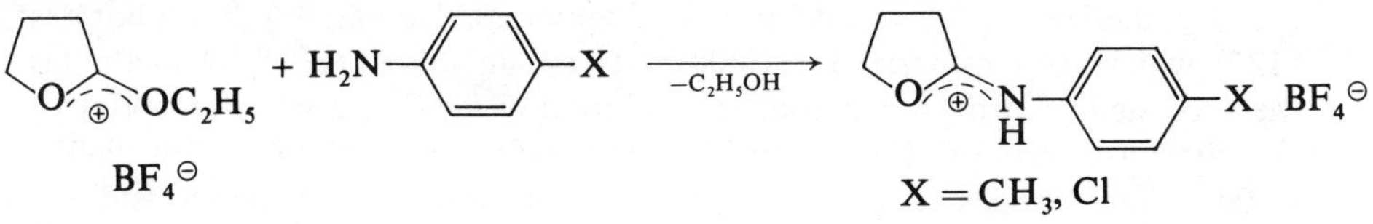

[Reaction of tetrahydrofuran-2-ylidene(OC$_2$H$_5$) cation BF$_4^\ominus$ (118) + H$_2$N–C$_6$H$_4$–X → (N-aryl) product (119) + C$_2$H$_5$OH; X = CH$_3$, Cl]

(118)   (119)

H. VIA ADDITION OF HYDROGEN HALIDES AND ALCOHOLS TO NITRILES

The addition of hydrogen chloride or hydrogen bromide and alcohols to nitriles is the oldest synthetic method for alkoxymethyleniminium salts **120** (135). The range of application of this extremely versatile reaction has already been reviewed (121, 136).

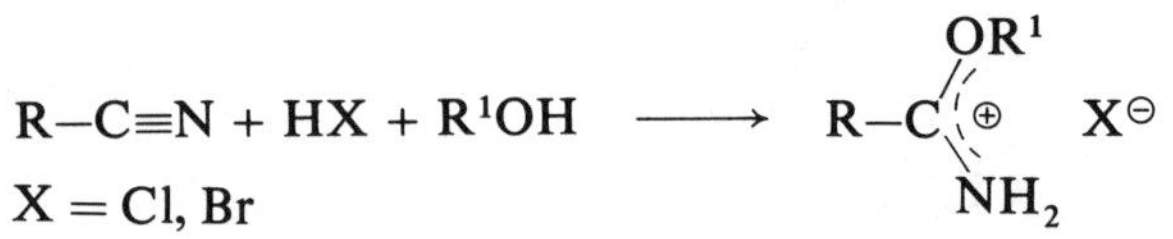

$$R-C{\equiv}N + HX + R^1OH \longrightarrow R-\overset{OR^1}{\underset{NH_2}{C^{\oplus}}} \quad X^{\ominus}$$
$$X = Cl, Br$$
(120)

In the presence of acids, alcohols also add to cyanamide via formation of iminium salts **(120a, 120b)**. Recently a method suitable for the industrial production of these salts **(120a)** has been developed (136a).

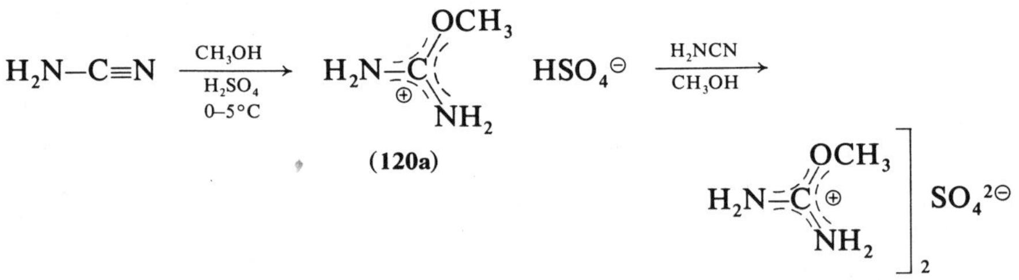

$$H_2N-C{\equiv}N \xrightarrow[H_2SO_4]{CH_3OH} H_2N{=}\overset{OCH_3}{\underset{NH_2}{C^{\oplus}}} \quad HSO_4^{\ominus} \xrightarrow[CH_3OH]{H_2NCN} \left[H_2N{=}\overset{OCH_3}{\underset{NH_2}{C^{\oplus}}} \right]_2 SO_4^{2\ominus}$$

(120a)   **(120b)**

I. VIA ADDITION OF ALCOHOLS TO NITRILIUM SALTS

The easily accessible nitrilium salts **(121)** (137) add alcohols forming alkoxymethyleniminium salts **(122)** (42, 138, 139). By this route, Pilotti et al.

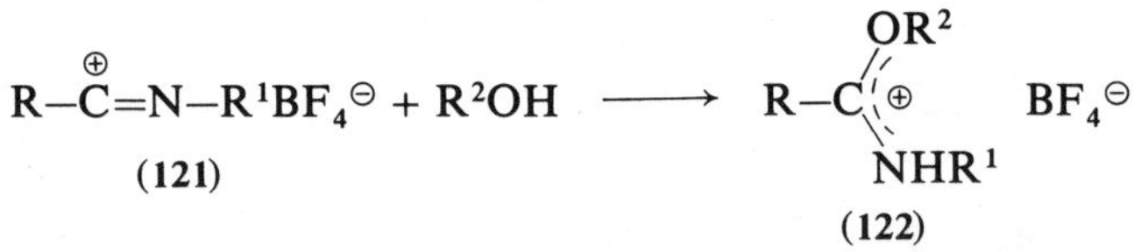

$$R-\overset{\oplus}{C}{=}N-R^1 BF_4^{\ominus} + R^2OH \longrightarrow R-\overset{OR^2}{\underset{NHR^1}{C^{\oplus}}} \quad BF_4^{\ominus}$$
(121)   **(122)**

(42) synthesized quite a number of alkoxymethyleniminium fluoroborates **(122)**, which are summarized in Table X. The main advantage of this method is the accessibility of those iminium salts in good yields that carry isopropoxy or cyclohexyloxy groups. They cannot be obtained from oxonium salts or dialkyl sulfates and amides, because these oxonium salts are unknown and the corresponding dialkyl sulfates are hard to obtain.

II. STRUCTURE OF ALKOXYMETHYLENIMINIUM SALTS

TABLE X

Alkoxymethyleniminium Salts from Alcohols and Nitrilium Salts (42)

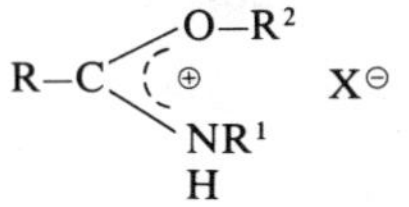

$$R-C\begin{smallmatrix}O-R^2\\ \oplus\\ NR^1\\ H\end{smallmatrix} \quad X^\ominus$$

R	R^1	R^2	X	Yield, %	m.p., °C
C_2H_5	C_2H_5	C_2H_5	BF_4	46	104–105
C_2H_5	C_2H_5	CH_3	BF_4	60	88–90
C_2H_5	C_2H_5	i-C_3H_7	BF_4	80	116–118
C_2H_5	C_2H_5	C_6H_{11}	BF_4	70	90.5–91
CH_3	C_2H_5	CH_3	BF_4	85	Oil
CH_3	C_2H_5	C_2H_5	BF_4	87	Oil
CH_3	C_2H_5	i-C_3H_7	BF_4	92	Oil
CH_3	C_2H_5	i-C_6H_{11}	BF_4	74	Oil

Phenol also adds to nitrilium salts; the resulting aryloxymethyleniminium salt (**123**) was not isolated, but rather reacted with alkali to give the corresponding imino ester (**124**) (139).

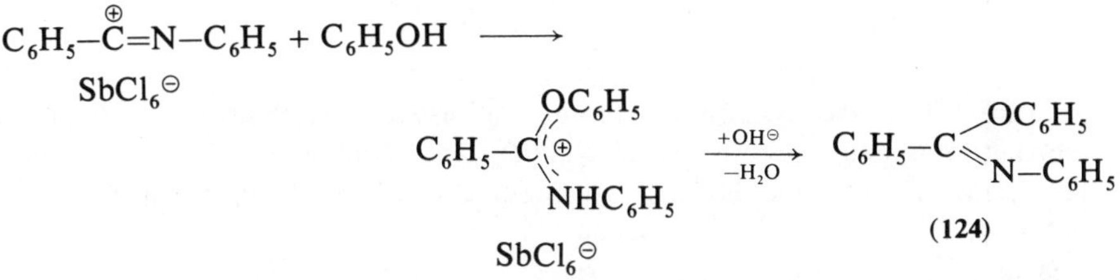

$$C_6H_5-\overset{\oplus}{C}=N-C_6H_5 + C_6H_5OH \longrightarrow$$
$$SbCl_6^\ominus$$

$$C_6H_5-C\begin{smallmatrix}OC_6H_5\\ \oplus\\ NHC_6H_5\end{smallmatrix} \xrightarrow[-H_2O]{+OH^\ominus} C_6H_5-C\begin{smallmatrix}OC_6H_5\\ \\ N-C_6H_5\end{smallmatrix}$$
$$SbCl_6^\ominus \qquad\qquad (\mathbf{124})$$
$$(\mathbf{123})$$

II. Structure of Alkoxymethyleniminium Salts

The structure of alkoxymethyleniminium salts has been established by various investigations. On acidic hydrolysis of the adducts from N,N-dimethyl amides and dimethyl sulfate only dimethylamine hydrochloride forms, not trimethylamine hydrochloride (65). The latter compound is the expected product resulting from N-alkylation:

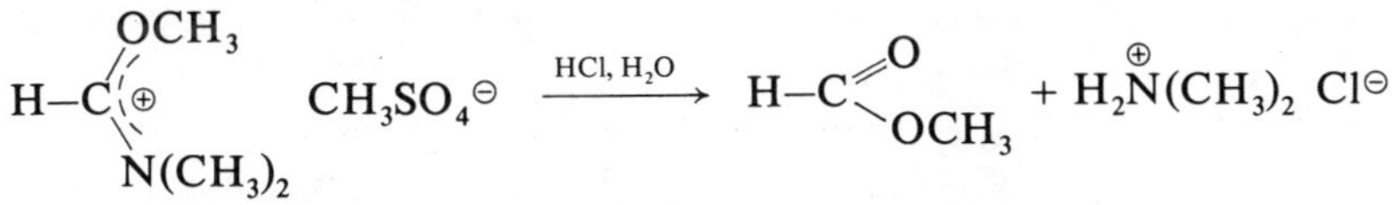

$$H-C\begin{smallmatrix}OCH_3\\ \oplus\\ N(CH_3)_2\end{smallmatrix} \quad CH_3SO_4^\ominus \xrightarrow{HCl, H_2O} H-C\begin{smallmatrix}O\\ \\ OCH_3\end{smallmatrix} + H_2\overset{\oplus}{N}(CH_3)_2 \quad Cl^\ominus$$

Furthermore, strongly nucleophilic reagents like alkoxide ions (13–15, 64, 68, 71) and cyanide ions (84, 140, 141) add to N,N-dialkylalkoxymethyleniminium salts like **125** to give amide acetals (**126**) or O,N-acetals of acylcyanides (**127**).

$$\underset{(126)}{\underset{N(CH_3)_2}{H-\overset{OR}{\underset{|}{C}}-OR}} \xleftarrow[-X^\ominus]{RO^\ominus} \underset{(125)}{\underset{N(CH_3)_2}{H-\overset{OR}{\underset{|}{C}}{}^\oplus}} \quad X^\ominus \xrightarrow[-X^\ominus]{+CN^\ominus} \underset{(127)}{\underset{N(CH_3)_2}{H-\overset{OR}{\underset{|}{C}}-CN}}$$

$$X = BF_4, CH_3SO_4$$

On the other hand, amide acetals like **105** and BF_3 yield N,N-dialkylalkoxymethyleniminium fluoroborates like **107** (13, 14).

$$\underset{(105)}{\underset{N(CH_3)_2}{3H-\overset{OC_2H_5}{\underset{|}{C}}-OC_2H_5}} + 4BF_3 \longrightarrow \underset{(107)}{\underset{N(CH_3)_2}{3H-\overset{OC_2H_5}{\underset{|}{C}}{}^\oplus}} \quad BF_4^\ominus + B(OC_2H_5)_3$$

The physical properties of alkoxymethylene iminium salts point to a polar structure. Thus alkoxymethyleniminium chlorides, fluoroborates, and hexachloroantimonates generally are crystalline solids, soluble in polar solvents like dimethylformamide, acetonitrile, and methylene chloride; they are insoluble in apolar solvents like ether and benzene.

Exhaustive conductivity measurements of nearly 50 alkoxymethyleniminium alkylsulfates and fluoroborates (16) in acetonitrile have established the polar structure of these compounds. Conductivity data may be represented with Kohlrausch's law of the square roots. Concentrations lower than 10^{-3} mole/liter give a linear $\lambda/\sqrt{c}$ plot, as expected for dissociated salts. The slope deviates from that calculated according to Debye–Hückel: more with the methylsulfates, less with the fluoroborates. Using the procedure by Fuoss and Kraus, and by Shedlovsky on the other hand, the degree of ion-pair formation was calculated from these measurements (16). For the dissociation of the salts into free ions the dissociation constants of the fluoroborates, in the range 10^{-2}–10^{-1}, were three to four times greater than those of the corresponding methylsulfates. The degree of dissociation of the salts covers a range of 80–99%.

From IR spectroscopic data one can conclude that the salts are completely ionized. Thus the IR spectra of alkoxymethyleniminium salts reveal a strong band in the range 1640–1710 cm^{-1} (41, 42, 44, 65, 68), which is attributed to

II. STRUCTURE OF ALKOXYMETHYLENIMINIUM SALTS

TABLE XI
Position of the $\overset{\oplus}{C=N}$ Band of Alkoxymethyleniminium Salts

$$R-C\overset{OR^3}{\underset{NR^1R^2}{\langle\oplus}}\quad X^{\ominus}$$

R	R^1	R^2	R^3	X	$\overset{\oplus}{C=N}$, cm^{-1}	Ref.
H	CH_3	CH_3	CH_3	CH_3SO_4	1710	65
CH_3	CH_3	CH_3	CH_3	CH_3SO_4	1675, 1640	65
CH_3	H	C_6H_5	CH_3	CH_3SO_4	1660	42
CH_3	H	n-C_6H_{13}	C_2H_5	BF_4	1673	42
CH_3	H	C_6H_5	C_2H_5	BF_4	1658	42
CH_3	C_2H_5	C_2H_5	CH_3	CH_3SO_4	1651	68
C_2H_5	H	C_2H_5	C_2H_5	BF_4	1668	42
C_2H_5	H	C_6H_5	C_2H_5	BF_4	1650	42
C_2H_5	H	C_2H_5	CH_3	BF_4	1674	42
C_2H_5	H	C_2H_5	i-C_3H_7	BF_4	1665	42
C_2H_5	H	C_2H_5	c—C_6H_{11}	BF_4	1664	42
C_2H_5	H	C_3H_7	CH_3	CH_3SO_4	1655	42
C_2H_5	CH_3	CH_3	CH_3	CH_3SO_4	1669	65
n-C_3H_7	CH_3	CH_3	CH_3	CH_3SO_4	1658	65
n-C_3H_7	C_2H_5	C_2H_5	CH_3	CH_3SO_4	1647	65
n-C_5H_{11}	CH_3	CH_3	CH_3	CH_3SO_4	1658	65
n-C_5H_{11}	C_2H_5	C_2H_5	CH_3	CH_3SO_4	1645	65
C_6H_5	H	C_2H_5	CH_3	CH_3SO_4	1658	42
C_6H_5	CH_3	CH_3	CH_3	CH_3SO_4	1655, 1633	65
$C_6H_5CH_2$	H	C_2H_5	CH_3	CH_3SO_4	1663	42
$C_6H_5CH_2$	CH_3	CH_3	CH_3	CH_3SO_4	1660	68
$NC-CH_2$	CH_3	CH_3	CH_3	CH_3SO_4	1656	68
$(CH_3)_2N\overset{O}{\overset{\|}{C}}-CH_2$	CH_3	CH_3	CH_3	CH_3SO_4	1649	68
$(CH_3)_2N-\overset{O}{\overset{\|}{C}}-(CH_2)_2$	CH_3	CH_3	CH_3	CH_3SO_4	1641	68
$CH_3SO_4^{\ominus}\ \overset{OCH_3}{\underset{\|}{\oplus}}$						
$(CH_3)_2N-C-(CH_2)_4$	CH_3	CH_3	CH_3	CH_3SO_4	1660	68
$(CH_3)_2N$	CH_3	CH_3	CH_3	CH_3SO_4	1666/1658	92
$(CH_3)_2N$	CH_3	CH_3	C_2H_5	$C_2H_5SO_4$	1661	92
$(C_2H_5)_2H$	C_2H_5	C_2H_5	CH_3	CH_3SO_4	1628	92
—CH— $\|$ $C(CH_3)_3$	t-C_4H_9	C_2H_5		BF_4	1670	44
—$(CH_2)_3$—	H	C_2H_5		BF_4	1669	42
—$(CH_2)_3$—	CH_3	CH_3		CH_3SO_4	1691	68
—$(CH_2)_5$—	H	C_2H_5		BF_4	1663	42

the $-C=N^{\oplus}$ vibration. This band is found to be at most 30 cm^{-1} higher than that of the amide C=O band, which the salt is generated from (65). Additionally, the spectra of iminium alkylsulfates show S=O bands between 1225–1235 cm^{-1} and 1050–1056 cm^{-1}, which are also found in the spectrum of the methylsulfate anion (65, 68). Table XI summarizes the position of the $C=N^{\oplus}$ vibrations of several alkoxymethyleniminium salts.

IR spectra indicate the existence of strong hydrogen bonds between the halide anion and the NH$_2$ groups in *N,N*-unsubstituted compounds **128** (129). The markedly increased stability of the "imino ester hydrogen halides" compared with *N,N*-disubstituted compounds **129** is possibly based on these findings. Owing to hydrogen bonding the nucleophilicity of the halide anion in **128** is decreased, whereas *N,N*-disubstituted compounds (**129**) do not allow hydrogen bonding, and dealkylation takes place.

$$R^1-C\!\!\begin{array}{c}OR^2\\ \oplus\\ NH_2\end{array}\ Cl^{\ominus} \qquad R^1-C\!\!\begin{array}{c}OR^2\\ \oplus\\ NR^3_2\end{array}\ Cl^{\ominus} \longrightarrow R^1-C\!\!\begin{array}{c}O\\ \\ NR^3_2\end{array} + R^2Cl$$

(128) 　　　　　　　　R^3 = Alkyl, Aryl
　　　　　　　　　　　　　(129)

NMR data on alkoxymethyleniminium salts (16, 17, 142) represent a structural proof also. The resonances of the ligands bound to the nitrogen atom or to the acyl carbon atom are shifted downfield (0.5–1 ppm) relative to the starting amide.

III. Reactions of Alkoxymethyleniminium Salts with Nucleophiles

Depending on the reaction conditions, alkoxymethyleniminium ions may react with nucleophiles to yield various products, and for that reason are termed "ambident." Hünig has thoroughly discussed the chemistry of these ambident cations (143). According to him there are two different reaction pathways. In the first, the nucleophile $Y^{\ominus}$ attacks at the carbon atom, the site of the lowest electron density. This is in essence an ion-pair association requiring almost no activation energy, and thus leading to kinetically controlled reaction products. If the energy gained through bond formation is large enough, the reaction products thus formed may be isolated (path *a*).

If the bond energy of the newly formed C–Y bond is low, an equilibrium is established (path *b*). $Y^{\ominus}$ will then attack at the polarized O–R^3 bond leading to dealkylation of the cation. In this way the changes are neutralized and the amide resonance energy restored. Therefore this reaction is irreversible leading

III. ALKOXYMETHYLENIMINIUM SALT—NUCLEOPHILE REACTIONS

$$
\begin{array}{c}
R-C\underset{NR^1R^2}{\overset{OR^3}{\underset{\oplus}{=}}} \\
\updownarrow \\
R-C\underset{NR^1R^2}{\overset{OR^3}{-Y}} \xleftarrow[a]{+Y^\ominus} R-C\underset{NR^1R^2}{\overset{OR^3}{\underset{\oplus}{-}}} \xrightarrow[b]{+Y^\ominus} R-C\underset{NR^1R^2}{\overset{OR^3}{-Y}} \\
\updownarrow \qquad\qquad\qquad\qquad \updownarrow \\
R-C\underset{NR^1R^2}{\overset{\oplus OR^3}{=}} \qquad R-C\underset{NR^1R^2}{\overset{OR^3}{\underset{\oplus}{=}}}\ Y^\ominus \\
\downarrow \\
R-C\underset{NR^1R^2}{\overset{O}{=}} + R^3Y
\end{array}
$$

to thermodynamically controlled reaction products. The reaction profile for the reaction of those ambident cations is graphically displayed as follows (143):

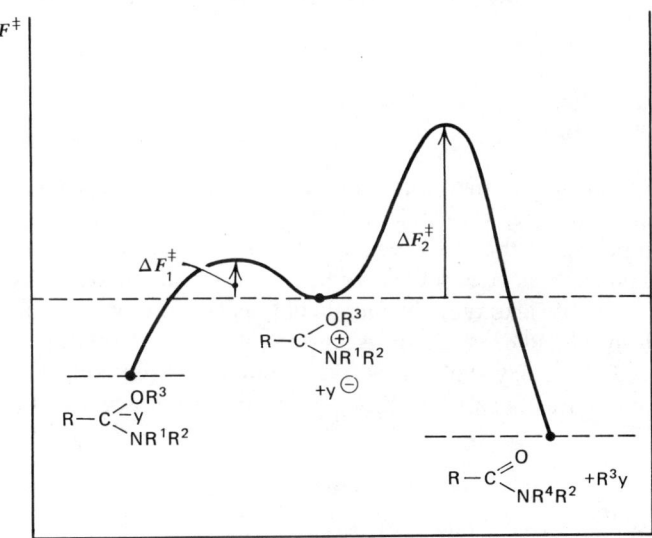

The actual course of the reaction is determined by the type of nucleophile, the cation stability, the temperature, and the reaction time as well as by the solvent used (143). The details are given with the particular reactions.

A. REACTIONS WITH HYDROXYL GROUPS

1. Hydrolysis

The mechanism of the hydrolysis of alkoxymethyleniminium salts **130** in acidic or neutral media has been studied only qualitatively (65, 78, 110, 125,

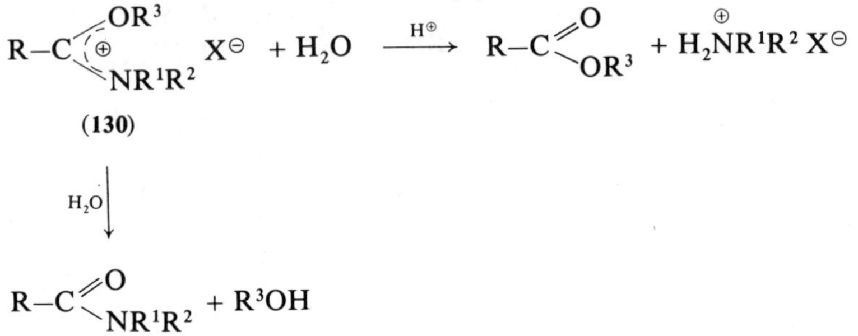

$$R-C{\overset{OR^3}{\underset{NR^1R^2}{\oplus}}} \; X^\ominus + H_2O \xrightarrow{H^\oplus} R-C{\overset{O}{\underset{OR^3}{\nwarrow}}} + H_2\overset{\oplus}{N}R^1R^2 \; X^\ominus$$

(**130**)

$$\downarrow H_2O$$

$$R-C{\overset{O}{\underset{NR^1R^2}{\nwarrow}}} + R^3OH$$

144). According to these studies, the C—N bond is cleaved preferentially, giving rise to esters and ammonium salts; amides and alcohols are also formed to a lesser extent. When the salts **131** and **132** are hydrolyzed, the C—N bond is cleaved exclusively (17).

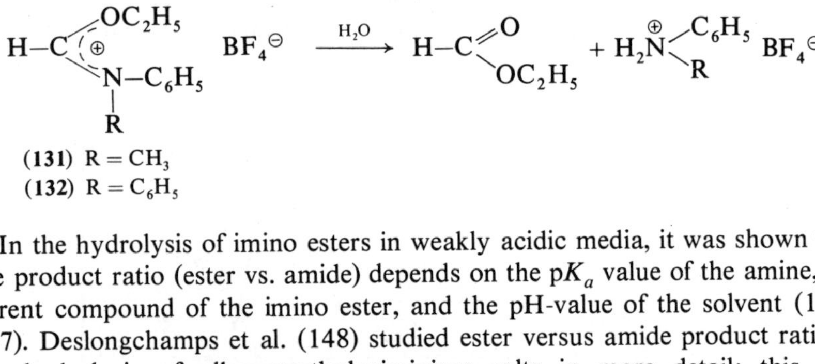

$$H-C{\overset{OC_2H_5}{\underset{\underset{R}{N-C_6H_5}}{\oplus}}} \; BF_4^\ominus \xrightarrow{H_2O} H-C{\overset{O}{\underset{OC_2H_5}{\nwarrow}}} + H_2\overset{\oplus}{N}{\overset{C_6H_5}{\underset{R}{\nwarrow}}} \; BF_4^\ominus$$

(**131**) R = CH$_3$
(**132**) R = C$_6$H$_5$

In the hydrolysis of imino esters in weakly acidic media, it was shown that the product ratio (ester vs. amide) depends on the pK_a value of the amine, the parent compound of the imino ester, and the pH-value of the solvent (145–147). Deslongchamps et al. (148) studied ester versus amide product ratio in the hydrolysis of alkoxymethyleniminium salts in more detail; this was performed in alkaline medium. According to this study the size of the substituent R strongly influences the product distribution. With alkoxymethyleniminium salts carrying bulky R groups at the acyl carbon atom, the

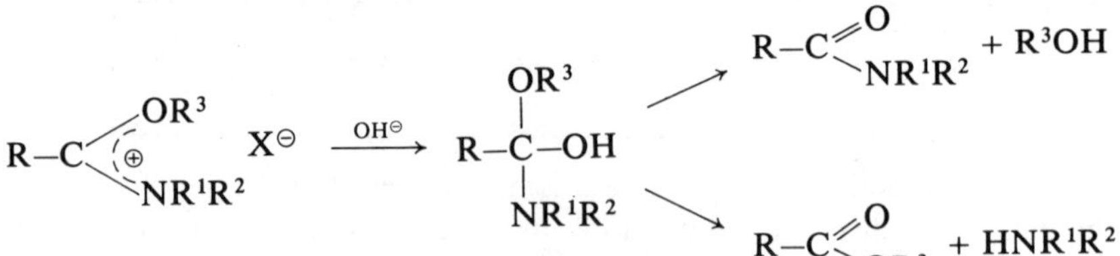

$$R-C{\overset{OR^3}{\underset{NR^1R^2}{\oplus}}} X^\ominus \xrightarrow{OH^\ominus} R-\underset{\underset{NR^1R^2}{|}}{\overset{\overset{OR^3}{|}}{C}}-OH \begin{array}{c} \nearrow R-C{\overset{O}{\underset{NR^1R^2}{\nwarrow}}} + R^3OH \\ \searrow R-C{\overset{O}{\underset{OR^3}{\nwarrow}}} + HNR^1R^2 \end{array}$$

(**133**)

III. ALKOXYMETHYLENIMINIUM SALT—NUCLEOPHILE REACTIONS

C—N bond is cleaved nearly exclusively, whereas small R groups equally well favor fission of the C—O and the C—N bond. For this reaction a tetrahedral intermediate **133** is assumed; its conformation determines the product ratio (148).

The easy hydrolysis of alkoxymethyleniminium salts is taken advantage of in several natural product synthesis. Thus acetamidosaccharide **134** is smoothly deacylated by first alkylating with triethyloxonium fluoroborate, then hydrolyzing the salt (149).

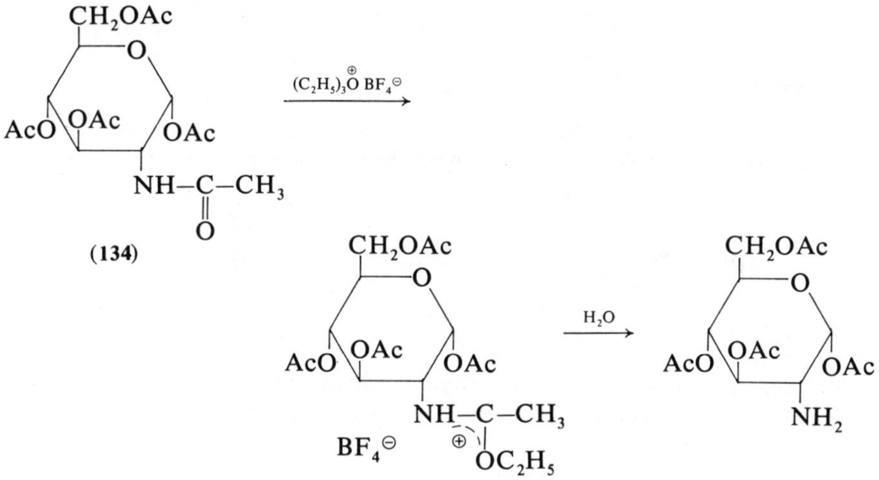

One advantage of this procedure is the conservation of ester and acetal functions as well as of glycosidic bonds. In the synthesis of tetracyclines a carboxamide function from **135** was successfully cleaved using this method (150).

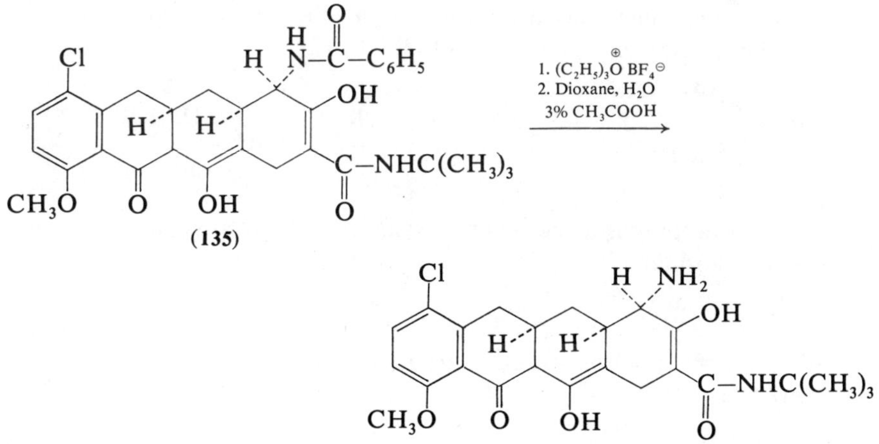

On alkaline hydrolysis of the salts **136** azobenzene derivatives (**137**) form (18).

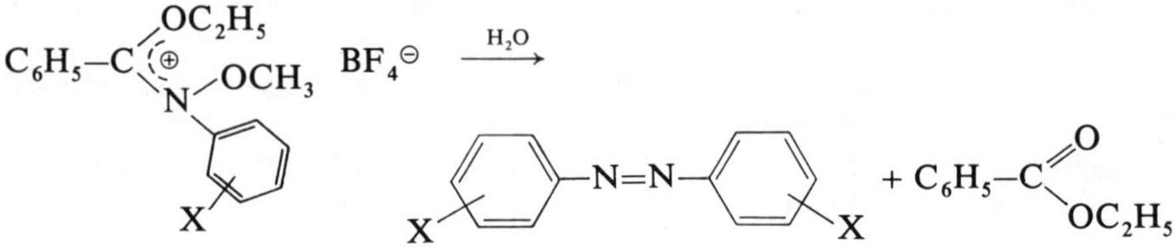

$$C_6H_5-C\overset{OC_2H_5}{\underset{N-OCH_3}{(+)}} \; BF_4^{\ominus} \xrightarrow{H_2O}$$

(**136**)

$$X\text{-}\underset{}{\bigcirc}\text{-}N=N\text{-}\underset{}{\bigcirc}\text{-}X + C_6H_5-C\overset{O}{\underset{OC_2H_5}{}}$$

(**137**)

2. Alcoholysis

Alcoholysis of imino ester hydrochlorides (**138**) with absolute alcohols yields ortho esters (**139**). This reaction has been reviewed frequently (121, 136, 151–153).

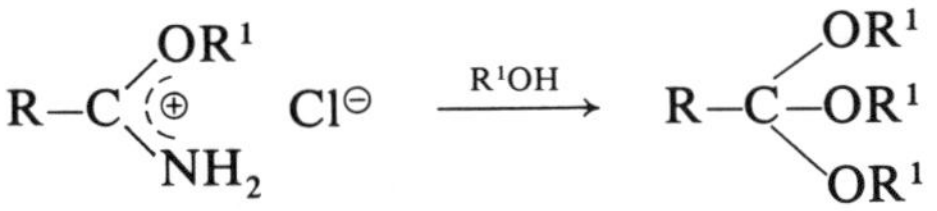

$$R-C\overset{OR^1}{\underset{NH_2}{(+)}} \; Cl^{\ominus} \xrightarrow{R^1OH} R-C\overset{OR^1}{\underset{OR^1}{-OR^1}}$$

(**138**) (**139**)

The reaction of the imino ester hydrochloride **140** with alcohols takes a different course, however, and leads to esters of β-alkoxypropionic acid (**141**) (154).

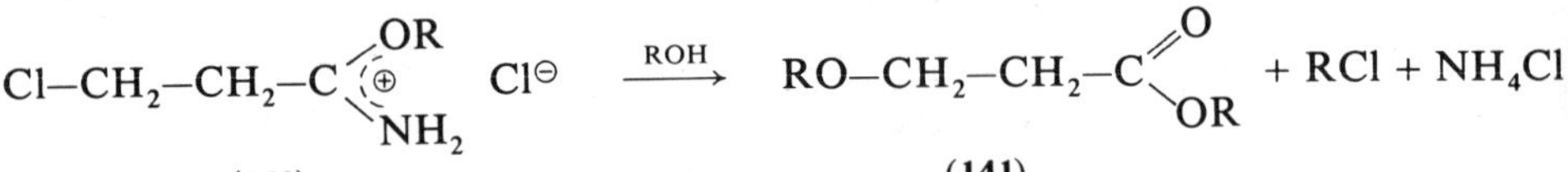

$$Cl-CH_2-CH_2-C\overset{OR}{\underset{NH_2}{(+)}} \; Cl^{\ominus} \xrightarrow{ROH} RO-CH_2-CH_2-C\overset{O}{\underset{OR}{}} + RCl + NH_4Cl$$

(**140**) (**141**)

Under thermodynamically controlled reaction conditions carboxylic acid esters and ammonium salts are formed on alcoholysis of N-alkyl- and N,N-dialkyl alkoxymethyleniminium salts (**142**) (20, 72).

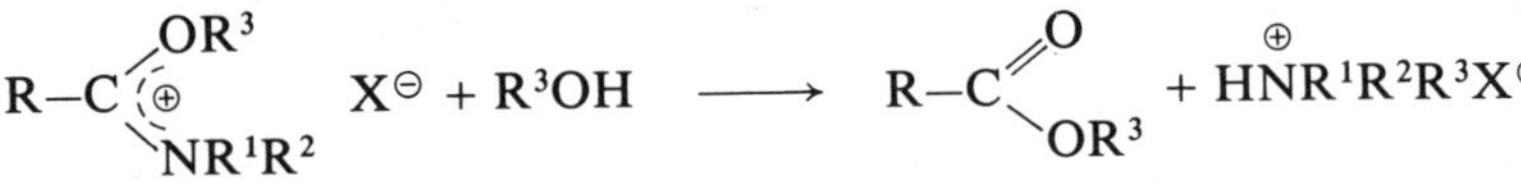

$$R-C\overset{OR^3}{\underset{NR^1R^2}{(+)}} \; X^{\ominus} + R^3OH \longrightarrow R-C\overset{O}{\underset{OR^3}{}} + \overset{\oplus}{H}NR^1R^2R^3X^{\ominus}$$

(**142**)

Methanol decomposes the iminium salts **143** into methyl chloride, phenol, and dimethylformamide (115).

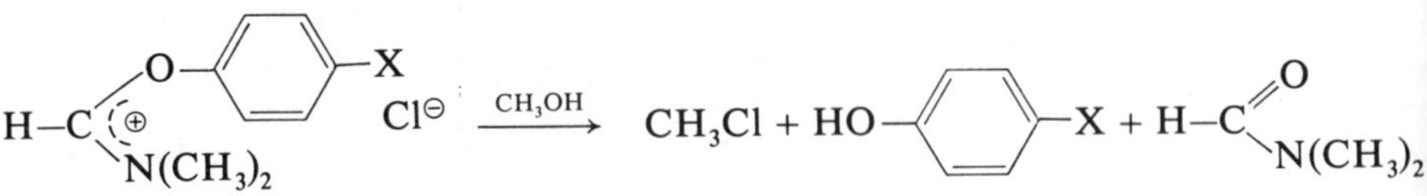

$$H-C\overset{O-\bigcirc-X}{\underset{N(CH_3)_2}{(+)}} \; Cl^{\ominus} \xrightarrow{CH_3OH} CH_3Cl + HO-\bigcirc-X + H-C\overset{O}{\underset{N(CH_3)_2}{}}$$

(**143**)

III. ALKOXYMETHYLENIMINIUM SALT—NUCLEOPHILE REACTIONS 227

The iminium ions **142** need not be isolated; to convert an amide into an ester it is sufficient to reflux the amide with an alkylating agent (e.g., dialkyl sulfate) and alcohol (20). In some cases the C—NHR as well as the C—OR bond cleaves, thus giving rise to carboxamides in addition to carboxylic acid esters (155). In his synthesis of anhydroaureomycin Muxfeld et al. (156) took advantage of this reaction to cleave an amide function from **144** under mild conditions.

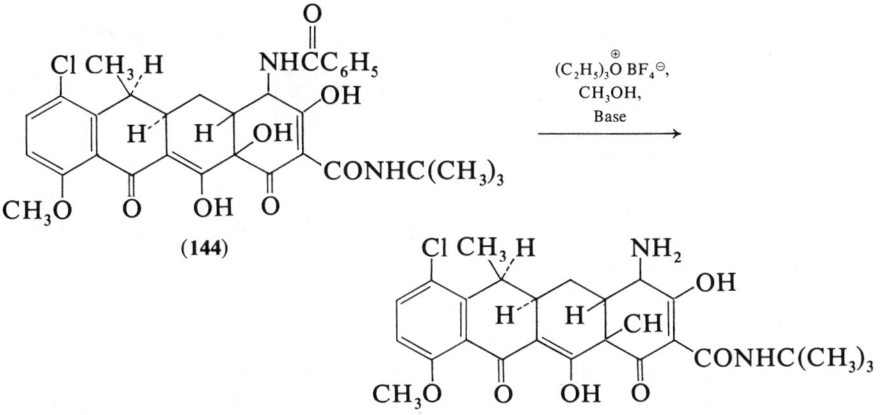

If salts **145** are alcoholyzed at low temperatures in an apolar solvent ortho esters **146** are obtained in yields of 3–50% (20). Dealkylation of the carboxonium ion (**147**) by the amine generated competes with orthoester formation; thus carboxylic acid esters are ubiquitous by-products. The basicity of the cleaved amine determines the orthoester–carboxylic acid ester ratio. Morpholinoalkoxycarbonium salts produce the highest yields of orthoesters.

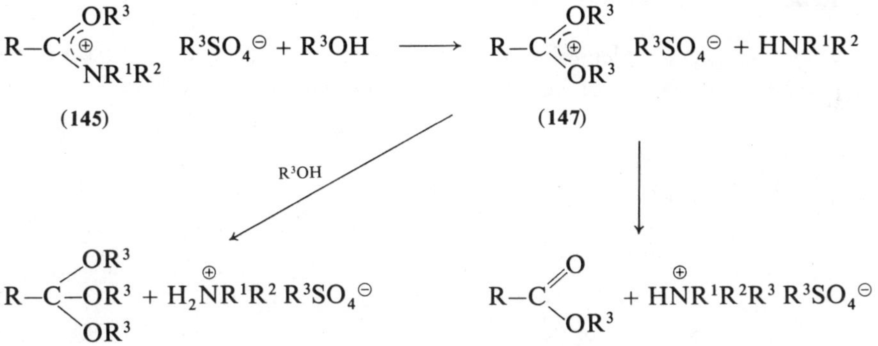

Alkoxymethyleniminium methylsulfates generally yield 30–50% orthoesters, whereas the corresponding fluoroborates give only 5–10% yields (20). Pinner's classical orthoester synthesis from alkoxymethyleniminium chlorides (**138**) and alcohols generates orthoesters in 50–80% yields (121, 136, 151–153).

Orthoesters **149** are also formed on alcoholysis of *N*-acylalkoxymethyleniminium chlorides (**148**) (91).

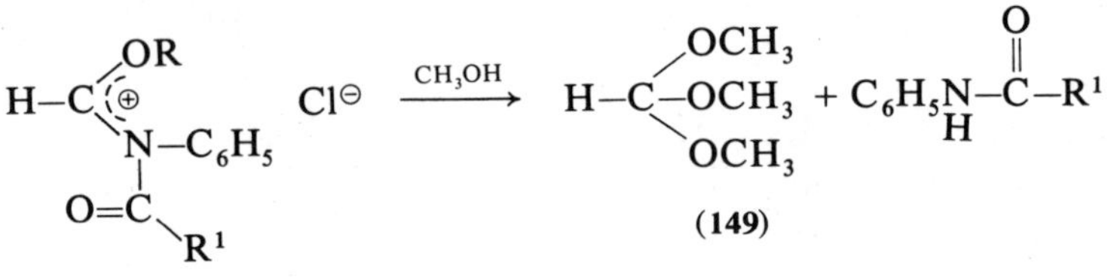

$$\text{H-C} \begin{matrix} \text{OR} \\ \oplus \\ \text{N-C}_6\text{H}_5 \\ | \\ \text{O=C-R}^1 \end{matrix} \quad \text{Cl}^{\ominus} \xrightarrow{\text{CH}_3\text{OH}} \text{H-C} \begin{matrix} \text{OCH}_3 \\ \text{-OCH}_3 \\ \text{OCH}_3 \end{matrix} + \text{C}_6\text{H}_5\overset{\text{H}}{\text{N}}-\overset{\text{O}}{\underset{\|}{\text{C}}}-\text{R}^1$$

(**148**) (**149**)

The weakly dissociated (dimethylamino)methoxyacetonitrile (**150**) behaves differently; it is transesterified with excess alcohol (157).

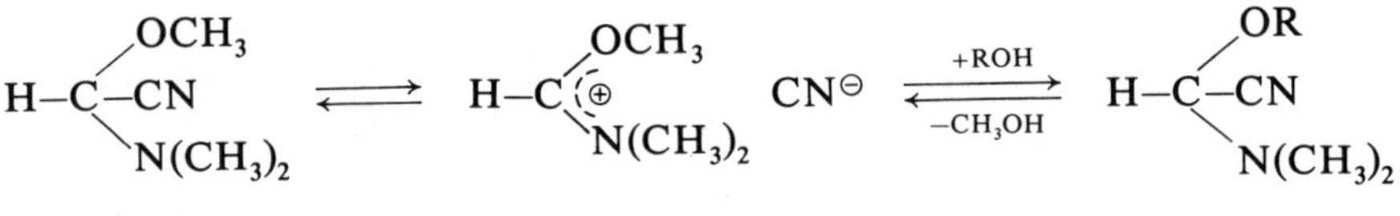

$$\text{H-C} \begin{matrix} \text{OCH}_3 \\ \text{-CN} \\ \text{N(CH}_3)_2 \end{matrix} \rightleftharpoons \text{H-C} \begin{matrix} \text{OCH}_3 \\ \oplus \\ \text{N(CH}_3)_2 \end{matrix} \text{CN}^{\ominus} \underset{-\text{CH}_3\text{OH}}{\overset{+\text{ROH}}{\rightleftharpoons}} \text{H-C} \begin{matrix} \text{OR} \\ \text{-CN} \\ \text{N(CH}_3)_2 \end{matrix}$$

(**150**)

3. Reactions with Hemiacetals

When alkoxymethyleniminium alkylsulfates **151** react with mixtures of alcohols and aldehydes or ketones in an equimolar ratio acetals or ketals are

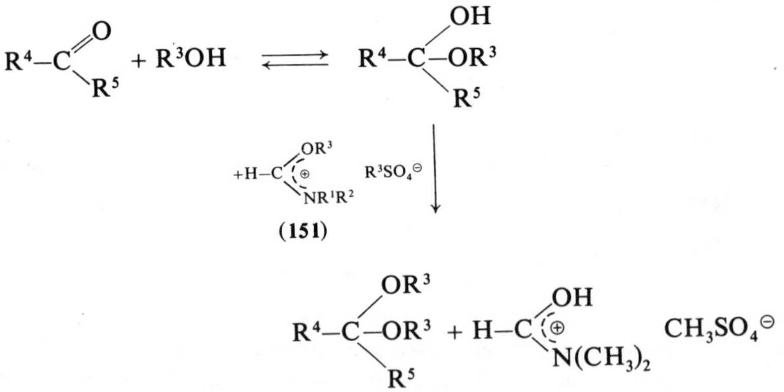

$$\text{R}^4-\text{C} \begin{matrix} \text{O} \\ \text{R}^5 \end{matrix} + \text{R}^3\text{OH} \rightleftharpoons \text{R}^4-\text{C} \begin{matrix} \text{OH} \\ -\text{OR}^3 \\ \text{R}^5 \end{matrix}$$

$$+\text{H-C}\begin{matrix}\text{OR}^3\\\oplus\\\text{NR}^1\text{R}^2\end{matrix} \quad \text{R}^3\text{SO}_4^{\ominus} \Bigg\downarrow$$

(**151**)

$$\text{R}^4-\text{C}\begin{matrix}\text{OR}^3\\-\text{OR}^3\\\text{R}^5\end{matrix} + \text{H-C}\begin{matrix}\text{OH}\\\oplus\\\text{N(CH}_3)_2\end{matrix} \quad \text{CH}_3\text{SO}_4^{\ominus}$$

(**152**)

formed in good yields (20, 158). During the course of the reaction hemiacetals are generated as intermediates, which are alkylated by **151** to give acetals **152**. This route allows the acetalization of aliphatic, aromatic, and α,β-unsaturated aldehydes and ketones. Cyclic acetals (1,3-dioxolanes, dioxanes, 1,3-dioxepanes, and 4,7-dihydro-1,3-dioxepines) are also obtained from aldehydes, ketones, and 1,2-, 1,3-, and 1,4-diols with N,N-dialkylmethoxymethyleniminium methylsulfates (**151**).

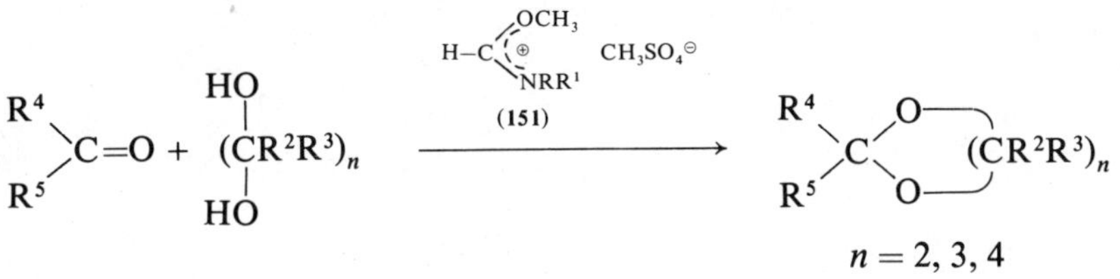

$$\begin{array}{c}R^4\\R^5\end{array}\!\!\!\!C\!=\!O + \begin{array}{c}HO\\|\\(CR^2R^3)_n\\|\\HO\end{array} \quad \xrightarrow{\begin{array}{c}H-C\overset{OCH_3}{\underset{NRR^1}{\oplus}}\quad CH_3SO_4^{\ominus}\\(\mathbf{151})\end{array}} \quad \begin{array}{c}R^4\\R^5\end{array}\!\!\!\!C\!\!\begin{array}{c}O\\\diagdown\\O\end{array}\!\!(CR^2R^3)_n$$

$$n = 2, 3, 4$$

For these acetalization reactions, morpholinomethoxymethylenecarbonium methylsulfate is the most reactive. These reactions parallel the acetal synthesis from alkoxymethyleniminium chlorides (**138**), aldehydes, and alcohols (136).

4. *Acidolysis*

When a mixture of a carboxylic acid and a NN-dialkylalkoxymethyleniminium-alkylsulfate (**152**) is heated, a carboxylic acid ester (**153**) is formed in good yields (159). Under these reaction conditions the iminium salts alkylate the carboxylic acids in a thermodynamically controlled reaction. By this route

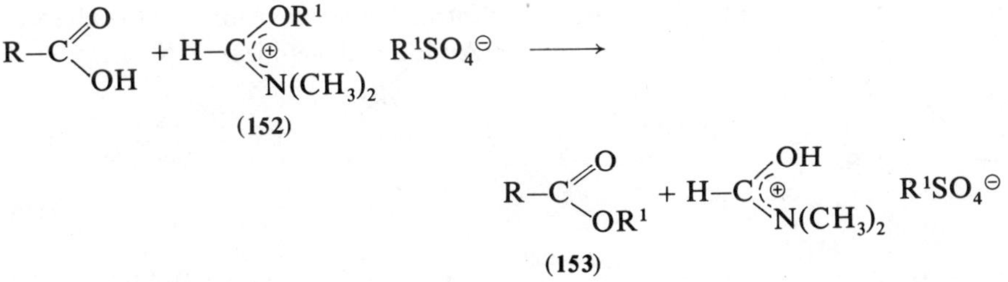

$$R-C\!\!\begin{array}{c}\diagup O\\\diagdown OH\end{array} + H-C\overset{OR^1}{\underset{N(CH_3)_2}{\oplus}} \quad R^1SO_4^{\ominus} \longrightarrow$$

$$(\mathbf{152})$$

$$R-C\!\!\begin{array}{c}\diagup O\\\diagdown OR^1\end{array} + H-C\overset{OH}{\underset{N(CH_3)_2}{\oplus}} \quad R^1SO_4^{\ominus}$$

$$(\mathbf{153})$$

methyl and ethyl esters of aliphatic and aromatic carboxylic acids were synthesized. Dicarboxylic acids may also be esterified in this manner.

N-Acylated amino acids (**154**) are converted into their esters by N,N-dimethyl(aryloxy)methyleniminium chlorides (113).

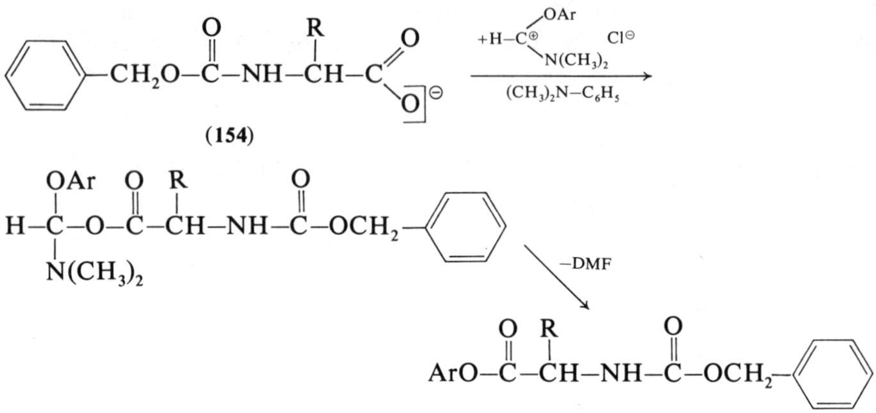

B. REACTIONS WITH THIOL GROUPS

1. Thiolysis

When alkoxymethyleniminium chlorides (**138**) react with hydrogen sulfide, *N*-unsubstituted thioamides are produced. These reactions have been known for a long time, and have been reviewed by Roger and Neilson (121).

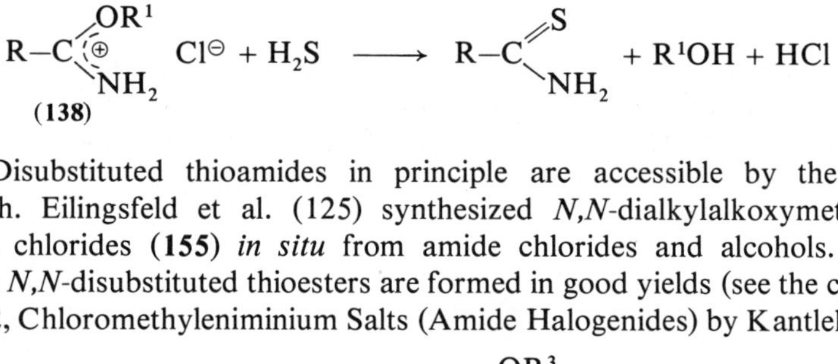

$$R-C\begin{matrix}OR^1\\ \oplus\\ NH_2\end{matrix} \; Cl^\ominus + H_2S \longrightarrow R-C\begin{matrix}S\\ \\ NH_2\end{matrix} + R^1OH + HCl$$
(**138**)

N,N-Disubstituted thioamides in principle are accessible by the same approach. Eilingsfeld et al. (125) synthesized *N,N*-dialkylalkoxymethyleniminium chlorides (**155**) *in situ* from amide chlorides and alcohols. Upon thiolysis *N,N*-disubstituted thioesters are formed in good yields (see the chapter in Part 2, Chloromethyleniminium Salts (Amide Halogenides) by Kantlehner.

$$R-C\begin{matrix}Cl\\ \oplus\\ NR^1R^2\end{matrix} \; Cl^\ominus + R^3OH \xrightarrow{-HCl} R-C\begin{matrix}OR^3\\ \oplus\\ NR^1R^2\end{matrix} \; Cl^\ominus \xrightarrow[-R^3OH\; -HCl]{+H_2S} R-C\begin{matrix}S\\ \\ OR^3\end{matrix}$$
(**155**) (**156**)

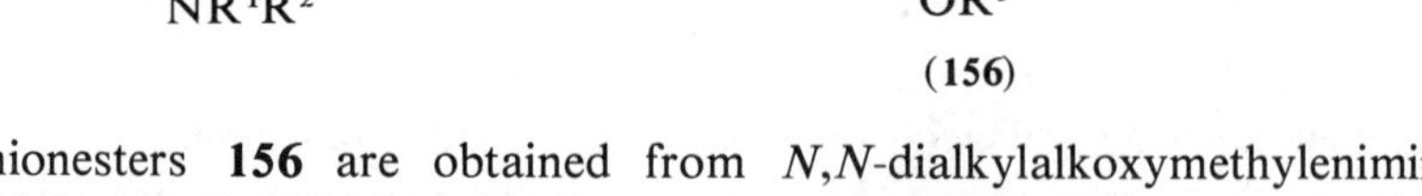

$$R-C\begin{matrix}OR^3\\ \oplus\\ NR^1R^2\end{matrix} \; BF_4^\ominus \xrightarrow{H_2S,\; Pyridine} R-C\begin{matrix}S\\ \\ OR^3\end{matrix} + H_2\overset{\oplus}{N}R^1R^2\; BF_4^\ominus$$
(**156**)

Thionesters **156** are obtained from *N,N*-dialkylalkoxymethyleniminium fluoroborates when thiolyzed in the presence of pyridine (19).

Table XII summarizes the thionesters generated in this way.

TABLE XII

Thionesters (**156**) from *N,N*-Dialkylalkoxymethyleniminium Fluoroborates and Hydrogen Sulfide (19)

Thionester

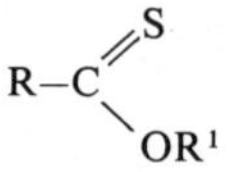

$$R-C\overset{\displaystyle S}{\underset{\displaystyle OR^1}{\diagup\!\!\!\diagdown}}$$

R	R^1	Yield, %	b.p./pressure, °C/torr	m.p., °C
C_6H_5	C_2H_5	84	125/20	—
$C_6H_5-CH_2$	C_2H_5	76.5	108–111/12	—
$C_6H_5-(CH_2)_2$	C_2H_5	70	119–123/12	—
$p\text{-}CH_3-C_6H_4$	C_2H_5	65	131–136/12	—
$p\text{-}CH_3O-C_6H_4$	C_2H_5	60	162–163/12	—
$p\text{-}Cl-C_6H_4$	C_2H_5	60	138–139/14	—
$p\text{-}NO_2-C_6H_4$	C_2H_5	84	—	108
C_6H_5	CH_3	56	104–106/12	—
C_6H_5	$n\text{-}C_3H_7$	62	121–130/12	—
C_6H_5	$n\text{-}C_4H_9$	85	134–141/10	—
C_6H_5	$i\text{-}C_3H_7$	33	115–120/12	—

2. Reactions with Mercaptans and Thiophenols

N,N-Dimethylalkoxymethyleniminium methylsulfates (**157**) react with sodium mercaptides or mercaptan/triethylamine mixtures to form the very reactive dimethylformamide mercaptals (**158**) (160, 161). Thiophenol and

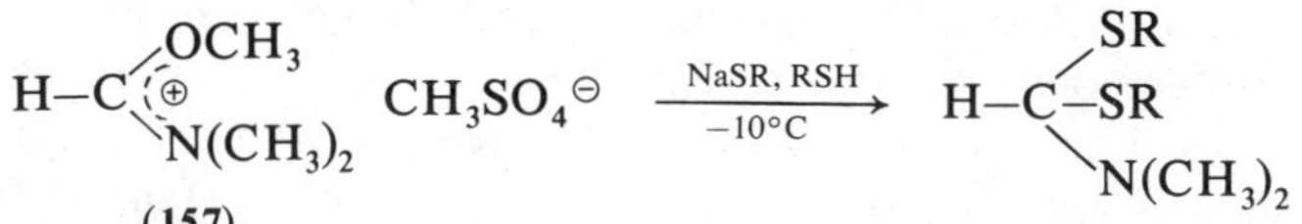

$$H-C\overset{\displaystyle OCH_3}{\underset{\displaystyle N(CH_3)_2}{\overset{\oplus}{\diagup\!\!\!\diagdown}}} CH_3SO_4^{\ominus} \xrightarrow[-10°C]{NaSR, RSH} H-C\overset{\displaystyle SR}{\underset{\displaystyle N(CH_3)_2}{\diagup\!\!\!\diagdown}}SR$$
(**157**) (**158**)

$R = C_2H_5$, $n\text{-}C_3H_7$, $i\text{-}C_3H_7$, C_4H_9, $t\text{-}C_4H_9$, C_6H_{13}, $t\text{-}C_6H_{13}$, C_7H_{15}, C_9H_{19}, $C_{10}H_{21}$, $CH_2-CH=CH_2$, $CH_2-C_6H_5$

$R-R = -CH_2-CH_2-$

thiophenethiol do not react as aliphatic mercaptans. Here the salts **157** behave as alkylating agents, giving rise to methyl aryl sulfides (**159**) (160).

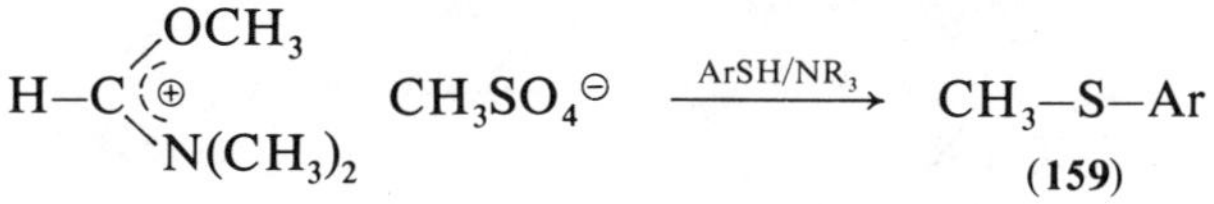

$$H-C\overset{\displaystyle OCH_3}{\underset{\displaystyle N(CH_3)_2}{\overset{\oplus}{\diagup\!\!\!\diagdown}}} CH_3SO_4^{\ominus} \xrightarrow{ArSH/NR_3} CH_3-S-Ar$$
(**157**) (**159**)

C. REACTIONS OF ALKOXYMETHYLENIMINIUM SALTS WITH SELENOL GROUPS

Only Hartmann (162) has studied the reaction of N,N-dimethylmethoxymethyleniminium methylsulfate (157) with sodium hydrogen selenide. This reaction represents a useful synthesis of selenodimethylformamide (160).

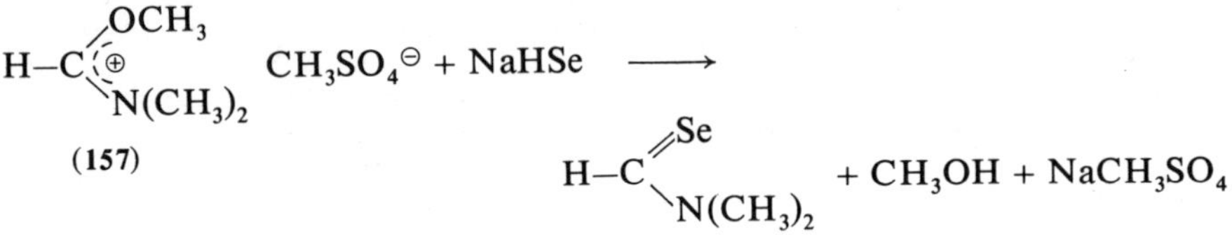

$$H-C \overset{OCH_3}{\underset{N(CH_3)_2}{\oplus}} \ CH_3SO_4^{\ominus} + NaHSe \longrightarrow H-C \overset{Se}{\underset{N(CH_3)_2}{}} + CH_3OH + NaCH_3SO_4$$

(157)   (160)

D. REACTIONS WITH NH-CONTAINING COMPOUNDS

1. Ammonia and Primary and Secondary Amines

The alkoxy group of alkoxymethyleniminium chlorides (138) exchanges with amino, alkylamino, and dialkylamino groups. This type of reaction of

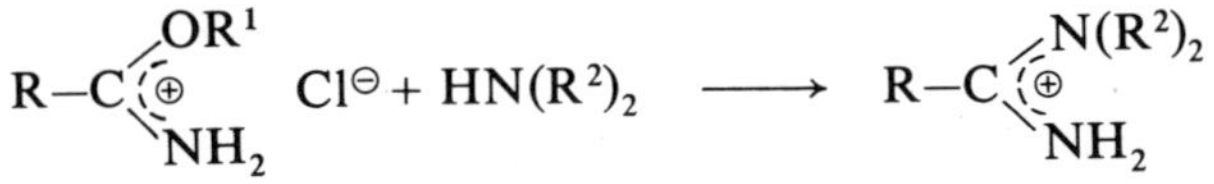

$$R-C \overset{OR^1}{\underset{NH_2}{\oplus}} \ Cl^{\ominus} + HN(R^2)_2 \longrightarrow R-C \overset{N(R^2)_2}{\underset{NH_2}{\oplus}} \ Cl^{\ominus}$$

(138)

N,N-unsubstituted compounds has been reviewed extensively (121); thus we summarize only more recent results here.

With N-mono and N,N-disubstituted alkoxymethyleniminium salts the alkoxy group is also smoothly substituted by amino, alkylamino, and dialkylamino groups (23, 41, 66, 144, 163–171, 171b), if reacted with ammonia or primary or secondary amines. Amidinium alkyl sulfates (161) are often

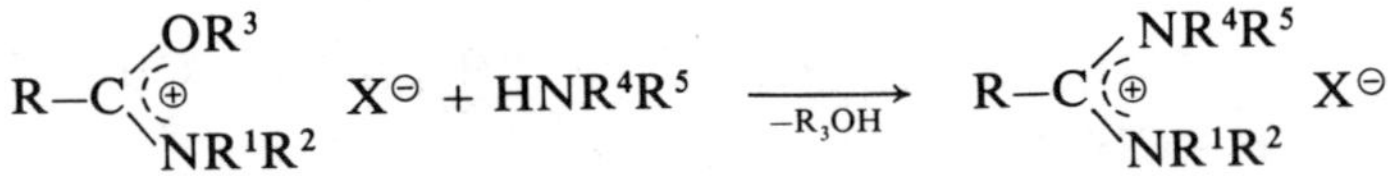

$$R-C \overset{OR^3}{\underset{NR^1R^2}{\oplus}} \ X^{\ominus} + HNR^4R^5 \xrightarrow[-R_3OH]{} R-C \overset{NR^4R^5}{\underset{NR^1R^2}{\oplus}} \ X^{\ominus}$$

(161)

obtained as hygroscopic oils. Sometimes they are not isolated, but immediately reacted with base to yield free amidines (162) (169).

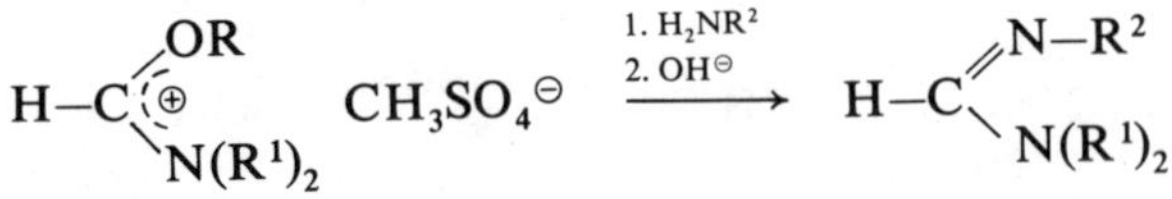

$$H-C \overset{OR}{\underset{N(R^1)_2}{\oplus}} \ CH_3SO_4^{\ominus} \xrightarrow[\text{2. OH}^{\ominus}]{\text{1. H}_2NR^2} H-C \overset{N-R^2}{\underset{N(R^1)_2}{}}$$

(162)

III. ALKOXYMETHYLENIMINIUM SALT—NUCLEOPHILE REACTIONS

With excess morpholine the salt **163** forms the salt **164**; partial aminolysis of **164** leads to **165** (34).

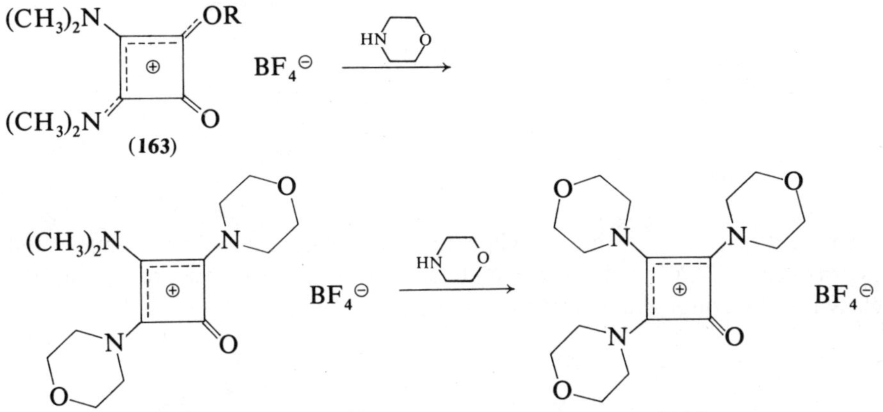

On reaction of the iminium salt **165b** with α-halo or β-hydroxy ketones in liquid ammonia, imidazoles (**165a**) are generated (171a).

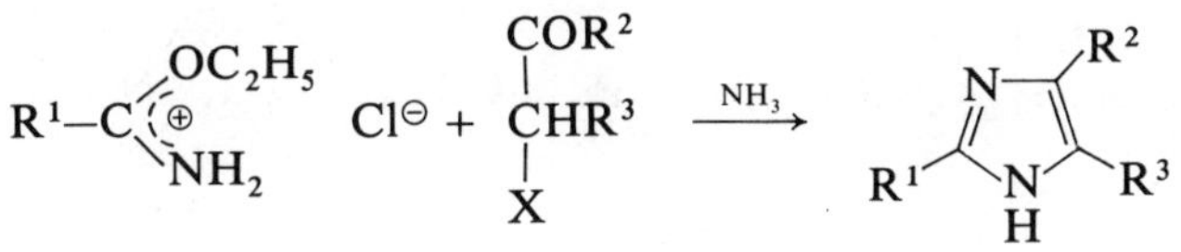

With the salt **165c** glycine or sarcosine yield the betaine **165d**; **165e** cyclizes to the compounds **165f** or **165g**, (171b).

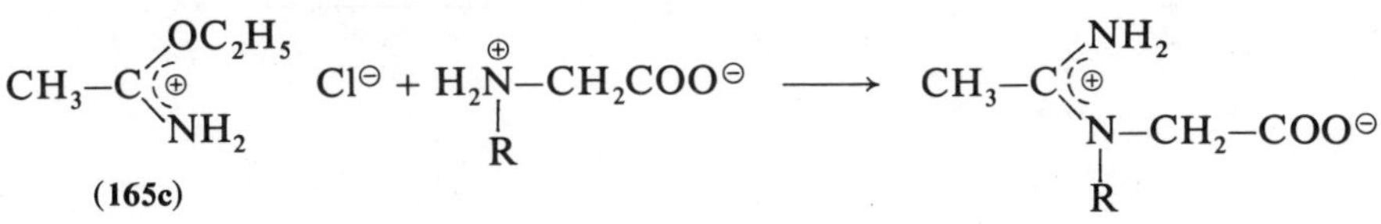

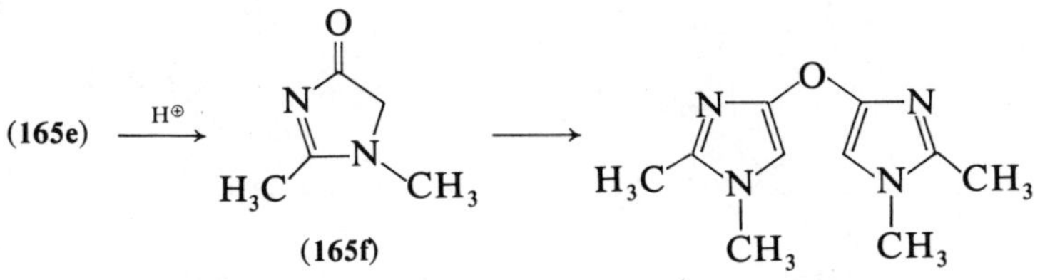

The alkoxy group of weakly dissociated (dialkylamino)alkoxyacetonitriles (**166**) exchanges with the amino group of secondary amines (157).

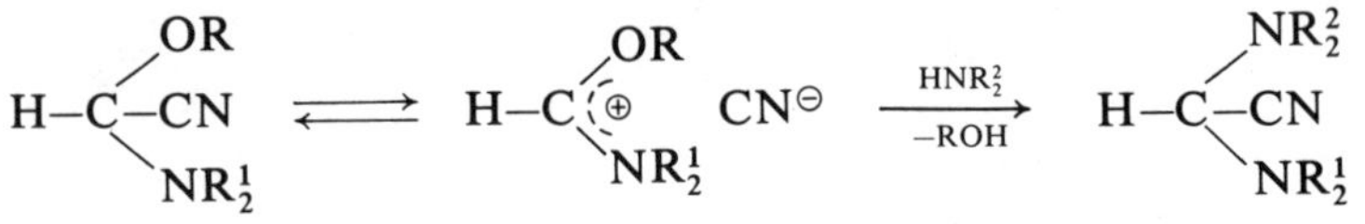

$$H-\underset{NR^1_2}{\overset{OR}{C}}-CN \rightleftharpoons H-\underset{NR^1_2}{\overset{OR}{C^{\oplus}}} \quad CN^{\ominus} \xrightarrow[-ROH]{HNR^2_2} H-\underset{NR^1_2}{\overset{NR^2_2}{C}}-CN$$

(**166**)

In contrast with primary aliphatic and cyclic aliphatic amines and hydrazines N,N,N'-trialkylformamidines (**167**) are formed from **166** via cleavage of alcohol and hydrogen cyanide (157).

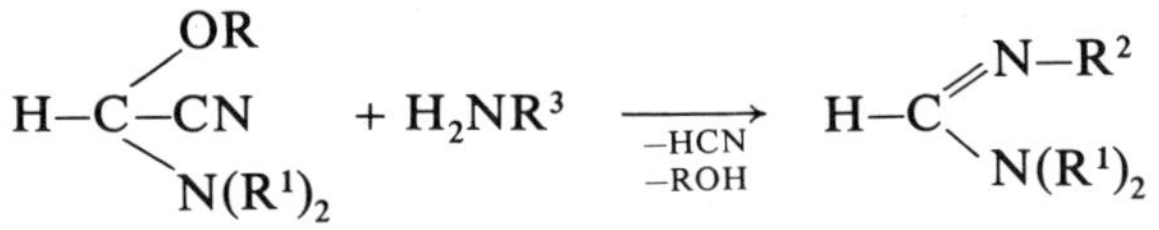

$$H-\underset{N(R^1)_2}{\overset{OR}{C}}-CN + H_2NR^3 \xrightarrow[-ROH]{-HCN} H-\underset{N(R^1)_2}{\overset{N-R^2}{C}}$$

(**166**)   (**167**)

$R^1 = CH_3$
$R^2 = CH_3, CH(CH_3)_2, n\text{-}C_3H_7, n\text{-}C_4H_9, c\text{-}C_6H_{11}, N(CH_3)_2$

If diamines are reacted with dialkylmethyleniminium salts like **168**, transamination may occur in addition to substitution of the alkoxy group (70).

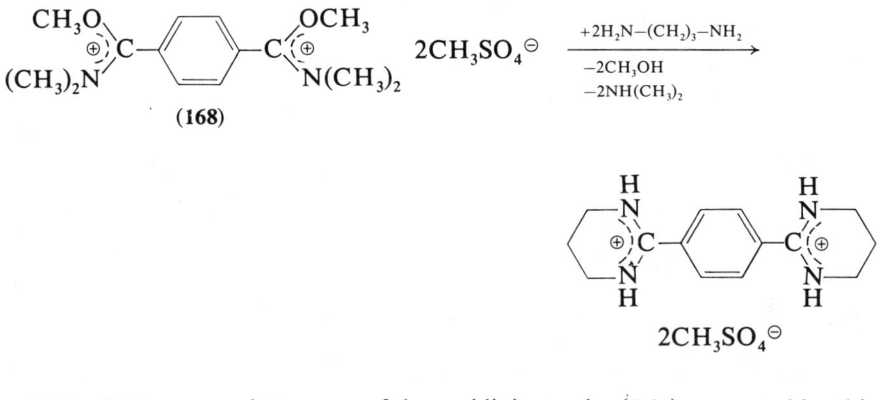

$$\underset{(CH_3)_2N}{\overset{CH_3O}{\diagdown}}C^{\oplus}-\underset{}{\bigcirc}-C^{\oplus}\underset{N(CH_3)_2}{\overset{OCH_3}{\diagup}} \quad 2CH_3SO_4^{\ominus} \xrightarrow[-2NH(CH_3)_2]{+2H_2N-(CH_2)_3-NH_2}$$

(**168**)

[structure: bis(tetrahydropyrimidinylidene) linked by phenylene, with 2CH$_3$SO$_4^{\ominus}$]

Table XIII summarizes some of the amidinium salts (**161**) generated by this route.

If appropriate workup conditions are employed, free amidines rather than amidinium salts are obtained (41, 168). Occasionally, it is the amino group that

TABLE XIII

Amidinium Salts from Alkoxymethyleniminium Salts and Ammonia or Primary or Secondary Amines

Amidinium salt

$$R-C\underset{NR^1R^2}{\overset{NR^3R^4}{\diagup}}\quad X^{\ominus}$$

R	R^1	R^2	R^3	R^4	X	Yield, %	n_D^{20}	m.p., °C	Ref.
H	H	CH_3	H	CH_3	CH_3SO_4	95	1.4665	—	67
H	H	CH_3	H	C_2H_5	CH_3SO_4	96	1.4554	—	166
H	H	C_2H_5	H	C_2H_5	CH_3SO_4	95	1.4583	—	67
H	H	CH_3	H	$n\text{-}C_4H_9$	CH_3SO_4	88	1.4554	—	166
H	H	CH_3	H	$c\text{-}C_6H_{11}$	CH_3SO_4	84	1.4831	—	166
H	H	C_2H_5	H	$c\text{-}C_6H_{11}$	CH_3SO_4	60	1.4813	—	166
H	H	CH_3	C_2H_5	C_2H_5	CH_3SO_4	80	1.4599	—	166
H	CH_3	CH_3	H	CH_3	CH_3SO_4	90	1.4635	—	165
H	CH_3	CH_3	H	C_6H_5	CH_3SO_4	90	—	178	165
H	CH_3	CH_3	CH_3	CH_3	CH_3SO_4	94	1.4640	—	165
H	C_2H_5	C_2H_5	C_2H_5	C_2H_5	CH_3SO_4	93	—	—	167
H	\multicolumn{2}{c	}{$-(CH_2)_2-O-(CH_2)_2-$}	\multicolumn{2}{c	}{$-(CH_2)_5-$}	CH_3SO_4	80	1.5024/30	30	171
H	\multicolumn{2}{c	}{$-(CH_2)_2-O-(CH_2)_2-$}	\multicolumn{2}{c	}{$-(CH_2)_2-O-(CH_2)_2-$}	CH_3SO_4	94.6	—	95	171
CH_3	H	H	H	H	Picrate	78	—	249–251	41
CH_3	H	CH_3	H	H	CH_3SO_4	98	1.4582	—	166
CH_3	H	C_2H_5	H	H	CH_3SO_4	81	1.4600	—	166
CH_3	H	CH_3	H	C_2H_5	CH_3SO_4	83	1.4650	—	166
CH_3	H	CH_3	H	$n\text{-}C_3H_7$	CH_3SO_4	88	1.4676	—	166
CH_3	H	CH_3	H	$c\text{-}C_6H_{11}$	CH_3SO_4	—	1.4892	97–104	166

TABLE XIII (cont.)

R	R^1	R^2	R^3	R^4	X	Yield, %	n_D^{20}	m.p., °C	Ref.
CH_3	H	C_2H_5	H	C_2H_5	CH_3SO_4	84	1.4680	48–53	166
CH_3	H	C_2H_5	H	n-C_4H_9	CH_3SO_4	83	1.4671	—	166
CH_3	H	C_2H_5	H	c-C_6H_{11}	CH_3SO_4	—	1.4872	98–101	166
CH_3	CH_3	CH_3	CH_3	CH_3	CH_3SO_4	91	1.4861	—	163
CH_3	C_2H_5	C_2H_5	C_2H_5	CH_3	CH_3SO_4	69	1.4467	—	163
C_2H_5	CH_3	CH_3	CH_3	CH_3	CH_3SO_4	85	1.4667	—	163
C_6H_5	H	H	H	H	Cl	71.5	—	166–168	41
C_6H_5	H	CH_3	H	CH_3	Cl	90	—	255–256	41
C_6H_5	H	CH_3	H	CH_3	Picrate	90	—	171–172	41
o-CH_3–C_6H_4	H	H	H	H	Cl	90	—	258–258.5	41
o-CH_3–C_6H_4	H	H	H	H	Picrate	90	—	235–236	41
o-$C_2H_5OC_6H_4$	H	H	H	H	Cl	91	—	195–196	41
o-$C_2H_5OC_6H_4$	H	H	H	H	Picrate	91	—	213–215	41
1-Naphthyl	H	H	H	H	Picrate	75	—	226.5	41
o-Cl–C_6H_4	H	H	H	H	Cl	90	—	280–282	41
o-Cl–C_6H_4	H	H	H	H	Picrate	90	—	218–220	41
![tolyl-dihydropyrimidinium]	H	—$(CH_2)_3$—		H	CH_3SO_4	54	—	257–261	70

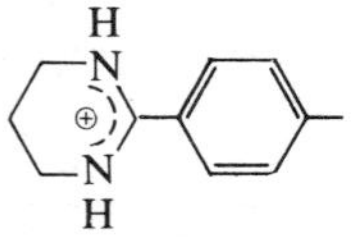

$CH_3SO_4^{\ominus}$

III. ALKOXYMETHYLENIMINIUM SALT—NUCLEOPHILE REACTIONS 237

is displaced while the alkoxy group remains. Thus the imino ester **170** is formed on reaction of ethoxyphenylmethyleniminium chloride (**169**) with aniline (**172**).

$$C_6H_5-C\overset{+}{\underset{NH_2}{\diagdown}}\overset{OC_2H_5}{\diagup} \quad Cl^\ominus + C_6H_5NH_2 \longrightarrow C_6H_5-C\overset{OC_2H_5}{\underset{NC_6H_5}{\diagdown}} + NH_4Cl$$

(**169**) (**170**)

Schlack and Richter (70) discovered an analogous reaction with the salt **168**. Lwowski (173) took advantage of this reaction to synthesize *N*-cyanoimino esters (**171**).

$$\underset{(CH_3)_2N}{\overset{CH_3O}{\diagdown}}\overset{+}{C}-\!\!\!\left\langle\;\;\right\rangle\!\!\!-\overset{+}{C}\overset{OCH_3}{\underset{N(CH_3)_2}{\diagup}} \quad 2CH_3SO_4^\ominus \quad \xrightarrow[\substack{-2H_2N(CH_3)_2 \\ CH_3SO_4^\ominus}]{+2H_2N-C_6H_5}$$

(**168**)

$$\underset{C_6H_5N}{\overset{CH_3O}{\diagdown}}C-\!\!\!\left\langle\;\;\right\rangle\!\!\!-C\overset{OCH_3}{\underset{N-C_6H_5}{\diagup}}$$

$$R-C\overset{OR^1}{\underset{NH_2}{\diagdown}} \quad Cl^\ominus + H_2N-CN + Na_2HPO_4 \xrightarrow[\substack{-NaH_2PO_4}]{-NaCl,\,-NH_3} R-C\overset{OR^1}{\underset{N-CN}{\diagdown}}$$

(**171**)

When imino ester fluoroborates **172** are treated with ammonia, they are transformed into amidinium fluoroborates via cleavage of alcohol. This type of reaction was performed with five-, six-, seven-, and eight-membered-ring imino ester fluoroborates (36).

$$\underset{R^1}{\overset{(H_2C)_n}{\diagup\!\!\!\diagdown}}\!\!\!\underset{N\overset{+}{\diagdown}OC_2H_5}{\overset{\diagup}{\diagdown}}\!\!\!R \quad BF_4^\ominus \xrightarrow[-C_2H_5OH]{NH_3} \underset{R^1}{\overset{(H_2C)_n}{\diagup\!\!\!\diagdown}}\!\!\!\underset{N\overset{+}{\diagdown}NH_2}{\overset{\diagup}{\diagdown}}\!\!\!R \quad BF_4^\ominus$$

$n = 2, 3, 4, 5$

(**172**)

Ethanolamine may also be used for these substitution reactions (52), for example in preparation of compound (**173**).

$$\text{(tetrahydroisoquinoline)}\overset{NH}{\underset{OC_2H_5}{\diagdown\!\!\!\!\!\!\!\!+}} \quad BF_4^\ominus \xrightarrow[-C_2H_5OH]{H_2N-CH_2-CH_2-X} \text{(tetrahydroisoquinoline)}\overset{NH}{\underset{\overset{|}{H}}{\diagdown\!\!\!\!\!\!\!\!+}}N-CH_2-CH_2-X$$

X = OH, Cl (**173**)

If a starting molecule like **174** contains a neighboring haloalkyl group, cyclization in addition to exchange may occur (52).

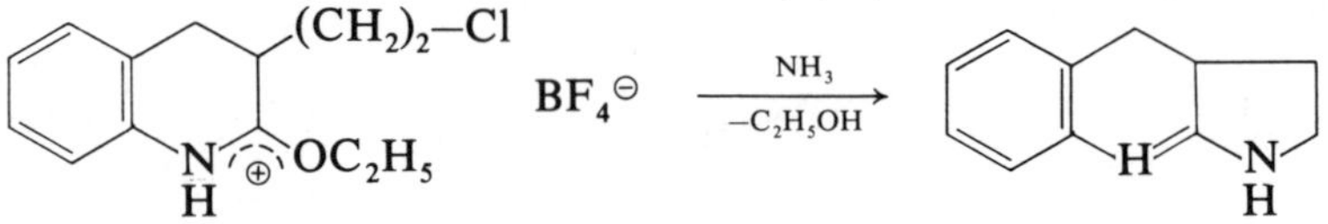

$$\underset{(\mathbf{174})}{\text{[tetrahydroquinoline with (CH}_2)_2\text{-Cl and }\overset{\oplus}{N}(H)\text{-OC}_2H_5]} \quad BF_4^{\ominus} \xrightarrow[-C_2H_5OH]{NH_3} \text{[fused bicyclic iminium]}$$

The iminium salt **120b** reacts with primary and secondary amines as well as with diamines to give guanidinium salts (**174a, 174b**) (136a).

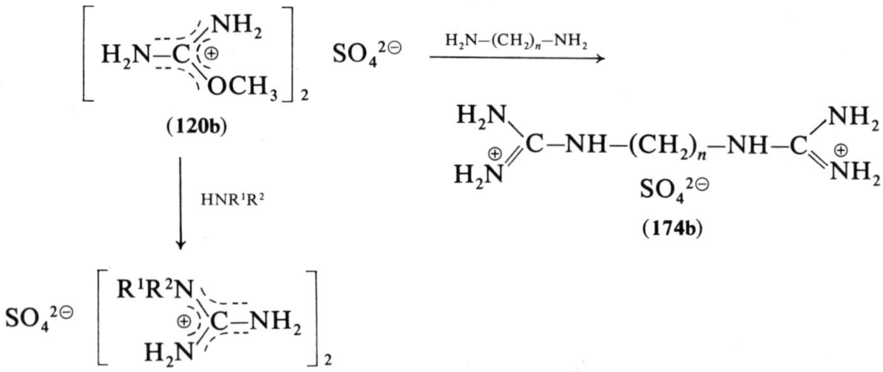

$$\left[H_2N-\overset{\oplus}{C}\!\!\begin{array}{c}NH_2\\OCH_3\end{array} \right]_2 SO_4^{2\ominus} \xrightarrow{H_2N-(CH_2)_n-NH_2}$$

(120b)

$$\underset{H_2N}{\overset{H_2N}{>}}\overset{\oplus}{C}-NH-(CH_2)_n-NH-\overset{\oplus}{C}\!\!\underset{NH_2}{\overset{NH_2}{<}} \quad SO_4^{2\ominus}$$

(174b)

↓ HNR¹R²

$$SO_4^{2\ominus} \left[\begin{array}{c} R^1R^2N\\ \overset{\oplus}{>}C-NH_2\\ H_2N \end{array} \right]_2$$

(174a)

$R^1 = H, R^2 = $ alkyl
$R^1 = R^2 = $ alkyl

2. N-Monoacylated Amines and Vinylogous Amides

When N-alkylalkoxymethyleniminium alkylsulfates **175** are heated with N-monosubstituted formamides, N,N'-dialkylformamidinium methylsulfates (**176**) result, mostly as oily products (67). The reaction ceases at the step of the

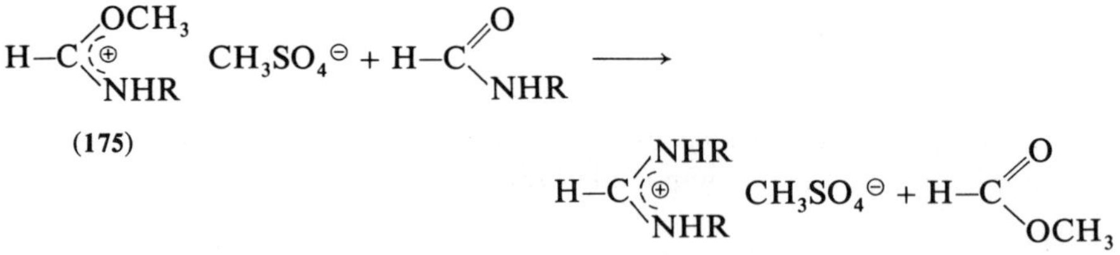

$$H-\overset{\oplus}{C}\!\!\begin{array}{c}OCH_3\\NHR\end{array} \quad CH_3SO_4^{\ominus} + H-C\!\!\begin{array}{c}O\\NHR\end{array} \longrightarrow$$

(175)

$$H-\overset{\oplus}{C}\!\!\begin{array}{c}NHR\\NHR\end{array} \quad CH_3SO_4^{\ominus} + H-C\!\!\begin{array}{c}O\\OCH_3\end{array}$$

(176)

amidinium salt **176** (67), whereas excess formamide promotes further reaction with dimethyl sulfate to give tris(formamido)methane (**45**) (61). N-Arylated formamides are converted with alkylating agents like dialkyl sulfates and

III. ALKOXYMETHYLENIMINIUM SALT—NUCLEOPHILE REACTIONS

sulfonic acid esters into N,N'-diarylformamidinium salts without isolation of intermediates (63). This reaction also presumably proceeds via N-arylalkoxymethyleniminium salts. There is only one case in which the amidinium salt is isolated; the free amides are obtained from the amidinium salts in all the other cases. Table XIV summarizes some of the amidinium salts or amidines generated by this route.

TABLE XIV

Formamidinium Salts or Formamidines from Alkoxymethyleniminium Salts and N-Monosubstituted Formamides

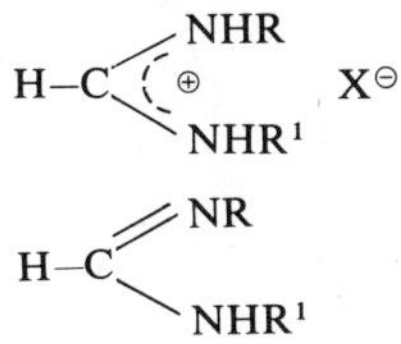

$$H-C\overset{NHR}{\underset{NHR^1}{\diagdown}}{}^{\oplus}\quad X^{\ominus}$$

$$H-C\overset{NR}{\underset{NHR^1}{\diagdown}}$$

$R = R^1$	X	Yield, %	m.p., °C	n_D^{20}	Ref.
CH_3	CH_3SO_4	95	—	1.4665	67
C_2H_5	CH_3SO_4	95	—	1.4583	67
C_6H_5	CH_3SO_4	78	181–182	—	63
p-CH_3O–C_6H_4	CH_3SO_4	92	178–179	—	63
C_6H_5	—	78	138	—	63
p-CH_3O–C_6H_4	—	92	110–112	—	63
p-CH_3–C_6H_4	—	95.3	140–141	—	63
p-NO_2–C_6H_4	—	98	234–235	—	63

Vinylogous amides (**177**) are converted by the salt **178** into vinylogous amidinium salts (**179**) (17).

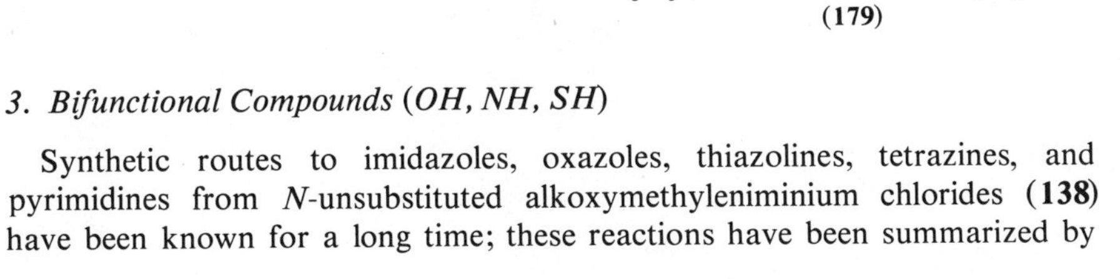

$$H-C\overset{OC_2H_5}{\underset{\underset{CH_3}{|}}{\diagdown}}{}^{\oplus}_{N-C_6H_5}\quad BF_4^{\ominus} + O=\!\!\!\diagup\!\!\!\diagdown_n\!\!\!N\overset{CH_3}{\underset{C_6H_5}{\diagdown}} \xrightarrow[-H-C\overset{O}{\diagdown OC_2H_5}]{}$$

(**178**)   (**177**)

$$\overset{CH_3}{\underset{C_6H_5}{\diagdown}}N\!\!-\!\!\diagup\!\!\!\diagdown_n\!\!\!\overset{\oplus}{N}\overset{CH_3}{\underset{C_6H_5}{\diagdown}}\quad BF_4^{\ominus}$$

(**179**)

3. Bifunctional Compounds (OH, NH, SH)

Synthetic routes to imidazoles, oxazoles, thiazolines, tetrazines, and pyrimidines from N-unsubstituted alkoxymethyleniminium chlorides (**138**) have been known for a long time; these reactions have been summarized by

Roger and Neilson (121). More recently Weidinger and Kranz (174–176) synthesized 1,3,4-oxadiazoles (**180**), 1,2,4-oxadiazoles (**181**), and 1,3,4-thiadiazoles (**182**) with the aid of alkoxymethyleniminium chlorides **138**.

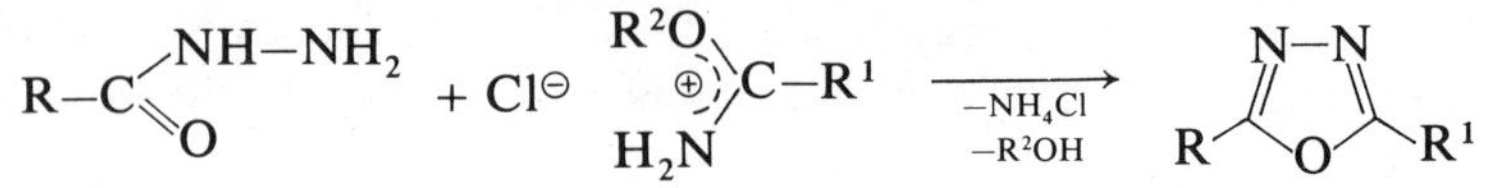

$$R-C\underset{O}{\overset{NH-NH_2}{\lessgtr}} + Cl^{\ominus} \quad \underset{H_2N}{\overset{R^2O}{\gtrless}}C-R^1 \quad \xrightarrow[-R^2OH]{-NH_4Cl} \quad R\underset{O}{\overset{N-N}{\diagdown\diagup}}R^1$$

(138)   (180)

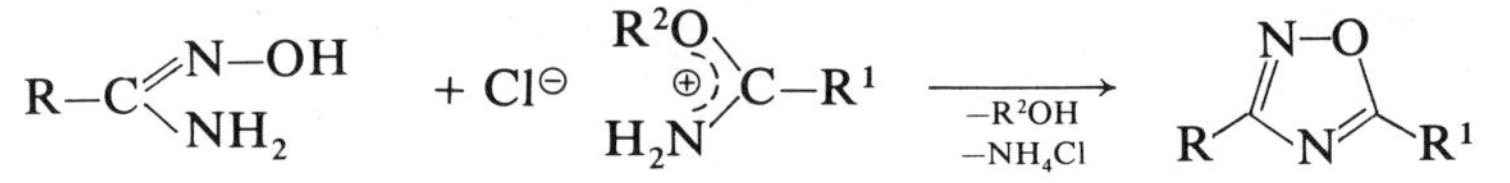

$$R-C\underset{NH_2}{\overset{N-OH}{\lessgtr}} + Cl^{\ominus} \quad \underset{H_2N}{\overset{R^2O}{\gtrless}}C-R^1 \quad \xrightarrow[-NH_4Cl]{-R^2OH} \quad R\underset{N}{\overset{N-O}{\diagdown\diagup}}R^1$$

(138)   (181)

$$R-C\underset{S}{\overset{NH-NH_2}{\lessgtr}} + Cl^{\ominus} \quad \underset{H_2N}{\overset{R^2O}{\gtrless}}C-R^1 \quad \xrightarrow[-NH_4Cl]{-R^2OH} \quad R\underset{S}{\overset{N-N}{\diagdown\diagup}}R^1$$

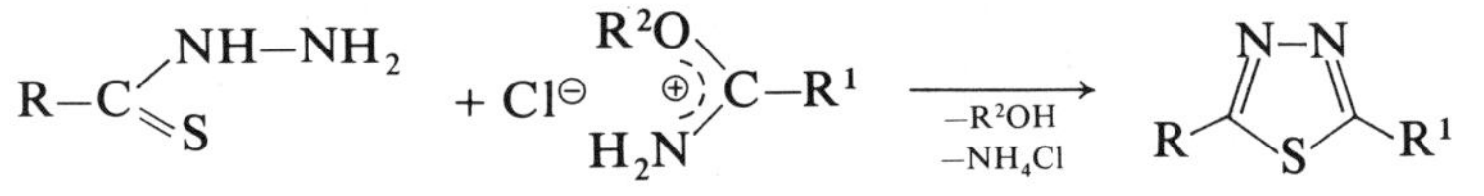

(138)   (182)

Eilingsfeld et al. (125) performed analogous cyclizations by reacting *N,N*-dialkylalkoxymethyleniminium chlorides **183** generated *in situ* from amide

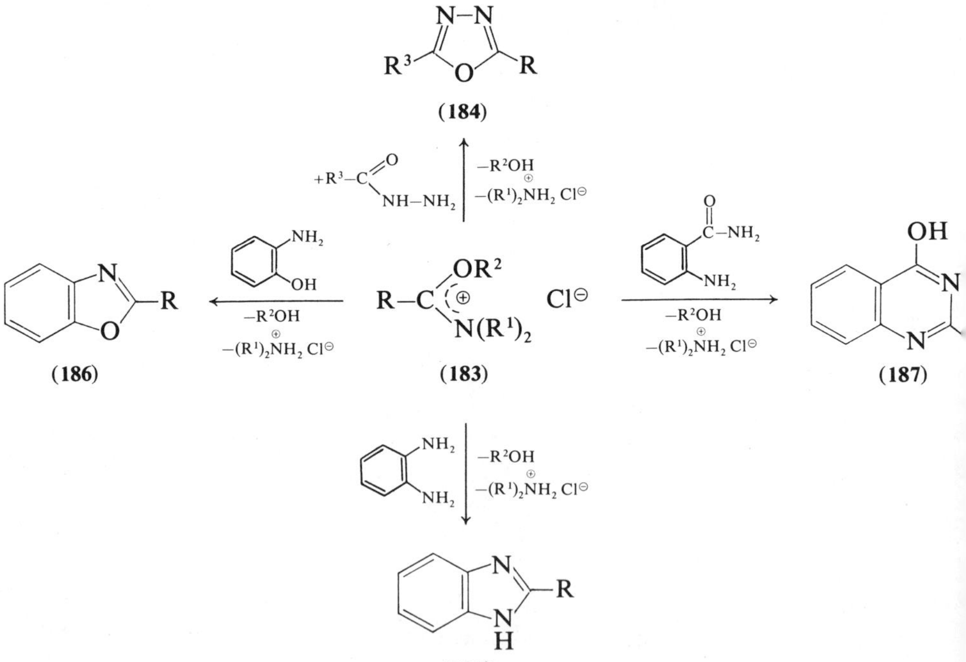

III. ALKOXYMETHYLENIMINIUM SALT—NUCLEOPHILE REACTIONS 241

chlorides and alcohols. Four disubstituted 1,3,4-oxadiazoles (**184**) were prepared from carboxylic acid hydrazides, 2-phenylbenzimidazole (**185**) from o-phenylenediamine, 2-phenylbenzoxazole (**186**) from o-aminophenol, and 2-phenyl-4-hydroxyquinazoline (**187**) from anthranilamide.

N-Unsubstituted salts **188** may also undergo this type of cyclization (25).

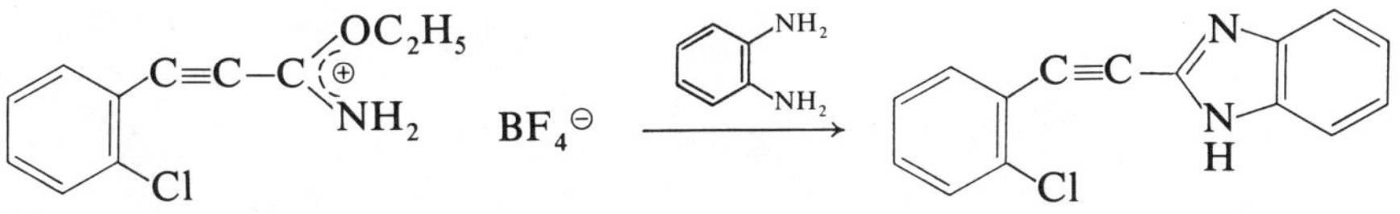

(**188**)

Uracil derivatives (**190**) are obtained from N-acylated ureas and N,N-dimethylmethoxymethyleniminium methylsulfates (**189**); the reaction is base-catalyzed (177).

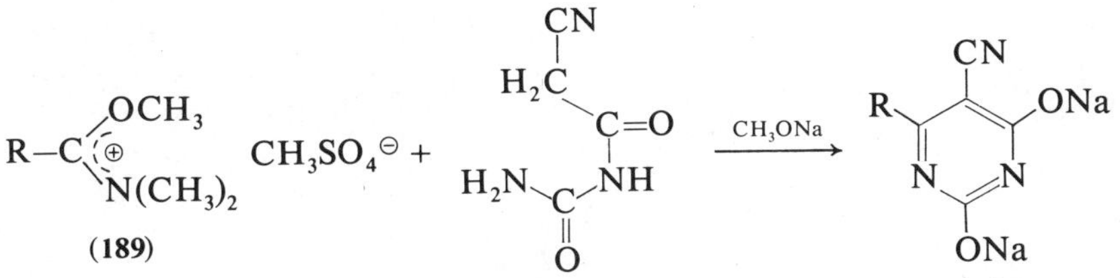

(**189**)   (**190**)

E. REACTIONS WITH BASE

From N-monosubstituted alkoxymethyleniminium salts (**191**) and base, imino esters are obtained. Tertiary amines (67, 178, 179), alkali hydroxides

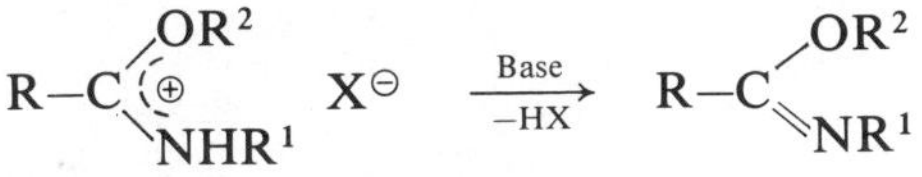

(**191**)

(42, 60), or alkaline cyanides (86) are the bases that have been used most often. The imino esters generated by this route are summarized in Table XV.

An exception to this pattern is found in the reaction of ethoxymethyleniminium chloride (**192**) and tributylamine, which gives s-triazine (**193**) in good yield (90).

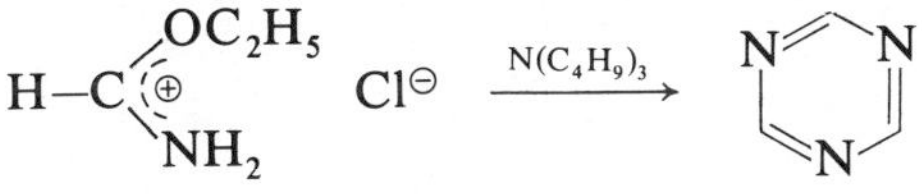

(**192**)   (**193**)

TABLE XV

Imino Esters from N-Monosubstituted Alkoxymethyleniminium Salts and Base

Imino ester

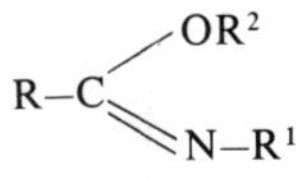

$$R-C\begin{array}{c}OR^2\\ \| \\ N-R^1\end{array}$$

R	R^1	R^2	Yield, %	b.p./pressure, °C/torr	m.p., °C	n_D^{20}	Ref.
H	CH₃	CH₃	51	47–49	—	1.3875	67
H	C₂H₅	CH₃	46	61–63	—	1.3881	67
C₂H₅	C₂H₅	i-C₃H₇	59	42–43/19	—	—	42
C₂H₅	C₂H₅	c-C₆H₁₁	69	100/19	—	—	42
C₂H₅	C₃H₇	CH₃	31	31–33/12	—	—	42
C₂H₅	C₆H₅	C₂H₅	61	97–98/12	—	—	42
CH₃	C₆H₅	C₂H₅	65	98–99/14	—	—	42
CH₃	CH₃	CH₃	69	77–78	—	1.4018	67, 86
CH₃	n-C₆H₁₃	C₂H₅	60	147–148/12	—	—	42
CH₃	C₂H₅	CH₃	25	92–93	—	1.4025	67
CH₃	C₂H₅	C₂H₅	6	113/760	—	—	
CH₃	C₆H₅	CH₃	36	91–92/17	—	—	42, 60
CH₃	C₂H₅	i-C₃H₇	37	125/760	—	—	42

CH_3	C_2H_5	*c*-C_6H_{11}	73	92–95/20	—	—	42
C_2H_5	CH_3	CH_3	56	96–98	—	1.4126	86
CH_3	*p*-$CH_3O–C_6H_4$	CH_3	52	115–116/10	—	—	42
C_2H_5	C_2H_5	C_2H_5	56	127–130/760	—	—	42
C_2H_5	C_2H_5	CH_3	32	113–115/760	—	—	42
C_6H_5	H	CH_3	37	95–97/14–15	—	—	42, 60
$NC–CH_2$	CH_3	C_2H_5	67	—	101–102	—	178
$NC–CH_2$	C_6H_5	C_2H_5	28	80–82/0.05	—	—	178
$NC–CH(CH_3)$	CH_3	C_2H_5	66	87–88/12	—	—	178
$NC–CH(CH_3)$	C_6H_5	C_2H_5	32	60–62/0.01	—	—	178
C_6H_5	H	CH_3	19	100–102/12	—	—	42
$C_6H_5–CH_2$	C_6H_5	C_2H_5	54	88–90/0.05	—	—	178
$C_6H_5–CH_2$	*p*-$NO_2–C_6H_4$	C_2H_5	35	—	57–58	—	178
p-$NO_2–C_6H_4$	C_6H_5	C_2H_5	37	—	75	—	178

The synthesis of lactim esters (**194**) from the corresponding salts has been studied extensively. Four- to 16-membered ring lactim esters have been

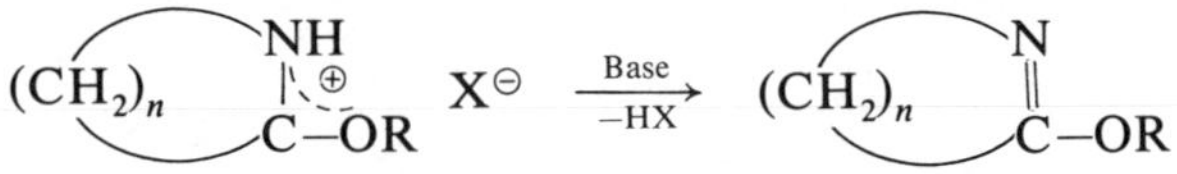

$$(CH_2)_n \overset{NH}{\underset{C-OR}{|\oplus}} X^\ominus \xrightarrow[-HX]{Base} (CH_2)_n \overset{N}{\underset{C-OR}{\|}}$$
(**194**)

prepared by this method. The reaction of three-membered ring lactim ester fluoroborates (**195**) with base shows different behavior. Here ring opening to amino acid esters (**196**) occurs upon treatment with aqueous potassium hydrogen carbonate (44). Table XVI lists some examples.

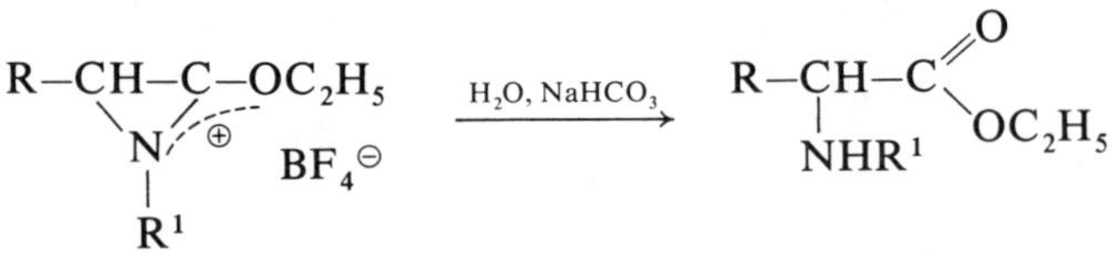

$$R-CH-C-OC_2H_5 \quad \xrightarrow{H_2O, NaHCO_3} \quad R-CH-C\overset{O}{\underset{OC_2H_5}{\diagdown}}$$
$$\underset{R^1}{\overset{N^\oplus}{|}} \quad BF_4^\ominus \qquad\qquad \underset{NHR^1}{|}$$
(**195**) (**196**)

Recently Glushkov and Granik (190) have reviewed the chemistry of lactim ethers. For this reason Table XVII contains only a few lactim ethers, which are generated from their salts.

TABLE XVII

Lactim Esters from Lactim Ester Salts and Base

Lactim ester 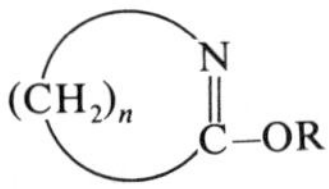 $(CH_2)_n \overset{N}{\underset{C-OR}{\|}}$					
n	R	Yield, %	b.p./pressure, °C/torr	n_D^{20}	Ref.
3	CH_3	57	117–118	1.4402	81, 86
3	C_2H_5	47	139/760	—	42
4	CH_3	71	143–144/710	1.4556	82
4	C_2H_5	—	86–88/70	—	49
5	CH_3	93.5	50/10	1.4620	82
5	C_2H_5	59	69/14	—	42
6	CH_3	88.8	56/10	1.4688	82
7	CH_3	79.5	82.5/10	1.4779	82
9	CH_3	96	54/0.2	1.4818	82
11	CH_3	92.5	101–103/0.3	1.4827	82

III. ALKOXYMETHYLENIMINIUM SALT—NUCLEOPHILE REACTIONS

TABLE XVI

Survey of Syntheses of Lactim Esters

Reaction	Ref.

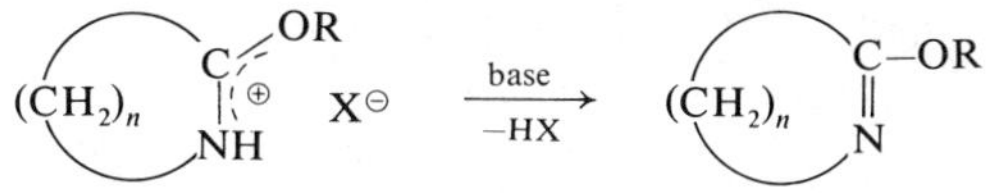

$$(CH_2)_n\underset{NH}{\overset{C-OR}{\diagup}}{}^{\oplus}\ X^{\ominus}\xrightarrow[-HX]{base}(CH_2)_n\underset{N}{\overset{C-OR}{\diagup}}$$

n	Ref.
n = 2	40
n = 3	40, 42, 46, 47, 53, 81, 86, 180
n = 4	40, 47, 49, 75, 82
n = 5	40, 42, 47, 73, 78, 82
n = 6	56, 82
n = 7	40, 76, 82
n = 9	40, 82
n = 10	40
n = 11	40, 82
n = 13	82
n = 14	40

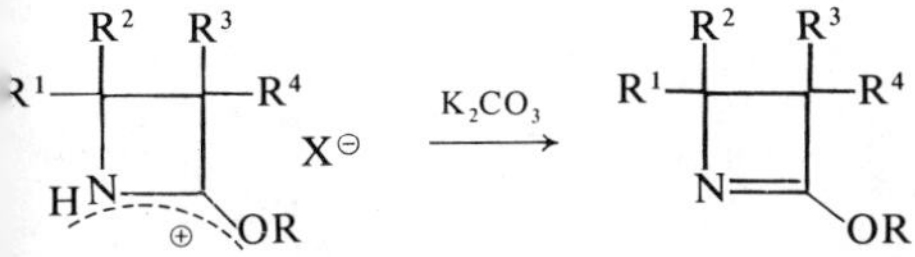

Reaction	Ref.
azetidinium salt → azetine (R¹, R², R³, R⁴ substituents) with K₂CO₃	50
pyrrolidinium salt → pyrroline (with COOC₂H₅ group) BF₄⁻, K₂CO₃	46
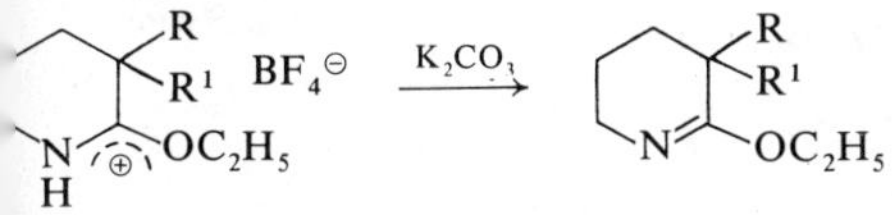 piperidinium salt → tetrahydropyridine, BF₄⁻, K₂CO₃	48

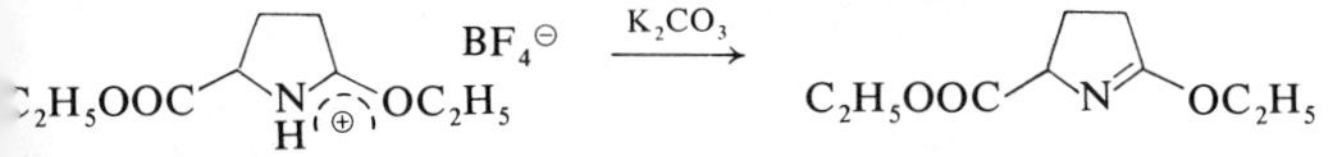

R	H	CH₃	C₂H₅	C₆H₅	COOC₂H₅	Cl	OCH₃	OCH₃	C₂H₅
R¹	H	H	H	H	H	H	H	COOC₂H₅	COOC₂H₅

R	C₆H₅	Cl	Br	Cl
R¹	COOC₂H₅	COOC₂H₅	COOC₂H₅	Cl

Reaction	Ref.
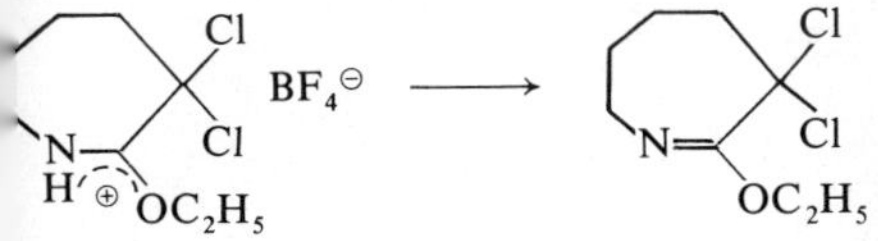 dichloro azepanium (BF₄⁻) → dichloro azepine with OC₂H₅	123
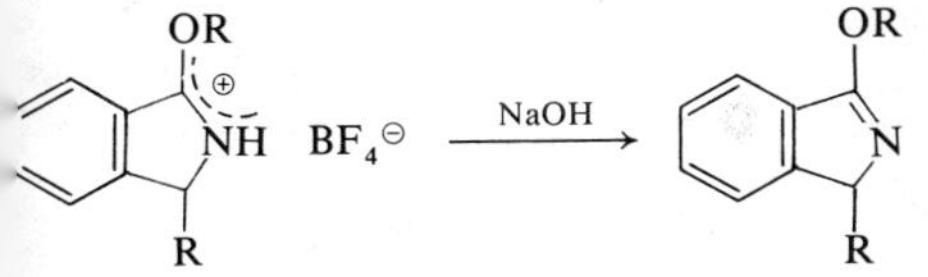 isoindolinium (OR, NH, BF₄⁻) → isoindole with NaOH	53, 54

TABLE XVI (cont.)

Reaction	Ref.
(structures) K₂CO₃	55
(structures) K₂CO₃	51
(structures) (CH₃)₃COK	56
(structures)	56
(structures)	56
(structures) (C₂H₅)₃O⁺ BF₄⁻, K₂CO₃	181
(structures) K₂CO₃	57

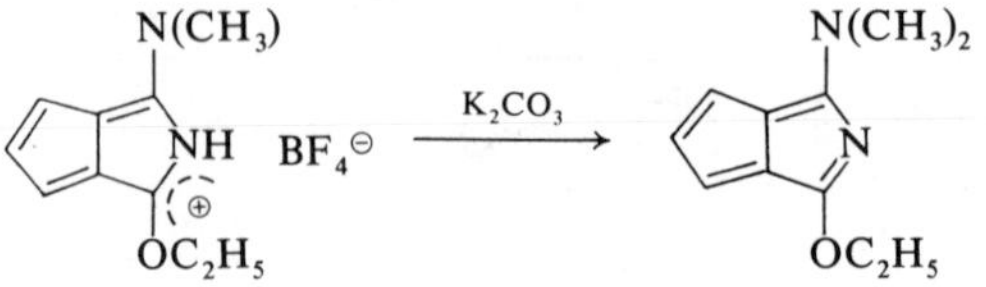

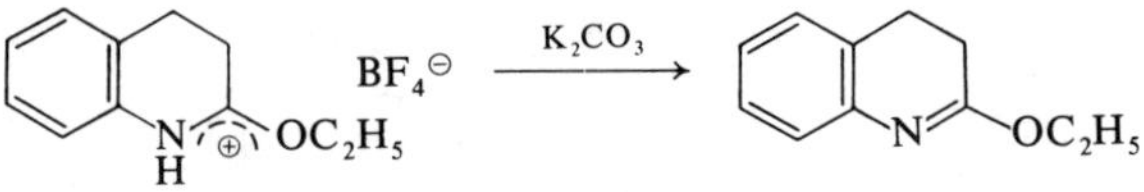

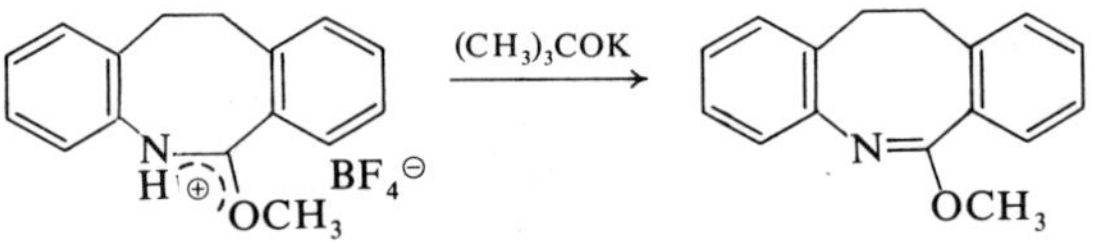

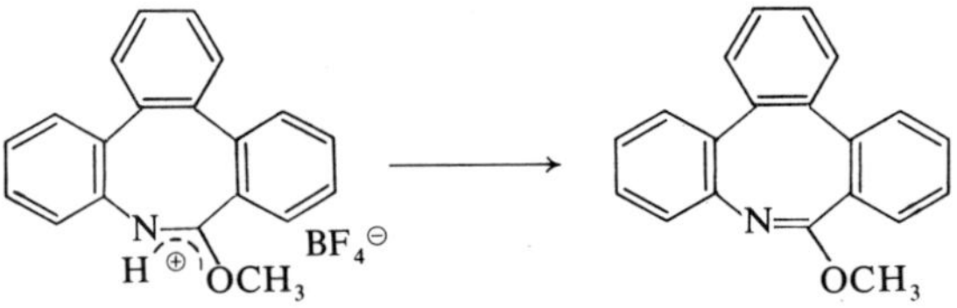

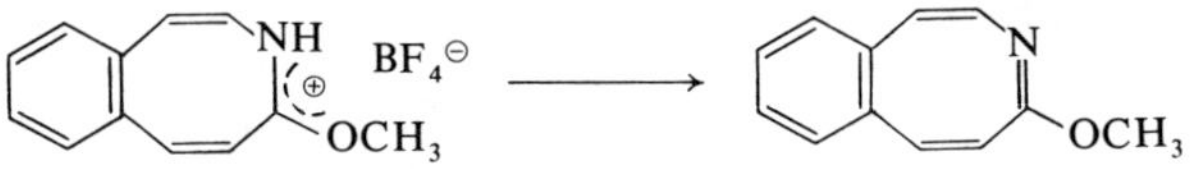

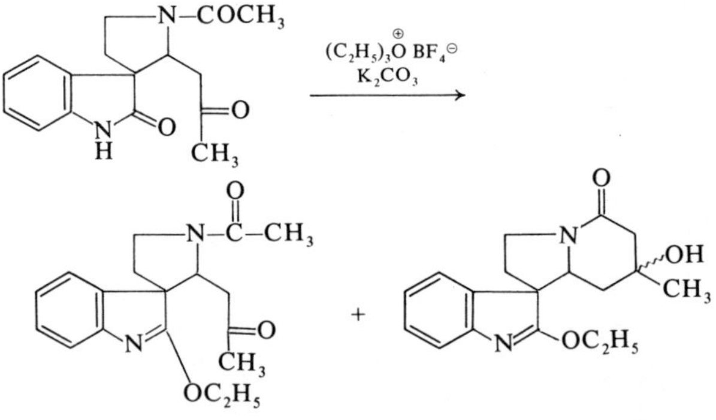

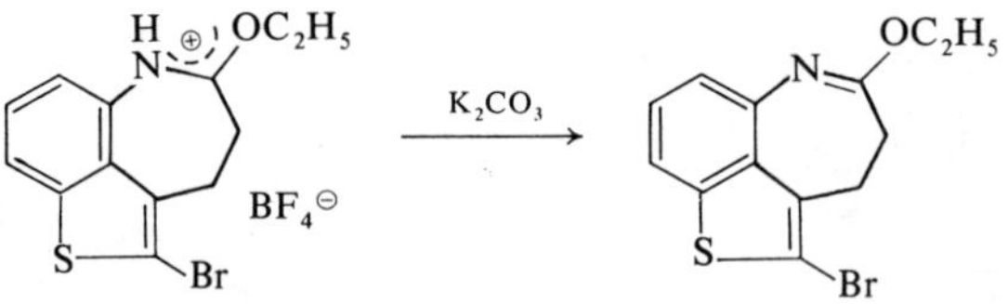

III. ALKOXYMETHYLENIMINIUM SALT—NUCLEOPHILE REACTIONS

TABLE XVI (*cont.*)

Reaction	Ref.
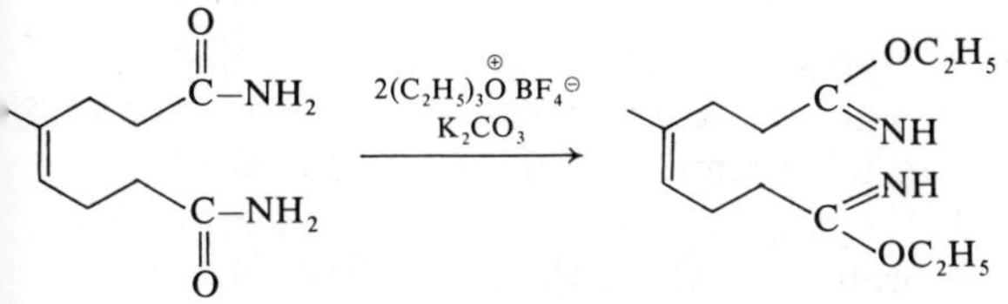	182
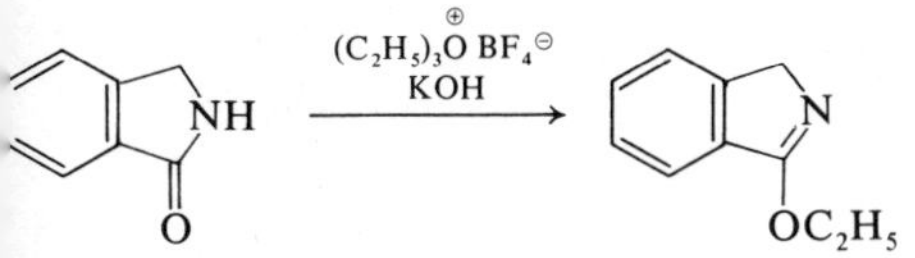	183
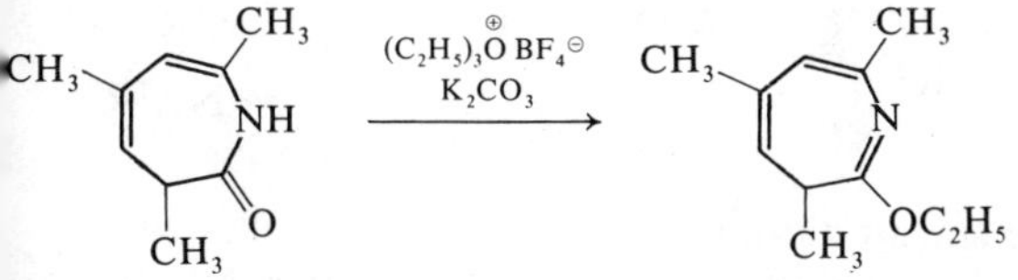	184
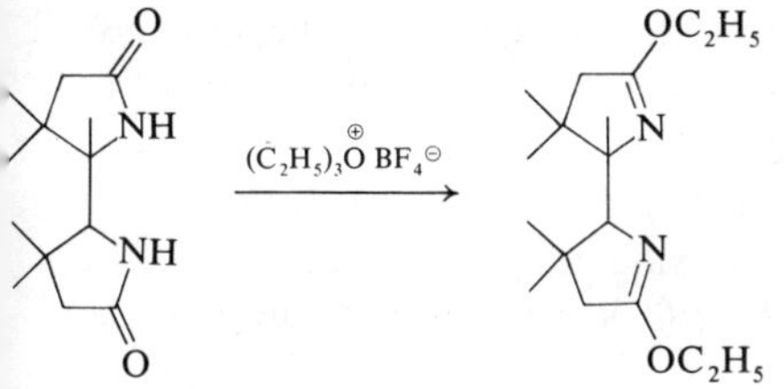	182, 185, 186
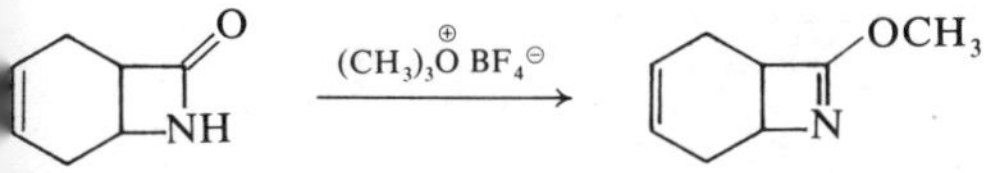	187
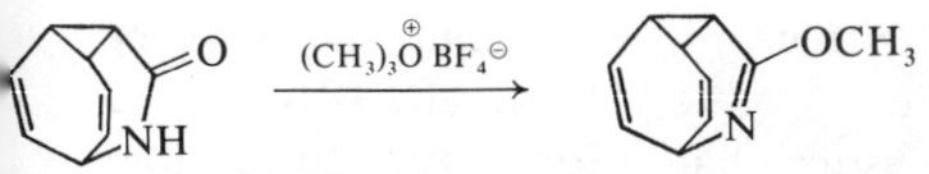	188, 189
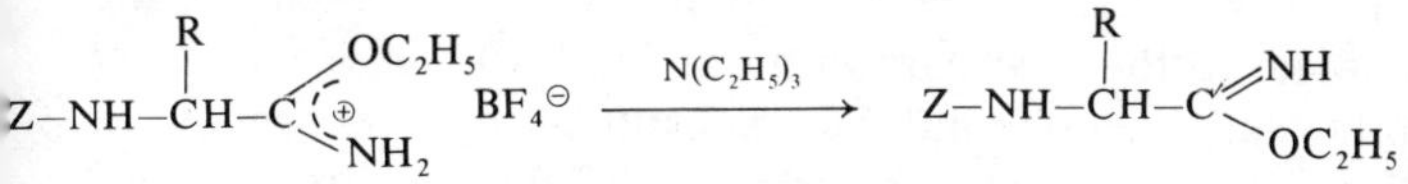 Z = carbobenzoxy, phthalyl, tosyl	155
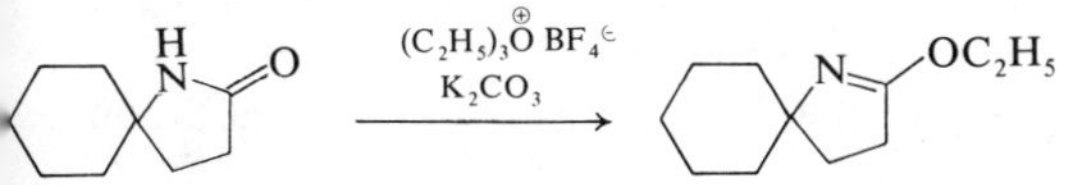	183

F. REACTIONS WITH ALKALI METAL ALKOXIDES

If alkali metal alkoxides act on N,N-dialkylalkoxymethyleniminium salts **197**, various products may be expected because of the ambident nature of these cations (15, 71, 143). At low temperatures nucleophilic addition of the alkoxide to the acyl carbon atom (reaction path a) occurs provided that there are no steric or electronic factors opposing this pathway. Acid amide acetals (**198**) form in this kinetically controlled reaction. This reaction was first performed by Meerwein et al. with iminium fluoroborates (13, 14), and by Bredereck et al. with iminium alkylsulfates (61, 64).

$$R-C\overset{O}{\underset{NR^1R^2}{\diagup}} + R^3OR^3 \xleftarrow[-X^\ominus]{b} R-C\overset{OR^3}{\underset{NR^1R^2}{\overset{\oplus}{\diagup}}} \quad X^\ominus + R^3O^\ominus \xrightarrow[-X^\ominus]{a} R-C\overset{OR^3}{\underset{NR^1R^2}{\diagup OR^3}}$$

$$(197) \hspace{5cm} (198)$$

To a lesser degree nucleophilic substitution at R^3 takes place as well, giving rise to ethers (71) (reaction path b). The reaction according to path b is the irreversible, thermodynamically controlled reaction, because charges are neutralized and furthermore the resonance energy of the amide function is gained. For the synthesis of amide acetals low temperatures are thus required. If the R group contains CH bonds in its α position they are made more acidic by the neighboring positive charge. The alkoxide may also attack at this position to abstract a proton, and to form ketene O,N-acetals (reaction path c).

$$R-\underset{H}{\overset{|}{C}}-C\overset{OR^3}{\underset{NR^1R^2}{\overset{\oplus}{\diagup}}} \quad X^\ominus + R^3O^\ominus \xrightarrow[-X^\ominus]{c} R-\overset{|}{C}=C\overset{OR^3}{\underset{NR^1R^2}{\diagup}}$$

The substituents attached to the iminium ion determine the extent of the alternative reaction paths a, b, and c. Some rules may be formulated, however, which help to predict the distribution of the products. Generally N,N-dialkyl-alkoxymethyleniminium ions are so energy-rich that at low temperatures, they follow the reaction path with the lowest activation energy; for example, the addition of the alcoholate to the acyl carbon atom (path a). This is a nearly irreversible reaction. Deviations are found if the substituent R is very bulky. In that case the reaction takes place at the periphery of the iminium ion, for example, at the polarized O—R^3 bond, leading to dealkylation. If the alkoxide adds to the acyl carbon atom, the acetal thus formed is strained, since owing to the change of hybridization ($sp^2 \rightarrow sp^3$) the substituents at the acetal are more densely packed than in the iminium ion. This makes the alkoxide addition

III. ALKOXYMETHYLENIMINIUM SALT—NUCLEOPHILE REACTIONS

reversible giving rise to the thermodynamically most stable products, namely, amide and ether (path b). Thus N,N-dimethylethoxyethylideniminium fluoroborate (**199**) yields dimethylacetamide diethyl acetal (**200**) with ethoxides (14), whereas the analogous reaction with N,N-dimethylethoxy(t-butyl)methyleniminium fluoroborate (**201**) only gives pivalic acid dimethylamide (**202**) and ether (15).

$$CH_3-C\begin{smallmatrix}OC_2H_5\\ \oplus\\ N(CH_3)_2\end{smallmatrix} \; BF_4^{\ominus} + NaOC_2H_5 \xrightarrow{-NaBF_4} CH_3-C\begin{smallmatrix}OC_2H_5\\ -OC_2H_5\\ N(CH_3)_2\end{smallmatrix}$$

(**199**)   (**200**)

$$(CH_3)_3C-C\begin{smallmatrix}OC_2H_5\\ \oplus\\ N(CH_3)_2\end{smallmatrix} \; BF_4^{\ominus} + NaOC_2H_5 \xrightarrow{-NaBF_4}$$

(**201**)

$$(CH_3)_3C-C\begin{smallmatrix}O\\ \diagup\\ \diagdown N(CH_3)_2\end{smallmatrix} + C_2H_5OC_2H_5$$

(**202**)

CH bonds in a position α to the carbonium carbon atom may react as carbon acids, and transfer a proton to the alkoxide, thus forming ketene O,N-acetals. If the CH bond is acidified by further electron-withdrawing groups capable of forming resonance structures (e.g., ester or cyano functions), amide acetal formation may be completely suppressed; ketene O,N-acetals are the only products. This rule is derived from a comparison of the products formed in the reactions of N,N-dimethylethoxyethylideniminium fluoroborate (**199**) (14), N,N-dimethylethoxy(cyanomethyl)methyleniminium fluoroborate (**203**) (15), and N,N-dimethyl-1-ethoxy-2-(ethoxycarbonyl)ethylideniminium methylsulfate (**204**) (163) with sodium ethoxide. In the first case (**199**) a mixture of amide acetal and ketene O,N-acetal is obtained, whereas in the other cases (**203, 204**) ketene O,N-acetals (**205, 206**) form exclusively.

$$CH_3-C\begin{smallmatrix}OC_2H_5\\ \oplus\\ N(CH_3)_2\end{smallmatrix} \; BF_4^{\ominus} \xrightarrow{NaOC_2H_5} CH_3-C\begin{smallmatrix}OC_2H_5\\ -OC_2H_5\\ N(CH_3)_2\end{smallmatrix} + CH_2=C\begin{smallmatrix}OC_2H_5\\ \diagdown N(CH_3)_2\end{smallmatrix}$$

(**199**)   (**200**)

$$NC-CH_2-C\begin{smallmatrix}OC_2H_5\\ \oplus\\ N(CH_3)_2\end{smallmatrix} \; BF_4^{\ominus} \xrightarrow{NaOC_2H_5} NC-CH=C\begin{smallmatrix}OC_2H_5\\ \diagdown N(CH_3)_2\end{smallmatrix}$$

(**203**)   (**205**)

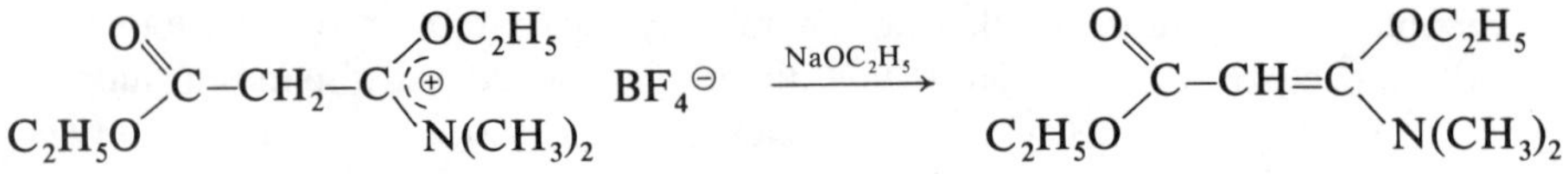

$$\underset{(204)}{\overset{\displaystyle C_2H_5O}{\underset{\displaystyle}{}}\overset{\displaystyle O}{\underset{\displaystyle}{\|}}C-CH_2-\overset{\displaystyle OC_2H_5}{\underset{\displaystyle N(CH_3)_2}{C^{\oplus}}}} \ BF_4^{\ominus} \xrightarrow{NaOC_2H_5} \underset{(206)}{\overset{\displaystyle C_2H_5O}{\underset{\displaystyle}{}}\overset{\displaystyle O}{\underset{\displaystyle}{\|}}C-CH=\overset{\displaystyle OC_2H_5}{\underset{\displaystyle N(CH_3)_2}{C}}}$$

Comparison of the reactions of N-alkyllactim ester salts with alkoxides is even more striking. From sodium ethoxide and N-methylbutyrolactim ester methylsulfate (207) only the lactam acetal (208) is generated (47, 163).

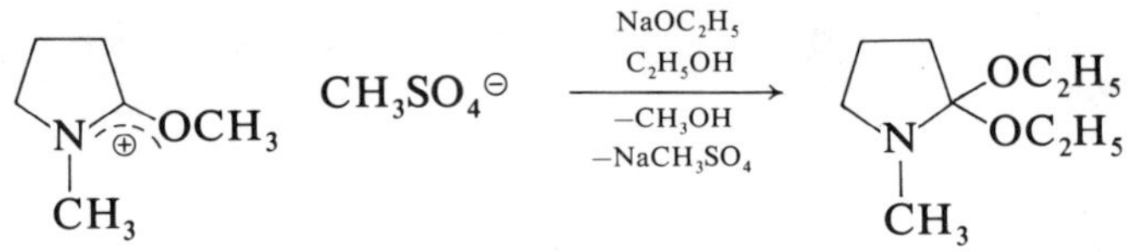

(207) pyrrolidine N-methyl with OCH₃, CH₃SO₄⁻ → (208) pyrrolidine N-methyl with OC₂H₅, OC₂H₅; reagents: NaOC₂H₅, C₂H₅OH, −CH₃OH, −NaCH₃SO₄

The cleavage of alcohol from the lactam acetal does not occur even with calcium, although alkali metals generally produce pure ketene O,N-acetals (210) from the corresponding acid amide acetals 209 (14, 163).

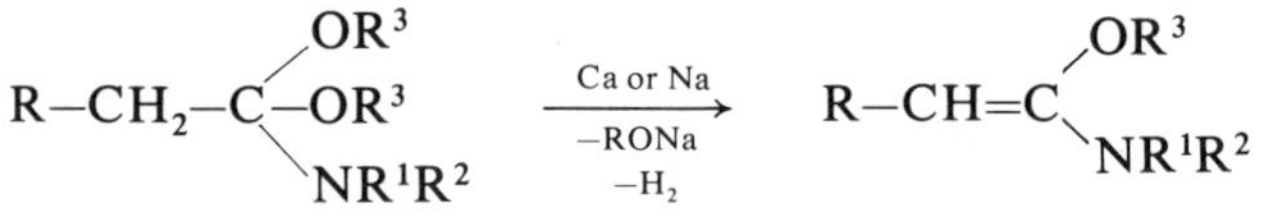

$$\underset{(209)}{R-CH_2-\overset{\displaystyle OR^3}{\underset{\displaystyle NR^1R^2}{C-OR^3}}} \xrightarrow[\substack{-RONa \\ -H_2}]{Ca\ or\ Na} \underset{(210)}{R-CH=\overset{\displaystyle OR^3}{\underset{\displaystyle NR^1R^2}{C}}}$$

On the other hand, the reaction of ethyl N-ethyl-(3-ethoxycarbonyl)-valerolactim ester fluoroborate (211) with sodium ethoxide smoothly yields the corresponding cyclic O,N-acetal 212 (122).

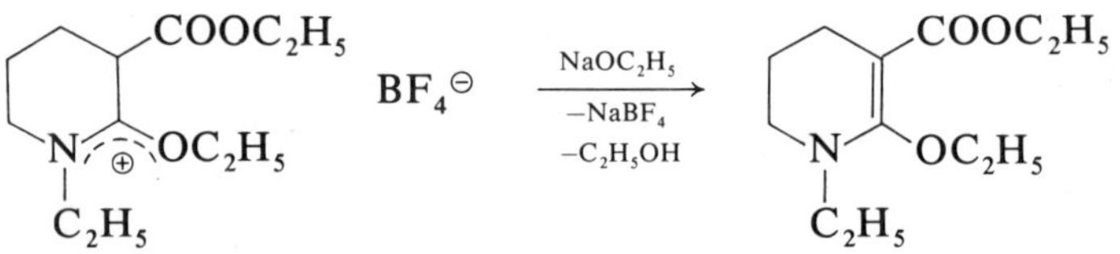

(211) → (212); reagents: NaOC₂H₅, −NaBF₄, −C₂H₅OH

It should be mentioned that a demonstration of the CH acidity by hydrogen/deuterium exchange of the parent compound of the alkoxymethyleniminium salts, namely, N,N-dimethylmethoxymethyleniminium methylsulfate (213), failed (191); however, such a demonstration with the structurally related thiazolium, benzthiazolium, imidazolium, and oxazolium salts was successful by base-catalyzed hydrogen/deuterium exchange (192–196).

III. ALKOXYMETHYLENIMINIUM SALT—NUCLEOPHILE REACTIONS

These results find further support in synthetic studies (197). Whereas thiazolium (**214**) and imidazolium (**215**) salts are deprotonated by triethylamine, and the nucleophilic carbene intermediates may be trapped with isocyanates and isothiocyanates, the analogous reaction of *N,N*-dimethylmethoxymethyleniminium methylsulfate (**214**) shows no acylation of the formyl carbon atom, but rather trimerization of the isocyanate (197).

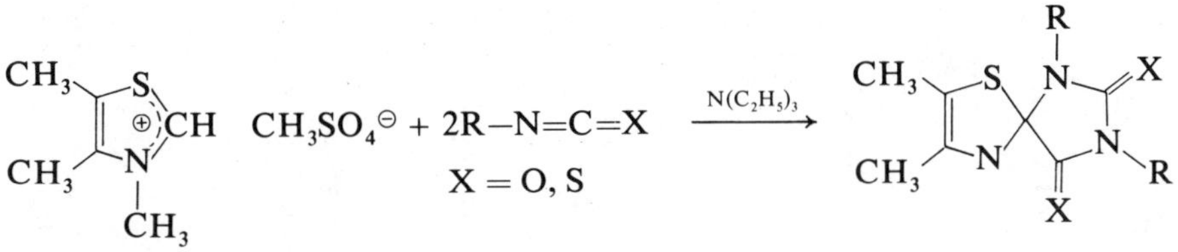

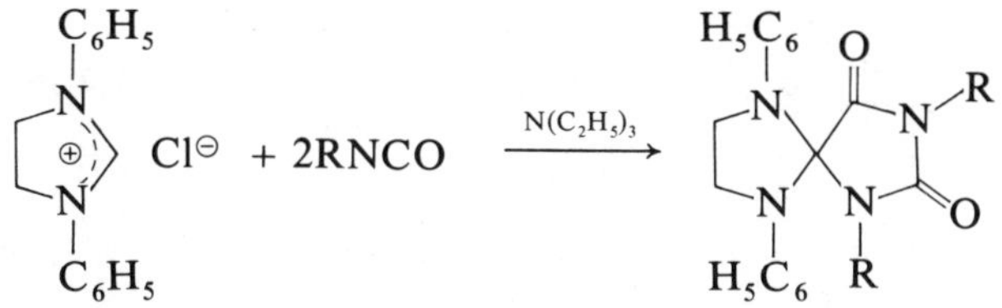

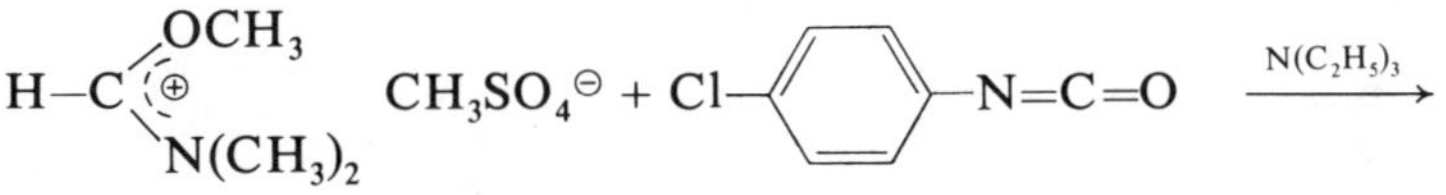

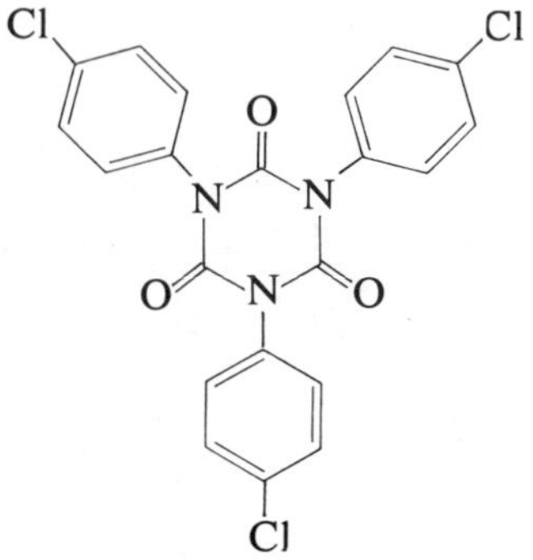

In the meantime a large number of carboxylic acid amide acetals and lactam acetals have been synthesized from iminium fluoroborates and methylsulfates; they are summarized in Table XVIII. Sometimes amide acetals are generated

TABLE XVIII

Amide Acetals from N,N-Dialkylalkoxymethyleniminium Salts and Alkali Metal Alkoxides

Amide acetals

$R-C{\underset{NR^1R^2}{\overset{O-R^3}{-}}}O-R^3$ from $R-C{\underset{NR^1R^2}{\overset{OR^3}{\oplus}}}$ $X^\ominus$ and $R^3O^\ominus$

R	R^1	R^2	R^3	$X^\ominus$	Yield, %	b.p./pressure, °C/torr	n_D^{20}	Ref.
H	CH_3	CH_3	CH_3	CH_3SO_4	75–87	104–105/731	1.3962	64, 71, 199–201
H	CH_3	CH_3	C_2H_5	BF_4	72	134–136	1.3987/18	13, 14, 200
H	CH_3	CH_3	C_2H_5	CH_3SO_4	63	128–132/740	1.4070	64, 71, 201
H	CH_3	CH_3	C_3H_7	CH_3SO_4	56	173–175/760	1.4083	71
H	CH_3	CH_3	i-C_3H_7	CH_3SO_4	42	69/50	1.4000	71
H	CH_3	CH_3	n-C_4H_9	CH_3SO_4	55	93/12	1.4164	71
H	CH_3	CH_3	t-C_4H_9	ClO_4	35	75–77/25	—	199
H	C_2H_5	C_2H_5	CH_3	CH_3SO_4	51	143–144/760	1.4070	71
H	n-C_3H_7	n-C_3H_7	CH_3	CH_3SO_4	77	73/12	1.4166	202
H	i-C_3H_7	i-C_3H_7	CH_3	CH_3SO_4	57	60–61/12	1.4160	71
H	i-C_3H_7	i-C_3H_7	C_2H_5	CH_3SO_4	50	73–76/12	1.4195	71
H	n-C_4H_9	n-C_4H_9	CH_3	CH_3SO_4	46	74–75/12	1.4248	71
H	—$(CH_2)_4$—		CH_3	CH_3SO_4	56	160–161/760	1.4320	71, 140
H	—$(CH_2)_5$—		CH_3	CH_3SO_4	60	83/15	1.4411	71, 140
H	—$(CH_2)_2$—O—$(CH_2)_2$—		CH_3	CH_3SO_4	57	87/15	1.4420	71, 140
CH_3	CH_3	CH_3	C_2H_5	BF_4	64	56–58/13	—	14
CH_3	CH_3	CH_3	C_2H_5	CH_3SO_4	65	32/8	1.4112/18	201
C_2H_5	CH_3	CH_3	CH_3	CH_3SO_4	25	30/12	1.4170/21	201
C_6H_5	CH_3	CH_3	C_2H_5	CH_3SO_4	40	32/10^{-5}	1.4896	201
p-NO_2-C_6H_4	CH_3	CH_3	C_2H_5	BF_4	53	85/0.01	—	15
$(C_2H_5O)_2CH$	CH_3	CH_3	C_2H_5	BF_4	76	71/0.1	—	15

				Yield, %	b.p./pressure, °C/torr	n_D^{20}	Ref.
$(CH_3)_2N-\overset{O}{\underset{\|}{C}}-$	CH_3	C_2H_5	BF_4	60	85/0.03	1.4489	15
$C_2H_5O-\overset{O}{\underset{\|}{C}}-$	CH_3	C_2H_5	BF_4	72	80/0.05	1.4278	15
$C_2H_5O-\overset{O}{\underset{\|}{C}}-$	$-(CH_2)_5-$	C_2H_5	BF_4	50	57/0.001	1.4530	21
NC	CH_3	C_2H_5	BF_4	56	68/12	—	15
$(C_2H_5O)_3C$	$-(CH_2)_4-$	C_2H_5	BF_4	67	52–53/0.001	1.4450	21

Amide acetals from diacids

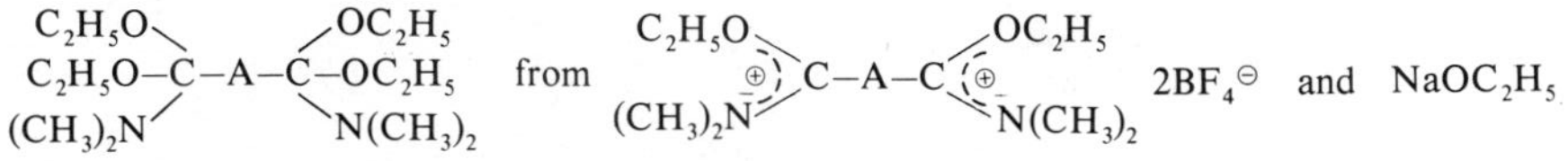

$$\begin{matrix}C_2H_5O\\C_2H_5O-C-A-C\\(CH_3)_2N\end{matrix}\begin{matrix}OC_2H_5\\OC_2H_5\\N(CH_3)_2\end{matrix} \quad \text{from} \quad \begin{matrix}C_2H_5O\\(CH_3)_2\overset{\ominus}{N}\end{matrix}\overset{\oplus}{C}-A-\overset{\oplus}{C}\begin{matrix}OC_2H_5\\N(CH_3)_2\end{matrix} \; 2BF_4^{\ominus} \quad \text{and} \quad NaOC_2H_5$$

A	Yield, %	b.p./pressure, °C/torr	m.p., °C	Ref.
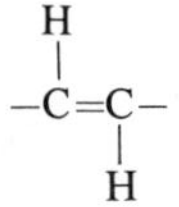 $\overset{H}{\underset{-C=C-}{\underset{\|}{H}}}$	68	122/0.05	65	15
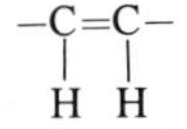 $-\overset{H}{\underset{H}{C}}-\overset{H}{\underset{H}{C}}-$	38	—	66–67	15
 para-phenylene	68	—	95	15

TABLE XVIII (cont.)

Lactam acetals	Yield,%	b.p./pressure, °C/torr	m.p., °C	n_D^{20}	Ref.
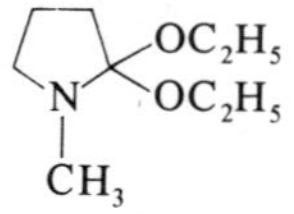 pyrrolidine-OC₂H₅/OC₂H₅, N-CH₃	73 60.5	57–58/10 59–60/11	— —	1.4373 1.4376	47, 164 13, 14
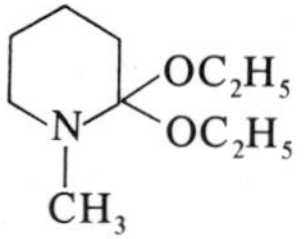 piperidine-OC₂H₅/OC₂H₅, N-CH₃	42	76–78/16	—	—	45, 47
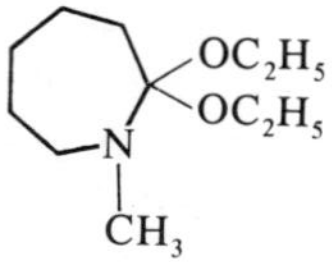 azepane-OC₂H₅/OC₂H₅, N-CH₃	67	88/22 27/0.02	— —	1.4758/22 1.4755/21	83 47, 203
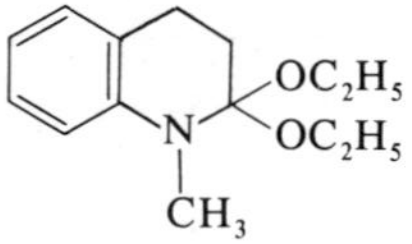 tetrahydroquinoline-OC₂H₅/OC₂H₅, N-CH₃	87	93/0.01	54–55	—	14
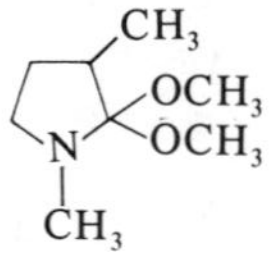 pyrrolidine-CH₃/OCH₃/OCH₃, N-CH₃	29	62/24	—	—	45
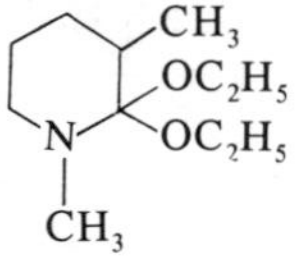 piperidine-CH₃/OC₂H₅/OC₂H₅, N-CH₃	40	77/21	—	—	45, 85

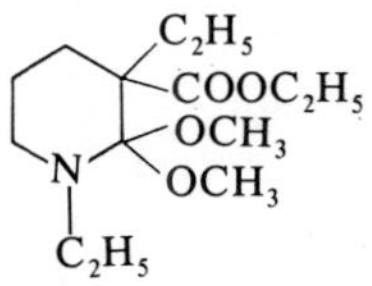

[structure: piperidine with C₂H₅, COOC₂H₅, OCH₃, OCH₃ substituents and N-C₂H₅] | 56 | 115–116/4–5 | — | 1.4631/24 | 122

Orthocarbamates

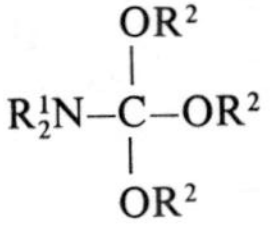

$$R^1_2N-C(OR^2)_3$$

R	R^1	R^2	Yield, %	b.p./pressure, °C/torr	n_D^{20}	Ref.
CH$_3$	CH$_3$	CH$_3$	41	34/12	1.4098	92
CH$_3$	CH$_3$	C$_2$H$_5$	48	70–72/17	1.4089	14,92
C$_2$H$_5$	C$_2$H$_5$	CH$_3$	12	48/12	1.4212	92
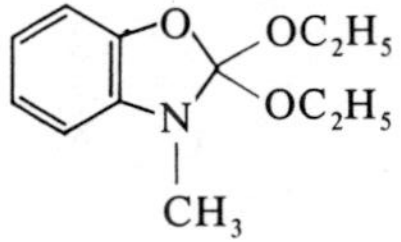 [benzoxazoline with OC$_2$H$_5$, OC$_2$H$_5$, N–CH$_3$]			56	97/11	—	14
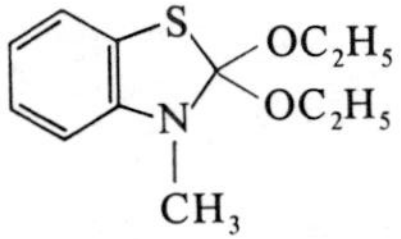 [benzothiazoline with OC$_2$H$_5$, OC$_2$H$_5$, N–CH$_3$]			75	126/1 m.p. 28	—	14

TABLE XVIII (cont.)

Urea acetals	Yield, %	b.p./pressure, °C/torr	n_D^{20}	Ref.
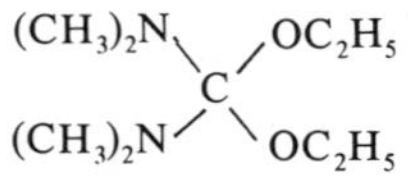 (CH$_3$)$_2$N–C(OC$_2$H$_5$)$_2$–N(CH$_3$)$_2$	56	35/2		14
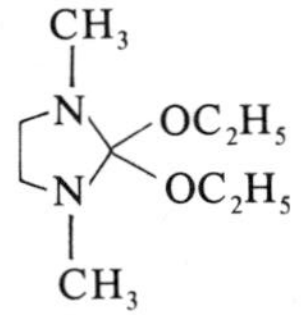 CH$_3$–N(–CH$_2$–)N(OC$_2$H$_5$)$_2$–CH$_3$ (cyclic)	65	30/0.01	—	14

Ketene O,N-acetals from iminium salts and alkoxides

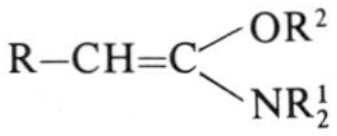

$$R-CH=C\begin{pmatrix}OR^2\\NR^1_2\end{pmatrix}$$

R	R^1	R^2	Yield, %	b.p./pressure, °C/torr	n_D^{20}	Ref.
H	CH$_3$	CH$_3$	58	104–105	1.4334	163
H	CH$_3$	C$_2$H$_5$	74	126	1.4367	14,163
H	C$_2$H$_5$	CH$_3$	48	142	1.4442	163
H	C$_2$H$_5$	C$_2$H$_5$	67	158–160	1.4402	163
CH$_3$	CH$_3$	CH$_3$	57	117–118	1.4250	163
CH$_3$	CH$_3$	C$_2$H$_5$	59	137	1.4331	163
C$_2$H$_5$	CH$_3$	C$_2$H$_5$	69	154–156	1.4330	163
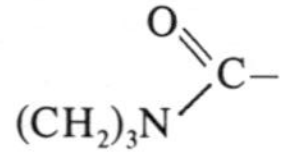 (CH$_2$)$_3$N–C(=O)–	CH$_3$	C$_2$H$_5$	62	141–142/10	1.5124	163
NC	CH$_3$	C$_2$H$_5$	51	85/0.01	—	15

III. ALKOXYMETHYLENIMINIUM SALT—NUCLEOPHILE REACTIONS

by this route, which are immediately reacted without isolation (23). Also some acid amides and lactams are alkylated at oxygen by triethyloxonium fluoroborate or dimethylsulfate, and the iminium salts thus generated are reacted without further characterization to amide or lactam acetals (45, 198).

Amide acetals (**217**) are also formed from (dialkylamino)methoxyacetonitriles **216** and sodium methoxide (140).

$$\underset{(216)}{\text{H}-\underset{\text{NR}^1\text{R}^2}{\overset{\text{OCH}_3}{\text{C}}}-\text{CN}} + \text{NaOCH}_3 \xrightarrow{-\text{NaCN}} \underset{(217)}{\text{H}-\underset{\text{NR}^1\text{R}^2}{\overset{\text{OCH}_3}{\text{C}}}-\text{OCH}_3}$$

$R^1 = R^2 = C_2H_5$
$R^1-R^2 = -(CH_2)_4-, -(CH_2)_5-, -(CH_2)_2-O-(CH_2)_2-$

From N,N-dialkyl-(dialkylamino)alkoxymethyleniminium fluoroborates like **218** and alkoxides free of alcohol the urea acetals **219** are obtained (14).

$$\underset{(218)}{\underset{(CH_3)_2N}{\overset{(CH_3)_2N}{\bigg\rangle}}\overset{\oplus}{\text{C}}-\text{OC}_2\text{H}_5 \ \ \text{BF}_4^{\ominus}} \xrightarrow[\text{CH}_3\text{CN}]{\text{NaOC}_2\text{H}_5} \underset{(219)}{\underset{(CH_3)_2N}{\overset{(CH_3)_2N}{\bigg\rangle}}\overset{\text{OC}_2\text{H}_5}{\underset{\text{OC}_2\text{H}_5}{\text{C}}}}$$

The same reaction in ethanol containing ethoxide leads to substitution of one dimethylamino group of the urea acetal **219** by an ethoxy group; triethyl orthocarbamic acid ester **220** is the product obtained (14). This ortho-

$$\underset{(219)}{\underset{(CH_3)_2N}{\overset{(CH_3)_2N}{\bigg\rangle}}\overset{\text{OC}_2\text{H}_5}{\underset{\text{OC}_2\text{H}_5}{\text{C}}}} \xrightarrow[-\text{HN(CH}_3)_2]{\text{C}_2\text{H}_5\text{OH}} \underset{(220)}{\underset{\text{C}_2\text{H}_5\text{O}}{\overset{(CH_3)_2N}{\bigg\rangle}}\overset{\text{OC}_2\text{H}_5}{\underset{\text{OC}_2\text{H}_5}{\text{C}}}}$$

carbamic acid ester (**220**) is also accessible from N,N-dialkyldiethoxymethyleniminium fluoroborate (**218**) and sodium ethoxide in ethanol (14).

$$\underset{(218)}{\underset{\text{OC}_2\text{H}_5}{\overset{(CH_3)_2N}{\bigg\rangle}}\overset{\oplus}{\text{C}}-\text{OC}_2\text{H}_5 \ \ \text{BF}_4^{\ominus}} \xrightarrow[\substack{-\text{NaBF}_4 \\ -\text{HN(CH}_3)_2}]{\substack{\text{NaOC}_2\text{H}_5 \\ \text{C}_2\text{H}_5\text{OH}}} \underset{(220)}{\underset{\text{C}_2\text{H}_5\text{O}}{\overset{(CH_3)_2N}{\bigg\rangle}}\overset{\text{OC}_2\text{H}_5}{\underset{\text{OC}_2\text{H}_5}{\text{C}}}}$$

The reaction of N,N-dialkyl-(dialkylamino)alkoxymethyleniminium alkylsulfates (**221**) and sodium ethoxide also generates orthocarbamic acid esters **223** via urea acetals **222** (92).

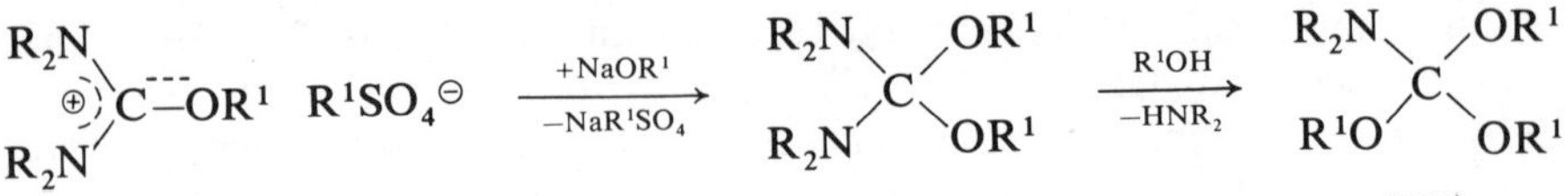

258   ALKOXYMETHYLENIMINIUM SALTS

$$\underset{(221)}{\overset{R_2N}{\underset{R_2N}{>}}\!\!\overset{\oplus}{C}\!\!-\!\!OR^1\ R^1SO_4^{\ominus}} \xrightarrow[-NaR^1SO_4]{+NaOR^1} \underset{(222)}{\overset{R_2N}{\underset{R_2N}{>}}\!\!C\!\!\overset{OR^1}{\underset{OR^1}{<}}} \xrightarrow[-HNR_2]{R^1OH} \underset{(223)}{\overset{R_2N}{\underset{R^1O}{>}}\!\!C\!\!\overset{OR^1}{\underset{OR^1}{<}}}$$

When ethoxide acts on cyclic O-alkylated ureas like **224**, no substitution of the amino function is observed, and the urea acetal **225** is isolated (14).

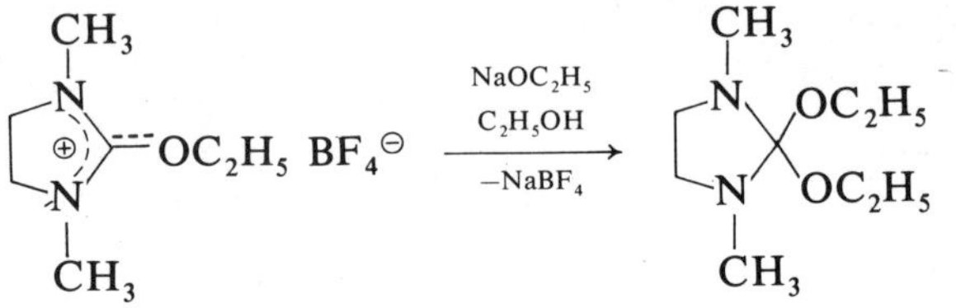

(224) → (225)

With **224** having an imidazolinium ring with N-CH_3, N-CH_3, $\oplus$, $=OC_2H_5\ BF_4^{\ominus}$ reacting with $NaOC_2H_5$/C_2H_5OH, $-NaBF_4$, to give **225**, the imidazolidine with N-CH_3, N-CH_3, and two OC_2H_5 groups.

These orthocarbamic acid esters and urea acetals are also listed in Table XVIII.

G. REACTIONS WITH ALKALI METAL HALIDES AND CYANIDES

When N,N-dimethylmethoxymethyleniminium methylsulfate (**213**) is treated with sodium halides, the halide anions are alkylated exclusively (71). This

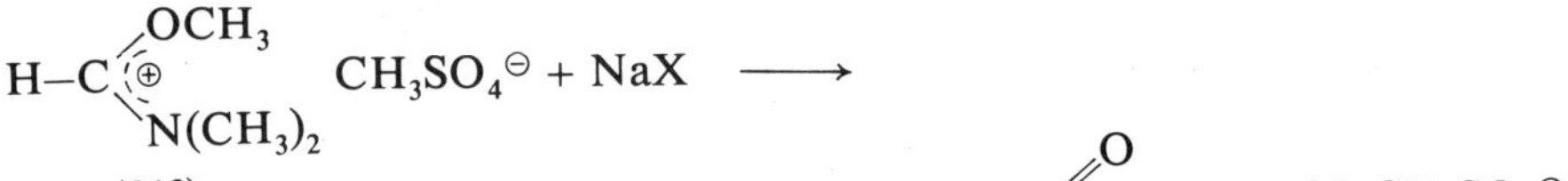

$$\underset{(213)}{H\!-\!\overset{OCH_3}{\underset{N(CH_3)_2}{\overset{\oplus}{C}\!\!\!<}}}\ CH_3SO_4^{\ominus} + NaX \longrightarrow CH_3X + H\!-\!C\!\!\overset{O}{\underset{N(CH_3)_2}{<}} + NaCH_3SO_4^{\ominus}$$

X = Cl, Br, I

reaction stresses the ambident character of the N,N-dialkylalkoxymethyleniminium ions (71, 143). The halide anion adds to the formyl carbonyl group in a kinetically controlled reaction. However, the bond formed is labile owing to the low nucleophilicity of the halide ion and it therefore easily dissociates. This is followed by nucleophilic substitution, leading to alkyl halides.

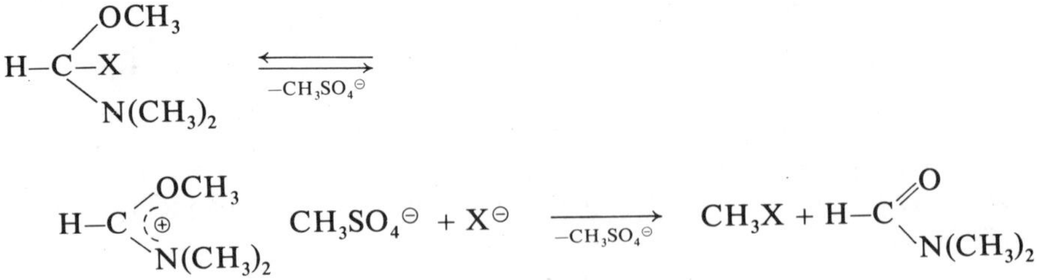

$$H\!-\!\overset{OCH_3}{\underset{N(CH_3)_2}{\overset{}{C}\!\!-\!\!X}} \underset{-CH_3SO_4^{\ominus}}{\rightleftarrows}$$

$$H\!-\!\overset{OCH_3}{\underset{N(CH_3)_2}{\overset{\oplus}{C}\!\!\!<}}\ CH_3SO_4^{\ominus} + X^{\ominus} \xrightarrow[-CH_3SO_4^{\ominus}]{} CH_3X + H\!-\!C\!\!\overset{O}{\underset{N(CH_3)_2}{<}}$$

III. ALKOXYMETHYLENIMINIUM SALT—NUCLEOPHILE REACTIONS

Owing to the higher nucleophilicity of cyanide ion, the reactions of N,N-dialkylalkoxymethyleniminium alkylsulfates and fluoroborates with alkali metal cyanides take a different route. In this case irreversible addition to the acyl carbon atom occurs in a kinetically controlled reaction giving rise to O,N-acetals of acyl cyanides (226) (86, 140, 141). The reactions may be performed in water, if the products are extracted with apolar solvents (benzene, cyclohexane) (86, 140). These reactions are also successful in acetonitrile (141).

$$R-C{\overset{OR^3}{\underset{NR^1R^2}{\oplus}}} \; X^\ominus + NaCN \xrightarrow{-NaX} R-C{\overset{OR^3}{\underset{NR^1R^2}{-CN}}}$$

(226)

Acyl cyanide O,N-acetals (226) generated by this method are summarized in Table XIX.

TABLE XIX

Acyl Cyanide O,N-Acetals from Dialkylalkoxymethyleniminium Salts and Alkali Cyanides

$$R-C{\overset{OR^3}{\underset{NR^1R^2}{-CN}}}$$

R	R^1	R^2	R^3	Yield, %	b.p./pressure, °C/torr	n_D^{20}	Ref.
H	CH_3	CH_3	CH_3	68	63–65/25	1.4110	86
H	CH_3	CH_3	C_2H_5	56.4	60/15	1.4141	141
H	C_2H_5	C_2H_5	CH_3	65	82–83/20	1.4233	86
H	$-(CH_2)_4-$		CH_3	50	99/19	1.4547	140
H	$-(CH_2)_5-$		CH_3	66.8	103–105/15	1.4580	140
H	$-(CH_2)_2-O-(CH_2)_2-$		CH_3	45.8	128/1	1.4666	140
CH_3	CH_3	CH_3	CH_3	70	48–51/15	1.4193	86
CH_3	CH_3	CH_3	C_2H_5	63	56/15	1.4200/25	141
C_2H_5	CH_3	CH_3	CH_3	73	58–60/12	1.4352	86
$-(CH_2)_3-$		CH_3	CH_3	66	72/12	1.4475	86

If excess sodium cyanide acts on N,N-dialkylethoxymethyleniminium fluoroborates (227) dissolved in acetonitrile for a prolonged period, two cyano groups are introduced and aminomalononitrile derivatives (228) (141) are thus formed.

$$R-C{\overset{OC_2H_5}{\underset{N(CH_3)_2}{\oplus}}} \; BF_4^\ominus \xrightarrow{NaCN} RC{\overset{CN}{\underset{N(CH_3)_2}{-CN}}}$$

(227)   (228)

H. REACTIONS WITH COMPLEX HYDRIDES

Borch (38) alkylated a series of *N*-mono- and *N,N*-disubstituted amides with the aid of triethyloxonium fluoroborate in methylene chloride.

The ethoxymethyleniminium fluoroborates **229** were reduced with sodium borohydride in absolute alcohol to the amines **230** without isolation; there are no isolable intermediates. The amines **230** are obtained in yields of 81–92%.

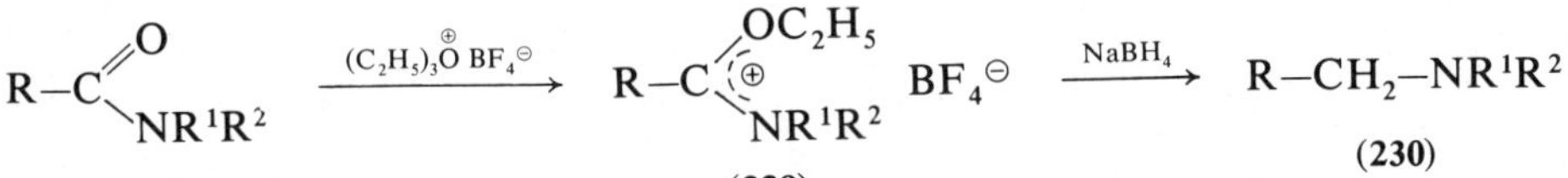

$$R-C(=O)NR^1R^2 \xrightarrow{(C_2H_5)_3\overset{\oplus}{O} BF_4^{\ominus}} \underset{(229)}{R-C(OC_2H_5)(\overset{\oplus}{})NR^1R^2} BF_4^{\ominus} \xrightarrow{NaBH_4} \underset{(230)}{R-CH_2-NR^1R^2}$$

Lactams are reduced the same way (38). Thus from ε-caprolactam hexamethylenimine (**231**) is formed in 92% yield.

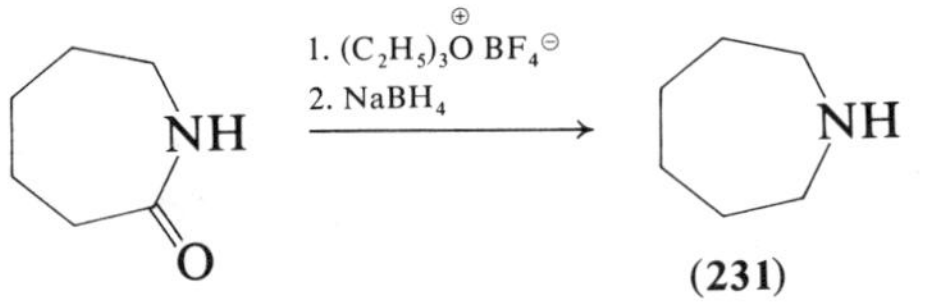

[caprolactam] $\xrightarrow[\text{2. NaBH}_4]{\text{1. }(C_2H_5)_3\overset{\oplus}{O} BF_4^{\ominus}}$ [hexamethylenimine] (**231**)

With NaBH$_4$, iminium salts like **232** react more readily than carboxyl groups, a reaction that can be taken advantage of in a simple synthesis of proline (**233**) (204).

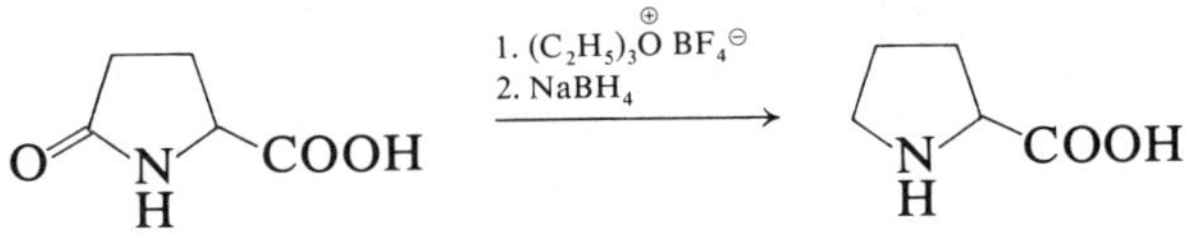

[structure **232**: 5-oxoproline] $\xrightarrow[\text{2. NaBH}_4]{\text{1. }(C_2H_5)_3\overset{\oplus}{O} BF_4^{\ominus}}$ [structure **233**: proline]

Primary amides are dehydrated to nitriles under the same reaction conditions (38):

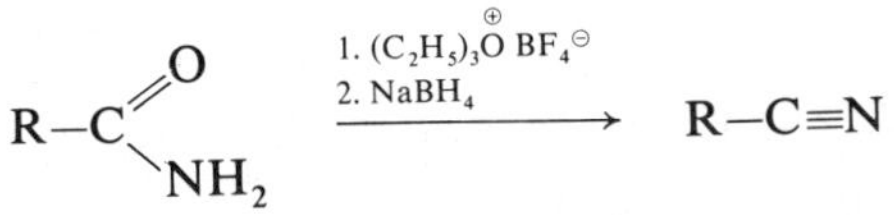

$$R-C(=O)NH_2 \xrightarrow[\text{2. NaBH}_4]{\text{1. }(C_2H_5)_3\overset{\oplus}{O} BF_4^{\ominus}} R-C\equiv N$$

Alkoxymethyleniminium fluoroborates (**234**) synthesized from nitrilium salts and alcohols are reduced with sodium borohydride to yield amines (138).

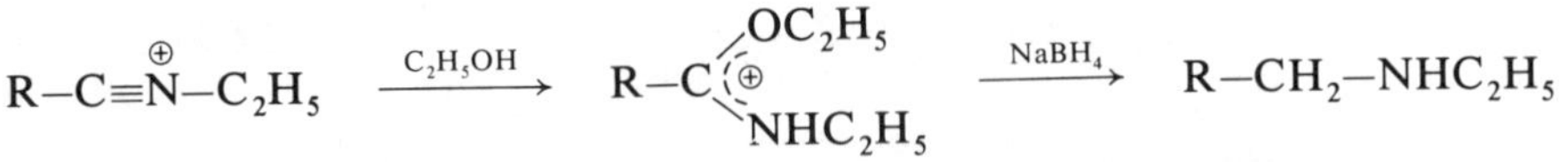

$$R-C\equiv\overset{\oplus}{N}-C_2H_5 \xrightarrow{C_2H_5OH} \underset{(234)}{R-C(OC_2H_5)(\overset{\oplus}{})NHC_2H_5} \xrightarrow{NaBH_4} R-CH_2-NHC_2H_5$$

III. ALKOXYMETHYLENIMINIUM SALT—NUCLEOPHILE REACTIONS

I. REACTIONS WITH ALKALI METAL PHOSPHIDES

Lithium diorganophosphides (**235**) will react with N,N-dimethylmethoxymethyleniminium methylsulfate (**213**) to give bis(diorganophosphino)-dimethylaminomethanes (**236**) (205). The reaction proceeds stepwise; initially the phosphide anion adds to the formyl carbon to yield (dimethylamino)-(diorganophosphino)methoxymethane (**237**), which reacts further with more phosphide to give bis(diorganophosphino)(dimethylamino)methane (**236**);

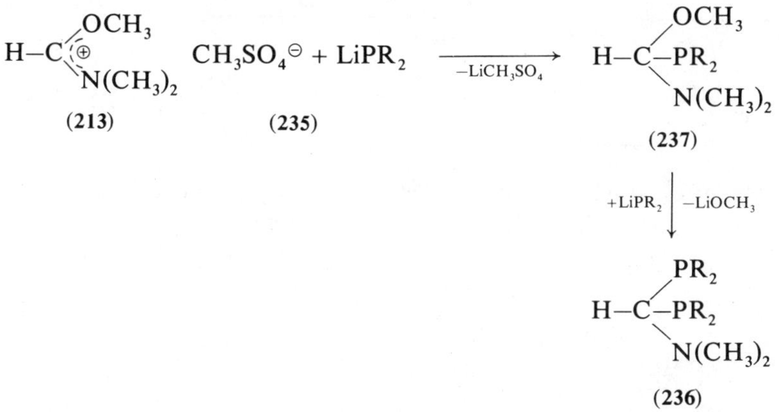

$$H-C\underset{N(CH_3)_2}{\overset{OCH_3}{\oplus}} \; CH_3SO_4^{\ominus} + LiPR_2 \xrightarrow[-LiCH_3SO_4]{} H-\underset{N(CH_3)_2}{\overset{OCH_3}{C-PR_2}}$$
(**213**) (**235**) (**237**)

$$\downarrow {+LiPR_2 \;\; -LiOCH_3}$$

$$H-\underset{N(CH_3)_2}{\overset{PR_2}{C-PR_2}}$$
(**236**)

lithium methoxide is eliminated. To a lesser extent dealkylation of the iminium salt **213** by the phosphide occurs, thus generating tertiary phosphines **238**.

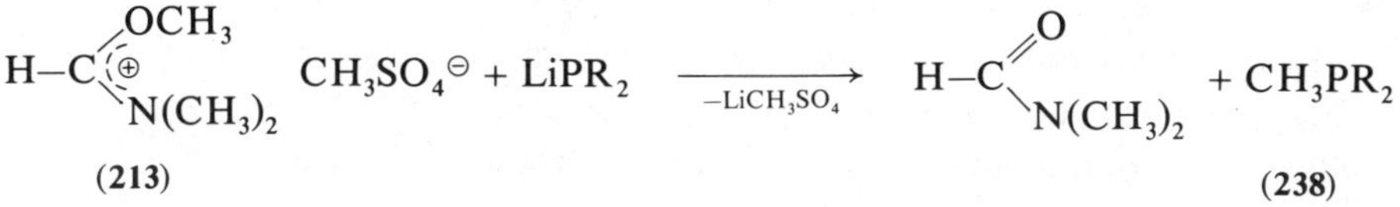

$$H-C\underset{N(CH_3)_2}{\overset{OCH_3}{\oplus}} \; CH_3SO_4^{\ominus} + LiPR_2 \xrightarrow[-LiCH_3SO_4]{} H-C\underset{N(CH_3)_2}{\overset{O}{\diagdown}} + CH_3PR_2$$
(**213**) (**238**)

(Dimethylamino)(diethylphosphino)methoxymethane (**239**) can be isolated only in an impure state, when lithium diethylphosphide is reacted with N,N-dimethylmethoxymethyleniminium methylsulfate in ether at low temperatures (205):

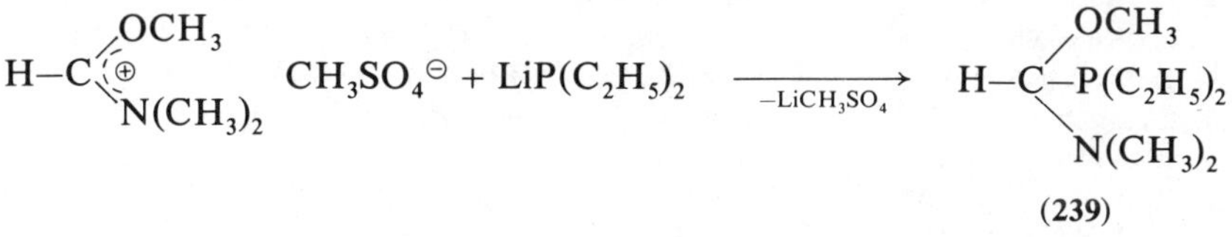

$$H-C\underset{N(CH_3)_2}{\overset{OCH_3}{\oplus}} \; CH_3SO_4^{\ominus} + LiP(C_2H_5)_2 \xrightarrow[-LiCH_3SO_4]{} H-\underset{N(CH_3)_2}{\overset{OCH_3}{C-P(C_2H_5)_2}}$$
(**213**) (**239**)

The properties of bis(diorganophosphino)(dimethylamino)methanes (**236**) are summarized in Table XX.

TABLE XX

Bis(diorganophosphino)(dimethylamino)methanes from *N,N*-Dimethylmethoxymethyleniminium Methylsulfate and Lithium Diorganophosphides

H–C(PR$_2$)(PR$_2$)N(CH$_3$)$_2$ R	Yield, %	b.p./pressure, °C/torr	m.p., °C	Ref.
CH$_3$	49	85–87/12	—	205
C$_2$H$_5$	52	119–120.5/6	—	205
n-C$_4$H$_9$	54	121–122/0.02	—	205
C$_6$H$_5$	57	—	82–84	205

Sodium diethylphosphite reacts with *N,N*-dimethylmethoxymethyleniminium methylsulfate (**213**) to yield the phosphite **240** (206).

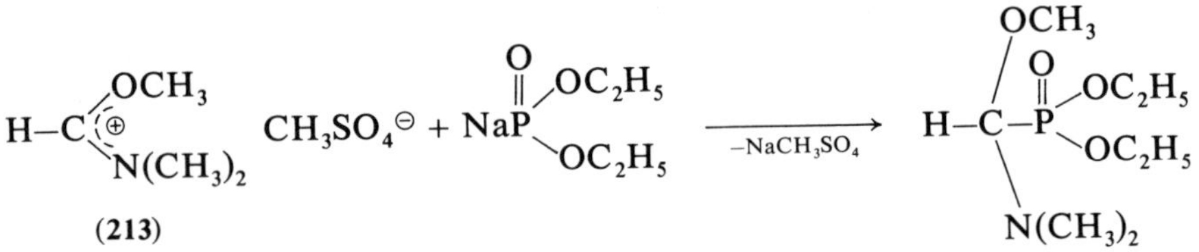

$$H-C^{\oplus}(OCH_3)(N(CH_3)_2) \cdot CH_3SO_4^{\ominus} + NaP(=O)(OC_2H_5)(OC_2H_5) \xrightarrow{-NaCH_3SO_4} H-C(OCH_3)(N(CH_3)_2)-P(=O)(OC_2H_5)(OC_2H_5)$$

(213) → (240)

ACKNOWLEDGMENT

To Dr. Heinz Eilingsfeld BASF AG, Ludwigshafen, and to Dr. Bernd Funke as well as Dr. Erwin Haug, both of Fachhochschule Aalen, Aalen, I am very indebted for fruitful discussions and critical revisions of the text.

REFERENCES

1. M. B. Robin, F. A. Bovey, and H. Basch, *The Chemistry of Amides*, J. Zabicky, Ed., John Wiley & Sons, New York, 1970.
2. R. Huisgen and H. Brade, *Chem. Ber.*, **90**, 1432 (1957).
3. R. Huisgen, H. Brade, H. Walz, and I. Glogger, *Chem. Ber.*, **90**, 1437 (1957).
4. G. Fraenkel and C. Niemann, *Proc. Natl. Acad. Sci. U.S.*, **44**, 688 (1958).
5. D. M. Brouwer and J. A. van Doorn, *Tetrahedron Letters*, **1971**, 3339.
6. G. A. Olah, A. N. White, and D. H. O'Brieu, *Chem. Rev.*, **70**, 580 (1970) and references cited therein.
7. R. A. McClelland and W. F. Reynolds, *J. Chem. Soc. Chem. Commun.*, **1974**, 824.
8. E. H. White, *J. Am. Chem. Soc.*, **77**, 6215 (1955).
9. R. Gompper and P. Altreuther, *Z. Anal. Chem.*, **170**, 205 (1959).
10. E. Allenstein and A. Schmidt, *Z. Anorg. Allgem. Chem.*, **344**, 113 (1966).
11. S. S. Hecht and E. S. Rothman, *J. Org. Chem.*, **38**, 395 (1973).

12. H. Meerwein, E. Battenberg, H. Gold, E. Pfeil and G. Willfang, *J. Prakt. Chem.*, **154**, 83 (1939).
13. H. Meerwein, P. Borner, O. Fuchs, H. J. Sasse, H. Schrodt, and J. Spille, *Chem. Ber.*, **89**, 2060 (1956).
14. H. Meerwein, W. Florian, N. Schön, and G. Stopp, *Ann. Chem.*, **641**, 1 (1961).
15. H. Bredereck, W. Kantlehner and D. Schweizer, *Chem. Ber.*, **104**, 3475 (1971).
16. P. Gross, Dissertation, University of Stuttgart, 1973.
17. H. G. Nordmann and F. Kröhnke, *Angew. Chem.*, **81**, 1004 (1969).
18. R. Kreher, A. Bauer, and H. Hennige, *Z. Naturforsch. B*, **29**, 231 (1974).
19. M. Mori, Y. Ban, and T. Oishi, *Intern. J. Sulfur Chem. A*, **2**, 79 (1972).
20. H. D. Gutbrod, Dissertation, University of Stuttgart, 1973.
21. U. Dinkeldein, Dissertation, University of Stuttgart, 1973.
22. Y. Ban, M. Kimura, and T. Oishi, *Heterocycles (Sendai Japan)*, **2**, 323 (1974); *Cheminform* **1974**, 46–438, p. 196.
23. R. F. Borch, C. V. Grudzinskas, D. A. Peterson, and L. D. Weber, *J. Org. Chem.*, **37**, 1141 (1972).
24. N. Kornblum and G. P. Coffey, *J. Org. Chem.*, **31**, 3447 (1966).
25. P. C. Unags and P. L. Southwick, *J. Heterocycl. Chem.*, **10**, 399 (1973).
26. H. v. Dobeneck and A. Uhl, *Ann. Chem.*, **1974**, 1550.
27. U. Mueller-Westerhoff, *Tetrahedron Letters*, **1972**, 4639.
28. W. Betz and J. Daub, *Angew. Chem.*, **83**, 289 (1971).
29. J. Daub and W. Betz, *Tetrahedron Letters*, **1972**, 3451.
30. W. Betz and J. Daub, *Chem. Ber.*, **107**, 2095 (1974).
31. K. Hafner, K. F. Bangert, and V. Orfanos, *Angew. Chem.*, **79**, 414 (1967).
32. H. Bredereck and K. Bredereck, *Chem. Ber.*, **94**, 2278 (1961).
33. H. E. Sprenger and W. Ziegenbein, *Angew. Chem.*, **80**, 541 (1968).
34. G. Seitz and H. Morck, *Chimia*, **26**, 368 (1972).
35. U. Schöllkopf and F. Gerhart, *Angew. Chem.*, **79**, 990 (1967); *Angew. Chem. Intern. Ed. Engl.*, **6**, 970 (1967).
36. T. Kato, A. Takada, and T. Ueda, *Chem. Pharm. Bull.*, **20**, 901 (1972); *C.A.*, **77**, 101514r (1972).
37. T. Kato, A. Takada, and T. Ueda, *Chem. Pharm. Bull.*, **22**, 984 (1974).
38. R. F. Borch, *Tetrahedron Letters*, **1968**, 61.
39. H. Ahlbrecht and C. Vonderheit, *Chem. Ber.*, **106**, 2009 (1973).
40. R. M. Moriarty, C. L. Yeh, K. C. Ramey, and P. W. Whitehurst, *J. Am. Chem. Soc.*, **92**, 6360 (1970).
41. L. Weintraub, S. R. Oles, and N. Kalish, *J. Org. Chem.*, **33**, 1679 (1968).
42. A. Piloti, A. Reuterhall, and K. Torsell, *Acta Chem. Scand.*, **23**, 818 (1969).
43. H. Meerwein, K. Bodenbrenner, P. Borner, F. Kunert, and K. Wunderlich, *Ann. Chem.*, **632**, 38 (1960).
44. J. C. Sheehan and M. Mehdi Nafissi, *J. Org. Chem.*, **36**, 4246 (1970).
45. T. Oishi, H. Nakakimura, M. Mori, and Y. Ban, *Chem. Pharm. Bull.*, **20**, 1735 (1972).
46. R. G. Glushkov, V. G. Granik, V. A. Volskova, V. A. Chernov, and S. M. Minakova, *Khim. Farm. Zh.*, **7** 14 (1973); *Cheminform*, **1974**, 7–299, p. 125.
47. V. G. Granik, A. M. Zhidkova, N. S. Kuryatov, V. P. Pakhomov, and R. G. Glushkov, *Khim. Geterotsikl. Soedin.* **1973**, 1532; *Cheminform*, **1974**, 10–298.
48. V. G. Granik, B. M. Paytin, J. V. Persianova, E. M. Peresleni, N. P. Kostyuclenko, R. G. Glushkov, and Y. N. Sheinker, *Tetrahedron*, **26**, 4367 (1970).
49. T. Oishi, M. Nagaı, T. Onuma, R. Moriyama, K. Tsutae, M. Ochiai, and Y. Ban, *Chem. Pharm. Bull.*, **17**, 2306 (1969).
50. D. Bormann, *Ann. Chem.*, **725**, 124 (1969).

51. B. M. Paytin, V. G. Granik, and R. G. Glushkov, *Pharm. Chem. J.*, **1971**, 683.
52. T. Jen, B. Dienel, F. Dowalo, H. Van Hoeven, P. Bender, and B. Loer, *J. Med. Chem.*, **16**, 633 (1973).
53. R. Kreher and H. Hennige, *Tetrahedron Letters*, **1973**, 1911.
54. R. Kreher and H. Hennige, *Tetrahedron Letters*, **1969**, 4695.
55. K. Hafner and F. Schmidt, *Angew. Chem.*, **85**, 450 (1973).
56. L. A. Paquette, L. B. Anderson, J. F. Hansen, S. A. Lang, Jr., and H. Berk, *J. Am. Chem. Soc.*, **94**, 4907 (1972).
57. R. Neidlein and M. Höhle, *Pharm. Ztg.*, **119**, 1651 (1974).
58. L. Kienitz, Dissertation, University of Stuttgart, 1973.
59. S. Kabuss, *Angew. Chem.*, **78**, 714 (1966).
60. A. Bühner, *Ann. Chem.*, **333**, 289 (1904).
61. H. Bredereck, R. Gompper, H. Rempfer, H. Keck, and K. Klemm, *Angew. Chem.*, **70**, 269 (1958).
62. H. Bredereck, R. Gompper, H. Rempfer, K. Klemm, and H. Keck, *Chem. Ber.*, **92**, 329 (1959).
63. H. Bredereck, R. Gompper, K. Klemm, and H. Rempfer, *Chem. Ber.*, **92**, 837 (1959).
64. H. Bredereck, F. Effenberger, and G. Simchen, *Angew. Chem.*, **73**, 493 (1961).
65. H. Bredereck, F. Effenberger, and G. Simchen, *Chem. Ber.*, **96**, 1350 (1963).
66. H. Bredereck, F. Effenberger, and E. Henseleit, *Angew. Chem.*, **75**, 790 (1963).
67. H. Bredereck, F. Effenberger, and E. Henseleit, *Chem. Ber.*, **98**, 2754 (1965).
68. H. Bredereck, F. Effenberger, and H. P. Beyerlin, *Chem. Ber.*, **97**, 3076 (1964).
69. H. Jaus, Dissertation, University of Stuttgart, 1970.
70. P. Schlack and W. Richter, *Chem. Ber.*, **103**, 3729 (1970).
71. H. Bredereck, G. Simchen, S. Rebsdat, W. Kantlehner, P. Horn, R. Wahl, H. Hoffmann, and P. Grieshaber, *Chem. Ber.*, **101**, 41 (1968).
72. H.-D. Gutbrod, Dissertation, University of Stuttgart, 1971.
73. R. E. Benson and T. L. Cairns, *J. Am. Chem. Soc.*, **70**, 2115 (1948).
74. R. E. Benson and T. L. Cairns, *Org. Synth.*, **31**, 72 (1951).
75. W. Z. Heldt, *J. Am. Chem. Soc.*, **80**, 5880 (1958).
76. H. Behringer and H. Meier, *Ann. Chem.*, **607**, 73 (1957).
77. S. Petersen and E. Tietze, *Chem. Ber.*, **90**, 909 (1957).
78. J. Körösi, *J. Prakt. Chem.*, **23** [4], 212 (1964).
79. K. H. Büchel, A. K. Bocz, and F. Korte, *Chem. Ber.*, **99**, 724 (1966).
80. S. W. Breuer and D. Ginsburg, *Israel J. Chem.*, **4**, 145 (1966); *C.A.*, **66**, 65382h (1967).
81. A. Etienne and Y. Correira, *Bull. Soc. Chim. France*, **1969**, 3704.
82. H. Lüssi, *Chimia*, **27**, 65 (1973).
83. V. G. Granik, M. K. Polievktov, and R. G. Glushkov, *Zh. Org. Khim.*, **7**, 1480 (1971).
84. V. G. Granik, M. K. Polievktov, and R. G. Gluskov, *Zh. Org. Khim.*, **7**, 1431 (1971); *C.A.*, **75**, 129299g (1971).
85. R. G. Gluskov, *Khim. Goterotsikl. Soedin*, **1973**, 954; *Cheminform*, **1973**, 43–299, p. 117.
86. H. Bredereck, G. Simchen and W. Kantlehner, *Chem. Ber.*, **104**, 924 (1971).
87. E. A. Werner, *J. Chem. Soc.*, **105**, 927 (1914).
88. P. A. Ongley, *Trans. Roy. Soc. New Zealand*, **77**, 10 (1948); *C.A.*, **42**, 8165 (1948).
89. J. W. Janus, *J. Chem. Soc.*, **1955**, 3551.
90. W. L. Hughes, H. A. Saroff, and A. L. Carney, *J. Am. Chem. Soc.*, **71**, 2476 (1949).
91. D. J. Brown and E. Hoerger, *J. Appl. Chem.*, **4**, 284 (1954).
92. H. Bredereck, F. Effenberger, and H. P. Beyerlin, *Chem. Ber.*, **97**, 1834 (1964).
93. H. Bredereck, R. Gompper, and G. Theilig, *Chem. Ber.*, **87**, 537 (1954).
94. H. Bredereck, R. Gompper, and D. Bitzer, *Chem. Ber.*, **92**, 1139 (1959).
95. R. Gompper and O. Christmann, *Chem. Ber.*, **92**, 1935 (1959).

96. H. Bredereck, R. Gompper, H. G. v. Schuh, and G. Theilig, *Angew. Chem.*, **71**, 753 (1959).
97. N. Kornblum and R. K. Blackwood, *J. Am. Chem. Soc.*, **78**, 4037 (1956).
98. S. D. Ross and M. M. Labes, *J. Am. Chem. Soc.*, **79**, 4155 (1957).
99. J. L. Neumeyer and J. G. Cannon, *J. Org. Chem.*, **26**, 4681 (1961).
100. M. G. Ahmed, R. W. Alder, G. H. James, M. L. Sinnott, and M. C. Whiting, *Chem. Commun.*, **1968**, 1533.
101. J. V. Paukstelis and M. Kim, *J. Org. Chem.*, **39**, 1494 (1974).
102. A. De Roocker and P. de Radzitzki, *Bull. Soc. Chim. Belges*, **79**, 531 (1970).
103. S. Masson, A. Tullier, and D. Villemin, *Compt. Rend. C*, **274**, 2092 (1972).
104. M. C. Lasne, S. Masson, and A. Thullier, *Bull. Soc. Chim. France*, **1973**, 1751.
105. S. Takemura, H. Niczato, and Y. Ueno, *Chem. Pharm. Bull.*, **19**, 1606 (1971).
106. M. G. Ahmed and R. W. Alder, *Chem. Commun.*, **1969**, 1389.
107. W. Seeliger, E. Aufderhaar, W. Diepers, R. Feinauer, R. Nehring, W. Thier, and H. Hellmann, *Angew. Chem.*, **78**, 913 (1966).
108. D. A. Tomalia and J. N. Paige, *J. Org. Chem.*, **38**, 422 (1973).
109. W. Hechelhammer, Ger. Pat. 948,973 (1956); *C.A.*, **53**, 6088 (1959).
110. F. H. Suydam, W. E. Greth, and N. R. Langerman, *J. Org. Chem.*, **34**, 292 (1969).
111. A. Pinner and F. Klein, *Ber. Deut. Chem. Ges.*, **10**, 1889 (1877).
112. K. Ikawa, F. Takami, Y. Fukui, and K. Tokuyama, *Tetrahedron Letters*, **1969**, 3279.
113. M. Itoh, *Chem. Pharm. Bull. (Tokyo)*, **18**, 784 (1970).
114. R. R. Koganty, M. B. Shambhu, and G. A. Digenis, *Tetrahedron Letters*, **1973**, 4511.
115. V. A. Pattison, J. G. Colson, and R. L. K. Carr, *J. Org. Chem.*, **33**, 1084 (1968).
116. R. Ohme and E. Schmitz, *Angew. Chem.*, **79**, 531 (1967).
117. H. L. Wheeler and T. B. Johnson, *Ber. Deut. Chem. Ges.*, **32**, 35 (1899).
118. H. L. Wheeler and T. B. Johnson, *J. Am. Chem. Soc.*, **23**, 135 (1900).
119. A. E. Arbuzov and V. E. Shishkin, *Dokl. Akad. Nauk. SSSR*, **141**, 349 (1961); *C.A.*, **56**, 11491.
120. A. Koda, K. Tokanobu, J. Isaka, T. Kashiwagi, K. Takahashi, S. Kawahara, and M. Murakami, *Yakugaku Zasshi*, **92**, 459 (1972); *C.A.*, **77**, 34411g (1972).
121. R. Roger and D. G. Neilson, *Chem. Rev.*, **61**, 179 (1961).
122. B. M. Paytin and R. G. Glushkov, *Pharm. Chem. J.*, **1969**, 256.
123. R. G. Glushkov, V. G. Smirnova, K. A. Zaitsewa, N. A. Novitskaya, M. D. Mashkovskii, and G. N. Pershin, *Khim. Farm. Zh.*, **8**, 14 (1974); *Cheminform*, **1974**, 25–341, p. 133.
124. H. Eilingsfeld, M. Seefelder, and H. Weidinger, *Angew. Chem.*, **72**, 836 (1960).
125. H. Eilingsfeld, M. Seefelder, and H. Weidinger, *Chem. Ber.*, **96**, 2671 (1963).
126. Z. Arnold and A. Holy, *Collection Czech. Chem. Commun.*, **27**, 2886 (1962).
127. E. Allenstein and A. Schmidt, *Z. Anorg. Allgem. Chem.*, **344**, 113 (1966).
128. E. Allenstein and A. Schmidt, *Chem. Ber.*, **97**, 1863 (1964).
129. E. Allenstein and P. Quis, *Chem. Ber.*, **97**, 1857 (1964).
130. Z. Arnold, *Collection Czech. Chem. Commun.*, **26**, 1723 (1961).
131. D. R. Hepburne and H. R. Hudson, *Chem. Ind., London*, **1974**, 664.
132. G. Ege and H. O. Frey, *Tetrahedron Letters*, **41**, 4217 (1972).
133. H. Meerwein, V. Hederich, H. Morschel, and K. Wunderlich, *Ann. Chem.*, **635**, 1 (1960).
134. E. Schaumann, S. Sieveking, and W. Walter, *Chem. Ber.*, **107**, 3589 (1974).
135. A. Pinner, *Die Iminoäther und ihre Derivate*, Verlag R. Oppenheim, Berlin, 1892.
136. H. Meerwein, Houben-Weyl, *Methoden der Organischen Chemie*, Vol. VI/3, Georg Thieme Verlag, Stuttgart, 1965, p. 300.
136a. S. Weiss and H. Krommer, *Chemiker-Ztg.*, **98**, 617 (1974).
137. H. Meerwein, P. Laasch, R. Mersch, and J. Spille, *Chem. Ber.*, **89**, 209 (1956).
138. R. F. Borch, *J. Org. Chem.*, **34**, 627 (1968).

139. J. E. Gordon and G. C. Turell, *J. Org. Chem.*, **24**, 269 (1959).
140. W. Kantlehner and P. Speh, *Chem. Ber.*, **105**, 1340 (1972).
141. H. Plieninger, Ramadan El-Berins, and Heduck Mah, *Chem. Ber.*, **104**, 3983 (1971).
142. T. D. Smith, *J. Chem. Soc. A*, **1966**, 841.
143. S. Hünig, *Angew. Chem.*, **76**, 400 (1964); *Angew. Chem. Intern. Ed. Engl.*, **3**, 548 (1964).
144. H. Bredereck, F. Effenberger, and G. Simchen, *Angew. Chem.*, **74**, 353 (1962); *Angew. Chem. Intern. Ed. Engl.*, **1**, 331 (1962).
145. T. Okuyama, T. C. Pletcher, D. J. Sahn, and G. L. Schmir, *J. Am. Chem. Soc.*, **95**, 1253 (1973).
146. T. Okayama, D. J. Sahn, and G. L. Schmir, *J. Am. Chem. Soc.*, **95**, 2345 (1973).
147. R. A. McClelland, *J. Am. Chem. Soc.*, **96**, 3690 (1974).
148. P. Deslongchamps, C. Lebreux, and R. Taillefer, *Can. J. Chem.*, **51**, 1665 (1973).
149. S. Hanessian, *Tetrahedron Letters*, **1967**, 1549.
150. H. Muxfeldt and W. Rogalski, *J. Am. Chem. Soc.*, **87**, 933 (1965).
151. S. R. Sandler and W. Karo, *Organic Functional Group Preparations*, Vol. II, Academic Press, New York, 1971.
152. R. H. De Wolfe, *Carboxylic Ortho Acid Derivatives*, Academic Press, New York, 1970.
153. R. H. De Wolfe, *Synthesis*, **1974**, 153.
154. C. L. Schilling, Jr., *J. Org. Chem.*, **39**, 1770 (1974).
155. W. Ried and E. Schmidt, *Ann. Chem.*, **695**, 217 (1966).
156. H. Muxfeldt, H. Döpp, J. E. Kaufman, J. Schneider, P. E. Hansen, A. Sasaki, and T. Geiser, *Angew. Chem.*, **85**, 508 (1973).
157. H. Bredereck, G. Simchen, and W. Kantlehner, *Chem. Ber.*, **104**, 932 (1971).
158. W. Kantlehner, H.-D. Gutbrod, and P. Gross, *Ann. Chem.*, **1974**, 690.
159. W. Kantlehner and B. Funke, *Chem. Ber.*, **104**, 3711 (1971).
160. A. I. Ivanova, B. P. Fedorov, and F. M. Stojanovic, *Izv. Akad. Nauk. SSSR*, **1965**, 2179; *C.A.*, **64**, 12538d (1966).
161. H. Bredereck, G. Simchen, and M. Hoffmann, *Chem. Ber.*, **106**, 3725 (1973).
162. H. Hartmann, *Z. Chem.*, **11**, 60 (1971).
163. H. Bredereck, F. Effenberger, and H. P. Beyerlin, *Chem. Ber.*, **97**, 3081 (1964).
164. H. Winberg, U.S. Pat. 3,239,534 (1966); *C.A.*, **64**, 17425 (1966).
165. H. Bredereck, F. Effenberger, and G. Simchen, *Chem. Ber.*, **98**, 1078 (1965).
166. H. Bredereck, F. Effenberger, and E. Henseleit, *Chem. Ber.*, **98**, 2887 (1965).
167. H. Bredereck, F. Effenberger, T. Brendle, and H. Muffler, *Chem. Ber.*, **101**, 1885 (1968).
168. K. Seckinger, *Helv. Chim. Acta*, **56**, 776 (1973).
169. R. D. Westland, M. M. Merz, S. M. Alexander, L. S. Newton, L. Bauer, T. T. Conway, J. M. Barton, K. K. Khullar, P. B. Devdhar, and M. M. Grenan, *J. Med. Chem.*, **15**, 1313 (1972).
170. H. Bredereck, G. Simchen, and H. U. Schenck, *Chem. Ber.*, **101**, 3058 (1968).
171. H. Bredereck, G. Simchen, and G. Beck, *Ann. Chem.*, **762**, 62 (1972).
171a. K. Wegner and W. Schunack, *Arch. Pharm. Weinheim*, **307**, 972 (1974); *Cheminform*, **1975**, 14–227.
171b. T. Wang, *J. Org. Chem.*, **39**, 3591 (1974).
172. W. Lossen, *Ann. Chem.*, **265**, 136 (1891).
173. W. Lwowski, *Synthesis*, **1971**, 263.
174. H. Weidinger and J. Kranz, *Chem. Ber.*, **96**, 1049 (1963).
175. H. Weidinger and J. Kranz, *Chem. Ber.*, **96**, 1059 (1963).
176. H. Weidinger and J. Kranz, *Chem. Ber.*, **96**, 1064 (1963).
177. H. Meindl and H. Ackermann, *Helv. Chim. Acta*, **55**, 1039 (1972).
178. H. Ahlbrecht and C. Vonderheid, *Chem. Ber.*, **106**, 2009 (1973).

REFERENCES

179. S. R. Sandler and W. Karo, *Organic Functional Group Preparations*, Vol. III, Academic Press, New York, 1972.
180. A. E. Wick, P. A. Bartlett, and D. Dolphin, *Helv. Chim. Acta*, **54**, 513 (1971).
181. T. Oishi, M. Ochiai, N. Nagai, and Y. Ban, *Tetrahedron Letters*, **1968**, 497.
182. M. Bertele, H. Boos, J. D. Dunitz, F. Elsinger, A. Eschenmoser, J. Felner, H. P. Gribi, H. Gschwend, E. F. Meyer, M. Pesaro, and R. Scheffold, *Angew. Chem.*, **76**, 393 (1964); *Angew. Chem. Intern. Ed. Engl.*, **3**, 490 (1964).
183. S. Petersen and E. Tietze, *Ann. Chem.*, **623**, 133 (1959).
184. L. A. Paquette, *J. Am. Chem. Soc.*, **86**, 4096 (1964).
185. J. Felner, A. Tischli, A. Wick, M. Pesaro, D. Bormann, E. L. Winnacker, and A. Eschenmoser, *Angew. Chem.*, **79**, 863 (1967); *Angew. Chem. Intern. Ed. Engl.*, **6**, 864 (1967).
186. Y. Yamada, D. Miljkovic, P. Wehrli, B. Golding, P. Löliger, R. Keese, K. Müller, and A. Eschenmoser, *Angew. Chem.*, **81**, 301 (1969); *Angew. Chem. Intern. Ed. Engl.*, **8**, 343 (1969).
187. L. A. Paquette and T. Kakihana, *J. Am. Chem. Soc.*, **90**, 3897 (1968).
188. L. A. Paquette and T. J. Barton, *J. Am. Chem. Soc.*, **89**, 5480 (1967).
189. P. Wegner, *Tetrahedron Letters*, **1967**, 4985.
190. R. G. Glushkov and V. G. Granik, *Advances in Heterocyclic Chemistry*, Vol. 12, Academic Press, New York, 1970, p. 185.
191. G. Simchen and W. Kantlehner, *Tetrahedron Letters*, **28**, 3535 (1972).
192. H. A. Staab, H. Irngartner, A. Mannschreck, and Mon-Tai Wu, *Ann. Chem.*, **659**, 55 (1966).
193. P. Haake and W. B. Miller, *J. Am. Chem. Soc.*, **85**, 4044 (1963).
194. R. A. Olofson and J. M. Landsberg, *J. Am. Chem. Soc.*, **88**, 4263 (1966).
195. R. A. Olofson, J. M. Landsberg, K. N. Honk, and J. S. Michelmann, *J. Am. Chem. Soc.*, **88**, 4264 (1966).
196. R. A. Olofson, W. R. Thompson, and J. S. Michelmann, *J. Am. Chem. Soc.*, **86**, 1865 (1964).
197. J. Hocker and R. Merten, *Ann. Chem.*, **751**, 145 (1971).
198. T. Oishi, M. Ochiai, T. Nakayama, and Y. Ban, *Chem. Pharm. Bull.*, **17**, 2314 (1969).
199. Z. Arnold and M. Kornilov, *Collection Czech. Chem. Commun.*, **29**, 645 (1964).
200. H. Brechbühler, H. Büchi, E. Hatz, J. Schreiber, and A. Eschenmoser, *Helv. Chim. Acta*, **48**, 1746 (1965).
201. C. Feugeas and D. Olschwang, *Bull. Soc. Chim. France*, **1968**, 4985.
202. W. Kantlehner, Dissertation, University of Stuttgart, 1968.
203. P. Cresson, *Compt. Rend. C*, **1972**, 1299.
204. H. J. Monteiro, *Synthesis*, **1974**, 137.
205. K. Issleib and M. Lischewski, *J. Organomet. Chem.*, **46**, 297 (1972).
206. H. Gross and B. Costisella, *J. Prakt. Chem.*, **311**, 925 (1969).

ADDENDUM

To Section I-A

Glyoxylic acid piperidides **241** react with HBF_4 to form the salts **242** (207).

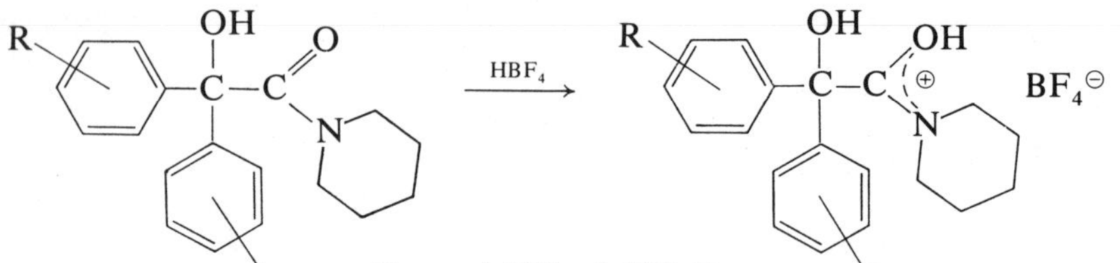

[Structure 241] →(HBF₄)→ [Structure 242] $BF_4^\ominus$

R = –4-CH_3, 3-CH_3O–, 4-F, 4-Br, 4-Cl–, 2-Cl

(241)   (242)

To Section I-A-1a

Further examples of alkylation of carboxylic acid amides by means of oxonium salts have been reported; some of these reaction are listed below.

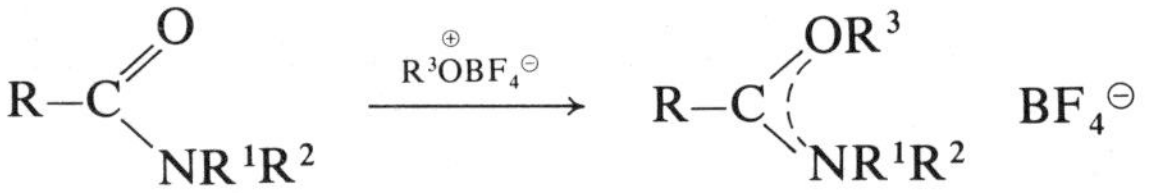

$$R-C(=O)NR^1R^2 \xrightarrow{R^3OBF_4^\ominus} R-C(OR^3)(NR^1R^2)\ BF_4^\ominus$$

R	R^1	R^2	R^3
R^4-C≡C–	H, CH_3	H, CH_3	C_2H_5 (refs. 208, 209)
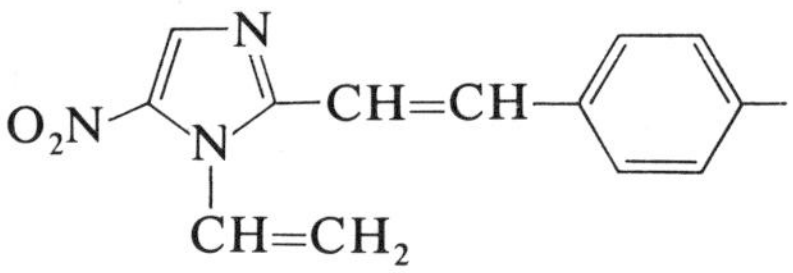 O_2N-[imidazole with N–CH=CH₂]-CH=CH-[C₆H₄]–	H, alkyl,	cycloalkyl	C_2H_5 (ref. 210)
CH_2CH–Fe(CO)$_4$	H, CH_3	H, CH_3	C_2H_5 (ref. 211)
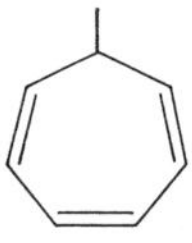 [cycloheptatrienyl]	CH_3	CH_3	C_2H_5 (ref. 212)
CH_3	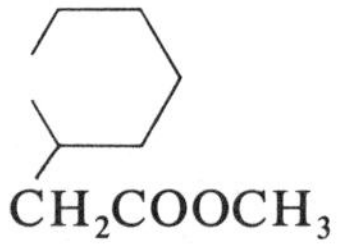 [cyclohexyl-CH₂COOCH₃]		C_2H_5 (ref. 213)
CH_3	H	–CH(COOH)–CH_3	CH_3* (ref. 214)

* As well as other *N*-acylated α-aminocarboxylic acids.

A series of amides have been reacted with oxonium salts; the resulting iminium salts, however, were not isolated but further reacted to form ketene aminals (215), iminoesters (216), and heterocycles (216).

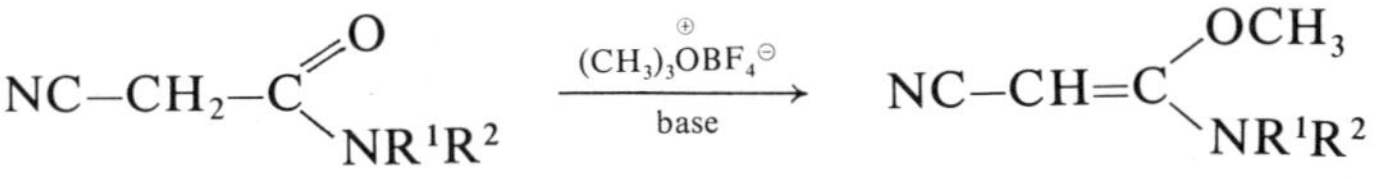

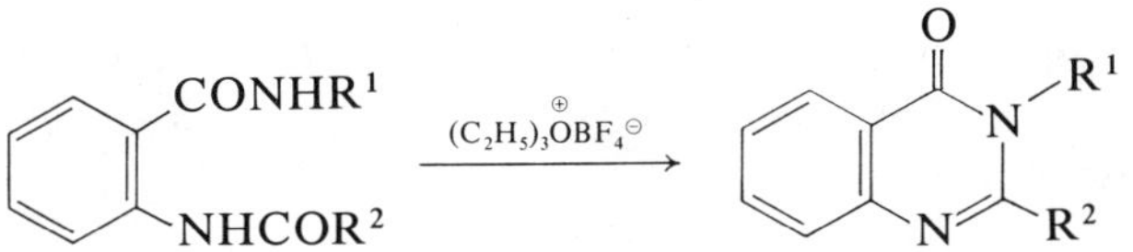

$$NC-CH_2-C(=O)NR^1R^2 \xrightarrow[\text{base}]{(CH_3)_3OBF_4^\ominus} NC-CH=C(OCH_3)NR^1R^2$$

[2-(CONHR^1)-C$_6$H$_4$-NHCOR^2] $\xrightarrow{(C_2H_5)_3OBF_4^\oplus}$ [quinazolinone with N-R^1 and C-R^2]

To Section I-A-1b

Lactams that were alkylated with oxonium salts to give iminium compounds (217, 218) are shown below. In a few examples (219–221) iminium salts have been reacted further without isolation.

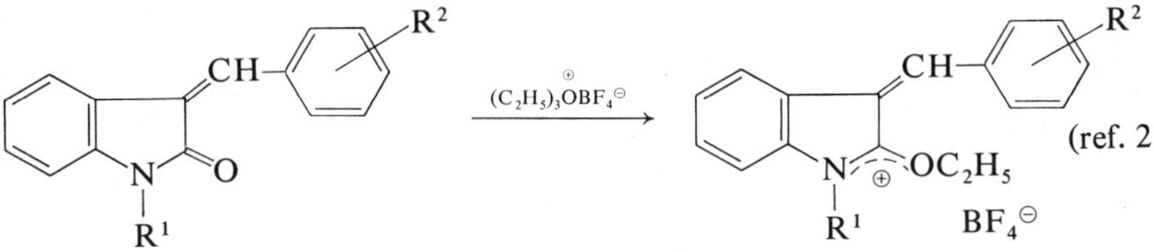

[indolinone with =CH-Ar(R^2), N-R^1] $\xrightarrow{(C_2H_5)_3OBF_4^\oplus}$ [iminium ether, OC$_2$H$_5$, BF$_4^\ominus$] (ref. 217)

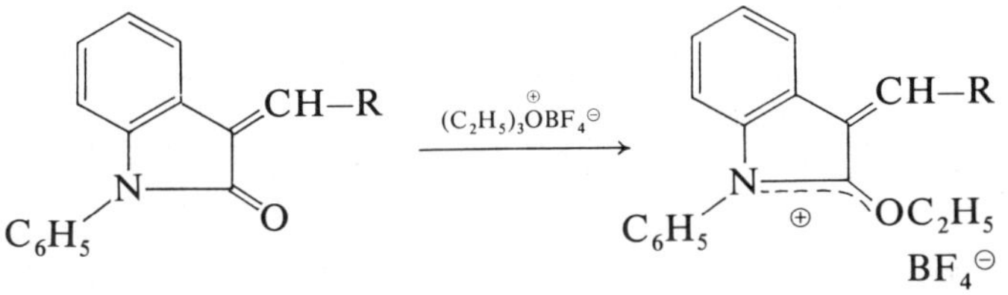

[indolinone with =CH-R, N-C$_6$H$_5$] $\xrightarrow{(C_2H_5)_3OBF_4^\oplus}$ [iminium ether, OC$_2$H$_5$, BF$_4^\ominus$] (ref. 218)

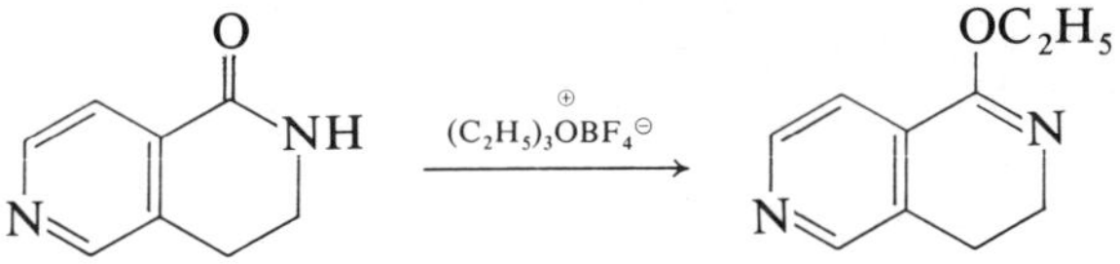

[isoquinolinone-NH] $\xrightarrow{(C_2H_5)_3OBF_4^\oplus}$ [OC$_2$H$_5$ imino ether] (ref. 219)

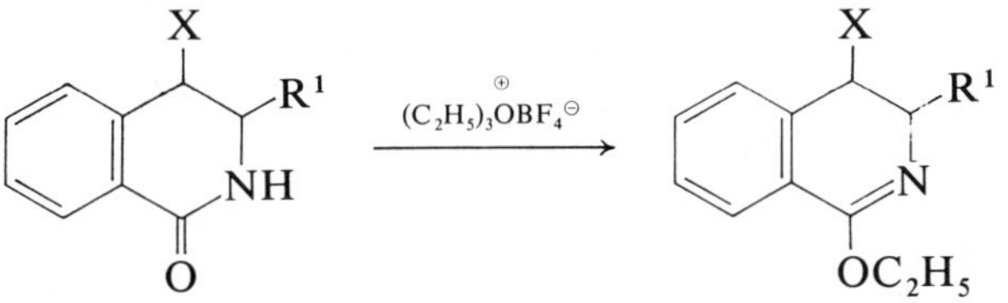

[3,4-dihydroisoquinolinone with X, R^1, NH] $\xrightarrow{(C_2H_5)_3OBF_4^\oplus}$ [imino ether OC$_2$H$_5$] (ref. 220)

$R^1 = H, CH_3; X = H, OH$

To Section I-A-3a

Alkylations of amides have also been achieved by use of dimethylsulfate (222).

To Section I-A-3b

By action of dimethylsulfate and lutidine on the lactam **243**, the lactimidicacid ester **244** was prepared (223); the *N*-methyl lactam **245** is formed as a byproduct.

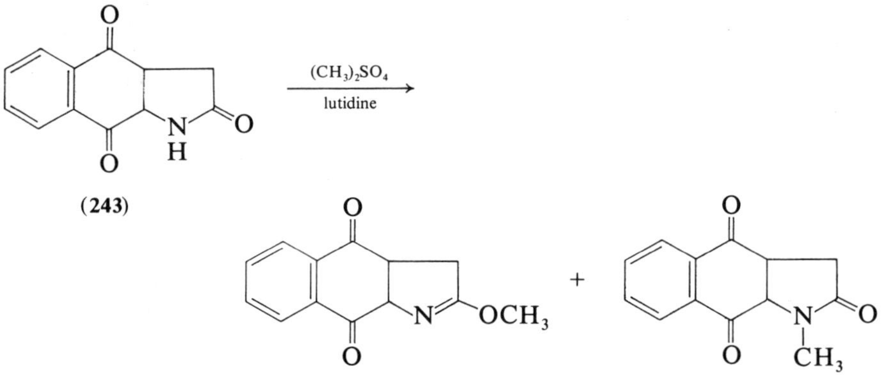

In a similar way lactimidic acid esters **246** have been prepared (224).

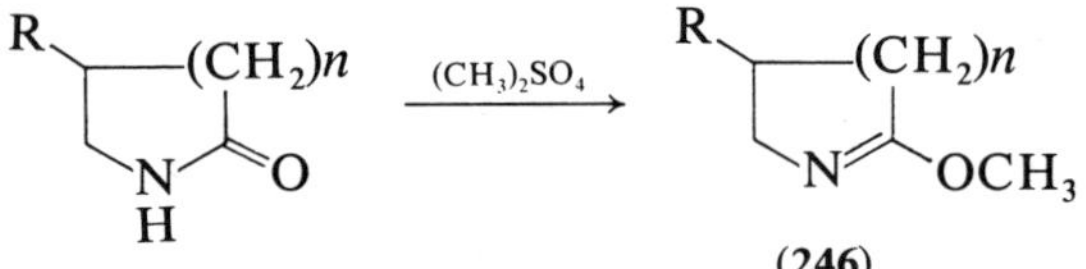

R = H, C$_2$H$_5$, n = 1–3,

To Section I-A-4

Fluorsulfonic acid esters and trifluormethanesulfonic acid silylester have proved to be suitable reagents for preparation of alkoxymethyleniminium salts and imino esters from carboxylic acid amides.

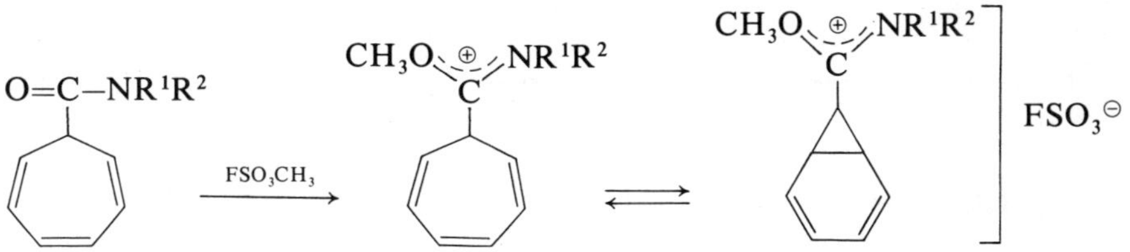

(ref. 212)

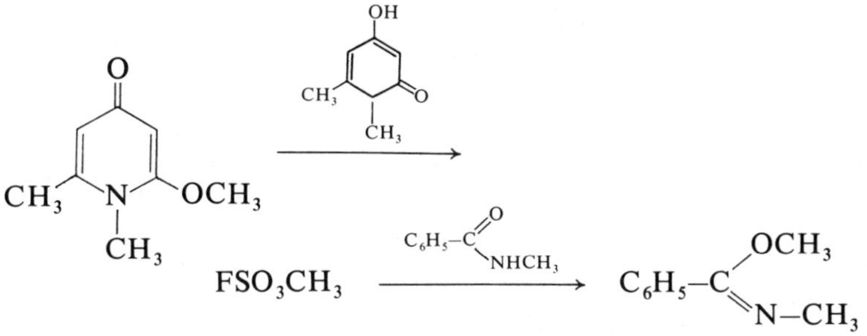

(ref. 225)

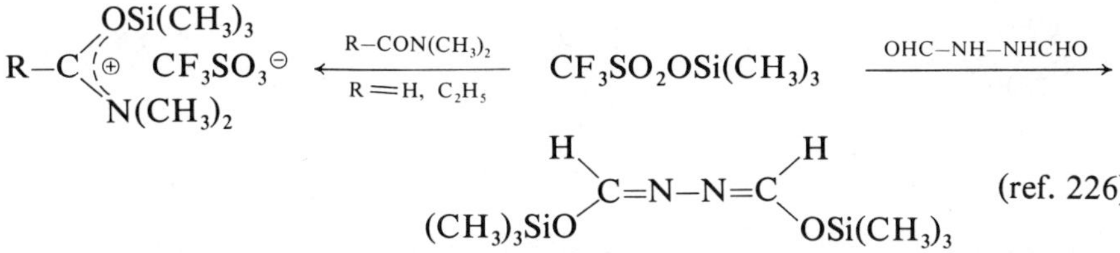

(ref. 226)

Sultones may also be used to alkylate amides on oxygen (227).

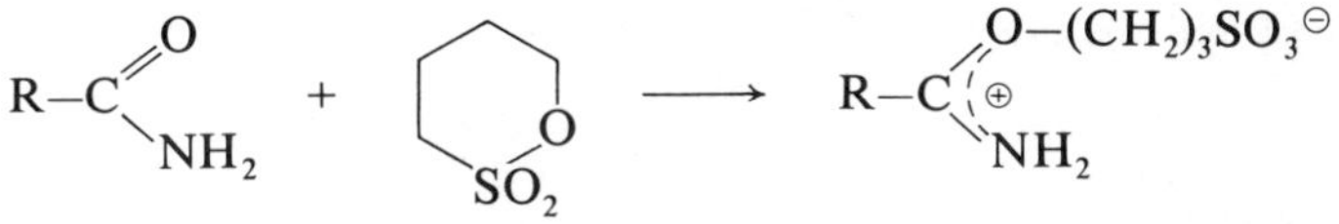

The protected glucopyranosyl bromide **246** reacts with secondary carboxylic acid amides in the presence of Ag_2O to form iminoesters **247** (228).

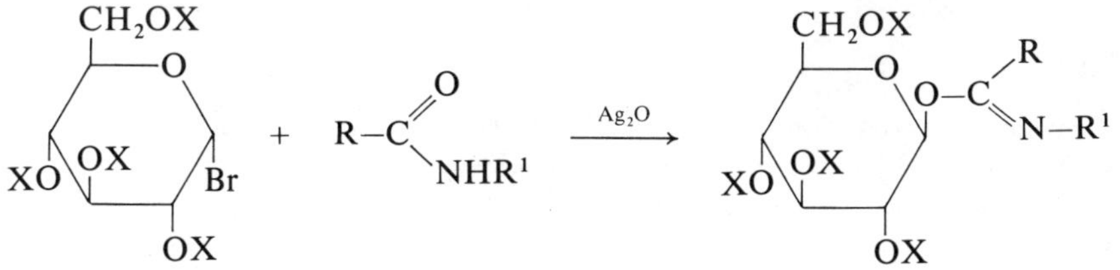

(**246**)
$x = C_6H_5-CH_2$

(**247**)

To Section I-C

Dimethylsulfate transforms the iminoester **248** to the iminium salt **249** (222).

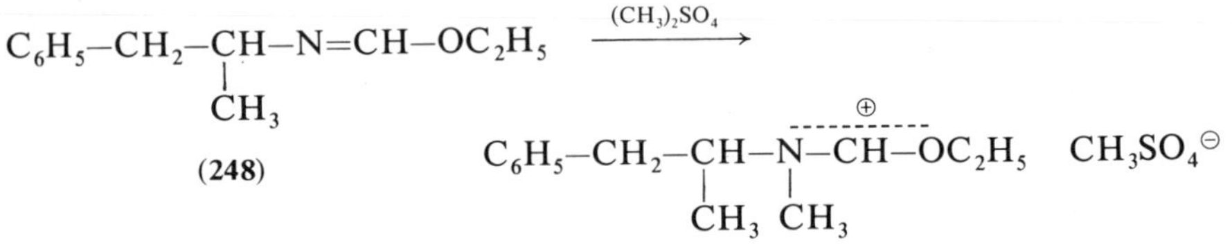

$$C_6H_5-CH_2-\underset{\underset{CH_3}{|}}{CH}-N=CH-OC_2H_5 \xrightarrow{(CH_3)_2SO_4}$$

$$C_6H_5-CH_2-\underset{\underset{CH_3}{|}}{CH}-\overset{\oplus}{\underset{\underset{CH_3}{|}}{N}}-CH-OC_2H_5 \quad CH_3SO_4^{\ominus}$$

(248)

(249)

By acylation of 5,5-dimethyloxazoline **250** iminium salts **251** are obtainable (229).

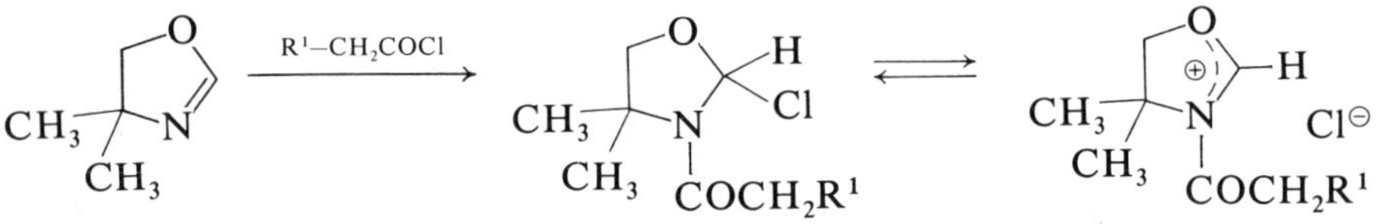

(250) → (251)

To Section I-H

A recently published very extensive review concerned with chemistry of iminoesters described the scope of Pinner synthethis for preparation of iminium salts (iminoesterhydrohalogenides) by action of alcohols on nitriles in the presence of hydrohalides (230).

To Section II

Reviews on physical, spectroscopic, and stereochemical properties of not *NN*-peralkylated alkoxymethyleniminium compounds have appeared (230).

To Section III-A-1

Hydrolysis reactions of *N*-unsubstituted and *N*-monosubstituted alkoxymethyleniminium salts have been reviewed recently (230).

A series of more recent investigations is concerned with the hydrolysis of *NN*-disubstituted alkoxymethyleniminium salts (influence of stereochemistry and pH dependence of product distribution-amid-versus ester formation (231–235).

*To Section III-A-2
and III-B-1*

A summary of thiolysis and alcoholysis reactions of *N*-unsubstituted and *N*-monosubstituted alkoxymethyleniminium compounds may be found in ref. 230.

To Section III-D

Reactions of *N*-unsubstituted and *N*-monosubstituted alkoxymethyleniminium salts with *NH*-containing compounds again recently have been reviewed (230).

To Section III-D-1

Aminolysis of iminium salts of type **252** yield amidines **253** (222).

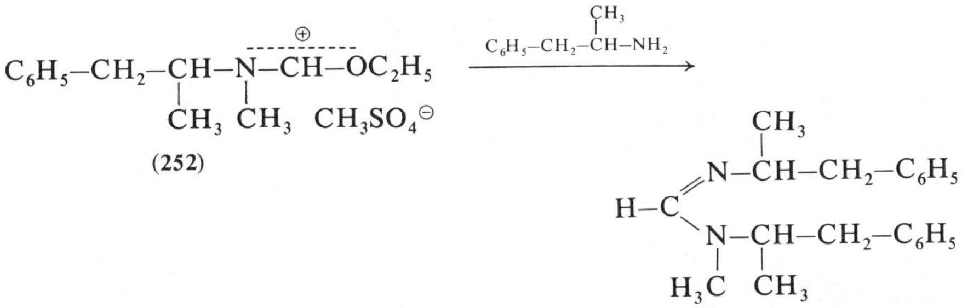

$$C_6H_5-CH_2-\overset{|}{\underset{CH_3}{C}}H-\overset{\oplus}{N}-\overset{|}{\underset{CH_3}{C}}H-OC_2H_5 \quad CH_3SO_4^{\ominus}$$
(252)

$$\xrightarrow{C_6H_5-CH_2-\overset{CH_3}{\underset{|}{C}}H-NH_2}$$

$$H-C\begin{cases} N=\overset{CH_3}{\underset{|}{C}}H-CH_2-C_6H_5 \\ N-\overset{|}{\underset{H_3C}{C}}H-CH_2-C_6H_5 \\ \quad\quad CH_3 \end{cases}$$
(253)

To Section III-E

For a recent review, see ref. 230.

To Section III-F

From alkalialcoholates or alcohol/triethylamine and from iminium salts **254** and **255** further orthoamide derivatives (217, 218, 229, 236) were obtained.

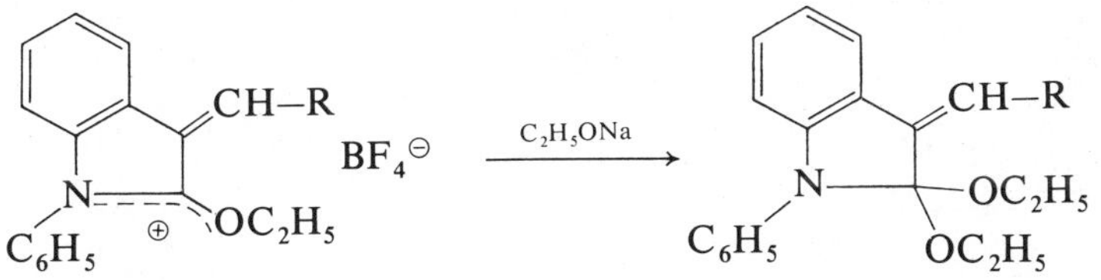

(254)

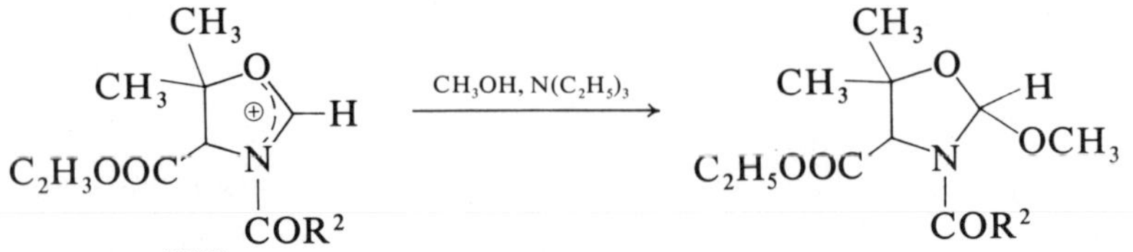

To Section III-H

Further examples for the reduction of iminium salts by means of $NaBH_4$ leading to amines have been reported (210, 219).

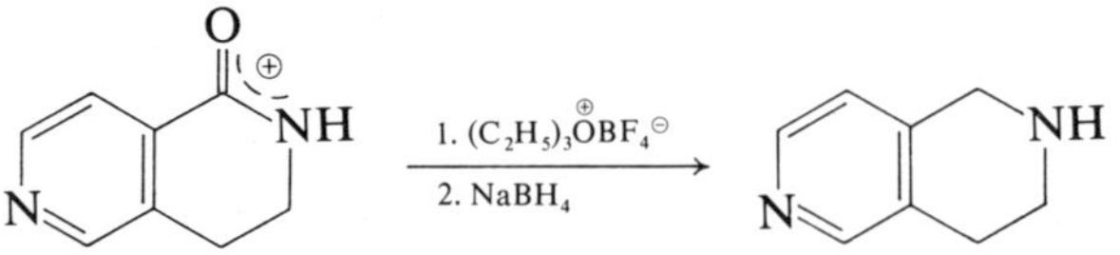

N-Ethylated aminocarbonic acids, such as N-ethylalamine **256**, have been prepared by reduction of iminium salts of type **257** (214).

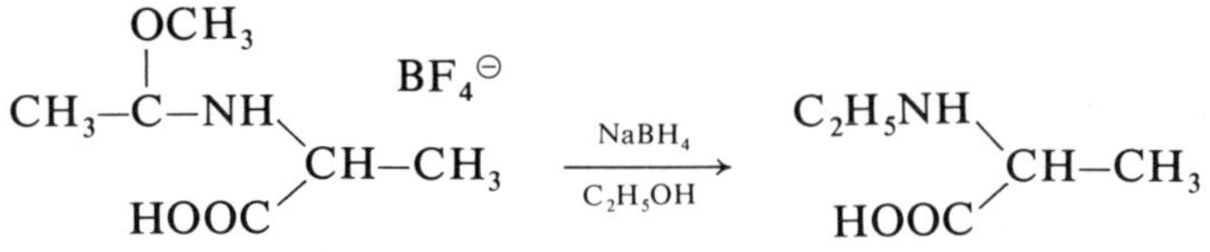

Section III-J

Miscellaneous reactions. Iminium salts, derived from unsaturated carboxylic acid amides, were shown to be suitable reagents for various cyclo addition reactions, as the following formulas show.

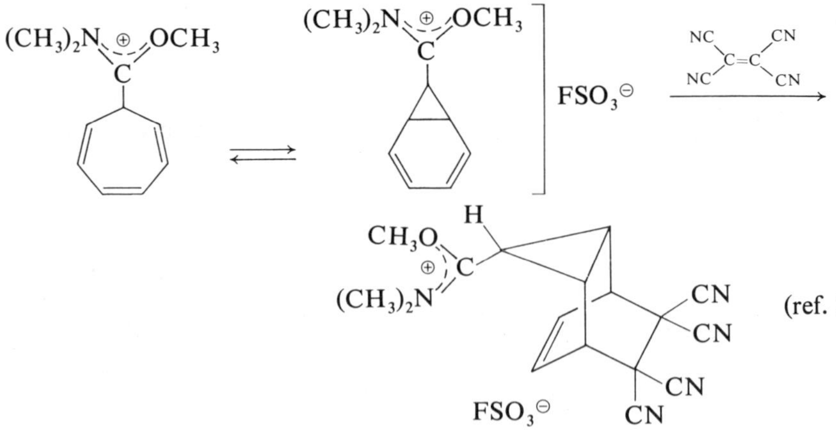

(ref. 212)

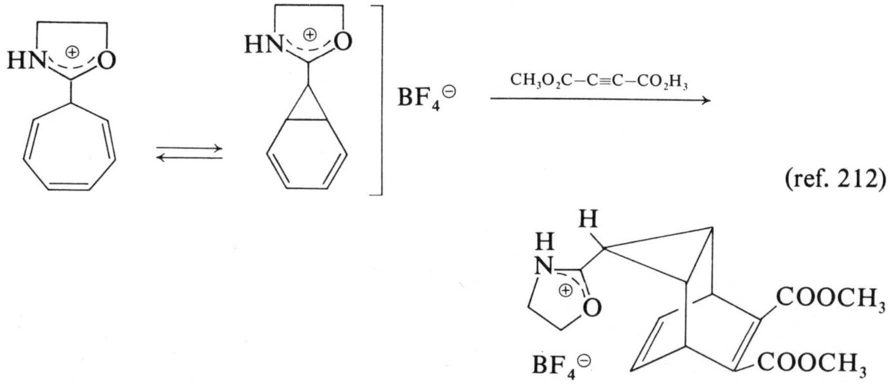

(ref. 212)

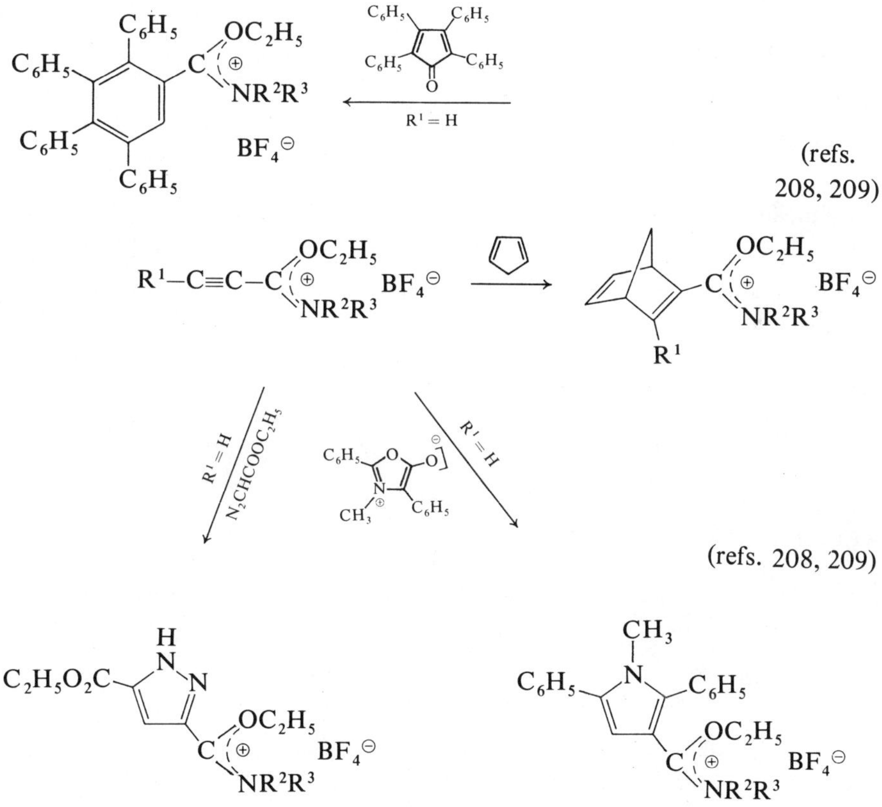

(refs. 208, 209)

(refs. 208, 209)

Reduction of iminium salt **258** with H_2/Pd leads to a mixture of iminium compounds with saturated carbon skeleton (212).

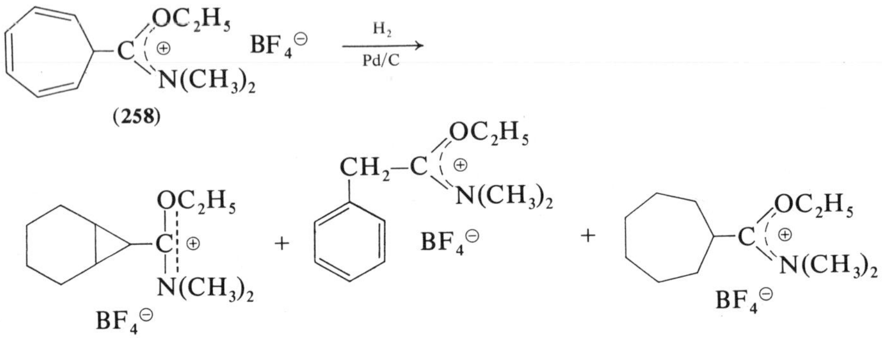

The complex iminium compounds **259** afford amidinium salts **260** on reaction with secondary amines (211).

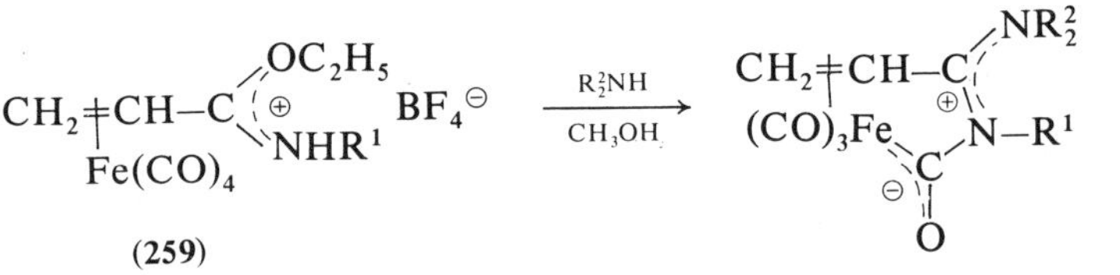

By means of BF_3-etherate the iminium salt **261** can be cyclized to form **262** (213).

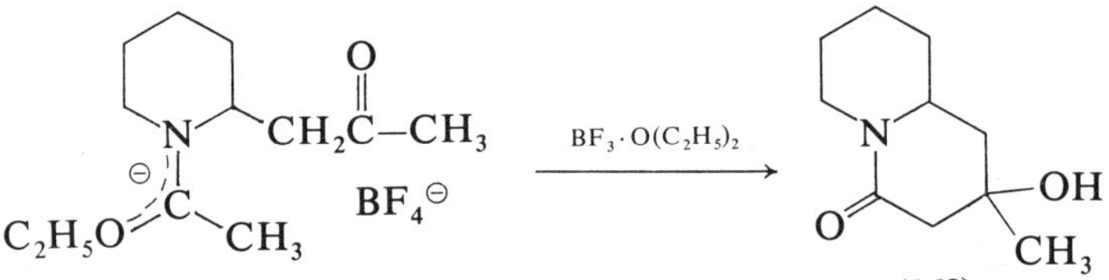

REFERENCES

207. Z. G. Kalugina and V. S. Shkaev, *Izo. Vyssh. Uchebn. Zaved Khim. Teknol.*, **19**, 328 (1976); *C. I.*, 26–220 (1976).
208. H. G. Viehe and J. S. Baum, *Chimia*, **29**, 300 (1975).
209. J. S. Baum and H. G. Viehe, *J. Org. Chem.*, **41**, 183 (1976).
210. W. J. Ross and W. B. Jamieson, *J. Med. Chem.*, **18**, 430 (1975).
211. A. N. Nesmeyanov, T. N. Sal'nikova, Yu. T. Struchkov, V. G. Andrianov, A. A. Pogrebnyak, L. V. Rybin, and M. I. Rybinskaya, *J. Organomet. Chem.*, **117**, C16 (1976).
212. W. Betz, J. Daub, and K. M. Rapp, *Liebigs Ann. Chem.*, **1974**, 2089.

213. Y. Ban, M. Kumura, and T. Oishi, *Chem. Pharm. Bull.*, **24,** 1490 (1976).
214. F. M F. Chen and N. L. Benoiton, *Can. J. Chem.*, **55,** 1433 (1977).
215. J. M. McCall, R. E. Ten Brink, and J. J. Ursprung, *J. Org. Chem.*, **40,** 3304 (1975).
216. T. Kato, A. Takada, and T. Ueda, *Chem. Pharm. Bull.*, **24,** 431 (1976).
217. V. P. Borovnik, L. S. Filatova, and V. P. Mamaev, *Izv. Sib. Otd. Akad. Nauk. SSSR, Ser. Khim. Nauk.*, **1975,** 137; *C. I.*, 44–253 (1975).
218. V. P. Borovnik, L. A. Gubanova, and V. P. Mamaev, *Izv. Sib. Otd. Akad. Nauk. SSSR, Ser. Khim. Nauk*, **1975,** 133; *C. I.*, 44–227 (1975).
219. F. Alhaique, R. G. E. Cozzani, and F. M. Riccieri, *Farm. Ed. Sci.*, **31,** 845 (1975); *C. I.*, 16–226 (1977).
220. G. D. Diana, W. B. Hinshaw, and H. E. Lape, *J. Med. Chem.*, **20,** 449 (1977).
221. T. Oishi, M. Fukui, Y. Ban, and M. Honda, *Heterocycles (Sendai Japan)*, **5** (1976), Spec. Issue 281; *C. I.*, 17–228 (1977).
222. N. B. Marchenko, V. G. Granik, R. G. Glushkov, L. I. Budanova, V. A. Kuzovkin, V. A. Parshin, and R. A. Al'tshuler, *Khim. Farm. Zh.*, **10,** 46 (1976); *C. I.*, 2–141 (1977).
223. B. V. Salov and A. I. Shakhrovich, *Khim. Geterotsikl. Soedin*, **1976,** 911; *C. I.*, 46–219 (1976).
224. T. Fujii, Sh. Yoshifuji, and K. Yamada, *Chem. Ind. (London)*, **1975,** 177.
225. P. Beak and J. K. Lee, *J. Org. Chem.*, **40,** 147 (1975).
226. G. Simchen and W. Kober, *Synthesis*, **1976,** 259.
227. W. Ried and E. Schmidt, *Liebigs Ann. Chem.*, **676,** 114 (1964).
228. J. R. Pougny and P. Sinaÿ, *Tetrahedron Lett.*, **1976,** 4073.
229. B. T. Golding and D. R. Hall, *J. Chem. Soc. Perkin I*, **1975,** 1302.
230. D. G. Neilson "Imidates including cyclic imidates," in S. Patei, Ed., *The Chemistry of Amidines and Imidates*, Chemistry of Functional Groups series, John Wiley and Sons, New York, 1975.
231. V. F. Smith, Jr., and G. L. Schmir, *J. Amer. Chem. Soc.*, **97,** 3171 (1975).
232. R. A. McClelland, *J. Amer. Chem. Soc.*, **97,** 3177 (1975).
233. P. Deslongchamps, S. Dubé, C. Lebreux, D. R. Patterson, and R. J. Taillefer, *Can. J. Chem.*, **53,** 2791 (1975).
234. P. Deslongchamps, *Tetrahedron*, **31,** 2463 (1975).
235. P. Deslongchamps and R. Taillefer, *Can. J. Chem.*, **53,** 3029 (1975).
236. R. C. Costin, C. J. Morrow, and H. Rapoport, *J. Org. Chem.*, **41,** 535 (1976).

ALKYLMERCAPTOMETHYLENIMINIUM SALTS

By W. KANTLEHNER, *D-7080-AALEN-Dewangen, Laachweg 14, Germany*

CONTENTS

I. Synthesis 279
   A. Via Alkylation of Thioamides 279
   B. Via Alkylation and Acylation of Thioimino Esters 287
   C. Via Acylation of Thioamides 289
   D. From Chloromethyleniminium Salts (Amide Chlorides) 290
   E. From Nitriles, Mercaptans, and Hydrogen Halides 290
   F. Via Rearrangement of Secondary Thioamides 291
   G. Via Reaction of Ketene *S,N*-Acetals with Electrophilic Reagents 291
   H. From Chloroalkylmercaptomethyleniminium Salts 293
II. Properties and Structure of Alkylmercaptomethyleniminium Salts 294
III. Reactions of Alkylmercaptomethyleniminium Salts 294
   A. Hydrolysis, Alcoholysis 294
   B. Reactions with Hydrogen Sulfide, Hydrogen Selenide, Mercaptides, and Alkoxides 297
   C. Proton Abstraction with Base 299
   D. Reactions with Amines and Amine Derivatives 304
   E. Reactions with Grignard Reagents 308
   F. Reactions with Inorganic Salts 308
   References 310
   Addendum 313

I. Synthesis

A. VIA ALKYLATION OF THIOAMIDES

In 1878 Wallach (1) discovered that thioacetanilide is ethylated at the sulfur atom by ethyl bromide; thus the salt **1** forms. Later it was shown that nearly all salts of types **4** and **5** are accessible by alkylation of thioamides (**2**) (2–55c) and thiolactams (**3**) (8, 26, 29, 30a, 31, 42–42b). The ligands R, R^1, and R^2 may be hydrogen, alkyl, and aryl groups.

Remarkably via the isolable monoalkylated intermediate **6**, a cyclobutene dication (**7**) was obtained by methylation of bis(dimethylamino)cyclobutenedithione (**8**) (52).

The alkylation of thioamides has been reviewed repeatedly (56–59).

Translated from German by Dr. W.-J. Richter, MPI für Kohleforschung, D 4330 Mühlheim.

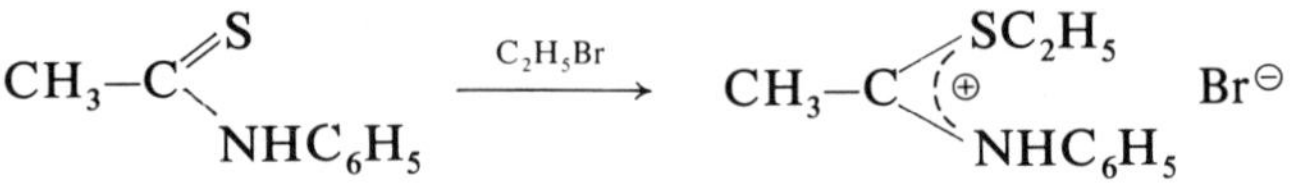

$$CH_3-C\underset{NHC_6H_5}{\overset{S}{\diagup}} \xrightarrow{C_2H_5Br} CH_3-C\underset{NHC_6H_5}{\overset{SC_2H_5}{\overset{\oplus}{\diagup}}} \quad Br^{\ominus}$$

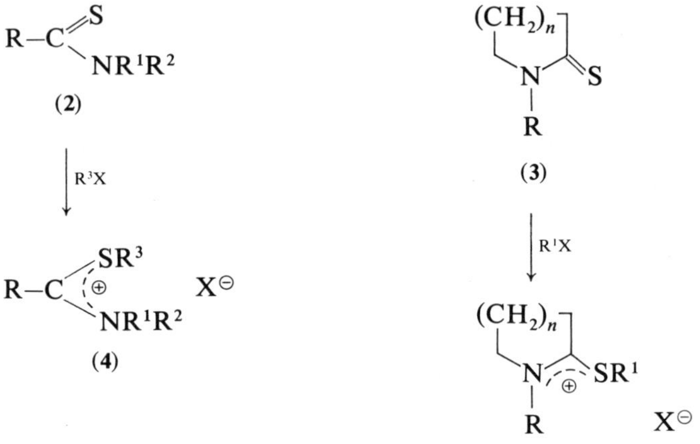

(1)

$$R-C\underset{NR^1R^2}{\overset{S}{\diagup}}$$

(2)

$\downarrow R^3X$

$$R-C\underset{NR^1R^2}{\overset{SR^3}{\overset{\oplus}{\diagup}}} \quad X^{\ominus}$$

(4)

structure (3): cyclic thioamide with $(CH_2)_n$, N–R, C=S

$\downarrow R^1X$

structure (5): cyclic with $(CH_2)_n$, N–R, $\overset{\oplus}{C}$–SR^1, $X^{\ominus}$

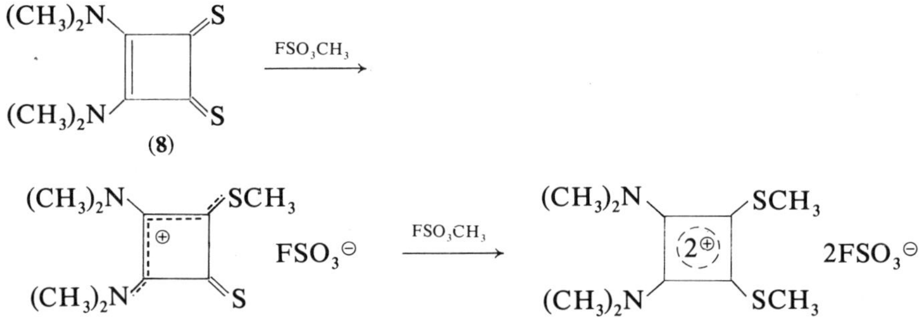

(8) four-membered ring with $(CH_3)_2N$ groups and two C=S

$\xrightarrow{FSO_3CH_3}$

(6) four-membered ring with $(CH_3)_2N$, SCH$_3$, C=S, $FSO_3^{\ominus}$

$\xrightarrow{FSO_3CH_3}$

(7) four-membered ring with $(CH_3)_2N$ groups, two SCH$_3$, $2\oplus$, $2FSO_3^{\ominus}$

Recently it was shown that selenoamides (**9**) may be alkylated with alkyl halides at the selenium atom to give the salts **10** (46).

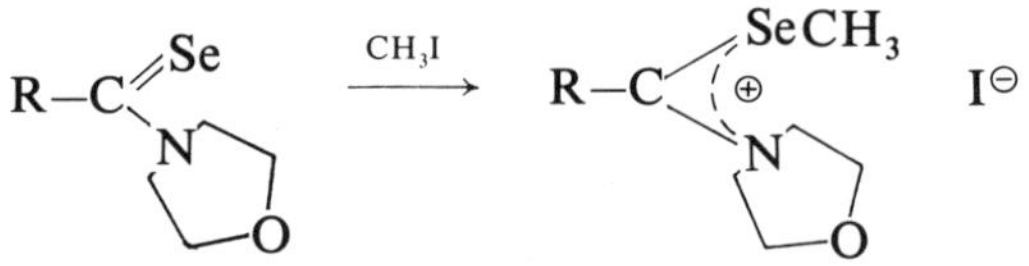

$$R-C\overset{Se}{\diagup}\underset{N}{} \xrightarrow{CH_3I} R-C\overset{SeCH_3}{\overset{\oplus}{\diagup}}\underset{N}{} \quad I^{\ominus}$$

(9)   (10)

The properties of some salts of type **4** are summarized in Table I. The alkylating agents used here are alkyl bromides (1, 2, 3, 30, 44, 49, 50), alkyl

TABLE I

Alkylmercaptomethyleniminium Salts (4, 5) by Alkylation of Thioamides

$$R-C\underset{NR^1R^2}{\overset{SR^3}{\diagup}}\ X^{\ominus}$$

R	R^1	R^2	R^3	X	Yield, %	m.p., °C	Ref.
H	H	H	CH_3	I	75	96	28
H	H	H	$C_6H_5CH_2$	Br	87	136	28
H	H	H	$(C_6H_5)_2CH$	Br	83	—	2
H	H	H	$(C_6H_5)_3C$	Cl	77	106–107	28
H	CH_3	CH_3	CH_3	I	—	122–123	5, 48
H	CH_3	CH_3	C_2H_5	$C_2H_5SO_4$	99	—	51
H	C_2H_5	C_2H_5	CH_3	I	—	111–112	5
H	—$(CH_2)_5$—		CH_3	I	—	119–120	5
H	C_6H_5	C_6H_5	CH_3	I	—	100 (dec.)	5
CH_3	H	H	$C_6H_5CH_2$	Br	95	174–176	49
CH_3	H	H	$(C_6H_5)_2CH$	Br	95	145–147	44
CH_3	H	H	p-CH_3O–$C_6H_4)_2CH$	Cl	95	101–109	44
CH_3	H	H	(mesityl)$_2$CH	Cl	95	185–195	44
CH_3	CH_3	CH_3	CH_3	I	98–100	85–97	42
CH_3	CH_3	CH_3	C_2H_5	I	74	139–141	40
CH_3	—$(CH_2)_4$—		CH_3	I	97	85.5–87	42
CH_3	—$(CH_2)_2$–O–$(CH_2)_2$—		CH_3	I	99	116–118 (131–132)	21, 46
CH_3	CH_3	C_6H_5	CH_3	I	95–97	135	42
CH_3	H	H	CH_3	I	95	95–100	49
C_2H_5	H	H	$C_6H_5CH_2$	Br	95	120–125	49
C_2H_5	C_2H_5	C_2H_5	CH_3	I	—	112–113	42
C_2H_5	CH_3	CH_3	CH_3	I	82	96.5–98	42
$(CH_3)_2CH$	H	H	CH_3	I	95	95–100	49
$(CH_3)_2CH$	H	H	$C_6H_5CH_2$	Br	90	150–155	49

TABLE I (cont.)

R	R¹	R²	R³	X	Yield, %	m.p., °C	Ref.
(CH₃)₂CH	CH₃	H	C₆H₅CH₂	Br	95	130–135	49
(CH₃)₂CH	CH₃	CH₃	CH₃	I	92	120–121	42
n-C₃H₇	CH₃	CH₃	CH₃	I	88	77–78	42
n-C₄H₉	CH₃	CH₃	CH₃	I	95–97	115–116.5	42
–(CH₂)₃–			CH₃	I	94–98	118–120	42
–(CH₂)₄–			CH₃	I	82	156–158	42
–(CH₂)₅–			CH₃	I	89	141–142.5	42
(CH₃)₃C	H	H	C₆H₅CH₂	Br	93	225–230	49
(CH₃)₃CCH₂	H	H	C₆H₅CH₂	Br	95	190–195	49
(CH₃)₃C–CHCH₃	H	H	C₆H₅CH₂	Br	95	165–167	49
c-C₆H₁₁	H	H	C₆H₅CH₂	Br	91	155–160	49
C₆H₅–CH₂	H	H	CH₃	I	—	138	4
C₆H₅–CH₂	H	H	C₂H₅	I	—	115–116	4
C₆H₅–CH₂	H	H	C₆H₅CH₂	Br	90	140–148	49
C₆H₅–CH₂	–(CH₂)₂–O–(CH₂)₂–		CH₃	I	95	161–166 158(dec.) 125–127	42, 21, 46
(C₆H₅)₂CH	H		C₆H₅CH₂	Br	95	155–165	49
1-naphthyl-CH₂	–(CH₂)₂–O–(CH₂)₂–		CH₃	I	86–96	150–160	42
2-naphthyl-CH₂	–(CH₂)₂–O–(CH₂)₂–		CH₃	I	95–98	154–157	42
C₆H₅	H	H	CH₃	I	80	167 (138–139)	3, 30
C₆H₅	H	H	C₂H₅	Br	40	185 (129.5–130.5)	3, 30
C₆H₅	H	H	C₂H₅	I	42	160 (142)	4, 30
C₆H₅	H	H	(C₆H₅)₂CH	Br	90	140–147	44
C₆H₅	H	H	C₆H₅CH–p-C₆H₄–OCH₃	Cl	90	110–115	44

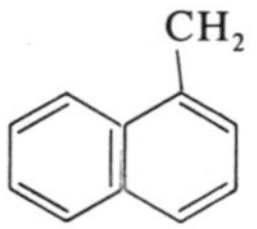

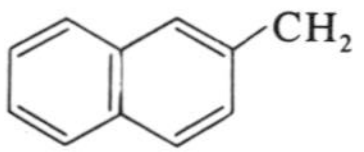

Ar	R	R'	R''	X	Yield (%)	mp (°C)	Refs.
C_6H_5	H	H	$(p\text{-}CH_3O\text{-}C_6H_4)_2CH$	Cl	90	106–114	44
C_6H_5	H	H	$(mesityl)_2CH$	Cl	80	100–110	44
C_6H_5	H	CH_3	CH_3	I	—	128–129	15
C_6H_5	H	C_2H_5	CH_3	I	—	121	15
C_6H_5	$-(CH_2)_5-$		CH_3	I	—	111	21
C_6H_5	$-(CH_2)_2-O-(CH_2)_2-$		CH_3	I	90	128–131 158 (dec.) 170	16, 21, 46
C_6H_5	$-(CH_2)_2-O-(CH_2)_2-$		C_2H_5	I	—	193, 136–137	16, 21
C_6H_5	CH_3	C_6H_5	CH_3	I	—	148–150	21
C_6H_5	C_6H_5	C_6H_5	CH_3	I	—	92	16
$o\text{-}HO\text{-}C_6H_4$	H	H	$C_6H_5CH_2$	Br	86	143–146	50
$o\text{-}HO\text{-}C_6H_4$	$-(CH_2)_2-O-(CH_2)_2-$		CH_3	I	—	189	16
$o\text{-}HO\text{-}C_6H_4$	$-(CH_2)_2-O-(CH_2)_2-$		C_2H_5	I	—	182	16
$o\text{-}CH_3\text{-}C_6H_4$	$-(CH_2)_2-O-(CH_2)_2-$		CH_3	I	—	185	16
$p\text{-}CH_3\text{-}C_6H_4$	H	H	CH_3	I	71	147	30
$p\text{-}CH_3\text{-}C_6H_4$	H	H	CH_3	CH_3SO_4	79	200	30
$p\text{-}CH_3\text{-}C_6H_4$	H	H	C_2H_5	I	37	188	30
$p\text{-}CH_3\text{-}C_6H_4$	H	H	C_2H_5	Br	38	178	30
$p\text{-}CH_3\text{-}C_6H_4$	H	H	CH_3	I	95	174	30
$p\text{-}CH_3O\text{-}C_6H_4$	H	H	CH_3	I	78	144	30
$p\text{-}CH_3O\text{-}C_6H_4$	H	H	CH_3	CH_3SO_4	64	162	30
$p\text{-}CH_3O\text{-}C_6H_4$	H	H	C_2H_5	Br	59	153	30
$p\text{-}CH_3O\text{-}C_6H_4$	H	CH_3	C_2H_5	I	—	114–115	16
$p\text{-}CH_3O\text{-}C_6H_4$	$-(CH_2)_2-O-(CH_2)_2-$		CH_3	I	86	142 (dec.)	16, 21, 46
$p\text{-}Cl\text{-}C_6H_4$	H	H	CH_3	I	93	123–125	30
$p\text{-}Cl\text{-}C_6H_4$	H	H	CH_3	CH_3SO_4	78	180	30
$p\text{-}Cl\text{-}C_6H_4$	H	H	C_2H_5	I	16	149	30
$p\text{-}Cl\text{-}C_6H_4$	H	H	CH_3	I	—	178	30
$p\text{-}(CH_3)_2N\text{-}C_6H_4$	$-(CH_2)_2-O-(CH_2)_2-$		CH_3	I	—	138–141	46
$p\text{-}CH_3CONH\text{-}C_6H_4$	$-(CH_2)_2-O-(CH_2)_2-$		CH_3	I	—	152–154	46
$HO-$	$-(CH_2)_2-O-(CH_2)_2-$		CH_3	I	—	136–138 (dec.)	21
	C_6H_5	H	CH_3	I	89	178–179	33

iodides (1–5, 15, 17, 21, 28, 30, 33, 36, 37, 42, 45, 49, 53, 54), alkyl chlorides (28, 32, 44, 49), dialkyl sulfates (6, 30, 30a, 34, 51), oxonium salts (28, 29, 31, 35, 47), and methyl fluorosulfonates (52, 55).

Thioamides are also alkylated by alkyl chlorides generated *in situ* from alcohols (60):

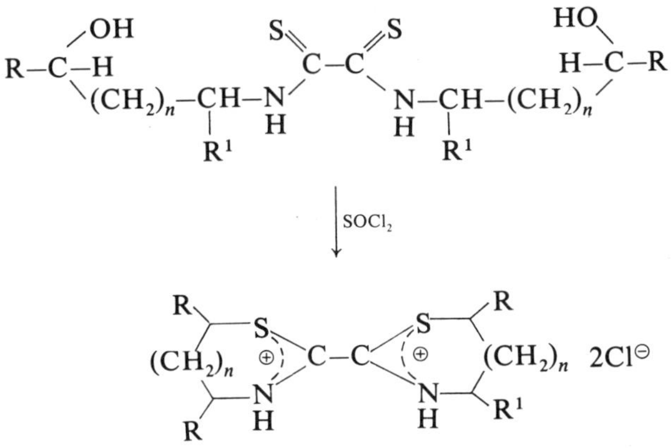

[Scheme showing two hydroxyalkyl thioamide structures connected, with SOCl$_2$ arrow leading to bicyclic dithiolium-type product with 2Cl$^\ominus$]

α-Halocarboxylic acids (7, 8, 12, 14, 28, 55a, 55b, 61–63), β-halocarboxylic acids (28), and α-haloketones (10, 11) are also suitable for the

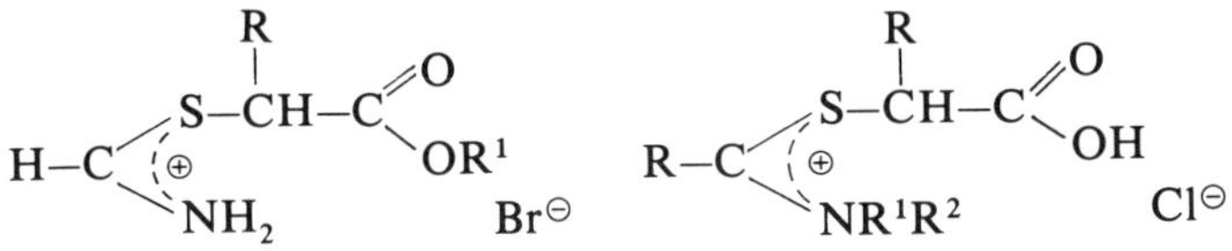

[Structures (11) and (12) showing S-alkylated thioamide salts with Br$^\ominus$ and Cl$^\ominus$ counterions]

(11)   (12)

alkylation of thioamides; here the salts **11** or **12** form. The salts **12** containing $R^1 = R^2 = H$ (**12a**) may be cyclized to thiazolone derivatives (**13**).

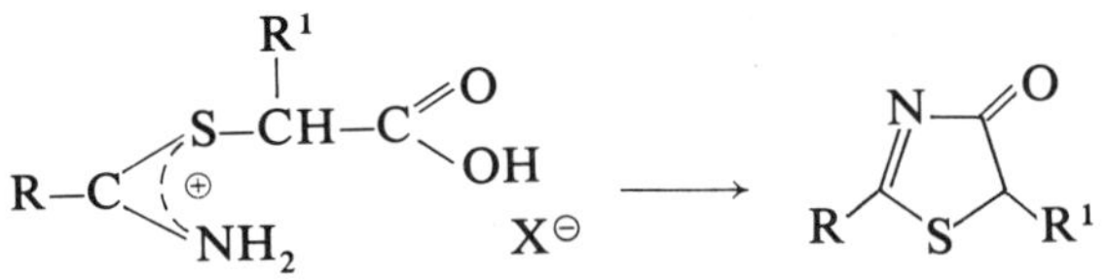

[Structure (12a) converting to thiazolone (13)]

(12a)   (13)

Thiazole derivatives (**14**) are also generated from α-haloketones and thioamides (10–12). This and similar reactions have been reviewed elsewhere (57, 59, 64).

$$R-C\overset{S-CH_2-\overset{O}{\overset{\|}{C}}-R^1}{\underset{NH_2}{\oplus}} \longrightarrow \underset{(14)}{\underset{S}{\overset{R^1}{\underset{\|}{\bigcirc}}}\overset{N}{R}}$$

If 1,2-dihalo compounds are reacted with thioamides, thiazolines or dications of the type **15** form. At lower reaction temperatures the salts **15**, and at higher temperatures the thiazolinium salts **16** or thiazolines, are preferentially obtained. This reaction has also been reviewed (57, 59, 64).

$$\underset{(16)}{CH_3-C\overset{\overset{H}{\overset{|}{N}}\rceil}{\underset{S\rfloor}{\oplus}}} \quad Br^{\ominus} \quad \xleftarrow{BrCH_2CH_2Br} \quad CH_3-C\overset{S}{\underset{NH_2}{\diagdown}} \quad \xrightarrow{BrCH_2CH_2Br}$$

$$\underset{(15)}{CH_3-C\overset{S-CH_2-CH_2-S}{\underset{NH_2}{\diagdown}}\overset{\oplus}{\underset{H_2N}{\diagdown}}C-CH_3}$$

There are some cases of base-catalyzed direct alkylation at the nitrogen atom of thioamides, namely, if the alkylating agent is capable of forming stable carbonium ions (44, 65, 66). Thus thiobenzamide reacts with trityl chloride in pyridine to yield *N*-tritylthiobenzamide (**17**) (65); with 4,4'-dimethoxydiphenylchloromethane in DMF in the presence of potassium carbonate it gives thioamide **18**, whereas in ether the salt **19** is obtained (44).

$$\underset{(17)}{C_6H_5-C\overset{S}{\underset{NHC(C_6H_5)_3}{\diagdown}}} \qquad \underset{(18)}{C_6H_5-C\overset{S}{\underset{NHCH(C_6H_4-p\text{-}OCH_3)_2}{\diagdown}}}$$

$$\underset{(19)}{C_6H_5-C\overset{SCH(C_6H_4-p\text{-}OCH_3)_2}{\underset{NH_2}{\oplus}}} \quad Cl^{\ominus}$$

Even those carbinols that easily form stable carbonium ions with acids, such as trityl alcohol, 9-hydroxyxanthene, and benzhydrol, alkylate primary thioamides at the nitrogen atom (44, 66). Thus, for example, thioformamide, *N*-monosubstituted thioformamides, acetamides, and benzamides are immediately alkylated at the nitrogen atom to give **20**, when reacted with 9-hydroxyxanthene in glacial acetic acid (44). *t*-Butanol, methyldiphenylcarbinol, and 4,4',4''-trimethoxytriphenylcarbinol do not react with thiobenzamide under the same reaction conditions (44).

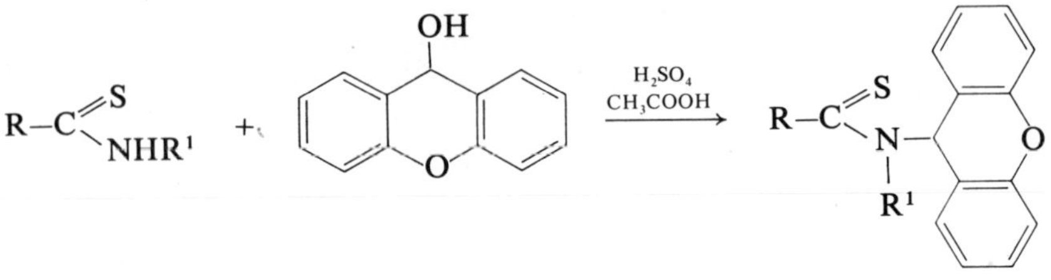

(20)

N-alkylation is the thermodynamically controlled reaction, whereas *S*-alkylation is considered to be kinetically controlled (66). *N*-Methyl-*N′*-thiosalicylpiperazine is also alkylated by alkyl halides at the nitrogen atom to give

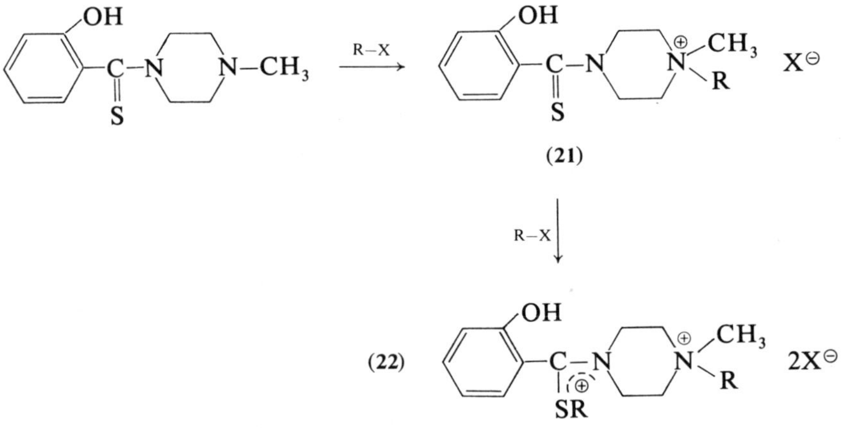

(21)

(22)

the ammonium salt **21**, which reacts with further alkyl halide at the sulfur atom to give **22** (32).

Methylation of indolethiocarboxylic acid amide (**23**) at the sulfur atom was also successful (55a).

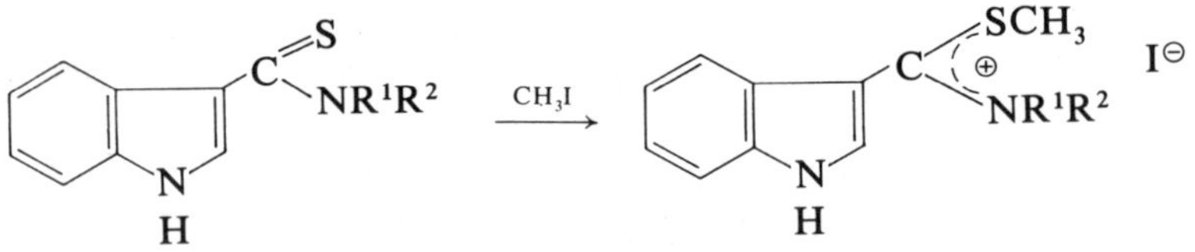

(23)

$R^1–R^2 = –(CH_2)_4–, –(CH_2)_5–, –(CH_2)_2–O–(CH_2)_2–$

I. SYNTHESIS

Thiohydrazides are methylated by methyl iodide at the sulfur atom; thus the salts (**24**) form in good yields (67).

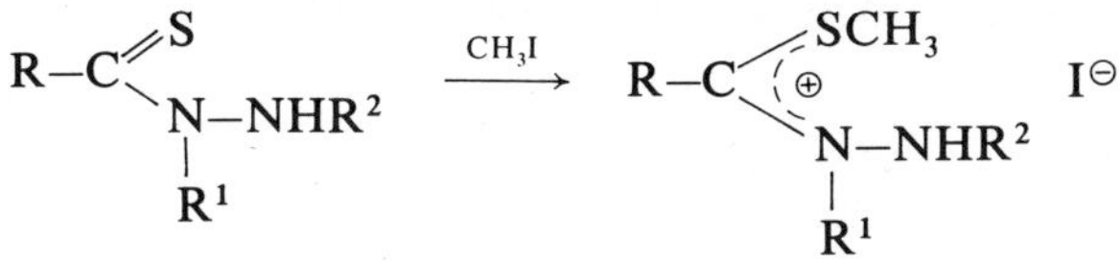

$$R-C(=S)(N(R^1)-NHR^2) \xrightarrow{CH_3I} [R-C(SCH_3)=N^+(R^1)-NHR^2] \; I^-$$

(**24**)

B. VIA ALKYLATION AND ACYLATION OF THIOIMINO ESTERS

Thioimino esters **25** are alkylated at the nitrogen atom with formation of the salts **26** (15, 33, 37, 68). Via this route the salts **27** and **28** have been obtained, which are of synthetic interest (33, 37).

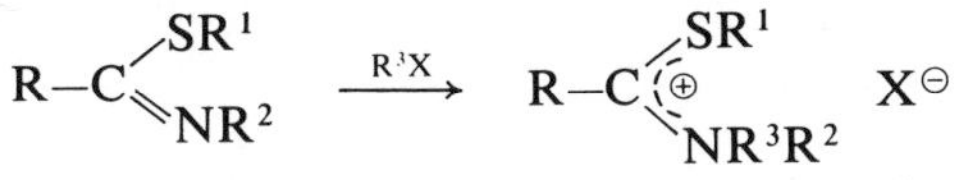

$$R-C(SR^1)=NR^2 \xrightarrow{R^3X} [R-C(SR^1)=N^+R^3R^2] \; X^-$$

(**25**) → (**26**)

[o,p-NHR-C_6H_4]-C(SR^3)=NR^2 $\xrightarrow{R^1I}$ [o,p-NHR-C_6H_4]-C(SR^3)=N^+R^1R^2 I^-

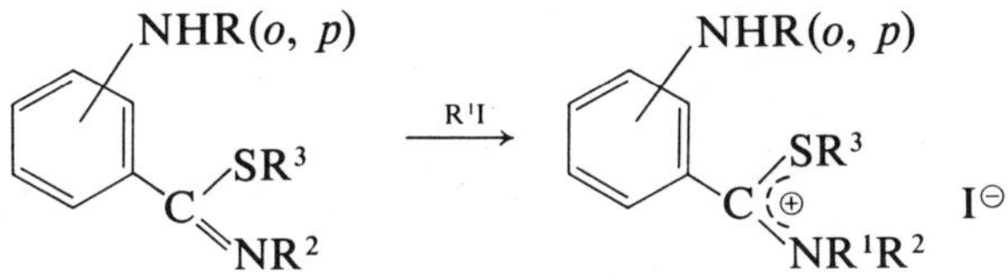

(**27**)

R = SO$_2$R′; 2,4-(NO$_2$)$_2$–C$_6$H$_3$; 2,4,6-(NO$_2$)$_3$–C$_6$H$_2$

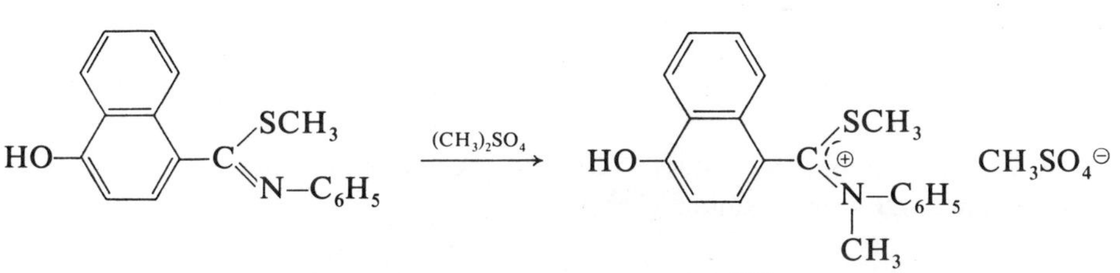

[HO-naphthyl]-C(SCH$_3$)=N-C$_6$H$_5$ $\xrightarrow{(CH_3)_2SO_4}$ [HO-naphthyl]-C(SCH$_3$)=N$^+$(CH$_3$)(C$_6$H$_5$) CH$_3$SO$_4^-$

(**28**)

Methyl fluorosulfonate has also been used for the alkylation of thioimino esters (55):

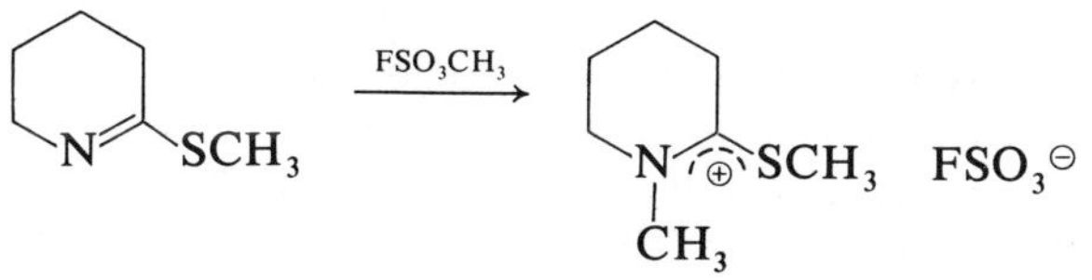

piperidine-N=C(SCH$_3$) $\xrightarrow{FSO_3CH_3}$ [piperidine-N$^+$(CH$_3$)=C(SCH$_3$)] FSO$_3^-$

If thioimino esters are acylated, iminium salts like **29** and **30** form as well; these are not isolated but rather are converted into pyrroline derivatives **31** with base or cyclized to the compounds **32** (54, 69, 70).

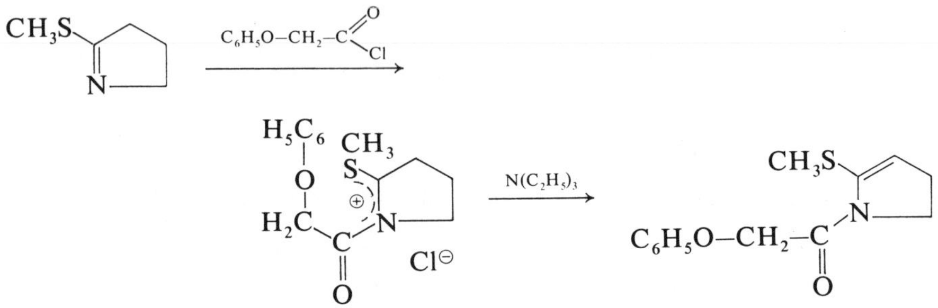

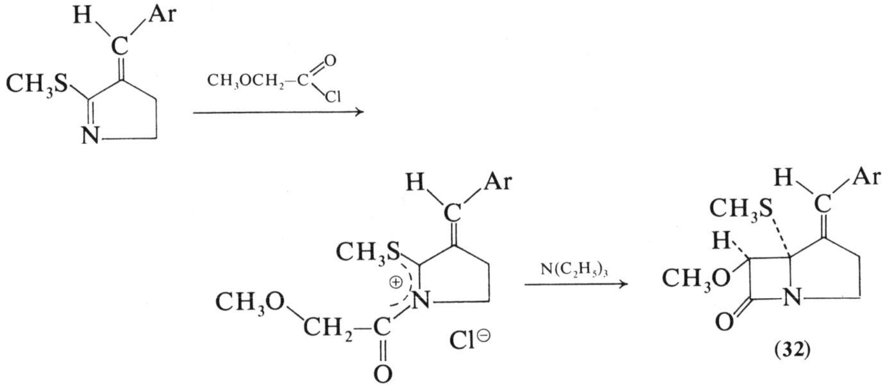

From the thioimino esters **32a** and acetic anhydride the mesoionic compounds **32b** were obtained (55b).

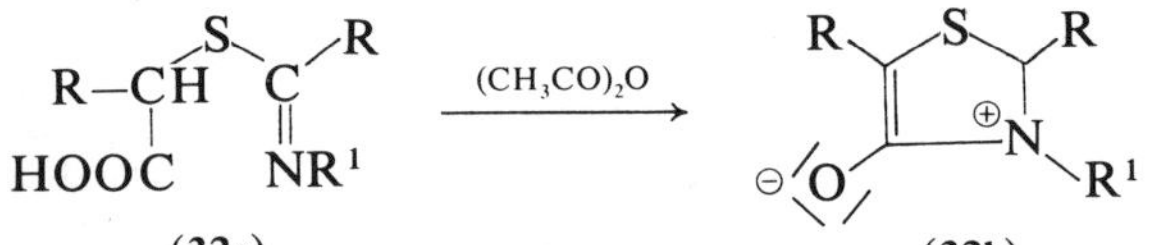

For a study of tautomerism in *N*-acylthioimino esters (**33**) and acylketene *S,N*-acetals (**34**), a series of *N*-acylthioimino esters (**33**) was obtained from thioimino esters generated *in situ* and acid chlorides; the iminium ions formed as intermediates were not isolated (49).

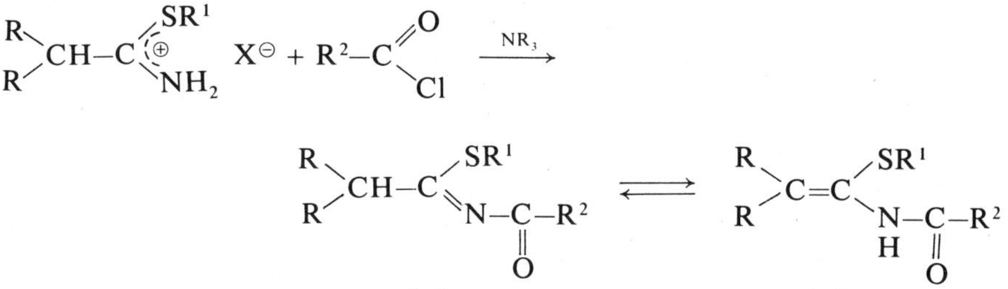

$$R_2CH-C(SR^1)(\overset{\oplus}{N}H_2)\ X^{\ominus} + R^2-C(=O)Cl \xrightarrow{NR_3}$$

$$R_2CH-C(SR^1)(N-C(=O)-R^2) \rightleftarrows R_2C=C(SR^1)(N(H)-C(=O)-R^2)$$

(33) (34)

To investigate the E/Z isomerism of N-(arylthio)thioimino esters (**35**), thioimino esters were reacted with arylsulfenyl chlorides to yield **35** (50). No intermediates were isolated here either.

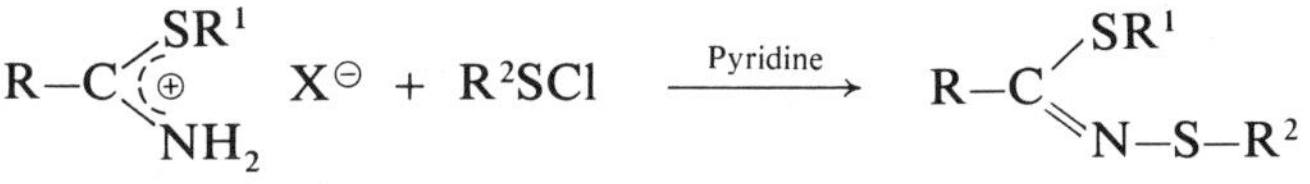

$$R-C(SR^1)(\overset{\oplus}{N}H_2)\ X^{\ominus} + R^2SCl \xrightarrow{\text{Pyridine}} R-C(SR^1)(=N-S-R^2)$$

(35)

C. VIA ACYLATION OF THIOAMIDES

There are several older studies concerning acylation of thioamides. According to them, carboxylic acid chlorides acylate thioamides at the nitrogen atom via formation of N-(acylthio)carboxylic acid amides. These reactions have been reviewed (57, 59). More recently the research groups of Goerdeler (71–73) and Walter (44, 74–78) have studied these reactions.

Thioamides are converted by ethyl-chlorothiolformate acid ester into the iminium salts **36** (79).

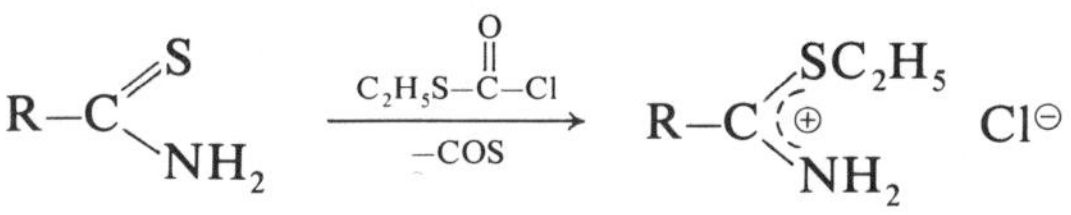

$$R-C(=S)(NH_2) \xrightarrow[\text{-COS}]{C_2H_5S-\overset{O}{\overset{\|}{C}}-Cl} R-C(SC_2H_5)(\overset{\oplus}{N}H_2)\ Cl^{\ominus}$$

(36)

Sulfenyl chlorides generally attack thioamides like **37** at the sulfur atom, primarily forming salts like **38** (74, 76, 80, 81). These have only rarely been isolated, but rather reacted with tertiary amines to give iminoalkyl disulfides **39**.

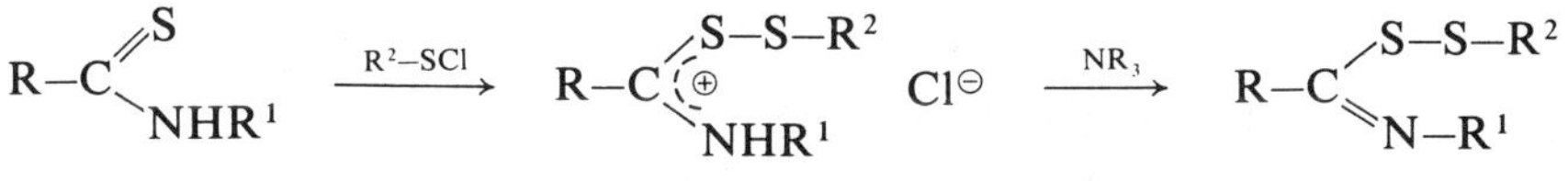

$$R-C(=S)(NHR^1) \xrightarrow{R^2-SCl} R-C(S-S-R^2)(\overset{\oplus}{N}HR^1)\ Cl^{\ominus} \xrightarrow{NR_3} R-C(S-S-R^2)(=N-R^1)$$

(37) (38) (39)

When β-chlorovinylmethyleniminium salts (**40**) act on *N,N*-disubstituted thioamides, 2-dialkylaminothiopyrylium salts (**41**) form (82).

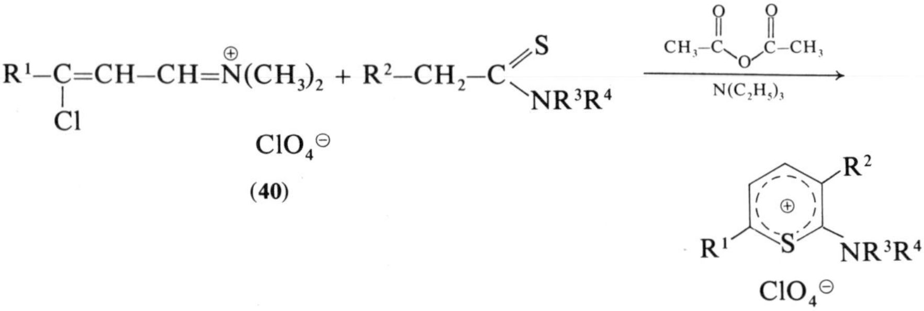

$$R^1-\overset{+}{C}=CH-CH=\overset{\oplus}{N}(CH_3)_2 \;+\; R^2-CH_2-C\overset{S}{\underset{NR^3R^4}{\diagdown}} \quad \xrightarrow[N(C_2H_5)_3]{CH_3-CO-O-CO-CH_3}$$

with Cl on the first carbon, $ClO_4^\ominus$

(**40**)

→ 2-dialkylaminothiopyrylium salt with R^1, R^2, NR^3R^4 substituents, $ClO_4^\ominus$

(**41**)

D. FROM CHLOROMETHYLENIMINIUM SALTS (AMIDE CHLORIDES)

The salts **42** are synthesized from mercaptoacetic acid and amide chlorides (83). This reaction, which in principle is very versatile, possesses the disadvantage that the salts **42** strongly absorb the HCl eliminated; that is why they are obtained as analytically pure samples only with great difficulty (83).

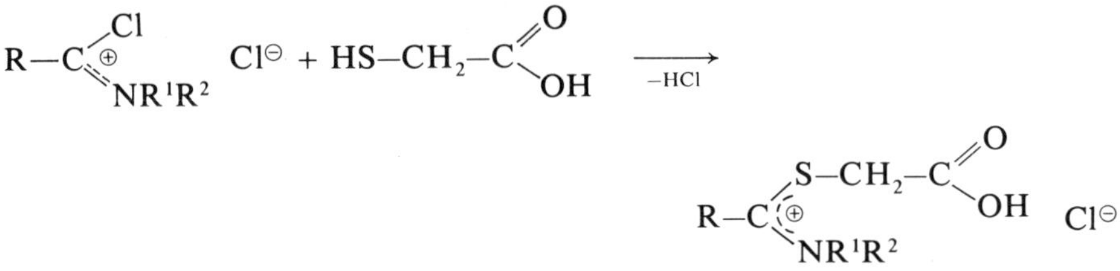

$$R-\overset{\oplus}{C}\overset{Cl}{\underset{NR^1R^2}{\diagdown}} Cl^\ominus \;+\; HS-CH_2-C\overset{O}{\underset{OH}{\diagdown}} \quad \xrightarrow{-HCl}$$

$$R-\overset{\oplus}{C}\overset{S-CH_2-C(=O)OH}{\underset{NR^1R^2}{\diagdown}} \; Cl^\ominus$$

(**42**)

E. FROM NITRILES, MERCAPTANS, AND HYDROGEN HALIDES

In the presence of hydrogen halides alkyl- or arylmercaptomethyleniminium salts **43** form via addition of mercaptans (4, 9, 15, 49, 62, 84–88) or thiophenols (89), respectively, to nitriles.

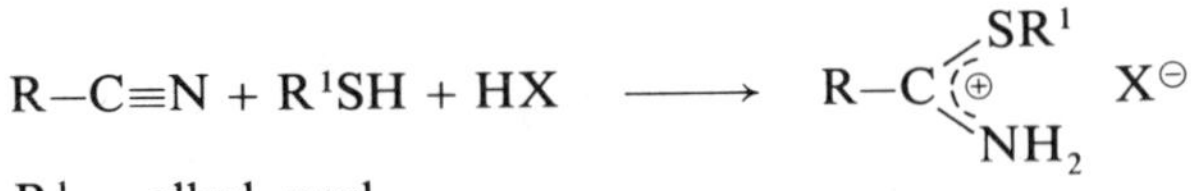

$$R-C\equiv N \;+\; R^1SH \;+\; HX \quad \longrightarrow \quad R-\overset{\oplus}{C}\overset{SR^1}{\underset{NH_2}{\diagdown}} \; X^\ominus$$

R^1 = alkyl, aryl

(**43**)

F. VIA REARRANGEMENT OF SECONDARY THIOAMIDES

Secondary thioamides like **44**, which carry substituents at their nitrogen atoms capable of forming stable carbonium ions, are rearranged with HCl in ether to give thioimino ester salts like **45** (44).

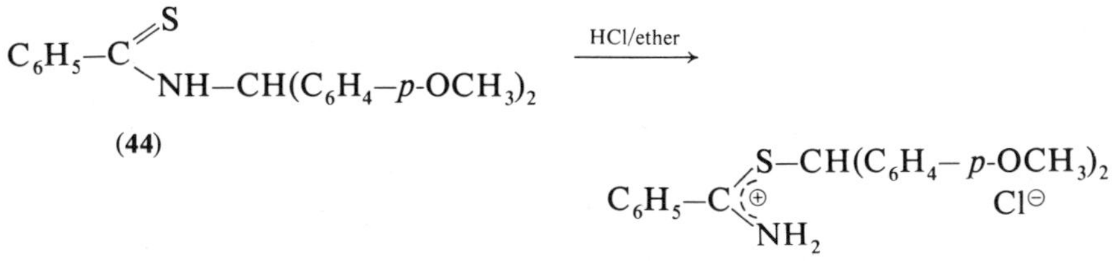

$$C_6H_5-C\underset{NH-CH(C_6H_4-p\text{-}OCH_3)_2}{\overset{S}{\diagup}} \xrightarrow{HCl/ether} C_6H_5-C\underset{NH_2}{\overset{S-CH(C_6H_4-p\text{-}OCH_3)_2}{\diagup}} \; Cl^{\ominus}$$

(44) → (45)

G. VIA REACTION OF KETENE S,N-ACETALS WITH ELECTROPHILIC REAGENTS

Ketene S,N-acetals **46** react with acids to give alkylmercaptomethyleniminium salts **47** (90). However, this reaction is important only in particular cases, since the salts **47** on the other hand are used as starting materials for the ketene S,N-acetal synthesis (see Section III-C).

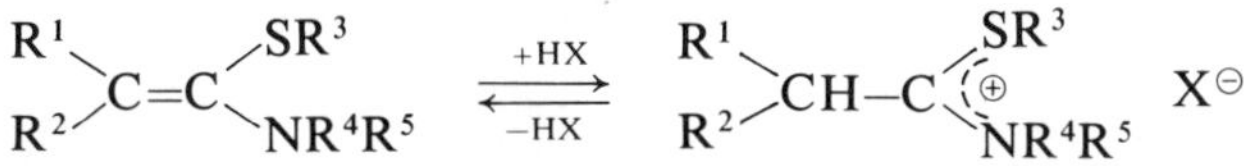

$$R^1\underset{R^2}{\diagdown}C=C\underset{NR^4R^5}{\overset{SR^3}{\diagup}} \underset{-HX}{\overset{+HX}{\rightleftarrows}} R^1\underset{R^2}{\diagdown}CH-C\underset{NR^4R^5}{\overset{SR^3}{\diagup}} \; X^{\ominus}$$

(46) (47)

Electrophilic reagents primarily form iminium salts (**49**) with ketene S,N-acetals (**48**), but they are transformed without isolation into ketene S,N-acetals (**50**) with base (90). Thus in the presence of triethylamine the ketene S,N-acetal

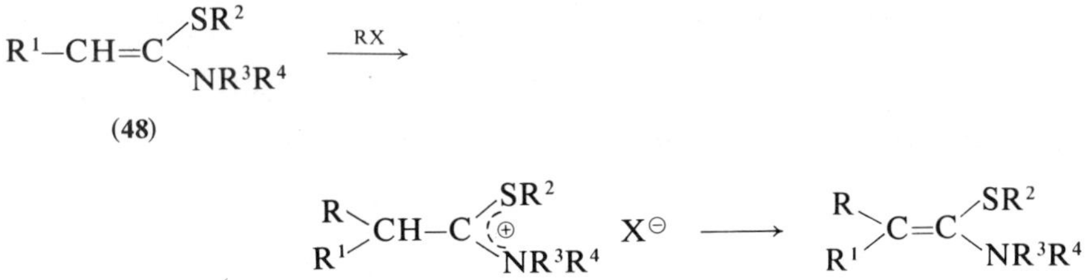

$$R^1-CH=C\underset{NR^3R^4}{\overset{SR^2}{\diagup}} \xrightarrow{RX}$$

(48)

$$\underset{R^1}{\overset{R}{\diagdown}}CH-C\underset{NR^3R^4}{\overset{SR^2}{\diagup}} \; X^{\ominus} \longrightarrow \underset{R^1}{\overset{R}{\diagdown}}C=C\underset{NR^3R^4}{\overset{SR^2}{\diagup}}$$

(49) (50)

52 forms from the acetal **51** and benzoyl chloride in DMF; with $HClO_4$ it gives the salt **53** (90).

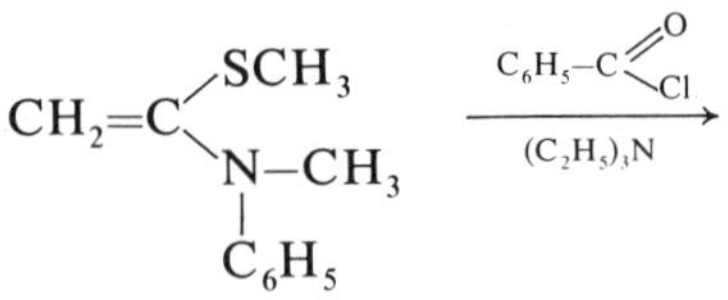

Ketene acetals **54** react with benzoyl chloride/triethylamine via the intermediate **55** to yield the β-ketothioamide derivatives **56** (90).

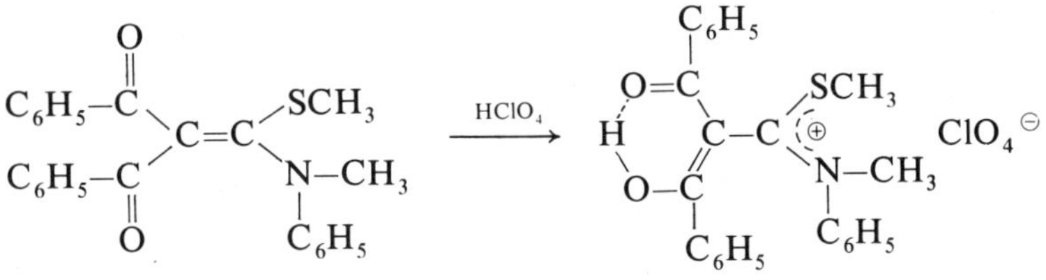

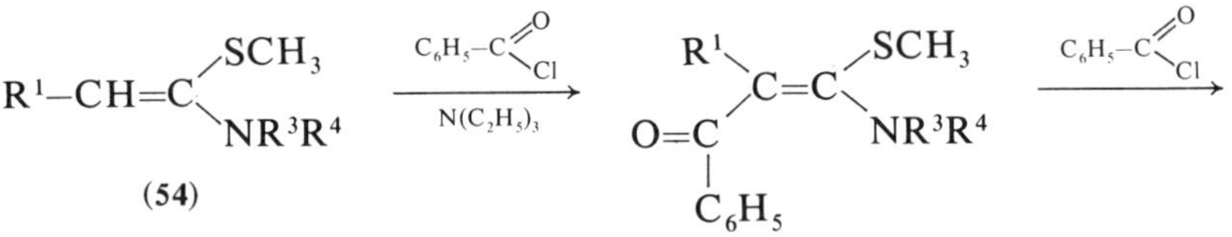

Acylation of **57** or **58** with ethyl chloroformate/triethylamine also generates the previously unknown ketene S,N-acetals **59** or **60**, respectively (90).

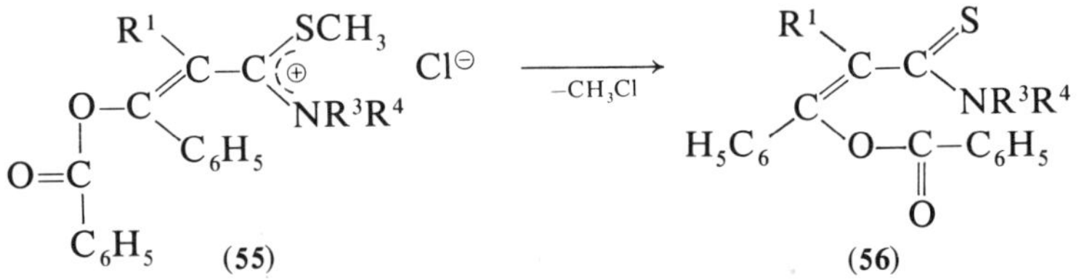

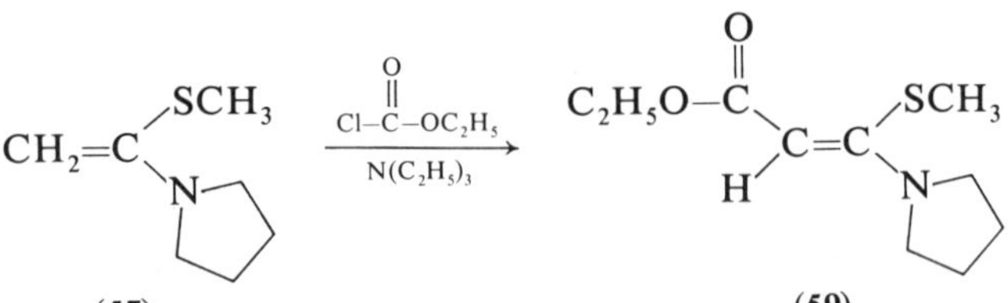

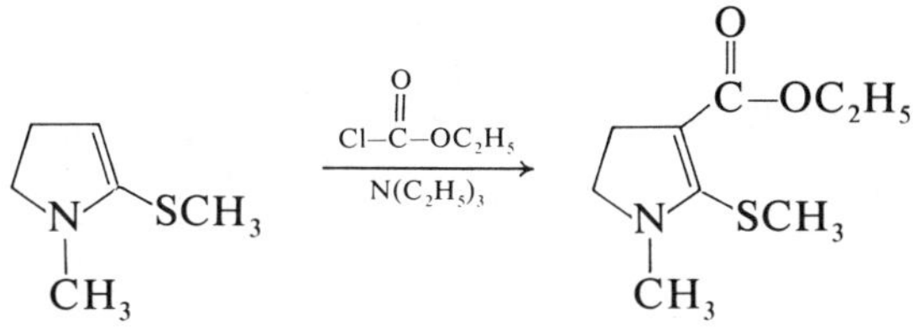

I. SYNTHESIS

(58) → (60)

Furthermore, a series of activated halogen compounds was reacted with ketene *S,N*-acetals without isolation of the initially formed iminium salts (90). Scheme 1 summarizes these reactions.

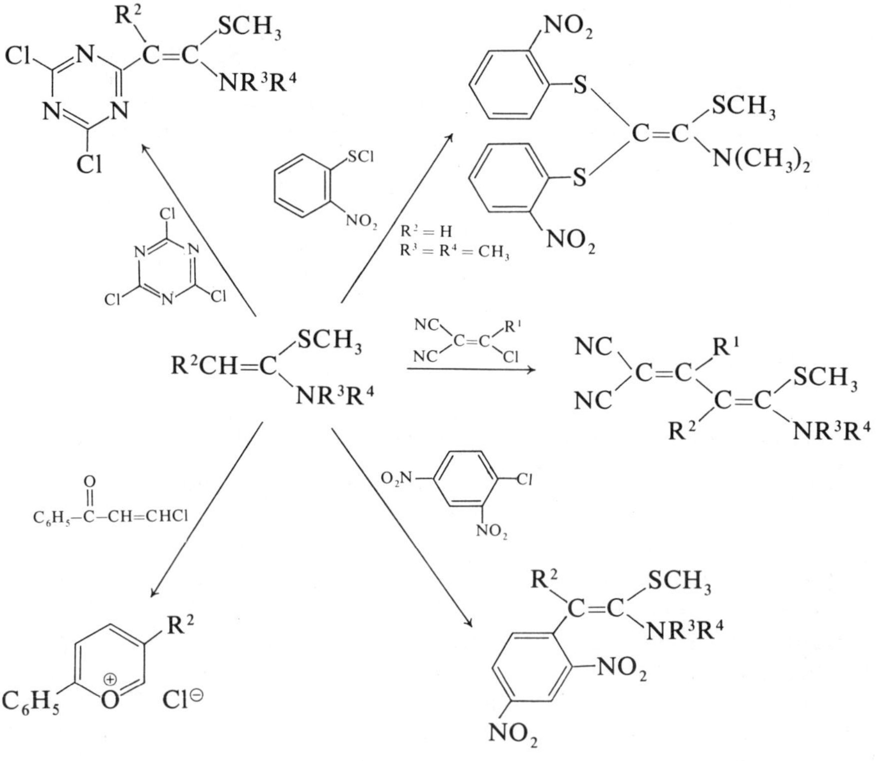

Scheme 1

H. FROM CHLOROALKYLMERCAPTOMETHYLENIMINIUM SALTS

When heated in benzene with chloroalkylmercaptomethyleniminium salts (**60a**), heterocycles like pyrrole and indole yield alkylmercaptomethyleniminium salts **60b**, **60c** (90a). In the presence of pyridine the salts **60a** react

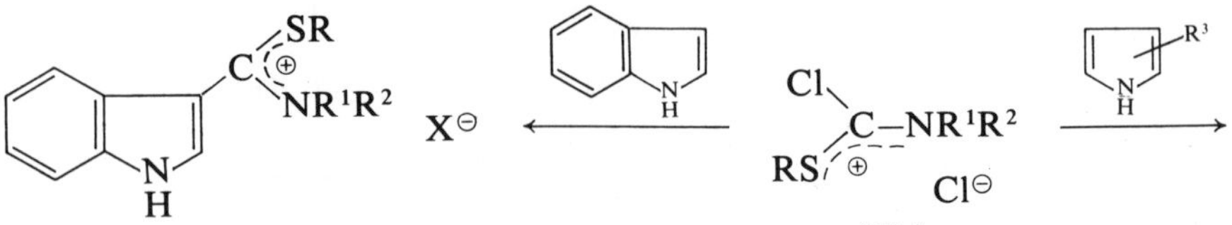

[Scheme showing conversion between (60a), (60b), and (60c)]

(60b): indole-substituted C(SR)=NR¹R² X⁻
(60a): Cl–C(SR)–NR¹R² Cl⁻ (from indole-H and pyrrole-R³)
(60c): pyrrole-substituted C(SR)=NR¹R² Cl⁻

For (60b): $R^1 = R^2 = CH_3$, X = Cl, ClO$_4$
$R^1-R^2 = -(CH_2)_4-$, X = Cl

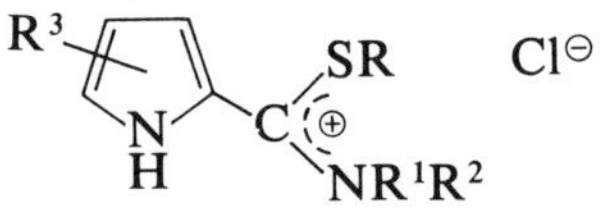

For (60c): $R^1 = R^2 = R = CH_3$
$R = CH_3$, $R^3 = -3COOC_2H_5$ -2,4-$(CH_3)_2$

with carbon acids like malononitrile in the same way. Here the initially formed iminium salts are deprotonated instantly, giving ketene S,N-acetals **60d** as the isolated products (90b).

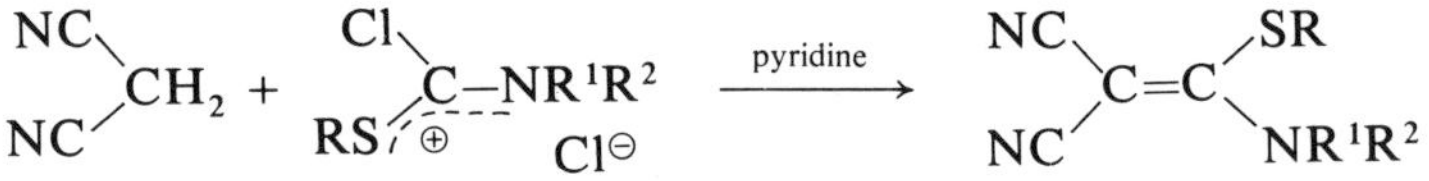

$$NC-CH_2-CN + Cl-C(SR)(NR^1R^2)\ Cl^- \xrightarrow{pyridine} (NC)_2C=C(SR)(NR^1R^2)$$

(60a) → (60d)

$R = CH_3$, $R^1 = R^2 = C_6H_5$
$R = CH_3$, $R^1-R^2 = -(CH_2)_5-$

II. Properties and Structure of Alkylmercaptomethyleniminium Salts

The physical properties of alkylmercaptomethyleniminium salts, such as their high melting points and their behavior toward solvents, point at a polar structure for this class of compounds. Their structure has been established by hydrolysis (1–4, 15, 52) as well as by transformation into imino thioesters (1, 2, 15, 21, 44). Furthermore the IR band at ~1580 cm^{-1} found in alkylmercaptomethyleniminium salts indicates the presence of a C=N$^\oplus$ group (21, 22, 27, 32, 40, 42).

With some salts derived from dithiosquaric acid amides, higher values (1772, 1620, and 1750, 1665 cm^{-1}) are also reported (52).

The NMR spectra point to hindered rotation around the C=N$^\oplus$ bond (40, 42, 52).

III. Reactions of Alkylmercaptomethyleniminium Salts

A. HYDROLYSIS, ALCOHOLYSIS

Alkylmercaptomethyleniminium salts (**4**) hydrolyze primarily to thioesters (**61**) in acidic media, whereas they give mercaptans (**62**) and amides in alkaline

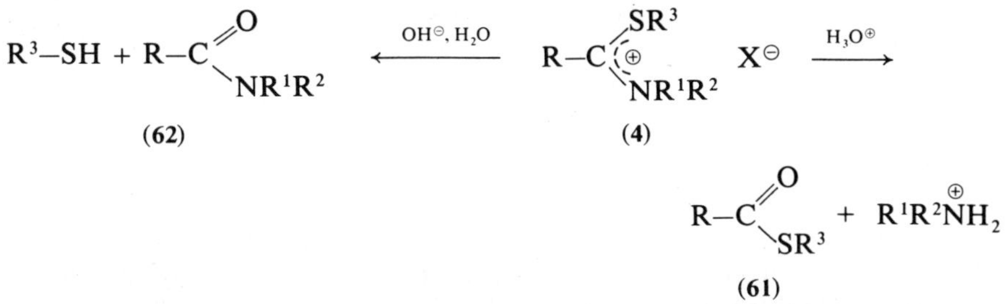

III. REACTIONS OF ALKYLMERCAPTOMETHYLENIMINIUM SALTS

$$R^3-SH + R-C(=O)NR^1R^2 \quad \xleftarrow{OH^\ominus, H_2O} \quad R-C^{\oplus}(SR^3)NR^1R^2 \;\; X^\ominus \quad \xrightarrow{H_3O^\oplus} \quad R-C(=O)SR^3 + R^1R^2\overset{\oplus}{N}H_2$$

(62) ← (4) → (61)

media (91). This reaction has already been summarized (57); kinetic studies have also been reported (40, 43, 92). Since the compounds **4** are hydrolytically cleaved, they may be used to synthesize mercaptans or amides (1–5, 15, 17, 28, 93) as well as thioesters (13–15, 44, 68, 83, 94, 95).

The iminium salt **62a** is converted to the amide **62b** or the ester **62c** by aqueous alkali or alcohol; on the other hand, aqueous pyridine yields the thioester **62d** (90a). The salt **62e** reacts the same way (90a).

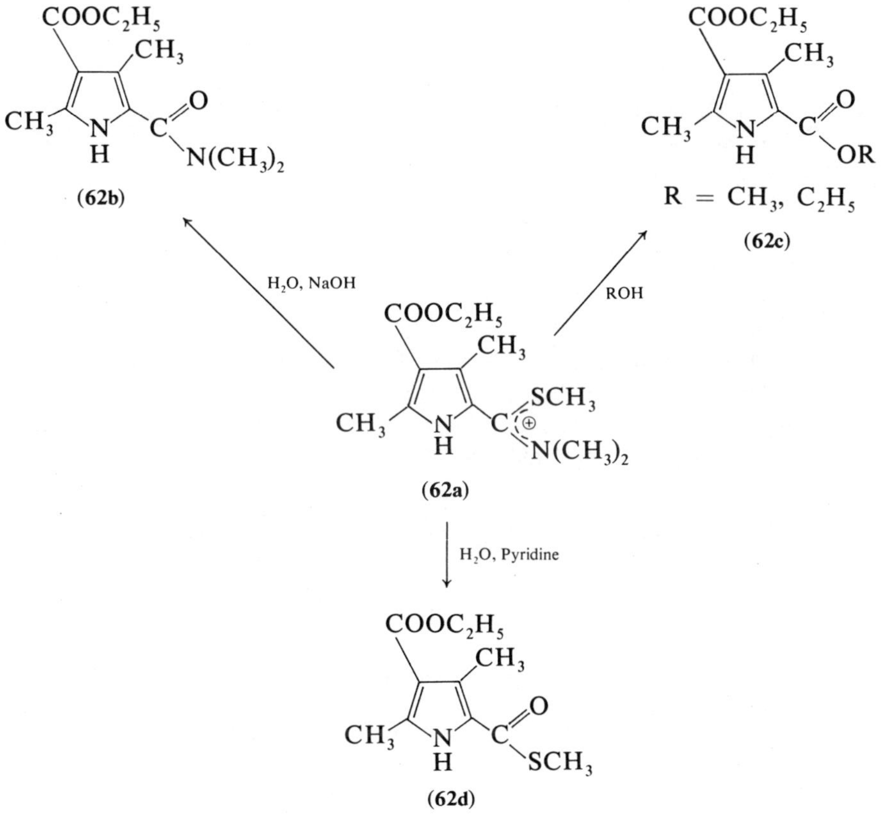

(62b) — pyrrole with COOC$_2$H$_5$, CH$_3$, CH$_3$, N–H, C(=O)N(CH$_3$)$_2$

(62c) — pyrrole with COOC$_2$H$_5$, CH$_3$, CH$_3$, N–H, C(=O)OR; R = CH$_3$, C$_2$H$_5$

(62a) — pyrrole with COOC$_2$H$_5$, CH$_3$, CH$_3$, N–H, C$^{\oplus}$(SCH$_3$)N(CH$_3$)$_2$

H$_2$O, NaOH ↖   ROH ↗

↓ H$_2$O, Pyridine

(62d) — pyrrole with COOC$_2$H$_5$, CH$_3$, CH$_3$, N–H, C(=O)SCH$_3$

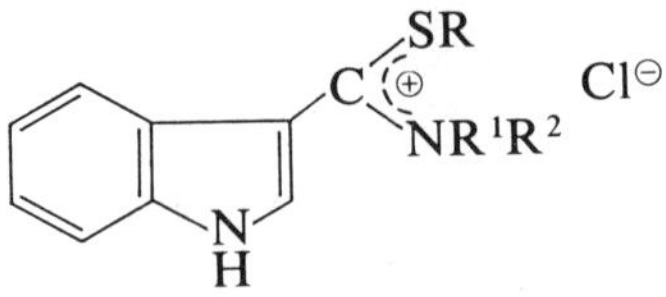

$$\text{(62e)} \quad \underset{\underset{H}{N}}{\text{indole-3-yl}}-C(\oplus)(SR)(NR^1R^2) \; Cl^\ominus$$

N,N-Disubstituted thioamides **2** are converted into *N,N*-disubstituted amides **63** if alkylated with trimethyloxonium fluoroborate (47) or dimethyl sulfate (6) and the salts obtained (**64**) are cleaved with alkali.

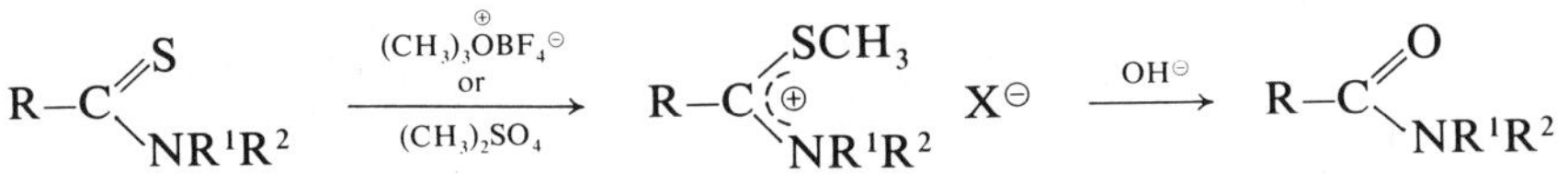

$$R-C(=S)(NR^1R^2) \xrightarrow[\text{(CH}_3)_2\text{SO}_4]{(CH_3)_3O^\oplus BF_4^\ominus \text{ or}} R-C(\oplus)(SCH_3)(NR^1R^2) \; X^\ominus \xrightarrow{OH^\ominus} R-C(=O)(NR^1R^2)$$
$$\text{(2)} \hspace{6cm} \text{(63)}$$

To synthesize α-mercaptocarboxylic acid esters (**64**) via hydrolysis, α-halocarboxylic acid esters are reacted with thioformamide to yield the salts **65**, followed by degradation with aqueous alcohol (28).

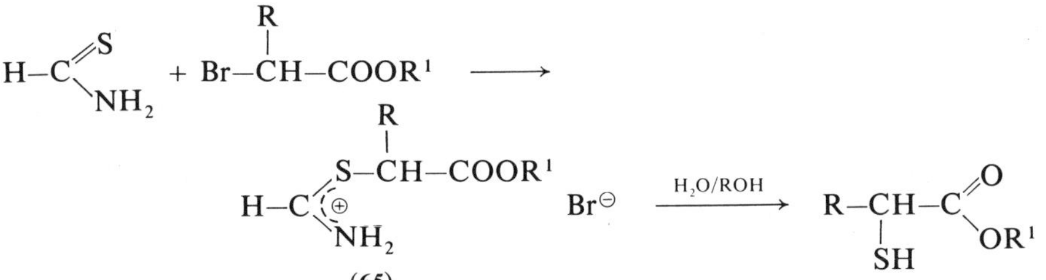

$$H-C(=S)(NH_2) + Br-CH(R)-COOR^1 \longrightarrow$$

$$H-C(\oplus)(S-CH(R)-COOR^1)(NH_2) \; Br^\ominus \xrightarrow{H_2O/ROH} R-CH(SH)-C(=O)(OR^1)$$
$$\text{(65)} \hspace{6cm} \text{(64)}$$

Salts of the type **66** carrying ligands R^1 at the sulfur atom capable of forming stable carbonium ions may also give thioamides **67** upon acidic hydrolysis; in addition, alkyl migration of ligand R^1 from sulfur to nitrogen sometimes occurs, giving thioamides **68** (44).

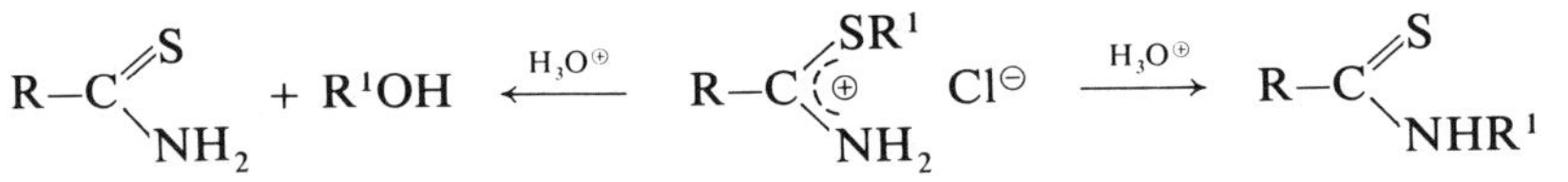

$$R-C(=S)(NH_2) + R^1OH \xleftarrow{H_3O^\oplus} R-C(\oplus)(SR^1)(NH_2) \; Cl^\ominus \xrightarrow{H_3O^\oplus} R-C(=S)(NHR^1)$$
$$\text{(67)} \hspace{4cm} \text{(66)} \hspace{4cm} \text{(68)}$$

(67) $R = CH_3, C_6H_5,$
$R^1 = HC(C_6H_5)_2$ \hspace{2cm} (68) $R = C_6H_5, \; R^1 = HC-(p-C_6H_4-OCH_3)_2$

Water causes the loss of dimethylamine from 3,4-bis(dimethylamino)-1-methylthio-2-thionocyclobutenylium fluorosulfate (**69**); thus dimethylamino-1-methylthio-3-oxo-4-thionocyclobutene **70** forms (52).

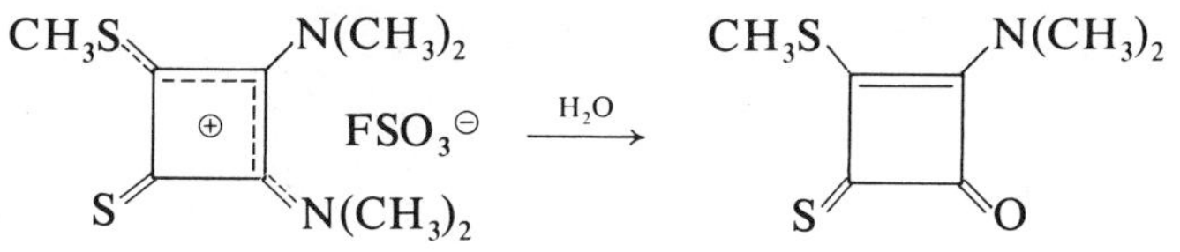

III. REACTIONS OF ALKYLMERCAPTOMETHYLENIMINIUM SALTS

(69) [CH₃S, N(CH₃)₂, S, N(CH₃)₂ cyclobutenyl cation] FSO₃⁻ →[H₂O] (70) [CH₃S, N(CH₃)₂, S, O cyclobutenone]

On hydrolysis of 3,4-bis(dimethylamino)-1,2-bis(methylthio)cyclobutenediylium bis(fluorosulfate) (**71**) both dimethylamino groups are lost to yield **72** (52).

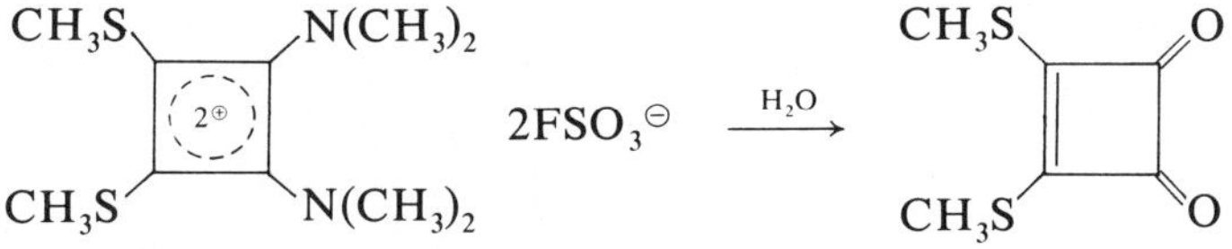

(71) [CH₃S, N(CH₃)₂, CH₃S, N(CH₃)₂ dication] 2FSO₃⁻ →[H₂O] (72) [CH₃S, O, CH₃S, O cyclobutenedione]

Alcoholysis of the salts **73** gives the *N,N*-disubstituted amide **74**, thiophenol, and ethyl iodide (68).

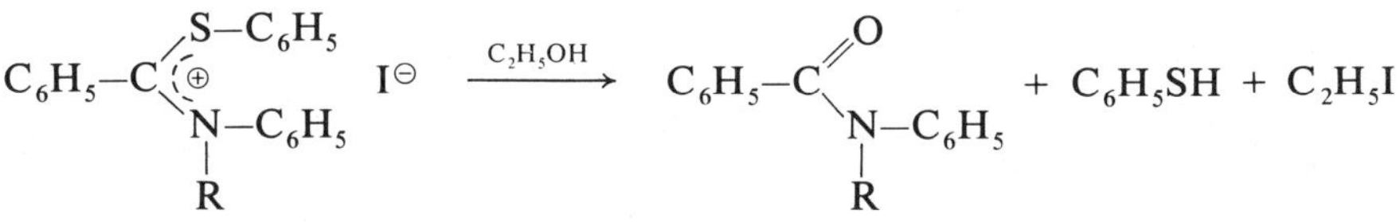

$$C_6H_5-C(\overset{S-C_6H_5}{\underset{N-C_6H_5}{\oplus}})\ I^\ominus \xrightarrow{C_2H_5OH} C_6H_5-C(\overset{O}{\underset{N-C_6H_5}{}}) + C_6H_5SH + C_2H_5I$$
$$\qquad\qquad\quad\;\, R \qquad\qquad\qquad\qquad\qquad\;\;\, R$$
$$\qquad\quad\; (73) \qquad\qquad\qquad\qquad\qquad (74)$$

B. REACTIONS WITH HYDROGEN SULFIDE, HYDROGEN SELENIDE, MERCAPTIDES, AND ALKOXIDES

The thiolysis of *N,N*-dialkylmercaptoalkylideniminium salts (**4**) yields dithioesters (**75**) via elimination of the dialkylamino groups (21, 52, 63, 83). A series of salts was converted with sodium hydrogen selenide in good yields into selenoamides (**76**), which are otherwise difficult to prepare (46).

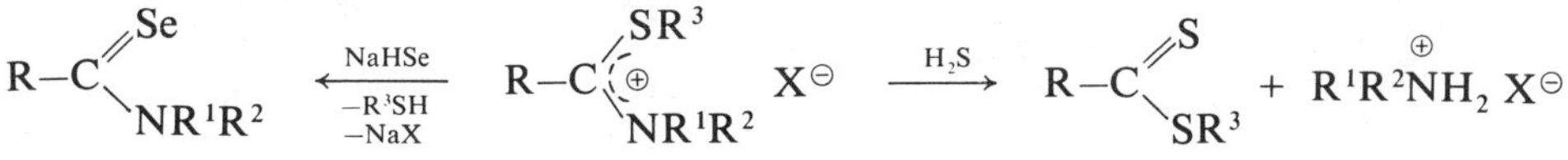

$$R-C(\overset{Se}{\underset{NR^1R^2}{}}) \xleftarrow[-R'SH \atop -NaX]{NaHSe} R-C(\overset{SR^3}{\underset{NR^1R^2}{\oplus}})\ X^\ominus \xrightarrow{H_2S} R-C(\overset{S}{\underset{SR^3}{}}) + R^1R^2\overset{\oplus}{N}H_2\ X^\ominus$$
$$\quad (76) \qquad\qquad\qquad\qquad (4) \qquad\qquad\qquad (75)$$

The reaction conditions of the thiolysis reaction seem to be of great importance, since under the influence of H_2O/Na_2S the salt **62a** forms the thioamide **76a**, whereas reaction with H_2S/pyridine yields the dithioester **76b** (90a).

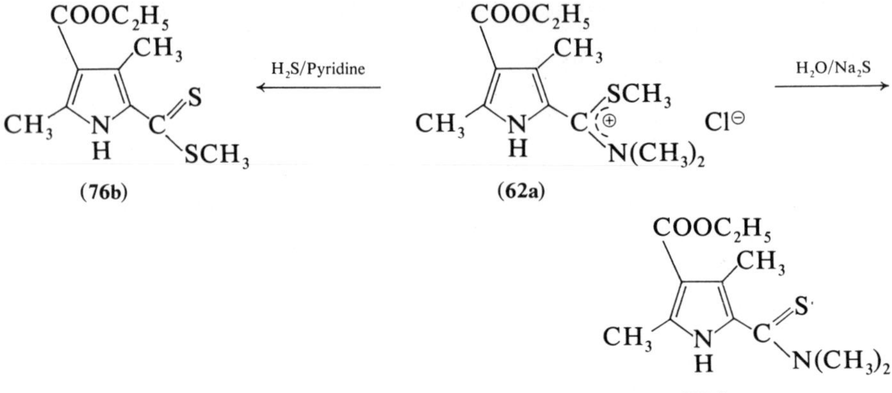

(76b) ⟵ H₂S/Pyridine ⟵ (62a) ⟶ H₂O/Na₂S ⟶ (76a)

Alkali mercaptides add to the salts **4** to give amide mercaptals (**77**) (96, 97). If the salts **4** contain a hydrogen atom in the position α to the acyl carbon atom, the mercaptide may also abstract a proton there, giving rise to ketene *S,N*-acetals **78** (98).

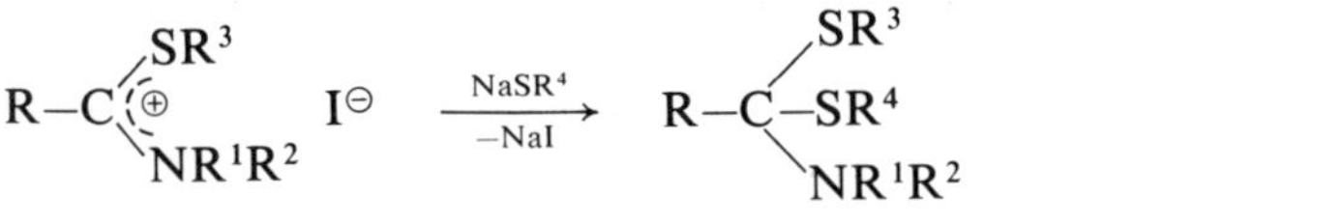

$$R-C(SR^3)(NR^1R^2)^{\oplus} \quad I^{\ominus} \xrightarrow[-NaI]{NaSR^4} R-C(SR^3)(SR^4)(NR^1R^2)$$

(4) (77)

$R^4 = CH_3, C_2H_5, \quad R = C_6H_5, \quad R^1 = R^2 = CH_3, \quad R^3 = CH_3$

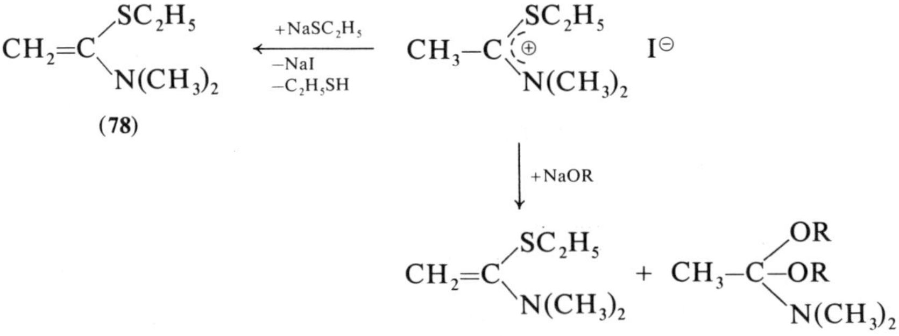

$$CH_2=C(SC_2H_5)(N(CH_3)_2) \xleftarrow[-NaI,\ -C_2H_5SH]{+NaSC_2H_5} CH_3-C(SC_2H_5)(N(CH_3)_2)^{\oplus} \quad I^{\ominus}$$

(78)

↓ +NaOR

$$CH_2=C(SC_2H_5)(N(CH_3)_2) + CH_3-C(OR)(OR)(N(CH_3)_2)$$

(78) (79)

Potassium *t*-butoxide behaves the same way; other alkoxides like sodium isopropoxide and sodium *n*-propoxide do yield the ketene *S,N*-acetal (**78**), but in addition the corresponding amide acetals (**79**) are formed (98).

If there is no α hydrogen, the alkoxide adds to give the mixed *O,S*-acetal (**80**), which decomposes into amide and dimethyl sulfide (96).

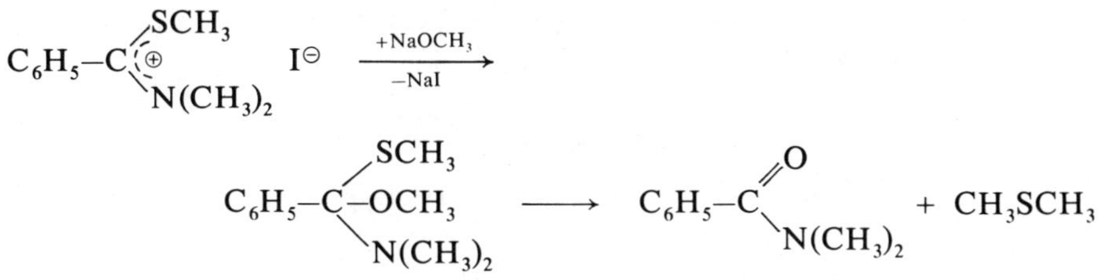

III. REACTIONS OF ALKYLMERCAPTOMETHYLENIMINIUM SALTS

$$C_6H_5-C(SCH_3)(\overset{\oplus}{N(CH_3)_2}) \;\; I^{\ominus} \xrightarrow[-NaI]{+NaOCH_3}$$

$$C_6H_5-C(SCH_3)(OCH_3)(N(CH_3)_2) \longrightarrow C_6H_5-C(=O)(N(CH_3)_2) + CH_3SCH_3$$

(80)

With sodium ethoxide the reaction of **81** takes a different course. A mixture of amide acetal **82** and amide thioacetal **83** is obtained (51).

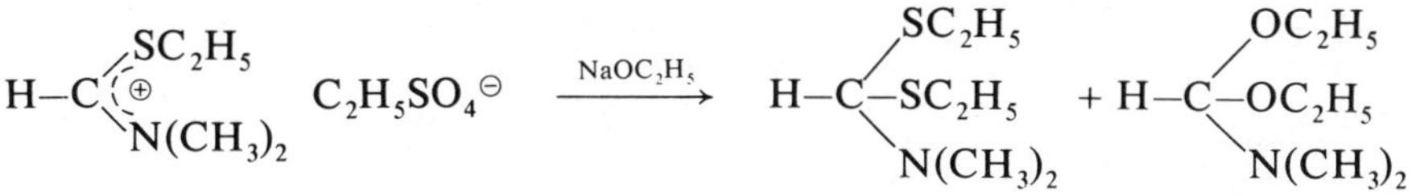

$$H-C(SC_2H_5)(\overset{\oplus}{N(CH_3)_2}) \;\; C_2H_5SO_4^{\ominus} \xrightarrow{NaOC_2H_5} H-C(SC_2H_5)(SC_2H_5)(N(CH_3)_2) + H-C(OC_2H_5)(OC_2H_5)(N(CH_3)_2)$$

(81)   (83)   (82)

If triethylamine and acetyl chloride act on the salt **84**, the thioacetal **85** forms. The same product results from the reaction of N-acetylthioacetamide (**86**) with benzylbromide and triethylamine (49).

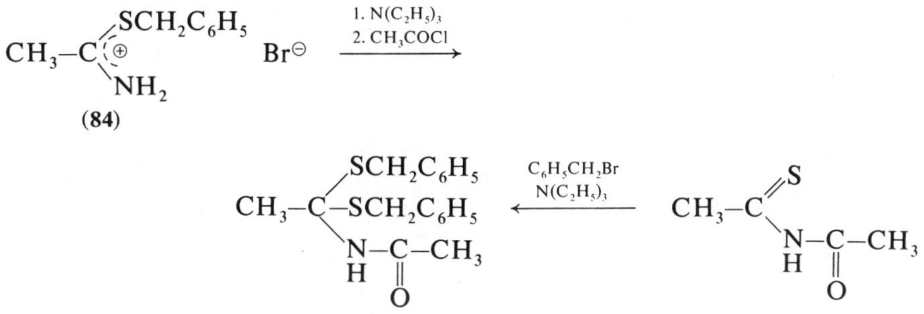

$$CH_3-C(SCH_2C_6H_5)(\overset{\oplus}{NH_2}) \;\; Br^{\ominus} \xrightarrow[2.\; CH_3COCl]{1.\; N(C_2H_5)_3}$$

$$CH_3-C(SCH_2C_6H_5)(SCH_2C_6H_5)(N(H)-C(=O)-CH_3) \xleftarrow[N(C_2H_5)_3]{C_6H_5CH_2Br} CH_3-C(=S)(N(H)-C(=O)-CH_3)$$

(85)   (86)

C. PROTON ABSTRACTION WITH BASE

Alkylmercaptomethyleniminium salts **87** not peralkylated at the nitrogen atom are deprotonated by bases like alkaline hydroxides, tertiary amines, or alkaline carbonates to yield thioimino esters **88** (1–4, 8, 15, 28, 30a, 33, 35, 42b, 44, 45, 49, 75).

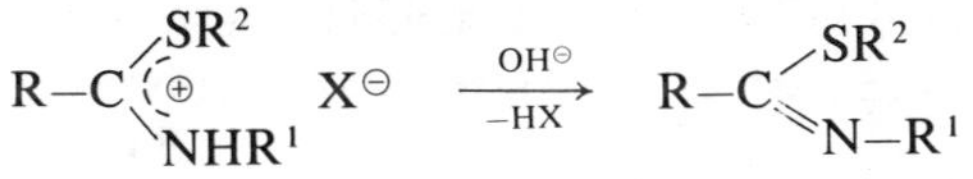

$$R-C(SR^2)(\overset{\oplus}{NHR^1}) \;\; X^{\ominus} \xrightarrow[-HX]{OH^{\ominus}} R-C(SR^2)=N-R^1$$

(87)   (88)

These reactions have already been reviewed (99). Thioimino ethers like **89–92** are accessible by this method (29, 31, 42b, 45, 54, 100).

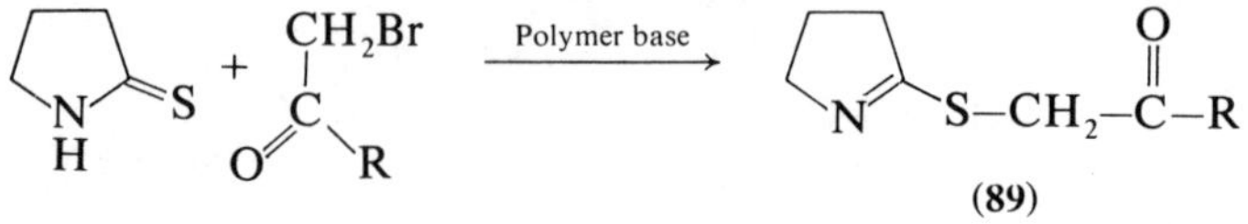

*[Reaction scheme showing pyrrolidine-2-thione + BrCH$_2$C(O)R with polymer base → compound (**89**)]*

*[Reaction scheme showing azepane-2-thione with (C$_2$H$_5$)$_3$O BF$_4$ / K$_2$CO$_3$ → compound (**89a**) with SC$_2$H$_5$ group]*

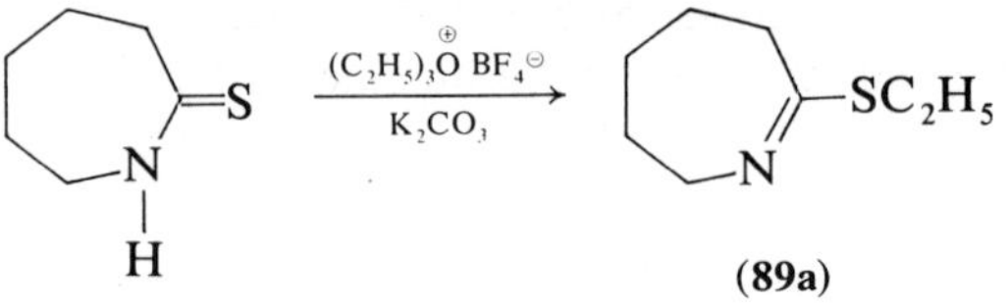

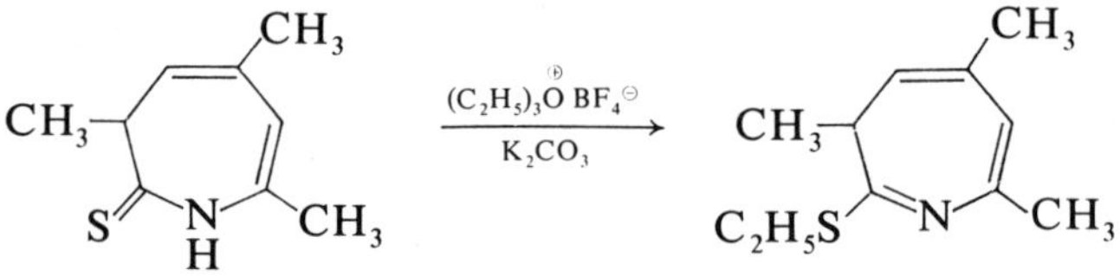

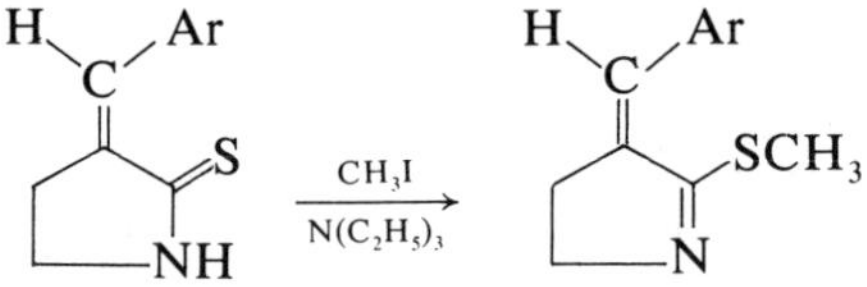

*[Reaction scheme of dimethyl-substituted azepinethione with (C$_2$H$_5$)$_3$O BF$_4$ / K$_2$CO$_3$ → compound (**90**)]*

*[Reaction scheme with CH$_3$I / N(C$_2$H$_5$)$_3$ → compound (**91**) with SCH$_3$ group]*

*[Reaction scheme of indoline-2-thione with CH$_2$=CH–CH$_2$Br / K$_2$CO$_3$ → compound (**92**)]*

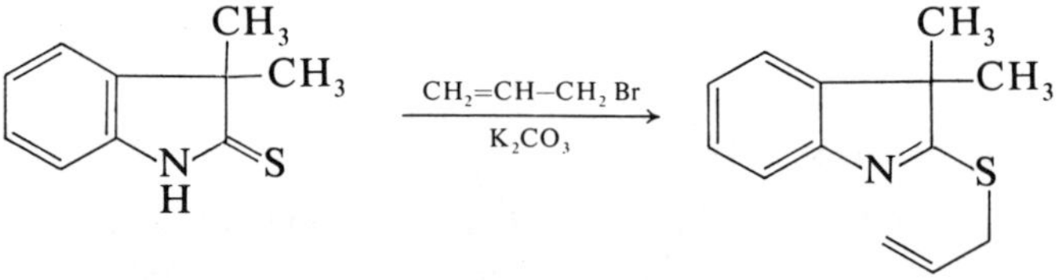

When sodium hydride in tetrahydrofuran acts on the salt **93**, the formyl hydrogen is abstracted to generate α-dimethylamino-α-methylthiocarbene (**94**) (101). **94** will react with further **93** to give a mixture of *cis* and *trans* isomers of 1,2-bis(dimethylamino)-1,2-bis(methylthio)ethylenes (**95**). As the main product the thioacetal **96** is isolated (52% yield) (101).

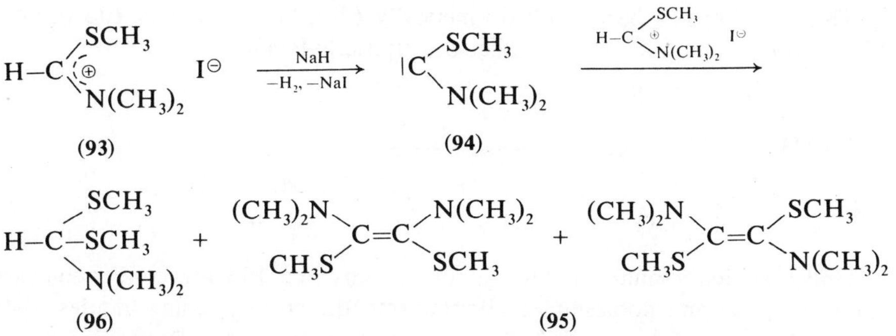

III. REACTIONS OF ALKYLMERCAPTOMETHYLENIMINIUM SALTS

[Structures **(93)**, **(94)**, **(96)**, **(95)** shown with NaH, -H₂, -NaI reaction]

Thiazolium salts **97** are deprotonated with bases like triethylamine to yield the nucleophilic carbenes **98**; they may be trapped with isocyanates or isothiocyanates as adducts of the type **99** (102).

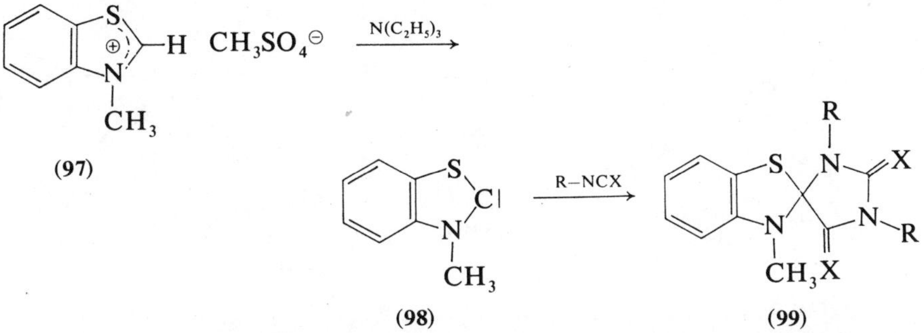

[Structures **(97)**, **(98)**, **(99)** shown with CH₃SO₄⁻, N(C₂H₅)₃, and R-NCX]

N,N-Persubstituted salts **4** and **5** having a CH bond in the position α to the acyl carbon atom may be deprotonated at the α position by bases like ethyl mercaptide (98) or alkoxides (21, 38, 39, 41–42a, 98, 103); they yield ketene S,N-acetals **100** or **101**.

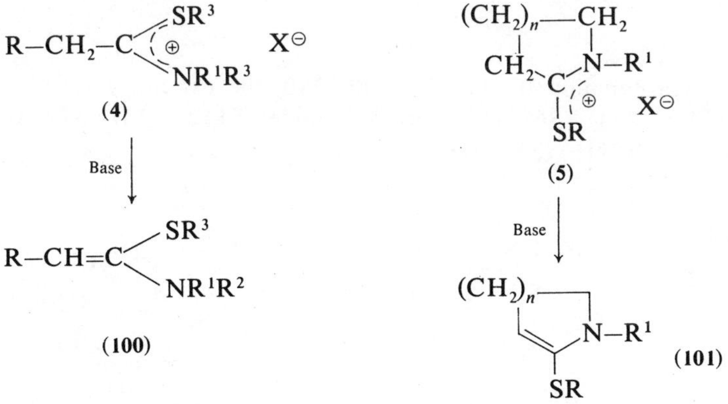

[Structures **(4)**, **(5)**, **(100)**, **(101)** shown with Base arrows]

This reaction has been studied kinetically (40). With the salts **102** proton abstraction can be successfully effected with triethylamine (54).

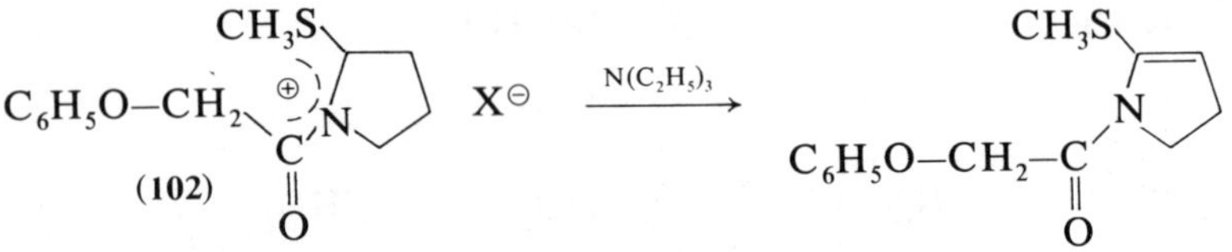

Under similarly mild conditions, thiolactams like **103** are converted with allyl bromides and potassium carbonate into the corresponding indoles **104**; on heating they undergo a thio-Claisen rearrangement (45). Even open-chain compounds undergo this Claisen rearrangement (39):

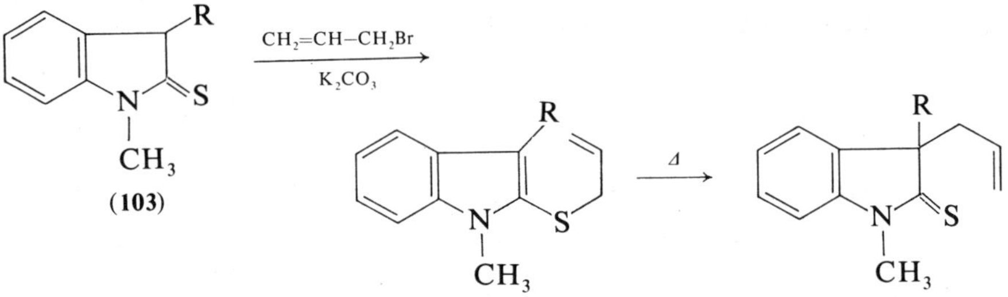

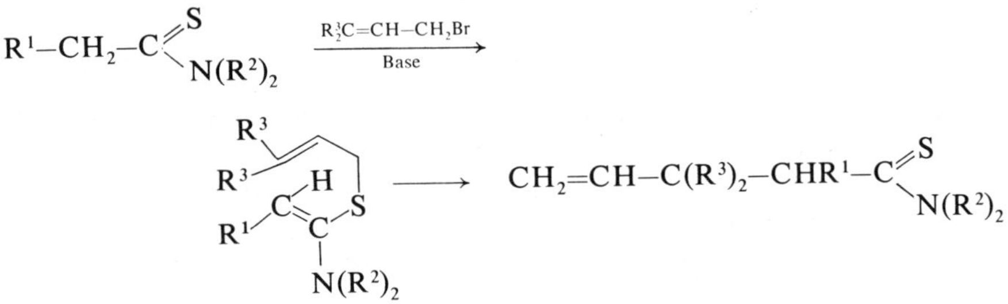

Via deprotonation of the salts **105–110**, the extremely reactive *o*-quinodimethides (**111**) (36), quinomethide imides (**112, 113**) (37), and quinomethides (**114–116**) (33, 104) form.

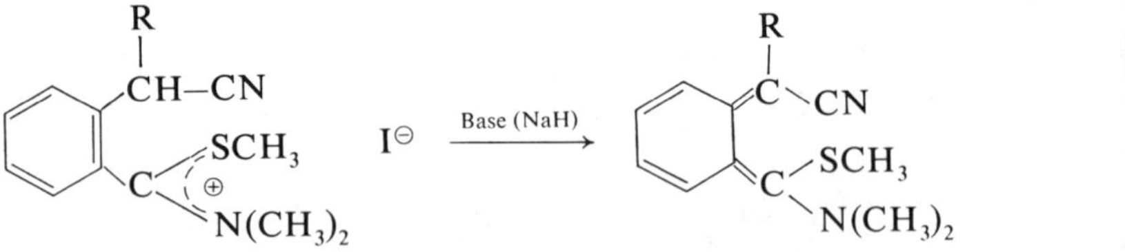

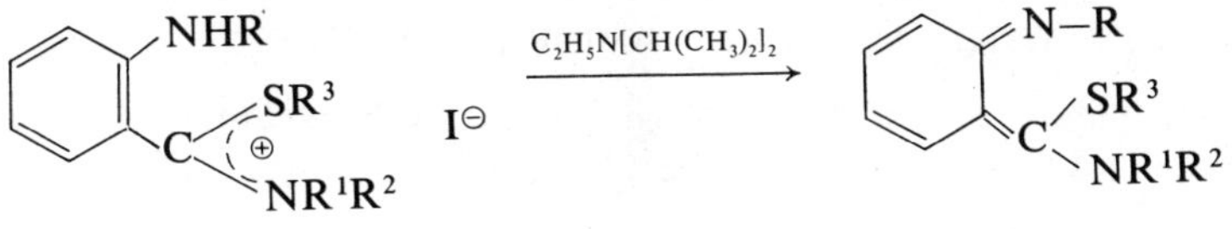

III. REACTIONS OF ALKYLMERCAPTOMETHYLENIMINIUM SALTS

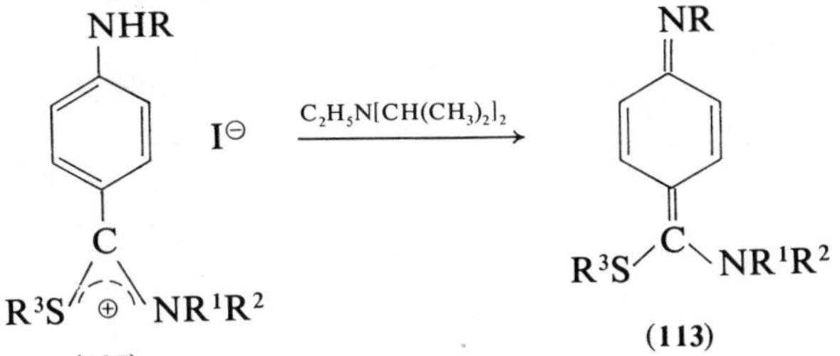

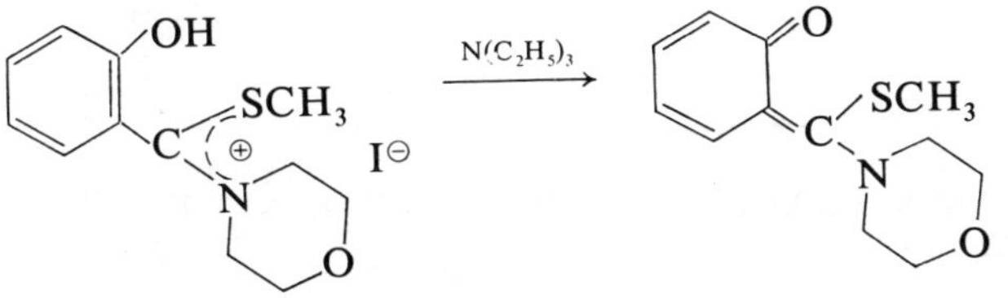

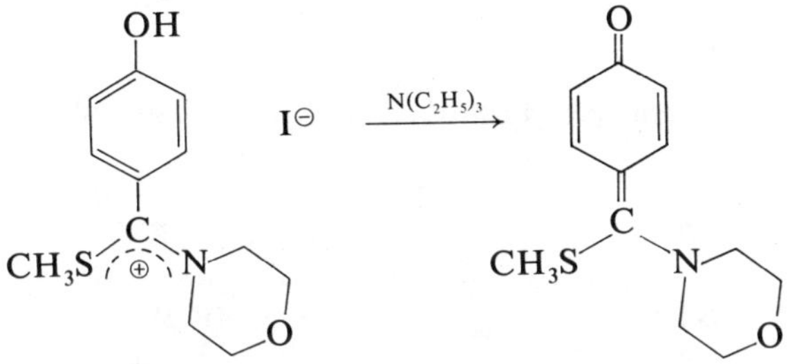

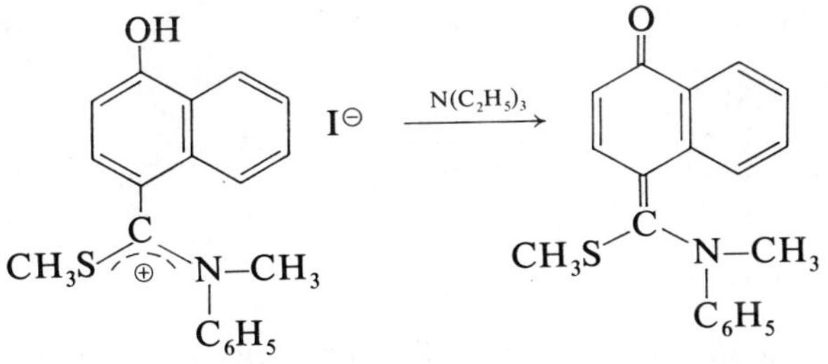

D. REACTIONS WITH AMINES AND AMINE DERIVATIVES

Primary aliphatic and aromatic amines as well as ammonia react with iminium salts **4** to yield amidines like **117** or amidinium salts **118** (21, 28, 90a, 96, 105–107).

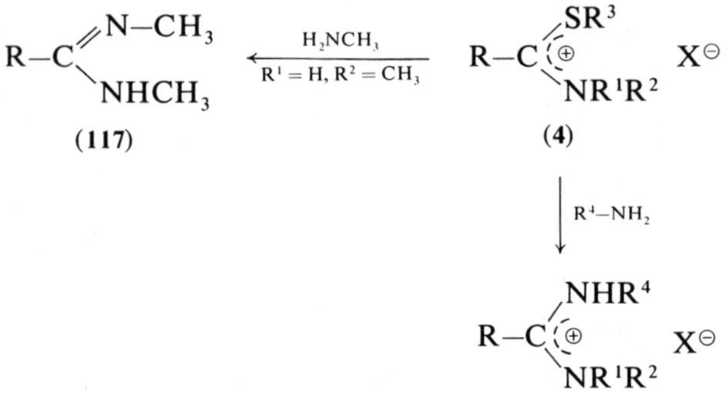

$$R-C\begin{matrix}N-CH_3\\NHCH_3\end{matrix} \xleftarrow[R^1=H,\ R^2=CH_3]{H_2NCH_3} R-C\begin{matrix}SR^3\\\oplus\\NR^1R^2\end{matrix} X^{\ominus}$$

(**117**)   (**4**)

$$\downarrow R^4-NH_2$$

$$R-C\begin{matrix}NHR^4\\\oplus\\NR^1R^2\end{matrix} X^{\ominus}$$

(**118**)

With N-unsubstituted salts nitriles sometimes form in addition to amidinium salts (106).

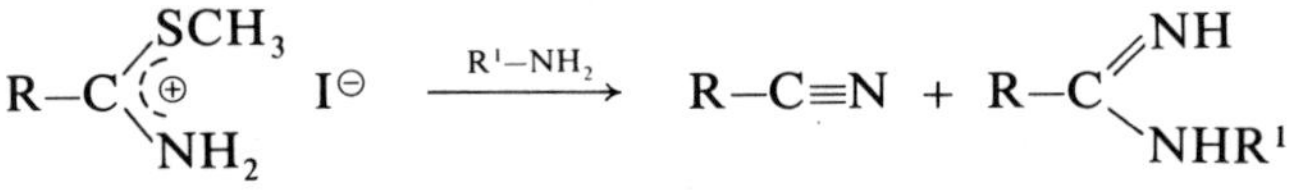

$$R-C\begin{matrix}SCH_3\\\oplus\\NH_2\end{matrix} I^{\ominus} \xrightarrow{R^1-NH_2} R-C\equiv N\ +\ R-C\begin{matrix}NH\\NHR^1\end{matrix}$$

Secondary amines convert **4** into N,N,N',N'-tetrasubstituted amidinium salts **119** (17, 21).

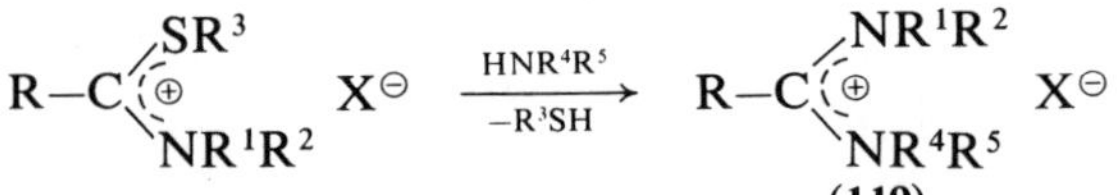

$$R-C\begin{matrix}SR^3\\\oplus\\NR^1R^2\end{matrix} X^{\ominus} \xrightarrow[-R^3SH]{HNR^4R^5} R-C\begin{matrix}NR^1R^2\\\oplus\\NR^4R^5\end{matrix} X^{\ominus}$$

(**119**)

Secondary and tertiary amines may dealkylate mercaptomethylenium salts to thioamides (105):

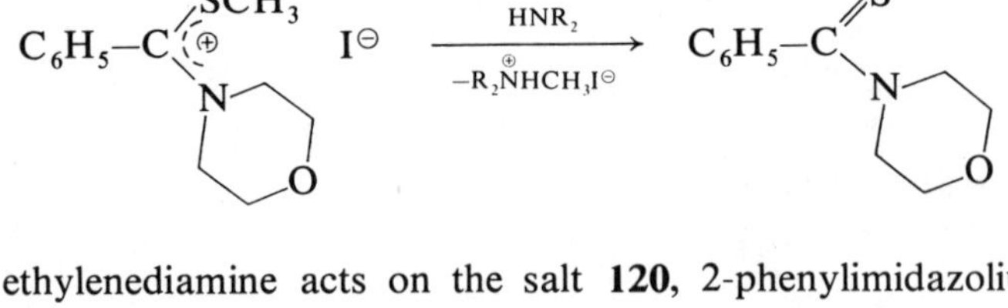

$$C_6H_5-C\begin{matrix}SCH_3\\\oplus\\N\end{matrix} I^{\ominus} \xrightarrow[-R_2\overset{\oplus}{N}HCH_3I^{\ominus}]{HNR_2} C_6H_5-C\begin{matrix}S\\N\end{matrix}$$

(morpholine rings on N)

When ethylenediamine acts on the salt **120**, 2-phenylimidazolium iodide (**121**) forms (96).

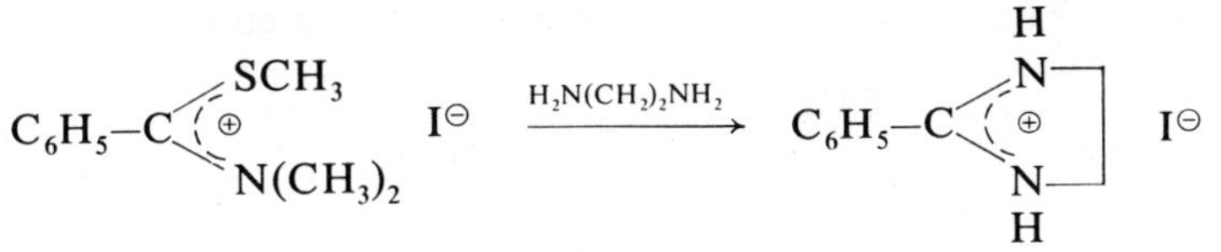

III. REACTIONS OF ALKYLMERCAPTOMETHYLENIMINIUM SALTS   305

(120) → **(121)**

The amidinium salts obtained this way may be deprotonated by further amine at the α positions to give ketene aminals **(122)** (17).

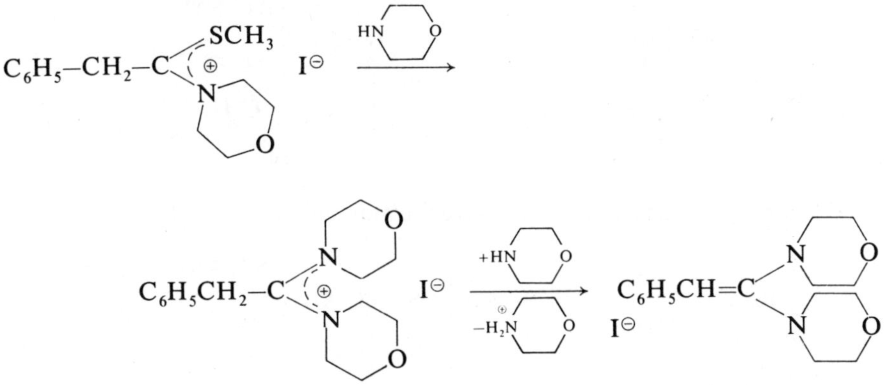

(122)

Mercaptomethyleniminium salts cyclize with o-phenylenediamine and o-aminophenol to 2-substituted benzimidazoles **123** (17, 96) or benzoxazoles **124** (96).

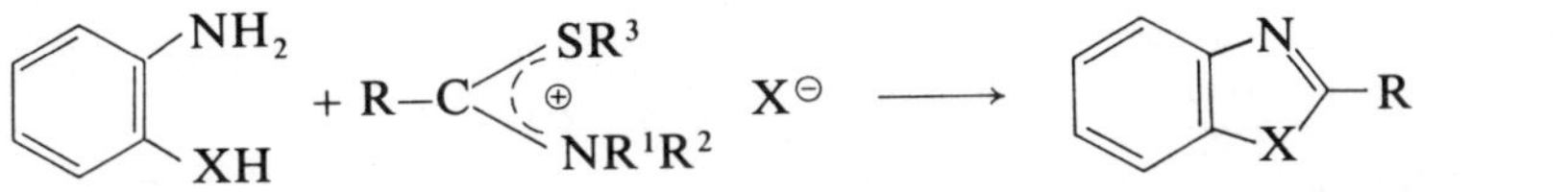

(123) X = NH   **(124)** X = O

2-Alkyl- or 2-aryl-4-ethoxycarbonyloxazolines **(125)** may be prepared from the iminium salts **126** and ethyl serine ester (108, 109).

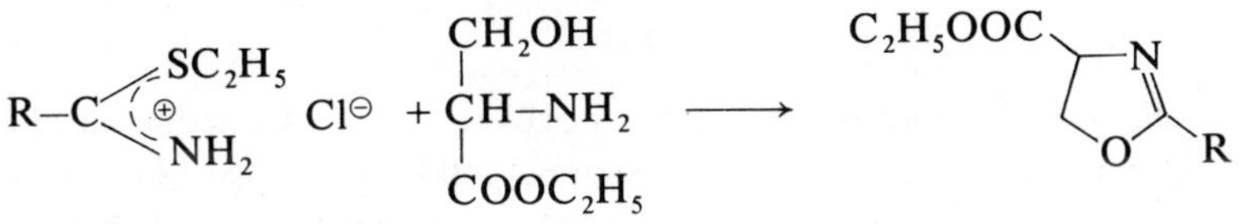

(126)   **(125)**

Hydrazine reacts with the salt **127** to give the dihydrotetrazine (**128**) (105).

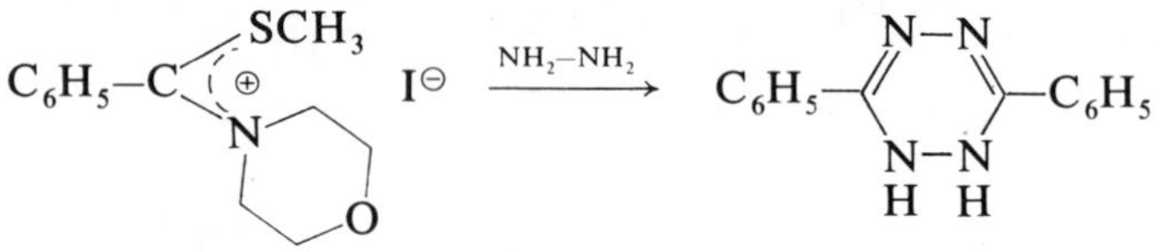

$$C_6H_5-C{\overset{SCH_3}{\underset{N}{\bigoplus}}}\quad I^{\ominus} \xrightarrow{NH_2-NH_2} C_6H_5-\underset{\underset{H\;H}{N-N}}{\overset{N-N}{\diagup\diagdown}}-C_6H_5$$

(**127**) (**128**)

The amidinium salts **130** are obtained from the salts **129** and hydrazine (53).

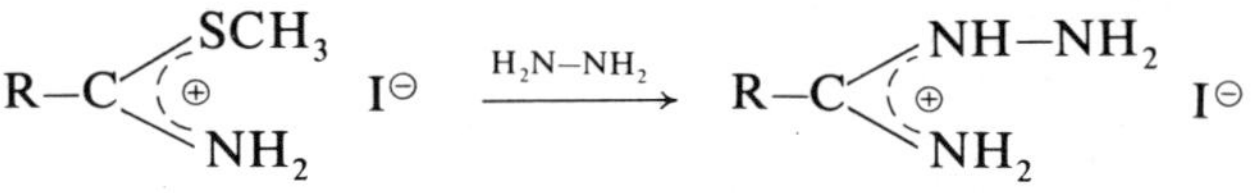

$$R-C{\overset{SCH_3}{\underset{NH_2}{\bigoplus}}}\quad I^{\ominus} \xrightarrow{H_2N-NH_2} R-C{\overset{NH-NH_2}{\underset{NH_2}{\bigoplus}}}\quad I^{\ominus}$$

(**129**) $R = C_6H_5, C_6H_5CH_2$

(**130**)

N,N-dimethylhydrazine and **131** give the amidinium salts **132** (107).

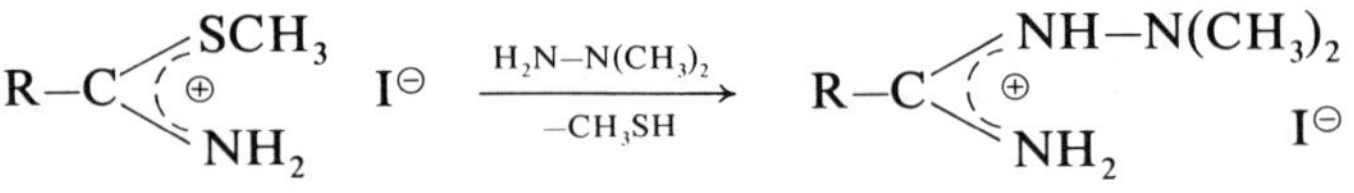

$$R-C{\overset{SCH_3}{\underset{NH_2}{\bigoplus}}}\quad I^{\ominus} \xrightarrow[-CH_3SH]{H_2N-N(CH_3)_2} R-C{\overset{NH-N(CH_3)_2}{\underset{NH_2}{\bigoplus}}}\quad I^{\ominus}$$

(**131**) (**132**)

Phenylhydrazine, benzhydrazide, ethyl carbonate hydrazide, thiosemicarbazide, and ethanolamine all react in a similar way (107).

Semicarbazide and **4** give thioimino esters (**133**) (21, 91), which are converted into 1,2,4-triazoles (**134**) on heating (18).

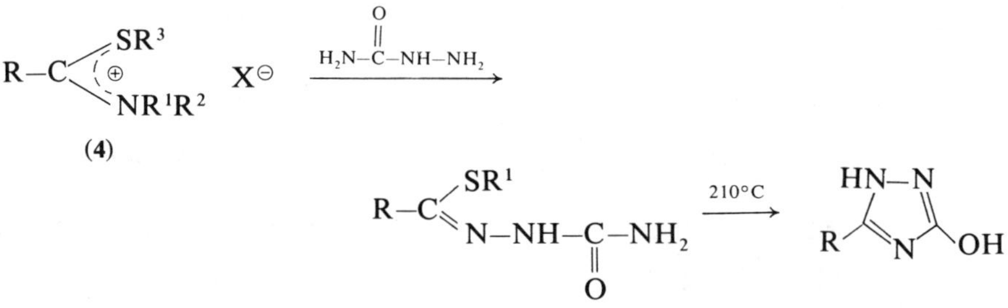

$$R-C{\overset{SR^3}{\underset{NR^1R^2}{\bigoplus}}}\quad X^{\ominus} \xrightarrow{H_2N-\overset{O}{\overset{\|}{C}}-NH-NH_2}$$

(**4**)

$$R-C{\overset{SR^1}{\underset{N-NH-\underset{\underset{O}{\|}}{C}-NH_2}{}}} \xrightarrow{210°C} R-\underset{N}{\overset{HN-N}{\diagup\diagdown}}-OH$$

(**133**) (**134**)

With the aid of **4** thiosemicarbazide is cyclized to 1,3,4-thiadiazoles (**135**) in acidic or neutral solution; no intermediates could be trapped (21, 107). If the condensation is performed in alkaline media, the intermediate **136** is isolable. This reaction may also be applied to alkylated thiolactams (21).

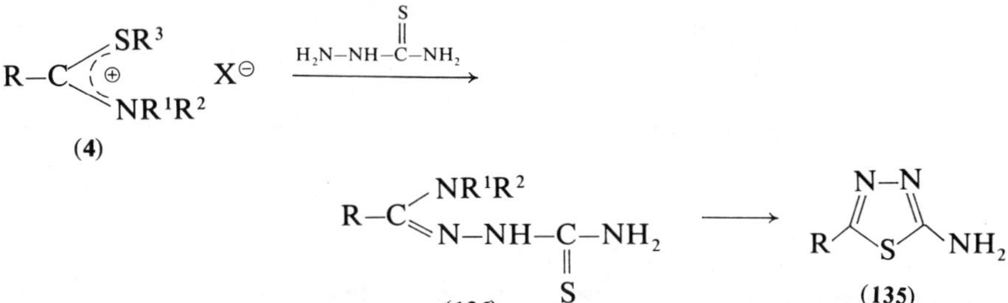

III. REACTIONS OF ALKYLMERCAPTOMETHYLENIMINIUM SALTS

$$R-C\begin{matrix}SR^3\\\oplus\\NR^1R^2\end{matrix} \quad X^\ominus \quad \xrightarrow{H_2N-NH-\underset{\underset{S}{\|}}{C}-NH_2}$$

(4)

$$R-C\begin{matrix}NR^1R^2\\\\N-NH-\underset{\underset{S}{\|}}{C}-NH_2\end{matrix} \longrightarrow$$

(136)

[structure 135: R-C on thiadiazole ring with NH₂]

(135)

Reacting bifunctional molecules of the type $H_2N-R-YH$ (Y = O, S, NH) with the iminium salts **136a** and **136b** gives heterocycles **136c–136g** (90a).

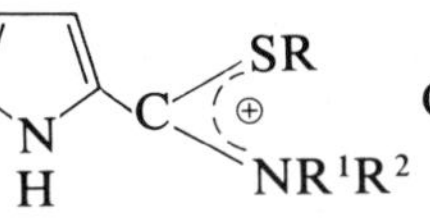

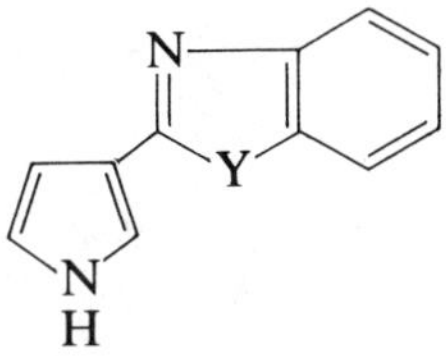

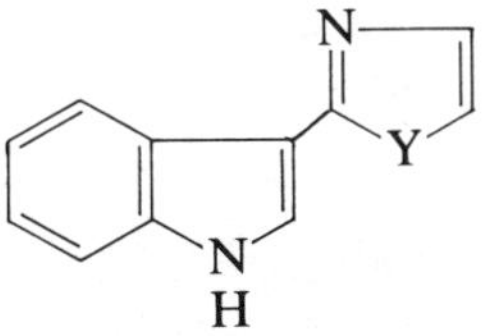

(136a) [pyrrole-C(SR)(NR¹R²) Cl⁻]

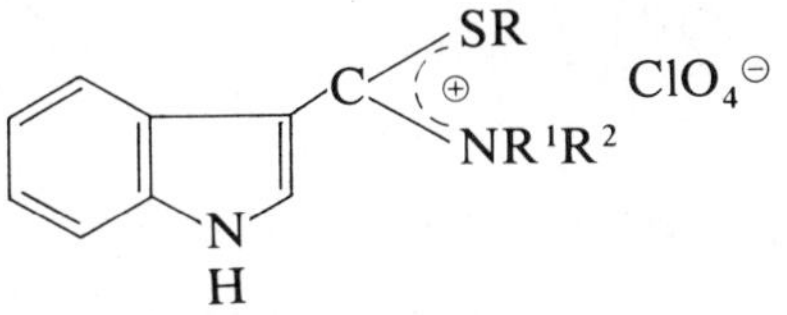

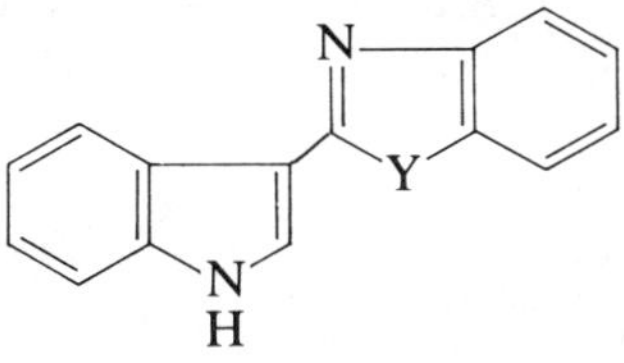

(136b) [indole-C(SR)(NR¹R²) ClO₄⁻]

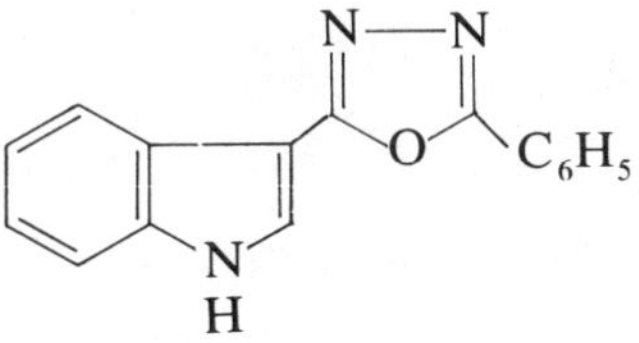

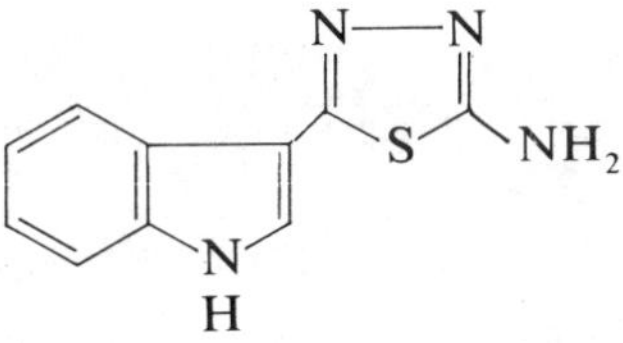

(136c) [pyrrole fused with benzazole, Y = O, S, NH]
Y = O, S, NH

(136d) [indole with 5-membered ring, Y = O, S, NH]
Y = O, S, NH

(136e) [indole with benzazole, Y = O, S, NH]
Y = O, S, NH

(136f) [indole with oxadiazole-C₆H₅]

(136g) [indole with thiadiazole-NH₂]

S-Alkylated thiohydrazides (**137**) give the amidinium salts **138** with hydrazines (67).

Hünig (30a) synthesized a series of amidrazones from alkylmercaptomethyleniminium salts and hydrazine or hydrazine derivatives.

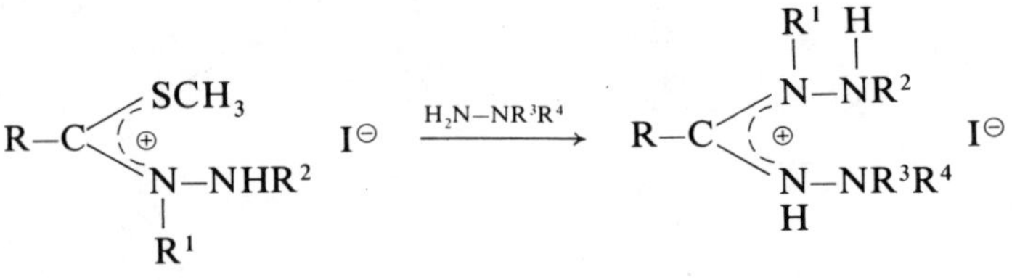

$$R-C\underset{N-NHR^2}{\overset{SCH_3}{\Bigg\langle_\oplus}} I^\ominus \quad \xrightarrow{H_2N-NR^3R^4} \quad R-C\underset{N-NR^3R^4}{\overset{N-NR^2}{\Bigg\langle_\oplus}} I^\ominus$$

(137)   (138)

Hydroxylamine displaces the amino function of thioimino ester salts with the formation of thiohydroxamic esters (**139**) (110).

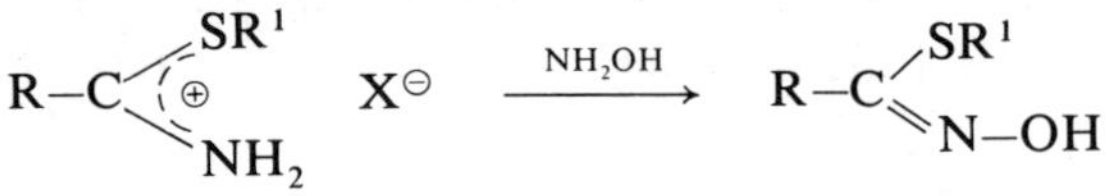

$$R-C\underset{NH_2}{\overset{SR^1}{\Bigg\langle_\oplus}} X^\ominus \quad \xrightarrow{NH_2OH} \quad R-C\underset{N-OH}{\overset{SR^1}{\Bigg\langle}}$$

(139)

E. REACTIONS WITH GRIGNARD REAGENTS

Grignard reagents add to the salts **4** primarily to form *S,N*-acetals of aldehydes or ketones (**140**) (48). *S,N*-Acetals having hydrogen atoms α to the acetal function lose mercaptans to form enamines (**141**) (48).

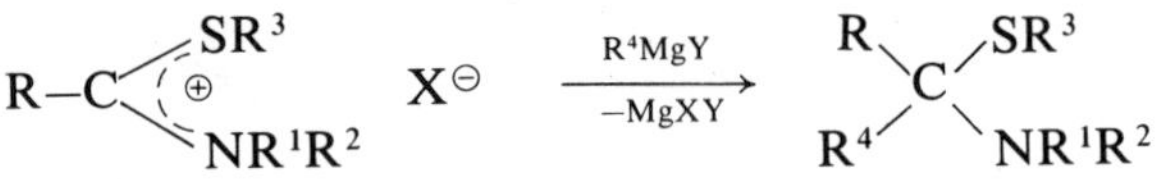

$$R-C\underset{NR^1R^2}{\overset{SR^3}{\Bigg\langle_\oplus}} X^\ominus \quad \xrightarrow[-MgXY]{R^4MgY} \quad \underset{R^4}{\overset{R}{\Bigg\rangle}}C\underset{NR^1R^2}{\overset{SR^3}{\Bigg\langle}}$$

$R = H, C_6H_5$

(4)   (140)

$$H-C\underset{\overset{|}{N}}{\overset{SCH_3}{\Bigg\langle_\oplus}} I^\ominus \quad \xrightarrow{C_6H_5CH_2MgCl}$$

$$\underset{H}{\overset{C_6H_5CH_2}{\Bigg\rangle}}C\underset{\overset{|}{N}}{\overset{SCH_3}{\Bigg\langle}} \quad \xrightarrow{-CH_3SH} \quad C_6H_5CH=CH-N\bigcirc O$$

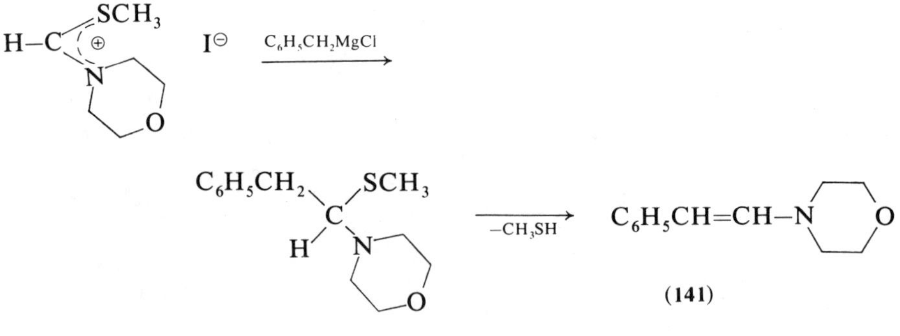

(141)

F. REACTIONS WITH INORGANIC SALTS

The iminium salt **142** is converted into dimethylbenzamide by mercury salts of carboxylic acids. In addition, carboxylic acid anhydrides are formed (111).

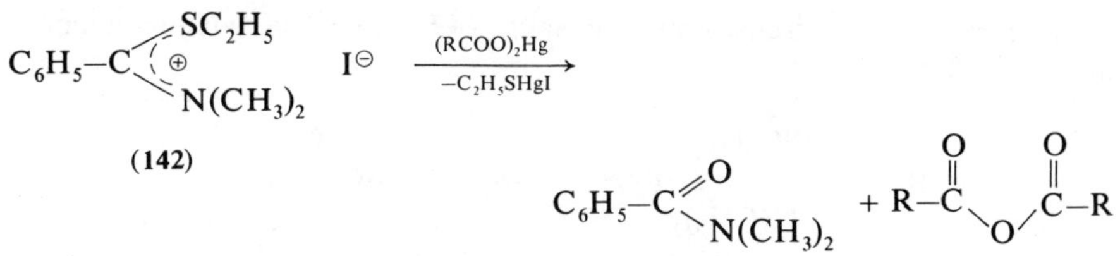

III. REACTIONS OF ALKYLMERCAPTOMETHYLENIMINIUM SALTS

$$C_6H_5-C\overset{SC_2H_5}{\underset{N(CH_3)_2}{\oplus}} \quad I^{\ominus} \quad \xrightarrow[-C_2H_5SHgI]{(RCOO)_2Hg}$$

(142)

$$C_6H_5-C\overset{O}{\underset{N(CH_3)_2}{\diagdown}} + R-\overset{O}{\underset{}{C}}-O-\overset{O}{\underset{}{C}}-R$$

Potassium cyanide adds to **143** yielding 2-dimethylamino-2-methylthiophenylacetonitrile (**144**) (96). The addition of silver cyanide to **145** proceeds analogously (111).

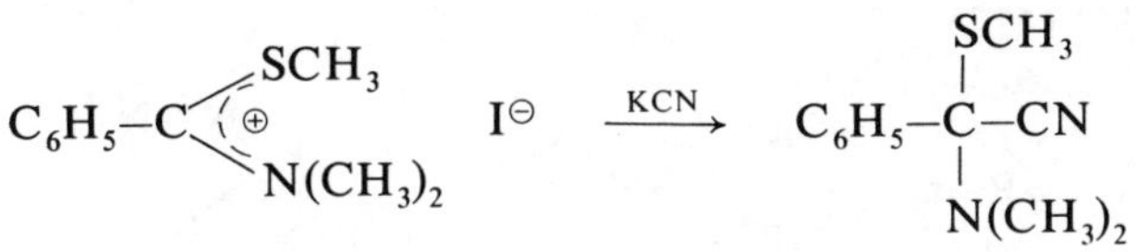

$$C_6H_5-C\overset{SCH_3}{\underset{N(CH_3)_2}{\oplus}} \quad I^{\ominus} \quad \xrightarrow{KCN} \quad C_6H_5-\overset{SCH_3}{\underset{N(CH_3)_2}{\overset{|}{C}}}-CN$$

(143) (144)

$$C_6H_5-C\overset{SC_2H_5}{\underset{\text{(morpholino)}}{\oplus}} \quad I^{\ominus} \quad \xrightarrow{AgCN} \quad C_6H_5-\overset{SC_2H_5}{\underset{\text{(morpholino)}}{\overset{|}{C}}}-CN$$

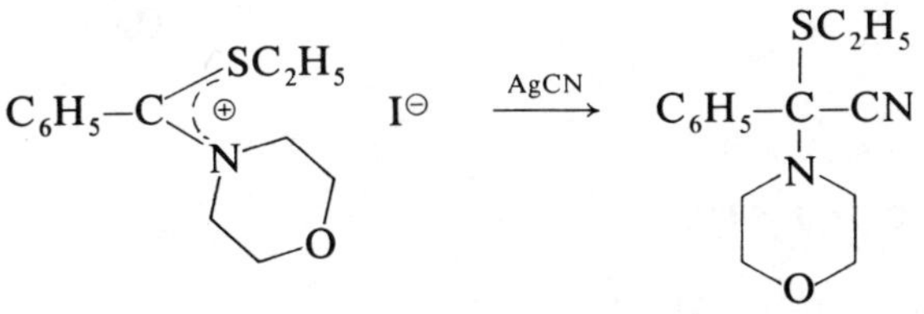

(145) (146)

The iminium salt **146** reacts with silver cyanide primarily to give 2-morpholino-2-ethylthiophenylpropionitrile (**147**), which is converted into 1-morpholino-1-cyanostyrene (**148**) by elimination of ethyl mercaptan (111).

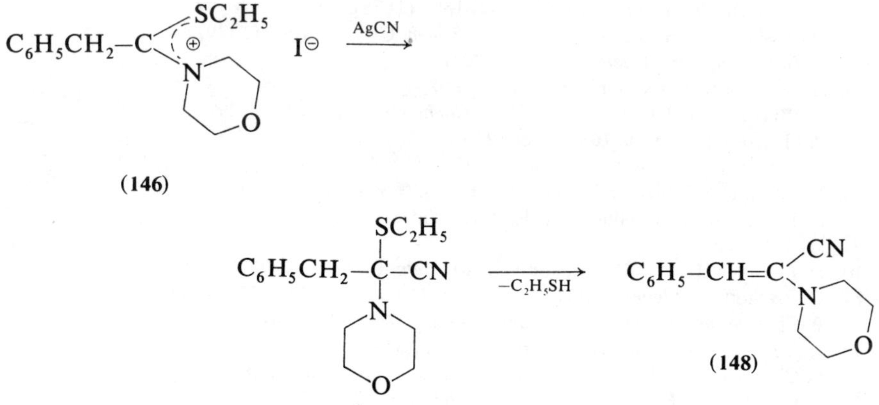

$$C_6H_5CH_2-C\overset{SC_2H_5}{\underset{\text{(morpholino)}}{\oplus}} \quad I^{\ominus} \quad \xrightarrow{AgCN}$$

(146)

$$C_6H_5CH_2-\overset{SC_2H_5}{\underset{\text{(morpholino)}}{\overset{|}{C}}}-CN \quad \xrightarrow{-C_2H_5SH} \quad C_6H_5-CH=C\overset{CN}{\underset{\text{(morpholino)}}{\diagdown}}$$

(147) (148)

Mercury cyanide transforms the salts **149** into dimethylaminomalononitriles (**150**) (111).

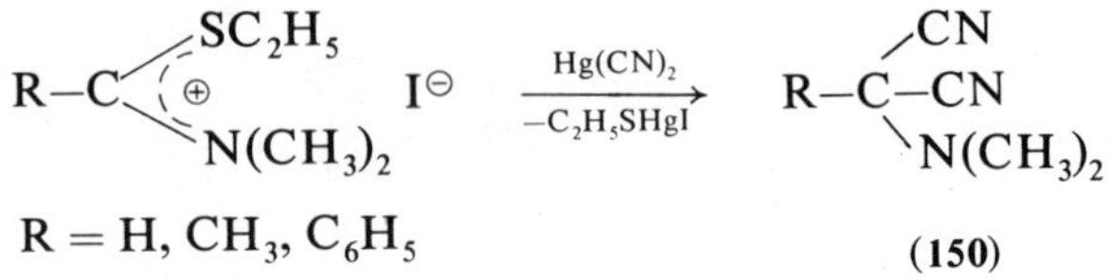

$$R-C\overset{SC_2H_5}{\underset{N(CH_3)_2}{(\oplus)}} \quad I^\ominus \quad \xrightarrow[-C_2H_5SHgI]{Hg(CN)_2} \quad R-C\overset{CN}{\underset{N(CH_3)_2}{-CN}}$$

$$R = H, CH_3, C_6H_5$$

(149)   (150)

Sodium azide converts the salts **4** into 1,5-dihydro-1,2,3,4-thia(S^{IV})-triazoles (**151**) (112).

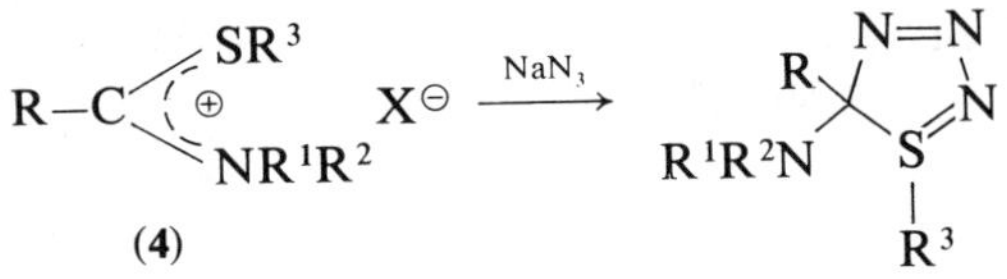

$$R-C\overset{SR^3}{\underset{NR^1R^2}{(\oplus)}} \quad X^\ominus \quad \xrightarrow{NaN_3} \quad \underset{R^1R^2N}{\overset{R}{\diagdown}}\overset{N=N}{\underset{S}{\diagup N}}$$

(4) R^3

(151)

ACKNOWLEDGMENT

To Dr. Heinz Eilingsfeld BASF AG, Ludwigshafen, and to Dr. Bernd Funke as well as Dr. Erwin Haug, both of Fachhochschule Aalen, Aalen, I am very indebted for fruitful discussions and critical revisions of the text.

REFERENCES

1. O. Wallach, *Ber. Deut. Chem. Ges.*, **11**, 1590 (1878).
2. O. Wallach and H. Bleibtreu, *Ber. Deut. Chem. Ges.*, **12**, 1061 (1879).
3. A. Bernthsen, *Ann. Chem.*, **192**, 56 (1878).
4. A. Bernthsen, *Ann. Chem.*, **197**, 341 (1879).
5. R. Willstätter und T. Wirth, *Ber. Deut. Chem. Ges.*, **42**, 1908 (1909).
6. P. May, *J. Chem. Soc.*, **103**, 2272 (1913).
7. J. J. Donleavy, *J. Am. Chem. Soc.*, **58**, 1004 (1936).
8. J. Tafel and P. Lawaczeck, *Ber. Deut. Chem. Ges.* **40**, 2842 (1907).
9. F. E. Condo, E. T. Hinkel, A. Fassero, and R. L. Shriner, *J. Am. Chem. Soc.*, **59**, 230 (1937).
10. H. Erlenmeyer, W. Büchner, and H. Lehr, *Helv. Chim. Acta*, **27**, 969 (1944).
11. G. Bischoff, O. Weber, and H. Erlenmeyer, *Helv. Chim. Acta*, **27**, 947 (1944).
12. P. Chabrier and S. H. Renard, *Compt. Rend.*, **226**, 582 (1948).
13. P. Chabrier and S. H. Renard, *Compt. Rend.*, **228**, 850 (1949).
14. P. Chabrier, S. H. Renard, und K. Smarzewska, *Bull. Soc. Chim. France*, **1949**, 237.
15. P. Böttcher and F. Bauer, *Ann. Chem.*, **568**, 218 (1950).
16. P. Chabrier, S. H. Renard, and K. Smarzweska, *Bull. Soc. Chim. France*, **1950**, 1167.

17. M. A. Rogers, *J. Chem. Soc.*, **1950**, 3350.
18. J. H. Renault, *Compt. Rend.*, **233**, 182 (1951).
19. R. Boudet, *Bull. Soc. Chim. France*, **18**, 377 (1951).
20. D. J. Fry and J. D. Kendall, *J. Chem. Soc.*, **1951**, 1716.
21. D. A. Peak and F. Stansfield, *J. Chem. Soc.*, **1952**, 4067.
22. J. D. S. Goulden, *J. Chem. Soc.*, **1953**, 997.
23. R. Boudet, *Compt. Rend.*, **239**, 1803 (1954).
24. K. Miyataka and T. Yoschikawa, *J. Pharm. Soc. Japan*, **75**, 1054 (1955) *C.A.*, **50**, 5633 (1956).
25. R. I. Meltzer, A. D. Lewis, and J. A. King, *J. Am. Chem. Soc.*, **77**, 4062 (1955).
26. J. Renault, *Ann. Chim.*, **10**, [12], 135 (1955) *C.A.*, **50**, 9408b (1956).
27. J. Fabian, M. Legrand, and P. Poirier, *Bull. Soc. Chim. France,* **1956**, 1499.
28. H. Bredereck, R. Gompper, and H. Seitz, *Chem. Ber.*, **90**, 3786 (1957).
29. S. Petersen and E. Tietze, *Ann. Chem.*, **623**, 166 (1959).
30. R. Reynaud, R. C. Moreau, und N. H. Thu, *Compt. Rend.*, **253**, 1968 (1961).
30a. S. Hünig and F. Müller, *Ann. Chem.*, **651**, 89 (1962).
31. L. A. Paquette, *J. Am. Chem. Soc.*, **86**, 4096 (1964).
32. P. Guepet, P. Pigamol, A. Carayon-Gentil, and P. Chabrier, *Bull. Soc. Chim. France*, **1965**, 224.
33. R. Gompper and R. R. Schmidt, *Chem. Ber.*, **98**, 1385 (1965).
34. H. Reinheckel, D. Jahnke, and G. Kretzschmar, *Chem. Ber.*, **99**, 11 (1966).
35. W. Ried and E. Schmidt, *Ann. Chem.*, **695**, 217 (1966).
36. R. Gompper, E. Kutter, and H. Kast, *Angew. Chem.*, **79**, 147 (1967).
37. R. Gompper and H. D. Lehmann, *Angew. Chem.*, **80**, 38 (1968).
38. P. J. W. Schuijl, H. J. T. Bos, and L. Brandsma, *Rec. Trav. Chim. Pays-Bas*, **87**, 123 (1968).
39. P. J. W. Schuijl and L. Brandsma, *Rec. Trav. Chim. Pays-Bas*, **87**, 927 (1968).
40. G. E. Lienhard and T. C. Wang, *J. Am. Chem. Soc.*, **90**, 3781 (1968).
41. R. Raap, *Can. J. Chem.*, **46**, 2255 (1968).
42. R. Gompper and W. Elser, *Ann. Chem.*, **725**, 64 (1969).
42a. R. Gompper and W. Elser, *Organic Synthesis*, Coll. Vol. 5, H. E. Baumgarten, Ed., John Wiley and Sons, New York, 1973, p. 780.
42b. M. Roth, P. Dubs, E. Götschi, and A. Eschenmoser, *Helv. Chim. Acta*, **54**, 710 (1971).
43. R. K. Chaturvedi and G. O. Schmir, *J. Am. Chem. Soc.*, **91**, 737 (1969).
44. W. Walter and J. Krohn, *Chem. Ber.*, **102**, 3786 (1969).
45. B. W. Bycroft and W. Landon, *Chem. Commun.*, **1970**, 168.
46. H. Hartmann, *Z. Chem.*, **11**, 60 (1971).
47. R. Mukerjee, *Chem. Commun.*, **1971**, 1113.
48. T. Yamaguchi, Y. Shimizu, and T. Suzuki, *Chem. Ind. London*, **1972**, 380.
49. W. Walter and J. Krohn, *Ann. Chem.*, **1973**, 443.
50. W. Walter and C. O. Meese, *Ann. Chem.*, **1973**, 832.
51. H. Bredereck, G. Simchen, and H. Hoffmann, *Chem. Ber.*, **106**, 3725 (1973).
52. G. Seitz, R. Schmiedel, and K. Maun, *Synthesis*, **1974**, 578.
53. K. N. Doyle and F. Kurzer, *Synthesis*, **1974**, 583.
54. A. K. Bose and J. D. Fahey, *J. Org. Chem.*, **39**, 115 (1974).
55. P. Beak, D. S. Müller, and J. Lee, *J. Am. Chem. Soc.*, **96**, 3867 (1974).
55a. L. Andrieu, J. Bitoun, M. Fatome, R. Granger, Y. Robbe, and A. Terol, *Chim. Thér.*, **9**, 449, 453 (1974); *Cheminform*, **1975**, 11–288, p. 111.
55b. K. T. Potts, E. Haughton, and U. P. Singh, *J. Org. Chem.*, **39**, 3627 (1974).
55c. R. A. Firestone, N. S. Maciejevicz, and G. Christensen, *J. Org. Chem.*, **39**, 3384 (1974).
56. P. Chabrier and S. H. Renard, *Bull. Soc. Chim. France*, **1949**, 272.

57. R. N. Hurd and G. De La Mater, *Chem. Rev.*, **61**, 45 (1961).
58. E. E. Reid, *Organic Chemistry of Bivalent Sulfur*, Vol. IV, Chemical Publishing Co., New York, 1962, p. 45.
59. W. Walter and J. Voss, "The Chemistry of Thioamides", *The Chemistry of Amides*, J. Zabicky, Ed., John Wiley & Sons, New York, 1970, p. 383.
60. D. A. Tomalia and J. N. Paige, *J. Org. Chem.*, **38**, 3949 (1973).
61. B. Holmberg, *Ark. Kemi, Mineral Geol.*, **24A** (3) (1947); *C.A.*, **45**, 581 (1951).
62. C. S. Marvel, P. de Radzitzky, and J. J. Brader, *J. Am. Chem. Soc.*, **77**, 5997 (1955).
63. K. A. Jensen and C. Pedersen, *Acta Chem. Scand.*, **15**, 1087 (1961).
64. J. M. Sprague and A. H. Lang, *Heterocyclic Compounds*, R. C. Elderfield, Ed., Vol. 5, John Wiley & Sons, New York, 1957, Chap. 8.
65. H. Bredereck, R. Gompper, and D. Bitzer, *Chem. Ber.*, **92**, 1139 (1959).
66. R. N. Hurd, G. De La Mater, and J. P. McDermott, *J. Org. Chem.*, **27**, 269 (1962).
67. R. Grashey, M. Baumann, and H. Bauer, *Chemiker-Ztg.*, **96**, 224 (1972).
68. A. E. Arbuzov, V. E. Šiškin, and S. S. Tjulenev, *Zh. Org. Chim.*, **1**, 1442 (1965).
69. A. K. Bose, M. S. Manhas, J. S. Chib, H. P. S. Chawla, and B. Dayal, *J. Org. Chem.*, **39**, 2877 (1974).
70. A. K. Bose, J. C. Kapur, S. G. Amin, and M. S. Manhas, *Tetrahedron Letters*, **1974**, 1917.
71. J. Goerdeler and H. Horstmann, *Chem. Ber.*, **93**, 663 (1960).
72. J. Goerdeler and H. Horstmann, *Chem. Ber.*, **93**, 671 (1960).
73. J. Goerdeler and K. Stadelbauer, *Chem. Ber.*, **98**, 1557 (1965).
74. W. Walter and P.-M. Hell, *Ann. Chem.*, **727**, 22 (1969).
75. W. Walter and J. Krohn, *Ann. Chem.*, **752**, 136 (1971).
76. W. Walter and H.-W. Meyer, *Ann. Chem.*, **1973**, 462.
77. W. Walter and J. Krohn, *Ann. Chem.*, **1973**, 476.
78. W. Walter and H.-W. Meyer, *Ann. Chem.*, **1974**, 776.
79. S. L. Razmak, E. M. Flagg, and F. Siebenthall, *J. Org. Chem.*, **38**, 2242 (1973).
80. W. Walter and P. M. Hell, *Angew. Chem.*, **77**, 720 (1965).
81. F. Fischer und R. Gottfried, *Z. Chem.*, **6**, 146 (1966).
82. H. Hartmann, *J. Prakt. Chem.*, **313**, 1113 (1971).
83. H. Eilingsfeld, M. Seefelder, and H. Weidinger, *Chem. Ber.*, **96**, 2671 (1963).
84. A. Pinner and F. Klein, *Ber. Deut. Chem. Ges.*, **11**, 1825 (1878).
85. W. Autenrieth and A. Brüning, *Ber. Deut. Chem. Ges.*, **36**, 3464 (1903).
86. H. Bader, J. D. Downer, and P. Driver, *J. Chem. Soc.*, **1950**, 2775.
87. H. Ulrich and J. Nuesslein, U.S. Pat. 2,008,649; *C.A.*, **29**, 5951.
88. Brit. Pat. 448,469 (1936); *C.A.*, **30**, 7584
89. R. T. Hartigan and J. B. Cloke, *J. Am. Chem. Soc.*, **67**, 709 (1945).
90. R. Gompper and W. Elser, *Ann. Chem.*, **725**, 73 (1969).
90a. R. L. N. Harris, *Australian J. Chem.*, **27**, 2635 (1974).
90b. H. Eilingsfeld and L. Möbius, *Chem. Ber.*, **98**, 1293 (1965).
91. R. K. Chaturvedi, A. E. McMahon, and G. L. Schmir, *J. Am. Chem. Soc.*, **89**, 6984 (1967).
92. R. B. Martin and A. Parcell, *J. Am. Chem. Soc.*, **83**, 4830 (1961).
93. A. A. Goldberg and W. Kelly, *J. Chem. Soc.*, **1948**, 1919.
94. W. Steinkopf and S. Müller, *Ber. Deut. Chem. Ges.*, **56**, 1931 (1923).
95. S. Gabriel and P. Heymann, *Ber. Deut. Chem. Ges.*, **24**, 783 (1891).
96. T. Mukaiyama, T. Yamaguchi, and H. Nohira, *Bull. Chem. Soc. Japan*, **38**, 2107 (1965).
97. T. Mukaiyama and T. Yamaguchi, *Bull. Chem. Soc. Japan*, **39**, 2005 (1966).
98. T. Mukaiyama, S. Aizawa, and T. Yamaguchi, *Bull. Chem. Soc. Japan*, **40**, 2641 (1967).
99. P. Chabrier and S. H. Renard, *Bull. Soc. Chim. France*, **1950**, 13.
100. K. Joshi, V. A. Rao, and N. Anaud, *Indian J. Chem.*, **11**, 1222 (1973).

101. T. Nakai and M. Okawara, *Chem. Commun.*, **1970**, 907.
102. J. Hocker and R. Merten, *Ann. Chem.*, **751**, 145 (1971).
103. R. Gompper and W. Elser, *Tetrahedron Letters*, **1964**, (1971).
104. R. Gompper and R. R. Schmidt, *Z. Naturforsch.*, **176**, 851 (1962).
105. P. Chabrier and S. H. Renard, *Compt. Rend.*, **230**, 1673 (1950).
106. R. Reynaud, R. C. Moreau, and Nguyen Hong Thu, *Compt. Rend.*, **253**, 2540 (1961).
107. R. Reynaud, R. C. Moreau, and T. Gousson, *Compt. Rend.*, **259**, 4067 (1964).
108. H. Bretschneider and G. Pickorski, *Monatsh. Chem.*, **85** 1110 (1954).
109. D. F. Elliot, *J. Chem. Soc.*, **1949**, 589.
110. J. B. Buchanan, U.S. Pat. 3,374,260 (1968).
111. T. Yamaguchi, K. Inomata, and T. Mukaiyama, *Bull. Chem. Soc. Japan*, **41**, 673 (1968).
112. S. I. Mathew and F. Stansfield, *J. Chem. Soc. Perkin Trans. I*, **1974**, 540.

ADDENDUM

To Section I-A

Many thioamides have been reacted with various alkylating reagents, but not in all cases were the primary formed iminium compounds isolated. More recent results are summarized in the following formulas.

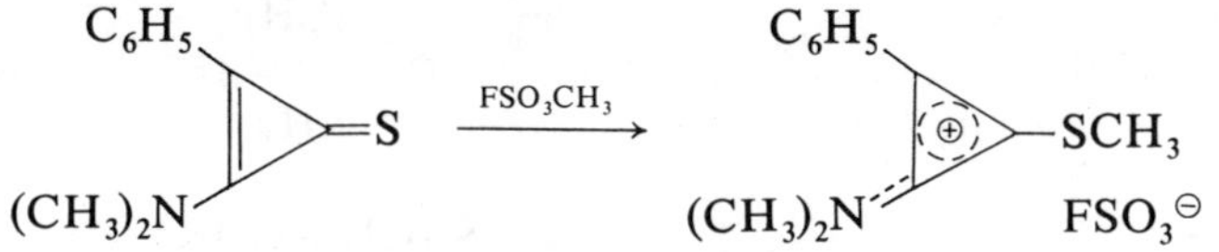

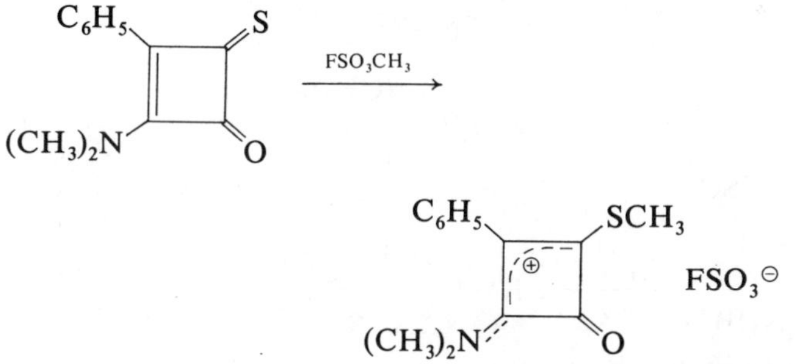

(refs. 113, 114)

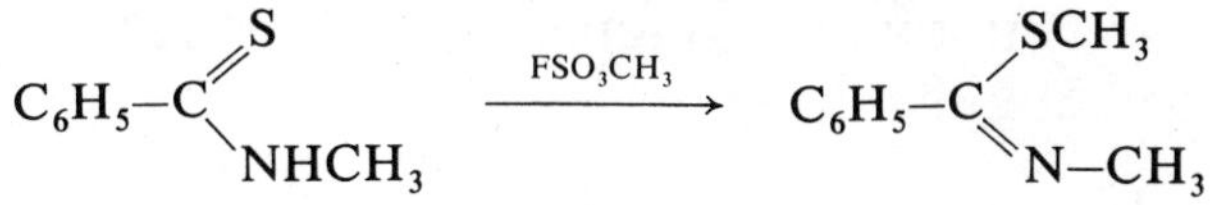

(ref. 115)

---

* Translated from German by W. Kantlehner, and by H. J. Benivschke, Fachhochschule Aalen, D-708-Aalen, Hohenstaufenstrasse, Germany.

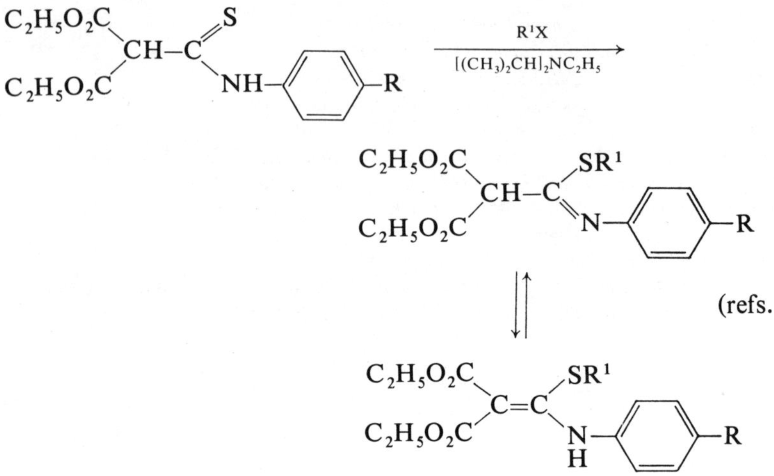

$$\underset{C_2H_5O_2C}{\overset{C_2H_5O_2C}{>}}CH-\overset{S}{\underset{NH}{C}}-\underset{}{\bigcirc}-R \xrightarrow[[(CH_3)_2CH]_2NC_2H_5]{R^1X}$$

$$\underset{C_2H_5O_2C}{\overset{C_2H_5O_2C}{>}}CH-\overset{SR^1}{\underset{N}{C}}-\underset{}{\bigcirc}-R$$

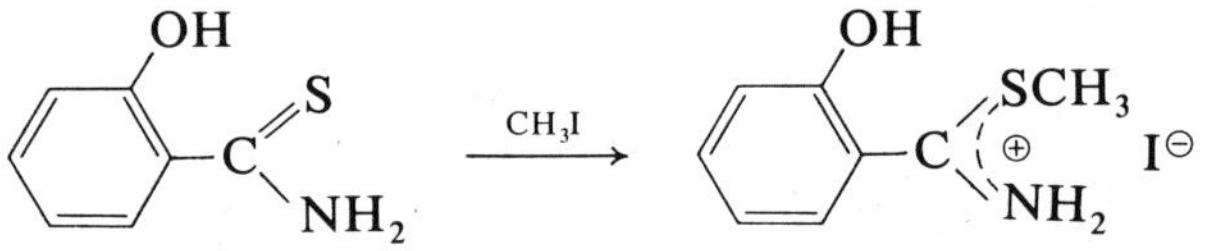

$$\updownarrow \qquad \text{(refs. 116, 117)}$$

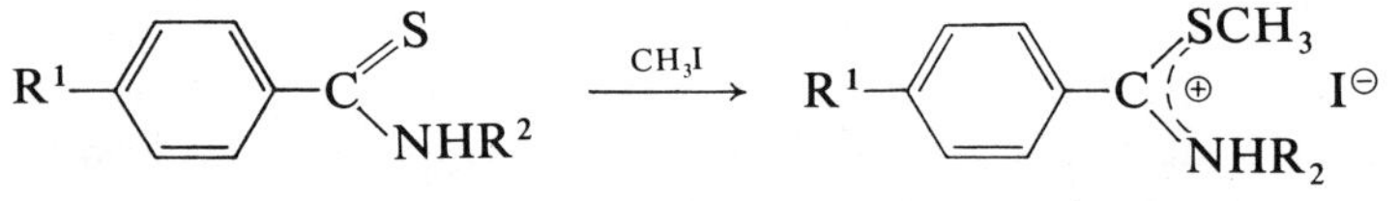

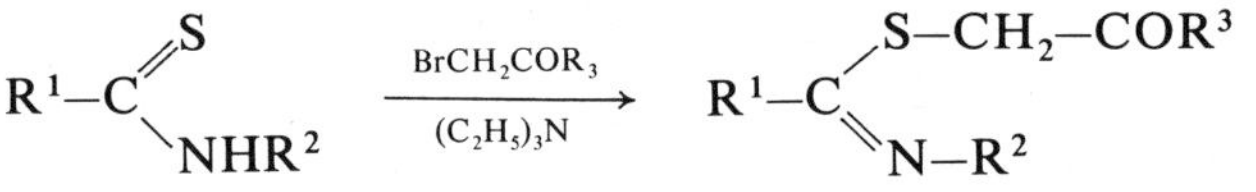

$$\underset{C_2H_5O_2C}{\overset{C_2H_5O_2C}{>}}C=\overset{SR^1}{\underset{\underset{H}{N}}{C}}-\underset{}{\bigcirc}-R$$

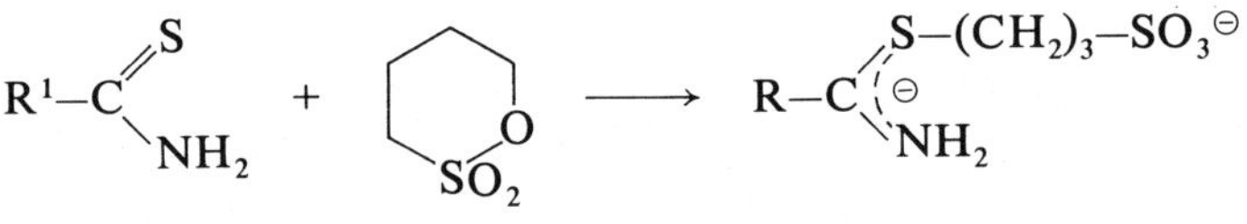

$$R^1-\bigcirc-\overset{S}{\underset{NHR^2}{C}} \xrightarrow{CH_3I} R^1-\bigcirc-\overset{SCH_3}{\underset{NHR_2}{C^\oplus}} \; I^\ominus \qquad \text{(ref. 118)}$$

$$\underset{}{\bigcirc}\overset{OH}{\underset{}{}}-\overset{S}{\underset{NH_2}{C}} \xrightarrow{CH_3I} \underset{}{\bigcirc}\overset{OH}{\underset{}{}}-\overset{SCH_3}{\underset{NH_2}{C^\oplus}} \; I^\ominus \qquad \text{(ref. 119)}$$

$$R^1-\overset{S}{\underset{NHR^2}{C}} \xrightarrow[(C_2H_5)_3N]{BrCH_2COR_3} R^1-\overset{S-CH_2-COR^3}{\underset{N-R^2}{C}} \qquad \text{(ref. 120)}$$

$$R^1-\overset{S}{\underset{NH_2}{C}} + \underset{SO_2}{\bigcirc}O \longrightarrow R-\overset{S-(CH_2)_3-SO_3^\ominus}{\underset{NH_2}{C^\ominus}} \qquad \text{(ref. 121)}$$

$$R^1-\overset{S}{\underset{NHR^3}{C}} \xrightarrow{R^2X} R^1-\overset{SR^2}{\underset{N-R^3}{C}}$$

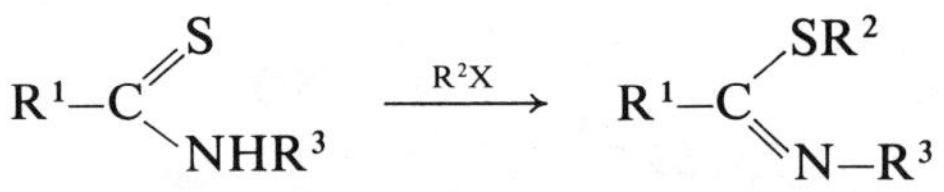

(a) R^1, R^2: CH_3, R^3: C_2-C_4-alkyl
(b) $R^1 = CH_3, CH(CH_3)_2, -C(CH_3)_3$ \qquad (ref. 122)
$R^3 = CH_3, R^3 = C_6H_5$

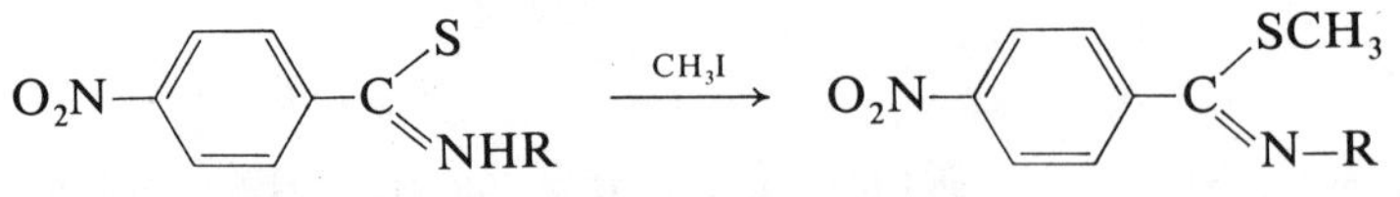

$$O_2N-\bigcirc-\overset{S}{\underset{NHR}{C}} \xrightarrow{CH_3I} O_2N-\bigcirc-\overset{SCH_3}{\underset{N-R}{C}} \qquad \text{(ref. 122)}$$

$R = CH_3, C_2H_5, CH(CH_3)_2, -C(CH_3)_3$

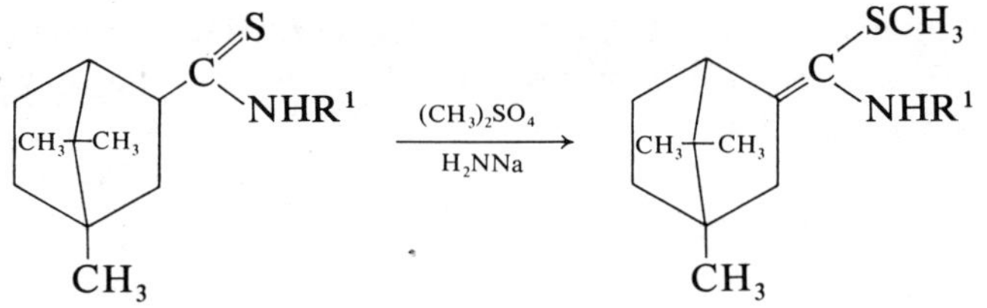

(ref. 123)

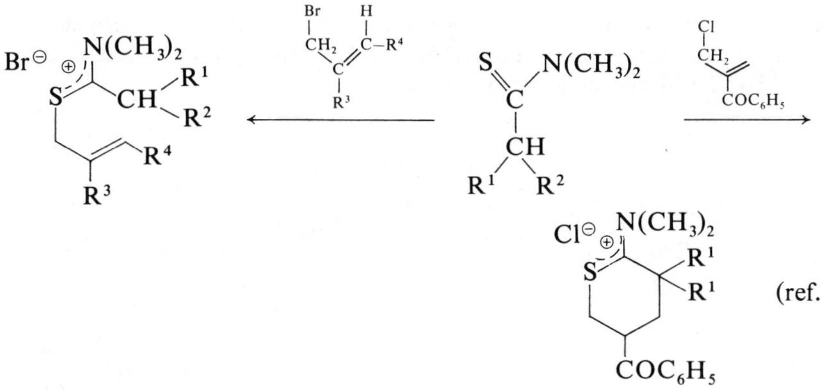

(ref. 124)

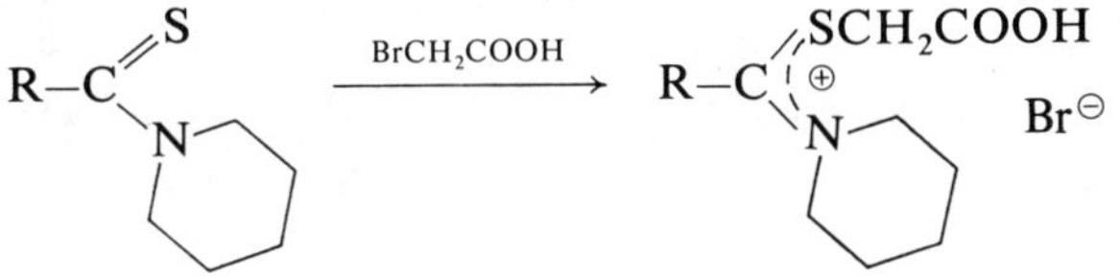

(ref. 125)

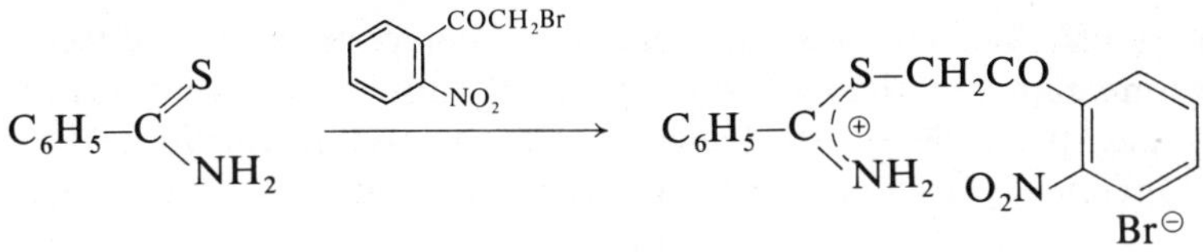

(ref. 126)

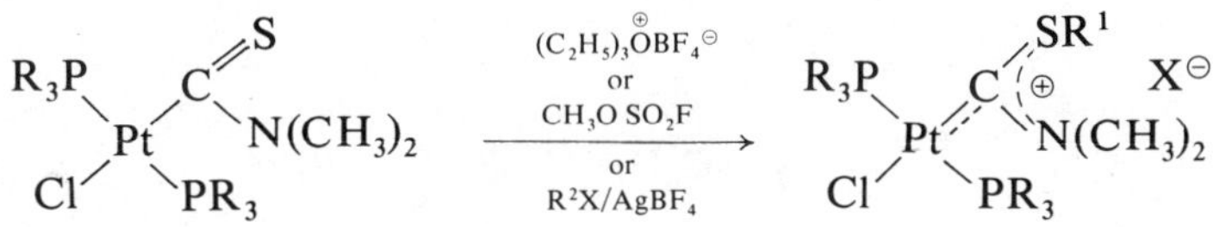

(ref. 127)

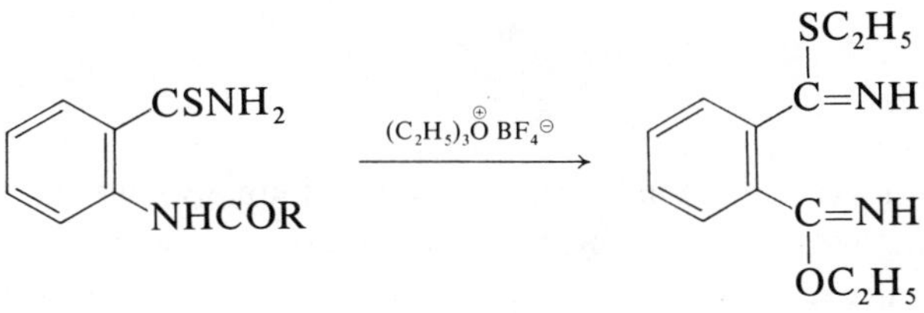

(ref. 128)

Kinetics of alkylation of *p*-substituted *N*-alkylthiobenzamides by methyljodide have been investigated (118). Two more recent reviews deal partly with alkylation of thioamides (129, 130).

To Section I-B

The action of phosgene or carboxylic acid chlorides, respectively on thioimidic acid esters, in the presence of tertiary amines affords ketene-*O*,*N*-acetals **152** and **153**, respectively (131).

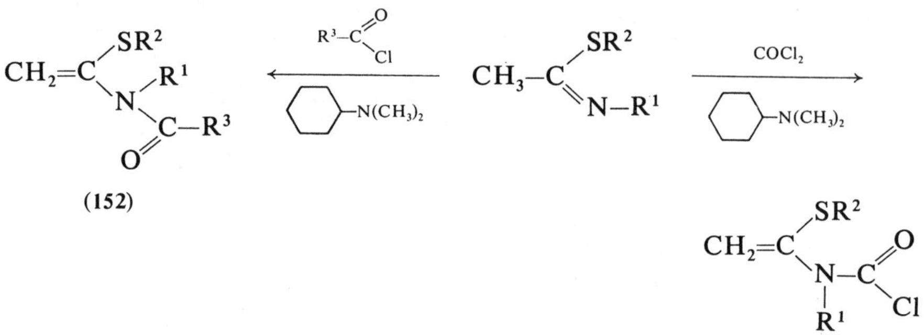

$$CH_2=C\begin{smallmatrix}SR^2\\N-R^1\\|\\C-R^3\\\|\\O\end{smallmatrix} \xleftarrow{\substack{R^3-C(=O)Cl \\ \bigcirc-N(CH_3)_2}} CH_3-C\begin{smallmatrix}SR^2\\N-R^1\end{smallmatrix} \xrightarrow{\substack{COCl_2 \\ \bigcirc-N(CH_3)_2}}$$

(152)

$$CH_2=C\begin{smallmatrix}SR^2\\N-C(=O)Cl\\|\\R^1\end{smallmatrix}$$

(153)

To Section I-E

The synthesis of alkylmercaptomethyleniminium compounds by treatment of nitriles with mercaptanes and hydrohalogenides is a topic of a recent review (130). Interestingly the nitrile group of acylcyanides can also be transformed to an iminium function by treatment with mercaptanes and hydrogenchloride (119).

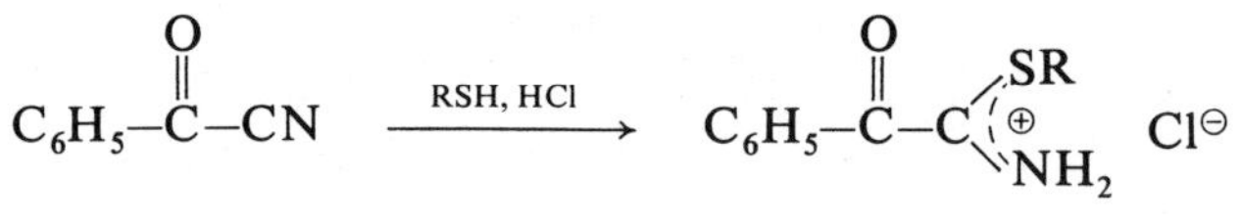

$$C_6H_5-\overset{O}{\underset{\|}{C}}-CN \xrightarrow{RSH, HCl} C_6H_5-\overset{O}{\underset{\|}{C}}-C\begin{smallmatrix}SR\\\oplus\\NH_2\end{smallmatrix} \;\; Cl^\ominus$$

To Section II

Restricted rotation around the partial C=N$^\oplus$ double bond of alkylmercaptomethyleniminium salt is subject of two more recent ^{1}H- and ^{13}C-NMR investigations (132, 133).

To Section III-A

Kinetic investigations on hydrolysis of iminium salts **154, 155** and **156** at different pH's in presence and absence of Hg and Ag ions have been done (134).

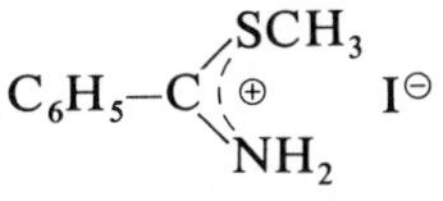

$$C_6H_5-C\overset{SCH_3}{\underset{NH_2}{\oplus}} \quad I^{\ominus}$$
(**154**)

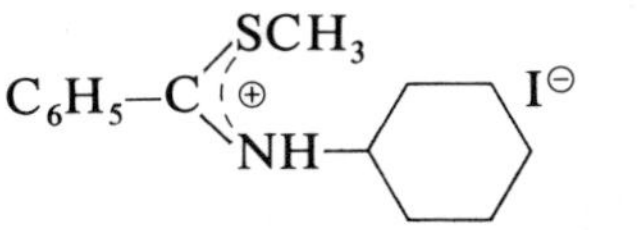

$$C_6H_5-C\overset{SCH_3}{\underset{NH-\text{cyclohexyl}}{\oplus}} \quad I^{\ominus}$$
(**155**)

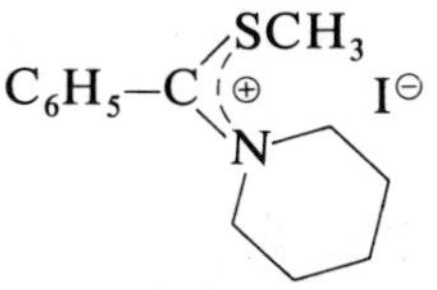

$$C_6H_5-C\overset{SCH_3}{\underset{N\text{-piperidyl}}{\oplus}} \quad I^{\ominus}$$
(**156**)

In the course of acidic hydrolysis of salts **157** and **158** the $C\overset{\oplus}{=}N$ bond is cleaved, yielding esters **159** and **160** (144, 125, 135).

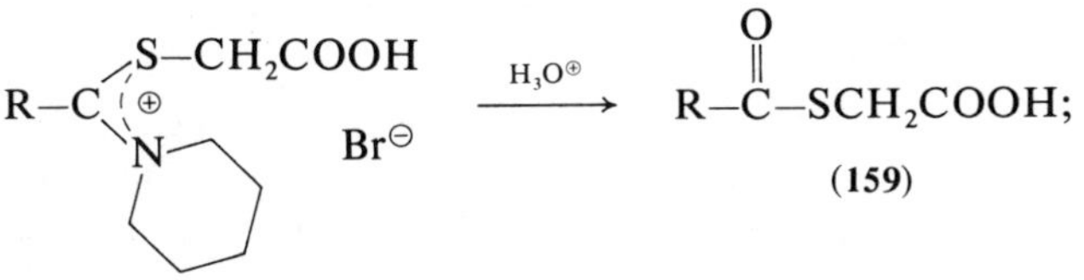

$$R-C\overset{S-CH_2COOH}{\underset{N\text{-piperidyl}}{\oplus}} Br^{\ominus} \xrightarrow{H_3O^{\oplus}} R-\overset{O}{\underset{\|}{C}}-SCH_2COOH;$$
(**157**) → (**159**)

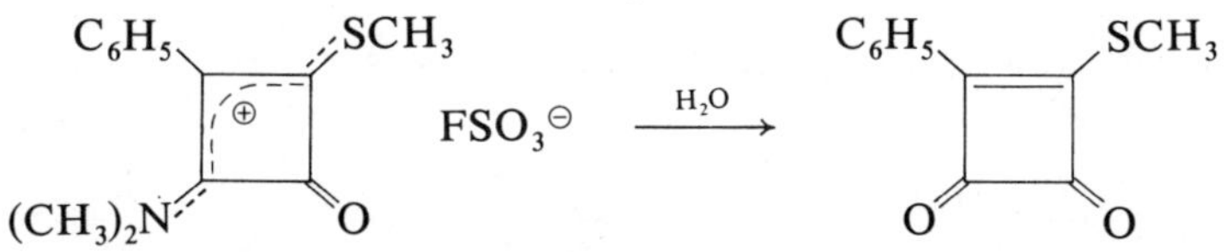

$$\underset{(CH_3)_2N}{\overset{C_6H_5}{\diagdown}}\square\overset{SCH_3}{\diagup}_{O} \quad FSO_3^{\ominus} \xrightarrow{H_2O} \underset{O}{\overset{C_6H_5}{\diagdown}}\square\overset{SCH_3}{\diagup}_{O}$$
(**158**) → (**160**)

To Section III-B

Thiolysis of compounds **157** affords dithiocarboxymethylesters **161** (125).

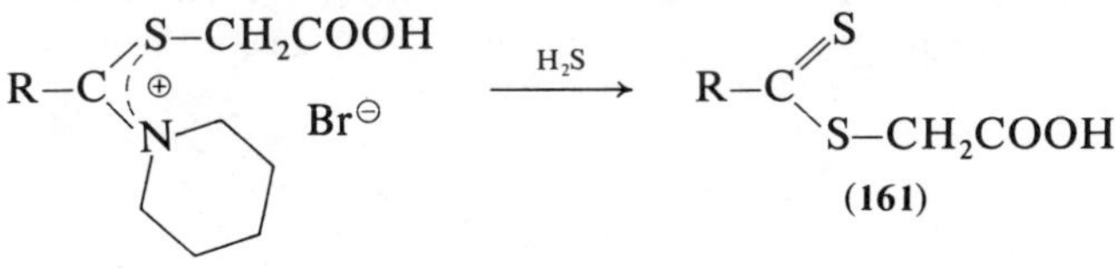

$$R-C\overset{S-CH_2COOH}{\underset{N\text{-piperidyl}}{\oplus}} Br^{\ominus} \xrightarrow{H_2S} R-C\overset{S}{\underset{S-CH_2COOH}{\diagdown}}$$
(**157**) → (**161**)

Treatment of the salt **162** with H_2S/pyridine leads to a mixture of compounds **163** and **164**, whereas thiolysis of **162** in the presence of acetic acid gives a mixture of **165** and **166** (136). Thiolysis of **167** with H_2S/pyridine proceeds selectively, yielding dithiosuccinimide (136).

(163) (164) (162) (165) (166)

(167)

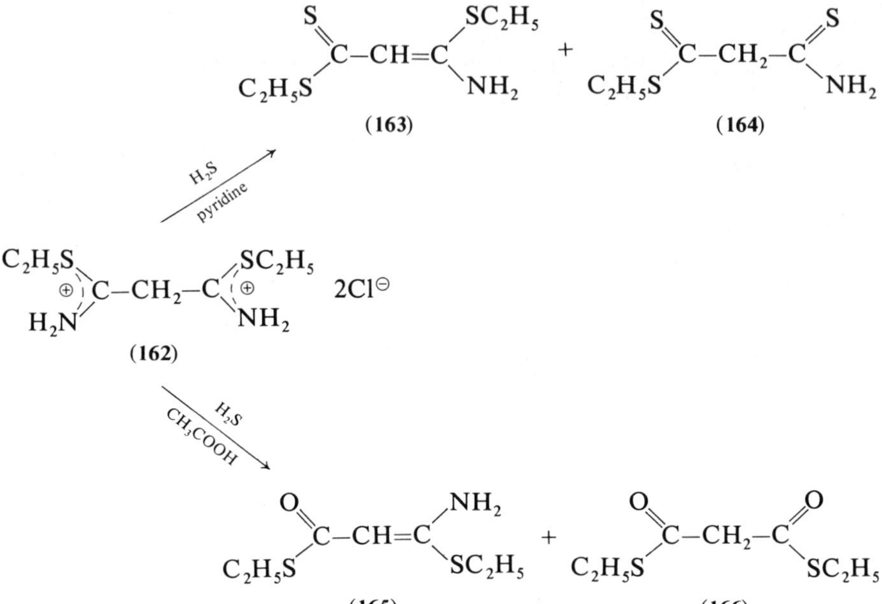

To Section III-C

Preparation of thioimidic acid esters from alkylmercaptomethyleniminium salts is the subject of a more recent review (130). Electron-rich olefines **168** can be prepared by action of the thiazolium salt **169** or orthoamides **170** (137). A mechanism involving nucleophilic carbenes has been discussed for these reactions.

(169) (170) (168)

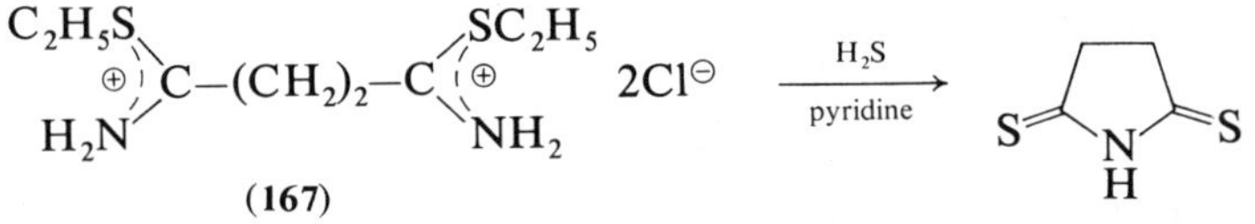

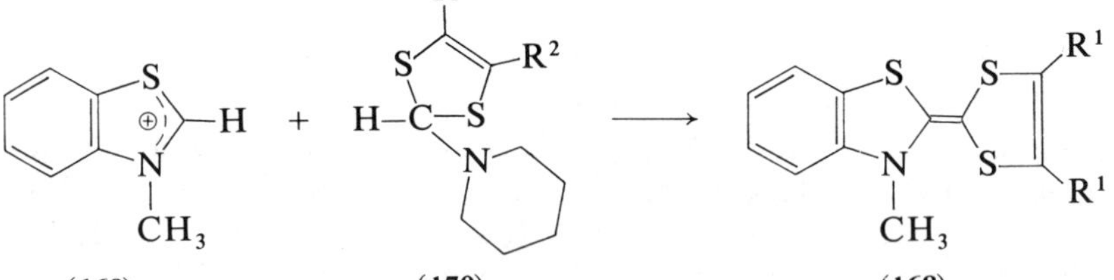

Treatment of salts **171** with strong bases induces a thia-Cope rearrangement, producing thioamides **172**, whereas salts **173** under similar conditions are deprotonated to form ketene-S,N-acetals **174** (124).

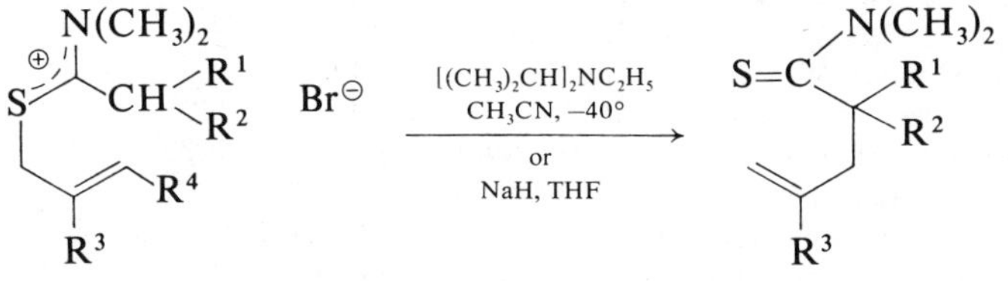

To Section III-D

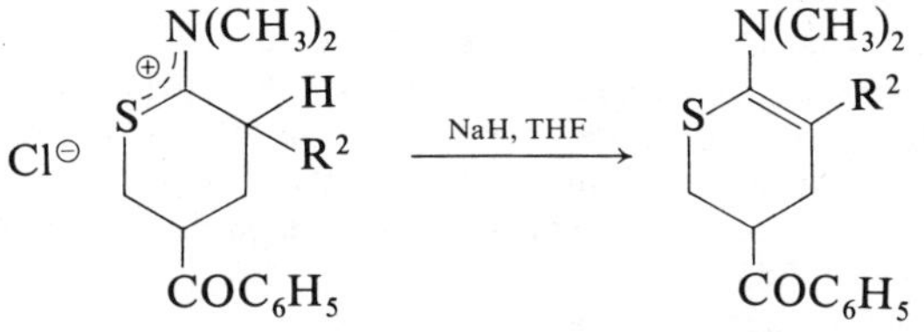

The salt **175** has been recommended for dehydration of oximes to nitriles (138).

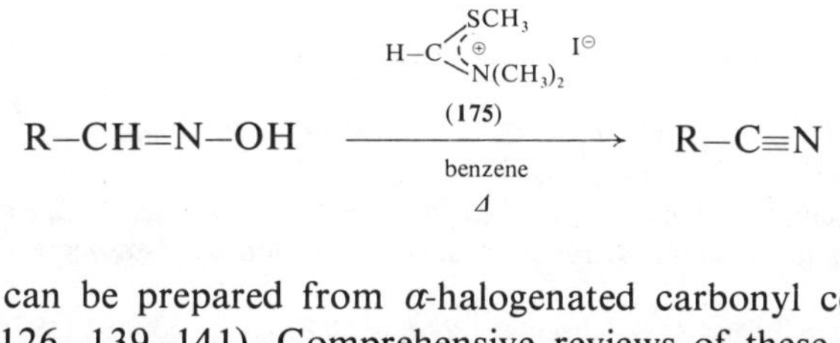

$$R-CH=N-OH \xrightarrow[\Delta]{\text{benzene}} R-C\equiv N$$

(via **175**)

Thiazoles can be prepared from α-halogenated carbonyl compounds and thioamides (126, 139–141). Comprehensive reviews of these reactions have been published (129, 130). Mechanistic studies of this thiazole synthesis, including isolation of intermediates, have been undertaken (143–145).

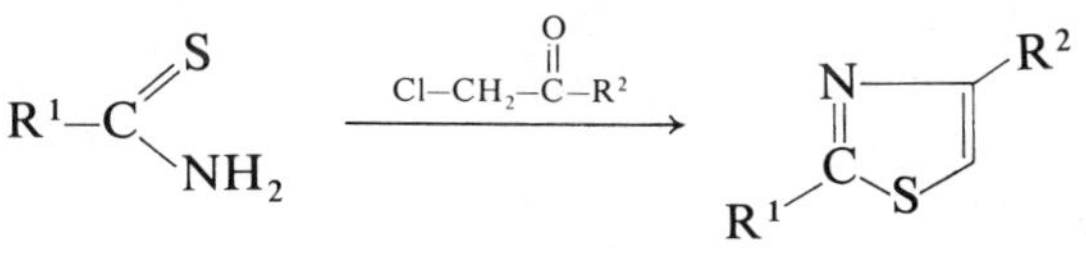

REFERENCES

113. G. H. R. Seitz and G. Arndt, *Synthesis*, **1976**, 692.
114. G. H. R. Seitz and G. Arndt, *Synthesis*, **1976**, 445.
115. P. Beak and J. K. Lee, *J. Org. Chem.*, **40**, 147 (1975).
116. W. Walter and H.-W. Meyer, *Liebigs Ann. Chem.*, **1975**, 19.
117. W. Walter and H.-W. Meyer, *Liebigs Ann. Chem.*, **1975**, 36.
118. R.-C. Moreau and P. Loiseau, *C. R. Acad. Sci.*, **C284**, 347 (1977).
119. P. Vinkler, K. Thimm, and J. Voß, *Liebigs Ann. Chem.*, **1976**, 2083.
120. W. Ried and L. Kaiser, *Synthesis*, **1976**, 535.
121. W. Ried and E. Schmidt, *Liebigs Ann. Chem.*, **676**, 114 (1964).
122. W. Walter and C. O. Meese, *Chem. Ber.*, **109**, 922 (1976).
123. A.-M. Lamazouère, F. Darré, and J. Sotiropoulos, *Bull. Soc. Chim. Fr.*, **1975**, 2269.
124. R. Gompper and W.-R. Ulrich, *Angew. Chem.*, **88**, 300 (1976).
125. N. H. Leon, *J. Pharm. Sci.*, **65**, 146 (1976).
126. K. T. Potts and J. L. Marshall, *J. Org. Chem.*, **41**, 129 (1976).
127. E. D. Dobrzynski and R. A. Angelici, *Inorg. Chem.*, **14**, 1513 (1975).
128. T. Kato, A. Takada, and T. Ueda, *Chem. Pharm. Bull.*, **24**, 431 (1976).
129. K. A. Petrov and L. N. Andreev, *Russian Chem. Rev.*, **40**, 505 (1971).
130. D. G. Neilson, in S. Patai, Ed., *The Chemistry of Amidines and Imidates*, John Wiley and Sons, New York, 1975, p. 385.
131. B. Zeeh and H. Kiefer, *Liebigs Ann. Chem.*, **1975**, 1984.
132. W. Walter and C. O. Meese, *Chem. Ber.*, **109**, 947 (1976).
133. C. Rabiller, Y. P. Renou, and G. J. Martin, *J. Chem. Soc. Perkin II*, **1977**, 536.
134. A. J. Hall and D. P. N. Satchell, *J. Chem. Soc. Perkin II*, **1976**, 1274.
135. K. I. Potts, J. Kane, E. Carnaham, and U.-P. Singh, *J. Chem. Soc., Chem. Commun.*, **1975**, 417.
136. G. Gattow and K. Hanewald, *Chem. Ber.*, **109**, 3243 (1976).
137. D. Buza and W. Karasuski, *Rocz. Chem.*, **49**, 2007 (1975); *C. I.*, 15–222 (1976).
138. T. L. Ho and C. M. Wong, *Synth. Commun.*, **5**, 299 (1975).
139. H. Schäfer and K. Gewald, *J. Prakt. Chem.* **317**, 771 (1975).
140. M. Fuertes, T. Garcia-López, G. Garcia-Munoz, and M. Stud, *J. Org. Chem.*, **41**, 4074 (1976).
141. V. A. Sedavkina, G. v. Bespalova, V. G. Garanzha, and L. K. Kulikova, *Khim. Farm. Zh.*, **10**, 66 (1976).
142. R. H. Wiley, D. C. England, and L. C. Behr, in *Organic Reactions*, Vol. 6, John Wiley and Sons, New York, 1951, p. 367.
143. A. Babadjamian, J. Metzger, and M. Chanon, *J. Heterocycl. Chem.*, **12**, 643 (1975).
144. A. Babadjamian, R. Gallo, J. Metzger, and M. Chanon, *J. Heterocycl. Chem.*, **13**, 1205 (1976).
145. H. Singh, S. Singh, and A. S. Cheema, *J. Indian Chem. Soc.*, **53**, 682 (1976); *C. I.*, 16–168 (1977).

AMIDINIUM SALTS

By W. KANTLEHNER, *D-7080-AALEN-Dewangen, Laachweg 14, Germany*

CONTENTS

I. Synthesis 322
   A. From Imide Halides or Amide Halides and Amines 322
   B. From Acid Amide–Acid Halide Adducts and Amines 325
      1. Acid Amide–Phosphorus Oxychloride Adducts 325
      2. Acid Amide Adducts with Other Acylating or Halogenating Agents 327
         a. Acid Amide–Phosphorus Trichloride Adducts 327
         b. Acid Amide Adducts with Sulfonyl Chlorides or Carboxylic Acid Chlorides 327
         c. Tertiary Carboxamides, Secondary Amines, and Titanium Tetrachloride . 327
   C. From Alkoxymethyleniminium Salts and Amines 328
   D. From Acid Amides and Acylating Agents 329
   E. From Acid Amides and Alkylating Agents 333
   F. From Nitriles and Ammonium Salts 333
   G. From Amines and Nitrilium Salts 333
   H. From Isonitriles and Amine Hydrochlorides or Amines 336
   I. From Orthoesters and Orthoamide Derivatives 337
   J. Via Alkylation and Acylation of Amidines 341
   K. Via Transamination of Amidinium Salts 344
   L. From Ketene Aminals and Ynamines 344
      1. Alkylation and Acylation of Ketene Aminals and Ynamines 344
      2. Oxidation of Ketene Aminals 351
   M. From Thioamides, Thioamide Derivatives, and Thioureas 353
II. Structure and Properties of Amidinium Salts 354
III. Reactions of Amidinium Salts 357
   A. Hydrolysis, Thiolysis 357
   B. Reactions with NH-Containing Compounds 360
      1. With Carboxylic Acid Amides 360
      2. With Ammonia, Primary and Secondary Amines, Hydroxylamine, and Hydrazines 361
   C. Reactions with Base 364
   D. Reactions of *N,N,N',N'*-Tetrasubstituted Amidinium Salts with Alkali Metal Alkoxides 367
      1. In the Presence of Alcohol 367
      2. In the Absence of Alcohol 369

Translated from German by Dr. W.-J. Richter, MPI für Kohleforschung, D 4330 Mühlheim.

E. Reactions with Mercaptides 370
F. Reactions of N,N,N',N'-Tetrasubstituted Formamidinium Salts with Alkali Amides 372
G. Reactions of N,N,N',N'-Tetrasubstituted Formamidinium Salts with Alkali Phosphides and Phosphonates 373
H. Anion Exchange 375
I. Miscellaneous Reactions 376
References 378
Addendum 382

I. Synthesis

Amidines (**1**) in general are strong bases and will react with acids to form amidinium salts (**2**). By this route chlorides, sulfates, acetates (1) nitrates,

$$R-C{\overset{NR^3}{\underset{NR^1R^2}{\diagup}}} + HX \longrightarrow R-C{\overset{NHR^3}{\underset{NR^1R^2}{\diagup}}}^{\oplus} X^{\ominus}$$

(**1**)   (**2**)

carbonates (2), picrates (3), chloroplatinates (3), and nitrates (4) have been synthesized. So it is well understood that amidinium salts are in principle accessible from various amidines. Predominantly N,N-unsubstituted, N-mono-, N,N-di-, and N,N,N'-trisubstituted amidinium salts are generated this way. Since the chemistry of the amidines has been repeatedly reviewed (5–7), this chapter mainly deals with reactions in which amidinium salts—especially N,N',N',N'-tetrasubstituted ones—are formed directly.

A. FROM IMIDE HALIDES OR AMIDE HALIDES AND AMINES

Carboxylic acid amides are converted into amide chlorides or imide chlorides by halide carriers such as PCl_5, $COCl_2$, $SOCl_2$, and oxalyl chloride. These reactions are covered in the chapter "Amide Halides." Ulrich (8) and Bonnett (9) published review articles on the chemistry of imide halides.

N-Monosubstituted amidinium salts (3,4), N,N'-disubstituted amidinium salts (**2a**) (10–18), and N,N,N'-trisubstituted amidinium salts (**2b**) (11, 19–21) are synthesized from imide halides (**3**) and ammonia or primary and secondary amines. However, in most cases the amidine was primarily set free, then converted into the corresponding hydrochloride, picrate, etc., and characterized as such.

$$R-C{\overset{O}{\underset{NHR^1}{\diagup}}} \xrightarrow{PCl_5} R-C{\overset{Cl}{\underset{NR^1}{\diagup}}} \xrightarrow{R^2R^3NH} R-C{\overset{NHR^1}{\underset{NR^2R^3}{\diagup}}}^{\oplus} Cl^{\ominus}$$

           (**3**)   (**2a**) $R^3 = H$
                   (**2b**) R = alkyl, aryl

The reaction of aliphatic amines with amide chlorides like **4** does lead to amidinium salts **5**, however, the separation from the amine hydrochloride simultaneously formed is hardly possible, as the example of the reaction of dimethylformamide chloride (**4**) with dimethylamine shows (22).

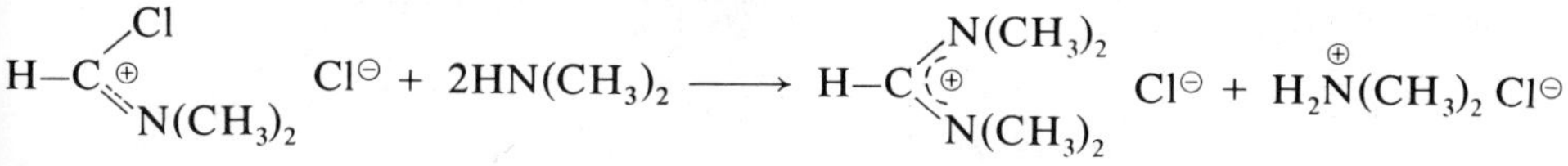

$$\text{H-C}^{\oplus}\begin{smallmatrix}\text{Cl}\\\text{N(CH}_3)_2\end{smallmatrix} \text{Cl}^{\ominus} + 2\text{HN(CH}_3)_2 \longrightarrow \text{H-C}^{\oplus}\begin{smallmatrix}\text{N(CH}_3)_2\\\text{N(CH}_3)_2\end{smallmatrix} \text{Cl}^{\ominus} + \text{H}_2\overset{\oplus}{\text{N}}(\text{CH}_3)_2\, \text{Cl}^{\ominus}$$

(4)   (5)

At higher reaction temperatures (150–180°C), ammonium salts such as the chloride, carbonate, or acetate may be reacted instead of the amines (23). Even ammonium chloride gives a reaction with dimethylformamide chloride (**4**) to yield *N,N*-dimethylformamidinium chloride (**6**) (23).

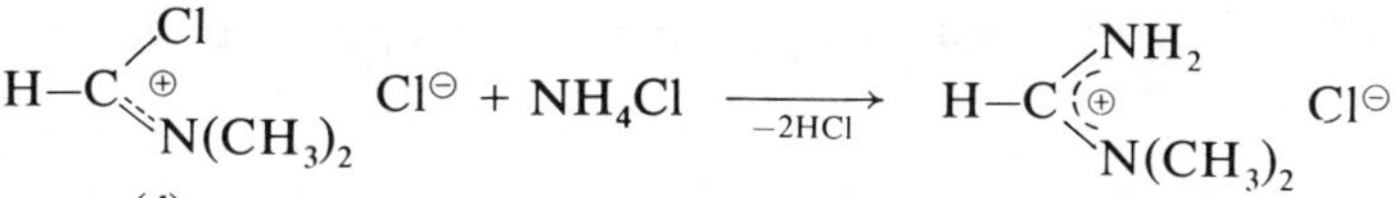

$$\text{H-C}^{\oplus}\begin{smallmatrix}\text{Cl}\\\text{N(CH}_3)_2\end{smallmatrix} \text{Cl}^{\ominus} + \text{NH}_4\text{Cl} \xrightarrow{-2\text{HCl}} \text{H-C}^{\oplus}\begin{smallmatrix}\text{NH}_2\\\text{N(CH}_3)_2\end{smallmatrix} \text{Cl}^{\ominus}$$

(4)   (6)

Owing to the lower basicity of aromatic amines their hydrochlorides will react with amide chlorides under much milder reaction conditions (23, 24).

The aminoanthraquinones are also easily converted with the aid of **4** into amidinium salts **7** (23, 25).

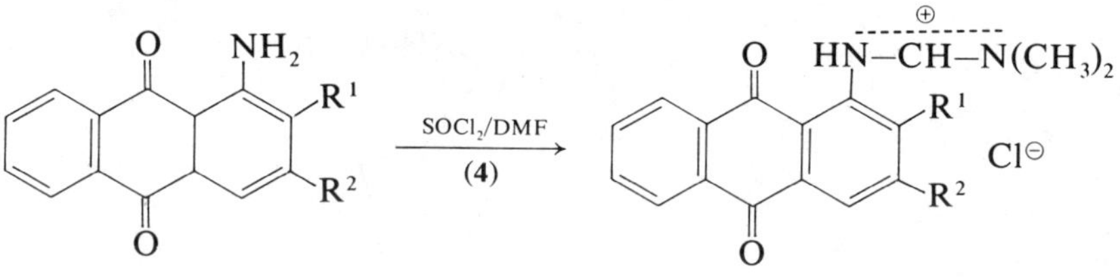

[Reaction scheme: aminoanthraquinone + SOCl₂/DMF (**4**) → amidinium salt **7** with HN=CH−N(CH₃)₂ group, Cl⁻]

Czŭrös et al. (26–28) synthesized a large number of amidinium salts **8** and **9** starting from amide chlorides **10** or **4** and *o*-substituted aromatic amines.

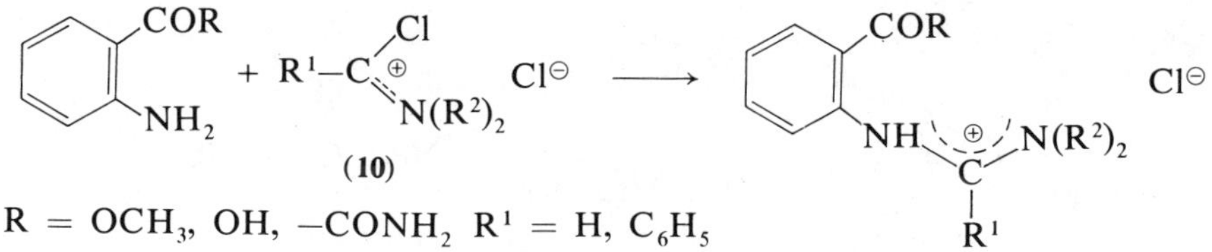

[Reaction scheme: o-COR-substituted aniline + R¹−C(Cl)=N(R²)₂ Cl⁻ (**10**) → amidinium salt **8**]

R = OCH₃, OH, −CONH₂   R¹ = H, C₆H₅
                                 R² = H, C₆H₅

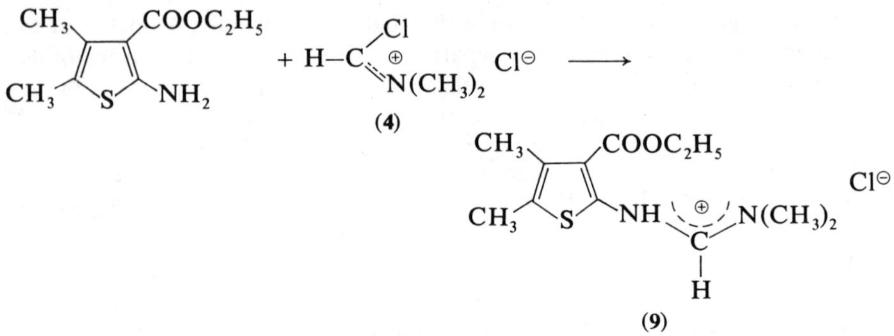

Amidinium salts like **11–14**, substituted at the nitrogen atom by amino-carbonyl, aminothiocarbonyl, or alkoxycarbonyl groups, may be obtained from amide chlorides and ureas, thioureas, or urethanes (23, 29–32). The chapter called "Amide Halides" summarizes the properties of some of these compounds.

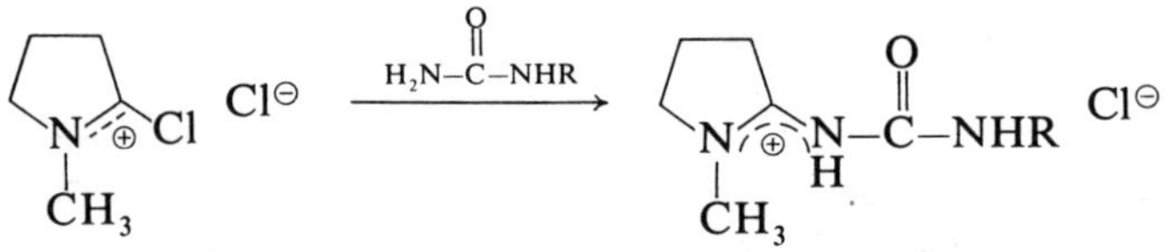

(11)

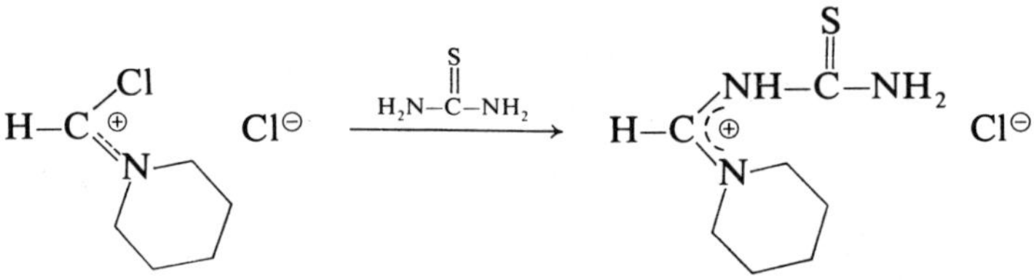

(12)

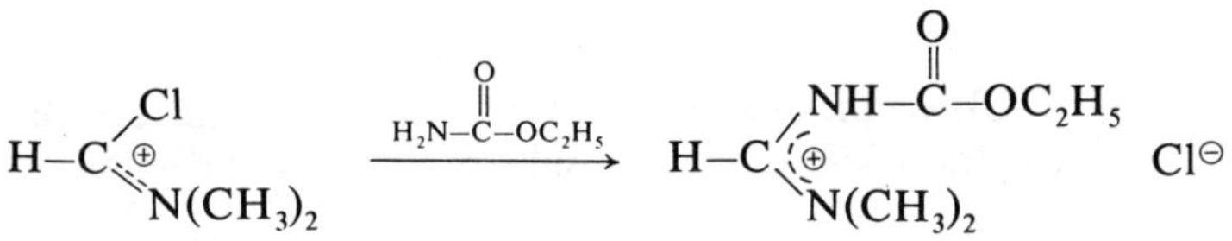

(13)

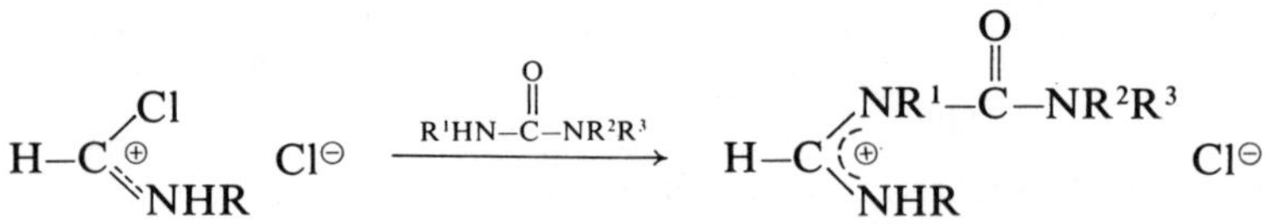

(14)

Amidinium salts are also accessible from bromomethyleniminium bromides (3), but this approach is of less importance owing to the complicated synthesis of the bromomethyleniminium salt.

B. FROM ACID AMIDE–ACID HALIDE ADDUCTS AND AMINES

1. Acid Amide–Phosphorus Oxychloride Adducts

If carboxylic acid amide–phosphorus oxychloride adducts **15** are reacted with aromatic amines, the amidinium salts **2a** or **2b** are formed (34–45) (see also the chapter "Adducts from Acid Amides and Acylation Reagents").

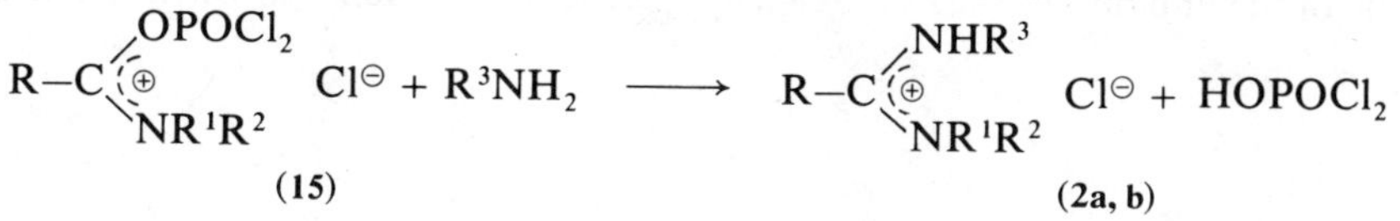

$$R-C\overset{\oplus}{\underset{NR^1R^2}{\diagdown}}{}^{OPOCl_2} \quad Cl^{\ominus} + R^3NH_2 \longrightarrow R-C\overset{\oplus}{\underset{NR^1R^2}{\diagdown}}{}^{NHR^3} \quad Cl^{\ominus} + HOPOCl_2$$

(15)   (2a, b)

In most cases, the amidinium salts are not isolated, but rather the amidines are generated from the salts with base. Raison (36) and Jutz (42) synthesized a large number of N,N,N',N'-tetrasubstituted amidinium salts like **16**; they started from acid amide–phosphorus oxychloride adducts and secondary aromatic amines or their hydrochlorides. The low solubility of the perchlorates (42) and the iodides (36) was sometimes utilized to isolate the products from the reaction mixture. Adducts from aliphatic carboxamides containing α-CH_2

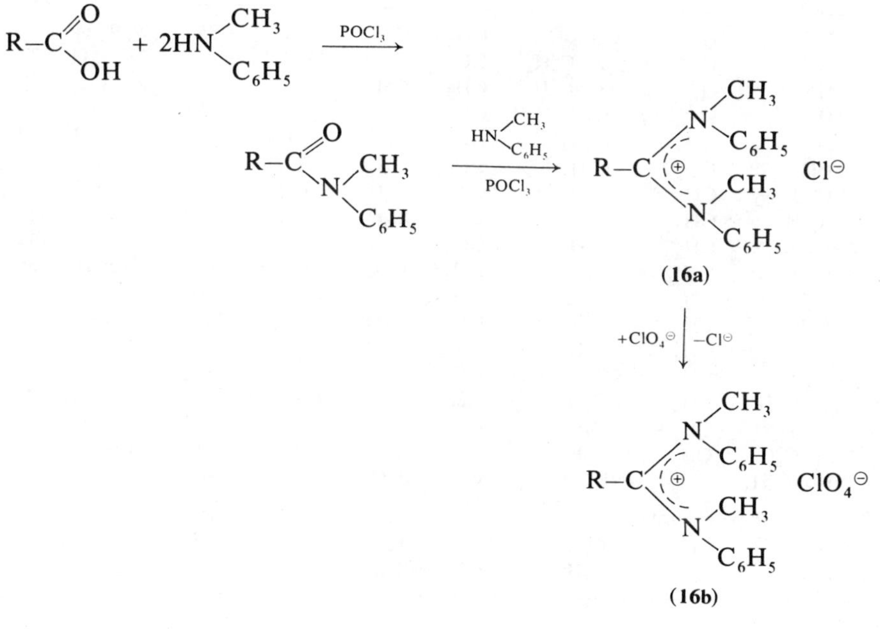

groups and phosphorus oxychloride show self-condensation (23, 38, 46, 47); this may interfere with the synthesis of amidinium salts from this kind of adducts. Jutz (42) elaborated two methods to suppress this self-condensation. In his first method phosphorus oxychloride is added dropwise to a mixture of amide and aromatic amine, then heated to 150°C. The second, even simpler, method involves direct reaction of phosphorus oxychloride with the free carboxylic acid and excess amine.

Bisamidinium perchlorates from glutaric acid and adipic acid are also generated by this route (42). The synthesis of N,N'-diphenylbutyramidine from aniline, butyric acid, and $POCl_3$ has already been reported in earlier times, (41). Some of the amidinium salts obtained by this method are summarized in Table I.

TABLE I

Amidinium Salts from Acid Amide–Phosphorus Oxychloride Adducts and Primary and Secondary Aromatic Amines

$$R-C\underset{NR^1R^2}{\overset{NR^3R^4}{\lessgtr}}\ \ X^\ominus$$

R	R^1	R^2	R^3	R^4	X	Yield, %	m.p., °C	Ref.
H	H	C_6H_5-	CH_3	CH_3	Cl	85	239–241	37
H	H	$p\text{-}CH_3\text{-}C_6H_4$	CH_3	CH_3	Cl	—	242–244	37
H	H	$p\text{-}NO_2\text{-}C_6H_4$	CH_3	CH_3	Cl	74	265 (dec.)	39
H	H	$p\text{-}CH_3O\text{-}C_6H_4$	CH_3	CH_3	Cl	85	214–215	39
H	CH_3	C_6H_5	CH_3	C_6H_5	ClO_4	98	122–133	42
CH_3	CH_3	C_6H_5	CH_3	C_6H_5	ClO_4	96	168	42
C_2H_5	CH_3	C_6H_5	CH_3	C_6H_5	ClO_4	93	145	42
$n\text{-}C_3H_7$	CH_3	C_6H_5	CH_3	C_6H_5	ClO_4	73	129	42
$i\text{-}C_3H_7$	CH_3	C_6H_5	CH_3	C_6H_5	ClO_4	35	—	42
C_6H_5	CH_3	$p\text{-}NH_2\text{-}C_6H_4$	CH_3	CH_3	I	—	261–262	36
C_6H_5	CH_3	C_6H_5	CH_3	CH_3	I	90	180–182	36
C_6H_5	CH_3	C_6H_5	CH_3	C_6H_5	I	50	170–172	36
C_6H_5	CH_3	C_6H_5	CH_3	C_6H_5	ClO_4	83	216–217	42
C_6H_5	CH_3	$p\text{-}NO_2\text{-}C_6H_4$	CH_3	CH_3	I	95	—	36
$p\text{-}CH_3\text{-}C_6H_4$	CH_3	C_6H_5	CH_3	C_6H_5	ClO_4	92	225–226	42
$p\text{-}Cl\text{-}C_6H_4$	CH_3	C_6H_5	CH_3	C_6H_5	ClO_4	72	165–166	42
$p\text{-}NO_2\text{-}C_6H_4$	CH_3	C_6H_5	CH_3	C_6H_5	ClO_4	49	199–200	42
$m\text{-}NH_2\text{-}C_6H_4$	CH_3	$p\text{-}NH_2\text{-}C_6H_4$	CH_3	CH_3	I	—	262	36
Bisamidinium perchlorates								
$-(CH_2)_3-$	CH_3	C_6H_5	CH_3	C_6H_5	ClO_4	48	185–186	42
$-(CH_2)_4-$	CH_3	C_6H_5	CH_3	C_6H_5	ClO_4	45	166–167	42

2. Acid Amide Adducts with Other Acylating or Halogenating Agents

(a) *Acid Amide–Phosphorus Trichloride Adducts.* N,N-Disubstituted and N,N,N'-trisubstituted amidinium salts, have been synthesized from primary, secondary, and tertiary amides as well as primary and secondary aromatic and aliphatic amines (12, 18, 35, 36, 48, 49); however, these compounds are converted directly into the free amidines **17** and **18** without isolation (see also the chapter "Adducts from Acid Amides and Acylation Reagents").

$$R-C\begin{smallmatrix}O\\NR^1R^2\end{smallmatrix} + H_2NR^3 \xrightarrow{PCl_3} R-C\begin{smallmatrix}NHR^3\\\oplus\\NR^1R^2\end{smallmatrix} Cl^{\ominus} \xrightarrow{OH^{\ominus}} R-C\begin{smallmatrix}N-R^3\\NR^1R^2\end{smallmatrix}$$
(**17**)

$$R-C\begin{smallmatrix}O\\NHR^1\end{smallmatrix} + HNR^2R^3 \xrightarrow{PCl_3} R-C\begin{smallmatrix}NR^2R^3\\\oplus\\NHR^1\end{smallmatrix} Cl^{\ominus} \xrightarrow{OH^{\ominus}} R-C\begin{smallmatrix}NR^1\\NR^1R^2\end{smallmatrix}$$
(**18**)

(b) *Acid Amide Adducts with Sulfonyl Chlorides or Carboxylic Acid Chlorides.* The syntheses of these adducts are dealt with in the appropriate chapters for the particular adducts.

(c) *Tertiary Carboxamides, Secondary Amines, and Titanium Tetrachloride.* Recently it was shown that secondary aliphatic amines and tertiary carboxamides combine to give amidinium salts **19** under the influence of stochiometric amounts of titanium tetrachloride (50). In addition, the amine hydrochloride forms, which is not always easily separated. This reaction is

$$2R-C\begin{smallmatrix}O\\N(R^1)_2\end{smallmatrix} + 4(R^1)_2NH \xrightarrow{TiCl_4} 2R-C\begin{smallmatrix}N(R^1)_2\\\oplus\\N(R^1)_2\end{smallmatrix} Cl^{\ominus} + 2(R^1)_2\overset{\oplus}{N}H_2Cl^{\ominus} + TiO_2$$

$R = H, C_2H_5$
$R^1 = CH_3, C_2H_5$

(**19**)

$$\downarrow \begin{smallmatrix}+NaPF_6\\-NaCl\end{smallmatrix}$$

$$R-C\begin{smallmatrix}N(R^1)_2\\\oplus\\N(R^1)_2\end{smallmatrix} PF_6^{\ominus}$$

successful with amides, the ligands of which must not be too bulky; for example, N,N-dimethylisobutyramide still condenses with dimethylamine, whereas N,N-dimethylpivalic acid amide does not undergo the corresponding reaction.

Large α substituents with-I effects again allow the reaction; thus N,N-dimethyl-α,α-dichloroacetamide is successfully aminated.

If excess amine is used for the reaction, often it is not the amidinium salts but rather their deprotonation products ketene aminals **20** (50) that are formed.

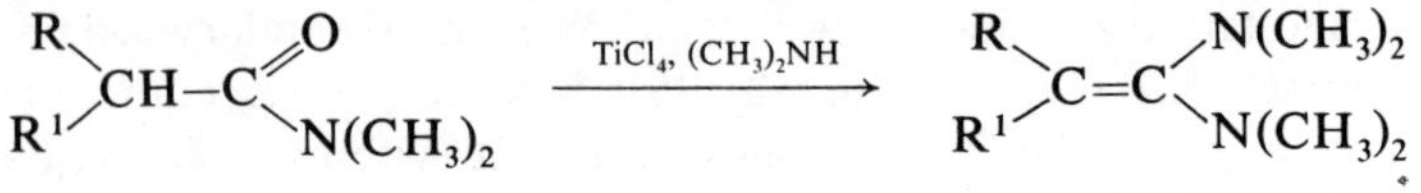

$$\begin{array}{c} R \\ R^1 \end{array}\!\!CH\!-\!C\!\!\begin{array}{c} O \\ N(CH_3)_2 \end{array} \xrightarrow{TiCl_4,\ (CH_3)_2NH} \begin{array}{c} R \\ R^1 \end{array}\!\!C\!=\!C\!\!\begin{array}{c} N(CH_3)_2 \\ N(CH_3)_2 \end{array}$$
(20)

R	H	Cl	C_6H_5
R¹	H	Cl	H

The imidazolium salt **21** or **22** is synthesized from benzoic acid, N,N'-dimethylethylenediamine, and titanium tetrachloride (47).

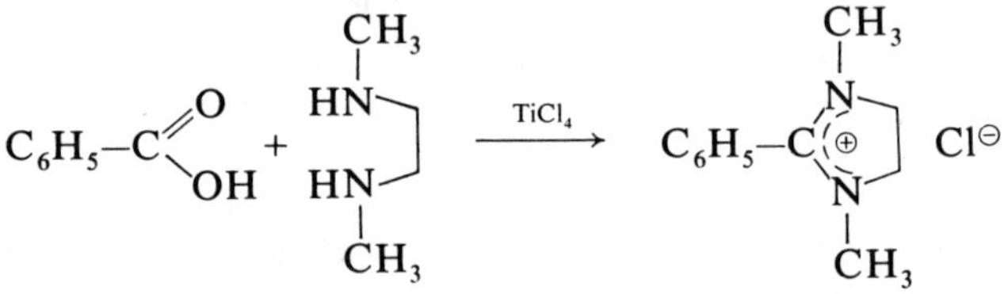

$$C_6H_5-C\!\!\begin{array}{c} O \\ OH \end{array} + \begin{array}{c} HN-CH_3 \\ HN-CH_3 \end{array} \xrightarrow{TiCl_4} C_6H_5-C\!\!\begin{array}{c} N(CH_3) \\ N(CH_3) \end{array}^{\oplus}\ Cl^{\ominus}$$
(21)

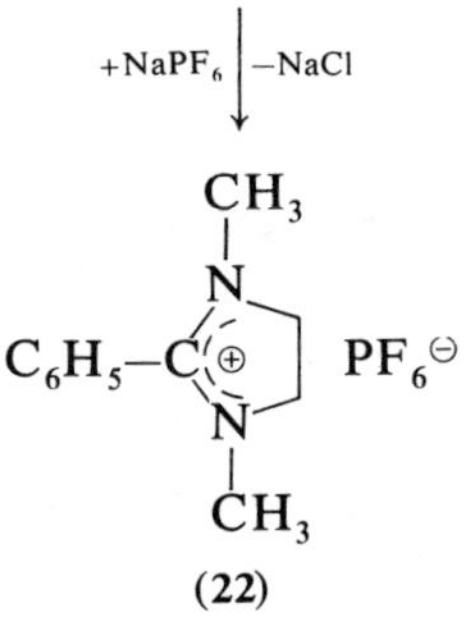

$$\downarrow {+NaPF_6\ |\ -NaCl}$$

$$C_6H_5-C\!\!\begin{array}{c} N(CH_3) \\ N(CH_3) \end{array}^{\oplus}\ PF_6^{\ominus}$$
(22)

C. FROM ALKOXYMETHYLENIMINIUM SALTS AND AMINES

N-Unsubstituted alkoxymethyleniminium salts like imino ester hydrochlorides (**23**) are capable of reacting with ammonia, primary and secondary amines, and with sulfonamides to yield amidinium salts (**24**); these are mostly not isolated as such but rather as free amidines (**25**).

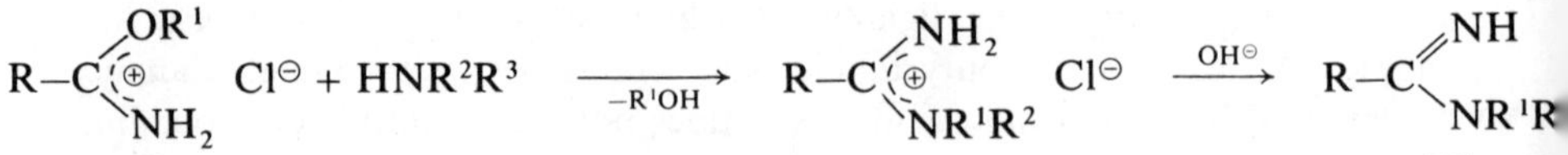

$$R-C\!\!\begin{array}{c} OR^1 \\ NH_2 \end{array}^{\oplus}\ Cl^{\ominus} + HNR^2R^3 \xrightarrow{-R^1OH} R-C\!\!\begin{array}{c} NH_2 \\ NR^1R^2 \end{array}^{\oplus}\ Cl^{\ominus} \xrightarrow{OH^{\ominus}} R-C\!\!\begin{array}{c} NH \\ NR^1R \end{array}$$
(23)   (24)   (25)

This reaction has been repeatedly reviewed (5–7, 51). The synthesis of amidinium salts **(26)** from *N*-mono- and *N,N*-disubstituted alkoxymethyleniminium salts **(27)** has been dealt with in the respective chapters about the reactions of these salts.

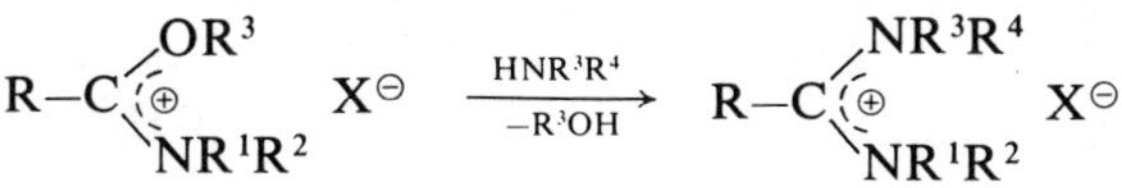

$$R-C{\overset{OR^3}{\underset{NR^1R^2}{\oplus}}} \ X^\ominus \xrightarrow[-R^3OH]{HNR^1R^4} R-C{\overset{NR^3R^4}{\underset{NR^1R^2}{\oplus}}} \ X^\ominus$$

(27)   (26)

D. FROM ACID AMIDES AND ACYLATING AGENTS

From *N-p*-tolylacetamide and phosphorus pentachloride Wallach (52) obtained the *N,N'*-di-*p*-tolylacetamidine hydrochloride **(28)**.

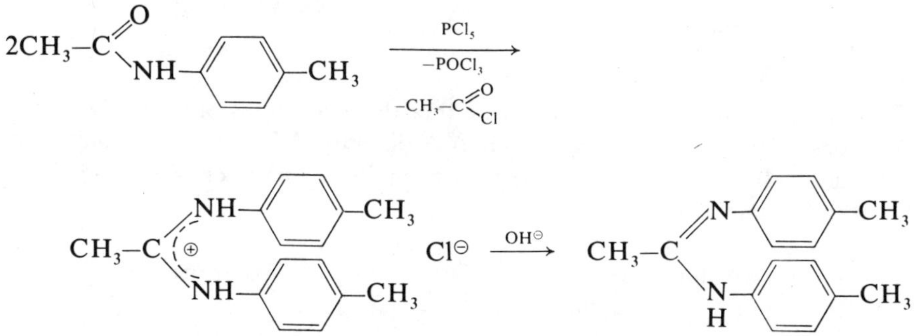

(28)

Later *N,N'*-diphenyl-acetamidine **(29)** was synthesized from acetanilide and phosgene (53), and the *N,N*-di-(*o*-tolyl)formamidinium chloride **(30)** from *N*-formyl-*o*-toluidine and phosgene (54).

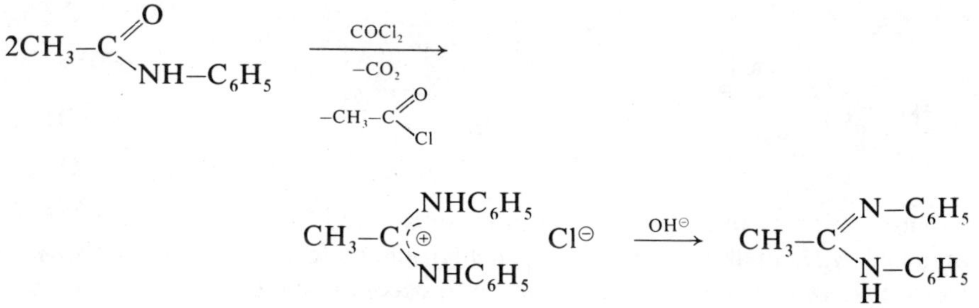

(29)

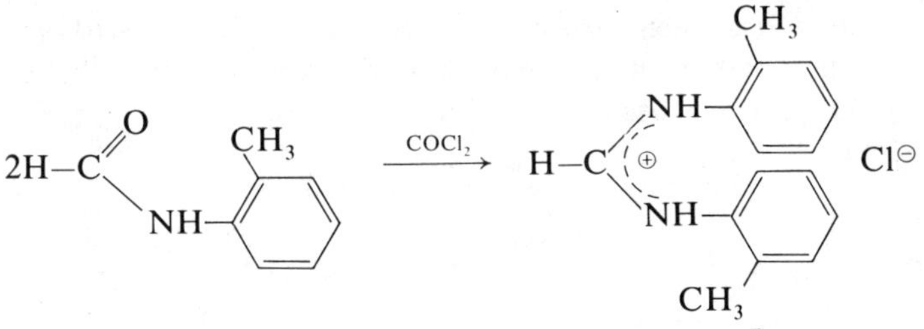

Warren and Wilson (55) converted formanilide and benzanilide with thionyl chloride into the corresponding amidinium salts (**31, 32**).

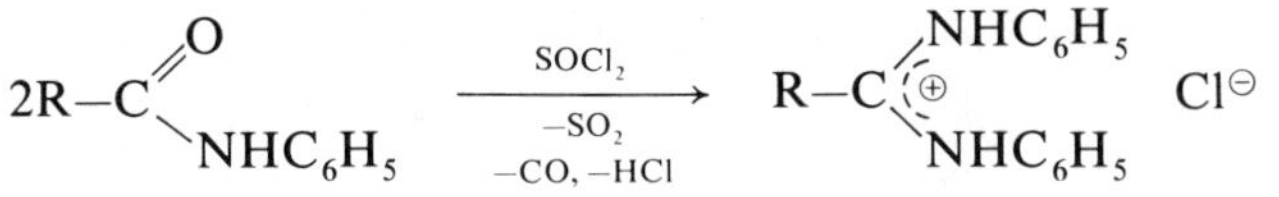

(**31**) R = H
(**32**) R = C$_6$H$_5$

Kühle (56) was able to demonstrate that these reactions proceed via amide chlorides. During these studies the *N,N*-dimethyl-*N'*-phenylformamidinium chloride (**34**) was generated from *N*-phenylformamide chloride (**33**) and dimethylformamide.

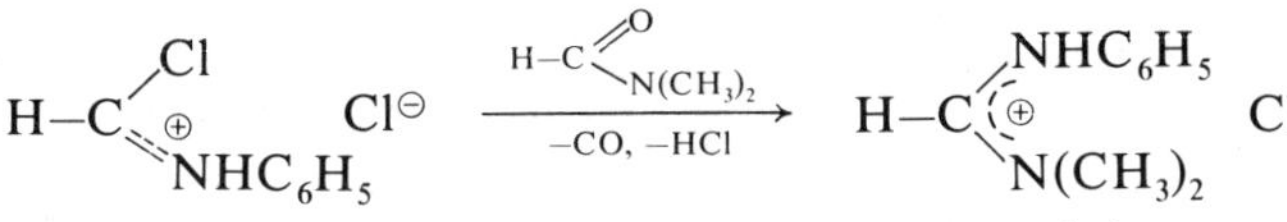

The reaction of thionyl chloride with excess DMF also has dimethylformamide chloride (**35**) as the intermediate; *N,N,N',N'*-tetramethylformamidinium chloride (**36**) is formed here (57).

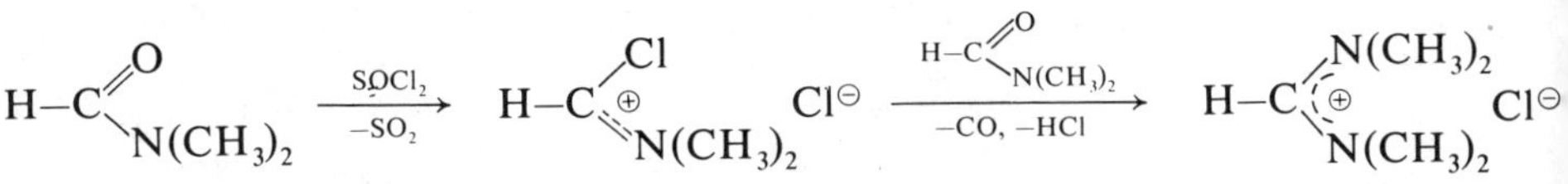

On reaction of phosgene with *N*-monosubstituted formamides, Jentsch (58) obtained *N,N'*-dialkyl(or diaryl)-*N'*-dichloromethylformamidinium salts (**37**), which take only a small amount of water to decompose into *N,N'*-disubstituted formamidinium salts (**38**).

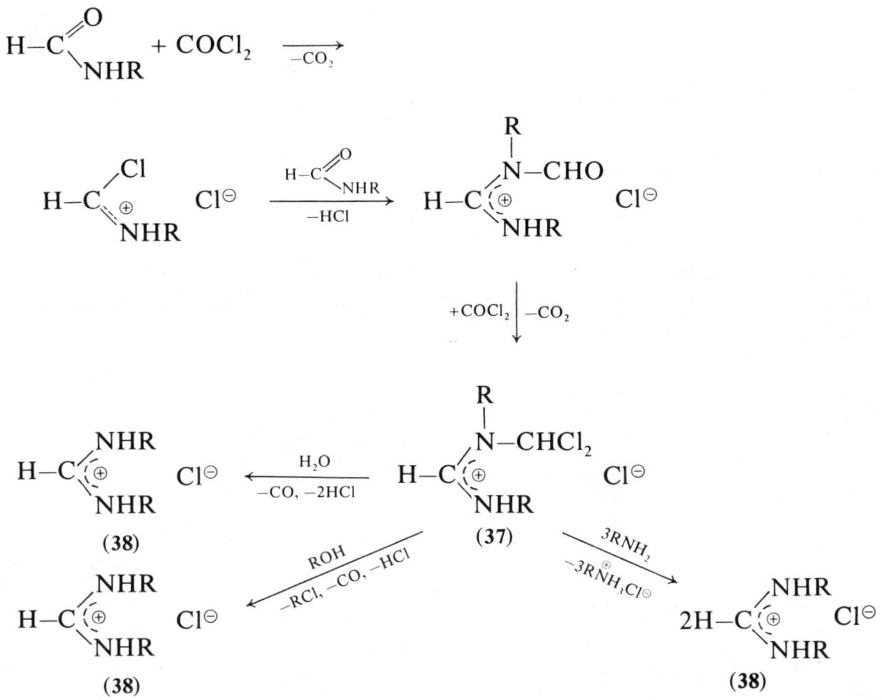

I. SYNTHESIS

$$H-C\underset{NHR}{\overset{O}{\diagup}} + COCl_2 \xrightarrow{-CO_2}$$

$$H-C\underset{NHR}{\overset{Cl}{\diagup}}{}^{\oplus} \;\; Cl^{\ominus} \xrightarrow[-HCl]{H-C\overset{O}{\diagup}_{NHR}} H-C\underset{NHR}{\overset{N-CHO}{\diagup}}{}^{\oplus} \;\; Cl^{\ominus}$$

$$\downarrow {+COCl_2 \;|\; -CO_2}$$

$$H-C\underset{NHR}{\overset{NHR}{\diagup}}{}^{\oplus} \;\; Cl^{\ominus} \xleftarrow[-CO,\,-2HCl]{H_2O} H-C\underset{NHR}{\overset{N-CHCl_2}{\diagup}}{}^{\oplus} \;\; Cl^{\ominus}$$
(38)   (37)

$$H-C\underset{NHR}{\overset{NHR}{\diagup}}{}^{\oplus} \;\; Cl^{\ominus} \qquad\qquad 2H-C\underset{NHR}{\overset{NHR}{\diagup}}{}^{\oplus} \;\; Cl^{\ominus}$$
(38)   (38)

Alcohols and amines are also capable of converting the N,N'-dialkyl- or N,N'-diaryl-N'-dichloromethylformamidinium salts into N,N'-disubstituted formamidinium salts (38) (59). N,N,N',N'-Tetramethylformamidinium arylsulfonates (39) form when dimethylformamide and arylsulfonic acid chlorides are heated (60–62). With this type of reaction the conversion is initiated by primary sulfonation at the amide oxygen atom.

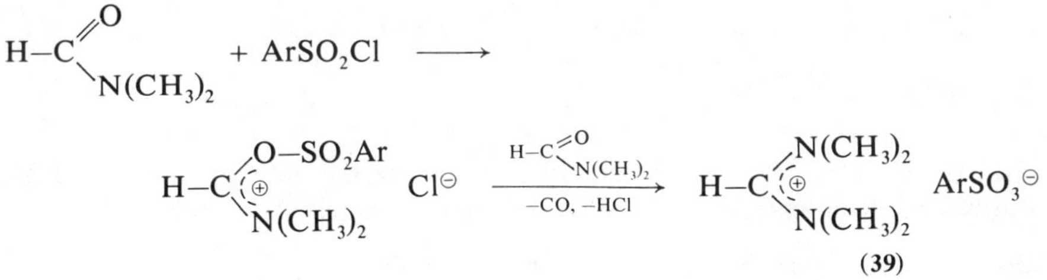

$$H-C\underset{N(CH_3)_2}{\overset{O}{\diagup}} + ArSO_2Cl \longrightarrow$$

$$H-C\underset{N(CH_3)_2}{\overset{O-SO_2Ar}{\diagup}}{}^{\oplus} \;\; Cl^{\ominus} \xrightarrow[-CO,\,-HCl]{H-C\overset{O}{\diagup}_{N(CH_3)_2}} H-C\underset{N(CH_3)_2}{\overset{N(CH_3)_2}{\diagup}}{}^{\oplus} \;\; ArSO_3^{\ominus}$$
(39)

N,N-Disubstituted formamides react with N,N-disubstituted carbamoyl chlorides to yield N,N,N',N'-tetrasubstituted formamidinium chlorides (40); CO_2 is eliminated (63–66).

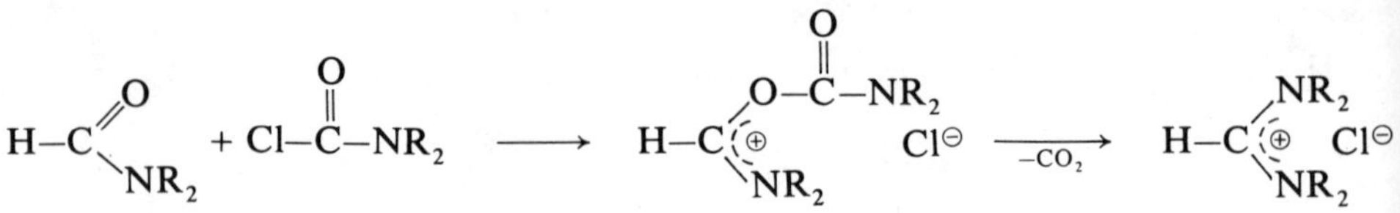

$$H-C\overset{O}{\underset{NR_2}{\diagdown}} + Cl-\overset{O}{\underset{\parallel}{C}}-NR_2 \longrightarrow H-C\overset{O-\overset{O}{\overset{\parallel}{C}}-NR_2}{\underset{NR_2}{\diagdown}^{\oplus}} Cl^{\ominus} \xrightarrow{-CO_2} H-C\overset{NR_2}{\underset{NR_2}{\diagdown}^{\oplus}} Cl^{\ominus}$$

(40)

R = CH$_3$, C$_2$H$_5$, CH$_2$Cl
R$_2$ = –(CH$_2$)$_2$O(CH$_2$)$_2$–

Adducts from *N,N*-disubstituted formamides and phenylisocyanide dichloride decompose on heating to 140°C giving bisformamidinium chlorides (**41**) via CO$_2$ loss (67).

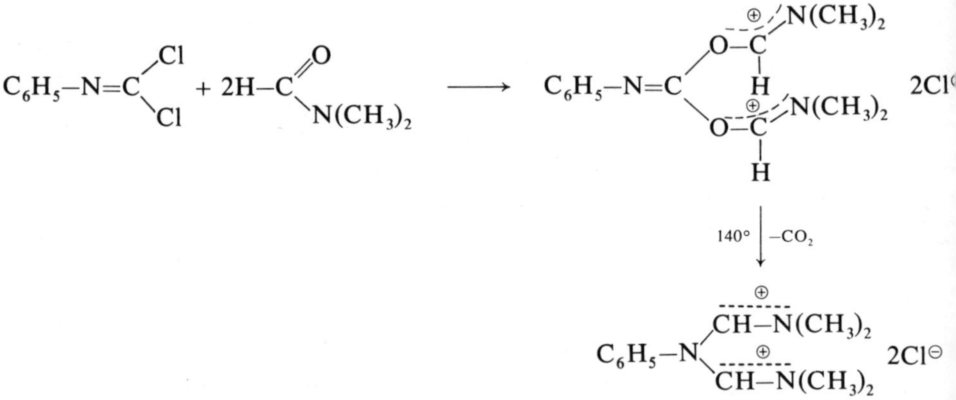

$$C_6H_5-N=C\overset{Cl}{\underset{Cl}{\diagdown}} + 2H-C\overset{O}{\underset{N(CH_3)_2}{\diagdown}} \longrightarrow C_6H_5-N=C\overset{O-\overset{\oplus}{C}\overset{N(CH_3)_2}{\diagdown}H}{\underset{O-\overset{\oplus}{C}\overset{N(CH_3)_2}{\diagup}H}{\diagdown}} \quad 2Cl^{\ominus}$$

$$\downarrow 140° \quad -CO_2$$

$$C_6H_5-N\overset{\overset{\oplus}{CH}-N(CH_3)_2}{\underset{\overset{\oplus}{CH}-N(CH_3)_2}{\diagdown}} \quad 2Cl^{\ominus}$$

(41)

The reaction of phenylisocyanide dichloride with formanilide does not proceed as uniformly as the aforementioned reactions; *N,N*-diphenylformamidinium chloride (**42**) forms, presumably via analogous intermediates (67).

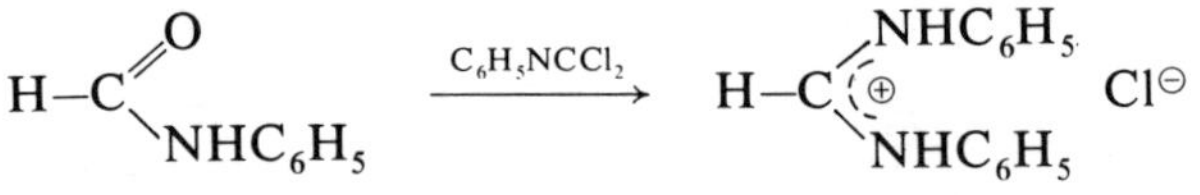

$$H-C\overset{O}{\underset{NHC_6H_5}{\diagdown}} \xrightarrow{C_6H_5NCCl_2} H-C\overset{NHC_6H_5}{\underset{NHC_6H_5}{\diagdown}^{\oplus}} Cl^{\ominus}$$

(42)

In the presence of silver hexafluoroantimonate the amidinium salts **43** are synthesized from chloroformic acid esters and tetramethylurea (68).

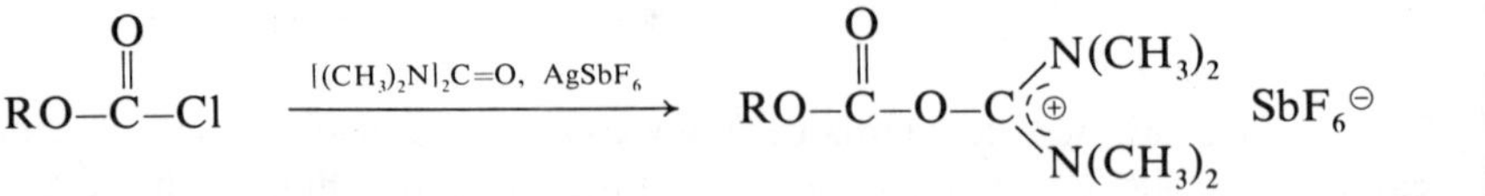

$$RO-\overset{O}{\underset{\parallel}{C}}-Cl \xrightarrow{[(CH_3)_2N]_2C=O, \, AgSbF_6} RO-\overset{O}{\underset{\parallel}{C}}-O-C\overset{N(CH_3)_2}{\underset{N(CH_3)_2}{\diagdown}^{\oplus}} \quad SbF_6^{\ominus}$$

R = C$_6$H$_5$, CH$_2$=CH–

(43)

I. SYNTHESIS

E. FROM ACID AMIDES AND ALKYLATING AGENTS

Alkylating agents like sulfonic acid esters and dialkylsulfates alkylate formanilides at the oxygen atom. These adducts react with further formanilide via elimination of formic acid ester to yield N,N'-diarylformamides (**44**) (37). Analogous reactions are observed on treatment of N-monoalkylated formamides with dimethyl sulfate (69).

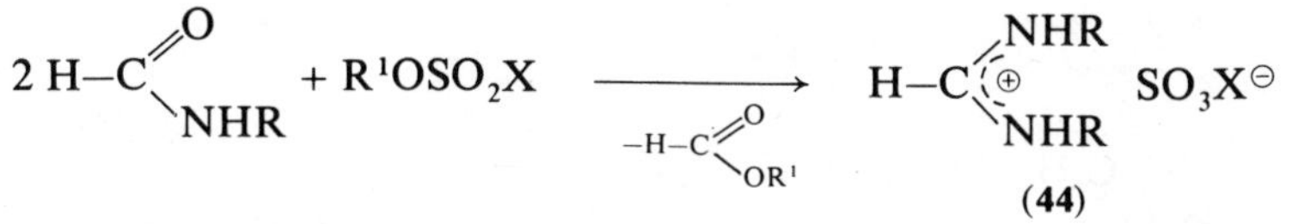

$$2\ H-C\underset{NHR}{\overset{O}{\diagdown}} + R^1OSO_2X \xrightarrow{-H-C\overset{O}{\diagdown}OR^1} H-C\underset{NHR}{\overset{NHR}{\diagdown}}{}^{\oplus}\ SO_3X^{\ominus}$$

(**44**)

There are examples of these reactions in the chapter "Alkoxymethyleniminium Salts," Sections 1-A-3 and 1-A-4.

F. FROM NITRILES AND AMMONIUM SALTS

Ammonium salts of primary and secondary aliphatic and aromatic amines add to nitriles to give N-mono- and N,N-disubstituted amidinium salts (**45**). Ammonium salts themselves, such as ammonium acetate, may also be used for this reaction.

$$R-C\equiv N + H_2\overset{\oplus}{N}R^1R^2\ X^{\ominus} \longrightarrow R-C\underset{NR^1R^2}{\overset{NH_2}{\diagdown}}{}^{\oplus}\ X^{\ominus}$$

(**45**)

The chemistry of these reactions has been reviewed often (5–7, 70–72). In the presence of hydrogen halides thioamides also add to nitriles, but the salt-like intermediates like **46** were not isolated, but rather hydrolyzed to yield N-acylthioamides (**47**) (73, 74).

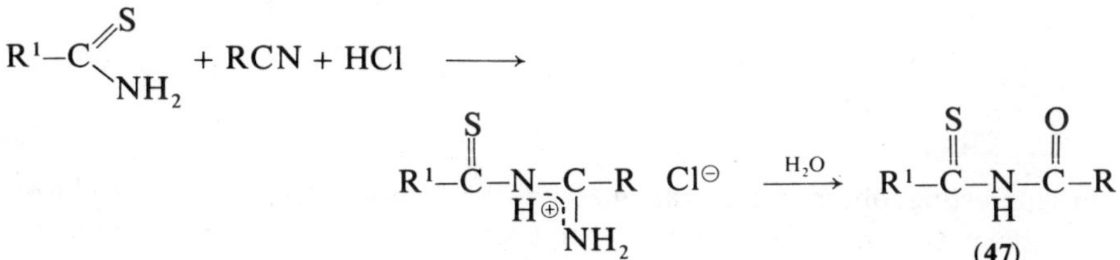

$$R^1-C\underset{NH_2}{\overset{S}{\diagdown}} + RCN + HCl \longrightarrow$$

$$\underset{\underset{NH_2}{\overset{H\oplus|}{}}}{R^1-\overset{\overset{S}{\|}}{C}-N=C-R}\ Cl^{\ominus} \xrightarrow{H_2O} R^1-\overset{\overset{S}{\|}}{C}-\underset{H}{N}-\overset{\overset{O}{\|}}{C}-R$$

(**46**) (**47**)

G. FROM AMINES AND NITRILIUM SALTS

Nitrilium salts add to aromatic and aliphatic amines via formation of amidinium salts like **48–50** (75–79). Generally, however, the amidinium salts are not isolated, but rather the free amidines like **51–53**. During these studies,

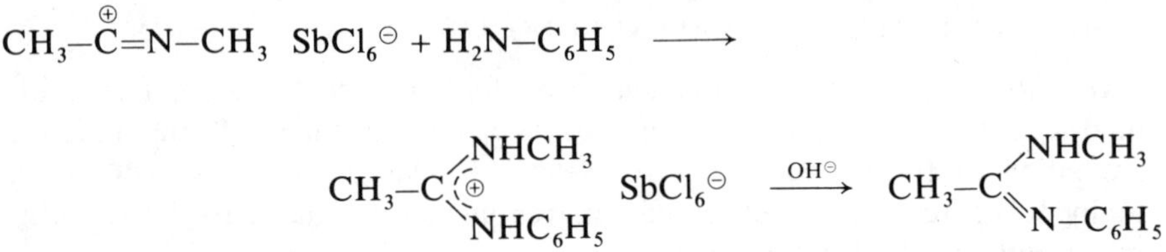

$$CH_3-\overset{\oplus}{C}=N-CH_3 \ SbCl_6^{\ominus} + H_2N-C_6H_5 \longrightarrow$$

$$CH_3-C\overset{NHCH_3}{\underset{NHC_6H_5}{\oplus}} SbCl_6^{\ominus} \xrightarrow{OH^{\ominus}} CH_3-C\overset{NHCH_3}{\underset{N-C_6H_5}{}}$$

(48)   (51)

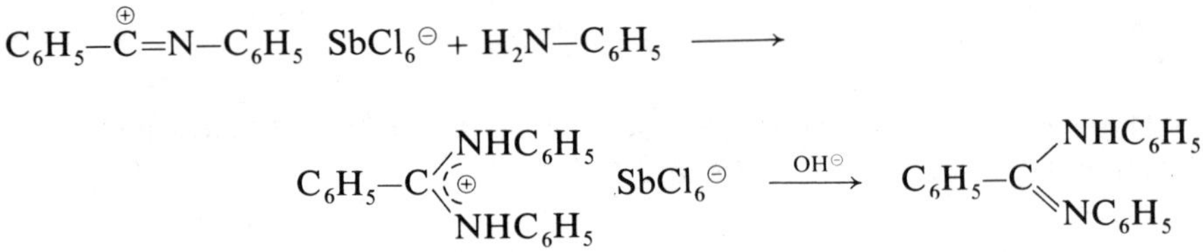

$$C_6H_5-\overset{\oplus}{C}=N-C_6H_5 \ SbCl_6^{\ominus} + H_2N-C_6H_5 \longrightarrow$$

$$C_6H_5-C\overset{NHC_6H_5}{\underset{NHC_6H_5}{\oplus}} SbCl_6^{\ominus} \xrightarrow{OH^{\ominus}} C_6H_5-C\overset{NHC_6H_5}{\underset{NC_6H_5}{}}$$

(49)   (52)

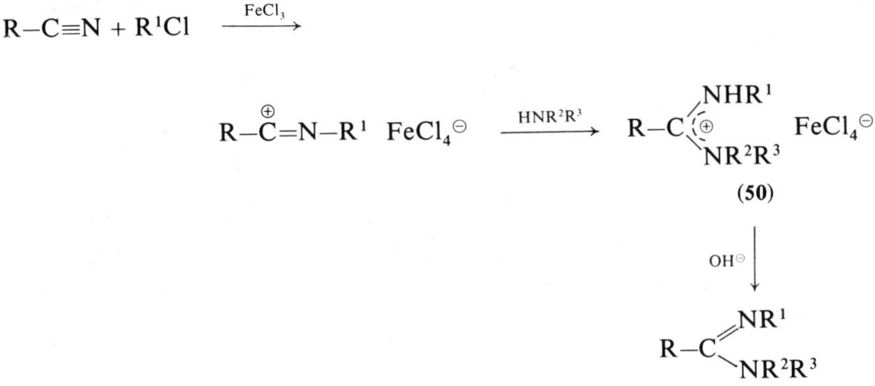

$$R-C\equiv N + R^1Cl \xrightarrow{FeCl_3}$$

$$R-\overset{\oplus}{C}=N-R^1 \ FeCl_4^{\ominus} \xrightarrow{HNR^2R^3} R-C\overset{NHR^1}{\underset{NR^2R^3}{\oplus}} FeCl_4^{\ominus}$$

(50)

$$\downarrow OH^{\ominus}$$

$$R-C\overset{NR^1}{\underset{NR^2R^3}{}}$$

(53)

an interesting observation was made (79): *N-isopropyl-N* phenyl-β-chloropropionic amidine (**54**) quickly rearranges to *N*-isopropyl-*N* phenylacrylamidinium chloride (**55**) on standing.

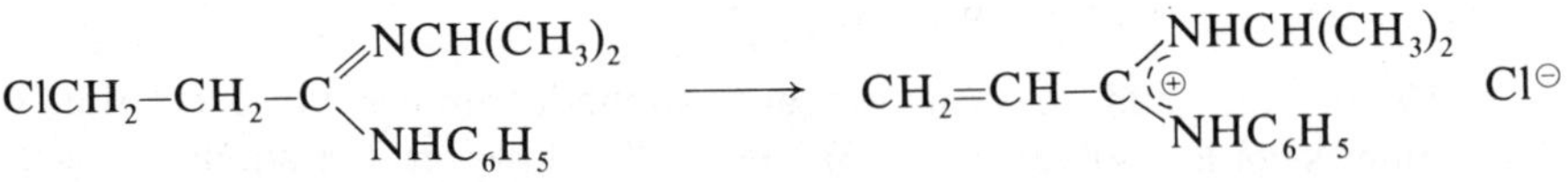

$$ClCH_2-CH_2-C\overset{NCH(CH_3)_2}{\underset{NHC_6H_5}{}} \longrightarrow CH_2=CH-C\overset{NHCH(CH_3)_2}{\underset{NHC_6H_5}{\oplus}} Cl^{\ominus}$$

(54)   (55)

I. SYNTHESIS

Nitrilium salts react with amino acid esters to yield heterocycles like **56–58** (78).

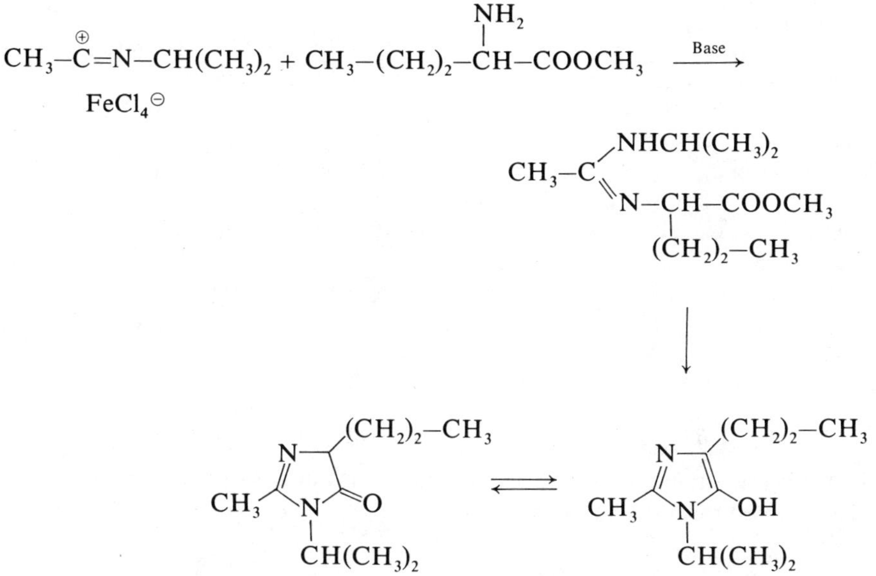

$$CH_3-\overset{\oplus}{C}=N-CH(CH_3)_2 \;\; FeCl_4^{\ominus} \;\; + \;\; CH_3-(CH_2)_2-\underset{NH_2}{\underset{|}{CH}}-COOCH_3 \xrightarrow{Base}$$

$$CH_3-C\begin{subarray}{l}\diagup NHCH(CH_3)_2 \\ \diagdown N-CH-COOCH_3 \\ \quad\quad\;\; | \\ \quad\quad\; (CH_2)_2-CH_3 \end{subarray}$$

(56) ⇌ tautomer (imidazolone / hydroxyimidazole with CH₃, (CH₂)₂–CH₃, N–CH(CH₃)₂ substituents)

$$CH_3-\overset{\oplus}{C}=N-C(CH_3)_3 \;\; FeCl_4^{\ominus} \;\; + \;\; \text{(2-aminobenzoate, COOCH}_3, NH_2) \xrightarrow{Base}$$

then Δ, o-dichlorobenzene → (**57**): 2-(*tert*-butylamino)-4(1H)-quinolinone (NHC(CH₃)₃)

or Δ, CH₃ONa / CH₃OH → (**58**): 3-*tert*-butyl-2-methyl-4(3H)-quinazolinone

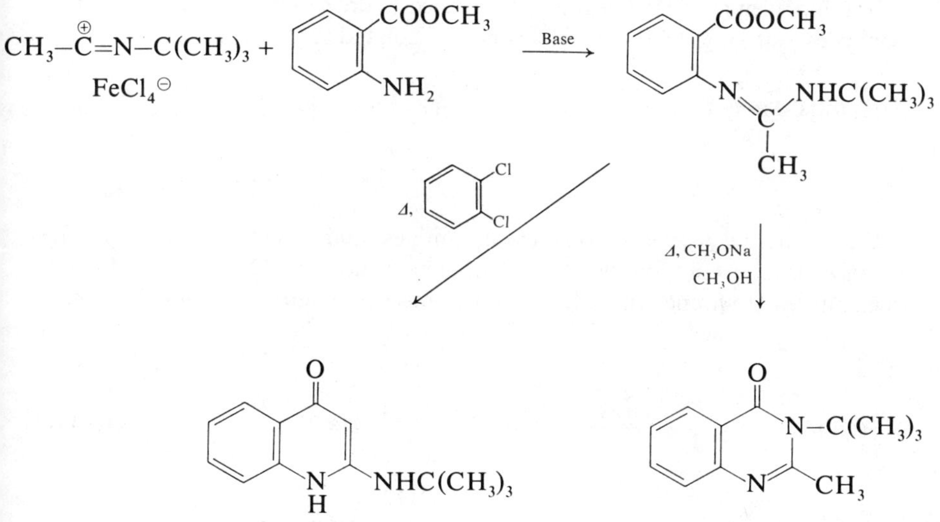

Propargylamine causes cyclization to the imidazole derivative **59** (78).

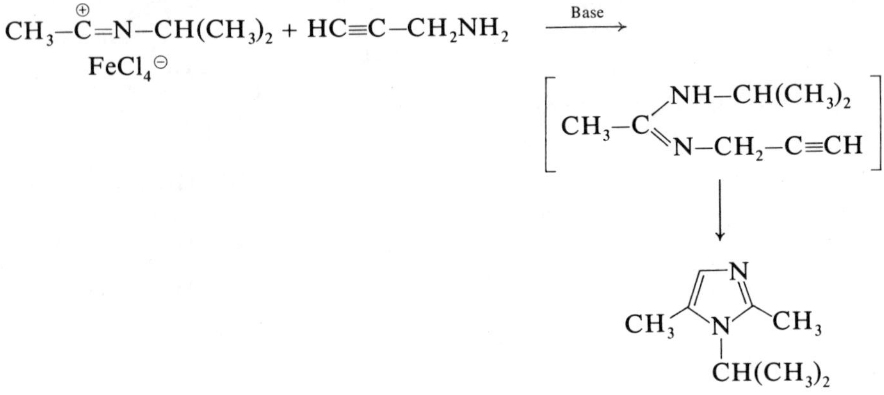

$$CH_3-\overset{\oplus}{C}=N-CH(CH_3)_2 + HC\equiv C-CH_2NH_2 \xrightarrow{\text{Base}}$$
$$FeCl_4^{\ominus}$$

$$\left[CH_3-C \underset{N-CH_2-C\equiv CH}{\overset{NH-CH(CH_3)_2}{\diagup}} \right]$$

↓

[imidazole ring with CH₃ at 4-position, CH₃ at 2-position, CH(CH₃)₂ on N-1]

(**59**)

H. FROM ISONITRILES AND AMINE HYDROCHLORIDES OR AMINES

Aromatic amine hydrochlorides add to isonitriles giving rise to N,N'-disubstituted formamidinium salts (**60**) (80).

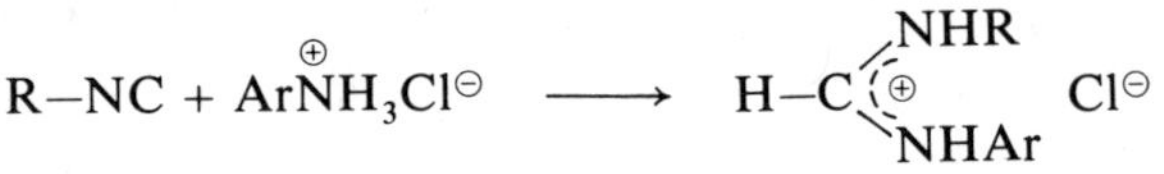

$$R-NC + Ar\overset{\oplus}{N}H_3Cl^{\ominus} \longrightarrow H-C\underset{NHAr}{\overset{NHR}{\diagup}}\ Cl^{\ominus}$$

(**60**)

In the presence of HCl N-monosubstituted hydroxylamines may also add to isonitriles, but only the free amidines **61** are isolated (81).

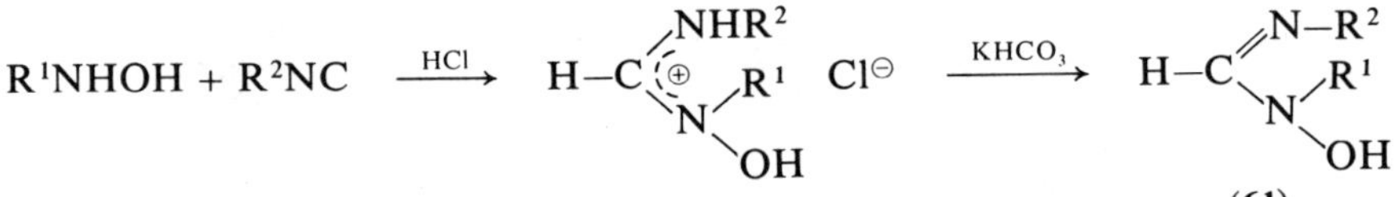

$$R^1NHOH + R^2NC \xrightarrow{HCl} H-C\underset{\underset{OH}{N}}{\overset{NHR^2}{\diagup}}R^1\ Cl^{\ominus} \xrightarrow{KHCO_3} H-C\underset{\underset{OH}{N}}{\overset{N-R^2}{\diagup}}R^1$$

(**61**)

When mercury acetate is present, amines and isonitriles react to give organomercuric compounds (**62**) of amidine structure (82). Similar additions to isonitriles are found with amines in the presence of auric compounds (83):

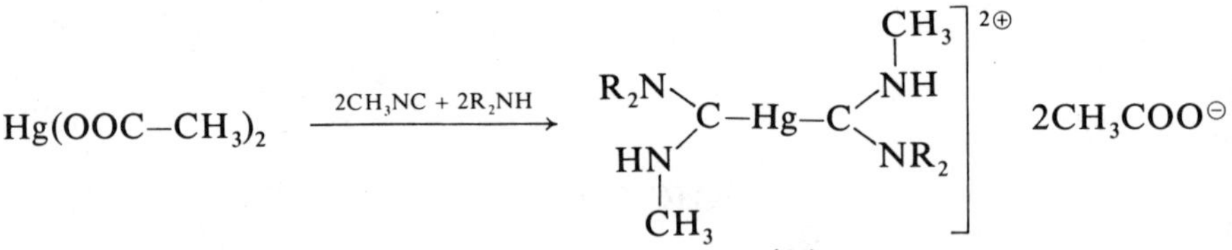

$$Hg(OOC-CH_3)_2 \xrightarrow{2CH_3NC\ +\ 2R_2NH} \left[\underset{HN}{\overset{R_2N}{\diagdown}}C-Hg-C\underset{NR_2}{\overset{NH}{\diagup}} \right]^{2\oplus} 2CH_3COO^{\ominus}$$
$$\overset{|}{CH_3}\qquad\qquad\overset{|}{CH_3}$$

(**62**)

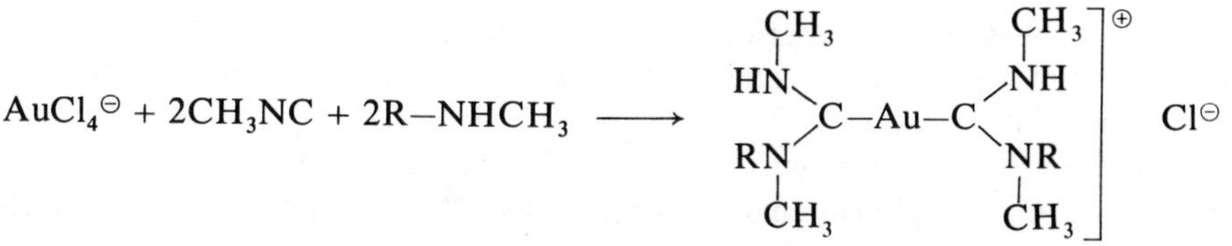

$$AuCl_4^{\ominus} + 2CH_3NC + 2R-NHCH_3 \longrightarrow \left[\begin{array}{c} \underset{RN}{\overset{HN}{\diagdown}}\underset{CH_3}{\overset{CH_3}{|}}C-Au-C\underset{\overset{|}{CH_3}}{\overset{\overset{|}{CH_3}}{\diagup NH}}\diagdown NR \end{array}\right]^{\oplus} Cl^{\ominus}$$

Isonitriles insert into the C–N bond of aminals, thus forming amidines (**63**) (84).

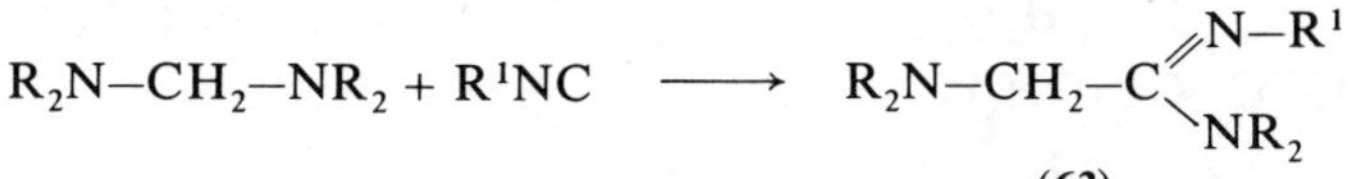

$$R_2N-CH_2-NR_2 + R^1NC \longrightarrow R_2N-CH_2-C\diagup\overset{N-R^1}{\diagdown NR_2}$$
(63)

I. FROM ORTHOESTERS AND ORTHOAMIDE DERIVATIVES

When dimethylammonium perchlorate, dimethylamine, and triethylorthoformic acid ester are heated, N,N,N',N'-tetramethylformamidinium perchlorate

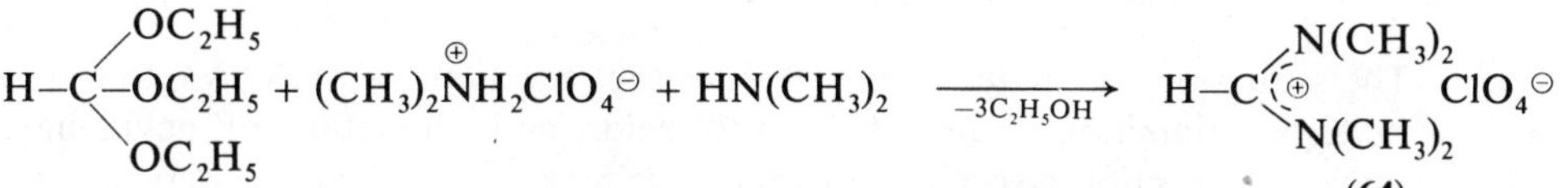

$$H-\underset{\underset{OC_2H_5}{|}}{\overset{\overset{OC_2H_5}{|}}{C}}-OC_2H_5 + (CH_3)_2\overset{\oplus}{N}H_2ClO_4^{\ominus} + HN(CH_3)_2 \xrightarrow{-3C_2H_5OH} H-C\overset{\diagup N(CH_3)_2}{\underset{\diagdown N(CH_3)_2}{\overset{\oplus}{}}} ClO_4^{\ominus}$$
(64)

(**64**) is obtained (85). N,N-Dimethyl-N'-t-butylformamidine (**65**) is synthesized the same way (86).

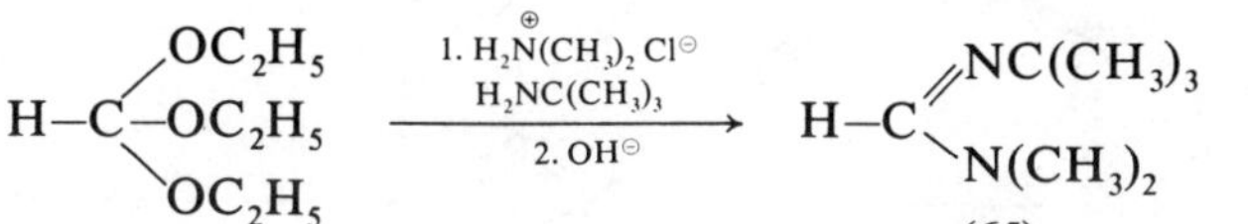

$$H-\underset{\underset{OC_2H_5}{|}}{\overset{\overset{OC_2H_5}{|}}{C}}-OC_2H_5 \xrightarrow[\text{2. OH}^{\ominus}]{\overset{1.\ H_2\overset{\oplus}{N}(CH_3)_2,\ Cl^{\ominus}}{H_2NC(CH_3)_3}} H-C\overset{\diagup NC(CH_3)_3}{\diagdown N(CH_3)_2}$$
(65)

Taylor and Ehrhart (87, 88) obtained from orthocarboxylic acid esters, glacial acetic acid, and ammonia or primary amines a series of N,N'-disubstituted formamidinium or acetamidinium acetates (**66**) in good yields. The free amidines obtained from those were partly converted into picrates or hydrochlorides. This synthesis of amidinium salts is very versatile, since aliphatic as well as aromatic amines may be reacted.

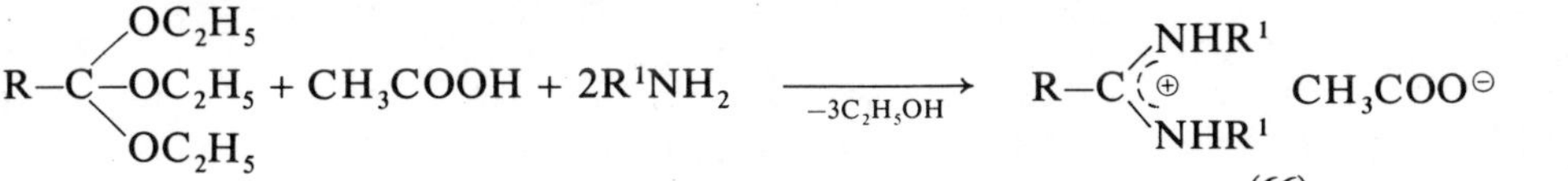

$$R-\underset{\underset{OC_2H_5}{|}}{\overset{\overset{OC_2H_5}{|}}{C}}-OC_2H_5 + CH_3COOH + 2R^1NH_2 \xrightarrow{-3C_2H_5OH} R-C\overset{\diagup NHR^1}{\underset{\diagdown NHR^1}{\overset{\oplus}{}}} CH_3COO^{\ominus}$$
(66)

Later studies mostly describe the isolation of the free amidines rather than the amidinium salts. These reactions have been reviewed recently by Sandler and Karo (7) as well as by De Wolfe (89). The reaction of aminal esters (**67**) (90) and tris(dialkylamino)methanes (**68**) (91) with dry hydrogen halides yields N,N,N',N'-tetraalkylformamidinium salts (**69, 70**); synthetically it is of little interest, since all reactions possible with tetraalkylformamidinium chlorides are also successful with tetraalkylformamidinium methylsulfates, the precursors of the aminal esters.

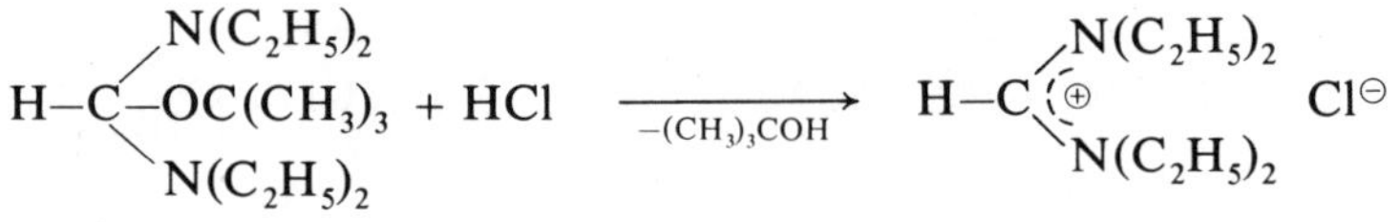

$$H-C \begin{matrix} N(C_2H_5)_2 \\ -OC(CH_3)_3 \\ N(C_2H_5)_2 \end{matrix} + HCl \xrightarrow{-(CH_3)_3COH} H-C^{\oplus} \begin{matrix} N(C_2H_5)_2 \\ \\ N(C_2H_5)_2 \end{matrix} Cl^{\ominus}$$

(**67**)   (**69**)

$$C_6H_5-C \begin{matrix} N(CH_3)_2 \\ -N(CH_3)_2 \\ N(CH_3)_2 \end{matrix} \xrightarrow{HCl} C_6H_5-C^{\oplus} \begin{matrix} N(CH_3)_2 \\ \\ N(CH_3)_2 \end{matrix} Cl^{\ominus}$$

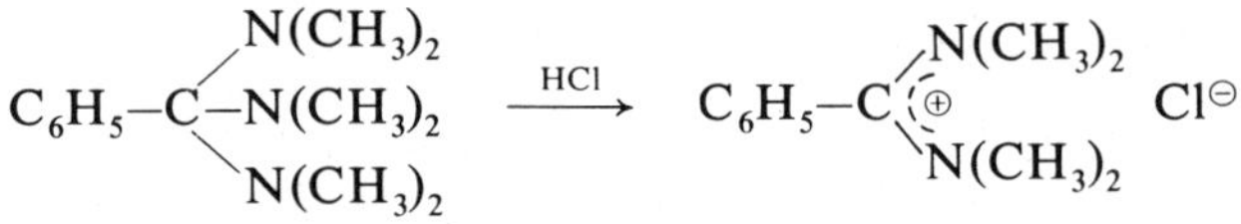

(**68**)   (**70**)

The situation is quite different with the reaction of N,N',N''-triaryl-substituted triaminomethanes (**71**) with acids and alkylating or acylating agents. This reaction also gives amidinium salts, but the starting materials are easily accessible (92–94) (e.g., from chlorodifluoromethane and sodium alkylarylamides, from orthoesters and N-alkyl-N-arylamines, and from sodium trichloroacetate and N-alkylanilides).

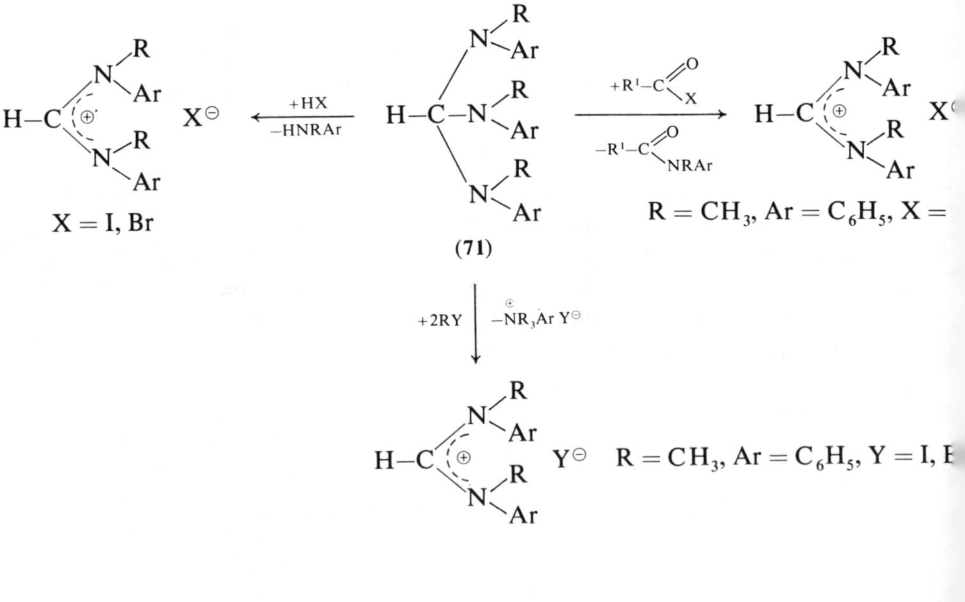

$X = I, Br$

(**71**)

$R = CH_3, Ar = C_6H_5, X =$

$+2RY \downarrow -\overset{\oplus}{N}R_3Ar\ Y^{\ominus}$

$Y^{\ominus} \quad R = CH_3, Ar = C_6H_5, Y = I, E$

N,N'-Diacyl-substituted formamidinium salts (**73**) are obtained from acids and tris(acylamino)methanes (**72**) (95).

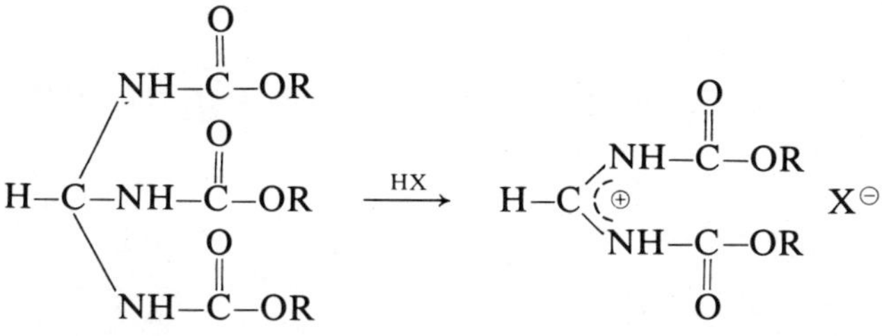

$$\underset{(72)}{H-C\begin{pmatrix}NH-C(=O)-OR\\NH-C(=O)-OR\\NH-C(=O)-OR\end{pmatrix}} \xrightarrow{HX} \underset{(73)}{H-C\begin{pmatrix}NH-C(=O)-OR\\NH-C(=O)-OR\end{pmatrix}^{\oplus} X^{\ominus}}$$

R = CH$_3$, C$_2$H$_5$
X = Cl, HSO$_4$

The synthesis of 1,1-bis(dimethylamino)acetonitrile (**75**) starts from aminal esters (**74**) and hydrogen cyanide or acyl cyanides (96–98). Also the triaminomethane **76** is converted into the nitrile **77** by HCN (99).

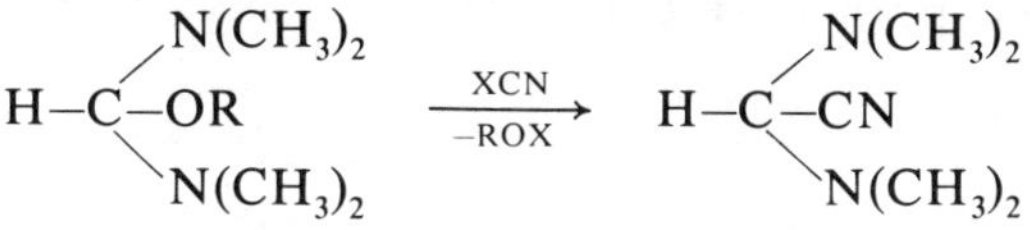

$$\underset{(74)}{H-C\begin{pmatrix}N(CH_3)_2\\OR\\N(CH_3)_2\end{pmatrix}} \xrightarrow[-ROX]{XCN} \underset{(75)}{H-C\begin{pmatrix}N(CH_3)_2\\CN\\N(CH_3)_2\end{pmatrix}}$$

$X = H, R^1-C{\Large\diagdown}\!\!\!{}^{O}$

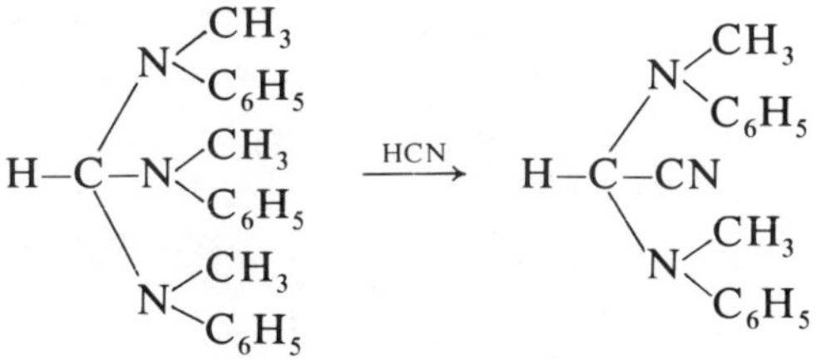

$$\underset{(76)}{H-C\begin{pmatrix}N(CH_3)(C_6H_5)\\N(CH_3)(C_6H_5)\\N(CH_3)(C_6H_5)\end{pmatrix}} \xrightarrow{HCN} \underset{(77)}{H-C\begin{pmatrix}N(CH_3)(C_6H_5)\\CN\\N(CH_3)(C_6H_5)\end{pmatrix}}$$

To obtain N,N^1-bis(dialkylamino)acetonitriles (**78**), one starts with the action of pyrrolidine, piperidine, and morpholine on hydrogen cyanide (99, 100), as well as of dibromoacetonitrile (**79**) on morpholine (101), and of (dialkylamino)-alkoxyacetonitriles (**80**), on secondary amines (98, 102).

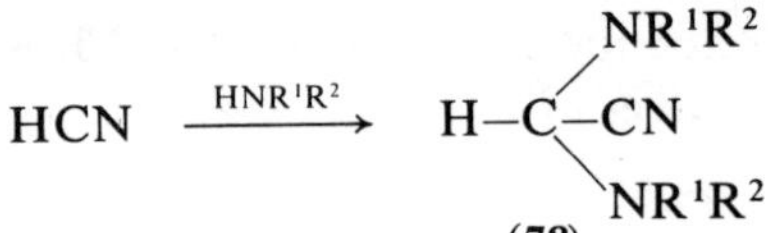

$$\text{HCN} \xrightarrow{\text{HNR}^1\text{R}^2} \underset{(78)}{\text{H}-\overset{\overset{\displaystyle \text{NR}^1\text{R}^2}{|}}{\underset{\underset{\displaystyle \text{NR}^1\text{R}^2}{|}}{\text{C}}}-\text{CN}}$$

$$\text{R}^1-\text{R}^2 = -(\text{CH}_2)_4-, -(\text{CH}_2)_5-, -(\text{CH}_2)_2-\text{O}-(\text{CH}_2)_2-$$

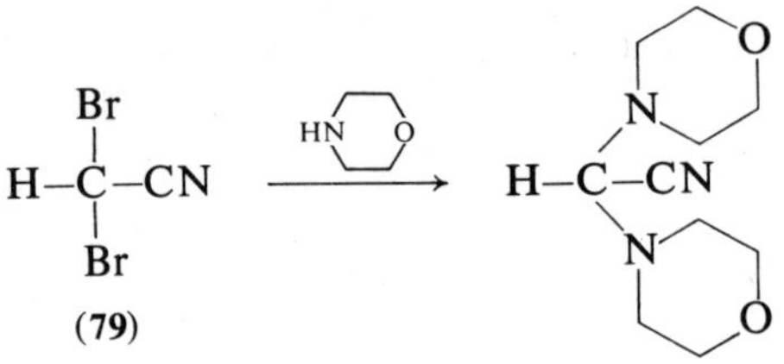

$$\underset{(79)}{\text{H}-\overset{\overset{\displaystyle \text{Br}}{|}}{\underset{\underset{\displaystyle \text{Br}}{|}}{\text{C}}}-\text{CN}} \xrightarrow{\text{HN}\diagdown\!\!\text{O}\diagup} \text{H}-\overset{\overset{\displaystyle \text{N}\diagup\diagdown\text{O}}{|}}{\underset{\underset{\displaystyle \text{N}\diagdown\diagup\text{O}}{|}}{\text{C}}}-\text{CN}$$

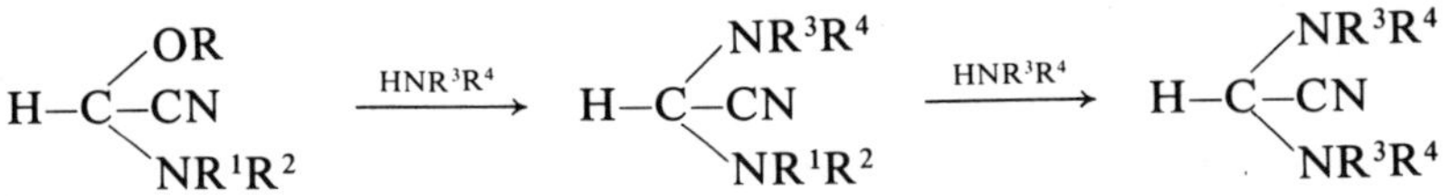

$$\underset{(80)}{\text{H}-\overset{\diagup\text{OR}}{\underset{\diagdown\text{NR}^1\text{R}^2}{\text{C}}}-\text{CN}} \xrightarrow{\text{HNR}^3\text{R}^4} \text{H}-\overset{\diagup\text{NR}^3\text{R}^4}{\underset{\diagdown\text{NR}^1\text{R}^2}{\text{C}}}-\text{CN} \xrightarrow{\text{HNR}^3\text{R}^4} \text{H}-\overset{\diagup\text{NR}^3\text{R}^4}{\underset{\diagdown\text{NR}^3\text{R}^4}{\text{C}}}-\text{CN}$$

Hydrogen cyanide reacts with a mixture of primary and secondary amines to yield the formamidines **81** (99). Primary amines give a reaction with

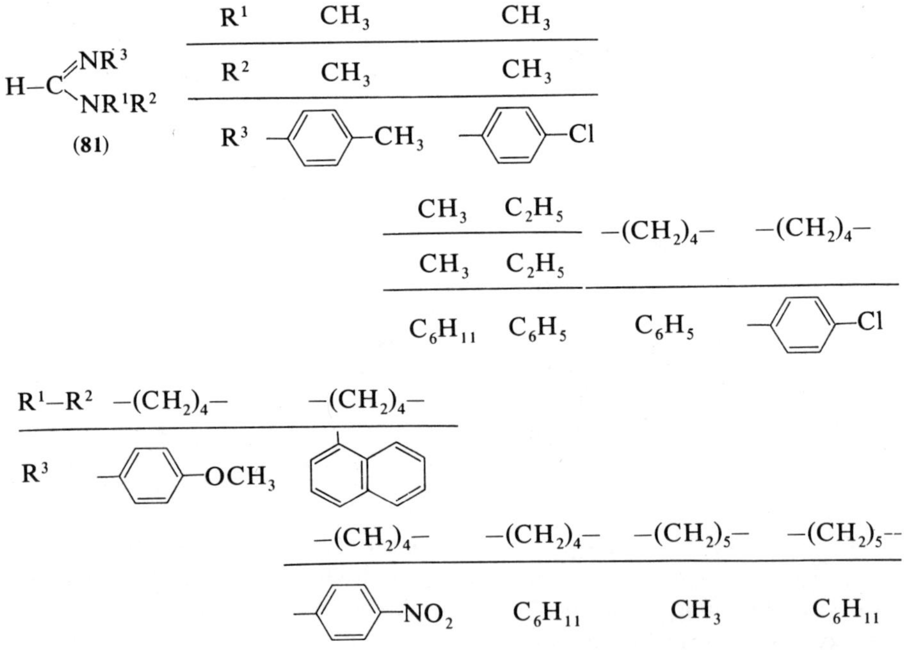

$$\underset{(81)}{\text{H}-\text{C}\overset{\diagup\text{NR}^3}{\diagdown\text{NR}^1\text{R}^2}}$$

R^1	CH_3	CH_3	
R^2	CH_3	CH_3	
R^3	–C$_6$H$_4$–CH$_3$	–C$_6$H$_4$–Cl	

CH_3	C_2H_5	$-(CH_2)_4-$	$-(CH_2)_4-$
CH_3	C_2H_5		
C_6H_{11}	C_6H_5	C_6H_5	–C$_6$H$_4$–Cl

R^1–R^2	$-(CH_2)_4-$	$-(CH_2)_4-$
R^3	–C$_6$H$_4$–OCH$_3$	naphthyl

$-(CH_2)_4-$	$-(CH_2)_4-$	$-(CH_2)_5-$	$-(CH_2)_5-$
–C$_6$H$_4$–NO$_2$	C_6H_{11}	CH_3	C_6H_{11}

hydrogen cyanide yielding N,N'-disubstituted formamidines (**82**), whereas with 1,2- or 1,3-diaminoalkanes Δ^2-imidazolines like **83** or 1,4,5,6-tetrahydropyrimidines like **84** are the products (103).

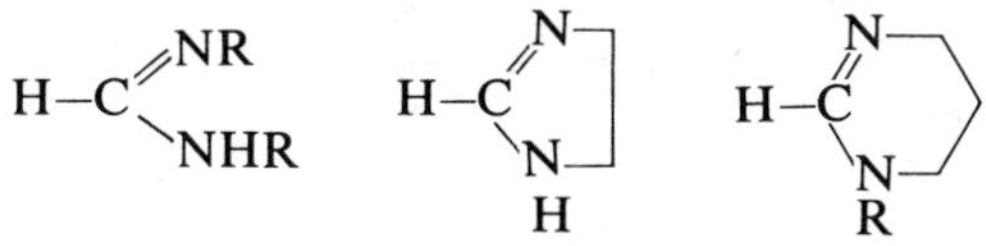

"Cyclic aminal esters" like **85** and **86** react with boron trifluoride etherate to give amidinium salts **87** or **88** (104).

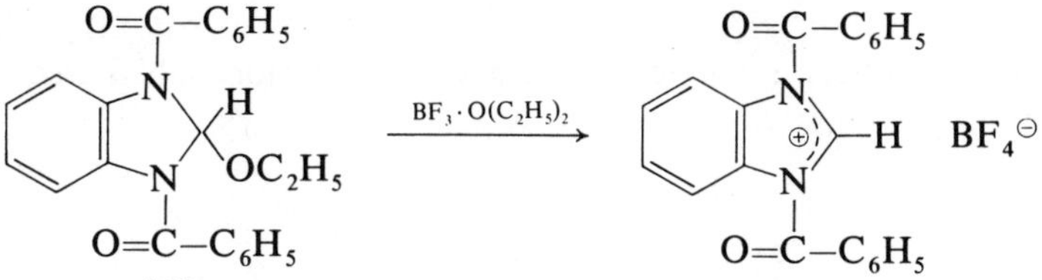

J. VIA ALKYLATION AND ACYLATION OF AMIDINES

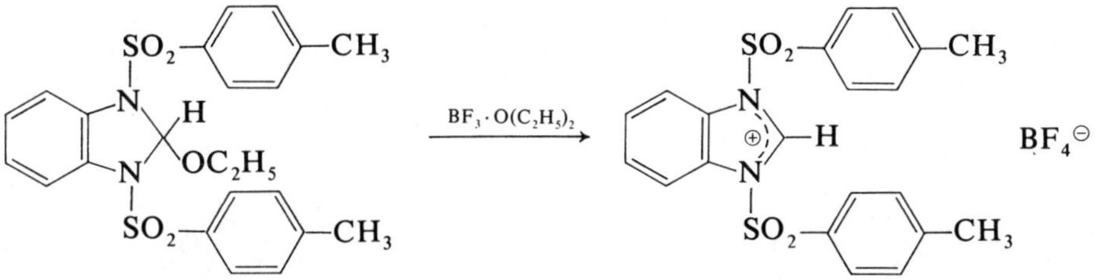

Unsubstituted amidines are transformed into N-monoalkylated amidinium salts, for example, **89**, by alkyl halides (105). Sometimes multiple alkylation occurs as well (106, 107). With alkylating agents, N-monosubstituted amidines (107) and N,N'-disubstituted amidines (106–108) form a mixture of products.

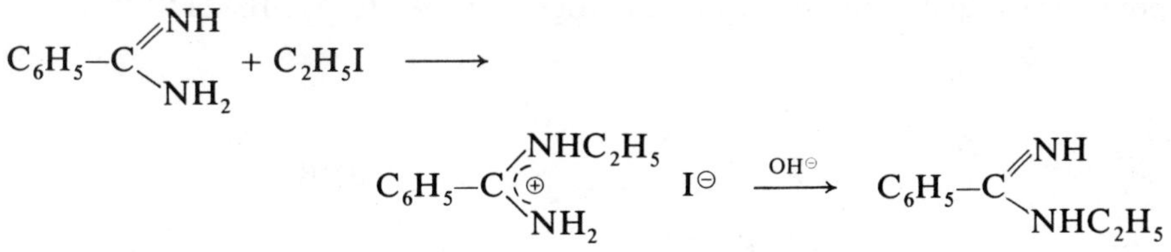

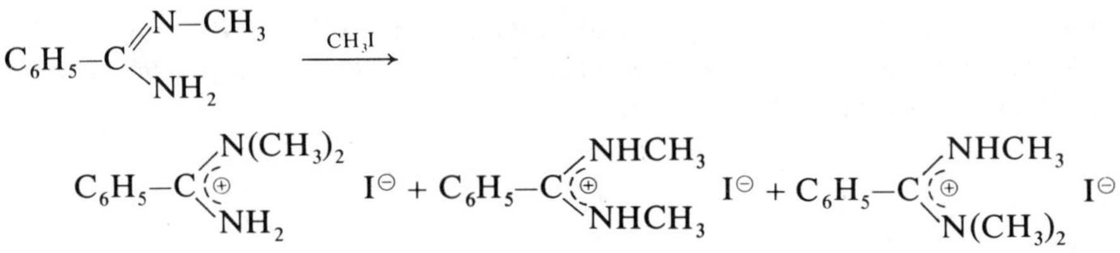

$$C_6H_5-C\underset{NH_2}{\overset{N-CH_3}{\diagup}} \xrightarrow{CH_3I}$$

$$C_6H_5-\overset{+}{C}\underset{NH_2}{\overset{N(CH_3)_2}{\diagup}} I^{\ominus} + C_6H_5-\overset{+}{C}\underset{NHCH_3}{\overset{NHCH_3}{\diagup}} I^{\ominus} + C_6H_5-\overset{+}{C}\underset{N(CH_3)_2}{\overset{NHCH_3}{\diagup}} I^{\ominus}$$

$$R-C\underset{NHR^2}{\overset{NR^1}{\diagup}} \xrightarrow{R^3X} R-\overset{+}{C}\underset{NHR^2}{\overset{NR^1R^3}{\diagup}} X^{\ominus} + R-\overset{+}{C}\underset{NR^2R^3}{\overset{NHR^1}{\diagup}} X^{\ominus}$$

A uniform reaction has been reported for the arylation of monosubstituted amidines with picryl chloride (109):

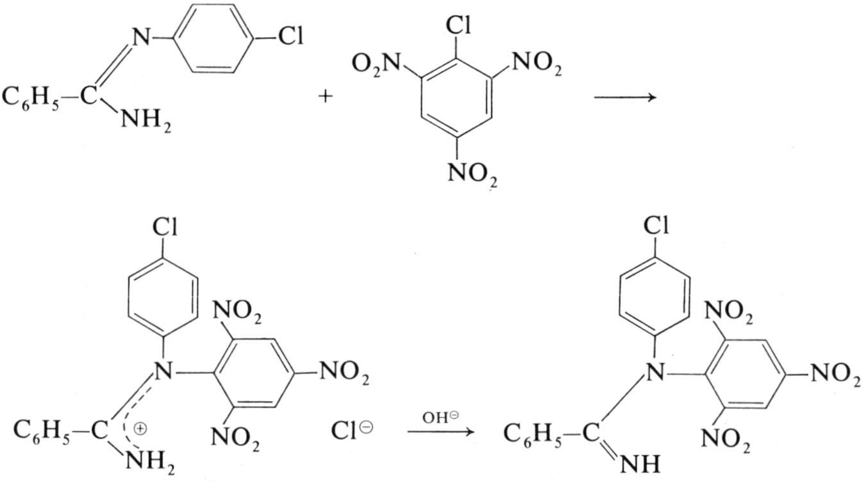

$$C_6H_5-C\underset{NH_2}{\overset{N-\text{Ar}-Cl}{\diagup}} + \text{picryl-Cl} \longrightarrow$$

$$C_6H_5-\overset{+}{C}\underset{NH_2}{\overset{N(\text{Ar})(\text{picryl})}{\diagup}} Cl^{\ominus} \xrightarrow{OH^{\ominus}} C_6H_5-C\underset{NH}{\overset{N(\text{Ar})(\text{picryl})}{\diagup}}$$

The alkylation of *N,N*-disubstituted and *N,N,N'*-trisubstituted amidines also proceeds unequivocally at the imide nitrogen (19, 20, 36, 92, 107, 110–112):

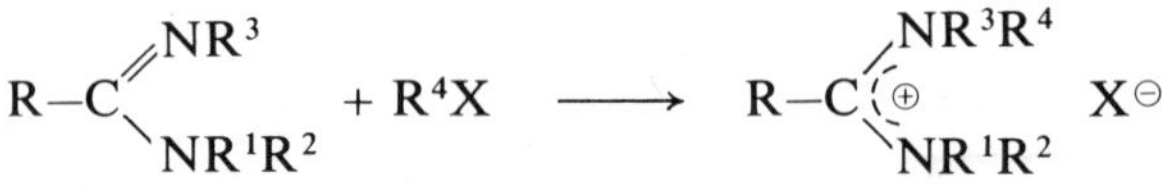

$$R-C\underset{NR^1R^2}{\overset{NR^3}{\diagup}} + R^4X \longrightarrow R-\overset{+}{C}\underset{NR^1R^2}{\overset{NR^3R^4}{\diagup}} X^{\ominus}$$

The properties of some of the amidinium salts obtained by this method are summarized in Table II.

Amidrazones and hydrazidimides may also be converted into amidinium salts (**90**) by alkylation (112, 113).

Amidinium Salts via Alkylation of Amidines

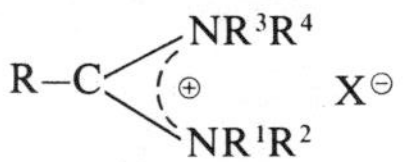

$$R-C{\overset{NR^3R^4}{\underset{NR^1R^2}{\oplus}}}\ X^{\ominus}$$

R	R^1	R^2	R^3	R^4	X$^\ominus$	m.p., °C	Ref.
H	CH$_3$	C$_6$H$_5$	CH$_3$	C$_6$H$_5$	I	162–164	92
C$_6$H$_5$	CH$_3$	CH$_3$	CH$_3$	p-NH$_2$–C$_6$H$_4$	I	261–262	36
C$_6$H$_5$	CH$_3$	CH$_3$	CH$_3$	m-NH$_2$–C$_6$H$_4$	I	—	36
m-NH$_2$–C$_6$H$_4$	CH$_3$	CH$_3$	CH$_3$	C$_6$H$_5$	I	164–165	36
C$_6$H$_5$	CH$_3$	CH$_3$	CH$_3$	pCH$_3$–C(=O)–NH–C$_6$H$_4$	I	247–248	36
p-CH$_3$–C(=O)–NH–C$_6$H$_4$	CH$_3$	CH$_3$	CH$_3$	pCH$_3$–C(=O)–NH–C$_6$H$_4$	I	256–257	36
C$_6$H$_5$	CH$_3$	CH$_3$	CH$_3$	m-NO$_2$–C$_6$H$_4$	I	184.5–185.5	36
m-NO$_2$–C$_6$H$_4$	CH$_3$	CH$_3$	CH$_3$	m-NO$_2$–C$_6$H$_4$	I	216	36
pC$_2$H$_5$–O–C(=O)–NH–C$_6$H$_4$	CH$_3$	CH$_3$	CH$_3$	p-C$_2$H$_5$–O–C(=O)–NH–C$_6$H$_4$	I	—	36
p-C$_2$H$_5$–O–C(=O)–NH–C$_6$H$_4$	CH$_3$	CH$_3$	CH$_3$	p-NH$_2$–C$_6$H$_4$	Br	275	36
p-NH$_2$–C$_6$H$_4$	CH$_3$	CH$_3$	CH$_3$	pC$_2$H$_5$–O–CO–NH–C$_6$H$_4$	I	145–150	36
p-CH$_3$O–C$_6$H$_4$	CH$_3$	CH$_3$	CH$_3$	p-NH$_2$–C$_6$H$_4$	I	222	36
p-CH$_3$O–C$_6$H$_4$	CH$_3$	CH$_3$	CH$_3$	m-NH$_2$–C$_6$H$_4$	I	183–184	36
p-NH$_2$–C$_6$H$_4$	CH$_3$	CH$_3$	CH$_3$	p-NH$_2$–C$_6$H$_4$	I	275–276	36
p-NH$_2$–C$_6$H$_4$	CH$_3$	CH$_3$	CH$_3$	m-NH$_2$–C$_6$H$_4$	I·H$_2$O	229	36
m-NH$_2$–C$_6$H$_4$	CH$_3$	CH$_3$	CH$_3$	p-NH$_2$–C$_6$H$_4$	I·H$_2$O	262	36
m-NH$_2$–C$_6$H$_4$	CH$_3$	CH$_3$	CH$_3$	m-NH$_2$–C$_6$H$_4$	I	195	36
p-NH$_2$–C$_6$H$_4$	–(CH$_2$)$_5$–		CH$_3$	p-NH$_2$–C$_6$H$_4$	I	243	36
p-NH$_2$–C$_6$H$_4$	C$_2$H$_5$	C$_2$H$_5$	CH$_3$	p-NH$_2$–C$_6$H$_4$	I	248	36

$$\text{C}_6\text{H}_5-\text{C}\begin{array}{c}\nearrow\text{N}-\text{C}_6\text{H}_5\\ \searrow\text{N}-\text{N}(\text{CH}_3)_2\\ |\\ \text{CH}_3\end{array} \xrightarrow{\text{CH}_3\text{I}} \text{C}_6\text{H}_5-\overset{\overset{\text{CH}_3}{|}}{\text{C}}\begin{array}{c}\nearrow\text{N}-\text{C}_6\text{H}_5\\ \oplus\\ \searrow\text{N}-\text{N}(\text{CH}_3)_2\\ |\\ \text{CH}_3\end{array}\ \text{I}^\ominus$$

(90)

When substituted amidines are acylated, the homogeneity of the reaction products is subject to the same limitations as in alkylation. A series of amidines was acylated with carboxylic acid chlorides and sulfonyl chlorides; the amidinium salts were not isolated as such, but mostly as the free amidines. This type of reaction has been reviewed by Shriner and Neumann (5).

$$\text{R}-\text{C}\begin{array}{c}\nearrow\text{NR}^2\\ \searrow\text{NHR}^1\end{array} + \text{R}^3-\text{C}\begin{array}{c}\nearrow\text{O}\\ \searrow\text{Cl}\end{array} \longrightarrow \text{R}-\text{C}\begin{array}{c}\nearrow\text{N}-\text{R}^2\\ \oplus\\ \searrow\text{NHR}^1\end{array}\ \text{Cl}^\ominus \xrightarrow{\text{OH}^\ominus} \text{R}-\text{C}\begin{array}{c}\nearrow\text{N}-\text{R}^2\\ \searrow\text{N}-\text{R}^1\end{array}$$

with $\text{O}=\text{C}\diagup\text{R}^3$ substituent on N.

K. VIA TRANSAMINATION OF AMIDINIUM SALTS

Amidinium salts may be transaminated by amines. These reactions are dealt with in Section III-B-2 of this chapter.

L. FROM KETENE AMINALS AND YNAMINES

1. Alkylation and Acylation of Ketene Aminals and Ynamines

Ketene aminals like **91–93** add acids HX (like HClO_4, HCl), forming amidinium salts **94–96** (114–116).

$$\text{R}-\text{NH}-\overset{\overset{\text{X}}{\|}}{\text{C}}-\text{CH}=\text{C}\begin{array}{c}\nearrow\text{N}\diagup\\ \searrow\text{N}\diagdown\end{array} + \text{HY} \longrightarrow \text{R}-\text{NH}-\overset{\overset{\text{X}}{\|}}{\text{C}}-\text{CH}_2-\text{C}\begin{array}{c}\nearrow\text{N}\diagup\\ \oplus\\ \searrow\text{N}\diagdown\end{array}\ \text{Y}^\ominus$$

(91) **(94)**

X = O, S
Y = Cl, ClO_4, picrate

$$\begin{array}{c}\text{R}_2\text{N}\\ \text{R}_2\text{N}\end{array}\text{C}=\text{C}\begin{array}{c}\text{NR}_2\\ \text{H}\end{array} + \text{HClO}_4 \longrightarrow \text{R}_2\text{N}-\text{CH}_2-\text{C}\begin{array}{c}\nearrow\text{NR}_2\\ \oplus\\ \searrow\text{NR}_2\end{array}\ \text{ClO}_4^\ominus$$

(92) **(95)**

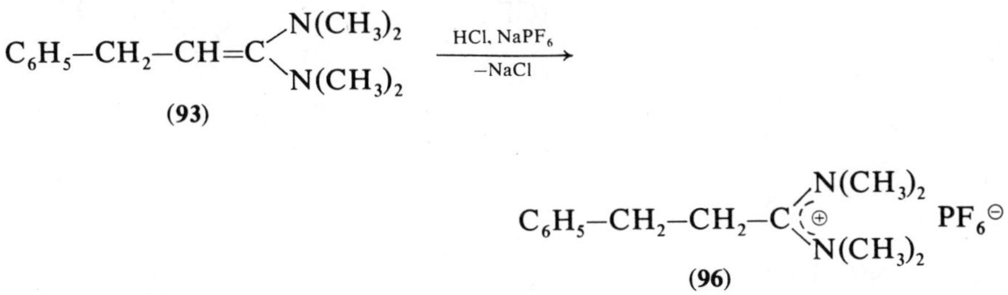

I. SYNTHESIS

$$C_6H_5-CH_2-CH=C{\overset{N(CH_3)_2}{\underset{N(CH_3)_2}{\diagup}}} \quad \xrightarrow[-NaCl]{HCl,\ NaPF_6}$$

(93)

$$C_6H_5-CH_2-CH_2-\overset{\oplus}{C}{\overset{N(CH_3)_2}{\underset{N(CH_3)_2}{\diagup}}} \quad PF_6^{\ominus}$$

(96)

With the aid of alkyl halides like benzyl bromide (114–116), allyl bromide (116), methyl iodide (114, 116), ethyl iodide (114), and butyl bromide (116) the ketene aminals **97, 92,** and **98** are converted into the amidinium salts **99–101**. The influence of α substituents on the course of the alkylation has been studied (114, 115).

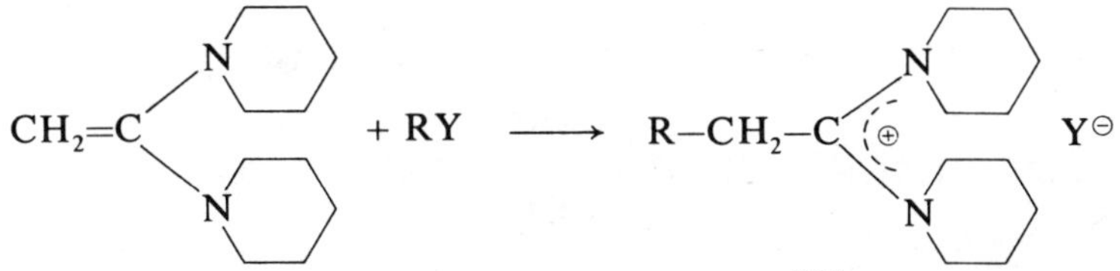

$$CH_2=C\begin{pmatrix}N\text{-piperidyl}\\N\text{-piperidyl}\end{pmatrix} + RY \longrightarrow R-CH_2-\overset{\oplus}{C}\begin{pmatrix}N\text{-piperidyl}\\N\text{-piperidyl}\end{pmatrix} Y^{\ominus}$$

(97)   (99)

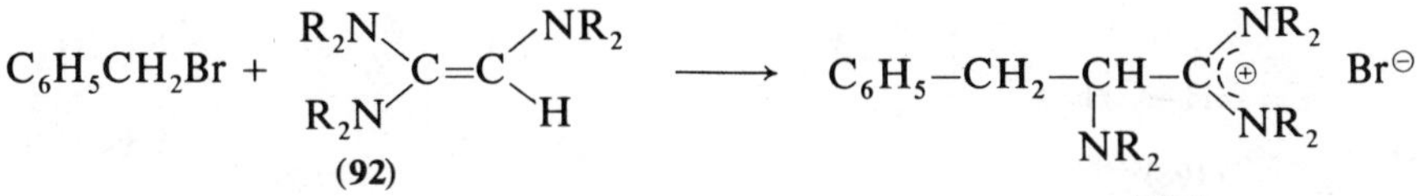

$$C_6H_5CH_2Br + {\overset{R_2N}{\underset{R_2N}{\diagup}}}C=C{\overset{NR_2}{\underset{H}{\diagup}}} \longrightarrow C_6H_5-CH_2-\underset{NR_2}{\overset{|}{CH}}-\overset{\oplus}{C}{\overset{NR_2}{\underset{NR_2}{\diagup}}} \quad Br^{\ominus}$$

(92)   (100)

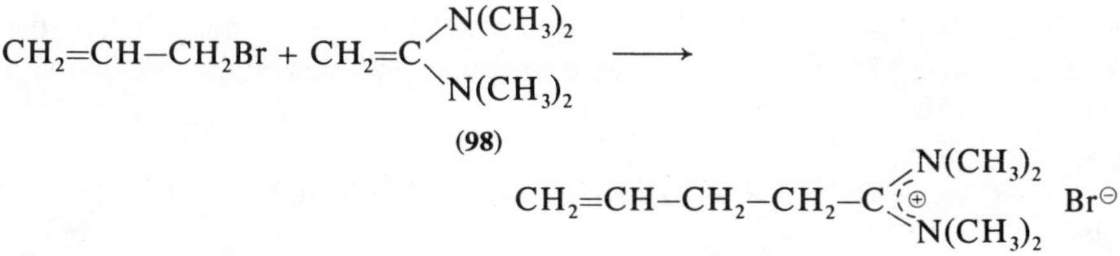

$$CH_2=CH-CH_2Br + CH_2=C{\overset{N(CH_3)_2}{\underset{N(CH_3)_2}{\diagup}}} \longrightarrow$$

(98)

$$CH_2=CH-CH_2-CH_2-\overset{\oplus}{C}{\overset{N(CH_3)_2}{\underset{N(CH_3)_2}{\diagup}}} \quad Br^{\ominus}$$

(101)

Ketene aminals like **102** react with α,ω-dihaloalkanes $X-(CH_2)_n-X$ (**103**) to yield linear diamidinium salts (**104, 105**) and cycloalkylamidinium salts (**106**). The cyclic components **106** dominate, if n equals 2, 4, or 5 in the dihaloalkanes **103**. 1,3-Diiodopropane reacts with **102** to form the tetrahydropyrimidinium salt **107** (117).

(107) (104) (102) (106) (105)

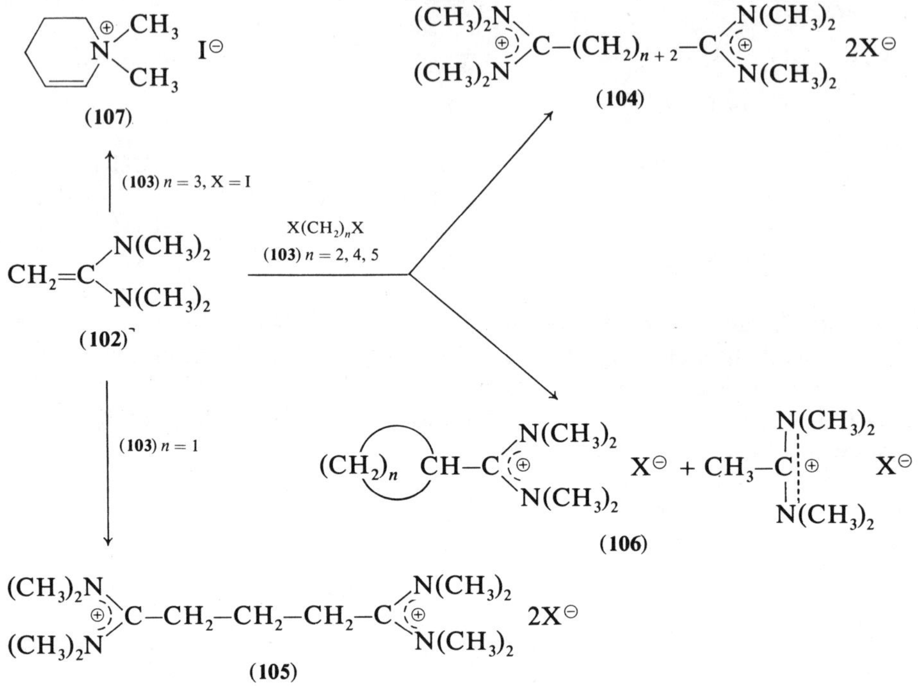

With methylen iodide 1,1,4,4-tetrakis(dimethylamino)-1,3-butadiene (**108**) reacts smoothly to give the cyclopropane derivative **109** (117, 118).

(108) (109)

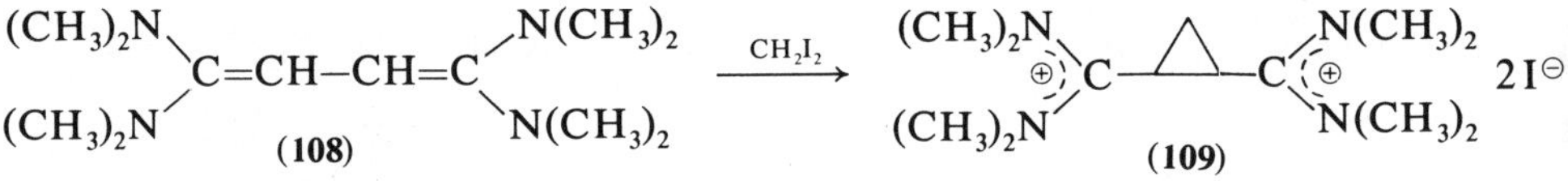

Other ketene aminals give very complex reaction mixtures with methylen iodide (117). *N,N*-Dialkylmethyleniminium salts (**110**) aminomethylate the ketene aminals **111** (115), which may also couple with diazonium salts in the *α* position (115).

(111) (110)

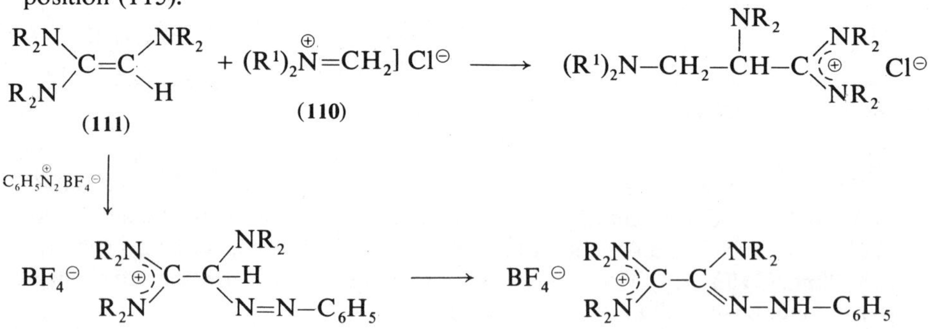

I. SYNTHESIS

The properties of these salts are summarized in Table III.

TABLE III

Amidinium Salts via HX Addition, Alkylation, or Acylation of Ketene Aminals

Amidinium salt

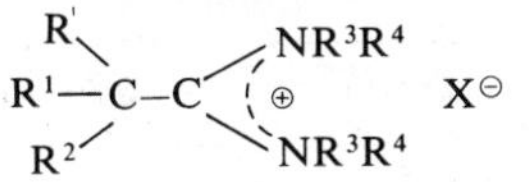

$$\begin{array}{c} R^1 \\ R^1-C-C \\ R^2 \end{array} \begin{array}{c} NR^3R^4 \\ \oplus \\ NR^3R^4 \end{array} \quad X^\ominus$$

R	R^1	R^2	R^3	R^4	X	Yield, %	m.p., °C	Ref.
H	H	H	$-(CH_2)_5-$		picrate	80	86–88	114
CH_3	H	H	$-(CH_2)_5-$		I	95	158–159	114
C_2H_5	H	H	CH_3	CH_3	I	84	225	116
C_2H_5	H	H	$-(CH_2)_5-$		I	92	212–212.5	114
CH_3	CH_3	CH_3	CH_3	CH_3	I	62	390 (subl)	116
$CH_2=CH-CH_2$	H	H	CH_3	CH_3	Br	77	153–154	116
$CH_2=CH-CH_2$	CH_3	CH_3	CH_3	CH_3	Br	92	212–213	116
n-C_4H_9	CH_3	CH_3	CH_3	CH_3	Br	69	263 (dec.)	116
$C_6H_5CH_2$	H	H	CH_3	CH_3	Cl	86	105–107	116
$C_6H_5CH_2$	H	H	CH_3	CH_3	PF_6	—	146–147.5	116
$C_6H_5CH_2$	H	H	$-(CH_2)_5-$		Br	75	185–186	114
$C_6H_5CH_2$	CH_3	CH_3	CH_3	CH_3	Br	63	181–183	116
$C_6H_5-NH-\overset{O}{\underset{\parallel}{C}}-$	H	CH_3	$-(CH_2)_5-$		I	89	197–199	114

Amidinium salt

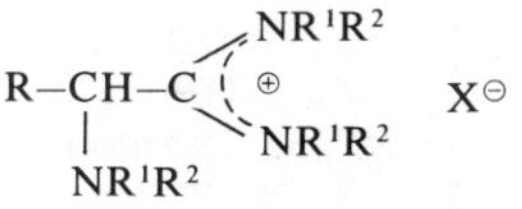

$$\begin{array}{c} R-CH-C \\ | \\ NR^1R^2 \end{array} \begin{array}{c} NR^1R^2 \\ \oplus \\ NR^1R^2 \end{array} \quad X^\ominus$$

R	R^1	R^2	X	Yield, %	m.p., °C	Ref.
H	$-(CH_2)_2-O-(CH_2)_2-$		ClO_4	93	249–249 (dec)	115
$C_6H_5CH_2$	$-(CH_2)_2-O-(CH_2)_2-$		Br	71	—	115
$-CH_2-N(CH_3)_2$	$-(CH_2)_5-$		Cl	71	—	115
$-CH_2-N\overset{\frown}{\underset{\smile}{}}O$	$-(CH_2)_5-$		Cl	90	—	115
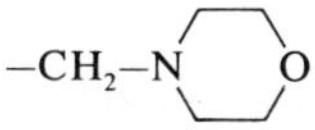 $-CH_2-N\overset{\frown}{\underset{\smile}{}}O$	$-(CH_2)_2-O-(CH_2)_2-$		Cl	88	—	115

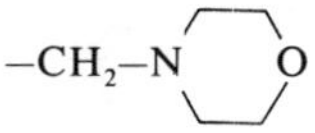

TABLE III (cont.)

Amidinium salt

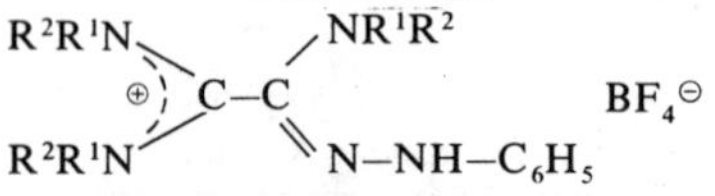

$$\begin{matrix} R^2R^1N & & NR^1R^2 \\ & \overset{\oplus}{C}-C & \\ R^2R^1N & & N-NH-C_6H_5 \end{matrix} \quad BF_4^{\ominus}$$

R¹	R²	Yield, %	m.p., °C	Ref.
$-(CH_2)_2-O-(CH_2)_2-$		87	—	115
$-(CH_2)_5-$		90	—	115

Amidinium salt

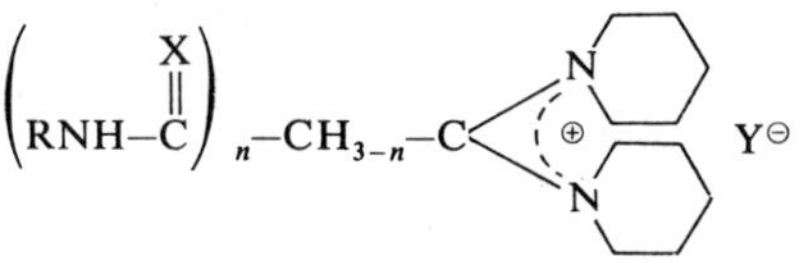

$$\left(RNH-\overset{X}{\underset{\|}{C}} \right)_n -CH_{3-n}-C\overset{N}{\underset{N}{\underset{\oplus}{\diagup}}} \quad Y^{\ominus}$$

R	X	Y	n	m.p., °C	Ref.
C_6H_5	S	ClO_4	1	209–211	114
C_2H_5	S	ClO_4	1	151–154	114
C_6H_5	O	ClO_4	1	223–226	114
$3,4-Cl_2-C_6H_3$	O	ClO_4	1	196–197	114
$2,5-Cl_2-C_6H_3$	O	ClO_4	1	241–243	114
C_6H_5	S	ClO_4	2	209–210	114
C_2H_5	S	ClO_4	2	148–149	114
$3,4-Cl_2-C_6H_3$	O	ClO_4	2	196–197	114
C_6H_5	O	Cl	1	195	114
C_6H_5	S	Cl	1	72–74	114
$2,5-Cl_2-C_6H_3$	O	Cl	1	177–180 (dec)	114
C_6H_5	O	Cl	2	200–202 (dec)	114
$3,4-Cl_2-C_6H_3$	O	Cl	2	185–187 (dec)	114
$2,5-Cl_2-C_6H_3$	O	Cl	2	175–177 (dec)	114
C_6H_5	S	Picrate	1	160–162	114
C_2H_5	S	Picrate	1	154–155	114
C_2H_5	S	Picrate	2	138–140 (dec)	114

With iminium salts (113) the ynamines 112 yield α,β-unsaturated amidinium salts (115); 114 is the likely intermediate (119).

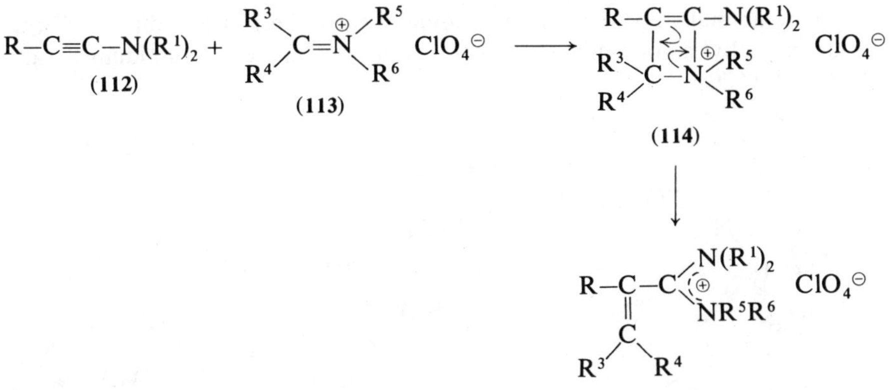

I. SYNTHESIS

R—C≡C—N(R^1)$_2$ + $\begin{matrix}R^3\\R^4\end{matrix}$C=N$^⊕\begin{matrix}R^5\\R^6\end{matrix}$ ClO$_4^⊖$ ⟶ R—C=C—N(R^1)$_2$ with $\begin{matrix}R^3\\R^4\end{matrix}$C—N$^⊕\begin{matrix}R^5\\R^6\end{matrix}$ ClO$_4^⊖$

(112)　(113)　(114)

↓

R—C—C$\begin{matrix}N(R^1)_2\\⊕\\NR^5R^6\end{matrix}$ ClO$_4^⊖$, with =C< bearing R^3, R^4

(115)

Isoquinolinium salts (**116**) also add to ynamines (**117**) via formation of amidinium salts (**118**) (119).

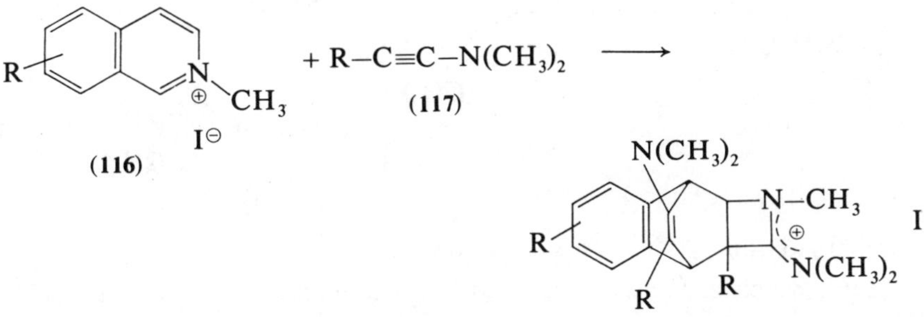

(116) + R—C≡C—N(CH$_3$)$_2$ ⟶ (118)

(117)

When ketene aminals (**119**) are acylated with acid chlorides, the amidinium salts **120** are generated (115).

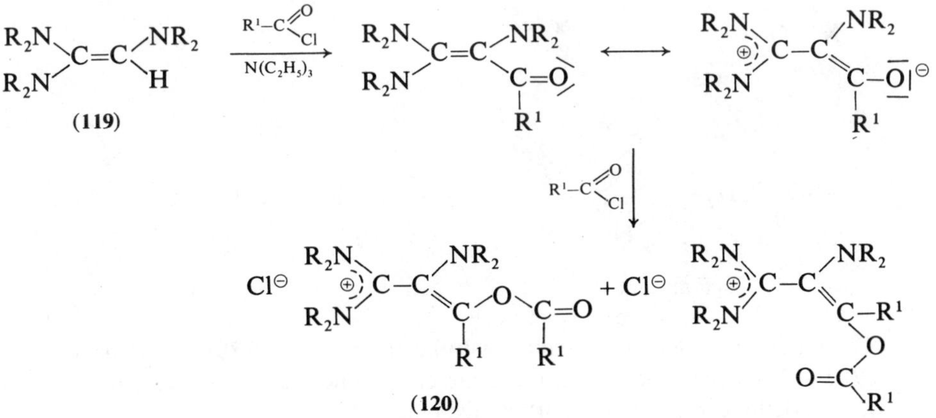

(119) $\xrightarrow[N(C_2H_5)_3]{R^1-C(O)Cl}$ ⟷

↓ R^1—C(O)Cl

(120) + Cl$^⊖$

Ketene aminals also react with amidinium salts like **121** and dimethylformamide–dimethyl sulfate adducts (**122**) to yield the vinylogous amidinium salts (**123, 124**) (120).

(**121**) (**123**)

(**122**) (**124**)

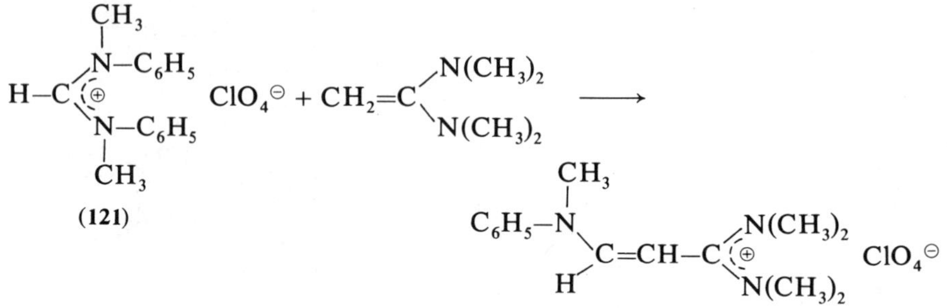

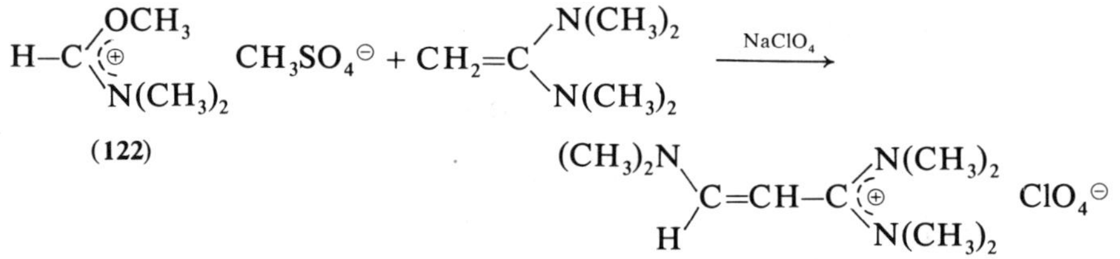

Sulfonyl chlorides acylate ketene aminals (**125**) via formation of amidinium salts (**126**), which are easily converted into sulfonated ketene aminals (**127**) (121).

(**125**)

(**126**) (**127**)

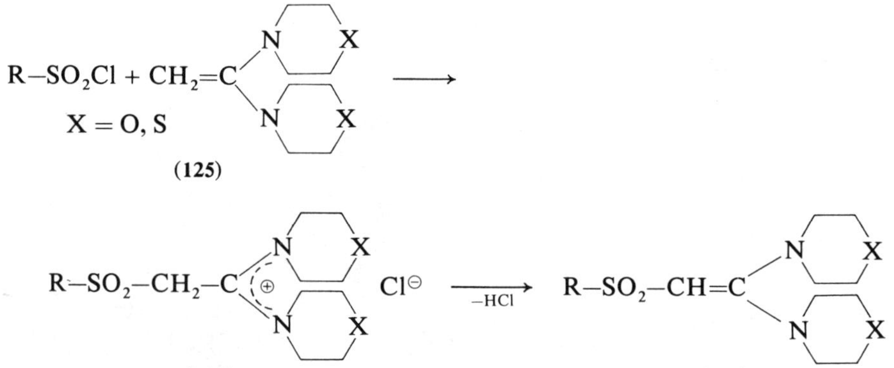

The reaction of elementary sulfur with ketene aminals (**128**) to give betaines (**129**) has been reported (122); the same compounds are also accessible from tetraaminoethylene (**130**) and carbon disulfide (123).

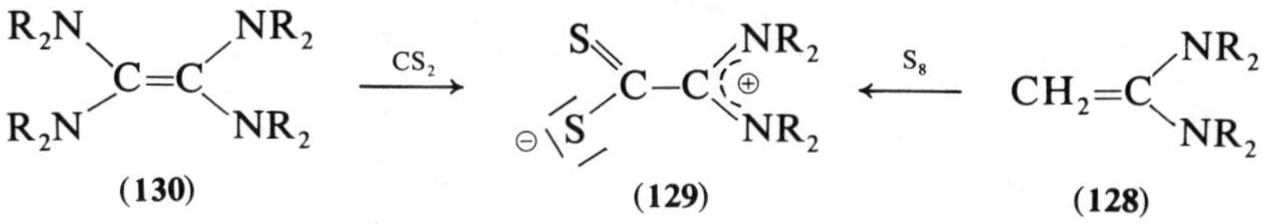

2. Oxidation of Ketene Aminals

Oxidizing agents convert ketene aminals into amidinium salts. There are careful and extensive studies on oxidation of tetrakis(dimethylamino)ethylene (**131**) leading to the octamethylbis(oxamidinium)cation (**132**) (124–133).

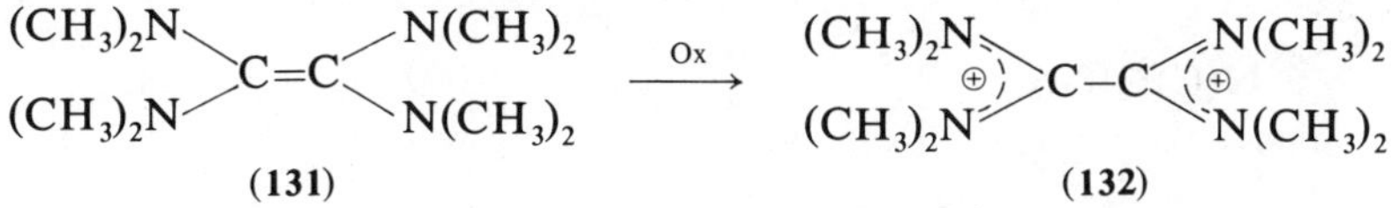

Among the oxidizing agents used were N_2O_4, Cu^I, Ag^I, and CCl_4. Reviews covering this reaction have been published (133, 134). Also other dications (**133, 134**) of this type are obtained by oxidizing tetraaminoethylene (**135**) (135). Triaminoethylenes like **136** react analogously with bromine (136).

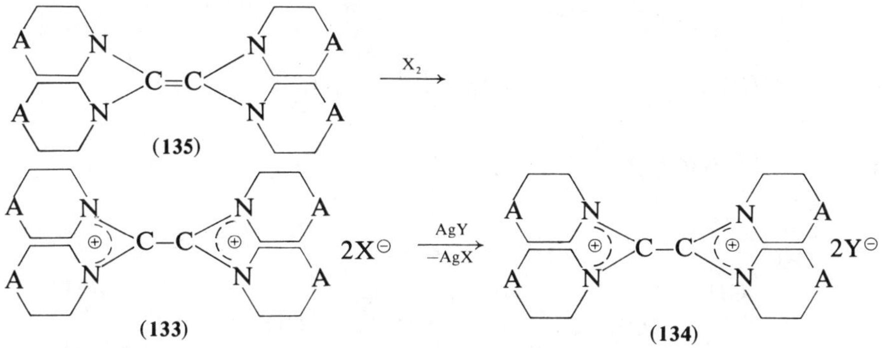

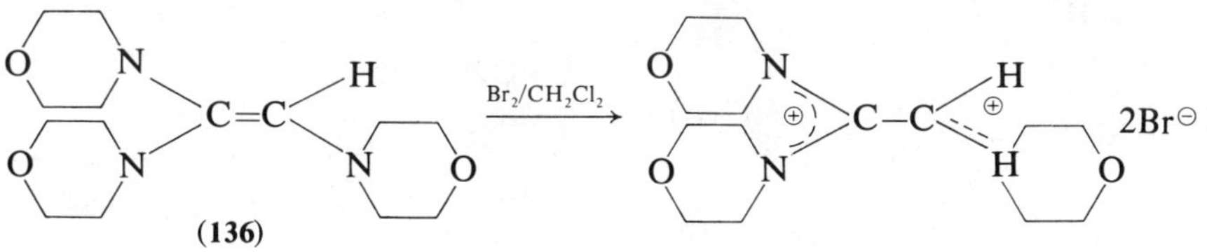

A = O, CH_2
X = Cl, Br, I
Y = NO_2, NO_3

The oxidation of ketene aminals like **137–141** with silver nitrate does not proceed so cleanly, because the reaction products occasionally dimerize, and these dimers may further react with the oxidizing agents (137, 138).

AMIDINIUM SALTS

I. SYNTHESIS

Bis(dimethylamino)fulvene (**142**) forms a 1:1 complex (**143**) with $FeCl_2$, which transforms on heating into a ferrocene derivative (**144**), which has an amidinium salt structure (139).

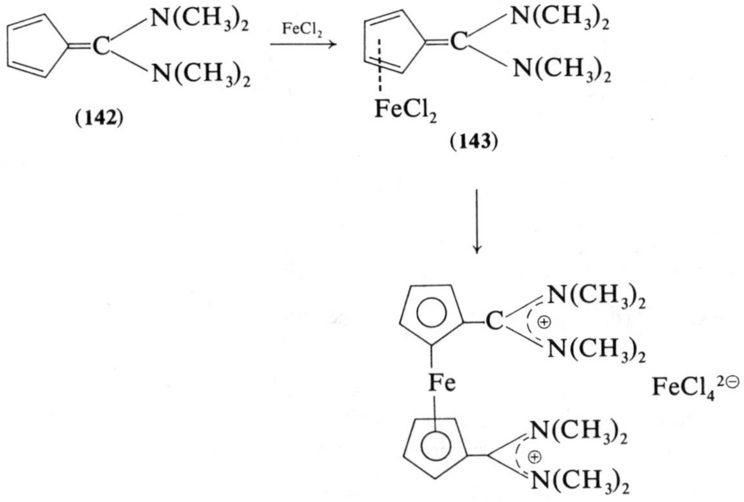

M. FROM THIOAMIDES, THIOAMIDE DERIVATIVES, AND THIOUREAS

Thioamides are alkylated by alkylating agents at the sulfur atom; the alkylmercaptomethyleniminium salts (**145**) thus formed give amidinium salts (**146**) or free amidines (**147**) with amines or amino derivatives. These reactions are dealt with in the chapter "Alkylmercaptomethyleniminium Salts."

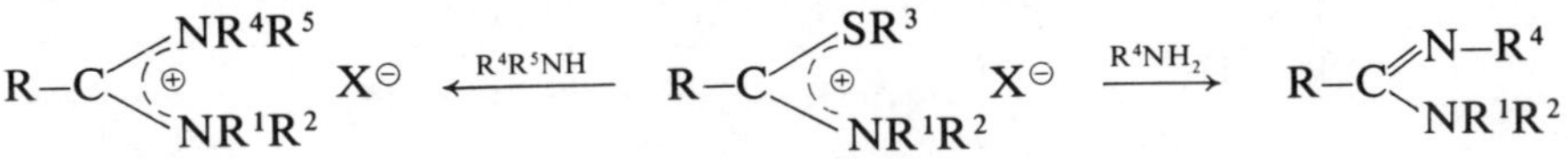

From the monothioamides of oxalic acid (**148**) the betaines **149** are obtained from primary amines in the presence of mercuric oxide (140).

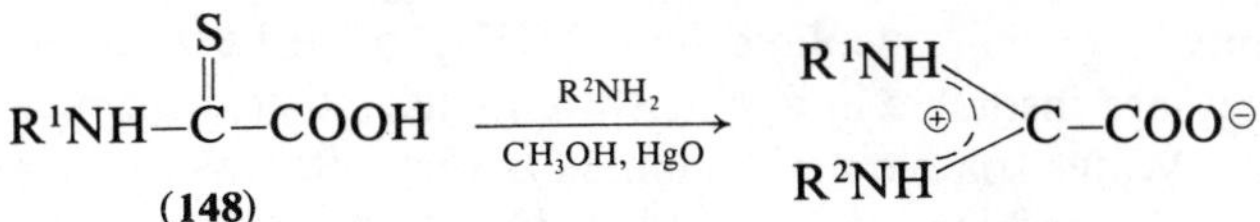

Walter and Ruess (141–144) introduced the synthesis of N,N,N'-trisubstituted formamidines (**150**) or formamidinium salts (**151**) by oxidizing N,N,N'-trisubstituted thioureas (**152**) with peracetic acid. This procedure is particularly suitable for the synthesis of compounds **150** or **151** carrying bulky substituents at the nitrogen atoms. Besides formamidinium salts, ureas (**153**) (141, 142) or thiourea trioxides (**154**) (143, 144) may be formed as by-products.

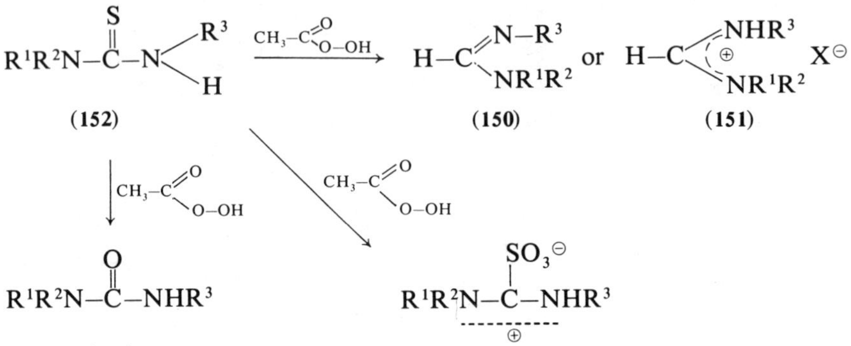

Via sulfide contraction (desulfurization) a series of amidines (**155**) or their amidinium salts was synthesized from β-ketocarboxylic acids according to the scheme shown (145).

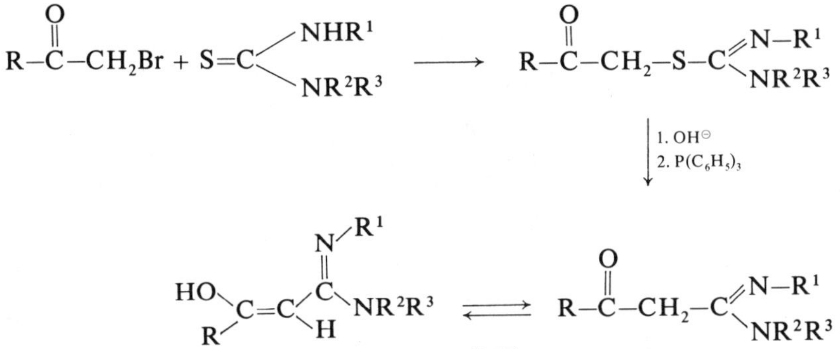

II. Structure and Properties of Amidinium Salts

Amidinium salts generally are well-crystallizing solids, which are soluble in polar solvents like water, alcohols, and DMF, provided certain anions are not present. They are insoluble in apolar solvents like ether, benzene, and carbon tetrachloride. Water-soluble amidinium salts very often can be precipitated as iodides (36, 42), perchlorates (42, 114, 115), tetrafluoroborates (92, 93) hexafluorophosphates (50, 116), tetraphenylborates (138), picrates (3), or chloroplatinates (4). The high melting points indicate a polar structure for these compounds. The positive charge of the cation is delocalized at the nitrogen atoms (5, 146):

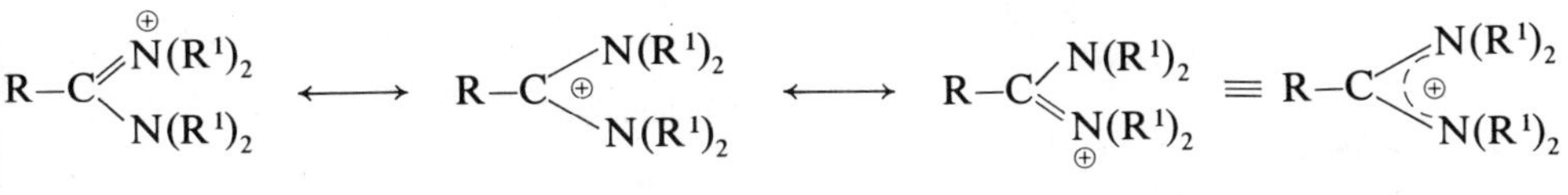

II. STRUCTURE AND PROPERTIES OF AMIDINIUM SALTS

$$R-C\overset{\oplus}{\underset{N(R^1)_2}{\diagup N(R^1)_2}} \longleftrightarrow R-C\overset{\diagup N(R^1)_2}{\underset{N(R^1)_2}{\oplus}} \longleftrightarrow R-C\overset{\diagup N(R^1)_2}{\underset{\underset{\oplus}{N(R^1)_2}}{}} \equiv R-C\overset{\diagup N(R^1)_2}{\underset{N(R^1)_2}{(\oplus}}$$

N,N'-Unsubstituted and N,N-disubstituted amidinium salts were studied by Raman and IR spectroscopy (147–150). The IR spectra of N,N'-tetrasubstituted amidinium salts show a band in the range 1590–1660 cm^{-1}, which is attributed to C=N$^\oplus$ valence vibration (42) (see Table IV). N,N,N'-Trisubstituted amidinium salts give similar data (151).

TABLE IV

Position of the C=N$^\oplus$ Valency Vibration of N,N'-Tetrasubstituted Amidinium Salts (42)

Amidinium salts 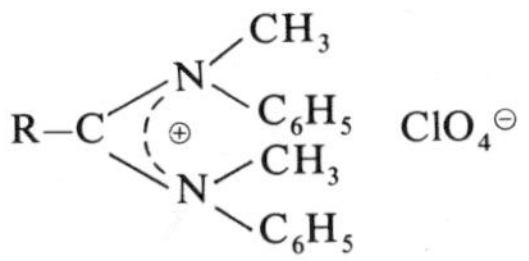 $R-C\overset{\diagup N(CH_3)(C_6H_5)}{\underset{N(CH_3)(C_6H_5)}{(\oplus}} \;ClO_4^\ominus$	$\nu_{C=N^\oplus}$ cm^{-1}
R = H	1652
CH$_3$	1607
C$_2$H$_5$	1605
n-C$_3$H$_7$	1605
i-C$_3$H$_7$	—
C$_6$H$_5$	1597
p-CH$_3$–C$_6$H$_4$	1598
p-Cl–C$_6$H$_4$	1597
p-NO$_2$–C$_6$H$_4$	1626
Bisamidinium perchlorates 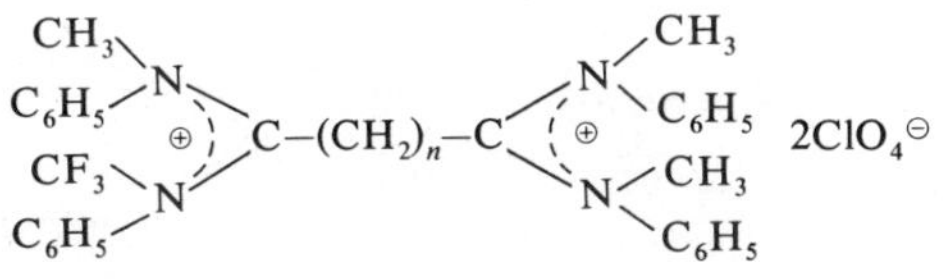 $\overset{CH_3}{\underset{C_6H_5}{}}N\diagdown\underset{CF_3}{\overset{}{}}\underset{C_6H_5}{}N\diagup C\overset{\oplus}{(}-(CH_2)_n-C\overset{\oplus}{)}\overset{N\diagdown CH_3/C_6H_5}{\underset{N\diagup CH_3/C_6H_5}{}} \; 2ClO_4^\ominus$	
n = 3	1605
n = 4	1605

The NMR spectra of amidinium salts (42, 91, 114, 116) support the proposed structure for the amidinium cations (NMR data, see Tables V and VI). The signals of the alkyl groups bound to nitrogen are shifted 0.5–0.8 ppm downfield compared with the free amine (114). Some NMR-spectroscopic studies deal with restricted rotation around the partial CN double bond (111,

TABLE V
NMR Data of some Amidinium Salts

Amidinium salt (solvent)	NMR data	Ref.
H–C^a(⊕)(N(CH$_3$)$_2$)$_2^b$ PF$_6^\ominus$ (CD$_3$CN)	$\tau_a = 7.1$(s), $\tau_b = 3.9$(s)	50
CH$_3$–C^c(⊕)(N(C$_6$H$_5$)(CH$_3^a$))$_2^b$ ClO$_4^\ominus$ (H$_2$O, HCl)	$\tau_a = 6.68$(s), $\tau_b = 2.63$(m), $\tau_c = 7.55$(s)	42
CH$_3^c$–CH$_2^b$–C(⊕)(N(CH$_3$)$_2$)$_2^a$ Cl$^\ominus$ (CD$_3$CN)	$\tau_a = 6.67$(s), $\tau_b = 6.7$(q), $\tau_c = 8.8$(t), $I_{bc} = 7.5$ Hz	50
CH$_3^c$–CH$_2^b$–C(⊕)(N(CH$_3$)$_2$)$_2^a$ I$^\ominus$ (CH$_2$Cl$_2$)	$\tau_a = 6.67$(s), $\tau_b = 7.13$(q), $\tau_c = 8.74$(t), $I_{bc} = 7.5$ Hz	116
CH$_2^e$=CH^d–CH$_2^c$–CH$_2^b$–C(⊕)(N(CH$_3$)$_2$)$_2^a$ Br$^\ominus$ (CH$_2$Cl$_2$)	$\tau_a = 6.62$(s), $\tau_b = 7.51$(t), $\tau_c = 6.95$(m), $\tau_d = 4.25$(m), $\tau_e = 4.83$(m)	116
(CH$_3$)$_3$C^b–C(⊕)(N(CH$_3$)$_2$)$_2^a$ I$^\ominus$ (CH$_2$Cl$_2$)	$\tau_a = 6.55$(s), $\tau_b = 8.40$(s)	116
C$_6$H$_5^d$–CH$_2^c$–CH$_2^b$–C(⊕)(N(CH$_3$)$_2$)$_2^a$ I$^\ominus$ (CH$_2$Cl$_2$)	$\tau_a = 6.78$(s), $\tau_{b,c} = 6.91$(m), $\tau_d = 2.69$(s)	116
CH$_2^e$=CH^d–CH$_2^c$–C(CH$_3$)(CH$_3$)^b–C(⊕)(N(CH$_3$)$_2$)$_2^a$ Br$^\ominus$ (CH$_2$Cl$_2$)	$\tau_a = 6.52$(s), $\tau_b = 8.37$(s), $\tau_c = 7.37$(d), $\tau_d = 4.2$(m), $\tau_e = 4.7$(m)	116

TABLE V (cont.)

Amidinium salt (solvent)	NMR data	Ref.
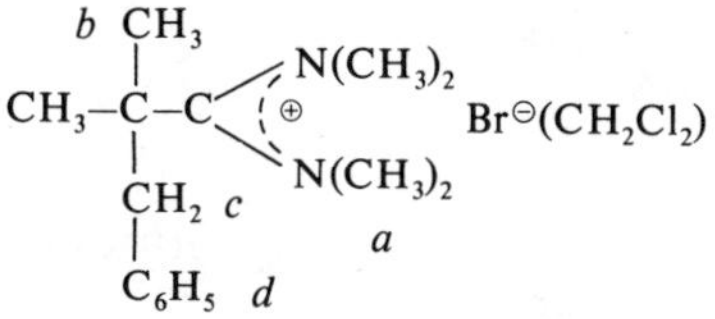 $\underset{d}{\overset{b}{\text{CH}_3}}\text{CH}_3-\underset{\underset{\underset{C_6H_5}{\vert}}{\overset{\vert}{\text{CH}_2}}\;c}{\overset{\vert}{\text{C}}}-\overset{\;\;\;\;\text{N(CH}_3)_2}{\underset{\;\;\;\;\text{N(CH}_3)_2}{\overset{\oplus}{\text{C}}}}\;\;\text{Br}^{\ominus}(\text{CH}_2\text{Cl}_2)$	$\tau_a = 6.70(s), \tau_b = 8.35(s), \tau_c = 6.88(s),$ $\tau_d = 2.7(s)$	116
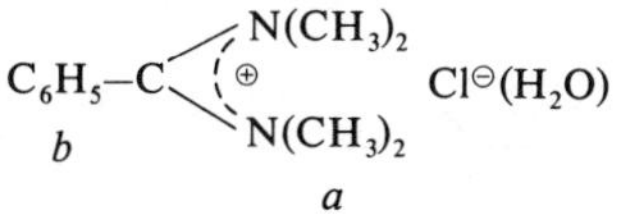 $\underset{b}{\text{C}_6\text{H}_5}-\overset{\;\;\;\;\text{N(CH}_3)_2}{\underset{\;\;\;\;\text{N(CH}_3)_2\;\;a}{\overset{\oplus}{\text{C}}}}\;\;\text{Cl}^{\ominus}(\text{H}_2\text{O})$	$\tau_a = 6.8(s), \tau_b = 2.4(s)$	116
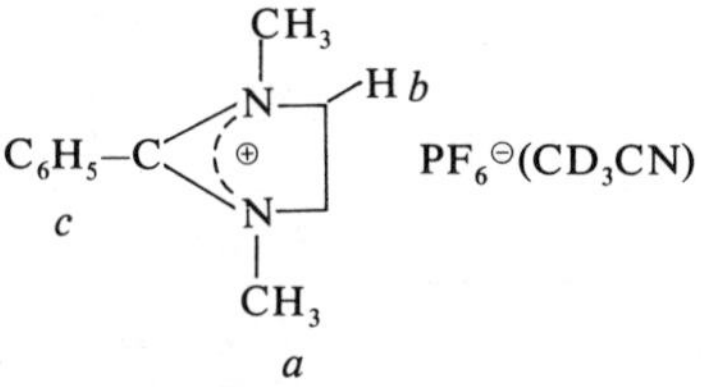 $\underset{c}{\text{C}_6\text{H}_5}-\text{C}\underset{\underset{\underset{a}{\text{CH}_3}}{\vert}}{\overset{\overset{\text{CH}_3}{\vert}}{\underset{\text{N}}{\overset{\text{N}}{\oplus}}}}\overset{\text{H }b}{\underset{}{}} \;\;\text{PF}_6^{\ominus}(\text{CD}_3\text{CN})$	$\tau_a = 7.0(s), \tau_b = 5.9(s), \tau_c = 2.3(m)$	116

152–160) and with the hydrogen/deuterium exchange of the NH protons (153, 161, 162). In the UV spectrum amidinium salts reveal an absorption band between 220 and 260 mμ, if only aliphatic or aliphatic and phenylic ligands are present in the molecule (50, 92, 114, 151) (see Table VII).

The structure of 1,1,3,3-tetrakis(dimethylamino)allyl perchlorate (**156**) was determined by X-ray methods (163); its electronic structure was determined by MO calculations (164).

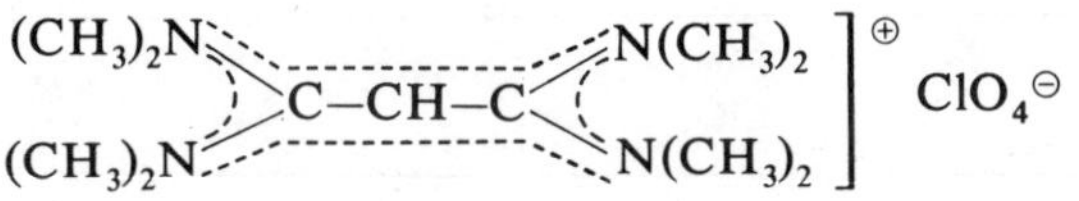

$$\left[\begin{array}{c}(\text{CH}_3)_2\text{N}\\(\text{CH}_3)_2\text{N}\end{array}\!\!\!\!\!\text{C}-\text{CH}-\text{C}\!\!\!\!\!\begin{array}{c}\text{N(CH}_3)_2\\\text{N(CH}_3)_2\end{array}\right]^{\oplus}\text{ClO}_4^{\ominus}$$
(**156**)

III. Reactions of Amidinium Salts

A. HYDROLYSIS, THIOLYSIS

Generally the hydrolysis of amidinium salts proceeds easily and yields acid amides (5). On alkaline hydrolysis of *N,N*-disubstituted amidines the stronger nucleophilic secondary amine remains in the molecule, as demonstrated by the

TABLE VI
Position of the α-CH Signals of Amidinium Salts in NMR Spectra (114)

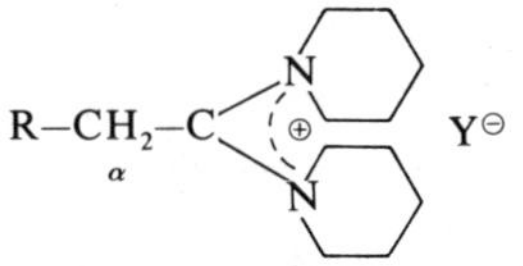

R—CH$_2$—C(⊕)(piperidine ring) Y$^\ominus$

R	Y	$\tau_{\alpha CH_2}$(CDCl$_3$)
H	Picrate	7.46
CH$_3$	I	7.23
C$_2$H$_5$	I	7.18

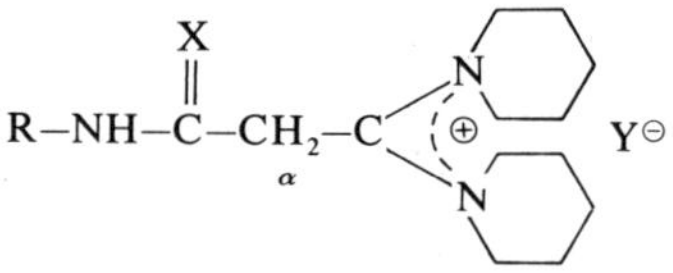

R—NH—C(=X)—CH$_2$—C(⊕)(piperidine) Y$^\ominus$

R	X	Y	$\tau_{\alpha CH_2}$(CDCl$_3$)
C$_2$H$_5$	S	Picrate	5.88
C$_2$H$_5$	S	ClO$_4$	5.81
C$_6$H$_5$	S	Picrate	5.33
C$_6$H$_5$	S	Cl	5.00
2,5 Cl$_2$—C$_6$H$_3$	O	Cl	5.20

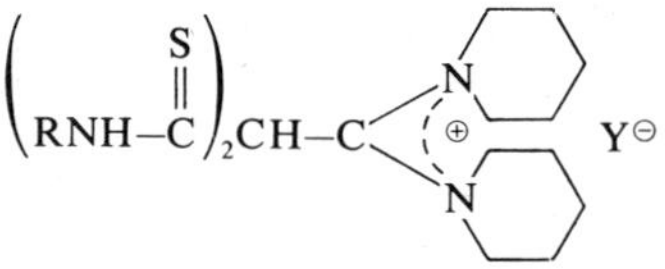

(RNH—C(=S))$_2$CH—C(⊕)(piperidine) Y$^\ominus$

R	Y	$\tau_{\alpha CH_2}$(CDCl$_3$)
C$_2$H$_5$	Picrate	3.45
C$_2$H$_5$	ClO$_4$	4.44

hydrolysis of N,N-diethylformamidinium methylsulfate (**157**) and of N,N-pentamethyleneformamidinium sulfate (**158**) (22).

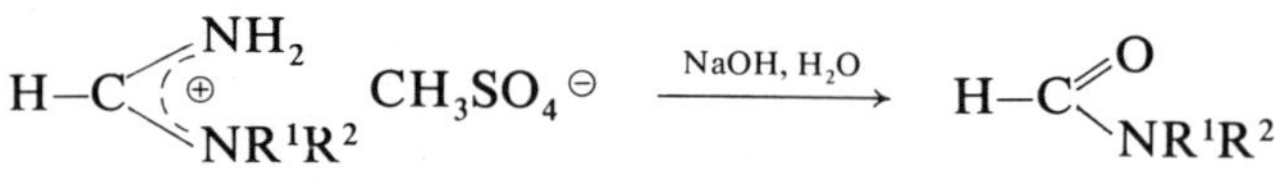

H—C(⊕)(NH$_2$)(NR^{1}R^2) CH$_3$SO$_4^\ominus$ $\xrightarrow{\text{NaOH, H}_2\text{O}}$ H—C(=O)(NR^{1}R^2)

(**157**) R^1 = R^2 = C$_2$H$_5$
(**158**) R^1—R^2 = —(CH$_2$)$_5$—

TABLE VII

UV Absorption Maxima of Amidinium Salts

Amidinium salt

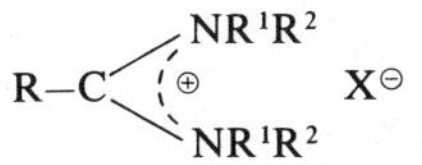

$$R-C\underset{NR^1R^2}{\overset{NR^1R^2}{\diagup}} \oplus \quad X^{\ominus}$$

R	R¹	R²	X	λ_{max}, nm		·log ε or ε	Ref.
H	CH_3	CH_3	PF_6	224 (H_2O)		4.076	50
H	CH_3	C_6H_5	I	245	} (C_2H_5OH)	21000	
				Shoulder 290		3500	92
H	CH_3	C_6H_5	Br	243	} (CH_3OH)	13100	
				Shoulder 290		1560	92
H	CH_3	C_6H_5	PF_6	245	} (C_2H_5OH)	24000	
				Shoulder 290		4310	92
H	C_2H_5	C_6H_5	I	245	} (C_2H_5OH)	19000	
				Shoulder 290		2720	92
CH_3	$-(CH_2)_5-$		Picrate	250		4.32	114
C_2H_5	$-(CH_2)_5-$		I	257		4.14	114
C_3H_7	$-(CH_2)_5-$		I	255		4.10	114
$C_6H_5-CH_2$	$-(CH_2)_5-$		Br	259		4.05	114

The alkaline hydrolysis of N,N,N'-trisubstituted amidinium salts **159** leads to loss of the secondary amine (23) at a pH of 5–8. Hydrolysis of the amidinium salt **160** proceeds similarly (165).

$$H-C\underset{NR^1R^2}{\overset{NHR}{\diagup}} \oplus \ Cl^{\ominus} + H_2O \longrightarrow H-C\underset{NHR}{\overset{O}{\diagup}} + H_2\overset{\oplus}{N}R^1R^2 \ Cl^{\ominus}$$

(**159**)

$$H-C\underset{NHCOC_6H_5}{\overset{N(CH_3)_2}{\diagup}} \oplus \ ClO_4^{\ominus} \xrightarrow{H_2O} H-C\underset{NH-\overset{O}{\overset{\|}{C}}-C_6H_5}{\overset{O}{\diagup}}$$

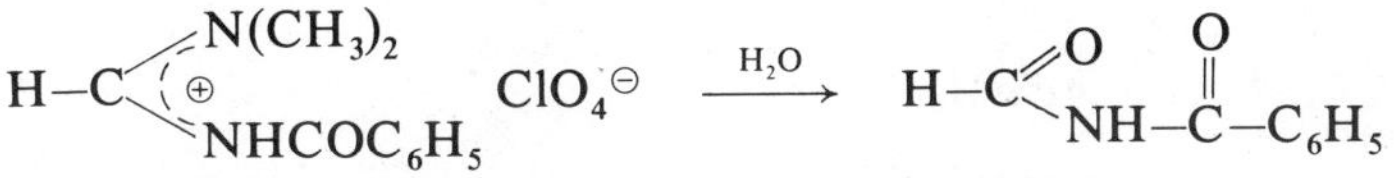

(**160**)

N,N,N',N'-Tetrasubstituted amidinium salts **161** or **162** react with water to give carboxamides (42, 91, 92, 114, 116–118).

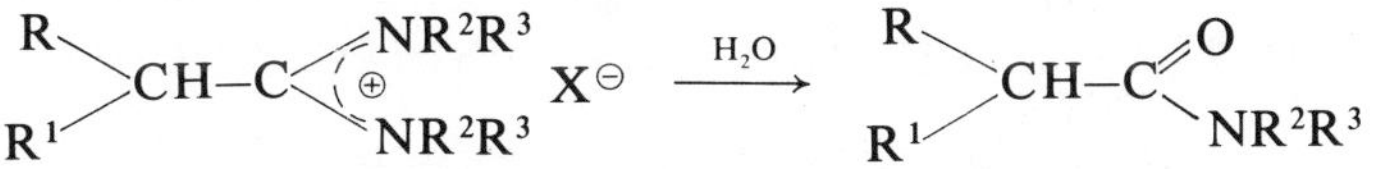

$$\underset{R^1}{\overset{R}{\diagdown}}CH-C\underset{NR^2R^3}{\overset{NR^2R^3}{\diagup}} \oplus \ X^{\ominus} \xrightarrow{H_2O} \underset{R^1}{\overset{R}{\diagdown}}CH-C\underset{NR^2R^3}{\overset{O}{\diagup}}$$

(**161**)

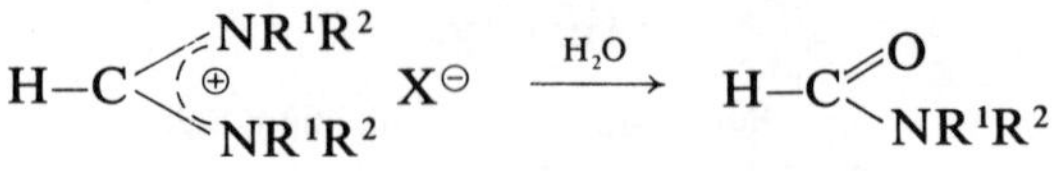

$$H-C\underset{NR^1R^2}{\overset{NR^1R^2}{\diagup}}\oplus \quad X^\ominus \xrightarrow{H_2O} H-C\underset{NR^1R^2}{\overset{O}{\diagup}}$$

(162)

Hydrolysis of the sesquihalides of hydrogen cyanide (163) is a stepwise process (166).

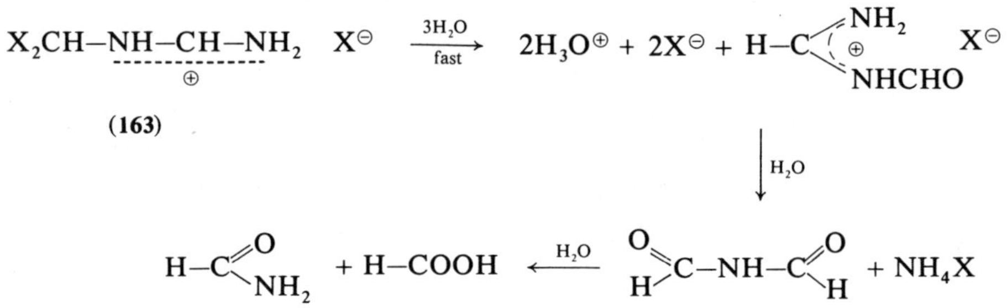

$$X_2CH-\underline{NH-CH-NH_2} \quad X^\ominus \xrightarrow[fast]{3H_2O} 2H_3O^\oplus + 2X^\ominus + H-C\underset{NHCHO}{\overset{NH_2}{\diagup}}\oplus \quad X^\ominus$$
$$\hspace{2cm}\oplus$$

(163)

$$\downarrow H_2O$$

$$H-C\underset{NH_2}{\overset{O}{\diagup}} + H-COOH \xleftarrow{H_2O} \underset{H}{\overset{O}{\diagdown}}C-NH-C\underset{H}{\overset{O}{\diagup}} + NH_4X$$

The action of H$_2$S on 1,1-diaminoacetonitriles (164) yields thioamides (165) (99).

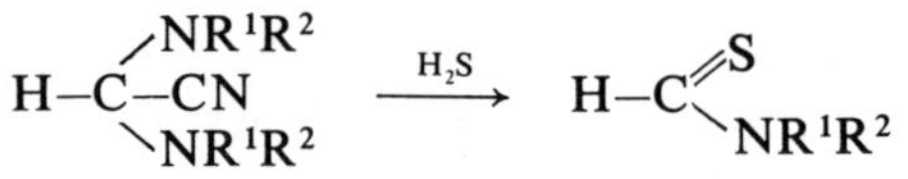

$$H-C\underset{NR^1R^2}{\overset{NR^1R^2}{\diagup}}-CN \xrightarrow{H_2S} H-C\underset{NR^1R^2}{\overset{S}{\diagup}} \quad\quad R^1-R^2 = -(CH_2)_4-, -(CH_2)_5-,$$
$$\hspace{6cm} -(CH_2)_2-O-(CH_2)_2-$$

(164) (165)

The synthesis of thioamides from amidines has been reviewed by Walter and Bode (167).

B. REACTIONS WITH NH-CONTAINING COMPOUNDS

1. *With Carboxylic Acid Amides*

Formamidinium methylsulfate reacts with formamide to yield tris(formamido)methane (165) in moderate yields (22%) (168). The analogous reaction of *N,N*-dialkylformamidinium methylsulfate with *N*-monosubstituted amides has not been observed (69).

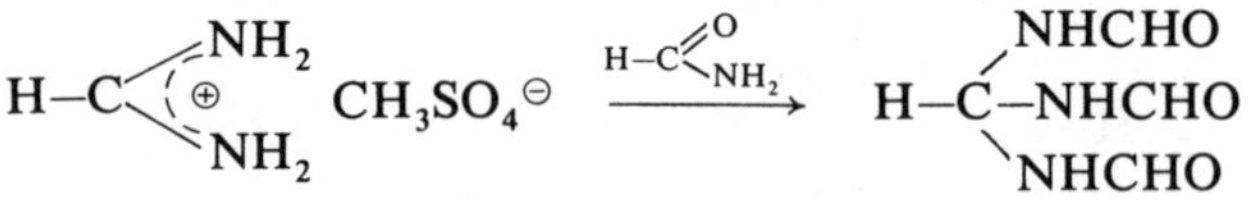

$$H-C\underset{NH_2}{\overset{NH_2}{\diagup}}\oplus \quad CH_3SO_4^\ominus \xrightarrow{H-C\underset{NH_2}{\overset{O}{\diagup}}} H-\underset{NHCHO}{\overset{NHCHO}{C}}-NHCHO$$

(165)

2. With Ammonia, Primary and Secondary Amines, Hydroxylamine, and Hydrazines

Amidinium salts are transaminated with ammonia and with primary and secondary amines (22, 23, 26, 27, 64, 102, 169). Thus the reaction of formamidinium methylsulfate (**166**) with diethylamine and morpholine gives N,N-diethylformamidinium methylsulfate (**167**) and N,N-(3-oxapentamethylene)formamidinium methylsulfate (**168**), respectively, in good yields (22). Excess amine does not give rise to N,N,N',N'-tetrasubstituted formamidinium salts (**169**).

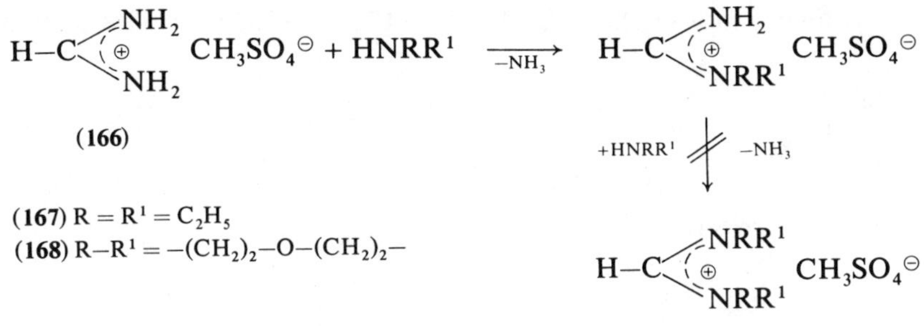

$$H-C(\overset{\oplus}{\underset{NH_2}{NH_2}}) \, CH_3SO_4^{\ominus} + HNRR^1 \xrightarrow{-NH_3} H-C(\overset{\oplus}{\underset{NRR^1}{NH_2}}) \, CH_3SO_4^{\ominus}$$

(**166**)

(**167**) $R = R^1 = C_2H_5$
(**168**) $R-R^1 = -(CH_2)_2-O-(CH_2)_2-$

$$+HNRR^1 \Big\updownarrow -NH_3$$

$$H-C(\overset{\oplus}{\underset{NRR^1}{NRR^1}}) \, CH_3SO_4^{\ominus}$$

(**169**)

It is possible with N,N,N'-trisubstituted amidinium salts to substitute successively both amino groups of the amidine with the aid of ammonia or primary amines. As in hydrolysis, the N,N-disubstituted ligand is eliminated first in the aminolysis reaction (23). This reaction may be applied to synthesize

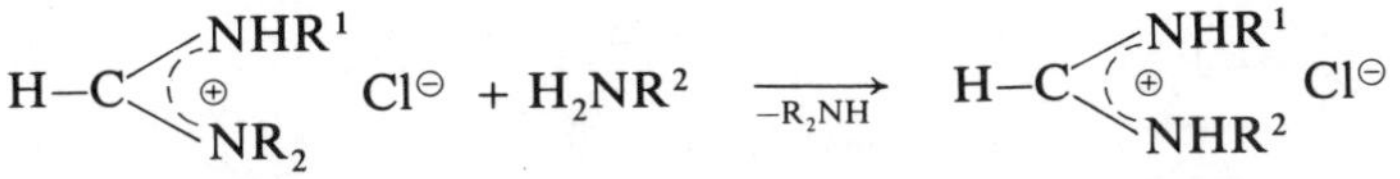

$$H-C(\overset{\oplus}{\underset{NR_2}{NHR^1}}) \, Cl^{\ominus} + H_2NR^2 \xrightarrow{-R_2NH} H-C(\overset{\oplus}{\underset{NHR^2}{NHR^1}}) \, Cl^{\ominus}$$

pyrimidines (23). Here the amidinium salts **170** carrying a β-carbonyl group cyclize with ammonium salts of weak acids, which function as ammonia

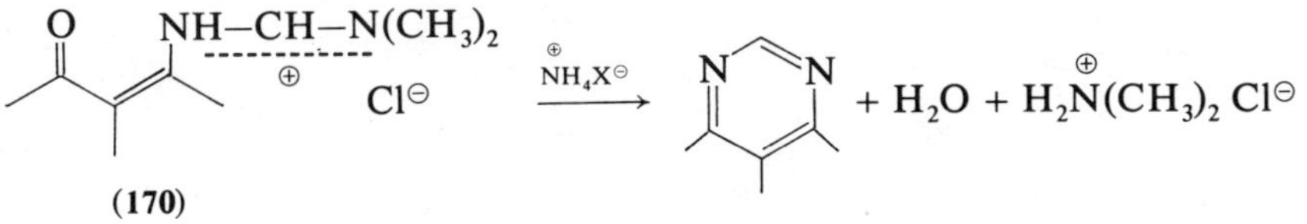

(**170**)

carriers. This reaction is particularly useful for the synthesis of anthrapyrimidines (**171**) (23).

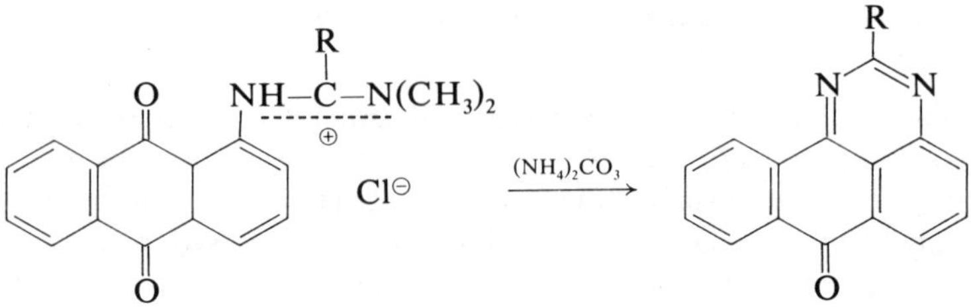

A series of *N,N,N'*-trisubstituted amidinium salts was converted with primary amines into the heterocycles **172–175** (26–28, 169). This particular

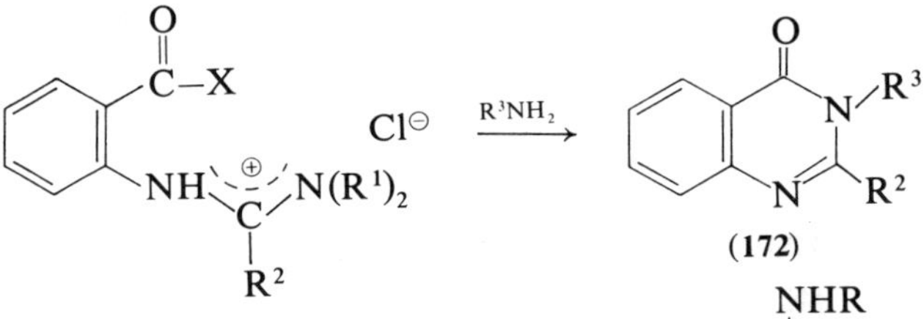

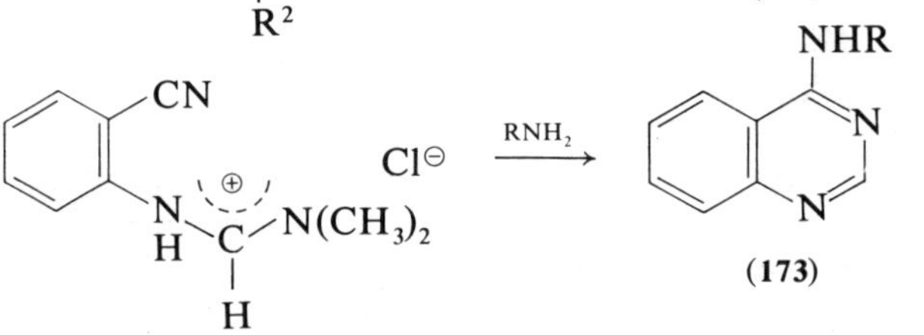

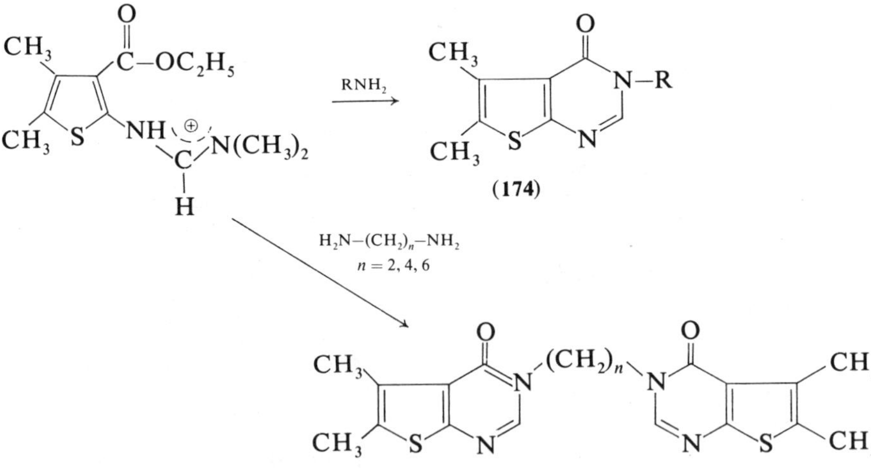

III. REACTIONS OF AMIDINIUM SALTS

aminolysis of amidinium salts can sometimes be taken advantage of to protect amino groups, for example, with the synthesis of *p*-aminobenzoic acid amides (**176**) (23). Here the aminobenzoic acid is reacted with DMF and thionyl chloride, where on the one hand the acid chloride forms, and the amino group is incorporated into an amidine structure (**177**) on the other hand. This acid chloride acts on the desired amine; finally the amidine structure is removed on heating with the acetate of a primary amine.

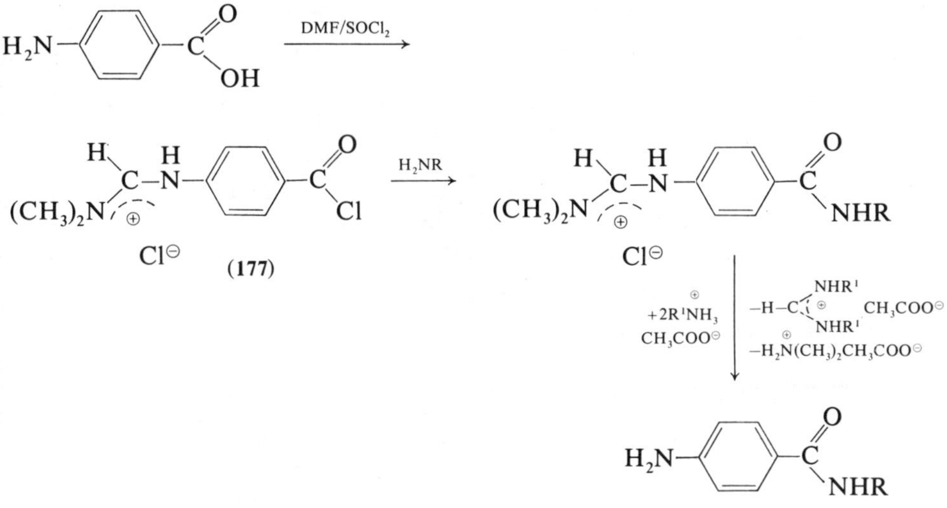

N,N,N',N'-Tetrasubstituted formamidinium salts like **178** may be transaminated with secondary amines (64).

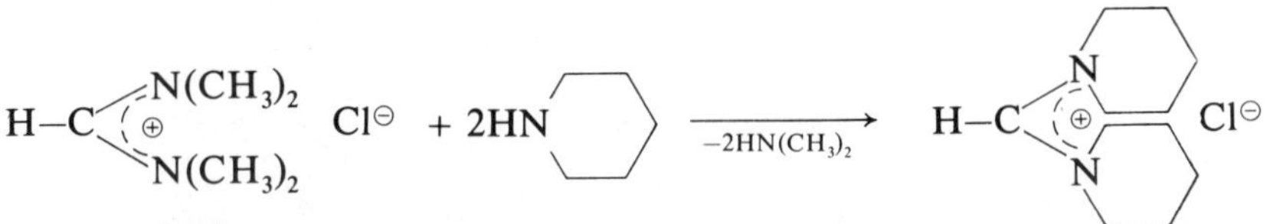

The same mechanism was proposed to rationalize the transamination of the weakly dissociated 1,1-bis(dialkylamino)acetonitriles (**179**) (102).

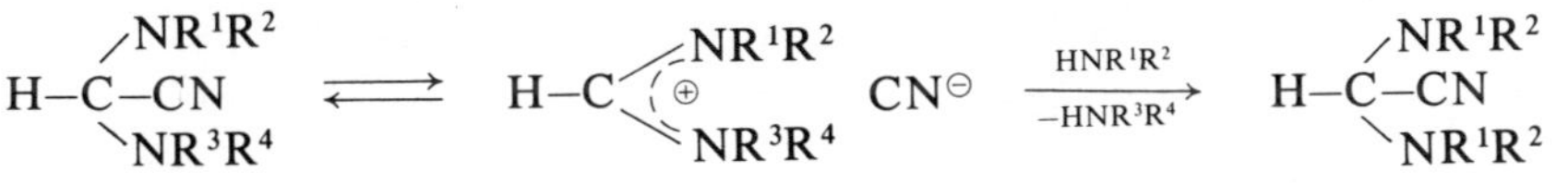

1,1-Dipiperidinoacetonitrile (**180**) reacts with aniline to yield the formamidine **181** (99).

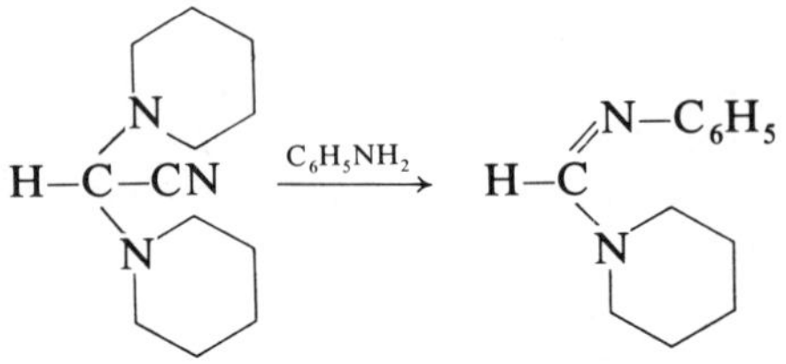

$$H-\underset{\underset{N(C_5H_{10})}{|}}{\overset{N(C_5H_{10})}{|}}{C}-CN \xrightarrow{C_6H_5NH_2} H-C\underset{N(C_5H_{10})}{\overset{N-C_6H_5}{\diagup}}$$

(**180**)   (**181**)

When treated with methoxyamine, the amidinium salt **182** gives **183**; the reaction proceeds via general base catalysis (170).

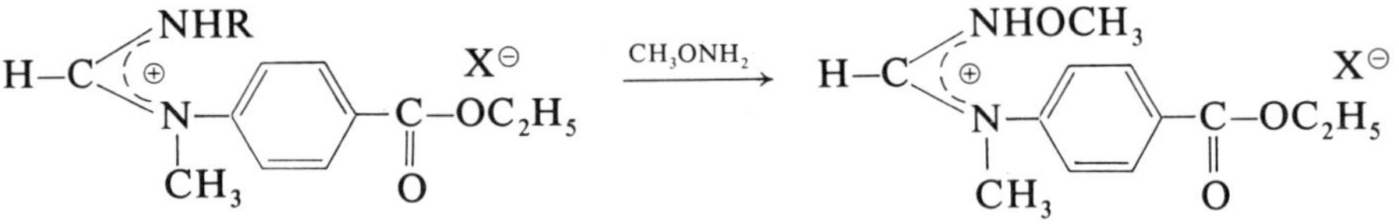

(**182**) $\xrightarrow{CH_3ONH_2}$ (**183**)

Amidrazones (**184**) are obtained from hydrazinolysis of the salts **185** (165).

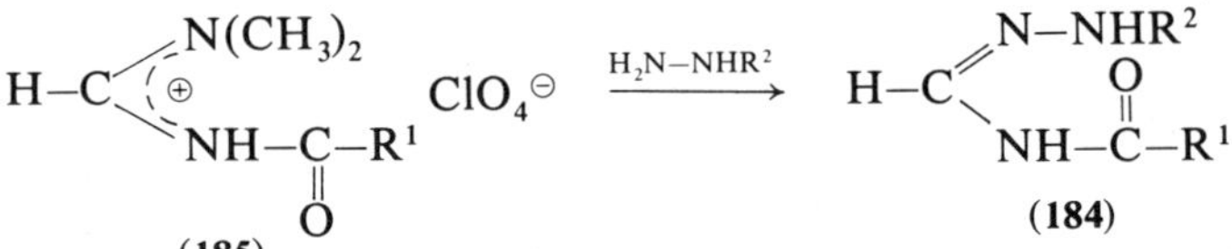

$$H-C\underset{NH-C(=O)-R^1}{\overset{N(CH_3)_2}{\diagup}}\ ClO_4^\ominus \xrightarrow{H_2N-NHR^2} H-C\underset{NH-C(=O)-R^1}{\overset{N-NHR^2}{\diagup}}$$

(**185**)   (**184**)

R^1	CH_3	C_6H_5	C_6H_5
R^2	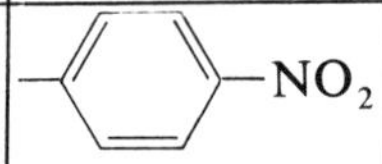 C_6H_4-NO_2	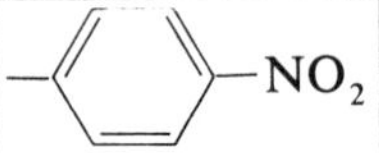 C_6H_4-NO_2	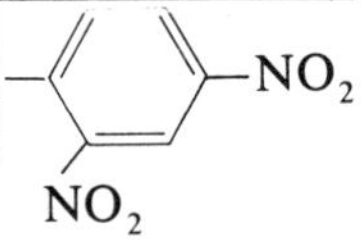 $C_6H_3(NO_2)_2$

C. REACTIONS WITH BASE

Upon treatment of amidinium salts **186**, which are not N,N,N',N''-persubstituted, with aqueous alkalies, generally the free amidines **186a** are obtained (5–7, 24, 77, 171, 172). By this route N,N,N'-trialkylformamidines

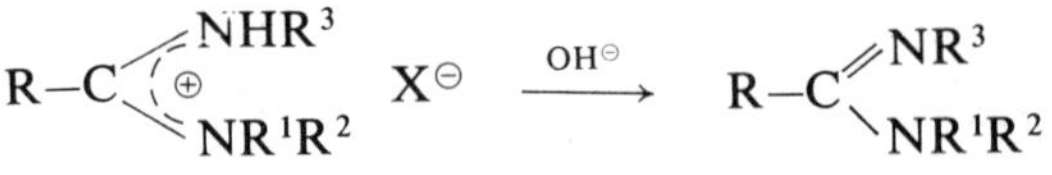

$$R-C\underset{NR^1R^2}{\overset{NHR^3}{\diagup}}{}^\oplus\ X^\ominus \xrightarrow{OH^\ominus} R-C\underset{NR^1R^2}{\overset{NR^3}{\diagup}}$$

(**186**)   (**186a**)

(24, 171, 172), trimethylformamidine (**187**) (171), and N,N,N'-trisubstituted acrylamidines (**188**) (77) were synthesized.

$$H-C{\overset{H}{\underset{N(CH_3)_2}{\overset{\mid}{\underset{\scriptstyle}{\overset{\displaystyle N-CH_3}{\oplus}}}}}} \quad CH_3SO_4^{\ominus} \quad \xrightarrow{K_2CO_3} \quad H-C{\overset{N-CH_3}{\underset{N(CH_3)_2}{}}}$$

(**187**)

$$CH_2=CH-C{\overset{H}{\underset{NR^1R^2}{\overset{\mid}{\underset{\scriptstyle}{\overset{\displaystyle N-CH(CH_3)_2}{\oplus}}}}}} \quad FeCl_4^{\ominus} \quad \xrightarrow{NaOH} \quad CH_2=CH-C{\overset{N-CH(CH_3)_2}{\underset{NR^1R^2}{}}}$$

(**188**)

If alcoholic alkoxides are used to synthesize the amidines from N,N-disubstituted amidinium salts, sometimes the carboxylic acid amide forms in addition to the amidine (22).

$$H-C{\overset{NH_2}{\underset{NR_2}{\overset{\oplus}{}}}} \quad CH_3SO_4^{\ominus} \quad \xrightarrow[C_2H_5OH]{NaOC_2H_5} \quad H-C{\overset{NH}{\underset{NR_2}{}}} + H-C{\overset{O}{\underset{NR_2}{}}}$$

When reacted with sodium hydride, N,N'-diphenyl-N,N'-dimethylformamidinium fluoroborate (**189**) adds hydride ion at the formyl carbon atom, giving rise to bis-(N-phenyl-N-methyl)-amino methane (**190**) (92).

$$H-C{\overset{CH_3}{\underset{\underset{CH_3}{N-C_6H_5}}{\overset{\mid}{\underset{\scriptstyle}{\overset{\displaystyle N-C_6H_5}{\oplus}}}}}} \quad BF_4^{\ominus} + NaH \quad \xrightarrow[-NaBF_4]{CH_3O-CH_2-CH_2-OCH_3} \quad {\overset{CH_3}{\underset{\underset{CH_3}{N-C_6H_5}}{\overset{\mid}{\underset{\scriptstyle}{\overset{\displaystyle N-C_6H_5}{CH_2}}}}}}$$

(**189**) (**190**)

On the other hand N,N,N',N'-tetrasubstituted amidinium salts (**191**) containing CH bonds in the α position are deprotonated by sodium hydride to ketene aminals (**192**) (173). If N,N,N',N'-tetraethylacetamidinium methylsulfate reacts with sodium hydride, no deprotonation could be observed (173).

$$R-CH_2-C{\overset{N(CH_3)_2}{\underset{N(CH_3)_2}{\overset{\oplus}{}}}} \quad CH_3SO_4^{\ominus} \quad \xrightarrow[-H_2,\ -NaCH_3SO_4]{NaH} \quad R-CH=C{\overset{N(CH_3)_2}{\underset{N(CH_3)_2}{}}}$$

(**191**) (**192**)

R = H, CH$_3$

Phenyllithium also converts amidinium salts (**193**) into ketene aminals (**194**) (42).

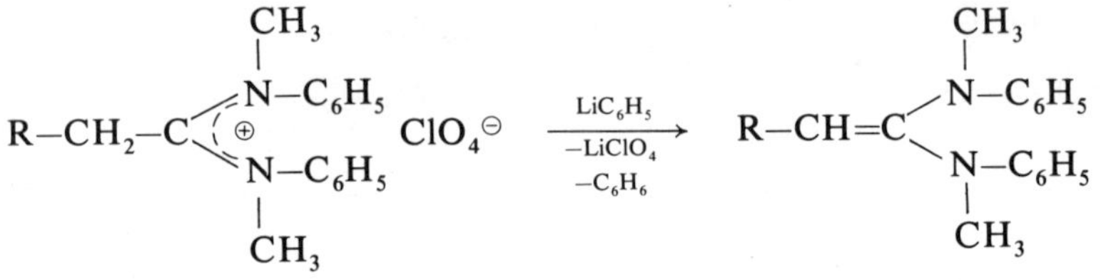

$$R-CH_2-C\underset{N-C_6H_5}{\overset{N-C_6H_5}{\oplus}}\begin{array}{c}CH_3\\|\\ \\|\\CH_3\end{array}\ ClO_4^{\ominus} \quad \xrightarrow[\substack{-LiClO_4\\-C_6H_6}]{LiC_6H_5} \quad R-CH=C\underset{N-C_6H_5}{\overset{N-C_6H_5}{}}\begin{array}{c}CH_3\\|\\ \\|\\CH_3\end{array}$$

(**193**)   (**194**)

The reaction of N,N,N',N',2,2-hexamethylpropionamidinium iodide (**195**) with t-butyllithium proceeds in a complex way (174). The aminal of pivalic aldehyde (**196**) is isolated in 20% yield.

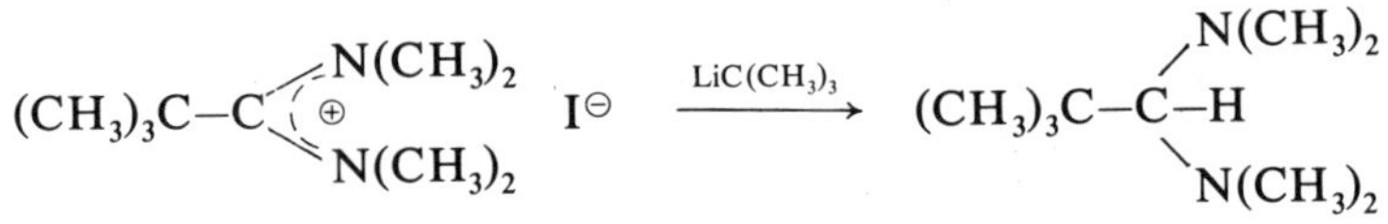

$$(CH_3)_3C-C\underset{N(CH_3)_2}{\overset{N(CH_3)_2}{\oplus}}\ I^{\ominus} \quad \xrightarrow{LiC(CH_3)_3} \quad (CH_3)_3C-\underset{N(CH_3)_2}{\overset{N(CH_3)_2}{C}}-H$$

(**195**)   (**196**)

If the N,N'-substituents of amidinium salts are aromatic, the salts are deprotonated to ketene aminals (**198**) even by aqueous alkalies (42). To suppress hydrolysis, the ketene aminals are taken up with a solvent not miscible with water (e.g., benzene) after the compounds have been liberated.

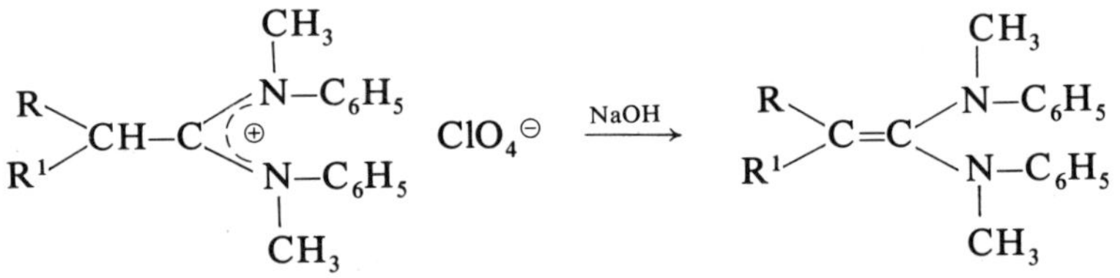

$$\underset{R^1}{\overset{R}{{>}}}CH-C\underset{N-C_6H_5}{\overset{N-C_6H_5}{\oplus}}\begin{array}{c}CH_3\\|\\ \\|\\CH_3\end{array}\ ClO_4^{\ominus} \quad \xrightarrow{NaOH} \quad \underset{R^1}{\overset{R}{{>}}}C=C\underset{N-C_6H_5}{\overset{N-C_6H_5}{}}\begin{array}{c}CH_3\\|\\ \\|\\CH_3\end{array}$$

(**197**)   (**198**)

R	H	H	H	CH_3
R^1	H	CH_3	C_2H_5	CH_3

When aqueous alkalies act on amidinium salts like **199**, which contain two aminocarbonyl or aminothiocarbonyl groups, the same observation could be made (114).

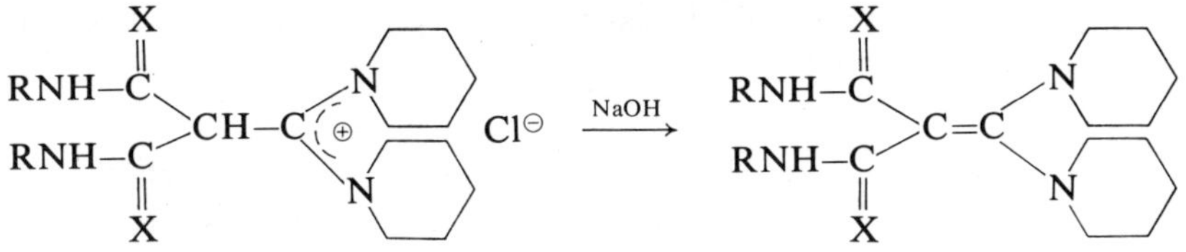

III. REACTIONS OF AMIDINIUM SALTS

(199)

X = O, S

If carboxylic acid dimethylamides condense with titanium tetrachloride in the presence of excess dimethylamine, the ketene aminals **201** are sometimes directly obtained via deprotonation of the amidinium salts **200**, which are formed initially (50).

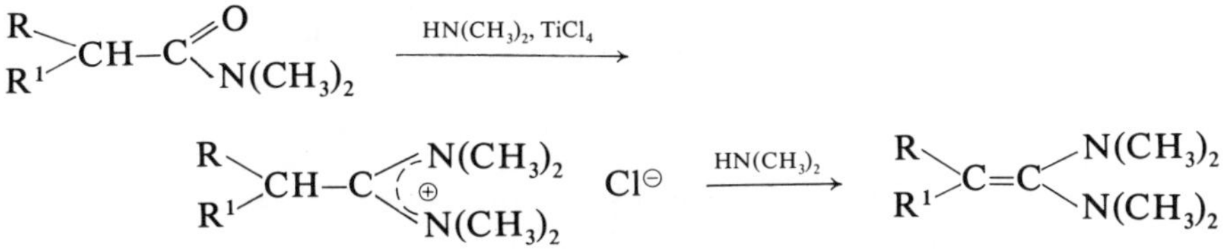

(200)　　　　　(201)

N,N,N',N'-Tetramethylacetamidinium chloride is smoothly deprotonated by dimethylamine, whereas the corresponding propionamidinium salts remain unreacted. This result emphasizes the varying acidity of the α hydrogens (50).

D. REACTIONS OF *N,N,N',N'*-TETRASUBSTITUTED AMIDINIUM SALTS WITH ALKALI METAL ALKOXIDES

1. In the Presence of Alcohol

If alcoholic alkoxides act on *N,N,N',N'*-tetramethylformamidinium chloride (**202**), acetals of dimethylformamide (**203**) form (65, 175, 176). This reaction definitely proceeds via aminal esters **204**, of which one dimethylamino group is displaced by the alcohol present.

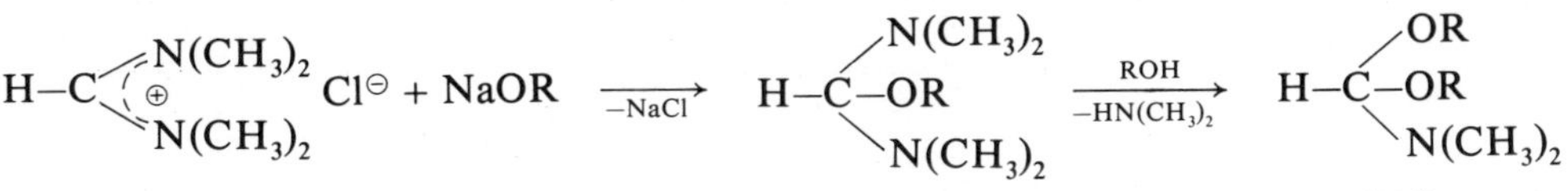

(202)　　　　　(204)　　　　　(203)

R = CH_3, C_2H_5, t-C_4H_9

Other N,N,N',N'-tetrasubstituted formamidinium salts (**205**) undergo this type of conversion (65, 92).

$$\text{H–C}\begin{array}{c}\diagup\text{NR}^1\text{R}^2\\\oplus\\\diagdown\text{NR}^1\text{R}^2\end{array}\text{Cl}^\ominus \xrightarrow[-\text{NaCl}, -\text{HNR}^1\text{R}^2]{\text{NaOCH}_3, \text{CH}_3\text{OH}} \text{H–C}\begin{array}{c}\diagup\text{OCH}_3\\-\text{OCH}_3\\\diagdown\text{NR}^1\text{R}^2\end{array}$$

(**205**)   (**206**)

$R^1 = CH_3$, $R^2 = C_6H_5$ $R^1-R^2 = -(CH_2)_2-O-(CH_2)_2-$

Starting from the weakly dissociated 1,1-bis-(dimethylamino)acetonitrile (**207**) and alcoholic alkoxide solutions, dimethylformamide acetals **203** are also synthesized in good yields (177). Table VIII summarizes the properties of the amide acetals **203** or **206** generated by this route.

$$\text{H–C–CN}\begin{array}{c}\diagup\text{N(CH}_3)_2\\\\\diagdown\text{N(CH}_3)_2\end{array} \xrightarrow[-\text{NaCN}, -\text{HN(CH}_3)_2]{\text{NaOR, ROH}} \text{H–C}\begin{array}{c}\diagup\text{OR}\\-\text{OR}\\\diagdown\text{N(CH}_3)_2\end{array}$$

(**207**)   (**203**)

$R = CH_3, C_2H_5, n\text{-}C_3H_7, i\text{-}C_3H_7, n\text{-}C_4H_9OH, t\text{-}C_4H_9OH$

TABLE VIII

Amide Acetals from N,N,N',N'-Tetrasubstituted Formamidinium Salts and Alcoholic Alkoxide Solutions

$$\text{H–C}\begin{array}{c}\diagup\text{OR}^3\\-\text{OR}^3\\\diagdown\text{NR}^1\text{R}^2\end{array} \text{ from } \text{H–C}\begin{array}{c}\diagup\text{NR}^1\text{R}^2\\\oplus\\\diagdown\text{NR}^1\text{R}^2\end{array}\text{X}^\ominus$$

R^1	R^2	R^3	X	Yield, %	b.p./pressure, °C/torr	n_D^{20}	Ref.
CH$_3$	CH$_3$	CH$_3$	Cl, CN	70	105/760	1.3970	64, 177
CH$_3$	CH$_3$	C$_2$H$_5$	CN	75	135–136/760	1.4000	65, 176, 177
CH$_3$	CH$_3$	n-C$_3$H$_7$	CN	78	57–59/12	1.4089	177
CH$_3$	CH$_3$	i-C$_3$H$_7$	CN	74	43–45/15	1.4012	177
CH$_3$	CH$_3$	n-C$_4$H$_9$	CN	82	96–98/20	1.4142	177
CH$_3$	CH$_3$	t-C$_4$H$_9$	Cl, CN	79	75–78/25	1.4160	175, 177
CH$_3$	C$_6$H$_5$	CH$_3$	Cl	—	123/15	—	65
−(CH$_2$)$_2$O(CH$_2$)$_2$−		CH$_3$	Cl	55	76.5–78/11	—	65

N,N,N',N'-Tetrasubstituted amidinium salts like **208** derived from higher carboxylic acids very likely react with alcoholic sodium alkoxides to give aminal esters (**209**) first, then amide acetals (**210**) with alcohol via cleavage of dimethylamine. The amide acetals **210** are in equilibrium with the ketene O,N-acetals **211** and alcohol (173).

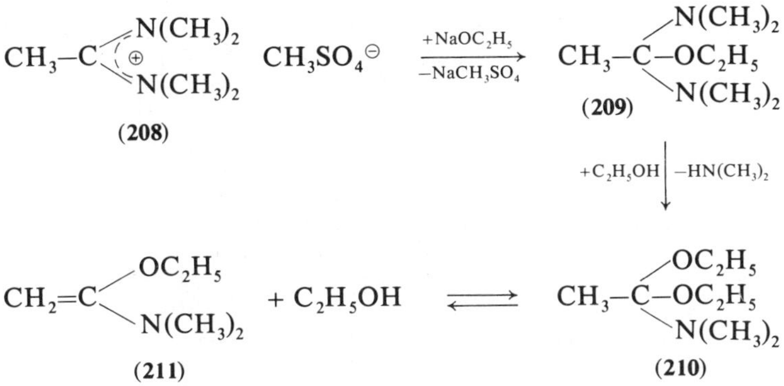

$$CH_3-C{\overset{\oplus}{\underset{N(CH_3)_2}{\diagup N(CH_3)_2}}} \; CH_3SO_4^{\ominus} \xrightarrow[-NaCH_3SO_4]{+NaOC_2H_5} CH_3-C{\overset{N(CH_3)_2}{\underset{N(CH_3)_2}{-OC_2H_5}}}$$

(**208**)   (**209**)

$+C_2H_5OH \; | \; -HN(CH_3)_2$

$$CH_2=C{\overset{OC_2H_5}{\underset{N(CH_3)_2}{\diagup}}} + C_2H_5OH \rightleftharpoons CH_3-C{\overset{OC_2H_5}{\underset{N(CH_3)_2}{-OC_2H_5}}}$$

(**211**)   (**210**)

2. In the Absence of Alcohol

N,N,N',N'-Tetraalkylsubstituted formamidinium salts (**212**) react with alcohol-free alkoxides to yield bis(dialkylamino)alkoxymethanes (aminal esters) (**213**); inert dry organic solvents like ether, cyclohexane, and n-hexane are used here (90, 171, 178–182).

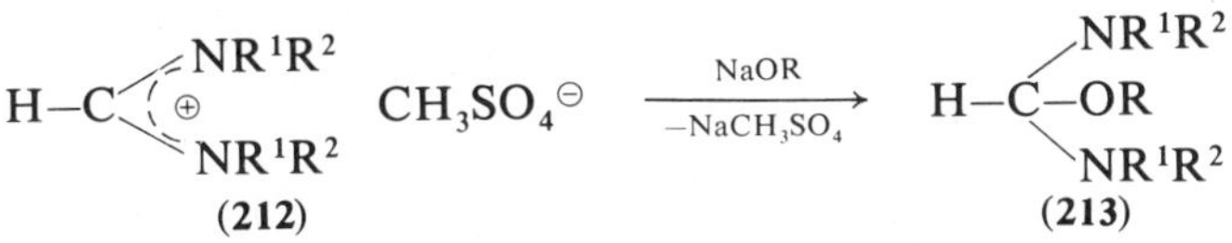

$$H-C{\overset{\oplus}{\underset{NR^1R^2}{\diagup NR^1R^2}}} \; CH_3SO_4^{\ominus} \xrightarrow[-NaCH_3SO_4]{NaOR} H-C{\overset{NR^1R^2}{\underset{NR^1R^2}{-OR}}}$$

(**212**)   (**213**)

Instead of the formamidinium methylsulfates, which are insoluble in apolar solvents, it is advantageous to use 1,1-bis(dialkylamino)acetonitriles like **214** because they dissolve in ether and cyclohexane (177).

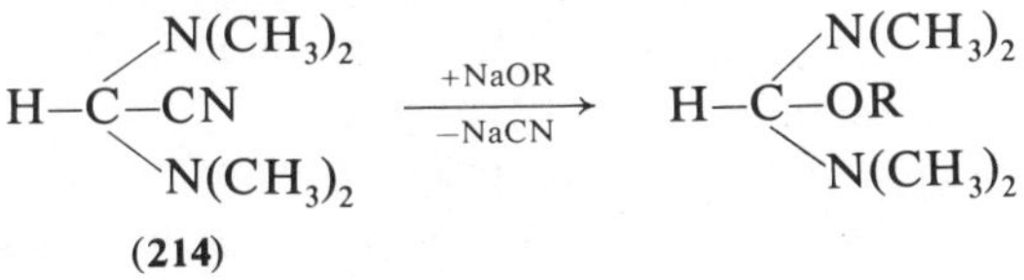

$$H-C{\overset{N(CH_3)_2}{\underset{N(CH_3)_2}{-CN}}} \xrightarrow[-NaCN]{+NaOR} H-C{\overset{N(CH_3)_2}{\underset{N(CH_3)_2}{-OR}}}$$

(**214**)

N,N'-Diphenyl-N,N'-dimethylformamidinium tetrafluoroborate (**215**) also reacts with alkali alkoxides to yield aminal esters (**216**) (92).

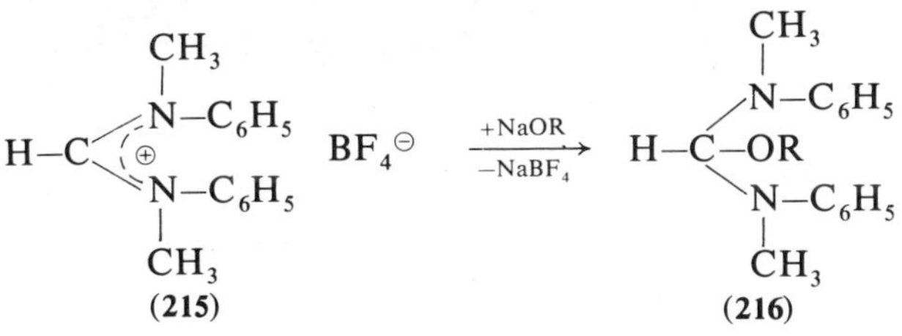

$$H-C{\overset{\oplus}{\underset{\underset{CH_3}{N-C_6H_5}}{\diagup \underset{CH_3}{N-C_6H_5}}}} \; BF_4^{\ominus} \xrightarrow[-NaBF_4]{+NaOR} H-C{\overset{\underset{CH_3}{N-C_6H_5}}{\underset{\underset{CH_3}{N-C_6H_5}}{-OR}}}$$

(**215**)   (**216**)

Table IX summarizes the properties of aminal esters.

TABLE IX

Aminal Esters from N,N,N',N'-Tetrasubstituted Formamidinium Salts and Alcohol-free Alkoxides

Aminal ester $H-C(NR^1R^2)(NR^1R^2)-OR$ from $H-C(NR^1R^2)(NR^1R^2)^{\oplus} \; X^{\ominus}$

R	R^1	R^2	X	Yield, %	b.p./pressure, °C/torr	n_D^{20}	Ref.
CH_3	CH_3	CH_3	CH_3SO_4	62	128/760	1.4158	180
CH_3	CH_3	CH_3	CN	82	32–33/25	1.4159	177
CH_3	$-(CH_2)_5-$		CH_3SO_4	35	56–59/10^{-3}	—	182
CH_3	CH_3	C_6H_5	BF_4	62	143–145/0.5	1.5909	92
C_2H_5	CH_3	CH_3	CH_3SO_4	79	142–144/736	1.4175	180
C_2H_5	CH_3	CH_3	CN	81.5	45–46/22	1.4176	177
n-C_3H_7	CH_3	CH_3	CH_3SO_4	72	50–52/12	1.4232	180
n-C_3H_7	CH_3	CH_3	CN	78	60–61/21	1.4208	177
i-C_3H_7	CH_3	CH_3	CH_3SO_4	62	41/10	1.4160	180
i-C_3H_7	CH_3	C_6H_5	BF_4	67	138–140/0.25	1.5696/25	92
i-C_3H_7	CH_3	CH_3	CN	76	43–45/15	1.4165	177
t-C_4H_9	CH_3	CH_3	CH_3SO_4	70	48–52/10–12	1.4260	180
t-C_4H_9	CH_3	CH_3	CN	83.5	50–52/12	1.4250	177
t-C_4H_9	C_2H_5	C_2H_5	CH_3SO_4	47	84–89/12	—	90
t-C_4H_9	CH_3	C_6H_5	BF_4	55	144–146/0.7	1.5609/25	92

The patent literature reports the reaction of tetramethylformamidinium methylsulfate (**217**) with sodium methoxide to yield a mixture of dimethylformamide dimethyl acetal (**218**), bis(dimethylamino)methoxymethane (**219**), and tetrakis(dimethylamino)ethylene (**220**) (183).

The occurrence of the three reaction products is definitely due to the manner of running the reaction or the workup. In this reaction primarily the aminal ester **219** forms, which dismutates in an equilibrium reaction to dimethylformamide dimethyl acetal (**218**) and tris(dimethylamino)methane (**221**) (179, 181, 184). The latter is known to convert into tetrakis(dimethylamino)ethylene (**220**) on heating with elimination of dimethylamine (185, 186).

E. REACTIONS WITH MERCAPTIDES

When N,N'-diphenyl-N,N'-dimethylformamidinium fluoroborate (**222**) acts on sodium methylmercaptide dissolved in methylmercaptan, a reaction takes

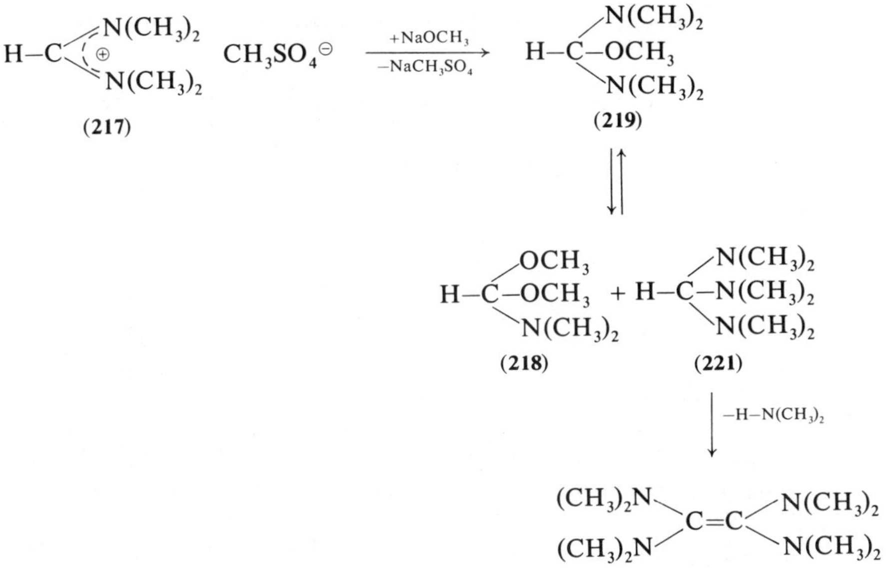

III. REACTIONS OF AMIDINIUM SALTS

$$H-C\underset{N(CH_3)_2}{\overset{N(CH_3)_2}{\oplus}} CH_3SO_4^{\ominus} \xrightarrow[-NaCH_3SO_4]{+NaOCH_3} H-C\underset{N(CH_3)_2}{\overset{N(CH_3)_2}{-}}OCH_3$$

(217)   (219)

$$\Updownarrow$$

$$H-C\underset{N(CH_3)_2}{\overset{OCH_3}{-}}OCH_3 \;+\; H-C\underset{N(CH_3)_2}{\overset{N(CH_3)_2}{-}}N(CH_3)_2$$

(218)   (221)

$$\Big\downarrow {-H-N(CH_3)_2}$$

$$\underset{(CH_3)_2N}{\overset{(CH_3)_2N}{\diagdown}}C=C\underset{N(CH_3)_2}{\overset{N(CH_3)_2}{\diagup}}$$

(220)

place which is analogous to the action of alcoholic alkoxides on formamidinium salts. The substitution of one amino function by mercaptide leads to the thioacetals of N-methylformanilide (223) (92).

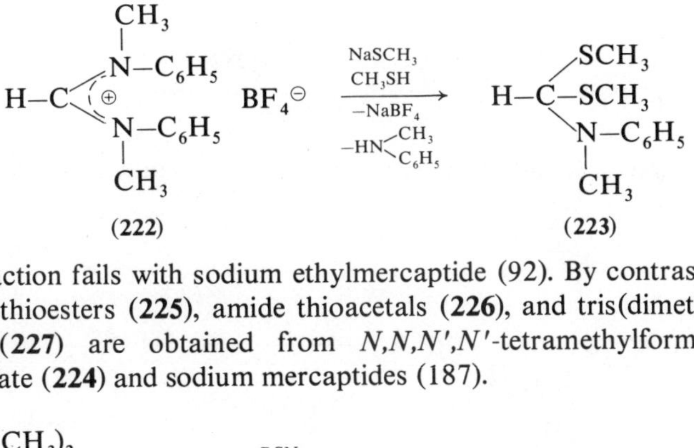

$$H-C\underset{\underset{CH_3}{|}}{\overset{\overset{CH_3}{|}}{\diagup}}\underset{N-C_6H_5}{\overset{N-C_6H_5}{\diagdown}} BF_4^{\ominus} \xrightarrow[\substack{-NaBF_4 \\ -HN\diagdown_{C_6H_5}^{CH_3}}]{NaSCH_3 \; CH_3SH} H-C\underset{\underset{CH_3}{|}}{\overset{SCH_3}{\diagup}}\underset{N-C_6H_5}{\overset{SCH_3}{\diagdown}}$$

(222)   (223)

This reaction fails with sodium ethylmercaptide (92). By contrast, mixtures of aminal thioesters (225), amide thioacetals (226), and tris(dimethylamino)-methane (227) are obtained from N,N,N',N'-tetramethylformamidinium methylsulfate (224) and sodium mercaptides (187).

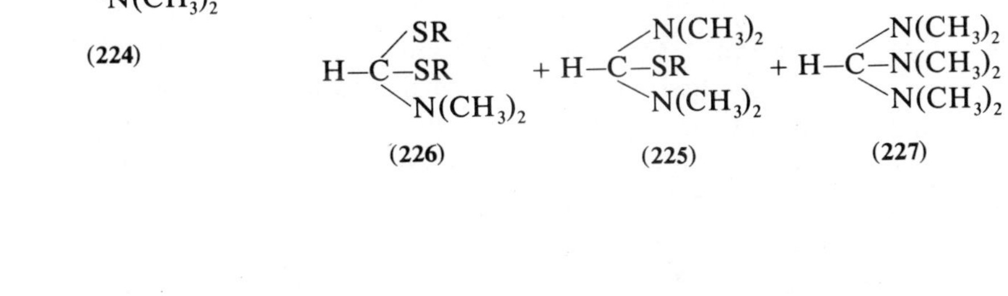

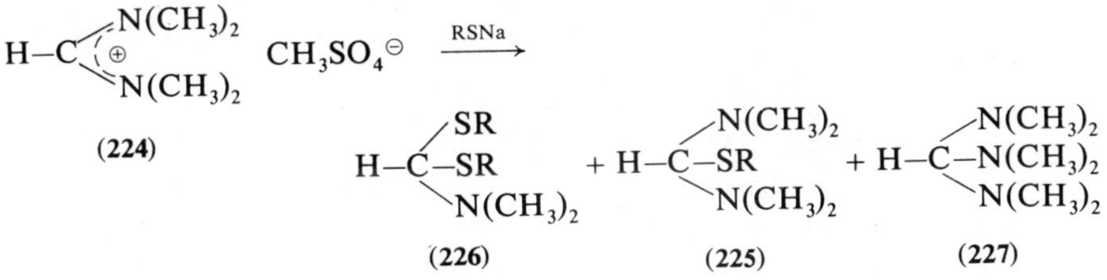

$$H-C\underset{N(CH_3)_2}{\overset{N(CH_3)_2}{\diagup}} CH_3SO_4^{\ominus} \xrightarrow{RSNa}$$

(224)

$$H-C\underset{N(CH_3)_2}{\overset{SR}{-}}SR \;+\; H-C\underset{N(CH_3)_2}{\overset{N(CH_3)_2}{-}}SR \;+\; H-C\underset{N(CH_3)_2}{\overset{N(CH_3)_2}{-}}N(CH_3)_2$$

(226)   (225)   (227)

F. Reactions of N,N,N',N'-Tetrasubstituted Formamidinium Salts with Alkali Amides

N,N'-diphenyl-N,N'-dimethyl-formamidinium iodide (**228**) is converted into N,N',N''-triphenyl-N,N',N''-trimethyltriaminomethane (**229**) with the aid of sodium hydride and N-methylaniline (92).

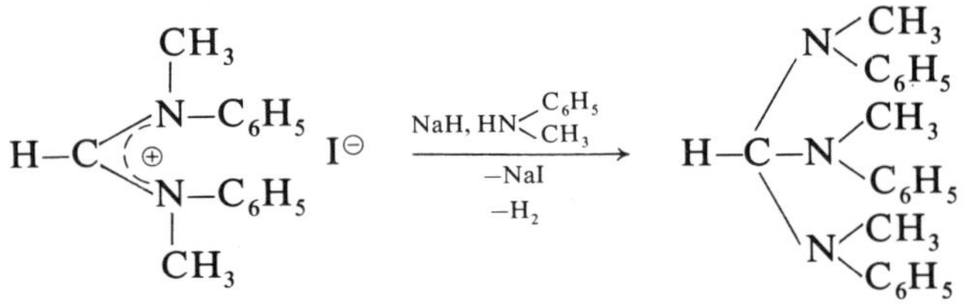

(228) → (229)

The reaction of N,N,N',N'-tetrasubstituted formamidinium salts like **230** with primary amines in the presence of sodium hydride yields formamidines (**231**) (92). The primary formation of a mixed triaminomethane such as **232** is very likely; this compound is stabilized by loss of secondary amine. The same

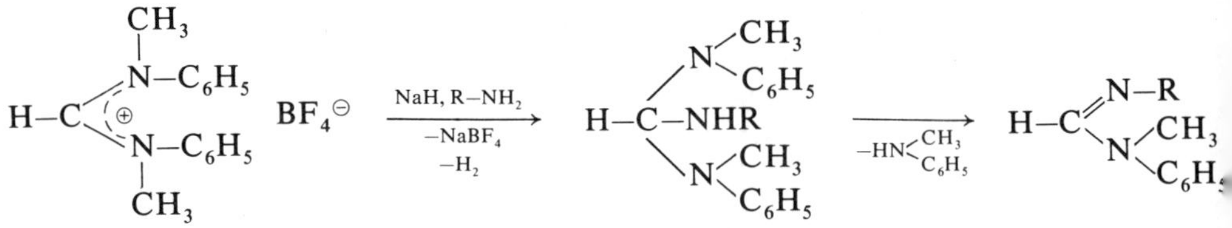

(230) → (232) → (231)

R = C_2H_5, n-C_4H_9, C_6H_5

reaction course is found in the reaction of the sodium salt of benzamide with **230** (92).

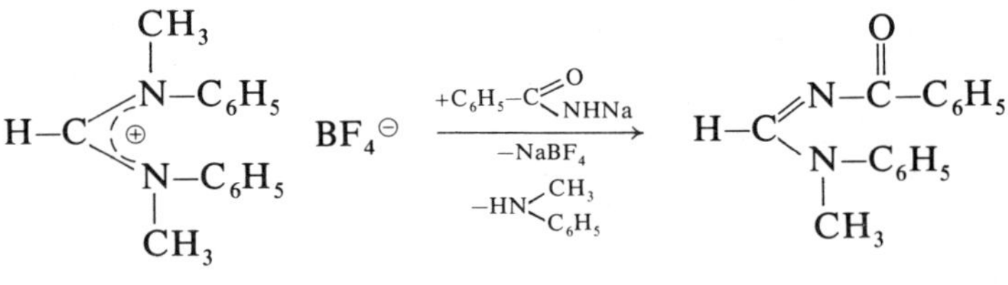

(230)

If triaminomethane **233** carries only aliphatic substituents at the nitrogen atom, it can be synthesized from alkali dialkylamides and N,N,N',N'-tetraalkylformamidinium salts (**234**) in ether (90, 188). The course of the reaction seems to be independent of the anion, since amidinium chlorides, perchlorates, and methylsulfates are capable of reacting this way. Even mixed substituted triaminomethanes are accessible by this route (90).

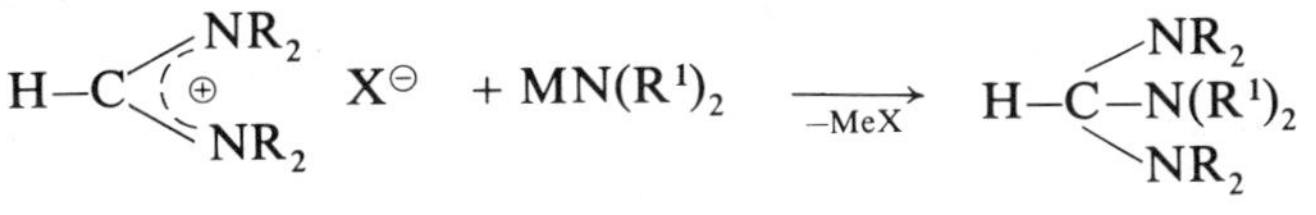

$$H-C{\overset{NR_2}{\underset{NR_2}{\oplus}}} \ X^{\ominus} \ + MN(R^1)_2 \xrightarrow{-MeX} H-C{\overset{NR_2}{\underset{NR_2}{-N(R^1)_2}}}$$

(234) (233)

X = Cl, ClO$_4$, CH$_3$SO$_4$
M = Li, Na

Table X summarizes the properties of the triaminomethanes (233) thus generated.

TABLE X

Tris(dialkylamino)methanes from N,N,N',N'-Tetraalkylformamidinium Salts and Alkali Dialkylamides

H-C(NR$_2$)(NR$_2$)(NR$_2^1$)	from	H-C(NR$_2$)(NR$_2$)$^{\oplus}$ X$^{\ominus}$	and	MeN(R^1)$_2$			
R	R^1	X	Me	Yield, %	b.p./pressure, °C/torr	n_D^{20}	Ref.
CH$_3$	CH$_3$	Cl	Li	63–67	42–43/12	1.4349/23	90
CH$_3$	CH$_3$	ClO$_4$	Li	63	145–146/760	—	90
CH$_3$	CH$_3$	Cl	Na	75	40–43/12	—	90
CH$_3$	CH$_3$	CH$_3$SO$_4$	Li	55	39–40/12	—	90
CH$_3$	C$_2$H$_5$	Cl	Li	42	66–68/12	1.4531	90
C$_2$H$_5$	C$_2$H$_5$	Cl	Li	26	106–107/12	1.4545	90
C$_2$H$_5$	CH$_3$	Cl	Li	37	86/12	1.4490	90

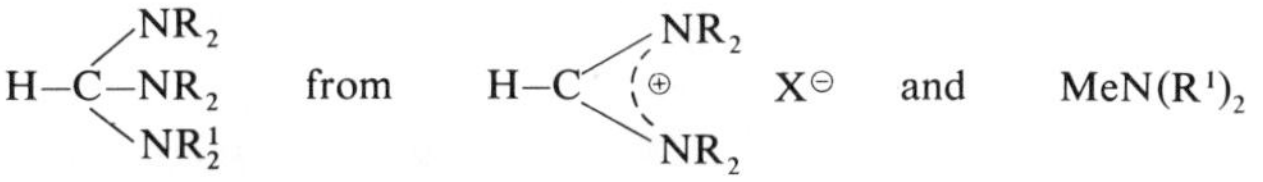

G. REACTIONS OF TETRAALKYLFORMAMIDINIUM SALTS WITH ALKALI PHOSPHIDES AND PHOSPHONATES

N,N,N',N'-Tetramethylformamidinium salts 235 and 236 react with lithium dialkylphosphides (237) to yield bis(dimethylamino)dialkylphosphinomethanes (238), which may undergo further reaction with lithium dialkylphosphide to generate bis(dialkylphosphino)dimethylaminomethane (239) (189).

Tetramethylformamidinium methylsulfate (236) gives somewhat better yields than the amidinium chloride 235 here. The compounds 238 are converted into the salts 240 by hydrogen chloride (189).

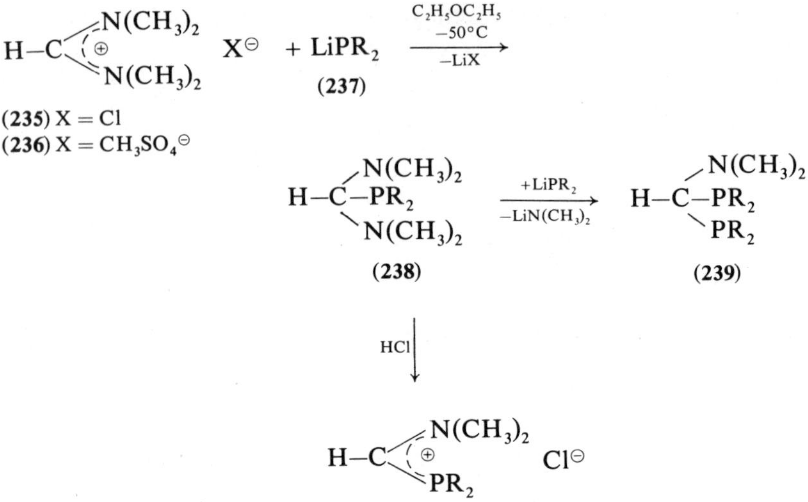

$$\underset{(235)\ X=Cl}{\underset{(236)\ X=CH_3SO_4^{\ominus}}{H-C{\overset{\oplus}{\underset{N(CH_3)_2}{\overset{N(CH_3)_2}{\diagup}}}}}}\ X^{\ominus}\ +\ \underset{(237)}{LiPR_2}\ \xrightarrow[-LiX]{C_2H_5OC_2H_5\ -50°C}$$

$$\underset{(238)}{H-C{\overset{N(CH_3)_2}{\underset{N(CH_3)_2}{\diagup}}}-PR_2}\ \xrightarrow[-LiN(CH_3)_2]{+LiPR_2}\ \underset{(239)}{H-C{\overset{N(CH_3)_2}{\underset{PR_2}{\diagup}}}-PR_2}$$

$$\downarrow HCl$$

$$\underset{(240)}{H-C{\overset{\oplus}{\underset{PR_2}{\overset{N(CH_3)_2}{\diagup}}}}}\ Cl^{\ominus}$$

Table XI summarizes the properties of the bis(dimethylamino)dialkylphosphinomethanes.

TABLE XI

Bis(dimethylamino)dialkylphosphinomethanes from N,N,N',N'-Tetramethylformamidinium Salts and Lithium Dialkylphosphides (189)

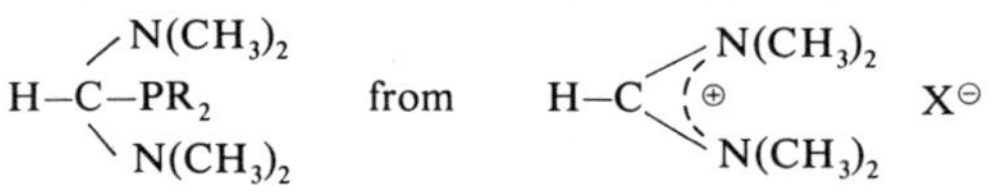

$H-C{\overset{N(CH_3)_2}{\underset{N(CH_3)_2}{\diagup}}}-PR_2$	from	$H-C{\overset{\oplus}{\underset{N(CH_3)_2}{\overset{N(CH_3)_2}{\diagup}}}}\ X^{\ominus}$		
R		X	Yield, %	b.p./pressure, °C/torr
CH_3		Cl	30	56–57/12
CH_3		CH_3SO_4	30	
C_2H_5		Cl	32	74–75/12
C_2H_5		CH_3SO_4	38	
n-C_4H_9		Cl	36	81–82/12
n-C_4H_9		CH_3SO_4	47	

When sodium diethylphosphite reacts with N,N,N',N'-tetramethylformamidinium methylsulfate (236), bis(dimethylamino)methanephosphonic acid diethyl ester (241) may be isolated (190).

$$\underset{(236)}{H-C\overset{\oplus}{\underset{N(CH_3)_2}{\diagdown}}^{N(CH_3)_2}} CH_3SO_4^\ominus + NaP\overset{O}{\underset{OC_2H_5}{\diagdown}}^{OC_2H_5} \xrightarrow[-NaCH_3SO_4]{} \underset{(241)}{H-C\overset{N(CH_3)_2}{\underset{N(CH_3)_2}{\diagdown}}-\overset{O}{P}(OC_2H_5)_2}$$

H. ANION EXCHANGE

If amidinium salts are not N,N,N',N'-persubstituted, anion exchange is easily achieved by liberation of the amidine and reaction with the corresponding acid (5):

$$R-C\overset{\oplus}{\underset{NR^1R^2}{\diagdown}}^{NHR^3} X^\ominus \xrightarrow{OH^\ominus} R-C\overset{NR^3}{\underset{NR^1R^2}{\diagdown}} \xrightarrow{HY} R-C\overset{\oplus}{\underset{NR^1R^2}{\diagdown}}^{NHR^3} Y^\ominus$$

Very often an anion exchange is accomplished, for example, if the generated amidinium salts show a low solubility. Using this method acetates, nitrates, sulfonates, thiosulfates, and chromates were synthesized, among others (5, 146).

An exchange by anion exchange is also often successful with N,N,N',N'-substituted amidinium salts taking advantage of the low solubility of the particular salts. Thus iodides (36, 42), perchlorates (42, 114, 115), tetrafluoroborates (92), hexafluorophosphates (50, 116), or tetraphenylborates (138) are obtained:

$$R-C\overset{\oplus}{\underset{N(R^1)_2}{\diagdown}}^{N(R^1)_2} X^\ominus \xrightarrow[-X^\ominus]{+Y^\ominus} R-C\overset{\oplus}{\underset{N(R^1)_2}{\diagdown}}^{N(R^1)_2} Y^\ominus$$

An alternative route starts from ortho-derivatives or ketene N,N-acetals (see Sections I–I and I–L), for example, reacting with alkoxides, and treating the ortho derivative formed with waterfree acids (90):

$$H-C\overset{\oplus}{\underset{NR_2}{\diagdown}}^{NR_2} X^\ominus \xrightarrow[-NaX]{+NaOR} H-C\overset{NR_2}{\underset{NR_2}{\diagdown}}-OR \xrightarrow[-ROH]{HY} H-C\overset{\oplus}{\underset{NR^2}{\diagdown}}^{NR^2} Y^\ominus$$

If salts with strongly nucleophilic anions are used for these double decompositions, weakly dissociated amidinium salts may also be generated. Thus from N,N,N',N'-tetramethylformamidinium methylsulfate (242) the 1,1-bis(dimethylamino)acetonitrile is synthesized, which is insoluble in water but

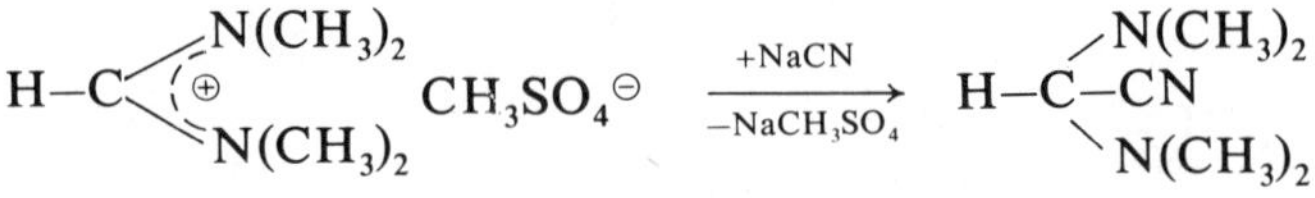

$$\text{H–C}\overset{\oplus}{\underset{N(CH_3)_2}{\overset{N(CH_3)_2}{\diagup}}} CH_3SO_4^{\ominus} \xrightarrow[-NaCH_3SO_4]{+NaCN} \text{H–C–CN}\underset{N(CH_3)_2}{\overset{N(CH_3)_2}{\diagup}}$$

(242)

soluble in benzene and cyclohexane (98). This class of bis(dialkylamino)acetonitriles is also accessible by addition of secondary amines to hydrogen cyanide (99, 100):

$$2\text{H–C≡N} + 2\text{HNR}^1\text{R}^2 \longrightarrow \text{H–C–CN}\underset{NR^1R^2}{\overset{NR^1R^2}{\diagup}} + \text{NH}_3$$

$$R^1–R^2 = –(CH_2)_4–, –(CH_2)_5–, –(CH_2)_2–O–(CH_2)_2–$$

Perchloric acid generates formamidinium perchlorates like **244** from nitriles like **243** (99).

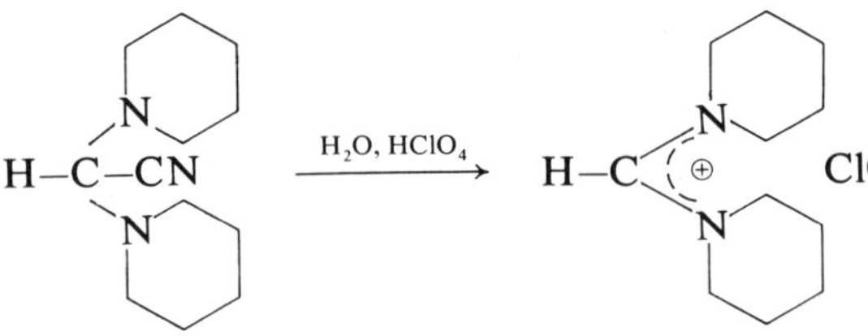

H–C–CN (piperidino)₂ →[H₂O, HClO₄] H–C⊕(piperidino)₂ ClO₄⁻

(243)   **(244)**

I. MISCELLANEOUS REACTIONS

If N,N'-di-n-butylformamidinium picrate (**245**) is thermally decomposed, or fused with sodium hydroxide, n-butylisonitrile (**246**) forms (35).

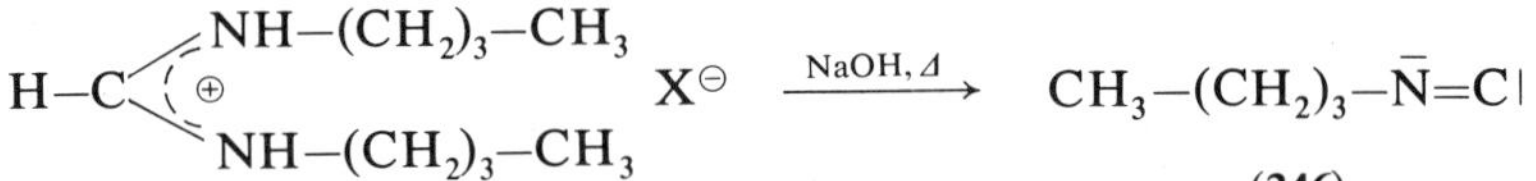

$$\text{H–C}\overset{\oplus}{\underset{NH–(CH_2)_3–CH_3}{\overset{NH–(CH_2)_3–CH_3}{\diagup}}} X^{\ominus} \xrightarrow{NaOH, \Delta} CH_3–(CH_2)_3–\bar{N}=C|$$

(245) → **(246)**

The formation of phenylisocyanate is observed when N,N-dimethyl-N'-phenyl-N'-(phenylcarbamoyl)formamidinium chloride (**247**) is heated (31).

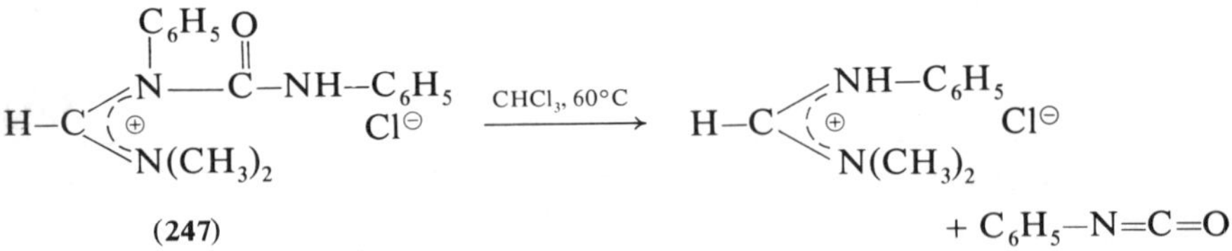

$$\text{H–C}\overset{\oplus}{\underset{N(CH_3)_2}{\overset{N(C_6H_5)-C(O)-NH-C_6H_5}{\diagup}}} Cl^{\ominus} \xrightarrow{CHCl_3, 60°C} \text{H–C}\overset{\oplus}{\underset{N(CH_3)_2}{\overset{NH–C_6H_5}{\diagup}}} Cl^{\ominus} + C_6H_5–N=C=O$$

(247)

The nitration of N,N'-dimethyl N,N'-diphenylformamidinium fluoroborate (**248**) occurs in both phenyl rings in the *para* position (92). Nitrations and

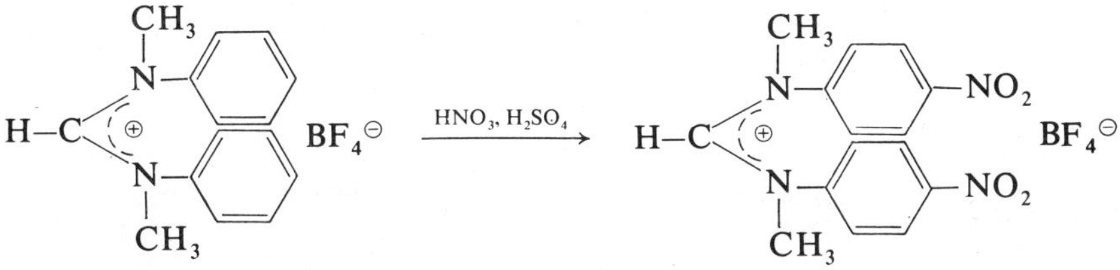

oxidations have also been performed with other amidinium salts carrying aromatic ligands at the nitrogen atom, such as **249** (25).

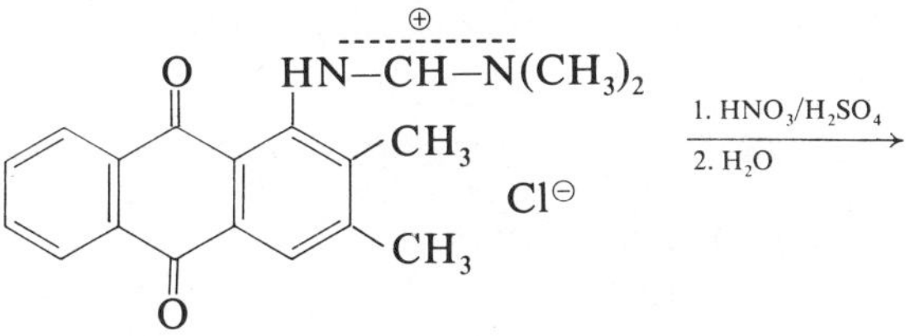

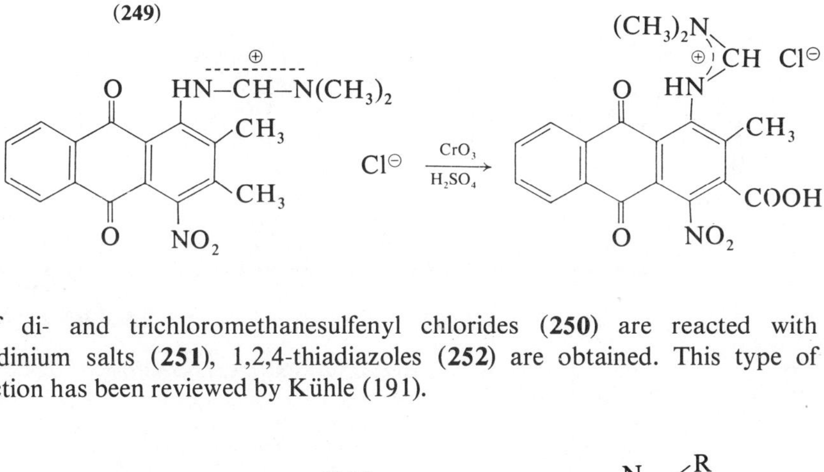

If di- and trichloromethanesulfenyl chlorides (**250**) are reacted with amidinium salts (**251**), 1,2,4-thiadiazoles (**252**) are obtained. This type of reaction has been reviewed by Kühle (191).

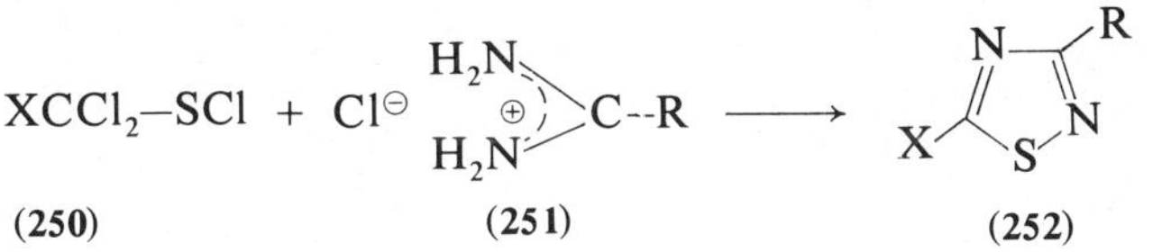

$X = H, F, Cl$

ACKNOWLEDGMENT

To Dr. Heinz Eilingsfeld BASF AG, Ludwigshafen, and to Dr. Bernd Funke as well as Dr. Erwin Haug, both of Fachhochschule Aalen, Aalen, I am very indebted for fruitful discussions and critical revisions of the text.

REFERENCES

1. A. Pinner, *Die Imidoäther und ihre Derivate*, Verlag Oppenheim, Berlin, 1892.
2. F. Kunckel and O. Sarfert, *Ber. Deut. Chem. Ges.*, **35**, 3169 (1902).
3. R. Walther and R. Grossmann, *J. Prakt. Chem.*, **78** (2), 478 (1908).
4. W. Lossen, F. Mierau, M. Kobbert, and G. Grabowski, *Ann. Chem.*, **265**, 129 (1891).
5. R. L. Shriner and F. W. Neumann, *Chem. Rev.*, **35**, 351 (1944).
6. H. Henecka and P. Kurtz, in Houben-Weyl *Methoden der organischen Chemie*, Vol. 8, Georg Thieme Verlag, Stuttgart, 1952, p. 702.
7. S. R. Sandler and W. Karo, *Organic Functional Preparations*, Vol. III, Academic Press, New York, 1972.
8. H. Ulrich, *The Chemistry of Imidoyl Halides*, Plenum Press, New York, 1968.
9. R. Bonnett, "Imidoyl Halides," *The Chemistry of the Carbon–Nitrogen Double Bond*, S. Patai, Ed., John Wiley & Sons, New York, 1970.
10. G. Gerhardt, *Ann. Chem.*, **108**, 219 (1858).
11. J. v. Braun and K. Weissbach, *Ber. Deut. Chem. Ges.*, **65B**, 1574 (1932).
12. E. Bamberger and J. Lorenzen, *Ann. Chem.*, **273**, 269 (1893).
13. H. v. Pechmann, *Ber. Deut. Chem. Ges.*, **28**, 869–879 (1895).
14. H. v. Pechmann, *Ber. Deut. Chem. Ges.*, **28**, 2362 (1895).
15. H. v. Pechmann, *Ber. Deut. Chem. Ges.*, **30**, 1779 (1897).
16. O. Wallach, *Ann. Chem.*, **184**, 1 (1877).
17. O. Wallach, *Ann. Chem.*, **214**, 202 (1882).
18. A. J. Hill and M. V. Cox, *J. Am. Chem. Soc.*, **48**, 3214 (1926).
19. E. Beckmann and E. Fellrath, *Ann. Chem.*, **273**, 1 (1893).
20. J. v. Braun, *Ber. Deut. Chem. Ges.*, **37**, 2678 (1904).
21. J. v. Braun, F. Jostes, and A. Heymons, *Ber. Deut. Chem. Ges.*, **60B**, 92–102 (1927).
22. H. Böhme and F. Soldan, *Chem. Ber.*, **94**, 3110 (1961).
23. H. Eilingsfeld, M. Seefelder, and H. Weidinger, *Angew. Chem.*, **72**, 836 (1960).
24. K. Seckinger, *Helv. Chim. Acta*, **56**, 776 (1973).
25. J. Arient and J. Podstata, *Collection Czech. Chem. Commun.*, **39**, 3117 (1974).
26. Z. Csuros, R. Soós, J. Palinkas, and I. Bitter, *Acta Chim. Acad. Sci. Hung.*, **63**, 215 (1970).
27. Z. Csuros, R. Soós, J. Palinkas, and I. Bitter, *Acta Chim. Acad. Sci. Hung.*, **68**, 397 (1971).
28. Z. Csuros, R. Soós, I. Bitter, and J. Palinkas, *Acta Chim. Acad. Sci. Hung.*, **72**, 59 (1972); *C.A.*, **76**, 153706d (1972).
29. Z. Arnold and J. Zemlicka, *Chem. Listy*, **52**, 459 (1958).
30. K. H. Beyer, H. Eilingsfeld, and H. Weidinger, (to BASF AG), Ger. Pat. 1,110,625; *C.A.*, **56**, 3363b (1962).
31. Z. Csuros, R. Soós, I. Bitter, and E. A. Karpati, *Acta Chim. Sci. Hung.*, **73**, 239 (1972); *C.A.*, **77**, 101082y (1972).
32. W. Jentsch and M. Seefelder, *Chem. Ber.*, **98**, 274 (1965).
33. B. A. Phillips, G. Fodor, J. Gal, F. Letourneau, and J. J. Ryan, *Tetrahedron*, **29**, 3309 (1973).
34. E. E. Bures and M. Kundera, *Cas. Cesk. Lek.*, **144**, 272 (1934); *C.A.*, **29**, 4750 (1935).

35. T. L. Davis and W. E. Yelland, *J. Am. chem. Soc.*, **59**, 1998 (1937).
36. C. G. Raison, *J. Chem. Soc.*, **1949**, 3319.
37. H. Bredereck, R. Gompper, and H. Rempfer, *Chem. Ber.*, **92**, 837 (1959).
38. H. Bredereck and K. Bredereck, *Chem. Ber.*, **94**, 2278 (1961).
39. H. Bredereck, F. Effenberger, and H. Botsch, *Chem. Ber.*, **97**, 3397 (1964).
40. H. Bredereck, F. Effenberger, H. Botsch, and H. Rehn, *Chem. Ber.*, **98**, 1081 (1965).
41. N. S. Drozdov and A. F. Bekhli, *J. Gen. Chem. USSR*, **14**, 472 (1944); *C.A.*, **39**, 4589[7].
42. C. Jutz and H. Amschler, *Chem. Ber.*, **96**, 2100 (1963).
43. K. Dierbach, East Ger. Pat. **26**, 918 (1964); *C.A.*, **61**, 13236h.
44. W. Klocker and M. Herbertz, *Monatsh. Chem.*, **96**, 1567 (1965).
45. E. C. Taylor and R. W. Morrison Jr., *Angew. Chem. Intern. Ed. Engl.*, **4**, 868 (1965).
46. H. Bredereck, R. Gompper, and K. Klemm, *Chem. Ber.*, **92**, 1456 (1959).
47. H. Eilingsfeld, M. Seefelder, and H. Weidinger, *Chem. Ber.*, **96**, 2899 (1963).
48. A. J. Hill and I. Rabinowitz, *J. Am. Chem. Soc.*, **48**, 732 (1926).
50. J. D. Wilson, C. F. Hobbs, and H. Weingarten, *J. Org. Chem.*, **35**, 1542 (1970).
51. R. Roger and D. G. Neilson, *Chem. Rev.*, **61**, 179 (1961).
52. O. Wallach, *Ann. Chem.*, **214**, 202 (1882).
53. Ger. Pat. 372,842 (1924); *C.A.*, **18**, 2176 (1924).
54. A. A. R. Sayigh and H. Ulrich, *J. Chem. Soc.*, **1963**, 3146.
55. W. H. Warren and F. E. Wilson, *Ber. Deut. Chem. Ges.*, **68**, 957 (1935).
56. E. Kühle, *Angew. Chem.*, **74**, 861 (1962).
57. W. Kantlehner and P. Speh, *Chem. Ber.*, **104**, 3714 (1971).
58. W. Jentsch, *Chem. Ber.*, **97**, 1362 (1964).
59. W. Jentsch, *Chem. Ber.*, **97**, 2756 (1964).
60. H. E. Ulery, *J. Org. Chem.*, **30**, 2464 (1965).
61. H. Schindelbauer, *Monatsh. Chem.*, **100**, 1590 (1969).
62. A. E. Kulikova and F. A. Ekstrin, *Z. Org. Chim.*, **4**, 1945 (1968); *C.A.*, **69**, 86541 (1969).
63. Z. Arnold, *Collection Czech. Chem. Commun.*, **24**, 760 (1959).
64. H. Gold and O. Bayer, *Chem. Ber.*, **94**, 2594 (1961).
65. H. Gold (to Farbenfabriken Bayer AG), Ger. Pat. 1,146,892 (1963); *C.A.*, **59**, 10070h (1963).
66. K. H. König and H. Pommer, Ger. Pat. 1,173,087 (1964); *C.A.*, **61**, 11899b.
67. E. Kühle, B. Anders, E. Klauke, H. Tarnow, and G. Zumach, *Angew. Chem.*, **81**, 18 (1969).
68. P. Beak and J. A. Borron, *J. Org. Chem.*, **38**, 2771 (1973).
69. H. Bredereck, F. Effenberger, and E. Henseleit, *Chem. Ber.*, **98**, 2754 (1965).
70. P. Oxley and W. F. Short, *J. Chem. Soc.*, **1946**, 147.
71. P. Oxley and W. F. Short, *J. Chem. Soc.*, **1950**, 859 and references cited therein.
72. F. C. Schäfer and A. P. Krapcho, *J. Org. Chem.*, **27**, 1255 (1962).
73. J. Goerdeler and H. Porrmann, *Chem. Ber.*, **94**, 2856 (1961).
74. W. Walter and J. Krohn, *Ann. Chem.*, **1973**, 476.
75. J. E. Gordon and G. C. Turrell, *J. Org. Chem.*, **24**, 269 (1959).
76. R. Fuks, *Tetrahedron*, **29**, 2147 (1973).
77. R. Fuks, *Eur. Polymer J.*, **9**, 835 (1973).
78. R. Fuks, *Tetrahedron*, **29**, 2153 (1973).
79. R. Fuks, *Bull. Soc. Chim. Belg.*, **82**, 571 (1973).
80. Y. V. Mitin, V. R. Glushenkova, and G. P. Vlasov, *Zh. Obshch. Khim.*, **32**, 3867 (1962); *C.A.*, **58**, 12440h.
81. G. Zinner, W. Heuer, and D. Moderhack, *Chemiker Ztg.*, **98**, 159 (1974).
82. U. Schöllkopf and F. Gerhart, *Angew. Chem.*, **79**, 990 (1967).
83. J. E. Parks and A. L. Balch, *J. Organomet. Chem.*, **57**, C 103 (1973).
84. G. Zinner, D. Moderhack, and W. Heuer, *Chemiker-Ztg.*, **98**, 112 (1974).

85. S. S. Malhotra and M. C. Whiting, *J. Chem. Soc.*, **1960**, 3812.
86. D. L. Harris and K. M. Wellman, *Tetrahedron Letters*, **1968**, 5225.
87. E. C. Taylor and W. A. Ehrhart, *J. Am. Chem. Soc.*, **82**, 3138 (1960).
88. E. C. Taylor and W. A. Ehrhart, *J. Org. Chem.*, **28**, 1108 (1963).
89. R. H. De Wolfe, *Carboxylic Ortho Acid Derivatives*, Academic Press, New York, 1970.
90. H. Bredereck, F. Effenberger, T. Brendle, and H. Muffler, *Chem. Ber.*, **101**, 1885 (1968).
91. C. F. Hobbs and H. Weingarten, *J. Org. Chem.*, **36**, 2885 (1971).
92. D. H. Clemens, E. Y. Shropshire, and W. D. Emmons, *J. Org. Chem.*, **27**, 3664 (1962).
93. D. H. Clemens and W. D. Emmons, *J. Am. Chem. Soc.*, **83**, 2588 (1961).
94. D. H. Clemens, U.S. Pat. 3,143,571; *C.A.*, **61**, 9435 (1964).
95. A. V. Stavrovskaya, V. T. Protopovova, and A. P. Skoldinov, *Zh. Org. Khim.*, **9**, 699 (1973).
96. H. Bredereck, G. Simchen, and P. Horn, *Angew. Chem.*, **77**, 508 (1965).
97. H. Bredereck, G. Simchen, and P. Horn, *Chem. Ber.*, **103**, 210 (1970).
98. H. Bredereck, G. Simchen, and W. Kantlehner, *Chem. Ber.*, **104**, 924 (1971).
99. M. Seefelder, *Chem. Ber.*, **92**, 2678 (1966).
100. J. G. Erickson, *J. Org. Chem.*, **20**, 1569 (1955).
101. M. Kerfanto, *Bull. Soc. Chim. France*, **1965**, 3544.
102. H. Bredereck, G. Simchen, and W. Kantlehner, *Chem. Ber.*, **104**, 932 (1971).
103. W. Jentzsch and M. Seefelder, *Chem. Ber.*, **98**, 1342 (1965).
104. G. Scherowsky, *Ann. Chem.*, **739**, 45 (1970).
105. A. Pinner and F. Klein, *Ber. Deut. Chem. Ges.*, **11**, 4 (1878).
106. F. L. Pyman, *J. Chem. Soc.*, **1923**, 367.
107. F. L. Pyman, *J. Chem. Soc.*, **1923**, 3359.
108. H. v. Pechmann and B. Heinze, *Ber. Deut. Chem. Ges.*, **30**, 1783 (1879).
109. R. Walter and R. Grossmann, *J. Prakt. Chem.*, **78** (2), 478 (1908).
110. C. Chew and L. Pyman, *J. Chem. Soc.*, **1927**, 2318.
111. J. S. McKennis and P. A. S. Smith, *J. Org. Chem.*, **37**, 4173 (1972).
112. R. F. Smith, D. S. Johnson, C. L. Hyde, T. C. Rosenthal, and A. C. Bates, *J. Org. Chem.*, **36**, 1155 (1971).
113. R. F. Smith, D. S. Johnson, R. A. Abgott, and M. J. Madden, *J. Org. Chem.*, **38**, 344 (1973).
114. D. H. Clemens, A. J. Bell, and J. L. O'Brien, *J. Org. Chem.*, **29**, 2932 (1964).
115. H. Böhme and W. Höver, *Ann. Chem.*, **748**, 59 (1971).
116. C. F. Hobbs and H. Weingarten, *J. Org. Chem.*, **33**, 2385 (1968).
117. C. F. Hobbs and H. Weingarten, *J. Org. Chem.*, **39**, 918 (1974).
118. C. F. Hobbs and H. Weingarten, *J. Am. Chem. Soc.*, **91**, 780 (1969).
119. R. Fuks, G. S. D. King, and H. G. Viehe, *Angew. Chem.*, **81**, 702 (1969).
120. C. Jutz and E. Müller, *Angew. Chem.*, **78**, 747 (1966).
121. W. E. Truce, D. J. Abraham, and P. Son, *J. Org. Chem.*, **32**, 990 (1967).
122. D. H. Clemens, A. J. Bell, and J. L. O'Brian, *Tetrahedron Letters*, **1965**, 3257.
123. H. E. Winberg and D. D. Coffman, *J. Am. Chem. Soc.*, **87**, 2776 (1965).
124. N. Wiberg and J. W. Buchler, *Z. Naturforsch.*, **19b**, 953 (1964).
125. N. Wiberg and J. W. Buchler, *Angew. Chem.*, **74**, 490 (1962).
126. N. Wiberg and J. W. Buchler, *Chem. Ber.*, **96**, 3223 (1963).
127. K. Kuwata and D. H. Geske, *J. Am. Chem. Soc.*, **86**, 2101 (1964).
128. W. H. Urry and J. Sheeto, *Photochem. Photobiol.*, **4**, 1067 (1965); *C.A.*, **64**, 12533a.
129. W. Carpenter, *J. Org. Chem.*, **30**, 3082 (1962).
130. W. Carpenter, A. Haymaker, and D. W. Moore, *J. Org. Chem.*, **31**, 789 (1966).
131. W. E. Thun, *J. Org. Chem.*, **32**, 503 (1967).
132. R. B. King, *Inorg. Chem.*, **4**, 1518 (1965).

133. N. Wiberg, *Angew. Chem.*, **80**, 809 (1968).
134. R. W. Hoffmann, *Angew. Chem.*, **80**, 823 (1968).
135. Tsuge Otohiko, Yanagi Kiyoshi, and Horic Masako, *Bull. Chem. Soc. Japan*, **44**, 2171 (1971).
136. H. Böhme, G. Auterhoff, and W. Höver, *Chem. Ber.*, **104**, 3350 (1971).
137. H. Weingarten and J. S. Wagner, *Tetrahedron Letters*, **1969**, 3267.
138. H. Weingarten and J. S. Wagner, *J. Org. Chem.*, **35**, 1750 (1970).
139. U. Mueller-Westerhoff, *Tetrahedron Letters*, **1972**, 4639.
140. J. Yoshimura, K. Fujimori, and Y. Sugiyama, *Nippon Kagaku Kaishi*, **1973**, 2242.
141. W. Walter and K.-P. Ruess, *Chem. Ber.*, **102**, 2640 (1969).
142. W. Walter and K.-P. Ruess, *Ann. Chem.*, **1974**, 225.
143. W. Walter and K.-P. Ruess, *Ann. Chem.*, **1974**, 243.
144. W. Walter and K.-P. Ruess, *Ann. Chem.*, **1974**, 253.
145. W. Heffe, R. W. Balsiger, and U. Thoma, *Helv. Chim. Acta*, **57**, 1242 (1974).
146. J. Walker, *J. Chem. Soc.*, **1949**, 1996.
147. D. N. Shigorin, *Zh. Fiz. Khim.*, **25**, 798 (1951).
148. J. Fabian, M. Legrand, and P. Poirier, *Bull. Soc. Chim. France*, **1956**, 1499.
149. J. C. Grivas and A. Taurins, *Can. J. Chem.*, **37**, 1260 (1959); *C.A.*, **54**, 2162 (1960).
150. G. R. Pettit and L. R. Garson, *Can. J. Chem.*, **43**, 2640 (1965).
151. Z. B. Kiro, Y. A. Teterin, L. N. Nikolenko, and B. I. Stepanov, *Zh. Org. Khim.*, **8**, 2573 (1972).
152. G. S. Hammond and R. C. Neumann, Jr., *J. Phys. Chem.*, **67**, 1655 (1963).
153. R. C. Neumann, G. S. Hammond, and T. J. Dougherty, *J. Am. Chem. Soc.*, **84**, 1506 (1962).
154. J. Sandström, *J. Phys. Chem.*, **71**, 2318 (1967).
155. R. C. Neumann and Y. Jonas, *J. Phys. Chem.*, **67**, 3532 (1971).
156. J. P. Marsh, Jr., and L. Goodman, *Tetrahedron Letters*, **1967**, 683.
157. J. Rauft and S. Dahne, *Helv. Chim. Acta*, **47**, 1160 (1964).
158. K. M. Wellman and D. L. Harris, *Chem. Commun.*, **1967**, 256.
159. R. C. Neumann, Jr., and V. Jones, *J. Org. Chem.*, **39**, 929 (1974).
160. R C. Neumann, Jr., and L. B. Young, *J. Phys. Chem.*, **69**, 2570 (1965).
161. R. C. Neumann, Jr., and G. S. Hammond, *J. Phys. Chem.*, **67**, 1659 (1963).
162. C. L. Perrin, *J. Am. Chem. Soc.*, **96**, 5631 (1974).
163. E. Oeser, *Chem. Ber.*, **107**, 627 (1974).
164. H. U. Wagner, *Chem. Ber.*, **107**, 634 (1974).
165. J. Liebscher and H. Hartmann, *Zh. Chem.*, **14**, 358 (1974).
166. E. Allenstein, A. Schmidt, and V. Beyl, *Chem. Ber.*, **99**, 431 (1966).
167. W. Walter and K.-D. Bode, *Angew. Chem.*, **78**, 517 (1966).
168. H. Bredereck, R. Gompper, H. Rempfer, K. Klemm, and H. Keck, *Chem. Ber.*, **92**, 329 (1959).
169. Z. Csuros, R. Soós, I. Bitter, and J. Palinkas, *Acta Chim. Acad. Sci. Hung.*, **69**, 361 (1971); *C.A.*, **75**, 129751y (1971).
170. W. P. Bullard, L. J. Farina, P. R. Farina, and S. J. Benkovic, *J. Am. Chem. Soc.*, **96**, 7295 (1974).
171. H. Bredereck, F. Effenberger, and G. Simchen, *Chem. Ber.*, **98**, 1078 (1965).
172. H. Bredereck, F. Effenberger, and E. Henseleit, *Chem. Ber.*, **98**, 2887 (1965).
173. H. Bredereck, F. Effenberger, and H. P. Bayerlin, *Chem. Ber.*, **97**, 3081 (1964).
174. C. F. Hobbs and H. Weingarten, *J. Org. Chem.*, **36**, 2881 (1971).
175. Z. Arnold and M. Kornilov, *Collection Czech. Chem. Commun.*, **29**, 645 (1964).
176. H. Gold, *Angew. Chem.*, **72**, 956 (1960).
177. W. Kantlehner and P. Speh, *Chem. Ber.*, **105**, 1340 (1972).

178. H. Bredereck, F. Effenberger, and G. Simchen, *Angew. Chem.*, **74**, 353 (1962).
179. H. Bredereck, G. Simchen, H. Hoffmann, P. Horn, and R. Wahl, *Angew. Chem.*, **79**, 311 (1967).
180. H. Bredereck, G. Simchen, S. Rebsdat, W. Kantlehner, P. Horn, R. Wahl, H. Hoffmann, and P. Grieshaber, *Chem. Ber.*, **101**, 41 (1968).
181. H. Bredereck, G. Simchen, and H. U. Schenck, *Chem. Ber.*, **101**, 3058 (1968).
182. H. Bredereck, G. Simchen, and G. Beck, *Ann. Chem.*, **762**, 62 (1972).
183. H. Winberg, U.S. Pat. 3,239,534 (1966); *C.A.*, **64**, 17425 (1966).
184. G. Simchen, H. Hoffmann, and H. Bredereck, *Chem. Ber.*, **101**, 51 (1968).
185. H. Bredereck, F. Effenberger, and H. J. Bredereck, *Angew. Chem.*, **78**, 984 (1966).
186. H. Weingarten and W. A. White, *J. Org. Chem.*, **31**, 3427 (1966).
187. H. Bredereck, G. Simchen, and H. Hoffmann, *Chem. Ber.*, **106**, 3725 (1973).
188. H. Bredereck, F. Effenberger, and T. Brendle, *Angew. Chem.*, **78**, 147 (1966).
189. M. Lischewski, K. Issleib, and H. Tille, *J. Organomet. Chem.*, **54**, 195 (1973).
190. H. Gross and B. Costisella, *J. Prakt. Chem.*, **311**, 925 (1969).
191. E. Kühle, *Synthesis*, **1971**, 617.

ADDENDUM*

To Section I

Preparation and reaction of amidines (192, 193), vinylogous amidines, and vinylogous amidinium salts (vinamidinium salts) (194) are subjects of recently published reviews.

To Section I-D

N-Monosubstituted formamides, acetamides, and benzamides were also successfully reacted with NN-dialkylcarbamidic acid chlorides to give amidinium salts, which have been, without isolation, transformed to free amidines by treatment with aqueous bases (KOH, K_2CO_3) (195).

$$R-C\overset{O}{\underset{NHR_1}{\diagdown}} \xrightarrow[110°]{Cl-\overset{O}{\overset{\|}{C}}-NR_2^2} R-C\overset{NHR^1}{\underset{NR_2^2}{\diagdown}}{}^{\oplus} Cl^{\ominus} \xrightarrow{OH^{\ominus}} R-C\overset{NR^1}{\underset{NR_2^2}{\diagdown}}$$

$R = H, CH_3, C_6H_5; R^2 = CH_3, C_2H_5;$
$R^1 = CH_3, C_2H_5, CH(CH_3)_2, n\text{-}C_3H_7, C(CH_3)_3$

A similar reaction has been reported to occur between caprolactam and dimethylcarbamic acid chloride (196).

* Translated from German by H. J. Benirschke, and by W. Kantlehner. Fachhochschule Aalen, 708 Aalen, Hohenstaufenstrasse, Germany.

To Section I-H

In the presence of cuprous-I-chloride amines add to isonitriles to give formamidines. A review on these reactions has appeared (197). Instead of amines hydrazines can be used in this reaction (198).

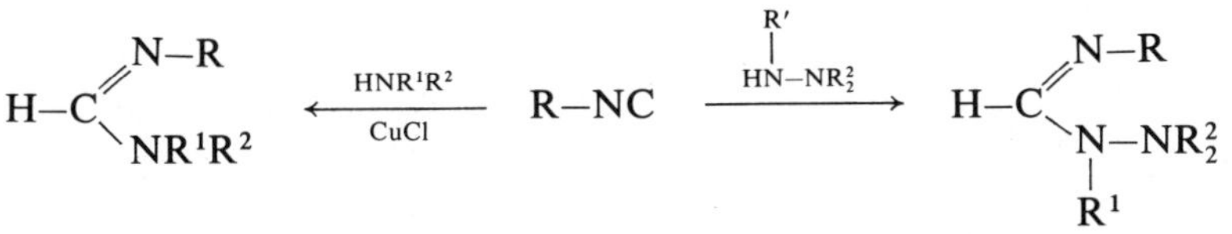

$$H-C{\overset{N-R}{\underset{NR^1R^2}{\Big\backslash}}} \xleftarrow{\underset{CuCl}{HNR^1R^2}} R-NC \xrightarrow{\underset{HN-NR_2^2}{\overset{R'}{|}}} H-C{\overset{N-R}{\underset{\underset{R^1}{\overset{|}{N-NR_2^2}}}{\Big\backslash}}}$$

To Section I-J

By stepwise alkylation $NNN'N'$-tetrasubstituted formamidinium salts with different N-alkyl substituents have been prepared, starting from NN'-disubstituted formamidinium salts (199).

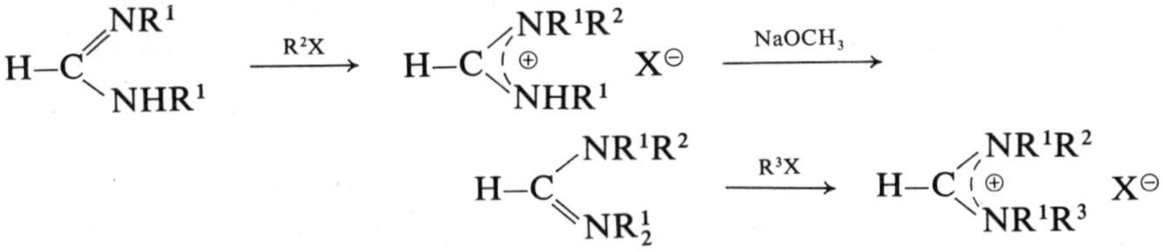

$$H-C{\overset{NR^1}{\underset{NHR^1}{\Big\backslash}}} \xrightarrow{R^2X} H-C^\oplus{\overset{NR^1R^2}{\underset{NHR^1}{\Big\backslash}}} X^\ominus \xrightarrow{NaOCH_3}$$

$$H-C{\overset{NR^1R^2}{\underset{NR_2^1}{\Big\backslash}}} \xrightarrow{R^3X} H-C^\oplus{\overset{NR^1R^2}{\underset{NR^1R^3}{\Big\backslash}}} X^\ominus$$

Amidines add to polynitro substituted aromatic compounds (benzene, naphthalene) producing adducts (**253**) or salts (**254**) respectively (200, 201).

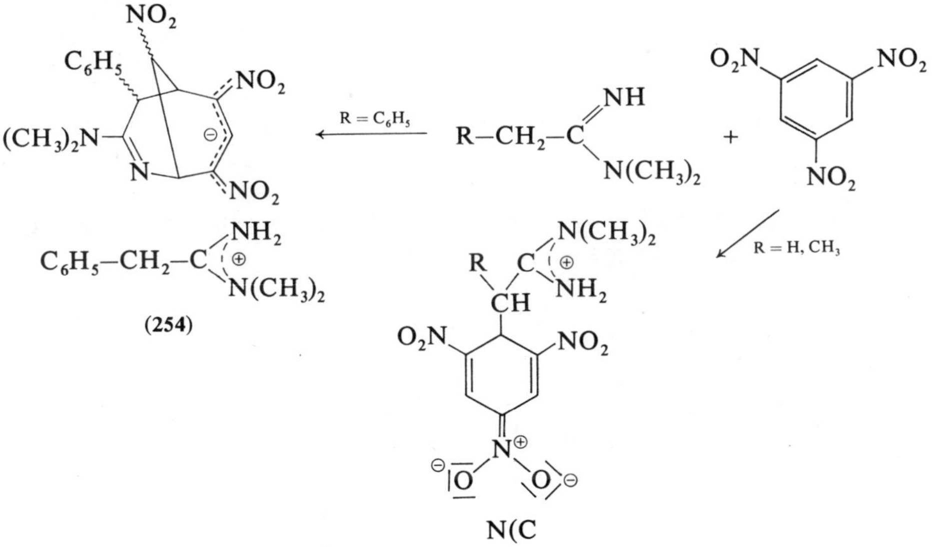

[Structural diagrams showing the reactions between amidines $R-CH_2-C(=NH)N(CH_3)_2$ and polynitro aromatics, giving adduct (253) and salt (254); $R = C_6H_5$ for (254), $R = H, CH_3$ for (253).]

(**254**) $C_6H_5-CH_2-C^\oplus{\overset{NH_2}{\underset{N(CH_3)_2}{\Big\backslash}}}$

(**253**)

To Section I-L-1

Action of arylisocyanates on the keteneaminale **255** yields dipoles **256**, which may be transformed by protonation or alkylation to, for example, amidinium salts **257** (202). Sulfonylisocyanates can also be used to prepare analogous dipoles (**258**) (202).

(258) (255) (256) (257)

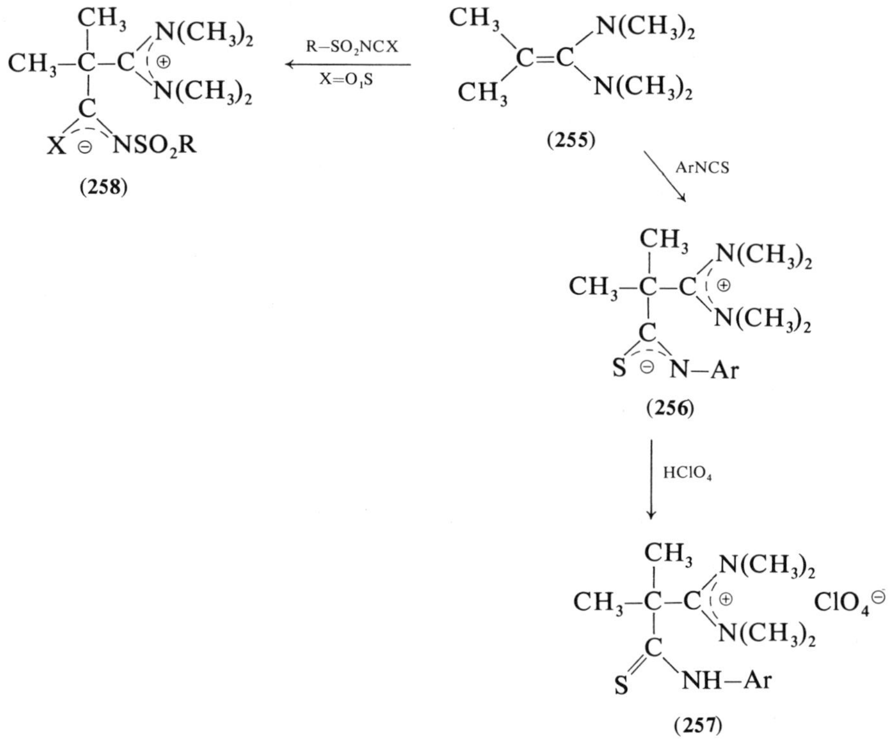

The ynamine **260** reacts with the cyclic iminium salt **259** to form the amidinium compound **261** (203).

(259) (260) (261)

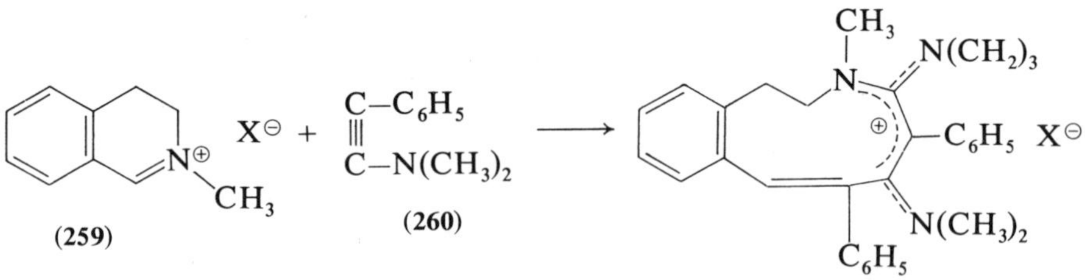

By alkylation and bromination of ynamines amidinium salts **262** and **263** were obtained among other products (204).

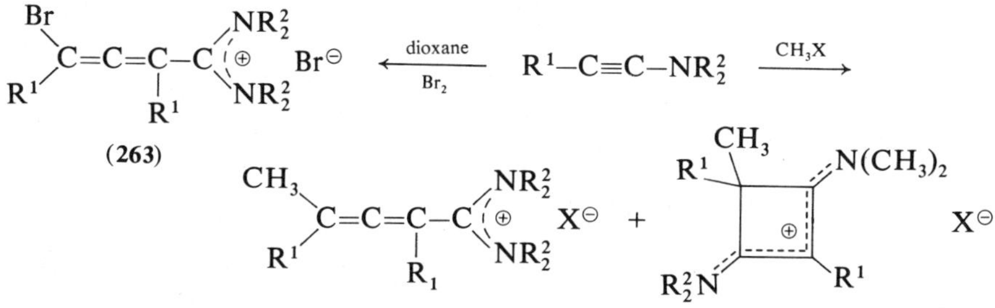

$$R^1-C\equiv C-NR_2^2 \xrightarrow{CH_3X}$$

(263), (262) formed via dioxane/Br$_2$ and CH$_3$X routes respectively, along with a cyclobutene-type bisamidinium salt.

To Section I-L-2

Halides, **264** derived from elements that are members of the fourth group of the periodic table, yield on photolysis in the presence of tetrakisdimethylaminoetylene besides radicals **265** and bisamidinium salts, such as **266** (205).

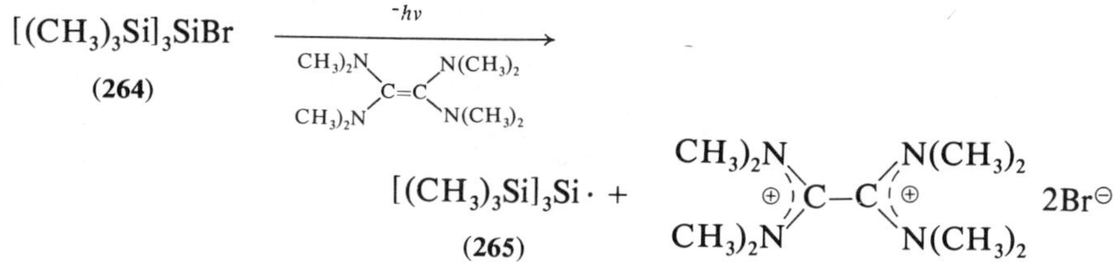

$$[(CH_3)_3Si]_3SiBr \xrightarrow{h\nu}$$

with (CH$_3$)$_2$N–C=C–N(CH$_3$)$_2$ derivatives giving [(CH$_3$)$_3$Si]$_3$Si· (**265**) and the bisamidinium salt (**266**) · 2Br$^\ominus$.

Tetrapiperidinoethylene **267** reacts with excessive malondinitrile to give the salt **268** (206).

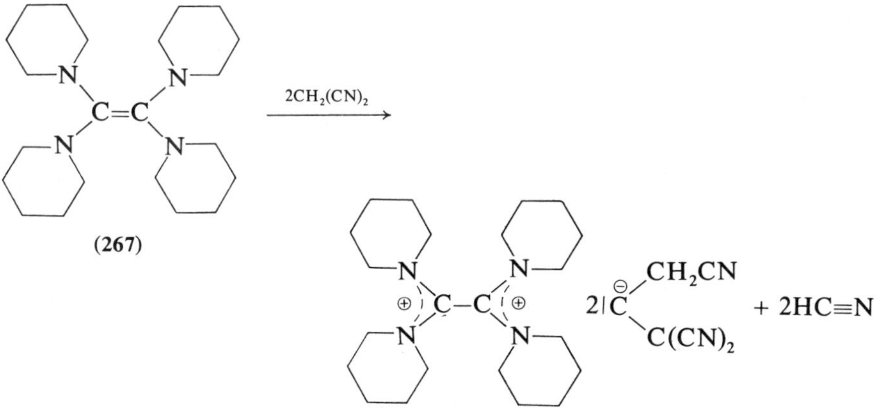

$$\xrightarrow{2CH_2(CN)_2}$$

(268) + 2 [CH(CN)–C(CN)$_2$]$^\ominus$ + 2HC≡N

To Section I-M

Via dioxides of thiourea further formamidinium salts (**207**) have been prepared.

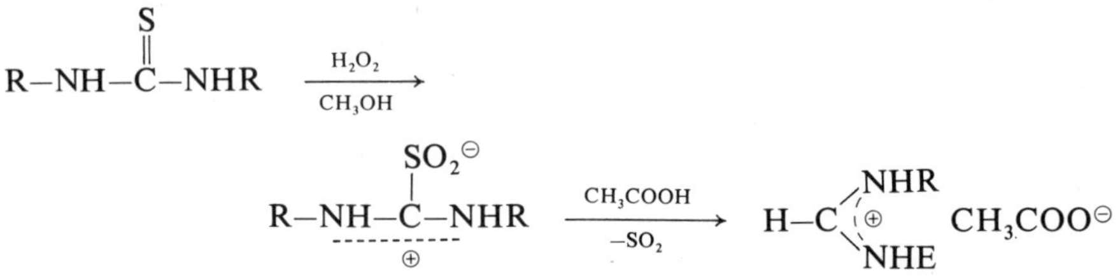

$$R-NH-\overset{S}{\overset{\|}{C}}-NHR \xrightarrow[CH_3OH]{H_2O_2} R-NH-\underset{\oplus}{\overset{SO_2^{\ominus}}{\underset{|}{C}}}-NHR \xrightarrow[-SO_2]{CH_3COOH} H-\overset{\oplus}{C}\begin{smallmatrix}NHR\\NHE\end{smallmatrix} \quad CH_3COO^{\ominus}$$

Oxidation of thiocarbazides affords trioxides **269** and **270**, and formamidrazones **271** are sometimes byproducts in this reaction (208, 209).

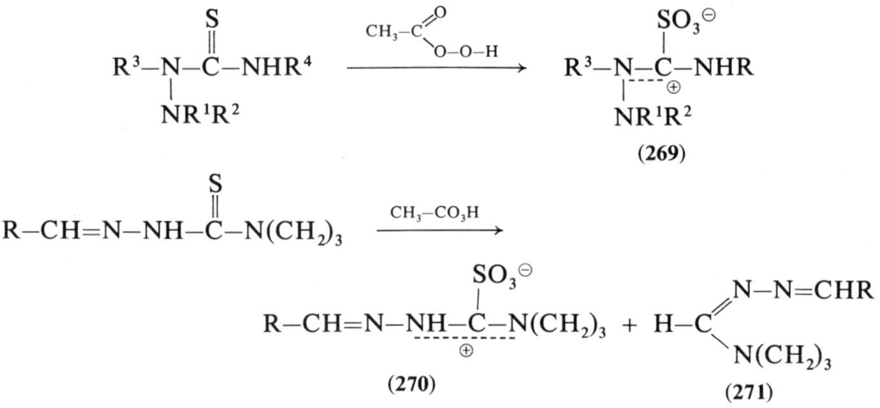

$$R^3-\underset{NR^1R^2}{\overset{S}{\underset{|}{\overset{\|}{N-C-NHR^4}}}} \xrightarrow{CH_3-C\overset{O}{\underset{O-O-H}{\diagdown}}} R^3-\underset{NR^1R^2}{\underset{|}{\overset{SO_3^{\ominus}}{\underset{|}{N-\overset{\oplus}{C}-NHR}}}}$$

(**269**)

$$R-CH=N-NH-\overset{S}{\overset{\|}{C}}-N(CH_2)_3 \xrightarrow{CH_3-CO_3H}$$

$$R-CH=N-NH-\underset{\oplus}{\overset{SO_3^{\ominus}}{\underset{|}{C}}}-N(CH_2)_3 \quad + \quad H-C\begin{smallmatrix}N-N=CHR\\N(CH_2)_3\end{smallmatrix}$$

(**270**) (**271**)

Dithiodiglycolic acid **272** and glycolic acid amidinium salt **273** may be oxidized to form the zwitterionic compound **274** (210).

$$(HOOC-CH_2-S)_2 \xrightarrow[NH_3, H_2O]{O_2}$$
(**272**)

$$\begin{smallmatrix}H_2N\\\oplus\\H_2N\end{smallmatrix}C-COO^{\ominus} \xleftarrow[H_2O]{KMnO_4} \begin{smallmatrix}H_2N\\\oplus\\H_2N\end{smallmatrix}C-CH_2OH \quad Cl^{\ominus}$$

(**274**) (**273**)

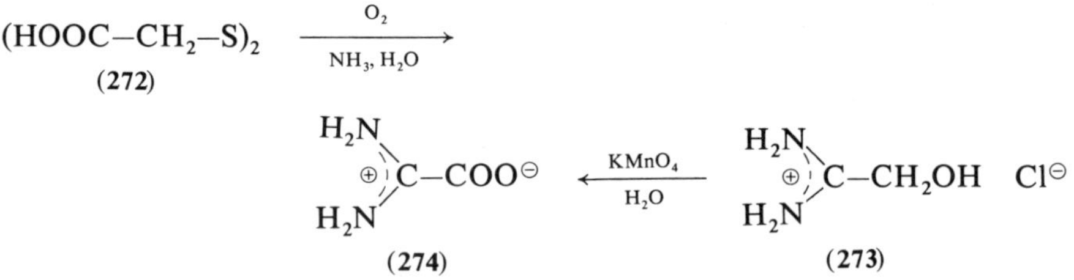

Section I-N

Amidinium salts from 1,1-dihalogenolefines and 1,1,1-trihalogenalkanes. Aminolysis of 1,1-dihalogenolefines, such as trichlorethylene **275** (211), trifluorochloroethylene (212, 213), trichlorovinylamines **276** (214), β,β-dichlorvinylketones **277** (215), *N*-(perchlorovinyl)benzimidchlorides **278** (216) as well as phenyltrichlorocyclopropen **279** (217) leads to amidines or amidinium salts, respectively.

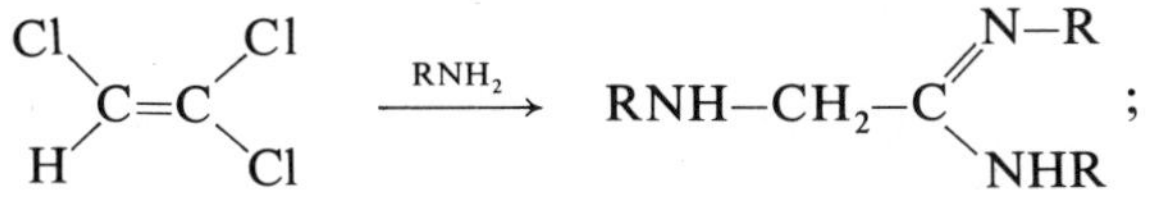

$$\underset{(275)}{\overset{Cl}{\underset{H}{>}}C=C\overset{Cl}{\underset{Cl}{<}}} \xrightarrow{RNH_2} RNH-CH_2-C\overset{N-R}{\underset{NHR}{\nwarrow}} ;$$

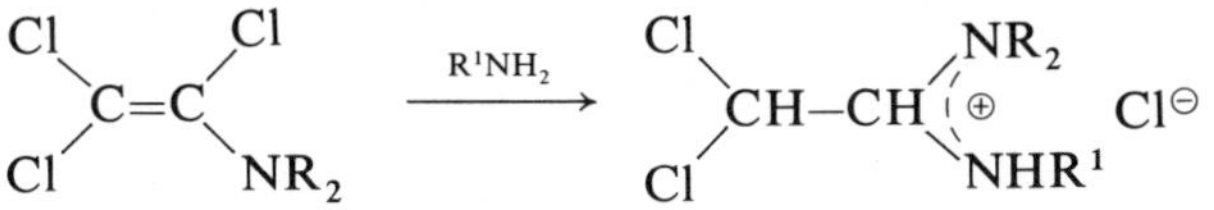

$$\underset{(276)}{\overset{Cl}{\underset{Cl}{>}}C=C\overset{Cl}{\underset{NR_2}{<}}} \xrightarrow{R^1NH_2} \underset{Cl}{\overset{Cl}{>}}CH-CH\overset{NR_2}{\underset{NHR^1}{\nwarrow}}^{\oplus} \quad Cl^{\ominus}$$

$$\underset{(277)}{R-\overset{O}{\underset{\|}{C}}-CH=C\overset{Cl}{\underset{Cl}{<}}} \xrightarrow{C_6H_5NH_2} R-\overset{\overset{\oplus}{H}NC_6H_5}{\underset{\|}{C}}-CH=C\overset{NHC_6H_5}{\underset{NHC_6H_5}{\nwarrow}}^{\oplus} \quad Cl^{\ominus}$$

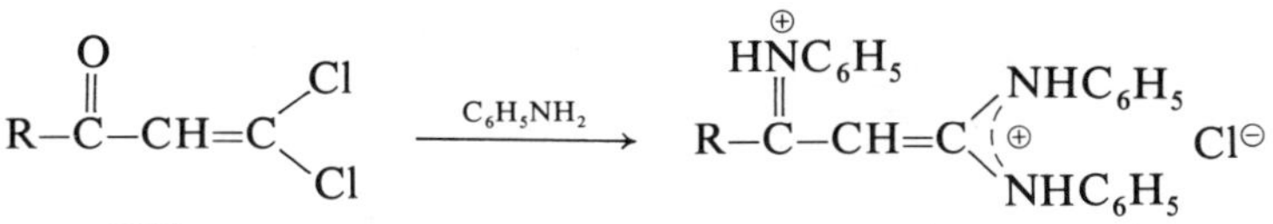

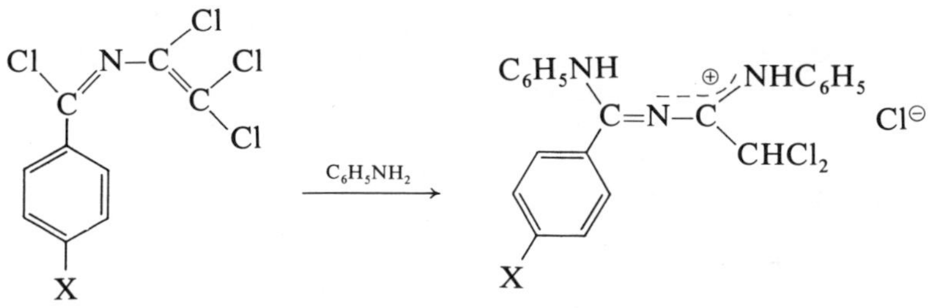

(278) + C₆H₅NH₂ → product with CHCl₂, Cl⁻

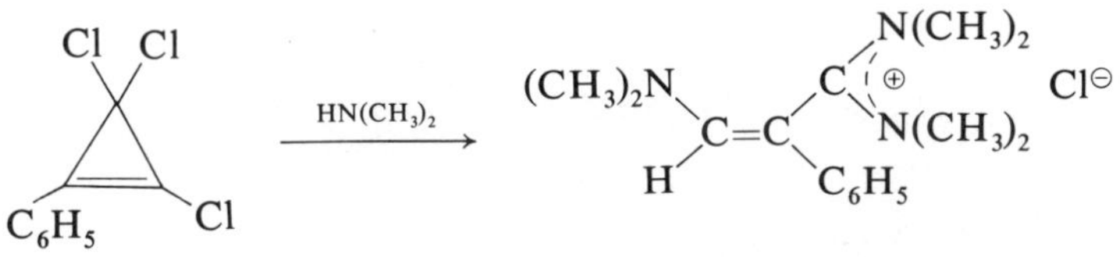

(279) + HN(CH₃)₂ → amidinium salt

From benzotrichloride and ammonia (218) or aromatic amines (213, 219–221), amidines, and amidinium salts have been prepared.

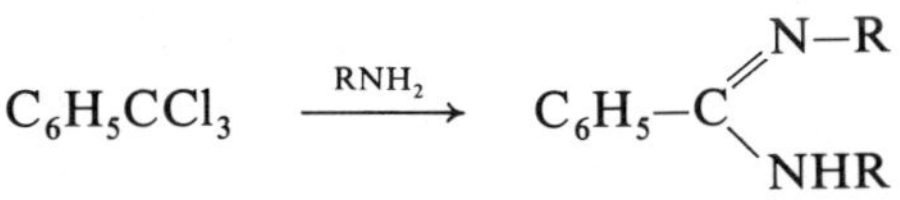

$$C_6H_5CCl_3 \xrightarrow{RNH_2} C_6H_5-C\overset{N-R}{\underset{NHR}{\nwarrow}}$$

Section I-O

Amidinium salts from chloromethyleniminium salts and nitriles. Treatment of nitriles with chloromethylene iminium salts in the presence of HCl leads to amidinium salts **280**, **281**, and **282**, which afford further amidinium compounds **283** and **284** on hydrolysis or aminolysis, respectively (222–224).

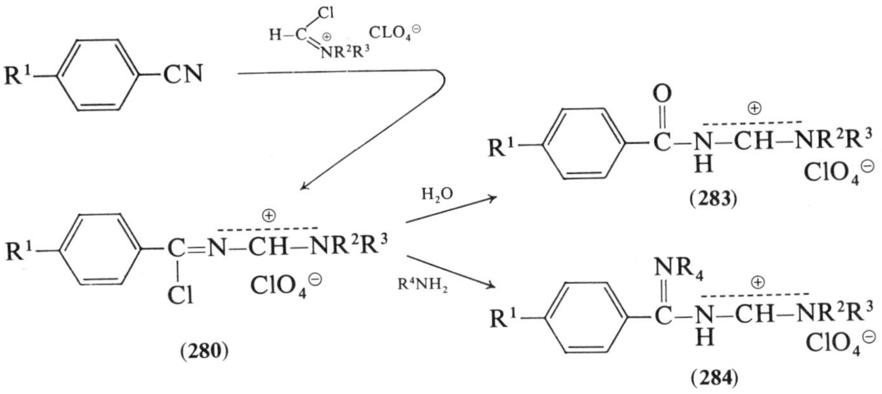

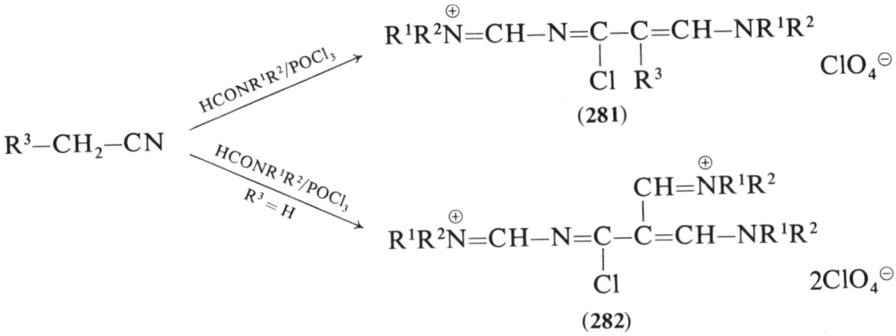

Similar reactions were performed with akylrhodanides (225).

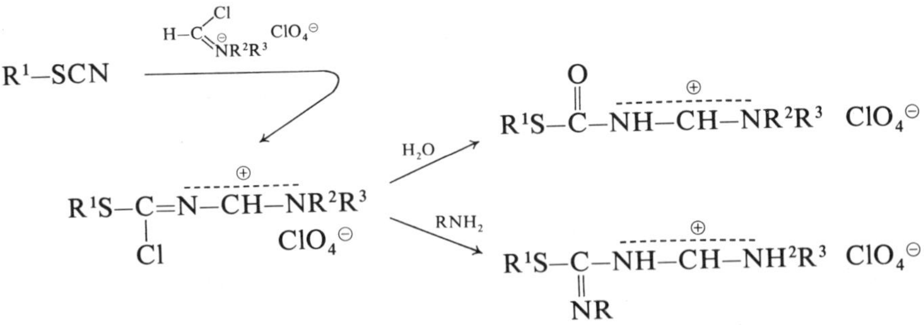

Section I-P

Miscellaneous methods. Reaction between aminhydrochlorides and thionoesters gives rise to formation of amidinhydrochlorides (226).

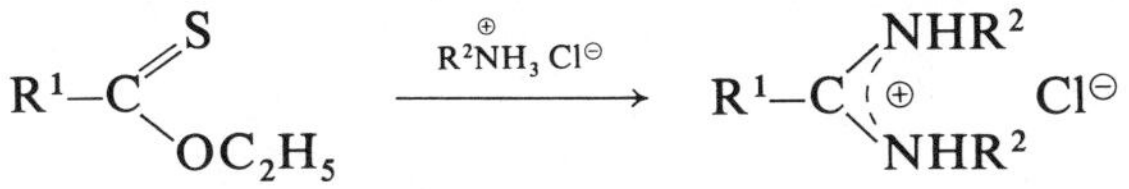

$$R^1-C\overset{S}{\underset{OC_2H_5}{\diagdown}} \xrightarrow{R^2NH_3\ Cl^\ominus} R^1-C\overset{NHR^2}{\underset{NHR^2}{\diagdown}}\oplus\quad Cl^\ominus$$

To Section II

An extensive discussion on physical properties and structural problems of amidinium compounds may be found in ref. 192.

To Section III-A

Further hydrolytic reactions of such amidinium-compounds as **285** leading to amides have been reported (222). Sodium sulfide and sodium thiosulfate react with **285** to form *N*-thioacylformamidines **286** (223, 227), whereas β-ketothiols **287** are cyclized to give thiazoles **288** (223).

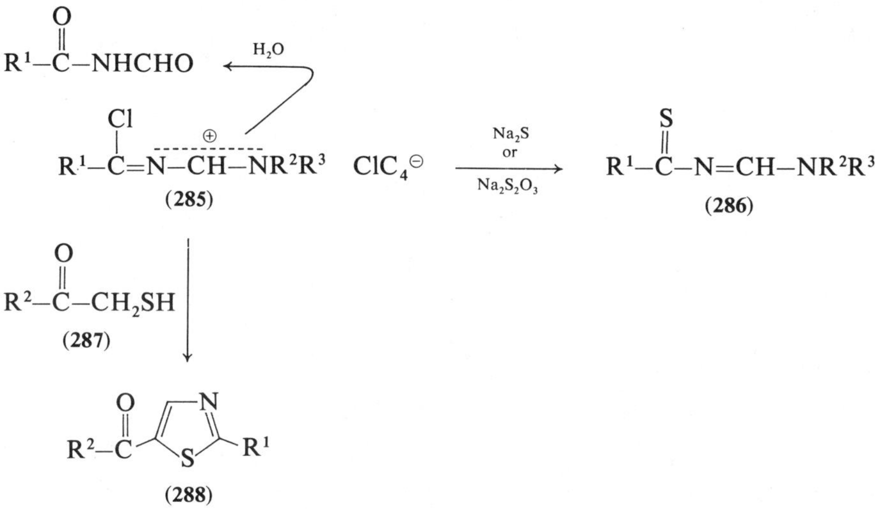

$$R^1-\overset{O}{\underset{\|}{C}}-NHCHO \xleftarrow{H_2O}$$

$$R^1-\overset{Cl}{\underset{|}{C}}=N-CH-NR^2R^3\ \oplus\quad ClC_4^\ominus \xrightarrow[Na_2S_2O_3]{Na_2S\ or} R^1-\overset{S}{\underset{\|}{C}}-N=CH-NR^2R^3$$

(285)   (286)

$$R^2-\overset{O}{\underset{\|}{C}}-CH_2SH$$

(287)

(288): thiazole with $R^2-\overset{O}{\underset{\|}{C}}-$ and R^1 substituents

To Section III-B-1

By treatment of amidinium salts **289**, **290**, **291**, and **292**, with thioamides, vinylogous thioamides, and *N*-thioacylformamidines, various heterocycles have been synthesized (222, 227, 228, 229).

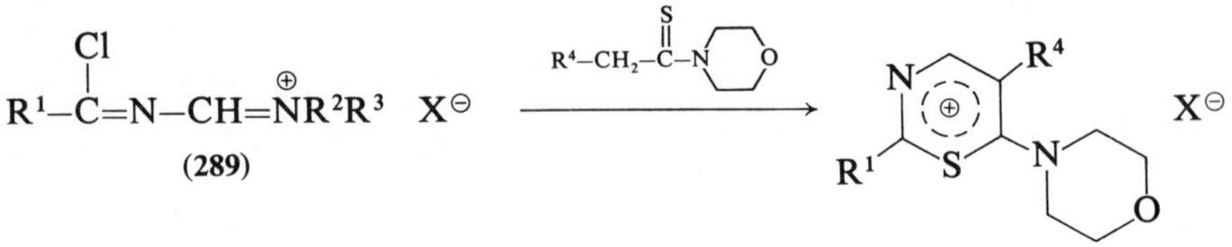

To Section III-J

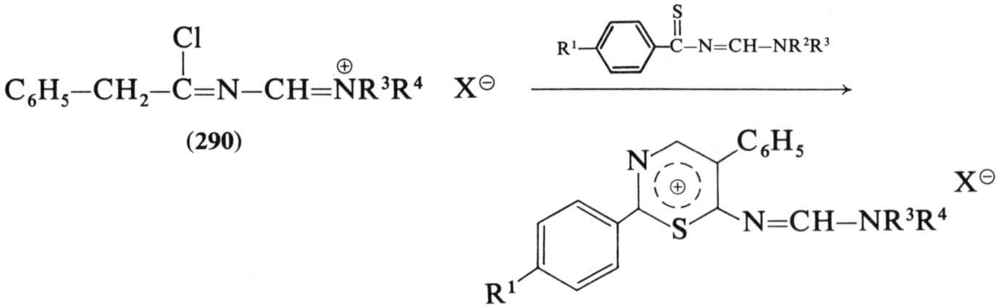

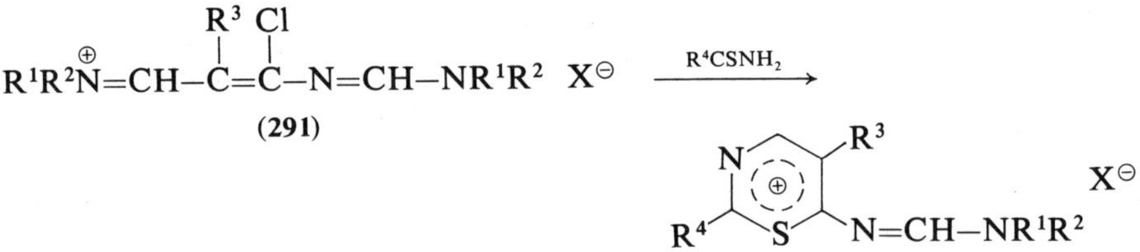

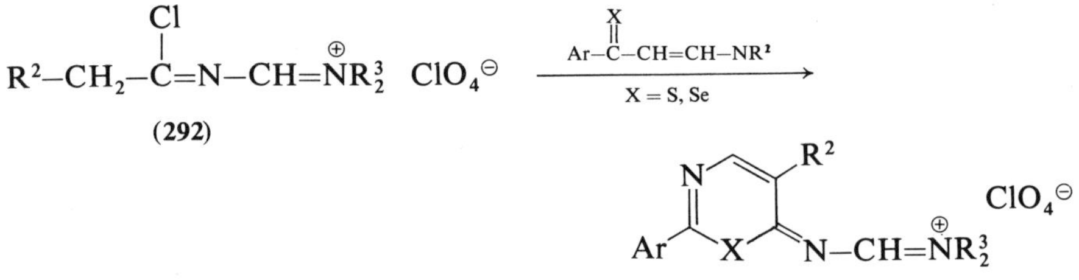

Pyrolysis of **261** affords compounds **293** (203).

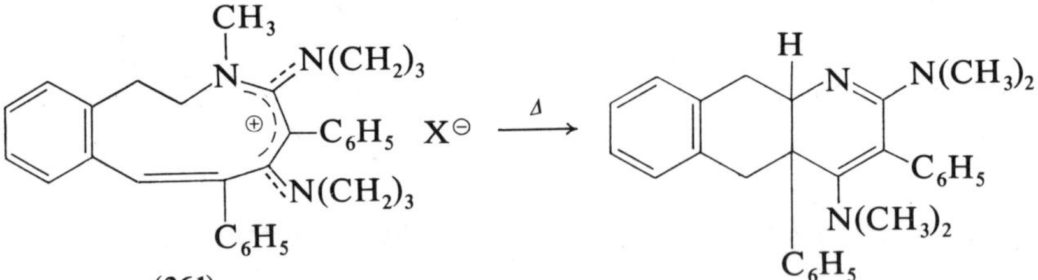

Azidoformamidinium chloride reacts with triphenylphosphane to form the adduct **294**, which on heating looses N_2 to give **295** (230).

$$\underset{H_2N}{\overset{H_2N}{>}}\!\!\overset{\oplus}{C}\!-\!N_3 \quad Cl^{\ominus} \xrightarrow[-30°]{P(C_6H_5)_3}$$

$$(C_6H_5)_3P=N-N=N-\overset{\oplus}{C}\!\!\underset{NH_2}{\overset{NH_2}{<}} \quad Cl^{\ominus} \xrightarrow{\underset{-N_2}{\Delta}} (C_6H_5)_3P=N-\overset{\oplus}{C}\!\!\underset{NH_2}{\overset{NH_2}{<}} \quad Cl^{\ominus}$$
$$\quad\quad\quad (294) \quad\quad\quad\quad\quad\quad\quad\quad\quad\quad\quad\quad\quad (295)$$

Further examples on electrophilic substitution of aromatic compounds, bound to an amidinium system, have been published (231).

$X = COCH_3, NO_3$

REFERENCES

192. S. Patai, Ed., *The Chemistry of Functional Groups*. Chemistry of Amidines and Imidates, John Wiley and Sons, New York, 1975.
193. F. Zymalkowski, Ed., *Methodicum Chimicum*, Vol. 6, *C-N-Verbindungen*, Georg Thieme Verlag, Stuttgart, 1974.
194. D. Lloyd and H. McNab, *Angew. Chem.*, **88**, 496 (1976).
195. E. Haug and W. Kantlehner, unpublished results.
196. V. P. Arya and S. Shenoy, *Indian J. Chem.*, **B14**, 763 (1976).
197. T. Saegusa and Y. Ito, *Synthesis*, **1975**, 291.
198. P. Jacobsen, *Acta Chem. Scand.*, **B30**, 847 (1976).
199. S. C. Wicherink, J. W. Scheeren, and R. J. F. Nivard, *Synthesis*, **1977**, 273.
200. R. R. Bard and M. J. Strauss, *J. Amer. Chem. Soc.*, **97**, 3789 (1975).
201. R. R. Bard and M. J. Strauss, *J. Org. Chem.*, **42**, 435 (1977).
202. E. Schaumann and S. Sieveking, *Liebigs Ann. Chem.*, **1976**, 2284.
203. R. Fuks, R. Merenyi, and H. G. Viehe, *Bull. Soc. Chim. Belg.*, **85**, 892 (1976).
204. R. Fuks and H. G. Viehe, *Bull. Soc. Chim. Belg.*, **86**, 219 (1977).
205. M. J. S. Gynane and M. F. Lappert, *J. Organomet. Chem.*, **114**, C4 (1976).
206. O. Tsuge, T. Iimure, and M. Horie, *Heterocycles (Sendai Japan)*, **4**, 13 (1976); *C. I.*, 20–232 (1976).
207. J. J. Havel and R. Q. Kluttz, *Synth. Commun.*, **4**, 389 (1974).
208. W. Walter and Ch. Rohloff, *Liebigs Ann. Chem.*, **1977**, 463.
209. W. Walter and Ch. Rohloff, *Liebigs Ann. Chem.*, **1977**, 477.
210. T. Wieland and A. Seeliger, *Liebigs Ann. Chem.*, **1976**, 820.

211. P. Ruggli and I. Morszak, *Helv. Chim. Acta*, **11,** 180 (1928).
212. R. L. Pruett, J. T. Barr, K. E. Rapp, C. T. Bahner, J. D. Gibron, and R. H. Lafferty, Jr., *J. Amer. Chem. Soc.*, **72,** 3646 (1950).
213. D. C. England, L. R. Melby, M. A. Dietrich, and R. V. Lindsay, Jr., *J. Amer. Chem. Soc.*, **82,** 5116 (1960).
214. A. J. Speziale and R. C. Freeman, *J. Amer. Chem. Soc.*, **82,** 909 (1960).
215. G. G. Levkovskaya, A. N. Mirskova, A. S. Atavin, and I. D. Kalikhuman, *Zh. Org. Khim.*, **10,** 2293 (1974); *C. I.*, 6–203 (1975).
216. B. S. Drach, V. A. Kovalev, and V. A. Kirsanov, *Zh. Org. Khim.*, **11,** 122 (1975); *C. I.*, 14–187 (1975).
217. M. T. Wu, D. Taub, and A. A. Patchett, *Tetrahedron Lett.*, **1976,** 2405.
218. Y. Takikawa, Ch. Takahashi, R. Sat, and S. Takizawa, *Nippon Kagaku Kaishi*, **1976,** 637; *C. I.*, 32–175 (1976).
219. O. Döbner, *Ber. Dtsch. Chem. Ges.*, **15,** 232 (1882).
220. O. Döbner, *Ann. Chem.*, **217,** 223 (1883).
221. S. P. Joshi, A. P. Khanolkar, and T. S. Wheeler, *J. Chem. Soc.*, **1936,** 793.
222. J. Liebscher and H. Hartmann, *Z. Chem.*, **15,** 16 (1975).
223. J. Liebscher and H. Hartmann, *Synthesis*, **1976** 403.
224. J. Liebscher and H. Hartmann, *Collect. Czech. Chem. Commun.*, **41,** 1565 (1976).
225. J. Liebscher and H. Hartmann, *Tetrahedron Lett.*, **1975,** 2977.
226. P. Reynaud, J.-D. Brion, and G. Ménard, *Bull. Soc. Chim. Fr.*, **1976,** 301.
227. J. Liebscher and H. Hartmann, *Z. Chem.*, **15,** 438 (1975).
228. J. Liebscher and H. Hartmann, *Z. Chem.*, **16,** 268 (1976).
229. J. Liebscher and H. Hartmann, *Tetrahedron*, **33,** 731 (1977).
230. W. Buder and A. Schmidt, *Z. Naturforsch*, **B30,** 503 (1975).
231. I. v. Borlev, A. F. Pozharskii, and V. N. Koroleva, *Khim. Geterotskl. Soedin*, **1975,** 1692; *C. I.*, 15–257 (1976).

ORTHOAMIDES: PROPERTIES AND REACTIONS

By G. SIMCHEN, *Institut für Organische Chemie, Biochemie und Isotopenforschung der Universität, D-7-Stuttgart 80, Pfaffenwaldring 55, Germany*

CONTENTS

I. Survey and Limitations 395
II. Properties of Orthoamides 396
   A. Electrolytic Dissociation 396
   B. Basicity 398
   C. Thermal Stability 398
   D. Pyrolysis 399
   E. Dismutation and Commutation 401
III. Reactions of Orthoamides with Nucleophiles 403
   A. Reactions with XH-Acidic Compounds 403
   B. Reactions with CH-Acidic Compounds 406
      1. Formic Acid Orthoamides with Active Methyl- and Methylene Groups . . . 406
      2. Carboxylic Acid Orthoamides and Lactam Acetals with Active Methyl and Methylene Groups 409
      3. Carbamic Acid Acetals, Urea Acetals, Alkoxytris(dialkylamino)methanes, and Tetrakis(dimethylamino)methane with Active Methyl and Methylene Groups 426
      4. Dialkoxymethylammonium Salts with Active Methyl and Methylene Groups . 434
      5. Orthoamides with Active Methyne Groups 435
   C. Reactions with NH-Acidic Compounds 439
      1. Carboxylic Acid Orthoamides with NH_2-Acidic Compounds 439
      2. Carbon Acid Orthoamides with NH_2-Acidic Compounds 444
      3. Tris(acylamino)methanes with NH_2-Acidic Compounds 444
      4. Orthoamides with Secondary Amines and Derivatives 445
      5. (Dialkoxymethyl)ammonium Salts with NH-Acidic Compounds . . . 453
   D. Reactions with OH-Acidic Compounds 454
      1. Carboxylic Acid Amide Acetals and Thioacetals with Alcohols . . 454
      2. Aminal Esters and Tris(dialkylamino)methanes with Alcohols . . 457
      3. Alkoxy(dialkylamino) and Bis(dialkylamino)acetonitriles with Alcohols . . 458
      4. Carbamic Acid Orthoamides with Alcohols 460
      5. Orthoamides with Water (Hydrolysis) 460

Translated from German by Dr. W.-J. Richter, MPI für Kohleforschung, D 4330 Mühlheim.

	6. Orthoamides with Phenols	461
	7. *N,N*-Dialkylformamide Acetals with Carboxylic Acids	465
	8. Orthoamides with Strong Acids and Hydrogen Chloride	468
E.	Reactions with PH-Acidic Compounds	470
F.	Reactions with SH-Acidic Compounds	471
G.	Reactions with Organometallic Compounds	473
H.	Synthesis of Heterocycles via Orthoamides	473
	1. Via Tris(acylamino)methanes	473
	a. By Transfer of Formamidine Carbon Atom	474
	b. By Transfer of Imidoformyl Group (HC=NH)	474
	c. By Transfer of Formylaminomethylene Group [HC(=)NHCHO]	475
	d. By Transfer of Formylformamidine Group	476
	2. Via Carboxylic Acid Amide Acetals, Aminal Esters, Bis(dialkylamino)acetonitriles	478
	a. Indoles	478
	b. Pyridines	478
	c. 4-Pyrones	479
	d. Pyrazoles, Isoxazoles	480
	e. Imidazoles, Oxazoles	480
	f. Pyrimidines, and Purines	481
	g. *s*-Triazines	485
IV.	Reactions of Orthoamides with Electrophiles	487
A.	At Alkoxy (Alkylmercapto) and Dialkylamino Groups	487
	1. Alkylation	487
	2. Reactions with Aldehydes and Ketones	487
	3. Acylation	492
	a. With Acyl Halides-, Cyanides-, Azides, and Acid Anhydrides	492
	b. With Heterocumulenes	496
B.	At α-Methyl (Methylene) Groups of Carboxylic Acid Amide Acetals	500
	1. With Activated Alkenes and Alkynes	500
	2. With Alkyl and Acyl Halides	502
	3. Via Allylic Rearrangement	503
C.	At Formyl Carbon of Orthoamides	506
	1. With Isocyanates, Isothiocyanates, and Carbodiimides	506
	a. Reaction Mechanism	509
	2. With *N*-Haloacylamides	510
	a. Reaction Mechanism	512
	3. With Nitrenes, Carbenes, Sulfur, and Selenium	513
	4. With Carbonyl Compounds and Their Derivatives	514
	a. Schiff Bases of Aromatic Aldehydes	514
	b. Azines	515
	c. Aromatic Aldehydes	517
	5. H–D Exchange Reactions at Dialkylformamide Dialkyl Acetals	518
V.	Elimination Reactions	519
A.	1,2-Elimination	519
B.	Elimination of Dimethylformamide from Cyclic Dimethylformamide Acetals	519
C.	Degradation of Cyclic Dimethylformamide Acetals to Alkenes	520
	References	521

I. Survey and Limitations

This chapter deals with the chemistry of orthoamides, which are compounds characterized by substitution at the carbonyl group of acyclic and cyclic carboxylic and carbonic acid amides, the substituents being two monovalent oxygen, sulfur, or nitrogen functions. Alkoxy(dialkylamino)- and Bis(dialkylamino)carboxylic acid nitriles are treated as special cases.

Generally we do not consider bicyclic (1), spirocyclic, and special heterocyclic orthoamides (2). The following nomenclature is used:

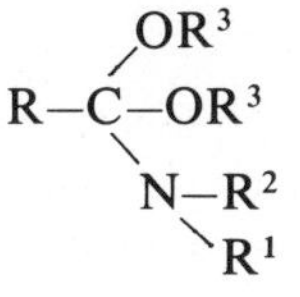

Amide acetals (**1**)

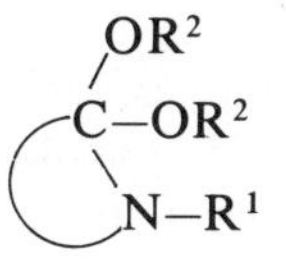

Lactam acetals (**2**)

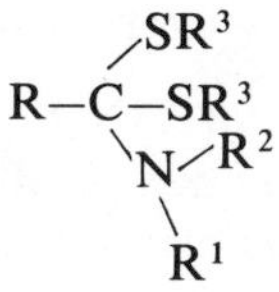

Amide thioacetals (**3**)

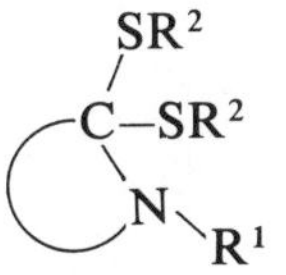

Lactam thioacetals (**4**)

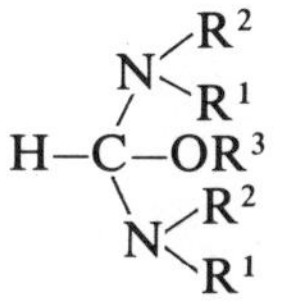

Alkoxybis(dialkylamino)methanes (**5**)

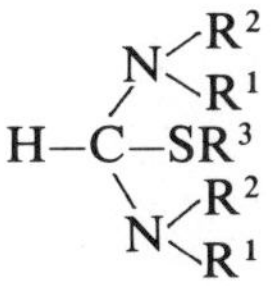

Alkylthiobis(dialkylamino)methanes (**6**)

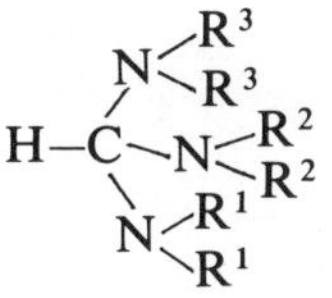

Tris(dialkylamino)methanes, Tris(alkylarylamino)methanes (**7**)

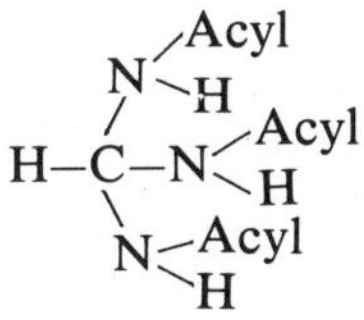

Tris(acylamino)methanes (**8**)

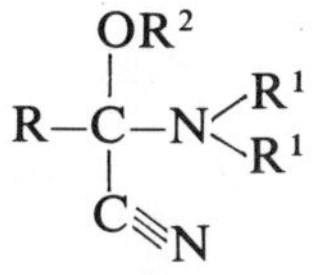

Alkoxy(dialkylamino)carboxylic acid nitriles (**9**)

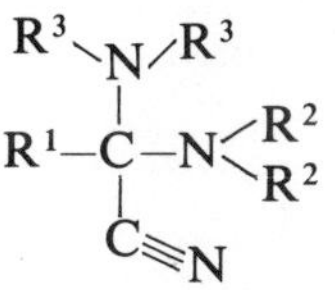

Bis(dialkylamino)carboxylic acid nitriles (**10**)

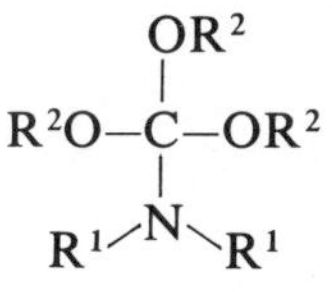

Carbamic acid acetals (**11**)

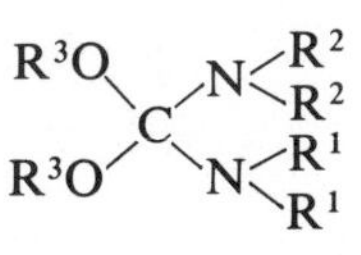

Urea acetals (**12**)

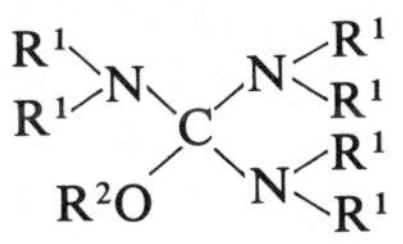

Alkoxytris(dialkylamino)methanes (**13**)

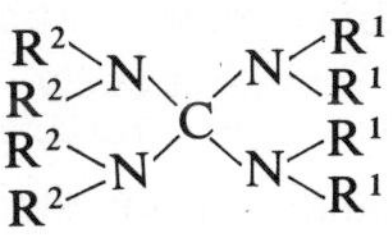

Tetrakis(dialkylamino)methanes (**14**)

where R = alkyl, substituted alkyl, aryl, or carboalkoxy group; R^1, R^2, R^3 = alkyl, aryl. The oxygen, sulfur, and nitrogen functions may also be bridged.

Carboxylic acid, lactam, carbamic acid, and urea acetals (**1, 2, 11, 12**) containing different alkoxy groups (mixed acetals) have not yet been described, but the successful synthesis of tris(dialkylamino)methanes with different dialkylamino groups has been reported (3).

II. Properties of Orthoamides

A. ELECTROLYTIC DISSOCIATION

Except for Tris(acylamino)methanes, orthoamides are liquids that smell like amines and are usually very soluble in apolar solvents. This behavior indicates only slight dissociation into carbiminium cations according to the following scheme:

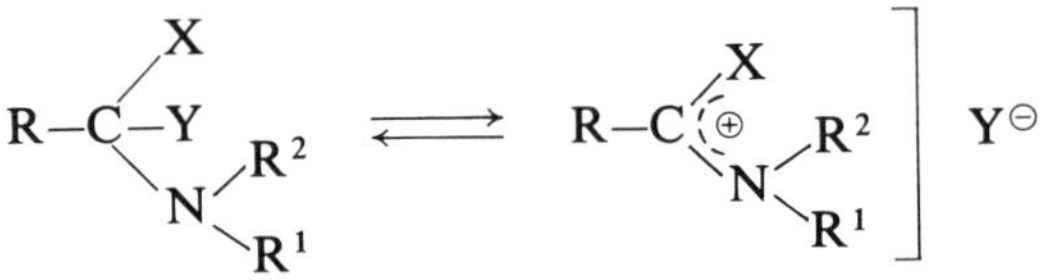

$$R-\underset{\underset{\underset{R^1}{N}}{|}}{\overset{\overset{X}{|}}{C}}-Y \cdot \overset{R^2}{} \rightleftharpoons \left[R-\underset{\underset{\underset{R^1}{N}}{|}}{\overset{\overset{X}{|}}{C}}{}^{\oplus}\overset{R^2}{} \right] Y^{\ominus}$$

	R	X	Y
1	H, alkyl	OR^3	OR^3
2	R^1–R	OR^3	OR^3
5	H	$N{<}_{R^1}^{R^2}$	OR^3
7	H	$N{<}_{R^2}^{R^1}$	$N{<}_{R^2}^{R^1}$
10	H	$N{<}_{R^2}^{R^1}$	CN

H. Meerwein et al. (4) have demonstrated the dissociation of cyclic amide acetals, but obtained inconclusive results in the case of simple carboxylic acid amide acetals. Determination of the electric conductivity (5) of N,N-dimethylformamide dimethyl acetal (**1a**, R = H, $R^1 = R^2 = R^3 = CH_3$), methylaminal ester (**5a** $R^1 = R^2 = R^3 = CH_3$), t-butylaminalester (**5c** $R^1 = R^2 = CH_3$, $R^3 = C(CH_3)_3$), and tris(dimethylamino)methane (**7a** $R^1 = R^2 = R^3 = CH_3$) showed that orthoamides are weak electrolytes (see Table I).

TABLE I

Specific (χ) and Equivalent Conductivity (λ) of 0.1 M Orthoamide Solutions

Orthoamide	In Dimethylformamide				In 1,2-Dimethoxyethane			
	After 10 min		After 60 min		After 60 min		After 720 min	
	$\chi \times 10^6$	$\lambda \times 10^1$	$\chi \times 10^6$	$\lambda \times 10^1$	$\chi \times 10^8$	$\lambda \times 10^4$	$\chi \times 10^8$	$\lambda \times 10^4$
1a	6.05	0.61	8.2	0.82	2.8	2.8	2.8	2.8
7a	12.5	1.25	15.0	1.5	9.3	9.3	11.2	11.2
5a	16.2	1.62	19.0	1.9	—	—	—	—
5c	18.5	1.85	22.2	2.22	18.0	18.0	71.0	71.0

The dissociation according to the above scheme occurs slowly; the greater the dielectric constant and the solvating ability of the solvent, the faster the dissociation. The conductivity increases going from amide acetal (**1**) to tris-(dialkylamino)methanes (**7**) to aminal esters (**5**). The rate and degree of dissociation at time t thus depend on the carbonium ions and anions that are formed.

Aminal esters (**5**) dissociate faster and to a greater extent than tris-(dimethylamino)methane (**7a**) because of the greater stability of the alkoxide ions. Since N,N-tetramethylformamidinium ions are especially stabilized, the dissociation to these compounds occurs much more rapidly and to a greater extent. Bis(dialkylamino)carboxylic acid nitriles (**10**) show an equivalence conductivity in DMF similar to those of aminal esters (**6, 7**). According to more recent measurements of compounds **10** in acetonitrile, the increased dissociation parallels the increasing stability of the amidinium cations being formed:

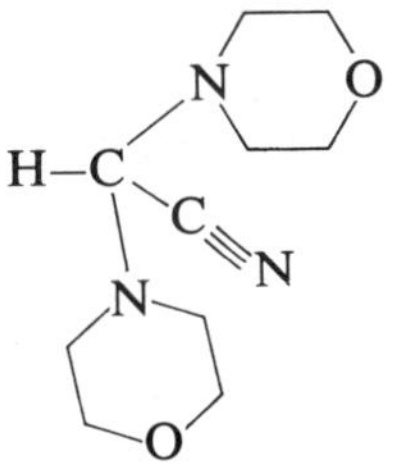

(**10d**) $\lambda = 3.6$

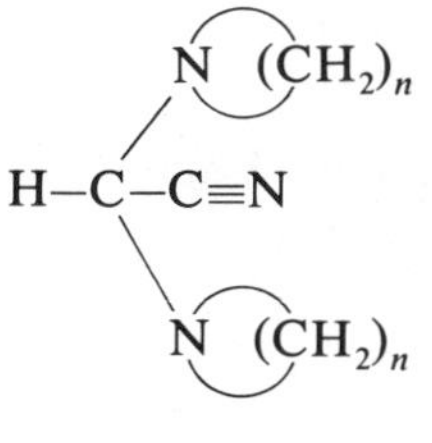

(**10b**) $n = 4$; $\lambda = 4.9$
(**10c**) $n = 5$; $\lambda = 4.3$

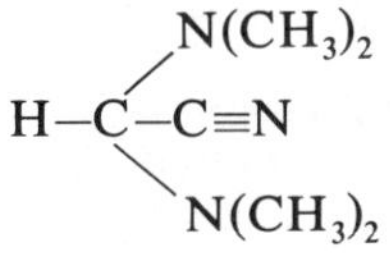

(**10a**) $\lambda = 8.5$

The constants of electrolytic dissociation are 1.0×10^{-4} and 7.7×10^{-4} for **10a** and **b**, respectively.

Lactam acetals are also weak electrolytes (**8**). The formation of carbiminium cations has been demonstrated polarographically.

B. BASICITY (9)

The basic properties of orthoamides are closely related to their reactivity. The usual methods of determining basicity are inadequate, because these compounds are sensitive to hydrolysis and react irreversibly with the titrating solution (see Section III-D-8).

Estimation of pK_b values (Table II) by NMR techniques is possible by observing the change of the chemical shift of the OH signal of t-butanol on addition of an orthoamide. This method is based on the formation of hydrogen bonds between t-butanol and the dialkylamino groups of an orthoamide. Since hydrogen bonding with alkoxy groups in orthoamides is also possible, the accuracy of this method is limited; this is especially true for cyclic derivatives. The limit of error is estimated to be ± 0.5 pK units.

TABLE II

pK_b Values of N,N-Dialkylcarboxylic Acid Amide Acetals (1)[a]

	pK_b
N,N-Dimethylformamide dimethyl acetal	6.25
N,N-Diethylformamide dimethyl acetal	7.6
N,N-Di-n-propylformamide dimethyl acetal	6.25
N,N-Diisopropylformamide dimethyl acetal	7.4
Pyrrolidinodimethoxymethane	7.4
Piperidinodimethoxymethane	6.9
2-Dimethylamino-1,3-dioxolane	7.25
2-Dimethylamino-1,3-dioxane	6.2
Triethylamine	10.65
Triallylamine	8.31
N-Methylmorpholine	7.41

[a] pK_b Values of some tertiary amines are also included. These data reflect the lower basicity of amide acetals as compared with trialkylamines.

The pK_b values of orthoamides **5** and **7** cannot be measured owing to their reaction with hydrogen bond donors (see Section III-D-2).

C. THERMAL STABILITY

Amide acetals generally may be distilled at temperatures up to 150°C with little or no decomposition. The more stable the carbonium ions formed from the alkoxy groups, the greater the possibility of observing decomposition to an N,N-dialkylcarboxylic acid amide and a dialkyl ether. The synthesis of N,N-dimethylformamide bis(4-methoxybenzyl) acetal (**15**) via mixed acetals fails (10) (see Section III-D-1).

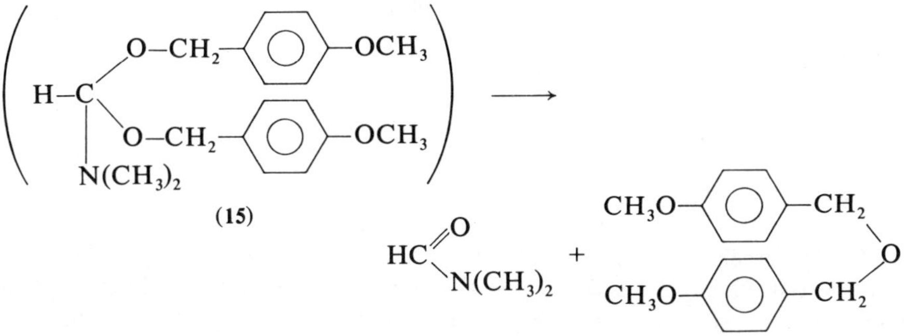

$$\left(\underset{N(CH_3)_2}{\underset{|}{H-C}} \underset{O-CH_2-\text{[C}_6\text{H}_4\text{]}-OCH_3}{\overset{O-CH_2-\text{[C}_6\text{H}_4\text{]}-OCH_3}{\diagup}} \right) \longrightarrow$$

(15)

$$HC\overset{O}{\underset{N(CH_3)_2}{\diagdown}} + \underset{CH_3O-\text{[C}_6\text{H}_4\text{]}-CH_2}{\overset{CH_3O-\text{[C}_6\text{H}_4\text{]}-CH_2}{\diagdown}} O$$

N,N-DMF didodecyl acetal can be isolated only as a crude product but cannot be purified by distillation (10).

Analogously, carbamide acetals (11) (**11**) decompose on heating at 100°C; alkoxytris(dialkylamino)methanes (12) (**13**) decompose at 80°C.

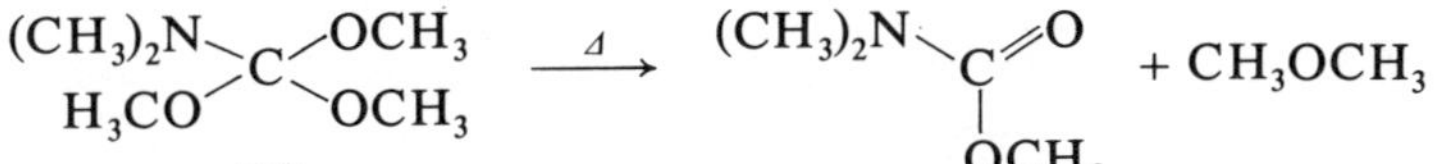

$$\underset{H_3CO}{\overset{(CH_3)_2N}{\diagdown}}C\underset{OCH_3}{\overset{OCH_3}{\diagup}} \xrightarrow{\Delta} \underset{\underset{OCH_3}{|}}{\overset{(CH_3)_2N}{\diagdown}}C\overset{O}{\diagup} + CH_3OCH_3$$

(11)

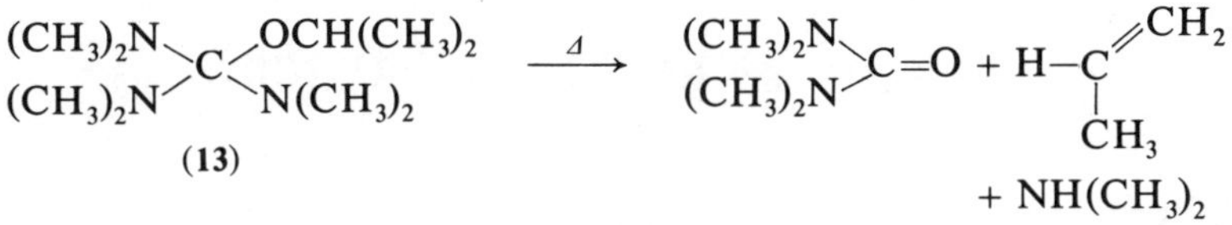

$$\underset{(CH_3)_2N}{\overset{(CH_3)_2N}{\diagdown}}C\underset{N(CH_3)_2}{\overset{OCH(CH_3)_2}{\diagup}} \xrightarrow{\Delta} \underset{(CH_3)_2N}{\overset{(CH_3)_2N}{\diagdown}}C=O + H-\underset{\underset{CH_3}{|}}{\overset{CH_2}{C}}$$

(13)

$$+ NH(CH_3)_2$$

Whereas alkoxy(dialkylamino)acetonitriles (**9**) slowly yield N,N-dialkyl-formamide and carboxylic acid (iso)-nitriles at 130–140°C (13) 1,1-bis-(dimethylamino)-2,2,2-trichloropropionitrile (14) (**10**, $R^1 = CCl_3$, $R^2 = R^3 = CH_3$) explodes at 100°C.

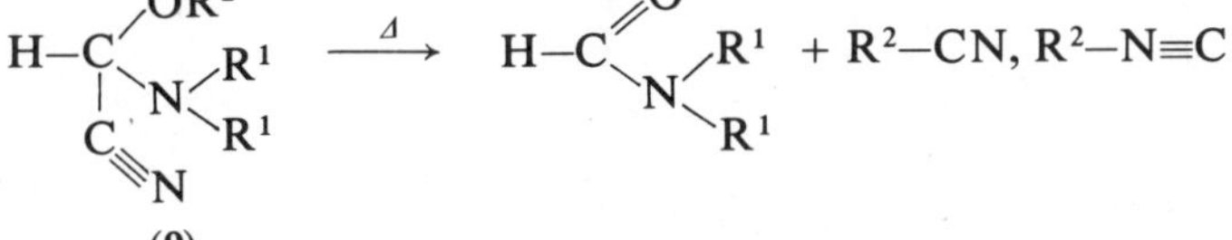

$$\underset{\underset{N}{\overset{|}{C}}}{\overset{H-C}{\diagdown}}\underset{N\diagdown R^1}{\overset{OR^2}{\diagup}} \xrightarrow{\Delta} H-C\underset{N\diagdown R^1}{\overset{O}{\diagup}}R^1 + R^2-CN, R^2-N\equiv C$$

(9)

Distillation of the homologous amide acetals **1** (R = CH_3, C_2H_5, etc.) under normal pressure gives some ketene O,N acetals (see Section V-A).

D. PYROLYSIS

Interestingly, tetrakis(dialkyl[or alkylaryl]amino)ethylenes are formed in good yields from tris(dialkyl[or alkylaryl]amino)methanes (**7**) via dialkyl[or alkylaryl]amine cleavage during pyrolysis.

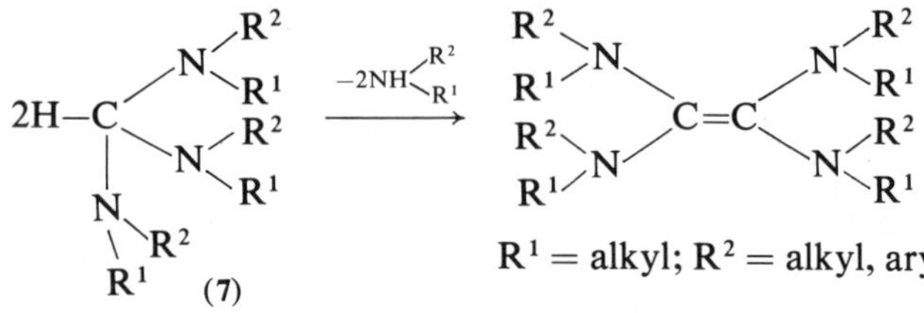

$$2H-C\begin{pmatrix}N(R^1)(R^2)\\N(R^1)(R^2)\\N(R^1)(R^2)\end{pmatrix} \xrightarrow{-2NH(R^1)(R^2)} (R^2)(R^1)N\underset{(R^1)(R^2)N}{\overset{(R^2)(R^1)N}{>}}C=C\underset{N(R^1)(R^2)}{\overset{N(R^1)(R^2)}{<}}$$

(7) R^1 = alkyl; R^2 = alkyl, aryl

H. Bredereck et al. (15) obtained tetrakis(dimethylamino)ethylene by heating tris(dimethylamino)methane (**7a**, $R^1 = R^2 = CH_3$) to 160°C. Several other compounds of this kind were similarly synthesized by heating tris(dialkylamino)methanes at 180–250°C (16). H. W. Wanzlick (17) has given an unambiguous demonstration of the dimerization of orthoamides (**5**);

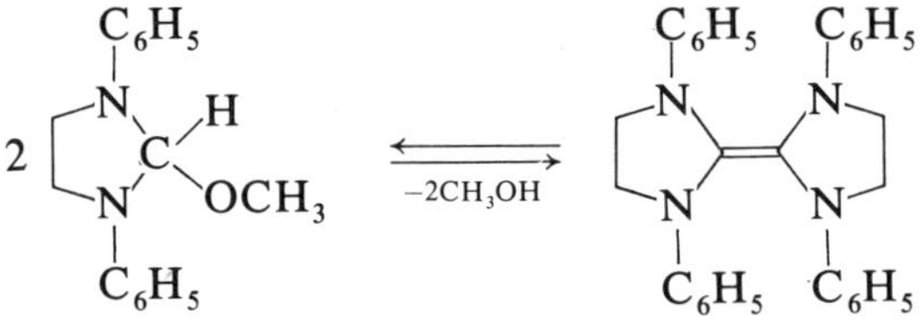

[Structure: 2 equivalents of imidazolidine with N-C$_6$H$_5$, C-H, C-OCH$_3$ ⇌ (−2CH$_3$OH) bis(imidazolidinylidene) dimer]

$R^1 = C_6H_5$; $R^2-R^2 = -(CH_2)_2-$; $R^3 = CH_3$

(**5**)

dimerization of orthoamides (**10**) via HCN elimination can also be achieved (18, 19).

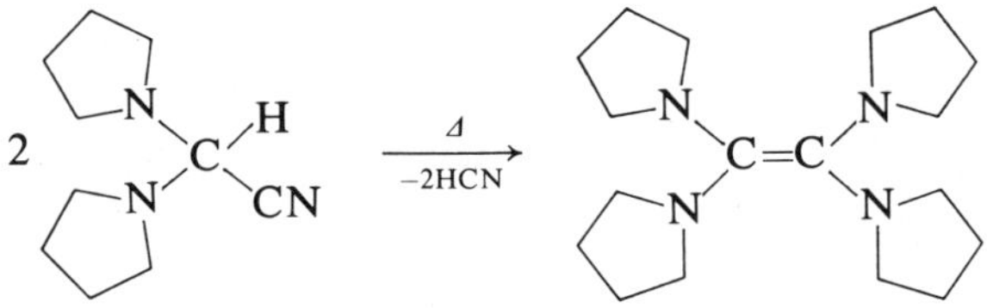

[Structure: 2 equivalents of dipyrrolidinyl-CH-CN $\xrightarrow{\Delta, -2HCN}$ tetrapyrrolidinyl ethylene]

$R^1 = H$, $R^2-R^2 = R^3-R^3 = -(CH_2)_4-$

(**10**)

The synthesis of tetrakis(dialkylamino)alkenes by aminolysis of *N,N*-dialkylformamide acetals must at least partially go through **7**, because the intermediate aminal esters (**5**) undergo dismutation (see Section II-E). The course of the reaction is still uncertain. Owing to the high temperature (160–250°C), either a 50 or 100% carbene mechanism seems to be likely.

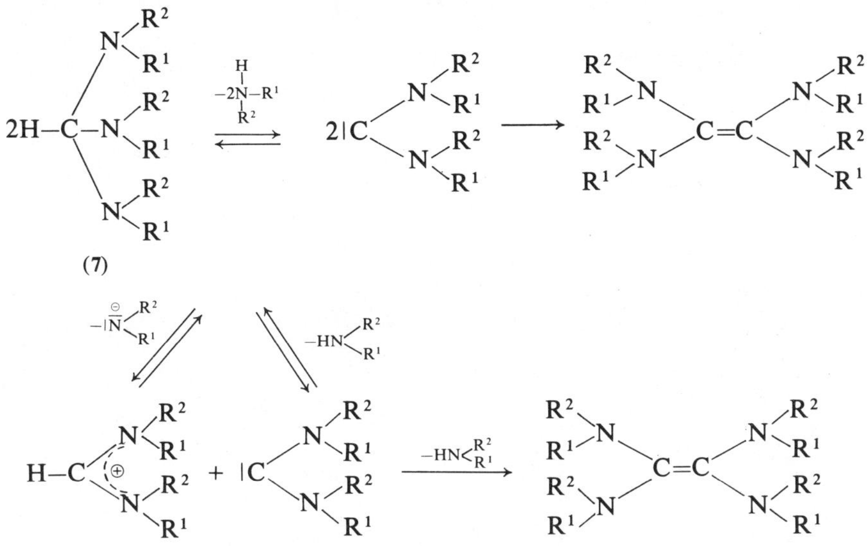

II. PROPERTIES OF ORTHOAMIDES

When heated at 160–250°C tris(formylamino)methane (**8a**) expectedly does not afford tetrakis(formylamino)ethylene, but rather eliminates CO and formamide, giving *N*-formylformamidine as an intermediate, which then dimerizes to *s*-triazine (20–22). The pyrolysis of **8a** represents an excellent synthesis of *s*-triazine.

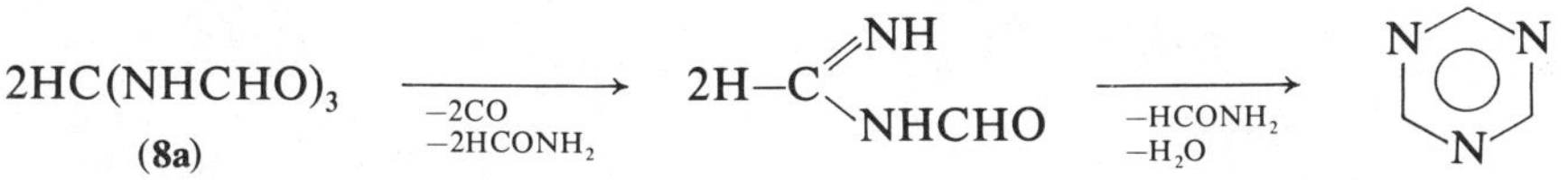

$$2HC(NHCHO)_3 \xrightarrow[-2HCONH_2]{-2CO} 2H-C\begin{smallmatrix}NH\\NHCHO\end{smallmatrix} \xrightarrow[-H_2O]{-HCONH_2} \text{[s-triazine]}$$

(**8a**)

The pyrolysis of tetrakis(dimethylamino)methane (**14**) apparently leads to 1,1-bis(dimethylamino)-2-methylaminoethylene via radicals (22a). The increase of reaction rate by the addition of radical initiators supports the proposed mechanism.

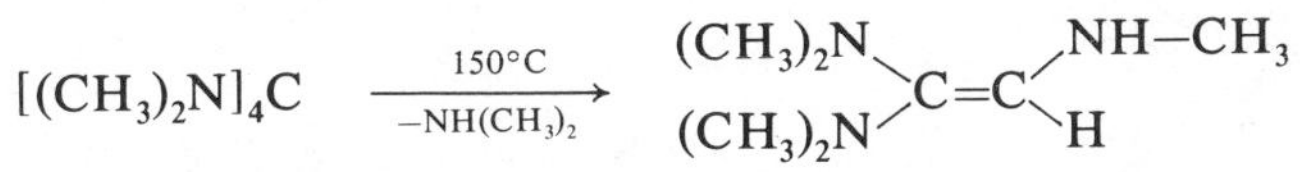

$$[(CH_3)_2N]_4C \xrightarrow[-NH(CH_3)_2]{150°C} (CH_3)_2N\text{-}C=C\text{-}(NH\text{-}CH_3)(H) / (CH_3)_2N$$

$R^1 = R^2 = CH_3$ (**14**)

E. DISMUTATION AND COMMUTATION

Dismutation to *N,N*-dialkylformamide acetals (**1**, R = H) and tris(dialkylamino)methanes (**7**) is a characteristic of aminal esters (**5**); this reaction occurs at 100°C or at lower temperatures when traces of alcohols are added. On the

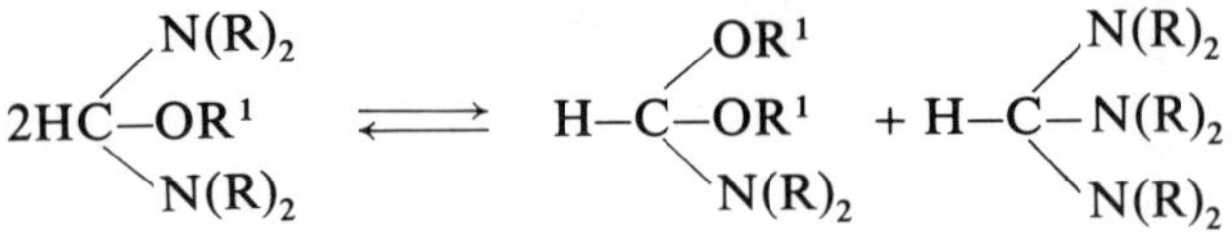

$$2H\overset{\displaystyle N(R)_2}{\underset{\displaystyle N(R)_2}{C-OR^1}} \rightleftharpoons H-\overset{\displaystyle OR^1}{\underset{\displaystyle N(R)_2}{C-OR^1}} + H-\overset{\displaystyle N(R)_2}{\underset{\displaystyle N(R)_2}{C-N(R)_2}}$$

(5)　　　　　　　　(1)　　　　　　(7)

$R = CH_3;\ R^1 = CH_3,\ C_2H_5,\ i\text{-}C_3H_7,\ tC_4H_9$

Dismutation　$H-\overset{N(R)_2}{\underset{N(R)_2}{C-OR^1}} \rightleftharpoons \left[H-C\overset{N(R)_2}{\underset{N(R)_2}{\oplus}} \right] OR^{1\ominus}$

$H-C\overset{N(R)_2}{\underset{N(R)_2}{\oplus}} + |\overset{R}{\underset{R}{N}}-\overset{H}{\underset{N(R)_2}{C-OR^1}} + OR^{1\ominus} \rightleftharpoons H-\overset{N(R)_2}{\underset{N(R)_2}{C-N(R)_2}} + H-\overset{OR^1}{\underset{N(R)_2}{C-OR^1}}$

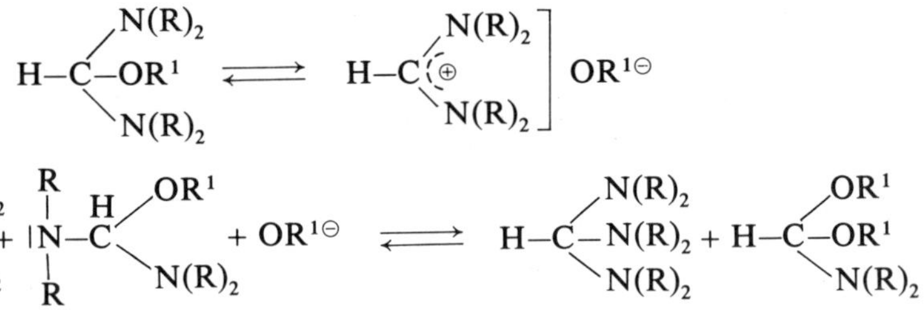

Commutation

$H-\overset{N(R)_2}{\underset{N(R)_2}{C-N(R)_2}} \rightleftharpoons H-C\overset{N(R)_2}{\underset{N(R)_2}{\oplus}} + {}^\ominus|\overline{N}(R)_2$

$H-\overset{OR^1}{\underset{N(R)_2}{C-OR^1}} \rightleftharpoons H-C\overset{OR^1}{\underset{N(R)_2}{\oplus}} + {}^\ominus|\overline{O}R^1$

$\Updownarrow \quad \Updownarrow$

$H-\overset{N(R)_2}{\underset{N(R)_2}{C-OR^1}} + H\overset{N(R)_2}{\underset{N(R)_2}{C-OR^1}}$

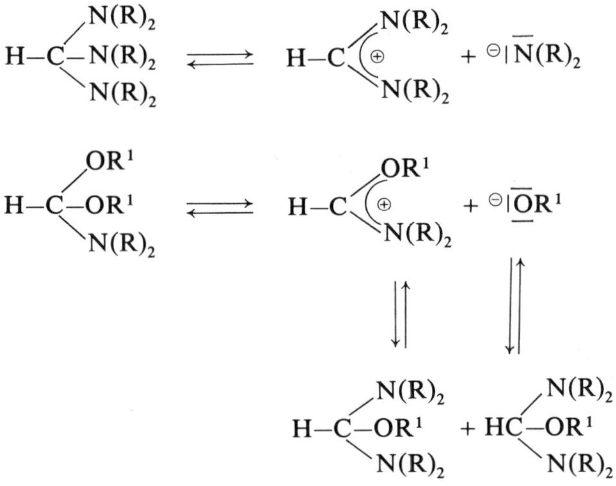

other hand, **1** (R = H) and **7** partially commute to **5** (5, 5a). Electrolytic dissociations initiate these reactions. The rate of dismutation is determined by nucleophilic substitution; that of commutation is determined by electrolytic dissociation. Since the latter is faster, the equilibrium is reached earlier by commutation than by dismutation. The greater rate of commutation causes aminal esters **5** to react as if they were single compounds.

When heated, aminal thioesters (**6**) reach an analogous equilibrium. In this case, however, the commutation is slower than the dismutation (24).

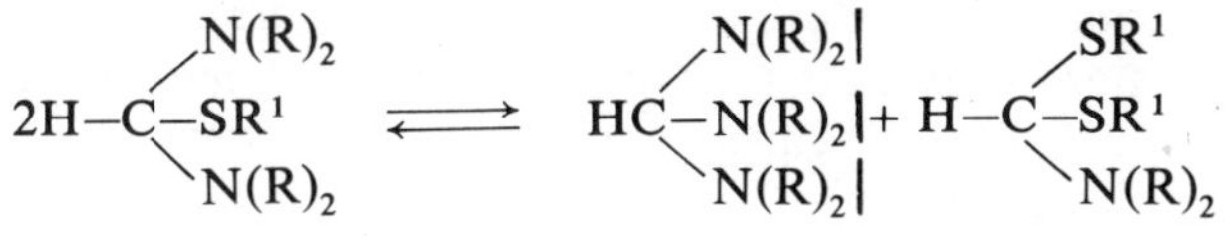

III. REACTIONS OF ORTHOAMIDES WITH NUCLEOPHILES

$$2H-\underset{N(R)_2}{\overset{N(R)_2}{C}}-SR^1 \rightleftharpoons H\underset{N(R)_2}{\overset{N(R)_2}{C}}-N(R)_2 + H-\underset{N(R)_2}{\overset{SR^1}{C}}-SR^1$$

(6)   (7)   (3)

$R = CH_3; R^1 = C_2H_5, i\text{-}C_3H_7, t\text{-}C_4H_9$

Tris(dialkylamino)methanes (7) undergo exchange of dialkylamino groups at room temperature (3). Equilibrium may also be reached by commutation.

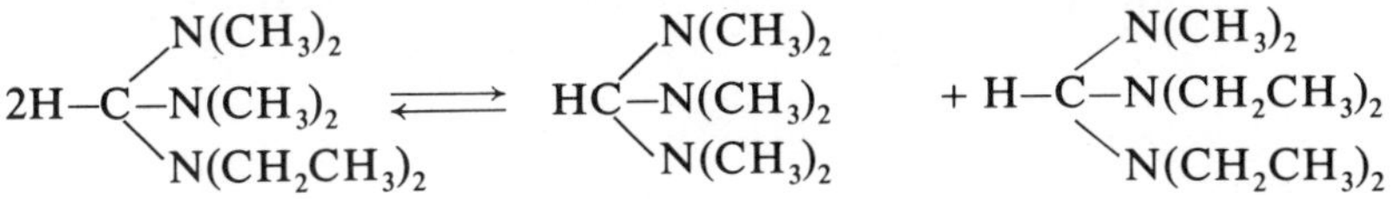

$$2H-\underset{N(CH_2CH_3)_2}{\overset{N(CH_3)_2}{C}}-N(CH_3)_2 \rightleftharpoons H\underset{N(CH_3)_2}{\overset{N(CH_3)_2}{C}}-N(CH_3)_2 + H-\underset{N(CH_2CH_3)_2}{\overset{N(CH_3)_2}{C}}-N(CH_2CH_3)_2$$

(7)

$$2H-\underset{N(CH_2CH_3)_2}{\overset{N(CH_3)_2}{C}}-N(CH_2CH_3)_2 \rightleftharpoons H\underset{N(CH_2CH_3)_2}{\overset{N(CH_3)_2}{C}}-N(CH_3)_2 + H-\underset{N(CH_2CH_3)_2}{\overset{N(CH_2CH_3)_2}{C}}-N(CH_2CH_3)_2$$

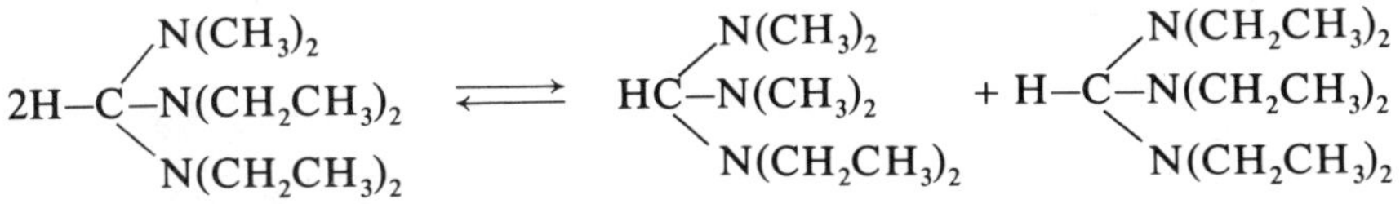

Thus mixed tris(dialkylamino)methanes must be stored at low temperatures (3).

III. Reactions of Orthoamides with Nucleophiles

A. REACTIONS WITH XH ACIDIC COMPOUNDS

The reactions of orthoamides with XH-acidic compounds may be depicted as follows if thermodynamically stable C—X bonds are formed.

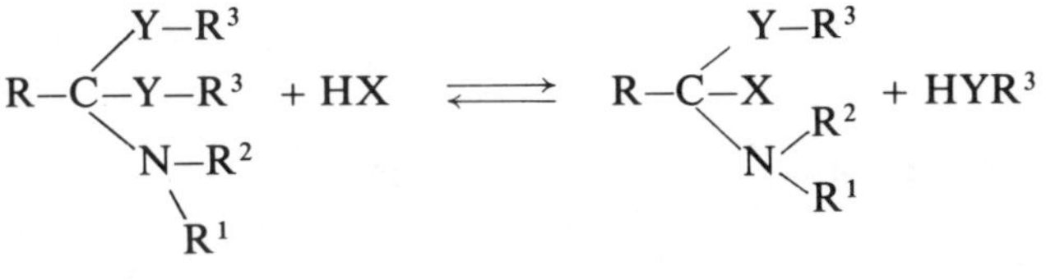

$$R-\underset{N-R^2}{\overset{Y-R^3}{C}}-Y-R^3 + HX \rightleftharpoons R-\underset{N(R^1)(R^2)}{\overset{Y-R^3}{C}}-X + HYR^3$$
$$\overset{|}{R^1}$$

(1–3, 5–7, 11–14)

$$R-\underset{N(R^1)(R^2)}{\overset{Y-R^3}{C}}-Y-R^3 + H-X-H \longrightarrow R-\underset{N(R^1)(R^2)}{\overset{X}{C}} + 2HY-R^3$$

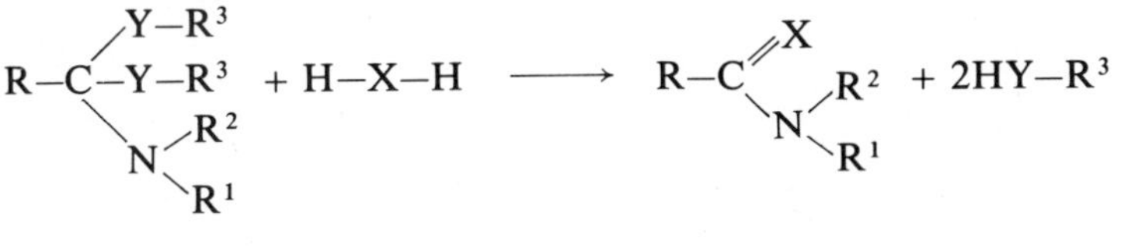

$Y = O, S, NR^3$
$R = H$, alkyl substituted alkyl, aryl, OR^3, $N(R^1)(R^2)$
$R^1, R^2, R^3 = $ alkyl

Reactions with HX give orthoamides by simple exchange. Reactions with XH$_2$ lead to quasi-substitution of the carbonyl oxygen of carboxylic acid amides by X, owing to subsequent β-elimination. Two extreme possibilities of the first reaction step must be discussed: The orthoamide dissociates to give the ambident carbiminium cation and the anion YR$^{3\ominus}$, which is protonated by HX to yield the nucleophile X$^{\ominus}$. Finally the carbiminium cation and X$^{\ominus}$ combine to give the orthoamide (Path a). The formation of the ambident cation may also

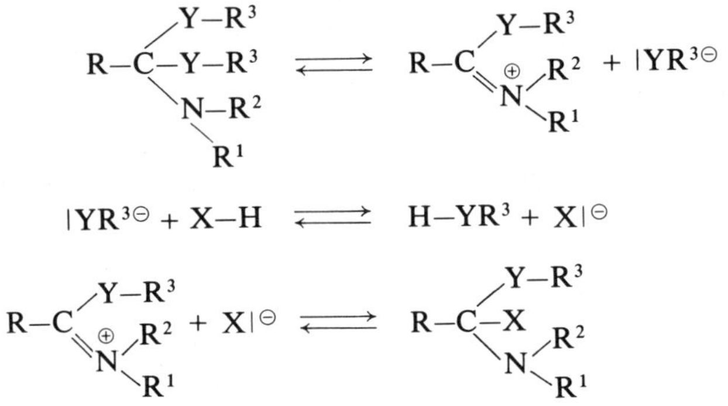

$$R-\underset{\underset{R^1}{\overset{}{N-R^2}}}{\overset{Y-R^3}{C-Y-R^3}} \rightleftharpoons R-\underset{\underset{R^1}{\overset{\oplus}{N}}}{\overset{Y-R^3}{C}}R^2 + |YR^{3\ominus}$$

$$|YR^{3\ominus} + X-H \rightleftharpoons H-YR^3 + X|^{\ominus}$$

$$R-\underset{\underset{R^1}{\overset{\oplus}{N}}}{\overset{Y-R^3}{C}}R^2 + X|^{\ominus} \rightleftharpoons R-\underset{\underset{R^1}{\overset{}{N}}}{\overset{Y-R^3}{C-X}}R^2$$

Path a

occur by an acid–base raction of the orthoamide with HX (Path b). The orthoamide, the acidity of HX, and the solvent all influence the choice of alternate paths.

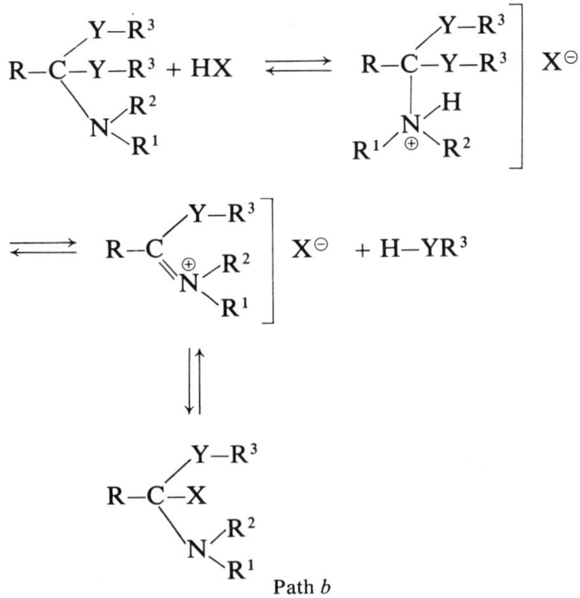

$$R-\underset{\underset{R^1}{\overset{}{N}}}{\overset{Y-R^3}{C-Y-R^3}}R^2 + HX \rightleftharpoons \left[R-\underset{\underset{R^{1\oplus}}{\overset{}{N}}\overset{H}{\underset{R^2}{}}}{\overset{Y-R^3}{C-Y-R^3}} \right] X^{\ominus}$$

$$\rightleftharpoons \left[R-\underset{\underset{R^1}{\overset{\oplus}{N}}}{\overset{Y-R^3}{C}}R^2 \right] X^{\ominus} + H-YR^3$$

$$\Updownarrow$$

$$R-\underset{\underset{R^1}{\overset{}{N}}}{\overset{Y-R^3}{C-X}}R^2$$

Path b

The reaction path via dissociation of the orthoamide is to be preferred when very stable, ambident cations are formed (e.g., from **5, 7, 10,** or **13**). Low kinetic or thermodynamic acidity of the XH-acidic compound (like fluorene, xanthene) as well as high solvent polarity or solvation power (like DMF) favor this path. This equilibrium is followed by an irreversible β-elimination of HYR^3, when the reaction involves compounds of the general formula XH$_2$.

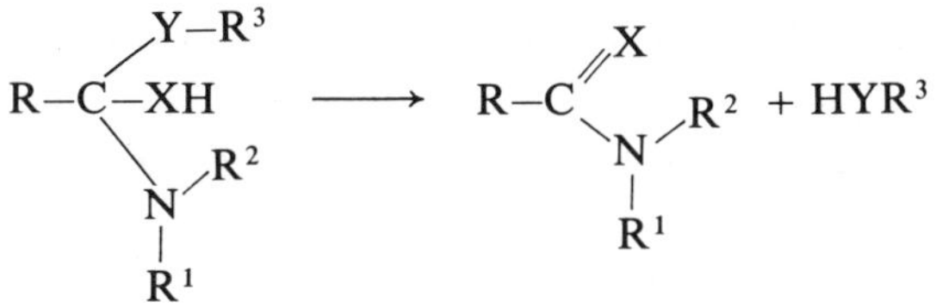

$$R-\underset{\underset{R^1}{\overset{|}{N}-R^2}}{\overset{Y-R^3}{\underset{|}{C}-XH}} \longrightarrow R-\underset{\underset{R^1}{\overset{|}{N}-R^2}}{C\overset{X}{\diagup}} + HYR^3$$

A striking feature of orthoamides (as compared with orthoesters) is their enormously increased reactivity. This is partly caused by the basicity of the medium, partly by the formation of more stable carbiminium cations. If the newly formed CX bond is labile (owing to stable anions X$^\ominus$ or XH$^\ominus$), the reaction ends either by simple salt formation (in **5, 7, 13**) or by competition with thermodynamically controlled S_N reactions at Y–R^3 (Y = O).

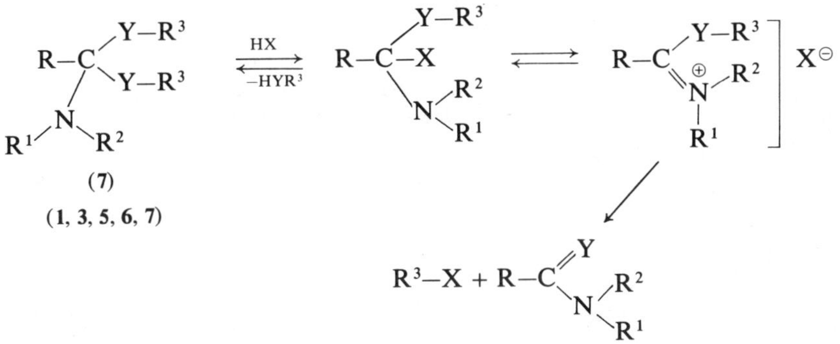

$$R-C\underset{\underset{R^1 \diagup N \diagdown R^2}{}}{\overset{Y-R^3}{\diagup Y-R^3}} \underset{-HYR^3}{\overset{HX}{\rightleftarrows}} R-\underset{\underset{R^1}{N-R^2}}{\overset{Y-R^3}{C-X}} \rightleftarrows \left[R-\underset{\underset{R^1}{\overset{|}{N}-R^2}}{\overset{Y-R^3}{C \oplus}}\right] X^\ominus$$

(7)
(1, 3, 5, 6, 7)

$$\downarrow$$

$$R^3-X + R-\underset{\underset{R^1}{N-R^2}}{C\overset{Y}{\diagup}}$$

Alkylation of X$^\ominus$ also restricts the yields of orthoamides **1–4, 9, 11,** and **12** as they are synthesized from ambident cations and nucleophiles.

Reactivity of orthoamides to XH-acidic compounds varies over a great range. If deprotonation of the weak acid HX or H$_2$X is the rate-determining step (e.g., recombination of the ions is much faster) the reactivity increases with basicity and with extent of electrolytic dissociation (which parallels basicity) (5).

Thus formic acid orthoamides show the following order of increasing reactivity (5, 25–28a):

N,N-Dimethylformamide dimethyl acetal (**1a**)
N,N-Dimethylformamide diethyl acetal (**1b**)
N,N-Dimethylformamide di-t-butyl acetal (**1c**)
Bis(dimethylamino)methoxymethane (**5a**)
Bis(dimethylamino)ethoxymethane (**5b**)
Bis(dimethylamino)-t-butoxymethane (**5c**)
Tris(dimethylamino)methane (**7a**)

Thio analogues (24) as well as carbamic acid acetals (11) and carboxylic acid orthoamides with electron-withdrawing groups (29) at the α position react more slowly with XH-acidic compounds owing to less electrolytic dissociation and basicity.

In carbonic acid orthoamides reactivity increases with the number of dialkylamino groups in the following way (11, 12, 30):

N,N-Dialkylcarbamic acid acetal
N,N,N',N'-Tetraalkylurea acetals
Alkoxytris(dialkylamino)methane
Tetrakis(dialkylamino)methane

All these acetals are derivatives of carbonic acid

$$O=C{\overset{OH}{\underset{OH}{\diagdown}}}$$

The increased reactivity caused by greater dissociation (and basicity) is opposed by slower recombination owing to greater stability of the carbiminium cations. Steric effects may also play a role (see Section III-B). For example, this is the reason for the lower reactivity of dipiperidinomethoxymethane and trimorpholinomethane as compared with bis(dimethylamino)methoxymethane and tris(dimethylamino)methane (23).

The decreased reactivity of amide acetals of long-chain carboxylic acids may be explained in the same way.

It is noteworthy that N-quaternization activates cyclic N,N-dialkylformamide acetals, which otherwise react very slowly (31).

B. REACTIONS WITH CH-ACIDIC COMPOUNDS

1. Formic Acid Orthoamides with Active Methyl and Methylene Groups

Formic acid orthoamides react with CH_2- and CH_3-acidic compounds to give dialkylaminomethylene compounds (4, 7, 10, 23–28a, 32–48, 51–53, 56–61). Since an aminomethylene functionality is introduced here, this enamine synthesis may be considered as a formylation reaction.

III. REACTIONS OF ORTHOAMIDES WITH NUCLEOPHILES

$$\underset{B}{\overset{A}{\diagdown}}CH_2 + H-\underset{\underset{R^2}{\overset{R^1}{\diagup}}}{\overset{YR^3}{\overset{\diagup}{C}-YR^3}} \longrightarrow \underset{B}{\overset{A}{\diagdown}}C=C\underset{\underset{R^2}{\diagdown}}{\overset{H}{\diagup}R^1}$$

$$+\ 2H-YR^3$$

A = activating group
B = activating group, H, alkyl
YR^3 = alkoxy, alkylmercapto, dialkylamino

Diformylation of activated vicinal methylene groups is also possible (23, 51). Deprotonation of the activated methylene compound is the first reaction step,

$$\underset{B}{\overset{A}{\diagdown}}\underset{\underset{CH_2}{|}}{\overset{CH_2}{\ }} \xrightarrow[(5,\ 7)]{HC-YR^3 \atop N_{R^1}^{R^2}} \underset{B}{\overset{A}{\diagdown}}\underset{\underset{H}{|}}{\overset{H \atop |}{\underset{C-N}{C-N}\diagdown\overset{R^2}{R^1}}}\diagdown\overset{R^1}{\underset{R^2}{\diagup}}$$

thus forming the carbanion and the ambident cation (see Section III-A); elimination of the corresponding alcohol or thiol always occurs. Ionic recombination and β-elimination to give a dialkylaminomethylene derivative are generally much faster processes. Alkoxy(alkylmercapto)methylene compounds could never be isolated. Which of the numerous orthoamides **1**, **5**, or **7** should be used depends on the kinetic acidity of the active methylene compound.

In the case of strongly CH-acidic compounds like nitromethane or 1,3-dicarbonyl compounds, the use of a *N,N*-dimethylformamide dialkyl acetal is sufficient.

Carbon acids of low acidity such as nitriles and carboxylic acid esters react only with much more reactive aminal esters such as *t*-butyl aminal ester (**5c**) or tris(dimethylamino)methane. Table III gives a survey.

Formation of stable tetraalkylformamidinium ions is responsible for the greater reactivity of aminal esters (23, 25, 26, 32, 39, 52). Increased reactivity which parallels increasing bulk of the alkoxy groups in series **1** and **5** is presumably due to increased dissociation and the parallel increase of basicity of the alkoxides (26, 27).

In comparison with **1** and **5** the thio analogues **3** and **6** are less reactive and so are seldom used (24, 48).

The reactivity of the bis(dialkylamino)acetonitrile **10** (R = H) is comparable to that of the aminal ester **5**. However, since the HCN formed during the first reaction step polymerizes, isolation of the products is sometimes difficult (7).

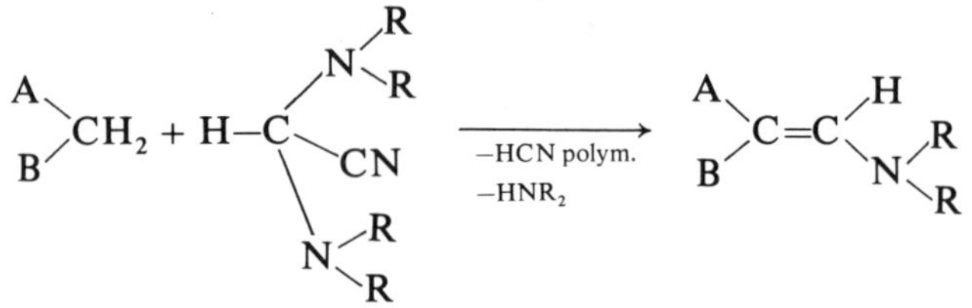

ORTHOAMIDES: PROPERTIES AND REACTIONS

$$\underset{B}{\overset{A}{>}}CH_2 + H-C\underset{N(R)_2}{\overset{N(R)_2}{<}}CN \quad \xrightarrow[-HNR_2]{-HCN\ polym.} \quad \underset{B}{\overset{A}{>}}C=C\underset{N(R)_2}{\overset{H}{<}}$$

(10)

R = alkyl
A, B = activating groups

Reactions of active methylene compounds with **1, 3, 5–7,** and **10** is effected by heating at 80–160°C while the more volatile products (alcohols, amines) are distilled off. Dipolar aprotic solvents like DMF strongly assist the reaction (26). Very often orthoformamides may be reacted in situ by adding first an alkoxydialkyliminium salt and subsequently the methylene compound to the alkoxide in alcohol (35, 53).

Tris(acylamino)methane (**8**) reacts with active methylene compounds at 150–200°C giving acylaminomethylene derivatives (54, 55). In the reaction with tris(formylamino)methane, *N*-formylformamidine was shown to be the intermediate and the actual formylating agent (21, 22). These reactions are important in the synthesis of pyrimidines (see Section III-H-2-f).

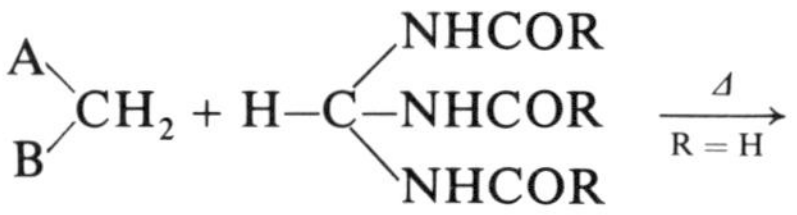

$$\underset{B}{\overset{A}{>}}CH_2 + H-C(NHCOR)_3 \quad \xrightarrow[R=H]{\Delta}$$

(8)

$$H-C\underset{NH_2}{\overset{N-COR}{<}} \quad \xrightarrow[-NH_3]{\overset{A}{\underset{B}{>}}CH_2} \quad \underset{B}{\overset{A}{>}}C=C\underset{}{\overset{H}{|}}-NH-CRO$$

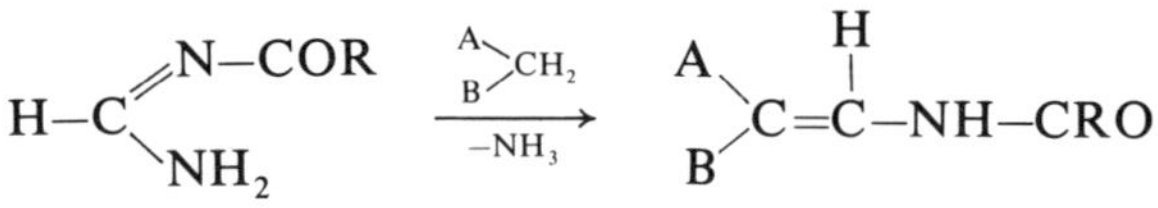

Tables III through IX give a survey of the reaction conditions, scope, and limitations of the enamine synthesis using orthoformamides.

2. Carboxylic acid orthoamide and Lactam Acetals with Active Methyl and Methylene Groups

Like N,N-dialkylformamide acetals, their homologues N-alkyllactam acetals react with carbon acids to give enamines (4, 62–64). Here also the reactivity is analogous to that of orthoformic acid amides. First of all, increased reactivity because of better stabilized ambident cations is observed. Thus N,N-dimethylacetamide diethyl acetal (1e) reacts with acidic compounds faster than dimethylformamide diethyl acetal.

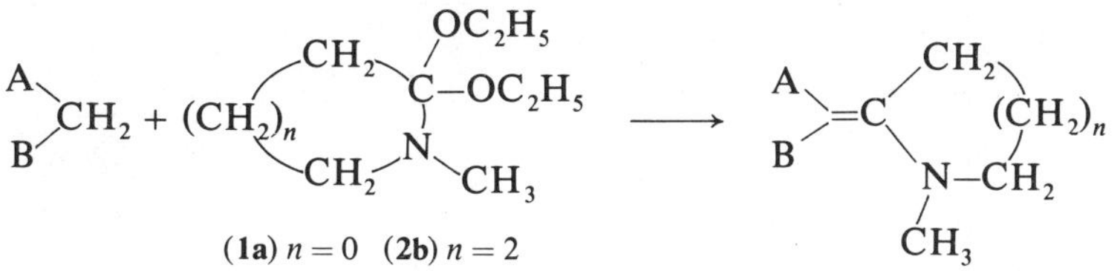

(1a) $n = 0$ (2b) $n = 2$
(2a) $n = 1$ (2c) $n = 3$

The effect of the carbiminium stabilization is clearly demonstrated by the case of the relatively reactive acetal 2e (4). On the other hand, destabilization

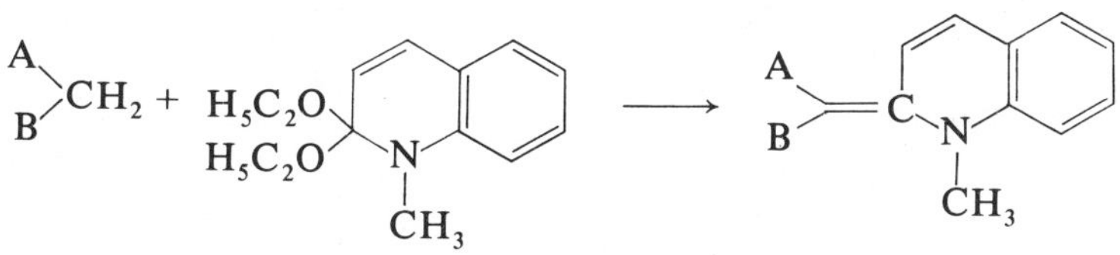

(2e)

of the carbiminium ions by electron-withdrawing substituents causes a remarkable decrease in reactivity (29).

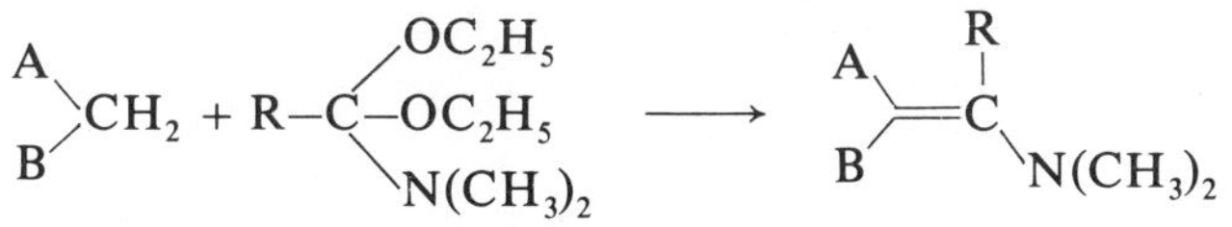

(1e–1g)

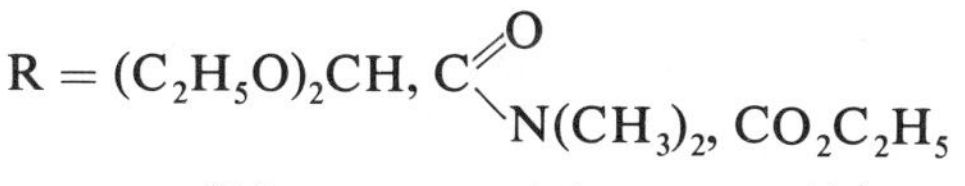

$R = (C_2H_5O)_2CH$, $C\begin{smallmatrix}O\\N(CH_3)_2\end{smallmatrix}$, $CO_2C_2H_5$
   (1e)   (1f)   (1g)

TABLE III
Enamines from N,N-DMF Dialkyl Acetals **1** (R = H) and Active Methylene Compounds

Starting compound	Acetal[a]	Reaction conditions	Reaction product	Yield, %	Ref.
Nitro compounds					
CH_3-NO_2	**1b**	60–80°C, ethanol	$(CH_3)_2N-CH=CHNO_2$	60–85	4, 33, 35
Nitriles					
$R-CH_2-CN$ R = CN, CO_2CH_3, $CONH_2$	**1b**	20–80°C, 1–4 hr, ethanol	$R-C=CN$ $\quad\ \ \|$ $\quad\ \ CH$ $\quad\ \ \|$ $\quad\ \ N(CH_3)_2$	60–93	4, 25, 56
$HOOC-CH_2-CN$	**1a** N,N-DMF dineopentyl acetal	80°C, 2 hr, benzene	$\quad\ \ H\ \ H$ $\quad\ \ \|\ \ \ \|$ $(CH_3)_2N-C=C-CN$	96	10
1,3-Dicarbonyl compounds					
$R-\underset{\underset{O}{\|\|}}{C}-CH_2COR$	**1a, b**	80°C, 1–2 hr.	$R-\underset{\underset{O}{\|\|}}{C}-\underset{\underset{CHN(CH_3)_2}{\|}}{C}-COR$	68–77	4, 44
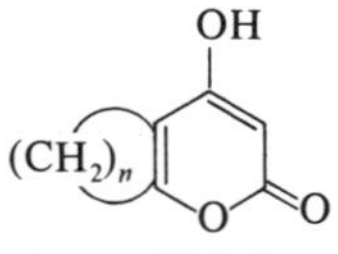 ![pyranone with OH and $(CH_2)_n$] $n = 0, 1, 2$	**1a**	20°C, 1,4-dioxane	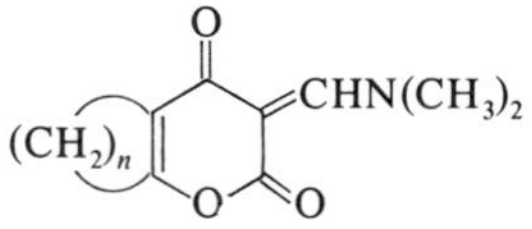 ![pyranone product with $CHN(CH_3)_2$ and $(CH_2)_n$]	30–80	45

Ketone	Reagent	Conditions	Product	Yield (%)	Refs.
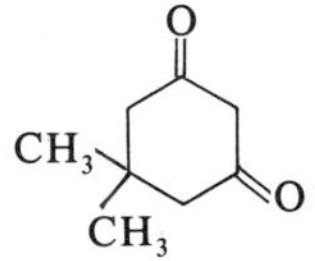 ![ketone1] O=C-CH₃ with CH₃, CH₃ cyclohexanedione	1a	80°C, 1 min, benzene	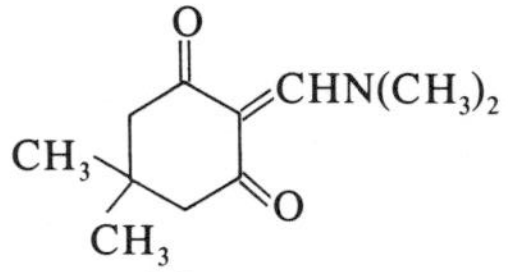 dimethylaminomethylene cyclohexanedione	100	27
Ketones					
$R-\overset{O}{\underset{\|}{C}}-CH_3$ $R = C_6H_5, 2\text{-}C_6H_4OH,$ $C_6H_5\text{-}CH=CH, CH_3$ $C_6H_5\text{-}COCH_2\text{-}C_6H_5$	1a, b	80–120°C	$R-\overset{O}{\underset{\|}{C}}-CH=C\overset{H}{\underset{N(CH_3)_2}{}}$	30–90	4, 25, 38
	1a, c	80–100°C, 30 min	$C_6H_5COC\text{-}C_6H_5$ with $=CHN(CH_3)_2$ and H	60	27
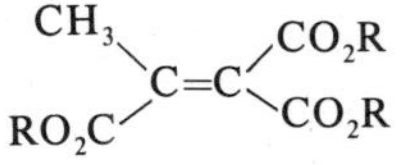 $\overset{CH_3}{\underset{RO_2C}{}}C=C\overset{CO_2R}{\underset{CO_2R}{}}$	1b	80°C, 3 hr, DMF	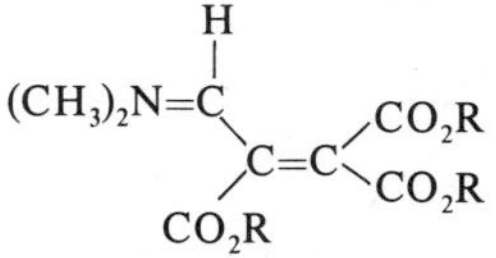 $(CH_3)_2N\text{-}CH=C\overset{CO_2R}{\underset{CO_2R}{}}$ with CHN(CH₃)₂ and H	83–87	41
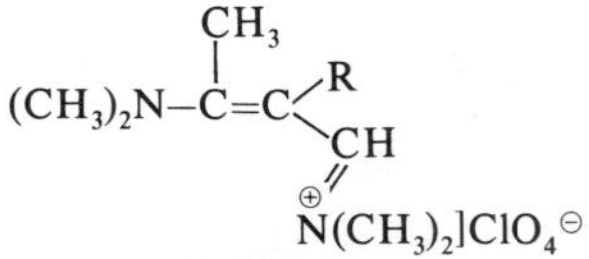 $(CH_3)_2N-C=C\overset{CH_3}{\underset{R}{}} \overset{\oplus}{CH}$ $[N(CH_3)_2]ClO_4^{\ominus}$ $R = H, CH_3, C_6H_5$	1a, b	80°C, 4 hr, DMF	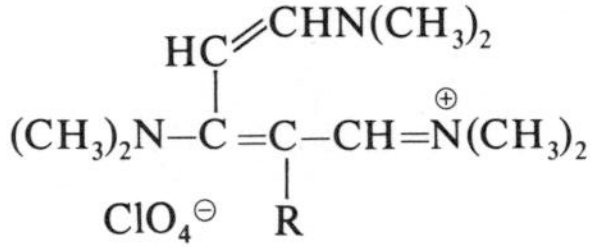 $(CH_3)_2N-C=C-CH=\overset{\oplus}{N}(CH_3)_2$ with $HC=CHN(CH_3)_2$, R, $ClO_4^{\ominus}$	35–88	36
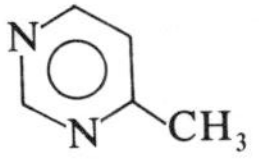 2-methylpyrazine	1b	160°C, 30 hr	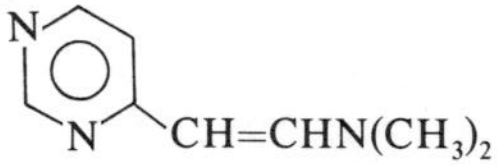 2-(dimethylaminovinyl)pyrazine CH=CHN(CH₃)₂	70	26, 34
 cyclopentadiene	1b	~130°C	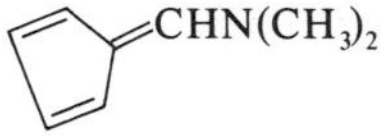 cyclopentadienylidene =CHN(CH₃)₂	32	4

[a] **1a** = *N,N*-DMF dimethyl acetal; **1b** = *N,N*-DMF diethyl acetal; **1c** = *N,N*-DMF di-*t*-butyl acetal.

TABLE IV

Enamines from Aminal Esters (5) and Active Methylene Compounds

Starting material	5[a]	Reaction conditions	Reaction product	Yield, %	Ref.
Nitriles					
R—CH$_2$—CN R = CN, CO$_2$—CH$_3$	5a	20°C, 30–60 min	R—C(CN)=CHN(CH$_3$)$_2$	86–89	25, 56
R—CH$_2$—CN	5a, c	80–160°C, 3–24 hr	R—C(CN)=CH—CH=N(CH$_3$)$_2$	50–85	56, 57, 58
R = H, CH$_3$, C$_2$H$_5$, C$_6$H$_5$, N(CH$_3$)$_2$, OCH$_3$, SC$_2$H$_5$, (C$_2$H$_5$O)$_2$—P=O					
CH$_3$—CH=CHCN	5c	120°C, 20 hr.	NC—CH=CH—CH=CH—N(CH$_3$)$_2$	45	56
Ketones					
R—C(=O)—CH$_2$R^1	5a, c	80–140°C, 1–12 hr	R—C(=O)—C(R^1)=CH—N(CH$_3$)$_2$	64–92	25, 57

R	CH$_3$	CH$_3$	C$_6$H$_5$	C$_6$H$_5$	subst. C$_6$H$_5$
R^1	H	CH$_3$	H	N(CH$_3$)$_2$	N(CH$_3$)$_2$

$CH_3C-CH_2N(CH_3)_2$ $\parallel$ O	5c	140°C, 3 hr	$(CH_3)_2N-CH=CH-CHC-CH_2$ $\|$ $N(CH_3)_2$ $\|\|$ O	64 57
![cyclobutanone] $R\ R^1$ (O= on cyclobutane with R, R¹)	5c	50°C	![cyclobutanone with CHN(CH₃)₂] $R\ R^1$	~95 43
$R=H, R^1 = $ (4-methylcyclohexenyl)				
$R-R^1 = $ (octahydrophenanthrene / decalin structures)				

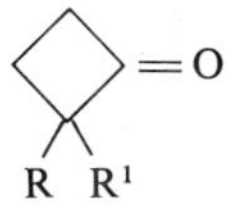

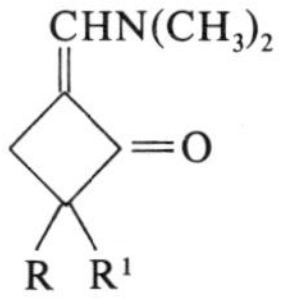

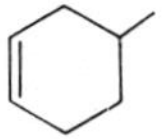

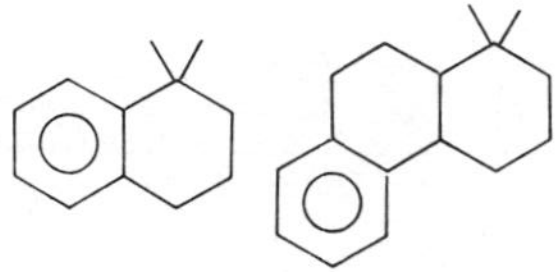

Carboxylic acid esters

$R-CH_2C\overset{O}{\underset{}{\diagdown}}_{OR^1}$	5c	160°C, 3–20 hr	$R-C\overset{CO_2R^1}{\underset{CHN(CH_3)_2}{\diagup}}$	48–88 32, 57
R H H CH₃ N(CH₃)₂ N(CH₂)₅ C₂H₅				
R¹ C₂H₅ c-C₆H₁₁ C₂H₅ C₂H₅ C₂H₅				

R   OC₂H₅   SC₂H₅   CH₂CO₂C₂H₅   P(O)   NHCHO
                                                    $\|\|$
                                                    O				

R C₂H₅ C₂H₅ C₂H₅ C₂H₅ C₂H₅				
R $H_5C_6\overset{O}{\underset{}{C}}-N$ (piperidine with vinyl and methyl substituents)				
R¹ CH₃				

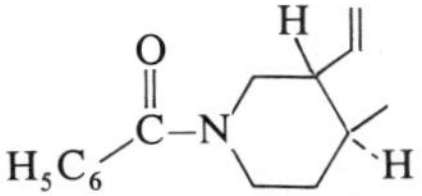

TABLE IV (cont.)

Starting material	5^a	Reaction conditions	Reaction product	Yield, %	Ref.
$CH_3CH=CHCO_2C_2H_5$	5c	140°C, 1 hr	$(CH_3)_2N-CH=CH-CH=CH-\overset{\mid}{C}HCO_2C_2H_5$	69	32
$(CH_3)_3Si-CH_2CO_2C_2H_5$	5c	160°C, 6 hr	$(CH_3)_2N-CH=CHCO_2C_2H_5$	77	58
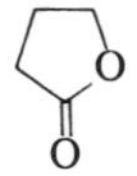 (butyrolactone)	5c	160°C, 2–3 min	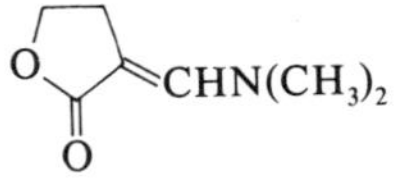 =CHN(CH_3)_2 (α-dimethylaminomethylene-γ-butyrolactone)	64	32

Carboxylic acid thioesters

Starting material	5^a	Reaction conditions	Reaction product	Yield, %	Ref.
$R-CH_2-C(=O)SC_2H_5$ R = H, CH_3	5c	110–140°C, ½–1 hr	$R-C(=O)-C(SC_2H_5)$ $\;\;\;\;\;\;\;\;\;\;$ CHN(CH_3)_2	44–48	32

Thiocarboxylic acid esters

Starting material	5^a	Reaction conditions	Reaction product	Yield, %	Ref.
$CH_3-CH_2-C(=S)R$ R = OC_2H_5, SC_2H_5	5c	130–140°C, 20–30 min	$CH_3-C(=S)-C\cdot R$ $\;\;\;\;\;\;\;\;\;\;$ CHN(CH_3)_2	23–58	32

Imidates

Starting material	5^a	Reaction conditions	Reaction product	Yield, %	Ref.
$C_6H_5-CH_2-C(=N-R)OC_2H_5$	5c	160°C, 16 hr	$C_6H_5-C(=N-R)-C(OC_2H_5)$ $\;\;\;\;\;\;\;\;\;\;$ CHN(CH_3)_2	23–83	58
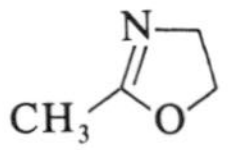 (2-methyl-4,5-dihydrooxazole)	5c	150°C, 48 hr	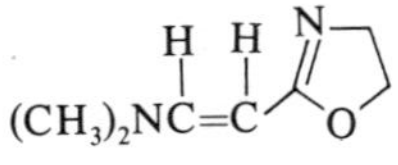 H H N O ring with $(CH_3)_2NC=C$	7–24	58

414

Isonitriles

Structure	Reagent	Conditions	Yield	Ref
C≡N–CH$_2$–CO$_2$C$_2$H$_5$	5c	20°C, 24 hr	74	58
C≡N–CH$_2$–C$_6$H$_5$	5c	160°C, 3 hr	91	58
C≡N–C(CO$_2$C$_2$H$_5$)=CHN(CH$_3$)$_2$			58	
(CH$_3$)$_2$NCH=N–C(C$_6$H$_5$)=CHN(CH$_3$)$_2$			58	

Carboxamides

Structure	Reagent	Conditions	Yield	Ref
CH$_3$CON(CH$_3$)R; R = H, CH$_3$	5c	160°C, 4 hr	51–63	32
R–CH$_2$–C(=O)N(CH$_3$)R^1	5c	160°C, 24 hr	5–15	32
(CH$_3$)$_2$N–CH=CH–CON(CH$_3$)R				
R, CH$_3$, CH$_3$ / R^1, H, CH$_3$; R–R^1 = (CH$_2$)$_{3-5}$				
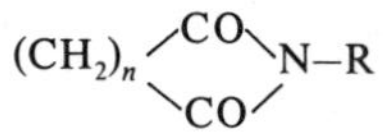 (CH$_2$)$_n$(CO)$_2$N–R; R = H, CH$_3$; n = 2, 3	5c	110–120°C, 1–15 hr	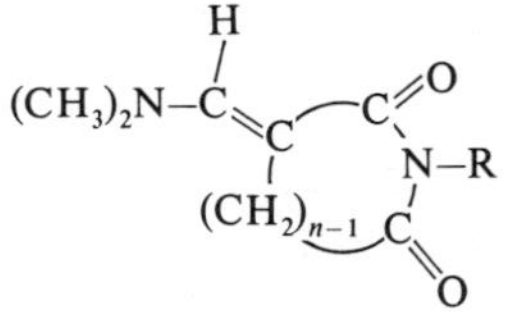 46–78	23, 5
cyclic imide with (CH$_2$)$_{n-1}$, =CHN(CH$_3$)$_2$ substituent			49	
Dipiperidinomethoxymethane		140°/2 hr 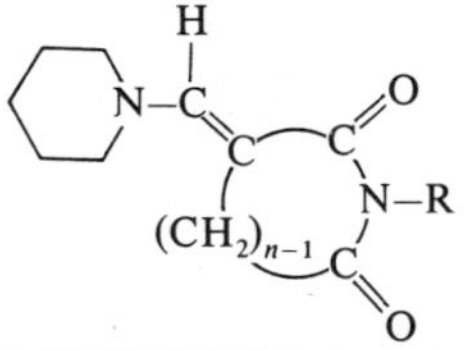		23
cyclic imide with (CH$_2$)$_{n-1}$, =CH-piperidino substituent				

TABLE IV (cont.)

Starting material		Reaction conditions	Reaction product	Yield, %	Ref.
Thioamides					
R–CH$_2$–C(=S)–N(CH$_3$)(R^1)	5^a	120–140°C, 30–60 min	(CH$_3$)$_2$N–C(H)=C(R)–C(=S)–N(R^1)(CH$_3$)	72–94	32
R H H CH$_3$ CH$_3$					
R^1 H CH$_3$ H CH$_3$					
R–R′ = –(CH$_2$)$_3$–, –(CH$_2$)$_4$–					
Amidines					
CH$_3$C(=NR)–N(CH$_3$)$_2$	5c	160°C, 16–50	(CH$_3$)$_2$N–C(=NR)–CH=CHN(CH$_3$)$_2$	5–52	58
R = CH$_3$, C$_6$H$_5$					
Phosphonium salts					
(C$_6$H$_5$)$_3$P$^⊕$–CH$_2$–C$_6$H$_4$R	5c	140–160°C, ½–1 hr	(C$_6$H$_5$)$_3$P=C(C$_6$H$_4$R)–CHN(CH$_3$)$_2$	60–90	39
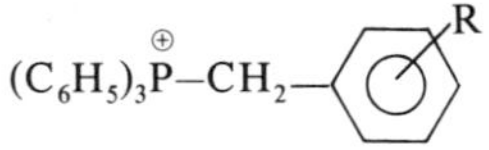 R = H, 4-OCH$_3$, 4-F, 4-Cl, 4-Br, 4-CH$_3$, 4-N(CH$_3$)$_2$			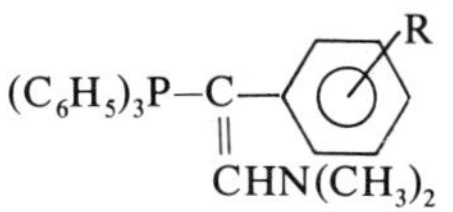		
Phosphonic acid esters					
(H$_5$C$_2$O)$_2$P(=O)–CH$_2$C$_6$H$_5$	5c	160°C, 4 hr	(H$_5$C$_2$O)$_2$P(=O)–C(C$_6$H$_5$)=	60	58

Substituted toluenes

Substrate	Reagent	Conditions	Product	Yield (%)	Refs
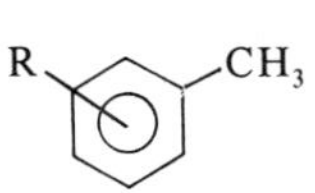 R—C$_6$H$_4$—CH$_3$ (R = 2-NO$_2$, 4-NO$_2$, 2-CN, 4-CN, 4-SO$_2$N(CH$_3$)$_2$, 2,6-dichloro, 3-methyl-2-nitro)	5c	160°C, 2–24 hr	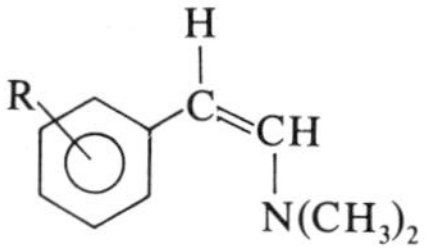 R—C$_6$H$_4$—CH=CH—N(CH$_3$)$_2$	45–75	26, 28, 61
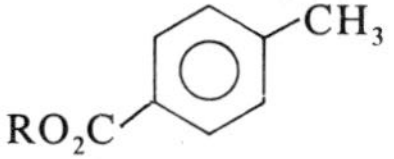 4-RO$_2$C—C$_6$H$_4$—CH$_3$	5c	160°C, 6–24 hr	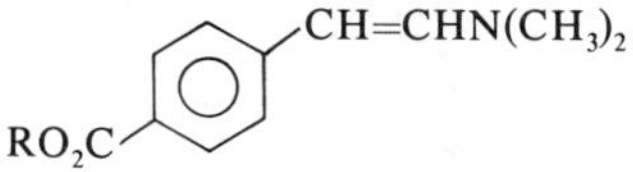 RO$_2$C—C$_6$H$_4$—CH=CHN(CH$_3$)$_2$	10–15	26
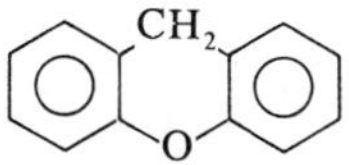 xanthene-CH$_2$ (9H-xanthene)	5c	160°C, 5 hr	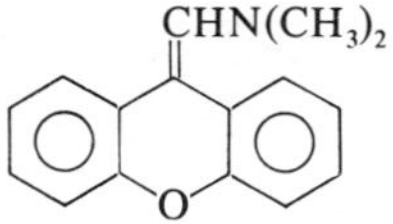 9-[=CHN(CH$_3$)$_2$]-xanthene	17	26

Methyl-substituted heterocycles

Substrate	Reagent	Conditions	Product	Yield (%)	Refs
Het—CH$_3$ (Het. = 4-pyridyl-, 2-pyridyl N-oxide-, 4-quinolyl-, 2-quinolyl-, 9-acridyl-, 4-(3-cyano)-pyridyl-, 2-benzoxazolyl-, 2-benzimidazolyl-)	5c	160°C, 1–12 hr	Het—CH=CHN(CH$_3$)$_2$	32–84	26, 42
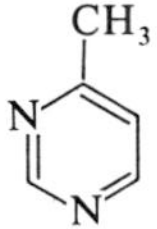 4-methylpyrimidine	5c	160°C, 1 hr	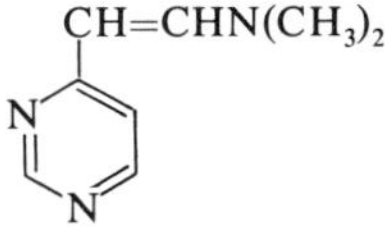	70	26
4-methylpyrimidine	5c	130°C, ½ hr, DMF	4-pyrimidinyl—CH=CHN(CH$_3$)$_2$	94	26
4-methylpyrimidine	5b	130°C, 1 hr, DMF		62	26

[a] **5a** = bis(dimethylamino)methoxymethane; **5b** = bis(dimethylamino)ethoxymethane; **5c** = bis(dimethylamino)-t-butoxymethane.

TABLE V

Bisenamines from Aminal Esters (5) and Active Methylene Compounds

Starting material	5[a]	Reaction conditions	Reaction product	Yield, %	Ref.
$NC-CH_2-CH_2CN$	5c	65°C, 1 hr	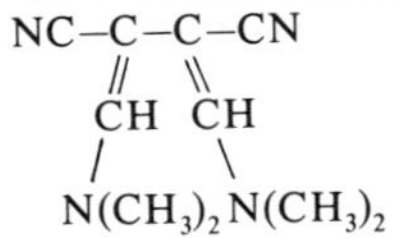 $NC-C=C-CN$ with CH CH and $N(CH_3)_2$ $N(CH_3)_2$	85	23
$(C_6H_5-SO_2)_2(CH_2)_2$	5c	140–160°C, 20 min	$(C_6H_5-SO_2)_2[C=CHN(CH_3)_2]_2$	95	40, 57
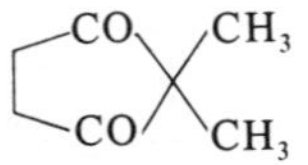 (cyclic diester, CO_2CH_3 / CO_2CH_3 on ring)	5c	70–80°C, 2 hr	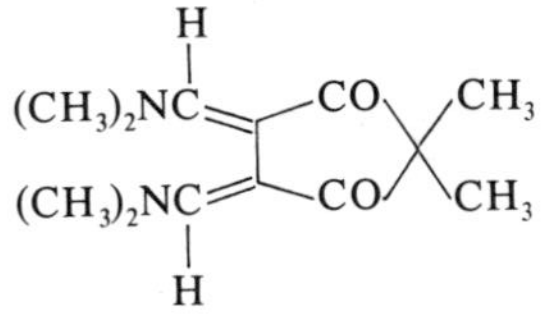 $(CH_3)_2NC$ and $(CH_3)_2NC$ substituted ring with $CO-CH_3$, $CO-CH_3$	91	23
$CH_3-COCH_2-CH_2-CO_2C_6H_5$	5c	120°C, 3 hr	$(CH_3)_2NCH=CH-C-CH_2-C-CO_2C_2H_5$ with $\overset{\|}{O}$ and $\overset{CH}{\|}$, $N(CH_3)_2$	50	23
$H_5C_2O_2CCH_2-CH_2-CO_2H_5$	5c	160°C, 5 hr	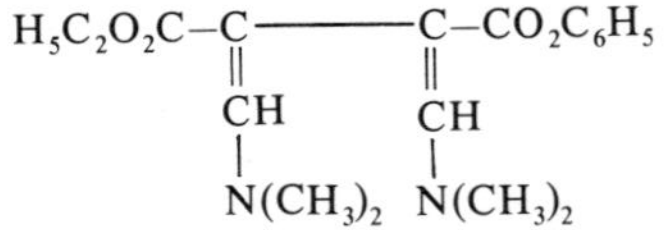 $H_5C_2O_2C-C=C-CO_2C_6H_5$ with CH CH and $N(CH_3)_2$ $N(CH_3)_2$	6	23
$H_5C_2O_2C-(CH_2)_4-CO_2C_2H_5$	5c	170°C, 4 hr	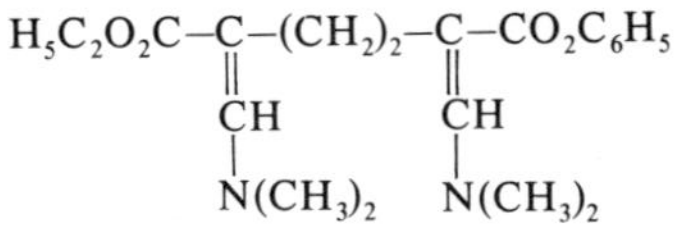 $H_5C_2O_2C-C=C-(CH_2)_2-C=C-CO_2C_6H_5$ with CH CH and $N(CH_3)_2$ $N(CH_3)_2$	16	23

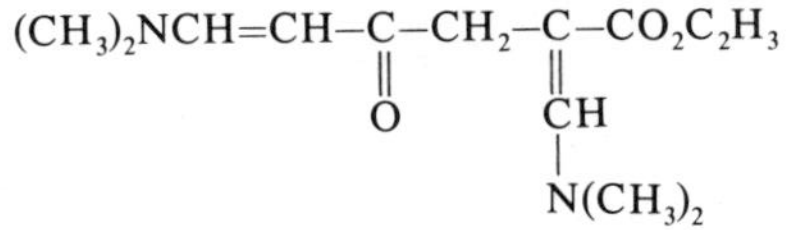

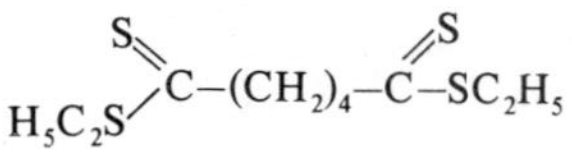

Structure	Reagent	Conditions	Product	Yield (%)	Ref.
$H_5C_2S\overset{S}{\overset{\|}{C}}-(CH_2)_4-\overset{S}{\overset{\|}{C}}-SC_2H_5$	5c	140°C, 30 min	$H_5C_2SC(=S)-\overset{CH-N(CH_3)_2}{\underset{CH-N(CH_3)_2}{C}}-(CH_2)_2-\overset{CS_2H_5}{\underset{S}{C}}=CH-N(CH_3)_2$	48	23
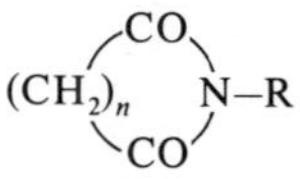 $\overset{CO}{\underset{CO}{(CH_2)_n}}\!\!>\!\!N\text{—}R$ $n = 2, 3;\ R = H, CH_3, C_2H_5$	5c	120°C, 1–4 hr	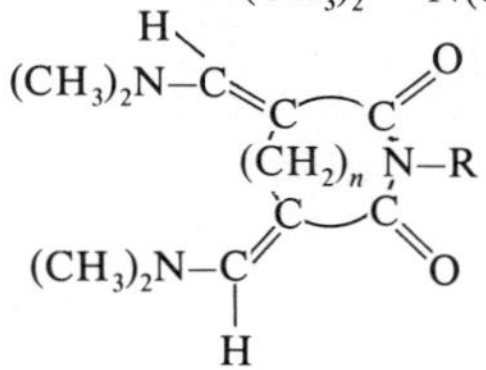 $(CH_3)_2N-\overset{H}{\underset{}{C}}=\!\!\overset{CO}{\underset{CO}{(CH_2)_n}}\!\!N\text{—}R$ $(CH_3)_2N-\overset{H}{\underset{}{C}}=$	64–85	23
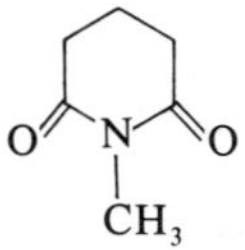	Dipiperidino-methoxymethane	150°C, 2 hr	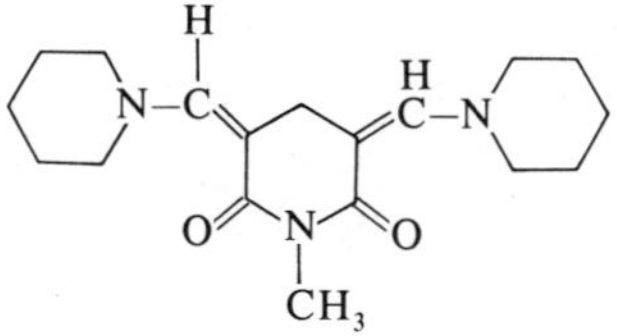 piperidino=CH—(glutarimide N-CH$_3$)—CH=piperidino	60	23
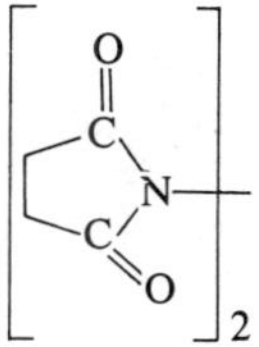 $\left[\overset{O}{\underset{O}{}}\!\!\!>\!\!N-\right]_2$	5c	150°C, 2 hr	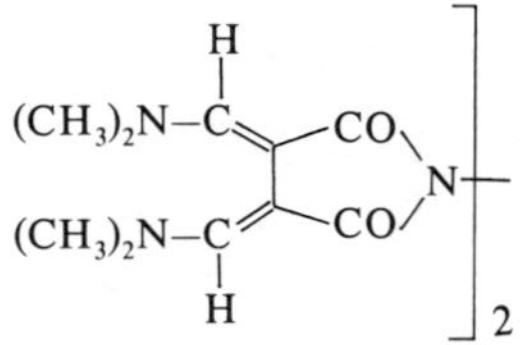 $\left[(CH_3)_2N-CH=\overset{CO}{\underset{CO}{}}\!\!>\!\!N-\right]_2$	52	23

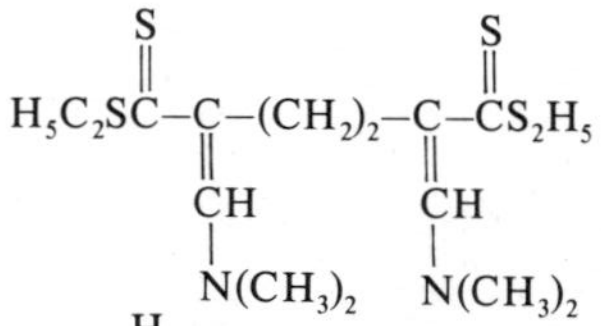

[a] **5a** = bis(dimethylamino)methoxymethane; **5b** = bis(dimethylamino)ethoxymethane; **5c** = bis(dimethylamino)-*t*-butoxymethane.

TABLE VI
Enamines from Tris(dialkylamino)methanes (7) and Active Methylene Compounds

Starting material	7[a]	Reaction conditions	Reaction product	Yield, %	Ref.
$R-CH_2-CN$ $R = CN, CO_2CH_3, C_6H_5$	7a	20°C, 15–20 min, diethyl ether	$R-C(=CHN(CH_3)_2)-CN$	73–88	46, 60
$R-COCH_3$ $R = CH_3, C_6H_5$	7a	20–35°C, 2 hrs diethyl ether	$R-C(=O)-CH=CHN(CH_3)_2$	60–80	46, 60
 O_2N-C$_6$H$_4$-CH$_3$	7a	150°C, 5 hrs	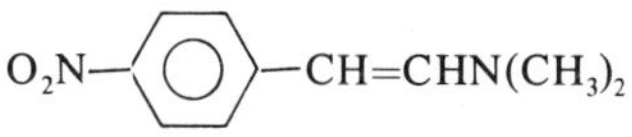 O_2N-C$_6$H$_4$-CH=CHN(CH_3)_2	100	46, 60
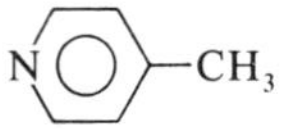 4-methylpyridine	7a	150°C, 5 hr	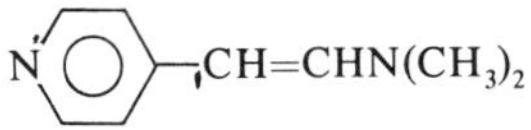 4-(CH=CHN(CH_3)_2)pyridine	52	46, 60
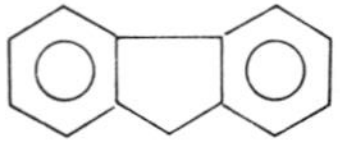 fluorene	7a	150°C, 5 hr	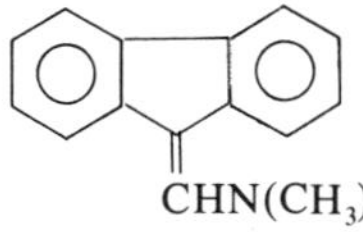 9-(CHN(CH_3)_2)-fluorene	95	30
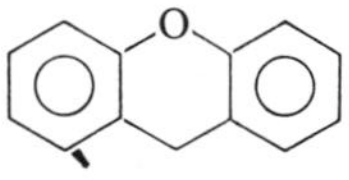 xanthene	7a	150°C, 48 hr	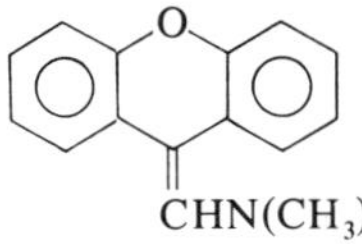 9-(CHN(CH_3)_2)-xanthene	71	30
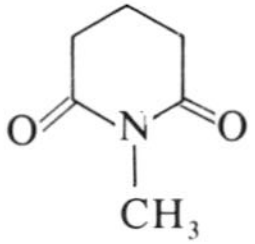 N-methylglutarimide	Trismorpholinomethane	140°C, 2 hr	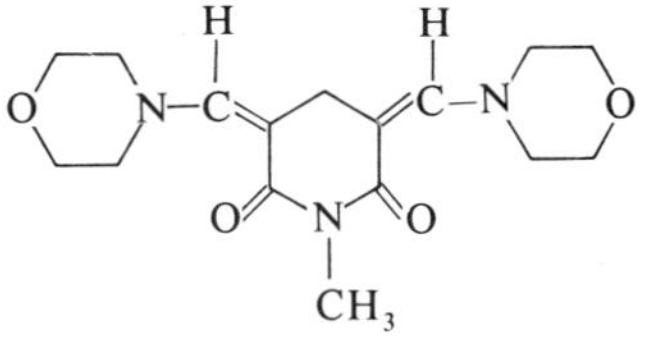 bis-morpholinomethylene-N-methylglutarimide	38	23

TABLE VII

Enamines from N,N-DMF Dialkyl Dithioacetals (**3**, R = H) or Aminal Thioesters (**6**) and Active Methylene Compounds

Starting material	3 or 6	Reaction conditions	Reaction product	Yield, %	Ref.
CH_3-NO_2	$HC(SC_6H_5)_2$ $\mid$ $N(CH_3)_2$		$O_2N-CH=CHN(CH_3)_2$	77	48
$R-CH_2CN$ $R = CN, CO_2CH_3$	$HC(SR^1)_2$ $\mid$ $N(CH_3)_2$ $R^1 = C_2H_5, C(CH_3)_3$	80°C	$NC-C=CHN(CH_3)_2$ $\mid$ R	50–64	24
$C_6H_5COCH_3$	$HC(SC_2H_5)_2$ $\mid$ $N(CH_3)_2$	70–80°C, 2 hr	$C_6H_5COCH=CH$ $\mid$ $N(CH_3)_2$	72	48
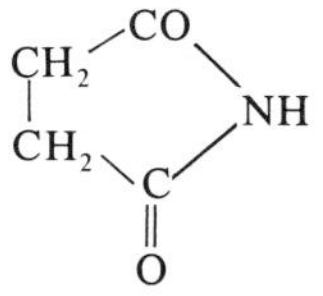 ![oxazolidinone structure] CH$_2$–CO / NH \ CH$_2$–C=O	$HC(SC_2H_5)_2$ $\mid$ $N(CH_3)_2$	170°C, 3 hr	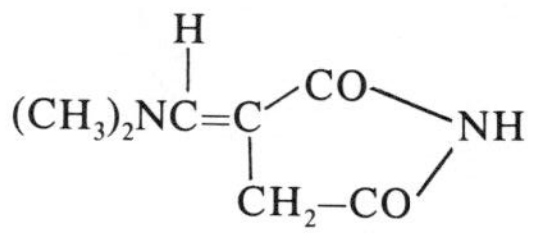 $(CH_3)_2NC=C\underset{H}{\overset{CO-NH}{\underset{CH_2-CO}{\diagup\diagdown}}}$	50	48
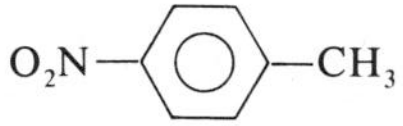 O_2N–C$_6H_4$–CH_3	SC_2H_5 $\mid$ $HC[N(CH_3)_2]_2$	160°C, 30 min	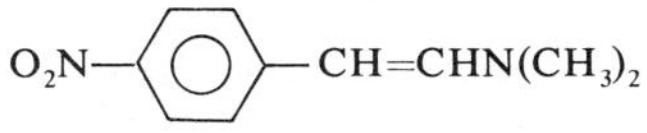 O_2N–C$_6H_4$–$CH=CHN(CH_3)_2$	47	24
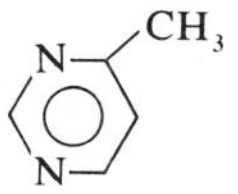 2-(CH_3)-pyrazine	$SC(CH_3)_3$ $\mid$ $HC[N(CH_3)_2]_2$	170–180°C, 1 hr	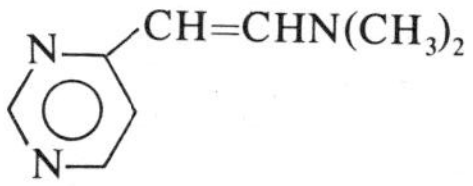 pyrazinyl–$CH=CHN(CH_3)_2$	63	24

421

TABLE VIII
N-Acylenamines from Tris(acylamino)methanes (8) and Active Methylene Compounds

Starting material	HC(NHCOR)$_3$ (8)	Reaction conditions	Reaction product	Yield, %	Ref.
R^{1}CH$_2$CO$_2$CH$_3$ R^1 = CO$_2$CH$_3$, CN	R = H	150–160°C, 1–2 hr	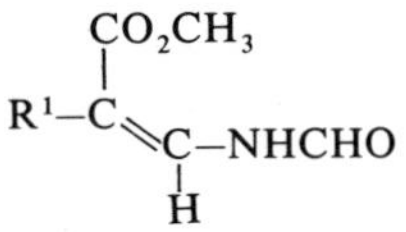 R^1–C(CO$_2$CH$_3$)=C(H)–NHCHO	18–48	55
C$_6$H$_5$COCH$_2$R^1 R^1 = H, CH$_3$, C$_2$H$_5$	R = CH$_3$	200°C, 1–2 hr	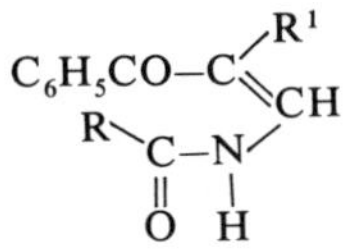 C$_6$H$_5$CO–C(R^1)=CH–N(CH)–C(R)=O (enamide)	15–32	54
	R = C$_2$H$_5$	160–200°C, 1½ hr	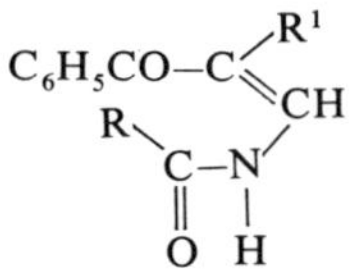 C$_6$H$_5$CO–C(R^1)=CH–N(CH)–C(R)=O	11–37	54
R^1 = H	R = n-C$_3$H$_7$	160°C, 1½ hr	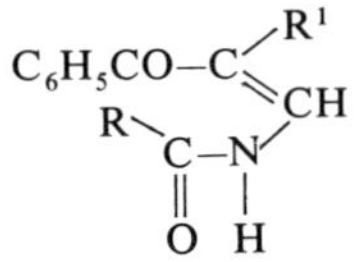 C$_6$H$_5$CO–C(R^1)=CH–N(CH)–C(R)=O	37	54
R^1 = H	R = C$_6$H$_5$, 4-CH$_3$O–C$_6$H$_4$	160°C, 1½ hr	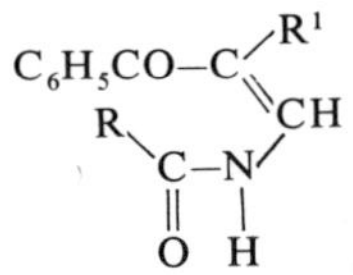 C$_6$H$_5$CO–C(R^1)=CH–N(CH)–C(R)=O	21–22	54
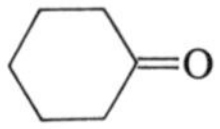 cyclohexanone	R = CH$_3$	190°C	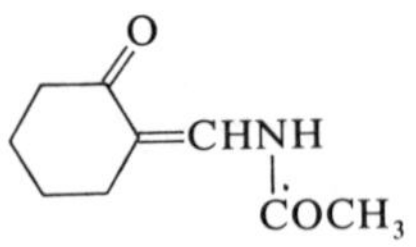 2-(α-acetamidomethylene)cyclohexanone (=CHNH–COCH$_3$)	15	54

TABLE IX

Enamines from Carboxylic Acid Orthoamides (1) or Lactam Acetals (2) and Active Methylene Compounds

Starting material	Carboxylic acid orthoamide or lactam acetal	Reaction conditions	Reaction product	Yield, %	Ref.
$R-CH_2-R^1$	$CH_3-C(OC_2H_5)_2$ $\|$ $N(CH_3)_2$	20–130°C[a]	$R-C-R^1$ $\|\ \ \|$ $CH_3\ N(CH_3)_2$	30–51	4, 65
R H CN CO_2CH_3 R^1 NO_2 CN CN R H H H R^1 —⟨NO_2,NO_2⟩ C_6H_5CO					
⟨cyclopentadiene⟩	$CH_3-C(OC_2H_5)_2$ $\|$ $N(CH_3)_2$	130°C,[a] 30 min	⟨cyclopentadienylidene⟩=C(CH_3)N(CH_3)_2	40	4, 65
CH_3NO_2	$R-C(OC_2H_5)_2$ $\|$ $N(CH_3)_2$ $R = CH(OC_2H_5)_2, CO_2C_6H_5,$ $CON(CH_3)_2$	80°C, 1–5 hrs	$R-C=CHNO_2$ $\|$ $N(CH_3)_2$	37–96	29
CH_3NO_2	$(H_5C_2O)_2-C$—⟨C_6H_4⟩—$C(OC_2H_5)_2$ $\|\ \|$ $N(CH_3)_2\ \ \ \ \ \ \ \ \ \ \ \ \ \ \ \ N(CH_3)_2$	80°C, 1 hr	$O_2N-C=C$—⟨C_6H_4⟩—$C=CNO_2$ $\ \ \ \ \ |\ |\ \ \ \ \ \ $ H N(CH_3)_2 $\ \ \ $ N(CH_3)_2 H	60	29

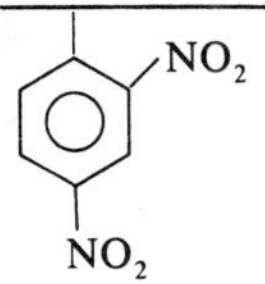

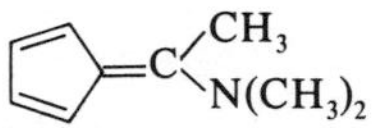

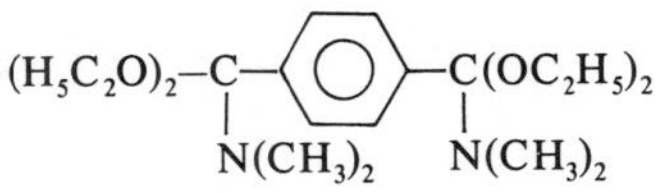

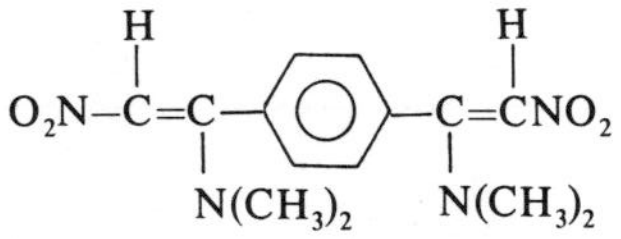

TABLE IX (cont.)

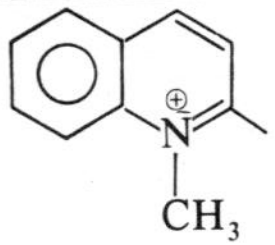

Starting material	Carboxylic acid orthoamide or lactam acetal	Reaction conditions	Reaction product	Yield, %	Ref.
R–CH$_2$–R^1	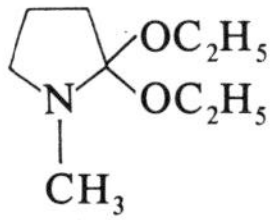 [pyrrolidine with OC$_2$H$_5$, OC$_2$H$_5$, N–CH$_3$]	75–80 °C [a]	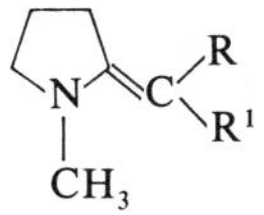 [pyrrolidinylidene with =C(R)(R^1), N–CH$_3$]	31–88	4, 65
R H CN CO$_2$CH$_3$ H					
R^1 NO$_2$ CN CN O$_2$N–C$_6$H$_3$–NO$_2$					
R H H					
 R^1 C$_6$H$_5$CO [quinolinium salt with CH$_3$, N–CH$_3$]	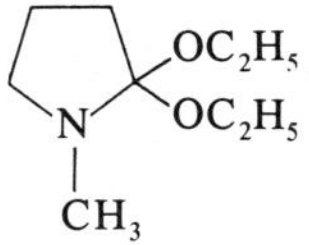 [pyrrolidine with OC$_2$H$_5$, OC$_2$H$_5$, N–CH$_3$]	70 °C	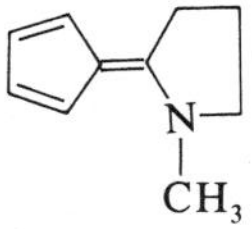 [cyclopentadienylidene-pyrrolidine with N–CH$_3$]	75	4, 65
[cyclopentadiene]	[dihydroquinoline with OC$_2$H$_5$, OC$_2$H$_5$, N–CH$_3$]				
R–CH$_2$–R^1	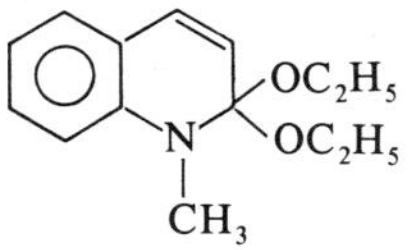	20–100 °C [a]	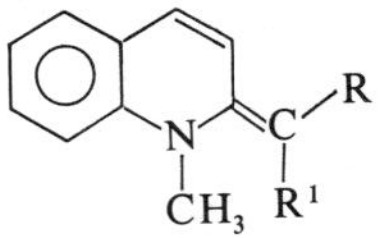 [quinoline with =C(R)(R^1), N–CH$_3$]	89–96	4, 66

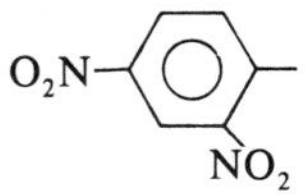

R	R¹	Starting material	Conditions	Product	Yield	Ref.

| R | H | | | | | |
| R¹ | O₂N-C₆H₃-NO₂ | quinolinium salt with N-CH₃ | 80°C | | 94 | 4, 66 |

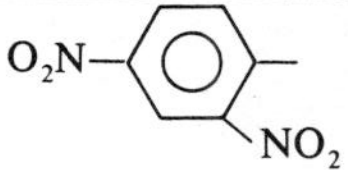

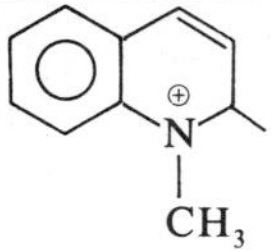

R–CH₂–R¹

| R | H | CO₂C₂H₅ | CONH₂ |
| R¹ | NO₂ | CN | CN |

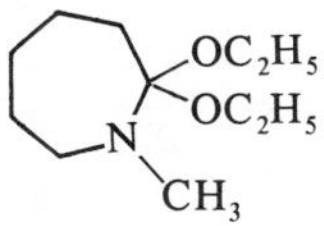

azepane with 2,2-(OC₂H₅)₂, N–CH₃ → 20°C, 30 min–24 hr → azepane with =C(R)(R¹), N–CH₃ ; 36–80 ; 62

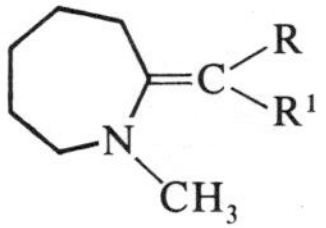

R–CH₂–R¹

R	CO₂C₂H₅	H	CN
R¹	CO₂C₂H₅	COC₆H₅	C₆H₅
R	CO₂CH₃	CO₂C₂H₅	
R¹	C₆H₅	C₆H₅	

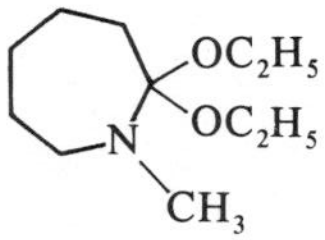

azepane with 2,2-(OC₂H₅)₂, N–CH₃ → 120–150°C, 3–6 hr → azepane with =C(R)(R¹), N–CH₃ ; 37–60 ; 62, 63

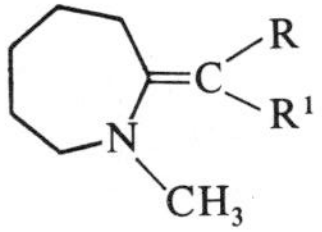

R–CH₂–R¹

R	H	CN	CONH₂
R¹	NO₂	CN	CN
R	CO₂C₂H₅	CN	H
R¹	CO₂C₂H₅	C₆H₅	CO-aryl

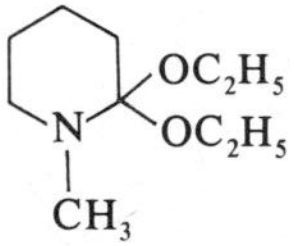

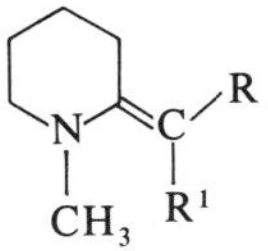

piperidine with 2,2-(OC₂H₅)₂, N–CH₃ → piperidine with =C(R)(R¹), N–CH₃

[a] Exothermic reaction.

The steric effects of a long carbon chain or increasing ring size slow down the recombination of the ions and thus the overall reaction rate. Consequently, thermodynamically controlled alkylation becomes a competitive reaction (63).

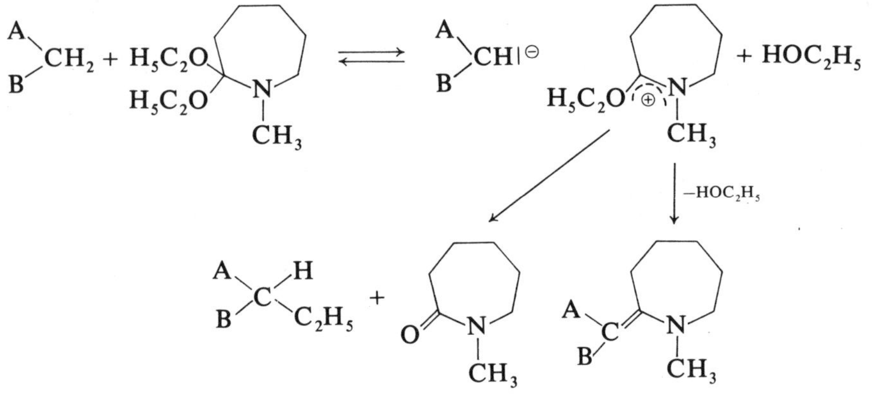

Thioacetals like N,N-dimethylbenzamide diethyl mercaptals react with stronger CH-acidic compounds such as malononitrile in the same way (49, 50).

3. Carbamic Acid Acetals, Urea Acetals, Alkoxytris(dialkylamino)methanes, and Tetrakis(dimethylamino)methane with Active Methyl and Methylene Groups

N,N-dialkylcarbamic acid acetals (11) are much less reactive than N,N-dialkylformamide acetals owing to their decreased dissociation and basicity.

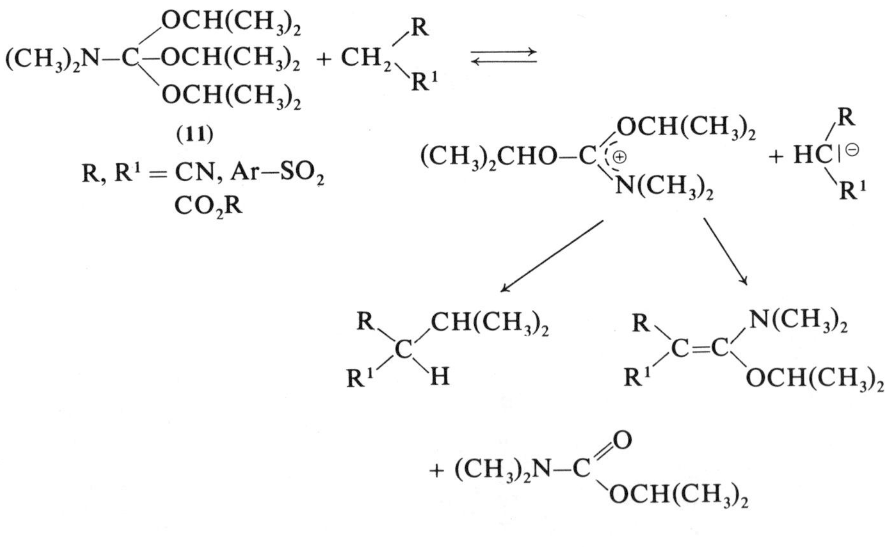

III. REACTIONS OF ORTHOAMIDES WITH NUCLEOPHILES 427

Only reactions with strong carbon acids (see Table X) yield enamines. The same holds true for cyclic compounds (4, 66). Moreover, alkylation becomes the competitive reaction, especially when the ambident cation is treated with weak nucleophiles (11).

Substitution of an alkoxy group by an alkylthio group apparently increases the reactivity (4) (see Table X).

Because of the formation of bis(dialkylamino)alkoxycarbonium ions, N,N'-tetraalkylurea acetals (12) are more reactive, and react with CH-acidic compounds in ether at room temperature (11) (see Table XI). The overall reaction becomes more complex, since dialkylamino groups easily exchange with the alkoxy groups of the urea acetal, and β-elimination of ethanol or dialkylamine from the intermediate may occur. Alkoxydialkylaminomethylene compounds are formed as well as bis(dialkylamino)methylene compounds. Introduction of bulky alkoxy substituents does not change the results (11), but the use of N,N-bridged urea acetals does (4).

As in the case of formic acid orthoamides, the reactivity of carboxylic acid and carbonic acid orthoamides increases with increasing replacement of alkoxy groups by dialkylamino groups. Thus alkoxytris(dialkylamino)methanes (13) are extremely reactive (11,12) (see Table XII).

These rules, based on the corresponding stability of the carbiminium ion intermediates, apply only as long as ionic recombination and β-elimination to

$$\underset{R^1}{\overset{R}{\diagdown}}CH_2 + (R^2)_2N-\underset{\underset{N(R^2)_2}{|}}{\overset{\overset{OR^3}{|}}{C}}-OR^3 \rightleftharpoons \underset{R^1}{\overset{R}{\diagdown}}\underset{\underset{N(R^2)_2}{|}}{\overset{\overset{OR \ H}{| \ |}}{C-C}}-N(R^2)_2$$

(12)

$$\swarrow_{-HOR^3} \qquad \downarrow_{-H(NR^2)_2}$$

$$\underset{R^1}{\overset{R}{\diagdown}}C=C\underset{\diagdown N(R^2)_2}{\diagup N(R^2)_2} \qquad \underset{R}{\overset{R^1}{\diagdown}}C=C\underset{\diagdown OR^3}{\diagup N(R^2)_2}$$

$$(R^2)_2N-\underset{\underset{N(R^2)_2}{|}}{\overset{\overset{OR^3}{|}}{C}}-OR^3 \underset{-HN(R^2)_2}{\overset{HOR^3}{\rightleftharpoons}} R^3O-\underset{\underset{N(R^2)_2}{|}}{\overset{\overset{OR^3}{|}}{C}}-OR^3 \quad \text{etc.}$$

$$\underset{R^1}{\overset{R}{\diagdown}}CH_2 + \underset{}{\overset{R^2}{\diagdown}}\underset{\underset{R^2}{N}}{\overset{\overset{OR^3}{|}}{C}}\underset{OR^3}{\diagdown} \longrightarrow \underset{R^1}{\overset{R}{\diagdown}}C=C\underset{\underset{R^2}{|}}{\overset{\overset{R^2}{|}}{\diagdown N}}$$

TABLE X

Enamines from *N,N*-Dialkylcarbamic Acid (Thio) Acetals (**11**) and Active Methylene Compounds

Starting materials	Orthoamide	Reaction conditions	Reaction product	Yield, %	Ref.
R–CH$_2$CN	(CH$_3$)$_2$NC[OCH(CH$_3$)$_2$]$_3$	60–80°C, 2 hr	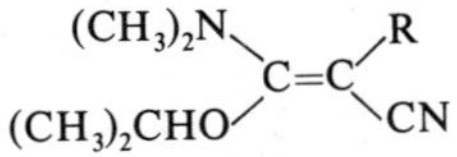 (CH$_3$)$_2$N–C(R)=CN / (CH$_3$)$_2$CHO	69–72	11
R = CN, CO$_2$C$_2$H$_5$ R–CH$_2$–R^1	[benzoxazoline with OC$_2$H$_5$, OC$_2$H$_5$, N–CH$_3$]	100–150°C	[benzoxazolinylidene =C(R)(R^1), N–CH$_3$]	37–65	4, 66
R CO$_2$CH$_3$ H R^1 CN 2,4-dinitrophenyl [rhodanine: H$_2$C–S, O=C, N–C$_2$H$_5$, C=S]	[benzoxazoline with OC$_2$H$_5$, OC$_2$H$_5$, N–CH$_3$]	110°C	[benzoxazolinylidene =C linked to rhodanine S–C=S, C=O, N–C$_2$H$_5$, N–CH$_3$]	72	4, 66

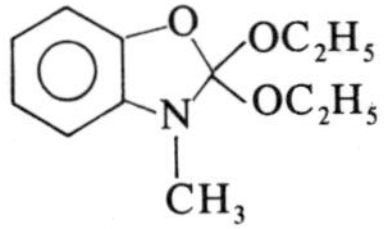

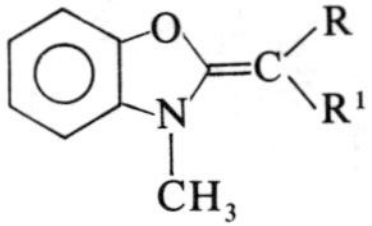

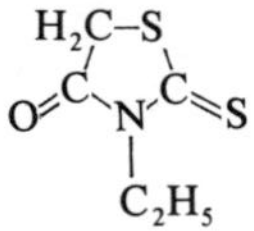

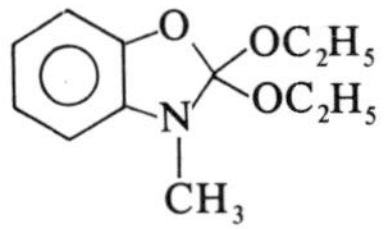

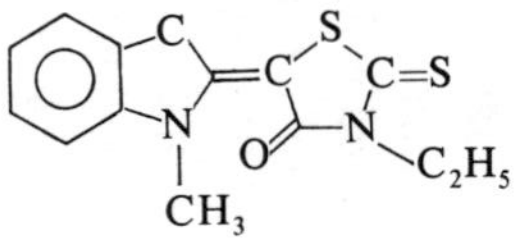

		100°C		85	4, 66
$R{-}CH_2{-}R^1$ R: CN, CN, CO_2CH_3, H R^1: CN, CN, 2,4-dinitrophenyl		20–100°C		50–98	4, 66
	,,	115°C		88	4, 66
	,,	140°C		92	4, 66

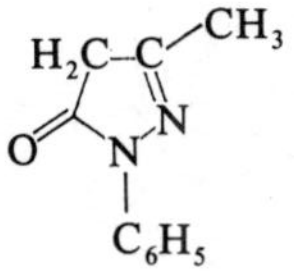

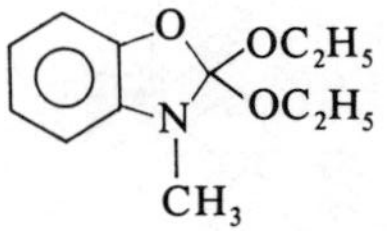

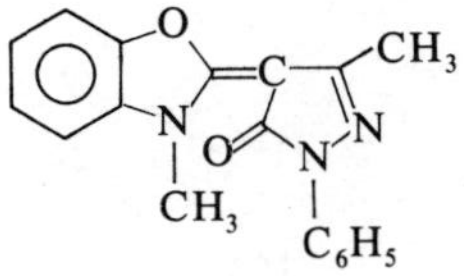

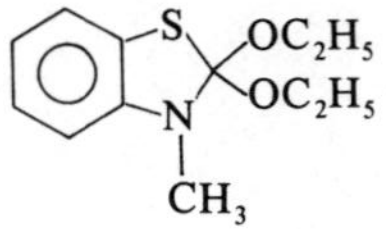

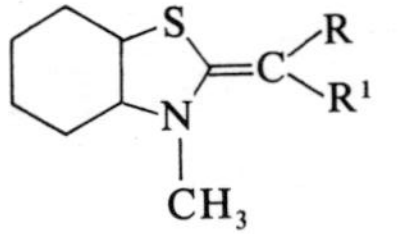

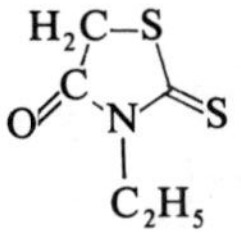

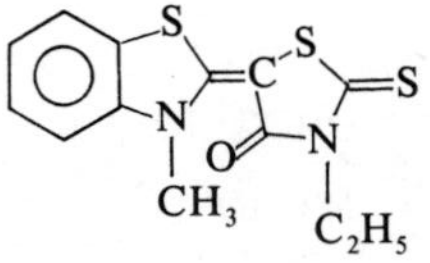

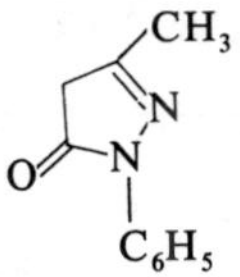

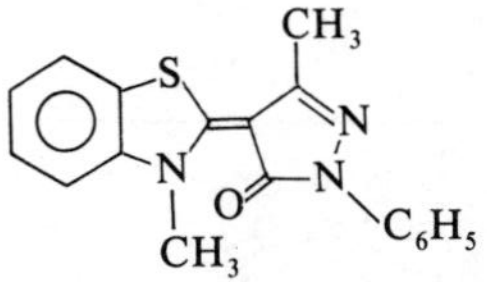

TABLE XI

Enamines from N,N'-Tetraalkylurea Acetals (**12**) and Active Methylene Compounds

Starting materials	Orthoamide	Reaction conditions	Reaction product	Yield, %	Ref.
CH_3NO_2	$(CH_3)_2N-C[OCH(CH_3)_2]_2$	20°C	$(CH_3)_2N-C=C{<}^H_{NO_2}$	82	11
$CH_2{<}^{CN}_{CN}$	"	20°C	$(CH_3)_2N-C=C{<}^{CN}_{CN}$; $(CH_3)_2CHO$	67	11
$R-CH_2-R^1$	(cyclic orthoamide with OC_2H_5, OC_2H_5, $N-CH_3$, $N-CH_3$)	20–180°C	(cyclic =C(R)(R¹) with N–CH₃, N–CH₃)	70–81	4, 65

R	H	CN	H
R^1	NO_2	CN	2,4-dinitrophenyl

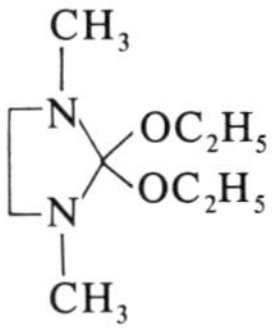

 (cyclopentadienyl)	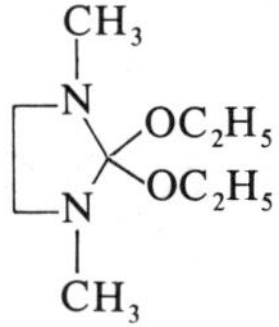 (cyclic orthoamide with OC_2H_5, OC_2H_5, $N-CH_3$, $N-CH_3$)	100°C	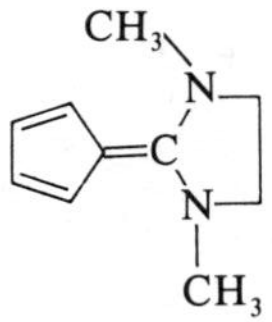 (cyclopentadienylidene imidazolidine with $N-CH_3$, $N-CH_3$)	94	4, 65

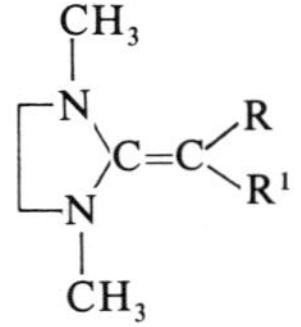

R–CH$_2$–R^1 → [diethoxy intermediate, 20°C] → [alkylidene product, 61–91%] [4, 66]

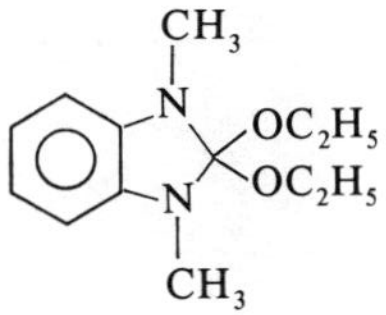

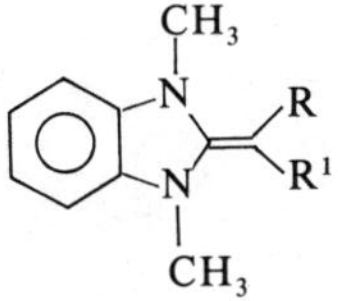

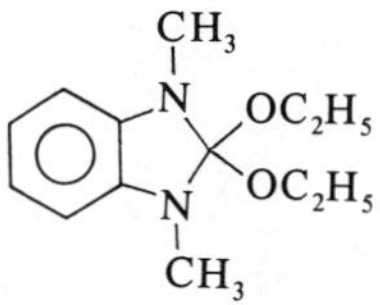

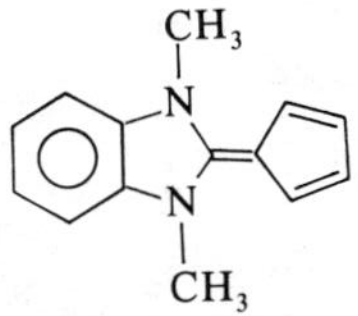

R	H	CN	CO$_2$CH$_3$	H
R^1	NO$_2$	CN	CN	COC$_6$H$_5$
R	H			
R^1	2,4-dinitrophenyl			

[N-methylquinolinium iodide] → [diethoxy intermediate, 80°C] → [cyclopentadienylidene product]

[cyclopentadiene]

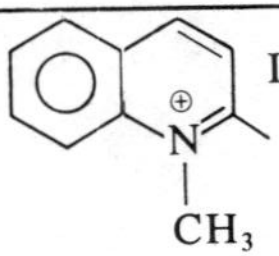

TABLE XII

Enamines from Alkoxytris(dimethylamino)methanes (13) or (14) and Active Methylene Compounds

Starting material	Orthoamide	Reaction conditions	Reaction product	Yield, %	Ref.
R—CH$_2$—CN R = H, CH$_3$, C$_2$H$_5$, C$_3$H$_7$	[(CH$_3$)$_2$N]$_3$COCH(CH$_3$)$_2$	~60–80°C[a]	[(CH$_3$)$_2$N]$_2$C=C(R)—CN	20–78	11
R—CH$_2$—CN R = OCH$_3$, N(CH$_3$)$_2$		~60–80°C[a]	(CH$_3$)$_2$N_C=C(CN[b])(R) (CH$_3$)$_2$N/	~20	11
R—CH$_2$—CN R = CN, CO$_2$CH$_3$		0–20°C[a]	[(CH$_3$)$_2$N]$_2$C=C(R)—C≡N	67–73	11
CH$_3$—C(=O)—R		120–170°C 5–7 hr	[(CH$_3$)$_2$N]$_2$C=C(H)—C(=O)R	23–25	12
R = O—CH(CH$_3$)$_2$ N(CH$_3$)$_2$					
H$_2$NOC—CH$_2$CONH$_2$	[(CH$_3$)$_2$N]$_3$COC$_2$H$_5$	~60–80°C	[(CH$_3$)$_2$N]$_2$C=C(CONH$_2$)$_2$	81	12
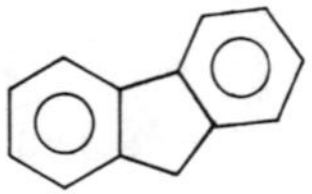 fluorene (9H-fluorene)	C[N(CH$_3$)$_2$]$_4$	140°C, 6 hr	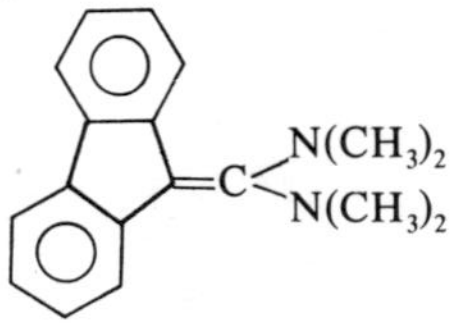 9-[bis(dimethylamino)methylene]fluorene	87	30

[a] Exothermic reaction.
[b] Unstable.

III. REACTIONS OF ORTHOAMIDES WITH NUCLEOPHILES

form enamines proceed very fast. The rules fail in the case of stable carbanions [e.g., from bis(p-toluenesulfonyl)methane]; here the reaction stops after the formation of hexalkylguanidinium salts (11,12).

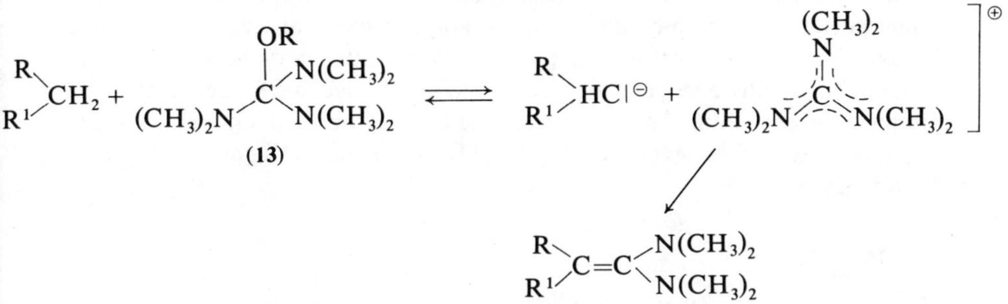

$$\underset{R^1}{\overset{R}{>}}CH_2 + \underset{(13)}{(CH_3)_2N-\underset{N(CH_3)_2}{\overset{OR}{\overset{|}{C}}-N(CH_3)_2}} \rightleftharpoons \underset{R^1}{\overset{R}{>}}HCl^{\ominus} + \left[(CH_3)_2N-\overset{N(CH_3)_2}{\underset{}{\overset{|}{C}}}-N(CH_3)_2 \right]^{\oplus}$$

$$\downarrow$$

$$\underset{R^1}{\overset{R}{>}}C=C\overset{N(CH_3)_2}{\underset{N(CH_3)_2}{<}}$$

Unfortunately thermal instability (> 100°C) limits the use of alkoxytris-(dialkylamino)methanes (11,12).

In connection with the aforementioned reactions, the in situ reaction of **13** greatly extends their applicability, and also gives a much better yield of enamines. Up to now this variation was applied only to nitriles. Here dimethylamine is added to an N,N'-tetramethylcarbamide halide (**16**) solution along with the corresponding nitrile; then sodium isopropoxide is added. 3,3-Bis(dimethylamino)acrylonitriles are formed in exothermic reactions (11).

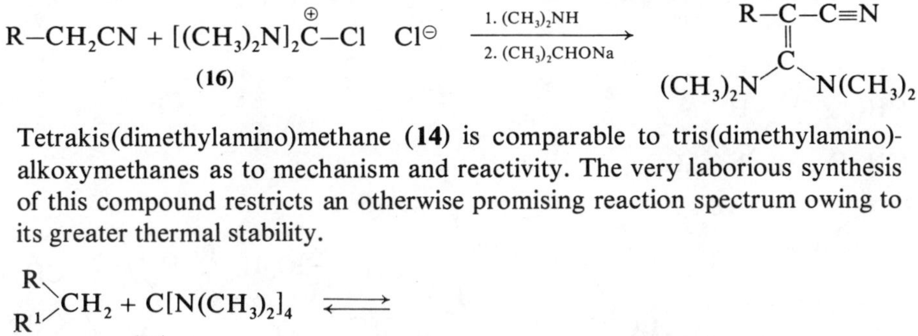

$$R-CH_2CN + \underset{(16)}{[(CH_3)_2N]_2\overset{\oplus}{C}-Cl} \; Cl^{\ominus} \xrightarrow[\text{2. }(CH_3)_2CHONa]{\text{1. }(CH_3)_2NH} \underset{(CH_3)_2N}{\overset{R-C-C\equiv N}{\underset{N(CH_3)_2}{\overset{\|}{C}}}}$$

Tetrakis(dimethylamino)methane (**14**) is comparable to tris(dimethylamino)-alkoxymethanes as to mechanism and reactivity. The very laborious synthesis of this compound restricts an otherwise promising reaction spectrum owing to its greater thermal stability.

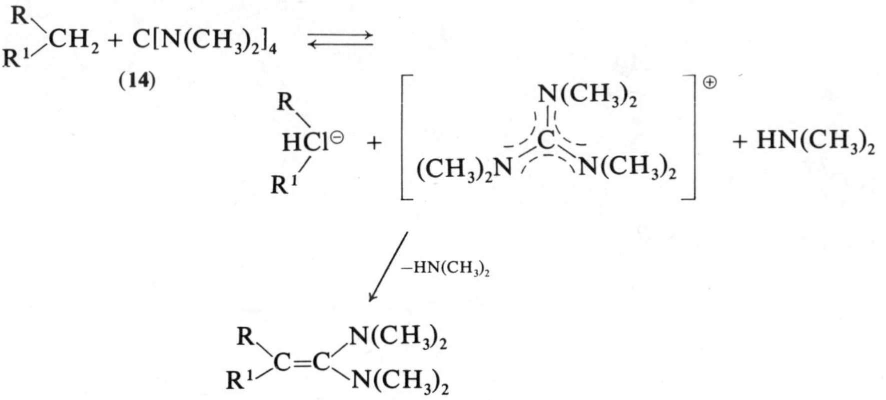

$$\underset{R^1}{\overset{R}{>}}CH_2 + C[N(CH_3)_2]_4 \rightleftharpoons$$
$$(14)$$

$$\underset{R^1}{\overset{R}{>}}HCl^{\ominus} + \left[(CH_3)_2N-\overset{N(CH_3)_2}{\underset{}{\overset{|}{C}}}-N(CH_3)_2 \right]^{\oplus} + HN(CH_3)_2$$

$$\downarrow -HN(CH_3)_2$$

$$\underset{R^1}{\overset{R}{>}}C=C\overset{N(CH_3)_2}{\underset{N(CH_3)_2}{<}}$$

4. Dialkoxymethylammonium Salts with Active Methyl and Methylene Compounds

Electrophilic properties (especially those of O,O-bridged N,N-dialkylformamide acetals) can be increased considerably by N-quaternization (31). Such N-quaternized amide acetals react with carbon acids via dialkoxymethylation in refluxing methylene chloride. The initially formed aldehyde acetals may eliminate to enol ethers under these conditions. Double dialkoxymethylation was also observed in several examples (31). See Table XIII for a summary of these reactions.

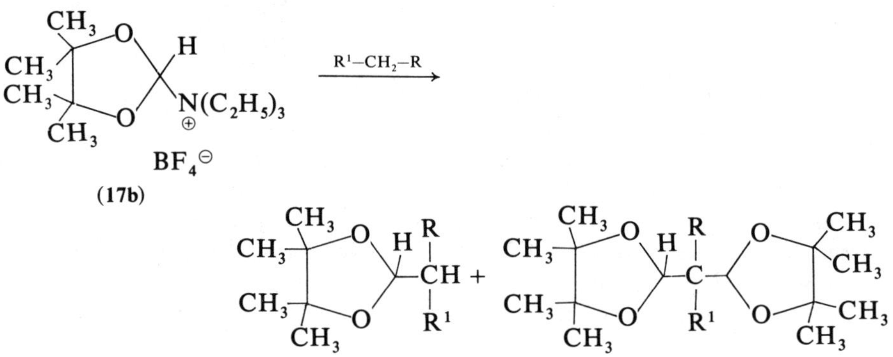

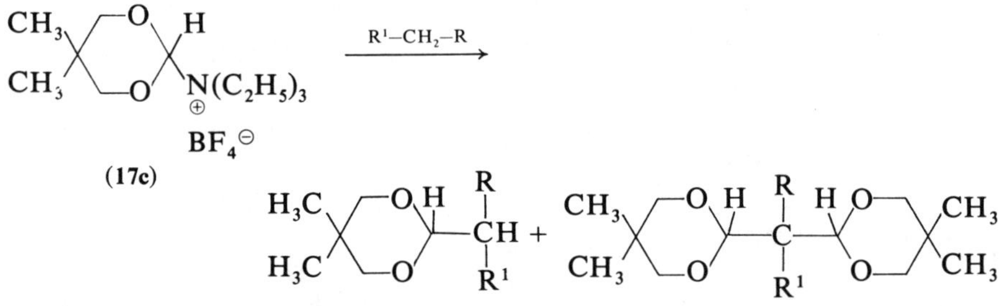

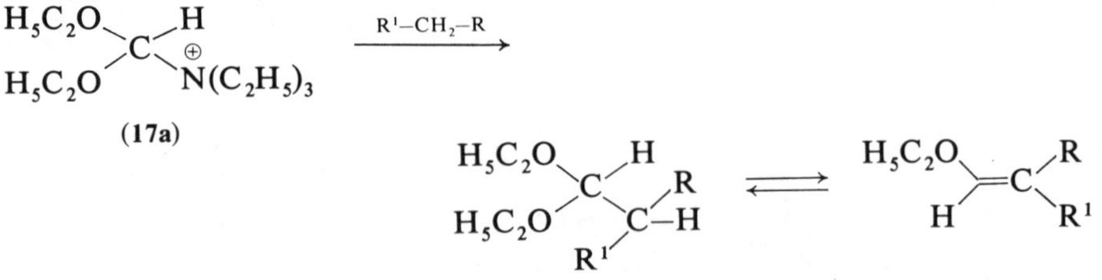

TABLE XIII

Enol Ethers and Aldehyde Acetals from Active Methylene Compounds and Dialkoxy Methylammonium Salts (17) (31)

Starting material R^1-CH_2-R		Dialkoxymethylammonium salts	Yield, %	
R^1	R		Dialkoxymethyl derivative	Bis(dialkoxy)methyl derivative
H	NO_2	17a	44	10
$CO_2C_2H_5$	$CO_2C_2H_5$	17a	75	—
$CO_2C_2H_5$	CN	17a	—	41
$CO_2C_2H_5$	$COCH_3$	17a	56	—
$COCH_3$	$COCH_3$	17a	—	77
CN	CN	17a	—	52
H	NO_2	17b	51	29
$COCH_3$	$COCH_3$	17b	88	—
$CO_2C_2H_5$	CN	17b	54	—
H	NO_2	17c	26	44
CH_3	NO_2	17c	38	—
C_6H_5	NO_2	17c	61	—
$CO_2C_2H_5$	$CO_2C_2H_5$	17c	83	—
$CO_2C_2H_5$	CN	17c	—	86
$CO_2C_2H_5$	$COCH_3$	17c	89	—
CN	CN	17c	—	73

5. Orthoamides with Active Methyne Groups

The first step in the reaction of orthoamides (e.g., **1**) with tertiary carbon acids parallels the aforementioned double decomposition; along with the ambident cation, the tertiary carbanion is generated by alcohol elimination.

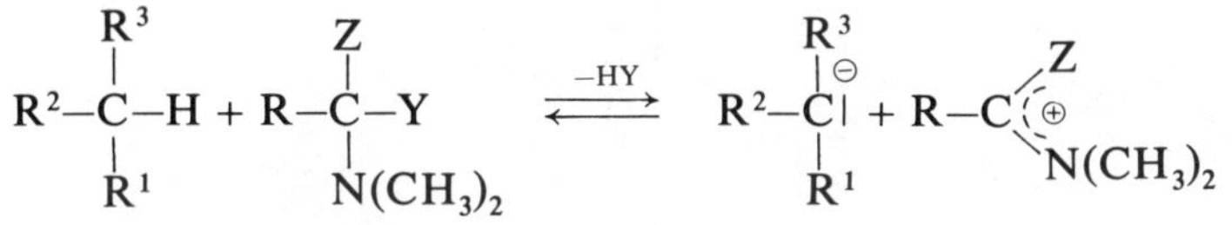

$$R^2-\underset{R^1}{\underset{|}{\overset{R^3}{\overset{|}{C}}}}-H + R-\underset{N(CH_3)_2}{\underset{|}{\overset{Z}{\overset{|}{C}}}}-Y \underset{-HY}{\rightleftarrows} R^2-\underset{R^1}{\underset{|}{\overset{R^3}{\overset{|}{C}}}}{}^{\ominus} + R-C{\overset{Z}{\underset{N(CH_3)_2}{\lessgtr}}}{}^{\oplus}$$

(1)

R = H, CH_3, C_2H_5, C_3H_7, C_6H_5
R^1, R^2, R^3 = alkyl, aryl, CO_2-alkyl, CN
$R^1-R^2-R^3$ = ≡C—aryl, ≡C—alkyl, ≡C—H, ≡N|
Y, Z = O—alkyl

The type of secondary reactions depends on the structure of both the anions and the cations to a much greater extent than in the case of active methylene compounds. The following rules may be formulated (52).

TABLE XIV
Aldehyde (Ketone) O,N- and N,N-Acetals from Orthoamides and Active Methyne Compounds

Starting material	Orthoamide	Reaction conditions	Reaction product	Yield, %	Ref.
$H-C\equiv C-H$	$HC[N(CH_3)_2]_2$ $OC(CH_3)_3$ **5c**	140°C, 5 hr	$[(CH_3)_2N]_2\overset{H}{\underset{\vert}{C}}-C\equiv C-\overset{H}{\underset{\vert}{C}}[N(CH_3)_2]_2$	80	67
$R-C\equiv C-H$ $R = n$-butyl, Phenyl, 4-chlorophenyl	**5c, 7a**	90–140°C, 4–7 hr	$R-C\equiv C-\overset{H}{\underset{\vert}{C}}[N(CH_3)_2]_2$	55–87	52
$C_6H_5-C\equiv C-H$	**1b**	140°C	$C_6H_5-C\equiv C-\overset{H}{\underset{\underset{N(CH_3)_2}{\vert}}{C}}-OC_2H_5$		68
$C_6H_5-C\equiv CH$	$H-C(OCH_3)_2$ $\underset{R}{\overset{\vert}{N}}{-}R$	140°C, 6 hr	$C_6H_5-C\equiv C-\overset{H}{\underset{\underset{R}{\overset{\vert}{N}}-R}{\overset{\vert}{C}}}-O-CH_3$		67

R = CH$_3$, CH(CH$_3$)$_2$, CH$_2$CH(CH$_3$)$_2$, R—R=(CH$_2$)$_4$

H–C≡N	R–C(OC$_2$H$_5$)$_2$–N(CH$_3$)$_2$ R = H, CH$_3$, C$_2$H$_5$, C$_3$H$_7$	20°C, 20–80 hr; and 50–80°C, 1–2 hr	N≡C–C(OC$_2$H$_5$)(R)–N(CH$_3$)$_2$	58–86	52
H–C≡N	C$_6$H$_5$–C(OC$_2$H$_5$)$_2$–N(CH$_3$)$_2$	20°C, 16 hr and 90°C, 2 hr	NC–C(OC$_2$H$_5$)(C$_6$H$_5$)–N(CH$_3$)$_2$	19	52, 68
HCN	**5c, 7a**	20°C, 12 hr; and 50–60°C, 1 hr	NC–C(N(CH$_3$)$_2$)(H)–N(CH$_3$)$_2$ [a]	71–95	52

[a] For a simple synthesis see the chapter by W. Kantlehner in this volume called Alkoxymethylene Iminium Salts.

If the charge of the carbanion is neither delocalized by resonance nor inaccessible owing to steric effects, the kinetically controlled ion recombination gives aldehyde (ketone) O,N- or N,N-acetals, starting from **1**, **5**, and **7**. This is exemplified by the reactions with hydrogen cyanide and acetylenes. The introduction of a cyano group by displacement of Mercaptide succeeds with mercury-II-cyanide (see Table XIV).

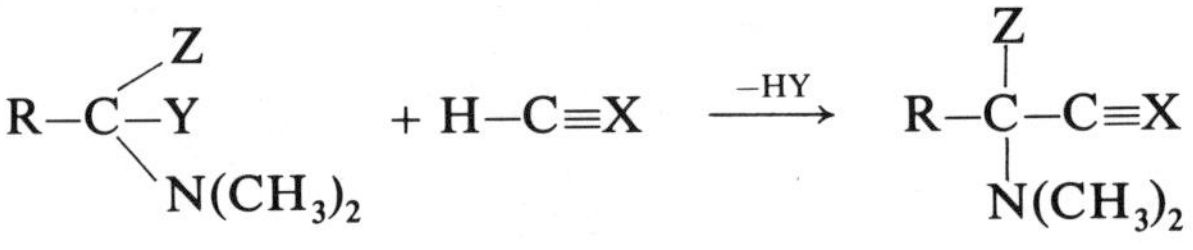

$$R-\underset{N(CH_3)_2}{\overset{Z}{C}}-Y \;+\; H-C\equiv X \xrightarrow{-HY} R-\underset{N(CH_3)_2}{\overset{Z}{C}}-C\equiv X$$

$$X = CH, CR, N$$

If the carbanion is only a weak nucleophile, owing to delocalization, inductive, and/or steric effects (as with trisubstituted methane derivatives), the following reaction paths are possible. On heating, alkylation of the carbanion with carboxylic amide acetals (**1**) occurs as a thermodynamically controlled

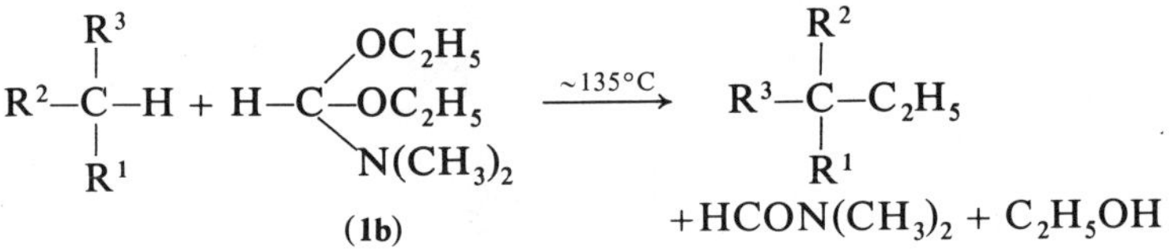

$$R^2-\underset{R^1}{\overset{R^3}{C}}-H \;+\; H-\underset{N(CH_3)_2}{\overset{OC_2H_5}{C}}-OC_2H_5 \xrightarrow{\sim 135°C} R^3-\underset{R^1}{\overset{R^2}{C}}-C_2H_5$$
$$(\mathbf{1b}) \qquad\qquad +HCON(CH_3)_2 + C_2H_5OH$$

R^1	R^2	R^3
CH_3	$CO_2C_2H_5$	$CO_2C_2H_5$
C_6H_5	$CO_2C_2H_5$	$CO_2C_2H_5$
CH_3	$CO_2C_2H_5$	CN
C_6H_5	$CO_2C_2H_5$	CN
C_6H_5	CN	CN
$CO_2C_2H_5$	$CO_2C_2H_5$	$CO_2C_2H_5$

reaction. This reaction is also favored by bulky ambident cations, and by increased polarizability of the O-alkyl bond of the carbiminium ion.

In the reaction of aminal esters (**5**) and tris(dimethylamino)methane under mild conditions, the stable amidinium ions do not react further with

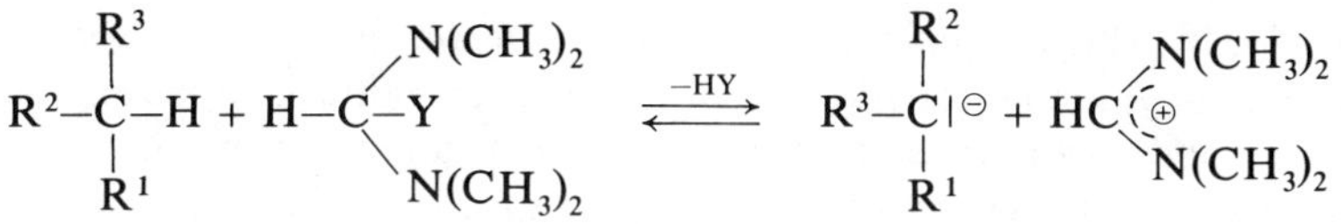

III. REACTIONS OF ORTHOAMIDES WITH NUCLEOPHILES

$$R^2-\underset{R^1}{\underset{|}{C}}-H + H-\underset{N(CH_3)_2}{\underset{|}{C}}-Y \xrightleftharpoons{-HY} R^3-\underset{R^1}{\underset{|}{C}}|^\ominus + HC\overset{\oplus}{\underset{N(CH_3)_2}{\diagdown}}^{N(CH_3)_2}$$

(5)

$Y = O\text{-alkyl}, N(CH_3)_2$

carbanions. More drastic reaction conditions cause aminolysis when **5** is reacted with carboxylic acid esters.

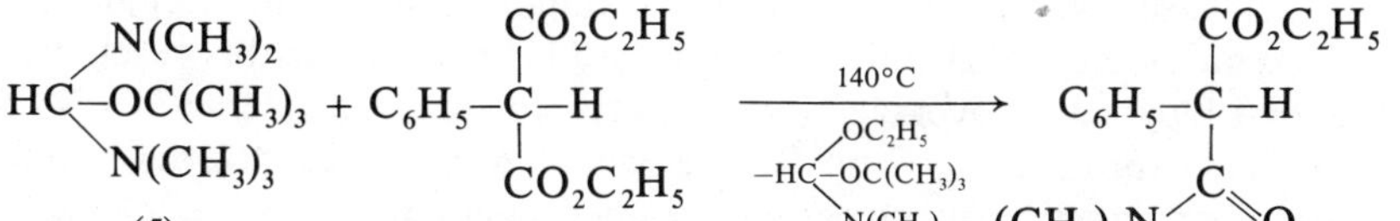

$$HC\overset{N(CH_3)_2}{\underset{N(CH_3)_3}{\diagdown}}-OC(CH_3)_3 + C_6H_5-\underset{CO_2C_2H_5}{\underset{|}{C}}-H \xrightarrow[\substack{OC_2H_5 \\ -HC-OC(CH_3)_3 \\ N(CH_3)_2}]{140°C} C_6H_5-\underset{\underset{(CH_3)_2N}{C}\diagdown O}{\underset{|}{C}}-H$$

(5)

These types of reactions are also observed to a lesser extent if active methylene compounds are used. This is particularly true if strong resonance stabilization of carbanions coincides with great stability of the ambident cations (11,30,40) or with steric hindrance (63).

C. REACTIONS WITH NH-ACIDIC COMPOUNDS

1. Carboxylic Acid Orthoamides with NH-acidic Compounds

The reaction proceeds analogously to that of active methylene compounds. For example, when formic acid orthoamides are reacted, N,N,N'-trisubstituted formamidines are the isolated products.

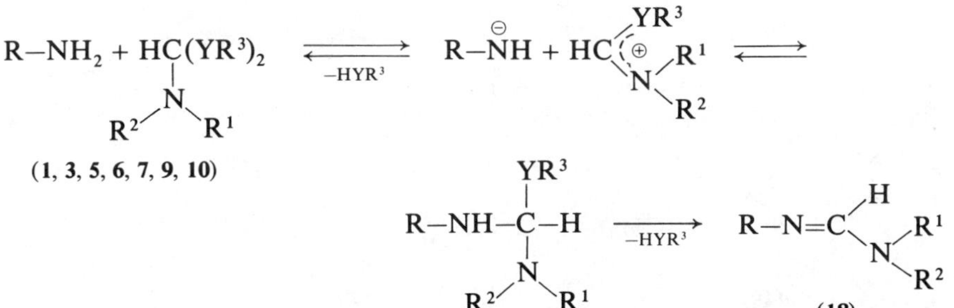

$$R-NH_2 + HC(YR^3)_2 \xrightleftharpoons[-HYR^3]{} R-\overset{\ominus}{N}H + HC\overset{\oplus}{\underset{N}{\diagdown}}\overset{YR^3}{\underset{R^2}{\diagup R^1}} \rightleftharpoons$$
$$\underset{R^2}{\underset{|}{N}}\underset{R^1}{\diagdown}$$

(1, 3, 5, 6, 7, 9, 10)

$$R-NH-\underset{\underset{R^2}{\underset{|}{N}}\underset{R^1}{\diagdown}}{\underset{|}{C}}-H \xrightarrow{-HYR^3} R-N=C\overset{H}{\underset{N}{\diagdown}}\overset{R^1}{\underset{R^2}{\diagup}}$$

(18)

YR^3 = alkoxy, alkylmercapto, dialkylamino, CN
R = alkyl, aryl, hetaryl, dialkylamino, arylamino, acyl

Dialkylformamide acetals (**1**), aminal esters (**5**), tris(dimethylamino)-methane (**7a**), and alkoxy(dimethylamino)acetonitriles (**9**) are reacted between

20 and 80°C* with aliphatic (69, 73), aromatic (4, 70, 71), heteroaromatic (70, 71), and acyl amino functions (4, 72, 73). Trisubstituted formamidines are formed in good yield, especially if the primary amine shows little nucleophilicity (high NH acidity). We believe this method to be the most general and most elegant formamidine synthesis known. Among others, this method has been used in oligonucleotide synthesis. 2'-Desoxy-ribonucleosides containing primary amino groups on the pyrimidine or purine ring react with N,N-dimethylformamide dimethyl acetal in DMF giving N-dimethylaminomethylene derivatives at room temperature.

In the case of ribonucleosides, the amidination is accompanied by acetalization of the 2',3'-cis-diol of the ribose (**19**). To avoid this reaction, the diol is first blocked by acetalization with orthoformic acid esters (74,75) (**20**). This selective protection of nucleoside amino groups (**21, 22, 23**) exhibits considerable synthetic advantages over simple acylation (74,75).

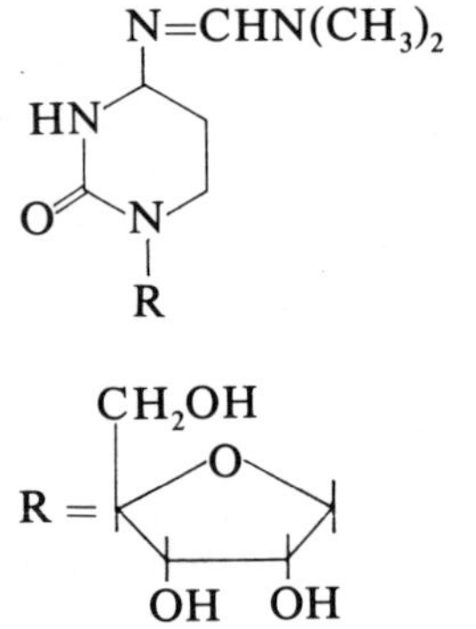

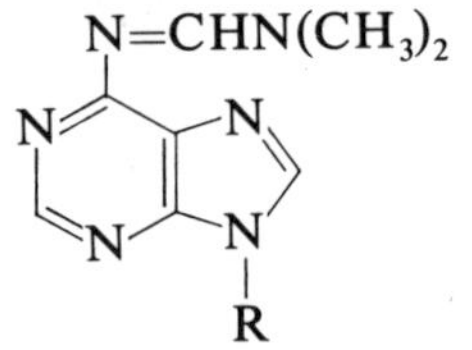

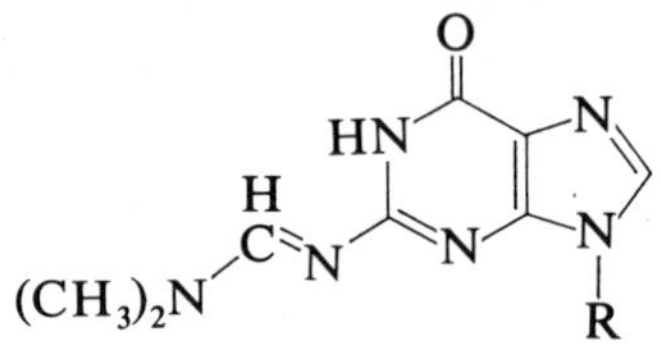

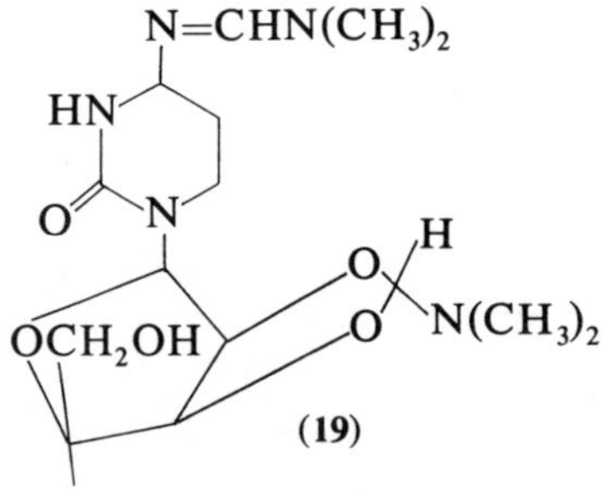

Reactions of amide acetals (**1**) with aliphatic amines proceed slowly owing to their low dissociation and basicity. Orthoamides **5** or **9** which dissociate to a greater extent, smoothly yield N,N,N'-trialkylformamidines (32,69,76).

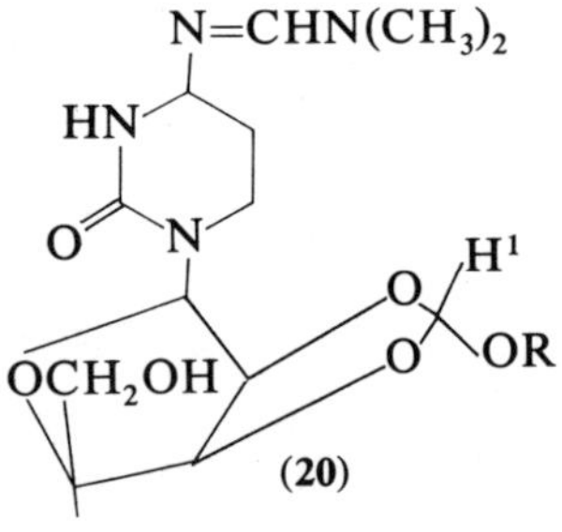

* For the reaction of orthoformamides with 1,3-bifunctional amines see Section III-H-2-c.

TABLE XV

Amidines and Amidrazones from NH_2-Acidic Compounds and Carboxylic Acid Orthoamides

Starting material	Orthoamide	Reaction conditions	Reaction product	Yield, %	Ref.
RNH_2 $R = CH_3, n\text{-}C_3H_7,$ $n\text{-}C_4H_9, CH(CH_3)_2,$ $c\text{-}C_6H_{11}$	$H-C\begin{smallmatrix}-OR^1\\-CN\\-N(CH_3)_2\end{smallmatrix}$ (9a) $R^1 = CH_3$ (9b) $R^1 = C_2H_5$	20–65°C, 2–24 hr	$R-N\overset{H}{=}CN(CH_3)_2$	52–61	69
CH_3NH_2	5a	35°C, 7 hr	$CH_3-N\overset{H}{=}CN(CH_3)_2$	35	76
$Ar-NH_2$ Ar = 4-chlorophenyl	9a	20°C, 24 hr	$Ar-N\overset{H}{=}C-N(CH_3)_2$	77	69
$Ar-NH_2$ Ar = 4-CH_3-, 3,4-dichloro-, 2-nitro-, 3-nitro-, 4-nitro-, 2,4-dinitro-, 2,4,6-trinitrophenyl	1a, b	70–80°C, 2–3 hr	$Ar-N\overset{H}{=}C-N(CH_3)_2$	89–95	70, 71 4, 73
$Ar-NH_2$ Ar = 2-nitro-, 3-nitro-4-nitro-, 2,4-dinitrophenyl	7a	–10–20°C, 2 hr	$ArN=CHN(CH_3)_2$	83–93	46, 60
Hetarylamines: 2-Aminopyrimidine, 2-Aminothiazole, 2-Aminobenzthiazole, 2-Aminopyridine	1a	60°C, 2 hr	Hetaryl–$N=CHN(CH_3)_2$	89–95	77
4-Aminopyrimidine, 2-Aminopurine, 6-Aminopurine	1a	20°C, 12–48 hr, DMF	Hetaryl–$N=CHN(CH_3)_2$	~100	74, 75

TABLE XV (cont.)

Starting material	Orthoamide	Reaction conditions	Reaction product	Yield, %	Ref.
2-, 3-, 4-Aminopyridine, 3-, 4-Amino-1,2,4-triazole, Amino-s-triazine, 5-Amino-1,2,3,4-tetrazole	1b	50–80°C, 1–2¼ hr	Hetaryl–N=CHN(CH$_3$)$_2$	78–95	70, 71
2-Amino-1,3,4-oxadiazole	1b	50°C	1,3,4-Oxadiazolyl–N=CHN(CH$_3$)$_2$	83–95	78
(CH$_3$)$_2$N–NH$_2$	9a	65°C, 2 hr	(CH$_3$)$_2$N–N=CHN(CH$_3$)$_2$	82	69
(CH$_3$)$_2$N–NH$_2$	1a		(CH$_3$)$_2$N–N=CHN(CH$_3$)$_2$		73
	Fumaric acid bis(dimethylamide) diethyl acetal	80°C, 4hr	(CH$_3$)$_2$N–C(N(CH$_3$)$_2$)=C–C=C(H)(NN(CH$_3$)$_2$)–H, NN(CH$_3$)$_2$	47	29
R–NH–NH$_2$ R = 4-nitrophenyl, 2,4-dinitrophenyl, 4-methylphenylsulfonyl	1a, b	80°C, benzene	R–N(H)–N=C(H)–N(CH$_3$)$_2$		4, 73
R–C(=O)–NH$_2$	1a	80°C, benzene	RCON=CHN(CH$_3$)$_2$	~80	73
	5c	20°C		90	32
HCONH$_2$	5c	0–20°C	(CH$_3$)$_2$N–C(H)=N–CHO	100	79

Substrate	Reagent	Conditions	Product	Yield (%)	Ref.
R–CH$_2$CONH$_2$	5c (2 mol)	120–160°C, $\frac{1}{2}$–7 hr	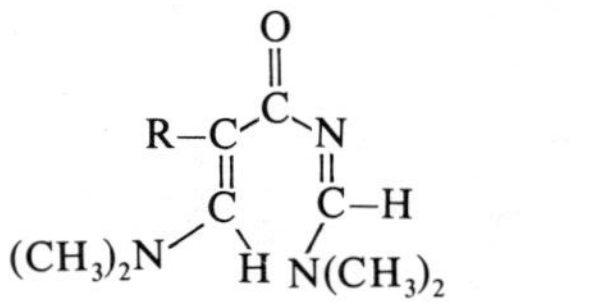 R–C(=O)–C(N(CH$_3$)$_2$)=C(–C–H)–N=CH–N(CH$_3$)$_2$ (H)	83–90	32
CH$_3$–CH=CHCONH$_2$	5c (2 moles)	140°C, 2 hr	(CH$_3$)$_2$N(CH=CH)$_2$–C(=O)–N=CH–N(CH$_3$)$_2$ (H)	65	32
H$_2$N=CH–N(CH$_3$)$_2$ $^\oplus$	1b	~80°C	(CH$_3$)$_2$N–C(H)=N–CH=N(CH$_3$)$_2$ $^\oplus$	80	80
ROCONH$_2$	1a		ROCN=CHN(CH$_3$)$_2$ ‖ O	73	73
				55	55
H$_2$N–C(=X)–NH$_2$	1a, b	60–80°C, 3 hr	X=C(N=CHN(CH$_3$)$_2$)(N=CHN(CH$_3$)$_2$)	41–93	71, 73
X = O, S	7a	0–20°C		64–73	46
R–SO$_2$NH$_2$	1a	~80°C	R–SO$_2$–N=CH–N(CH$_3$)$_2$	73	73
	7a	0–20°C		84	46, 60
HC(SR)(SR)N(CH$_3$)$_2$				30	48
X–N=C(NH$_2$)(NH$_2$)	1b	60°C	X–N=C(N=CHN(CH$_3$)$_2$)(N=CHN(CH$_3$)$_2$)	64–87	70, 72
X = NO$_2$, SO$_3$H					
H$_2$N–C(=O)–NHNH$_2$	1b	80°C	H$_2$N–C(=O)–NH–N=CHN(CH$_3$)$_2$	69	4

Because of its extreme reactivity, **5** reacts with carboxylic acid amides giving N,N-dialkyl-N'-[3-(dimethylamino)acryloyl]formamidines via N-acyl-N',N'-dialkylformamidine intermediates (32). Generally, the reaction rate increases with the acidity of the NH_2 protons.

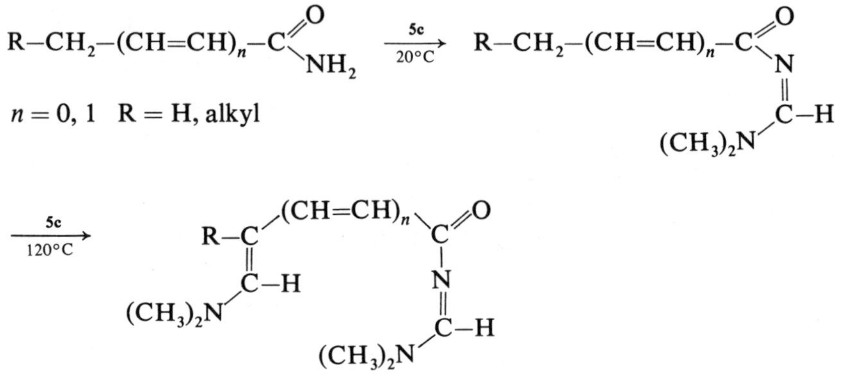

$$R-CH_2-(CH=CH)_n-C\begin{smallmatrix}O\\NH_2\end{smallmatrix} \xrightarrow[20°C]{5c} R-CH_2-(CH=CH)_n-C\begin{smallmatrix}O\\N\\\|\\C-H\\(CH_3)_2N\end{smallmatrix}$$

$n = 0, 1 \quad R = H, \text{alkyl}$

$$\xrightarrow[120°C]{5c} R-C\begin{smallmatrix}(CH=CH)_n\\\|\\C-H\\(CH_3)_2N\end{smallmatrix}\begin{smallmatrix}O\\C\\|\\N\\\|\\C-H\\(CH_3)_2N\end{smallmatrix}$$

Table XV lists some products of the reactions of carboxylic acid orthoamides with NH_2-acidic compounds.

Acetals of higher carboxylic acid amides (73), lactam acetals (62), and thioacetals (49) react with NH_2 acidic compounds to afford the corresponding products.

2. Carbon Acid Orthoamides with NH_2-Acidic Compounds

Reactions of alkoxytris(dimethylamino)methanes (**13**) with primary carboxylic acid amides have been well studied (11,12). N-Acyl-N',N''-tetramethylguanidines are formed in exothermic reactions, in which the orthoamides are preferentially used in situ.

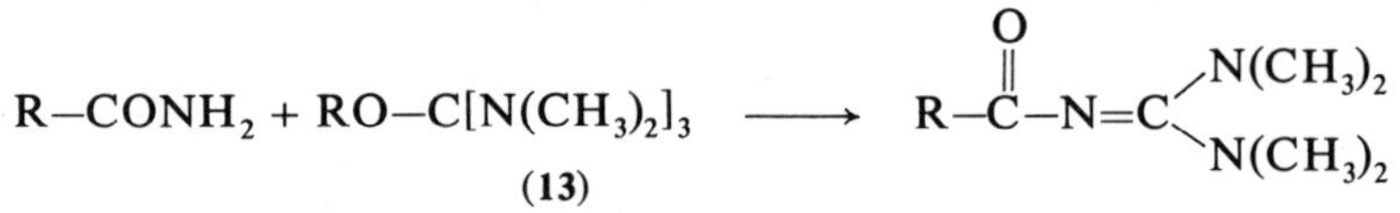

$$R-CONH_2 + RO-C[N(CH_3)_2]_3 \longrightarrow R-\underset{\|}{\overset{O}{C}}-N=C\begin{smallmatrix}N(CH_3)_2\\N(CH_3)_2\end{smallmatrix}$$
(13)

Surprisingly, malonamide reacts only at its methylene group (12) (see Table XII). Aliphatic amines like α-aminoacetonitrile yield N-alkyl-N',N''-tetramethylguanidines (11).

3. Tris(acylamino)methanes with NH_2-Acidic Compounds

Tris(formylamino)methane and primary aromatic amines give N,N'-diarylformamidines, generally in low yields (70, 81), whereas primary aliphatic amines yield N-formylamines (81). The synthesis of heterocycles containing bifunctional amines is of much greater preparative importance (see Section III-H).

4. Orthoamides with Secondary Amines and Derivatives

Orthoamides (1, 5, 7) exchange their alkoxy and/or dialkylamino groups with secondary amines; this reaction was already exemplified by their disproportionation (5). These reactions are successful with N,N-dialkylformamide acetals and higher-boiling aliphatic, cycloaliphatic, and aromatic amines (4,16,82).

The aminal ester 5 is formed as an intermediate (82,84). The type of reaction product isolated depends on the amount of the starting amine and the properties of the alcohol formed. Using an equimolar amount of the amine and keeping the alcohol in the reaction mixture leads only to transamination of the amide acetals 1 (4).

Use of an excess of secondary amine and removal of the alcohol from the equilibrium convert the extremely reactive compounds 5 into tris(dimethylamino)methanes (7) (16,82). For the synthesis of 7, compounds 5 may be reacted *in situ* (16); under these conditions, exchange reactions with low molecular weight aliphatic amines are also successful:

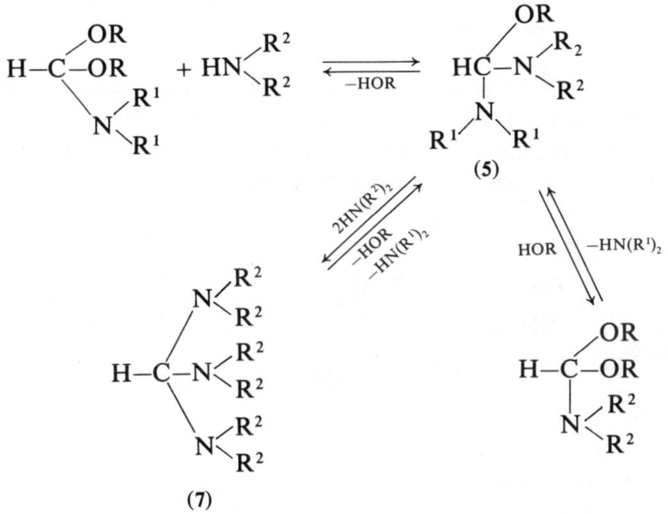

The yield of 7 decreases with increasing size of the alkyl groups on the secondary amine (16,82). Owing to the thermal lability of 7—formation of tetrakis(dialkylamino)ethylenes (15,16,84)—the temperature of the amination reaction may not exceed 100°C, especially with cycloaliphatic amines (82).

Tris(dimethylamino)methane **(7a)** reacts with secondary amines to give mixed compounds **7** (3). This exchange reaction is not useful in synthesis, since the resulting compounds **7** of varied composition cannot be separated by distillation.

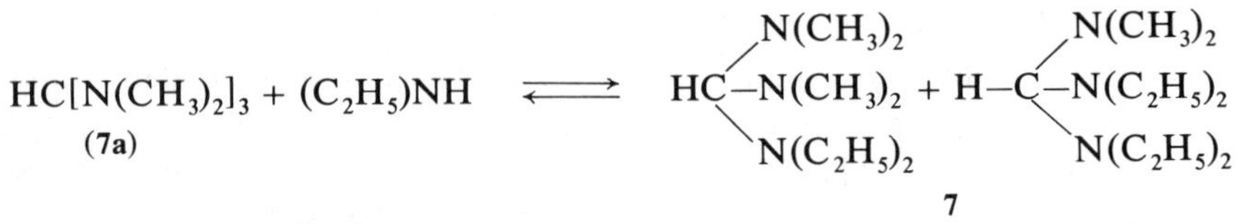

$$HC[N(CH_3)_2]_3 + (C_2H_5)NH \rightleftharpoons HC\begin{subarray}{l}N(CH_3)_2\\-N(CH_3)_2\\N(C_2H_5)_2\end{subarray} + H-C\begin{subarray}{l}N(CH_3)_2\\-N(C_2H_5)_2\\N(C_2H_5)_2\end{subarray}$$
(7a)   **7**

Mixed tris(dialkylamino)methanes can be made from **5** and aromatic amines or from amidinium salts and lithium dialkylamides (3,82).

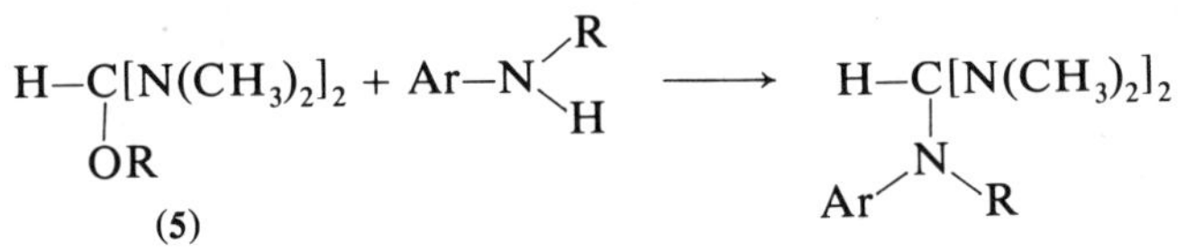

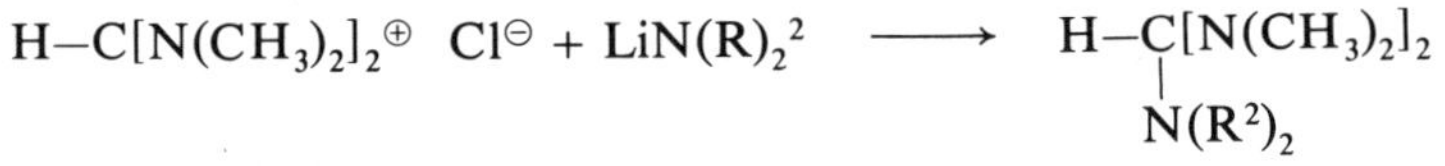

$$H-C[N(CH_3)_2]_2\begin{subarray}{l}\\OR\end{subarray} + Ar-N\begin{subarray}{l}R\\H\end{subarray} \longrightarrow H-C[N(CH_3)_2]_2\begin{subarray}{l}\\N\\Ar\quad R\end{subarray}$$
(5)

$$H-C[N(CH_3)_2]_2^\oplus\ Cl^\ominus + LiN(R)_2^2 \longrightarrow H-C[N(CH_3)_2]_2\begin{subarray}{l}\\N(R^2)_2\end{subarray}$$

The substitution of alkoxy by dialkylamino groups is of special importance in the synthesis of bis(dialkylamino)acetonitriles from **9** (69,85). The exchange proceeds exothermically and in high yield. The use of excess secondary amine also allows substitution of the dialkylamino groups.

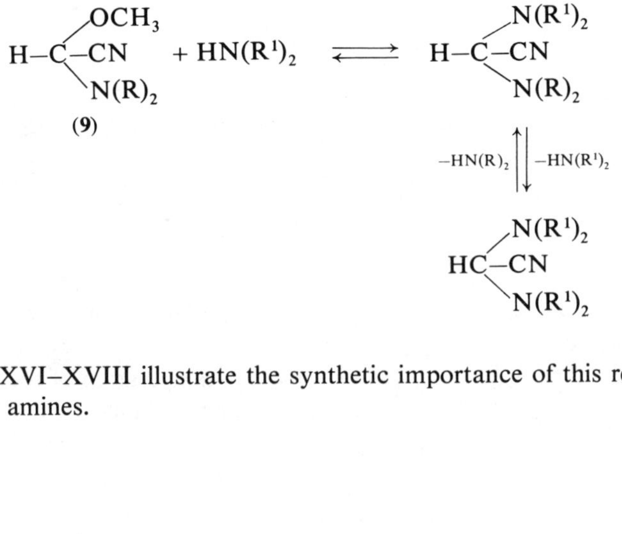

$$H-C\begin{subarray}{l}OCH_3\\-CN\\N(R)_2\end{subarray} + HN(R^1)_2 \rightleftharpoons H-C\begin{subarray}{l}N(R^1)_2\\-CN\\N(R)_2\end{subarray}$$
(9)

$$\updownarrow \;-HN(R)_2 \;\;-HN(R^1)_2$$

$$HC\begin{subarray}{l}N(R^1)_2\\-CN\\N(R^1)_2\end{subarray}$$

Tables XVI–XVIII illustrate the synthetic importance of this reaction with secondary amines.

TABLE XVI
Tris(dialkylamino)methanes from N,N-Dialkylformamide Acetals (1) or Aminal Esters (5) and Secondary Amines

Orthoamide	sec-Amine	Reaction conditions	Substituted methane product	Yield, %	Ref.
1a	$NH(CH_3)_2$	15–100°C, 50 hr	Tris(dimethyl-amino)	33	82
1a	$(n\text{-}C_3H_7)_2NH$	100–130°C	Tris(di-n-propyl-amino)	6	16
1a	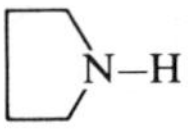 pyrrolidine (N–H)	100°C	Trispyrrolidino	10	16
1a	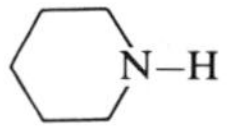 piperidine (N–H)	100–130°C	Trispiperidino	80	16
1a	 morpholine (O, NH)	100–130°C	Trismorpholino	75	16
1b	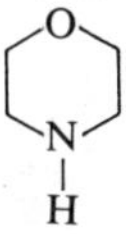 morpholine	150°C, 48 hr	Trismorpholino	31	83
1a	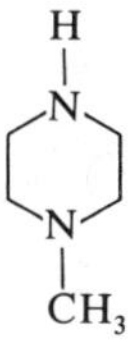 N-methylpiperazine	100–130°C	Tris-(4-methyl-piperazino)	60	16
1a	$C_6H_5NHCH_3$	100–130°C	Tris(N-methyl-anilino)	70	16
1a	$C_6H_5NHC_2H_5$	100–130°C	Tris(N-ethyl-anilino)	10	16
5c	$NH(CH_3)_2$	100°C, 10 hr	Tris(dimethyl-amino)	41	82
$[N(C_2H_5)_2]_2$ $HC-OC(CH_3)_3$	$NH(C_2H_5)_2$	120°C, 12 hr	Tris(diethyl-amino)	20	82
$[N(C_3H_7)_2]_2$ $HC-OC(CH_3)_3$	$NH(C_3H_7)_2$	130°C, 4 hr	Tris(di-n-propyl-amino)	11	82
5c	$C_6H_5NHCH_3$	20°C, 10 hr	Bis(dimethyl-amino)(N-methyl-anilino)	76	82
5c	$C_6H_5NHC_2H_5$	80°C, 22 hr	(N-Ethylanilino)-bis(dimethylamino)	45	82

TABLE XVII

Bis(dialkylamino)acetonitriles from *N,N*-(Dimethylamino)-methoxyacetonitrile (**9a**) and Secondary Amines[a] (Molar Ratio 1:1) (69)

sec Amine	Reaction conditions	Products	Yield, %
$NH(CH_3)_2$	0°C, 3–4 hr; 20°C, 1 hr	$HC\begin{smallmatrix}[N(CH_3)_2]_2\\CN\end{smallmatrix}$	86
$NH(C_2H_5)_2$	20°C, 3 hr	$H-C\begin{smallmatrix}N(C_2H_5)_2\\-CN\\N(CH_3)_2\end{smallmatrix}$	41
		$\begin{smallmatrix}[N(C_2H_5)_2]_2\\|\\H-C-CN\end{smallmatrix}$	7
$NH(C_3H_7)_2$	20°C, 3 hr	$H-C\begin{smallmatrix}N(CH_3)_2\\-CN\\N(C_3H_7)_2\end{smallmatrix}$	44
		$H-C\begin{smallmatrix}[N(C_3H_7)_2]_2\\CN\end{smallmatrix}$	12
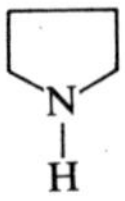 pyrrolidine (NH)	20°C	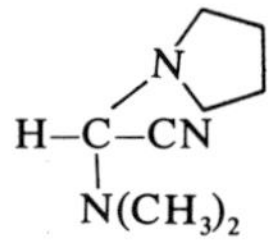 $H-C\begin{smallmatrix}\text{(pyrrolidinyl)}\\-CN\\N(CH_3)_2\end{smallmatrix}$	45
		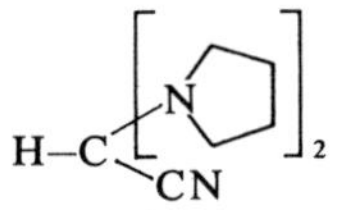 $H-C\begin{smallmatrix}[\text{pyrrolidinyl}]_2\\CN\end{smallmatrix}$	28
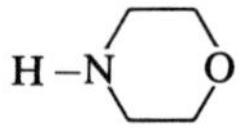 morpholine (H–N)	20°C	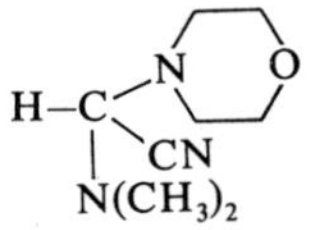 $H-C\begin{smallmatrix}\text{(morpholinyl)}\\CN\\N(CH_3)_2\end{smallmatrix}$	47
		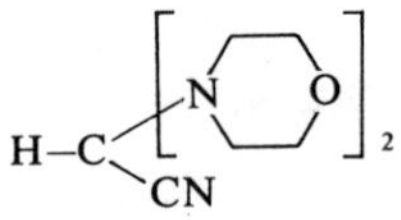 $H-C\begin{smallmatrix}[\text{morpholinyl}]_2\\CN\end{smallmatrix}$	20
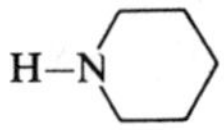 piperidine (H–N)	20°C	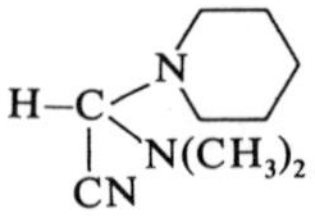 $H-C\begin{smallmatrix}\text{(piperidinyl)}\\N(CH_3)_2\\CN\end{smallmatrix}$	51
		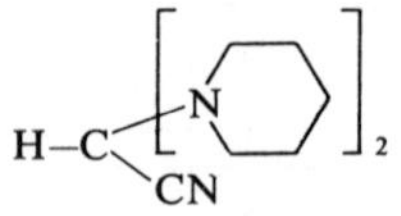 $H-C\begin{smallmatrix}[\text{piperidinyl}]_2\\CN\end{smallmatrix}$	

[a] Homologous acetonitriles react analogously (69).

TABLE XVIII

Bis(dialkylamino)acetonitriles from *N,N*-(dimethylamino)methoxyacetonitrile (**9a**) and Secondary Amines (Molar Ratio 1:2)

sec Amine	Reaction conditions	Products	Yield, %	Ref.
$HN(C_2H_5)_2$	100°C, 48 hr	$H-C(CN)[N(C_2H_5)_2]_2$	14	69
		$H-C(CN)[N(CH_3)_2][N(C_2H_5)_2]$	25	
$HN(C_3H_7)_2$	140°C, 24 hr	$H-C(CN)[N(C_3H_7)_2]_2$	25	69
		$H-C(CN)[N(C_3H_7)_2][N(CH_3)_2]$	28	
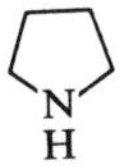 pyrrolidine (N-H)	140°C, 14 hr	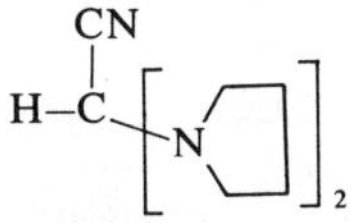 $H-C(CN)[\text{pyrrolidinyl}]_2$	87	69, 84
 piperidine (N-H)	140°C, 14 hr	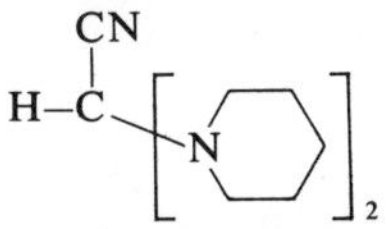 $H-C(CN)[\text{piperidinyl}]_2$	90	69, 84
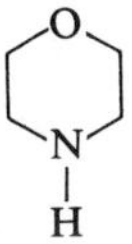 morpholine (N-H)	140°C, 14 hr	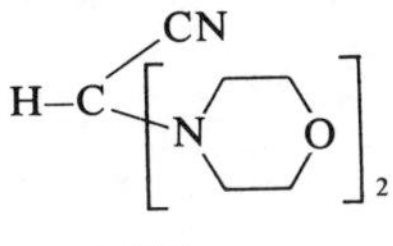 $H-C(CN)[\text{morpholinyl}]_2$	85	69
$H-N(CH_3)-(CH_2)_3-N(CH_3)-H$	80°C	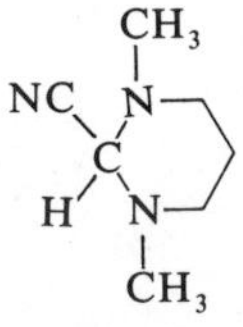 1,3-dimethyl-2-cyano-hexahydropyrimidine	65	84
$H_5C_2-NH-(CH_2)_2-NH-C_2H_5$	80°C	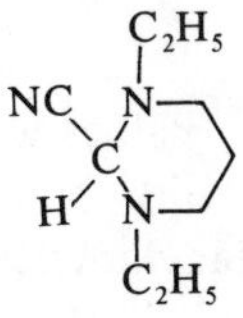 1,3-diethyl-2-cyano-imidazolidine derivative	65	84
$CH_3-NH-C_6H_{11}\text{-}c$	80°C	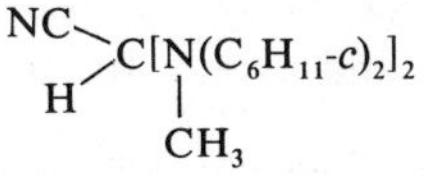 $NC-C(H)(CH_3)[N(C_6H_{11}\text{-}c)_2]_2$	60	84

There have been only a few experiments with homologous alkoxybis-(dialkylamino)acetonitriles. Interestingly, 1-pyrrolidino-1,2,2,2-tetraethoxy-propionitrile reacts with pyrrolidine in both possible ways (14):

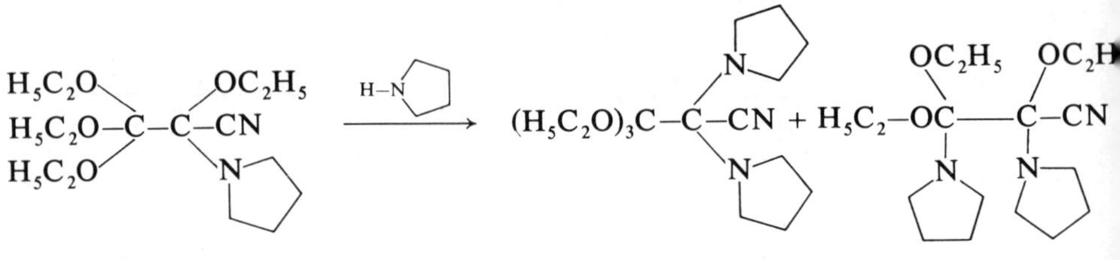

Orthoamides form an acid—base equilibrium with secondary amino compounds like N-monosubstituted acylamines or diacylamines (23). The kinetically controlled addition products are in equilibrium with the ion pair formed as a result of the instability of the newly formed C–N bond. With amide acetals (**1**), alkylation of the acylamide nitrogen follows in an S_N2 reaction (86).

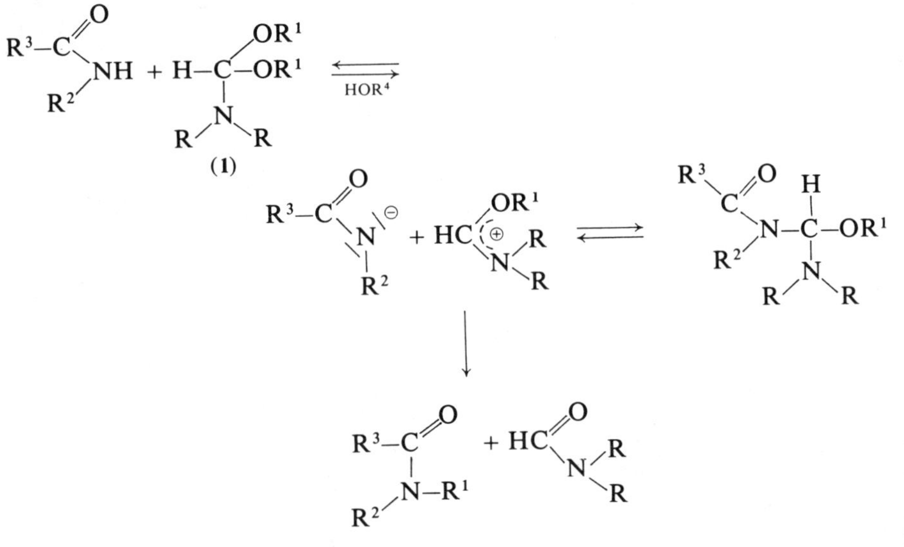

The correlation of pK_a values with the course of the reaction was established for the reaction of **1a** with NH-acidic compounds (87).* Thus NH-acidic compounds like N-alkylacylamides, N-diacylamides, and N-alkylsulfonamides react as described above. If the acidity increases, the nucleophilicity of the anion decreases and excess **1a** competes with the anion in the reaction with the

* This correlation cannot be easily transferred to higher N N-dimethylformamide acetals.

III. REACTIONS OF ORTHOAMIDES WITH NUCLEOPHILES 451

ambident cation. Therefore, the formation of trimethyl orthoformate and tetramethylammonium salt is observed in addition to methylation.

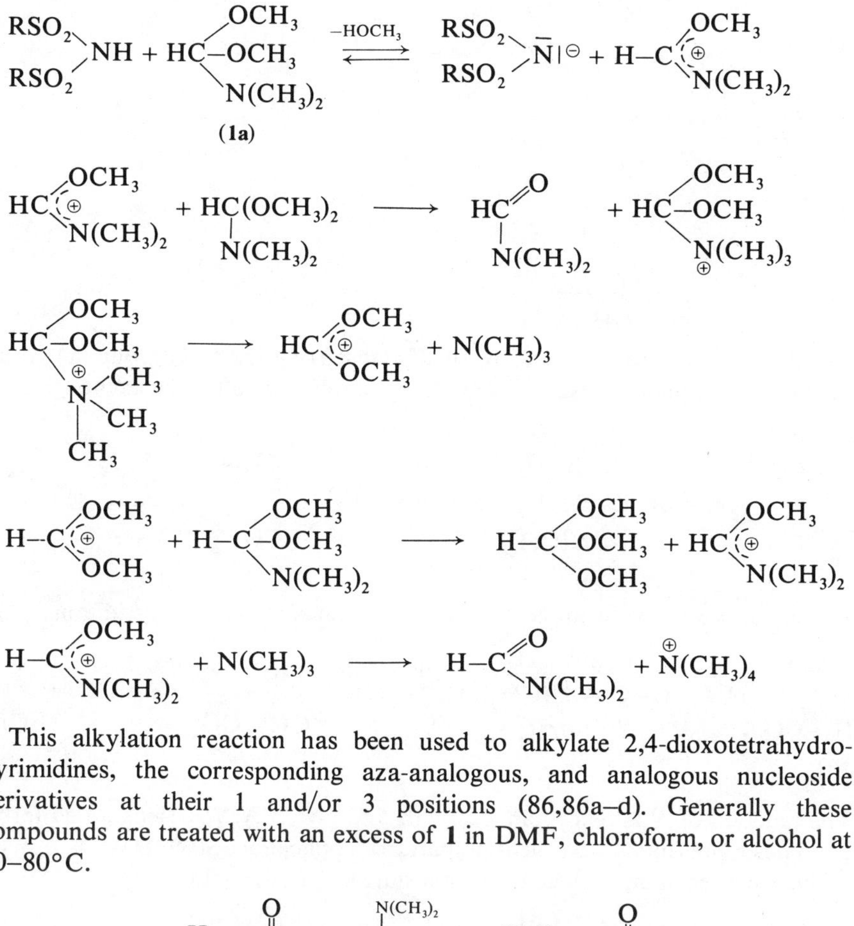

$$\begin{array}{c}RSO_2\\RSO_2\end{array}\!\!>\!\!NH + HC\!\!<\!\!\begin{array}{c}OCH_3\\OCH_3\\N(CH_3)_2\end{array} \xrightleftharpoons{-HOCH_3} \begin{array}{c}RSO_2\\RSO_2\end{array}\!\!>\!\!\bar{N}|^{\ominus} + H-C\!\!<\!\!\begin{array}{c}OCH_3\\\oplus\\N(CH_3)_2\end{array}$$

(1a)

$$HC\!\!<\!\!\begin{array}{c}OCH_3\\\oplus\\N(CH_3)_2\end{array} + HC(OCH_3)_2 \longrightarrow HC\!\!<\!\!\begin{array}{c}O\\N(CH_3)_2\end{array} + HC\!\!<\!\!\begin{array}{c}OCH_3\\OCH_3\\\stackrel{\oplus}{N}(CH_3)_3\end{array}$$

$$HC\!\!<\!\!\begin{array}{c}OCH_3\\OCH_3\\\stackrel{\oplus}{N}\!\!<\!\!\begin{array}{c}CH_3\\CH_3\end{array}\\CH_3\end{array} \longrightarrow HC\!\!<\!\!\begin{array}{c}OCH_3\\\oplus\\OCH_3\end{array} + N(CH_3)_3$$

$$H-C\!\!<\!\!\begin{array}{c}OCH_3\\\oplus\\OCH_3\end{array} + H-C\!\!<\!\!\begin{array}{c}OCH_3\\OCH_3\\N(CH_3)_2\end{array} \longrightarrow H-C\!\!<\!\!\begin{array}{c}OCH_3\\OCH_3\\OCH_3\end{array} + HC\!\!<\!\!\begin{array}{c}OCH_3\\\oplus\\N(CH_3)_2\end{array}$$

$$H-C\!\!<\!\!\begin{array}{c}OCH_3\\\oplus\\N(CH_3)_2\end{array} + N(CH_3)_3 \longrightarrow H-C\!\!<\!\!\begin{array}{c}O\\N(CH_3)_2\end{array} + \stackrel{\oplus}{N}(CH_3)_4$$

This alkylation reaction has been used to alkylate 2,4-dioxotetrahydropyrimidines, the corresponding aza-analogous, and analogous nucleoside derivatives at their 1 and/or 3 positions (86,86a–d). Generally these compounds are treated with an excess of **1** in DMF, chloroform, or alcohol at 60–80°C.

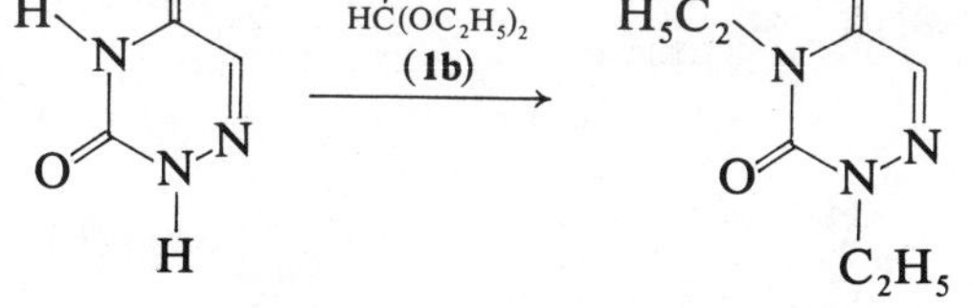

As experiments with numerous substrates have shown (86a), this alkylation method gives no by-products. *O*-Alkylation has not been reported at the CONH group or at the hydroxy group of a sugar, or at the N_7 atom of a purine ring. Owing to its high selectivity, this procedure is superior to the classical ones.

The alkylation is also successful with 2-dimethylamino-1,3-dioxolane in DMF at ~100°C (86e):

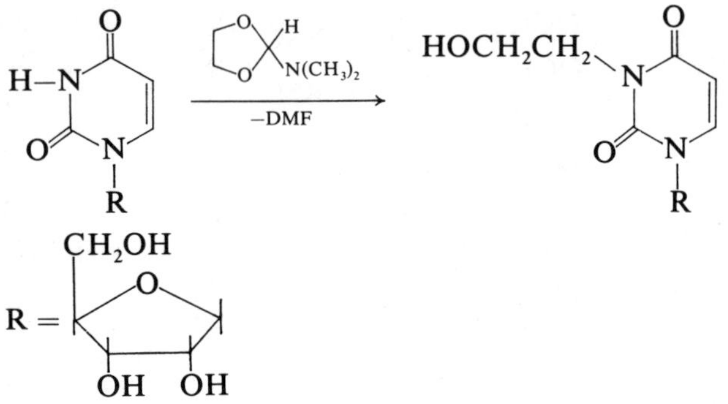

If the ambident cation is incapable of transferring the alkyl group, as in the reaction with amide thioacetals (**3**), the reaction ends after the exchange step (48).

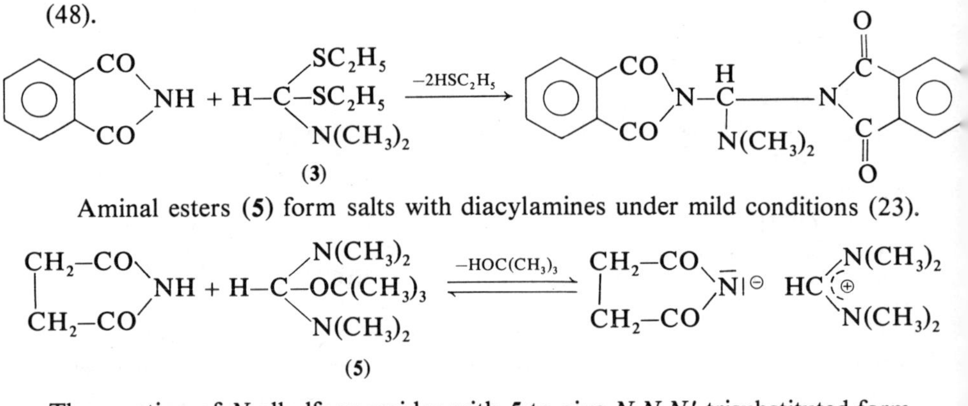

Aminal esters (**5**) form salts with diacylamines under mild conditions (23).

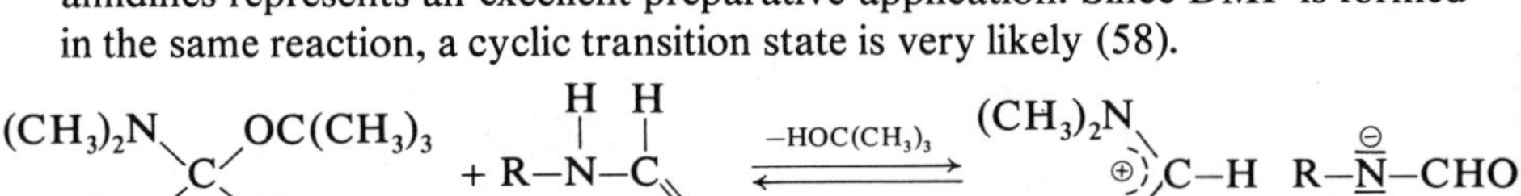

The reaction of *N*-alkylformamides with **5** to give *N,N,N'*-trisubstituted formamidines represents an excellent preparative application. Since DMF is formed in the same reaction, a cyclic transition state is very likely (58).

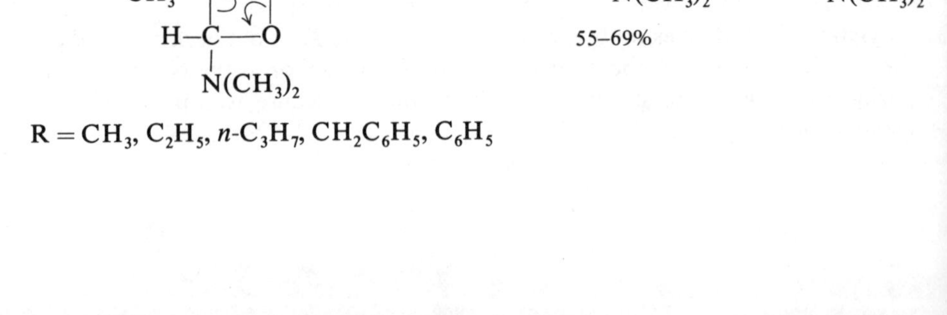

$R = CH_3, C_2H_5, n\text{-}C_3H_7, CH_2C_6H_5, C_6H_5$

5. (Dialkoxymethyl)ammonium salts with NH-acidic compounds

As does N-protonation (4,87,89), N-alkylation enormously increases the reactivity of carboxylic acid amide acetals. Primary aromatic amines react with the adduct formed from 5,5-dimethyl-1,3-dioxonium tetrafluoroborate and triethylamine, resulting in double dialkoxymethylation under mild conditions (90).

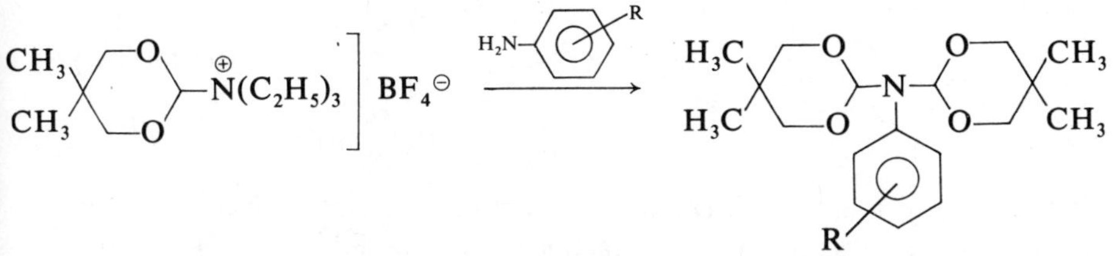

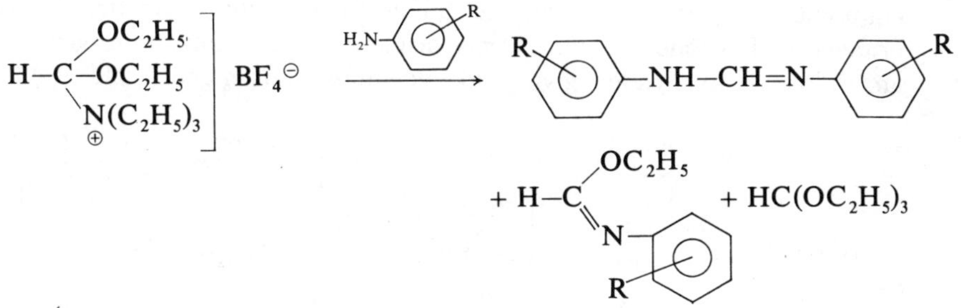

N,N'-Diarylformamidines result from reactions with excess aromatic amines as well as with acyclic N-quaternized formamide acetals (90). Cyclic (dialkoxymethyl)ammonium salts give N,N-disubstituted formamide acetals with secondary amines. This reaction is of some importance in the synthesis of N-aryl substituted formamide acetals (91).

Acyclic(dialkoxymethyl)ammonium salts give formamide acetals in good yields with secondary aromatic amines. Tetraalkylformamidinium salts are formed as a by-product if aliphatic amines are used, as in the reaction of **1a** with strong acids (91) (see Section III-D-8).

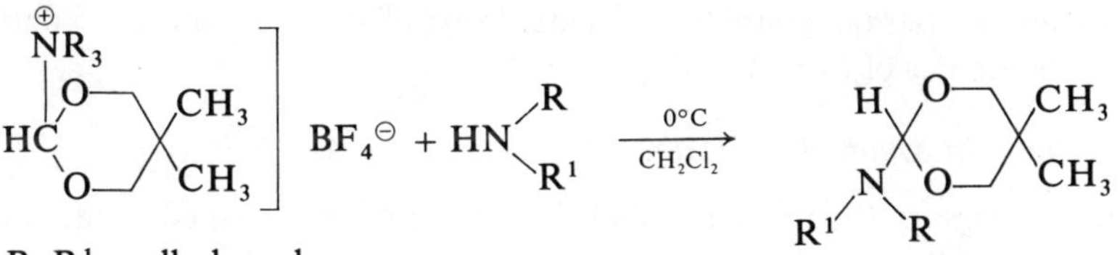

R, R^1 = alkyl, aryl

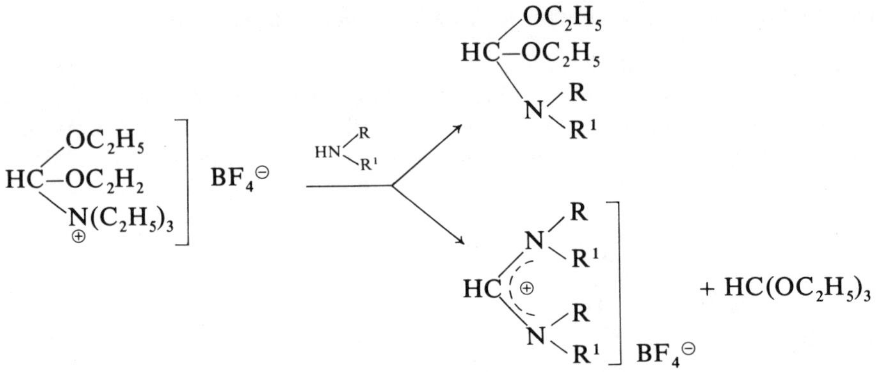

[Reaction scheme showing:

$\left[\begin{array}{c} \text{OC}_2\text{H}_5 \\ \text{HC}-\text{OC}_2\text{H}_2 \\ \text{N}(\text{C}_2\text{H}_5)_3 \end{array} \right]$ BF$_4^\ominus$ + HN$\begin{array}{c}\text{R}\\\text{R}^1\end{array}$

leading to:

HC$\begin{array}{c}\text{OC}_2\text{H}_5\\\text{OC}_2\text{H}_5\\\text{N}\begin{array}{c}\text{R}\\\text{R}^1\end{array}\end{array}$

and

$\left[\text{HC}^\oplus \begin{array}{c}\text{N}\begin{array}{c}\text{R}\\\text{R}^1\end{array}\\\text{N}\begin{array}{c}\text{R}\\\text{R}^1\end{array}\end{array} \right]$ BF$_4^\ominus$ + HC(OC$_2$H$_5$)$_3$]

Like secondary amines, secondary carboxylic acid amides yield *N*-acyl-*N*-alkylformamide acetals (92). Reactions of 1,3-dithiolan-2-yl-trimethylammonium iodide with aromatic amines are fairly interesting. Whereas primary and secondary amines only displace trimethylamine, tertiary amines and indole are formylated, giving aldehyde thioacetals in good yield (93).

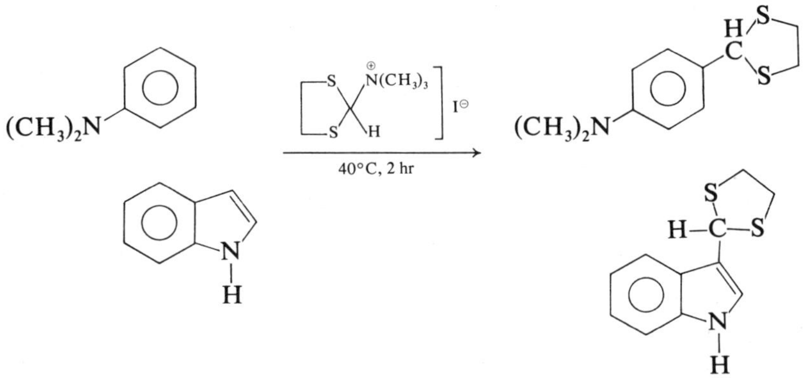

[Reaction scheme: (CH$_3$)$_2$N-C$_6$H$_5$ and indole react with 1,3-dithiolan-2-yl-trimethylammonium iodide at 40°C, 2 hr to give the corresponding 4-(dimethylamino)phenyl 1,3-dithiolane and 3-indolyl 1,3-dithiolane products.]

D. REACTIONS WITH OH-ACIDIC COMPOUNDS

Generally speaking, the nature of the final products depends on the pK_a value of the acids (87). Strongly nucleophilic agents with pK_a values > 15 displace alkoxy or dialkylamino groups in orthoamides. Acids with pK_a values between 2 and 15 are alkylated or amidated. Strong acids (pK_a < 2) form tetraalkylammonium or tetraalkylamidinium salts. Most of the experiments to date deal with orthoamides of formic acid.

1. *Carboxylic Acid Amide Acetals and Thioacetals with Alcohols*

In protic solvents, a dissociation equilibrium is rapidly established, causing a fast exchange of alkoxy groups with the solvent, for example, alcohol. This

TABLE XIX
N,N-dialkylcarboxylic Acid Amide Acetals via Interchange of Acetals

Starting material	Alcohol	Yield	Ref.
1a	Ethanol	53–69	96, 27
	n-Propanol	80	96
	Isopropanol	57	96
	n-Butanol	89	96
	Isobutanol	69	10
	sec-Butanol	67	10
	Neopentanol	81–84	10, 27
1a	Cyclohexanol	48	27
	Allyl alcohol	78	27
	Benzyl alcohol	61–88	10, 27
	1,2-Ethanediol	94	96
	2,3-Dimethyl-2,3-butanediol	84	96
	1,3-Propanediol	84	96
	2,2-Dimethyl-1,3-propanediol	92	27
1b	n-Butanol		4
	1,2-Ethanediol		4
$\mathrm{CH_3\underset{\underset{N(CH_3)_2}{|}}{C}(OC_2H_5)_2}$	1,2-Ethanediol	86	4

substitution is essential in the very simple synthesis of N,N-dialkylcarboxylic acid amide acetals from N,N-dialkylcarboxylic acid amide, dimethyl sulfate, and alcoholic solutions (94–96). Table XIX summarizes the reactions that have been performed with various alcohols.

This interchange of acetals at the same time represents one of the most important and simplest syntheses of higher amide acetals (diols yield cyclic amide acetals) (4,10,27,96). This method usually consists in heating a dimethyl acetal with excess alcohol, whereas methanol is distilled over a column.

The applicability is limited by the thermal stability of the acetals (10) and by bulky substituents adjacent to the hydroxyl group of the alcohol. Thus isopropanol (96) and neopentanol (10) react smoothly with **1**, whereas t-butanol (27) shows no reaction at all. Mixed acetals are not isolated; using too

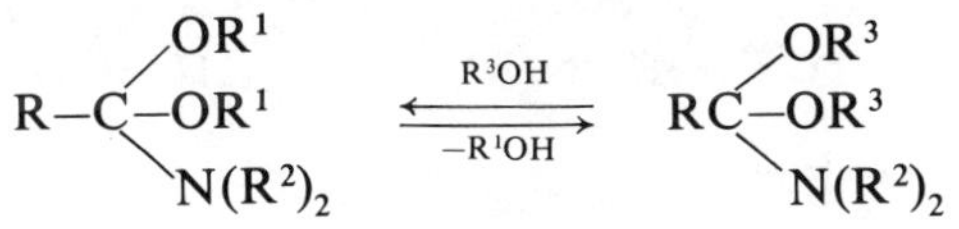

$$\mathrm{R{-}C}\genfrac{}{}{0pt}{}{\diagup OR^1}{\diagdown N(R^2)_2}\!\!{-}OR^1 \underset{-R^1OH}{\overset{R^3OH}{\rightleftarrows}} \mathrm{RC}\genfrac{}{}{0pt}{}{\diagup OR^3}{\diagdown N(R^2)_2}\!\!{-}OR^3$$

(1)

little alcohol results in recovery of the starting material along with some product showing complete interchange of the alkoxy groups. For the reaction of N,N-dimethylacetamide acetals with α,β-unsaturated alcohols see Section IV-B-3.

Protecting *cis*-vicinal hydroxyl groups of sugars by acetal formation is of some importance (74,75,97). Acetalizations using N,N-dimethylacetamide and N,N-dimethylbenzamide dialkyl acetals (**1**) (97) have been applied to this purpose.

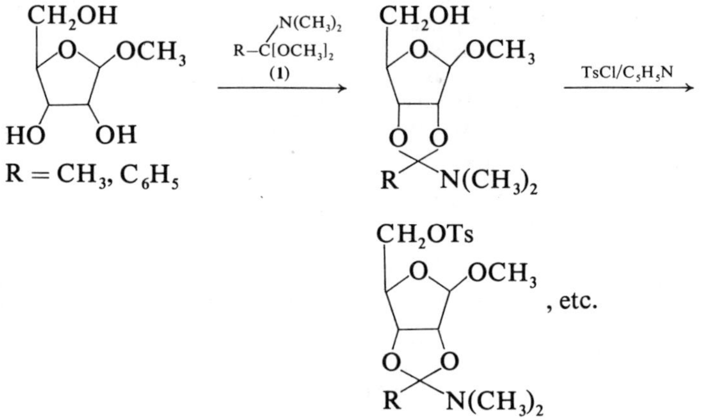

$$\text{structures shown: starting diol} \xrightarrow{R-C(OCH_3)_2 N(CH_3)_2 \;(\mathbf{1})} \text{cyclic acetal} \xrightarrow{\text{TsCl}/C_5H_5N} \text{tosylate, etc.}$$

$R = CH_3, C_6H_5$

In connection with these experiments, the following observation is rather important: trialkylammonium salts of ribonucleoside-2′,3-phosphates react with **1** giving cyclic 2′,3′-phosphates (75e,97a).

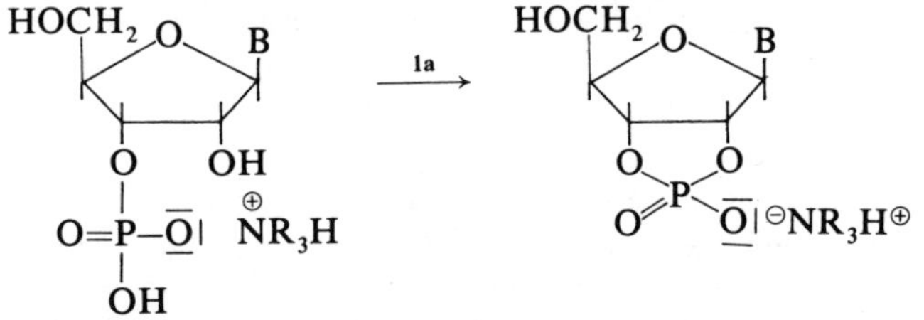

If tetraalkylammonium salts are reacted, dephosphorylation occurs, the mechanism of which has not yet been studied (97a).

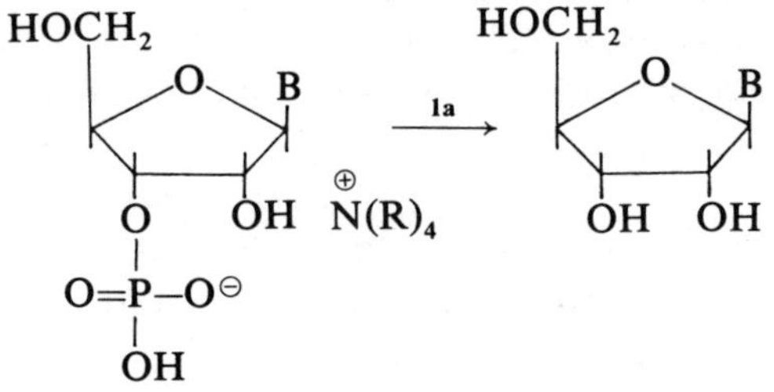

The alcoholysis of carboxylic acid amide thioacetals (**3**), which occurs only with higher-boiling alcohols under harsh conditions, is of minor preparative interest. The greater nucleophilicity of mercaptans is evidenced by an alkylation by benzyl alcohol (48).

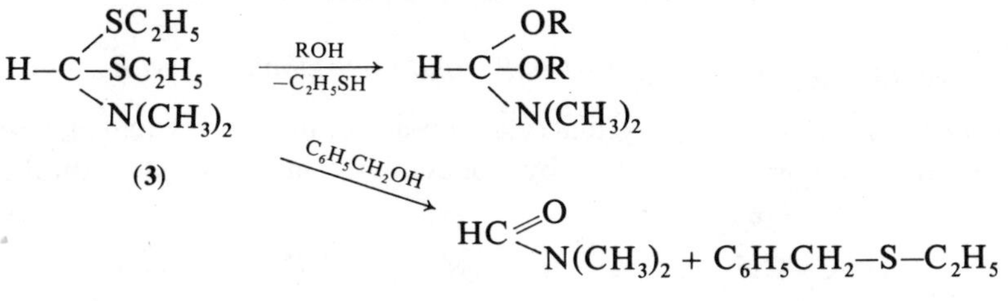

$$\underset{(3)}{\text{H--C}(SC_2H_5)_2 N(CH_3)_2} \xrightarrow[-C_2H_5SH]{ROH} \text{H--C}(OR)_2 N(CH_3)_2$$

$$\xrightarrow{C_6H_5CH_2OH} \text{HC}(=O)N(CH_3)_2 + C_6H_5CH_2\text{--S--}C_2H_5$$

R = *n*-hexyl, *n*-heptyl, cyclohexyl

2. *Aminal esters and Tris(dialkylamino)methanes with Alcohols*

Aminal esters establish a dissociative equilibrium as well as an acid–base relationship with alcohols. This represents the first step in the alcoholysis of tris(dialkylamino)methanes (**7**). Through an acid–base reaction, substitution of one dialkylamino group occurs in the presence of excess alcohol (5,96).

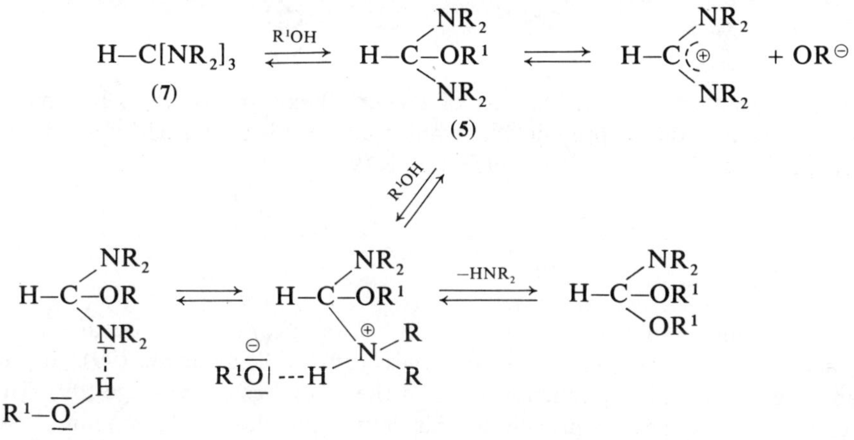

$$\underset{(7)}{\text{H--C}[NR_2]_3} \xrightleftharpoons{R'OH} \text{H--C}(OR^1)(NR_2)_2 \rightleftharpoons \text{H--C}^{\oplus}(NR_2)_2 + OR^{\ominus}$$
(5)

$$\Updownarrow R'OH$$

$$\underset{\substack{| \\ R^1\text{--}\bar{O}}}{\text{H--C}(OR)(NR_2)_2\text{--}H} \rightleftharpoons \underset{R^1\bar{O}|\text{---}H\text{--}N^{\oplus}R_2}{\text{H--C}(NR_2)(OR^1)} \xrightarrow{-HNR_2} \text{H--C}(NR_2)(OR^1)_2$$

Double reaction of **5** with alcohols (27,98) can be utilized to synthesize amide acetals with bulky alkoxy groups. Compounds of type **5** are reacted only *in situ.**

---

* *N,N*-dimethylformamide di-*t*-butyl acetal is formed by heating a mixture of 10 moles *t*-butanol, 1 mole potassium *t*-butoxide, and 1 mole tetramethylformamidine methylsulfate for 7 hr using a column; *t*-butanol is distilled off (98). For a better synthesis see the following section.

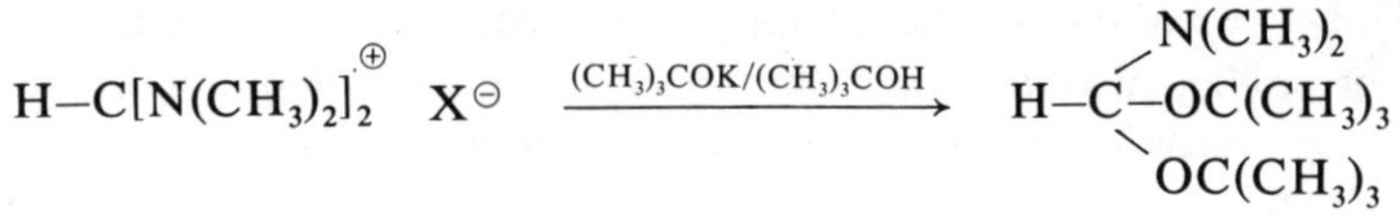

$$H-C[N(CH_3)_2]_2^{\oplus} \ X^{\ominus} \xrightarrow{(CH_3)_3COK/(CH_3)_3COH} H-C\begin{smallmatrix} N(CH_3)_2 \\ -OC(CH_3)_3 \\ OC(CH_3)_3 \end{smallmatrix}$$

X = Cl, ClO$_4$ (ref. 27), CH$_3$SO$_4$ (ref. 38)

3. Alkoxy(dialkylamino)- and Bis(dialkylamino) acetonitriles with Alcohols

In alcoholic solution alkoxy(dialkylamino)acetonitriles (**9**) exchange alkoxy groups via equilibria established by conductivity measurements (6,7). This

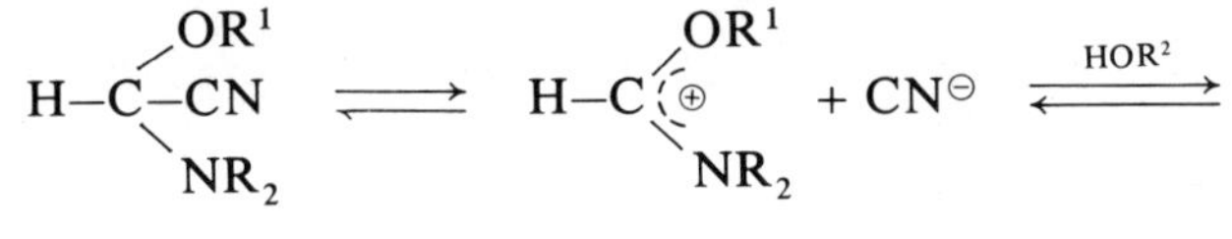

$$H-C\begin{smallmatrix} OR^1 \\ -CN \\ NR_2 \end{smallmatrix} \rightleftharpoons H-C\begin{smallmatrix} OR^1 \\ \oplus \\ NR_2 \end{smallmatrix} + CN^{\ominus} \xrightarrow{HOR^2} \rightleftharpoons$$

(**9**)

$$H-C\begin{smallmatrix} OR^1 \\ -O^{\oplus}-R^2 \\ NR_2 \ \ H \end{smallmatrix} \rightleftharpoons H-C\begin{smallmatrix} OR^2 \\ \oplus \\ NR_2 \end{smallmatrix} + HOR^1$$

$$H-C\begin{smallmatrix} OR^2 \\ \oplus \\ NR_2 \end{smallmatrix} + CN^{\ominus} \rightleftharpoons H-C\begin{smallmatrix} OR^2 \\ -CN \\ NR_2 \end{smallmatrix}$$

procedure allows the introduction of higher alkoxy groups (69). If a molar amount of alkoxide is present, the cyanide ion of **9** is displaced by an alkoxy group, giving *N,N*-dialkylformamide acetals (99).

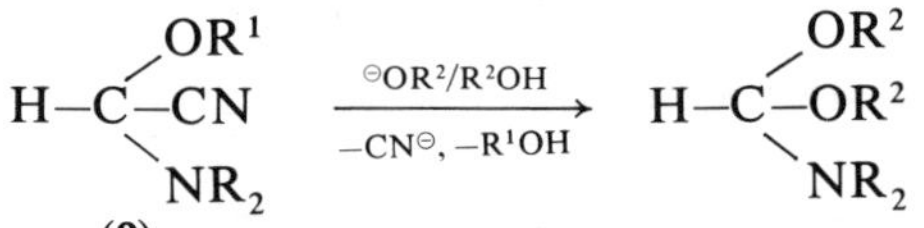

$$H-C\begin{smallmatrix} OR^1 \\ -CN \\ NR_2 \end{smallmatrix} \xrightarrow[-CN^{\ominus},\ -R^1OH]{^{\ominus}OR^2/R^2OH} H-C\begin{smallmatrix} OR^2 \\ -OR^2 \\ NR_2 \end{smallmatrix}$$

(**9**)

Analogously, substitution of cyanide by an alkoxy group occurs in the reaction of bis(dialkylamino)acetonitriles (**10**) with alkoxide (99). In the absence of alcohol, the reaction stops at the aminal ester (**5**) stage (99). This reaction is the best approach to this important class of compounds and represents one of the few known S_N reactions of orthoamides in the absence of protic solvents.

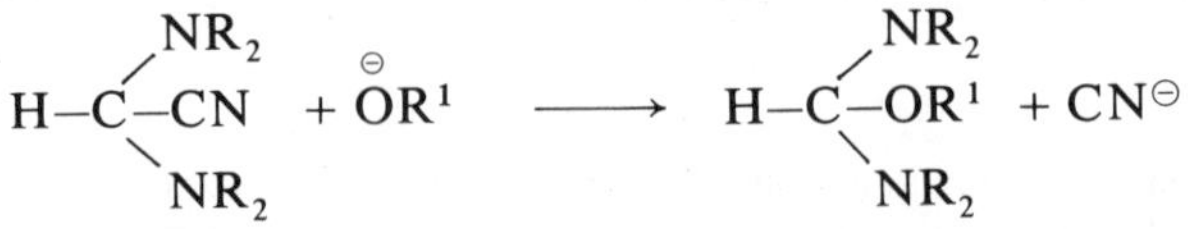

$$H-C\begin{smallmatrix} NR_2 \\ -CN \\ NR_2 \end{smallmatrix} + {}^{\ominus}OR^1 \longrightarrow H-C\begin{smallmatrix} NR_2 \\ -OR^1 \\ NR_2 \end{smallmatrix} + CN^{\ominus}$$

(**10**) \hspace{4cm} (**5**)

III. REACTIONS OF ORTHOAMIDES WITH NUCLEOPHILES

When the reaction occurs in excess alcohol, amide acetals (1) are formed in good yields; this procedure is especially recommended for the synthesis of compounds 1 containing bulky alkoxy groups (99).

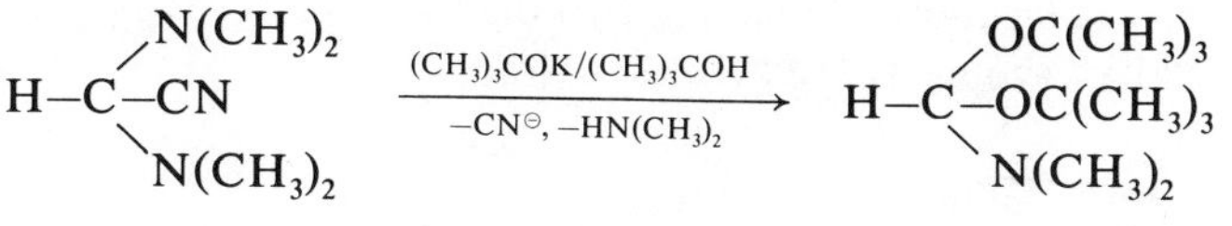

$$H-C\begin{pmatrix}N(CH_3)_2\\-CN\\N(CH_3)_2\end{pmatrix} \xrightarrow[-CN^\ominus, -HN(CH_3)_2]{(CH_3)_3COK/(CH_3)_3COH} H-C\begin{pmatrix}OC(CH_3)_3\\-OC(CH_3)_3\\N(CH_3)_2\end{pmatrix}$$

(10)   (1)

A summary of these reactions is given in Table XX.

TABLE XX

Alkoxy(dialkylamino)acetonitriles (9), Amide Acetals (1), and Aminal Esters (5) from Alkoxy(dialkylamino)- and Bis(dialkylamino)acetonitriles (9, 10)

Starting material, orthoamide	Product	Yield	Ref.
$H-C(OR)(CN)-N(CH_3)_2$ $R = CH_3, C_2H_5$	$H-C(OR^1)(CN)-N(CH_3)_2$ $R^1 = n\text{-}C_3H_7$, $i\text{-}C_3H_7$	61–65	69
$H-C(OCH_3)(CN)-N(R)_2$ $R = CH_3, C_2H_5$	$H-C(OCH_3)_2-N(R)_2$	63–70	99
$H-C(OCH_3)(CN)-X$ $X = 1$-pyrrolidino, 1-piperidinyl, 4-morpholino	$H-C(OCH_3)_2-X$ with $N(R)_2$	73–76	99
$H-C[N(CH_3)_2]_2-CN$ $R = CH_3, C_2H_5, n\text{-}C_3H_7, i\text{-}C_3H_7, t\text{-}C_4H_9$	$H-C[N(CH_3)_2]_2-OR$	76–84	99
$H-C[N(CH_3)_2]_2-CN$ $R = CH_3, C_2H_5, n\text{-}C_3H_7, i\text{-}C_3H_7, n\text{-}C_4H_9, t\text{-}C_4H_9$	$H-C(OR)_2-N(CH_3)_2$	70–82	99

4. Carbonic Acid Orthoamides with Alcohols

In principle, these compounds react similarly to formic acid orthoamides. Orthoamides containing one dialkylamino group (**11**) react by exchange of alkoxy groups (11); those containing two dialkylamino groups (**12**) exchange one for an alkoxy group (4,11,95,100).

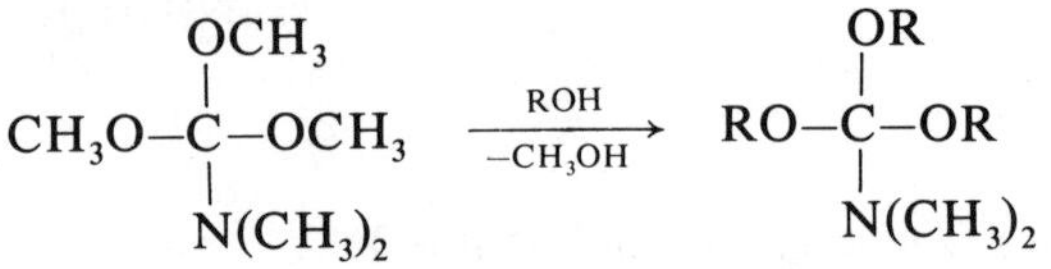

$$\underset{(\mathbf{11})}{\underset{N(CH_3)_2}{\overset{OCH_3}{CH_3O-\underset{|}{\overset{|}{C}}-OCH_3}}} \xrightarrow[-CH_3OH]{ROH} \underset{N(CH_3)_2}{\overset{OR}{RO-\underset{|}{\overset{|}{C}}-OR}}$$

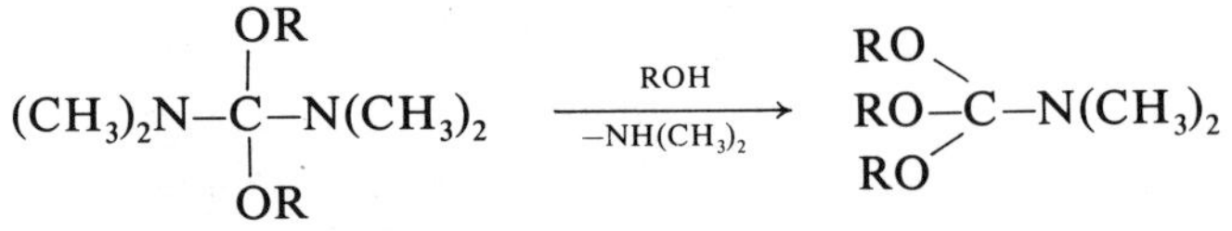

$$\underset{(\mathbf{12})}{\underset{OR}{\overset{OR}{(CH_3)_2N-\underset{|}{\overset{|}{C}}-N(CH_3)_2}}} \xrightarrow[-NH(CH_3)_2]{ROH} \underset{RO}{\overset{RO}{RO-\underset{/}{\overset{\backslash}{C}}-N(CH_3)_2}}$$

Alkoxytris(dimethylamino)methanes (**13**) undergo solvolysis with excess alcohol, eventually giving tetraalkylurea acetals (**11**). Because of the low thermal stability of both the starting material and the products, a rather low yield is obtained.

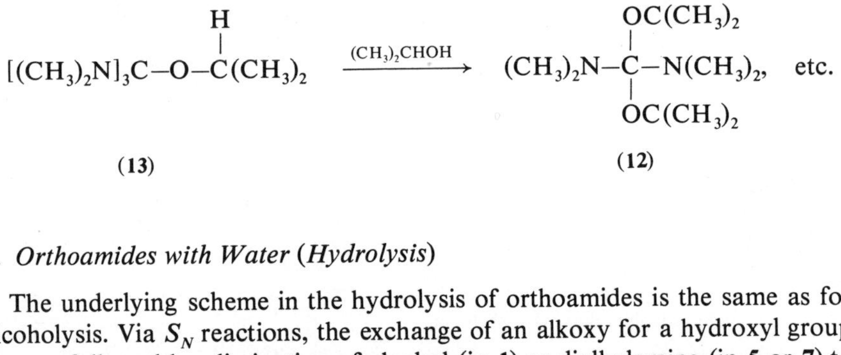

$$\underset{(\mathbf{13})}{[(CH_3)_2N]_3C-O-\overset{H}{\underset{|}{C}}(CH_3)_2} \xrightarrow{(CH_3)_2CHOH} \underset{(\mathbf{12})}{\underset{OC(CH_3)_2}{\overset{OC(CH_3)_2}{(CH_3)_2N-\underset{|}{\overset{|}{C}}-N(CH_3)_2}}}, \text{ etc.}$$

5. Orthoamides with Water (*Hydrolysis*)

The underlying scheme in the hydrolysis of orthoamides is the same as for alcoholysis. Via S_N reactions, the exchange of an alkoxy for a hydroxyl group occurs, followed by elimination of alcohol (in **1**) or dialkylamine (in **5** or **7**) to give *N,N*-dialkylcarboxylic acid amides (4,87,101). These results hold for both

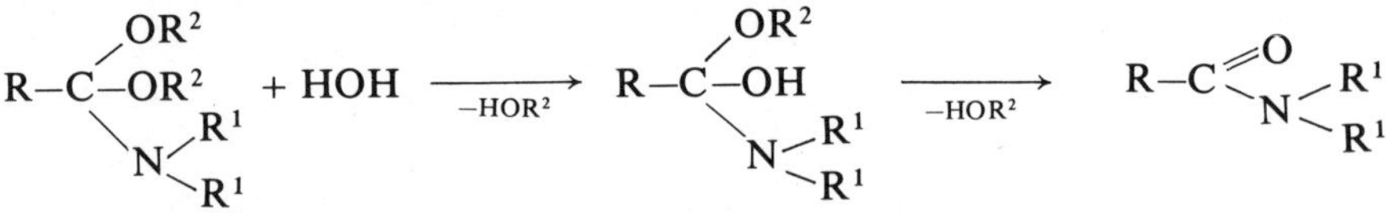

$$\underset{(\mathbf{1})}{R-\overset{OR^2}{\underset{N\diagdown R^1}{\overset{\diagup}{\underset{\diagdown}{C}}-OR^2}}} + HOH \xrightarrow{-HOR^2} R-\overset{OR^2}{\underset{N\diagdown R^1}{\overset{\diagup}{\underset{\diagdown}{C}}-OH}} \xrightarrow{-HOR^2} R-C\overset{\diagup\!\!\!=O}{\underset{N\diagdown R^1}{\diagdown R^1}}$$

neutral and weakly alkaline media (101). The competitive *N*-protonation, followed by C–N bond cleavage and formation of carboxylic acid esters, occurs in strongly acidic media (4).

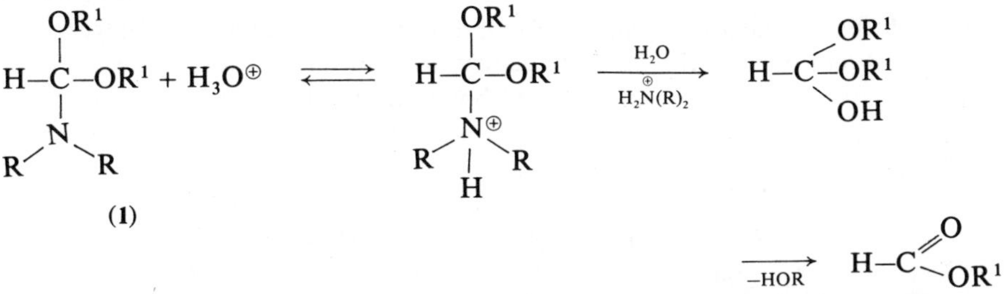

$$\underset{(1)}{\underset{R}{\overset{OR^1}{\underset{N}{\overset{|}{\text{H–C–OR}^1}}}}\text{R}} + H_3O^{\oplus} \rightleftharpoons \underset{R\underset{H}{\overset{|}{N^{\oplus}}}R}{\overset{OR^1}{\overset{|}{\text{H–C–OR}^1}}} \xrightarrow[H_2N(R)_2]{H_2O} \text{H–C}\underset{OH}{\overset{OR^1}{\overset{\diagup}{\diagdown}}} OR^1 \xrightarrow{-HOR} \text{H–C}\underset{OR^1}{\overset{O}{\diagdown}}$$

According to the hydrolysis steps formulated above* the rate of saponification corresponds to the general reactivity of orthoamides with XH-acidic compounds.

The sensitivity toward hydrolysis increases as follows:

> carboxylic acid amide acetals (1)
> aminal esters (5)
> tris(dialkylamino)methane (7)

The same is true for orthocarbonic acid amides. The hydrolysis is slowed down with increasing chain length of the alkoxy group in **1** (R = H).

Cyclic amide acetals are somewhat more stable than acyclic compounds of the same molecular weight owing to their decreased dissociation (4, 74, 75, 97). *O*-Arylamide acetals hydrolyze very rapidly, since dissociation occurs more easily (102).

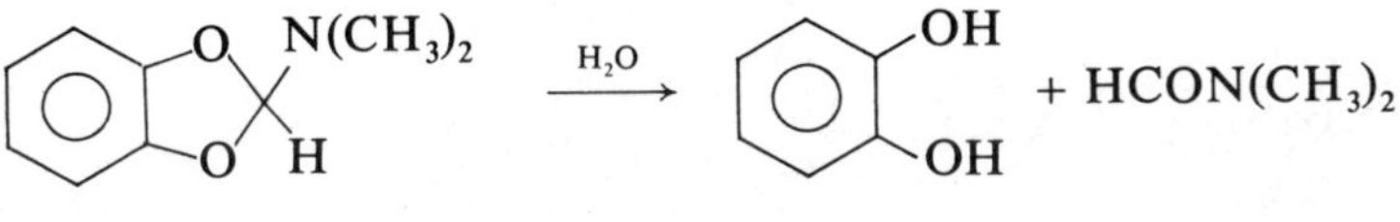

Aminal esters (**5**) and tris(dialkylamino)methanes (**7**) generally yield carboxylic acid amides in a fast reaction sequence. Only in acidic media could the amidinium salts, as the first intermediates of hydrolysis, be isolated in a few cases (103,104). As expected, orthocarbamic acid acetals, especially cyclic acetals, are marked by their special stability (4).

6. Orthoamides with Phenols

Acetals of dimethylformamide react with phenols via (1) *O*-alkylation (106,107); (2) aromatic substitution (108); or (3) salt formation, for example, tetraalkylammonium or tetraalkylformamidinium salts (87,108).

---

* Hydrolysis—C–N bond cleavage—up to 10% is observed for bicyclic amide acetals (105).

The wide range of pK_a values of phenols (~0–10) requires discussion. If the pK_a value of the phenol is between 3 and 10, alkylation by an S_N2 mechanism is generally observed (87,106,107). Only if there is strong steric hindrance by substituents in the 2 and 6 positions does the slower substitution of the aromatic ring of dialkylphenols occur (108). Interchange of acetals has never been observed, not even in the case of 1,2-dihydroxybenzene; this can be traced to the low stability of the resulting C–O bond and the high concentration of the phenoxide.

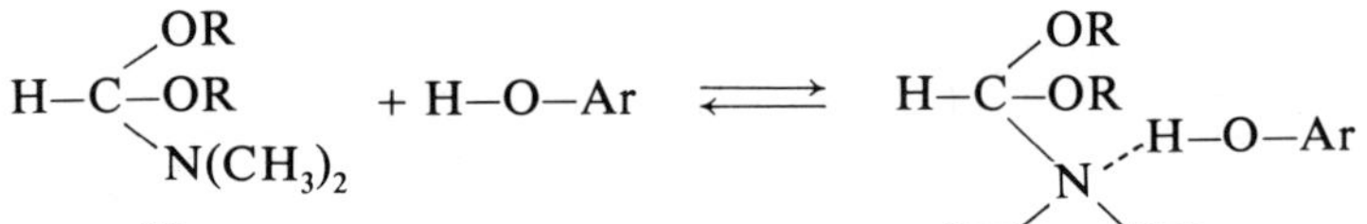

$$\text{H-C(OR)}_2\text{N(CH}_3\text{)}_2 + \text{H-O-Ar} \rightleftharpoons \text{H-C(OR)}_2\cdots\text{H-O-Ar}\cdot\text{N(CH}_3\text{)}_2$$
(1)

$R = CH_3, C_2H_5, C_6H_5-CH_2$

$$\text{H-C(OR)}_2\text{-N}^+\text{H(CH}_3\text{)}_2 \cdot |\bar{O}\text{Ar} \rightleftharpoons \text{H-C}^+\text{(OR)}_2\text{N(CH}_3\text{)}_2 + |\bar{O}\text{Ar} + \text{ROH}$$

$\downarrow S_N2$

$$\text{H-C(=O)N(CH}_3\text{)}_2 + \text{ArO-R}$$

$$\text{H-C}^+\text{(OR)N(CH}_3\text{)}_2 + \text{[Ar-}\bar{O}\text{]} \longrightarrow \text{(CH}_3\text{)}_2\text{N-CH=[cyclohexadienone]=O} + \text{HOR}$$

$\downarrow H_2O$

$$\text{O=CH-[Ar]-OH}$$

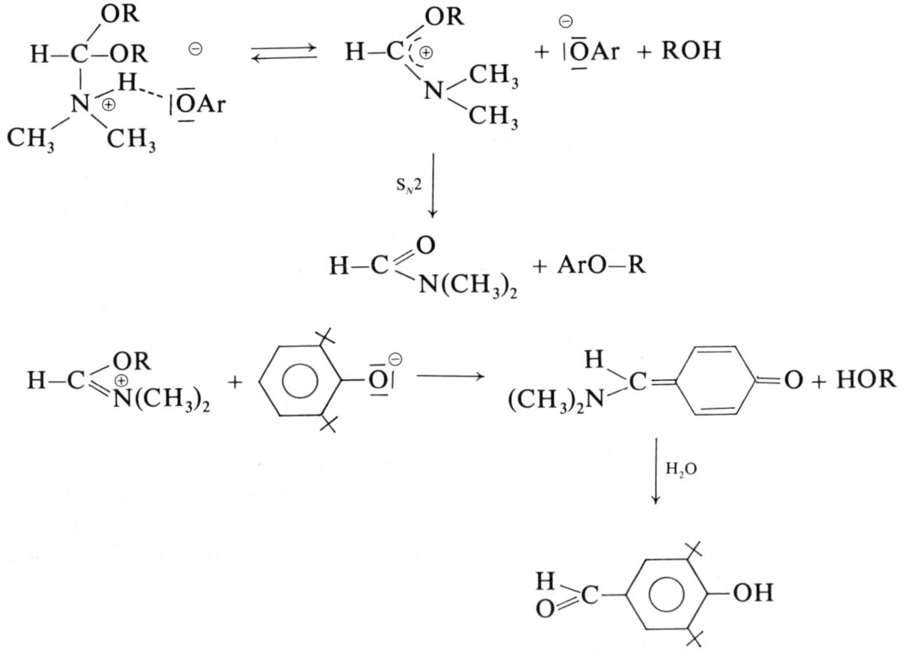

With phenols having a pK_a value < 3 salt formation is observed (87, 108). The nucleophilicity of the phenoxide is too low for alkylation. The composition of the salts (tetraalkylammonium or tetraalkylformamidinium salts) is closely related to the structure of the acetals. Tetraalkylammonium salts are formed by

III. REACTIONS OF ORTHOAMIDES WITH NUCLEOPHILES

reaction of N,N-dimethylformamide acetal (**1**), which is more nucleophilic than the phenoxide ion, with the ambident cation. The formation of tetraalkylformamidinium salts starts with the cleavage of the dialkylamino group. Therefore they are the main products if strongly acidic phenols are reacted. The interaction of weakly acidic phenols with the dialkylamino group of **1** has the character of a hydrogen bond, whereas strongly acidic phenols protonate nitrogen to form strong N–H bonds.

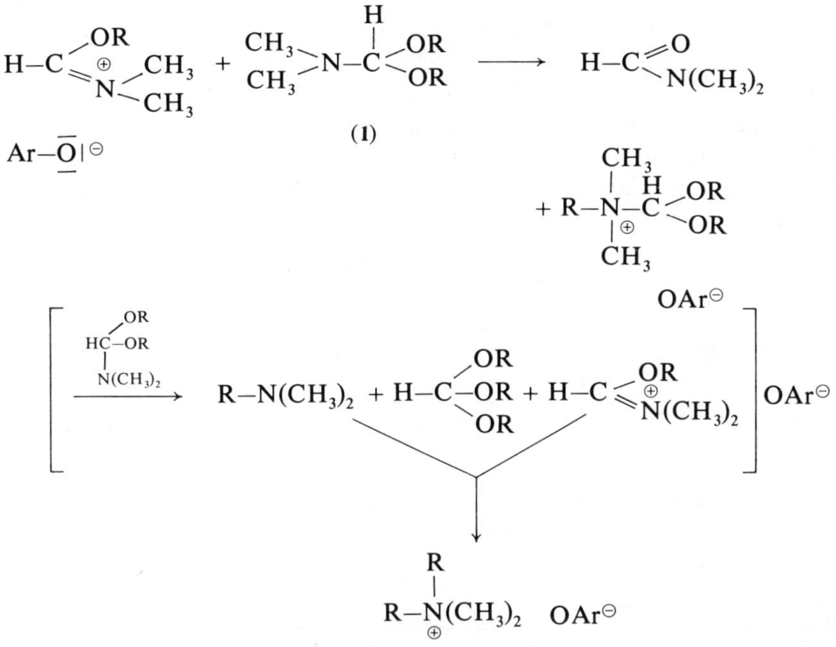

$$H-C \overset{\oplus}{\underset{N(CH_3)_2}{\underset{|}{\overset{OR}{\diagup}}}} + (CH_3)_2N-CH(OR)_2 \longrightarrow H-C \underset{N(CH_3)_2}{\overset{O}{\diagdown}}$$

$$Ar-\bar{O}|^{\ominus} \qquad (1)$$

$$+ R-\underset{CH_3}{\overset{CH_3}{\underset{|}{\overset{|}{N}}}}-\overset{H}{\underset{\oplus}{C}}\underset{OR}{\overset{OR}{\diagup}}$$

$$\left[\xrightarrow{\underset{N(CH_3)_2}{\overset{HC-OR}{\underset{|}{\overset{OR}{\diagup}}}}} R-N(CH_3)_2 + H-C(OR)_3 + H-C\underset{N(CH_3)_2}{\overset{OR}{\underset{\oplus}{\diagup}}} \middle| OAr^{\ominus} \right] \overset{OAr^{\ominus}}{}$$

$$\downarrow$$

$$R-\underset{\oplus}{\overset{R}{\underset{|}{N}}}(CH_3)_2 \quad OAr^{\ominus}$$

If the N–H bond is labile, the proton is easily transferred to the alkoxy groups, followed by a kinetically controlled cleavage of the alcohol. With strong N–H bonds, cleavage of the dialkylamino group is faster. The increasing size of the alkoxy groups augments the amount of tetraalkylformamidinium salt. This is caused either by the lower stability of the carbiminium cation (nonplanar) or by steric hindrance of the N-alkylation of the acetal.

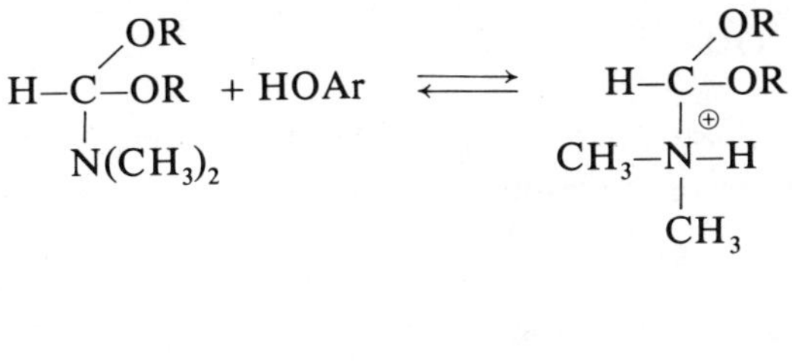

$$H-\underset{N(CH_3)_2}{\overset{OR}{\underset{|}{C}-OR}} + HOAr \rightleftharpoons H-\underset{\underset{CH_3}{\overset{|}{CH_3-N-H}}}{\overset{OR}{\underset{\oplus}{C-OR}}}$$

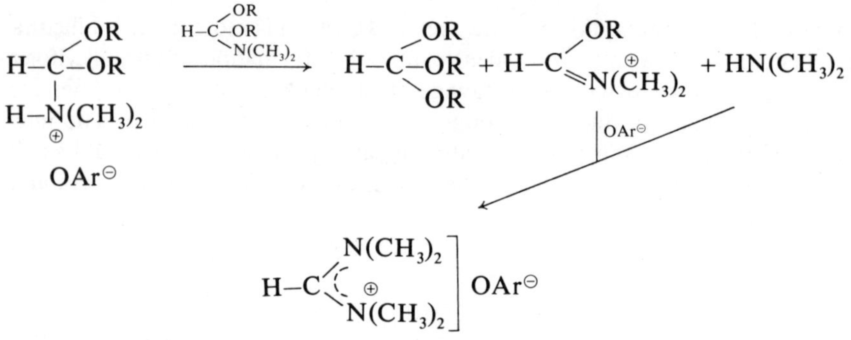

464   ORTHOAMIDES: PROPERTIES AND REACTIONS

$$\underset{\underset{\underset{OAr^{\ominus}}{\oplus}}{H-\overset{|}{\underset{N(CH_3)_2}{C}}-OR}}{H-C\overset{OR}{\underset{OR}{\diagdown}}} \longrightarrow H-\overset{OR}{\underset{N(CH_3)_2}{C}}-OR \longrightarrow H-C\overset{OR}{\underset{OR}{\diagdown}}OR + H-C\overset{OR}{\underset{\oplus N(CH_3)_2}{\diagup}} + HN(CH_3)_2$$

$$\left[H-C\overset{N(CH_3)_2}{\underset{N(CH_3)_2}{\diagdown}}\oplus \right] OAr^{\ominus}$$

Nucleophilic attack of **1** may occur more easily at a less planar (hence more reactive) carbiminium ion. This reaction mechanism could not be established for **1a** (87), but may play an important role for compound **1b** and others.

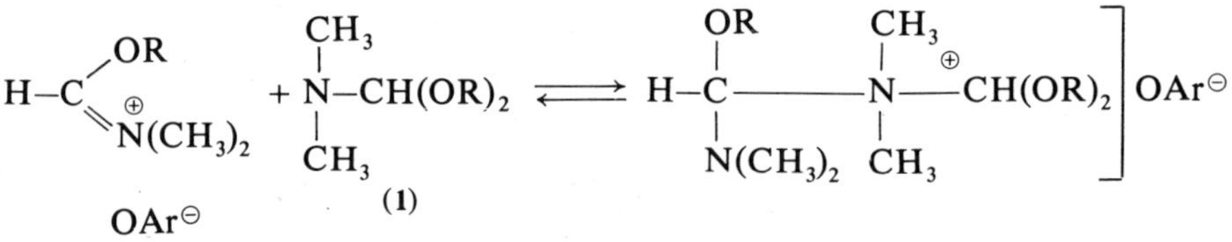

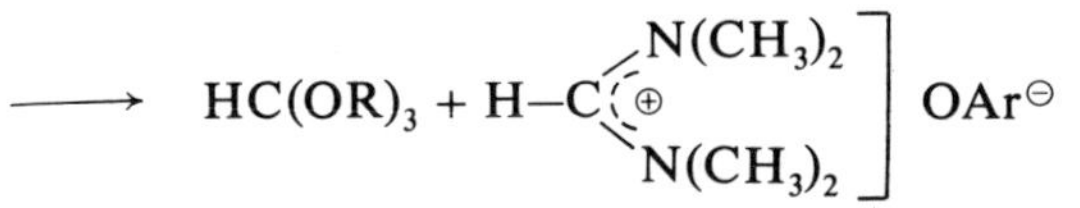

$$H-C\overset{OR}{\underset{\oplus N(CH_3)_2}{\diagup}} + \underset{\underset{CH_3}{|}}{\overset{CH_3}{\underset{|}{N}}-CH(OR)_2} \rightleftharpoons \left[H-\overset{OR}{\underset{N(CH_3)_2}{\overset{|}{C}}}-\overset{CH_3}{\underset{CH_3}{\overset{|}{N}\oplus}}-CH(OR)_2 \right] OAr^{\ominus}$$

$$\longrightarrow HC(OR)_3 + \left[H-C\overset{N(CH_3)_2}{\underset{N(CH_3)_2}{\diagdown}}\oplus \right] OAr^{\ominus}$$

Salt and orthoester formation are favored by high concentrations of **1**, as the reaction conditions indicate. In order to get a high yield of phenol ethers, it is advisable to maintain a low stationary concentration of **1** (107).

Aminal esters **5** and bis(dimethylamino)acetonitrile (**10**) yield, with phenols, amidinium salts of the following structure (7,108):

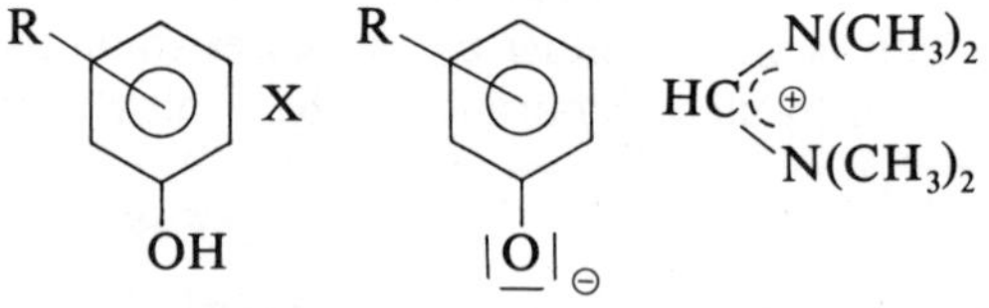

With nucleophilic phenols like β-naphthol, 2,6-di-*t*-butyl- and 2,6-dimethylphenol, formylation of the aromatic ring, giving aldehydes, occurs at temperatures of 100–130°C.

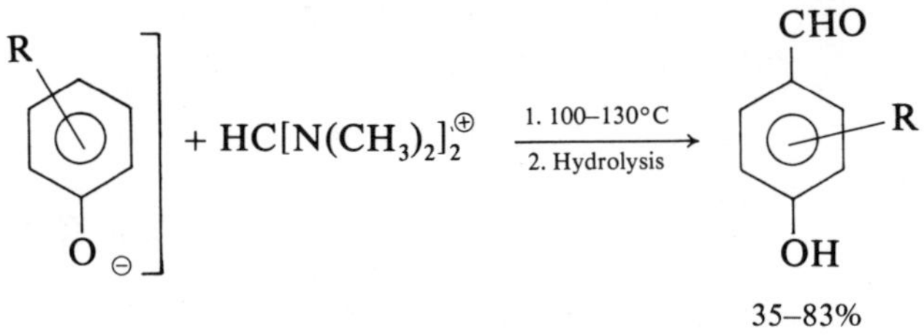

III. REACTIONS OF ORTHOAMIDES WITH NUCLEOPHILES

[Phenolate with R substituent] + HC[N(CH$_3$)$_2$]$_2^\oplus$ $\xrightarrow[\text{2. Hydrolysis}]{\text{1. 100–130°C}}$ [2-hydroxy-5-R-benzaldehyde]

35–83%

7. N,N-dialkylformamide Acetals with Carboxylic Acids

The reaction of carboxylic acids with pK_a values >2.5 and N,N-dialkylformamide acetals is a valuable method of esterification (10,86e,87,106, 107,109). To obtain high yields, the compounds are best treated at room temperature (sometimes at 80°C) in an inert solvent like benzene, ether, or methylene chloride maintaining a low, stationary concentration of **1** (10,107). High concentration of **1** favors the formation of by-products [see Section III-A-4(f)] (107). Thus the reaction of acetic acid with excess **1** yields only trialkyl orthoformate (110).

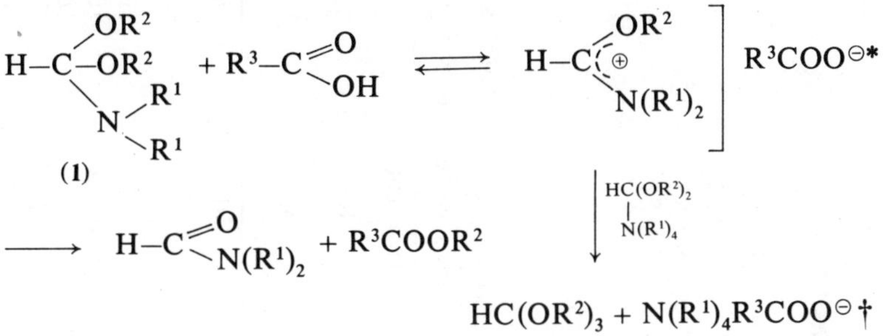

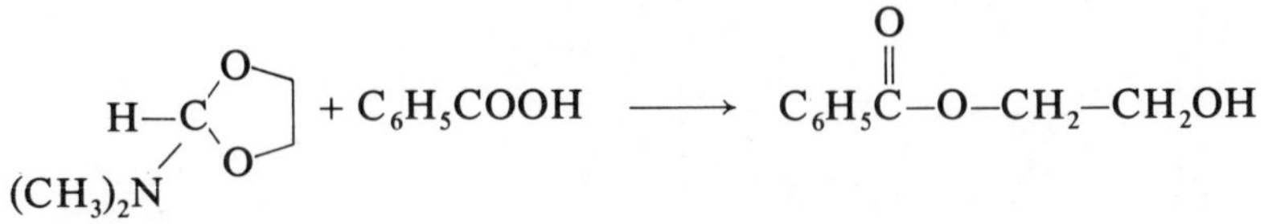

$$\underset{(1)}{H-C(OR^2)_2 N(R^1)_2} + R^3-C(=O)OH \rightleftharpoons \left[H-C(OR^2)_2^{\oplus} N(R^1)_2 \right] R^3COO^{\ominus *}$$

$$\longrightarrow H-C(=O)N(R^1)_2 + R^3COOR^2$$

$$\downarrow \begin{array}{c} HC(OR^2)_2 \\ | \\ N(R^1)_4 \end{array}$$

$$HC(OR^2)_3 + N(R^1)_4 R^3COO^{\ominus} \dagger$$

$$\underset{(CH_3)_2N}{H-C\langle O\text{-}O\rangle} + C_6H_5COOH \longrightarrow C_6H_5\overset{O}{\underset{\|}{C}}-O-CH_2-CH_2OH$$

Phenolic hydroxyl groups do not react under these conditions; therefore phenolcarboxylic acid can be selectively esterified (10,107).

In order to esterify carboxylic acids with alcohols, for which the amide acetals are either difficultly or not accessible, a procedure using N,N-dimethylformamide dineopentyl acetal is recommended (10). This acetal alkylates

---

* Oily products, which are sometimes observed at the beginning of the reaction, have been assigned to this structure.

† By-products in the case of high concentration of **1**.

TABLE XXI

Esterification of Carboxylic Acids with N,N-Dimethylformamide Dialkyl Acetals (1)

$$\underset{N(CH_3)_2}{H-C(OR)_2}$$

Carboxylic acid	R	Reaction conditions	Yield, %	Ref.
Benzoic acid	CH_3	Benzene, 80°C, 1 hr	83	10
Benzoic acid	C_2H_5	Benzene, 80°C, 1 hr	94	10, 109
Benzoic acid	i-C_4H_9	Benzene, 80°C, 1 hr	81	10
Benzoic acid	± sec-C_4H_9	Benzene, 80°C, 2 hr	94	10
Benzoic acid	$C_6H_5CH_2$	Benzene, 80°C, 2 hr	95	10
2,4,6-Trimethylbenzoic acid	CH_3	Benzene, 80°C, 1 hr	82	10
2,4,6-Trimethylbenzoic acid	C_2H_5	Benzene, 80°C, 1 hr	93	10
2,4,6-Trimethylbenzoic acid	$C_6H_5CH_2$	Benzene, 80°C, 2 hr	94	10
4-Nitrobenzoic acid	CH_3	1,2-Dichloroethane, 24°C, 71 hr	90	107
4-Nitrobenzoic acid	CH_3	1,2-Dichloroethane, 30°C, 4 hr	65	107
4-Nitrobenzoic acid	CH_3	1,2-Dichloroethane, 30°C, 25 hr	70	107

Acid	R	Conditions	Yield	Ref
4-Nitrobenzoic acid	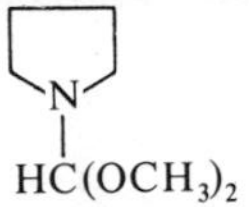 pyrrolidine–HC(OCH$_3$)$_2$	1,2-Dichloroethane, 30°C, 4 hr	92	107
4-Nitrobenzoic acid	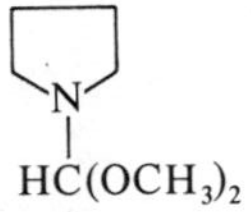 pyrrolidine–HC(OCH$_3$)$_2$	1,2-Dichloroethane, 30°C, 25 hr	100	107
4-Nitrobenzoic acid	CH$_3$	20°C, 24 hr	32	87
3,5-Dinitrobenzoic acid	CH$_3$	20°C, 30 hr	21	87
2-Naphthoic acid	C$_6$H$_5$CH$_2$	1,2-Dichloroethane, 84°C, 3 hr	92	106, 107
2-Hydroxybenzoic acid	C$_2$H$_5$	Methylene chloride, 22°C, 36 hr	93	106, 107
3-Pyridinecarboxylic acid	C$_2$H$_5$	Methylene chloride, 22°C, 16 hr	81	106, 107
Adipic acid	C$_2$H$_5$	Methylene chloride, 22°C, 24 hr	75[a]	106, 107
Fumaric acid	C$_6$H$_5$CH$_2$	Methylene chloride, 22°C, 35 hr	81[a]	106, 107
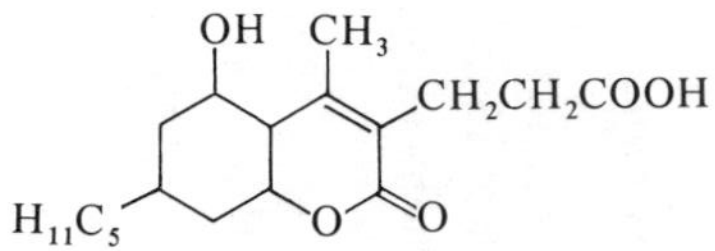 (bicyclic lactone structure with OH, CH$_3$, CH$_2$CH$_2$COOH, H$_{11}$C$_5$ substituents)	C$_2$H$_5$	Benzene, 80°C, 3 hr	72	111

[a] Diester.

carboxylic acids very slowly, whereas the interchange of the acetal proceeds rapidly. Thus one may esterify carboxylic acids with an alcohol in the presence of the dineopentyl acetal as an "alkyl group carrier" (for example, R^1 = 4-methoxybenzyl or 4-dodecylbenzyl).

$$HC\underset{N(CH_3)_2}{\overset{OCH_2C(CH_3)_3}{<}}\!\!\!{OCH_2C(CH_3)_3} + 2R^1OH \rightleftharpoons HC\underset{N(CH_3)_2}{\overset{OR^1}{<}}\!\!\!{OR^1} + 2HOCH_2(CH_3)_3$$

(1)

$$H-\underset{N(CH_3)_2}{\overset{OR^1}{C}}\!\!\!{OR^1} + R^2-COOH \longrightarrow H-C\underset{N(CH_3)_2}{\overset{O}{\diagup}} + R^2-COOR^1 + R^1-OH$$

The synthesis of methyl esters may be improved by using 1,1-bis(methoxymethyl)pyrrolidine instead of 1a. With this reagent the esterification proceeds much faster, and fewer by-products (tetraalkylammonium salts) are observed (107).

According to Eschenmoser et al. (10) these esterifications are S_N2 reactions, as indicated by the complete inversion of the chiral center of (+)(S)-sec-butanol when (+)-dimethylformamide di-sec-butyl acetal is used. The alkylation is the rate-determining step. This inversion of a chiral center was used to invert the configuration of axial 3α-hydroxy groups in steroids (107).

The reactions are summarized in Tables XXI–XXIII.

Anions of carboxylic acids with pK_a values of 0–1 are not alkylated because their nucleophilicity is too low. Reaction with 1a yields tetraalkylammonium salts, as discussed in Section III-D-4 (87).

Reactions of carboxylic acids with aminal esters (5) and tris(dialkylamino)-methanes (7) are rarely mentioned, since such reactions lack synthetic interest. 5 and 7 form tetraalkylformamidinium salts at room temperature. If the temperature is raised, carboxylic acids may be amidated.

8. Orthoamides with Strong Acids and Hydrogen Chloride

In reactions with strong acids, for example, perchloric acid, the similar results to those discussed earlier (Section III-D-6) are obtained. After N-protonation tetramethylformamidinium salts are formed with excess 1a (87).

$$2HC(OCH_3)_2 + HX \longrightarrow \left[HC\underset{N(CH_3)_2}{\overset{N(CH_3)_2}{<}}\!\!\!{\oplus}\right] X^\ominus + HC(OCH_3)_3 + CH_3OH$$
$$|$$
$$N(CH_3)_2$$

(1a)

$X = ClO_4, ArSO_3$

III. REACTIONS OF ORTHOAMIDES WITH NUCLEOPHILES

TABLE XXII

Esterfication of Amino Acid and Peptide Derivatives with Dimethylformamide Dibenzyl Acetal (A) and Dimethylformamide Bis(4-dodecylbenzyl) Acetal (B)

RCOOH[a]	Acetal	Reaction conditions	Yield, %	Ref.
N-DOBC-L-Val	A	Benzene, 80°C, 1.5 hr	97	10, 109
N-DOBC-L-Phe	A	Benzene, 80°C, 48 hr	90	10
N-DOBC-L-Try	A	Benzene, 80°C, 2 hr	78	10, 109
N-DOBC-Gly-L-Leu	A	Benzene, 80°C, 1 h	73	10,109
N-DOBC-Gly-L-Tyr	A	Methylene chloride, 25°C, 48 hr	80	10
N-BOC-Gly-L-Ph	A	Methylene chloride, 25°C, 90 hr	80	10
N-DOBC-L-Phe	B	Methylene chloride, 20°C, 18 hr	84	109
N-DOBC-C-Try	B	Methylene chloride, 20°C, 50 hr	68	109
N-DOBC-L-Phe-L-Ala	B	Methylene chloride, 20°C, 120 hr	75	109

[a] DOBC = 4-Decyloxybenzyloxycarbonyl; BOC = t-butoxycarbonyl.

TABLE XXIII

Esterification of Amino Acids and Dipeptide Derivatives with Benzyl (A), 4-Dodecyl (B), and 4-Methoxybenzyl Alcohol (C) via Dimethylformamide Dineopentyl Acetal in Methylene Chloride at Room Temperature

RCOOH[a]	Alcohol	Reaction time, hr	Yield, %	Ref.
N-BOC-Gly-L-Phe	A	72	85	10
N-DOBC-L-Phe	B	48	86	10
N-DOBC-L-Ala	B	43	83	10
N-DOBC-Gly	B	47	86	10
N-DOBC-L-Phe-L-Ala	B	53	78	10
N-DOBC-L-Phe	C	41	80	10
N-DOBC-L-Val	C	47	73	10
N-DOBC-Gly	C	46	73	10
N-Z-Gly	C	47	74	10
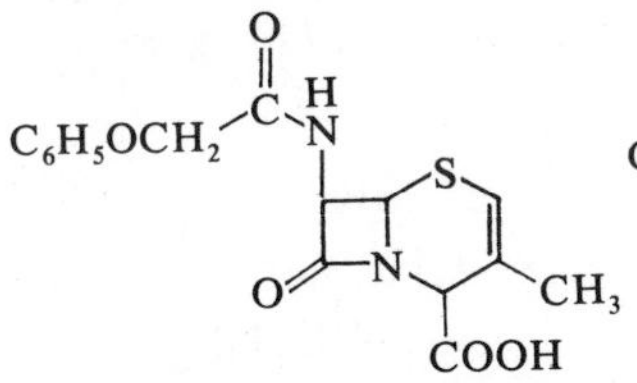 $C_6H_5OCH_2$-C(=O)-NH-[β-lactam-dihydrothiazine with CH₃ and COOH]	C		90	112

[a] BOC = t-Butoxycarbonyl; DOBC = 4-decyloxybenzyloxycarbonyl; Z = benzyloxycarbonyl.

Owing to the stronger nucleophilic character of chloride ion, hydrogen chloride maintains a special position. In this case it is possible to isolate a product* with the following structure, if **1b** is used (4):

* This hydrolyzes to ethyl formate and dimethylammonium chloride.

$$\left[\begin{array}{c} H_2C_2O \\ C_2H_5O \end{array}\!\!>\!\!C\!\!<\!\!\begin{array}{c} H \\ \overset{\oplus}{N}(CH_3)_2 \\ | \\ H \end{array}\right] Cl^{\ominus}$$

Excess **1b** forms tetramethylformamidinium chloride (89). **1a** forms tetramethylammonium chloride, as expected; at the same time an S_N reaction partly occurs, giving methyl chloride (87,89).*

Aminal esters and tris(dialkylamino)methanes react with strong acids and hydrogen chloride giving tetraalkylformamidinium salts as the sole products (3,46,113).

E. REACTIONS OF ORTHOAMIDES WITH PH-ACIDIC COMPOUNDS

Phosphoric acid dialkyl esters react as acetylene or hydrogen cyanide with N,N-dimethylformamide acetals under mild conditions, giving phosphono O,N-acetals (37,114–117). If the temperature is raised a little, these compounds react with further phosphoric acid dialkyl esters to give 1,1-diphosphonotrimethylamines (37,114,117).

$$\begin{array}{c} H-C(OR)_2 \\ | \\ N(CH_3)_2 \end{array} + (R^1O)_2P\!\!<\!\!\begin{array}{c} O \\ H \end{array} \longrightarrow (R^1O)_2\overset{\overset{O}{\|}}{P}-\underset{H}{\overset{}{C}}\!\!<\!\!\begin{array}{c} OR \\ N(CH_3)_2 \end{array}$$

$$\xrightarrow{(R^1O)_2\overset{\overset{O}{\|}}{P}H} (R^1O)_2\!-\!\overset{\overset{O}{\|}}{P}\!-\!\underset{\underset{N(CH_3)_2}{|}}{\overset{\overset{H}{|}}{C}}\!-\!\overset{\overset{O}{\|}}{P}(OR^1)_2$$

Cyclic phosphoric acid esters form only high-molecular-weight products with dimethylformamide acetals. This is also true in reactions of cyclic dimethylformamide acetals with phosphoric acid dialkyl esters (114).

$$\begin{array}{c} HC(SR)_2 \\ | \\ N(CH_3)_2 \end{array} + (R^1O)_2\overset{\overset{O}{\|}}{P}H \longrightarrow (R^1O)_2\overset{\overset{O}{\|}}{P}\!-\!CH\!\!<\!\!\begin{array}{c} SR \\ N(CH_3)_2 \end{array}$$

At 100°C also, N,N-dimethylformamide thioacetals and phosphoric acid dialkyl esters react to give phosphono S,N-acetals (114). The reaction of dibutyl- and diphenylphosphine oxide with **1a** gives dimethylaminobis(dibutyl-[or diphenyl]phosphinyl)methanes (114).

---

* Amide thioacetals form more stable salts with compounds like picric acid (37).

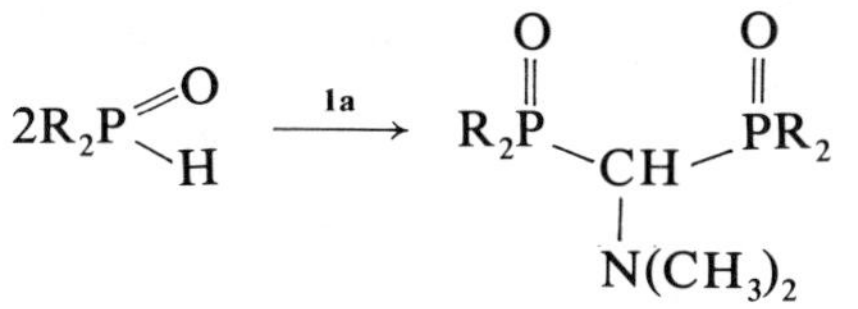

$$2R_2P\overset{O}{\underset{H}{\diagdown}} \xrightarrow{1a} R_2\overset{O}{\overset{\|}{P}}-\underset{\underset{N(CH_3)_2}{|}}{CH}-\overset{O}{\overset{\|}{P}}R_2$$

F. REACTIONS WITH SH-ACIDIC COMPOUNDS

N,N-dialkylcarboxylic acid amide acetals yield the corresponding thioacetals (**3**) with mercaptans. This exchange occurs in the synthesis of **3** from N,N-dialkylcarboxylic acid amide–dimethyl sulfate complexes and sodium mercaptide/mercaptan (24,37,48).

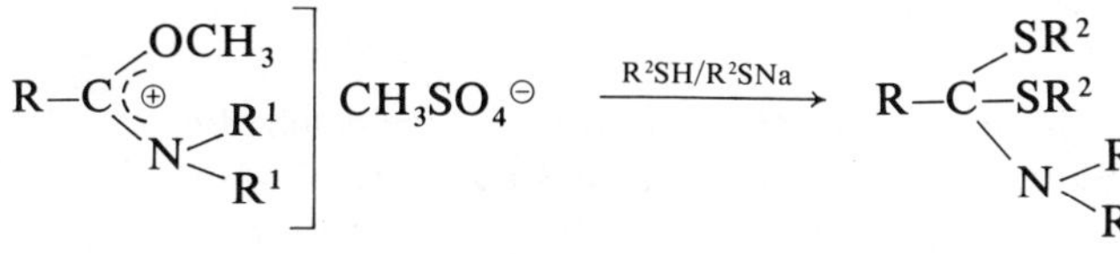

$$\left[R-C\underset{N{\diagdown}{R^1 \atop R^1}}{\overset{OCH_3}{\diagup}\oplus} \right] CH_3SO_4^{\ominus} \xrightarrow{R^2SH/R^2SNa} R-\underset{N{\diagdown}{R^1 \atop R^1}}{\overset{SR^2}{\overset{|}{C}}-SR^2}$$

(**3**)

The mercaptonolysis of **1** avoids the use of mercaptides, and proceeds at 100–120°C in good yields (24,48,118–120a).

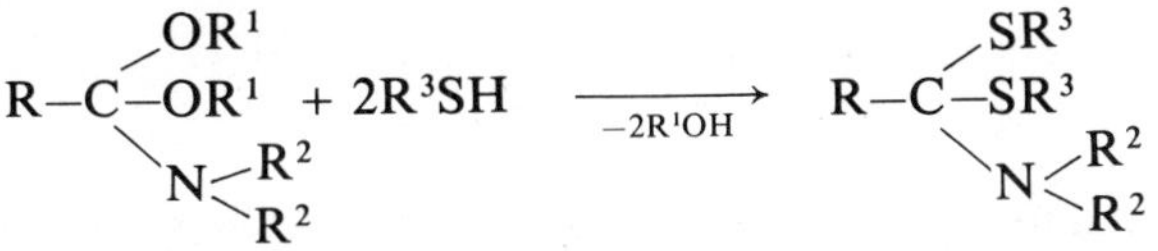

$$R-\underset{N{\diagdown}{R^2 \atop R^2}}{\overset{OR^1}{\overset{|}{C}}-OR^1} + 2R^3SH \xrightarrow{-2R^1OH} R-\underset{N{\diagdown}{R^2 \atop R^2}}{\overset{SR^3}{\overset{|}{C}}-SR^3}$$

(**1**)

The synthesis of O,S,N-acetals from the N,N-dimethylformamide–dimethyl sulfate complex and sodium mercaptide fails because of disproportionation, as does the synthesis from the N,N-dimethylthioformamide–diethyl sulfate adduct and sodium ethoxide. As expected, the reaction of **1b** with ethyl mercaptan also fails.*

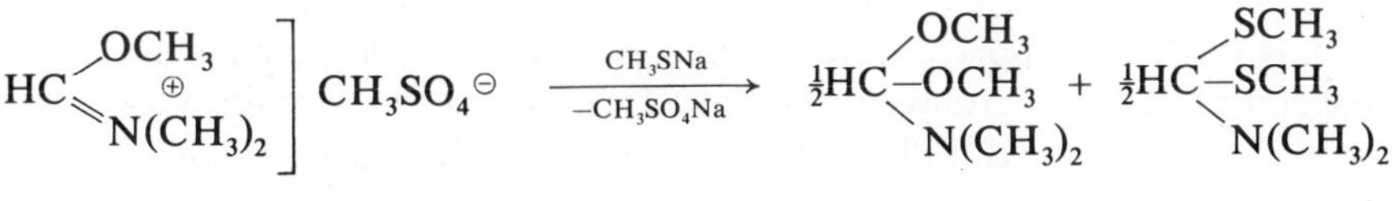

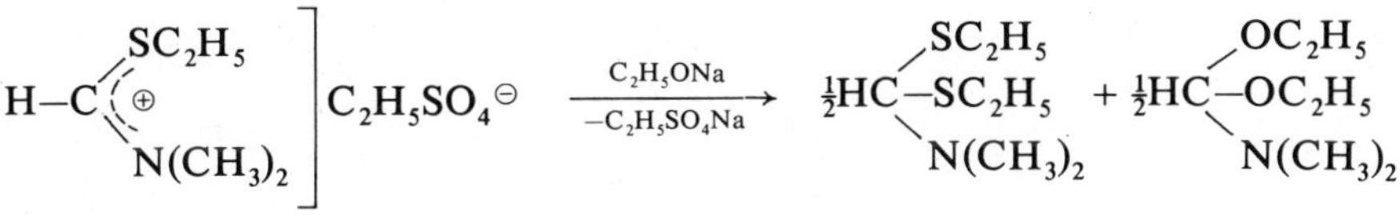

$$\left[HC\underset{N(CH_3)_2}{\overset{OCH_3}{\diagup}\oplus}\right] CH_3SO_4^{\ominus} \xrightarrow[-CH_3SO_4Na]{CH_3SNa} \tfrac{1}{2}HC\underset{N(CH_3)_2}{\overset{OCH_3}{\diagup}}-OCH_3 + \tfrac{1}{2}HC\underset{N(CH_3)_2}{\overset{SCH_3}{\diagup}}-SCH_3$$

$$\left[H-C\underset{N(CH_3)_2}{\overset{SC_2H_5}{\diagup}\oplus}\right] C_2H_5SO_4^{\ominus} \xrightarrow[-C_2H_5SO_4Na]{C_2H_5ONa} \tfrac{1}{2}HC\underset{N(CH_3)_2}{\overset{SC_2H_5}{\diagup}}-SC_2H_5 + \tfrac{1}{2}HC\underset{N(CH_3)_2}{\overset{OC_2H_5}{\diagup}}-OC_2H_5$$

* The ethylthio(dimethylamino)methoxymethane described in the literature (120) is presumably a mixture of **1a** and DMF diethyl thioacetal.

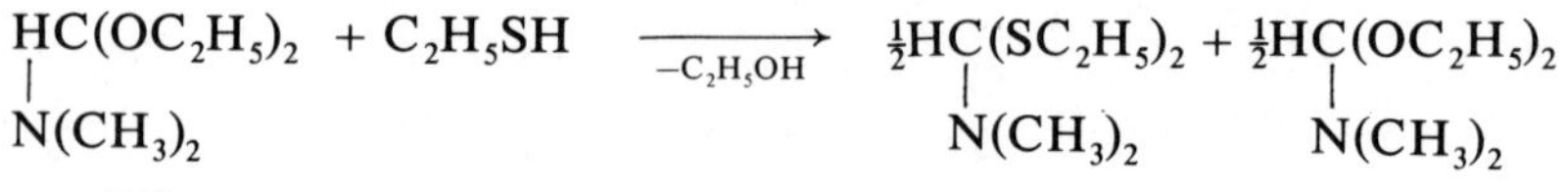

$$\underset{(1b)}{\underset{N(CH_3)_2}{\overset{HC(OC_2H_5)_2}{|}}} + C_2H_5SH \xrightarrow{-C_2H_5OH} \tfrac{1}{2}\underset{N(CH_3)_2}{\overset{HC(SC_2H_5)_2}{|}} + \tfrac{1}{2}\underset{N(CH_3)_2}{\overset{HC(OC_2H_5)_2}{|}}$$

The action of equimolar amounts of mercaptans with aminal esters (5) yields aminal thioesters at room temperature (24).

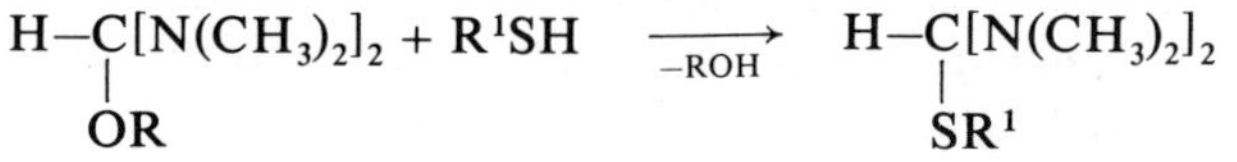

$$\underset{(5)}{\underset{OR}{\overset{H-C[N(CH_3)_2]_2}{|}}} + R^1SH \xrightarrow{-ROH} \underset{SR^1}{\overset{H-C[N(CH_3)_2]_2}{|}}$$

"SH-acidic compounds," characterized by delocalization of the negative charge on sulfur, are methylated at sulfur by **1a**. This method allows the transformation of compounds like 2-thioxo- or 4-thioxodihydropyrimidines into the corresponding methylthiopyrimidines (121a).

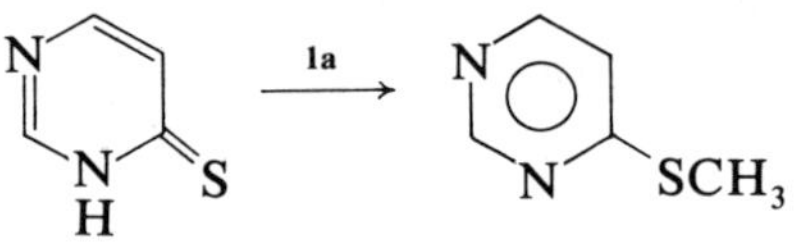

2-Dimethylamino-1,3-dioxolane yields the analogous 2-hydroxyethyl derivatives when heated for a longer time:

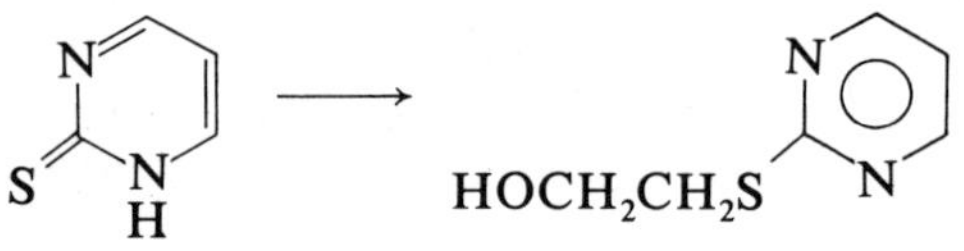

If this method is applied to nucleosides using dimethylformamide dineopentyl acetal as the alkyl group carrier (see Section III-D-7) the primary hydroxyl group of ribose is successfully exchanged for an alkylthio group (121a).

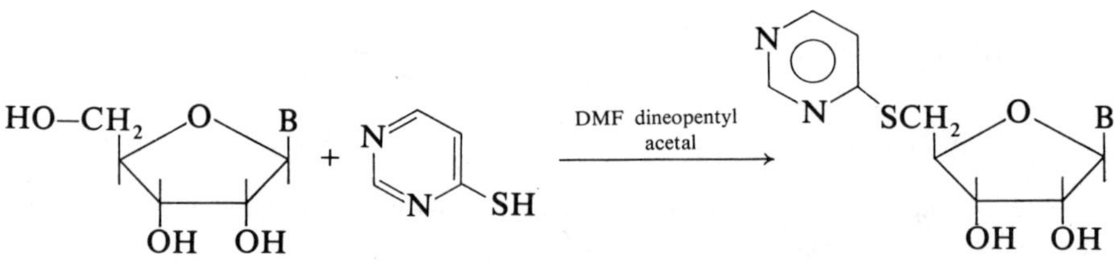

III. REACTIONS OF ORTHOAMIDES WITH NUCLEOPHILES

There are few examples of the thiolysis of orthoamides. But we believe such a reaction represents an adequate synthesis of N,N-disubstituted thioamides, as shown in the example of bis(dialkylamino)acetonitriles (10) (121).

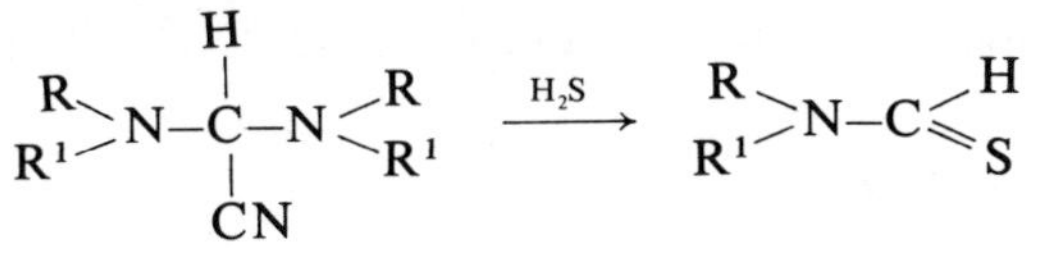

$$\begin{array}{c} R \\ R^1 \end{array} N - \underset{\underset{CN}{|}}{\overset{\overset{H}{|}}{C}} - N \begin{array}{c} R \\ R^1 \end{array} \xrightarrow{H_2S} \begin{array}{c} R \\ R^1 \end{array} N - C \begin{array}{c} H \\ S \end{array}$$

(10)

G. REACTIONS WITH ORGANOMETALLIC COMPOUNDS

The direct displacement of the alkoxide or mercaptide of orthoamides is possible only with strong nucleophiles, since these are poor leaving groups.* Thus N,N-dimethylformamide acetals (1) give α-substituted methylamines with Grignard reagents (65,37). Attempts to trap the aldehyde O,N-acetals with 1 mole Grignard reagent failed (37).

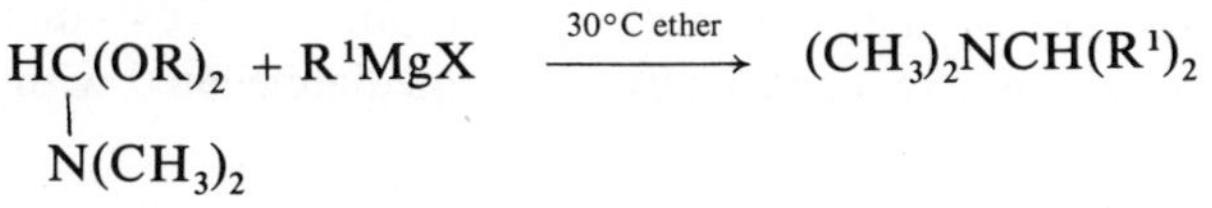

$$\underset{\underset{N(CH_3)_2}{|}}{HC(OR)_2} + R^1MgX \xrightarrow{30°C \text{ ether}} (CH_3)_2NCH(R^1)_2$$

(1)

$$R = C_2H_5, \quad R^1 = C_2H_5, C_6H_5$$

H. SYNTHESIS OF HETEROCYCLES WITH ORTHOAMIDES

1. With Tris(acylamino)methanes

Tris(formylamino)methane (8a) is extensively and widely used for the synthesis of heterocycles (22). At 140–160°C 8a decomposes into the N-formylformamidine (8b), which is important in the synthesis of heterocycles (21).

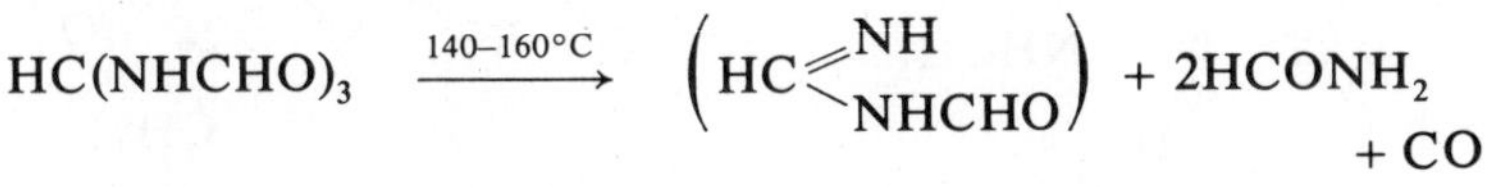

$$HC(NHCHO)_3 \xrightarrow{140-160°C} \left(HC \begin{array}{c} NH \\ NHCHO \end{array} \right) + 2HCONH_2 + CO$$

(8a) \hspace{3cm} (8b)

N-Formylformamidine may react by transfer of the formamidine carbon atom, the imidoformyl group, the formylaminomethylene group (HC(=)NHCHO), or of itself. We may classifiy the reactions of 8 in heterocyclic synthesis according to these transfers.

---

* For the reactions of 10 with alkoxides see Section III-D-3.

(a) *By Transfer of the Formamidine Carbon Atom.* Because of its minor importance, there are only a few publications on this reaction. At 180°C benzimidazole is formed from *o*-phenylenediamine in 90% yield, whereas ethylenediamine gives only 1,2-bis(formylamino)ethylene (81). Anilinoguanidinium sulfate yields 3-formylamino-1-phenyl-1,2,4-triazole (122).

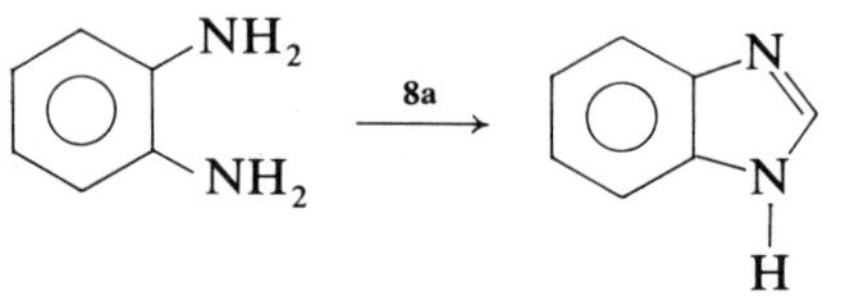

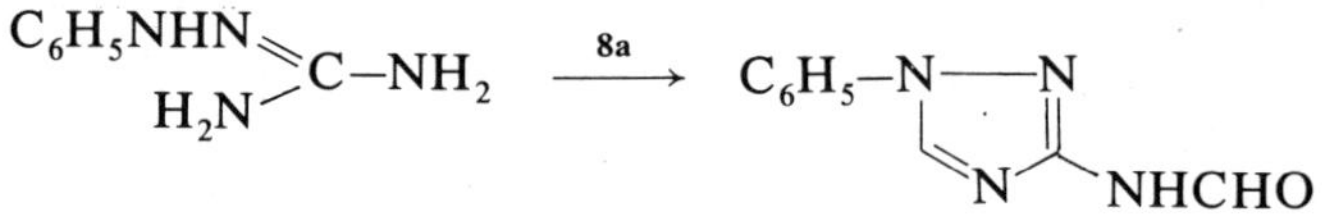

α,β-Unsaturated β-aminovinylcarbonyl compounds like 4-aminouraciles or 3-amino-5-pyrazolones react to yield pyridines diannellated with heterocycles (123–125).

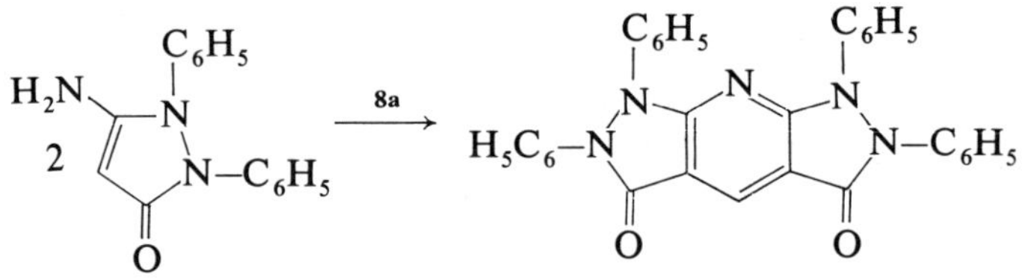

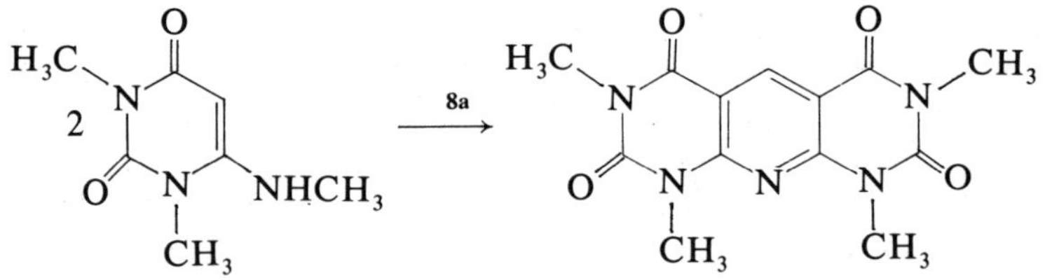

(b) *By Transfer of the Imidoformyl Group (HC=NH).* α-Hydroxy- and α-haloketones react with **8a** at 140°C neat or in DMF giving 1,3-oxazolones in 55–87% yield (22). Partial ring closure giving imidazole derivatives is also observed; here **8a** formally behaves as a formamidine.

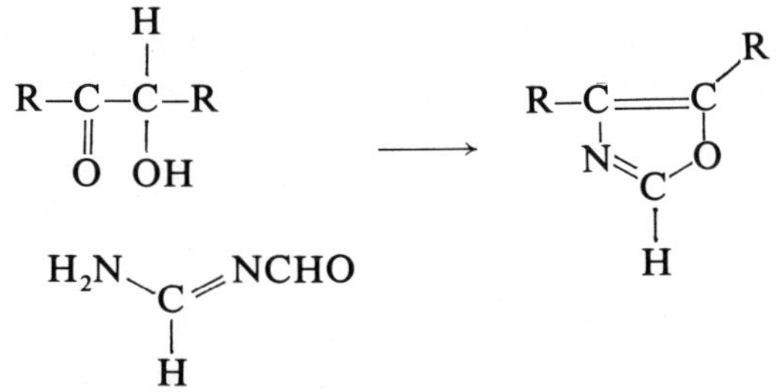

(8b)

R = aryl, alkyl

(*c*) *By Transfer of the Formylaminomethylene Group* [*HC*(=)*NHCHO*]. The incorporation of a C–N–C element leads to a series of simple and important heterocyclic syntheses. Thus the reaction of **8a** with hydrazine in formamide gives 1,2,4-triazole at 150°C in 54% yield (81).

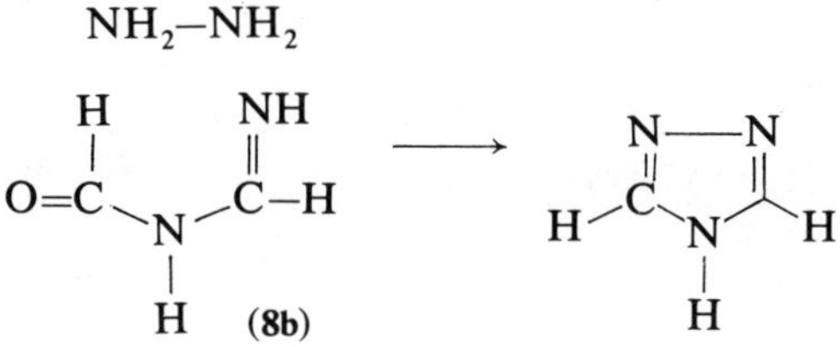

(8b)

1,3-Diamines of the guanidine type react smoothly with **8a** in boiling DMF, giving monoamino-*s*-triazines in 40–100% yield (22,70,71,122,126).

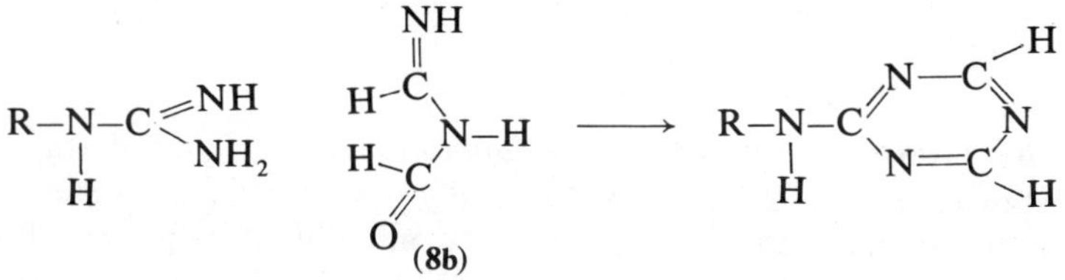

(8b)

R = H, alkyl, aryl, acyl

The simple, high-yield synthesis of *s*-triazine by thermolysis of **8a** at 165–170°C also fits into this class (20–22, 122).

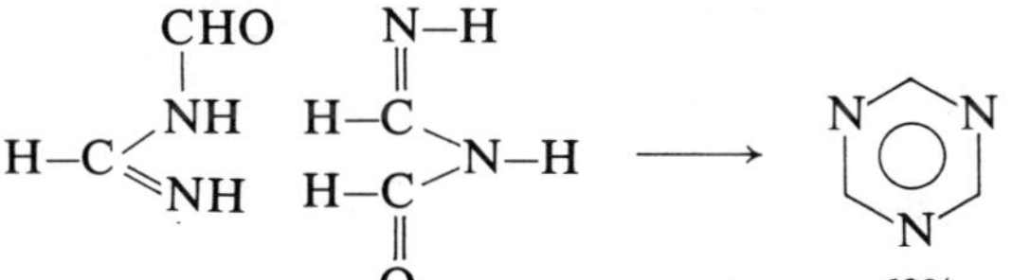

62%

476   ORTHOAMIDES: PROPERTIES AND REACTIONS

If guanidine is replaced by urea, 2,4-dihydroxy-s-triazine is surprisingly isolated in 43% yield (70,71). Supposedly urea and **8a** form formylbiuret, which cyclizes.

$$2H_2NCONH_2 \xrightarrow{8a} H_2NCONHCONHCHO \longrightarrow$$

Heterocycles with an enamine structure can be treated with **8a** resulting in annellation of a pyrimidine ring. New pyrimido[4,5-d]pyrimidine and pyrazolo[3,4-d]pyrimidine syntheses are based on this principle (123–125).

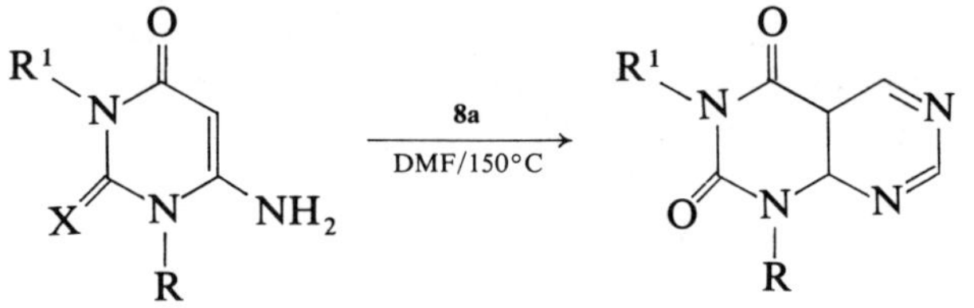

R, R^1 = H, alkyl, aryl
X = O, S

40–62%

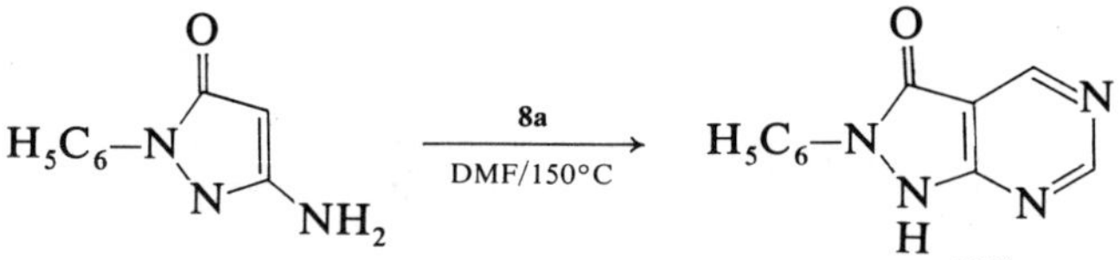

60%

(d) *By Transfer of the Formylformamidine Group.* Pyrimidines, which are unsubstituted in their 2 positions, are easily prepared from **8a** and ketones in formamide in the presence of a catalytic amount of p-toluenesulfonic acid (22,127). Aldehydes also give pyrimidines, but only in low yields.

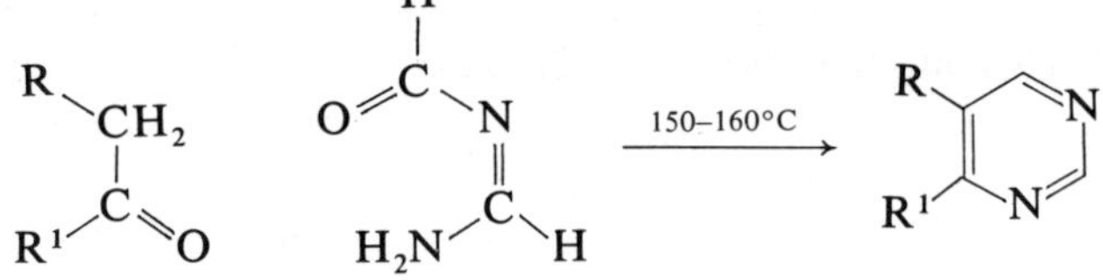

R, R^1 = alkyl, aryl
R–R^1 = $(CH_2)_3$, $(CH_2)_4$

III. REACTIONS OF ORTHOAMIDES WITH NUCLEOPHILES

This synthetic principle may be extended to arylacetonitriles (128), arylacetamides (128), and malonic and cyanoacetic acid derivatives, among others.

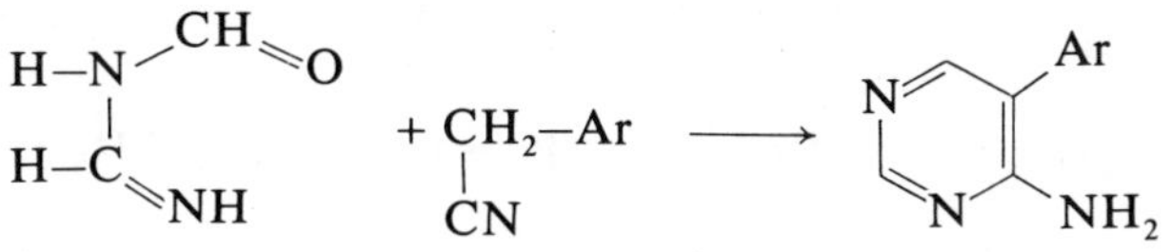

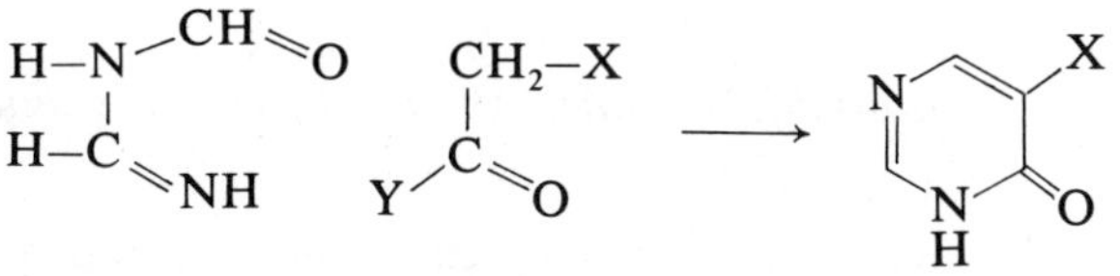

$$\begin{array}{c} H-N\diagdown^{CH=O} \\ H-C\diagup_{NH} \end{array} + \begin{array}{c} CH_2-Ar \\ | \\ CN \end{array} \longrightarrow \begin{array}{c} \text{pyrimidine with Ar at 5, NH}_2 \text{ at 4} \end{array}$$

$$\begin{array}{c} H-N\diagdown^{CH=O} \\ H-C\diagup_{NH} \end{array} \quad \begin{array}{c} CH_2-X \\ | \\ Y^{C\diagdown}{}_O \end{array} \longrightarrow \begin{array}{c} \text{4-oxo-pyrimidine with X at 5} \end{array}$$

$X = CO_2R, CN, CONH_2$
$Y = OR, NH_2$

The latter compounds yield especially interesting pyrimidines substituted in the 4 and 5 positions (55,129). The mechanism of this pyrimidine synthesis has been established (55). The authors contend that formylaminomethylene derivatives are intermediates which cyclize on reaction with ammonia or ammonium formate. This synthetic principle applied to aminoacetonitrile and its derivatives represents the most elegant synthetic approach to purine presently known (130–133):

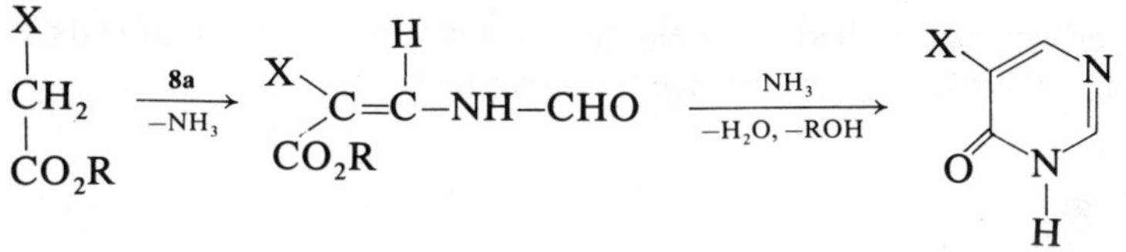

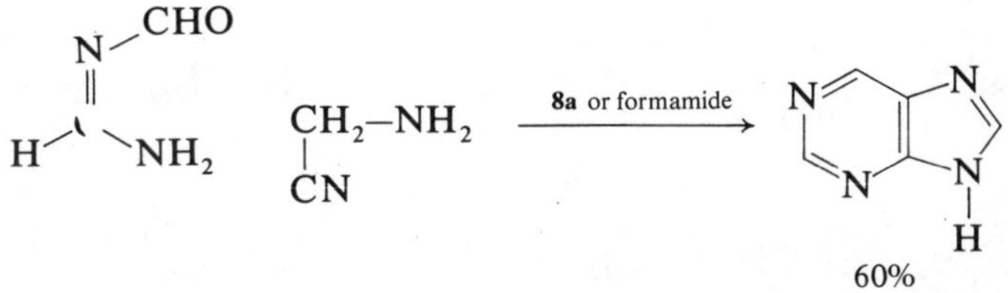

$$\begin{array}{c} X \\ | \\ CH_2 \\ | \\ CO_2R \end{array} \xrightarrow[-NH_3]{8a} \begin{array}{c} X\diagdown \quad H \\ C=C-NH-CHO \\ CO_2R \end{array} \xrightarrow[-H_2O, -ROH]{NH_3} \text{4-oxo-pyrimidine with X at 5}$$

$$\begin{array}{c} N\diagdown^{CHO} \\ \| \\ H\diagup \diagdown NH_2 \end{array} \quad \begin{array}{c} CH_2-NH_2 \\ | \\ CN \end{array} \xrightarrow{\text{8a or formamide}} \text{purine}$$

60%

Analogous syntheses open the way to 7-substituted purines, and also 4,5-diaminopyrimidines generated by their hydrolysis (132). Homologues, with respect to their acyl groups, of **8a** are less reactive by far; but it is possible to force the reactions with ketones at 150–200°C. β-(Acylamino)vinylketones are the main products. Hence a ring-forming reaction may occur on heating with formamide, ammonia, or formamidinium acetate (54). By using formamide as a solvent, pyrimidines are prepared in a "one-pot process" (54).

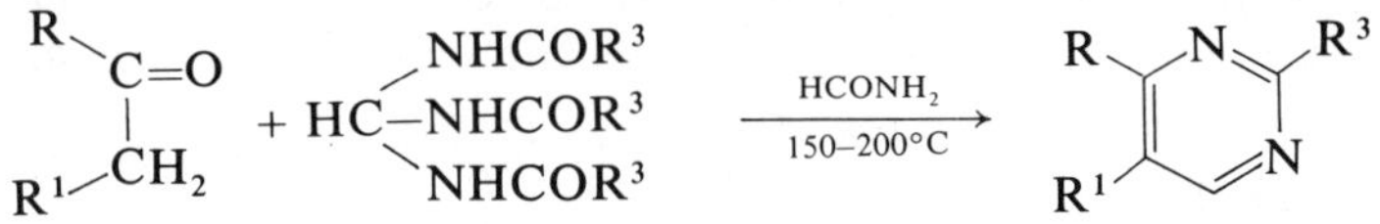

$$R\!\!-\!\!\underset{R^1\diagdown CH_2}{\overset{\diagup C=O}{|}} + HC\!\!-\!\!\underset{\diagdown NHCOR^3}{\overset{\diagup NHCOR^3}{NHCOR^3}} \xrightarrow[150-200°C]{HCONH_2} \underset{R^1}{\overset{R}{\diagdown}}\!\!\underset{\diagdown N}{\overset{\diagup N\diagdown R^3}{|}}$$

$R = C_6H_5$, $R^1 = H$, CH_3, C_2H_5; $R^3 = CH_3$, C_2H_5, C_3H_7, C_6H_5
$R\!-\!R^1 = (CH_2)_4$

2. Via Carboxylic Acid Amide Acetals, Aminal Esters, Bis(dialkylamino)-acetonitriles

Syntheses of heterocycles have been based on the reaction of XH-acidic compounds to give dialkylaminomethylene compounds. These may cyclize intramolecular with a functional group, if present in a suitable position, or with corresponding bifunctional components. The classification of these syntheses is made according to final products, irrespective of whether the dialkylaminomethylene compound is only an intermediate or an isolated product.

(a) *Indoles.* 1-Alkyl-2-nitrobenzenes react with **1a**, **1b**, or **5** to afford the corresponding dimethylaminomethylene derivatives (26). After the reduction of the *o*-nitro group to the amino group, indoles are formed by cyclization (26a). This new indole synthesis was first discovered by Bredereck et al. (26a,28,28a) and later widely applied by Leimgruber (61); normally it affords excellent yields and shows little dependence on substituents.

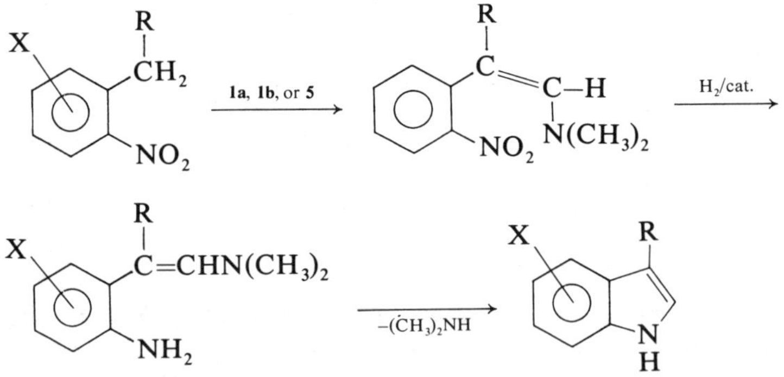

(b) *Pyridines.* Dialkylaminomethylene derivatives of CH_2-acidic compounds, which carry a carbonyl group in their 4 positions or which is generated *in situ*, yield *N*-alkyl-4(2)-pyridones with amines (41,45).

III. REACTIONS OF ORTHOAMIDES WITH NUCLEOPHILES

Cyclization of 4-amino substituted dialkylaminomethylene derivatives represents an alternative pathway to pyridines (42).

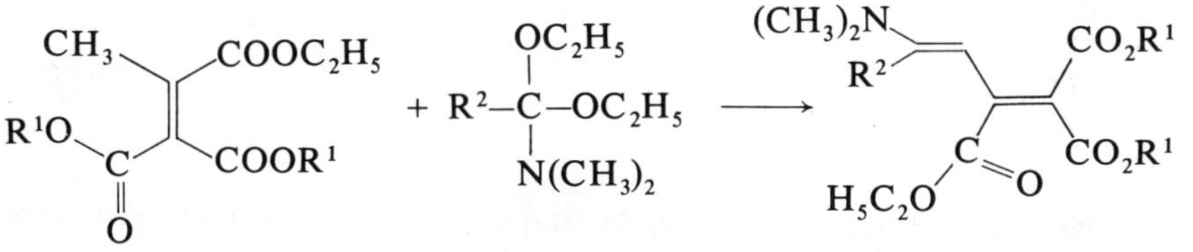

$R^1 = C_2H_5, CH_3$

$R^2 = $ H, 3-pyridyl, 3-cyanophenyl, 3-cyano-2-quinolyl

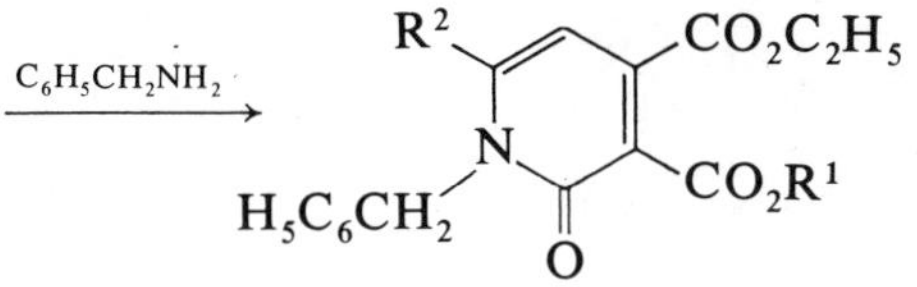

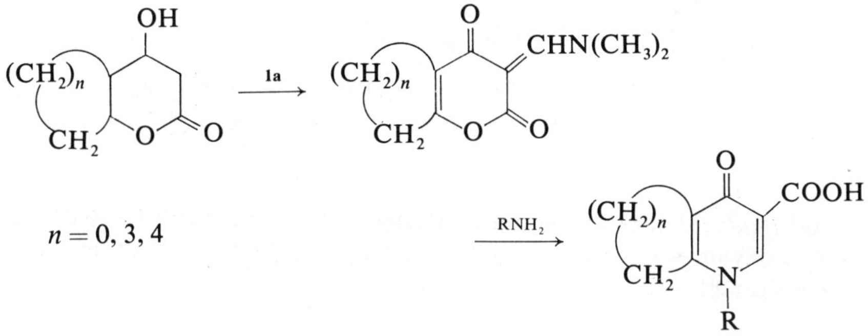

$n = 0, 3, 4$

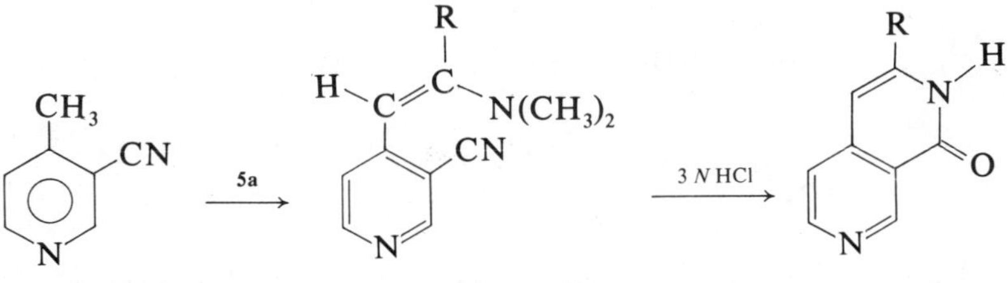

$R = $ H, CH_3

(c) *4-Pyrones.* B. Föhlisch (38) discovered a new chromone synthesis by reacting 2-hydroxy-acetophenone with **1a** and subsequently cyclizing the intermediate in dilute sulfuric acid.

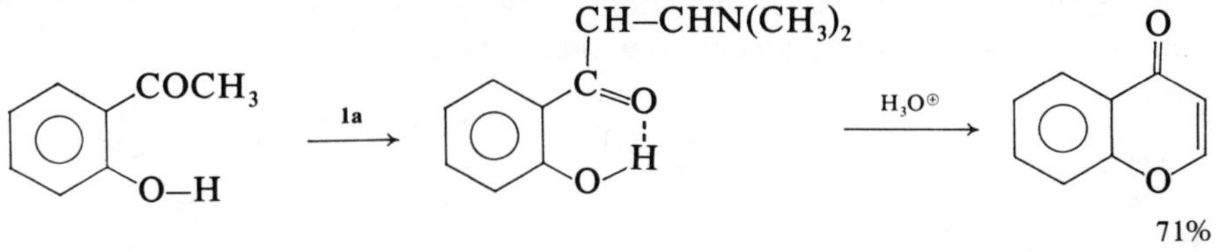

[Scheme: o-hydroxyacetophenone → (via 1a) intermediate with CH–CHN(CH₃)₂ and C=O···H–O → (H₃O⁺) chromone, 71%]

(d) Pyrazoles, Isoxazoles. Dialkylaminomethylene derivatives, generated from ketones or nitriles with **1** or **5**, react with hydrazine or hydroxylamine to give pyrazoles and isoxazoles (40,51,57,134,135); thus they react as derivatives of β-dicarbonyl compounds do.

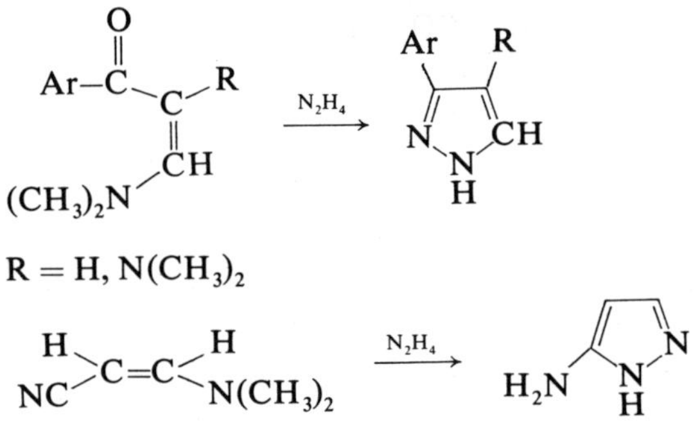

[Scheme: Ar–C(=O)–C(R)=CH–N(CH₃)₂ + N₂H₄ → pyrazole with Ar, R substituents; R = H, N(CH₃)₂]

[Scheme: NC–CH=CH–N(CH₃)₂ + N₂H₄ → 3-aminopyrazole (H₂N-)]

(e) Imidazoles, Oxazoles. Orthoamides **1** and **9** react with 1,2-diamines and 2-hydroxyamines giving imidazoles and oxazoles (27,49,69), but there are only a few examples known.

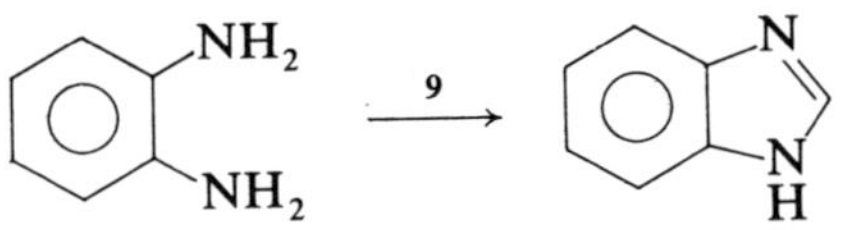

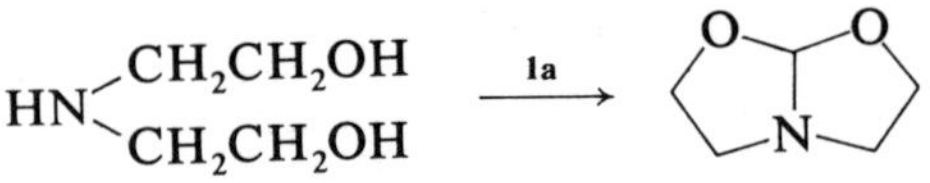

[Scheme: o-phenylenediamine + 9 → benzimidazole]

[Scheme: HN(CH₂CH₂OH)₂ + 1a → bicyclic bis-oxazolidine]

Benzoylhydrazine and dimethylbenzamide diethyl thioacetal form a 1,2,4-oxadiazole:

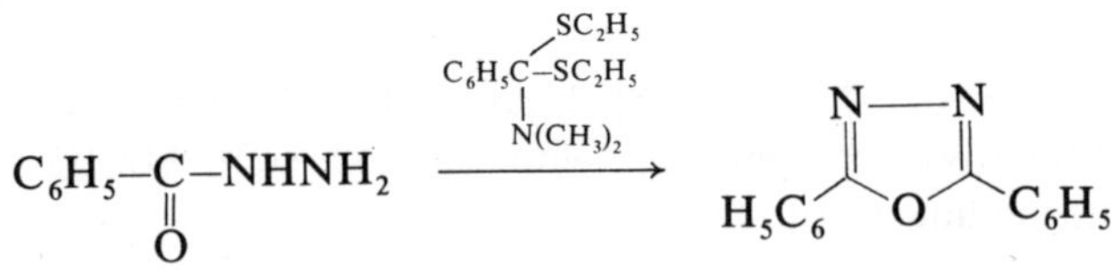

[Scheme: C₆H₅–C(=O)–NHNH₂ + C₆H₅C(SC₂H₅)₂N(CH₃)₂ → 2,5-diphenyl-1,3,4-oxadiazole (H₅C₆–/N=N\–C₆H₅ with O)]

III. REACTIONS OF ORTHOAMIDES WITH NUCLEOPHILES

(f) *Pyrimidines and Purines*. There are several examples that illustrate the synthesis of pyrimidines from β-dialkylaminomethylenecarbonyl compounds or the corresponding nitriles with 1,3-diamines (see Table XXIV). These reactions are successful between 80 and 130°C, generally in good yields. The cyclization of N-(2-pyridyl)-3-dimethylaminoacrylamides to pyrido[1,2-a]-pyrimidines may be attributed to a similar mechanism.

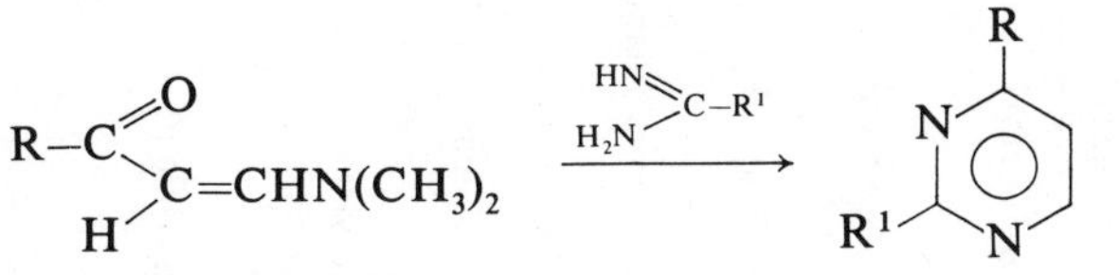

R, R¹: see Table XXIV

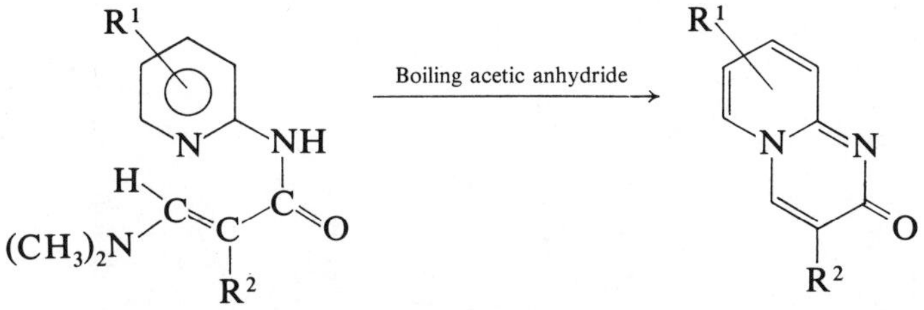

R^1 = H, 6-CH$_3$, 6,8-(CH$_3$)$_2$, 8-CH$_3$, 7-Cl
R^2 = COCH$_3$, CO$_2$C$_2$H$_5$

32–89%

β-Dialkylaminoacrylic acid amides are obtained by reaction of carboxylic acid amides with **1** or **5**. After displacement of the dialkylamino group by the amino group, cyclization with **1** may occur (33).

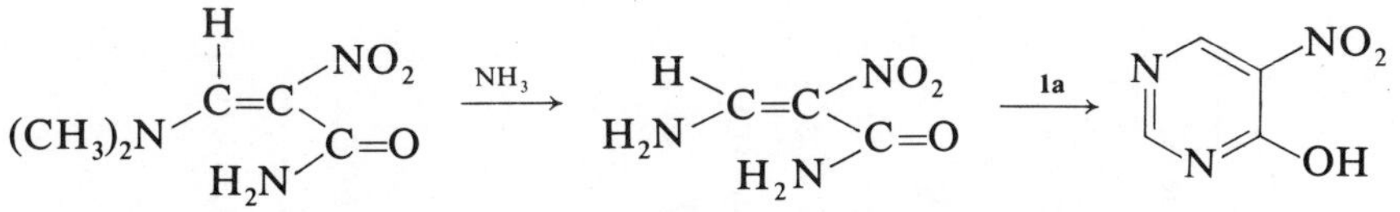

A recent synthesis of pyrimidines utilizes β-dialkylaminomethylene compounds, which react, for example, with acylisocyanates to give N-acyl-β-dialkylaminoacrylic acid amides. These are cyclized with reagents such as ammonia to yield pyrimidines (33).

TABLE XXIV
Pyrimidines from Dialkylaminomethylene Compounds and 1,3-Diamines

Starting material	1,3-Diamine	Reaction conditions	Pyrimidine substitutent	Yield, %	Ref.
$CH_3COCH=CHN(CH_3)_2$	Guanidine carbonate	180°C	2-Amino-4-methyl		135
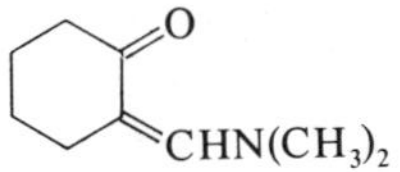 (2-oxocyclohexylidene)methyl-$CHN(CH_3)_2$	Guanidine carbonate	180°C	2-Amino-4,5-tetramethylene		135
$C_6H_5COCH=CHN(CH_3)_2$	Guanidine	78°C, 2 hr, ethanol	2-Amino-4-phenyl	84	135
$C_6H_5CO-C(N(CH_3)_2)=CHN(CH_3)_2$	Guanidine	78°C, 3 hr, ethanol	2-Amino-5-dimethylamino-4-phenyl	77	40, 57
$C_6H_5CO-C(N(CH_3)_2)=CHN(CH_3)_2$	Benzamidine	110°C, 3 hr	5-Dimethylamino-2,4-diphenyl	56	40, 57
$C_6H_5CO-C(N(CH_3)_2)=CHN(CH_3)_2$	Acetamidine	25°C, 100 hr, ethanol	5-Dimethylamino-2-methyl-4-phenyl	42	40, 57
$C_6H_5CO-C(N(CH_3)_2)=CHN(CH_3)_2$	Formamidine acetate	160°C, 0.75 hr	5-Dimethylamino-4-phenyl	39	40, 57
$H_5C_2O-C(=O)-C(H)=C-CHN(CH_3)_2$	Guanidine	78°C, 3 hr, ethanol	2-Amino-4-hydroxy	66	56
$H_5C_2O-C(=O)-C(H)=C-CHN(CH_3)_2$	Thiourea	100°C, 2 hr, water, NaOH	4-Hydroxy-2-mercapto	78	56

Starting material	Reagent	Conditions	Product	Yield (%)	Refs.
$H_5C_2O-\overset{O}{C}-\overset{H}{C}=CHN(CH_3)_2$	S-Methylisothiourea	100°C, 20 hr, water, NaOH	2-Cyanamino-4-hydroxy	40	56
$H_5C_2O-\overset{O}{C}-\overset{N(CH_3)_2}{C}=CHN(CH_3)_2$	Thiourea	140°C, 2 hr	5-Dimethylamino-4-hydroxy-2-mercapto	66	40, 57
$N\equiv C-CH=CH-N(CH_3)_2$	Formamidine acetate	78°C, 15 hr, ethanol	4-Amino	47	51
$N\equiv C-CH=CH-N(CH_3)_2$	Guanidine	78°C, 62 hr, ethanol	2,4-Diamino	51	51
$N\equiv C-\overset{R}{C}=CHN(CH_3)_2$	Guanidine	78°C, 4 hr, ethanol	2,4-Diamino-5-alkyl	72–74	56
$R = CH_3, C_2H_5$					
$N\equiv C-\overset{R}{C}=CHN(CH_3)_2$	Thiourea	130°C, 3 hr	5-Alkyl-4-amino-2-mercapto	72–82	56
$R = H, CH_3, C_2H_5, n-C_3H_7$					
$NC-\overset{R}{C}=CHN(CH_3)_2$	Phenylacetamidine	110°C, 5 hr, n-butanol	5-Alkyl-4-amino-2-benzyl	52–60	56
$R = CH_3, C_2H_5$					
$NC-\overset{N(CH_3)_2}{C}=CHN(CH_3)_2$	Formamidine acetate	120°C, 1 hr	4-Amino-5-dimethylamino	28	40, 57
$NC-\overset{N(CH_3)_2}{C}=CHN(CH_3)_2$	Guanidine	78°C, 2 hr, ethanol	2,4-Diamino-5-dimethylamino	30	40, 57

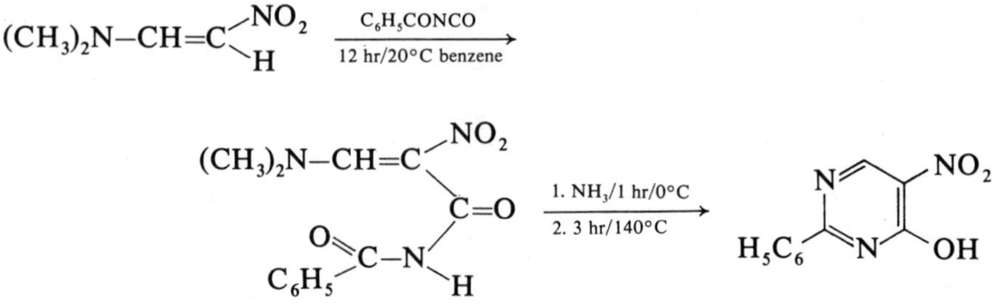

[Reaction scheme: $(CH_3)_2N-CH=C(NO_2)H$ + C_6H_5CONCO, 12 hr/20°C benzene →]

[Scheme showing intermediate with $(CH_3)_2N-CH=C(NO_2)(C=O)$ linked to $O=C-N(H)(C_6H_5)$, then: 1. NH_3/1 hr/0°C; 2. 3 hr/140°C → pyrimidine product with H_5C_6, NO_2, OH substituents.]

The synthesis of 5-(dimethylaminomethylene)hydantoins with **5** makes the purines accessible (136):

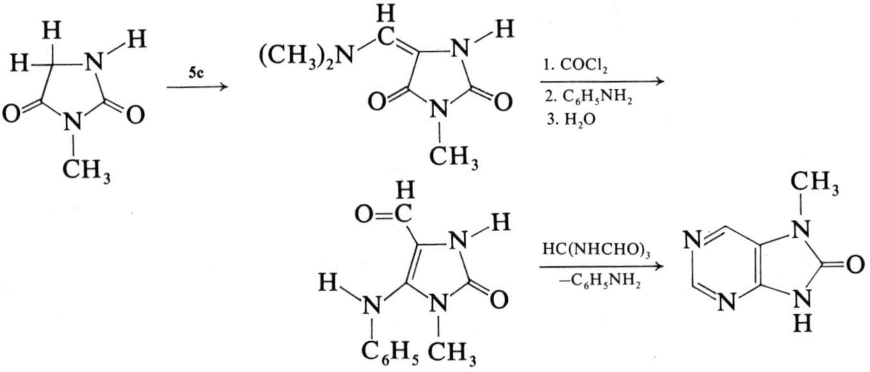

[Scheme: N-methyl hydantoin → (with **5c**) → 5-(dimethylaminomethylene)-N-methylhydantoin → 1. $COCl_2$; 2. $C_6H_5NH_2$; 3. H_2O → imidazolone intermediate with C_6H_5 and CH_3 substituents → $HC(NHCHO)_3$, $-C_6H_5NH_2$ → 1-methyl-1H-purin-2(3H)-one.]

Recently, a new synthesis of purine was reported. 4-Amino-5-arylazouracils and **1a** initially yielded dimethylaminomethylene derivatives as the primary product. If this intermediate is heated to 220°C, 8-dimethylaminopurines are formed after thermolytic cleavage of the N—N bond (137).

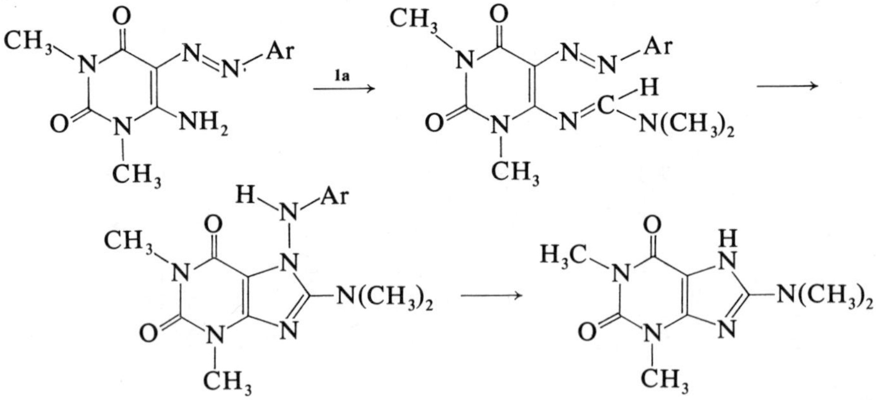

[Scheme: 1,3-dimethyl-6-amino-5-arylazouracil —**1a**→ dimethylaminomethylene derivative with $-N=C(H)N(CH_3)_2$ group → 7-arylamino-8-dimethylamino-1,3-dimethylxanthine → 8-dimethylamino-1,3-dimethylxanthine (N-H).]

(g) *s-Triazines*. The simplest synthesis of 2,4-diamino-*s*-triazines starts from guanidine and **1b** (70,71,72a). Dimethylaminomethyleneguanidines are formed as intermediates, which lead to ring closure to the *s*-triazine with excess guanidine. If substituted guanidines are used, *N,N'*-disubstituted or *N*-monosubstituted 2,4-diamino-*s*-triazines result (see Table XXV). It is still not entirely clear which structural features determine the alternative reaction pathways.

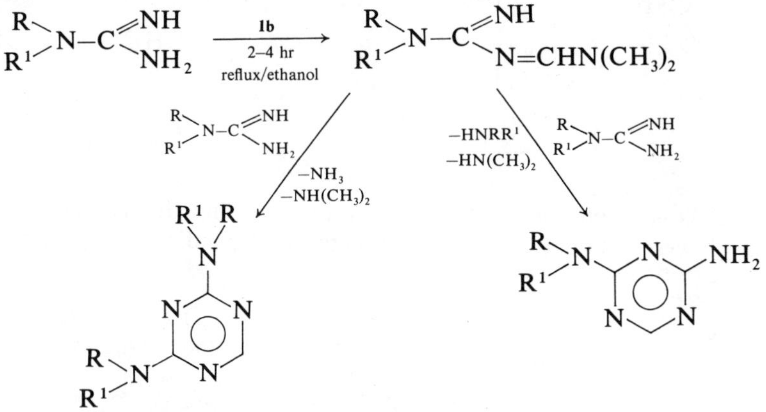

Amidines react with **1b**, giving 2,4-disubstituted *s*-triazines (56,70). *S*-Methylisothiourea similarly affords 2-amino-4-methylmercapto-*s*-triazines (70). This technique may be greatly facilitated by reacting **1b** in situ (71).

TABLE XXV

2,4-Diamino-*s*-triazines from Guanidines and **1b** (71)

Guanidine substituent	*s*-Triazine substituent	Yield, %
—	2,4-Diamino	86
Methyl	2-Amino-4-methylamino	66
Ethyl	2-Amino-4-ethylamino	63
n-Butyl	2,4-Bis(*n*-butylamino)	18
Cyclohexyl	2,4-Bis(cyclohexylamino)	14
	2-Amino-4-cyclohexylamino	48
Piperidyl	2-Amino-4-piperidyl	68
Phenyl	2,4-Dianilino	53
N-Methyl-*N*-phenyl	2-Amino-4-(*N*-methylanilino)	16
m-Tolyl	2,4-Bis-(*m*-tolylamino)	31
p-Tolyl	2,4-Bis-(*p*-tolylamino)	46
Benzyl	2,4-Bis(benzylamino)	43
p-Methoxyphenyl	2,4-Bis-(*p*-methoxyphenyl)amino	33

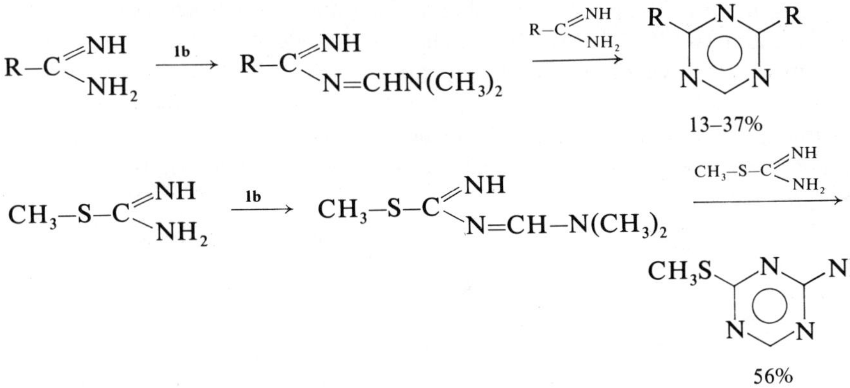

The cycloaddition of isocyanates with N,N-dimethyl-N'-hetarylformamidines gives different s-triazine derivatives, depending on the reaction conditions (77).

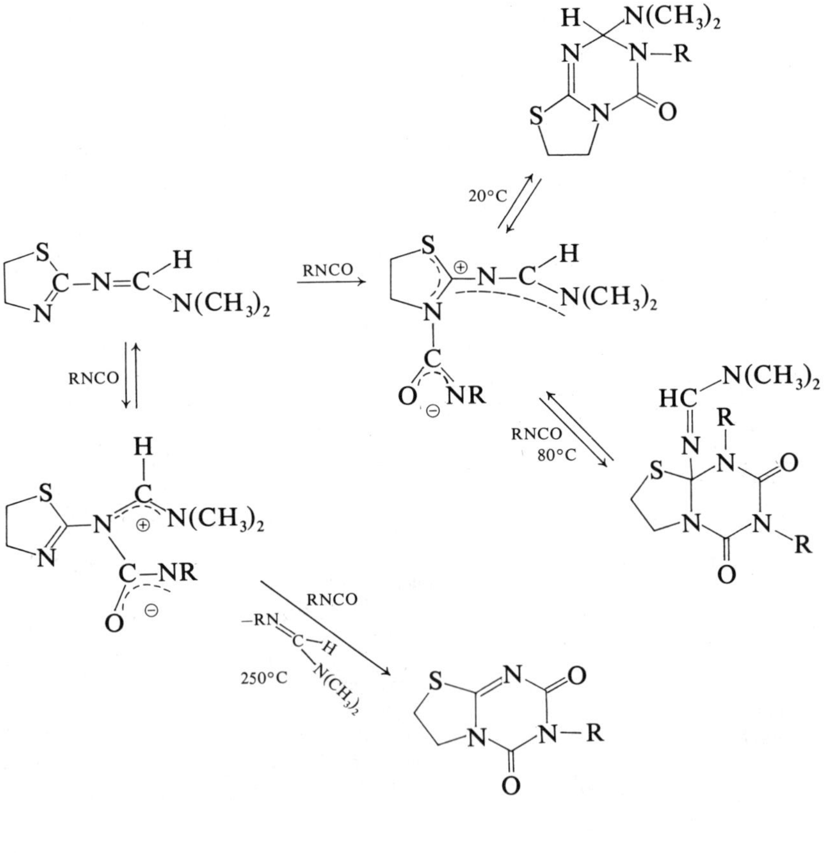

IV. Reactions of Orthoamides with Electrophiles

A. AT ALKOXY (ALKYLMERCAPTO) AND DIALKYLAMINO GROUPS

1. Alkylation

If the reaction follows an S_N2 mechanism, alkylating agents such as alkyl halides (4) or arylsulfonic acid esters (138) react with N,N-dialkylformamide acetals (N,N-dialkylformamide thioacetals) at room temperature to yield 1,1-(dialkoxy[1,1-dialkylthio]methyl)trialkylammonium salts. The alkylated products of cyclic thioacetals exhibit considerable thermal stability (93):

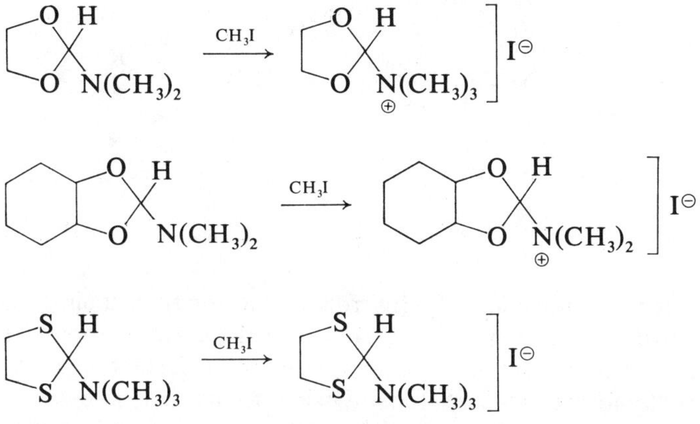

Dialkoxymethylammonium salts of acyclic acetals are inaccessible by this method. See (31) for its preparation. In a complex reaction **1b** and methyl iodide give DMF, tetramethylammonium iodide, ethyl orthoformate, and ethyl iodide (4).*

$$2(CH_3)_2NCH(OC_2H_5)_2 + 2CH_3I \longrightarrow (CH_3)_2NCHO + [N(CH_3)_4]^{\oplus}I^{\ominus} +$$
$$\text{(1b)} \hspace{4cm} C_2H_5I + HC(OC_2H_5)_3$$

Alkylating agents that react by an S_N1 mechanism attack N,N-dialkylformamide acetals (**1**) at their alkoxy groups (140).

Tris(alkylarylamino)methanes form alkylarylformamidinium salts after N-alkylation (113). Tris(dialkylamino)methanes are expected to react analogously.

---
* The synthesis of amide acetals from DMF–dimethyl sulfate complex and sodium alkoxide reveals an analogous competitive reaction if the reaction conditions (excess alkoxide, low temperatures) are not maintained (139).

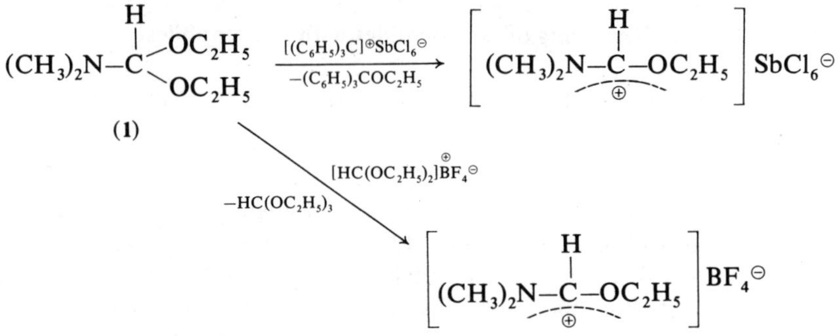

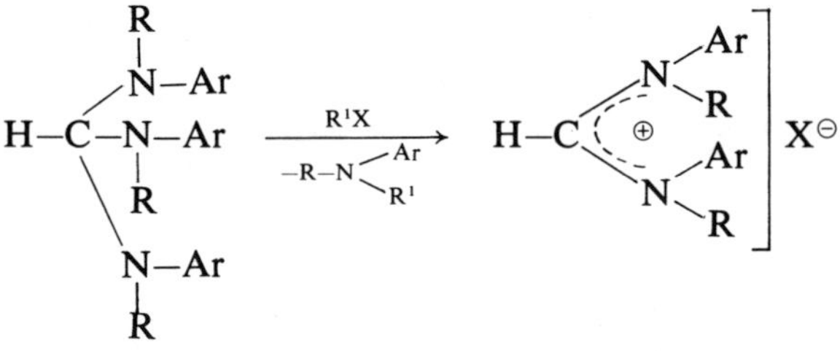

ORTHOAMIDES: PROPERTIES AND REACTIONS

$$(CH_3)_2N-\underset{\underset{OC_2H_5}{|}}{\overset{\overset{H}{|}}{C}}-OC_2H_5 \quad \xrightarrow[-(C_6H_5)_3COC_2H_5]{[(C_6H_5)_3C]^{\oplus}SbCl_6^{\ominus}} \quad \left[(CH_3)_2N-\underset{\oplus}{\overset{\overset{H}{|}}{C}}-OC_2H_5\right] SbCl_6^{\ominus}$$

(1)

$$\xrightarrow[-HC(OC_2H_5)_3]{[HC(OC_2H_5)_2]^{\oplus}BF_4^{\ominus}} \quad \left[(CH_3)_2N-\underset{\oplus}{\overset{\overset{H}{|}}{C}}-OC_2H_5\right] BF_4^{\ominus}$$

$$H-\underset{\underset{N-Ar}{|}}{\overset{\overset{N-Ar}{|}}{C}}-\underset{R}{\overset{R}{N}}-Ar \quad \xrightarrow[-R-N\diagdown^{Ar}_{R^1}]{R^1X} \quad \left[H-C \underset{N\diagdown_{R}^{Ar}}{\overset{N\diagup^{Ar}_{R}}{\oplus}}\right] X^{\ominus}$$

(7)

Bis(dimethylamino)- and Bis(pyrrolidino)acetonitrile maintain a special position owing to their marked dissociation. In ether at room temperature, alkylation with alkyl tosylates occurs, leading to the formation of alkyl cyanide and tetraalkylformamidinium salts. Owing to its much lower dissociation Bis(piperidino)-acetonitrile is alkylated "the normal way" at the piperidine nitrogen atom.

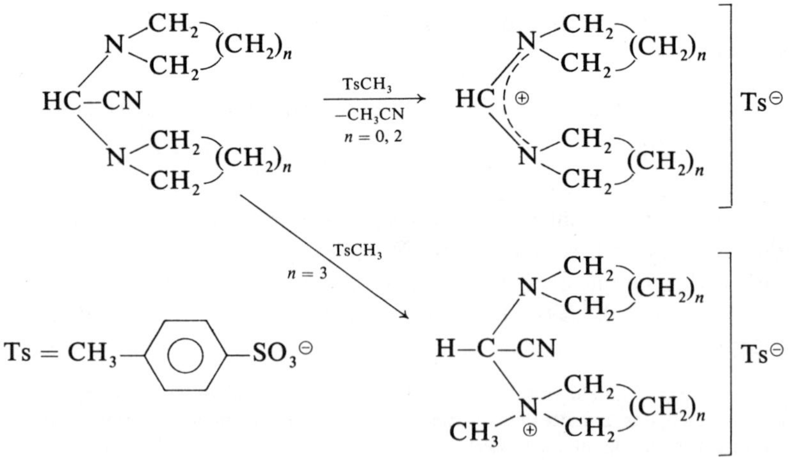

$$HC-CN \xrightarrow[-CH_3CN]{TsCH_3} \left[HC^{\oplus}\right] Ts^{\ominus} \quad n = 0, 2$$

$$\xrightarrow[n=3]{TsCH_3} \quad H-C-CN \quad Ts^{\ominus}$$

$$Ts = CH_3-\!\!\left\langle\!\!\!\bigcirc\!\!\!\right\rangle\!\!-SO_3^{\ominus}$$

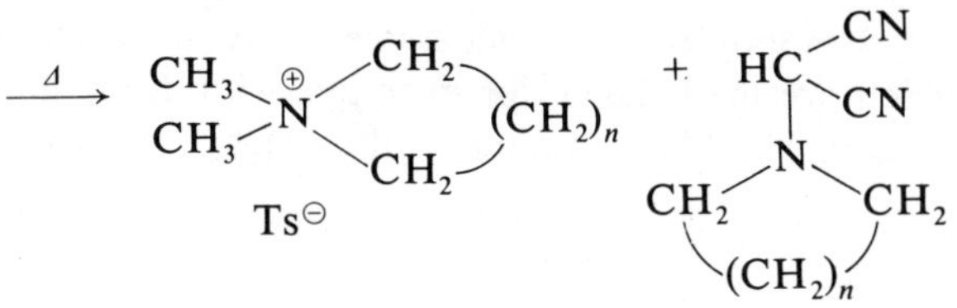

2. Reactions with Aldehydes and Ketones

The treatment of simple aliphatic aldehydes and ketones with N,N-dialkylformamide acetals either yields aldol-like products or it fails to react at all.

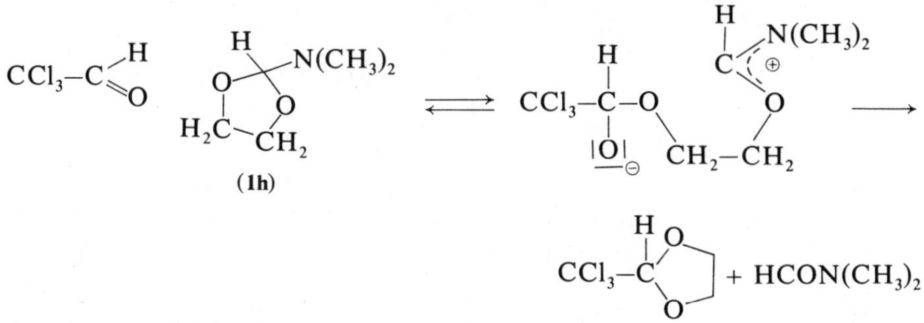

Aliphatic aldehydes containing electron-withdrawing groups at their α positions react with cyclic acetals (**1b**) (presumably via O-adducts) to form aldehyde acetals in very low yields (107). The first step of the reaction is believed to be the same as in the reaction of 4-nitrobenzaldehyde* with **1a** (141). From a Cannizzaro reaction, 4-nitrobenzyl alcohol and, from an S_N reaction, methyl anisate are finally isolated.

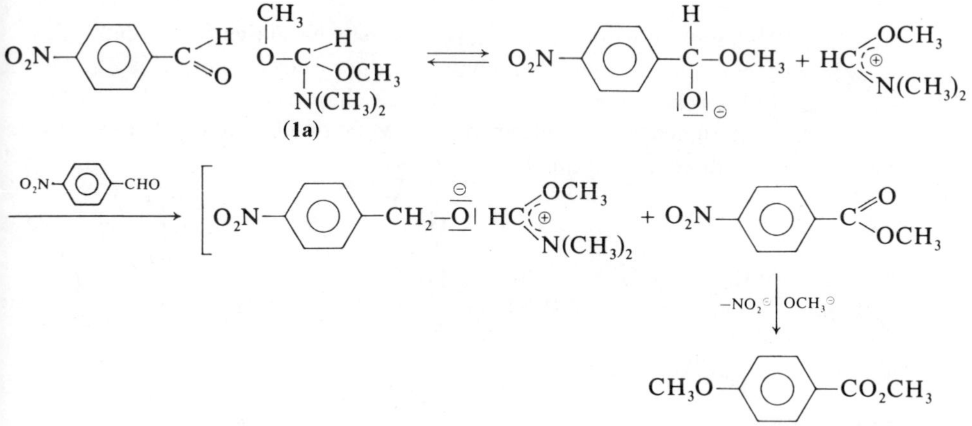

---

* For the reactions of other aromatic aldehydes with **1** or **5**, giving C-substituted products, see Section IV-C-4(c).

In a relatively smooth reaction, aldehydes may be acetalized via their hemiacetals, which are methylated, for example, with **1a** as carboxylic acids are (141). The higher the boiling points of the aldehydes, the better the yield (141).

$$R-\underset{OH}{\underset{|}{\overset{H}{\overset{|}{C}}}}-OCH_3 \xrightarrow[-CH_3OH]{1a} R-\underset{OCH_3}{\underset{|}{\overset{H}{\overset{|}{C}}}}-\overset{\ominus}{O} + \underset{(CH_3)_2N}{\overset{CH_3O}{\diagdown}}\overset{\oplus}{C}-H$$

$$\downarrow$$

$$R-\underset{OCH_3}{\underset{|}{\overset{H}{\overset{|}{C}}}}-OCH_3 + HCON(CH_3)_2$$

Under mild conditions, aliphatic aldehydes react with the much more reactive **7a** to give 1-dimethylaminoalkenes in good yields; the aldehyde aminals are the proposed intermediates (141a).

$$\underset{R^2}{\overset{R^1}{\diagdown}}\overset{H}{\underset{|}{C}}-CHO \xrightarrow{7a} \underset{R^2}{\overset{R^1}{\diagdown}}\overset{H}{\underset{|}{C}}\diagdown\underset{\underset{H}{\overset{|}{C}}\diagdown N(CH_3)_2}{\overset{O^\ominus}{\diagup}} + H-\overset{\oplus}{C}\underset{N(CH_3)_2}{\overset{N(CH_3)_2}{\diagup}} \xrightarrow{-DMF}$$

$$\underset{R^2}{\overset{R^1}{\diagdown}}\overset{H}{\underset{|}{C}}-\underset{H}{\overset{|}{C}}\underset{N(CH_3)_2}{\overset{N(CH_3)_2}{\diagup}} \xrightarrow{-NH(CH_3)_2} \underset{R^2}{\overset{R^1}{\diagdown}}C=C\underset{N(CH_3)_2}{\overset{H}{\diagup}}$$

Cyclic *N,N*-dimethylformamide acetals (**1b**) can be activated by catalytic amounts of acetic acid. The open-chain product selectively yields the 3-ethyleneketal of 5α-cholestan-3-one in methylene chloride in good yield. Owing to the steric requirement of the intermediate (see below) the keto groups in the 17 and 20 positions remain untouched (107, 142).

$$(CH_3)_2N-\underset{O}{\overset{H}{\underset{|}{C}}}\diagup\overset{O}{\diagdown}\diagdown \underset{\text{(1f)}}{\rightleftharpoons} \xrightarrow{H^\oplus} (CH_3)_2N-\overset{H}{\underset{O}{\underset{|}{\overset{\oplus}{C}}}}\diagup\overset{O}{\diagdown} \rightleftharpoons (CH_3)_2\overset{\oplus}{N}=\underset{O}{\overset{H-O}{\underset{|}{C}}}\diagdown \overset{H}{\underset{CH_2}{\underset{|}{CH_2}}}$$

Alkoxy(dimethylamino)- and bis(dialkylamino)acetonitriles (**10**) show a similar behavior toward aromatic aldehydes as in the alkylation reaction. Via

IV. REACTIONS OF ORTHOAMIDES WITH ELECTROPHILES

TABLE XXVI
(Dialkylamino)arylacetonitriles from Aromatic Aldehydes with Alkoxy(dimethylamino)- and Bis(dialkylamino)acetonitriles

Aldehyde, ArCHO Ar	Orthoamide	Reaction conditions	Yield, %	Ref.
Phenyl	Methoxy(dimethylamino)-acetonitrile	2 days, 20°C 4 hr, 80°C	37	6, 85
4-Chlorophenyl	Methoxy(dimethylamino)-acetonitrile	3 hr, 80°C	33	85
2-Furyl	Methoxy(dimethylamino)-acetonitrile	4 hr, 80°C	46	85
Phenyl	Bis(dimethylamino)-acetonitrile	15 hr, 50°C	72	7
4-Methylphenyl	Bis(dimethylamino)-acetonitrile	36 hr, 60°C	49	7
4-Methoxyphenyl	Bis(dimethylamino)-acetonitrile	12 hr, 60°C	64	7
4-Chlorophenyl	Bis(dimethylamino)-acetonitrile	4 hr, 60°C	65	7
Phenyl	Bis(piperidino)-acetonitrile	36 hr, 80°C	67	7
4-Methoxyphenyl	Bis(piperidino)-acetonitrile	24 hr, 60°C	67	7
4-Methoxyphenyl l,	Bis(pyrrolidino)-acetonitrile	24 hr, 60°C	60	7

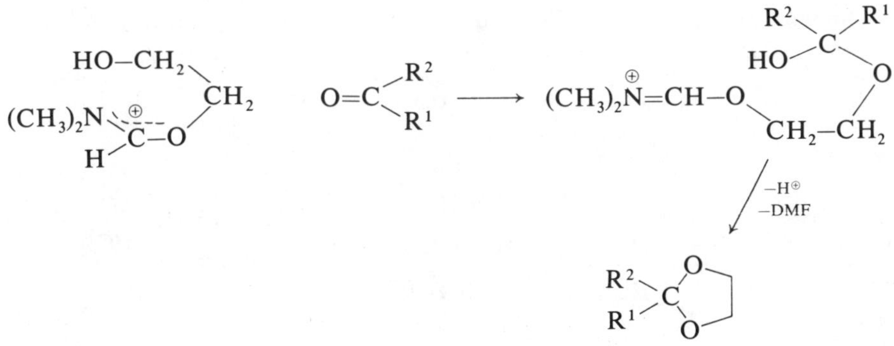

cyanide transfer the cyanohydrin anions are formed first. These generally react further to give α-(dialkylamino)arylacetonitriles. Only in the case of strongly-acidic cyanohydrins is benzoin condensate observed.

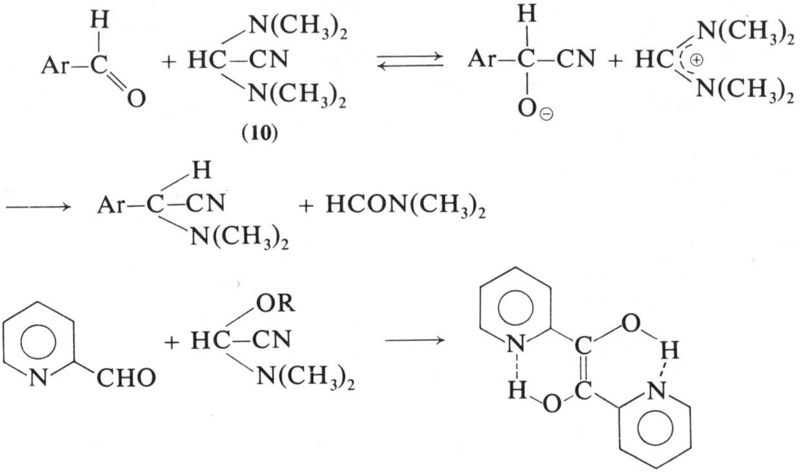

492 ORTHOAMIDES: PROPERTIES AND REACTIONS

$$Ar-\underset{O}{\overset{H}{C}} + HC\underset{N(CH_3)_2}{\overset{N(CH_3)_2}{-}}CN \rightleftharpoons Ar-\underset{O^{\ominus}}{\overset{H}{C}}-CN + HC\underset{N(CH_3)_2}{\overset{N(CH_3)_2}{\overset{\oplus}{-}}}$$

(10)

$$\longrightarrow Ar-\underset{N(CH_3)_2}{\overset{H}{C}}-CN + HCON(CH_3)_2$$

[Reaction of 2-pyridinecarboxaldehyde + HC(OR)(N(CH₃)₂) giving bicyclic product]

The reactions with aldehydes are summarized in Table XXVI.

3. Acylation

Based on the experimental evidence, one may conclude that strongly electrophilic acylating agents react with *N,N*-dialkylformamide acetals at the alkoxy groups and form esters, whereas weakly nucleophilic agents react at the dialkylamino group and form amides.

(a) *With Acyl halides-, Cyanides-, Azides, and Acid Anhydrides.** **1a** reacts with 4-dimethylaminobenzoyl chloride via *N*-acylation giving 4-dimethylaminobenzamide (101). 2-Chlorobenzoxazole reacts analogously, whereas

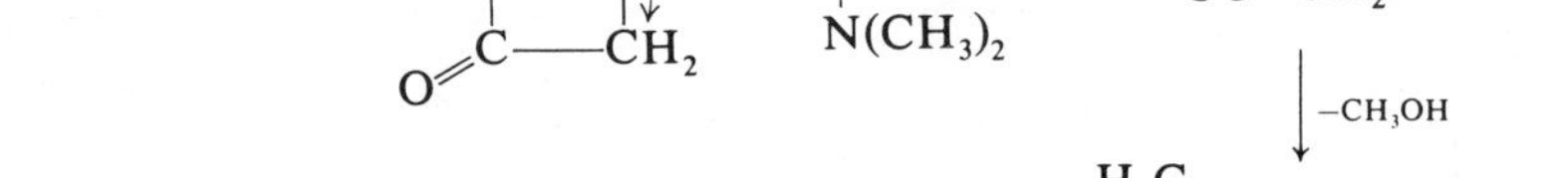

$$(CH_3)_2N-\langle\bigcirc\rangle-COCl \xrightarrow[-CH_3Cl]{\overset{1a}{-HCO_2CH_3}} (CH_3)_2N-\langle\bigcirc\rangle-CON(CH_3)_2$$

$$2CH_2=C=O \longrightarrow \underset{O=C-CH_2}{\overset{CH_2-C\overset{O}{\underset{\downarrow}{\diagdown}}}{|}} + HC\underset{N(CH_3)_2}{\overset{OCH_3}{-}}OCH_3 \longrightarrow \underset{CO-CH_2|^{\ominus}}{\overset{CH_2-CO_2CH_3}{|}} \quad HC\underset{N(CH_3)_2}{\overset{OCH_3}{\overset{\oplus}{\diagup}}}$$

$$\downarrow -CH_3OH$$

$$\underset{H}{\overset{H_3C}{\diagdown}}\underset{\underset{C}{\overset{\|}{C}}}{C}-C-CO_2CH_3 \\ \underset{N(CH_3)_2}{|}$$

* For the reaction of cyclic acetals with acetic anhydride see Section V-C.

IV. REACTIONS OF ORTHOAMIDES WITH ELECTROPHILES

ketene forms methyl dimethylaminomethylenemethylacetoacetate via O-acylation of diketene (101).

Cyanogen bromide and arylcyanates react first at the dimethylamino group of **1b**, as indicated by analysis of the reaction products (143).

Dialkoxychlorophosphines and **1a** form dialkylmethylphosphites. With cyclic dimethylformamide acetals O-acylation is also the initial reaction step. The reaction of **1a** with phosphorus trichloride to give trimethylphosphite also proceeds according to the same principles (144). The final product of this reaction is dimethylaminomethylenebis(phosphonic acid dimethyl ester), which arises from a complex reaction sequence.

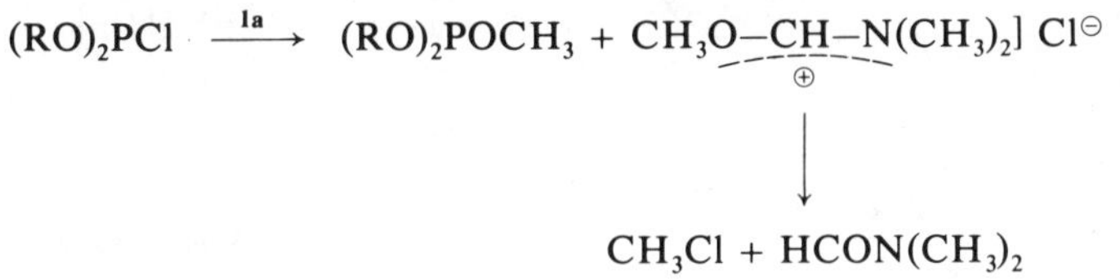

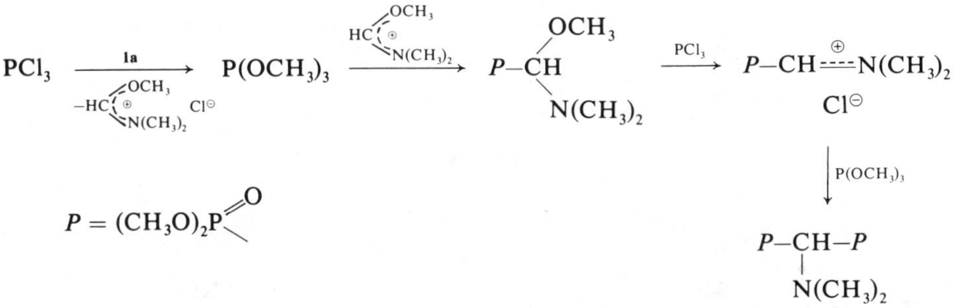

$$(RO)_2PCl \xrightarrow{1a} (RO)_2POCH_3 + CH_3O\text{-}CH\text{-}N(CH_3)_2]\ Cl^{\ominus}$$

$$\downarrow$$

$$CH_3Cl + HCON(CH_3)_2$$

$$PCl_3 \xrightarrow[\substack{1a \\ -HC(OCH_3)(N(CH_3)_2)\ Cl^{\ominus}}]{} P(OCH_3)_3 \xrightarrow{HC(OCH_3)(N(CH_3)_2)} P\text{-}CH(OCH_3)(N(CH_3)_2) \xrightarrow{PCl_3} P\text{-}CH\text{---}N(CH_3)_2\ Cl^{\ominus}$$

$$P = (CH_3O)_2P{=}O$$

$$\downarrow P(OCH_3)_3$$

$$P\text{-}CH\text{-}P\ |\ N(CH_3)_2$$

Acylcyanides, which are less electrophilic, initially form N-adducts with N,N-dimethylformamide acetals (**1**), to finally yield carboxylic acid esters and alkoxy(dimethylamino)acetonitriles; this reaction sequence may be used to synthesize the latter compounds (85).

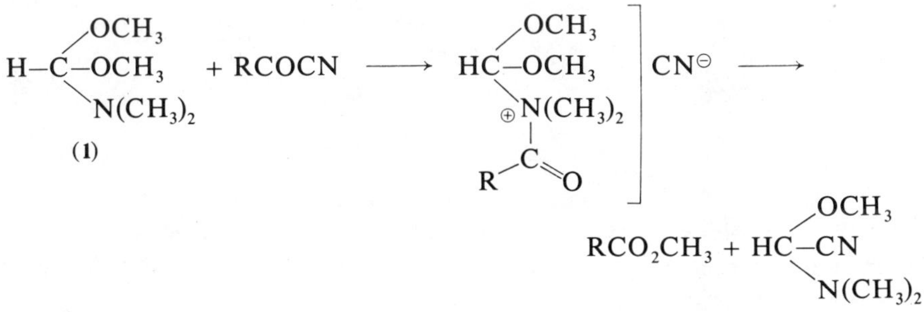

$$H\text{-}C(OCH_3)(OCH_3)(N(CH_3)_2) + RCOCN \longrightarrow [HC(OCH_3)(OCH_3)(\overset{\oplus}{N}(CH_3)_2\text{-}C(R)=O)]\ CN^{\ominus} \longrightarrow$$
(1)

$$RCO_2CH_3 + HC(OCH_3)(CN)(N(CH_3)_2)$$

Carboxylic acid azides reacting with **1a** and **1b** generally yield the products arising from the corresponding isocyanates (see Section IV-C-1). Only azides such as 4-nitrobenzoyl (145) or sulfonyl azides (138), which do not generally rearrange to isocyanates, give acid esters via *O*-acylation. Sulfonic acid esters subsequently undergo alkylation with **1a** and **1b**, as has been previously described. The extremely reactive azidinium salts exhibit analogous behavior (138).

(**1a**)

$$Ar{-}SO_2N_3 \xrightarrow{\mathbf{1b}} Ar{-}SO_3C_2H_5 + HCON(CH_3)_2 + C_2H_5N_3$$

$$\downarrow \mathbf{1b}$$

$$(CH_3)_2\overset{\oplus}{N}(C_2H_5)_2]ArSO_3^{\ominus}$$

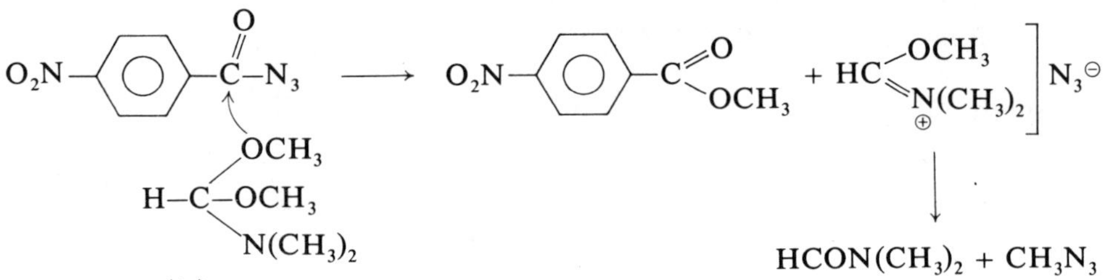

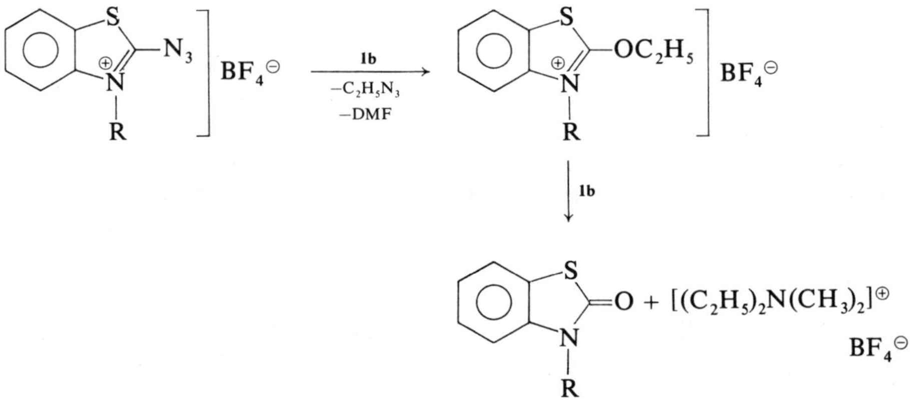

In the photolysis of weakly electrophilic carboxylic acid azides, the intermediate nitrene may be trapped by **1a**, giving *N,N*-dimethyl-*N'*-acylhydrazines (145).

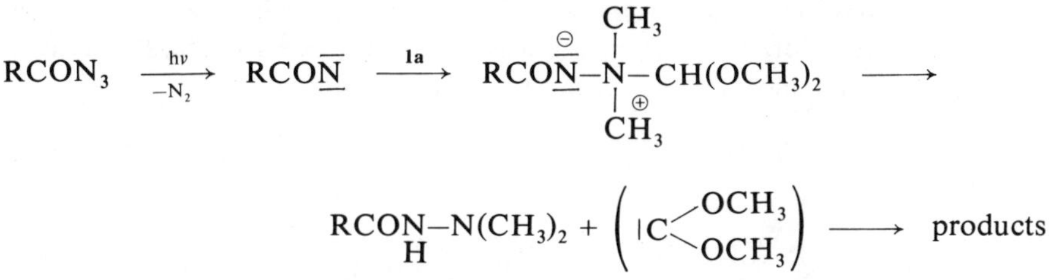

IV. REACTIONS OF ORTHOAMIDES WITH ELECTROPHILES

$$RCON_3 \xrightarrow[-N_2]{h\nu} RCO\overline{N} \xrightarrow{1a} RCO\overline{N}-\overset{\overset{CH_3}{|}}{\underset{\underset{CH_3}{|}}{N}}-CH(OCH_3)_2 \longrightarrow$$

$$RCON-N(CH_3)_2 + \left(IC\underset{OCH_3}{\overset{OCH_3}{\diagup}} \right) \longrightarrow \text{products}$$
$$H$$

$R = C_6H_5-, \ 4\text{-}CH_3O\text{-}C_6H_4-$

Aminal esters (**5**) are converted into bis(dimethylamino)acetonitriles (**10**) with acylcyanides (85). These proceed further via N-adducts with excess acylcyanide to N,N-dimethylcarboxylic acid amides and (dimethylamino)malononitrile (**7**).

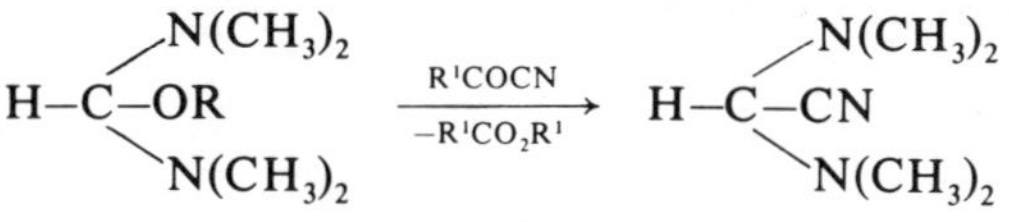

$$H-\underset{\underset{N(CH_3)_2}{|}}{\overset{\overset{N(CH_3)_2}{|}}{C}}-OR \xrightarrow[-R'CO_2R']{R'COCN} H-\underset{\underset{N(CH_3)_2}{|}}{\overset{\overset{N(CH_3)_2}{|}}{C}}-CN$$
$$(\mathbf{5}) (\mathbf{10})$$

$$H\overset{\overset{N(CH_3)_2}{|}}{\underset{\underset{N(CH_3)_2}{|}}{C}}-CN \xrightleftharpoons{R'COCN} H\overset{\overset{N(CH_3)_2}{|}}{\underset{\underset{\underset{NC}{|}}{\overset{R}{C}}}{C}}-\overset{R}{\underset{O^\ominus}{C}}-\overset{+}{N(CH_3)_2} \longrightarrow RCON(CH_3)_2 + H\overset{\overset{N(CH_3)_2}{|}}{\underset{\underset{CN}{|}}{C}}-CN$$
$$(\mathbf{10})$$

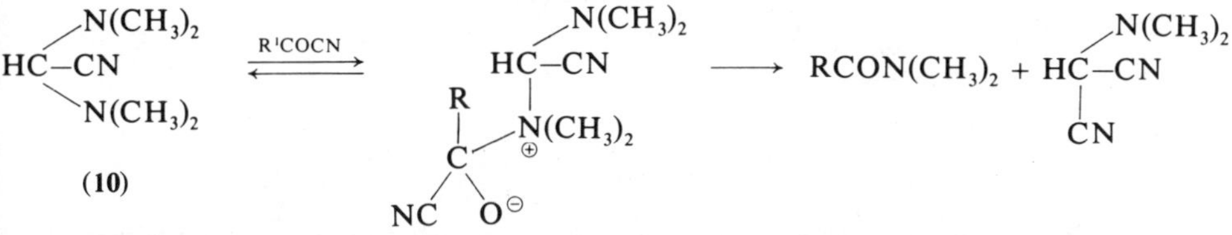

On the other hand, tetraalkylformamidinium hexachloroantimonate is formed in good yield from **10** with carboxylic acid chlorides in the presence of antimony pentachloride (7).

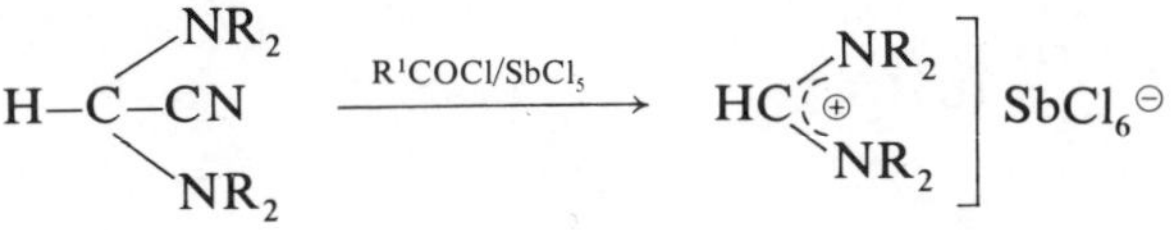

$$H-\underset{\underset{NR_2}{\diagdown}}{\overset{\overset{NR_2}{\diagup}}{C}}-CN \xrightarrow{R'COCl/SbCl_5} \left[HC\underset{NR_2}{\overset{NR_2}{\diagup}}^{\oplus} \right] SbCl_6^\ominus$$

Tris(alkylarylamino)methanes (**7**) similarly yield formamidinium salts with acyl halides (113). Tris(acylamino)methanes (**8**) exhibit entirely different

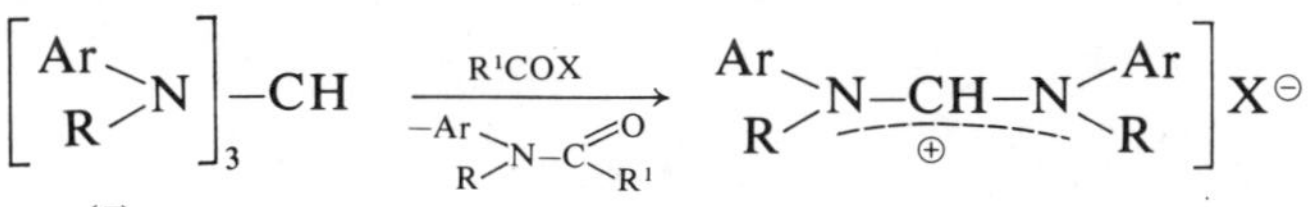

$$\left[\underset{R}{\overset{Ar}{\diagdown}}N \right]_3 -CH \xrightarrow[-Ar\diagdown N-C\diagup^{O}_{R'}]{R'COX} \left[\underset{R}{\overset{Ar}{\diagdown}}N-\overset{+}{CH}-N\underset{R}{\overset{Ar}{\diagup}} \right] X^\ominus$$
$$(\mathbf{7})$$

behavior. They undergo exchange of acyl groups with carboxylic acid anhydrides at ~140°C. This reaction is primarily used for the synthesis of higher tris(acylamino)methanes from tris(formylamino)methane (22, 146).

$$HC(NHCHO)_3 \xrightarrow{(RCO)_2O} HC(NHCOR)_3$$
(8)

$$R = CH_3, C_2H_5, C_3H_7, C_6H_5$$

(b) *With Heterocumulenes.* Aliphatic and aromatic iso[thio]cyanates form adducts at the dialkylamino groups of acyclic *N,N*-dialkylcarboxylic acid amide acetals in the first reaction step (101,147,148). This has been proved by hydrolysis of the reaction mixture as well as by other methods (101,147,149).* Whereas the hydrolysis of **1a** or **1b** yields 98.5% DMF, the addition of a molar amount of phenyl iso[thio]cyanate gives *N,N*-dimethyl-*N'*-phenyl(thio)urea.

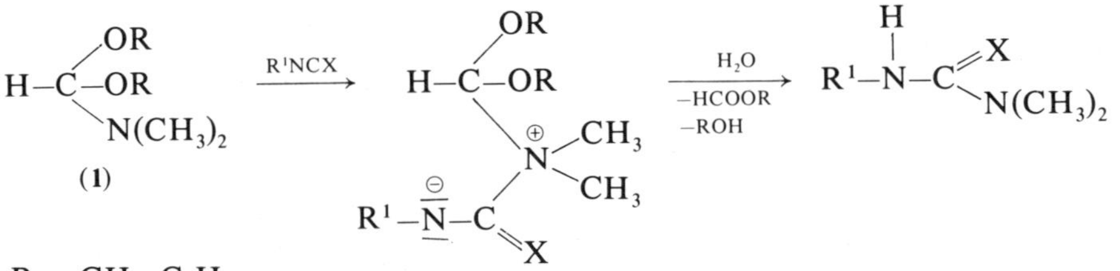

$$\underset{(1)}{\overset{OR}{\underset{N(CH_3)_2}{H-C-OR}}} \xrightarrow{R^1NCX} \underset{\underset{R^1-\overset{\ominus}{N}-C\overset{CH_3}{\underset{X}{\diagdown}}}{H-C-OR}}{\overset{OR}{\underset{\overset{\oplus}{N}\diagup}{\text{   }}}} \xrightarrow[\substack{-HCOOR \\ -ROH}]{H_2O} R^1-\overset{H}{\underset{}{N}}-C\overset{X}{\underset{N(CH_3)_2}{\diagdown}}$$

$R = CH_3, C_2H_5$
$R^1 = C_6H_5, 4\text{-}NO_2\text{-}C_6H_4\text{-}$
$X = O, S$

Isolation and characterization of heterocumulene adducts has been achieved in the reaction of bis(dialkylamino)acetonitriles (**10**) with phenyl isothiocyanate (7).

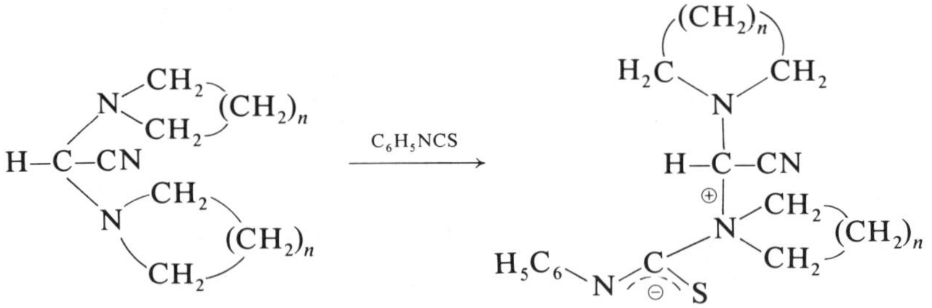

(**10**)

$n = 0, 2, 3$

---

*For subsequent formation of *C*-acylated products with *N,N*-dialkylformamide acetals see Section IV-C.

IV. REACTIONS OF ORTHOAMIDES WITH ELECTROPHILES

The reactions of isothiocyanates with **5** and **7** proceed analogously. In these cases the adducts are less stable, so isothioureas are the isolated products. The trimers of isothiocyanates are interesting by-products (101,150), which become the main product when **5** and **7** are reacted with aryl isothiocyanates (149,150).

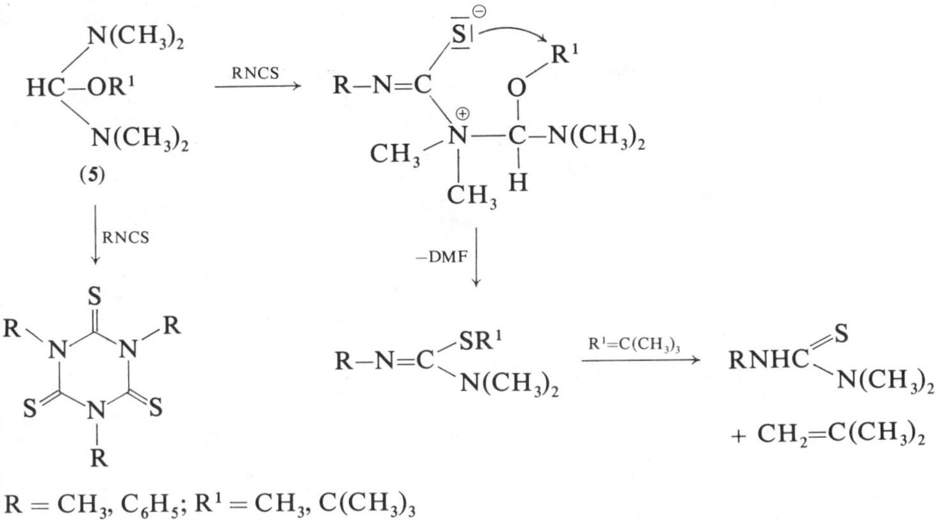

$R = CH_3, C_6H_5; R^1 = CH_3, C(CH_3)_3$

The strongly electrophilic 4-nitrophenyl isothiocyanate displays little regiospecificity. Addition may occur at the alkoxy groups as well as at the dimethylamino groups of **1a** or **1b**.

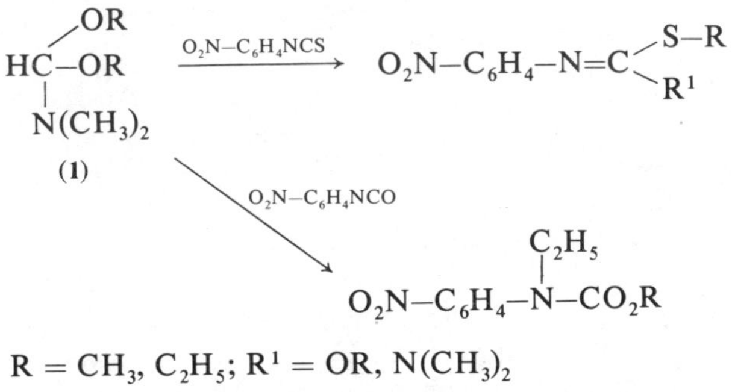

$R = CH_3, C_2H_5; R^1 = OR, N(CH_3)_2$

For conformational reasons (strong steric interaction between the axial protons in the ring of the *N*-adduct) cyclic *N,N*-dimethylformamide acetals are attacked by heterocumulenes at their alkoxy groups. Only when the oxygen atoms in the ring are strongly shielded can the reaction occur at the dimethylamino group (101,145):

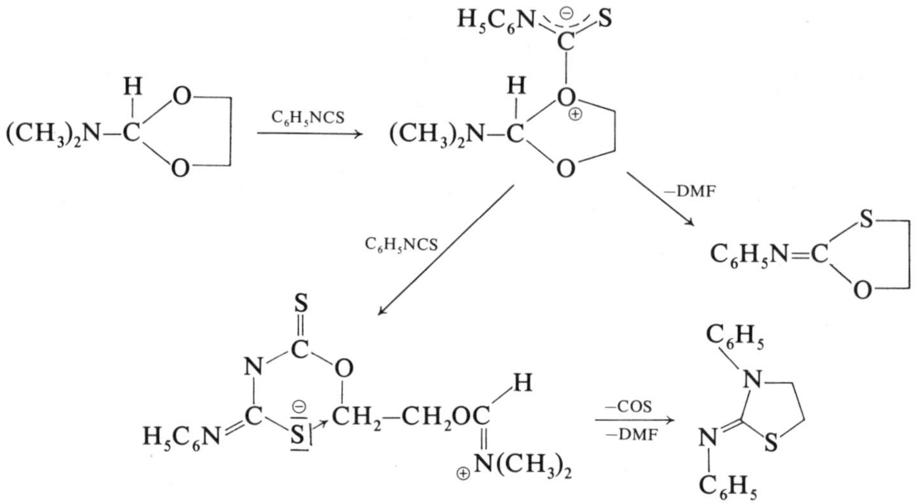

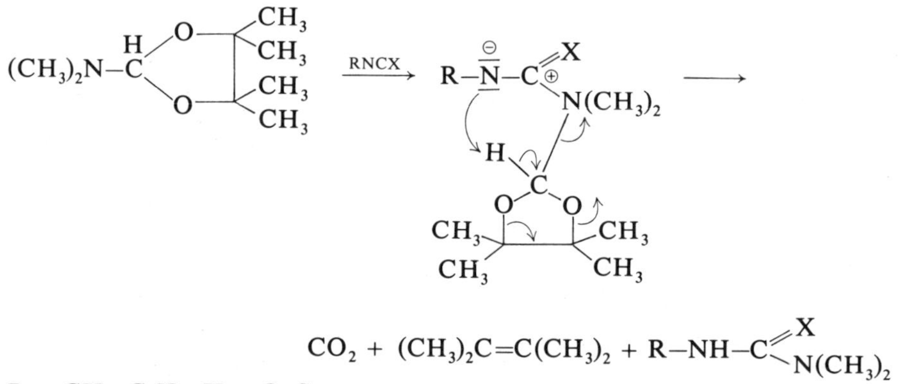

$$R = CH_3, C_6H_5; \quad X = O, S$$

When **1b** is reacted with less reactive heterocumulenes like carbon disulfide, carbon oxysulfide, or carbon dioxide, the formation of adducts competes with the dissociation of **1b**, which is then followed by nucleophilic addition of ethoxide to the heterocumulene. The product distribution clearly indicates the decreasing importance of the adduct-forming mechanism in reactions of weak electrophiles (151); see Table XXVII.

IV. REACTIONS OF ORTHOAMIDES WITH ELECTROPHILES   499

TABLE XXVII

Distribution of Products when CO_2, COS, and CS_2 are treated with **1b**

Heterocumulene	Reaction conditions	Product distribution	Overall yield, %
CO_2	180°C, 100 hr	Ethyl carbonate/ ethyl N,N-dimethyl carbamate = 96/4	62
COS	140°C, 50 hr	Monothiocarbonic acid O,S-diethyl ester/N,N-dimethyl-thiocarbamic acid S-ethyl ester = 58/42	69
CS_2	100°C, 80 hr	Dithiocarbonic acid O,S-diethyl ester/N,N-dimethyl-dithiocarbamic acid ethyl ester = 60/40	74

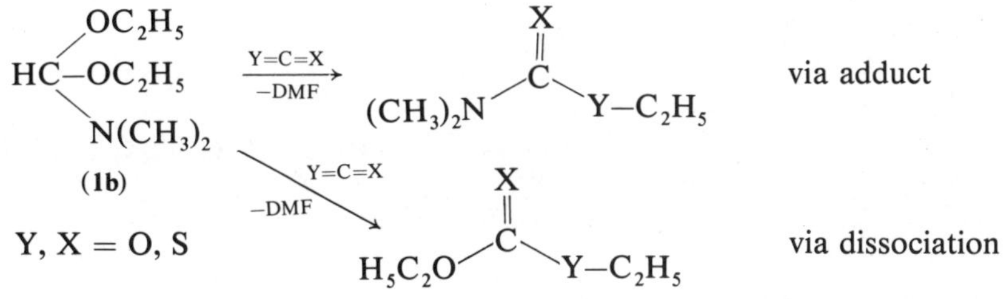

The extreme reactivity of the orthoamides **5** and **7** is again evident (46,151). They react at room temperature with carbon disulfide to give tetramethylformamidinium dithiocarbamates. Carbon oxysulfide and carbon dioxide behave analogously (151).

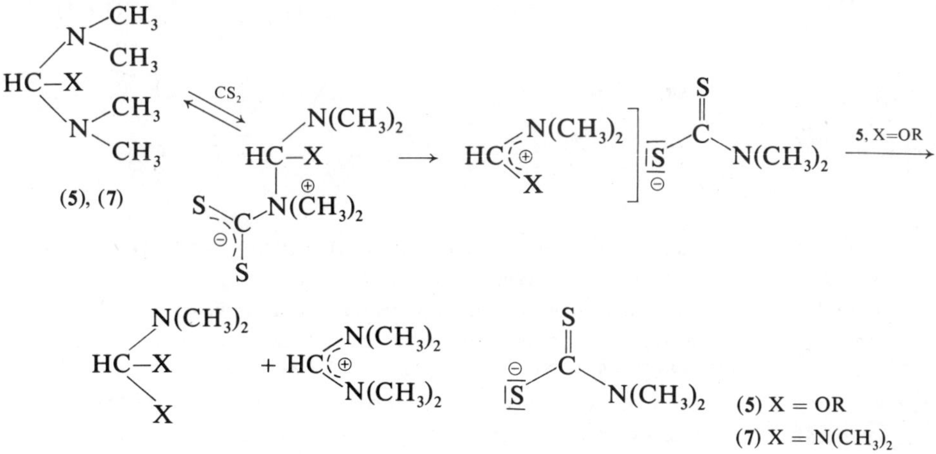

B. AT α-METHYL (METHYLENE) GROUPS OF CARBOXYLIC ACID AMIDE ACETALS

Since homologues of *N,N*-dialkylformamide acetals are in equilibrium with ketene *O,N*-acetals, they undergo electrophilic reactions at the α-methylene group.

1. With Activated Alkenes and Alkynes

Dimethylacetamide acetals (**1**) react with electron-deficient alkenes in a Michael-type addition to give α-alkoxyenamines. Excess alkene gives rise to cyclic products (152).

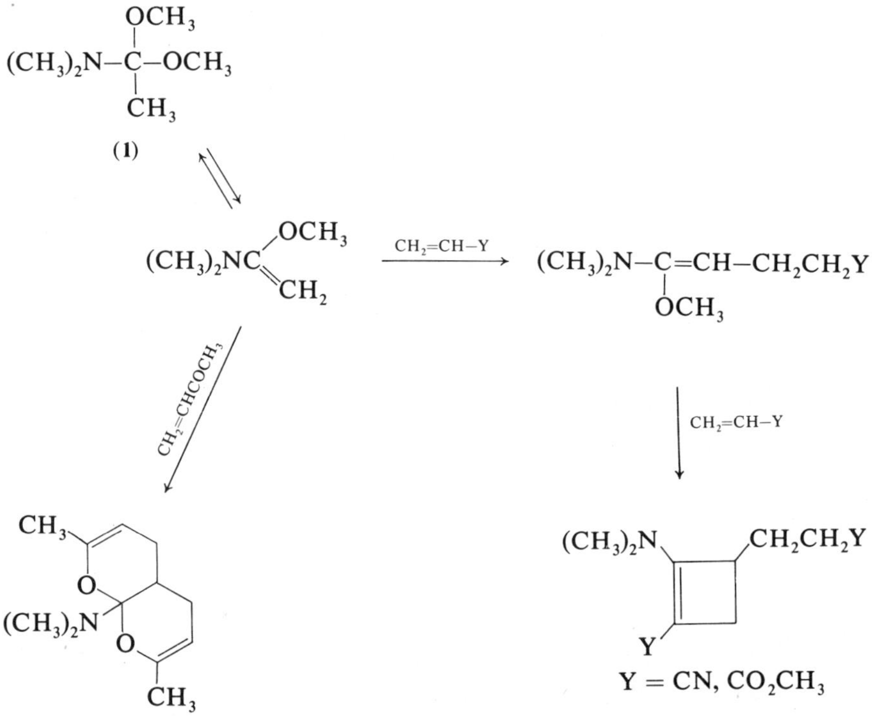

Whereas 2,2-diethoxy-1-methylpiperidine (**2c**) nearly exclusively yields 3,3-disubstituted products with excess alkene, the reaction with 2,2-dimethoxy-1-methylpyrrolidine stops once 3-monosubstituted products have been formed. Remarkably, lactams (via an S_N reaction), not α-alkoxyenamines, are the isolated products (153, 154). This result leads to the conclusion that with five-membered rings the transformation of the alkoxyenaminium ion to the lactam successfully competes with deprotonation to the alkoxyenamine. This

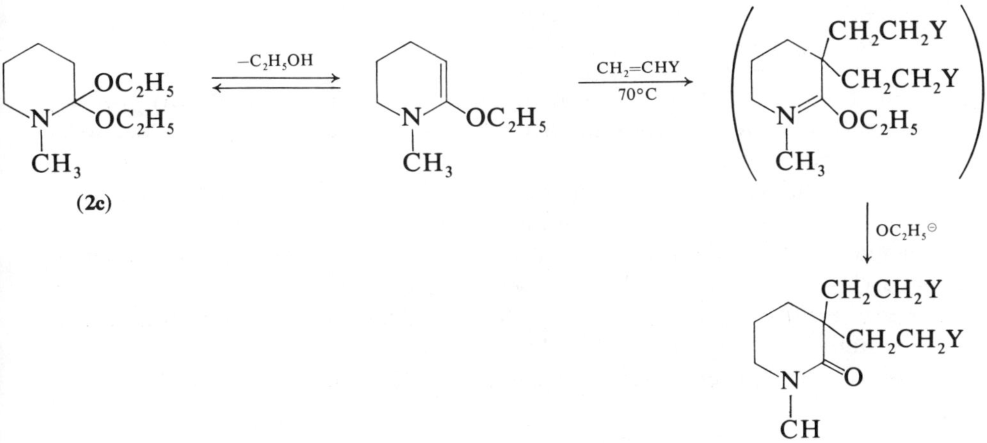

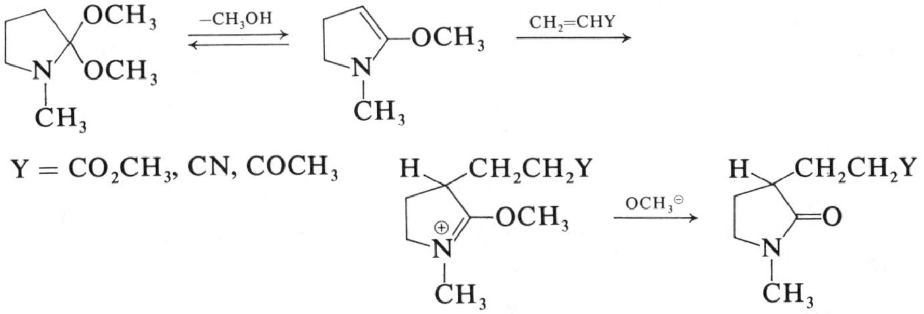

IV. REACTIONS OF ORTHOAMIDES WITH ELECTROPHILES

Y = CO$_2$CH$_3$, CN, COCH$_3$

assumption is supported by the finding that the elimination of alcohol from five-membered lactam acetals is extremely difficult (95).

Dimethyl acetylenedicarboxylate and 2,2-dimethoxy-1-methyl-pyrrolidine (**2**) yield a 1:1 adduct in benzene, whereas in dioxane a 2:1 adduct is also formed (154).

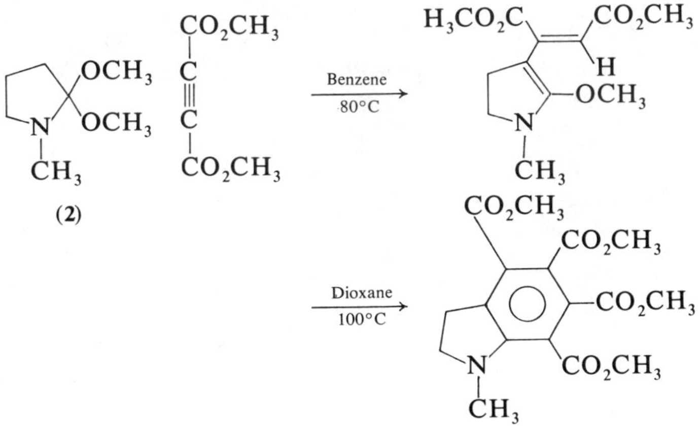

2. With Alkyl and Acyl Halides

Benzyl halides yield alkoxyiminium salts with the intermediately formed ketene O,N-acetals; the salts are then transformed to the carboxylic acid amide by the action of halide ion (152,153,155). The acylation reaction has been studied in more detail using N-methylcaprolactam diethyl acetal (**2d**) as a model (155). According to this study a carboxylic acid ester and an ambident iminium cation are formed first. The latter may yield the caprolactam upon attack by halide ions, or, after removal of an acidic proton, it may be acylated in that position giving a 2-alkoxy-3-acyliminium halide. This is then either deprotonated by excess acetal or desalkylated to give a 3-acyllactam. The free 2-alkoxyenamino ketone reacts immediately with additional acyl halide.

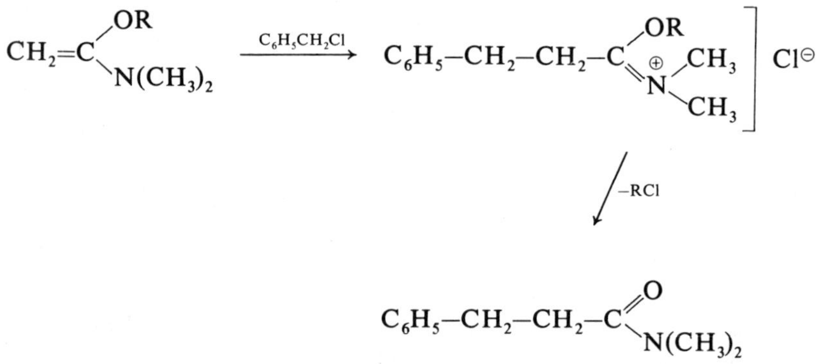

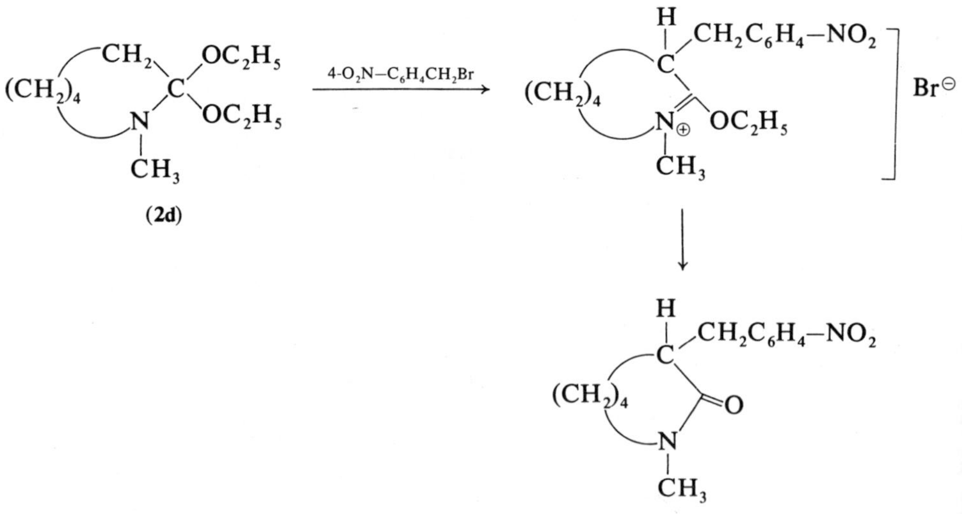

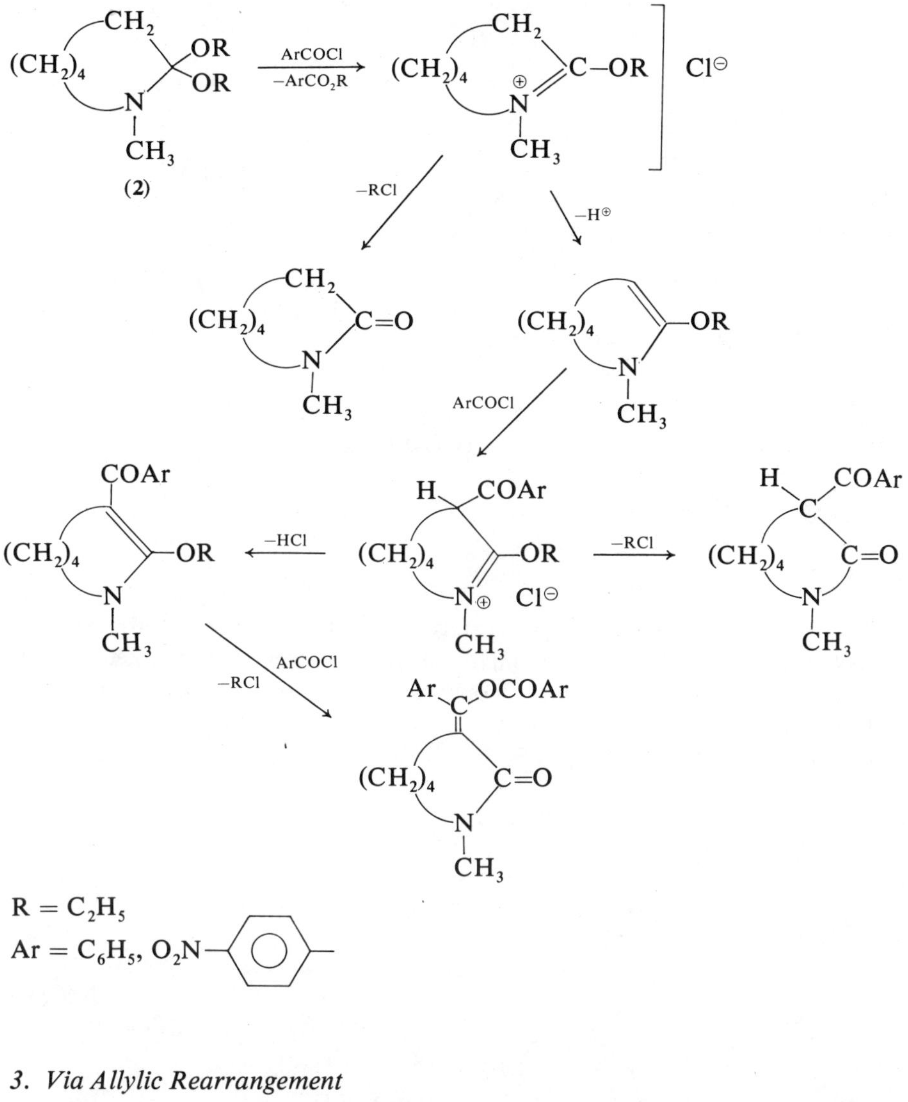

IV. REACTIONS OF ORTHOAMIDES WITH ELECTROPHILES

R = C₂H₅
Ar = C₆H₅, O₂N—⟨C₆H₄⟩—

3. Via Allylic Rearrangement

A modification of the Claisen rearrangement of vinyl ethers, consisting of the reaction of N,N-dialkylcarboxylic acid amide acetals (**1**) with allyl and benzyl alcohols, is more important than the aforementioned substitution reactions. This rearrangement, first discovered by Meerwein et al. (4) and then developed as a preparative method by Eschenmoser et al. (156), allows the stereospecific introduction of an alkane acid unit into the 3 position of an allyl alcohol (156–159).

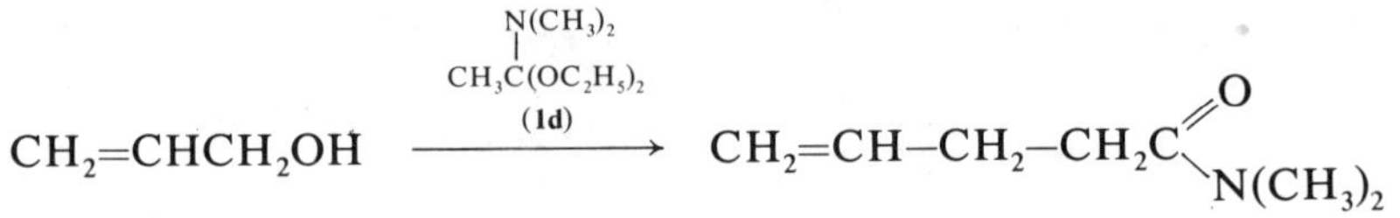

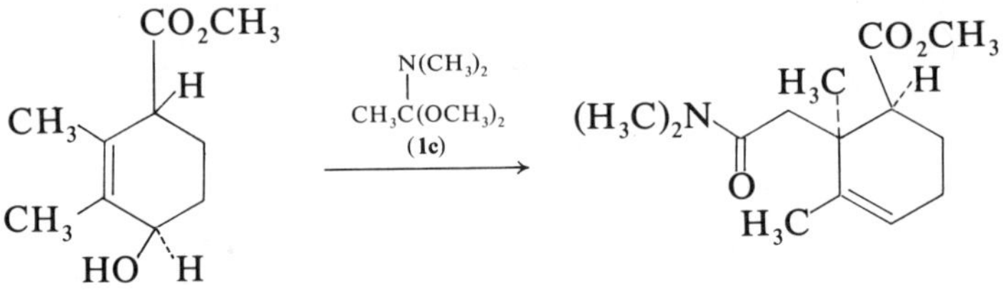

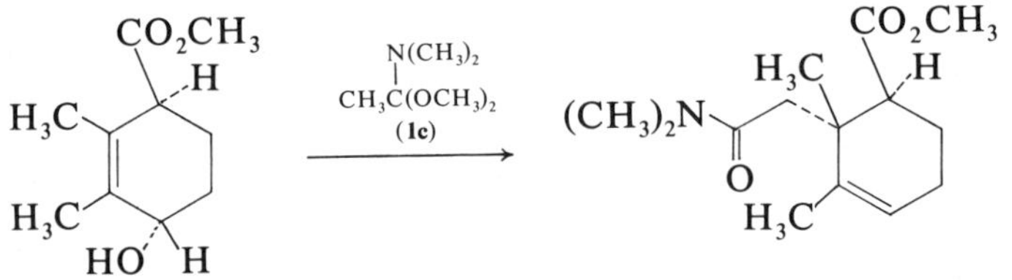

$$CH_2=CHCH_2OH \quad \xrightarrow{\substack{N(CH_3)_2 \\ | \\ CH_3C(OC_2H_5)_2 \\ (1d)}} \quad CH_2=CH-CH_2-CH_2C{\overset{O}{\underset{N(CH_3)_2}{\diagup}}}$$

[Cyclohexene with CO$_2$CH$_3$, H, CH$_3$, CH$_3$, HO, H substituents] $\xrightarrow{\substack{N(CH_3)_2 \\ | \\ CH_3C(OCH_3)_2 \\ (1c)}}$ [rearranged product with (H$_3$C)$_2$N–C(=O)–CH$_2$– group on cyclohexene bearing CO$_2$CH$_3$, H, H$_3$C, H$_3$C]

[Cyclohexene with CO$_2$CH$_3$, H, H$_3$C, H$_3$C, HO, H substituents] $\xrightarrow{\substack{N(CH_3)_2 \\ | \\ CH_3C(OCH_3)_2 \\ (1c)}}$ [rearranged product with (CH$_3$)$_2$N–C(=O)–CH$_2$– group on cyclohexene bearing CO$_2$CH$_3$, H, H$_3$C, H$_3$C]

Dimethylacetamide dialkyl acetals (156–161), dimethylpropionamide dialkyl acetals (162,163), their mixtures with the corresponding ketene O,N-acetals, and ketene O,N-acetals alone have been used in the rearrangement. This reaction is successful not only with α,β-unsaturated alcohols, but also with 1,3-dienols (162,163):

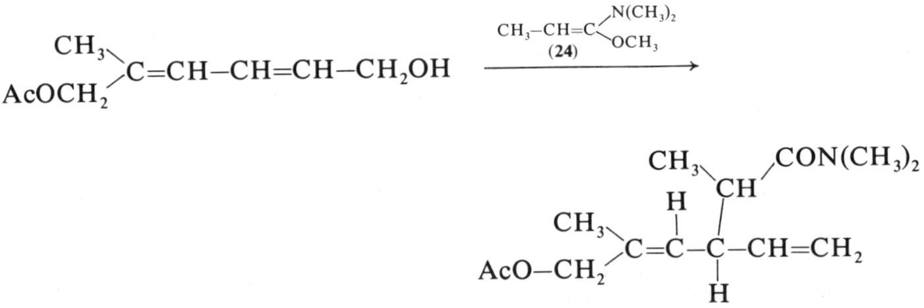

$$\underset{AcOCH_2}{\overset{CH_3}{\diagdown}}C=CH-CH=CH-CH_2OH \quad \xrightarrow{\substack{CH_3-CH=C{\overset{N(CH_3)_2}{\underset{OCH_3}{\diagdown}}} \\ (24)}}$$

$$\underset{AcO-CH_2}{\overset{CH_3}{\diagdown}}C=\underset{H}{\overset{CH_3}{\underset{|}{C}}}-\underset{H}{\overset{\overset{\displaystyle CH_3 \diagdown \diagup CON(CH_3)_2}{CH}}{\underset{|}{C}}}-CH=CH_2$$

Remarkably enough, benzyl alcohols which do not undergo a normal Claisen rearrangement yield arylacetic acid dialkylamides at 120–180°C. Moreover, condensation with excess orthoamide at the orthomethyl group is sometimes observed (156).

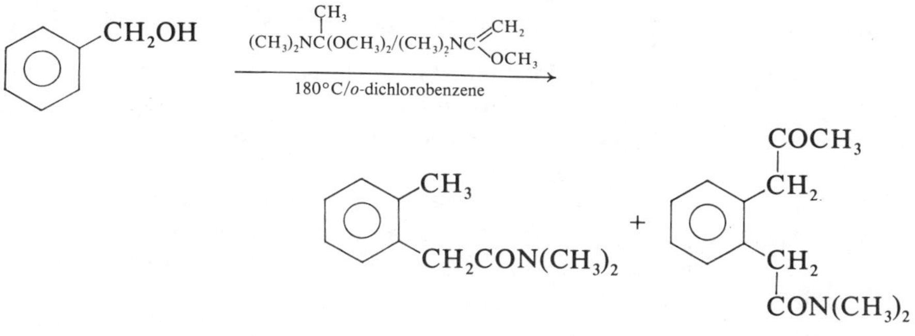

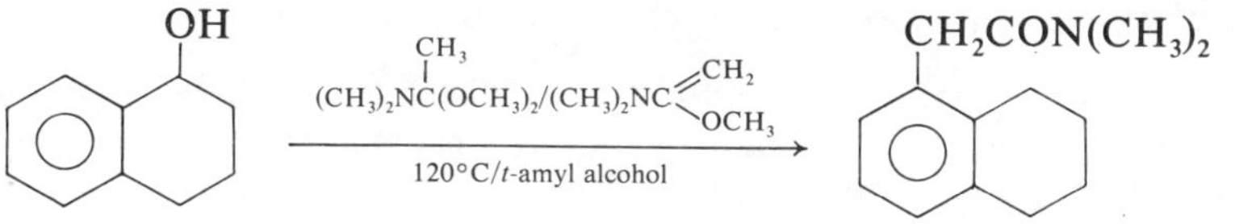

IV. REACTIONS OF ORTHOAMIDES WITH ELECTROPHILES

[Scheme: Benzyl alcohol (PhCH₂OH) reacts with (CH₃)₂NC(OCH₃)₂/(CH₃)₂NC(CH₃)(CH₂)(OCH₃) at 180°C/o-dichlorobenzene to give o-methyl-benzyl-CH₂CON(CH₃)₂ and o-(COCH₃)-C₆H₄-CH₂-CH₂-CON(CH₃)₂]

[Scheme: 1-hydroxy-tetrahydronaphthalene reacts with (CH₃)₂NC(OCH₃)₂/(CH₃)₂NC(CH₃)(CH₂)(OCH₃) at 120°C/t-amyl alcohol to give the corresponding CH₂CON(CH₃)₂ substituted product]

The mild reaction conditions (120–140°C) are due to the thermodynamics of the rearrangement, since an amide function is formed during the formation of a C=O bond from a C–O bond (156). The reaction is initiated either by interchange of the acetals with the unsaturated alcohol or by the addition of the alcohol to the ketene O,N-acetal:

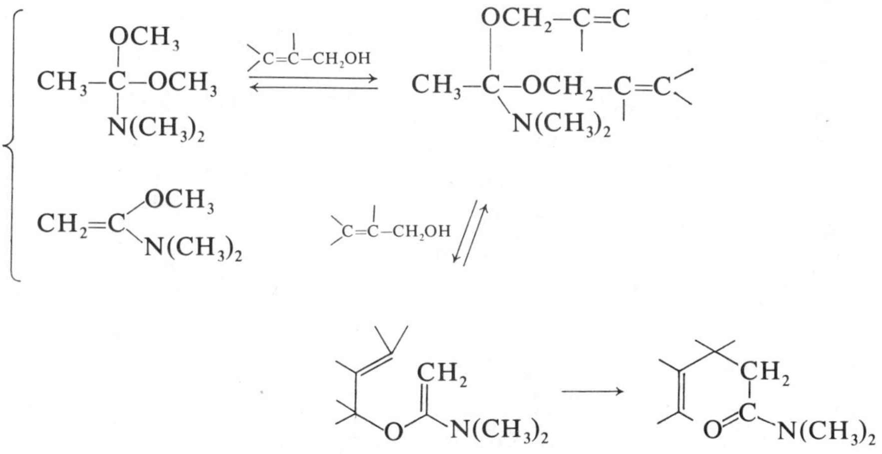

[Mechanism scheme showing equilibria between CH₃–C(OCH₃)₂–N(CH₃)₂, CH₂=C(OCH₃)–N(CH₃)₂, and the allyl-substituted intermediate CH₃–C(OCH₂–C=C)(N(CH₃)₂)(OCH₂–C=C), leading via [3,3]-sigmatropic rearrangement to the ketene N,O-acetal and then to the γ,δ-unsaturated amide with N(CH₃)₂]

Products of more than 90% optical purity have been obtained from chiral allyl alcohols via inversion. The degree of asymmetric induction by far exceeds that of the common Claisen rearrangement (159).

C. AT THE FORMYL CARBON OF ORTHOFORMAMIDES

Heterocumulenes and *N*-haloamides (as nitrene precursors) form adducts with dialkylformamide acetals (**1**) and thioacetals (**3**) in a reversible reaction. These adducts may undergo electrophilic substitution of the formyl carbon atom.

With aminal esters (**5**) and tris(dimethylamino)methane (**7a**) the adduct formation is usually followed by irreversible cleavage of the C–N bond.

Less electrophilic reagents like carbonyl compounds also react at the formyl carbon of **7a**.

1. With Isocyanates, Isothiocyanates, and Carbodiimides

Acyclic dialkylformamide acetals (**1**) yield 5-alkoxy-5-dialkylamino-2,4-dioxo[dithio or diarylimino]imidazolidines (parabanic acid N,O-acetals) with

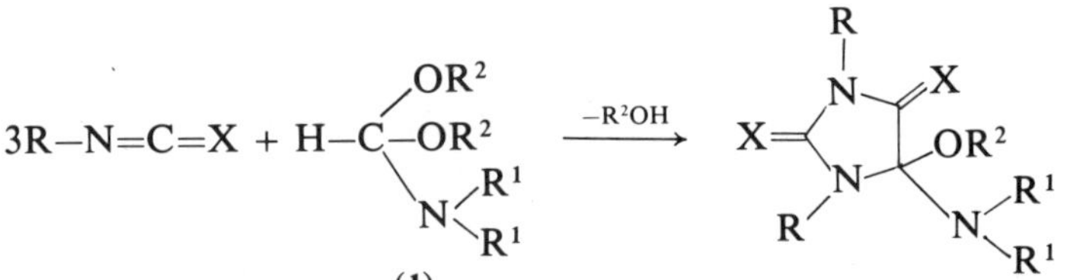

$$3R-N=C=X + H-C(OR^2)(OR^2)(NR^1R^1) \xrightarrow{-R^2OH} \text{imidazolidine}$$
(**1**)

R	X	
aryl	O	
alkyl	O	
aryl	N-aryl	R^1, R^2 = alkyl

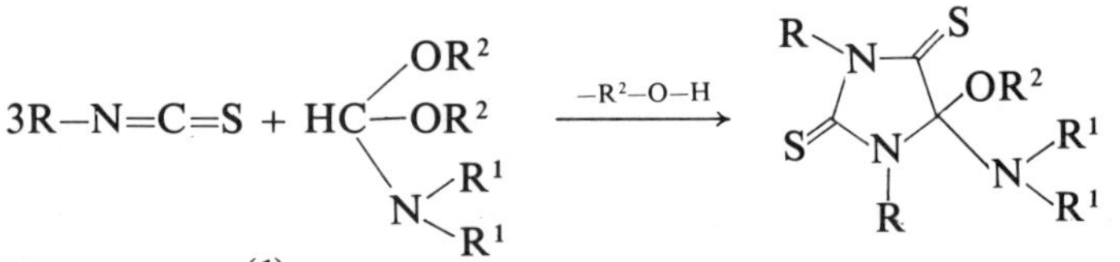

$$3R-N=C=S + HC(OR^2)(OR^2)(NR^1R^1) \xrightarrow{-R^2-O-H} \text{imidazolidine}$$
(**1**)

R: see Table XXIX

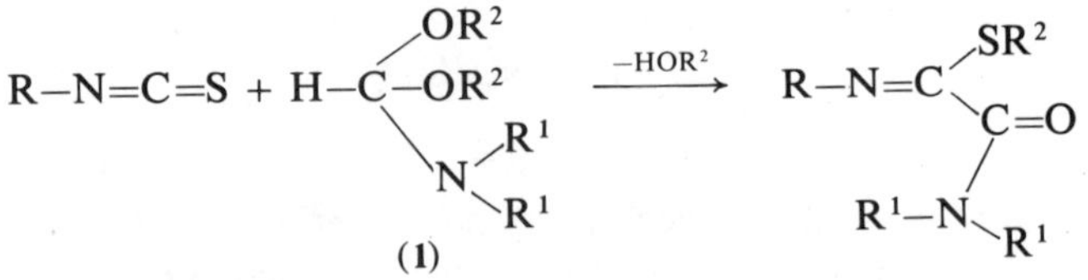

$$R-N=C=S + H-C(OR^2)(OR^2)(NR^1R^1) \xrightarrow{-HOR^2} R-N=C(SR^2)-C(=O)-NR^1R^1$$
(**1**)

R: see Table XXX

TABLE XXVIII

2,4-Dioxoimidazolidines from Orthoformamides and Isocyanates

Orthoamide	Isocyanate, RNCO	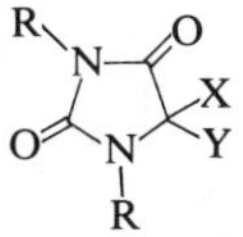 X, Y	Reaction conditions	Yield, %	Ref.
1b	Cyclohexyl, t-butyl, isopropyl	$X = OC_2H_5$, $Y = N(CH_3)_2$	120–140°C, 2–6 hr	48–56	147, 165
1b	Phenyl, 4-chlorophenyl, 4-methylphenyl, 2,6-dimethylphenyl, 4-ethoxyphenyl, β-naphthyl	$X = OC_2H_5$, $Y = N(CH_3)_2$	70–100°C, 1–4 hr, petroleum ether or toluene	96–99	147, 165
1b	2-Phenethyl	$X = OC_2H_5$, $Y = N(CH_3)_2$	80°C, 4 hr, n-hexane	94	147
HC(SC$_2$H$_5$)$_2$N(CH$_3$)$_2$	Phenyl	$X = SC_2H_5$, $Y = N(CH_3)_2$	Exothermic	84	151
HC[N(CH$_3$)$_2$]$_3$	Cyclohexyl	$X = Y = N(CH_3)_2$	Ether, 20°C	57	46
H–C(OCH$_3$)(CN)N(CH$_3$)$_2$	4-Methylphenyl	$X = OCH_3$, $Y = N(CH_3)_2$	Exothermic	59	9
HC(CN)[N(CH$_3$)$_2$]$_2$	n-Propyl, n-Butyl	$X = CN$, $Y = N(CH_3)_2$	35°C, 1 hr, ether	98	7
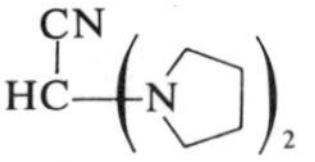 HC(CN)(N⟨pyrrolidinyl⟩)$_2$	n-Butyl	$X = CN$, $Y = N$⟨pyrrolidinyl⟩ 	35°C, 8 hr, ether	54	7

TABLE XXIX

5-Alkoxy-5-dimethylamino-1,3-dialkyl-2,4-dithioxoimidazolidines from Dimethylformamide Diethyl Acetal (1b) and Isothiocyanates (166, 167)

Isothiocyanate	Imidazolidine yield, %	Reaction conditions
Methyl	65	120–140°C, 16 hr
Ethyl	63	130–140°C, 14 hr
n-Butyl	64	130–140°C, 20 hr
Isopropyl	53	130–140°C, 15 hr
Cyclohexyl	42	130–140°C, 14 hr
2-Phenethyl	46	140°C, 7 hr

TABLE XXX

(Alkyl[aryl]imino)(ethylthio)-N,N-dimethylacetamides from Dimethylformamide Diethyl Acetal and Isothiocyanates (166, 167)

$$R-N=C{\overset{SC_2H_5}{\underset{CON(CH_3)_2}{}}}$$

Isothiocyanate	yield, %	Reaction conditions
Cyclohexyl	23	130–140°C, 14 hr
Phenyl	83	130–140°C, 12 hr toluene
4-(Dimethylamino)phenyl	66	130–140°C, 12 hr toluene
4-Ethoxyphenyl	62	130–140°C, 12 hr toluene
1-Naphthyl	43	140–150°C, 14 hr toluene
Benzyl	28	130–140°C, 18 hr

TABLE XXXI

5-Ethoxy-1,3-diaryl-2,4-diarylimino-5-dimethylaminoimidazolidines from Diarylcarbodiimides and Dimethylformamide Diethyl Acetal in Toluene (164)

Carbodiimide	Imidazolidine yield, %	Reaction conditions
Diphenyl	77	120°C, 5 hr
Bis(4-methylphenyl)	66	120°C, 5 hr
Bis(4-methoxyphenyl)	61	120°C, 5 hr
Bis(4-chlorophenyl)	84	120°C, 5 hr

isocyanates* (147,165) and N,N'-diarylcarbodiimides (164); see Tables XXVIII–XXXI. Dialkylformamide thioacetals react analogously (151). Isothiocyanates with small alkyl groups yield the corresponding derivatives of 2,4-dithioxoimidazolidines, whereas those with bulky alkyl (or aryl) groups give (alkyl[aryl]imino)(alkylmercapto)-N,N-dialkylacetamides (166,167).

Parabanic acid O,N-acetals may also be obtained from the reaction of methoxy(dimethylamino)acetonitrile and isocyanates (9). Reactions with bis-(dialkylamino)acetonitriles (10) yield 5-dimethylamino-5-cyano-2,4-dioxoimidazolidines (7).

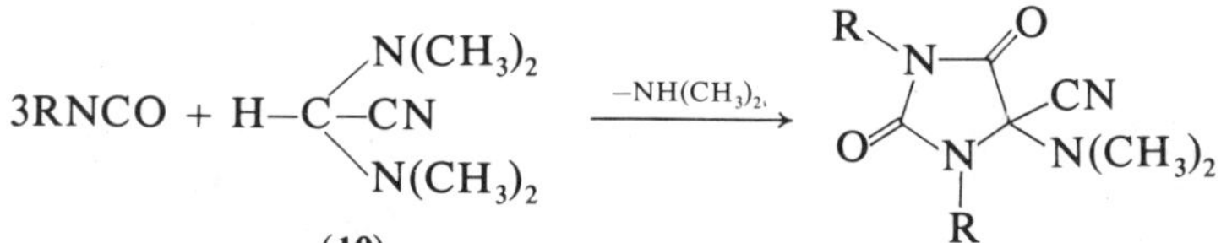

$$3RNCO + H-C{\overset{N(CH_3)_2}{\underset{N(CH_3)_2}{-CN}}} \xrightarrow{-NH(CH_3)_2}$$ [imidazolidine product]

(10)

The alcohols, dialkylamines, or mercaptans released during these reactions are completely (147,165) converted to [thio]carbamic acid derivatives by

---

* These are either used directly or generated *in situ* by Curtius reaction of the corresponding acid azides (145).

IV. REACTIONS OF ORTHOAMIDES WITH ELECTROPHILES

isocyanates, but only partially when isothiocyanates are used (167). The reaction conditions are determined by the electrophilic properties of the heterocumulenes.

Aromatic isocyanates react rapidly with **1** at 20–80°C in benzene; aliphatic isocyanates, isothiocyanates, and diarylcarbodiimides require 120–140°C.

The nucleophilic properties of orthoamides are important, as would be expected. With bis(dialkylamino)acetonitriles, isocyanates react exothermically, whereas with other orthoamides, the reaction must be heated.

(a) Reaction Mechanism. The reversible formation of adducts with the dialkylamino group is the key step in electrophilic substitution at the formyl carbon atom. If the newly formed C–N bonds are stable (e.g., in the reaction with **5**) elimination of the dialkylamino groups will occur instead of substitution. The intermediate formation of these adducts has been repeatedly (101,145,147,149) established, unequivocally in the reaction of bis(dimethylamino)acetonitrile with heterocumulenes (7). These adducts are quaternary ammonium salts which may form ylides through an intramolecular proton shift. When 2-dimethylamino-4,5-tetramethyl-1,3-dioxolane is treated with iso-[thio]cyanates, the isolation of [thio]urea is considered to be proof of a proton shift (101,145,149) (see Section IV-A-3(b)).

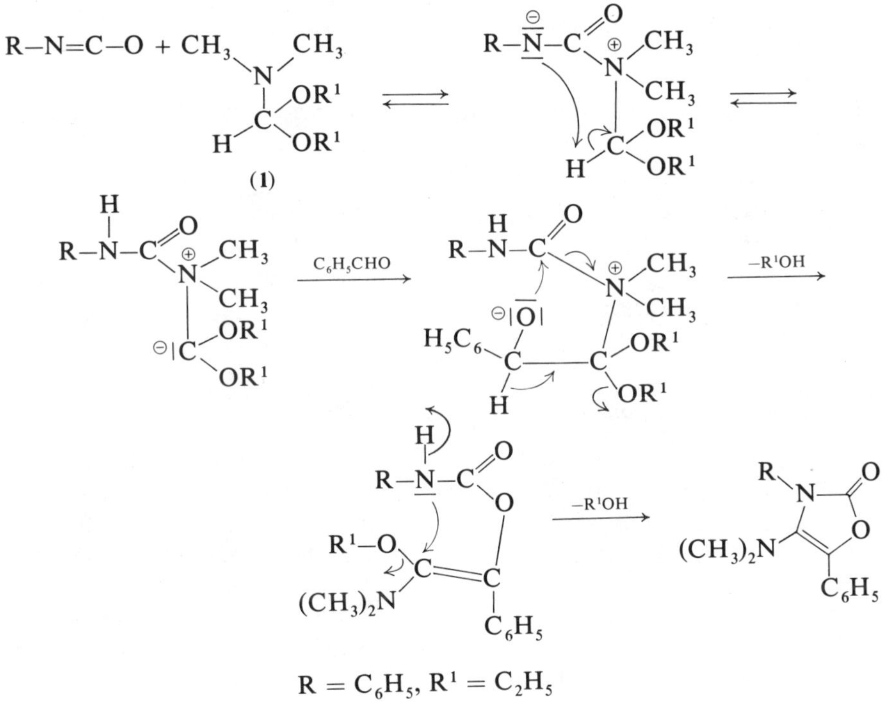

$$R = C_6H_5, \quad R^1 = C_2H_5$$

Moreover, during the reaction of **1b** and phenyl isocyanate, the ylide was trapped with benzaldehyde (147). The ylides undergo a Stevens-type rearrangement. Via 1,2 shifts, intermediates are formed which cyclize with further iso[thio]cyanate or carbodiimide to yield imidazolidine derivatives (101,147).

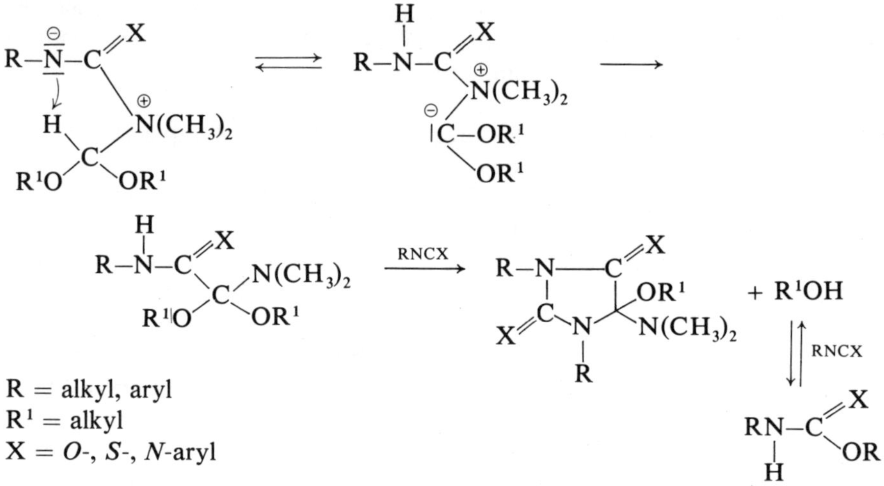

R = alkyl, aryl
R^1 = alkyl
X = O-, S-, N-aryl

In isothiocyanates containing bulky R groups, the attack of a second molecule of isothiocyanate is more difficult for steric reasons. In these cases, the interchange of alkyl groups, normally a slow reaction, predominates (101):

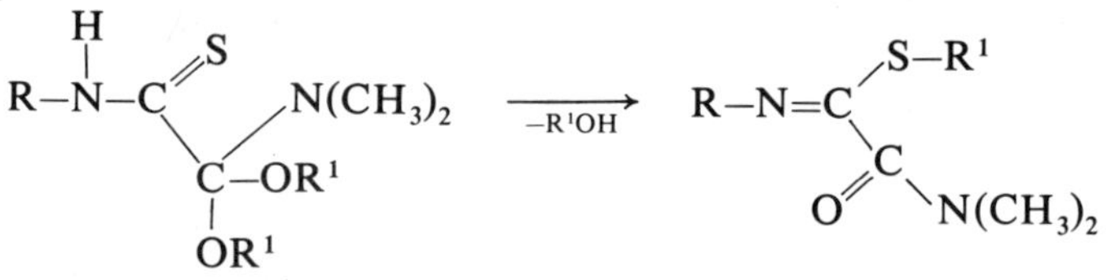

2. With N-Haloacylamides

If an *N*-haloacylamide is added to excess dimethylformamide dialkyl acetal at 0°C, adducts precipitate and are then exothermically transformed into acylisocyanate *O,N*-acetals at room temperature (see Table XXXII). *N*-Acyl-*N',N'*-dimethylurea acetals may be regarded as the intermediates, since they can be isolated from reactions of **1a**.

In addition to orthoformic acid trialkyl esters, tetramethylammonium halides are formed in reactions of **1a**; tetramethylformamidinium halides are formed with **1b** (168, 169). During this reaction, hydrogen chloride is formally

IV. REACTIONS OF ORTHOAMIDES WITH ELECTROPHILES

TABLE XXXII

Acylisocyanate O,N-Acetals from Dimethylformamide Dialkyl Acetals and N-Haloacylamides (169)

RCOHNX		Dimethylformamide dialkyl acetal	Acylisocyanate O,N-acetal yield, %
R	X		
CH_3	Cl	1b	69
CH_3	Br	1b	69
C_2H_5	Cl	1b	39
C_2H_5	Br	1b	61
n-C_3H_7	Cl	1b	53
i-C_3H_7	Cl	1b	41
i-C_3H_7	Br	1b	57
CH_3	Cl	DMF di-n-propyl acetal	64
CH_3	Cl	1b	70[a]
C_6H_5	Br	1b	54
NH_2	Cl	1b	40
$C_6H_5CH_2NH$	Cl	1b	22
$N(CH_3)_2$	Cl	1b	42
OC_2H_5	Cl	1b	76

[a] N-Acetyl-N,N'-dimethylurea dimethyl acetal.

generated and consumes some of the orthoamide. However, triethylamine has been used effectively as a base in this reaction (169).

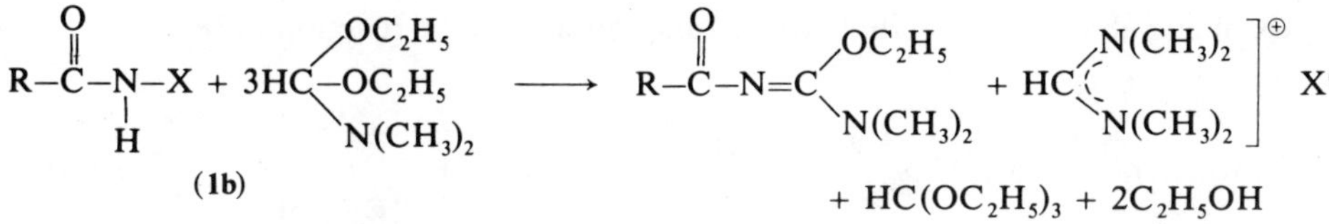

$$R-\overset{O}{\underset{H}{C}}-N-X + 3HC\overset{OC_2H_5}{\underset{N(CH_3)_2}{\diagdown OC_2H_5}} \longrightarrow R-\overset{O}{C}-N=C\overset{OC_2H_5}{\diagdown N(CH_3)_2} + \left[HC\overset{N(CH_3)_2}{\diagdown N(CH_3)_2}\right]^{\oplus} X^{\ominus}$$

(1b)

$$+ HC(OC_2H_5)_3 + 2C_2H_5OH$$

R = alkyl, aryl, NH_2, NH-alkyl, N(alkyl)$_2$, O-alkyl

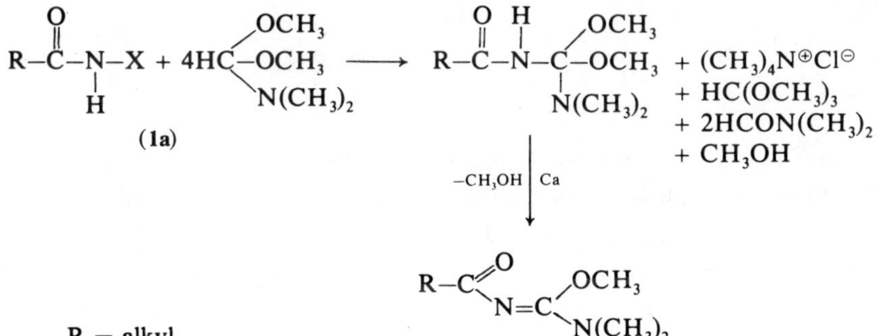

$$R-\overset{O}{\underset{H}{C}}-N-X + 4HC\overset{OCH_3}{\underset{N(CH_3)_2}{\diagdown OCH_3}} \longrightarrow R-\overset{O}{C}-\overset{H}{N}-\overset{}{\underset{N(CH_3)_2}{C}}-OCH_3 + (CH_3)_4N^{\oplus}Cl^{\ominus}$$

(1a)

$$+ HC(OCH_3)_3$$
$$+ 2HCON(CH_3)_2$$
$$+ CH_3OH$$

$$\downarrow -CH_3OH \mid Ca$$

$$R-C\overset{O}{\underset{N=C}{\diagdown}}\overset{OCH_3}{\diagdown N(CH_3)_2}$$

R = alkyl
X = Cl, Br

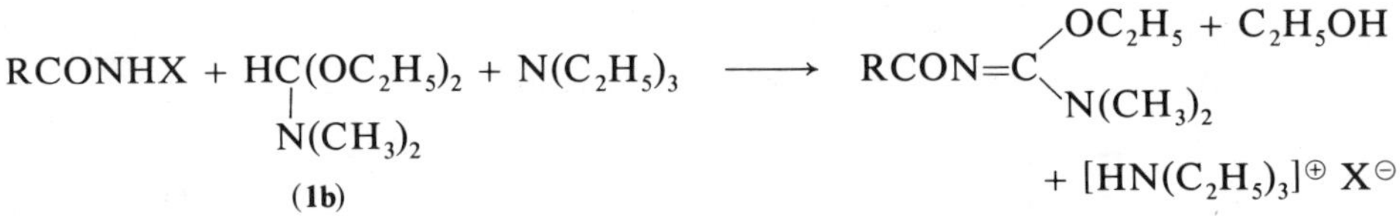

$$\text{RCONHX} + \underset{\underset{\text{(1b)}}{|}}{\text{HC}(\text{OC}_2\text{H}_5)_2} + \text{N}(\text{C}_2\text{H}_5)_3 \longrightarrow \text{RCON}=\text{C}\begin{array}{l}\diagup \text{OC}_2\text{H}_5 \\ \diagdown \text{N}(\text{CH}_3)_2\end{array} + \text{C}_2\text{H}_5\text{OH}$$
$$\text{N}(\text{CH}_3)_2$$
$$+ [\text{HN}(\text{C}_2\text{H}_5)_3]^{\oplus}\ X^{\ominus}$$

Finally, the synthesis of acylisocyanate O,N-acetals is successful if the dimethylformamide–dimethyl sulfate adduct and N-haloacylamides are reacted in the presence of a sodium alkoxide and triethylamine without prior isolation of the orthoamides **1** (169).

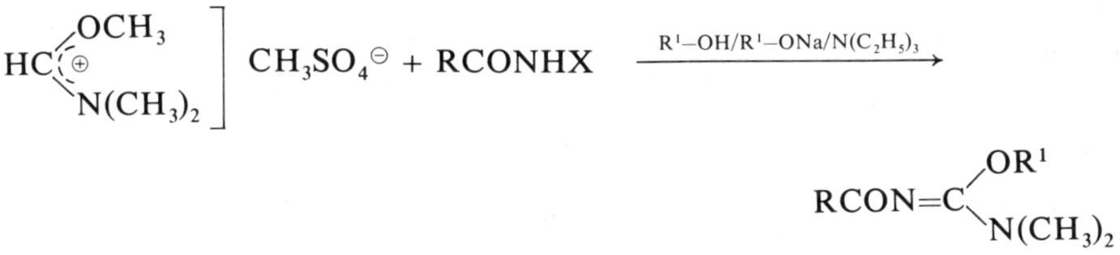

$$\left[\text{HC}\overset{\oplus}{\diagup}\begin{array}{l}\text{OCH}_3\\ \text{N}(\text{CH}_3)_2\end{array}\right] \text{CH}_3\text{SO}_4^{\ominus} + \text{RCONHX} \xrightarrow{\text{R}^1-\text{OH}/\text{R}^1-\text{ONa}/\text{N}(\text{C}_2\text{H}_5)_3}$$

$$\text{RCON}=\text{C}\begin{array}{l}\diagup \text{OR}^1 \\ \diagdown \text{N}(\text{CH}_3)_2\end{array}$$

Acylisocyanate N,N-acetals are obtained from orthoamides **5** in low yields.

Tetramethylformamidinium halides are either the main products, or the only isolable product in reactions with **7a** (169).

Sodium salts of N-halosulfonamides apparently do not form adducts. The reaction yields N-arylsulfonylformamidines as well as aldehydes. Here the oxidizing ability of the N-haloamide function directs the reaction (138):

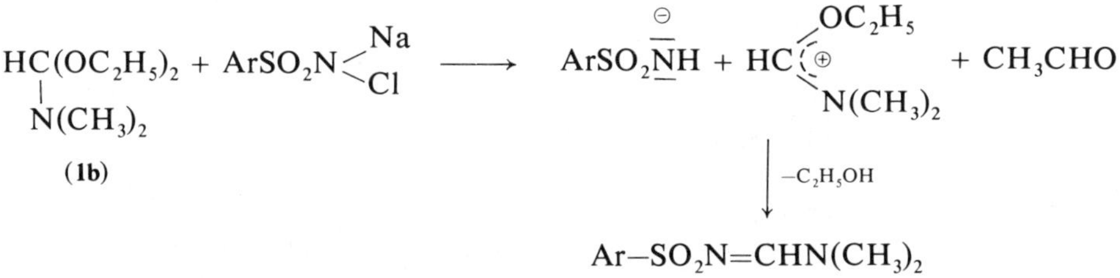

$$\underset{\underset{\text{(1b)}}{|}}{\text{HC}(\text{OC}_2\text{H}_5)_2} + \text{ArSO}_2\text{N}\begin{array}{l}\diagup \text{Na}\\ \diagdown \text{Cl}\end{array} \longrightarrow \text{ArSO}_2\underline{\overset{\ominus}{\text{N}}}\text{H} + \text{HC}\overset{\oplus}{\diagup}\begin{array}{l}\text{OC}_2\text{H}_5\\ \text{N}(\text{CH}_3)_2\end{array} + \text{CH}_3\text{CHO}$$
$$\text{N}(\text{CH}_3)_2$$

$$\downarrow {-\text{C}_2\text{H}_5\text{OH}}$$

$$\text{Ar}-\text{SO}_2\text{N}=\text{CHN}(\text{CH}_3)_2$$

(a) *Reaction Mechanism* (89). Isolation and characterization of an N-adduct has been accomplished, and the following structure was assigned:

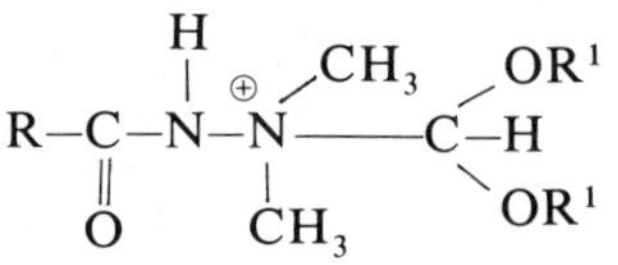

$$\underset{\text{O}}{\overset{\text{H}}{\underset{\|}{\text{R}-\text{C}}}}-\overset{\oplus}{\text{N}}-\underset{\text{CH}_3}{\overset{\text{CH}_3}{\text{N}}}-\text{C}\begin{array}{l}\diagup\text{OR}^1\\ -\text{H}\\ \diagdown\text{OR}^1\end{array}$$

IV. REACTIONS OF ORTHOAMIDES WITH ELECTROPHILES

The acidity of the formyl proton was proved by H–D exchange (−5°C, 1 min), and also by deprotonation (sodium alkoxide) and subsequent alkylation. This is a proof for ylide formation in basic media:

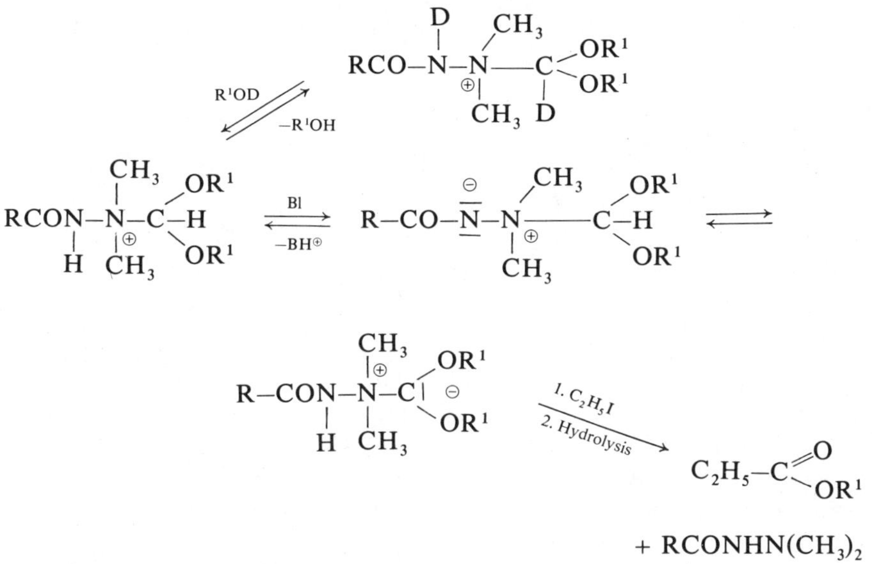

$$\text{RCON}-\underset{H}{\overset{CH_3}{N}}-\underset{CH_3}{\overset{OR^1}{C}}-H \underset{-BH^{\oplus}}{\overset{B|}{\rightleftharpoons}} R-CO-\underset{CH_3}{\overset{CH_3}{N}}-\underset{\oplus}{\overset{\ominus}{N}}-\underset{OR^1}{\overset{OR^1}{C}}-H \rightleftharpoons$$

$$\underset{R^1OD}{\overset{-R^1OH}{\rightleftharpoons}} \quad RCO-\underset{\oplus}{\overset{D}{N}}-\underset{CH_3}{\overset{CH_3}{N}}-\underset{D}{\overset{OR^1}{C}}-OR^1$$

$$R-CON-\underset{H}{\overset{CH_3}{\overset{|\oplus}{N}}}-\underset{CH_3}{\overset{OR^1}{C}}\ominus \xrightarrow{\substack{1.\ C_2H_5I \\ 2.\ Hydrolysis}} C_2H_5-C\overset{O}{\underset{OR^1}{\diagdown}}$$

$$+ \ RCONHN(CH_3)_2$$

The C–N bond formation finally occurs via an intramolecular S_E1 or S_N mechanism. The S_E1 mechanism via separated ion pairs was ruled out by crossover experiments.

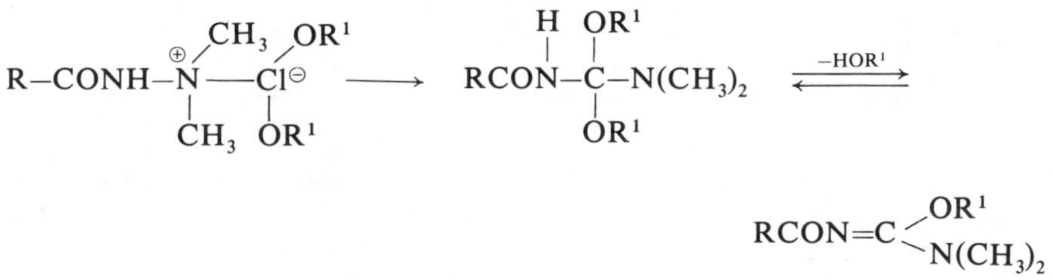

$$R-CONH-\overset{\oplus}{\underset{CH_3}{\overset{CH_3}{N}}}-\underset{OR^1}{\overset{OR^1}{C}}Cl^{\ominus} \longrightarrow RCON-\underset{OR^1}{\overset{H}{\underset{|}{C}}}-N(CH_3)_2 \xrightarrow{-HOR^1} \rightleftharpoons$$

$$RCON=C\overset{OR^1}{\underset{N(CH_3)_2}{\diagdown}}$$

3. With Nitrenes, Carbenes, Sulfur, and Selenium

Nitrenes photolytically generated from carboxylic acid azides may be trapped by **1a**. Addition to the dimethylamino group gives rise to an intermediate, which under these reaction conditions, yields a carboxylic acid hydrazide and dimethoxycarbene, which reacts further (145) (see Section IV-A-3a).

Phenyldiazomethane does not react with **1a**, even at 100°C. Benzaldazin was found as decomposition product (141). The reaction of phenyldiazomethane and **5c** contrasts by yielding a mixture of bis(dimethylamino)-alkenes, the formation of which is not necessarily due to reaction of the phenylcarbene with **5c** (141, 170) (see Section IV-C-4b). The reaction of **1b** with sulfur (selenium) yields thio(seleno)urethanes:

$$\underset{\mathbf{1b}}{\text{H}-\overset{\text{OC}_2\text{H}_5}{\underset{\text{N(CH}_3)_2}{\text{C}-\text{OC}_2\text{H}_5}}} \xrightarrow[-\text{C}_2\text{H}_5\text{OH}]{\text{X}} \text{X}=\underset{\text{N(CH}_3)_2}{\overset{\text{OC}_2\text{H}_5}{\text{C}}}$$

$$\text{X}=\text{S, Se}$$

4. With Carbonyl Compounds and Their Derivatives

(a) *Schiff Bases of Aromatic Aldehydes.* Aminal esters (**5**) give arylamino-2,2-bis(dimethylamino)ethylenes with Schiff bases of aromatic aldehydes at 160–170°C in good yields (136,170) (see Table XXXIII). The reaction is strongly influenced by substituents: +I (+M) donor substituents in the $o(p)$ position of Ar or Ar' slow down the reaction; bulky substituents in the o position have the same effect. −I (−M) substituents increase the reactivity.

$$\underset{\text{Ar}^1}{\overset{\text{Ar}-\text{N}}{\diagdown}}\text{C}-\text{H} + \text{HC}\underset{\text{N(CH}_3)_2}{\overset{\text{OC(CH}_3)_3}{\diagdown}}\text{N(CH}_3)_2 \xrightarrow{-(\text{CH}_3)_3\text{COH}} \underset{\text{Ar}^1}{\overset{\text{ArNH}}{\diagdown}}\text{C}=\text{C}\underset{\text{N(CH}_3)_2}{\overset{\text{N(CH}_3)_2}{\diagdown}}$$

(**5c**)

The weakly electrophilic Schiff bases from aliphatic amines and aromatic

TABLE XXXIII

1-Aryl-1-arylamino-2,2-bis(dimethylamino)ethylenes from **5c** and Schiff Bases (136, 170)

Ar	Ar¹	Reaction conditions	Product yield, %
C_6H_5	C_6H_5	160°C, 4 hr	65
4-CH_3-C_6H_4	C_6H_5	160°C, 5 hr	45
3-CH_3-C_6H_4	C_6H_5	160°C, 5 hr	55
2-CH_3-C_6H_4	C_6H_5	160°C, 5 hr	35
4-Cl-C_6H_4	C_6H_5	150°C, 5 hr	70
3-Cl-C_6H_4	C_6H_5	150°C, 5 hr	65
2-Cl-C_6H_4	C_6H_5	130°C, 5 hr	10
C_6H_5	4-$CH_3C_6H_4$	160°C, 6 hr	40
C_6H_5	4-ClC_6H_4	160°C, 4 hr	55
C_6H_5	2-Cl-C_6H_4	160°C, 4 hr	55
C_6H_5	$C_6H_5CH=CH$	140°C, 6 hr	50
C_6H_5	2-Pyridyl	150°C, 4 hr	65

IV. REACTIONS OF ORTHOAMIDES WITH ELECTROPHILES   515

aldehydes do not react at all. In the first reaction step **5c** and the Schiff base are supposed to form an intermediate of the following structure:

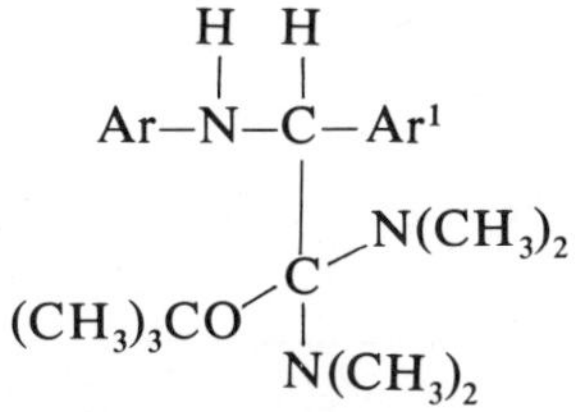

$$\begin{array}{c} H \quad H \\ | \quad\; | \\ Ar-N-C-Ar^1 \\ | \\ (CH_3)_3CO-C-N(CH_3)_2 \\ | \\ N(CH_3)_2 \end{array}$$

This intermediate may be formed either by addition of bis(dimethylamino)-carbene (from **5c**) to the iminocarbonyl bond, or by adduct formation at one of the dimethylamino groups of **5c** and subsequent Stevens rearrangement. Since the reaction does not proceed in polar solvents (DMF), but does so rapidly in solvents of low dielectric constant, we prefer the adduct mechanism (136).

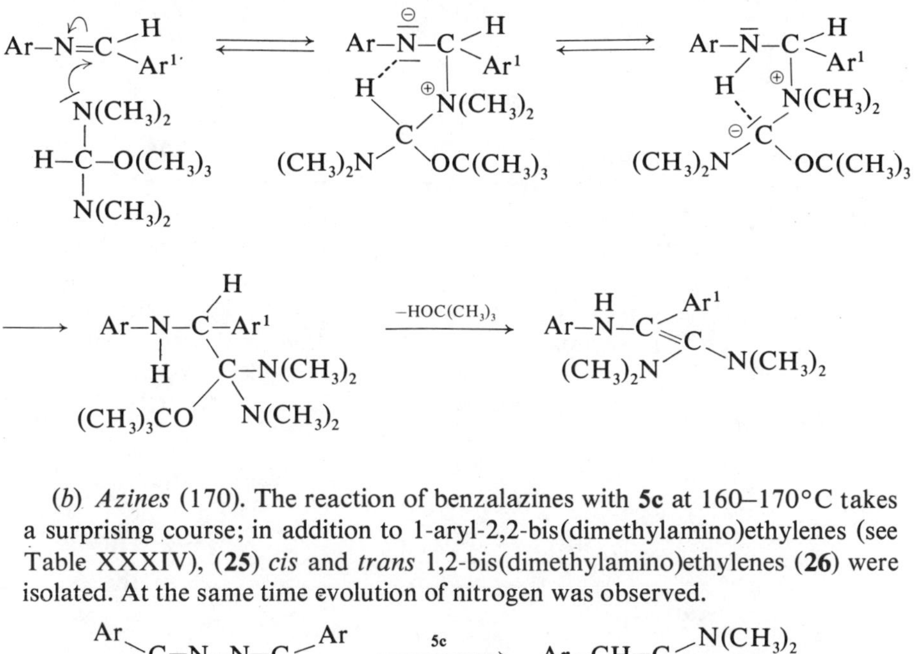

$$Ar-N=C\begin{matrix}H\\Ar^1\end{matrix} \quad \rightleftarrows \quad Ar-\bar{N}-C\begin{matrix}H\\Ar^1\end{matrix} \quad \rightleftarrows \quad Ar-\bar{N}-C\begin{matrix}H\\Ar^1\end{matrix}$$

with substituents $N(CH_3)_2$, $H-C-O(CH_3)_3$, $N(CH_3)_2$ on left; middle: H, $\overset{\oplus}{C}$, $(CH_3)_2N$, $OC(CH_3)_3$; right: H, $\overset{\ominus}{C}$, $(CH_3)_2N$, $OC(CH_3)_3$.

$$\longrightarrow \quad \begin{matrix} H \\ | \\ Ar-N-C-Ar^1 \\ | \quad\; \backslash \\ H \quad\; C-N(CH_3)_2 \\ \quad (CH_3)_3CO \quad N(CH_3)_2 \end{matrix} \quad \xrightarrow{-HOC(CH_3)_3} \quad Ar-N-C\begin{matrix}H \quad Ar^1\\ \diagdown\; \diagup \\ C \\ \diagup\; \diagdown \\(CH_3)_2N \quad N(CH_3)_2\end{matrix}$$

(b) *Azines* (170). The reaction of benzalazines with **5c** at 160–170°C takes a surprising course; in addition to 1-aryl-2,2-bis(dimethylamino)ethylenes (see Table XXXIV), (**25**) *cis* and *trans* 1,2-bis(dimethylamino)ethylenes (**26**) were isolated. At the same time evolution of nitrogen was observed.

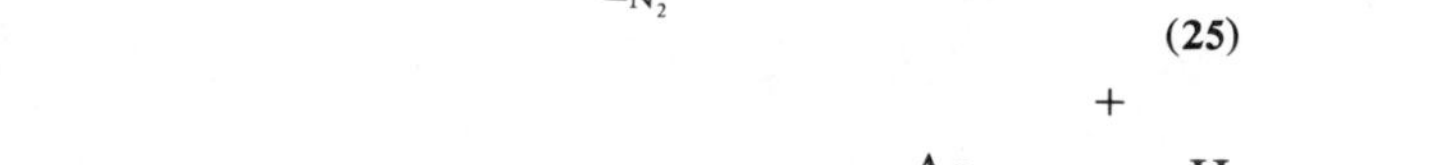

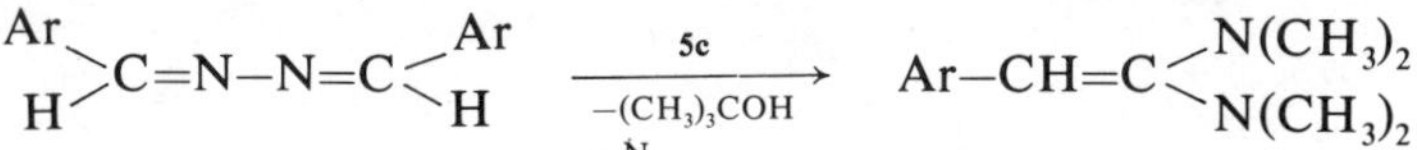

$$\begin{matrix}Ar\\ \diagdown \\ H\end{matrix} C=N-N=C \begin{matrix}Ar\\ \diagup \\ H\end{matrix} \quad \xrightarrow[-N_2]{\underset{-(CH_3)_3COH}{5c}} \quad Ar-CH=C\begin{matrix}N(CH_3)_2\\ \\ N(CH_3)_2\end{matrix}$$

$$(25)$$

$$+$$

$$\begin{matrix}Ar\\ \diagdown \\ (CH_3)_2N\end{matrix} C=C \begin{matrix}H\\ \diagup \\ N(CH_3)_2\end{matrix}$$

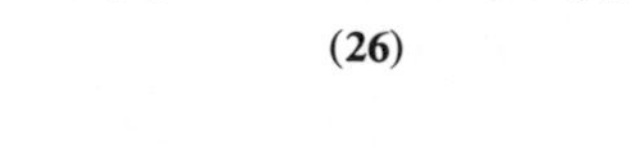

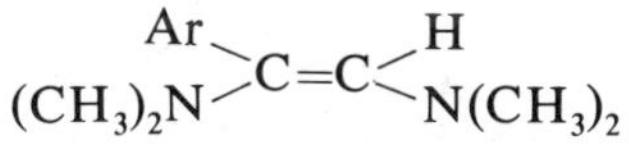

$$(26)$$

TABLE XXXIV

1-Aryl-2,2-bis(dimethylamino)ethylenes (**25**) from **5c** and Azines[a] (170)

Starting material

$$\text{ArCH=N-N=C} \begin{smallmatrix} \text{Ar}^1 \\ \text{R} \end{smallmatrix}$$

Ar	Ar^1	R	Ar–CH=C[N(CH$_3$)$_2$]$_2$ yield, %
2-pyridyl	2-pyridyl	H	32
2-furyl	2-furyl	H	33
C$_6$H$_5$	C$_6$H$_5$	C$_6$H$_5$	90
4-CH$_3$C$_6$H$_4$	C$_6$H$_5$	C$_6$H$_5$	81
4-CH$_3$OC$_6$H$_4$	C$_6$H$_5$	C$_6$H$_5$	88
4-(CH$_3$)$_2$NC$_6$H$_4$	C$_6$H$_5$	C$_6$H$_5$	69
2-naphthyl	C$_6$H$_5$	C$_6$H$_5$	77
C$_6$H$_5$CH=CH	C$_6$H$_5$	C$_6$H$_5$	60

[a] At 160–170°C for 4–6 hr.

Compounds (**26**) are not formed if azines from furfural or pyridine-2-carboxaldehyde are used. Moreover, 1-aryl-2,2-bis(dimethylamino)ethylenes (**25**) are the sole products if azines from aromatic aldehydes and benzophenone are reacted with **5c**. The reaction course may be rationalized by invoking an elimination of aryldiazomethane from an intermediate to yield a ketenaminal:

$$\begin{smallmatrix}\text{Ar}\\\text{H}\end{smallmatrix}\!\!>\!\!\text{C=N-N=C(C}_6\text{H}_5)_2 \xrightarrow[-(\text{CH}_3)_3\text{COH}]{\textbf{5c}} \begin{smallmatrix}\text{Ar}\\\text{H}\end{smallmatrix}\!\!>\!\!\text{C=C}\!\!<\!\!\begin{smallmatrix}\text{N(CH}_3)_2\\\text{N(CH}_3)_2\end{smallmatrix}$$
(**25**)

$$\text{Ar-}\underset{\underset{\underset{(\text{CH}_3)_3\text{C-O}}{|}}{\underset{\text{N(CH}_3)_2}{\text{C-N(CH}_3)_2}}}{\overset{\overset{\text{H}}{|}}{\text{C}}}\text{-}\underset{}{\overset{\text{H}}{\text{N}}}\text{-N=C}\!<\!\begin{smallmatrix}\text{Ar}\\\text{R}\end{smallmatrix} \xrightarrow{-(\text{CH}_3)_3\text{COH}} \text{Ar-}\underset{\underset{(\text{CH}_3)_2\text{N}}{\overset{\overset{\text{C}}{\parallel}}{}}\text{N(CH}_3)_2}{\overset{\overset{\text{H}}{|}}{\text{C}}}\text{-N-N=C}\!<\!\begin{smallmatrix}\text{Ar}\\\text{R}\end{smallmatrix}$$

$$\rightleftharpoons \text{Ar-}\underset{\underset{(\text{CH}_3)_2\overset{\oplus}{\text{N}}}{\overset{\overset{\text{C}}{}}{}}\text{N(CH}_3)_2}{\overset{\overset{\text{H}}{|}}{\text{C}}}\text{-}\overset{\ominus}{\text{N}}\text{-N=C}\!<\!\begin{smallmatrix}\text{Ar}\\\text{R}\end{smallmatrix} \longrightarrow \begin{smallmatrix}\text{Ar}\\\text{H}\end{smallmatrix}\!\!>\!\!\text{C=C}\!<\!\begin{smallmatrix}\text{N(CH}_3)_2\\\text{N(CH}_3)_2\end{smallmatrix}$$
(**25**)

R = H, Ar

$$+ \text{N}_2\text{C}\!<\!\begin{smallmatrix}\text{Ar}\\\text{R}\end{smallmatrix}$$

IV. REACTIONS OF ORTHOAMIDES WITH ELECTROPHILES

If thermal decomposition is not too rapid, the aryldiazomethanes react partially with **5c** giving *cis* and *trans* 1,2-bis(dimethylamino)ethylenes, which could be obtained also by direct reaction of aryldiazomethane with **5c** (141).

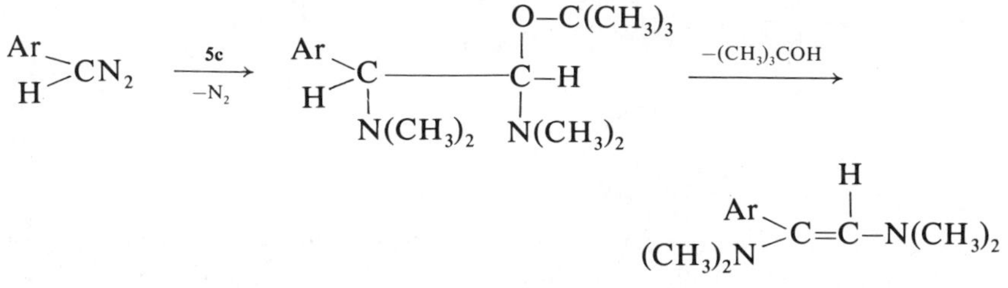

$$\underset{H}{\overset{Ar}{>}}CN_2 \xrightarrow[-N_2]{5c} \underset{H}{\overset{Ar}{>}}\underset{N(CH_3)_2}{\overset{}{C}}-\underset{N(CH_3)_2}{\overset{O-C(CH_3)_3}{\overset{|}{C}}}-H \xrightarrow{-(CH_3)_3COH}$$

$$\underset{(CH_3)_2N}{\overset{Ar}{>}}C=\overset{H}{\underset{}{\overset{|}{C}}}-N(CH_3)_2$$

cis, trans

Diphenyldiazomethane (from arylidenediphenylmethylenehydrazines) is inert to **5c** (141). The thermal decomposition gives mainly tetraphenylethylene (170).

$$2(C_6H_5)_2CN_2 \xrightarrow[-N_2]{5c} (C_6H_5)_2C=C(C_6H_5)_2$$

(c) *Aromatic Aldehydes*. The reactions of **5c** with aromatic aldehydes generally yield a variety of products that are difficult to separate. From the reactions of *p*-tolyl- and anisaldehyde, the products could be isolated in 30% yield (5a,171). Further investigations showed that an aminal of an arylglyoxal is initially formed and then undergoes disproportionation:

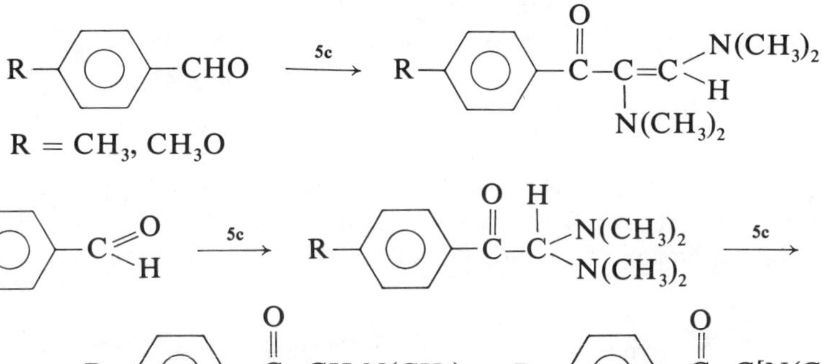

$$R-\langle O \rangle-CHO \xrightarrow{5c} R-\langle O \rangle-\overset{O}{\overset{\|}{C}}-\underset{N(CH_3)_2}{\overset{|}{C}}=C\underset{H}{\overset{N(CH_3)_2}{<}}$$

R = CH$_3$, CH$_3$O

$$R-\langle O \rangle-C\overset{O}{\underset{H}{\lessgtr}} \xrightarrow{5c} R-\langle O \rangle-\overset{O}{\overset{\|}{C}}-\overset{H}{\underset{}{\overset{|}{C}}}\overset{N(CH_3)_2}{\underset{N(CH_3)_2}{<}} \xrightarrow{5c}$$

$$R-\langle O \rangle-\overset{O}{\overset{\|}{C}}-CH_2N(CH_3)_2 + R-\langle O \rangle-\overset{O}{\overset{\|}{C}}-C[N(CH_3)_2]_3$$

The dimethylaminomethyl aryl ketone reacts with **5c** to give the final product; the acyltris(dimethylamino)methane was characterized by its hydrolysis product.

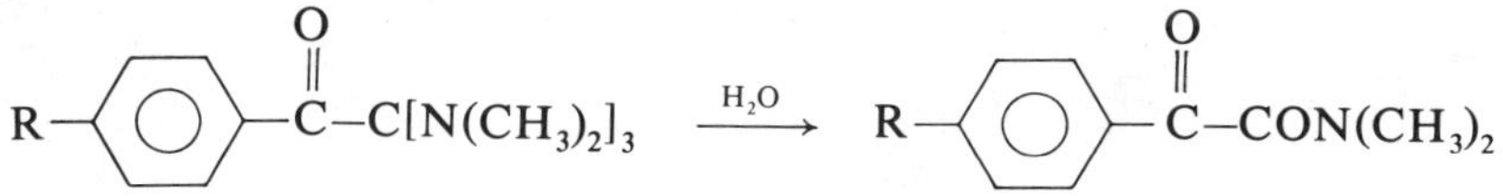

$$R-\langle O \rangle-\overset{O}{\overset{\|}{C}}-C[N(CH_3)_2]_3 \xrightarrow{H_2O} R-\langle O \rangle-\overset{O}{\overset{\|}{C}}-CON(CH_3)_2$$

1a reacts with 4-chlorobenzaldehyde with 18% yield (141).

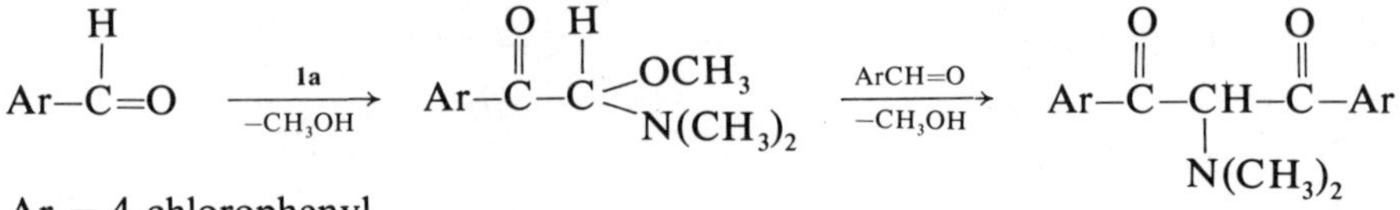

$$\text{Ar-CH=O} \xrightarrow[-\text{CH}_3\text{OH}]{\mathbf{1a}} \text{Ar-}\underset{\text{H}}{\overset{\text{O}}{\text{C}}}-\underset{\text{H}}{\text{C}}\underset{\text{N(CH}_3)_2}{\overset{\text{OCH}_3}{<}} \xrightarrow[-\text{CH}_3\text{OH}]{\text{ArCH=O}} \text{Ar-}\overset{\text{O}}{\overset{\|}{\text{C}}}-\underset{\text{N(CH}_3)_2}{\text{CH}}-\overset{\text{O}}{\overset{\|}{\text{C}}}-\text{Ar}$$

Ar = 4-chlorophenyl

5. H–D Exchange at Dialkylformamide Dialkyl Acetals

Acyclic *N,N*-dialkylformamide acetals (**1**) exchange their formyl protons with deuterium in *O*-deutero alcohol, usually within a few minutes at room temperature (172,173). The rate of the exchange decreases with the increase of the steric requirements of the alkoxy groups and with the increasing pK_a value of the alcohol. The basicity of the nitrogen atom as well as the size of the *N*-alkyl groups significantly influence the reaction rate. If the alkyl groups are comparable in size, the fastest exchange occurs at the most basic acetal. An increase in the basicity of the reaction media gives rise to a decrease of the reaction rate. These results as well as kinetic data are in accordance with an ylide mechanism for the exchange reaction. Initially, an acid–base equilibrium is established. The more acidic the alcohol, the more the equilibrium is shifted to the side of the ammonium compound.

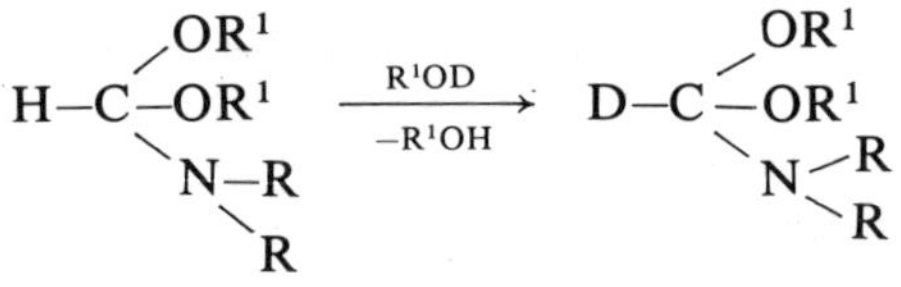

$$\text{H-C}\begin{smallmatrix}\text{OR}^1\\-\text{OR}^1\\\text{N-R}\\\text{R}\end{smallmatrix} \xrightarrow[-\text{R}^1\text{OH}]{\text{R}^1\text{OD}} \text{D-C}\begin{smallmatrix}\text{OR}^1\\-\text{OR}^1\\\text{N}<\text{R}\\\text{R}\end{smallmatrix}$$

(**1**)

In the *N*-protonated acetal the formyl hydrogen can be replaced by deuterium. The exchange rate increases with the acidity of the alcohol and the basicity of the acetal, as mentioned previously. Foreign bases (triethylamine)

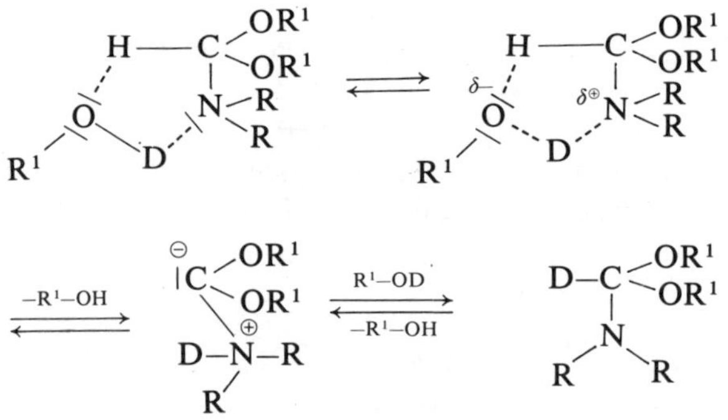

compete with the acetal for the deuteron, leading to a decrease of the reaction rate.

The alternative carbene mechanism can be ruled out, since the precursors of the alkoxy(dialkylamino)carbenes do not undergo H–D exchange (173).

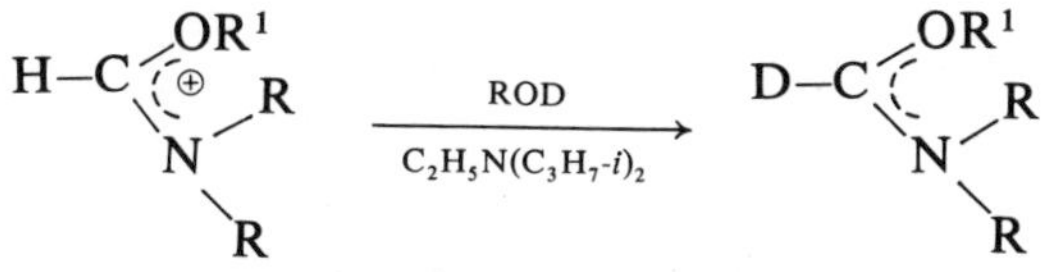

$$H-C{\overset{OR^1}{\underset{N(R)(R)}{\oplus R}}} \xrightarrow[C_2H_5N(C_3H_7\text{-}i)_2]{ROD} D-C{\overset{OR^1}{\underset{N(R)(R)}{R}}}$$

V. Elimination Reactions

A. 1,2-ELIMINATION

Homologues of dialkylformamide acetals (**1**) form a temperature-dependent equilibrium with ketene O,N-acetals. Dimethylacetamide acetals partially eliminate alcohols on distillation; the same holds true for acetals of higher carboxylic acid amides (4,95). Cyclic acetals are exceptions, of course.

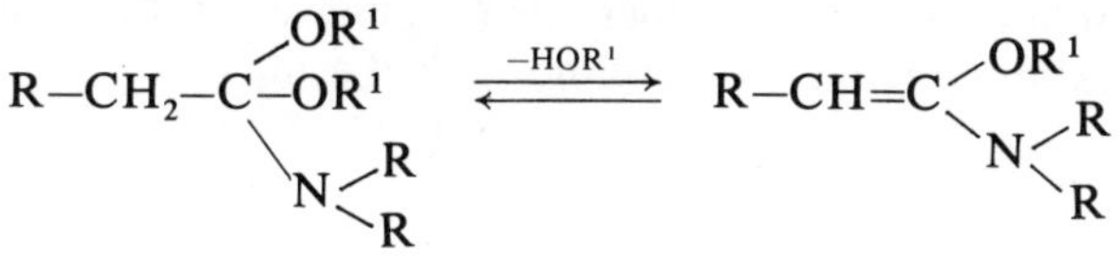

$$R-CH_2-C{\overset{OR^1}{\underset{N(R)(R)}{OR^1}}} \underset{}{\overset{-HOR^1}{\rightleftarrows}} R-CH=C{\overset{OR^1}{\underset{N(R)(R)}{}}}$$

(**1**)

The complete transformation into ketene O,N-acetals is accomplished by treatment with sodium (4), calcium (95), or sodium hydride (98) at reflux. If either O-alkylated N,N-dialkylcarboxylic acid amides with electron-withdrawing groups in the α-position or homologous amidinium salts are treated with alkoxides, ketene O,N-acetals (29) or aminals (95) are obtained respectively.

B. ELIMINATION OF DIMETHYLFORMAMIDE FROM CYCLIC DIMETHYLFORMAMIDE ACETALS

If cyclic dimethylformamide acetals are heated to 130–150°C for one or more days, they form epoxides (174). The cyclic acetals do not have to be isolated; it is sufficient to heat vicinal diols with **1a**. The mechanism of the reaction involves a ring opening and subsequent S_Ni substitution to give the epoxide.

Thus *trans*-cyclohexanediol yields *cis*-cyclohexene oxide, and *meso*-hydrobenzoin yields *trans*-stilbene oxide, whereas *cis*-cyclohexanediol and D,L-hydrobenzoin of course just form acetals.

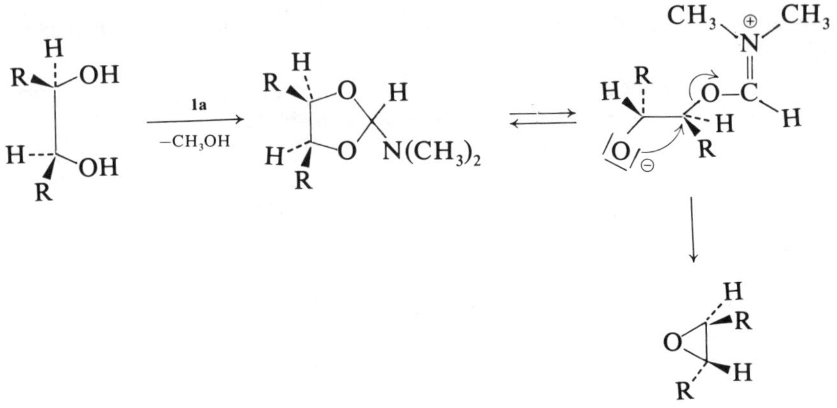

The acid-catalyzed elimination of DMF from 2-dimethylamino-4-alkylidene-1,3-dioxolane at room temperature is extremely interesting since it represents a new route to formerly inaccessible enols (175–177).

The rate of elimination increases with the acidity of the "acid." These acids include *t*-butanol, methanol, and acetic acid. Enols are obtained in highly concentrated solutions. The product distribution (enol or dienol) can be regulated by the amount and nature of the applied acid.

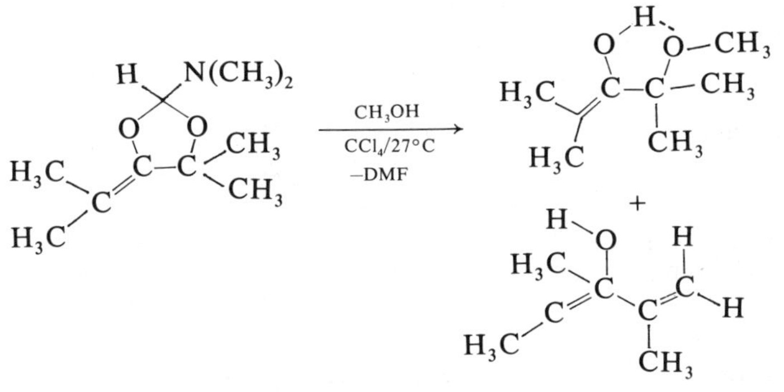

C. DECOMPOSITION OF CYCLIC DIMETHYLFORMAMIDE ACETALS TO ALKENES

It has been shown that the reaction of iso[thio]cyanates with 2-dimethylamino-4,5-tetramethyl-1,3-dioxolane yields tetramethylethylene. The intermediate was formulated as a cyclic dialkoxycarbene which fragments to carbon dioxide and the alkene (145,167).

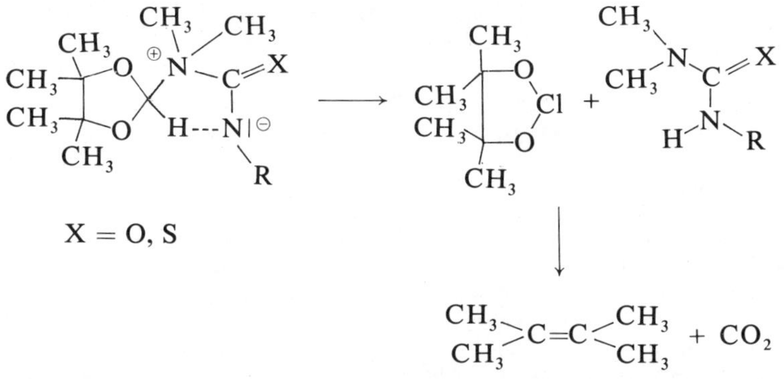

The fragmentation of 2-dimethylamino-*trans*-4,5-diphenyl-1,3-dioxolane with acetic anhydride at 165–180°C presumably proceeds via a similar mechanism. The *cis* compound yields predominantly the *cis* olefin (178).

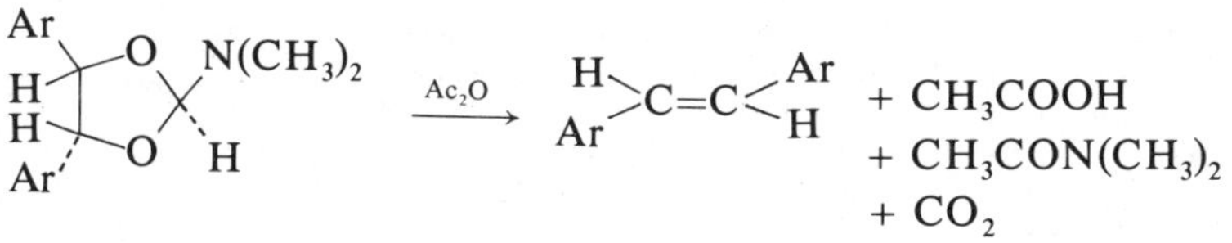

REFERENCES

1. R. Feinauer, *Synthesis*, **1971**, 16.
2. R. De Wolfe, *Organic Chemistry, Carboxylic Ortho Derivatives*, Academic Press, New York, 1969.
3. H. Bredereck, F. Effenberger, T. Brendle, and H. Muffler, *Chem. Ber.*, **101**, 1885 (1968); H. Muffler, Dissertation, University of Stuttgart, 1969.
4. H. Meerwein, W. Florian, N. Schön, and G. Stopp, *Ann. Chem.*, **641**, 1 (1961).
5. G. Simchen, H. Hoffmann, and H. Bredereck, *Chem. Ber.*, **101**, 51 (1968).
5a. H. Bredereck, G. Simchen, P. Horn, and R. Wahl, *Angew. Chem.*, **79**, 311 (1967).
6. P. Horn, Dissertation, University of Stuttgart, 1967.
7. R. Baur, Dissertation, University of Stuttgart, 1973.
8. V. G. Granik, M. K. Polievktov, and R. G. Glushkov, *Zh. Org. Khim.*, **7**, 1431 (1971).
9. W. Kantlehner, Dissertation, University of Stuttgart, 1968.
10. H. Brechbühler, H. Büchi, E. Hatz, J. Schreiber, and A. Eschenmoser, *Helv. Chim. Acta*, **48**, 1746 (1965).
11. J. Jaus, Dissertation, University of Stuttgart, 1972.
12. L. Kienitz, Dissertation, University of Stuttgart, 1973.
13. G. Simchen, unpublished results.
14. U. Dinkeldein, Dissertation, University of Stuttgart, 1973.
15. H. Bredereck, F. Effenberger, and H. J. Bredereck, *Angew. Chem.*, **78**, 984 (1966).
16. J. W. Scheeren and R. J. F. Nivard, *Rec. Trav. Chim. Pays-Bas*, **88**, 289 (1969).
17. H. W. Wanzlick, *Angew. Chem.*, **74**, 129 (1962).
18. H. E. Winberg, J. E. Carnahan, D. D. Coffmann, and M. Brown, *J. Am. Chem. Soc.*, **87**, 2055 (1965); H. E. Winberg, U.S. Pat 3,239,534 (1966).

19. M. Brown, U.S. Pat. 3,214,428 (1965); *C.A.*, **64**, 3501 (1966).
20. H. Bredereck, R. Gompper, H. Rempfer, H. Keck, and K. Klemm, *Angew. Chem.*, **70**, 269 (1958).
21. H. Bredereck, F. Effenberger, and A. Hofmann, *Chem. Ber.*, **96**, 3260 (1963).
22. H. Bredereck, R. Gompper, H. G. v. Schuh, and G. Theilig, *Angew. Chem.*, **71**, 753 (1959).
22a. H. Weingarten, *J. Am. Chem. Soc.*, **89**, 3713 (1967).
23. H. Bredereck, G. Simchen, and G. Beck, *Ann. Chem.*, **762**, 62 (1972).
24. H. Bredereck, G. Simchen, and H. Hoffmann, *Chem. Ber.*, **106**, 3725 (1973).
25. H. Bredereck, F. Effenberger, and H. J. Botsch, *Chem. Ber.*, **97**, 3397 (1964).
26. H. Bredereck, G. Simchen, and R. Wahl, *Chem. Ber.*, **101**, 4048 (1968).
26a. R. Wahl, Dissertation, University of Stuttgart, 1967.
27. Z. Arnold and M. Kornilov, *Collection Czech. Chem. Commun.*, **29**, 645 (1964).
28. H. Weber, Diplomarbeit University of Stuttgart, 1967.
28a. H. Bredereck, *Pharm. Ztg.*, **116**, 780 (1971).
29. H. Bredereck, W. Kantlehner, and D. Schweizer, *Chem. Ber.*, **104**, 3475 (1971).
30. H. Weingarten and N. K. Edelmann, *J. Org. Chem.*, **32**, 3293 (1967).
31. S. Kabuss, *Synthesis*, **1971**, 312.
32. H. Bredereck, G. Simchen, and B. Funke, *Chem. Ber.*, **104**, 2709 (1971).
33. G. Simchen, *Angew. Chem.*, **76**, 860 (1964).
34. H. Bredereck and G. Simchen, *Angew. Chem.*, **75**, 1102 (1963).
35. Th. Severin and B. Brück, *Chem. Ber.*, **98**, 3848 (1965).
36. Z. Arnold and A. Holý, *Collection Czech. Chem. Commun.*, **28**, 2040 (1963).
37. J. Gloede, L. Haase, and H. Gross, *Z. Chem.*, **9**, 201 (1969).
38. B. Föhlisch, *Chem. Ber.*, **104**, 348 (1971).
39. H. Bredereck, G. Simchen, and W. Griebenow, *Chem. Ber.*, **106**, 3732 (1973).
40. W. Griebenow, Dissertation, University of Stuttgart, 1969.
41. R. F. Borch, C. V. Grudzinskas, D. A. Peterson, and L. D. Weber, *J. Org. Chem.*, **37**, 1141 (1972).
42. E. Wenkert, E. B. Spitzner, and R. L. Webb, *Australian J. Chem.*, **25**, 433 (1972).
43. B. M. Trost and M. Preckel, *J. Am. Chem. Soc.*, **95**, 7862 (1973).
44. M. C. Seidel, *J. Org. Chem.*, **37**, 600 (1972).
45. E. E. Kilbourn and M. C. Seidel, *J. Org. Chem.*, **37**, 1145 (1972).
46. T. Brendle, Dissertation, University of Stuttgart, 1967.
47. M. Weigele, S. L. De Bernardo, J. P. Tengi, and W. Leimgruber, *J. Am. Chem. Soc.*, **94**, 5927 (1972).
48. J. A. Ivanova, B. P. Fedorov, and F. W. Stojanovic, *Nachr. Akad. Wiss. UdSSR, Abt. Chem. Wiss.*, **1965**, 2179; *C.A.*, **64**, 12538 (1966).
49. T. Mukayama and T. Yamaguchi, *Bull. Chem. Soc. Japan*, **39**, 2005 (1966).
50. T. Yamaguchi and T. Mukayama, *Bull. Chem. Soc. Japan*, **40**, 1952 (1967).
51. J. Kapasakalidis, Dissertation, University of Stuttgart, 1972.
52. H. Bredereck, G. Simchen, and P. Horn, *Chem. Ber.*, **103**, 210 (1970).
53. G. Simchen, Dissertation, University of Stuttgart, 1962.
54. H. Bredereck, F. Effenberger and H. J. Treiber, *Chem. Ber.*, **96**, 1505 (1963).
55. H. Bredereck, G. Simchen, and H. Traut, *Chem. Ber.*, **98**, 3883 (1965).
56. H. Saur, Dissertation, University of Stuttgart, 1971.
57. H. Bredereck, G. Simchen, and W. Griebenow, *Chem. Ber.*, **107**, 1545 (1974).
58. F. Wagner, Dissertation, University of Stuttgart, 1974.
59. B. Funke, Diplomarbeit, University of Stuttgart, 1967.
60. H. Bredereck, F. Effenberger, and T. Brendle, *Angew. Chem.*, **78**, 147 (1966).
61. A. D. Batcho and W. Leimgruber, Ger. Offen. 2057840 (1971); C.A., 75, 63605 v (1971).

62. V. Granik, A. N. Akalaev, and R. G. Glushkov, *Zh. Org. Khim.*, **7**, 1146 (1971).
63. V. G. Granik, A. N. Akelaev, and R. G. Glushkov, *Zh. Org. Khim.*, **7**, 2429 (1971).
64. V. G. Granik, A. G. Sukhoruskin, N. S. Kuryatov, V. P. Pakhomov, and R. G. Glushkov, *Khim. Geterotsikl. Soedin.*, **1973**, 954.
65. N. Schön, Dissertation, University of Marburg, 1959.
66. G. Stopp, Dissertation, University of Marburg, 1959.
67. H. Riese, Dissertation, University of Stuttgart, 1969.
68. H. Eilingsfeld, M. Seefelder, and H. Weidinger, *Angew. Chem.*, **72**, 836 (1960).
69. H. Bredereck, G. Simchen, and W. Kantlehner, *Chem. Ber.*, **104**, 932 (1971).
70. H. Bredereck, F. Effenberger, and A. Hofmann, *Angew. Chem.*, **75**, 825 (1963).
71. H. Bredereck, F. Effenberger, and A. Hofmann, *Angew. Chem.*, **97**, 61 (1964).
72. H. Bredereck, F. Effenberger, and M. Hajek, *Chem. Ber.*, **101**, 3062 (1968).
72a. H. Bredereck, F. Effenberger, and M. Hajek, *Chem. Ber.*, **98**, 3178 (1965).
73. H. E. Winberg, U.S. Pat. 3,121,084 (1964); *C.A.*, **60**, 13197 (1964).
74. J. Zemlicka, S. Chladek, A. Holý, and J. Smrt, *Collection Czech. Chem. Commun.*, **31**, 3198 (1966).
75. J. Zemlicka and A. Holý, *Collection Czech. Chem. Commun.*, **32**, 3159 (1967).
75a. J. Smrt, *Collection Czech. Chem. Commun.*, **32**, 3380 (1967).
75b. J. Smrt, *Collection Czech. Chem. Commun.*, **32**, 3958 (1967).
75c. J. Smrt, *Collection Czech. Chem. Commun.*, **33**, 1462 (1968).
75d. A. Holý, J. Smrt, and F. Sorm, *Collection Czech. Chem. Commun.*, **32**, 2980 (1967).
75e. A. Holy, S. Chladek, and J. Zemlicka, *Collection Czech. Chem. Commun.*, **34**, 253 (1969).
76. T. Brendle, Diplomarbeit University of Stuttgart, 1964.
77. R. Richter and H. Ulrich, *Chem. Ber.*, **103**, 3525 (1970).
78. P. Henklein and G. Westphal, *Z. Chem.*, **12**, 103 (1972).
79. R. Hildebrand, Diplomarbeit University of Stuttgart, 1967.
80. H. Gold, *Angew. Chem.*, **72**, 956 (1960).
81. W. Bauer, Dissertation, University of Stuttgart, 1962.
82. H. Bredereck, G. Simchen, and H. U. Schenck, *Chem. Ber.*, **101**, 3058 (1968).
83. H. Porkert, Dissertation, University of Stuttgart, 1966.
84. H. E. Winberg, U.S. Pat. 3,239,519 (1966); *C.A.*, **64**, 15854 (1966).
85. H. Bredereck, G. Simchen, and W. Kantlehner, *Chem. Ber.*, **104**, 924 (1971).
86. J. Zemlicka, *Collection Czech. Chem. Commun.*, **28**, 1060 (1963).
86a. A. Holý, S. Chladek, and J. Zemlicka, *Collection Czech. Chem. Commun.*, **34**, 232 (1969).
86b. J. Zemlicka, *Collection Czech. Chem. Commun.*, **33**, 3796 (1968).
86c. A. Holý and J. Zemlicka, *Collection Czech. Chem. Commun.*, **34**, 3921 (1969).
86d. J. Zemlicka, *Collection Czech. Chem. Commun.*, **35**, 3572 (1970).
86e. A. Holý, R. W. Bald, and Ng. D. Hong, *Collection Czech. Chem. Commun.*, **36**, 2658 (1971).
87. J. Gloede and B. Costisella, *J. Prakt. Chem.*, **313**, 277 (1971).
88. H. Eilingsfeld, M. Seefelder, and H. Weidinger, *Chem. Ber.*, **96**, 2671 (1963).
89. H. Bredereck, G. Simchen, and H. Porkert, *Chem. Ber.*, **103**, 256 (1970).
90. W. Tritschler and S. Kabuss, *Synthesis*, **1973**, 423.
91. W. Tritschler and S. Kabuss, *Synthesis*, **1972**, 32.
92. S. Kabuss and W. Tritschler, *Synthesis*, **1972**, 418.
93. K. Hiratani, T. Nakai, and M. Okawara, *Bull. Chem. Soc. Japan*, **46**, 3510 (1973).
94. H. Bredereck, F. Effenberger, and G. Simchen, *Angew. Chem.*, **73**, 493 (1961).
95. H. Bredereck, F. Effenberger, and H. P. Beyerlin, *Chem. Ber.*, **97**, 3081 (1964).
96. H. Bredereck, G. Simchen, S. Rebsdat, W. Kantlehner, P. Horn, R. Wahl, H. Hoffmann, and P. Grieshaber, *Chem. Ber.*, **101**, 41 (1968).

97. S. Hanessian and E. Moraliogeu, *Tetrahedron Letters*, **1971**, 813.
97a. J. Zemlicka and S. Chladek, *Tetrahedron Letters*, **1969**, 715.
98. G. Simchen, unpublished data.
99. W. Kantlehner and P. Speh, *Chem. Ber.*, **105**, 1340 (1972).
100. H. Eilingsfeld, G. Neubauer, M. Seefelder, and H. Weidinger, *Chem. Ber.*, **97**, 1232 (1964).
101. H. Bredereck, G. Simchen, and S. Rebsdat, *Chem. Ber.*, **101**, 1872 (1968).
102. G. Ege and H. O. Frey, *Tetrahedron Letters*, **1972**, 4217.
103. C. F. Hobbs and H. Weingarten, *J. Org. Chem.*, **36**, 2885 (1971).
104. J. Hocker and R. Merten, *Chem. Ber.*, **105**, 1651 (1972).
105. R. Feinauer and E. Henckel, *Ann. Chem.*, **716**, 135 (1968).
106. H. Vorbrüggen, *Angew. Chem.*, **75**, 296 (1963).
107. H. Vorbrüggen, *Ann. Chem.*, **1974**, 821.
108. W. Rütz, Dissertation, University of Stuttgart, 1968.
109. H. Brechbühler, H. Büchi, E. Hatz, J. Schreiber, and A. Eschenmoser, *Angew. Chem.*, **75**, 296 (1963).
110. See ref. 88.
111. K. E. Fahrenholtz, M. Lurie, and R. W. Kierstead, *J. Am. Chem. Soc.*, **89**, 5934 (1967).
112. J. A. Webber, E. M. Van Heyningen, and R. T. Vasileff, *J. Am. Chem. Soc.*, **91**, 5674 (1969).
113. D. H. Clemens, E. Y. Shropshire, and W. D. Emmons, *J. Org. Chem.*, **27**, 3664 (1962).
114. H. Gross, B. Costisella, and L. Haase, *J. Prakt. Chem.*, **311**, 577 (1969).
115. H. Gross and B. Costisella, *Angew. Chem..*, **80**, 364 (1968).
116. H. Gross and B. Costisella, *Angew. Chem.*, **80**, 445 (1968).
117. H. Gross and B. Costisella, *J. Prakt. Chem.*, **311**, 925 (1969).
118. F. M. Stoyanovitch, I. A. Ivanova, and B. P. Fedorov, *Bull. Soc. Chim. France*, **1970**, 2013.
119. F. M. Stoyanovitch and I. A. Ivanova, UdSSR Pat. 234,399 (1969); *C.A.*, **70**, 105963g (1969).
120. D. Olschwang, *Bull. Soc. Chim. France*, **1971**, 3354.
120a. C. Fengeas and D. Olschwang, *Bull. Soc. Chim. France*, **1969**, 332.
121. M. Seefelder, *Angew. Chem.*, **78**, 339 (1966); M. Seefelder, *Chem. Ber.*, **99**, 2678 (1966).
121a. A. Holý, *Tetrahedron Letters*, **1972**, 585.
122. H. Bredereck, O. Smerz, and R. Gompper, *Chem. Ber.*, **94**, 1883 (1961).
123. H. Bredereck, F. Effenberger, and R. Sauter, *Angew. Chem.*, **72**, 77 (1960).
124. H. Bredereck, F. Effenberger, and R. Sauter, *Chem. Ber.*, **95**, 2049 (1962).
125. H. Bredereck, F. Effenberger, and W. Resemann, *Chem. Ber.*, **95**, 2796 (1962).
126. H. Bredereck, F. Effenberger, and A. Hofmann, *Angew. Chem.*, **74**, 354 (1962).
127. H. Bredereck, R. Gompper, and B. Geiger, *Chem. Ber.*, **93**, 1402 (1960).
128. G. Tsatsaronis and F. Effenberger, *Chem. Ber.*, **94**, 2876 (1961).
129. H. Bredereck, F. Effenberger, and E. H. Schweizer, *Chem. Ber.*, **95**, 803 (1962).
130. H. Bredereck, F. Effenberger, and G. Rainer, *Angew. Chem.*, **73**, 63 (1961).
131. H. Bredereck, F. Effenberger, G. Rainer, and H. P. Schosser, *Ann. Chem.*, **659**, 133 (1962).
132. H. Bredereck, F. Effenberger, and G. Rainer, *Ann. Chem.*, **673**, 88 (1964).
133. H. Bredereck, F. Effenberger, and G. Rainer, *Ann. Chem.*, **673**, 88 (1964).
134. H. J. Botsch, Dissertation, University of Stuttgart, 1962.
135. H. Bredereck, F. Effenberger, H. J. Botsch, and H. Rehn, *Chem. Ber.*, **98**, 1081 (1965).
136. G. Lang, Dissertation, University of Stuttgart, 1974.
137. F. Yoneda, M. Higuchi, and T. Nagamatsu, *J. Am. Chem. Soc.*, **96**, 5607 (1974).
138. H. J. Kretschmar, Diplomarbeit, University of Stuttgart, 1967.
139. G. Simchen, unpublished data.

140. H. Meerwein, V. Hederich, H. Morschel, and K. Wunderlich, *Ann. Chem.*, **635**, 1 (1960).
141. F. Wagner, Dissertation, University of Stuttgart, 1972.
141a. G. Hauthal and D. Schied, *Z. Chem.*, **9**, 62 (1969).
142. H. Vorbrüggen, *Steroids*, **1**, 45 (1963).
143. D. Martin and A. Weise, *Chem. Ber.*, **99**, 3367 (1966).
144. H. Gross and B. Costisella, *Z. Chem.*, **10**, 404 (1970).
145. H. Bredereck, G. Simchen, and G. Beck, *Chem. Ber.*, **104**, 3794 (1971).
146. H. Bredereck, R. Gompper, F. Effenberger, H. Keck, and H. Heise, *Chem. Ber.*, **93**, 1398 (1960).
147. H. Bredereck, G. Simchen, and E. Göknel, *Chem. Ber.*, **103**, 236 (1970).
148. S. Rebsdat, Dissertation, University of Stuttgart, 1967.
149. E. Göknel, Dissertation, University of Stuttgart, 1967).
150. H. J. Bredereck, Diplomarbeit University of Stuttgart, 1968; H. J. Bredereck, Dissertation, University of Stuttgart, 1970.
151. H. Hoffmann, Dissertation, University of Stuttgart, 1968.
152. T. Oishi, M. Ochiai, M. Nagai, and Y. Ban, *Tetrahedron Letters*, **1968**, 497.
153. T. Oishi, H. Nakakimura, M. Mori, and Y. Ban, *Chem. Pharm. Bull.*, **20**, 1735 (1972).
154. T. Oishi, S. Murakami, and Y. Ban, *Chem. Pharm. Bull.*, **20**, 1740 (1972).
155. V. G. Granik, N. S. Kuryatov, V. P. Pakhomov, E. M. Granik, I. V. Persianova, and R. G. Gluskhov, *Zh. Org. Khim.*, **8**, 1521 (1972).
156. A. E. Wick, D. Felix, K. Steen, and A. Eschenmoser, *Helv. Chim. Acta*, **47**, 2425 (1964).
157. H. Muxfeldt, R. S. Schneider, and J. B. Mooberry, *J. Am. Chem. Soc.*, **88**, 3670 (1966).
158. I. J. Bolton, R. G. Harrison, and B. Lythgoe, *Chem. Commun.*, **1970**, 1512.
159. R. K. Hill, R. Soman, and S. Sawada, *J. Org. Chem.*, **37**, 3737 (1972).
160. F. E. Ziegler and J. G. Sweeny, *Tetrahedron Letters*, **1969**, 1097.
161. D. F. Morrow, T. P. Culbertson, and R. M. Hofer, *J. Org. Chem.*, **32**, 361 (1967).
162. W. Sucrow, *Angew. Chem.*, **80**, 626 (1968).
163. W. Sucrow, *Tetrahedron Letters*, **1970**, 4725.
164. D. Schweizer, Diplomarbeit University of Stuttgart, 1967.
165. H. Bredereck, G. Simchen, and E. Göknel, *Angew. Chem.*, **76**, 861 (1964).
166. H. Bredereck, G. Simchen, and S. Rebsdat, *Angew. Chem.*, **77**, 507 (1965).
167. H. Bredereck, G. Simchen, and S. Rebsdat, *Chem. Ber.*, **101**, 1863 (1968).
168. H. Bredereck, G. Simchen, and H. Porkert, *Angew. Chem.*, **78**, 826 (1966).
169. H. Bredereck, G. Simchen, and H. Porkert, *Chem. Ber.*, **103**, 245 (1970).
170. H. Bredereck, G. Simchen, and G. Kapaun, *Chem. Ber.*, **104**, 792 (1971).
171. H. Bredereck, G. Simchen, G. Kapaun, and R. Wahl, *Chem. Ber.*, **103**, 2980 (1970).
172. G. Simchen, S. Rebsdat, and W. Kantlehner, *Angew. Chem.*, **79**, 869 (1967).
173. G. Simchen and W. Kantlehner, *Tetrahedron*, **28**, 3535 (1972).
174. H. Neumann, *Chimia*, **23**, 267 (1969).
175. H. M. R. Hoffmann and E. A. Schmidt, *J. Am. Chem. Soc.*, **94**, 1373 (1972).
176. E. A. Schmidt and H. M. R. Hoffmann, *J. Am. Chem. Soc.*, **94**, 7832 (1972).
177. H. M. R. Hoffmann and E. A. Schmidt, *Angew. Chem.*, **85**, 227 (1973).
178. F. W. Eastwood, K. J. Harrington, J. S. Josan, and J. L. Pura, *Tetrahedron Letters*, **1970**, 5223.

IMINIUM AND NITRILIUM SALTS IN THE DIMERIZATION OF CYANO COMPOUNDS IN ACIDS

By SHOZO YANAGIDA and MITSUO OKAHARA, *Department of Applied Chemistry, Faculty of Engineering, Osaka University, Yamadakami, Suita, Osaka, Japan*; and SABURO KOMORI, *Nara Technical College, Yata-cho-22, YamatoKooriyama-shi, Nara, Japan*

CONTENTS

I. Introduction 528
II. Dimerization of Nitriles with Sulfuric Acid or Sulfur Trioxide 529
   A. Dimerization with Sulfuric Acid 529
   B. Dimerization with Sulfur Trioxide 530
   C. Reactions of 1,2,3,5-Oxathiadiazine 2,2-Dioxides 530
III. Dimerization of Cyano Compounds with Hydrogen Halides 531
   A. Dimeric Iminium Salts from Cyano Compounds with No α-Hydrogen 531
      1. Hydrogen Cyanide 531
      2. Cyanamides 532
      3. Thiocyanates 535
      4. Cyanates 537
      5. Cyanogen Halides 538
      6. Aromatic Nitriles 538
      7. α-Halogenated Nitriles 540
   B. Dimeric Iminium Salts from Cyano Compounds with α-Hydrogen 541
      1. Chloroacetonitrile 541
      2. Formation of N-(α-Chloroalkenyl)alkylamidinium Chlorides 542
      3. Codimerization of Cyano Compounds 544
   C. Reactions of N-(α-Chloroalkenyl)alkylamidinium Chlorides 548
      1. Reactions with Water 548
      2. Thermal Decomposition 549
      3. Phosgenation 550
      4. Miscellaneous Reactions 551
   D. Intra- and Intermolecular Dimerization of Dicyano Compounds 552
      1. Intramolecular Dimerization 552
      2. Cyclization of Malononitrile 554
      3. Cyclization of Dichloromalononitrile 556
IV. Dimerization of Cyano Compounds with Lewis Acids 557
   A. Formation and Reactions of Diazapyrylium Salts 557
   B. Miscellaneous Reactions 560
V. Mechanism of Dimerization 562
   References 568

I. Introduction

Cyano groups exhibit some nucleophilicity at nitrogen and consequently undergo various reactions catalyzed by acid. The interactions of nitriles with protic acids have received much attention, and the resulting reactions have been comprehensively reviewed (1–10). In general, the association of a proton with the lone pair of electrons on the nitrogen has been recognized as the first step in the following equilibrium:

$$R-C\equiv N \underset{}{\overset{nHX}{\rightleftarrows}} RC\equiv N\cdot nHX \rightleftarrows \underset{X^{\ominus}}{R-C\equiv \overset{\oplus}{N}H} \rightleftarrows R-C\underset{X}{\overset{NH}{\diagdown}} \quad (2)$$

$$\underset{(1)}{} \overset{HX}{\rightleftarrows} R-\overset{\oplus}{C}\underset{X}{\overset{NH_2}{\diagdown}} X^{\ominus} \quad (3)$$

Nitrilium ion **1** is much more susceptible to nucleophilic attack than the iminium ion **3** and plays an important role in the reactions of cyano compounds with nucleophiles in the presence of acids. Some Lewis acids accelerate the formation of both the nitrilium and iminium ions **1** and **3**, and also enhance their stability. The Pinner imidate synthesis and the Houben–Hoesch reaction can be understood as proceeding via attack on nitrilium ions by alcohols and by electron-rich aromatic rings, respectively. As exemplified by the Ritter reaction (11), cyano groups also show strong affinity toward carbonium ions, giving *N*-substituted nitrilium intermediates. A novel reaction involving such nitrilium intermediates is the formation of amidinium salts by the dimerization of nitriles under the influence of acid. In general, nitrilium species are subject to further nucleophilic attack by cyano or imidoyl (**2**) nitrogen, leading to dimerization. Some dimeric nitrilium intermediates undergo further reaction such as cyclization. This chapter deals with the dimerization of cyano compounds in acid and the reactions of the resulting dimeric iminium products. The mechanism of the dimerization is also discussed.

$$R-\bar{C}\equiv N \xrightarrow{Y^{\oplus}} \underset{(1a, c, d, g, f)}{R-\overset{\oplus}{C}\equiv N-Y} \xrightarrow{RCN \text{ or } 2} \underset{(1b, e, 1'b)}{R-C\underset{N\equiv C-R}{\overset{N-Y}{\diagdown}}^{\oplus}}$$

$$\xrightarrow{HX} R-\overset{\oplus}{C}\underset{N-C-R}{\overset{H}{\diagdown}}\underset{X}{\overset{N-Y}{\diagdown}} X^{\ominus}$$

Cyclization etc.

II. Dimerization of Nitriles with Sulfuric Acid or Sulfur Trioxide

In the reaction of nitriles with sulfuric acid, the presence of alkenes or alcohols yields carbonium ions which alkylate the nitrogen atom of the cyano group (Ritter reaction). Without alkenes or alcohols, nitriles generally form 1:1 complexes, possibly nitrilium salts, which produce amides on hydrolysis (3). Some nitriles, however, undergo dimerization with sulfuric acid and sulfur trioxide.

A. DIMERIZATION WITH SULFURIC ACID

Eitner (12) reported that in the reaction of acetonitrile with fuming sulfuric acid (28% SO_3), a crystalline product of the composition $2CH_3CN \cdot H_2SO_4$ was isolated. Recently, a similar 2:1 adduct was observed in the reaction of trichloroacetonitrile with sulfuric acid (13). Moreover, the dimerization of benzonitrile, chloroacetonitrile, and fluoronitrile under comparable conditions was also observed. It is worth noting that employment of just a small quantity of sulfuric acid is most favorable for dimerization.

Structure **4** was proposed for the isolated 2:1 adducts. However, the IR spectrum of the 2:1 adduct of acetonitrile reported by Oku (14), with characteristic absorptions at 1557, 1638, and 1739 cm^{-1}, is quite similar to those of nitrile–HCl 2:2 adducts, suggesting that structures **5** or **6** are more likely. Further investigation is needed before a definite structure can be written.

$$R-C\begin{matrix}N-SO_3H\\N-COR\\H\end{matrix}$$

(**4a**) R = CH_3
(**4b**) R = CCl_3

$$CH_3-C\begin{matrix}NH_2\\\oplus\\HN-C=CH_2\\OSO_3^\ominus\end{matrix}$$

(**5**)

$$CCl_3-C\begin{matrix}NH_2\\\oplus\\N-C-CCl_3\\OSO_3^\ominus\end{matrix}$$

(**6**)

The treatment of benzonitrile with fuming sulfuric acid affords N-benzoylbenzamidine (**7**) and 4,6-diphenyl-1,2,3,5-oxathiadiazine 2,2-dioxide (**8a**), as well as a trace of 2,4,6-triphenyl-s-triazine (13). Their respective yields depend on the concentration of sulfur trioxide; higher sulfur trioxide concentration decreases the yield of **7** and increases that of **8a**. The former may form through **8a**.

$$Ph-C\equiv N \xrightarrow[SO_3]{H_2SO_4} Ph-C\begin{matrix}NH\\N-CO-Ph\\H\end{matrix} + Ph-C\begin{matrix}N\\\\N\end{matrix}\begin{matrix}SO_2\\\\O\end{matrix}\begin{matrix}\\\\C\\Ph\end{matrix} + Ph-C\begin{matrix}N\\\\N\end{matrix}\begin{matrix}\\\\C\\Ph\end{matrix}C-Ph$$

(**7**)   (**8a**)   trace

B. DIMERIZATION WITH SULFUR TRIOXIDE

Aromatic nitriles react with sulfur trioxide at low temperature, giving 4,6-diaryl-1,2,3,5-oxathiadiazine 2,2-dioxides (**8a–8e**) (15, 16). The mass spectrum of **8a** has been reported to be in agreement with structure **8** (17). It has been noted that **8a** is formed in low yield in the reaction of benzonitrile with chlorosulfonic acid in carbon disulfide at 0°C (17).

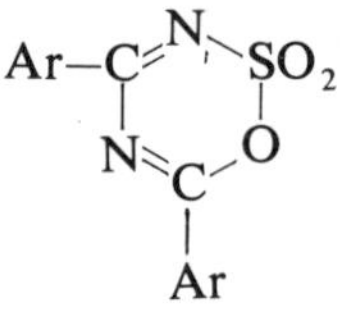

a: Ar = Ph
b: Ar = p-MePh
c: Ar = m-MePh
d: Ar = (thienyl)
e: Ar = $(CH_3)_2N$

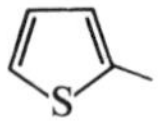

(8)

Aryl cyanates were incorrectly reported to give the corresponding cyclic products **8** (Ar = ArO) (18). It has recently been demonstrated that aryl cyanates give the symmetrical oxathiadiazines **9a** as well as cyanogen chloride (19). Dimethylcyanamide gives a mixture of isomers **8e** and **9c**. The interesting reactions of **9a** have been reviewed (19).

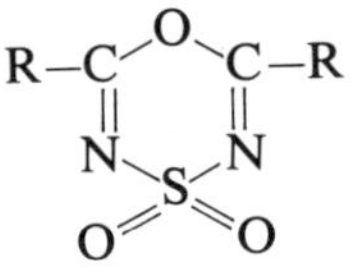

a: R = ArO
b: R = Cl
c: R = $(CH_3)_2N$

(9)

Acetonitrile has been reported to react abnormally with sulfur trioxide to yield a product of unknown structure, $3CH_3CN \cdot 2SO_3$, (12). Monosulfonation of nitriles by sulfur trioxide has also been observed (20).

C. REACTIONS OF 1,2,3,5-OXATHIADIAZINE 2,2-DIOXIDES

Hydrolysis of **8** gives N-aroylarylamidines (**10**) or diacylamines (**11**), depending on the reaction conditions as shown in Scheme 1 (15). With ethanol, **8a** gives benzamidinesulfonic acid (**12**) and ethyl benzoate (15). Pyrolysis of **8a** results in the formation of a brittle, brown, glassy polymer as well as benzonitrile (17).

It has recently been shown that oxathiadiazines (**8**) are useful intermediates for heterocyclic synthesis (16). Reactions with imidates, amidines, trichloroacetonitrile, urea, thiourea, and guanidine give the respective triazine derivatives **13–17**, shown in Scheme 2 in fair to good yields. These reactions have been interpreted as an initial Diels–Alder reaction followed by elimination

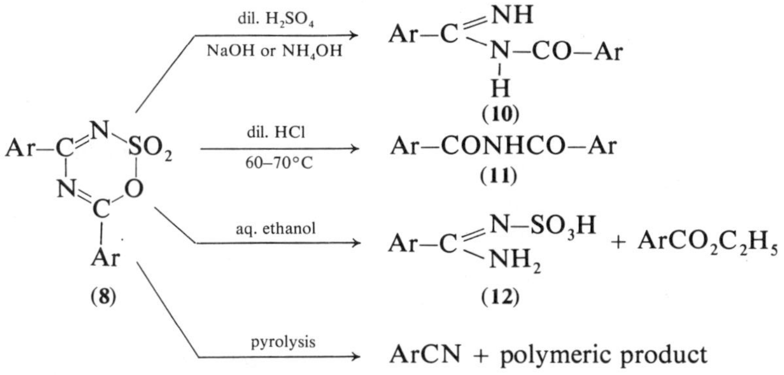

III. Dimerization of Cyano Compounds with Hydrogen Halides

[Scheme 1 showing reactions of compound (8) Ar-C(=N)-SO2-N=C(-O)-Ar with various reagents:
- dil. H₂SO₄ / NaOH or NH₄OH → Ar-C(=NH)-N(H)-CO-Ar (10)
- dil. HCl, 60-70°C → Ar-CONHCO-Ar (11)
- aq. ethanol → Ar-C(=N-SO₃H)(NH₂) (12) + ArCO₂C₂H₅
- pyrolysis → ArCN + polymeric product]

Scheme 1

of alcohol, water, or ammonia from the intermediate adducts (21). With hydrazine derivatives or hydroxylamine, the five-membered heterocycles **18–20** shown in Scheme 3 are formed (22).

The treatment of **8** with active methylene compounds or enamines affords pyrimidine derivatives **21–24** shown in Scheme 4 (23). Reaction with sulfur ylides followed by treatment with hot aqueous potassium chloride solution gives the pyrimidine derivative **25** (24).

III. Dimerization of Cyano Compounds with Hydrogen Halides

It has long been known that nitriles and hydrogen halides form addition products of varying compositions such as $RC\equiv N \cdot nHX$ ($n = 1, 5$), and $2RC\equiv N \cdot nHX$ ($n = 1, 2, 3$), depending especially on the reaction temperature (1, 5). Although dimeric adducts have been proposed as intermediates to explain side reactions in the Stephen reduction (25) and the Houben–Hoesch reaction (4), isolation of these compounds was quite difficult (26). At that time the dimerization of cyano compounds with hydrogen halides was of limited interest, but more recently it has been established that most nitriles form dimeric amidinium salts, especially with hydrogen chloride.

A. DIMERIC IMINIUM SALTS FROM CYANO COMPOUNDS WITH NO α-HYDROGEN

1. Hydrogen Cyanide

Hydrogen cyanide reacts slowly with excess hydrogen chloride and without external heating to give the so-called "sesquichloride" having the composition $2HC\equiv N \cdot 3HCl$ (27). Hydrogen bromide and hydrogen iodide afford even

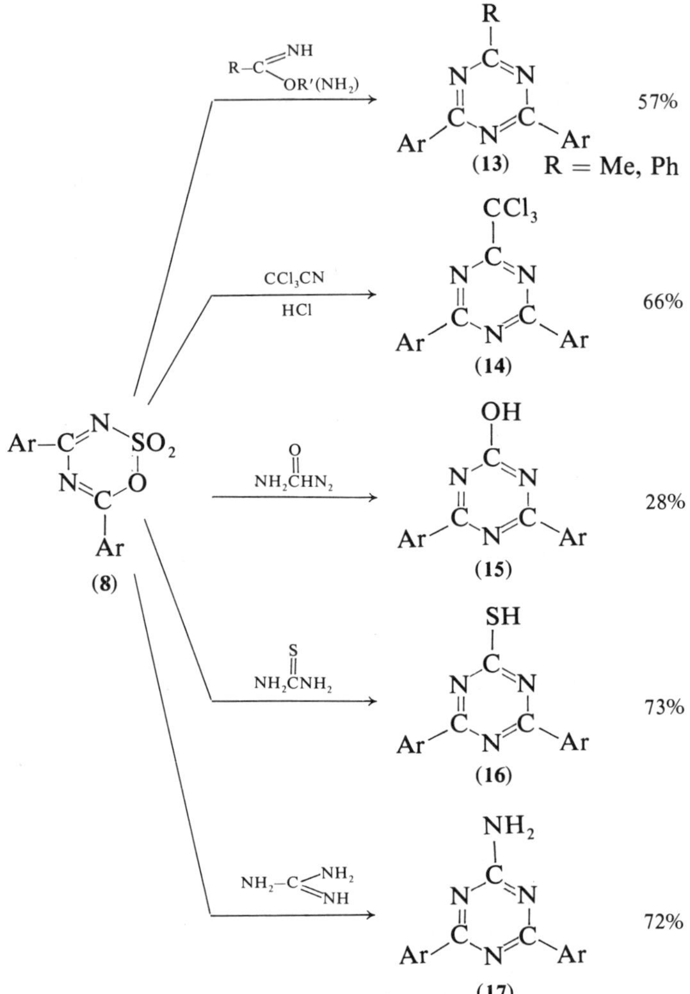

DIMERIZATION OF CYANO COMPOUNDS IN ACIDS

Scheme 2

more readily their corresponding sesquihalides. Recently their structure was shown to be that of N-(dihalomethyl)formamidinium halides (**26a, b, c**) on the basis of IR spectroscopy (28). The amidinium chloride **26a** reacts with antimony pentachloride and aluminum trichloride, yielding the rather stable salts **26d** and **26e**, respectively (28).

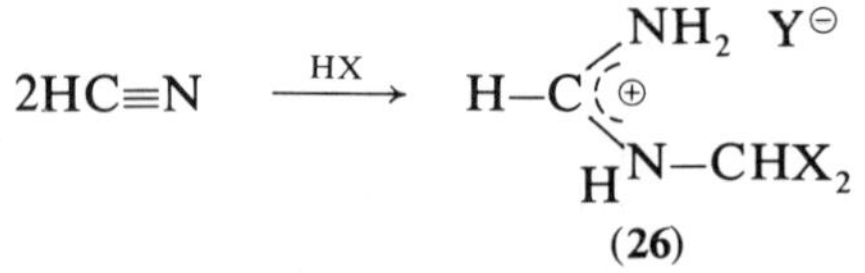

$$2HC\equiv N \xrightarrow{HX} H-C\underset{HN-CHX_2}{\overset{NH_2}{\lessgtr}}\ Y^{\ominus}$$

(26)

a: $X = Y = Cl$
b: $X = Y = Br$
c: $X = Y = I$
d: $X = Cl, Y = SbCl_6$
e: $X = Cl, Y = AlCl_4$

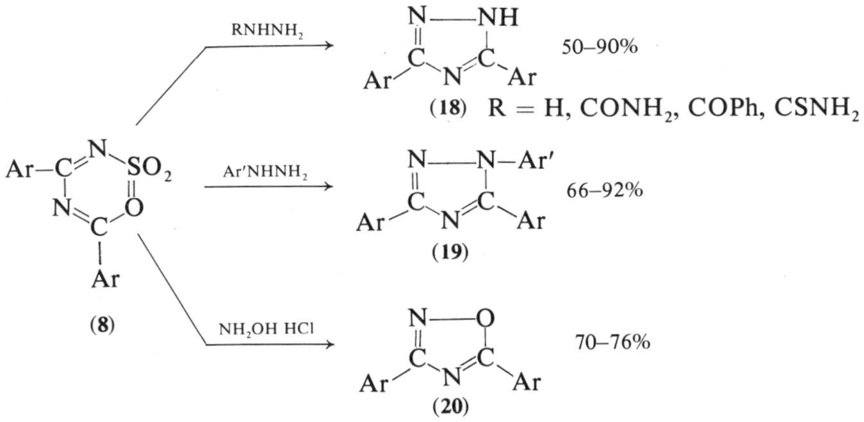

III. DIMERIZATION OF CYANO COMPOUNDS WITH HYDROGEN HALIDES 533

Scheme 3

The amidinium chloride **26a** is hygroscopic and also loses hydrogen chloride under the influence of bases or on heating to produce s-triazine hydrochloride (**27**) (29), which had originally been erroneously regarded as chloromethyleneformamidine (**28**) (30).

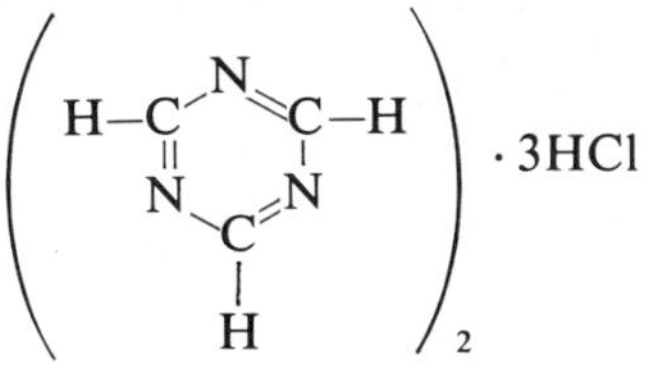

ClCH=N—CH=NH
(**28**)

Hydrolysis of **26a** gives N-formylformamidinium chloride (**31**). Treatment with ethanol gives formamidinium chloride, ethyl orthoformate, and ethyl chloride. Similar reactions are to be expected for **26b** and **26c** (29). It has been reported that dimethyl sulfoxide reacts with **26b** in dichloromethane, yielding N-formylformamidinium bromide and dimethyl sulfide dibromide (28).

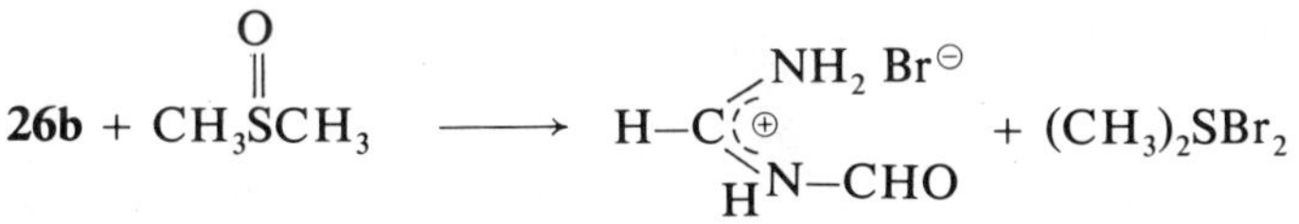

$$\mathbf{26b} + CH_3SCH_3 \longrightarrow H-C\substack{NH_2 \\ \oplus \\ HN-CHO} Br^{\ominus} + (CH_3)_2SBr_2$$

2. Cyanamides

Cyanamides have long been known to react with hydrogen chloride at low temperatures to give 1:2 addition products, the structures of which have

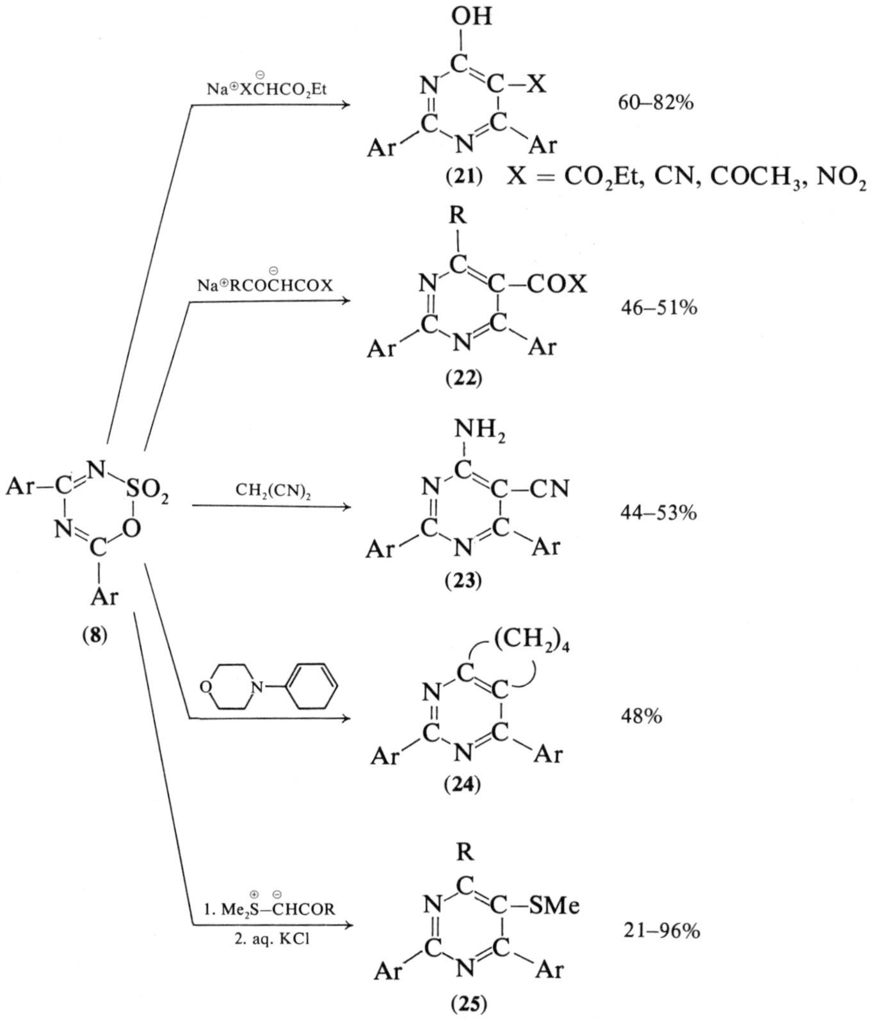

Scheme 4

recently been established to be chloroformamidinium chlorides (**3a, b, c**) (*32*). The resonance structure **3** was supported by the observation of nonequivalent methyl proton signals in the NMR spectrum of **3c** in liquid SO_2 (*33*). Hydrogen bromide also reacts with cyanamide yielding **3d**. The chloride **3a** is converted into **3e** on treatment with stannic tetrachloride (*32*).

III. DIMERIZATION OF CYANO COMPOUNDS WITH HYDROGEN HALIDES   535

$$\underset{R^2}{\overset{R^1}{\diagdown}}N-C\equiv N \xrightarrow{HX} \underset{R^2}{\overset{R^1}{\diagdown}}N-\overset{\oplus}{C}\underset{X}{\overset{NH_2}{\diagup}} \; X^{\ominus} \quad \left(\underset{H}{\overset{H}{\diagdown}}N-\overset{\oplus}{C}\underset{Cl}{\overset{NH_2}{\diagup}}\right)_2 \cdot SnCl_6{}^{2\ominus}$$

(3)   (3e)

a: X = Cl, R¹ = R² = H
b: X = Cl, R¹ = H, R² = Me
c: X = Cl, R¹ = R² = Me
d: X = Br, R¹ = R² = H

A United States patent (34) has disclosed that chloroformamidinium chlorides **3** derived from *N,N*-disubstituted or *N-t*-butylcyanamides may be converted into guanylchloroformamidinium chlorides (**29**) in quantitative yields when they are heated above their melting points or when they are reacted with the starting cyanamides above their melting points. Differential thermal analysis revealed that the dimerization of **3c** occurs endothermically at about 130°C.

$$\underset{R^2}{\overset{R^1}{\diagdown}}N-\overset{\oplus}{C}\underset{Cl}{\overset{NH_2}{\diagup}} \; Cl^{\ominus} \xrightarrow{\Delta \text{ or } R^1R^2N-C\equiv N} \left(\underset{R^2}{\overset{R^1}{\diagdown}}N-C\underset{N-\underset{Cl}{C}-N\underset{R^2}{\overset{R^1}{\diagup}}}{\overset{NH_2}{\diagup}}\right)^{\oplus} Cl^{\ominus}$$

(3)   (29)

a: R¹ = R² = alkyl or aryl
b: R¹ = H, R² = *t*-butyl
c: R¹ = R² = Me

The NMR spectrum of **29c** revealed that all its methyl groups are magnetically nonequivalent owing to the partial double bond character of each carbon–nitrogen bond, thus supporting the resonance structure **29** (33).

The bonded chlorine atom in **29** undergoes substitution reactions with nucleophiles. Treatment with aqueous sodium hydroxide at room temperature gives tetraalkylguanylureas (**30**) (35). Reaction with alkali metal alkoxides affords alkylguanylisoureas (**31**) (36), and treatment with ammonia, primary amines, or secondary amines produces biguanide derivatives (**32**) (37).

3. Thiocyanates

Allenstein et al. (38) have shown that methyl thiocyanate reacts with hydrogen chloride at low temperature (below 0°C) to give chloromethylthio-iminium chloride (**3f**). With hydrogen bromide, methyl thiocyanate and phenyl thiocyanate give the bromides **3g** and **3h**, respectively. At 60°C in a sealed tube, however, aliphatic thiocyanates were found to afford dimeric amidinium salts **33** (39).

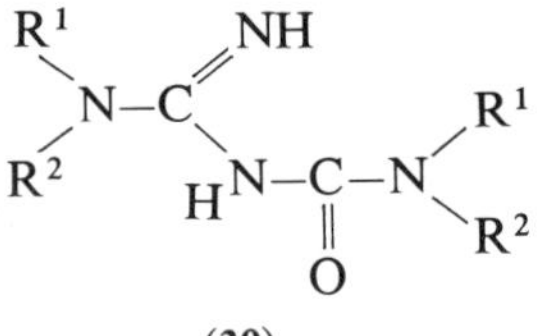

$$
\begin{array}{ccc}
\text{(30)} & \text{(31)} & \text{(32)}
\end{array}
$$

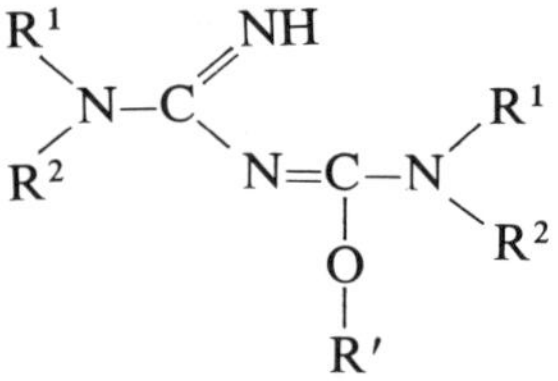

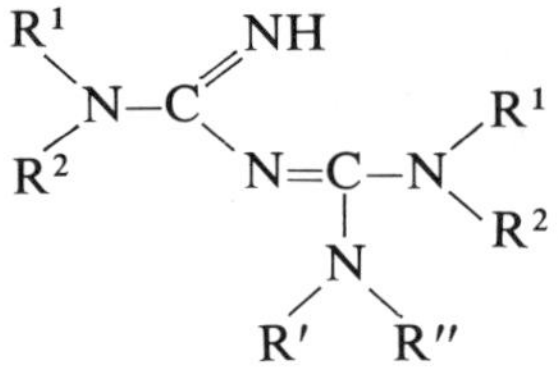

Structures (30), (31), (32): substituted guanidine/urea-type dimers with R^1, R^2 groups.

$$R^1-S-C\equiv N \xrightarrow[0°C]{HX} R^1-S-C\underset{X}{\overset{\oplus NH_2}{\diagdown}} \; X^{\ominus}$$

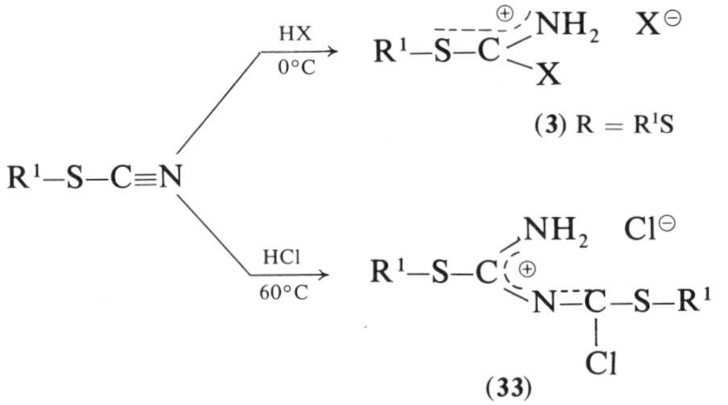

(3) R = R¹S

$$R^1-S-C\equiv N \xrightarrow[60°C]{HCl} R^1-S-C\underset{N=C-S-R^1}{\overset{NH_2}{\diagdown}} \; Cl^{\ominus}$$
                                                                    |
                                                                    Cl
(33)

f: X = Cl, R¹ = Me
g: X = Br, R¹ = Me
h: X = Br, R¹ = Ph

Analytically pure salts were reported only with ethyl thiocyanate and octyl thiocyanate. The other salts seemed to be contaminated with by-products, possibly 2,4,6-tris(alkylthio)-s-triazines.

Taking into account the stabilization of carbonium ions by electron donation from sulfur led to a proposal of the resonance structure **33′** (40). In the NMR spectrum of the dimeric salt from ethyl thiocyanate, however, the two sets of α-methylene protons are identical in chemical shift and are not deshielded as much as would be expected from the structure **33′** (39). This implies that the resonance form **33′** is only a minor contributor to the structure of **33**.

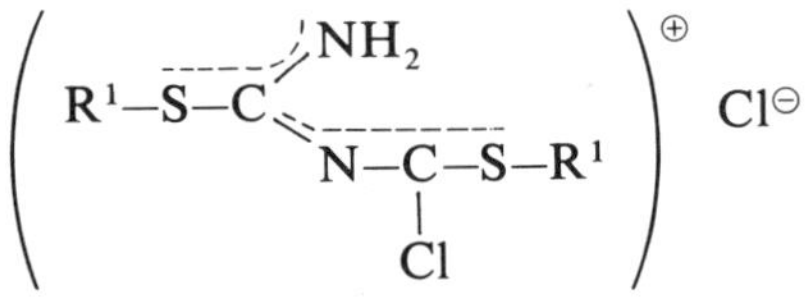

$$\left(R^1-S-C\underset{N-C-S-R^1}{\overset{NH_2}{\diagdown}} \right)^{\oplus} Cl^{\ominus}$$
                                    |
                                    Cl
(33′)

On hydrolysis with water, only the salt **33** from ethylthiocyanate gives the expected bis(ethylthioformyl)amine. Others afford only unidentified oily or tarry substances (39).

Recently, an unusual dimerization of thiocyanate anion has been reported in weakly acidic solution (41):

$$S=C=N^{\ominus} + H^{\oplus} \underset{}{\overset{pH\;5}{\rightleftharpoons}} S=C=NH$$

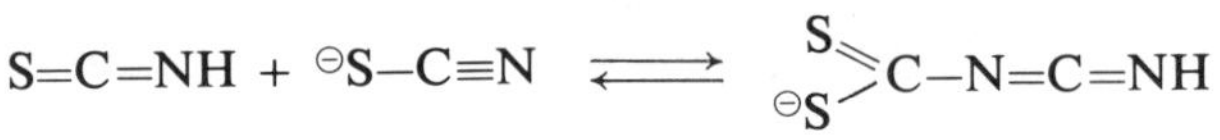

$$S=C=NH + {}^{\ominus}S-C\equiv N \rightleftharpoons \underset{{}^{\ominus}S}{\overset{S}{\diagdown}}C-N=C=NH$$

4. Cyanates

In the presence of acids, aryl cyanates trimerize to 2,3,5-triaryloxy-s-triazines, and alkyl cyanates isomerize to isocyanates, which then trimerize to trisubstituted isocyanurates (42, 43):

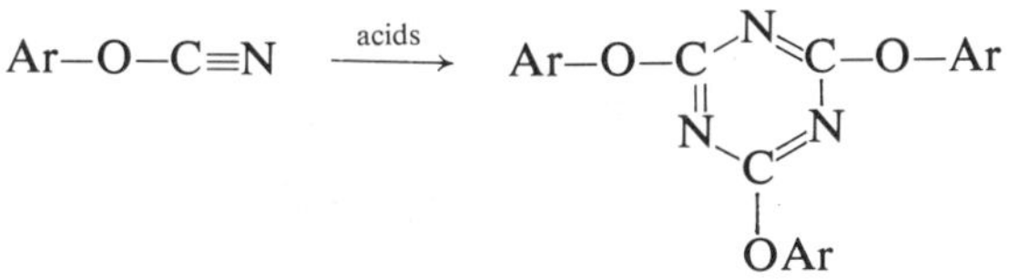

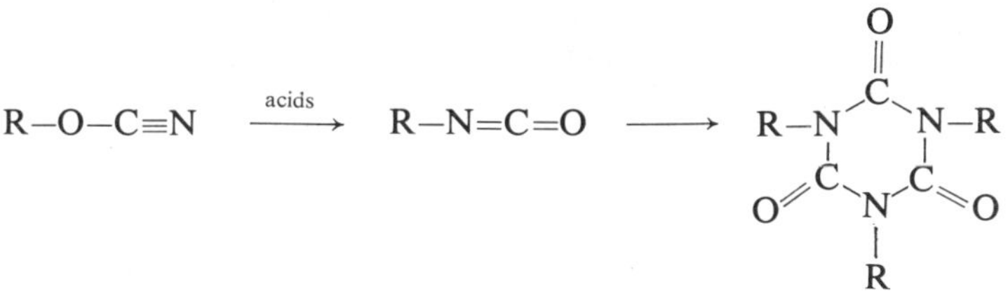

$$\text{Ar-O-C}\equiv\text{N} \xrightarrow{\text{acids}} \text{2,3,5-triaryloxy-s-triazine (OAr)}$$

$$\text{R-O-C}\equiv\text{N} \xrightarrow{\text{acids}} \text{R-N=C=O} \longrightarrow \text{trisubstituted isocyanurate}$$

Martin et al. (44) have shown that phenyl cyanate reacts with dry hydrogen chloride and with hydrogen bromide in ethyl ether at 0°C, giving the iminium salts **3i** and **3j**, respectively.

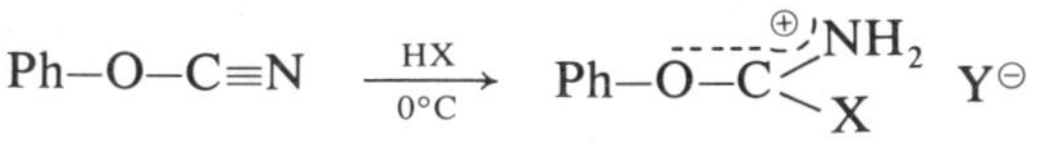

$$\text{Ph-O-C}\equiv\text{N} \xrightarrow[0°C]{HX} \text{Ph-O-C}(\overset{\oplus}{NH_2})(X) \; Y^{\ominus}$$

(3)

i: X = Y = Cl
j: X = Y = Br
k: X = Cl, Y = SbCl$_6$

The resonance structure seems reasonable in view of the greater stability of these compounds over the corresponding iminium salts **3** obtained from nitriles and thiocyanates. The more stable salt **3k** is obtainable from phenyl cyanate, hydrogen chloride, and antimony pentachloride under similar reaction conditions (44). The salts **3i** and **3j** may be converted slowly but readily on warming into 2,3,5-triphenoxy-s-triazine with evolution of the corresponding hydrogen halide.

No isolation of dimeric salts has been reported yet.

5. Cyanogen Halides

The well-known reactions of cyanogen chloride with hydrogen chloride and of cyanogen bromide with hydrogen bromide give 2,4,6-trichloro- and tribromo-s-triazines, respectively (45, 46):

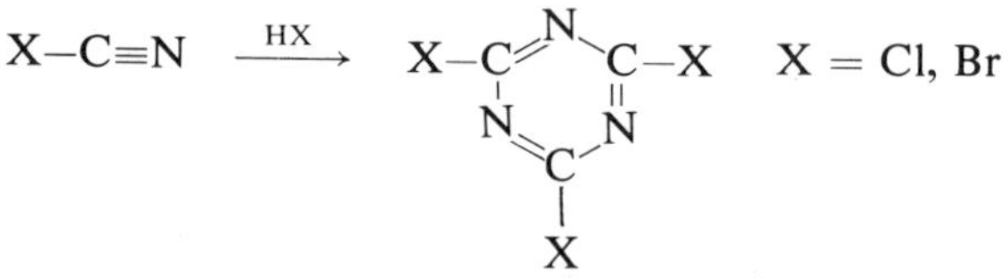

$$X-C\equiv N \xrightarrow{HX} \text{[2,4,6-trihalo-s-triazine]} \quad X = Cl, Br$$

Hydrogen iodide, however, reduces the cyanogens into hydrogen cyanide and hydrogen halides with the formation of iodine (47).

Allenstein and Schmidt (48) reported the isolation of haloiminium salts **3l–3n** by the reaction of cyanogen halide–antimony pentachloride complexes with hydrogen chloride.

$$X-C\equiv N \cdot SbCl_5 \xrightarrow{HCl} \left(\begin{array}{c} X \\ Cl \end{array} C-NH_2 \right)^{\oplus} SbCl_6^{\ominus}$$

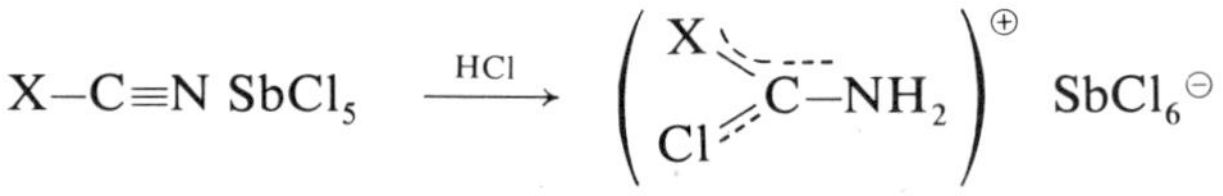

(3)

l: X = Cl
m: X = Br
n: X = I

No dimeric salts have been reported so far.

6. Aromatic Nitriles

Aromatic nitriles react with hydrogen halides at low temperatures to give arylhaloiminium halides (**3o**), which are stable only in the presence of hydrogen halides and decompose to the starting nitriles in air (1, 49–51). Some aromatic nitrile–Lewis acid complexes react with hydrogen chloride to give

TABLE I

N-(Arylchloromethylidene)arylamidinium Chlorides

34		Mole ratio HCl/ArCN	Reaction temp., °C	Reaction time, hr	Yield, %
a	Ph	0.9	30	192	23[a]
b	p-MePh	1.0	60	286	27
c	m-MePh	0.9	60	112	15

[a] 57% was recorded after 75 days.

III. DIMERIZATION OF CYANO COMPOUNDS WITH HYDROGEN HALIDES   539

stable iminium salts, for example, **3p** and **3q** (52). At high temperature, especially in the presence of hydrogen chloride and Lewis acids, trimerization to 2,4,6-triaryl-s-triazines occurs (53, 54).

It has recently been demonstrated that when some aromatic nitriles are reacted with approximately an equivalent of hydrogen chloride at 30–60°C in a sealed tube, the dimerization takes place very slowly, giving amidinium salts **34** (Table I) (39). The synthesis of the corresponding salts (**34**) from o-toluonitrile, p-nitrobenzonitrile, and p-chlorobenzonitrile did not succeed.

$$Ar-C\equiv N \xrightarrow{HX, 0°C} \left(Ar-C{\overset{NH_2}{\underset{X}{\diagup\!\!\!\diagdown}}}\right)^{\oplus} Y^{\ominus} \quad \left(Ar-C{\overset{NH_2}{\underset{Cl}{\diagup\!\!\!\diagdown}}}\right)_2 SnCl_6^{2\ominus}$$

$$\text{(3)} \qquad\qquad \text{(3q)}$$

o: X = Y = Cl, Br, or I
p: X = Cl, Y = SbCl$_6$

$$Ar-C\equiv N \xrightarrow[30-60°C]{HCl} Ar-C{\overset{NH_2\ Cl^{\ominus}}{\underset{N-C-Ar}{\diagup\!\!\!\diagdown}}}_{\ \ \ \ \ \ |\ \ \ Cl}$$

(34a, b, c)

Janz and Ahmad (55) reported that benzonitrile reacts with hydrogen bromide at room temperature, giving a solid substance having the composition PhCN · HBr. The absence of a peak around 2250 cm^{-1} and a new band at 1660 cm^{-1} in the IR spectrum would suggest a dimeric structure similar to **33** rather than the speculated nitrilium structure. Further studies are apparently required for confirmation.

The salts **34** are thermally unstable and are readily converted into 2,4,6-triaryl-s-triazines by heating. Differential thermal analysis showed that 2 moles of **34a** condensed at 123°C and then decomposed to 1 mole of 2,4,6-triphenyl-s-triazine, 1 mole of benzonitrile, and 4 moles of hydrogen chloride, as follows:

$$\mathbf{34a} \xrightarrow[-2HCl]{\Delta} \left(\begin{array}{c} Ar-C{\overset{NH_2}{\diagup\!\!\!\diagdown}}\cdots C-N-C-Ar \\ N\diagdown\ \ \diagup N\ \ \ |\ \ \ \ | \\ C\ \ \ \ \ Ar\ \ Cl \\ | \\ Ar \end{array}\right)^{\oplus} Cl^{\ominus} \xrightarrow[-4HCl]{-ArCN} \begin{array}{c} Ar-C{\overset{N}{\diagdown\!\!\!\diagup}}C-Ar \\ \|\ \ \ \ \ \ \ \ \ | \\ N\diagdown\ \ \diagup N \\ C \\ | \\ Ar \end{array}$$

(35)

The tetramer **35** appears to have been formed during the pyrolysis.

Hydrolysis products of **34a** depend on the reaction conditions; aqueous alkali yields *N*-benzoylbenzamidine, and water gives dibenzoylamine (39).

7. α-Halogenated Nitriles

Hantzsch (49) has reported that at low temperatures, trichloroacetonitrile reacts with hydrogen bromide to give either 1:1 or 1:2 nitrile–HBr adducts, depending on the solvent used. Recently Wakabayashi et al. (56) have shown that trichloroacetonitrile trimerizes into 2,4,6-tris(trichloromethyl)-*s*-triazine in good yield when the reaction temperature is raised in the presence of both hydrogen halides and Lewis acids. It is worth noting, however, that hydrogen chloride alone gave a 1.6% yield of the triazine, whereas hydrogen bromide gave a 34% yield.

Grundmann et al. (26) have reported an accidental success in the dimerization of trichloroacetonitrile with hydrogen chloride. This was ascertained by the isolation of bis(trichloroacetyl)amine on hydrolysis of the intermediate adduct. In our laboratory, however, confirmation attempts were unsuccessful.

Troeger (57) has reported that 2,2-dichloropropionitrile formed a dimeric adduct with hydrogen bromide. Later Grundmann et al. (26) reexamined the reaction and reported a product of composition $2CH_3CCl_2CN \cdot HBr$. The fact that the salt gives bis(2,2-dichloropropionyl)amine on hydrolysis indicates a structure similar to **34**, although this is still open to question.

Troeger (57) also has reported an unusual dimeric product in the reaction of 2,2-dichloropropionitrile with hydrogen chloride, but this was not observed by our group. Examination of the reported elemental analysis of the hydrolysis product indicates that Troeger's compound may be the codimerization product of 2,2-dichloropropionitrile and propionitrile present as a contaminant in the reaction (see Section III-B-3).

Trifluoroacetonitrile reacts with hydrogen chloride in ethyl ether below 0°C, giving an unidentified precipitate, which is observable only in the sealed reaction tube. The precipitate disappears with evolution of gas when the tube is opened. It is surprising that 2,4,6-tris(trifluoromethyl)-*s*-triazine was obtained in 69% yield from the reaction mixture (33).

$$CF_3C\equiv N \xrightarrow[0°C,\ \text{in ether}]{HCl} \underset{\underset{CF_3}{|}}{\underset{N\diagdown_C\diagup N}{CF_3-C\diagup^N\diagdown C-CF_3}}$$

B. DIMERIC IMINIUM SALTS FROM CYANO COMPOUNDS WITH α-HYDROGEN

Nitriles having an α-hydrogen atom also form iminium salts (3) with hydrogen halides at about 0°C (1, 49, 61). Some Lewis acid salts such as **3u** and **3v** have also been reported (51, 52). Their structure has been widely investigated by IR spectroscopy (62) and X-ray diffraction analysis (63, 64). IR spectroscopy suggests a contribution by the resonance structure **3**. On the other hand, the structure **3'** has been proposed based on the carbon–halogen and carbon–nitrogen bond lengths. This implies a minor contribution by the resonance form **3**.

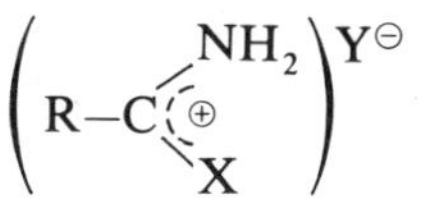

$$\left(R-C\underset{X}{\overset{NH_2}{\oplus}}\right) Y^{\ominus} \qquad \left(R-C\underset{Cl}{\overset{NH_2}{\lessgtr}}\right)_2^{\oplus} SnCl_6^{2\ominus}$$

(3) 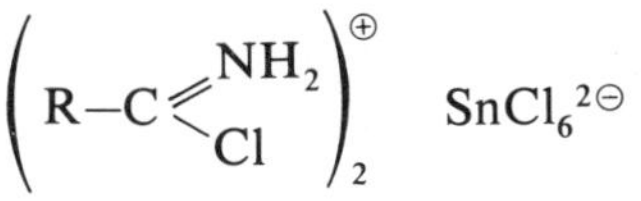   (3v) R = CH$_3$

r: R = CH$_3$, X = Y = Cl
s: R = CH$_3$, X = Y = Br
t: R = CH$_3$, X = Y = I
u: R = CH$_3$, X = Cl, Y = SbCl$_6$

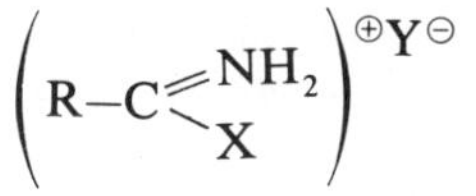

$$\left(R-C\underset{X}{\overset{NH_2}{\lessgtr}}\right)^{\oplus} Y^{\ominus}$$

(3') R = PhCH$_2$, BrCH$_2$, FCH$_2$

The iminium salt **3r** is very unstable in air, decomposing to starting material (61). Partly because of this, and partly because of the ambiguous and inconsistent reports on the stable salts of some nitriles with hydrogen chloride, the discovery of the general dimerization reaction of nitriles with hydrogen chloride was delayed until fairly recently.

1. Chloroacetonitrile

In 1904, Troeger et al. (58) first reported a stable salt from chloroacetonitrile and hydrogen chloride, which they incorrectly identified as chloromethylimidoyl chloride **2** (R = ClCH$_2$, X = Cl). Later, salts of varying compositions, for example, 2ClCH$_2$CN·HCl (26), 2ClCH$_2$CN·3HCl (59, 60), and 2ClCH$_2$CN·2HCl were also reported. It has been established that chloroacetonitrile reacts with hydrogen chloride in ether at 0°C to form a 2:3 adduct having the structure **36a**, and that the adduct gradually loses hydrogen chloride to form a rather stable 2:2 adduct (**37a**) by either path *a* or *b* as shown (65). The adduct **37a** further loses hydrogen chloride under prolonged reduced pressure to give 2,4,6-tris(chloromethyl)-*s*-triazine. It would seem that the 2 : 1 adduct reported by Grundmann et al. (26) was a mixture of **37a** and this triazine.

$$\text{ClCH}_2-\underset{\underset{a}{\overset{\overset{H\,Cl}{|}}{\underset{\text{(36a)}}{N-C-CHCl}}}{\overset{NH_2\ Cl^{\ominus}}{C\oplus}}} \xrightarrow[b]{\overset{a}{-HCl}} \begin{array}{c} \text{ClCH}_2-\underset{\underset{Cl}{|}}{\overset{\overset{NH_2\ Cl^{\ominus}}{C\oplus}}{N-C-CH_2Cl}} \\ \downarrow \\ \text{ClCH}_2-\underset{\underset{Cl}{|}}{\overset{\overset{NH_2\ Cl^{\ominus}}{C\oplus}}{\underset{H}{N}-C=CHCl}} \end{array}$$

(**37a**) (*cis* and *trans*)

The adduct obtained by heating the components in a sealed tube is the stable 2 : 2 adduct **37a**, not the 2 : 3 adduct **36a**. The 2 : 3 adducts from bromo- and fluoroacetonitriles probably have the same structure as **36a** (59).

It was reported that chloroacetonitrile reacts with hydrogen bromide to yield a 2 : 1 adduct whose structure and composition are worth reinvestigating as well as those of the adduct from 2,2-dichloropropionitrile and hydrogen bromide (58).

2. *Formation of N-(α-Chloroalkenyl)alkylamidinium Chlorides*

Since Gautier (66) first reported the stable salt from propionitrile and hydrogen chloride to be a 1 : 1 adduct, some stable 2 : 2 adducts have been incorrectly assigned imidoyl chloride structures (**2**) (58, 67, 68). Hinkel et al. (69) reported that chloromethyliminium chloride (**3r**) self-condenses at room temperature, or at 100°C in a sealed tube, to yield the dimeric salt **37'b**. It has now been shown that nitriles having α-hydrogen react slowly with more than an equivalent of hydrogen chloride, especially in a closed vessel at temperatures up to 65°C, to give a viscous solution of the dimeric amidinium salts (**36**), from which *N*-α-chloroalkenylalkylamidinium chlorides (**37**) arise, and either crystallize simultaneously or solidify in analytically pure form on treatment with dry ether (65). The NMR spectrum of the viscous solution from phenylacetonitrile did not show any olefinic proton signals (33). This fact, combined with the isolation of **36a** in the case of chloroacetonitrile, indicates that **37** is not the initial product but rather is formed later by dehydrochlorination of **36**, or prototropic rearrangement of **37'**.

III. DIMERIZATION OF CYANO COMPOUNDS WITH HYDROGEN HALIDES

TABLE II
N-(α-Chloroalkenyl)alkylamidinium chlorides[a]

37	R^1	R^2	Yield, %	m.p., °C
a	Cl	H	90[b]	122.0–123.0
b	H	H	67	–
c	CH$_3$	H	86	125.0–127.5
d	n-C$_6$H$_{13}$	H	37	115.0–116.5
e	Ph	H	53	135.0–138.0
f	CH$_3$	CH$_3$	76	193.0–194.0 (dec)
g	C$_2$H$_5$	CH$_3$	75	162.5–164.0 (dec)
h	ClCH$_2$	CH$_3$	82	98.0–100.0
i	Cl	Cl	99[b]	119.0–121.0 (dec)
j	CH$_3$	Cl	99[b]	175.0–176.0 (dec)
k	n-C$_4$H$_7$	Cl	38	145.0–148.0 (dec)
l	C$_2$H$_5$	Br	51	130.0–133.0

[a] Reaction conditions: molar ratio HCl/nitrile 1.1–1.9; reaction time 100–300 hr; reaction temperature, 60–65°C.
[b] Under cooling at 0°C in ether.

Typical amidinium salts (**37**) are listed in Table II along with reaction conditions. An excess of hydrogen chloride and prolonged reaction time are required for dimerization. Heating is not always necessary; dichloroacetonitrile and 2-chloropropionitrile dimerize quantitatively in ether even at 0°C.

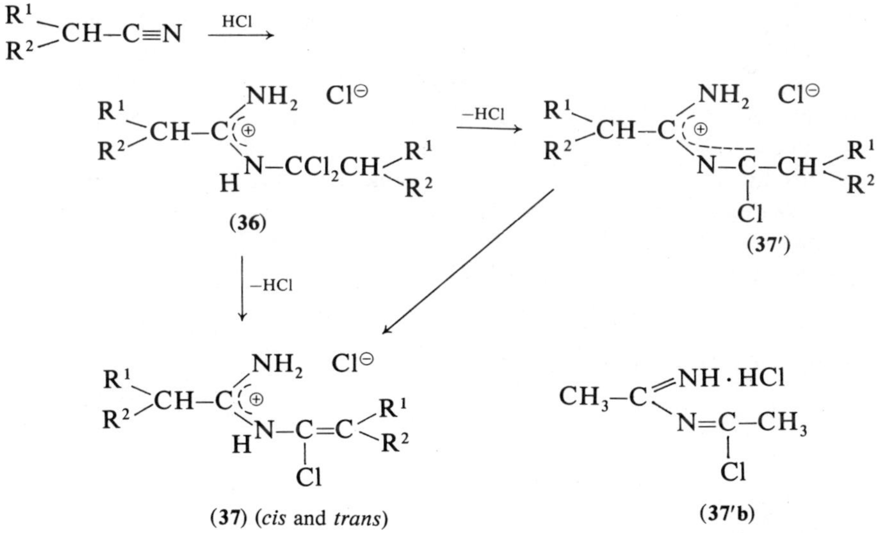

The structural assignment of **37** is based on spectral data. NMR spectra in liquid SO$_2$ afford substantial structural information, for example, the presence of an amidinium group and an alkenyl group consisting of *cis* and *trans* forms in almost equal proportions. The mass spectra exhibit peaks corresponding to M–HCl and M–HCl–Cl. Some spectra included the molecular ion and fragmentation pattern of the *s*-triazine of composition $(R^1R^2CN)_3$, suggesting contamination with the *s*-triazines, possibly during workup.

Isolation of the amidinium salts from methoxyacetonitrile, 3-alkoxypropionitrile, ethyl cyanoacetate, and diphenylacetonitrile did not succeed. Our attempts to dimerize acetonitrile with hydrogen bromide under various conditions were unsuccessful (33).

3. Codimerization of Cyano Compounds

The first example of a reaction between different cyano compounds in acid is, to our knowledge, the mixed trimerization to *s*-triazines reported by Grundmann et al. (26). More recently the trimerization reaction of trichloroacetonitrile with cyanogen chloride (70), nitriles (71), and thiocyanates (71) has been investigated.

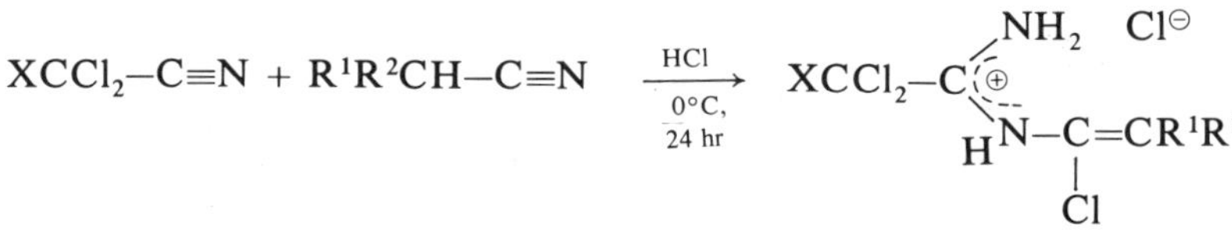

$$XCCl_2-C\equiv N + R^1R^2CH-C\equiv N \xrightarrow[0°C, 24\,hr]{HCl} XCCl_2-C \begin{matrix} NH_2 \\ \oplus \\ N-C=CR^1R^2 \\ | \\ Cl \end{matrix} \quad Cl^{\ominus}$$

(38) X = Cl (*cis* and *trans*)
(39) X = CH$_3$ (*cis* and *trans*)

In our laboratory, it was found that trichloroacetonitrile and 2,2-dichloropropionitrile react with aliphatic nitriles and hydrogen chloride at 0°C, affording fair yields of *N*-(α-chloroalkenyl)trichloroacetamidinium chlorides (**38**) and *N*-(α-chloroalkenyl)-2,2-dichloropropionamidinium chlorides (**39**), respectively (33). Ethyl thiocyanate also selectively codimerizes in a similar way, producing the codimer salts **40**. Table III gives the results of these codimerizations. An excess of the chlorinated nitriles and low reaction temperatures produce the best selectivity; otherwise, homodimerization of aliphatic nitriles or thiocyanates occurs.

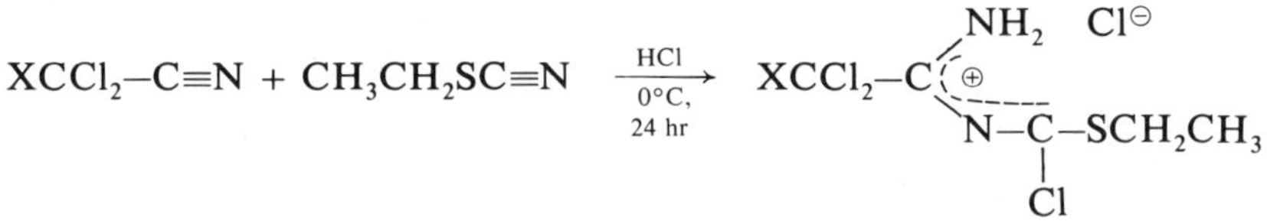

$$XCCl_2-C\equiv N + CH_3CH_2SC\equiv N \xrightarrow[0°C, 24\,hr]{HCl} XCCl_2-C \begin{matrix} NH_2 \\ \oplus \\ N-C-SCH_2CH_3 \\ | \\ Cl \end{matrix} \quad Cl^{\ominus}$$

(40) a: X = Cl
b: X = CH$_3$

III. DIMERIZATION OF CYANO COMPOUNDS WITH HYDROGEN HALIDES

TABLE III
Codimers 38–42

Codimers	Halogenated nitriles[a]	R^1	R^2	Selectivity, %[c]	Yield, %
38					
a	CCl_3CN	CH_3	H	93	50
b	CCl_3CN	CH_3	CH_3	100	53
c	CCl_3CN	CH_3	$n\text{-}C_6H_{11}$	100	71
d	CCl_3CN	CH_3CH_2	CH_3CH_2	100	23
39					
a	CH_3CCl_2CN	CH_3	H	74	46
b	CH_3CCl_2CN	CH_3	CH_3	99	36
c	CH_3CCl_2CN	CH_3	$n\text{-}C_6H_{11}$	100	12
d	CH_3CCl_2CN	CH_3CH_2	CH_3CH_2	100	25
40					
a	CCl_3CN	CH_3CH_2SCN		100	44
b	CH_3CCl_2CN	CH_3CH_2SCN		100	65
41	CF_3CN[b]	CH_3	H	~80[d]	~51[d]
42	$PhCCl_2CN$	CH_3	CH_3	~50[d]	~6[d]

[a] Molar ratio of nitrile to halogenated nitrile is 1 except in the case of trifluoronitrile.
[b] Molar ratio $CH_3CH_2CN/CF_3CN = 1.31$.
[c] Determined by NMR analysis.
[d] Because of unidentified by-products, the values are inaccurate.

Reactions of trichloroacetonitrile with benzonitrile and with acetonitrile, however, resulted in the exclusive isolation of 2,4-bis(trichloromethyl)-6-phenyl- and 6-methyl-s-triazines, respectively. They may be formed by further reaction of the intermediate codimer salts (**38**).

Trifluoroacetonitrile also codimerized with propionitrile to give **41**. The selectivity, however, was not as good as with trichloroacetonitrile (see Table III).

The success of these selective dimerizations may be explained as a result of the formation of reactive nitrilium ions from the halogenated nitriles. A low concentration of these nitrilium ions is to be expected because of their weaker basicity, but this is overcompensated by their strong electrophilic reactivity.

In the codimerization of α,α-dichlorophenylacetonitrile with isobutyronitrile, the yield of the codimer (**42**) and the selectivity were both unexpectedly low. The low reactivity may be due to the stabilization of the nitrilium intermediate by neighboring group participation of the phenyl ring (72). Similarly the slight drop in selectivity in the case of **41** and the easier trimerization to 2,4,6-tris(trifluoromethyl)-s-triazine, as compared with the case of trichloroacetonitrile, may be ascribed to the stabilization of trifluoroacetonitrilium ion by back-coordination by the trifluoromethyl group (73).

It appears now that the nitrogen atoms of imidoyl compounds (**2**) are also effective nucleophiles towards nitrilium ions **1** (see Section V). Their basicity exerts a slight influence on the direction of their codimerization. Evidence for this was obtained by examining the distribution of dimeric products of the following codimerization reactions:

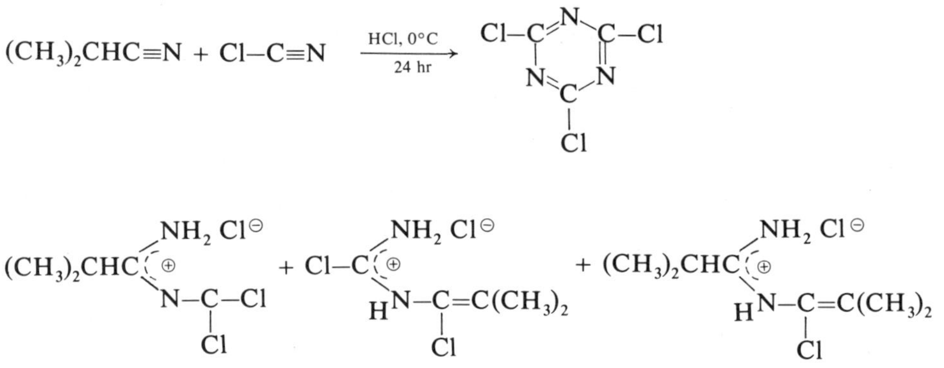

$$(CH_3)_2CHC\equiv N + Cl-C\equiv N \xrightarrow[24\ hr]{HCl,\ 0°C}$$ [triazine product: Cl–C, C–Cl, N, N, C–Cl ring]

$$(CH_3)_2CHC(NH_2\ Cl^\ominus)(\oplus)-N(H)-C(Cl)=CCl \quad + \quad Cl-C(NH_2\ Cl^\ominus)(\oplus)-N(H)-C(Cl)=C(CH_3)_2 \quad + \quad (CH_3)_2CHC(NH_2\ Cl^\ominus)(\oplus)-N(H)-C(Cl)=C(CH_3)_2$$

(43) (44) (37j) (1)

Mole ratios **43 : 44 : 37j** = 4.1 : 1.6 : 1 (by NMR).

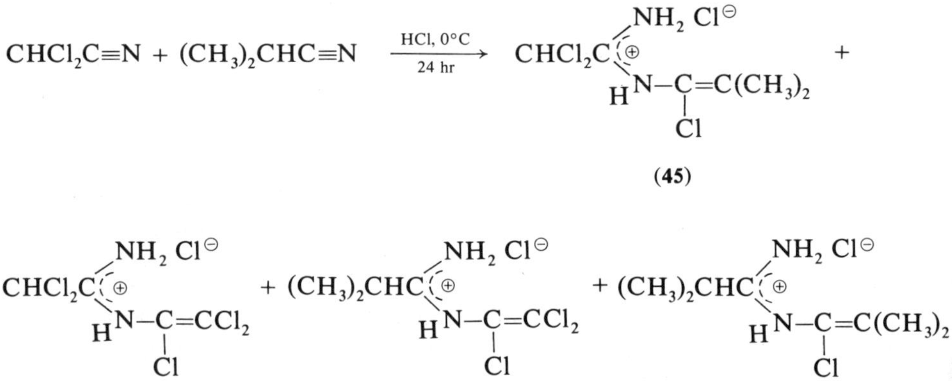

$$CHCl_2C\equiv N + (CH_3)_2CHC\equiv N \xrightarrow[24\ hr]{HCl,\ 0°C} CHCl_2C(NH_2\ Cl^\ominus)(\oplus)-N(H)-C(Cl)=C(CH_3)_2 \quad +$$

(45)

$$CHCl_2C(NH_2\ Cl^\ominus)(\oplus)-N(H)-C(Cl)=CCl_2 \quad + \quad (CH_3)_2CHC(NH_2\ Cl^\ominus)(\oplus)-N(H)-C(Cl)=CCl_2 \quad + \quad (CH_3)_2CHC(NH_2\ Cl^\ominus)(\oplus)-N(H)-C(Cl)=C(CH_3)_2$$

(37i) (46) (37f) (2)

Mole ratios **45 : 37i : 46** and/or **37f** = 4 ~ 5 : 18 ~ 19 : 1 (by NMR).

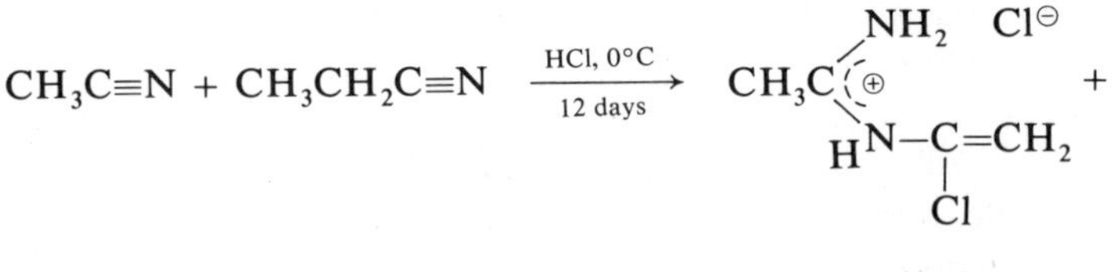

$$CH_3C\equiv N + CH_3CH_2C\equiv N \xrightarrow[12\ days]{HCl,\ 0°C} CH_3C(NH_2\ Cl^\ominus)(\oplus)-N(H)-C(Cl)=CH_2 \quad +$$

(37b)

III. DIMERIZATION OF CYANO COMPOUNDS WITH HYDROGEN HALIDES

$$CH_3-C\underset{HN-C=CHCH_3}{\overset{NH_2\ Cl^{\ominus}}{\underset{|}{\overset{\oplus}{\diagup}}}}\ +\ CH_3CH_2-C\underset{HN-C=CH_2}{\overset{NH_2\ Cl^{\ominus}}{\underset{|}{\overset{\oplus}{\diagup}}}}\ +\ CH_3CH_2-C\underset{HN-C=CHCH_3}{\overset{NH_2\ Cl^{\ominus}}{\underset{|}{\overset{\oplus}{\diagup}}}} \quad (3)$$

$$\overset{|}{Cl}\overset{|}{Cl}\overset{|}{Cl}$$

$$(47)(48)(37c)$$

Mole ratios **37b**:**47**:**48**:**37c** = 0.4 + α:1.4:1:1.6 (determined by gas-liquid phase chromatography after converting the resulting mixed dimeric salts into the corresponding dichloropyrimidines by phosgenation. α means a small quantity of 6-chloro-2-methyl-4(3H)-pyrimidinone hydrogen chloride salts. (See Section III-C-3.)

Codimerization 1 suggests that cyanogen chloride and/or the hypothetical dichloroimine are more nucleophilic toward nitrilium ions owing to the electron-donating effect of the chlorine atom:

$$\overset{\delta^{\oplus}}{\ddot{C}l}-C\equiv N{}^{\delta^{\ominus}} \qquad \overset{\delta^{\oplus}}{\underset{Cl}{\overset{\ddot{C}l}{\diagup}}}C=N\overset{\delta^{\ominus}}{H}$$

This tendency is quite consistent with the nucleophilic character of cyanogen chloride as observed in other reactions (74, 75).

Reactions 2 and 3 imply that dichloroacetonitrile is a better nucleophile than isobutyronitrile, and that propionitrile is slightly more nucleophilic than acetonitrile. These implications conflict with the expected electron-withdrawing effect of the chlorine atoms of the dichloromethyl group and with the electron-withdrawing ability of a methyl group bonded to a saturated carbon atom (76). There seem to be two different interpretations for these results.

1. Let us consider the following equilibrium between imidoyl chloride (**2**) and iminium salt (**3**) with their respective enamine forms (**2′** and **3′**). The

$$\underset{Y}{\overset{X}{\diagup}}CH-C\overset{\oplus}{\equiv}\overset{\cdot\cdot}{N}H \underset{-Cl^{\ominus}}{\overset{Cl^{\ominus}}{\rightleftarrows}} \underset{Y}{\overset{X}{\diagup}}CH-C\overset{NH}{\underset{Cl}{\diagdown}} \underset{-HCl}{\overset{HCl}{\rightleftarrows}} \underset{Y}{\overset{X}{\diagup}}CH-C\overset{NH_2}{\underset{Cl}{\overset{\oplus}{\diagdown}}}\ Cl^{\ominus}$$

$$(1)(2)(3)$$

$$\Updownarrow \Updownarrow$$

$$\underset{Y}{\overset{X}{\diagup}}C=C\overset{NH_2}{\underset{Cl}{\diagdown}} \underset{-HCl}{\overset{HCl}{\rightleftarrows}} \underset{Y}{\overset{X}{\diagup}}C=C\overset{NH_3^{\oplus}}{\underset{Cl}{\diagdown}}\ Cl^{\ominus}$$

$$(2')(3')$$

equilibrium should lie to the right owing to the resonance stabilization of the chlorine atom or the hyperconjugative stabilization of the methyl group when X and/or Y are chlorine atoms or methyl groups. The enamine forms would act as effective nucleophiles (see Section V).

2. We assume the following equilibrium:

$$\text{R}-\overset{\oplus}{\text{C}}\equiv\text{NH} + \text{R}^1-\text{C}\equiv\text{N} \rightleftarrows \text{R}-\text{C}\underset{\text{N}\equiv\text{C}-\text{R}^1}{\overset{\text{NH}}{\diagup}}$$

(1)   (1b)

The dimeric nitrilium intermediate **1b** would also be more reactive when R' is an electron-attracting group, and would then be subject to attack by chloride ion, irreversibly giving the dimeric salts after tautomerization. The kinetically controlled products in Ritter reactions of camphene with trichloroacetonitrile and with dichloroacetonitrile may support this interpretation (11, 78).

The codimerization of trichloroacetonitrile with *N,N*-dimethylcyanamide fails; *N,N*-dimethylchloroformamidinium chloride (**3c**) readily separates from the reaction mixture.

C. REACTIONS OF *N*-(α-CHLOROALKENYL)ALKYLAMIDINIUM CHLORIDES

1. Reactions with Water

Most of the dimeric salts **37** are readily hydrolyzed by water to give diacylamines (**49**) in satisfactory yields (65). However, **37i** and the codimer salts **38** give the *N*-acylenamines **50** and **51**, respectively, whereas the codimer **39** is completely hydrolyzed to **52**. Water readily attacks the electron-deficient carbon atom activated by electron-withdrawing groups, followed by immediate separation of the enamines **50** and **51** from the aqueous phase. Complete hydrolysis is thus interrupted (33).

$$\text{R}^1\text{R}^2\text{CH}-\underset{\underset{\text{H}}{\text{N}-\text{C}=\text{CR}^1\text{R}^2}}{\overset{\text{NH}_2}{\text{C}}}\ \text{Cl}^\ominus \xrightarrow{\text{H}_2\text{O}} \text{R}^1\text{R}^2\text{CH}-\overset{\text{O}}{\overset{\|}{\text{C}}}-\underset{\text{H}}{\text{N}}-\overset{\text{O}}{\overset{\|}{\text{C}}}-\text{CHR}^1\text{R}^2$$

(37)   (49)

Hydrolysis of **37h** afforded **53** through partial dehydrochlorination of **49h** (65).

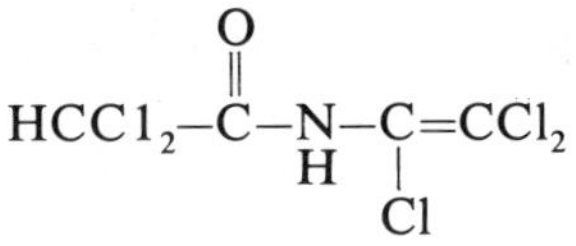

III. DIMERIZATION OF CYANO COMPOUNDS WITH HYDROGEN HALIDES

$$\text{HCCl}_2-\underset{\text{H}}{\overset{\overset{\text{O}}{\|}}{\text{C}}}-\underset{\text{Cl}}{\overset{}{\text{N}}}-\text{C}=\text{CCl}_2$$
(50)

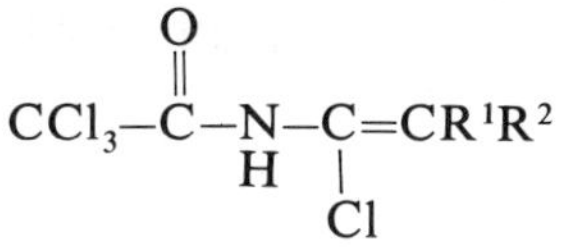

$$\text{CCl}_3-\underset{\text{H}}{\overset{\overset{\text{O}}{\|}}{\text{C}}}-\underset{\text{Cl}}{\overset{}{\text{N}}}-\text{C}=\text{CR}^1\text{R}^2$$
(51)

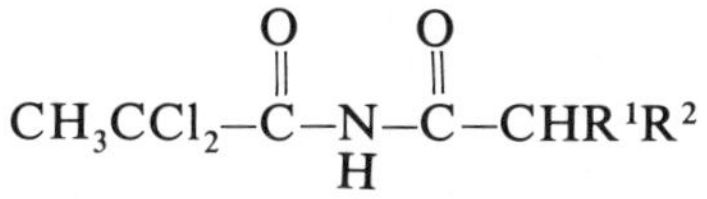

$$\text{CH}_3\text{CCl}_2-\overset{\overset{\text{O}}{\|}}{\text{C}}-\underset{\text{H}}{\text{N}}-\overset{\overset{\text{O}}{\|}}{\text{C}}-\text{CHR}^1\text{R}^2$$
(52)

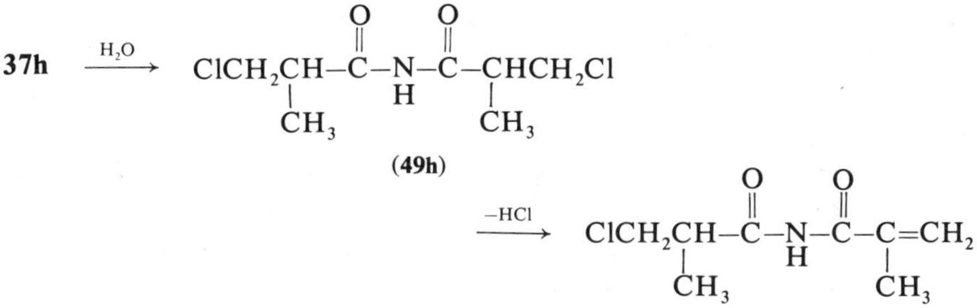

$$\mathbf{37h} \xrightarrow{\text{H}_2\text{O}} \text{ClCH}_2\underset{\text{CH}_3}{\text{CH}}-\overset{\overset{\text{O}}{\|}}{\text{C}}-\underset{\text{H}}{\text{N}}-\overset{\overset{\text{O}}{\|}}{\text{C}}-\underset{\text{CH}_3}{\text{CH}}\text{CH}_2\text{Cl}$$
(49h)

$$\xrightarrow{-\text{HCl}} \text{ClCH}_2\underset{\text{CH}_3}{\text{CH}}-\overset{\overset{\text{O}}{\|}}{\text{C}}-\underset{\text{H}}{\text{N}}-\overset{\overset{\text{O}}{\|}}{\text{C}}-\underset{\text{CH}_3}{\text{C}}=\text{CH}_2$$
(53)

2. Thermal Decomposition

Heating the salts **37** above their melting points causes decomposition to the starting nitriles and hydrogen chloride along with formation of a trace of s-triazines, for example, the trimers of the starting nitriles (33). However, the dimeric salt **37i** gives an almost quantitative yield of 2,4,6-tris(dichloromethyl)-s-triazine, and **34a** similarly gives 2,4,6-triphenyl-s-triazine (see Section III-A-6). Furthermore, pyrolysis of the codimers **38–40** affords good yields of 6-substituted 2,4-bis(trichloromethyl)- and 2,4-bis(1,1-dichloroethyl)-s-triazines **54–56**, respectively (33).

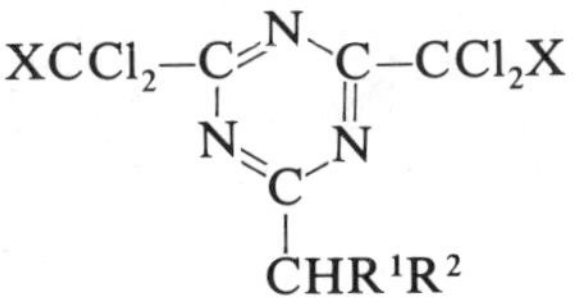

(54) X = Cl
(55) X = CH$_3$

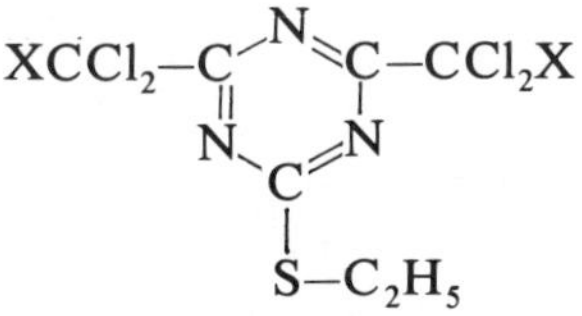

(56) X = Cl, CH$_3$

The failure to observe any reaction between nitriles and the dimeric salts **37** or **38** (33) implies that the cyclization probably proceeds through a stepwise reaction:

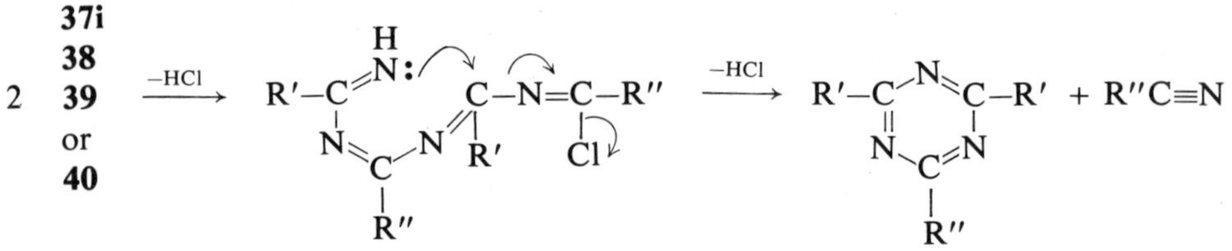

37i
38
2 **39** $\xrightarrow{-HCl}$ [intermediate structure] $\xrightarrow{-HCl}$ [pyrimidine structure] + R″C≡N
or
40

The selective formation of 6-substituted 2,4-bis(trichloromethyl)-*s*-triazines in the mixed trimerization with trichloroacetonitrile may be explained by the formation of codimers **38** (70, 71).

3. Phosgenation

Reaction with phosgene in an appropriate solvent at 100°C under pressure gives various pyrimidine derivatives (9, 79). The salts **37a–e**, prepared from nitriles having α-methylene groups, give good yields of 2,5-disubstituted 4,6-dichloropyrimidines (**57**). The salts **37f–k**, from nitriles having only one α-hydrogen, afford 5,5-disubstituted-2-alkylidene-4,6-dichloro-2,5-dihydropyrimidines (**58**) in high yield with but one exception. The salt **37i** gives **59** without further chlorination.

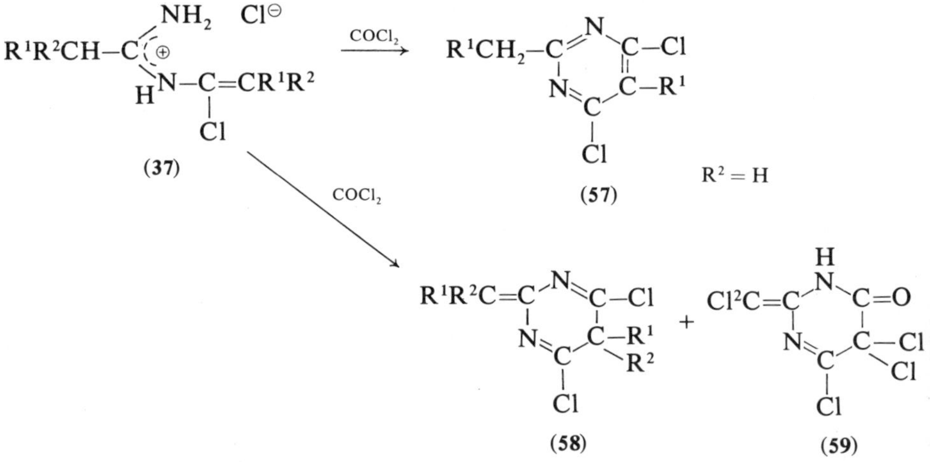

[Reaction scheme showing compound (37) converting via COCl₂ to compounds (57), (58), and (59)]

The codimers **38** and **39** react with phosgene in a similar fashion, giving the pyrimidine derivatives **60** and **61**, respectively. In the latter case, structure **61′** is a possible alternative, since some of these compounds are unstable and decompose to give dialklcyano amides (**62**) during workup (33).

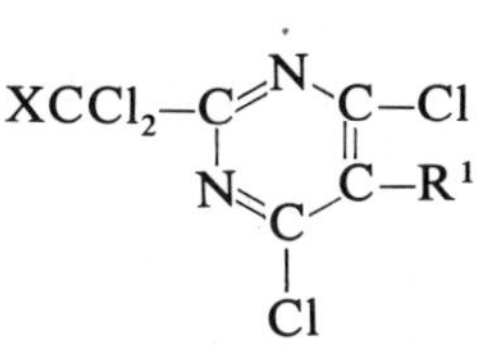

III. DIMERIZATION OF CYANO COMPOUNDS WITH HYDROGEN HALIDES

(60) X = Cl, CH$_3$

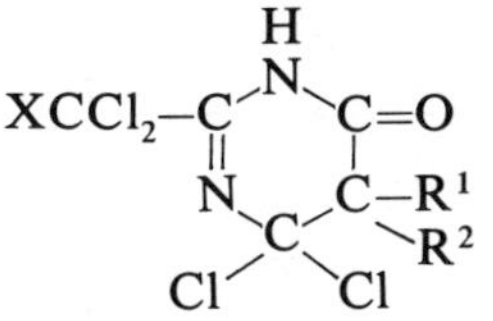

(61)

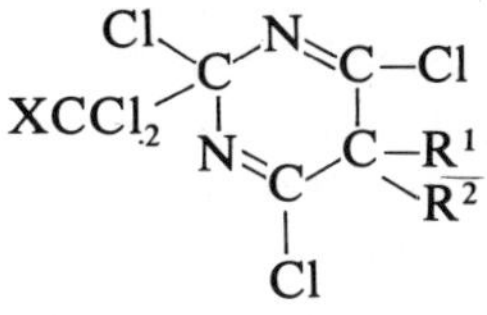

(61')

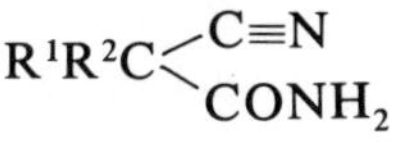

(62)

4. Miscellaneous Reactions

Michael et al. (67) reported the isolation of N-phenylpropionamidine in the reaction of **37c** with aniline, which led to an incorrect structure assignment for the amidinium salts **37c**.

Reaction with ammonia in ether under pressure at 60°C gives good yields of 2,4,6-trisubstituted s-triazines (**63**). With phenylhydrazine or hydroxylamine, triazole **64** and oxadiazole **65** are obtained (33).

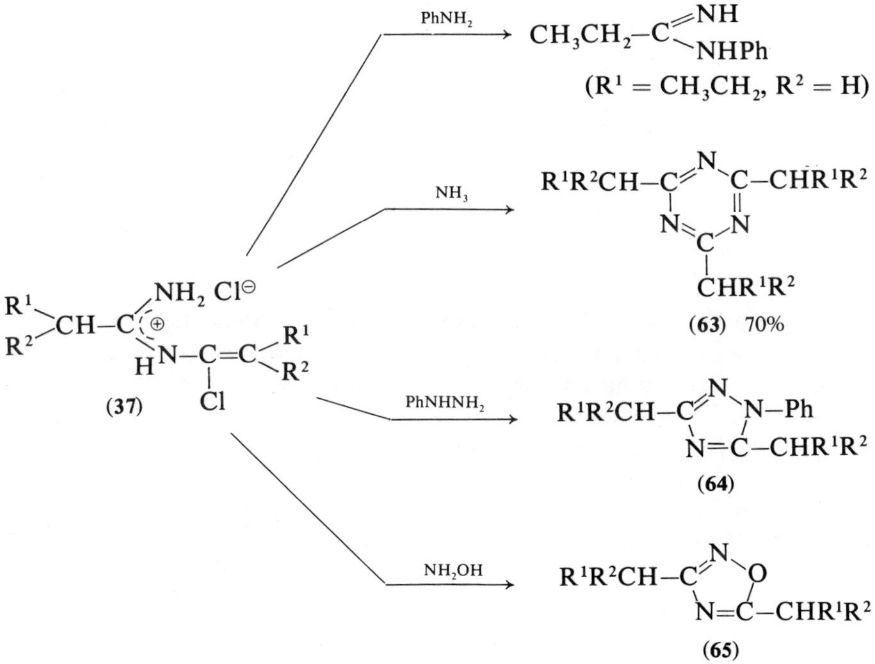

(R^1 = CH$_3$CH$_2$, R^2 = H)

(63) 70%

(64)

(65)

Zil'berman et al. (80) reported that the 2:3 adduct from chloroacetonitrile (possibly **36a**) reacts with anhydrous methanol to give a crystalline product with the composition $2ClCH_2CN \cdot 2CH_3OH \cdot HCl$. This product decomposes into chloroacetamidinium chloride, methyl chloroacetate, and ethyl ether on treatment with water. The other 2:3 adducts from bromo- and fluoroacetonitriles react in a similar way. Most of the salts **37**, however, are stable in methanol.

D. INTRA- AND INTERMOLECULAR DIMERIZATION OF DICYANO COMPOUNDS

1. Intramolecular Dimerization

Compounds with two appropriately positioned cyano groups undergo intramolecular dimerization on treatment with hydrogen halides to give five-, six-, and seven-membered ring heterocycles. Johnson et al. (7) and Meyer (4) have already reviewed these types of reactions.

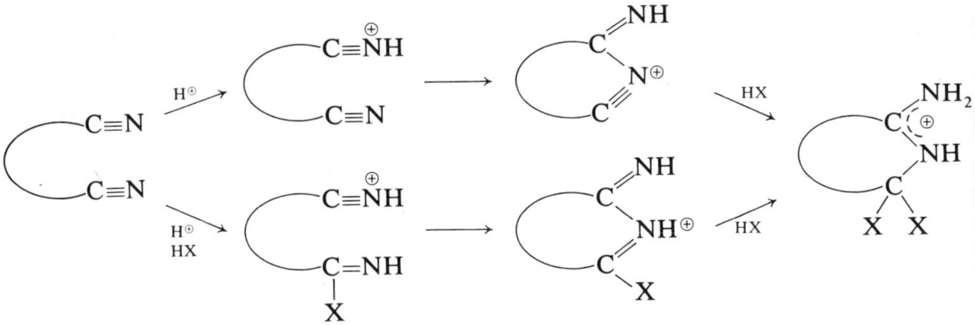

In this section, some important facts are summarized from a mechanistic point of view.

1. Substituents bonded to the cyano group determine the direction of cyclization. As shown in the schematic equations, a nitrile carbon atom bonded to sulfur, nitrogen, or unsaturated carbon will always act as the last nucleophile and bear the halogen atom in the cyclized product (7, 81). In view of this tendency, the following reaction is worth noting (82):

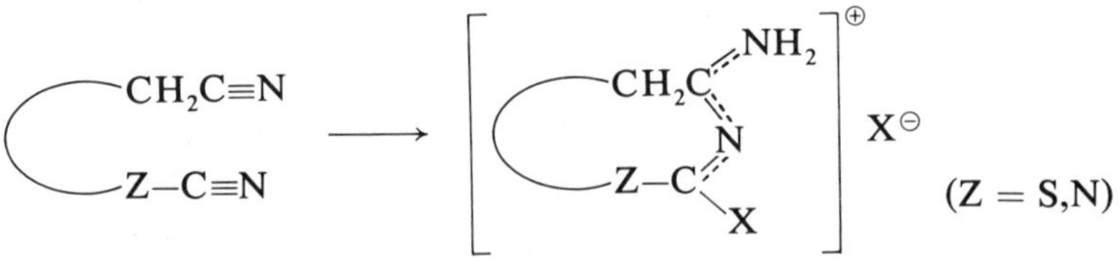

(Z = S, N)

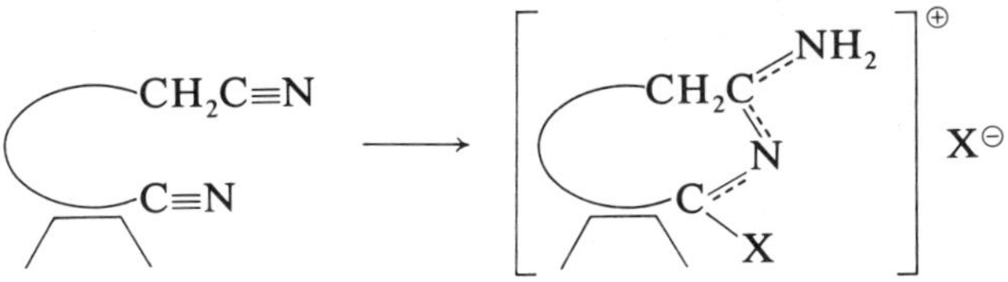

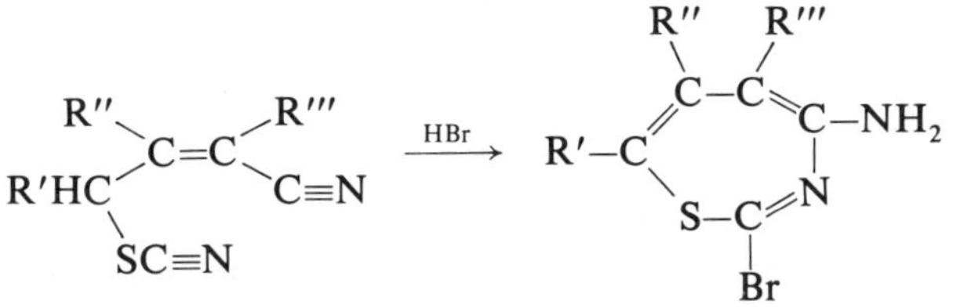

III. DIMERIZATION OF CYANO COMPOUNDS WITH HYDROGEN HALIDES

2. The two cyano groups involved in the reaction must be positioned close enough to each other for the intramolecular cyclization to occur. The *cis* dinitrile **66** gives the azepine derivative **67**, but neither **68** nor **69** gives any cyclic products, since in these latter compounds the two cyano groups are not close enough and also do not readily isomerize to the *cis* isomer under the experimental conditions (83).

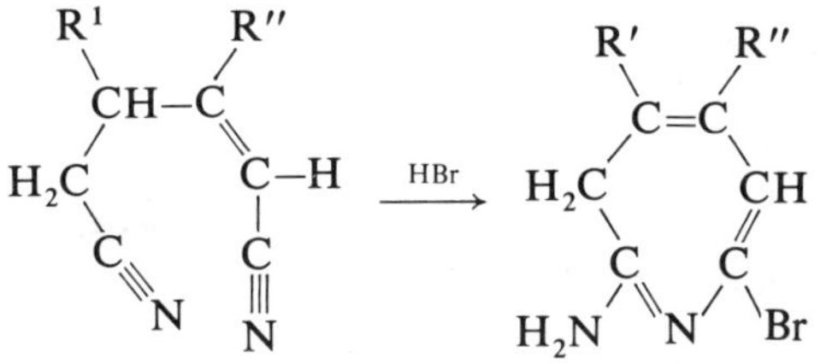

(66)   (67) 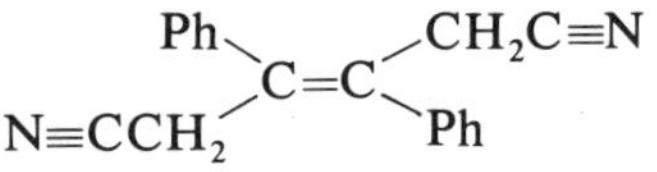   (68)

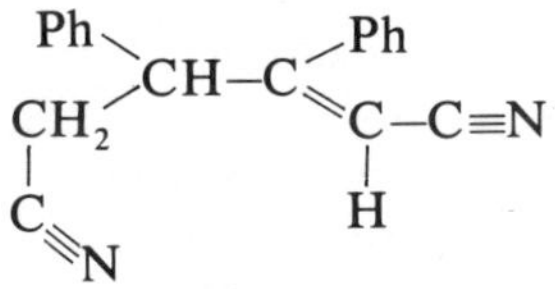

(69)

3. Effectiveness of hydrogen halides is proportional to their acid strength; for example, hydrogen bromide and hydrogen iodide are more effective than hydrogen chloride.

4. Succinonitrile and glutaronitrile give initially the pyrrolidinium halides **70** and piperidinium halides **71** (84). The ratio of hydrogen halides to cyano groups

in the product is the same as that of the initial dimeric salt from hydrogen cyanide and chloroacetonitrile.

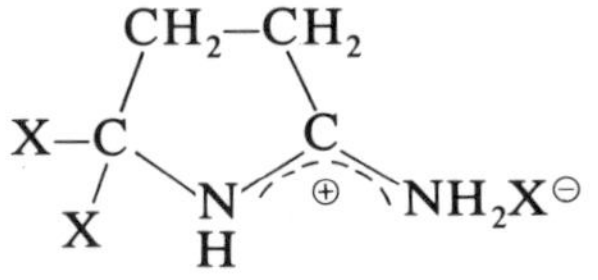

$$\text{(70)} \qquad \text{(71)}$$

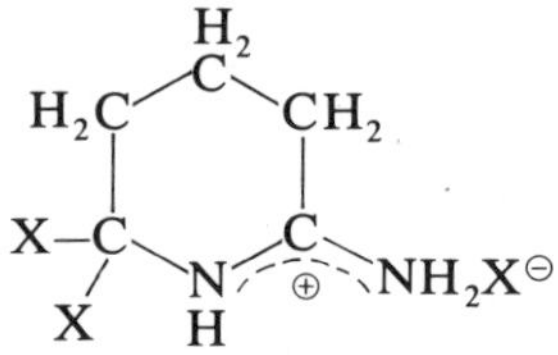

In view of these facts, intramolecular dimerization reactions are identical in mechanism to the dimerization of cyano compounds, and it may therefore be concluded that the reactions proceed through an intramolecular attack on the most reactive nitrilium ion by either another cyano group or the corresponding imidoyl halide.

Johnson et al. (7) noted the following reactions in their review article:

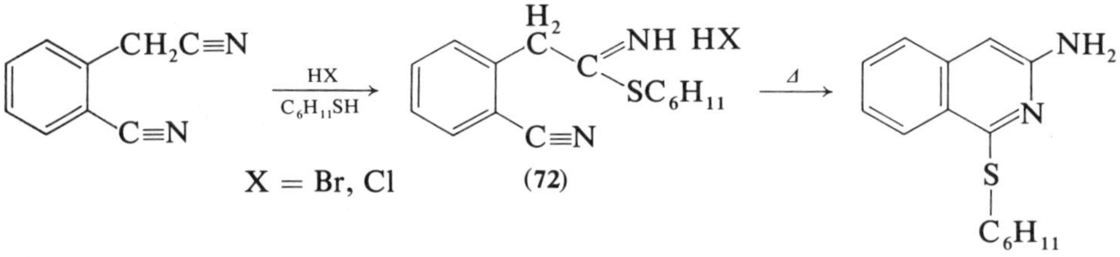

$$\text{X = Br, Cl} \qquad \text{(72)}$$

The isolation of **72** fully substantiates the difference in reactivity between the two possible nitrilium intermediates, and the subsequent cyclization reaction may be explained by the symbiotic effect on the iminium carbon atom (77, 85).

2. Cyclization of Malononitrile

Malononitrile reacts with hydrogen bromide or with hydrogen iodide at low temperature ($-78°C$), giving iminium salts (86) represented by the resonance structure **73**. On the other hand, with hydrogen chloride no reaction occurs under the same conditions. At higher reaction temperature ($35°C$), however, an unusual dimerization occurs to yield 2-amino-1,1,3-tricyano-1-propene (**74**) (86). This Claisen type condensation probably proceeds by nucleophilic attack on the nitrilium carbon atom by either the conjugate base of **73** (a kind of imidoyl intermediate), or malononitrile. Some Claisen type dimerizations have also been observed with amidinium chlorides (2).

The cyanopropene **74** reacts with hydrogen bromide, giving either **75** or **76**. The direct reaction of malononitrile with hydrogen bromide also leads to the same product at high temperature (87). Between the two possible paths, path *a*

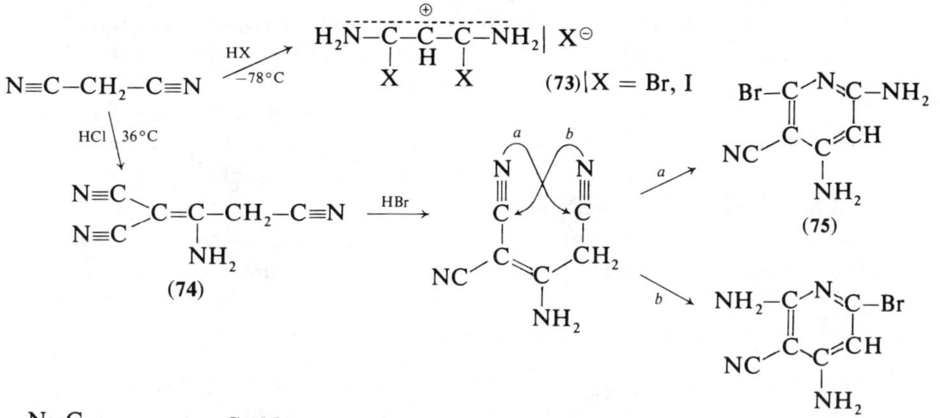

III. DIMERIZATION OF CYANO COMPOUNDS WITH HYDROGEN HALIDES

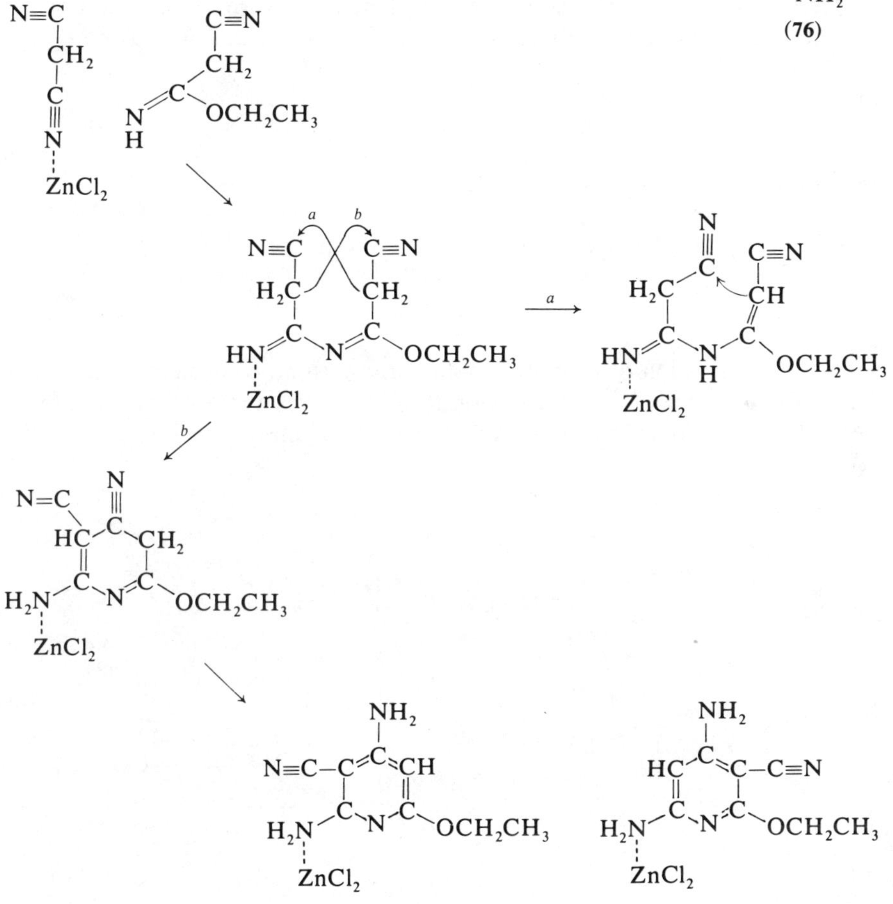

seems more probable in view of the higher reactivity of a nitrilium ion from a cyano group which is attached to a methylene group. Thus **75** is more likely than **76**.

Recently Silver et al. (88) showed that the reaction with ethanol in the presence of zinc chloride or cadmium halides produces the ethoxypyridine **77** or **78**. A mechanism involving both inter- and intramolecular condensations has been proposed (89). Zinc chloride must play an important role not only in the intermolecular dimerization step but also in the intramolecular cyclization. It can not be predicted which is more likely, path *a* or *b*, at present.

3. Cyclization of Dichloromalononitrile

Dichloromalononitrile reacts with hydrogen chloride in ether at 0°C to yield a rather stable adduct. The adduct was assigned structure **79** on the basis of elemental analysis and its methanolysis products (**80** and chlorine gas) (33).

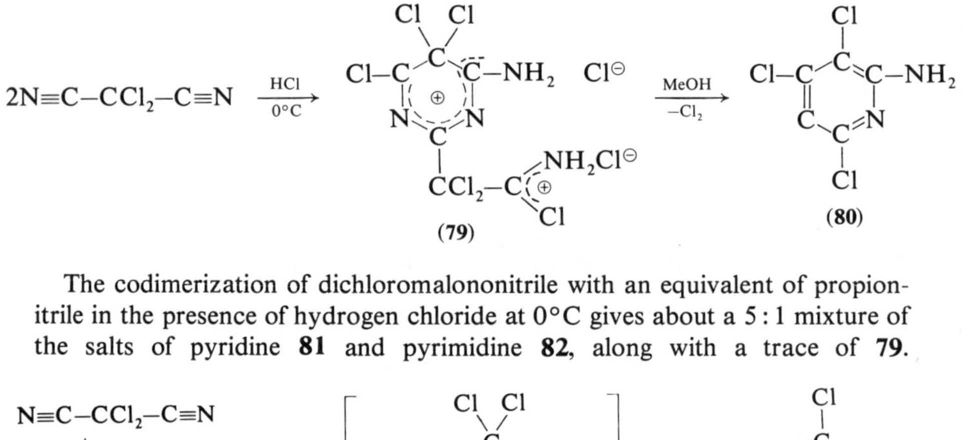

The codimerization of dichloromalononitrile with an equivalent of propionitrile in the presence of hydrogen chloride at 0°C gives about a 5:1 mixture of the salts of pyridine **81** and pyrimidine **82**, along with a trace of **79**.

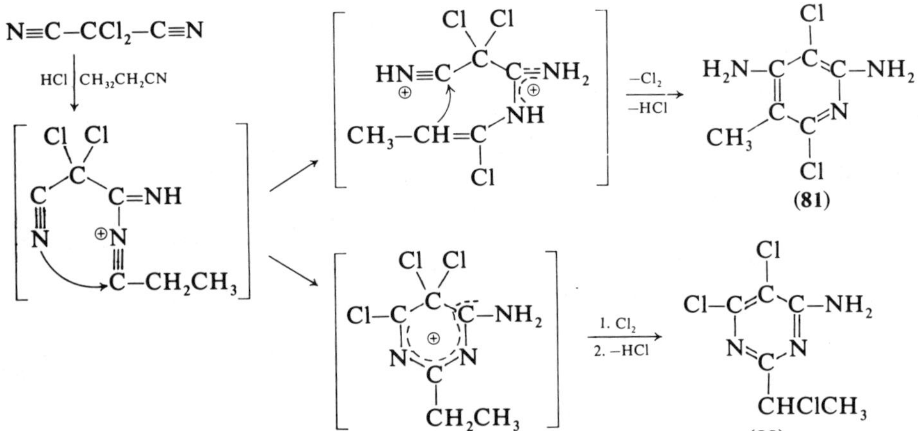

Determination of the scope and limitations of these reactions are now in progress.

IV. Dimerization of Cyano Compounds with Lewis Acids

Nitriles form various complexes with Lewis acids, the compositions of which range from simple 1:1 adducts to more complicated structures. Their physical properties and reactions have been comprehensively reviewed (5, 90). It can be concluded that nitrile–Lewis acid complexes react with electrophiles such as acyl halides and alkyl halides to give active nitrilium salts, which undergo nucleophilic attack by a second mole of nitrile to yield dimeric salts. These reactions usually require vigorous conditions, partly because the nitrilium–Lewis acid salts are rather stable.

A. FORMATION AND REACTIONS OF DIAZAPYRYLIUM SALTS

Aromatic nitrile–Lewis acid ($SnCl_4$, $AlCl_3$, $SbCl_5$) complexes react with acyl chlorides at about 150°C to produce diazapyrylium salts which can be depicted as resonance structures (**83**) (91). The methylthiocyanate–antimony pentachloride complex also reacts with acetyl chloride in a similar fashion, yielding the corresponding salt (**83b**). The benzonitrile–zinc chloride complex, however, reacts with benzoyl chloride to yield the 1:1 adduct **1c** (R = Ph, M = Zn) N-acylnitrilium salts (**1c**) are generally considered to be intermediates in the formation of the pyrylium salts (**83**) (92).

Aryl cyanates react with benzoyl chloride in the presence of aluminum chloride or antimony pentachloride under milder conditions, yielding the corresponding diazapyrylium salts (**83c**) in good yields (19).

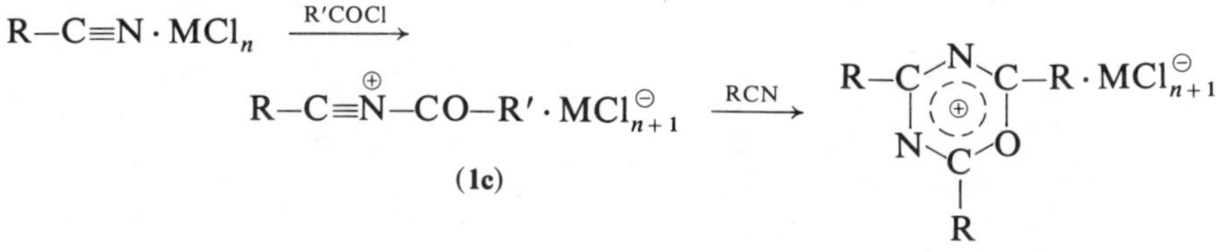

$$R-C\equiv N \cdot MCl_n \xrightarrow{R'COCl} R-C\equiv \overset{\oplus}{N}-CO-R' \cdot MCl_{n+1}^{\ominus} \xrightarrow{RCN}$$

(**1c**)

(**83**)

a: R = Ar, R' = Ar
b: R = MeS, R' = Me
c: R = PhO, R' = Ph

There are some reports of the formation of diazapyrylium salts in the absence of Lewis acids; N,N-disubstituted cyanamides react with benzoyl chloride to form the salts **83d** (93), and phenyl cyanate and phosgene form the

1:1 product **84a**, which reacts further with additional phenyl cyanate to yield the pyrylium salt **83e** (93).

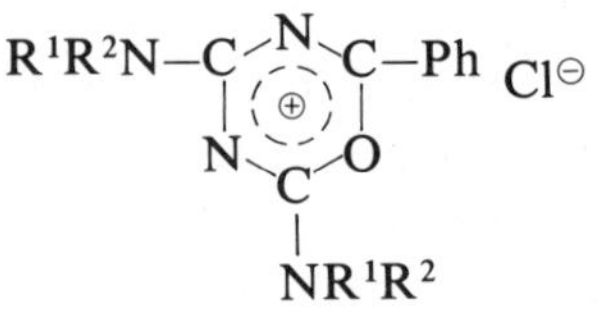

$$R^1R^2N-\underset{\underset{NR^1R^2}{\underset{|}{C}}}{\underset{N\diagdown C\diagup O}{C\diagdown N\diagup C}}-Ph \quad Cl^\ominus$$
(83d)

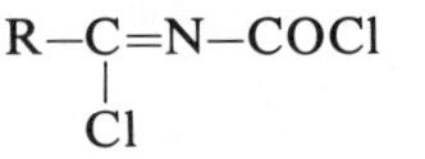

$$R-C=N-COCl$$
$$\underset{Cl}{|}$$
(84)
a: R = PhO
b: R = Ph

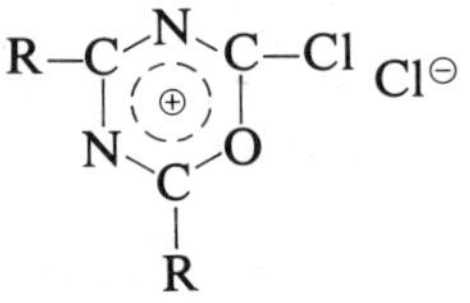

$$R-\underset{\underset{R}{\underset{|}{C}}}{\underset{N\diagdown C\diagup O}{C\diagdown N\diagup C}}-Cl \quad Cl^\ominus$$
(83)
e: R = PhO
f: R = Ph

It is worth noting that N-(α-chlorobenzylidene)carbamoyl chloride (**84b**), obtainable from benzonitrile, phosgene, and hydrogen chloride, undergoes nucleophilic attack by nitriles only in the presence of hydrogen chloride. Aromatic nitriles produce 2-chloro-4,6-diaryl-s-triazine, possibly through the pyrylium intermediate **83f** (95). Aliphatic nitriles having two α-hydrogen atoms afford the pyrimidine derivatives **86** and **87** through the dimeric intermediate **85** (96). In the latter case the initial attack on the carbonyl carbon in **84b** by imidoyl chloride and/or nitrile was also confirmed by the isolation of **88**. However, this is a minor reaction.

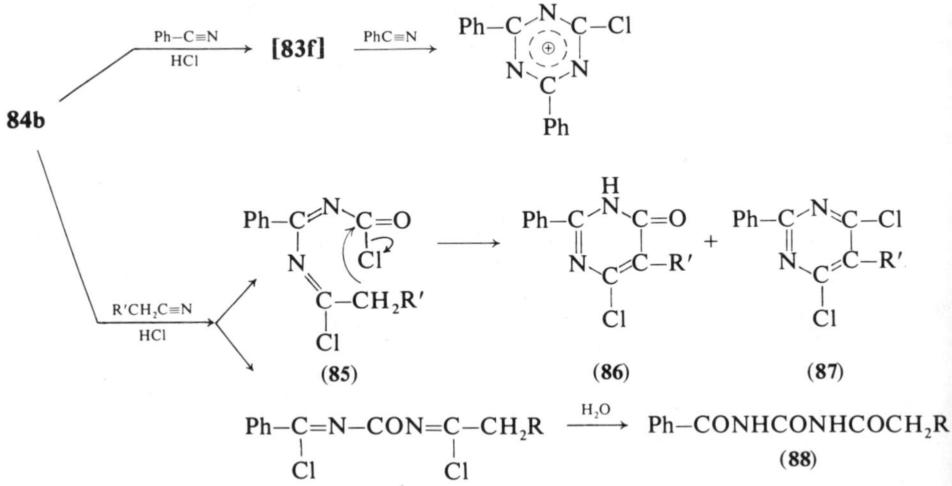

Furthermore, in the reaction of acetonitrile with acetyl chloride, the formation of trimethyldiazapyrylium chloride (**89**) has been proposed on the basis of a conductivity measurement (97).

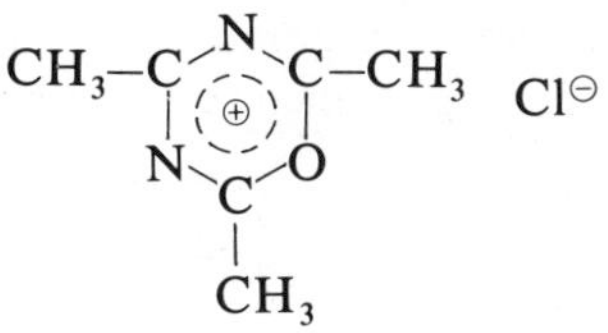

IV. DIMERIZATION OF CYANO COMPOUNDS WITH LEWIS ACIDS

(89)

The pyrylium salts **83a** react with ammonia, urea, and thiourea, giving s-triazine derivatives **90–92**, respectively (98). The reactions with malononitrile, methyl cyanoacetate, and benzoylacetonitrile give pyrimidine derivatives (**93**) (99). Phenylhydrazine and hydroxylamine give triazoles (**94**) and oxadiazoles (**95**), respectively, and hydrolysis affords mixtures of diacylamines and amides (**93**). The metal halides are readily exchanged by perchloride anion on treatment with sodium perchlorate in acetonitrile (98).

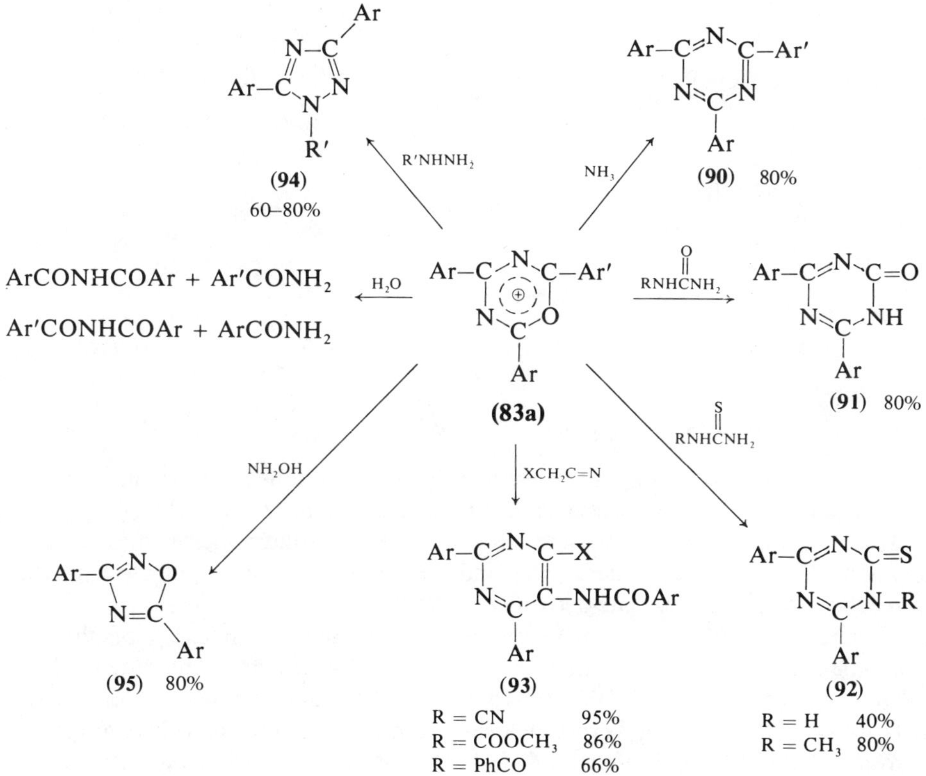

(94) 60–80%

(90) 80%

ArCONHCOAr + Ar'CONH$_2$
Ar'CONHCOAr + ArCONH$_2$

(83a)

(91) 80%

(95) 80%

(93)
R = CN   95%
R = COOCH$_3$ 86%
R = PhCO   66%

(92)
R = H   40%
R = CH$_3$ 80%

Diazapyrylium salts **83c** react with ammonia in ethanol, giving the s-triazine **96**, and hydrolysis affords the ring-opened compound **97** (18).

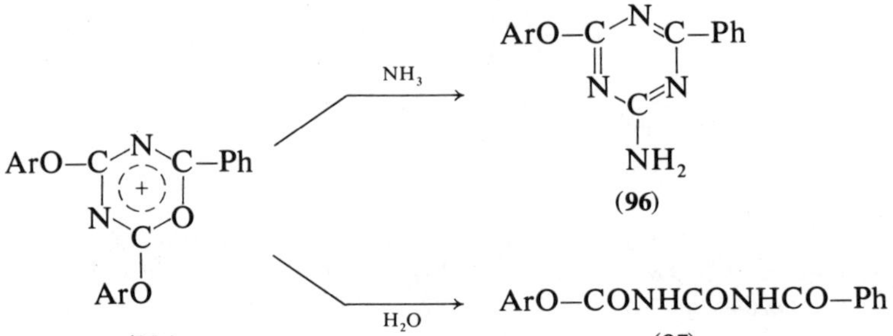

Diazapyrylium salts **83d** also react with ammonia to give s-triazines **98** with one exception (94). The salt from N,N-dimethylcyanamide gives 2-dimethylamino-4-hydroxy-6-phenyl-s-triazine **99**. Pyrolysis of **83d** produces benzonitrile, N,N-disubstituted cyanamide, and N,N-disubstituted carbamoyl chloride (94).

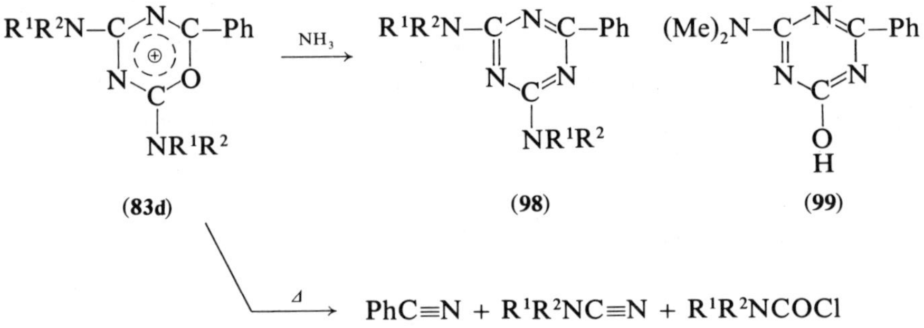

$$\text{(83d)} \xrightarrow{\Delta} PhC\equiv N + R^1R^2NC\equiv N + R^1R^2NCOCl$$

B. MISCELLANEOUS REACTIONS

Nitriles react with alkylating agents to produce N-substituted nitrilium salts (**1**) (100, 101). Meerwein et al. (102) have shown that the N-arylnitrilium salt **1d** reacts with cyano compounds to give dimeric nitrilium intermediates (**1e**), which subsequently undergo a kind of intramolecular Houben–Hoesch reaction to give quinazoline derivatives (**100**).

Martin and Weise (43, 44) recently reported that the reaction of phenyl cyanate–stannic chloride complex with t-butyl chloride gives pseudourea (**103**) on hydrolysis. Dimeric complex **102** was first assumed to arise from **101**. However, in view of the known tendency of nitrile–stannic chloride complexes to adopt a cis configuration (**104**) (90), the formation of **102** from **101** seems

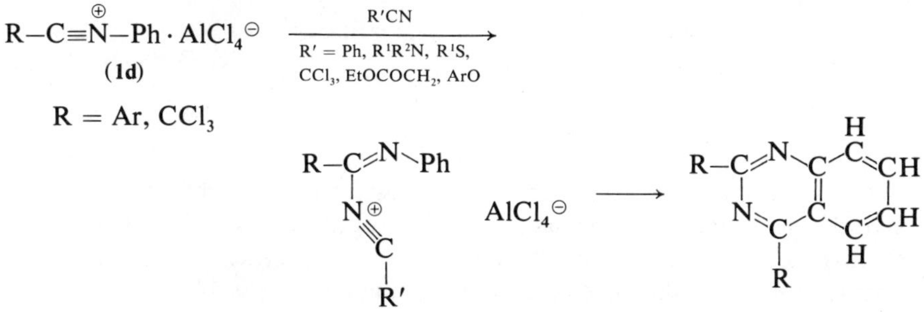

IV. DIMERIZATION OF CYANO COMPOUNDS WITH LEWIS ACIDS

$$R-\overset{\oplus}{C}=N-Ph \cdot AlCl_4^{\ominus} \xrightarrow[\substack{R' = Ph, R^1R^2N, R^1S, \\ CCl_3, EtOCOCH_2, ArO}]{R'CN}$$

(1d)

R = Ar, CCl$_3$

[Structure (1e): R–C(=N–Ph)–N(⊕)≡C–R'] · AlCl$_4^{\ominus}$ ⟶ [Structure (100): quinazoline-type ring with R and N–R]

unlikely. The intermediate **102** probably arises from attack of cyanate on the initially formed nitrilium salt, possibly **1f**.

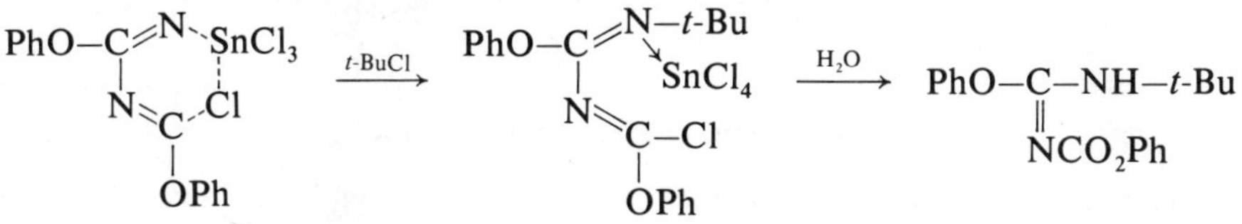

[Structure (101): PhO–C(=N–SnCl$_3$)–N=C(Cl)–OPh] $\xrightarrow{t\text{-BuCl}}$ [Structure (102): PhO–C(=N–t-Bu·SnCl$_4$)–N=C(Cl)–OPh] $\xrightarrow{H_2O}$ [Structure (103): PhO–C(=O)–NH–t-Bu, NCO$_2$Ph]

(101)   (102)   (103)

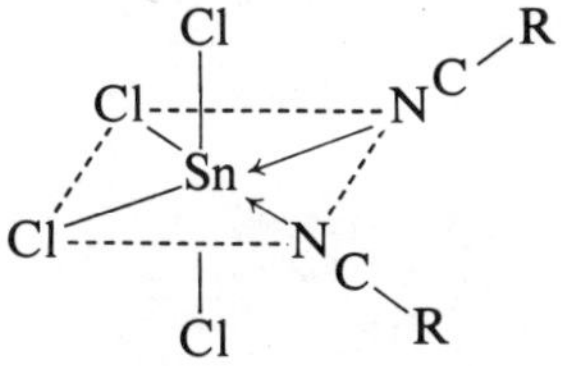

[Structure (104): Sn with Cl ligands coordinated to two N=C–R groups]

$[PhO-\overset{\oplus}{C}\equiv N-t\text{-Bu}]_2 SnCl_6^{2\ominus}$

(1f)

(104)

Klages et al. (52) succeeded in isolating *N*-unsubstituted nitrilium salts (**1g–j**) from the reaction of the corresponding nitrile–Lewis acid complexes with hydrogen chloride at an appropriately low temperature. It is worth noting that the reaction temperature is critical and since disproportionation occurs at higher temperatures, giving iminium salts (**3**) and the starting complexes:

$$R-C\equiv N \cdot SbCl_5 \xrightarrow{HCl} R-\overset{\oplus}{C}=NH \cdot SbCl_6^{\ominus}$$

(1g) R = CH$_3$
(1h) R = *p*-NO$_2$Ph

$$[R-C\equiv N]_2 \cdot SnCl_4 \xrightarrow{HCl} [R-\overset{\oplus}{C}=NH]_2 SnCl_6^{2\ominus}$$

(1i) R = CH$_3$
(1j) R = Ph

$$\text{R-C}\overset{\oplus}{\equiv}\text{NH}\cdot\text{SbCl}_6^{\ominus} \xrightarrow{\Delta} \text{R-C}\underset{\text{Cl}}{\overset{\text{NH}_2\cdot\text{SbCl}_6^{\ominus}}{\langle\oplus}} + \text{R-C}\equiv\text{N}\cdot\text{SbCl}_5$$

(3)

It has recently been found (54) that phosphorus pentachloride–hydrogen chloride, and antimony pentachloride–hydrogen chloride co-catalysts are effective not only in the trimerization of aromatic nitriles to 2,4,6-triaryl-s-triazines but also in the dimerization to the amidinium salts **34**. As Fialkov and Buryanov (103) have pointed out, the following equilibrium, involving nitrilium salt **1k**, probably plays an important role in these reactions:

$$\text{R-C}\equiv\text{N} + \text{HCl} + \text{PCl}_5 \rightleftarrows \text{R-C}\overset{\oplus}{\equiv}\text{NH}\ \text{PCl}_6^{\ominus}$$

(1k)

V. Mechanism of Dimerization

The following equilibrium depicts the interactions and reactions of cyano compounds with hydrogen halides. The direction of the equilibrium depends on reaction temperature, concentration of acids, and the types of cyano compounds and acids involved.

$$\text{R-C}\equiv\text{N} + \text{HX} \rightleftarrows \text{RCN}\cdot n\text{HX} \rightleftarrows$$

$$\underset{(1)}{\text{R-C}\overset{\oplus}{\equiv}\text{NHX}^{\ominus}} \rightleftarrows \underset{(2)}{\text{RC}\underset{X}{\overset{\text{NH}}{\langle}}} \rightleftarrows \underset{(3)}{\text{RC}\underset{X}{\overset{\text{NH}_2\ X^{\ominus}}{\langle\oplus}}}$$

$$R = R^1R^2CH \qquad \underset{(2')}{\underset{R^2}{\overset{R^1}{\rangle}}C=C\underset{X}{\overset{\text{NH}_2}{\langle}}} \qquad \underset{(3')}{\underset{R^2}{\overset{R^1}{\rangle}}C=C\underset{X}{\overset{\overset{\oplus}{\text{NH}_3}X^{\ominus}}{\langle}}}$$

The dimerization reactions can be simply understood as proceeding by attack of a nucleophile, such as the nitrile itself, **2**, or **2′**, on the most reactive electrophile, either nitrilium ion **1** or iminium ion **3**.

Some other mechanisms for the dimerization reactions have been proposed. Lazaris et al. (104) originally proposed the mechanism involving attack by nitriles on the nitrile–halogen acid complexes. Later Zil'berman modified this mechanism by proposing a cyclic electron transfer reaction (1).

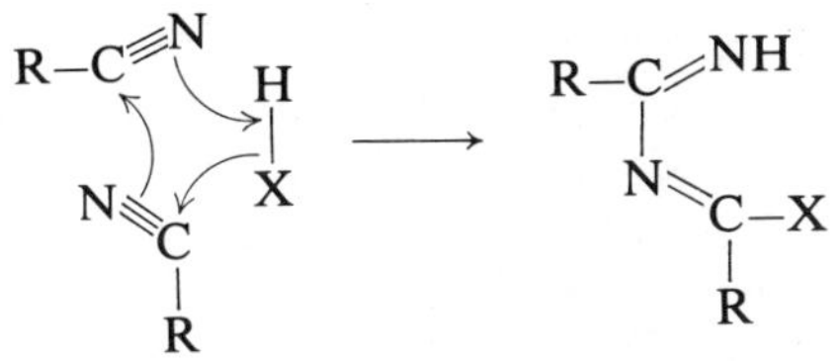

V. MECHANISM OF DIMERIZATION

$$\text{R-C≡N} \cdots \text{H} \quad \text{N≡C-R} \quad \text{X} \longrightarrow \text{R-C(=NH)-N=C(X)-R}$$

Johnson and Madronero (7) proposed a mechanism that proceeds via iminium ion **3** in the intramolecular dimerization of cyano compounds. Our group previously proposed a similar mechanism (39). Ruske, however, proposed a mechanism via nitrilium ion **1** (6).

Recently, iminium intermediates have received much attention as electrophiles. Speziale and Freeman (105) have demonstrated that *N,N*-diethyl-1,2,2-trichlorovinylamine (**105**) reacts with amines to afford the dichloroacetamidine (**107**) through the iminium intermediate **106**. Similarly, the reactions of iminium salts **108** and **110** with aryl cyanates (106) and *N,N*-dialkylcyanamides (107) have been reported to yield amidinium salts **109** and **111**, respectively:

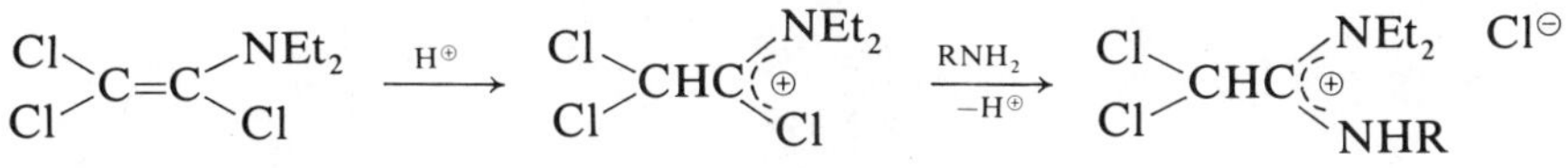

$$Cl_2C=C(Cl)(NEt_2) \xrightarrow{H^\oplus} Cl_2CHC^\oplus(Cl)(NEt_2) \xrightarrow[-H^\oplus]{RNH_2} Cl_2CHC^\oplus(NEt_2)(NHR) \; Cl^\ominus$$

(**105**) \quad (**106**) \quad (**107**)

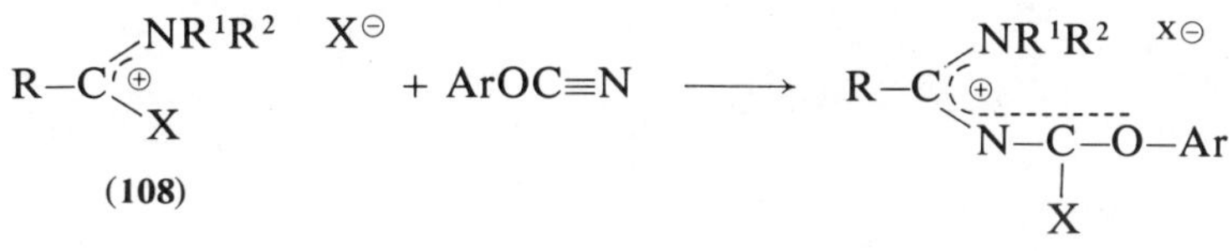

$$R-C^\oplus(NR^1R^2)(X) \; X^\ominus + ArOC≡N \longrightarrow R-C^\oplus(NR^1R^2)(N-C(X)-O-Ar) \; X^\ominus$$

(**108**) \quad (**109**)

$$Cl_2\overset{\oplus}{C}-NR^1R^2 \; Cl^\ominus + R^3R^4N-C≡N \longrightarrow Cl-C^\oplus(NR^1R^2)(N-C(Cl)-NR^1R^2) \; Cl^\ominus$$

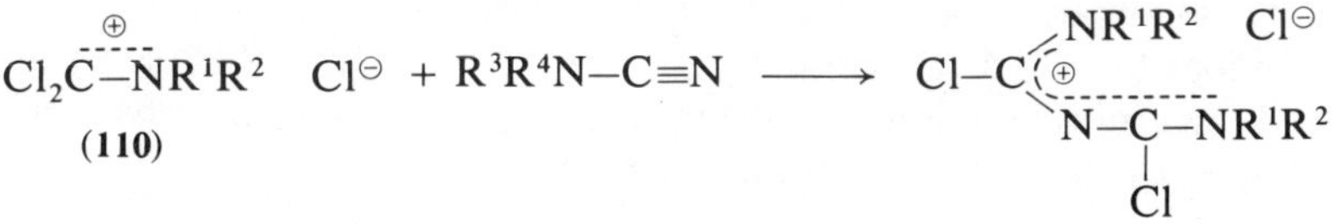

(**110**) \quad (**111**)

On the basis of these reactions, it was once thought that iminium ions **3** might be the most electrophilic species (39). However, on the basis of the following results, it must be concluded that nitrilium intermediates **1** play the decisive role in the dimerization of cyano compounds.

First, the reaction of *N*,*N*-dimethylchloroformamidinium chloride (**3c**) with *N*,*N*-diethylcyanamide resulted in isolation of a mixture of dimeric salts, and no preference for the codimer **29′** was observed. Dimethylcyanamide was also detected in trace amounts. Second, *N*-phenyltrichloromethylimidoyl chloride (**112**) did not react with benzonitrile in the presence of hydrogen chloride alone at either 0 or 100°C. In the presence of Lewis acids, however, the quinazoline derivative **100** was reported to be formed (102). These facts suggest that the two above reactions proceed through nitrilium intermediates **1l** and **1m**, respectively.

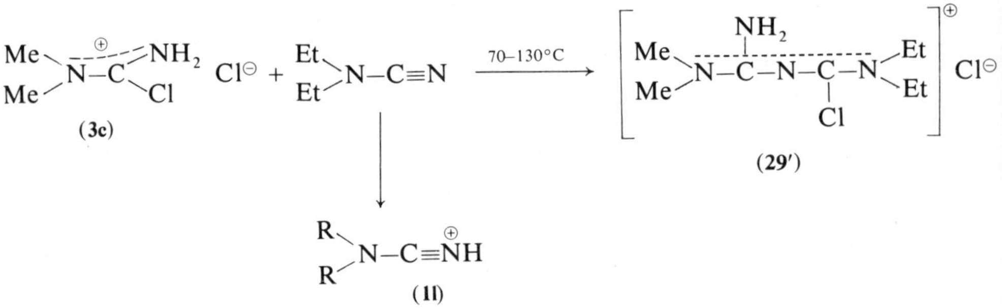

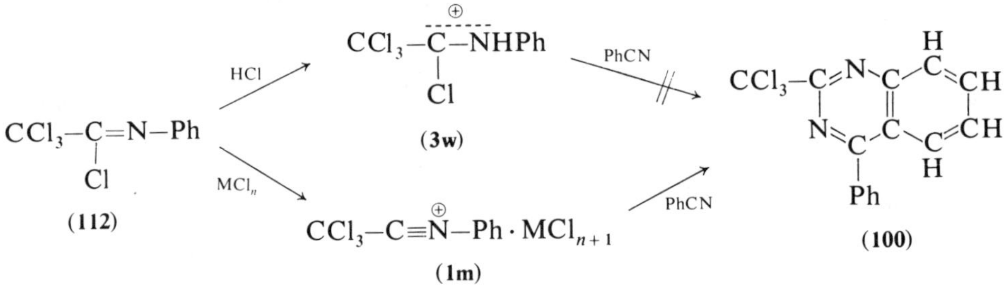

$$\underset{(\mathbf{3c})}{\overset{Me}{\underset{Me}{>}}N\overset{\oplus}{-}C\overset{NH_2}{\underset{Cl}{<}}}\; Cl^{\ominus} + \underset{Et}{\overset{Et}{>}}N-C\equiv N \xrightarrow{70-130°C} \left[\underset{Me}{\overset{Me}{>}}N-\underset{\underset{Cl}{|}}{C}-\overset{\overset{NH_2}{|}}{N}-C-N\underset{Et}{\overset{Et}{<}} \right]^{\oplus} Cl^{\ominus}$$

$$\underset{(\mathbf{1l})}{\overset{R}{\underset{R}{>}}N-C\equiv \overset{\oplus}{N}H}$$

$$\underset{(\mathbf{112})}{CCl_3-\underset{\underset{Cl}{|}}{C}=N-Ph} \xrightarrow[MCl_n]{HCl} \underset{(\mathbf{3w})}{\overset{\oplus}{CCl_3-\underset{\underset{Cl}{|}}{C}-NHPh}} \xrightarrow{PhCN} \underset{(\mathbf{1m})}{CCl_3-C\equiv \overset{\oplus}{N}-Ph \cdot MCl_{n+1}} \xrightarrow{PhCN} \underset{(\mathbf{100})}{CCl_3-\underset{\underset{N}{\parallel}}{\overset{\underset{C}{\parallel}}{C}}\overset{N}{\underset{\underset{Ph}{|}}{\diagdown}}\underset{\underset{H}{|}}{\overset{H}{\underset{C}{|}}}\overset{C}{\underset{C}{\diagup}}\underset{H}{\overset{CH}{\diagdown}}}$$

Klages et al. (52) succeeded in isolating *N*-unsubstituted nitrilium intermediates as their Lewis acid salts **1g–j**, as already noted (see Section IV-B). Recently, Olah and Kiovsky (108) have observed the presence of various nitrilium ions by NMR spectroscopy. Proton, ^{13}C, and ^{15}N NMR spectra have revealed that nitrilium ions have linear configurations with only a limited contribution by the imino carbonium ion form:

$$R-C\equiv \overset{\oplus}{N}H \rightleftarrows R-\overset{\oplus}{C}=NH$$

It was further shown that resonance form **1n′** is a major contributor to the structure of the nitrilium ion **1n**. This means that nitrilium ions bonded to unsaturated carbon atoms are stabilized and thus are less reactive than those

V. MECHANISM OF DIMERIZATION

bonded to saturated carbon atoms. The specificity in the cyclization of unsymmetric dinitriles partially bonded to unsaturated carbons and the relatively slow dimerization of benzonitrile are quite consistent with this interpretation.

$$CH_2=CH-\overset{\oplus}{C}\equiv NH \rightleftarrows {}^{\oplus}CH_2-CH=C=NH$$
$$(1n) \hspace{3cm} (1n')$$

The carbonium ions are stabilized by adjacent electron-donating atoms, such as nitrogen, oxygen, and sulfur, and functional groups such as aromatic rings, or by hyperconjugation (40, 109). The more a carbonium ion is stabilized, the less reactive it is. This is particularly apparent in the case of the nitrilium ions **1** and **1o'**. The results of intramolecular dimerization and codimerization are also quite consistent with this principle.

$$R-X-\overset{\oplus}{C}\equiv NH \rightleftarrows R-\overset{\oplus}{X}=C=NH \quad X = N, O, S$$
$$(1o) \hspace{3cm} (1o')$$

As noted in Section III-B-C, imidoyl halides (**2**), that is, the conjugate bases of iminium salts (**3**), act as nucleophiles toward nitrilium salts. The isolation of imidoyl chlorides **113** (110) and **114** (111) have been reported. However, both reactions need reinvestigation, since isolation of even N-substituted imidoyl chlorides having α-hydrogen atoms are very rare (10).

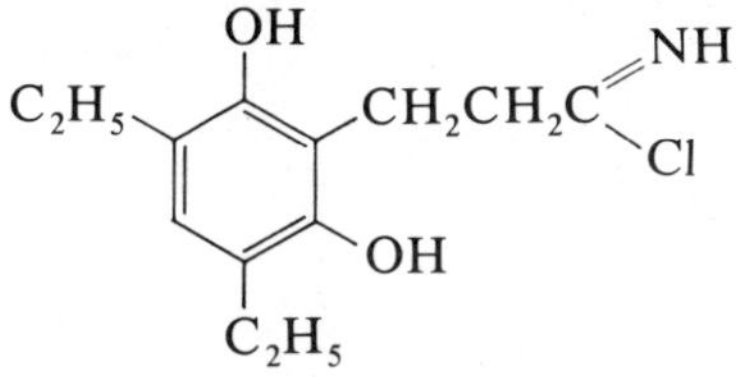

$$R-NH-CH_2C\overset{\displaystyle \nearrow NH}{\searrow Cl}$$

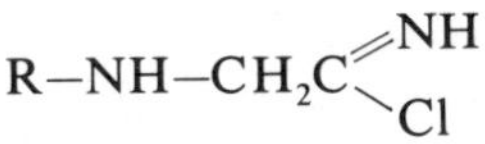

(114)

R = ArCO, ArSO$_2$

(113)

Recently, Ohoka et al. (112) have demonstrated that when nitriles are treated with three equivalents of deuterium chloride for each α-hydrogen in ethyl ether at 0°C for 4 days, they usually undergo deuterium exchange of the α-hydrogen atoms (this was not true for 2-ethylbutyronitrile) (Table IV). Simchen and Kramer (113) also noted the H–D exchange reaction in arylacetonitriles.

The following two proton tautomerizations can explain the exchange reactions:

$$R^1R^2CH-C\equiv N \rightleftarrows R^1R^2C=C=NH$$

$$R^1R^2CH-C\underset{X}{\overset{NH}{\diagdown}} \rightleftarrows R^1R^2C=C\underset{X}{\overset{NH_2}{\diagdown}}$$
(2) (2′)

TABLE IV

H–D Exchange of α-Hydrogens of Nitriles

$$R^1R^2CHC\equiv N + DCl \underset{}{\overset{0°C}{\rightleftarrows}} R^1R^2CDC\equiv N$$

R^1	R^2	% Deuteration[a]
CH_3	CH_3	45
CH_3	CH_3CH_2	18
CH_3CH_2	CH_3CH_2	0
CH_3	$ClCH_2$	82
Ph	CH_3	76
Ph	H	85
CH_3	H	80
$ClCH_2$	H	86

[a] Determined by NMR.

It is generally recognized that the nitrile and the enamine structures are more stable than the ketenimine and the imine tautomers, respectively (114). Thus, it is reasonable to assume that the imidoyl halide (2) once formed could easily be in equilibrium with an α-haloenamine (2′). Accordingly the H–D exchange reaction probably proceeds by the attack of a proton on the β-carbon atom of the α-chloroenamine rather than on that of the ketenimine tautomer:

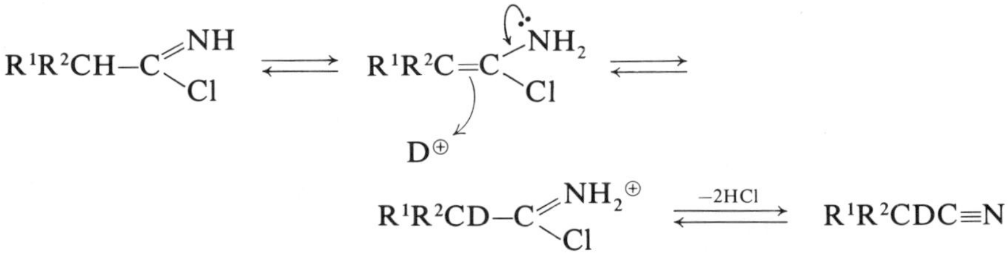

$$R^1R^2CH-C\underset{Cl}{\overset{NH}{\diagdown}} \rightleftarrows R^1R^2C=C\underset{Cl}{\overset{\ddot{N}H_2}{\diagdown}} \rightleftarrows$$

$$D^\oplus$$

$$R^1R^2CD-C\underset{Cl}{\overset{NH_2^\oplus}{\diagdown}} \underset{}{\overset{-2HCl}{\rightleftarrows}} R^1R^2CDC\equiv N$$

This is strong evidence for the contribution of imidoyl halides and α-haloenamines in the reaction of cyano compounds with hydrogen halides.

On the basis of the information obtained so far, the mechanism shown in Scheme 5 may be advanced. The formation of 2:3 adducts such as **36** and **70** must be explained by path b, since 2:2 adducts such as **37** never give the corresponding 2:3 adduct **36** on treatment with hydrogen chloride.

The trimerization to s-triazines observed in the reactions with cyanogen chloride, phenyl cyanate, and trifluoroacetonitrile probably proceed through

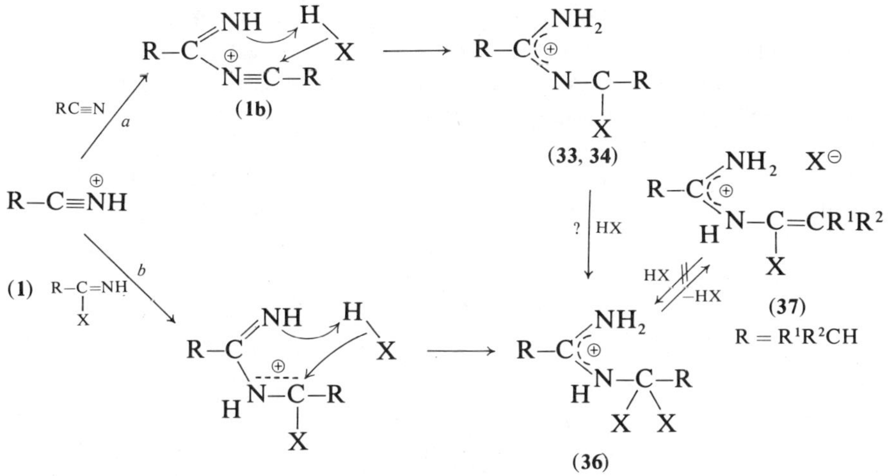

V. MECHANISM OF DIMERIZATION

Scheme 5

further attack by the nitrile, the imidoyl halide **2**, or the conjugate base of dimer salts such as **34** on nitrilium intermediate **1b** in the same fashion.

The dissociation constants pK_{HX} of hydrogen chloride and hydrogen bromide in acetonitrile have been determined to be 8.9 and 5.5, respectively (115), suggesting that hydrogen halides are incompletely ionized and retain their covalent bonding in dilute solution. On the other hand, at low temperature and at high acid concentration, nitriles form 1:1 or 1:5 complexes with hydrogen chloride (116). These are quite similar to nitrile–Lewis acids complexes and nitrile–iodine monochloride complexes (117), and are not ionic. Thus low concentrations and low temperatures are not favorable for dimerization reactions.

Hydrogen iodide and hydrogen bromide are stronger acids than hydrogen chloride, and are preferably used in the formation of nitrilium ion **1**. However, iodide and bromide ions are better nucleophiles than chloride ion. Furthermore, as melting points of the iminium salts **3r–3t**, indicate, the iminium iodide and bromide are more stable than the chloride. This reflects the order of the electron donating ability of halogen atoms (1, 109). Thus the reactions of nitriles with hydrogen bromide lead exclusively to precipitation of the iminium bromides with no intermolecular dimerization.

It has been shown that in 100% sulfuric acid, half the nitriles are protonated and the corresponding amides are formed by subsequent attack of sulfate anion (118). It should be noted that acetonitrile, propionitrile, and benzonitrile all hydrate at comparable rates, whereas chloroacetonitrile, which is less basic, hydrates more rapidly by factors of 3 to 10. This fact is a result of the high reactivity of the chloroacetonitrilium intermediate (118).

The dissociation constant ($pK_{HX} = 7.25$ in acetonitrile), polarographic analysis (119), and Raman spectra (120) indicate that dilute sulfuric acid is also incompletely ionized in nitrile solution. Accordingly nitriles, especially those bonded to electron-withdrawing groups, undergo dimerization in the appropriate concentration of sulfuric acid.

It is reasonable to assume that the dipolar nitrilium ion **116** is an intermediate in the reaction of nitriles with SO_3 (19).

$$R-C\overset{\oplus}{\equiv}N-\underset{\underset{O}{\|}}{\overset{\overset{O}{\|}}{S}}-O^{\ominus}$$

(116)

Although perchloric acid is more completely ionized in nitrile solutions than hydrogen halides or sulfuric acid (121), nitriles do not dimerize readily under these conditions, presumably owing to the lack of nucleophilicity of the perchloride anion. It should be noted that dipropionylamine was isolated by electrolysis using propionitrile as solvent (122). The dimerization of propionitrile may occur under the experimental conditions.

In conclusion, the dimerization of cyano compounds to amidinium salts may be regarded as concurrent Houben–Hoesch and Ritter type reactions. Activation of nitrilium ions by introducing electron-withdrawing groups near the cyano group promises wide applications in synthesis.

REFERENCES

1. E. N. Zil'berman, *Russ. Chem. Rev.*, **31**, 615 (1962).
2. R. Bonnetti, *The Chemistry of the Carbon–Nitrogen Double Bond*, S. Patai, Ed., John Wiley & Sons, New York, 1970, p. 597.
3. F. C. Schafer, *The Chemistry of the Cyano Group*, Z. Rapport, Ed., John Wiley & Sons, New York, 1970, p. 239.
4. A. I. Meyers and J. C. Sircar: ref. 3, p. 341.
5. J. Grundnes and P. Klaboe, ref. 3, p. 123.
6. W. Ruske, *Friedel-Crafts and Related Reactions*, Vol. 3, part 1, G. A. Olah, Ed., John Wiley & Sons, New York, 1964, p. 383.
7. F. Johnson and R. Madronero, *Advances in Heterocyclic Chemistry*, Vol. 6, A. R. Katritzky, Ed., Academic Press, New York, 1966, p. 95.
8. G. Simchen and G. Entenmann, *Angew. Chem. Intern. Ed. Engl.*, **12**, 119 (1973).
9. S. Yanagida and S. Komori, *Synthesis*, **1973**, 189.
10. H. Ulrich, *The Chemistry of Imidoyl Halides*, Plenum Press, New York, 1968.
11. L. I. Krimen and D. J. Gota, *Organic Reactions*, Vol. 17, W. G. Dauben, Ed., John Wiley & Sons, New York, 1969, p. 213.
12. P. Eitner, *Chem. Ber.*, **26**, 2833 (1893).
13. E. N. Zil'berman, *Chem. Chem. Tech. USSR*, 653 (1967).

14. M. Oku, MA Thesis, Osaka University, Faculty of Science, Osaka, 1969.
15. P. Eitner, *Chem. Ber.*, **25**, 461 (1892).
16. H. Weidinger and J. Kranz, *Chem. Ber.*, **96**, 2070 (1963).
17. J. I. Jones and H. M. Paisley, *Chem. Commun.*, 128 (1967).
18. D. Martin and A. Weise, *Chem. Ber.*, **100**, 3736 (1967).
19. E. Grigat, *Angew. Chem. Intern. Ed. Engl.*, **11**, 949 (1972).
20. H. G. Gilbert, *Chem. Rev.*, **62**, 549 (1962).
21. M. Lora-Tamayo and R. Masronero, *1,4-Cycloaddition Reactions*, J. Hamer, Ed., Academic Press, New York, 1967, p. 127.
22. H. Weidinger and J. Kranz, *Chem. Ber.*, **96**, 2070 (1973).
23. H. Weidinger and H. J. Sturn, *Ann. Chem.*, **716**, 143 (1968).
24. S. Ueda, Y. Hayashi, and R. Oda, *Tetrahedron Letters*, 4967 (1969).
25. J. A. Knight and H. D. Zook, *J. Am. Chem. Soc.*, **74**, 4560 (1952).
26. C. Grundmann, G. Weisse, and S. Seide, *Ann. Chem.*, **577**, 77 (1952).
27. L. Claisen and W. Matthews, *J. Chem. Soc.*, **41**, 264 (1882).
28. E. Allenstein, A. Schmidt, and V. Beyl, *Chem. Ber.*, **99**, 431 (1966).
29. C. Grundmann and A. Kreutzberger, *J. Am. Chem. Soc.*, **76**, 5646 (1954).
30. L. E. Hinkel and R. T. Dunn, *J. Chem. Soc.*, **1930**, 1834.
31. L. Claisen and W. Matthews, *Chem. Ber.*, **16**, 309 (1883).
32. M. Kuhn and R. Mecke, *Chem. Ber.*, **94**, 3016 (1961).
33. S. Yanagida et al., unpublished work.
34. I. Hechenblikner, U.S. Pat. 2,768,204 (1956).
35. I. Hechenblikner, U.S. Pat. 2,719,174 (1955).
36. I. Hechenblikner, U.S. Pat. 2,704,297 (1955).
37. I. Hechenblikner, U.S. Pat. 2,768,205 (1956).
38. E. Allenstein and P. Quis, *Chem. Ber.*, **97**, 3162 (1964).
39. S. Yanagida, M. Yokoe, I. Katagiri, M. Ohoka, and S. Komori, *Bull. Chem. Soc. Japan*, **46**, 303 (1973).
40. G. A. Olah, *Science*, **168**, 1298 (1970).
41. L. Schriver, *Bull. Chim. France*, **6** (pt. 1), 1884 (1973).
42. E. Grigat and R. Putter, *Chem. Ber.*, **97**, 3012 (1964).
43. D. Martin and A. Weise, *Chem. Ber.*, **100**, 3747 (1967).
44. D. Martin and A. Weise, *Chem. Ber.*, **100**, 3736 (1967).
45. A. Hantzsch and L. Mai, *Chem. Ber.*, **28**, 2466 (1895).
46. A. Perret and R. Perret, *Bull. Soc. Chim. France*, **7** (5), 743 (1940).
47. R. B. Mooney and H. G. Reid, *J. Chem. Soc.*, **1933**, 1318.
48. E. Allenstein and A. Schmidt, *Chem. Ber.*, **97**, 1286 (1964).
49. A. Hantzsch, *Chem. Ber.*, **64**, 667 (1931).
50. H. Bilte, *Chem. Ber.*, **25**, 2533 (1892).
51. F. Klages and W. Grill, *Ann. Chem.*, **594**, 21 (1955).
52. F. Klages, R. Ruhnau, and W. Hauser, *Ann. Chem.*, **626**, 60 (1959).
53. W. J. Sandner and W. L. Fierce, U.S. Pat. 3,071,586 (1963).
54. S. Yanagida, M. Yokoe, I. Katagiri, M. Ohoka, and S. Komori, *Bull. Chem. Soc. Japan*, **46**, 306 (1973).
55. G. J. Janz and I. Ahmad, *Electrochim. Acta*, **9**, 1539 (1964).
56. K. Wakabayashi, M. Tsunoda, and Y. Suzuki, *Bull. Chem. Soc. Japan*, **42**, 2924 (1969).
57. J. Troeger, *J. Prakt. Chem.*, **46**, 353 (1892).
58. J. Troeger, *J. Prakt. Chem.*, **69**, [2], 347 (1904).
59. A. y. Lazaris, E. N. Zilberman, and O. D. Strizhakov, *J. Gen. Chem USSR*, **32**, 890 (1962).
60. S. Yanagida, M. Ohoka, M. Okahara, and S. Komori, *J. Org. Chem.*, **34**, 2972 (1969).

61. G. J. Janz and S. S. Danyluk, *J. Am. Chem. Soc.*, **81**, 3846, 3850 (1952).
62. E. Allenstein and A. Schmidt, *Spectrochim. Acta*, **20**, 1451 (1964).
63. S. W. Peterson and J. M. Williams, *J. Am. Chem. Soc.*, **88**, 2866 (1966).
64. B. Matkovic, S. W. Peterson, and J. W. Williams, *Croat. Chem. Acta*, **39**, 139 (1967).
65. S. Yanagida, T. Fujita, M. Ohoka, I. Katagiri, and S. Komori, *Bull. Chem. Soc. Japan*, **46**, 292 (1973).
66. A. Gautier, *Compt. Rend.*, **63**, 920 (1866).
67. A. Michael and J. F. Wing, *Am. Chem. J.*, **7**, 71 (1885).
68. E. W. Shand, U.S. Pat. 2,411,064 (1946).
69. L. E. Hinkel and G. J. Treharne, *J. Chem. Soc.*, 866 (1945).
70. H. Herlinger, *Angew. Chem.*, **76**, 437 (1964).
71. K. Wakabayashi, M. Tsunoda, and Y. Suzuki, *Bull. Chem. Soc. Japan*, **42**, 2931 (1969).
72. Y. Pocker, *Molecular Rearrangements*, Part 1, P. Mayo, Ed., John Wiley & Sons, New York, 1963, p. 10.
73. G. A. Olah and C. U. Pittman, Jr., *J. Am. Chem. Soc.*, **88**, 3310 (1966).
74. Y. Kodama, *J. Syn. Org. Chem. Japan*, **21**, 525 (1963).
75. R. Fuks and M. Hartemink, *Tetrahedron*, **29**, 297 (1973).
76. V. W. Laurie and J. S. Muenter, *J. Am. Chem. Soc.*, **88**, 2883 (1966).
77. T.-L. Ho, *Chem. Rev.*, **75**, 1 (1975).
78. N. K. Kochetkov, A. Ya. Khorlin, and K. I. Lopatina, *J. Gen. Chem. USSR*, **29**, 77 (1959).
79. S. Yanagida, T. Fujita, M. Ohoka, R. Kumagai, and S. Komori, *Bull. Chem. Soc. Japan*, **46**, 299 (1973).
80. E. N. Zil'berman and A. Y. Lazaris, *J. Gen. Chem. USSR*, **31**, 1224 (1961).
81. A. Taurin and R. Tanli, *Can. J. Chem.*, **52**, 843 (1974).
82. F. Johnson, private communication.
83. W. A. Nasutavicus, S. W. Tobey, and F. Johnson, *J. Org. Chem.*, **32**, 3325 (1967).
84. L. G. Duquette and F. Johnson, *Tetrahedron*, **23**, 4517, 4539 (1967).
85. R. C. Pearson and J. Sonstag, *J. Am. Chem. Soc.*, **89**, 1827 (1967).
86. E. Allenstein and P. Quis, *Chem. Ber.*, **97**, 1857 (1964).
87. R. A. Carboni, D. D. Coffman, and E. G. Howard, *J. Am. Chem. Soc.*, **80**, 2838 (1958).
88. J. L. Silver, M. Y. Al-Janabi, R. M. Johnson, and J. L. Burmeister, *Inorg. Chem.*, **10**, 994 (1971).
89. F. Freeman, *Chem. Rev.*, **19**, 126 (1965).
90. R. A. Walton, *Quart. Rev.*, **19**, 126 (1965).
91. H. Meerwein, P. Laasch, R. Mersch, and J. Spille, *Chem. Ber.*, **89**, 209 (1956).
92. R. R. Schmidt, *Angew. Chem. Intern. Ed. Engl.*, **12**, 212 (1973).
93. K. Bredereck and R. Richter, *Chem. Ber.*, **95**, 2454 (1966).
94. E. Grigat and R. Putter, DOS 1,668,108 (1968).
95. S. Yanagida, H. Hayama, M. Yokoe, and S. Komori, *J. Org. Chem.*, **34**, 4125 (1969).
96. S. Yanagida, M. Yokoe, M. Ohoka, and S. Komori, *Bull. Chem. Soc. Japan*, **44**, 2182 (1971).
97. R. A. Slavinskoya, L. V. Levchenko, T. N. Sumarokova, and A. V. Karelora, *J. Gen. Chem. USSR*, **39**, 465 (1969).
98. R. R. Schmidt, *Chem. Ber.*, **98**, 334 (1965).
99. I. Shibuya and M. Kurabayashi, *Bull. Chem. Soc. Japan*, **42**, 2382 (1969).
100. R. Fuks, *Tetrahedron*, **29**, 2147 (1973).
101. F. Johnson, L. G. Duquette, W. L. Parker, and W. A. Nasutavicus, *J. Org. Chem.*, **39**, 1434 (1974).
102. H. Meerwein, P. Laasch, R. Mersh, and J. Nentwig, *Chem. Ber.*, **89**, 224 (1956).

103. Y. A. Fialkov and Y. B. Buryanov, *Zh. Obshch. Khim.*, **26**, 1003 (1956); *C.A.*, **50**, 16511 (1956).
104. A. Y. Lazaris, E. N. Zil'berman, and O. D. Strizhakov, *Zh. Obshch. Khim.*, **32**, 900 (1962).
105. A. J. Speziale and R. C. Freeman, *J. Am. Chem.*, **82**, 909 (1960).
106. E. Grigat and R. Putter, *Angew. Chem. Intern. Ed. Engl.*, **6**, 206 (1967).
107. Z. Janousek and H. G. Viehe, *Angew. Chem. Intern. Ed. Engl.*, **12**, 74 (1973).
108. G. A. Olah and T. E. Kiovsky, *J. Am. Chem. Soc.*, **90**, 4666 (1968).
109. T. Nakai, *J. Synthetic Chem. Japan*, **28**, 708 (1970).
110. E. M. Philbin and T. S. Wheeler, *Chem. Ind.*, **1952**, 449.
111. E. Ronwin, *Can. J. Chem.*, **35**, 1031 (1957).
112. M. Ohoka, T. Kojitani, S. Yanagida, M. Okahara, and S. Komori, *J. Org. Chem.*, **40**, 3540 (1975).
113. G. Simchen and W. Kramer, *Chem. Ber.*, **102**, 3656 (1969).
114. J. B. Hendrickson, D. J. Cram, and G. S. Hammond, *Organic Chemistry*, 3rd ed., McGraw-Hill Book Co., 1970, p. 164.
115. I. M. Kolthoff, S. Bruckenstein, and M. K. Chantooni, *J. Am. Chem. Soc.*, **83**, 3927 (1961).
116. F. E. Murray and W. G. Scheider, *Can. J. Chem.*, **33**, 797 (1955).
117. R. S. Mulliken, *J. Phys. Chem.*, **56**, 801 (1952).
118. N. C. Deno, R. W. Gaugler, and M. J. Wisotsky, *J. Org. Chem.*, **31**, 1967 (1966).
119. J. F. Coetzee and I. M. Kolthoff, *J. Am. Chem. Soc.*, **79**, 6110 (1957).
120. E. M. Arnett, *Progress in Physical Organic Chemistry*, Vol. 1, S. G. Cohen et al., Eds., John Wiley & Sons, New York, 1963, p. 279.
121. J. F. Coetzee and D. K. McGuire, *J. Phys. Chem.*, **69**, 1810 (1963).
122. Y. Matsuda, K. Kimura, C. Iwakura, and H. Tamura, *Bull. Chem. Soc. Japan*, **46**, 430 (1973).

ADDENDUM

To Section V

3-Halopropionitriles have now been found by NMR to undergo rapid H–D exchange of the α-hydrogen, compared with aceto and propionitriles, in spite of any PMR spectroscopic evidence suggesting the formation of either the nitrilium or the iminium ions (123). These facts imply that the equilibrium reactions forming nitrilium (**1**) or iminium (**2**) ions might be more rapid at room temperature than the time scale of PMR, or that the H–D exchange path through the ketenimine might be prevailing due to the increased contribution of 3-halo ketenimine tautomers owing to the heterovalent resonance as follows;

$$X-CH_2CH_2-C\equiv N \rightleftarrows X-CH_2CH=C=NH \longleftrightarrow \overset{-}{X}\overset{+}{C}H_2CH=C=NH$$

REFERENCE

123. S. Yanagida, Unpublished work.

N-ACYLPYRIDINIUM SALTS

By A. N. KOST and S. I. SUMINOV, *Department of Chemistry, Moscow State University, Moscow 117234, USSR*; and A. K. SHEINKMAN, *Department of Chemistry, Donets State University, Donetsk, USSR*

CONTENTS

I. Introduction 573
II. Synthesis of Salts of *N*-Acylheterocyclic Cations 574
III. Structure of Salts of *N*-Acylheterocyclic Cations 582
IV. Reactivity of *N*-Acylheterocyclic Cations 587
   A. Acylation with Pyridine Bases 589
   B. Kinetic Studies of Reactions Involving *N*-Acylheterocyclic Cations. Nucleophilic and General Basic Catalysis with Pyridinium Bases 598
   C. Reactions of *N*-Acylpyridinium Cations Involving Ring Opening 608
   D. Reissert Compounds 612
   E. Hetarylation Reactions 616
      1. Hetarylation of Activated Aromatic Rings 617
      2. π-Excessive Heterocycles 619
      3. CH-Acidic Compounds 624
      4. Compounds Containing Lone Electron Pairs 627
      5. Mechanism of the Hetarylation Reactions in the Presence of Acyl Halides . . 629
   F. Single-Electron Transfer Reactions of *N*-Acylheterocyclic Cations 632
   References 634

I. Introduction

This chapter is a systematic review of the available data on synthesis, structure, and chemistry of *N*-acyl derivatives of aromatic six-membered nitrogen heterocyclic compounds, such as pyridine (**1**) and condensed structures **2** and **3**, where X is an electron-withdrawing group such as RCO,

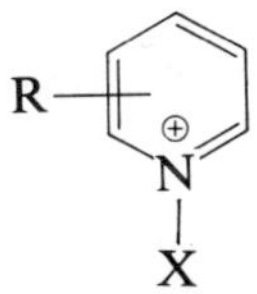

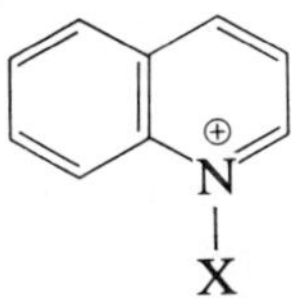

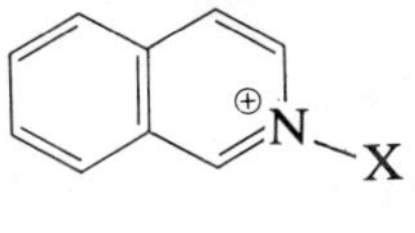

CON(R′)$_2$, COOR′, less frequently CN, P(O)(O$^\ominus$)$_2$, P(O)(OR′)$_2$, and P(O)(NH$_2$)O$^\ominus$.

A recent review (1) provides a brief outline of the structure and reactivity features of these compounds. As for the new branch of chemistry dealing with vinylogues of N-acylpyridinium salts, one should consult the original papers, for example, Ref. 2.

N-Acyl heterocyclic cations are very similar in structure and chemical behavior to their more familiar analogues containing four-coordinate nitrogen atoms [quaternary ammonium salts, (3–5) N-oxides (6, 7), etc.]. Also, they behave similarly to such hetero analogues as imidazolium, pyrylium, and thiapyrylium salts and their benz-fused derivatives (8–12), at least toward some reagents. On the other hand, there are reactions unique to these compounds that justify discussing them as a separate class. The positively charged nitrogen atom adjacent to the carbonyl group facilitates transfer of the acyl group (hydrolysis, acylation) and solvolysis in polar media (with decomposition to the components). During the last 10–15 years, synthetic procedures have been developed that utilize acylation and phosphorylation of nucleophiles with salts of N-acylheterocyclic cations (including *in situ* reactions), and techniques for the introduction of heterocyclic rings into such CH acids as ketones, dialkylanilines, pyrrole, furan, and indole have been proposed [such a reaction is named "hetarylation" (1, 13)]. Thus it became possible to correlate nucleophilic reactivity of pyridine bases and electrophilic reactivity of acylating and phosphorylating agents on the one hand, and ease of the formation and stabilities or reactivities in various reactions of salts of acylheterocyclic cations on the other.

II. Synthesis of Salts of N-Acylheterocyclic Cations

Reactions of pyridine and analogous nitrogen bases with acyl halides may follow different pathways leading to solvates (*a*), N-acylheterocyclic cations (*b*), and ketenes (*c*).* The product ratio and the preferential formation of some of the products depend on electronic and structural factors and on the reaction

$$R\text{-Py} + R'COX \rightleftharpoons \begin{array}{c} \text{solvate, } nB \cdot mR^1COX \\ \text{ion pair} \\ R^1CH{=}C{=}O + B \cdot HCl \end{array} \rightleftharpoons R\text{-Py}^+\text{-COR}^1 \; X^\ominus$$

(with R^{1}CH$_2$COX)

---

*For preparation of ketenes by dehydrochlorination of acyl chlorides in the presence of tertiary amines, see refs. 14 and 15. The same reaction has been recommended for the preparation of pure hydrochlorides of tertiary amines. These can be isolated in quantitative yields under certain conditions (16, 17).

II. SYNTHESIS OF SALTS OF N-ACYLHETEROCYCLIC CATIONS

conditions (including *temperature, solvent, molar ratio of the reactants, and the order of introducing them*). In many cases, stabilities of N-acylammonium salts also depend on the nature of the counterion (see below).*

The first reports pointing to the formation of crystalline adducts of unknown structure from heterocyclic nitrogen bases and acyl halides appeared as early as the 1860s (22–25). The adducts were described as unstable hygroscopic products which rapidly darkened in air. For a long time, acylpyridinium salts were believed nonexistent (26–28). Later, pyridine was shown to react exothermally with acetyl chloride in petroleum ether to produce crystalline acetylpyridinium chloride (29–31), with pyridinium hydrochloride and ketene formed as by-products (32–34). Ketene partially polymerized to dehydroacetic acid (16, 22, 33). Quinoline, quinaldine, and 2,6-lutidine yielded mixtures of their hydrochlorides and the corresponding adducts under the same conditions (32, 35).

At present N-acylpyridinium chlorides are prepared at -60 to $-65°C$ by addition of pyridine to an equimolar quantity of acetyl chloride in waterfree ether under dry nitrogen (36–41), and the product is washed repeatedly with waterfree benzene and ether. The snow-white crystals thus obtained can be stored in sealed ampuls in the cold for a time, and even recrystallized provided certain precautions are taken. The same procedure proves useful for the synthesis of salts of quinoline, quinaldine, and acridine (43) (Table I).

Picolines reacted with acetyl chloride in the cold to yield unstable colored products (31). In the presence of $TiCl_4$, the reaction of quinoline and CH_3COCl gave two complexes: $(C_9H_7NCOCH_3)_2 \cdot TiCl_6$ and $(C_9H_7NCOCH_3) \cdot TiCl_6(COCH_3)$. A precipitate of the composition $(C_6H_7NCOCH_3) \cdot SnCl_6(COCH_6)CH_3COCl$ resulted from mixing equimolar quantities of 2-picoline, CH_3COCl, and $SnCl_4$ (55). One paper (56) reported the formation of similar salts from pyridine bases, benzoyl chloride, and Lewis acids. As one would expect, the stability of acylpyridinium salts increased when substituents occurred that could participate in delocalization of the

(4)

---

* Recently, systems comprising a tertiary amine and an acyl chloride have been shown to give rise to charge-transfer complexes; see, for example, refs. 18–20. This can hardly be the case with pyridine derivatives. Thus according to ref. 21, pyridine and its methyl homologues are likely to play the role of π donors in complexes with tetrahalophthalic anhydrides.

TABLE I
Some Salts of Pyridine Bases and Acyl Chlorides Isolated as Crystalline Products

Base	Acylating agent	Reaction conditions		m.p., °C	Ref.
		Solvent	Temperature, °C		
1	2	3	4	5	6
Pyridine	Acetyl chloride	Chloroform	Room	—	29
		None	Room	100 (dec)	30
		Excess acetyl chloride	—	97–99 (dec)	31, 37, 39
		Ether	−65	104–107	36, 42, 44–46
	Propionyl chloride	Ether	−65	108–109	36, 42, 44
	Butyryl chloride	Ether	−65	110–114	36
	Lauryl chloride	None	Room	—	47
	Furoyl chloride	Petroleum ether	−20	60 (dec)	30
	Cinnamoyl chloride	Ether	Slightly cooled	122–124	48
	5-(Nitrofuryl)acryloyl chloride	Ether	−65	189–191	49
	o-Nitrobenzoyl chloride	Benzene	50–60	149–150	50
	m-Nitrobenzoyl chloride	Benzene	50–60	124–125	47, 50
	p-Nitrobenzoyl chloride	Benzene	55–60	200–202	34, 36
	p-Chlorobenzoyl chloride	Petroleum ether	Slightly cooled	185–187	36
	3,5-Dinitrobenzoyl chloride	Benzene	100	110–112	36

Base	Acyl halide	Solvent	Temperature	mp	Ref
	Trifluoroacetic anhydride	Ether	−78	38.5−40 (dec)	37, 39
2-Picoline[a]	Acetyl chloride	Excess acetyl chloride	Cooled	61−63	31
3-Picoline	Acetyl chloride	Excess acetyl chloride	Cooled	77−78	31
4-Picoline	Acetyl chloride	Excess acetyl chloride	Cooled	125−127	31
4-(Dimethyl-amino)pyridine	Acetyl chloride	—	—	106	51
4-Vinylpyridine	Acetyl chloride	Ether	0	151	37, 39
	Trifluoroacetic anhydride	Ether	−78	75−89	37, 39
Quinoline[b]	Acetyl chloride	Petroleum ether	−60 to −75	86−87	53
	p-Nitrobenzoyl chloride	Petroleum ether	−60 to −75	124−125 (dec)	53
2-Methylquinoline	Benzoyl chloride	Petroleum ether	−60 to −75	207−208 (dec)	52
Isoquinoline[b]	Acetyl chloride	Benzene or ether	0	—	53
	Benzoyl chloride	Benzene or ether	0	—	53
	3,4-Dimethoxybenzoyl chloride	Benzene or ether	0	—	53
	p-Nitrobenzoyl chloride	Tetrahydrofuran	0−5	210−212 (dec)	54
Acridine[b]	Acetyl chloride	Ether	−5	238−239	32, 53
	p-Nitrobenzoyl chloride	Tetrahydrofuran	0−5	230−232	54
Benzo[f]quinoline	p-Nitrobenzoyl chloride	Tetrahydrofuran	0−5	208−210 (dec)	54
Phenanthridine	p-Nitrobenzoyl chloride	Tetrahydrofuran	0−5	194−195 (dec)	54
4,9-Diazapyrene[c]	p-Nitrobenzoyl chloride	Tetrahydrofuran	0−5	340−342 (dec)	54

[a] Adduct with composition $2C_6H_7N \cdot 3CH_3COCl$.
[b] Holeček reported the isolation of crystalline bromides of N-acetylpyridinium, -quinolinium, -isoquinolinium, and -acridinium, but without giving any experimental details[43].
[c] 4,9-Diacylium salt was isolated.

positive charge (51, 57, 58). Thus 4-aminopyridine reacted in ether with acetyl chloride (or benzoyl or toluenesulfonyl chlorides) to yield the corresponding salts (**4**).

However, the replacement of chlorine with the acetate anion in (**4**) resulted in a rapid conversion of the salt to 4-acetamidopyridine. 1-Acetyl-4-(dimethylamino)pyridinium chloride seemed to be even more stable (51, 58), but 4-acetamidopyridine reacted with CH_3COCl in tetrahydrofuran to give an unstable quaternary salt (58).

The interaction of benzoyl chloride with quinoline gave the disolvate $C_9H_7N \cdot 2C_6H_5COCl$ (56), whereas with a 1:1 molar ratio of the reactants a mixture of N-benzoylhetaryl chloride and the hydrochloride of the base was usually formed (50).* So far, attempts to isolate N-benzoylpyridinium chloride have been unsuccessful (34–36, 59), but aroyl halides bearing electron-withdrawing substituents in the ring formed stable acylpyridinium salts (34, 36). With p-nitrobenzoyl chloride, the reaction went at elevated temperatures (50–60°C), and with 3,5-dinitrobenzoyl chloride at as high as 100°C (36). Cinnamoyl chloride also yielded a readily crystallizing pyridinium salt (48). A recent paper by Olah and Szilagyi (61) reported the acylation of pyridine with solutions of acylium fluoroantimonates in SO_2 to result in N-acetyl-, propionyl-, and benzoylpyridinium fluoroantimonates:

$$\langle\text{Py}\rangle N + RCO^{\oplus}SbF_6^{\ominus} \xrightarrow{-60°C} \left[\langle\text{Py}\rangle\overset{\oplus}{N}-COR\right] SbF_6^{\ominus}$$

N-Nitro- and N-nitrosopyridinium fluoroborates were prepared in a similar way (62).

Various carbamoyl chlorides gave very stable adducts with pyridine, quinoline, and isoquinoline (53, 63–67). Diphenylcarbamoylpyridinium chloride could be recrystallized from 10% hydrochloric acid or converted, under appropriate conditions, to the corresponding iodide, chloroplatinate, and picrate (64, 65). Dimethylcarbamoyl chloride reacted with pyridine to give adduct (**5**) which sublimed in vacuum without decomposition (66). Similar salts were reported for 4-picoline, 3,5-lutidine, and nicotinamide (66, 68);

$$\left[\langle\text{Py}\rangle\overset{\oplus}{N}-CON(CH_3)_2\right] Cl^{\ominus}$$

(**5**)

---

* Reissert (60) failed to isolate the product of the reaction between benzoyl chloride and quinoline. His attempts to apply the Schotten–Baumann technique (using aqueous alkalis) led to the isolation of "pseudobases" of a number of N-acylheterocyclic cations.

however, no reaction occurred in a mixture of α-substituted pyridines (2,6-dimethyl- and 2,6-dimethoxypyridines, 2,4,6-collidine) and $(CH_3)_2NCOCl$, perhaps because of steric hindrances.

Pyridine and its homologues reacted with such imidoyl chlorides as (o-$CH_3C_6H_4N=CCl)_2$, $CH_3COCCl=NNHAr$, and $ArNHN=CClO_2R$ (69–73) to produce stable compounds **6–8**. When heated above 160–180°C, adducts **8** underwent conversion to new heterocyclic structures (72).

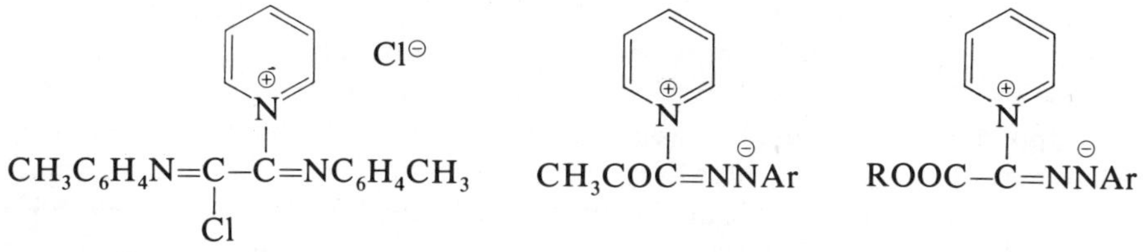

$$\text{Py}^+\text{—N}(CH_3C_6H_4N=C(Cl)-C=NC_6H_4CH_3)\quad Cl^- \qquad \text{Py}^+\text{—N}(CH_3COC=NNAr)^- \qquad \text{Py}^+\text{—N}(ROOC-C=NNAr)^-$$

(6) \qquad (7) \qquad (8)

According to the patent literature (24, 25, 74), addition of methyl chlorocarbonate to solutions of pyridine or 2-picoline in benzene led readily to precipitation. The precipitates contained 2 moles of the base per 1 mole of the chlorocarbonate and yielded symmetric or nonsymmetric carbonates when treated with water or aliphatic alcohols (75). Later, the formation of crystalline adducts of ethyl carbonate with pyridine, 2-picoline, or quinoline was repeatedly reported (25, 76–82); however, attempted isolation of the pure salts failed. Increasing temperature caused their decomposition to pyridine, pyridine hydrochloride, CO_2, and ethyl chloride or ethylpyridinium chloride (74, 80, 81), and trace amounts of moisture promoted the formation of diethyl carbonate. However, dihydro derivatives of some of the adducts could be isolated (83):

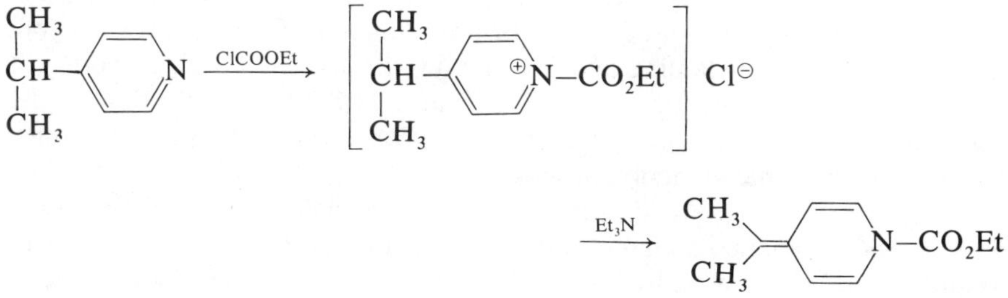

$$\text{(CH}_3)_2\text{CH-Py-N} \xrightarrow{\text{ClCOOEt}} [\text{(CH}_3)_2\text{CH-Py}^+\text{-N-CO}_2\text{Et}]\,Cl^- \xrightarrow{\text{Et}_3\text{N}} (CH_3)_2C=\text{Py}=N-CO_2Et$$

For the reaction of pyridine bases with phosgene, see refs. 84 and 85.

Like acetyl chloride, *t*-butoxycarbonyl chloride reacts with 4-dimethylaminopyridine in waterfree ether at 0°C to give the salt **9** (86). Since this salt is much more stable than the initial ester (it remains unchanged during several

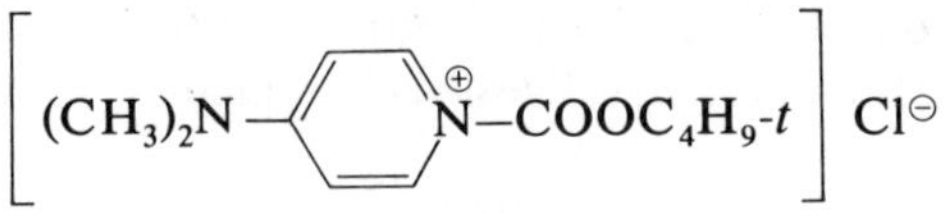

$$[(CH_3)_2N-\langle\rangle-\overset{\oplus}{N}-COOC_4H_9\text{-}t]\ Cl^{\ominus}$$

(9)

months in the cold) it has been recommended in the patent literature (87–89) for the introduction of the *t*-butoxycarbonyl function into amines and amino acids in aqueous solutions.

Pure adducts of pyridine bases and cyanogen bromide were not isolated. In the case of quinoline, isoquinoline, and acridine, the reaction went to the corresponding pseudobases in wet ether, for example, 1-cyano-2-oxydihydroquinolines (**10**) (90–92); ethers of type **11** occurred as by-products. Compound **10** was converted to 1-cyanoquinolinium chloride with hydrochloric acid (89). Johnson (93, 94) utilized this scheme in syntheses of quite a number of 1-cyanoquinolinium and isoquinolinium fluoroborates with various substituents in the rings.

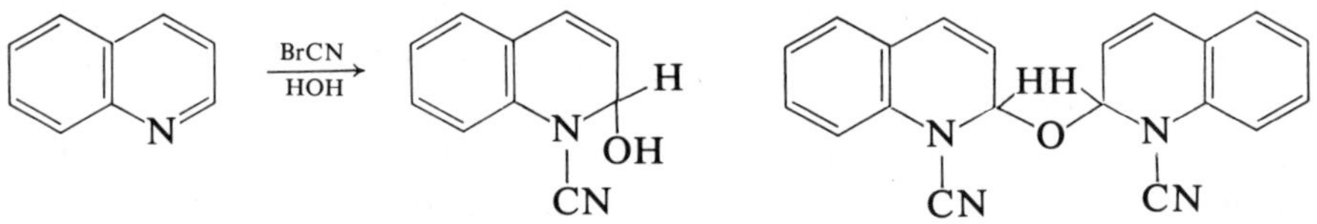

(10)   (11)

Pyridine, quinoline, isoquinoline, and 4-picoline readily formed adducts with various substituted arylsulfonyl chlorides $RC_6H_4SO_2Cl$ (95–100). Stabilities of the adducts toward hydrolysis decreased on going from isoquinoline to pyridine and quinoline (96); in the case of polysulfonyl chlorides, the hydrolysis could be blocked at the step of substitution of only one SO_2Cl group (97).

On the other hand, *p*-acetamidobenzenesulfonyl chloride proved non-reactive toward triethylamine and pyridine in dry acetone (101). A pure 1:1 adduct resulted from mixing ether solutions of $C_6H_5SO_2Cl$ and pyridine, under conditions excluding the presence of water. Otherwise, pyridinium benzenesulfonate was the main reaction product (99).

A higher stability of *N*-alkyl- or *N*-arylsulfonylpyridinium salts compared to their *N*-acyl analogues manifested itself in the fact that adducts **12** could be isolated starting from anhydrides of methane- and benzenesulfonic acids (102, 103).

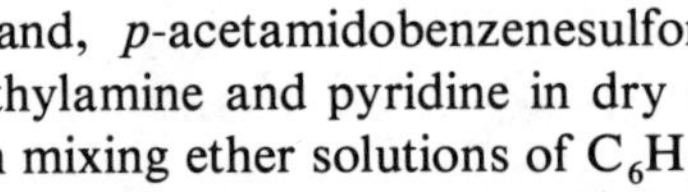

$$\langle\rangle\overset{\oplus}{N}-SO_2R \quad RSO_2O^{\ominus}$$

(12)

II. SYNTHESIS OF SALTS OF N-ACYLHETEROCYCLIC CATIONS

Buncel's review (104) gives a detailed account of reactions of pyridine and other tertiary amines with chlorosulfonates, $ROSO_2Cl$. Usually the products were alkyl halides and pyridine sulfotrioxide, resulting from the decomposition of intermediate quaternary salts $ROSO_2N^\oplus C_5H_5Cl^\ominus$ at temperatures that depended on the nature of the group R (thus the decomposition occurred at $-80°C$ with n-butyl but at 0 to $-20°C$ with iso or tertiary radicals) (105). However, $C_6H_5OSO_2Cl$ did not react with pyridine and quinoline in the cold (106). Addition of pyridine to cooled solutions of alkylchlorosulfinates gave a mixture of 1-acyl and 1-alkyl salts, indicating that pyridine attacked both the sulfur atom and the S—OR bond simultaneously (106, 107):

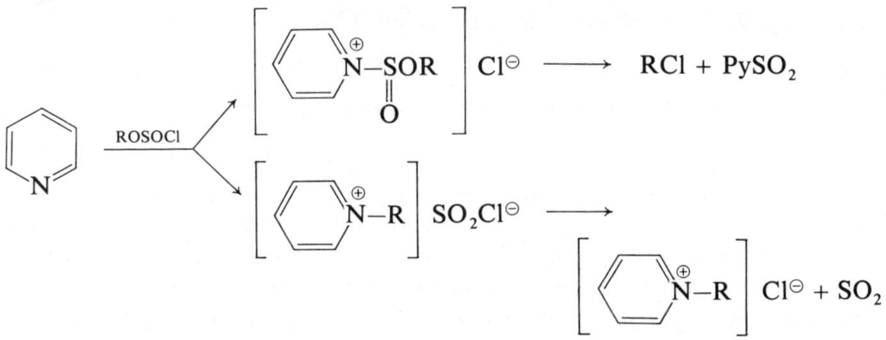

$$Py \xrightarrow{ROSOCl} \begin{cases} [Py^\oplus\text{-SOR}(=O)] Cl^\ominus \longrightarrow RCl + PySO_2 \\ [Py^\oplus\text{-R}] SO_2Cl^\ominus \longrightarrow [Py^\oplus\text{-R}] Cl^\ominus + SO_2 \end{cases}$$

Interactions of pyridine and its derivatives with $SOCl_2$, SO_2Cl_2, $ClSO_3H$, $POCl_3$ and other inorganic chlorides were also studied (108–112). The crystalline adducts that could be isolated in some cases (e.g., with $POCl_3$ or $POBr_3$) (110) should be regarded as molecular complexes.

In recent years, many pyridinium quaternary salts with phosphoric acid residues attached to the nitrogen atom were synthesized. For instance, 4-dimethylamino- and 4-morpholinopyridines reacted with an excess of $POCl_3$ and triethylamine to give, after alkaline hydrolysis, amorphous salts **13** (113).

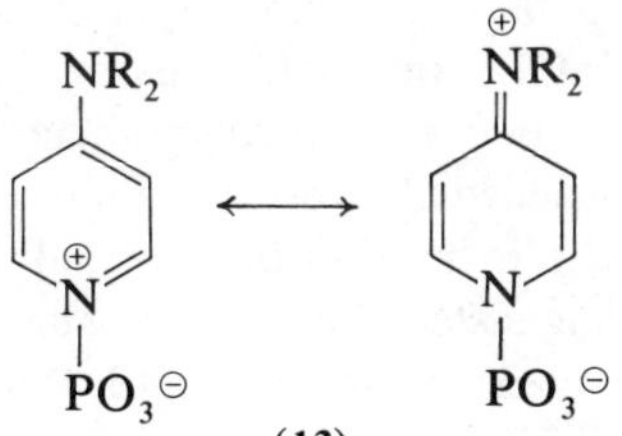

(13)

Similar salts of 4-amino, 4-isopropylamino-, and 4-dimethylaminopyridines were prepared from N-phosphoryl salts of 1-methylimidazole (51, 114) by exchange reactions (51, 114). A fresh approach to pyridine N-phosphonium salts (e.g., **14**, **15**) and their alkyl homologues (115–119) suggested

$$\text{pyridine} + (RO)_2P(O)H \xrightarrow{HgCl_2} \underset{(14)}{\begin{array}{c}\text{Py}^+\\ {}^-O-P(Cl)(OR)_2\end{array}} \xrightarrow{XH} \underset{(15)}{\begin{array}{c}\text{Py}^+\\ {}^-O-P(X)(OR)_2\end{array}}$$

X = OR', NHR', OCOR'

cooxidation of $P^{3\oplus}$ esters and pyridinium bases. Adducts of type (15) were obtained by treating the oxidation products with alcohols, amines, or acids (115, 116). For dithiophosphoric salts, see ref. 120.

III. Structure of Salts of *N*-Acylheterocyclic Cations

Even in the earliest papers (23, 29) adducts of pyridine and acyl chlorides were treated as salt-like structures. The arguments for this were as follows: insolubility of these compounds in most nonpolar solvents, the fact that the most stable of them could be recrystallized, rather high melting points of the adducts, their ability to enter into anionic exchange reactions, and the specific nature of the acylation reactions that these adducts participated in, which always went rapidly and in one direction. Structures of complex and even of molecular compounds also were considered.

During the last 10 years, data have been collected that demonstrate the validity of the ionic structure, with the acyl group bonded to the positively charged nitrogen atom and a halogeno counterion. This structure was substantiated by measurements of conductivity, heats of mixing, UV, IR, and NMR spectra, kinetic studies, and results obtained with polarographic and acid–base titration techniques.

Paul and co-workers (31, 35, 55, 56, 121) carried out titration of solutions of pyridine, picolines, quinoline, and dimethylaniline in acetyl and benzoyl chlorides with $TiCl_4$, $SnCl_4$, or $AsCl_5$ using visual or conductometric techniques. The reactions between nitrogen heterocycles and Lewis acids (chloride ion acceptors) proceeded to completion via one or two neutralization points corresponding to the formation of "normal" and "acid" salts (35, 55, 121, 122). These results suggested the initial formation of solvates of the type $R_3N \cdot R'COCl$ which behaved as ion pairs and underwent dissociation to the *N*-acylium ion and chloride ion during the titration (thus representing unsolvated bases):

$$B + AcCl \longrightarrow B \cdot AcCl \rightleftarrows BAc^\oplus + Cl^\ominus \quad \text{(formation of base)}$$
$$AcCl \rightleftarrows Ac^\oplus + Cl^\ominus$$

$$2AcCl + MCl_4 \rightleftharpoons 2Ac^\oplus + MCl_6{}^{2\ominus} \rightleftharpoons Ac^\oplus + AcMCl_6$$
(formation of acid)

$$2BAc^\oplus + MCl_6{}^{2\ominus} + 2Cl^\ominus + 2Ac^\oplus \longrightarrow (BAc)_2MCl_6 + 2AcCl$$
(neutralization reactions)

or

$$BAc^\oplus + MCl_6{}^{2\ominus} + Cl^\ominus + 2Ac^\oplus \longrightarrow (BAc)_2MCl_6 + AcCl$$

The same authors observed a rather high conductivity of solutions of 2-, 3-, and 4-picolines in acetyl chloride, though somewhat lower than in the case of benzyltrimethylammonium chloride (31, 123). Solutions of crystalline N-benzoyltriethylammonium and N-acetylpyridinium chlorides in SO_2 at $-30°C$ showed similar conductivities, corresponding to those of standard ionic salts such as $(C_2H_5)_3N^\oplus HSbCl_6{}^\ominus$ and $R_3O^\oplus SbCl_6{}^\ominus$ (124). Of course these results depended to a certain degree on the highly ionic nature of the solvent used.

Holeček (125) studied conductivities of acetyl bromide solutions of pyridine, isoquinoline, quinoline, and acridine to show that in these systems the adducts $(B \cdot CH_3CO)^\oplus Br^\ominus$ occurred which behaved as 1:1 electrolytes. A rapid increase of the salt conductivity with decreasing concentrations indicated an increased amount of dissociated ion pairs, in agreement with Paul's results (see below) (126, 127). Determination of the ionization constants gave values of 0.7×10^{-4} for pyridine and $1.8-1.9 \times 10^{-4}$ for quinoline, isoquinoline, and acridine salts.

More recently, Bogatkov et al. (128) performed potentiometric determinations of thermodynamic equilibrium constants for reactions of benzoyl chloride and pyridine or triethylamine in acetonitrile, and obtained a higher value for pyridine (1.2×10^{-4}) compared to triethylamine (2×10^{-5}) despite a higher basicity of the latter. The authors attributed their result to the resonance stabilization of the N-benzoylpyridinium salt.*

Mesomerism might hardly be expected with unsubstituted N-acylpyridinium salts (124, 130) and hence these should have low stabilities, as mentioned in the preceding section. The stabilizing effect of changes of the 1-acyl substituent and of introduction of the 4-amino substituents has already been indicated (58, 66, 86). Kinetics of hydrolysis and nucleophilic substitution reactions of N-acylpyridinium salts (e.g., ref. 46) indicated that the 4-methoxy group and (to a lesser degree) some other substituents are directly conjugated to the reaction centre.

1-Dimethylcarbamoylpyridinium chloride (5) behaved as a strong electrolyte in aqueous solutions and showed time-independent linear conductivity

---

* For solid N-acyltrialkylammonium salts, see refs. 44, 124, and 129. Crystalline adducts with arylsulfonyl chlorides were described in several papers which appeared at different times (see, e.g., ref. 101).

versus $\sqrt{C}$ dependence. The same pattern was observed in acetonitrile solutions though complicated by partial solvolysis with decomposition of the salt to the components (66). Recently, the conductivity of nitrobenzene solutions of N-diphenylcarbamoylisoquinolinium chloride was measured (53). It may be noted also that a benzoyl chloride–pyridine mixture dissolved in aqueous acetone (89% by weight) exhibited conductivity and could benzoylate butylamine (131). Arylsulfonyl chlorides were nonconductive in nitromethane or liquid SO_2 but became rather conductive in the presence of pyridine or isoquinoline. Pyrrole did not give ionic salts nor cause ionization of arylsulfonyl chlorides (99). Electrolysis of a waterfree mixture of $ArSO_2Cl$ and pyridine (the latter in excess) led to salts of arylsulfonic acids (99).

Paul (126, 127) performed a calorimetric determination of heats of solution of 2-picoline, quinoline, and dimethylaniline in acetyl and benzoyl chlorides and also determined heats of neutralization of the bases with Lewis acids. All the plots of ΔH versus concentration of the base showed a characteristic break in the region of low concentration, indicating a higher degree of dissociation of the solvates under such conditions (responsible for enhancement of heats of ionization); the same feature was characteristic of molar conductivity curves (125, 127). Thus graphical methods could be used to provide both the enthalpy of dissociation (ΔH) and the enthalpy of solvate formation (ΔH_1). An independent determination of the latter parameter was made in CCl_4 solution where ionization was hindered.

Polarographic reduction of acylpyridinium salts in pyridine gave characteristic two-wave patterns ($E_{1/2}$ of 0.7–0.9 and 1.3–1.7 V), indicating a stepwise reduction, first to 1-acyl-1,4-dihydropyridines and then to 1-acylpiperidine (36, 53, 132, 133). Mixtures of benzoyl chloride or p-methoxybenzoyl chloride and pyridine behaved similarly, which suggested the *in situ* formation of acylpyridinium salts in these systems also (36).

A wealth of IR spectral data were collected (using KBr disk or Nujol and fluorinated oil mull techniques) for N-acetyl- and N-propionylpyridinium chlorides (36, 44), acetylpyridinium, quinolinium, isoquinolinium, and acridinium bromides (134) as well as adducts of acetyl or benzoyl chlorides with pyridine or 3- or 4-picolines (135). Usually the spectra were discussed in reference to the spectra of free bases and acyl halides, pyridinium salts, acylium complex salts ($CH_3CO^{\oplus}AlCl_4^{\ominus}$, $2C_6H_5CO^{\oplus}TlCl_6^{\ominus 2}$) and amides of the corresponding acids. Data exist concerning the position of certain bands in the IR spectra of some other salts (56, 57, 66, 136, 137).

Most authors reported a higher frequency shift of the carbonyl stretching frequencies in the spectra of N-acylpyridinium halides compared to the reference compounds (1790–1804 cm^{-1} compared to 1760–1800 cm^{-1}; Table II), indicating the occurrence of a positively charged nitrogen atom adjacent to the carbonyl group. Lower frequencies were observed in the spectra whenever

III. STRUCTURE OF SALTS OF N-ACYLHETEROCYCLIC CATIONS

TABLE II

Carbonyl Stretching Frequencies in the IR Spectra of N-Acylpyridinium Chlorides

Chloride	v, cm^{-1}	Ref.
1-Acetylpyridinium	1804	36, 44
1-Propionylpyridinium	1796	44
1-(p-Chlorobenzoyl)pyridinium	1790	36
1-(p-Nitrobenzoyl)pyridinium	1800	36
1-(3,5-Dinitrobenzoyl)pyridinium	1802	36
1-Acetyl-4-dimethylaminopyridinium	1755	57, 58
1-Acetyl-4-acetamidopyridinium	1800, 1730	57
1-Acetyl-4-aminopyridinium	1780	57, 58
1-Acetyl-4-methylaminopyridinium	1770	58
1-Acetyl-4-isopropylaminopyridinium	1772	58
1-Acetyl-4-morpholinopyridinium	1770	58
1-Acetyl-4-piperidylpyridinium	1755	58
1-Dimethylcarbamoylpyridinium	1753	66
1-Dimethylcarbamoyl-4-picolinium	1750	66
1-Dimethylcarbamoyl-3,5-lutidinium	1757	66
1-Diphenylcarbamoylpyridinium	1745	66
1-Dimethylcarbamoylnicotinamidinium	1757	68
1-(t-butoxycarbonyl)-4-dimethylaminopyridinium	1770	86
1-Acetylquinolinium (bromide)	1800	53, 134
2-Carbamidoisoquinolinium (fluoroborate)	1795	137

there were factors favoring charge delocalization over the ring, such as the occurrence of 4-amino or 4-dialkylamino substituents. It might well be that the same factors were responsible for a low-intensity band at 1800 cm^{-1} in the spectrum of the salt of acetyl bromide and quinoline (134), as well as for the absence of high-frequency bands in the spectra of isoquinoline and acridine salts. On the other hand, the cited spectra might correspond to mixtures of N-acylium salts and hydrobromides of bases rather than pure compounds, and this would explain why the band intensities in the region 1700–1800 cm^{-1} appeared decreased (53).

The paper (135) was unique for quite a different approach to N-acylpyridinium salts which were treated there in terms of structure (**16**) involving carbonium ions and the pyridine molecule as an electron-donating group.

$$\left[\underset{}{\underset{}{\bigcirc}}\text{N:} \longrightarrow \text{O}\!\!=\!\!\!=\!\!\text{CR} \right] \text{Cl}^{\ominus}$$

(**16**)

Bands at 2000–2100 cm^{-1} in the spectra of salts of acetyl and benzoyl chlorides with pyridine and picoline were assigned to vibrations of the carbonium ions, whereas bands at 1322 cm^{-1} ($C_5H_5N \cdot CH_3COCl$) and 1380 cm^{-1} ($C_5H_5N \cdot C_6H_5COCl$) to the N–O charge transfer. At present, it is difficult to discuss the validity of this point of view because the spectra cited by other authors show bands at higher than 1800 cm^{-1} to be weak or nonexistent (36, 44, 134). A medium-intensity band at 1335 cm^{-1} was observed also in the spectrum of the adduct of pyridine with acetyl bromide (134); the authors of ref. 134 associate it with a 1355 cm^{-1} band of the initial CH_3COBr (CH bending). It is difficult to reconcile the NMR data with the idea of the dative nature of the interaction between the ring and the acyl group.

Pyridine, triethylamine, or dimethylaniline reacted with terephthaloyl chloride in dichloroethane with a rapid redistribution of band intensities at 1760 and 1770 cm^{-1}, though no frequency shift of the carbonyl stretches occurred (136). At the same time, in the NQR spectrum the chlorine signal at 31.1 MHz disappeared, indicating an ionic structure. These observations argued against so-called "Hantzsch isomerism" (at least in the ground state) which was thought to occur in N-acylpyridinium halide systems (124):

(17)

The IR spectra of N-phosphorylpyridinium salts were discussed in refs. 116, 117, and 119.

UV spectral data concerning solid N-acylpyridinium salts are limited to those cited for 1-acetyl-4-dialkylaminopyridinium chlorides (57, 58), 1-dimethylcarbamoylpyridinium chloride (66, 87), and 1-phosphoryl-4-alkylaminopyridinium salts (51, 113, 116). The spectra of these N-acyl salts resemble those of the aromatic bases, disregarding a certain bathochromic shift. Kinetic studies of acyl transfer and hydrolysis of acetic anhydride (e.g., refs. 46 and 138) furnished data on the UV spectra of N-acyl derivatives of substituted pyridines *in situ*; corrections for absorptions due to pyridine and its salts were made and the differential spectrum thus obtained revealed that acetylpyridinium acetate absorbs at λ_{max} of 272 (ε 4.3 × 10^3) and 225 nm (ε 7 × 10^3). Differences in the UV spectra permitted spectrophotometric control of solvolysis, hydrolysis, and transacylation reactions.

Only fragmentary data on the NMR spectra of the compounds in question were reported (51, 57, 58, 61, 66, 86, 113). The spectra of 1-acetyl, 1-

propionyl, and 1-benzoylpyridinium fluoroantimonates measured in SO_2 solution at $-60°C$ showed that α-, β-, and γ-proton chemical shifts increased with the electron-withdrawing power of the acyl substituent in the order $CH_3CO < C_2H_5CO < C_6H_5CO$ (61). Compare it with conclusions drawn from hetarylation experiments (139). A good correlation of the observed and expected δ values was also found in the case of $C_5H_5N^\oplus NO_2$, $C_5H_5N^\oplus NO$, and $C_5H_5N^\oplus H$ fluoroborates (62).

Signals from α and β protons underwent the same downfield shift on going from 1-acetyl-4-dimethylaminopyridinium chloride to its 1-(t-butoxycarbonyl) analogue with a weaker electron acceptor at nitrogen (58, 86), though according to the IR data the extent of charge delocalization was greater in the first case (v_{CO} was found at 1755 cm^{-1} in the former and at 1770 cm^{-1} in the latter compound). NMR spectra of 1-phosphoryl-4-aminopyridinium salts were rather difficult to rationalize (51, 113). It is likely that iminium forms contributed more to the resonance stabilization of 4-dimethylamino derivatives compared to the isopropylamino substituted species (cf. **13**). As a result, the deshielding of α protons was markedly greater in the first case, which accounted for the difference in the chemical shifts (8.2 and 8.6 ppm, respectively). Shifts of the β-proton signals were opposite to those observed for α protons.

Quite a number of IR and UV studies were performed in search of evidence for the occurrence of 1-acylpyridinium salts in solutions and mixtures containing pyridine and anhydrides of carboxylic acids, in view of their possible participation in acylation reactions (45, 140–143). In nonpolar media, such as carbon tetrachloride, mixing pyridine and acetic anhydride caused no spectral changes (140, 141, 143), and no solid residue was left on evaporating the solvent (37, 40). Similarly, UV spectra of mixtures of acetic anhydride and pyridine in cyclohexane were additive with respect to the spectra of the components (140). However, trifluoroacetic anhydride gave stable salts with pyridine and 4-vinylpyridine (37–40) which should be attributed to the counterion effect. On the other hand, data were obtained that demonstrated an important role played by solvents of high ionizing power in heterolysis of acid anhydrides in the presence of tertiary amines. Pyridine–acetic anhydride mixtures in acetone solutions exhibited conductivity just a little higher than the additive value (140); still, the conductivity increased with addition of pyridine to acetic anhydride until a 1:1 molar ratio was reached (143). As mentioned above, there is spectral evidence for the formation and subsequent hydrolysis of 1-acetylpyridinium acetate in aqueous solutions (46).

IV. Reactivity of N-Acylheterocyclic Cations

In the preceding section, we have shown that the addition products of acyl halides (and, possibly, acid anhydrides if in ionizing solvents) to six-membered nitrogen heterocyclic compounds are as a rule quaternary salts. Rather few compounds of this type have been isolated in the crystalline form; however, their occurrence in solutions and participation in reactions *in situ* are quite evident. Thus it appears justified to discuss the chemical behavior of mixtures of acylating agents and nitrogen bases in terms of reactions of the corresponding *N*-acylcycloammonium salts, even when the latter could not be detected as intermediates.

N-Acylpyridinium salts contain several reaction centers both in the ring and the acyl group. Accordingly, their reactions can be arranged into the following six main groups.

1. The attack by nucleophiles at the off-ring carbon atom bound to the nitrogen atom. The result is the elimination of the base and transfer of the acyl group to one of the components of the environment (acylation of water, alcohols, amines, acids) or to the counterion (solvolysis):

$$X^\ominus \; [\text{Py}^+\text{-COR}] \xrightarrow{\text{Nuc (H)}} \text{Py} + \text{RCONuc} + \text{HX}$$

Here also belong reactions involving transformations of the acyl group: hydrolysis CN → CONH$_2$ (137); decomposition of acyl salts of chlorocarbonates, chlorosulfonates, etc., via attack of the counterion (Cl$^\ominus$) on the C–O–R bond (81, 104, 144); and the formation of ketenes (attack by the pyridine ring on the proton in the position α to the carbonyl group) (14, 15).

Recently an interesting reaction between aryl ketones and the pyridine complex with SOCl$_2$ which leads to *N*-(α-styryl)pyridinium salts has been described. It is likely that the initial step of this reaction is attack by the S atom of the *N*-acyl intermediate on the carbonyl oxygen (145, 146).

2. The attack of nucleophiles (OH$^-$, arylamines) at the ring α-carbon atom resulting in so-called "pseudobases," or, more frequently, in ring opening (Zincke–Koenig reactions) (147, 148):

$$[\text{Py}^+\text{-COR}] \; X^\ominus \xrightarrow[-\text{HX}]{\text{Nuc (H)}} [\text{dihydro-Nuc-COR}] \rightleftharpoons [\text{ring-opened-Nuc-COR}]$$

3. Addition of nucleophiles to the heterocyclic ring at the α and γ positions. The resulting products are *N*-acyl-1,2 (or 1,4)-dihydro derivatives or, when the latter are unstable, substituted heterocycles (after elimination of the acyl group

and aromatization). These reactions, which can be included in the term "hetarylation" (1, 13), are exceedingly interesting thanks to many synthetic applications. Often they yield products otherwise inaccessible (including naturally occurring compounds), which in turn can be used as starting materials for further syntheses.

4. Reactions involving single electron transfer to N-acyl salts followed by radical dimerization (the Dimroth reduction) or radical hetarylation.

5. Reactions involving methyl, methylene, or vinyl side groups activated by the presence of the N-acyl function (37–40, 53, 149–153).

6. Reactions of N-acylheterocyclic cations involving electrophilic substitution (see, e.g., refs. 154 and 155).

Reactions of types 3–5 are known for various aromatic and heterocyclic cations (tropylium, cyclopropenylium, thiapyrylium, etc.), a fact that reveals the similarity of the π systems of these species. The positively charged nitrogen atom participates in the transfer of the substituent effect (156) though probably through an inductive rather than conjugative mechanism. Cooksey and Johnson (94) have measured pK_{ROH} values for a number of pseudobases to show that the electron-withdrawing power of the $\geq \overset{\oplus}{N}-CN$ group in N-cyanoquinolinium salts approximates that of the $\geq O^{\oplus}$ atom in benzopyrilium and exceeds that of the $\geq S^{\oplus}$ atom.

A. ACYLATION WITH PYRIDINE BASES

The first reports on the use of pyridine in acylation of alcohols and phenols date back to the 1890s (see bibliographies in Refs. 157 and 158). However, originally pyridine was believed to play the role of alkali. Einhorn and Hollandt (157, 159) studied reactions involving quite a number of acylating agents and hydroxy compounds to examine the advantages and limitations of this technique, which thus became known as "Einhorn acylation."

Acylation in pyridine can be recommended for compounds sensitive to alkali. Pyridine is a useful solvent and, in addition, the reaction does not necessarily require acyl halides if in pyridine and can proceed with acids in the presence of chlorinating agents (e.g., $COCl_2$, $SOCl_2$, benzoyl chloride) or, more conveniently with acid anhydrides (especially acetic anhydride). Benzoylpyridinium chloride is likely to be formed under such conditions using benzoyl chloride (157). The reaction can be run in acid media as well. Thus eugenol and isoeugenol can be acylated in acetic acid. Recently, a technique has been reported which suggests formylation of aliphatic and cycloaliphatic alcohols with formylacetic anhydride, or a mixture of formic acid and acetic anhydride (160). Pyridine compounds are known for the important role they play in syntheses of chlorocarbonates, carbonates, chlorosulfonates, and a number of other compounds with mixed functions (74, 104, 144, 161). With metal alkoxides instead of hydroxy compounds, only trace amounts of pyridine are necessary, since the exchange process provides regeneration of pyridine.

It has been shown that 4-dimethylaminopyridine can be used as an effective co-catalyst (together with pyridine or triethylamine) in the esterification of alcohols (see below) (88, 161–165). Application of this technique makes it possible to run reactions at low temperatures and to use such acylating agents as anhydrides of aromatic acids; it also provides successful acylation of quite a number of sterically hindered alcohols and phenols (165). Interesting syntheses can be performed with acyl halides fixed on a polymer matrix (166–168). This method can be used to block one of the two functions of symmetric diols, in step-by-step building of oligomeric chains, and so on, for example:

$$\text{(P)}-C_6H_4CH_2COCl \xrightarrow[\text{Py}]{\text{HO(CH}_2)_n\text{OH}} \text{(P)}-C_6H_4CH_2COO(CH_2)_n OH \longrightarrow$$

$$\xrightarrow[\text{Py}]{\text{TrCl}} \text{(P)}-C_6H_4CH_2COO(CH_2)_n OC(C_6H_5)_3 \xrightarrow{\text{HOH}} HO(CH_2)_n OC(C_6H_5)_3$$

For applications of the pyridine method to the chemistry of carbohydrates and cellulose see refs. 49 and 169–171, and for polyesterification reactions in the presence of tertiary amines see refs. 136 and 172–175.

Experimental data exist that provide direct demonstration of high acylating reactivity of pure N-acylpyridinium salts toward alcohols and phenols (48, 49, 73–75, 176). Though 1-nitropyridinium fluoroborate cannot be used in the nitration of aromatic compounds, it effects O-nitration of water (hydrolysis) and alcohols (62). Indirect evidence for participation of acylammonium salts in acylation reactions has been reported (177, 178).

The reaction of aliphatic acyl halides and CH_3OD results in a mixture of deuterated and nondeuterated esters, suggesting a competition between two reaction pathways (179). Path a is the predominant, if not the only, one when

$$RR'CHCOCl + (C_2H_5)_3N \begin{array}{c} \xrightarrow{a} RR'C=C=O + (C_2H_5)_3N \cdot HCl \xrightarrow{CH_3OD} \\ \longrightarrow RR'CDCOOCH_3 \\ \xrightarrow{b} RR'CHCO\overset{\oplus}{N}(C_2H_5)_3 Cl^{\ominus} \xrightarrow{CH_3OD} \end{array}$$

$$\longrightarrow RR'CHCOOCH_3 + (C_2H_5)_3N \cdot DCl$$

the acyl chlorides involved are sterically hindered at the reaction center, because steric factors play a more important role at the step of the formation of the quaternary salt than at the step of α-hydrogen elimination (acid–base equilibrium). The extent of deuterosubstitution for acyl halides decreases in the order chloride > bromide > iodide > acetate, that is, with "improvement" of the leaving group (180, 181).

There exist no data of this kind concerning reactions in pyridine, but one should expect that in this case the reaction must follow path *b* exclusively, at least with sterically nonhindered acyl halides. This conclusion is based on the lower basicity of pyridine than that of triethylamine and on the fact that, as mentioned above, *N*-benzoylpyridinium is characterized by a stronger resonance stabilization than *N*-benzoyltriethylammonium.

Most syntheses of arylsulfonates of alcohols, phenols, and polyols are run in pyridine (182). Experiments with solid adducts demonstrate the intermediate formation of *N*-arylsulfonylammonium salts (96, 97, 101–104). In most cases this reaction goes in higher yields than that involving aqueous solutions of sodium hydroxide. However, in some cases the resulting esters (especially esters of lower alcohols) react with pyridine to give alkylpyridinium salts as by-products. To avoid this, one may use 2,6-lutidine rather than pyridine, though the former reacts much more slowly (183). This reaction offers a means for detecting cinnamic alcohol fragments in lignin (182, 184).

$$C_6H_5-CH=CHCH_2OH \xrightarrow[Py]{TsCl} C_6H_5CH=CH-CH_2\overset{\oplus}{N}\diagup\!\!\!\diagdown \longrightarrow$$

$$\xrightarrow[KCN]{p\text{-}ONC_6H_4N(CH_3)_2} C_6H_5CH=CH-\underset{CN}{C}=N-\diagup\!\!\!\diagdown-N(CH_3)_2$$

Usually, it is difficult to interrupt tosylation of diols at the step of monosubstitution. Mesitylene sulfonyl chloride in pyridine is more selective (185).

Arylsulfonation of 1,4- and 1,5-diols, some amino alcohols, and γ-hydroxyamides proceeds via the intermediate formation of monosulfonates which can undergo cyclization to various oxa and azaheterocycles (186, 187).

Esters of carboxylic (but not of arylsulfonic) acids and primary or secondary alcohols, phenols, or polyphenols occur in carboxylic acid–pyridine–arylsulfonyl chloride systems (188). This has been used most widely in acylation of tertiary alcohols unstable to dehydration. The reaction is believed to proceed via *in situ* formation of acid anhydrides and regeneration of arylsulfonyl chlorides:

$$(RCO)_2O \xrightarrow{R'OH} RCOOH + RCOOR'$$
$$\underset{TsCl + Py}{\nwarrow \qquad \nearrow}$$

Recently, a stereospecific synthesis of olefins via thermolysis of β-lactones has been suggested. This is a favorable alternative to the Wittig olefin synthesis (189):

$$\begin{array}{c} R \\ R'-\overset{|}{C}-COOH \\ R''-\overset{|}{C}-OH \\ R''' \end{array} \quad \xrightarrow[\substack{Py \\ 0-5°C}]{PhSO_2Cl} \quad \begin{array}{c} R \\ R'-\overset{|}{C}-CO \\ R''-\overset{|}{C}-O \\ R''' \end{array} \quad \xrightarrow{140-160°C} \quad \underset{R''}{\overset{R'}{>}}C=C\underset{R'''}{\overset{R}{<}}$$

Analytical applications of pyridine acylation of hydroxyl groups are described in refs. 141, 158, and 190–194. Varying reactivities of various species toward acylation with acetic anhydride in pyridine solutions provide a means for determining the total amount of hydroxy compounds as well as the amount of primary and secondary alcohols in the presence of tertiary ones, of lower alcohols in the presence of higher ones, and of simple phenols in the presence of sterically hindered species. Also, techniques have been developed for analysis of mixtures containing isomeric primary and secondary alcohols or of primary and secondary hydroxy groups in polyglycols, by means of plotting differential second-order kinetic curves (192, 193).

N-Acylpyridinium salts may react with carboxylic acids to give anhydrides or mixed anhydrides:

$$\left[R-\underset{O}{\overset{\|}{C}}-\overset{\oplus}{N}\diagup\hspace{-0.2em}\diagdown \right] \overset{\ominus}{Cl} + R'COOH \rightarrow RCOOCOR' + Py \cdot HCl$$

This type of reaction, discovered in 1892 (23, 27, 195, 196), retains its value as a useful synthetic route to aliphatic, aromatic, and heterocyclic anhydrides (30, 63, 197–199). Symmetric products can be obtained by treating a mixture of acyl chloride and pyridine with water (better in an inert organic solvent). Techniques involving the use of carboxylic acids mixed with acyl chlorides or acyl chlorides in the presence of alkali agents (Na_2CO_3, $BaCO_3$, $NaHCO_3$) (to remove hydrogen chloride) have no advantages as a rule. Then one can use reactions involving acyl halides *in situ*, when a solution of pyridine and acid is treated with a chlorinating agent (phosgene or thionyl chloride) (176, 200, 201) or the process is run with a mixture of arylsulfonyl chloride, pyridine, and acid as starting material (187). Such reactions can be carried out with the use of solid adducts of pyridine and acetyl chlorides, as has been shown by numerous experiments (30, 47, 48, 64, 176, 202). Mixed anhydrides are usually obtained by addition of acetyl or benzoyl chlorides or chloroformates to ether or benzene solutions of pyridine containing nonreactive acids and by treating the mixture with ice (196, 203–206).

IV. REACTIVITY OF N-ACYLHETEROCYCLIC CATIONS

Recently an even simpler approach to anhydrides has been found (207, 208): one can simply add acyl chlorides to aqueous solutions of calculated quantities of sodium carboxylates in the presence of catalytic amounts of pyridine, 3-aminopyridine, or pyridine N-oxide, provided the compounds are stable to water. Replacing pyridine with picolines decreases the yields (to ~80%), and 2- and 4-amino- and 2-fluoropyridines as well as all the pyridinecarboxylic acids have no catalytic action (207). Another synthetic procedure is based on the use of free acids, cyanogen bromide, and pyridine in approximately 1:1:1 molar ratio as starting material. The reaction takes about 15–20 min at room temperature (208). Pyridine–acetic anhydride mixture can be used in syntheses of anhydrides of homophthalic acids, opening the way to a number of naturally occurring lactones (209, 210).

N-Benzoylpyridinium chloride reacts in aqueous solution with some inorganic salts. Thus it causes acylation of sodium azide giving an 85% yield of benzoyl azide (30, 211). A similar reaction occurs with hydrogen sulfide in petroleum ether at $-20°C$ (30, 212). On the other hand, shaking a 1:1:1 mixture of benzoyl chloride, aqueous potassium cyanide, and pyridine results in the formation of so-called "dimeric benzoyl cyanide" in addition to benzoic anhydride (207, 213). Substituted benzoyl chlorides react to give similar mixtures, and o-chloro-, bromo-, methoxy-, or p-nitrobenzoyl chlorides yield exclusively dimeric products of type **18**. The reaction is thought to proceed via the following scheme (213):

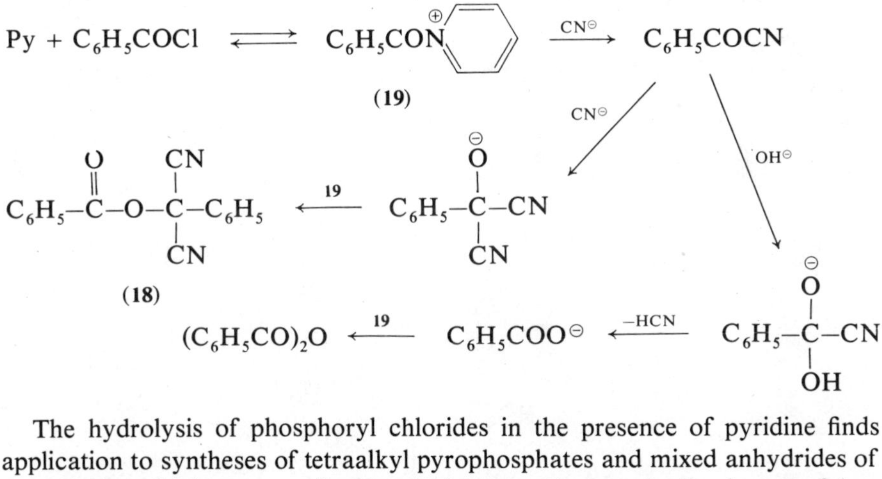

$$Py + C_6H_5COCl \rightleftharpoons C_6H_5CON^{\oplus}\langle\rangle \xrightarrow{CN^{\ominus}} C_6H_5COCN$$

(19)

$$\begin{array}{c} O \quad CN \\ \| \quad | \\ C_6H_5-C-O-C-C_6H_5 \\ | \\ CN \end{array} \xleftarrow{19} \begin{array}{c} O^{\ominus} \\ | \\ C_6H_5-C-CN \\ | \\ CN \end{array} \xleftarrow{CN^{\ominus}} \quad \xrightarrow{OH^{\ominus}}$$

(18)

$$(C_6H_5CO)_2O \xleftarrow{19} C_6H_5COO^{\ominus} \xleftarrow{-HCN} \begin{array}{c} O^{\ominus} \\ | \\ C_6H_5-C-CN \\ | \\ OH \end{array}$$

The hydrolysis of phosphoryl chlorides in the presence of pyridine finds application to syntheses of tetraalkyl pyrophosphates and mixed anhydrides of phosphoric acid (see, e.g., refs. 214–217). As a rule, isoquinoline is as useful as pyridine, whereas 2-picoline gives lower yields and less pure products, and quinoline does not enter the reaction at all. The reaction can be run in water–alkali media in the presence of 1–10% admixtures of pyridine, conditions that favor regeneration of the N-phosphoryl intermediate (215).

Acylpyridinium salts and acyl derivatives of other tertiary amines can split off the acyl residue when treated with ethers (218), just as in the case of alkylpyridinium salts. Thus veratrole, anisole, and diphenyl ether can react with acylpyridinium chlorides at 220°C to give phenol esters. In the absence of pyridine, acyl halides cause no such splitting of ethers. The synthesis of acyloin from benzothiazole-2-aldehyde, pyridine, and acetic anhydride offers an interesting example. The reaction probably goes via *O*-acylation by the *N*-acylpyridinium ion. Under the same conditions, chloral reacts to give a quantitative yield of chloral hydrate diacetate (219).

Of little importance are synthetic applications of amine acylation with acylpyridinium salts, since acylation with acid anhydrides usually gives satisfactory results without any catalyst. Still, there are reasons favoring aroylation of amines of higher NH acidity in pyridine solutions (158, 220). In the case of polyfunctional compounds, the technique finds wider application, especially when there are small differences in nucleophilic reactivities of the functional groups present (221–223). The acylation of phenol **20** gives the cyclic compound **21** (223), which is a model structure for a "tetrahedral intermediate" of acyl transfer reactions (180, 181). Various *N*-phosphorylpyridinium salts have been successfully used both in *N*-acylation of amino acids and in syntheses of polypeptides (87–89, 115, 118, 119).

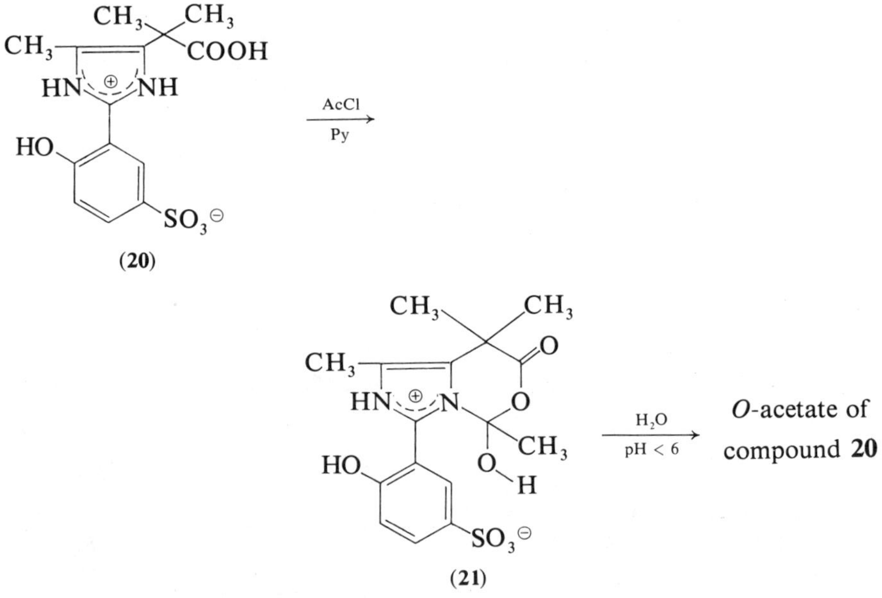

There are quite a few papers dealing with *N*-acylation of aza heterocycles by acyl chloride–pyridine base mixtures in spite of the fact that the reaction involves a complicated equilibrium of competing processes, including C-acylation and hetarylation (41, 42, 224–229):

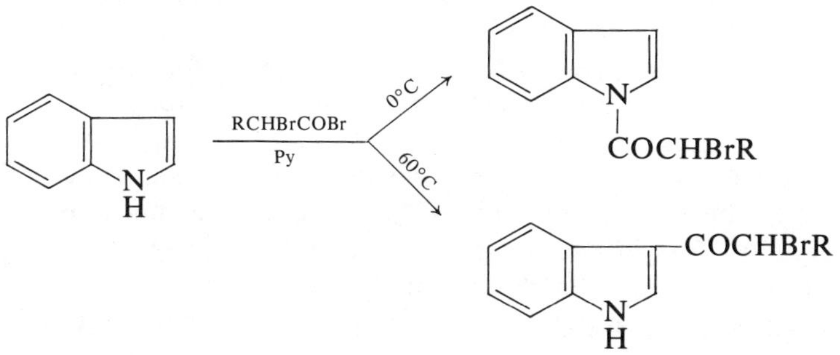

Cyclization of the intermediate may make acylation of pyridine amino derivatives even more complicated (228):

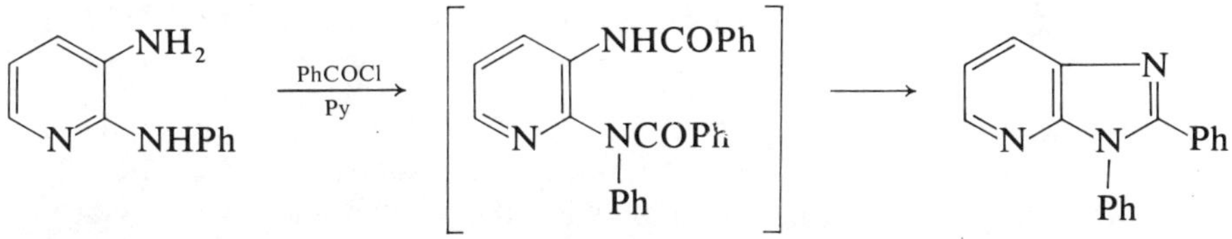

Heating a mixture of nicotine or N-methylanabasine and acetic anhydride or acetyl chloride leads to cleavage of the hydrogenated ring and results in β-PyCH=CH(CH$_2$)$_n$CH$_2$N(CH$_3$)COCH$_3$ (229).

There are examples of N-acylation of polymers containing heterocyclic rings (e.g., refs. 230 and 231). For the use of pyridine bases in analysis of amines and imines, including naturally occurring compounds, see refs. 158, 191, 220, 232–234.

Polycondensation of diacyl dichlorides with diamines, tetramines, and diaminodiphenols have been studied extensively (230, 235–241); the tertiary amines present in the systems influence the reaction rate, relative reactivities of particular functions, molecular weights of the products, and the extent of cross-polymerization. These effects have not found a clear-cut explanation. The synthesis of sulfonamides is discussed in refs. 242 and 243. Arylsulfonylation of heterocycles containing the amino function alongside the hydroxy group or halide atom sometimes involves side reactions of ring opening favored by betaine-type stabilization of the acylated product (244, 245). For participation of particular N-acyl- and N-arylsulfonylpyridinium salts in acylation of amines and aminoacids, see refs. 47, 63, 86–88, and 176.

Reactions of amides of carboxylic acids and acylpyridinium chlorides most clearly demonstrate the high acylating reactivity of the latter. Amides undergo mono- or diacylation or dehydration to nitriles depending on the nature of the pyridinium base, the acyl halide, and the amide, as well as on the conditions (246, 247). Acyl halides of weaker acids usually cause monoacylation, whereas derivatives of stronger ones react to yield diacylated products, and acyl halides of strong acids cause dehydration of amides (158, 248). The reaction goes most smoothly in the presence of benzene- and toluenesulfonyl chlorides (187, 249).*

There are numerous examples of N-acylation of sulfonamides, especially as used in the synthesis of pharmaceuticals (see, e.g., refs. 255 and 256).

Reactions of C-acylation of organic compounds with acyl halides in the presence of pyridine bases are less familiar. Thus benzoyl chloride reacts with dimethyl ketene acetal in triethylamine to yield a mixture of three compounds (257):

$$CH_2=C(OCH_3)_2 \xrightarrow{C_6H_5CO\overset{\oplus}{N}R_3Cl^\ominus} C_6H_5COCH_2\overset{\oplus}{C}(OCH_3)_2 \; Cl^\ominus \longrightarrow$$

$$(C_6H_5CO)_2CHCO_2CH_3 + C_6H_5COCH_2CO_2CH_3 + C_6H_5COCH=C(OCH_3)_2$$

Recently, methods have been developed for C-acylation of pyrrole, indole, furan, and their alkyl derivatives at reduced temperature (41, 42, 224, 225, 258). However, they allow the use of only a rather limited number of acylating agents ($COCl_2$, $ClCOCO_2C_2H_5$, CCl_3COCl, chlorides, and bromides of some α-halogen acids). The intermediate formation of N-acylium salts has been demonstrated (42, 258). Only hetarylation occurs in reactions of these substrates with most acyl and benzoyl chlorides in the presence of pyridine bases (see below). One more reaction involving acylation of the CH unit is exemplified by the following scheme (259):

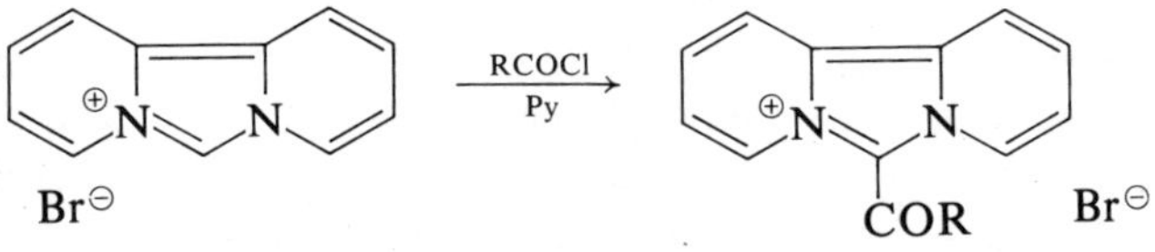

A well-known synthesis of acylamino ketones (the Dakin–West reaction) includes an intermediate C-4 acylation of substituted azlactones (162, 260, 261).

---

* Note that various reagents of the type "pyridine base–acyl chloride" (with $COCl_2$, ClCN, BrCN, $POCl_3$, $SOCl_2$, $ArSO_2Cl$, chloroformates for the latter) have been traditionally used in organic synthesis as soft dehydrating agents. Cf. dehydration of alcohols (250, 251), oximes (252, 253), and N-substituted formamides (254). However, there are no data as to whether N-acyl salts occur as intermediates in these systems.

IV. REACTIVITY OF N-ACYLHETEROCYCLIC CATIONS

Acylation of some organometallic derivatives of alkylpyridines first occurs at the nitrogen atom (83):

$$\text{pyridine-C(CH}_3)_2\text{-CH}_2\text{M} \xrightarrow{\text{RCOCl}}$$

M = MgCl, HgCl, Li

$$\left[\text{RCON}^\oplus\text{-C}_6\text{H}_4\text{-C(CH}_3)_2\text{-CH}_2\text{M} \cdot \text{Cl}^\ominus \right] \longrightarrow \text{RCON=}\langle\text{ring}\rangle\text{(CH}_3)(\text{CH}_3)$$

R = Me, Ph, OMe, OEt

As a rule, various β-dicarbonyl compounds as well as ketones characterized by keto–enol tautomerism yield O-esters rather than C-acyl derivatives when treated with N-acylpyridinium salts *in situ*, in contrast to the case of acylation of sodium enolates (262–265). The reactions proceed under very mild conditions, sometimes at room temperature or even in the cold. However Dieckmann et al. (266, 267) point to the occurrence of C-acyl derivatives in acylation of cyanoacetate and dihydroresorcinol in pyridine; also in the case of β-aminocrotonic ester and related compounds, N- and/or C-acylated products can occur, depending on the acyl chloride used (268). The use of other tertiary amines gives rise to high yields of acylation products (130, 257, 269, 270). Thus benzoyl chloride reacts with benzoyl acetate in triethylamine to give a 49% yield of the C-acyl and a 29% yield of the O-acyl derivatives; in dimethylaniline, the same reaction goes exclusively in the direction of the C-acyl derivative, and 2.6% of the C- and 75.8% of the O-acylated products are formed in pyridine (270). Doering and McEwen were the first to offer an explanation of the role played by pyridine in such reactions (130, 270, 271). These authors obtained a mixture of 1-benzoyl-4-phenacyl-1,4-dihydropyridine (**22**) and O-benzoylacetophenone from benzoyl chloride, pyridine, and acetophenone. They suggested a simultaneous attack by the N-benzoylpyridinium salt at both the C-2 and C-4 atoms of acetophenone, yielding the stable compound **22** and 1-benzoyl-2-phenacyl-1,2-dihydropyridine. The latter underwent rearrangement via a geometrically favored quasi-six-membered transition state (**23**) to the O-adduct. The same scheme provides an explanation of the kinetics of the reaction (272).

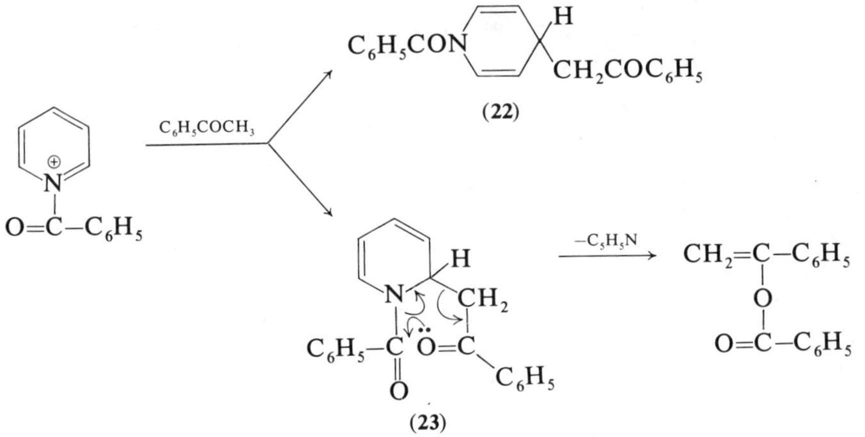

According to recent studies (264, 273), ethyl acetoacetate reacts with benzoyl chloride to yield the *O*-benzoate in pyridine, triethylamine, 2-picoline, and 2,6-lutidine. The yields decrease in going from pyridine to 2,6-lutidine, owing to an increase in steric hindrances. The reaction is first order in both components, and the reaction rate is inversely proportional to the concentration of $Cl^\ominus$. One obtains only the *O*-acetate using acetyl chloride or a mixture of acetyl chloride with pyridine. However, the sodium derivative of ethyl acetoacetate reacts with acetyl chloride to give a 98% yield of the *C*-acetyl derivative and with *N*-acetylpyridinium chloride to give the *O*-ester in addition. In nonpolar media, all the reactions appear slowed down. The process is believed to involve the keto form of acetoacetate rather than adducts to the pyridinium nucleus. In the case of benzoylation, the rate-determining step is that of the formation of benzoylammonium salts, the most unfavorable process with species sterically hindered at nitrogen. Clearly, further investigations are needed to solve the problem of the mechanism of *O*-acylation of β-keto–enols. In view of the above discussion, it appears that the idea of the intermediate formation of *C*-adducts in acylation of phenols, naphthols, and similar compounds must be dismissed (44, 274). In all these cases, the replacement of pyridine with tertiary amines does not cause any significant decrease in yield; 2,6-dimethylphenol, with no vacant α positions, readily undergoes acylation in pyridine.

B. KINETIC STUDIES OF REACTIONS INVOLVING *N*-ACYLHETEROCYCLIC CATIONS. NUCLEOPHILIC AND GENERAL BASIC CATALYSIS WITH PYRIDINIUM BASES

During the last 15 years, kinetics of acyl (and phosphoryl) transfer in *N*-acylpyridinium ion–nucleophile systems has been studied for a variety of

IV. REACTIVITY OF N-ACYLHETEROCYCLIC CATIONS

nucleophiles (water, alcohols, amines, anions of organic and inorganic acids, components of buffer solutions, etc.). These studies provided data on kinetic and thermodynamic factors determining stabilities of salts of N-acylpyridinium cations *in situ* (though such data are insufficient as yet) and on the reactivity of nucleophiles, depending on their structure and steric characteristics. To a certain degree our understanding of mechanisms of nucleophilic reactions has been improved. The theoretical and experimental approaches employed and the results obtained in these works are closely related to the problem of nucleophilic catalysis in reactions involving hydrolysis of acyl halides, anhydrides, and esters of carboxylic (and phosphoric) acids, or transacylation and transphosphorylation (180, 181, 275, 276). It should be emphasized that many such processes play important roles in biochemical reactions (275, 276).

According to the data of refs. 67, 132, and 277, solid N-dimethylcarbamoyl pyridinium chloride (**5**) is stable in protic solvents, but undergoes rapid solvolysis when dissolved in acetonitrile, chloroform, dimethyl sulfoxide, or ethylene carbonate. UV spectra of "aged" solutions in such aprotic solvents are strongly sensitive to small changes in the component ratio, evidencing the occurrence of a dynamic equilibrium:

$$\left[\text{Py}^{\oplus}\text{—CON(CH}_3)_2 \right] \text{Cl}^{\ominus} \rightleftarrows \text{Py} + (\text{CH}_3)_2\text{NCOCl}$$
(**5**)

Disappearance of N-dimethylcarbamoylpicolinium chloride in a similar reaction with 4-picoline follows pseudo-first-order kinetics, though the dissociation constant (30 ± 10) usually depends on the initial concentration of the salt. A possible explanation of solvent-dependent variations of stabilities of the salts can be given in terms of Pearson's concept of hard and soft acids and bases: although the chloride ion is stabilized by hydrogen bonding in alcoholic media, it behaves as a "hard" base in the absence of hydroxyl groups and attacks the hardest acidic center of ion **5**, that is, the carbonyl group (132).

Likewise, N-dimethylcarbamoylpyridinium chloride (**5**) is stable in acidic aqueous solutions (pH 1–5) and rapidly decomposes in alkaline media (67,

$$\mathbf{5} + \text{AcO}^{\ominus} \longrightarrow \text{CH}_3\text{COOCN(CH}_3)_2 + \text{Py}$$
$$\underset{\text{O}}{\parallel}$$
$$\downarrow \text{H}_2\text{O}$$
$$\text{CO}_2 + \text{AcO}^{\ominus} + (\text{CH}_3)_2\text{NH}$$

132). A direct nucleophilic substitution at the nitrogen atom of ion **5** occurs on the addition of $NH_2OH \cdot HCl$, imidazole, NaN_3, and formate ion, as evidenced by the product composition.

Components of buffer solutions (e.g., acetate ion) also favor hydrolysis of salt **5**.

Note that with few exceptions the rate constants of the second-order reactions involving the nine nucleophiles studied correlate with those of the corresponding reactions of *p*-nitrophenyl acetate (278); one may conclude that both compounds react according to the same hydrolysis mechanism, notwithstanding the occurrence of a positively charged nitrogen atom in *N*-dimethylcarbamoylpyridinium chloride (**5**). In the case of the similarly structured compound **24**, the influence of the electron-withdrawing 3-carbamoyl function (68) proved unexpectedly weak. The reaction rate increases sevenfold in going from pure water solutions to solutions containing acetate, $HPO_4^{2\ominus}$, imidazole, tris(hydroxymethylamino)methane, and $OH^{\ominus}$. In the same series, the rate constants of ring-opening reactions involving nicotinamide derivatives (see Section IV-C, p. 560) become five orders higher.

$$\underset{(24)}{\overset{\displaystyle \text{pyridinium-CONH}_2}{\underset{\displaystyle CON(CH_3)_2}{}}} \xrightarrow{\text{Nuc (H)}} \text{pyridine-CONH}_2 + (CH_3)_2NCONuc$$

Jencks and Fersht (46, 138, 279) compared the behavior of solid *N*-acetylpyridinium chloride and acetic anhydride–pyridine base mixtures in aqueous media to show that considerable amounts of *N*-acylpyridinium salts and their homologues can accumulate in buffer solutions. In fact, one can observe the appearance and disappearance of the characteristic absorption in the region 280–290 nm on mixing aqueous solutions of pyridine and acetic anhydride using a "flow technique" accessory. An increase of the acetate ion concentration lowers the quantity of acylpyridinium formed and increases the rate of the inverse reaction (k_{-1}). Pyridine has the opposite effect.

$$C_5H_5N + Ac_2O \underset{k_{-1}}{\overset{k_1}{\rightleftarrows}} C_5H_5\overset{\oplus}{N}COCH_3 + Ac^{\ominus}$$

Calculations performed for water solutions with an ionic power $\mu = 1.0$ yield values of 84 mole^{-1} sec^{-1} for k_1, 900 mole^{-1} sec^{-1} for k_{-1}, and 9.2×10^{-2} for the equilibrium constant (K_p) (compare the data of refs. 125 and 128).

A marked dependence of the kinetic parameters on basicity of the nitrogen atom was observed in a series of 4-substituted pyridines (Table III). A

IV. REACTIVITY OF N-ACYLHETEROCYCLIC CATIONS

TABLE III
Reactivities of N-Acetylpyridinium Ions (46)

Acetylpyridinium ion from:	pK_a of base	k_1, mole^{-1} sec^{-1}	K_p	Rate constants of hydrolysis, mole^{-1} sec^{-1}		$-\Delta F°$ of hydrolysis
				Water	Acetate ion	
Pyridine	5.51	84	0.092	6.9	910	18.31
4-Picoline	6.33	490	1.86	2.3	262	17.67
3,4-Lutidine	6.79	1160	9.67	1.5	120	17.33
4-Methoxypyridine	6.82	935	31.2	0.4	30	16.65
Acetylimidazole	—	—	2950	—	—	14.05

satisfactory linear correlation exists between the values of log k_1 and log K_p on the one hand and pK_a of the base on the other, with the slopes (or the Brönsted parameters β) of 0.87 and 1.6, respectively, indicating the effective electronegativity of the acyl group to be higher than that of a proton (46). These data are consistent with the above-mentioned NMR results, namely, a stronger deshielding of the ring protons in N-acylium salts compared to pyridine hydrochloride. The plot of log K_p versus pK_a shows positive deviations for N-acetyl-4-methoxypyridinium and N-acetylimidazolinium, which implies that these tertiary amines react with acetic anhydride to yield products of higher stability than can be predicted on the basis of their basicity, probably because of a resonance stabilization effect (46, 58):

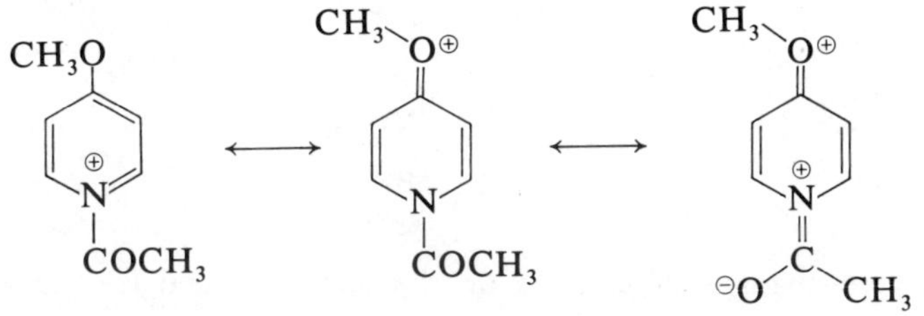

It is important that acylation of anisidine and p-toluidine with acetic anhydride in aqueous solution in the presence of pyridine follows pseudo-first-order kinetics with the rate constant independent of the type and concentration of arylamine. It shows that the acyl transfer involves the formation of acetylpyridinium ion as the rate-determining step succeeded by a rapid reaction with the nucleophile (46, 138). Thus the mentioned amines successfully compete with water in concentrations of 5×10^{-4} to 5×10^{-3} M; this reveals the high selectivity of acetylpyridinium ions to the type of the nucleophile (46,

280–282). The corresponding data on competing reactions involving water and the acetate ion (Table III) follow the Brönsted equation rather well, with β 0.5 and 0.68, respectively, and with expected negative deviations in the case of N-acetyl-4-methoxypyridinium ion. Pyridine furnishes a k_{AcO}/k_{H_2O} reactivity ratio two orders higher than that observed in the case of acetylimidazole, another compound frequently used in kinetic studies.

Low reactivity of acetic anhydride–pyridine mixtures toward water explains the possibility of using commercial-grade reagents in acylation reactions, of syntheses of mixed anhydrides in aqueous media, and so on.

The reactions involving the above-mentioned four acetylpyridinium salts and a variety of O- and N-nucleophiles (alkoxide and phenoxide ions, aliphatic and alicyclic amines, hydrazines) give k_2 values indicative of a decrease of sensitivity of acetylpyridinium ions toward basicity of the attacking nucleophile ("leveling") with stronger bases (279). In a series of O-nucleophiles, the Brönsted constants decrease from 0.9 ($AcO^{\ominus}$) to 0.2 (metal alkoxides). Less reactive acetylimidazolium ion exhibits no such "leveling."

Thus N-acylpyridinium ions behave as specific derivatives of acetic acid characterized by a reduced sensitivity to nucleophilic reactivity of the reagents involved and highly sensitive to the basicity of the leaving group. Collation of the related reactions (say, with p-nitrophenoxide ion or 4-picoline) of aliphatic acetates, phenyl acetate and its derivatives, acetic anhydride, and N-acylpyridinium ion reveals that such sensitivity is characteristic of reactions involving "unfavorable" leaving groups ($RO^{\ominus}$).

Recently, Jencks (279, 282) suggested a classification of acyl transfer reactions according to nucleophilic reactivities of nucleophiles, and β values characteristic of nucleophiles and leaving groups:

$$+0.3 \quad \overset{\displaystyle O}{\underset{\displaystyle \|}{} } \quad -0.3 \qquad +0.9 \quad \overset{\displaystyle O}{\underset{\displaystyle \|}{} } \quad -0.9 \qquad +1.4 \quad \overset{\displaystyle O}{\underset{\displaystyle \|}{} } \quad -1.4$$
$$Y \cdots C \text{——} X \qquad\qquad Y \cdots C \cdots X \qquad\qquad Y \text{——} C \cdots X$$
$$\text{Class I} \qquad\qquad\qquad \text{Class II} \qquad\qquad\qquad \text{Class III}$$

(numbers give changes of charges on the attacking and leaving groups as compared to the total reaction charge change of 1.7).

The reactions that belong to the first class are those of strongly basic nucleophiles and acylating agents bearing favorable leaving groups (including N-acylpyridinium ions); they are characterized by a nonsymmetric transition state containing few newly formed or broken bonds (the "early" transition state).

Conversely, the third class embraces the reactions of weakly basic nucleophiles with compounds containing "unfavorable" leaving groups. In terms of the tetrahedral transition adduct concept [according to Bender (180)], the decomposition of such an adduct represents the rate-determining step of the reaction.

A variety of reactions of the second class are characterized by a rather symmetric transition state with respect to both the attacking and leaving groups (β 0.7–1.0). Here belong reactions between acetylpyridinium ions and amines characterized by similar pK_a values. One can hardly expect a tetrahedral transition state in this case.

$$R-\underset{(25)}{\langle\rangle}\overset{\oplus}{N}-PO_3^-$$

The behavior of aliphatic and heterocyclic amines and imines with pK_a 2.6–10.8 in the aminolysis of 4-substituted N-phosphorylpyridinium ions (25) in aqueous solution shows low sensitivity toward basicity of the attacking group (the Brönsted nucleophilic constants fall in the range 0.2–0.3) (113). The anions carrying positively charged atoms—pyridines, hydrazine, and hydroxylamine—exhibit unusually high reactivity (the so-called α effect); however, they are almost nonreactive toward cyclohexyl- and isopropylamines and diethylamine, perhaps because of steric hindrance. Values of k_2 for nucleophiles strongly depend on the nature of the leaving group in compound 25. The correlation plots are characterized by greater slopes for less basic nucleophiles (thus water gives β −1.13, morpholine −0.96, methylamine −0.88). On the other hand, the plot of log $k_{CH_3NH_2}$ versus log $K_{pyrazole}$ is a rather straight line for all the four N-phosphorylpyridinium ions. So an increase in nucleophilic reactivity does not cause changes in the aminolysis mechanism.

Other literature data on stabilities of N-phosphorylpyridinium ions (25) and some N-acyl derivatives toward hydrolysis are summarized in Table IV. Both types of compounds demonstrate a decrease in the hydrolysis rate with increasing basicity of the substituent at C-4 favoring charge delocalization over the conjugation chain. Naturally, contributions of onium or iminium forms should increase in the series HO < NH_2 < RNH < R_2N. Accordingly, resonance-stabilized N-acetyl-4-dimethylaminopyridinium undergoes a 2000 times slower hydrolysis than unsubstituted N-acetylpyridinium chloride.

It is not our aim to discuss all the cases of nucleophilic catalysis in reactions involving carboxylic or phosphoric acid derivatives (for such discussions see refs. 1, 275, and 276). We therefore restrict ourselves to papers providing additional data on systems that undoubtedly contain N-acyl- and N-phosphorylpyridinium salts.

For instance, pyridine bases were shown to have a catalytic effect (of nucleophilic origin) on hydrolysis of acetic anhydride in water (140, 283), acetate buffer (280, 284), and water–organic solvent mixtures (283, 285, 286),

TABLE IV

Hydrolysis of N-Phosphorylpyridinium Ions and their Derivatives

Ion	Hydrolysis in 0.1 N HCl, 27°C		Ref.
	$t_{1/2}$, min	k_{exp}, min^{-1}	
25, R = OH	2.1	3.3×10^{-1}	51
25, R = NH$_2$	184	3.75×10^{-3}	51
25, R = NHPr-i	451	1.53×10^{-3}	51
25, R = N(CH$_3$)$_2$	455	1.52×10^{-3}	51, 113
N-Phosphoryl-N-methylimidazolium	57	1.2×10^{-2}	51
N-Acetylpyridinium	—	450^a	46
N-Acetyl-4-aminopyridinium	—	3.2×10^{-1a}	50, 58
N-Acetyl-4-methylaminopyridinium	—	1.9×10^{-1}	58
N-Acetyl-4-dimethylaminopyridinium	3.9^a	1.7×10^{-1a}	51, 57, 58
N-Acetyl-4-piperidylpyridinium	—	1.3×10^{-1}	58
N-Acetyl-4-acetamidopyridinium	—	3^a	57
N-(t-butoxycarbonyl)-4-dimethylaminopyridinium	2.1^b	—	86, 87

^{a} Solution in acetate buffer, pH 5.5, 25°C.
^{b} $t_{1/3}$.

and on hydrolysis of some mixed anhydrides (140, 287), various acyl chlorides in aqueous acetone (131, 288), acetyl fluoride (281), and arylsulfonyl chlorides in water (289, 290) and in water–polar solvent mixtures (291). Data exist on nucleophilic catalysis of methanolysis of 2,4-dinitrophenyl acetate, picryl acetate, and picryl benzoate, and arylsulfonyl chlorides with various pyridine derivatives (292–296). The results obtained in these works agree rather well with a general scheme proposed in refs. 180 and 181:

$$\text{RCOX} + \text{B} \underset{k_{-1}}{\overset{k_1}{\rightleftarrows}} \begin{bmatrix} \text{O}^{\ominus} \\ | \\ \text{RC--X} \\ | \\ \text{B}^{\oplus} \end{bmatrix} \rightleftarrows \begin{bmatrix} \text{RCB}^{\oplus} \\ \| \\ \text{O} \end{bmatrix} \text{X}^{\ominus} \xrightarrow{\text{Nuc (H)}}_{k_2} \text{RCONuc}$$

where Nuc (H) stands for water, MeOH, etc.

Hydrolysis of acetic anhydride proved very sensitive to acetate ion concentration (280, 284, 286). Thus the catalytic effect by pyridine was nearly suppressed in the presence of the buffer, 1.06 M in AcO$^{\ominus}$ and 0.54 M in acetic acid (a 40% dioxane solution) (280). On the other hand, an unusually high isotopic effect was observed in these reactions, with k_{H_2O}/k_{D_2O} 5 ± 1 (284). This phenomenon was explained on the basis of the assumption that the rate-determining step was solvolysis of the acylpyridinium ion (297–299). Next, assuming $k_{-1}(OAc^{\ominus}) \gg k_2(H_2O)$ (279, 282) provided a ready explanation of why the values of k_B for pyridine bases were inversely proportional to the concentration of acetate ions. Change of the limiting step depending on the structure of the pyridine base involved was described in ref. 288.

Solvolysis of acetic anhydride underwent a 10^5-fold acceleration on going from methanol to aqueous solutions (285). At the same time, pyridine had a negligibly higher catalytic activity in methanol than acetate ion. The rate of pyridine-promoted hydrolysis gradually decreased with replacing water by dioxane in water–dioxane mixtures (280, 286). It was assumed that the equilibrium constant K_p of the formation of the N-acylpyridinium ion rapidly decreased in less polar solvents (owing to both a decrease in k_1 and simultaneous increase in k_{-1}). Changes in solvent composition usually caused less pronounced changes in such terms of the kinetic equation that directly referred to water (k_{H_2O}) and acetate ion (k_{AcO^-}) (282).

Reference 131 dealt with hydrolysis of a number of acyl halides in 89% aqueous acetone solutions in the presence of n-butylamine, trimethyl- and triethylamines, pyridine, and collidine. Some of the acyl halides were found to undergo S_N2 reactions (involving the formation of N-acyl salts) whereas others underwent monomolecular ionization (an S_N1 reaction). The rate of saponification of N-piperidyl- and N-(diisopropylamino)carbonyl chlorides and 2,4,6-trimethylbenzoyl chloride remained practically the same whether in the presence or absence of any amine, whereas the process involving acetyl-, pivalyl-, benzoyl-, and 2,4,6-trimethyl-3,5-dinitrobenzoyl chlorides went at a rate 10^5–10^7 times higher in the presence of pyridine. The resulting ordering of the catalytic activity of tertiary amines fitted the nucleophilic reactivity series: pyridine > $(CH_3)_3N$ > $(C_2H_5)_3N$ > collidine.

The first-order equation $k_{exp} = k_{H_2O} + k_B[B]$ (289, 296) provided a good description of the hydrolysis of benzenesulfonyl chloride in aqueous solution in the presence of 10 derivatives of pyridine bearing various electron-donor and electron-acceptor substituents at the 3 and 4 positions of the ring. The reaction rate increased in the presence of donor and decreased in the presence of acceptor substituents. It proved possible to rationalize the obtained k_B values in terms of the Brönsted equation ($\beta = 0.45$) or correlate them with the Hammett parameters σ° ($\rho = -2.68 \pm 0.10$). Thus this reaction exhibited a lower sensitivity toward basicity of nucleophiles than that characteristic of carbonyl compounds (acetic anhydride, $\beta = 0.87$) and a higher sensitivity compared to phosphoric acids (hydrolysis of N-phosphorylpyridinium ions, $\beta = 0.3$). It was assumed that in the case of sulfur the transition state would involve longer-range interactions than in the case of carbon, because of the higher polarizability of sulfur. The validity of the Hammett equation also pointed to the development of a positively charged reaction center in the transition state (however, see ref. 290).

An extensive series of publications by Litvinenko and co-workers (300–320) was devoted to studies of kinetic patterns of benzoylation and arylsulfonation of arylamines in aprotic solvents in the presence of tertiary amines. High catalytic activity of various substituted pyridines was demonstrated. These compounds proved much more reactive than triethylamine and dimethylaniline (300, 301). The relative catalytic activity series followed the basicity series based on the pK_a values for pyridine bases in aqueous solutions: 3-NO_2 < 3-Cl < H < 4-CH_3 < 4-$(CH_3)_2$N. Application of the Hammett equation yielded $\rho = -3.74$, which showed the catalytic activity to depend strongly on structural changes in the catalyst; 2-picoline and 2,6-lutidine were not effective in this reaction. In ref. 302, a detailed correlation was drawn between the rate constants of catalytic (k_B) and noncatalytic (k_0) reactions involving arylamine $RC_6H_4NH_2$ on the one hand and the structural characteristics of the catalyst on the other; the obtained ρ values of -1.35 and 2.67, respectively, indicated an increase in the effective charge at the reaction center of the transition state in the case of pyridine-catalyzed reactions. A k_B/k_0 ratio increased with a decrease in acylated amine basicity, a feature characteristic of nucleophilic reactions. The occurrence of m- and p-substituents in benzoyl chlorides caused less pronounced changes ($\rho = +1.24$); p-chloro-, p-methoxy-, and p-dimethylaminobenzoyl chlorides showed markedly lower reactivities than would be expected from the corresponding σ values. A suggested explanation of this phenomenon involved the assumption of a stronger polar conjugation between the substituent and the reaction center than that characteristic of the standard Hammett series [dissociation of carboxylic acids (304)].

The same authors reported kinetic data favoring the formation of N-acylium salts in the acylation of p-anisidine with m-tolyl- and (p-nitrophenyl)sulfonyl chlorides in nitrobenzene solutions in the presence of pyridine (309, 310). The value of $k_{\text{eff.}} = k_0 + k_{\text{Py}}[\text{Py}]$ was found to decrease as the reaction proceeded. This might be possible when the intermediate adduct was fully or partially ionized in a strongly ionizing solvent so that the reaction could follow either of the following two pathways: (a) interaction of the arylamine and the ion pair and (b) interaction of the arylamine and the acylpyridinium cation. The accumulation of counterions ($Cl^\ominus$, $Br^\ominus$) would shift the dissociation equilibrium and then the equilibrium involving the formation of the primary adduct, just as was the case with hydrolysis of acetic anhydride and phenyl acetates (282).

Bonner and co-workers (143, 321–323) carried out kinetic studies of trifluoroacetylation and acetylation of a number of phenols and alcohols with trifluoroacetic, acetic, and trifluoroacetylacetic anhydrides in CCl_4 solutions in the presence of pyridine and its homologues. In the absence of pyridine and at 25°C, acetic anhydride was nonreactive toward p-chlorophenol, p-cresol, o-

nitrophenol, and salicylic aldehyde; it reacted slowly with isopropanol and n-butanol (321). Additional pyridine had a pronounced catalytic action only in the case of phenols; the reaction rate rapidly increased to its maximum value with the catalyst concentration. The temperature factor of the reaction in the case of p-chlorophenol was very low. 3- and 4-picolines proved highly reactive catalysts, but 2-methyl- and 2,6-dimethylpyridines did not catalyze acylation of p-chlorophenol (322). Trifluoroacetylacetic anhydride reacted with isopropanol to yield a 1:2 mixture of acetate and trifluoroacetate; the addition of pyridine had little influence on either the reaction rate or the product ratio (143). Theoretically, the formation of the N-trifluoroacetylpyridinium intermediate would accelerate both reactions involving phenols and alcohols. It should be mentioned that the reaction was shown to be second order in the base concentration.

It was assumed that the role of these pyridines was twofold: (1) they form hydrogen-bonded solvates with phenols, thus increasing the nucleophilicity of the oxygen atom, and (2) at the step of the transition state, they form solvates with the leaving acetoxy group (143, 322, 323). According to calculations, p-cresol and p-chlorophenol involved in hydrogen bonding had 300 and 700 times higher nucleophilic reactivities than the corresponding free molecules, respectively (321). Pyridine–aliphatic alcohol systems were characterized by a rather low level of association which explained weak catalytic effects at 25°C. Experiments with deuterated phenol and p-chlorophenol yielded 1.51 and 1.40 ratios of k_H/k_D, which demonstrated an important role of heterolysis of the O–H bonds in acylation reactions (322).

Korshak and co-workers studied low-temperature polyesterification in the presence of tertiary amines to arrive at the same conclusion of two catalytic mechanisms governing the acylation of phenols and bisphenols with acyl chlorides (136, 173, 324–329). General base catalysis was judged the more important in reactions involving strongly acidic phenols or strongly basic amines, such as triethylamine (325). Nevertheless, nucleophilic catalysis predominated, seemingly, in the case of sterically nonhindered pyridine bases (175, 326, 327). Thus experiments involving competitive benzoylation of phenol and methanol gave the amounts of phenol conversions rapidly decreasing together with pK_a's of the base across the series tributylamine, triethylamine, dimethylethylamine, sym-collidine, quinaldine, 2-picoline, isoquinoline, 3-picoline, pyridine, quinoline (173). The solvent nature and its propensity for the formation of solvates with the reactants had a pronounced effect on the relative contributions of nucleophilic and general base catalytic processes. The ratio $q = (1 - x)/x$ (x is a level of phenol conversion) reached its maximum value in dioxane, acetone, or nitrobenzene, whereas it was smallest in benzene or dichloroethane.

C. REACTIONS OF N-ACYLPYRIDINIUM CATIONS INVOLVING RING OPENING

Reactions involving cleavage of the pyridine ring (the Zincke–Koenig reactions) (147, 148, 245) yielding glutaconic dialdehyde and its anils and enamines are characteristic of quaternary pyridinium and benzopyridinium salts containing electron-withdrawing substituents at the nitrogen atom that cannot undergo transfer to the nucleophiles present in the reaction mixture. The first reports describing this reaction, which appeared at nearly the same time (330, 331), dealt with reactions of N-(2,4-dinitrophenyl)- and N-cyanopyridinium salts in the presence of alkali or aromatic amines. Koenig (332–335) extended the reaction to arylamines bearing various substituents in the rings, to α- and β-naphthylamines and other polycyclic amines, N-alkylanilines, substituted tetrahydroquinolines, and dihydroindoles. In addition to pyridine, 4-picoline and 4,4'-dipyridyl (but not 2-picoline) were successfully used in this reaction. More recently, Vompe (336, 337) reported cleavage of 4-alkoxy, 4-aryloxy, and 4-arylmercaptopyridines with cyanogen halides. These reactions can have synthetic applications thanks to the possibility of recyclization of the products to pyridine or pyrone rings with new 4-substituents (338). A procedure was reported which involved an interesting reaction of ring opening associated with attack by cyanogen bromide and diethylamine perchlorate leading to the simplest polymethylene dye (**26**) (339).

$$[\text{Py}\overset{\oplus}{N}-CN]\ Br^{\ominus} + (C_2H_5)_2NH \cdot HClO_4 \longrightarrow$$

$$\longrightarrow [Et_2\overset{\oplus}{N}=CHCH=CHCH=CH\overset{\oplus}{N}\text{Py}]\ (ClO_4^{\ominus})_2$$
$$(\mathbf{26})$$

In addition to cyanogen chloride and cyanogen bromide, SO_3 and ethyl chlorosulfate (334, 340, 341), $COCl_2$, $POCl_3$, PCl_5, thiophosgene, and other thioacyl chlorides (332, 342) were successfully used in the Koenig reaction. Acyl chlorides were commonly thought unable to initiate pyridine ring opening (332, 343). However, such a reaction was reported to occur with diphenylcarbamoyl chloride and pyridine in the presence of alkali (63). Nearly at the same time, pyridine adducts with bis(o-tolyloxalyl)imidoyl chloride and N-phenylbenzimidoyl chloride were shown to react in this way with various aromatic amines in alcohol solutions (69). The same reaction was claimed to go in the benzoyl chloride–alkali system; however, the yields of glutaconic aldehydes were not reported (344). Under alkaline conditions, cleavage of N-nitro- and N-nitrosopyridinium salts went in nearly quantitative yield to glutaconic dialdehyde derivatives (62). The formation of anils of glutaconic dialdehyde results in colored solutions with a characteristic fluorescence (99, 345). This is likely the origin of intense reddish-violet colors observed in aqueous and water–alcohol solutions of pyridine and its homologues in the presence of alkali and various arylsulfonyl chlorides. The reaction was recommended as a sensitive test for detecting pyridine and its derivatives (345–351).

Compounds containing activated methyl or methylene groups facilitate opening of the 1-acylpyridinium ring; they undergo condensation with the products to yield polymethine dyes. Thus aroyl chloride–pyridine mixtures react rather readily with 2-cumarone, oxindole, or 1-methyloxindole (352). Sequentially treating 2-picoline with ethyl chloroformate and dilute acid (79) results in the formation of a yellow oil of the composition of a 2 : 1 adduct less one molecule of HCl. The initially formed N-acyl salt is believed to undergo cleavage by the action of another molecule of picoline, which behaves as a methylene base under these conditions. Similar products have been obtained with 2,4- and 2,6-lutidines, 4-ethylpyridine, and a pyridine–2-picoline mixture (79).

Johnson has carried out a detailed kinetic study of ring-opening reactions of N-dimethylcarbamoylpyridinium cations in alkali solutions under widely varying pH conditions (67, 68, 353, 354). He has shown that the equilibrium formation of pseudobases by $OH^\ominus$ attack at the ring C-2 is a faster process than ring opening. The rate of the latter reaction is proportional to $[OH^\ominus]^2$ or $[B]\cdot[OH^\ominus]$ where B is the common base (e.g., $CO_3{}^{2\ominus}$). The resulting enoles of 5-dimethylureido-2,4-pentadienales (27) are relatively stable and can be isolated from the solution.

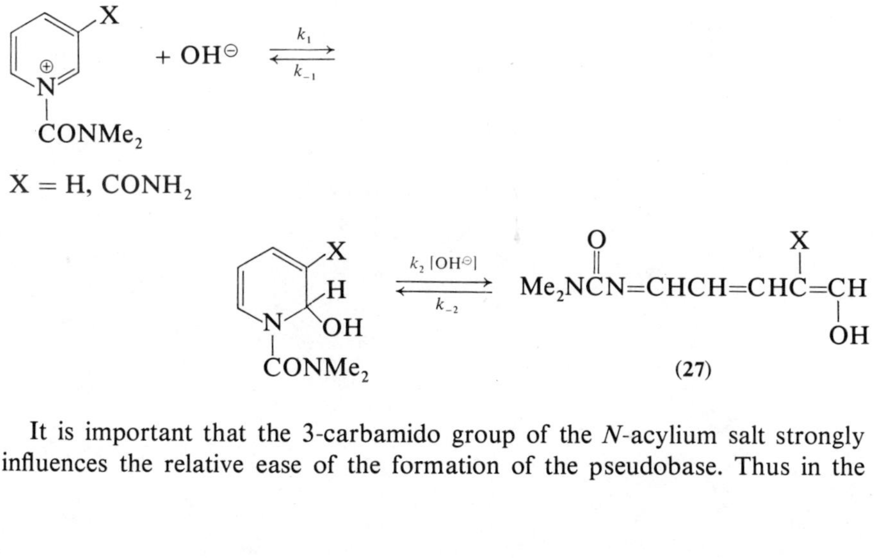

$X = H, CONH_2$

(27)

It is important that the 3-carbamido group of the N-acylium salt strongly influences the relative ease of the formation of the pseudobase. Thus in the presence of this substituent, the equilibrium between the open form and the ring form occurs at pH as low as 8.6, compared to pH 10.6 for the unsubstituted compound; $k_2^{CONH_2}$ and k_2^H differ by five orders of magnitude, compared to a 3- to 14-fold difference observed in the case of acyl transfer reactions (68) (see above).

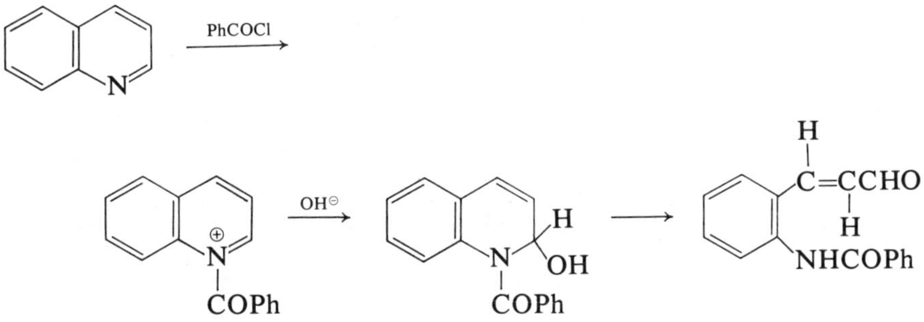

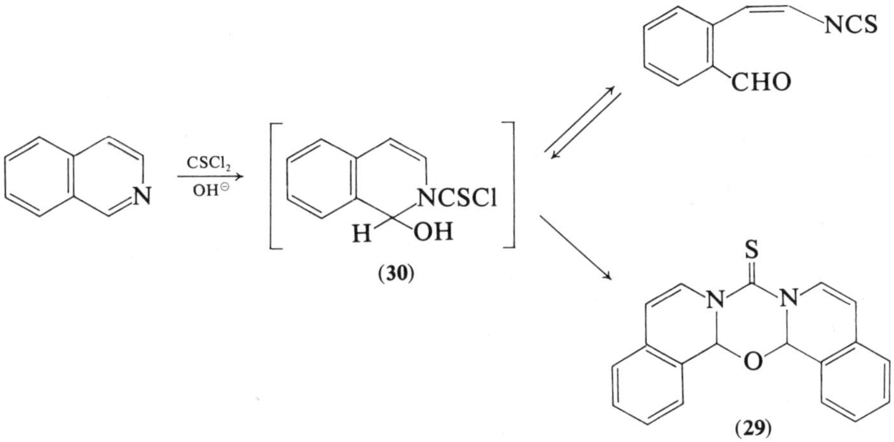

When treated with bases, N-alkyl salts of quinoline and isoquinoline readily yield pseudobases that are exceedingly stable and occur only in the ring-closed form (355). This is why quinoline was originally thought to form the pseudobase **28** with benzoyl chloride (60). More recently the adduct was shown to be the ring-opened structure *trans-o*-benzamidocinnamaldehyde (356–358). It has been pointed out (357) that the Reissert method was of practical importance for syntheses of a number of polyfunctional compounds which were hardly accessible otherwise. Thus substituted cinnamaldehydes were obtained (though in low yields) from 6-methyl, 6-methoxy, and 6-halo-quinolines. However, other quinolines carrying electron-donor substituents did not undergo cleavage. Isoquinoline reacted with thiophosgene in alkaline medium to yield the ring-opened product and **29**, probably as a result of dehydration of pseudobase **30** (359).

The *N*-benzoyl salt of 5-nitroquinoline reacted to yield a stable pseudobase which was shown to have a cyclic structure. It was demonstrated that in this case no acyclic tautomeric forms occurred (358).

The series of papers by Johnson and co-workers (93, 94, 137, 155, 360) dealing with *N*-cyanoquinolinium and *N*-cyanoisoquinolinium salts demonstrates the important role played by pseudobases in most reactions of such compounds (acid–base equilibria of two types, hydrolysis, ring cleavage, the formation of *O*-esters, bromination, nitration).

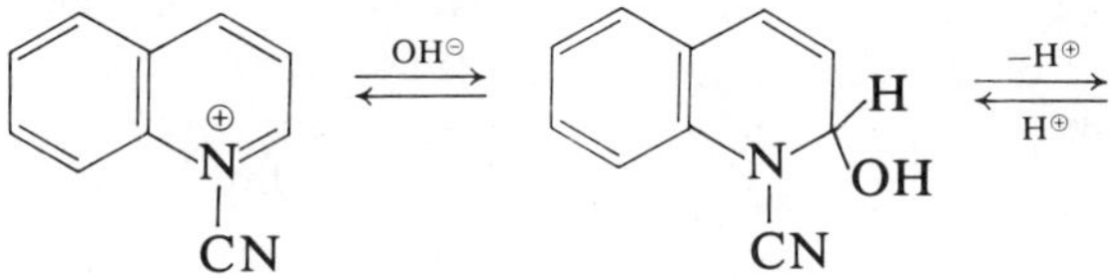

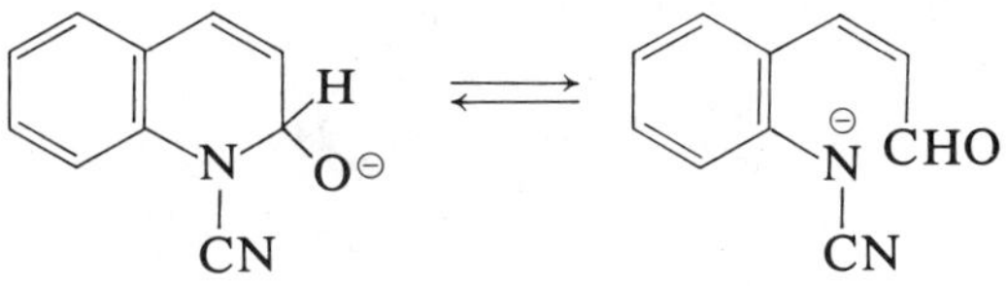

Pseudobases **31** are very stable and give equilibrium mixtures of ions **32** and **33** even under acidic conditions. The equilibria can be quantitatively described with the help of pK_{ROH} values (cf. dissociation of triarylmethylcarbinols) (94, 137). The Hammett equation provides a good description of the acidity versus the nature of 3-, 5-, 6-, and 7-substituent dependence in the *N*-cyanoquinolinium series, though the value of ρ depends on the substituent position, which indicates a dependence on conjugation with the $>\overset{\oplus}{\underset{|}{N}}CN$ group.

Electron-donating substituents increase the pK_{ROH} value (and hence the stability of the pseudobase) compared with the unsubstituted ion, whereas the introduction of Cl, Br, NO_2, especially in position 6, results in a rapid decrease of this parameter. The same approach proves applicable to evaluating the electron-withdrawing effects of substituents on the *N* atom. *N*-Methyl and *N*-cyanoquinolinium salts differ in pK_{ROH} by 17–18 units provided all the other substituents are the same. pK_{ROH} increases by seven or more units on fusion of a benzene ring to the pyridine nucleus (i.e., on going to quinoline and isoquinoline).

Thus the formation of pseudobases and pyridine ring opening are two different steps involved in the same process of nucleophilic attack (HO$^\ominus$, arylamines) at the α-carbon atom of the heteroaromatic ring. This process frequently competes with reactions of transfer of the acyl group to other nucleophiles.

D. REISSERT COMPOUNDS

Reactions of N-acylbenzopyridinium salts with CN$^\ominus$ ion are of great importance for organic syntheses. In contrast to hydroxide ion, cyanide ion causes no cleavage of the heteroaromatic ring and, with few exceptions, no elimination of the acyl group. In this case, the attack at the α-carbon atom results in exceedingly stable 1-acyl-2-cyano-1,2-dihydroquinolines or corresponding 2-acyl-1-cyano-1,2-dihydroisoquinolines [so-called Reissert compounds (361, 362)].* N-Acylpyridinium salts do not enter this reaction (60, 356), and it is likely that the only pyridine adduct of this type (**34**) (90, 362) owes its stability to the formation of an additional ring. However, it is generally assumed that pyridine Reissert compounds could be obtained under properly chosen reaction conditions.

Several techniques have been proposed for the preparation of quinoline Reissert compounds. One of them consists of addition of aqueous potassium cyanide to a mixture of quinoline and an acid halide (60, 356). Naturally, it is applicable only in the case of relatively nonreactive acid halides, primarily in the aromatic series. Aliphatic and some aromatic acid halides undergo hydrolysis in aqueous media. The method works better with potassium than with sodium cyanide (364).

The use of water–methylene chloride medium finds wider application to syntheses of Reissert compounds (365–370). Though the system is a heterogeneous one, all the components of the reaction mixture are soluble in one of the two phases, which make it more convenient than the previous technique. In addition, with the use of small amounts of water, reactions involving more reactive acyl halides become possible (369). In some cases, however, the use of waterfree solvents (mainly benzene) is required to exclude hydrolysis of acyl halides. In this case hydrogen cyanide must be employed rather than the potassium salt, which is an obvious disadvantage, considering

---

* See refs. 90 and 358 for the reaction of 1-cyanoquinolinium with waterfree HCN.

the highly toxic character of gaseous HCN. Still, this modification makes accessible a much wider range of Reissert compounds (371, 372, 372a). On the other hand, attempts to prepare benzoyl chloride adducts using potassium cyanide and such solvents as waterfree acetonitrile, benzonitrile, ether, dioxane, chloroform, or excess quinoline were a failure (373). The use of dimethylformamide and liquid SO_2 media is described in refs. 373 and 374.

There remains some uncertainty as to the nature of the effect of substituents in the quinoline ring. Thus the authors of refs. 374 and 375 were unable to prepare Reissert compounds from derivatives bearing the following substituents in the nucleus: 2-methyl; 5-, 6-, 7-, and 8-nitro; 5-amino; 5-acetamido; 6-dimethylamino; 8-hydroxy; 8-methoxy; and 8-acetoxy. On the other hand, the following derivatives readily enter the reaction with benzoyl chloride in aqueous KCN: 6-methyl; 6-chloro; 6-bromo; 6-methoxy, and 7-methoxyquinolines (371, 374, 376); 3-acetamidoquinoline (367); various 4-substituted quinolines (377); 5,6-benzoquinoline (361); and 2,3′- and 6,6′-diquinolines (378, 379).

Recently, Reissert compound **35** was prepared in high yield from thieno-[2,3-*f*]quinoline (380).

In a methylene chloride–water system, the Reissert reaction goes with many 3-, 4-, 5-, and 7-substituted as well as disubstituted quinolines (367), including compounds that would not react in aqueous medium (373). The above examples show that the ability of quinolines to yield Reissert compounds depends first on the electronic characteristics of the substituents. As a rule, reactions involving quinolines containing electron-donating 3-, 4-, 5-, and 7-substituents give high yields of products, whereas electron-acceptor groups on the ring decrease the yields (367). It is likely that electron-donating substituents facilitate the reaction with acyl halides yielding 1-acylquinolinium chlorides by increasing the nucleophilicity of the quinoline. With 2- and 8-substituted compounds, steric hindrance at the reaction center predominates. To check this supposition, seven 2-substituted and nine 8-substituted quinolines were studied with the result that none of them underwent the reaction (367, 375).

The substituent effect on the reactivity of *N*-acylisoquinolinium salts *in situ* in the Reissert reaction has been studied still less. Benzene ring substituents were reported to have little effect on the reaction. Thus 6,7-dimethoxyisoquinoline readily gave the corresponding 1-cyano-2-benzoyl-1,2-

dihydro derivative which was utilized in syntheses of a number of isoquinoline alkaloids (381). Likewise, 3-methyl and 3-methoxy groups had no hindering action (361, 381). The Reissert reaction would not go with 1-methoxyquinoline, though it did with 2-azafluoranthene; the compounds have similar steric characteristics (362). The Reissert reaction was successfully applied to some other polysubstituted isoquinolines during syntheses of alkaloids and their analogues (382–384).

As mentioned above, 5-nitroisoquinoline reacted with benzoyl chloride and potassium cyanide in aqueous solution under Reissert reaction conditions to yield, unexpectedly enough, a pseudobase together with only traces of the Reissert compound (358). The authors explained their results in terms of Pearson's theory. The enhanced acidity of the C-1 center in 5-nitroisoquinoline predetermined the attack by the "hard" $OH^\ominus$ base rather than "soft" cyanide ion.

All the earlier attempts to carry out the Reissert reaction with acridine were a failure (60, 356). Later 9-acridinecarbonitrile (**36**) and the nitrile of amygdalic acid were obtained by treating acridine and potassium cyanide with benzoyl chloride (385). The latter product might result from interaction of benzaldehyde (a standard product of decomposition of Reissert compounds) with excess potassium cyanide.

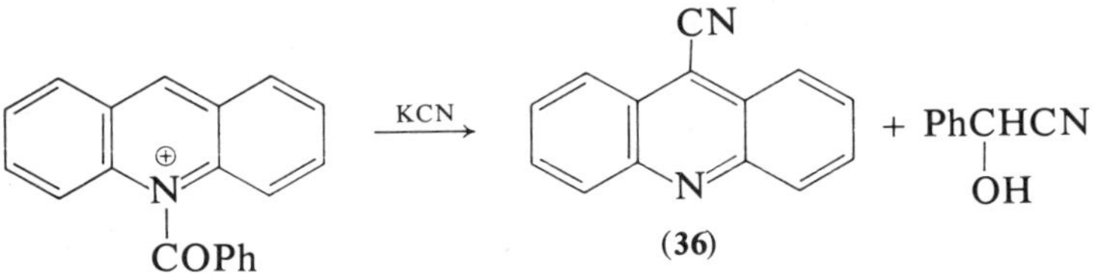

It proved possible to extend the Reissert reaction to phenanthridine (386), *m*-phenanthroline (362) [*o*-phenanthroline does not react (387)], and phthalazine (370). It is characteristic of all the aromatic diazines that they give exclusively monocyano derivatives. The behavior of 1,6-naphthyridine is somewhat unusual (388). The latter reacts with aliphatic acyl halides and KCN in a water–methylene chloride system to yield typical Reissert compounds. On the other hand, it gives a stable pseudobase when treated with benzoyl chloride, whereas in the case of methyl- and benzenesulfonylchlorides the only product is 1-naphthyridinecarbonitrile (**37**).

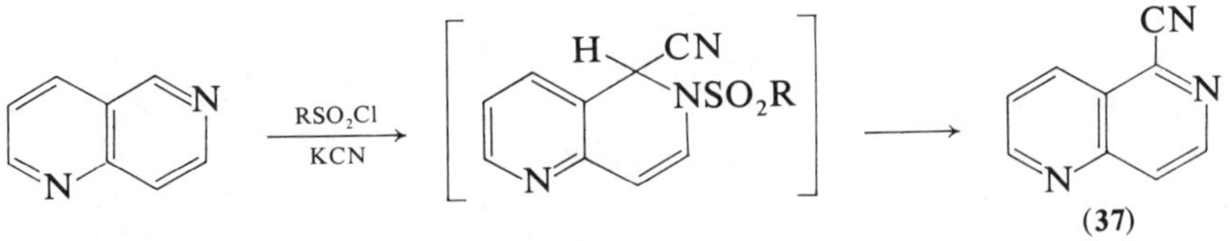

In all these cases the reaction occurs at position 5. This fits well with the above observation of the higher reactivity of N-acylisoquinolinium salts compared with quinolinium salts.

Electronic and structure characteristics of acyl substituents at the ring nitrogen atom and their effects on the reactivity of the corresponding pyridinium and benzopyridinium salts in the Reissert reaction are of great interest with regard to the problem of the transfer of electronic effects of substituents through a positively charged heteroatom. A variety of chlorides of aliphatic, aromatic, and heterocyclic acids, as well as diacyl dichlorides, readily undergo this reaction (369). Ortho-substituted benzoyl chlorides were found to readily yield Reissert compounds irrespective of the electronic properties of the substituent, whereas *m*-nitro- and *m*-chlorobenzoyl chlorides reacted to give low yields of the corresponding nitriles, and attempts to react them with 4-nitrobenzoyl and 3,5-dinitro- and 2,4-dinitrobenzoyl chlorides were a complete failure (389). Experiments with cyclopropane- and cyclohexanecarbonyl chlorides (369) and chloroformates (359, 390) proved successful. Acyl bromides and anhydrides of carboxylic acids, however, gave unsatisfactory yields (369), probably because of a less favorable reaction rate ratio between formation and hydrolysis of the intermediate N-acylium salts.

The Reissert reaction occurred also with diphenylcarbamoyl chloride (391) and arylsulfonyl chlorides (392), though only when isoquinoline and phthalazine were involved. According to the data of ref. 362, quinoline did not undergo the reaction under such conditions.

A usual Reissert condensation occurred with quinolines and isoquinolines by the action of various chlorothiocarbonyl derivatives (359, 393). However, quinoline reacted with thiophosgene and potassium cyanide to yield, as the final products, the ring-opened product **38** and 3-oxoimidazo[1,5-*a*]quinoline (**39**). The following scheme was proposed to explain the formation of the latter product (359):

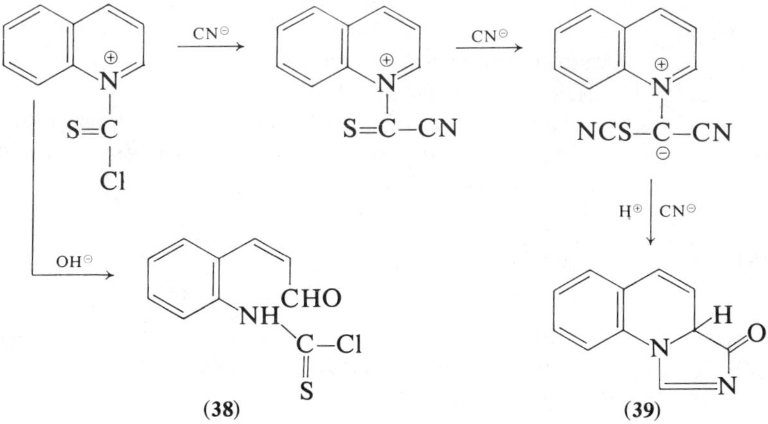

A marked polarizability of the sulfur atom is responsible for the fact that the quaternary salt formed from $CSCl_2$ and quinoline has its softest acid center at the thiocarbonyl function. So the harder base $OH^\ominus$ attacks the C-2 position of the ring to yield **38** whereas the softer base $CN^\ominus$ reacts at the C=S bond, eventually giving **39**. The yield of aldehyde **38** increases with the $OH^\ominus$ concentration, in conformity with the cited scheme.

A great deal of attention has been given to the reaction of N-acylium salts of six-membered nitrogen heterocycles with cyanide ion because of the wide applications of the resulting Reissert compounds to organic syntheses (361, 362). First of all, these compounds give aldehydes in good yields. This method has been recommended for the preparation of aldehydes from acyl chlorides containing readily reducible and other labile groups. Thus o-nitrobenzoyl chloride cannot be reduced to the aldehyde by the Rosenmund reduction, but application of the Reissert technique leads to a 58% yield of o-nitrobenzaldehyde.

Other, still more numerous applications utilize the ability of Reissert compounds to yield carbanions (**40**) by the action of strong bases (C_6H_5Li, NaH) in polar media at -10 to $-20°C$. These so-called "Reissert anions" react with various electrophiles to yield, after splitting off the acyl group, substituted quinolines and isoquinolines. Alkyl, allyl and benzyl halides, aldehydes, lactones, and unsaturated compounds containing activated double bonds (vinylpyridines, acrylonitrile, etc.) were successfully used in this reaction as electrophiles. The latest papers (394) claim that some of the anions can be obtained by the action of aqueous alkali (in the presence of trialkylbenzylammonium chlorides). Dimethylformamide medium and the use of NaH provide the most favorable conditions for arylation of Reissert compounds (395, 395a). Reissert anions can undergo intramolecular rearrangement to ketones in both the quinoline and isoquinoline series (370, 384, 396).

E. HETARYLATION REACTIONS

Generally the term "hetarylation," or "heteroarylation," includes reactions that lead to the introduction of aromatic heterocyclic residues in a nonsubstituted or partially hydrogenated form into organic species (1, 13). In a wider sense, this includes all the reactions of N-acyl heterocyclic cations with nucleophiles (including CH acids in the presence of basic catalysts, organometallic compounds, and compounds bearing lone electron pairs), provided the products retain the heterocyclic nucleus. Reactions involving $OH^\ominus$ and $CN^\ominus$ ions also fall under this wider definition, and it is merely because of their important applications and because they have been studied independently from other reactions of this kind that we discuss them in a separate section.

1. Hetarylation of Activated Aromatic Rings

Like a similar reaction of azo coupling with aryldiazonium salts, C-hetarylation naturally requires that compounds subjected to hetarylation be rather strong nucleophiles, such as dialkylanilines, phenols, and other activated nuclei, and also π-excessive heterocycles (pyrrole, indole, furan, etc.).

In 1952, McEwen and co-workers (397) observed the formation of small quantities of 1-benzoyl-2-(p-dimethylaminophenyl)-1,2-dihydroquinoline in a mixture of quinoline, dimethylaniline, and benzoyl chloride which was allowed to stand for 3 days. Likewise, pyridine attacked the *para* position of the dimethylaniline nucleus in the presence of benzoyl chloride and copper metal powder to give 4-(p-dimethylaminophenyl)pyridine (398).

Further studies (1, 13, 53) demonstrated that these reactions exemplify a more general process. When heated, various bases (pyridine, quinoline, isoquinoline, acridine, variously annellated benzoquinolines and isoquinolines) nearly always reacted with dialkylanilines (see Scheme 1) in the presence of acyl halides to introduce the heterocyclic residue into the aromatic nucleus.* Reactivities of bases toward dimethylaniline measured under standard conditions increased across the series pyridine, phenanthridine, quinoline, isoquinoline, acridine. Reaction with pyridine required a catalyst—a Lewis acid [$TiCl_4$ proved the best one (139)] or copper metal powder (398). In most cases, aromatic aroyl chlorides were more reactive than aliphatic acyl halides, in agreement with the higher electron-withdrawing power of the former (139).

Collating data on reactions involving various acyl halides is very important for an understanding of the reaction mechanism (see below). For instance, in the presence of CH_3COX, where X is Cl, Br, or I, pyridine reacted to give 25, 26, and 43.5% yields, respectively, of pyridyldimethylaniline (139). A similar increase of yields was observed in reactions involving mixtures of C_6H_5COX and phenanthridine, quinoline, or isoquinoline in going from F to Cl, to Br, and to I (402–404, 407).

In addition to acyl halides, such acylating agents as chlorides of dicarboxylic

---

*Recently, "pyridylation" of various organic compounds mixed with pyridine (anthracene derivatives, Schiff bases) was reported to occur during the anodic oxydation process (399–401). In this case, however, incorporation of the pyridine ring involved the nitrogen atom; that is, quaternary salts were formed, similarly to the case of the Ortoleva–King reaction.

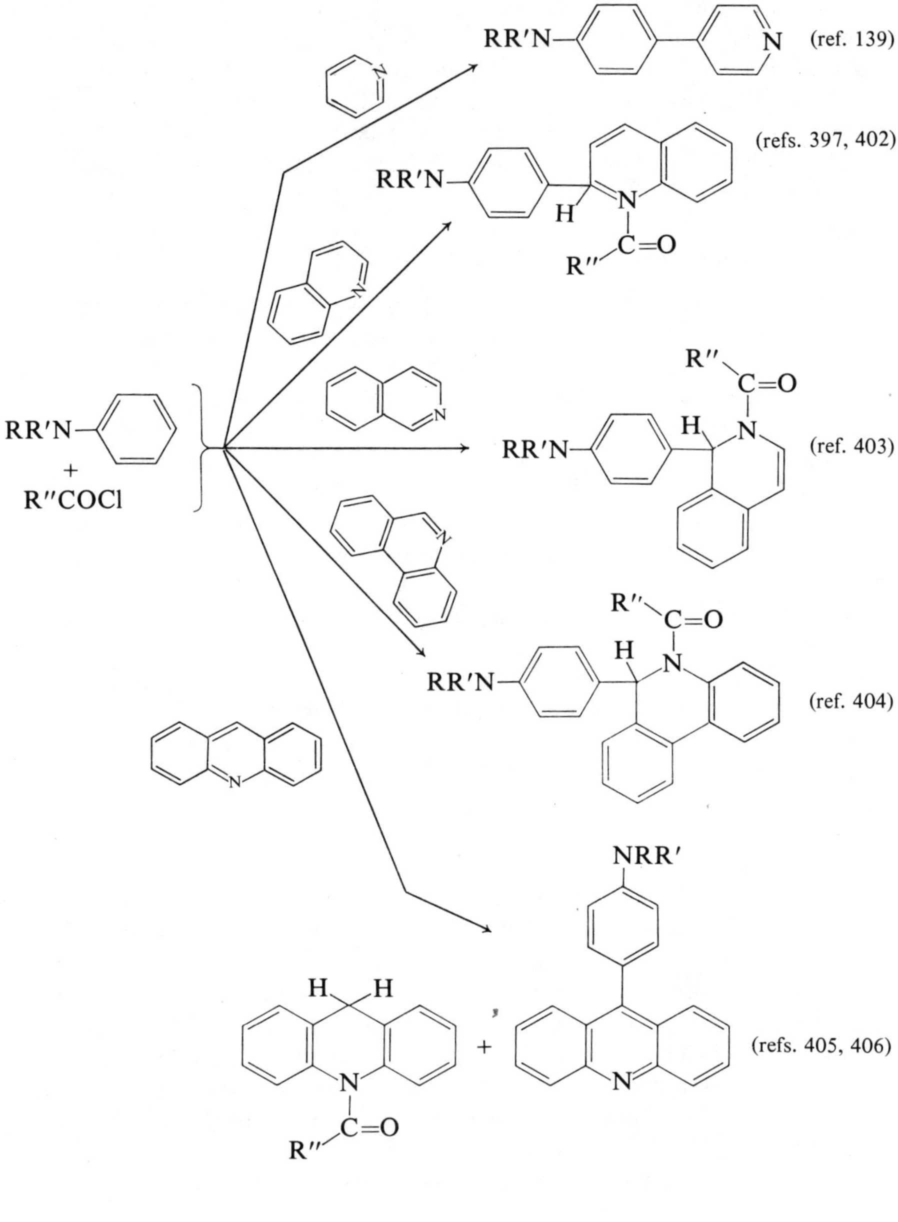

Scheme 1

acids (402), some acyl chlorides of phosphoric acids (408), β-chlorovinyl ketones [which represent vinylogues of acyl halides (409)], and cyanuric chloride (409, 410) were shown to undergo this reaction.

As for substituted pyridine bases, only those with substituents remote from the reaction center could effect hetarylation, such as 6-nitro-, 6-, 7-, and 8-methyl-, or 6-bromoquinoline (407). The reaction followed an unusual pathway when there were 2- or 4-methyl substituents in the quinoline nucleus [see below (52)].

The effect of alkyl groups in the dialkylamines manifested itself by a gradual decrease in the yields of the pyridylation products with increasing length of the alkyl chain (139). Heteroanalogues of anilines such as N-phenylpyrrolidine, N-phenylmorpholine, N,N'-diphenylpiperazine (139, 403, 406), N-alkylindolines (403, 406, 411, 412) and N-alkyl-1,2,3,4-tetrahydroquinolines (403, 404, 406, 407, 413, 414) also were reactive. The reaction was subject to strong steric limitations. In all cases, substrates incorporated heterocyclic substituents at the position *para* to the dialkylamino group; application of a TLC technique failed to detect any *ortho* isomers. Only with 1-ethyl-6-methyl-1,2,3,4-tetrahydroisoquinoline (occupied *para* position) did isoquinolylation occur at position 7; even then the yield was not more than 7% (53). In the phenol series, it proved possible to introduce the quinoline residue into the rings of the dimethyl ether of resorcinol and 2,6-di-(*t*-butyl)phenol (where usual O-acylation was impossible because of steric hindrance at the OH group). Anisole, benzamide, and N-methylbenzamide would not enter the reaction even when under severe conditions, perhaps because of the lower nucleophilicity of the *para* carbon atoms (53). Anisole itself underwent demethylation at 220°C (218). The main process for phenol is O-acylation.

2. π-Heterocycles

Reference 415 reported on the formation of a colorless pyridine-containing compound in the reaction of pyrrole with pyridine and BrCN, providing the first example of hetarylation of the pyrrole ring. More recently Treibs studied this reaction in detail using a number of substituted pyrroles (416–419) and extended the technique to such acylating agents as acetyl and benzoyl chlorides and ethyl chlorocarbonate (418). 2,4-Dimethylpyrrole reacted with C_6H_5COCl and pyridine (419) to give the 2-benzoyl derivative as the main product (see also ref. 258), and 1-benzoyl-2,4-dimethyl-5-(N-benzoyl-1,4-dihydro-4-pyridyl)pyrrole in only 5% yield. Reactions involving 2-ethoxypyrroles followed a still more complicated pathway (420):

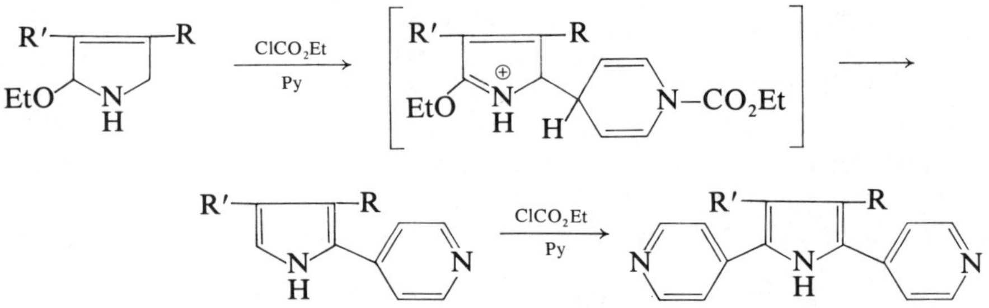

Hydroxypyrroles (pyrrolinones) reacted, depending on a number of factors (417, 420–424), in the direction of either the acylation–pyridylation products of dihydro structure **41** or 2-(4-pyridyl)-, 2,5-di-(4-pyridyl)pyrroles (after elimination of the OH group). It is interesting that 2,4-dimethylpyrrole reacted with acridine in the absence of catalysts to give molecular compound **42** (418, 425), formally representing the acridylation product. Later, the reaction of hetarylation of the ring was shown to be more generally applicable, and to proceed with various heterocyclic systems in the presence of acylating agents of many classes (417, 426, 427). Pyrroles proved the most reactive ones. The reactivity of heterocyclic compounds decreased across the series pyrrole > indole > 2-methylfuran > furan > 2-methylselenophene > 2-methylthiophene > thiophene (41, 42, 422, 428–435). Similar trends were observed for other electrophilic substitution reactions involving five-membered heteroaromatic systems (436).

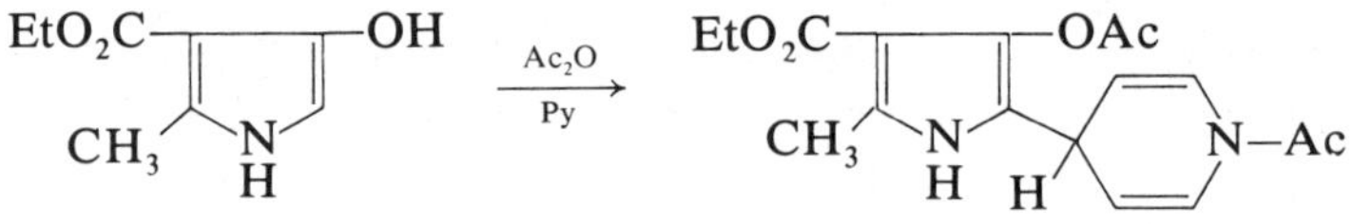

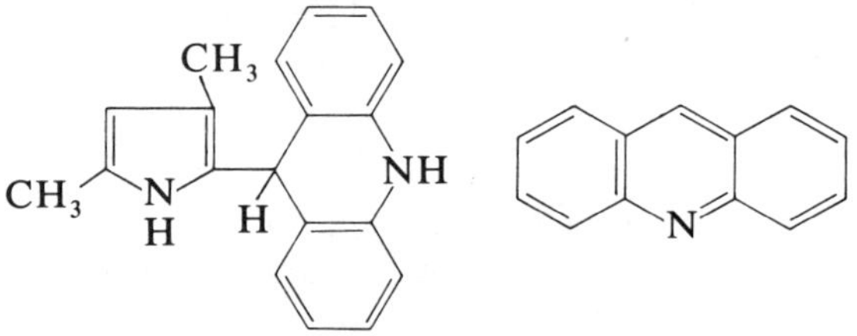

Of pyridine bases, isoquinoline and acridine were the most reactive species (426); then reactivity decreased in going to quinoline, phenanthridine, and

IV. REACTIVITY OF N-ACYLHETEROCYCLIC CATIONS

benzo[f]quinoline (54). As a rule, isoquinoline reacted with unsubstituted or N-alkylpyrroles in the presence of acyl halides to give a mixture of three compounds (43–45). The product ratio depended on both C-nucleophilicity of the pyrrole nucleus and electrophilicity of the N-acylisoquinolinium salt: the most reactive salts gave mainly bis-adducts (45), whereas the least active 2-diphenylcarbamoylisoquinolinium chloride reacted to yield exclusively compound 43. 1-Phenylpyrrole gave adducts containing only one heterocyclic residue (at the α position of the pyrrole ring) even when treated with the most active N-acylium salts. Monohetarylpyrroles (43) could in turn undergo hetarylation to 2,5-dihetarylpyrroles, incorporating either different heterocyclic residues, or different acyl groups.

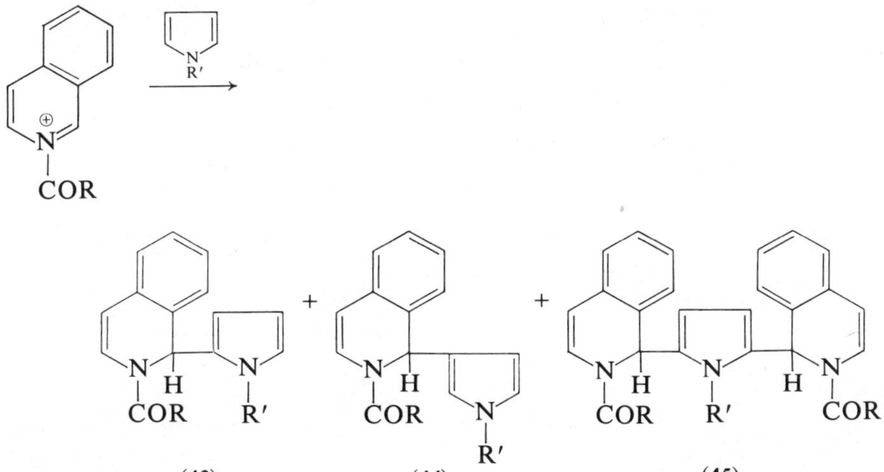

The same technique provided one-step routes to various heterocyclic furan derivatives (433, 434). Thus acylisoquinolinium salts reacted to yield, depending on the electronic nature of the acyl halides, mono- or bis-hetarylfurans (46 and 47, respectively); the formation of β isomers never occurred provided there was a vacant α position. When boiled with alcoholic solutions of alkalis, the products were converted to 1-furylisoquinolines, as one would expect (434).

Acridine immediately gave fully aromatized acridylpyrroles, the corresponding furans (e.g., 48), or indoles (426, 432). As a by-product acridane was separated.

Hetarylation of indoles with N-acylium salts (pyridinium, quinolinium, isoquinolinium) resulted in substitution of the hydrogen atom at C-3 of the pyrrole ring with the formation of compounds of types 49–51 (422, 428–432, 437, 438). The nature of the acylating agent and of the solvent plays an

(46) (47) (48) (49) (50) (51)

important though not yet completely understood role (41, 42, 231). Thus various authors reported 1-benzoyl-4-(indol-3-yl)-1,4-dihydropyridine (**49**, R' = H, R = C_6H_5) to be the product of the reaction between pyridine, benzoyl chloride, and indole (428, 429, 437), whereas with *p*-toluenesulfonyl chloride the corresponding compound was obtained in low yield and quite readily underwent conversion to pyridylindole (**50**, R' = H). As already mentioned, indole and 2-methylindole gave exclusively the acylation products (at N or C-3, depending on the temperature) when treated with acyl chlorides or bromides of halocarboxylic acids in the presence of pyridine bases (41, 42). Pyridylation of indole occurred in a mixture of $C_2H_5CHBrCOBr$, pyridine, and indole, whereas in the presence of 2,4,6-trimethylpyridine, 3-(α-bromobutyryl)indole was obtained (41).

Treatment of 2-(tosylimino)indoline and its *N*-methyl homologue with acetic anhydride and pyridine led to both acylation and pyridylation simultaneously (439, 440):

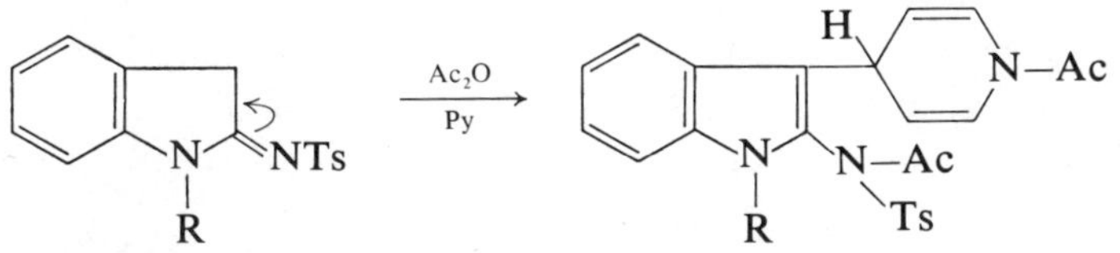

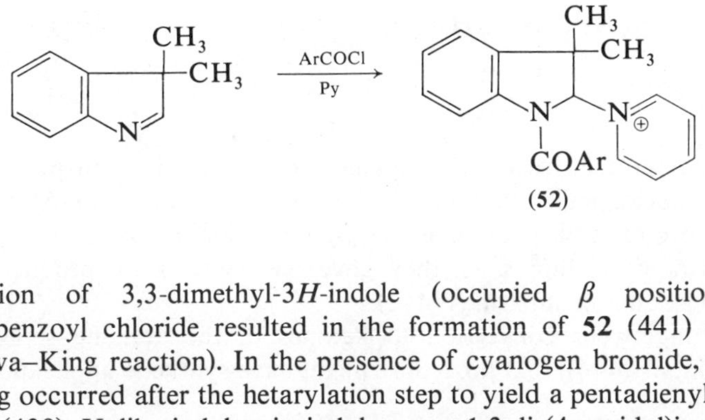

Acylation of 3,3-dimethyl-3*H*-indole (occupied β position) with *p*-chlorobenzoyl chloride resulted in the formation of **52** (441) (i.e., like the Ortoleva–King reaction). In the presence of cyanogen bromide, pyridine ring opening occurred after the hetarylation step to yield a pentadienyl derivative of indole (430). Unlike indoles, isoindoles gave 1,3-di-(4-pyridyl)isoindoles when treated with pyridine and benzoyl chloride (423).

Of other π-excessive heterocycles, selenophene and 2-methylthiophene could be hetarylized, but only with *N*-acylisoquinolinium salts (435). In the case of weakly nucleophilic thionaphthene, the only products were *N*,*N*'-diacyl-1,1',2,2'-tetrahydro-1,1'-diisoquinolines, which might result from recombination of intermediate *N*-acylium radicals.

3. CH-Acidic Compounds

Interaction of pyridine bases and ketones, β-dicarbonyl compounds, and some other classic CH acids in the presence of acyl halides resulted in C- and O-acylation (see above) or hetarylation of substrates, depending on the nature of the nitrogen base and, apparently, the pK_a of the enol form (130, 397, 429, 442–444). In the latter case, ketones reacted with N-acylpyridinium salts *in situ* to form 1-acyl-1,4-dihydropyridines (**53**) (130, 442), with quinolinium salts to give 2-substituted 1-acyl-1,2-dihydroquinolines of type **54** (130, 429), and with N-acylisoquinolinium or phenanthridinium salts to give 1-substituted 2-acyl-1,2-dihydroisoquinolines (**55**) or 6-substituted 5-acyl-5,6-dihydrophenanthridines (444), respectively.

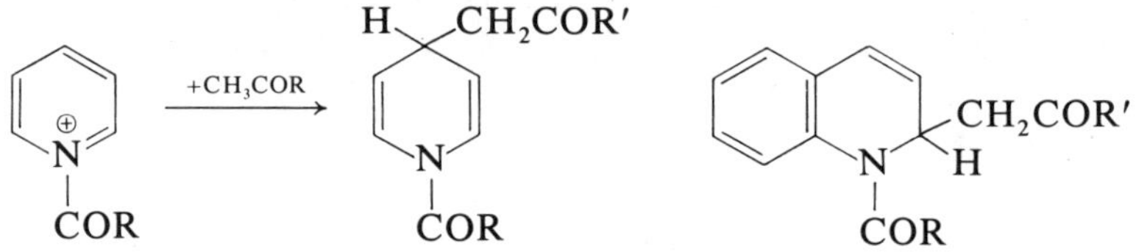

Acylpyridinium salts react with cyanacetic acid esters with pyridine ring opening after nucleophilic addition at C-2 (429). Nitroalkanes (445) reacted to give unusual products also containing no pyridine ring. However, when treated with N-acylisoquinolinium salts, they gave the hetarylation products incorporating two isoquinolinium rings (445).

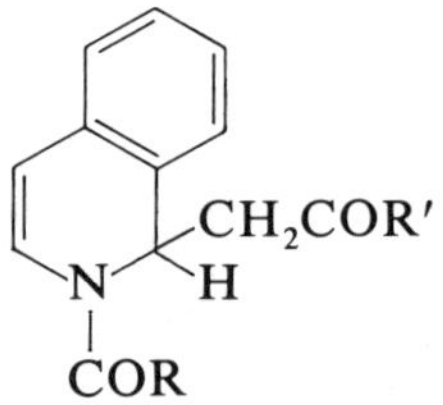

Acridine underwent conversion to ketones of the acridine series under similar conditions (444); that procedure might have synthetic applications:

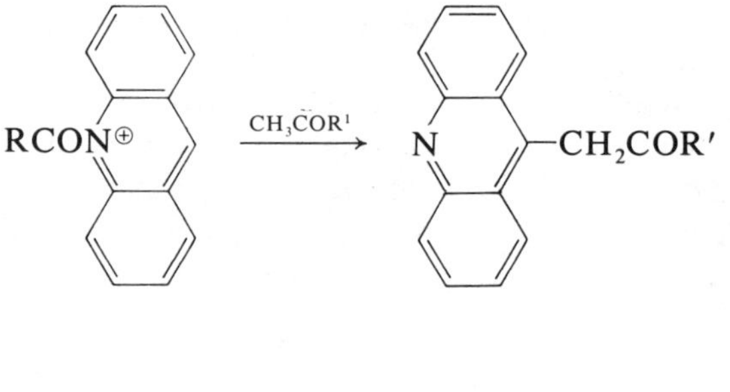

IV. REACTIVITY OF N-ACYLHETEROCYCLIC CATIONS

Hetarylation reactions were utilized to introduce heterocyclic residues into the *A* ring of 3-ketosteroids and the *D* ring of 17-ketosteroids as well as into some steroid lactones (446, 447). It is worth mentioning that the secondary OH groups remained unaffected in these reactions (446).

Hetarylation of *β*-dicarbonyl compounds (diketones, malonic and acetoacetic esters) went even more readily (444); the resulting adducts underwent transhetarylation reactions involving other CH acids (271, 448, 449).

Here belong also reactions involving *N*-acylium salts and partially hydrogenated heterocycles containing CH_2 or CH groups activated by the presence of neighboring carbonyl functions, including pyrazolones, thiazoline-2,4-diones, rhodanines, isorhodanines, and azlactones (422, 450–452):

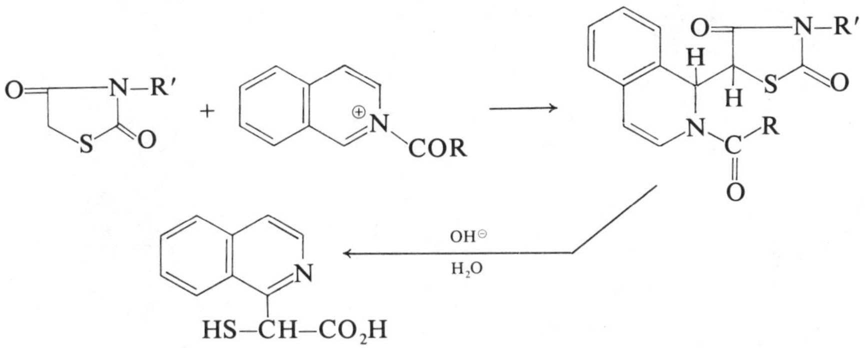

The products of hetarylation of thiazolidines are of interest, since they converted to thioglycolic acids of the heterocyclic series after alkaline hydrolysis (451).

The Dakin–West reaction was shown to give rise to products of pyridylation of oxazolones (261, 453), though these probably resulted from some side process (261).

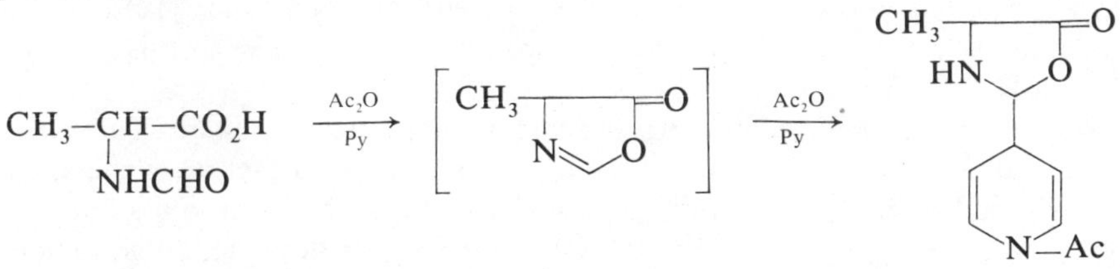

This observation aroused interest in hetarylation of Δ^2-oxazol-5-ones and pseudooxazol-5-ones with various pyridine derivatives (162, 261, 454) in the presence of anhydrides (261, 454), familiar in the Dakin–West reaction, as well as in the presence of acyl chlorides and chlorides of dicarboxylic acids (454–

457). When catalyzed with 4-dialkylaminopyridines, most α-amino acids underwent the Dakin–West reaction even at room temperature to give high yields of acylamino ketones (162, 454). In other cases, replacement of pyridine with 4-picoline was advantageous (457). For more details, see ref. 260. The authors of ref. 261 supposed that the pyridylation reaction might be characteristic of other heterocyclic enols as well. Thus according to their data, pyridine reacted exothermally with 2-methoxycarbonyl-5-phenylfuran to yield 2-phenyl-4-(1-methoxycarbonyl-1,4-dihydro-4-pyridyl-4)-4,5-dihydrofuran-5-one.

Homophthalic anhydride and its tetrahydro derivative reacted with aroyl chlorides in the presence of pyridine under mild conditions to give acylation–hetarylation products, for example, **56** (458, 459); with picoline, the usual *O*-esters were mainly obtained. Reference 460 reported the isolation of 1-acetyl-2-carboxymethyl-1,2-dihydropyridine (**57**) from an attempt to acylate niacytine with acetic anhydride in pyridine solution.

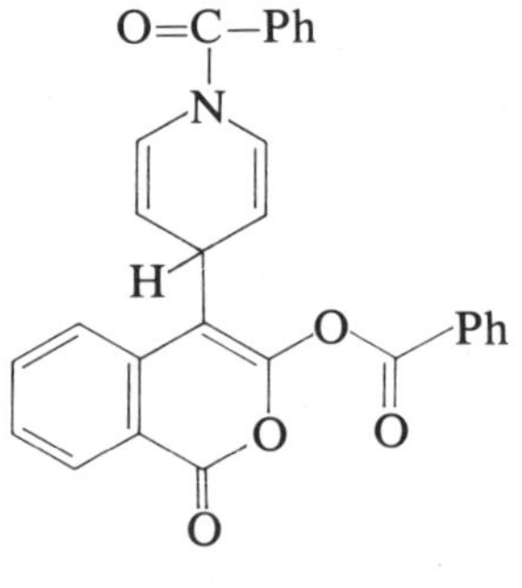

(56) 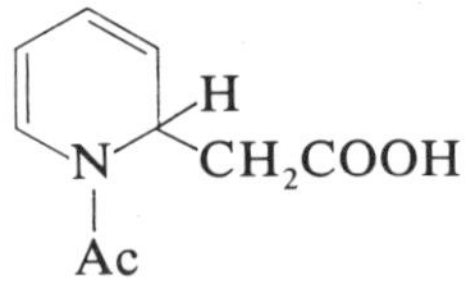   (57)

Data exist that claim the successful hetarylation of other organic derivatives characterized by pK_a values of 15–21 (at least with the most active *N*-acylium salts), including cyclopentadiene, azulene, indene, and phenylacetylene; on the other hand, fluorene (pK_a 22.9) and ferrocene did not undergo this reaction (53, 461).

Many authors have pointed to the fact that *N*-acylium salts of quinaldine and lepidine could not be used as hetarylation agents. In fact, acyllepidinium cations underwent conversions to 1-acyl-4-methyl-1,2-dihydroquinolines (at 50°C) or "dimeric species" (at 100°C), depending on the reaction conditions (52). Examination of the mass spectrum of the dimer obtained in the reaction with acetyl chloride suggested structure **58**; the dimer could be formed according to the scheme shown (similar to the scheme of dimerization of *N*-alkyllepidinium salts). No dimeric units were observed in the case of quinaldine, perhaps because the methyl group of salt **59** was less activated than

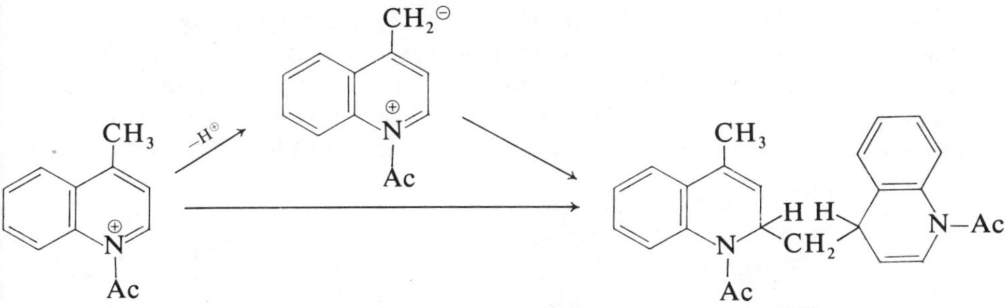

IV. REACTIVITY OF N-ACYLHETEROCYCLIC CATIONS

that of its lepidine analogue. Instead, 1-acyl-1,2,3,4-tetrahydroquinaldines were formed under various conditions. Their formation probably involved disproportionation (accompanied by hydride transfer) of 1-acyl-2-methyl-1,2-dihydroquinoline (**60**):

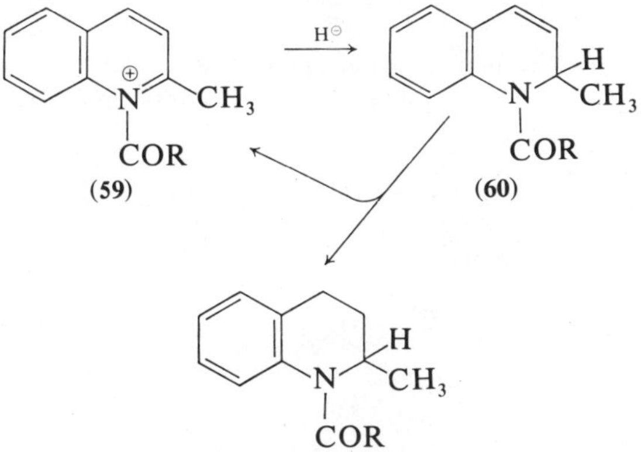

4. Compounds Containing Lone Electron Pairs

Reactions of N-acylheteroaromatic cations with compounds having lone electron pairs at the heteroatom most frequently involve attack on the carbonyl carbon atom, resulting in elimination of the heterocycle and acyl transfer (acylation of amines, alcohols, thiols, etc.). However, there are several examples of such reactions which follow the hetarylation pathway. These include the well-known synthesis of N-(4-pyridyl)pyridinium chloride in the chlorination of pyridine with thionyl chloride (462–464). It is likely that the reaction involves an attack by pyridine, acting as a nucleophile, on the N-acylium salt at the C-4 atom.

There exist a very limited number of examples of intramolecular hetarylation reactions. Here belong the formation of dyes in the reaction of quinaldic acid with acetic anhydride, or quinaldyl chloride with quinoline (465–467).

Recently, quinoline was shown to add at position 2 not only cyanide ions, but also alkoxide and mercaptide ions in the presence of acylating agents (ethyl chlorocarbonate anhydrides and halides of carboxylic and sulfonic acids) (468, 469). The reaction gives high yields of 1,2-dihydro derivatives **61** and **62**. The products are used for the synthesis of polypeptides (469, 470) and are of pharmacological interest (468).

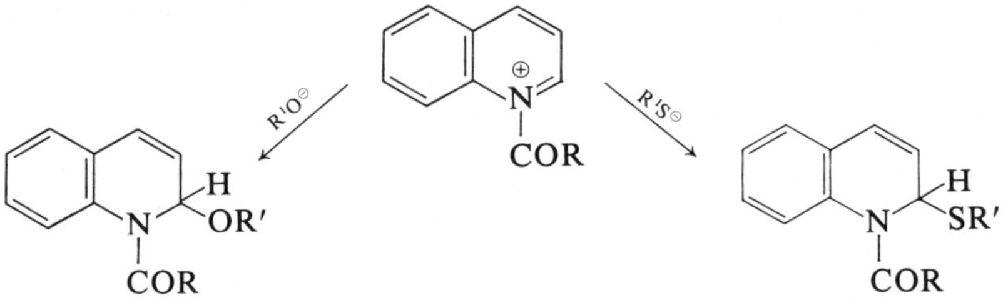

Arbuzov's rearrangement of trialkylphosphites under the action of nitrogen heterocycles in the presence of acyl halides offers an interesting example of such reactions (471, 472):

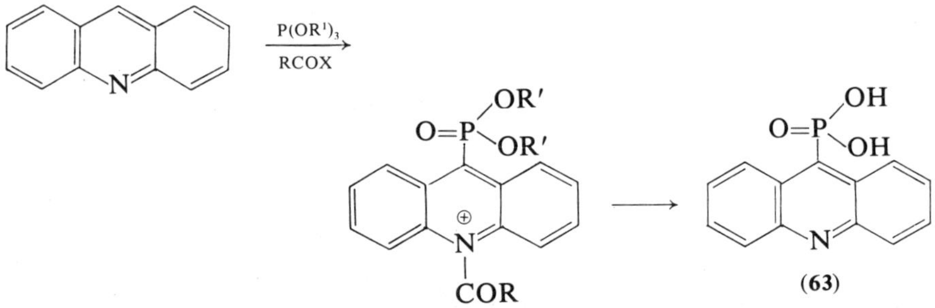

Reactivity of bases decreases here in going from acridine to isoquinoline and pyridine and then to quinoline. The reaction allows a one-step synthesis of heterocyclic phosphoric acids (e.g., **63**). The same synthetic procedure proves even simpler when acridine, chlorophosphite (or dialkylchlorophosphine), and higher boiling alcohols are used as starting materials (13).

Reactions of N-acylium salts and some organometallic compounds (attack at the R–Me bond) are discussed next. Pyridine was reported (361) to react with Grignard reagents in the presence of acylating agents to give the corresponding 1,4-dihydropyridines which readily underwent conversion to 4-substituted pyridines. However, more recently, Grignard reagents were shown to attack N-acylpyridinium salts at position 2 rather than 4 (but never at the acyl group!) (83, 473). Likewise (at C-2) went the reaction of N-

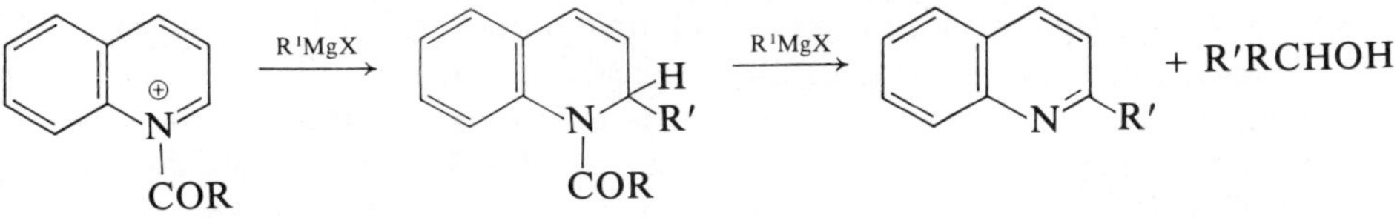

acylpyridinium salts *in situ* with silver acetylenides (474). As mentioned above, the C-2 atom behaves as a hard acid center, which, in terms of the HSAB concept, makes it subject to attack by $R^{\ominus}$.

Addition of Grignard reagents to a mixture of quinoline and an acyl halide resulted in the corresponding 2-alkyl-1,2-dihydro derivative (52, 449). The reverse order of mixing the reactants (providing an excess of the organomagnesium compound) led to aromatization to give a 2-substituted quinoline (449, 475). In this case, the acyl residue underwent conversion to the corresponding tertiary alcohol.

5. *Mechanisms of the Hetarylation Reactions in the Presence of Acyl Halides*

Initially, it had been assumed (139) that pyridylation of dialkylanilines with pyridine and acyl halides in the presence of $AlCl_3$ proceeded via the formation of the *N*-acylium salt with subsequent replacement of the anion with $AlCl_4^{\ominus}$, thus concentrating the positive charge on C-4 (1):

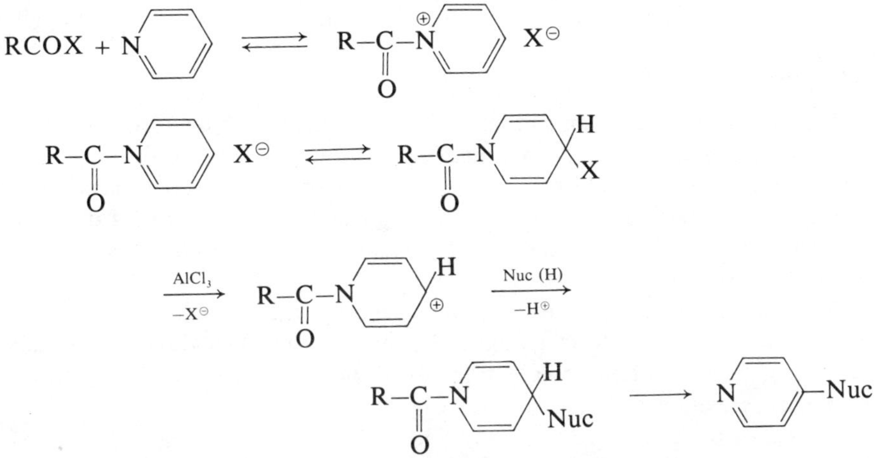

However, this scheme does not appear to be quite adequate (13, 53). Variable product composition (fully aromatized products in the case of quinoline, isoquinoline, and their benzene-condensed homologues; dimeric species in hetarylation of compounds of lower nucleophilic reactivity), together with the fact that many reaction mixtures give rise to EPR signals when dihydro structures are the final products, lead one to believe that two reaction mechanisms can operate, an ionic and a radical one (13, 379, 407).

Cationic hetarylation with acridine leads to fully aromatized heterocycles resulting from hydride transfer from intermediate dihydro adducts to the initial N-acylium ion **64**. The hydride transfer scheme has been substantiated by experiments involving 9-morpholinyl-1-methylacridane and 1-ethylacridinium methoiodide (406). Also, data exist that point to the possibility of aromatizing the intermediate N-acyldihydro structures via elimination of RCHO (139).

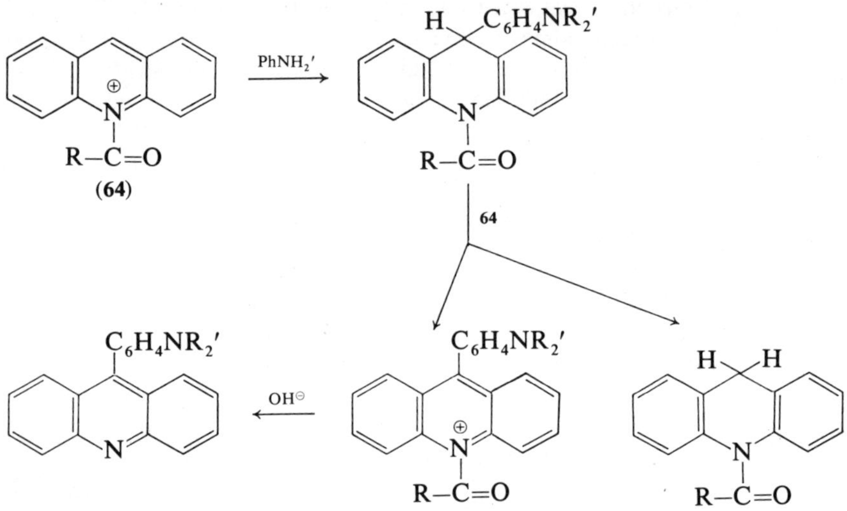

The pyridylation reaction can follow both a radical (with the formation of 4-substituted 1-acyl-1,4-dihydropyridines) and a cationic scheme (leading to 4-substituted pyridines). We have mentioned already that pyridine reacts with indoles in waterfree benzene to yield 1-acyl-4-(indol-3-yl)-1,4-dihydropyridines (429), whereas in more polar media, the same reaction results in 1-acyl-3-(4-pyridyl)indoles (42).

In weakly ionizing solvents, we believe, the initially formed N-acylpyridinium salts behave as tight ion pairs so that N-acylpyridyl radicals can occur because of electron transfer from the anion to the lowest vacant π orbital of the cation:

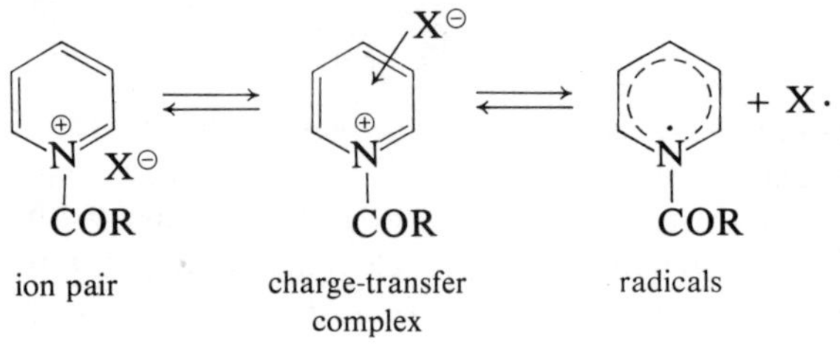

IV. REACTIVITY OF N-ACYLHETEROCYCLIC CATIONS

Such electron transfer (giving rise to charge-transfer complexes) is very common with pyridinium quaternary salts (476–478). The formation of radicals must be even more facile for benzopyridines which offer additional possibilities for charge delocalization.

The radicals represent electrophiles reactive enough to hetarylate most substrates though they cannot participate in hydride transfer reactions (13). The occurrence of radicals manifests itself most plainly in the formation of dimeric structures of type **65** (435), if some reagent to accept positive charge is used. These arguments find support in the fact that, unlike reactions involving acylpyridinium salts, the yields of quinolylation and isoquinolylation products are weakly influenced by the electronic nature of the acylating agent but depend strongly on the nature of the anion (and decrease along the series I > Br > Cl > F). However, the latter observation is also consistent with a cationic mechanism. Besides, UV irradiation accelerates these reactions (403).

The charge distribution patterns for salts of N-acylhetero-aromatic cations in the stationary state resemble those characteristic of the corresponding quaternary salts and can be predicted, although only qualitatively, on the basis of their IR and NMR spectra (479). Thus both alkyl and acylpyridinium salts give proton NMR spectra indicating that the maximum proton deshielding caused by electron density redistribution over the ring occurs at the α position (61, 62). However, this observation does not explain the preferred formation of 1,4-, 1,9-, and 1,2-dihydrostructures from pyridine, acridine, and quinoline, and isoquinoline, respectively.

According to Kosower and Klinedinst (476), substitution at C-4 of the pyridine ring occurs when the reactants (electron donors) form charge-transfer complexes with N-methylpyridinium salts; in case the formation of such complexes is hindered, 2-substitution is favored. In the systems we are discussing, the acyl residues at the nitrogen atom exhibit an enhanced affinity for pyridine ring electrons compared to the methyl group, and thus should facilitate complex formation, hence favoring 4-substitution. Thus according to Kosower (480), the benzoylpyridinium cation reacts with acetophenone enolate anion via the formation of an adduct in which the nucleophilic atom of the ambident group is located near the C-4 atom and the positively charged nitrogen near the anionic center. However, Kosower himself points to the fact that this rule does not hold with reactions involving cyanide ion, which attacks pyridinium salts exclusively at C-4, though spectra of the intermediates give no evidence for a charge-transfer complex.

In light of the HSAB theory by Pearson and Songstad (481), the direction of the attack by nucleophiles (generalized Lewis bases) on N-acylheterocyclic cations (ambident ions) should depend on the relatively "hard" or "soft" character of certain ring atoms. This explains the preference for attack at position 4 of pyridinium salts, but again fails to clarify the situation with quinoline and isoquinoline. It is likely that thermodynamic factors play the determining role (482): these favor the 1,2-dihydroquinoline structure over the 1,4-, since the C-3–C-4 double bond in the former is conjugated to the benzene ring.

F. SINGLE-ELECTRON TRANSFER REACTIONS OF N-ACYLHETEROCYCLIC CATIONS

In addition to the occurrence of radical states, at least with certain classes of N-acylheterocyclic cations in nonpolar solvents, the compounds in question facilitate reactions involving electron transfer from external electron donors. Thus single electron reduction potentials decrease across the pyridine cation series with increasing electron-withdrawing power of the substituent at the heteroatom (133). In fact, reduction of N-alkylpyridinium salts requires sodium amalgam or hydride complexes, whereas N-acylpyridinium derivatives can be reduced even by treatment with zinc powder at slightly elevated temperatures; also addition of potassium iodide to pyridine–arylsulfonyl chloride systems has been reported to result in a vigorous iodine release (99).

Weitz and co-workers (483) were the first to observe the formation of radicals in treating pyridine with benzoyl chloride and zinc powder. These rather stable radicals occurred in the crystalline form and recombined to 1,1'-diacyl-1,1',4,4'-tetrahydro-4,4'-dipyridyls (82). Nearly at the same time, Dimroth (484, 485) observed the formation of compound **66** in reducing pyridine with zinc powder in acetic anhydride (the structure of this compound was described in detail in ref. 486).

(66)

Reduction of pyridine with zinc in acetic acid (the Wibaut reaction) provided a synthetic route to 4-ethylpyridine (463, 487–490); an industrial process was worked out on the basis of this reaction (491–493). The reaction mechanism was discussed in refs. 492 and 494.* The same technique proved applicable to the synthesis of 2- and 3-methyl-4-ethylpyridines from 2- and 3-picolines (499), though at first many attempts to carry out this reaction proved a failure (497, 498). In the case of chloroformate this reaction may be a pathway to the ester of isonicotinic acid (500):

$$\text{[Pyridinium}-\text{N}-\text{COOEt]} \xrightarrow{\text{Zn}}$$

$$\text{EtO}_2\text{C}-\text{N}\underset{\text{H}}{\overset{\text{H}}{\diagup\diagdown}}\text{N}-\text{CO}_2\text{Et} \xrightarrow[195-230^\circ\text{C}]{\text{S}} \text{N}\diagup\diagdown\text{COOEt}$$

Oxidation of the reaction mixture resulted in 4,4′-dipyridyl (501).

Conditions were found (494, 502–505) which made it possible to synthesize 1,1′-diacyl-1,1′,4,4′-tetrahydro-4,4′-dipyridyls bearing various substituents at nitrogen. 4-Substituted pyridines reacted with acid anhydrides and zinc to yield 1,4-diacetyl-4-alkyl-1,4-dihydropyridines (505–507). It is likely that here again a radical process was involved:

$$\underset{\underset{\text{COOEt}}{|}}{\overset{\text{COOEt}}{\underset{\oplus}{\text{N}}}} \xrightarrow{\text{Zn}} \underset{\underset{\text{COOEt}}{|}}{\overset{\text{EtOOC}\diagdown\diagup\text{COOEt}}{\text{N}}}$$

Recently, an elegant solution for the synthesis of the alkaloid ellipticine based on reductive acylation of 3-(1-methoxyethyl)pyridine has been suggested (508). The "reductive dimerization" reaction is also characteristic of N-acyl derivatives of benzopyridines (509–511). Thus acetylisoquinolinium salts undergo reduction to a 1:1 mixture of epimers of bis(2-acetyl-1,2-dihydro-1-isoquinoline) (511).

However, in the case of acridine, N-acetylacridane rather than the dimer is formed (512), probably because of the greater stability of the N-acetylacridyl radical. Instead of recombining, the latter undergoes reduction to the anion and then adds a proton.

---

* Reduction of mixtures of pyridine (495) or quinoline (496) and chloroalkylcarbonates with $NaBH_4$ found application as a means for preparation of 1-methoxycarboxyl-1,4- and/or -1,2-dihydropyridines and 1-ethoxycarbonyl-1,2-dihydroquinoline.

N-Acylpyridinium radicals formed in the Dimroth reaction not only can recombine but also can cause hetarylation of aromatic or heterocyclic nucleophiles when the latter are introduced into the reaction mixture (513–515):

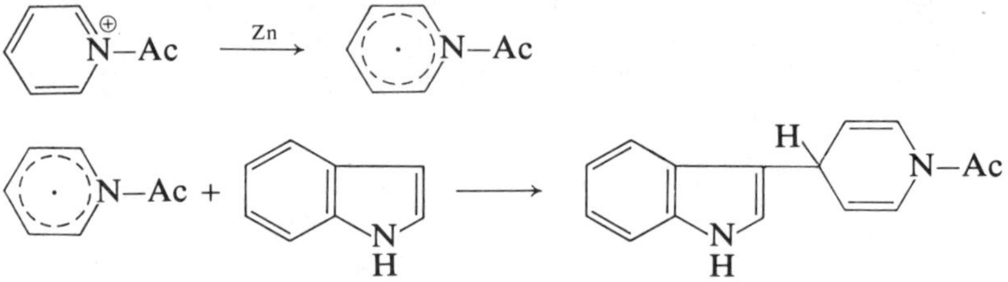

Quinoline, isoquinoline, and acridine undergo a similar reaction with acetic anhydride and aromatic or heterocyclic nucleophiles in the presence of some other active metals. In such a way 2-aryl[hetaryl]-*N*-acetyl-1,2-dihydroquinolines and isoquinolines, and also 4-aryl[hetaryl]pyridines and 9-aryl[hetaryl]acridines have been obtained in high yields.

It appears that reductive hetarylation may have useful synthetic applications, since it goes under rather mild conditions compared to the process involving Lewis acids, and sometimes gives different products from the latter technique.

We conclude our discussion by pointing out that the reactivity of the pyridine ring activated by the formation of *N*-acylium salts should be analyzed with regard to properties of protonated bases, *N*-alkylium salts, *N*-oxides, *N*-amino compounds, etc., which exhibit more or less analogous chemical behavior. However, to follow this principle would greatly exceed the scope of our review.

REFERENCES

1. A. K. Sheinkman, S. I. Suminov, and A. N. Kost, *Chem. Rev. USSR*, **42**, 1415 (1973).
2. G. W. Fischer, *Chem. Ber.*, **103**, 3470 (1970).
3. R. A. Abramovitch and J. G. Saha, *Advances in Heterocyclic Chemistry*, Vol. 6, 1966, p. 22.
4. E. Shaw, *Pyridine and its Derivatives*, E. Klingsberger, Ed., Part II, Interscience Publishers, New York, 1960, p. 1.
5. D. Beke, *Advances in Heterocyclic Chemistry*, Vol. 1, 1963, p. 167.
6. A. R. Katritzky and J. M. Lagowski, *Chemistry of the Heterocyclic N-Oxides*, Academic Press, New York, 1971.
7. M. Hamana, *Lectures in Heterocyclic Chemistry*, Vol. 1, Utah, 1972, p. 51.
8. A. Treibs and H. Bader, *Chem. Ber.*, **90**, 789 (1957).
9. F. Kröhnke and K. Dickoré, *Chem. Ber.*, **92**, 46 (1959).
10. H. A. Staab, *Angew. Chem.*, **74**, 407 (1962).

REFERENCES

11. J. Bergman, *Tetrahedron Letters*, **1972**, 4723.
12. J. Bergman and L. Renström, *4th Intern. Congr. Heterocycl. Chem., Salt Lake City, July, 1973, Abstr. Papers*, p. 259.
13. A. K. Sheinkman, *Khim. Geterotsik. Soedin.*, **1974**, 3.
14. W. E. Hanford and J. C. Sauer, *Organic Reactions*, Vol. 3, New York, 1946, p. 108.
15. F. I. Luknitzkij and B. A. Wovsi, *Chem. Rev. USSR*, **38**, 1072 (1969).
16. E. Wedekind, J. Häussermann, W. Weisswange, and M. Miller, *Ann. Chem.*, **378**, 261 (1911).
17. S. V. Bogatkov, R. I. Kruglikova, A. D. Lavrukhina, and E. M. Tcherkasova, *J. Org. Chem. USSR*, **1**, 885 (1965).
18. S. G. Entelis and O. V. Nesterov, *Dokl. Akad. Nauk SSSR*, **148**, 1323 (1963).
19. S. D. Stavrova, G. V. Peregudov, S. B. Goldstein, and S. S. Medvedev, *Dokl. Akad. Nauk. SSSR*, **169**, 630 (1966).
20. I. E. Kardash, N. P. Gluchoedov, A. N. Pravednikov, and S. S. Medvedev, *Dokl. Akad. Nauk SSSR*, **169**, 876 (1966).
21. S. Chakrabarti, *Spectrochim. Acta*, **24A**, 790 (1968).
22. M. Dennstedt and J. Zimmerman, *Chem. Ber.*, **19**, 75 (1886).
23. G. Minunni, *Gazz. Chim. Ital.*, **22**, *II*, 113 (1892).
24. Ger. Pat. 109,933 (1900); *Chem. Zentralbl.*, **1900**, II, 460.
25. Ger. Pat. 116, 386 (1901); *Chem. Zentralbl.*, **1901**, I, 287.
26. E. Wedekind, *Ann. Chem.*, **318**, 90 (1901).
27. E. Wedekind, *Chem. Ber.*, **34**, 2070 (1901).
28. A. E. Tshitshibabin, *Zh. Russ. fiz. Khim. Obstsch.*, **33**, 404 (1901).
29. V. Prey, *Chem. Ber.*, **75**, 537 (1942).
30. H. Adkins and Q. E. Thompson, *J. Am. Chem. Soc.*, **71**, 2242 (1949).
31. R. C. Paul, D. Singh, and S. S. Sandhu, *J. Chem. Soc.*, **1959**, 315.
32. W. M. H. Dehn, *J. Am. Chem. Soc.*, **34**, 1399 (1912).
33. P. Mesnard and M. Bertucat, *Bull. Soc. Pharm. Bordeaux*, **98** (2), 57 (1959).
34. K. Freudenberg and D. Peters, *Chem. Ber.*, **52**, 1463 (1919).
35. B. S. Manhas, R. C. Paul, and S. S. Sandhu, *J. Chem. Soc.*, **1959**, 325.
36. A. K. Sheinkman, S. A. Portnova, Yu. N. Sheinker, and A. N. Kost, *Dokl. Akad. Nauk SSSR*, **157**, 1416 (1964).
37. J. A. Moore and J. A. Goldstein, *Am. Chem. Soc. Polym. Prepr.*, **13**, 1273 (1972).
38. J. A. Moore and J. A. Goldstein, *J. Polymer Sci. A-1*, **10**, 2103 (1972).
39. J. A. Moore and J. A. Goldstein, *Prepr. Intern. Symp. Macromol., Helsinki, 1972*, Vol. 1, Sect. 1, p. 177.
40. J. A. Moore and J. A. Goldstein, *Nuova Chim.*, **49** (4), 79 (1973); *C.A.*, **79**, 79230 (1973).
41. J. Bergmann, J-E. Bäckvall, and J.-O. Lindström, *Tetrahedron*, **29**, 971 (1973).
42. J. Bergman, *J. Heterocycl. Chem.*, **7**, 1071 (1970).
43. J. Holeček and J. Klikorka, *Sb. Ved. Prac. Vysokej Skola Chem. Technol. Pardubice*, **2**, 15 (1964); *C.A.*, **64**, 5815 (1966).
44. D. Cook, *Can. J. Chem.*, **40**, 2362 (1962).
45. D. E. Koshland, Jr., *J. Am. Chem. Soc.*, **74**, 2286 (1962).
46. A. R. Fersht and W. P. Jencks, *J. Am. Chem. Soc.*, **92**, 5432 (1970).
47. C. I. Lurje, *J. Gen. Chem. USSR*, **18**, 1517 (1948).
48. H. E. Baumgarten, *J. Am. Chem. Soc.*, **75**, 1239 (1957).
49. P. Krkocšca, J. Kováč, and F. Herrmann, *Cellul. Chem. Technol.*, **7**, 563 (1973).
50. B. M. Bogoslovski, *J. Gen. Chem. USSR*, **7**, 255 (1937).
51. M. Wakselman and E. Guibé-Jampel, *Tetrahedron Letters*, **1970**, 1521.
52. A. K. Sheinkman, A. N. Kost, A. N. Prilepskaja, and N. A. Kluev, *Khim. Geterotsikl. Soedin.*, **1972**, 1105.

53. A. K. Sheinkman, Doctoral Thesis, Rostov-Don, 1972.
54. A. P. Kucherenko, Doctoral Thesis, Donetzk, 1973.
55. K. Goyal, R. C. Paul, and S. S. Sandhu, *J. Chem. Soc.*, **1959**, 322.
56. R. Paul, M. S. Bains, and G. Singh, *J. Indian Chem. Soc.*, **35**, 489 (1958).
57. M. Wakselman and E. Guibé-Jampel, *Tetrahedron Letters*, **1970**, 4715.
58. E. Guibé-Jampel and M. Wakselman, *Bull. Soc. Chim. France*, **1971**, 2554.
59. W. M. Dehn and A. A. Ball, *J. Am. Chem. Soc.*, **36**, 2091 (1914).
60. A. Reissert, *Chem. Ber.*, **38**, 1603 (1905).
61. G. A. Olah and P. J. Szilagyi, *J. Am. Chem. Soc.*, **91**, 2949 (1969).
62. G. A. Olah, J. A. Olah, and N. A. Overchuk, *J. Org. Chem.*, **30**, 3373 (1965).
63. J. Herzog, *Chem. Ber.*, **40**, 1831 (1907).
64. J. Herzog and K. Budy, *Chem. Ber.*, **44**, 1584 (1911).
65. E. v. Meyer and A. Nicolaus, *J. Prakt. Chem.*, **82** (2), 521 (1910).
66. S. L. Johnson and K. A. Rumon, *J. Phys. Chem.*, **68**, 3149 (1964).
67. S. L. Johnson and H. M. Giron, *J. Org. Chem.*, **37**, 1383 (1972).
68. S. L. Johnson and D. L. Morrison, *Biochemistry*, **9**, 1460 (1970).
69. F. Reitzenstein and W. Breuning, *J. Prakt. Chem.*, **83** (2), 97 (1911).
70. P. W. Neber and H. Wörner, *Ann. Chem.*, **526**, 173 (1936).
71. R. Huisgen, E. Aufderhaar, and G. Wallbillich, *Chem. Ber.*, **98**, 1476 (1965).
72. R. Fusco, P. D. Croce, and A. Salvi, *Tetrahedron Letters*, **1967**, 3071.
73. D. S. Breslow, U.S. Pat. 3,717,560 (1973); *Ref. Zh. Khim.*, **1974**, 1S 355.
74. Ger. Pat. 118,537; *Friedl.*, **6**, 1166 (1900).
75. Ger. Pat. 117,625, 118,566; *Friedl.*, **6**, 1162, 1163 (1900).
76. H. S. Fry, *J. Am. Chem. Soc.*, **36**, 248 (1914).
77. J. Gadamer and F. Knoch, *Arch. Pharm.*, **259**, 135 (1921).
78. T. Hopkins, *J. Chem. Soc.*, **117**, 278 (1920).
79. J. F. Arens and D. A. Van Dorp, *Rec. Trav. Chim.*, **65**, 722 (1946).
80. P. Carré, *Bull. Soc. chim. France*, **3** (5), 1064 (1936).
81. P. W. Clinch and H. R. Hudson, *J. Chem. Soc.*, **1971**, B, 747.
82. J. E. Colchester, Brit. Pat. 1,189,084 (1970); *C.A.*, **73**, 25315 (1970).
83. G. Fraenkel and J. W. Cooper, *J. Am. Chem. Soc.*, **93**, 7228 (1971).
84. H. Babad and A. G. Zeiler, *Chem. Rev.*, **73**, 75 (1973).
85. A. F. Cockerill, G. L. O. Davies, and D. M. Rackham, *Tetrahedron Letters*, **1972**, 27.
86. E. Guibé-Jampel and M. Wakselman, *Chem. Commun.*, **1971**, 267.
87. E. Guibé-Jampel, Doctoral Thesis, University of Paris, 1971; *Ref. Zh. Khim.*, **1973**, 16 Zh 447.
88. E. Guibé-Jampel and M. Wakselman, Fr. Pat. 2,115,552; *C.A.*, 78, 84814 (1973).
89. M. Miyoshi and T. Onishi, Jap. Pat. 70-24766; *C.A.*, **74**, 64397 (1971).
90. O. Mumm and E. Herrendörfer, *Chem. Ber.*, **47**, 758 (1914).
91. T. Shimidzu, *J. Pharm. Soc. Japan*, **529**, 243; **537**, 942 (1926); *C.A.*, **20**, 2680 (1926).
92. T. Shimidzu, *J. Pharm. Soc. Japan*, **538**, 1043 (1926); *C.A.*, **21**, 2694 (1927).
93. M. D. Johnson, *J. Chem. Soc.*, **1962**, 283.
94. C. J. Cooksey and M. Johnson, *J. Chem. Soc.*, **1968**, B, 1191.
95. G. N. Schwartz and W. M. Dehn, *J. Am. Chem. Soc.*, **39**, 2444 (1917).
96. H. Majda-Grabowska and K. Okón, *Bull. Acad. Polon. Sci. Ser. Sci. Chim.*, **9**, 195 (1961); *Roczniki Chem.*, **36**, 141 (1962).
97. H. Majda-Grabowska and K. Okón, *Roczniki Chem.*, **37**, 379 (1963).
98. I. A. Orudgeva, A. A. Farkharov, P. S. Mamedova, Yu. A. Alekperova, Zh. I. Dzhafarov, and F. I. Khalilova, *Azerb. Neft. Khoz.*, **1970** (7), 32.
99. E. Gebauer-Fülnegg and F. Riesenfeld, *Monatsh. Chem.*, **47**, 185 (1926).
100. C. R. Gambill, T. D. Roberts, and H. Shechter, *J. Chem. Educ.*, **49**, 287 (1972).

101. W. Loop and E. Lührs, *Ann. Chem.*, **580**, 225 (1953).
102. L. Field, *J. Am. Chem. Soc.*, **74**, 394 (1952).
103. L. Field and P. H. Settlage, *J. Am. Chem. Soc.*, **76**, 1222 (1954).
104. E. Buncel, *Chem. Rev.*, **70**, 323 (1970).
105. J. Charalambous, M. J. Frazor, and W. Gerrard, *J. Chem. Soc.*, **1964**, 5480.
106. W. Gerrard, *J. Chem. Soc.*, **1940**, 218.
107. W. Gerrard, *J. Chem. Soc.*, **1936**, 688.
108. P. Ledus and P. Chabrier, *Bull. Soc. Chim. France*, **1963**, 2271.
109. J. C. Sheldon and S. Y. Tyree, *J. Am. Chem. Soc.*, **81**, 2290 (1959).
110. R. C. Paul, H. Khurana, S. K. Vasisht, and S. L. Chadha, *J. Indian Chem. Soc.*, **46**, 914 (1969).
111. R. G. Makitra, M. S. Makaruk, and M. N. Didich, *J. Gen. Chem. USSR*, **42**, 1877 (1972).
112. A. J. Banister, B. Bell, and L. J. Moore, *J. Inorg. Nucl. Chem.*, **34**, 1161 (1972).
113. G. W. Jameson and J. M. Lawlor, *J. Chem. Soc. B*, **1970**, 53.
114. E. Guibé-Jampel, M. Wakselman, and M. Vilkas, *Bull. Soc. Chim. France*, **1971**, 1308.
115. N. Yamazaki and F. Higashi, *Tetrahedron Letters*, **1972**, 415.
116. N. Yamazaki and F. Higashi, *Bull. Chem. Soc. Japan*, **46**, 1235 (1973).
117. N. Yamazaki and F. Higashi, *Bull. Chem. Soc. Japan*, **46**, 1239 (1973).
118. N. Yamazaki and F. Higashi, *Bull. Chem. Soc. Japan*, **46**, 3821 (1973).
119. N. Yamazaki and F. Higashi, *Bull. Chem. Soc. Japan*, **47**, 170 (1974).
120. E. Fluck, P. J. Retuert, and H. Binder, *Z. Anorg. Allgem. Chem.*, **397**, 225 (1973).
121. R. C. Paul, K. Chander, and G. Singh, *J. Indian Chem. Soc.*, **35**, 869 (1958).
122. J. Devynck, *Ann. Chim.*, **7**, 321 (1972).
123. M. Davies, *Trans. Faraday Soc.*, **31**, 1561 (1935).
124. F. Klages and E. Zange, *Ann. Chem.*, **607**, 35 (1957).
125. J. Holeček, *Sb. Ved. Prac. Vysokej Skola Chem. Technol. Pardubice*, **1965** (1), 19; *C.A.*, **64**, 5815 (1965).
126. J. Singh, R. Prashar, M. Lakhanpal, and R. C. Paul, *J. Sci. Ind. Res.*, **21b**, 450 (1962).
127. R. C. Paul, P. S. Gil, and J. Singh, *Indian J. Chem.*, **2**, 219 (1964).
128. S. V. Bogatkov, V. V. Korshak, T. I. Mitajshvili, V. A. Vasnev, S. V. Vinogradova, and E. M. Cherkasova, *Dokl. Akad. Nauk SSSR*, **194**, 328 (1970).
129. R. Gompper and R. Altreuter, *Z. Anal. Chem.*, **170**, 205 (1959).
130. W. E. Doering and W. E. McEwen, *J. Am. Chem. Soc.*, **73**, 2104 (1951).
131. I. Ugi and F. Beck, *Chem. Ber.*, **94**, 1839 (1961).
132. S. L. Johnson and K. A. Rumon, *J. Am. Chem. Soc.*, **87**, 4782 (1965).
133. M. K. Polievktov, A. K. Sheinkman, and L. N. Morozova, *Khim. Geterotsikl. Soedin.*, **1973**, 1067.
134. J. Holeček, J. Pavlik, and J. Klilorka, *Sb. Ved. Prac. Vysokej Skola chem. technol. Pardubice*, **1964** (2), 23; *C.A.*, **64**, 5815 (1965).
135. R. C. Paul and S. Chadra, *Spectrochim. Acta*, **22**, 615 (1966).
136. V. V. Korshak, V. A. Vasnev, S. V. Vinogradova, and T. I. Mitajshvili, *Vysokomol. soedin.*, **10A**, 2182 (1968).
137. B. J. Huckings and M. D. Johnson, *J. Chem. Soc.*, **1964**, 5371.
138. A. R. Fersht and W. P. Jencks, *J. Am. Chem. Soc.*, **91**, 2125 (1969).
139. A. N. Kost, A. K. Sheinkman, and N. F. Kazarinova, *J. Gen.Chem.* USSR, **34**, 2044 (1964).
140. V. Gold and E. G. Jefferson, *J. Chem. Soc.*, **1953**, 1409.
141. G. H. Schenk, P. Wines, and C. Mojzis, *Anal. Chem.*, **36**, 914 (1964).
142. R. C. Paul, R. N. Sawhney, and S. L. Chadha, *Indian J. Chem.*, **5**, 631 (1967).
143. T. G. Bonner, E. G. Gabb, and P. M. McNamara, *J. Chem. Soc., B*, **1968**, 72.
144. M. Matzner, R. P. Kurkjy, and R. J. Cotter, *Chem. Rev.*, **64**, 645 (1964).

145. H. M. Relles, *J. Org. Chem.*, **38**, 1570 (1973).
146. H. M. Relles, U.S. Pat. 3,696,158 (1972); *C.A.*, **78**, 16735 (1973).
147. Ch. Heideman, *Organic Reactions*, Vol. 7, John Wiley and Sons, New York, 1953, p. 279.
148. F. Kröhnke, *Angew. Chem.*, **75**, 317 (1963).
149. A. N. Kost and A. K. Sheinkman, *J. Gen. Chem. USSR*, **33**, 2077 (1963).
150. A. N. Kost, A. K. Sheinkman, and A. N. Rosenberg, *J. Gen. Chem. USSR*, **34**, 4046 (1964).
151. A. K. Sheinkman, A. N. Kost, V. I. Sheitshenko, and A. N. Rosenberg, *Ukr. Khim. Zh.*, **33**, 941 (1967).
152. G. N. Bogdanov, A. N. Rosenberg, and A. N. Sheinkman, *Khim. Geterotsikl. Soedin.*, **1971**, 1660.
153. S. M. Linch and M. Gordon, *J. Heterocycl. Chem.*, **9**, 789 (1972).
154. R. D. Brown and R. D. Harcourt, *Tetrahedron*, **8**, 23 (1960).
155. M. D. Johnson and J. H. Ridd, *J. Chem. Soc.*, **1962**, 291.
156. V. E. Kononenko, A. K. Sheinkman, T. N. Kashtanova, and S. N. Baranov, *Reakt. Sposobnost Org. Soedin.*, **8**, 185 (1971).
157. A. Einhorn and F. Hollandt, *Ann. Chem.*, **301**, 95 (1898).
158. Houben-Weyl *Methoden der organischen Chemie*, Vol. II, Stuttgart, 1953.
159. K. Auwers, *Chem. Ber.*, **37**, 3899 (1904).
160. A. Van Es and W. Stevens, *Rec. Trav. Chim.*, **84**, 704 (1965).
161. M. Hedayatullah, J.-C. Lévêque, and L. Denivelle, *Bull. Soc. Chim. France*, **1972**, 3808.
162. W. Steglich and G. Höfle, *Angew. Chem.*, **81**, 1001 (1969).
163. W. Steglich and G. Höfle, F. R. Ger. Pat. 1,958,954 (1971); *C.A.*, **75**, 34673 (1971).
164. H. Dryden, F. R. Ger. Pat. 2,137,856 (1972); *C.A.*, **76**, 127269 (1972).
165. G. Höfle and W. Steglich, *Synthesis*, **1972**, 619.
166. C. C. Leznoff and J. Y. Wong, *Can. J. Chem.*, **50**, 2892 (1972); **51**, 2452 (1973).
167. H. Seliger and G. Aumann, *Tetrahedron Letters*, **1973**, 2911.
168. M. B. Shambhu and G. A. Digenis, *Tetrahedron Letters*, **1973**, 1627.
169. N. K. Kochetkov, I. V. Torgov, and M. M. Botvinik, *Khimija Prirodnich soedinenij*, Akademizdat, Moscow, **1961**, pp. 65, 67, 77.
170. B. Jastorff and T. Krebs, *Chem. Ber.*, **105**, 3192 (1972).
171. A. A. Khidojatov, *Uzbeksk. Khim. Zh.*, **1973** (1), 55.
172. T. M. Frunze, V. V. Korshak, and L. B. Kozlov, *Chem. Rev. USSR*, **30**, 593 (1961).
173. V. V. Korshak, S. V. Vinogradova, and V. A. Vasnev, *Dokl. Akad. Nauk SSSR*, **191**, 614 (1970).
174. S. V. Vinogradova, V. A. Vasnev, and V. V. Korshak, *Vysokomol. soedin. B*, **13**, 600 (1971).
175. V. V. Korshak, S. V. Vinogradova, Ju. I. Perfilov, V. A. Vasnev, and A. I. Tarasov, *Tr. Mosk. khim.-Technol. Inst.*, **1972**, 150.
176. C. Sholtissek, *Chem. Ber.*, **89**, 2562 (1956).
177. R. Wegler, *Ann. Chem.*, **498**, 62 (1932); **506**, 77 (1933); **510**, 72 (1934).
178. G. Pracejus, *Ann. Chem.*, **622**, 10 (1959).
179. W. E. Truce and P. S. Bailey, Jr., *J. Org. Chem.*, **34**, 1341 (1969).
180. M. Bender, *Chem. Rev.*, **60**, 53 (1960).
181. S. Johnson, *Advances in Physical Organic Chemistry*, Vol. 5, 1967, p. 237.
182. F. Kröhnke, *Angew. Chem.*, **65**, 605 (1953).
183. W. F. Edgell and L. Parts, *J. Am. Chem. Soc.*, **77**, 4899 (1955).
184. B. O. Lindgren and H. Mikawa, *Acta Chem. Scand.*, **11**, 826 (1957).
185. S. E. Creasey and R. D. Guthrie, *Chem. Commun.*, **1971**, 801.
186. D. Klamann, *Monatsh. Chem.*, **84**, 54 (1953).
187. J. H. Brewster and C. J. Ciotti, Jr., *J. Am. Chem. Soc.*, **77**, 6214 (1955).

188. D. M. Smith and W. M. D. Bryant, *J. Am. Chem. Soc.*, **57**, 61 (1935).
189. W. Adam, J. Baeza, and J.-Ch. Liu, *J. Am. Chem. Soc.*, **94**, 2000 (1972).
190. S. Siggia, *Quantitative Analysis via Functional Groups*, 3rd ed., John Wiley & Sons, New York, 1963, p. 8.
191. G. H. Schenk and J. S. Fritz, *Anal. Chem.*, **32**, 987 (1960).
192. S. Siggia and J. G. Hanna, *Anal. Chem.*, **33**, 896 (1961).
193. J. G. Hanna and S. Siggia, *J. Polymer Sci.*, **56**, 297 (1962).
194. S. Huwyler, *Experientia*, **28**, 718 (1972).
195. Ger. Pat. 117,267 (1901); *Zbl.*, **1901**, I, 347.
196. Ger. Pat. 201,325 (1908); *Zbl.*, **1908**, II, 996.
197. H. G. Rule and T. R. Paterson, *J. Chem. Soc.*, **125**, 2155 (1924).
198. C. F. H. Allen, C. J. Kibler, D. M. McLachlin, and C. V. Wilson, *Organic Syntheses*, Coll. Vol. 3, John Wiley & Sons, New York, 1955.
199. P. Rambacher and S. Mäke, *Angew. Chem.*, **80**, 487 (1968).
200. F. H. Carpenter, *J. Am. Chem. Soc.*, **70**, 2964 (1948).
201. J. M. Adduci and R. S. Ramirez, *Org. Prep. Proced. Int.*, **2**, 321 (1970).
202. J. Herzog, *Ber. Deut. Pharm. Ges.*, **19**, 394 (1909); *Zbl.*, **1910**, I, 351.
203. A. Einhorn and R. Seuffert, *Chem. Ber.*, **43**, 2988 (1910).
204. J. M. Zeawin and A. M. Fisher, *J. Am. Chem. Soc.*, **54**, 3739 (1932).
205. A. E. Vasiliev, A. A. Khachaturjan, and A. Ya. Rosenberg, *Khim. Prir. Soedin.*, **1971**, 698.
206. A. Kallianos, J. Moldi, and M. Simpson, U.S. Pat. 3,646,201 (1972); *C.A.*, **76**, 153372 (1972).
207. R. K. Smalley and H. Suschitzky, *J. Chem. Soc.*, **1964**, 755.
208. T.-L. Ho and C. M. Wong, *Synth. Commun.*, **3**, 63 (1973).
209. M. Guyot and D. Molho, *Tetrahedron Letters*, **1973**, 3433.
210. M. V. Loseva and B. M. Bolotyn, *Khim. Geterotsikl. Soedin.*, **1972**, 1341.
211. J. W. van Reijendam and F. Baardman, *Synthesis*, **1973**, 413.
212. H. Böhme and H. Schran, *Chem. Ber.*, **82**, 453 (1949).
213. D. Westwood and R. K. Smalley, *Chem. Ind. London*, **1970**, 1408.
214. D. M. Brown, *Advances in Organic Chemistry*, Vol. 3, John Wiley & Sons, New York, 1963, p. 75.
215. G. Schrader, *Die Entwicklung neuer insektizider Phosporsäure-Ester*, Verlag Chemie, 1963.
216. R. Letters and A. M. Michelson, *J. Chem. Soc.*, **1962**, 71.
217. A. W. D. Avison, *J. Chem. Soc.*, **1955**, 732.
218. E. Wedekind, *Chem. Ber.*, **34**, 2070 (1901).
219. P. Baudet and C. Otten, *Helv. Chim. Acta*, **53**, 1330 (1970).
220. J. Markgraf, T. Easley, R. Katt, and M. Ruckman, *J. Chem. Eng. Data*, **17**, 268 (1972).
221. H. Yamamoto and M. Nakao, Jap. Pat. 72-26509; *C.A.*, **77**, 126437 (1972).
222. Y. Maki and T. Masugi, *Chem. Pharm. Bull.*, **21**, 685 (1973).
223. G. A. Rogers and T. C. Bruice, *J. Am. Chem. Soc.*, **95**, 4452 (1973).
224. J. W. Harbuck and H. Rapoport, *J. Org. Chem.*, **37**, 3618 (1972).
225. B. Lindström. *Chem. commun. Univ. Stockholm*, **1973**, (8), 32; *Ref. Zh. Khim.*, **1973**, 21 Zh 399.
226. M. Fukumura, K. Simaga, S. Okano, T. Nakatam, and K. Basaka, Jap. Pat. 48-15301 (1973); *Ref. Zh. Khim.*, **1974**, 13H331.
227. H. E. Foster and J. Hurst, *J. Chem. Soc., Perkin Trans. I*, **1973**, 2901.
228. R. A. Abramovitch and R. B. Rogers, *J. Org. Chem.*, **39**, 1802 (1974).
229. Ya. L. Goldfarb, R. M. Ispirjan, and L. I. Belenkij, *Izv. Akad. Nauk SSSR, Ser. Khim.*, **1969**, 923.

230. L. Crivetz and M. Brumă, *Rev. Roum. Chim.*, **11**, 1135 (1966); **12**, 1245 (1967).
231. W. Bracke, F. R. Ger. Pat. 2,219,002 (1972); *C.A.*, **78**, 17057 (1973).
232. V. R. Olson and H. B. Feldman, *J. Am. Chem. Soc.*, **59**, 2003 (1937).
233. C. A. Reynolds, S. F. H. Walker, and E. Cochran, *Anal. Chem.*, **32**, 983 (1960).
234. J. Brooks and W. Moore, *Can. J. Microbiol.*, **15**, 1433 (1969).
235. V. V. Korshak, T. M. Frunze, V. V. Kurashev, and K. L. Serova, *Vysokomol. Soedin.*, **3**, 205 (1961).
236. C. Giori, *Am. Chem. Soc. Polym. Prep.*, **11**, 1023 (1970).
237. A. Ja. Jakubovich, N. N. Voznesenskaja, and G. I. Braz, *Dokl. Akad. Nauk SSSR*, **194**, 116 (1970).
238. L. B. Sokolov and S. S. Medvedev, *Vysokomol. Soedin.*, **B10**, 514 (1968).
239. S. S. Medvedev, L. B. Sokolov, and D. F. Sokolova, *Vysokomol. Soedin.*, **A14**, 1313 (1972).
240. V. V. Korshak, N. I. Bekasova, V. V. Vagin, and A. A. Isineev, *Vysokomol. Soedin.*, **B15**, 6 (1973).
241. V. V. Korshak, V. V. Vagin, N. I. Bekasova, and A. A. Isineev, *Dokl. Akad. Nauk SSSR*, **212**, 638 (1973).
242. E. Morthey, *The Sulphonamides and Applied Compounds*, New York, 1948.
243. V. A. Zasosov, *Zh. Vses. Khim. Obschestva im. D. I. Mendeleeva*, **10**, 671 (1965).
244. K. Okui, K. Ito, M. Koizumi, K. Fukumoto, and T. Kametani, *J. Heterocycl. Chem.*, **9**, 1283 (1972).
245. E. P. Lira, *J. Heterocycl. Chem.*, **9**, 713 (1972).
246. Q. E. Thompson, *J. Am. Chem. Soc.*, **73**, 5841 (1951).
247. R. T. LaLonde and C. B. Davis, *J. Org. Chem.*, **35**, 771 (1970).
248. J. Mitchell, Jr. and C. E. Ashby, *J. Am. Chem. Soc.*, **67**, 161 (1945).
249. C. R. Stephens, E. J. Bianco, and F. J. Pilgrim, *J. Am. Chem. Soc.*, **77**, 1701 (1955).
250. J. Szmuszkovicz, *J. Am. Chem. Soc.*, **82**, 1180 (1960).
251. A. Guzman, P. Ortiz de Montellano, and P. Crabbé, *J. Chem. Soc., Perkin Trans. I*, **1973**, 91.
252. J. Klinot and A. Vystrčil, *Collection Czech. Chem. Commun.*, **27**, 377 (1962).
253. J. K. Chakrabarti and T. M. Hotten, *J. Chem. Soc. Chem. Commun.*, **1972**, 1226.
254. I. Ugi, U. Fetzer, U. Eholzer, H. Knupfer, and K. Offermann, *Angew. Chem.*, **77**, 492 (1965).
255. F. Seidel and O. Engelfried, *Chem. Ber.*, **69**, 2567 (1936).
256. C. Ziegler and J. Spraque, U.S. Pat. 3,663,615 (1971); *C.A.*, **77**, 61625 (1972).
257. S. M. McElvain and G. A. McKay, Jr., *J. Am. Chem. Soc.*, **78**, 6086 (1956).
258. D. Behr, S. Brandänge, and B. Lindström, *Acta Chem. Scand.*, **27**, 2411 (1973).
259. I. Calder and W. Sasse, *Australian J. Chem.*, **18**, 1023 (1965); **21**, 1023 (1968).
260. S. I. Sav'jalov, N. I. Aronova, and N. N. Makhova, in *Reaktsii i Metody Issled. Organ. Soedin.*, **22**, 9 (1971).
261. W. Steglich and G. Höfle, *Chem. Ber.*, **102**, 1129 (1969).
262. L. Claisen and E. Haase, *Chem. Ber.*, **33**, 1242, 3778 (1900).
263. I. V. Machinskaja and V. A. Barkhagh, *Reaktsii i Metody Issled. Organ. Soedin.*, **14**, 299 (1964).
264. M. Suama, Y. Nakav, and K. Ichikawa, *Bull. Chem. Soc. Japan*, **44**, 2811 (1971).
265. G. Kresze and H. Härtner, *Ann. Chem.*, **1973**, 650.
266. W. Dieckmann and R. Stein, *Chem. Ber.*, **37**, 3370 (1904).
267. W. Dieckmann and F. Breest, *Chem. Ber.*, **37**, 3384 (1904).
268. E. Benary, F. Reiter, and H. Soenderop, *Chem. Ber.*, **50**, 65 (1917).
269. L. Bouveault and A. Bongert, *Bull. Soc. Chim. France*, **27** (3), 1160 (1902).
270. P. E. Wright and W. E. McEwen, *J. Am. Chem. Soc.*, **76**, 4540 (1954).

271. R. L. Stutz, C. A. Reynolds, and W. E. McEwen, *J. Org. Chem.*, **26**, 1684 (1961).
272. W. R. Gilkerson, W. J. Argersinger, Jr., and W. E. McEwen, *J. Am. Chem. Soc.*, **76**, 41 (1954).
273. M. Suama, Y. Murata, and K. Ichikawa, *J. Chem. Soc. Japan*, **91**, 162, 168 (1970); *C.A.*, **73**, 24604, 24605 (1970).
274. C. A. Reynolds, S. F. H. Walker, and E. Cochran, *Anal. Chem.*, **32**, 983 (1960).
275. W. P. Jencks, *Catalysis in Chemistry and Enzymology*, McGraw-Hill Book Co., New York, 1969.
276. T. Bruice and S. Benkovic, *Bioorganic Mechanisms*, Vol. 2, W. A. Benjamin, New York, 1966.
277. S. L. Johnson and K. A. Rumon, *J. Phys. Chem.*, **68**, 3149 (1964).
278. W. P. Jencks and J. Carriuolo, *J. Am. Chem. Soc.*, **82**, 1778 (1960).
279. A. R. Fersht and W. P. Jencks, *J. Am. Chem. Soc.*, **92**, 5442 (1970).
280. S. L. Johnson, *J. Phys. Chem.*, **67**, 495 (1963).
281. C. A. Bunton and J. H. Fendler, *J. Org. Chem.*, **31**, 2307 (1966).
282. W. P. Jencks and M. Gilchrist, *J. Am. Chem. Soc.*, **90**, 2622 (1968).
283. S. Bafna and V. Gold, *J. Chem. Soc.*, **1953**, 1406.
284. A. R. Butler and V. Gold, *Proc. Chem. Soc.*, **1960**, 15; *J. Chem. Soc.*, **1961**, 4362.
285. J. Koskiallio, *Suomen Kemistilehti B*, **32**, 41 (1959).
286. J. Koskiallio, *Acta chem. Scand.*, **17**, 1417 (1963).
287. V. Gold and E. G. Jefferson, *J. Chem. Soc.*, **1953**, 1416.
288. E. A. Castro and R. B. Moodie, *J. Chem. Soc. Chem. Commun.*, **1973**, 828.
289. O. Rogne, *J. Chem. Soc. B*, **1970**, 727.
290. O. Rogne, *J. Chem. Soc. Perkin Trans. II*, **1972**, 489.
291. L. J. Stangeland, L. Senatore, and E. Ciuffarin, *J. Chem. Soc. Perkin Trans. II*, **1972**, 852.
292. A. M. Kirkien-Konasiewicz, *J. Chem. Soc.*, **1961**, 5430.
293. W. Ali and A. M. Kirkien-Konasiewicz, *Chem. Ind. London*, **1964**, 809.
294. A. Kirkien-Konasiewicz and A. Maccoll, *J. Chem. Soc.*, **1964**, 1267.
295. A. Kirkien-Konasiewicz, G. M. Sammy, and A. Maccoll, *J. Chem. Soc. B*, **1968**, 1364.
296. O. Rogne, *J. Chem. Soc. B*, **1971**, 1334.
297. C. A. Bunton and V. J. Shiner, Jr., *J. Am. Chem. Soc.*, **83**, 3207 (1961).
298. C. A. Bunton, N. Fuller, S. G. Perry, and V. J. Shiner, Jr., *Tetrahedron Letters*, **1961**, 458.
299. M. L. Bender, E. J. Pollock, and M. C. Neveu, *J. Am. Chem. Soc.*, **84**, 595 (1962).
300. L. M. Litvinenko, E. S. Rudakov, and A. I. Kirichenko, *Kinetika i Kataliz*, **3**, 651 (1962).
301. L. M. Litvinenko and A. I. Kirichenko, *Trudy po problemam primenenija korrelacionnykh uravnenij v organicheskoj khimii*, Vol. 1, Tartu, 1962, p. 151.
302. L. M. Litvinenko and A. I. Kirichenko, *Ukr. Khim. Zh.*, **31**, 67 (1965).
303. L. M. Litvinenko and A. I. Kirichenko, *Dokl. Akad. Nauk SSSR*, **176**, 97 (1967).
304. L. M. Litvinenko, A. I. Kirichenko, V. D. Berestetskaja, and I. V. Shpanko, *J. Org. Chem. USSR*, **4**, 462 (1968).
305. L. M. Litvinenko, A. I. Kirichenko, and A. S. Savchenko, *Reakt. Sposobnost org. soedin.*, **5**, 90 (1968).
306. L. M. Litvinenko, V. A. Savelova, and V. A. Shatskaja, *J. Gen. Chem. USSR*, **38**, 1028 (1968).
307. L. M. Litvinenko and A. I. Kirichenko, *Ukr. Khim. Zh.*, **34**, 1030 (1968).
308. L. M. Litvinenko, A. I. Kirichenko, A. S. Savchenko, and L. Ya. Galushko, *Reakt. Sposobnost Org. Soedin.*, **6**, 981 (1969).
309. L. M. Litvinenko, V. A. Savelova, V. A. Shatskaja, and T. N. Sadovskaja, *Dokl. Akad. Nauk SSSR*, **198**, 844 (1971).
310. V. A. Shatskaja, V. A. Savelova, and L. M. Litvinenko, *J. Gen. Chem. USSR*, **41**, 2256 (1971).

311. L. M. Litvinenko, A. S. Savchenko, A. I. Kirichenko, and L. Ya. Galushko, *Reakt. Sposobnost Org. Soedin.*, **8**, 523 (1971).
312. L. M. Litvinenko, A. I. Kirichenko, A. S. Savchenko, and L. Ya. Galushko, *Reakt. Sposobnost Org. Soedin.*, **8**, 1101 (1971).
313. N. M. Olejnik, M. N. Sorokin, and L. M. Litvinenko, *Ukr. Khim. Zh.*, **38**, 343 (1972).
314. L. M. Litvinenko, A. I. Kirichenko, A. S. Savchenko, L. Ya. Galushko, and O. E. Shumejko, *Ukr. Khim. Zh.*, **38**, 1024 (1972).
315. L. M. Litvinenko, A. I. Kirichenko, and A. S. Savchenko, *Ukr. Khim. Zh.*, **38**, 1136 (1972).
316. V. A. Savelova, T. N. Solomojchenko, and L. M. Litvinenko, *Reakt. Sposobnost Org. Soedin.*, **9**, 665 (1972).
317. V. A. Savelova, T. N. Solomojchenko, and L. M. Litvinenko, *J. Org. Chem. USSR*, **8**, 1011 (1972).
318. T. N. Solomojchenko, V. A. Savelova, and L. M. Litvinenko, *J. Org. Chem. USSR*, **10**, 534 (1974).
319. N. K. Vorobiev, E. A. Chizhova, and L. A. Malisheva, *Izv. Vysshikh Uchebn. Zavedenii, Khim. i Khim. Tekhnol.*, **16**, 1366 (1973).
320. A. P. Grekov, S. A. Sukhorukova, and G. V. Otroshko, *J. Org. Chem. USSR*, **10**, 526 (1974).
321. T. G. Bonner, P. M. McNamara, and B. Smethrust, *J. Chem. Soc. B*, **1968**, 114.
322. T. G. Bonner and P. M. McNamara, *J. Chem. Soc. B*, **1968**, 795.
323. T. G. Bonner and K. Hillier, *J. Chem. Soc., Perkin Trans II*, **1973**, 1828.
324. S. V. Vinogradova, V. V. Korshak, L. I. Komarova, V. A. Vasnev, and T. I. Mitajshvili, *Vysokomol. Soedin.*, **A14**, 2591 (1972).
325. S. V. Vinogradova, V. A. Vasnev, E. I. Vasnev, E. I. Fedin, and V. V. Korshak, *Izv. Akad. Nauk SSSR Ser. Khim.*, **1967**, 1620.
326. S. V. Vinogradova, V. A. Vasnev, V. V. Korshak, T. I. Mitajshvili, and A. V. Vasiljev, *Dokl. Akad. Nauk SSSR*, **187**, 1297 (1969).
327. S. V. Vinogradova, V. A. Vasnev, V. V. Korshak, A. V. Vasiljev, and Ju. I. Perfilov, *Izv. Akad. Nauk SSSR Ser. Khim.*, **1970**, 2138.
328. V. A. Vasnev, S. V. Vinogradova, A. I. Tarasov, and V. V. Korshak, *Zh. Vses. Khim. obshchestva im. D. I. Mendeleeva*, **17**, 472 (1972).
329. V. V. Korshak, V. A. Vasnev, S. V. Bogatkov, A. I. Tarasov, and S. V. Vinogradova, *Reakt. Sposobnost Org. Soedin.*, **10**, 375 (1973).
330. T. Zincke et. al., *Ann. Chem.*, **330**, 361 (1904); **333**, 296 (1904); **338**, 107 (1905); **339**, 193 (1905).
331. W. König, *J. Prakt. Chem.*, **69** (2), 105 (1904); **70**, 19 (1904).
332. W. König and R. Bayer, *J. Prakt. Chem.*, **83** (2), 325 (1911).
333. W. König and G. A. Becker, *J. Prakt. Chem.*, **85** (2), 353 (1912).
334. W. König, G. Ebert, and K. Centner, *Chem. Ber.*, **56**, 751 (1923).
335. I. L. Knunjanz and T. Ja. Kefeli, *J. Gen. Chem. USSR*, **15**, 678 (1945).
336. A. F. Vompe and N. F. Turicina, *Dokl. Akad. Nauk SSSR*, **114**, 1017 (1957).
337. A. F. Vompe, I. I. Levkoev, N. F. Turicina, V. V. Durmashkina, and L. V. Ivanova, *J. Gen. Chem. USSR*, **34**, 1758 (1964).
338. E. N. Marvell, G. Caple, and I. Shahidi, *J. Am. Chem. Soc.*, **92**, 5641 (1970).
339. G. Schwarzenbach and R. Weber, *Helv. Chim. Acta*, **25**, 1628 (1942).
340. P. Baumgarten, *Chem. Ber.*, **57**, 1622 (1924); **59**, 1166 (1926).
341. F. Klages and H. Trager, *Chem. Ber.*, **86**, 1327 (1953).
342. R. Graf, *Angew. Chem.*, **80**, 179 (1968).
343. T. Zincke, G. Heuser, and W. Möller, *Ann. Chem.*, **333**, 296 (1904).
344. H. Freytag, *Chem. Ber.*, **67**, 1995 (1934).

345. S. Zawadzki, *Prace Cent. Inst. Ochrony Pracy*, **21**, 127 (1971); *C.A.*, **76**, 89683 (1972).
346. P. Dounzon and A. Le Clerc, *Anal. Chim. Acta*, **12**, 239 (1955).
347. W. Ciusa and G. Barbiroli, *Ann. Chim.*, **53**, 1248 (1963); *Ref. Zh. Khim.*, 1964, 17 G 177.
348. J. Fuentes-Duchemin and E. Casassas, *Anal. Chim. Acta*, **44**, 462 (1969).
349. K. Gierschner and G. Baumann, *Z. Lebensm. Untersuch. Forsch.*, **139**, 132 (1969); *C.A.*, **70**, 95514 (1969).
350. J. Bartos, *Ann. Pharm. France*, **29**, 71 (1971); *C.A.*, **75**, 1474 (1972).
351. M. I. Bukovsky and S. A. Psaltyra, *Issled. v Obl. Tekh. Besopasnosti*, **1973**, 107; *Ref. Zh. Khim.*, **1974**, 12-i-600.
352. P. Pfeiffer and E. Enders, *Chem. Ber.*, **84**, 313 (1951).
353. S. L. Johnson and K. A. Rumon, *Tetrahedron Letters*, **1966**, 1721.
354. S. L. Johnson and K. A. Rumon, *Biochemistry*, **9**, 847 (1970).
355. J. W. Bunting and W. G. Meathrel, *Can. J. Chem.*, **51**, 1965 (1973).
356. A. Reissert, *Chem. Ber.*, **38**, 3415 (1905).
357. I. W. Elliot, *J. Org. Chem.*, **29**, 305 (1964).
358. R. Bramley and M. D. Johnson, *J. Chem. Soc.*, **1965**, 1372.
359. R. Hull, *J. Chem. Soc. C*, **1968**, 1777.
360. B. J. Huckings and M. D. Johnson, *J. Chem. Soc. B*, **1966**, 63.
361. W. E. McEwen and R. L. Cobb, *Chem. Rev.*, **55**, 511 (1955).
362. F. D. Popp, *Advances in Heterocyclic Chemistry*, Vol. 9, 1968, p. 1.
363. P. Davis and W. E. McEwen, *J. Org. Chem.*, **26**, 815 (1961).
364. H. Rupe and W. Frey, *Helv. Chim. Acta*, **22**, 673 (1939).
365. F. D. Popp and W. Blount, *Chem. Ind. London*, **1961**, 550.
366. F. D. Popp, W. Blount, and A. Soto, *Chem. Ind. London*, **1962**, 1022.
367. F. D. Popp, W. Blount, and P. Melvin, *J. Org. Chem.*, **26**, 4930 (1961).
368. F. D. Popp and W. Blount, *J. Org. Chem.*, **27**, 297 (1962).
369. F. D. Popp and A. Soto, *J. Chem. Soc.*, **1963**, 1760.
370. F. D. Popp and J. Wefer, *Chem. Commun.*, **1967**, 59.
371. F. D. Popp and W. E. McEwen, *J. Am. Chem. Soc.*, **79**, 3773 (1957).
372. H. Shirai and N. Oda, *Chem. Pharm. Bull.*, **8**, 744 (1960).
372a. J. M. Grosheintz and H. O. L. Fischer, *J. Am. Chem. Soc.*, **63**, 2021 (1941).
373. R. B. Woodward, *J. Am. Chem. Soc.*, **62**, 1626 (1940).
374. I. W. Elliott, Jr., *J. Am. Chem. Soc.*, **77**, 4408 (1955).
375. A. Gassmann and H. Rupe, *Helv. Chim. Acta*, **22**, 1241 (1939).
376. E. Späth and O. Brunner, *Chem. Ber.*, **57**, 1243 (1924).
377. M. Szafran and T. Dziembowska, *Roczniki Chem.*, **44**, 1805 (1970); *C.A.*, **74**, 87792 (1971).
378. M. Colonna and S. Fatutta, *Gazz. Chim. Ital.*, **83**, 622 (1953).
379. K. Ueda, *J. Pharm. Soc. Japan*, **57**, 825 (1937); *C.A.*, **32**, 1265 (1938).
380. N. B. Chapman, K. Clarke and K. S. Sharma, *J. Chem. Soc. C*, **1970**, 2334.
381. R. D. Haworth and W. H. Perkin, Jr., *J. Chem. Soc.*, **127**, 1434 (1925).
382. M. P. Cava and M. Srinivasan, *Tetrahedron*, **26**, 4649 (1970).
383. A. H. Jackson and G. W. Stewart, *Chem. Commun.*, **1971**, 149.
384. J. Knabe and A. Frie, *Arch. Pharm.*, **306**, 648 (1973).
385. K. Bauer, *Chem. Ber.*, **83**, 10 (1950).
386. G. Wittig, M. A. Jesaitis, and M. Glos, *Ann. Chem.*, **577**, 1 (1952).
387. E. J. Corey, A. L. Borror, and T. Foglia, *J. Org. Chem.*, **30**, 288 (1965).
388. Y. Hamada, J. Takeuchi, and H. Matsuoka, *Chem. Pharm. Bull.*, **18**, 1026 (1970).
389. G. L. Buchanan, J. W. Cook, and J. Loudon, *J. Chem. Soc.*, **1944**, 325.
390. F. D. Popp, L. E. Katz, C. W. Klinowski, and J. M. Wefer, *J. Org. Chem.*, **33**, 4447 (1968).

391. F. D. Popp, J. M. Wefer, and A. Catala, *J. Heterocycl. Chem.*, **2**, 317 (1965).
392. J. M. Wefer, A. Catala, and F. D. Popp, *Chem. Ind.*, **1965**, 140.
393. F. D. Popp and C. W. Klinowski, *J. Chem. Soc. C*, **1969**, 741.
394. M. Makosza, *Tetrahedron Letters*, **1969**, 677.
395. B. C. Uff and J. R. Kershaw, *J. Chem. Soc. C*, **1969**, 666.
395a. R. Piccirilli and F. D. Popp, *Can. J. Chem.*, **47**, 3261 (1969).
396. F. D. Popp and J. M. Wefer, *J. Heterocycl. Chem.*, **4**, 183 (1967).
397. W. E. McEwen, R. H. Terss, and I. W. Elliot, *J. Am. Chem. Soc.*, **74**, 3605 (1952).
398. E. Koenigs and E. Ruppelt, *Ann. Chem.*, **509**, 142 (1934).
399. V. D. Parker and L. Eberson, *Tetrahedron Letters*, **1969**, 2839.
400. M. Masui and H. Ohmori, *J. Chem. Soc., Perkin Trans. II*, **1973**, 1112.
401. H. Blount, *J. Electroanal. Chem.*, **42**, 271 (1973).
402. A. K. Sheinkman, A. N. Kost, A. N. Prilepskaja, and Zh. B. Shijan, *J. Org. Chem. USSR*, **4**, 1286 (1968).
403. A. K. Sheinkman, A. K. Tokarev, and S. N. Baranov, *Khim. Geterotsikl. Soedin.*, **1971**, 82.
404. A. K. Sheinkman, A. P. Kucherenko, and S. N. Baranov, *Khim. Geterotsikl. Soedin.*, **1970**, 1291; **1972**, 669.
405. A. K. Sheinkman, S. G. Potashnikova, and S. N. Baranov, *Khim. Geterotsikl. Soedin.*, **1969**, 563.
406. A. K. Sheinkman, S. G. Potashnikova, and S. N. Baranov, *J. Org. Chem. USSR*, **6**, 614 (1970).
407. A. K. Sheinkman, A. N. Prilepskaja, and A. N. Kost, *Khim. Geterotsikl. Soedin.*, **1970**, 1515.
408. A. K. Sheinkman, T. V. Samoilenko, and N. A. Kluev, *J. Gen. Chem. USSR*, **44**, 1472 (1974).
409. A. K. Sheinkman and A. N. Prilepskaja, *Khim. Geterotsikl. Soedin.*, **1971**, 860.
410. G. I. Migachev and B. I. Stepanov, *J. Gen. Chem. USSR*, **38**, 1368 (1968).
411. A. N. Kost, A. K. Sheinkman, and N. F. Kazarinova, *Khim. Geterotsikl. Soedin.*, **1966**, 722.
412. A. K. Sheinkman, A. N. Kost, and R. D. Bodnarchuk, *Khim. Geterotsikl. Soedin.*, **1967**, 183.
413. A. K. Sheinkman and A. K. Tokarev, *Khim. Geterotsikl. Soedin.*, **1969**, 955.
414. A. K. Sheinkman, A. N. Kost, and A. K. Prilepskaja, *Khim. Geterotsikl. Soedin.*, **1967**, 379.
415. H. Fischer and P. Ernst, *Chem. Ber.*, **59**, 138 (1926).
416. A. Treibs and A. Ohorodnik, *Ann. Chem.*, **611**, 149 (1958).
417. A. Treibs and A. Dietl, *Ann. Chem.*, **619**, 80 (1958).
418. A. Treibs and M. Fligge, *Ann. Chem.*, **652**, 176 (1962).
419. A. Treibs, K. Jacob, and R. Tribolett, *Ann. Chem.*, **739**, 27 (1970).
420. H. Plieninger, U. Lerch, and J. Kurze, *Angew. Chem.*, **75**, 724 (1963).
421. H. Plieninger, H. Bauer, W. Bühler, J. Kurze, and U. Lerch, *Ann. Chem.*, **680**, 69 (1964).
422. H. Deubel, D. Wolkenstein, H. Jokisch, T. Messerschmitt, S. Brodka, and H. v. Dobeneck, *Chem. Ber.*, **104**, 705 (1971).
423. H. v. Dobeneck, D. Wolkenstein, H. Deubel, and H. Reinhard, *Chem. Ber.*, **102**, 3500 (1969).
424. H. M. D. Deubel, D. Wolkenstein, and P. Doyle, Brit. Pat. 1,311,336 (1973); *Ref. Zh. Khim.* **1973**, 22 N 348.
425. A. Treibs, *Rev. chim. Roum.*, **7**, 1345 (1962).
426. A. K. Sheinkman and A. A. Deikalo, *Khim. Geterotsikl. Soedin.*, **1970**, 126; **1971**, 1654.
427. N. A. Kluev, G. A. Maltzeva, R. A. Khmelnitzky, A. K. Sheinkman, and A. A. Deikalo, *Izv. Timiryazev. Sel'skokhoz. Akad.*, **1974**, 200.

428. H. v. Dobeneck, H. Deubel, and F. Heichele, *Angew. Chem.*, **71**, 310 (1959).
429. H. v. Dobeneck and W. Goltzsche, *Chem. Ber.*, **95**, 1484 (1962).
430. J. C. Powers, *J. Org. Chem.*, **30**, 2534 (1965).
431. A. K. Sheinkman, S. G. Potashnikova, and G. N. Baranov, *Khim. Geterotsikl. Soedin.*, **1970**, 1292.
432. A. K. Sheinkman, A. N. Kost, S. G. Potashnikova, A. O. Ginzburg, and S. N. Baranov, *Khim. Geterotsikl. Soedin.*, **1971**, 648.
433. A. K. Sheinkman, A. A. Deikalo, A. P. Kucherenko, and S. N. Baranov, *Khim. Geterotsikl. Soedin.*, **1971**, 424.
434. A. K. Sheinkman, A. A. Deikalo, T. N. Stupnikova, N. A. Kluev, and G. A. Maltzeva, *Khim. Geterotsikl. Soedin.*, **1972**, 1099.
435. A. K. Sheinkman, T. V. Stupnikova, and A. A. Deikalo, *Khim. Geterotsikl. Soedin.*, **1973**, 1147.
436. G. Marino, *Chim. Ind. (Ital.)*, **55**, 349 (1973); *Khim. Geterotsikl. Soedin.*, **1973**, 579.
437. W. Fanshawe, W. Bauer, and S. Safir, U.S. Pat. 3,551,567 (1970); *Ref. Zh. Khim.*, **1971**, 18 N 427.
438. D. Beck and K. Schenker, *Helv. Chim. Acta*, **51**, 260 (1968).
439. A. S. Bailey, M. C. Chum, and J. J. Wedgwood, *Tetrahedron Letters*, **1968**, 5953.
440. A. S. Bailey, A. J. Buckley, W. A. Warr, and J. J. Wedgwood, *J. Chem. Soc. Perkin Trans. I*, **1972**, 2411.
441. K. Takayama, M. Isobe, K. Harano, and T. Taguchi, *Tetrahedron Letters*, **1973**, 365.
442. E. Ghigi, *Chem. Ber.*, **73**, 677 (1940); **75**, 764 (1942).
443. E. Ghigi, *Gazz. Chim. Ital.*, **76**, 352 (1946).
444. A. K. Sheinkman, A. K. Tokarev, S. G. Potashnikova, A. A. Deikalo, A. P. Kucherenko, and S. N. Baranov, *Khim. Geterotsikl. Soedin.*, **1971**, 643.
445. R. H. Terss and W. E. McEwen, *J. Am. Chem. Soc.*, **76**, 580 (1954).
446. A. K. Tokarev, L. N. Voloveiskij, A. K. Sheinkman, and S. N. Baranov, *J. Gen. Chem. USSR*, **42**, 460 (1972).
447. L. N. Voloveiskij, A. K. Sheinkman, M. Ja. Jakovleva, and A. K. Tokarev, *J. Gen. Chem. USSR*, **43**, 1414 (1973).
448. A. K. Sheinkman and A. K. Tokarev, *J. Org. Chem. USSR*, **7**, 855 (1971).
449. A. K. Sheinkman, A. K. Tokarev, and A. N. Prilepskaja, *Khim. Geterotsikl. Soedin.*, **1972**, 529.
450. A. K. Sheinkman, A. A. Deikalo, and S. N. Baranov, *Khim. Geterotsikl. Soedin.*, **1970**, 131.
451. A. A. Deikalo, A. K. Sheinkman, and S. N. Baranov, *Khim. Geterotsikl. Soedin.*, **1972**, 1359.
452. A. K. Sheinkman, A. A. Deikalo, T. V. Stupnikova, and S. N. Baranov, *Khim. Geterotsikl. Soedin.*, **1972**, 284.
453. S. Weber, H. L. Slates, and N. L. Wendler, *J. Org. Chem.*, **32**, 1668 (1967).
454. W. Steglich and G. Höfle, *Chem. Ber.*, **104**, 3644 (1971).
455. W. Steglich and G. Höfle, *Chem. Ber.*, **102**, 883 (1969).
456. W. Steglich and G. Höfle, *Chem. Ber.*, **102**, 899 (1969).
457. N. I. Aronova, N. N. Makhova, and S. I. Zav'jalov, *Izv. Akad. Nauk SSSR Ser. Khim.*, **1970**, 1835.
458. J. Schnekenburger, *Arch. Pharm.*, **298**, 722 (1965).
459. J. Schnekenburger and P. Kaiser, *Arch. Pharm.*, **304**, 482 (1971).
460. I. Fleming and J. B. Mason, *J. Chem. Soc. C*, **1969**, 2509.
461. A. K. Sheinkman, G. V. Samoilenko, and S. N. Baranov, *J. Gen. Chem. USSR*, **40**, 2339 (1970).
462. E. Koenigs and H. Greiner, *Chem. Ber.*, **64**, 1049 (1931).
463. K. Thomas and D. Jerchel, *Angew. Chem.*, **70**, 719 (1958).

464. R. F. Evans, H. C. Brown, and H. C. van der Plas, *Org. Syntheses*, **43**, 97 (1963).
465. E. Besthorn and J. Ibele, *Chem. Ber.*, **37**, 1236 (1904); **38**, 2127 (1905); **46**, 2762 (1913).
466. H. Wieland, O. Hettohe, and T. Hoshino, *Chem. Ber.*, **61**, 2371 (1928).
467. F. Knollpfeiffer and K. Schneider, *Ann. Chem.*, **530**, 34 (1937).
468. J. F. Muren and A. Weissman, *J. Med. Chem.*, **14**, 49 (1971).
469. Y. Kiso and H. Yajima, *J. Chem. Soc. Chem. Commun.*, **1972**, 942.
470. F. Sipos and D. W. Gaston, *Synthesis*, **1971**, 321.
471. A. K. Sheinkman, G. V. Samoilenko, and S. N. Baranov, *J. Gen. Chem. USSR*, **40**, 700 (1970).
472. A. K. Sheinkman, G. V. Samoilenko, and S. N. Baranov, *Dokl. Akad. Nauk SSSR*, **196**, 1377 (1971).
473. G. Fraenkel, J. W. Cooper, and C. M. Fink, *Angew. Chem.*, **82**, 518 (1970).
474. T. Agawa and S. I. Miller, *J. Am. Chem. Soc.*, **83**, 449 (1961).
475. A. K. Sheinkman and A. N. Prilepskaja, *Khim. Geterotsikl. Soedin.*, **1971**, 1148.
476. E. M. Kosower and P. E. Klinedinst, Jr., *J. Am. Chem. Soc.*, **78**, 3493 (1956).
477. E. M. Kosower, E. J. Land, and A. J. Swallow, *J. Am. Chem. Soc.*, **94**, 986 (1972).
478. J. Kuthan, M. Ferles, J. Volke, and N. V. Koshmina, *Tetrahedron*, **26**, 4361 (1970).
479. H. Zahradník and J. Koutecký, *Advances in Heterocyclic Chemistry*, Vol. 5, Academic Press, New York, 1965, p. 69.
480. E. M. Kosower, in *Progress in Physical Organic Chemistry*, Vol. 3, John Wiley & Sons, 1965.
481. R. G. Pearson and J. Songstad, *J. Am. Chem. Soc.*, **89**, 1827 (1967); *Chem. Rev. USSR*, **38**, 1222 (1969).
482. U. Eisner and J. Kuthan, *Chem. Rev.*, **72**, 1 (1972).
483. E. Weitz, A. Roth, and A. Nelken, *Ann. Chem.*, **425**, 161 (1921).
484. O. Dimroth and R. Heene, *Chem. Ber.*, **54**, 2934 (1921).
485. O. Dimroth and F. Frister, *Chem. Ber.*, **55**, 1223, 3693 (1922).
486. A. J. Nielsen, D. W. Moore, G. M. Muha, and K. H. Berry, *J. Org. Chem.*, **29**, 2175, 2898 (1964).
487. J. F. Arens and J. P. Wibaut, *Rec. Trav. Chim. Pays-Bas*, **61**, 59 (1942).
488. J. P. Wibaut, *Rec. Trav. Chim. Pays-Bas*, **60**, 119 (1941).
489. T. Urbanski, Z. Biernacki, D. Gürne, L. Halski, M. Mioduszewska, B. Serafinowa, J. Urbanski, and D. Zelazko, *Roczniki Chem.*, **27**, 161 (1953).
490. G. Wilbert, L. Reich, and L. E. Tenenbaum, *J. Org. Chem.*, **22**, 694 (1957).
491. L. Jankov and L. Nedeleva, *Farm. Bulg.*, **9** (1), 24 (1959).
492. N. F. Kazarinova, E. P. Babin, K. A. Solomko, M. I. Kotelenez, A. A. Artamonov, and A. K. Sheinkman, *J. Appl. Chem. USSR*, **36**, 649 (1963).
493. N. Gospodinov, Sh. Levi, and E. Mutafceva, *Farm Bulg.*, **17** (4), 12 (1967).
494. P. M. Atlani, J. F. Biellmann, R. Briere, H. Lemaire, and A. Rassat, *Tetrahedron*, **28**, 2827 (1972).
495. F. W. Fowler, *J. Org. Chem.*, **37**, 1321 (1972).
496. B. Bello, USSR Pat. 294,330 (1971); *Ref. Zh. Khim.*, **1972**, 9 N 301.
497. W. Solomon, *J. Chem. Soc.*, **1946**, 934.
498. J. P. Wibaut and D. van der Vennen, *Rec. Trav. Chim. Pays-Bas*, **66**, 236 (1947).
499. S. Wawzonek, M. F. Nelson, Jr., and P. J. Thelen, *J. Am. Chem. Soc.*, **74**, 2894 (1952).
500. D. A. van Dorp and J. F. Arens, *Rec. Trav. Chim. Pays-Bas*, **66**, 188 (1947).
501. A. E. Arbuzov, *Izv. Akad. Nauk SSSR, Otdel. Khim. Nauk*, **1945**, 451.
502. P. M. Atlani and J. F. Biellmann, *Tetrahedron Letters*, **1969**, 4829.
503. P. M. Atlani and J. F. Biellmann, *Compt. Rend.*, **271**, 688 (1970).
504. F. Lindwurm and J. Tomasz, Hung. Pat. 2,387 (1971); *C.A.*, **75**, 129672 (1971).
505. P. M. Atlani, J. F. Biellmann, R. Briere, and A. Rassat, *Tetrahedron*, **28**, 5805 (1972).

506. S. L. Johnson and S. A. Anthony, *J. Org. Chem.*, **37**, 2516 (1972).
507. P. M. Atlani, J. Biellmann, and J. Moron, *Tetrahedron*, **29**, 391 (1973).
508. K. N. Kilminster and M. Sainsbury, *J. Chem. Soc. Perkin. Trans.*, *I*, **1972**, 2264.
509. J. P. Wibaut and S. Vromen, *Rec. Trav. Chim.*, **67**, 547 (1948).
510. D. A. van Dorp and J. F. Arens, *Rec. Trav. Chim.*, **66**, 183 (1947).
511. A. T. Nielsen, *J. Org. Chem.*, **35**, 2498 (1970).
512. E. R. Blout and R. S. Corley, *J. Am. Chem. Soc.*, **69**, 763 (1947).
513. A. K. Sheinkman, V. A. Ivanov, and S. N. Baranov, *Dopovidi Akad. Nauk Ukr. RSR Ser. B*, **1970**, 619.
514. A. K. Sheinkman and Ju. N. Iljina, *Khim. Geterotsikl. Soedin.*, **1971**, 568.
515. A. K. Sheinkman, V. A. Ivanov, N. A. Kluev, and G. A. Maltzeva, *J. Org. Chem. USSR*, **9**, 2550 (1973).

ADDENDUM

To Section I

New reviews on heteroaromatic cations (516–520) including their N-acyl derivatives (521) have been published. This paper deals exclusively with the chemistry of N-acyl derivatives of pyridine and condensed structures based on pyridine. However, other heteroaromatic N-acylium salts, first of all imidazole derivatives, are now extensively studied. Of the works cited above, see refs. 11, 12, 275, 276, 279. For more recent works of N-acylimidazolium salts which are acylating agents of high selectivity see refs 522–529. The hetarylation reactions of N-acyl salts of imidazole, benzimidazole, and other condensed heteroaromatic structures are described in refs. 530–537. For N-acylimidazolium salts, see also refs. 538–545. Investigations of bifunctional catalysis in hydrolysis with polymers containing imidazole fragments have developed into a separate branch of study (517, 518, 546, 547).

To Section II

Stable N-acylpyridinium fluoborates can be prepared from the corresponding chlorides by anion exchange (548). N-Acylpyridinium, quinolinium, and isoquinolinium perchlorates are obtained by elimination of hydride ions from the corresponding 1,2-dihydro compounds under the action of triphenylmethyl perchlorate (532, 549, 550). 4-Benzylpyridines readily loose protons under the action of acyl chlorides giving anhydronium bases, 1-acyl-4-benzyl-1,4-dihydropyridines (551, 552). On the contrary, N-carbamoyl salts of polymethylene-bis-pyridines are exceedingly stable (553). For interaction of peroxocarboxylic acids and their acyl chlorides with pyridine see refs. 554, 555.

To Section IV-A

For the mechanism of acylation of alcohols in the presence of tertiary amines, see refs. 521, 556, 557. HN_3 may effectively be acylated in a pyridine/toluene mixture, which provides a means of preparation of mono- and dicarboxylic acids azides (558). Oxalyl chloride readily acylates phenols in the presence of 4-dimethylaminopyridine (559). For selective acylation of sugars, see refs. 560 and 561. Mixtures containing pyridine and acid anhydrides or acyl chlorides may be applied to quantitative determination of hydroxy functions in phenols, aliphatic polyols, polymers, and ionites (562, 563). Acylation of alcohols, phenols, thiols with acyloxyacids halides usually mechanically follows more than one way. Thus *o*-acetoxybenzoyl chloride predominantly gives the 1,3-benzodioxan derivative, while in pyridine, the usual acylation of the hydroxy function occurs (564).

Na-enolates of β-ketoesters undergo *O*-acylation in the presence of pyridine, predominantly to cis-isomers (565). Anhydrides of acids usually cause cyclization of ketone acylhydrazones to 2,2-dialkyl-4-acyl-1,2,4-oxadiazolines (see, e.g., refs. 566); the presence of pyridine increases reaction yields (567). Acylation of ketoximes in pyridine proceeds via rearrangements (see, e.g., refs. 568, 569).

A number of new examples of *C*-acylation in the Dakin–West syntheses have been described. It has been shown that the formation of azlactones is not the rate-determining step in these reactions (570, 571).

2-Cyanomethylpyridine and some of its heterocyclic analogues undergo acylation at the CH_2 group in heating together with acid anhydrides (572, 573).

Acyl chlorides of sterically hindered arylsulphonic acids or polymer acyl chlorides of this type are used in specific acylations, including the synthesis of peptides and polyoses (574–576). Works have been published on the action of arylsulphochlorides on aromatic and heteroaromatic amines in pyridine (577, 578), on the synthesis of tetrahydropyranes from 1,5-diols via tosylates (579), on the preparation of acid anhydrides from carboxylic acids (580), and of allenes from ketones (581) by use of trifluoromethanesulphonic acid.

In a number of works, mainly by Japanese scientists, the acylating action of *N*-phosphonopyridinium betaines has been utilized, for example, in the preparation of amides and esters from carboxylic acids (582) including syntheses of polypeptides or polyamides (524, 583, 584). See also ref. 585. For oxidation of the CH_3 group in acetophenone under the action of $SOCl_2$ in pyridine, see ref. 586.

To Section IV-B

Hydrolysis of *M*-methoxycarbonylpyridinium chlorides containing various 3- or 4-substituents, probably, goes via the formation of tetrahedral binary ions at high pK_a values (at low pK_a, these ions are not stable) (556, 587, 588).

A purely nucleophilic mechanism of hydrolysis and alcoholysis of *p*-nitrophenylacetates in the presence of pyridine, 3- and 4-picolines, 4-dimethylaminopyridine has been confirmed. 2-Picoline catalyzes hydrolysis as mere base owing to steric factors (587, 589). For aminolysis of these esters, see ref. 590. For the mechanism of hydrolysis of esters in the presence of polymers containing pyridine or imidazole units, see refs. †517, 518, and 591‡. Reactivity of *N*-acylpyridinium cations toward nucleophiles well fits the equation log $K = \log K_0 + N^+$ (where K_0 and N^+ are the characteristics of the nucleophile and electrophile, respectively) (592–594).

For hydrolysis and methanolysis of arylsulphochlorides in the presence of pyridine proceeding via the formation of arylsulphonylpyridinium ions as rate-limiting step, see ref. 595. A combined mechanism has been reported for sulphoacylation of phenols (596) as well as for the reaction of the anhydride of *p*—chlorobenzenesulphonic acid with methanol in the presence of pyridine as catalyst (597).

Kinetics of reactions between aromatic amines and acyl halides in aprotic media in the presence of trialkylamines (525, 526) and 2-hydroxypyridines (598) as well as of similar reactions of anhydrides or chlorides of arylsulphonic acids (597, 599–601) has been studied. Unlike pyridine and alkylpyridines, 2-hydroxypyridines act as bifunctional catalysts (598).

To Section IV-D

For the stoichiometry of the Reissert compounds from 5-nitroisoquinoline and 6-nitroquinoline, see ref. 602. The action of thiophosgene on pyridine in the presence of water and barium carbonate leads to ring opening and formation of 2-trans-4-cis-1-formyl-5-isothiocyano-2,4-butadiene (603). Under the same conditions, 4,7-dichloroquinoline gives β,4-dichloro-2-isothiocyanocinnamaldehyde (604), 4-methoxy-7-chloroquinoline reacts without ring opening to give 1,2-dihydro-4-chloro-3-formylquinoline-2-thione (605).

Ethyl chloroformate reacts with pyridine, sodium cyanate, and methylene chloride/water mixture to give 1-ethoxycarbonyl-2-cyano-1,2-dihydropyridine, the first Reissert compound in the pyridine series (606). If, however, a mixture of pyridine and methyl chloroformate is treated with lithium cuprates of the type R_2CuLi (R = alkyl, aryl), the reaction yields 1-methoxycarbonyl-4-R-1,4-dihydropyridines. With diethyl chlorophosphate instead of methyl chloroformate, the reaction goes in a similar way (607).

For synthesis of polymer structures based on Reissert compounds, see refs. 608 and 609.

To Section IV-E

Pyridine adds to triple bonds of dialkylalkinylborates in the presence of acetyl chloride to give a mixture of 4-substituted cis- and trans-1-acetyl-1,4-dihydropyridines (601).

$$R_3B-\underset{H}{\overset{R^1}{C}}=\overset{R^2}{C}-\underset{}{\underset{}{\bigcirc}}N-Ac$$

It has been shown, using the reaction of pyridine with tosyl chloride and cyclohexanone as example, that N-arylsulphacyl salts of pyridine bases can enter hetarylation reactions (however, giving low product yields) to form 4-substituted 1-arylsulphonyl-1,4-dihydropyridines (611).

For the preparation of 1-acetyl-2-carboxymethyl-1,2-dihydropyridine (pyridilation of acetic acid) in the presence of some phosphates, see ref. 612. Heating of isoquinoline together with acetic anhydride to 150–160°C also leads to 1-carboxymethyl-2-acetyl-1,2-dihydroisoquinoline. Phthalazine reacts in a similar way at one of its nitrogen atoms (613, 614). According to the same scheme but still more readily do isoquinoline, phthalazine, and 1,6-naphthyridine react with diketene. Quinoline and quinoxaline return unreacted from the latter reaction (613, 614). With isoquinoline, the main reaction product is the ketone rather than the acid, a reaction which corresponds to the addition of acetone instead of acetic acid (616, 617). The addition of acetophenone, malonic ester, phenylacetic acid goes in a similar way (617, 617a).

$$\text{isoquinoline} \xrightarrow[\text{Ac}_2\text{O, 100°}]{\text{RCOCH}_3} \text{N-Ac, CH}_2\text{COR} + \text{N-Ac, CH}_2\text{COOH}$$

Though N-arylsulphacilium salts are less reactive than the corresponding salts containing carboxylic acid residues, they can be applied to hetarylate such C–H-acids as substituted indoles and dimethylaniline. Acidic hydrolysis leads to decomposition of the adducts formed to the initial compounds (618). The reactions with various chlorides of phosphoric and phosphorous acids go readily (619). Hetarylations of various benzoquinones and 4,9-diazapyrene have been described in (620, 621). There exist data on hetarylation of indene, cyclopentadiene, azulene, fluorene (622), 2-pyrrolinones (623).

A mixture of an amide or lactam with $POCl_3$ can be used as acyl component in the interaction of quinoline or isoquinoline with indole (624, 625).

Reactions of various substituted indoles with acylating agents in pyridine not always go in only one way (626, 627). The process may be directed to acylation only. 3-α-Haloacylindoles obtained from these reactions rearrange to indolyl-3-acetic acids under the action of alkalis (628). For hetarylation with pyrimidine, see ref. 537. Condensation of aromatic aldehydes with participation of side chains of acridine and phenanthridine via N-acylium salts is described in ref. 629.

To Section IV-F

Phenanthrydine, benzo[f]- and benzo[h]quinolines react with anhydrides or acyl chlorides of even carboxylic acids in the presence of zinc metal powder to give dimeric N,N'-diacyl tetrahydroderivatives via the formation of anion radicals (529). Sodium borohydride reacts with 1-methoxycarbonyl-3-cyanopyridinium to give a mixture of 1,2-, 1,4-, and 1,6-tetrahydrocompounds (630). For the action of organomagnesium compounds, see refs. 631 and 632.

REFERENCES

516. S. V. Krivun, O. J. Alferova, and S. V. Sajapina, *Chem. Rev. USSR*, **43**, 1739 (1974).
517. H. Morawetz, *Proc. Roy. Austral. Chem. Inst.*, **1974**, 89; *Ref. Zh. Khim.*, **1974**, 23 S 269.
518. C. G. Overberger, Th. W. Smith, and K. W. Dixon, *J. Polym. Sci., Polym. Symp.*, **1975**, N50, 1.
519. A. Williams and K. T. Douglas. *Chem. Rev.*, **75**, 627 (1975).
520. O. N. Chupakhin and I. Ya. Postovskij, *Chem. Rev. USSR*, **45**, 908 (1976).
521. E. M. Cherkasova, S. V. Bogatkov, and Z. P. Golovina, *Chem. Rev. USSR*, **46**, 477 (1977).
522. J. Rebek, D. Brown, and S. Zimmerman, *J. Am. Chem. Soc.*, **97**, 454 (1975).
523. Yu. V. Mitin and O. V. Glinskaja, *J. Gen. Chem. USSR*, **41**, 1151 (1971).
524. N. Ogata and H. Tanaka, *Polymer J.*, **6**, 461 (1974).
525. V. A. Dadaly, Yu. S. Simanenko, S. A. Lapshin, L. M. Litvinenko, and V. T. Ribachenko, *J. Org. Chem. USSR*, **12**, 1483 (1976).
526. F. Ramirez, J. F. Marecek, and H. Okazaki, *J. Am. Chem. Soc.*, **97**, 7181 (1975).
527. Ch. Hauser and L. F. Theiling. *J. Org. Chem.*, **39**, 1134 (1974).
528. Ch. Dadaly, E. B. Titov, Yu. S. Simanenko, S. A. Lapshin, R. G. Semenova, and V. I. Ribachenko, *Ukr. Khim. Zh.*, **42**, 598 (1976).
529. A. K. Sheinkman, A. P. Kucherenko, I. V. Kukurina, N. A. Kluev, and E. N. Kurkutova, *Khim. Geterotsikl. Soed.*, **1977**, 229.
530. E. Abusharab, D.-Y. Lee, and L. Goodman, *J. Org. Chem.*, **40**, 3376 (1975).
531. E. Abusharab, A. Bindra, and L. Goodman, *J. Org. Chem.*, **40**, 3379 (1975).
532. T. V. Stupnikova, A. I. Serdjuk, E. Porcel, and A. K. Sheinkman, *Dokl. Akad. Nauk. Ukr. SSR, Ser. B*, **1977**, 47.
533. A. K. Sheinkman, V. I. Zherebchenko, T. V. Stupnikova, and L. A. Ribenko. *Khim. Geterotsikl. Soedin.*, **1976**, 1573.
534. A. K. Sheinkman, T. V. Stupnikova, N. A. Kluev, L. Yu. Petrovskaja, and V. G. Zhilnikov, *Khim. Geterotsikl. Soedin.*, **1977**, 238.
535. T. V. Stupnikova, A. K. Sheinkman, A. F. Pozharskij, N. A. Kluev, and E. N. Istratov, *Khim. Geterotsikl. Soedin.*, **1976**, 1682.
536. O. Gerngross, *Ber.*, **46**, 1913 (1913).

537. A. K. Sheinkman, A. F. Pozharskij, V. I. Sokolov, and T. V. Stupnikova, *Dokl. Akad. Nauk SSSR*, **226**, 1096 (1976).
538. E. Guibé-Jampel, G. Bram, and M. Vilkas, *Bull. Soc. Chim. France*, **1973**, 1021.
539. L. M. Litvinenko, V. A. Dadaly, S. A. Lapshin, and Yu. S. Simanenko, *Dokl. Akad. Nauk SSSR*, **219**, 1161 (1974).
540. D. G. Oakenfull, *Austral. J. Chem.*, **27**, 1423 (1974).
541. W. Palaitis and E. R. Thornton, *J. Am. Chem. Soc.*, **97**, 1193 (1975).
542. K. T. Douglas and A. Williams, *J. Chem. Soc. Perkin II*, **1976**, 515.
543. K. Martinek, A. P. Osipov, A. K. Yatsimiski, and I. V. Beresin. *Tetrahedron*, **31**, 709 (1975).
544. F. D'Andrea and U. Tonellato, *Chem. Commun.*, **1975**, 659.
545. V. I. Ribachenko, A. Yu. Chervinsky, L. M. Kapkan, R. G. Semenova, and E. V. Titov, *J. Org. Chem. USSR.*, **12**, 240 (1976).
546. S. Shinkai and T. Kunitake, *Polym. J.*, **7**, 387 (1975).
547. Y. Imanishi, Y. Amimoto, T. Sugihara, and T. Higashimura, *Macromol. Chem.*, **177**, 1401 (1976).
548. J. V. Paukstelis and M.-G. Kim, J. Org. Chem., **39**, 1503 (1974).
549. M. Wakselman and E. Guibé-Jampel, *Chem. Commun.*, **1976**, 21.
550. T. V. Stupnikova and A. K. Sheinkman, *Dokl. Akad. Nauk Ukr. SSR.*, in press (1977).
551. G. W. Fischer, *J. Prakt. Chem.*, **317**, 779 (1975).
552. B. M. Goldschmidt, B. L. van Duuren, and R. O. Goldstein, *J. Heterocycl. Chem.*, **13**, 517 (1976).
553. V. A. Patison, US Pat. 3849426, 3849427 (1974); *Ref. Zh. Khim.*, **1975**, 16T53, 55.
554. J. Y. Nedelec, J. Sorba, and D. Lefort, *Synthesis*, **1976**, 821.
555. V. I. Galibey, T. A. Tolpigina, and I. S. Voloshanovskij, *Isv. vyssh. Ucheb. Zaved., Khim. and Khim. Tekhnol.*, **19**, 1032 (1976).
556. G. A. Rogers and Th. C. Bruice, *J. Am. Chem. Soc.*, **96**, 2481 (1974).
557. S. V. Bogatkov, Z. P. Golovina, and E. M. Cherkasova, *Dokl. Akad. Nauk SSSR*, **229**, 98 (1976).
558. J. W. van Reijendam and F. Baardman, *Synthesis*, **1973**, 413.
559. D. J. Zwanenburg and W. A. Reynen, *Synthesis*, **1976**, 624.
560. M. Kugelman, A. K. Mallams, H. F. Vernay, D. F. Crowe, and M. Tanabe, *J. Chem. Soc., Perkin I*, **1976**, 1088.
561. J. F. Batey, C. Bullock, E. O'Brien, and J. M. Willams, *Carbohyd. Res.*, **43**, 43 (1975).
562. B. Haszczyc and W. Walczyk, *Wiad. Chem.*, **28**, 495 (1974).
563. N. G. Polanskij and G. V. Gorbunov, *Zh. Prikl. Khim.* (*Leningrad*), **47**, 2362 (1974).
564. Ch. Rüchardt and H. Brinkmann, *Chem. Ber.*, **108**, 3224 (1975).
565. G. Entenmann, *Tetrahedron Lett.* **1975**, 4241.
566. R. S. Sagitullin and A. N. Kost, *Vestn. Mosk. Univ., Ser. Khim.*, **1959**, N4, 187.
567. V. N. Yandovskij, *J. Gen. Chem. USSR.*, **12**, 1093 (1976).
568. M. V. Bhatt, C. G. Rao, and S. Rengaraju, *Chem. Commun.*, **1976**, 103.
569. A. N. Kost, L. N. Žukauskaite, and A. P. Stankevičius, *Khim. Geterotsikl. Soegin.*, **1971**, 504.
570. J. Lepschy, J. Höfle, L. Wilschwitz, and W. Steglich. *Ann. Chem.*, **1974**, 1753.
571. N. L. Allinger, G. L. Wang, and B. B. Dewhurst, *J. Org. Chem.*, **39**, 1730 (1974). *org. Chem.*, **39**, 1730 (1974).
572. E. V. Grishina, E. R. Sakhs, and V. P. Martinov, *Khim. Geterotsikl. Soedin.*, **1974**, 138.
573. F. S. Babichev and Yu. M. Volovenko, *Khim. Geterosikl. Soedin.*, **1975**, 1005.
574. S. E. Creasey and R. R. Gutrie, *J. Chem. Soc., Perkin I*, **1974**, 1373.
575. M. Mort, B. E. Stacey, and B. Tierney, *Carbohyd. Res.*, **43**, 183 (1975).

576. M. Rubinstein and A. Patchornik, *Tetrahedron Lett.*, **1972**, 2881.
577. K. Kurita, *Chem. and Ind.*, **1974**, 345.
578. J. W. Chittum, Ch. G. Tindall, R. D. Howells, V. Coates, and M. D. Shie. *J. Chem. Eng. Data*, **19**, 294 (1974).
579. P. Picard, D. Leclerq, and J. Moulines, *Tetrahedron Lett.*, **L975**, 2731.
580. Z. Zborucki, J. Wajcht, A. Walczak, and J. Kochman, Pat. Pol. 70928 (1974); *Ref. Zh. Khim.*, **1975**, 14 0 24.
581. P. J. Stang and R. J. Hargrove, *J. Org. Chem.*, **40**, 657 (1975).
582. N. Yamazaki and F. Higashi, *Synthesis*, **1974**, 436.
583. L. Le Guilly and G. Spach, *J. Polym. Sci. Polym. Symp.*, **1975**, 119.
584. N. Yamazaki, M. Niwano, J. Kawabata, and F. Higashi, *Tetrahedron*, **31**, 665 (1975).
585. N. Yamazaki and F. Higashi, *J. Macromol. Sci.*, **(A)9**, 761 (1975).
586. K. Oka and Sh. Hara, *Tetrahedron Lett.*, **1976**, 2783.
587. P. M. Bond, E. A. Castro, and R. B. Moodie, *J. Chem. Soc. Perkin II*, **1976**, 68.
588. P. M. Bond and R. B. Moodie, *J. Chem. Soc. Perkin II*, **1976**, 679.
589. A. R. Butler and T. H. Robertson, *J. Chem. Soc. Perkin II*, **1975**, 660.
590. N. M. Olejnik, L. M. Litvinenko, S. E. Terehova, S. Zvaric, and Kh.-Yu. Wagner, *Org. Chem. USSR*, **12**, 1732 (1976).
591. Y. Okahata and T. Kunitake, *Chemistry (Japan)*, **31**, 733 (1976); *Ref. Zh. Khim.*, **1977**, 5 S 251.
592. C. D. Ritchie, *J. Am. Chem. Soc.*, **97**, 1170 (1975).
593. A. Ross, *J. Am. Chem. Soc.*, **98**, 776 (1976).
594. L. M. Litvinenko. *J. Org. Chem. USSR*, in press.
595. O. Rogne, *Rept. Forsvar. Forskinginst.*, **1973**, N 63, 40, 46, 50. 66. *Ref. Zh. Khim.*, **1975**, 3-B-1239, 1240, 1241, 1243.
596. R. V. Visgert, E. V. Panov, Yu. G. Skripnik, and M. P. Starodubzeva, *J. Org. Chem. USSR*, **11**, 1894 (1975).
597. N. T. Maleeva, V. A. Savelova, L. M. Litvinenko, and L. G. Kurjakova, *J. Org. Chem. USSR*, **11**, 1015 (1975).
598. L. M. Litvinenko, V. A. Savelova, and A. V. Skripka, *Dokl. Akad. Nauk SSSR*, **216**, 1327 (1974).
599. T. N. Solomojchenko, V. A. Savelova, and L. M. Litvinenko, *J. Org. Chem. USSR*, **10**, 534 (1974).
600. V. A. Savelova, V. A. Shatsaja, L. M. Litvinenko, and N. I. Nikishina, *J. Gen. Chem. USSR*, **44**, 1124 (1974).
601. V. A. Shatskaja, V. A. Savelova, L. M. Litvinenko, and N. F. Korchachenko, *J. Org. Chem. USSR*, **12**, 589 (1976).
602. B. C. Uff, J. R. Kershaw, and S. R. Chabra, *J. Chem. Soc. Perkin I*, **1974**, 1146.
603. F. T. Boyle and R. Hull, *J. Chem. Soc. Perkin I*, **1974**, 1541.
604. R. Hull, P. J. van den Brock, and M. R. Swain, *J. Chem. Soc. Perkin I*, **1975**, 922.
605. R. Hull, P. J. van den Brock, and M. R. Swain, *J. Chem. Soc. Perkin I*, **1975**, 2271.
606. R. H. Reuss, N. G. Smith, and L. J. Winters, *J. Org. Chem.*, **39**, 2027 (1974).
607. E. Piers and M. Soucy, *Canad. J. Chem.*, **52**, 3563 (1974).
608. H. W. Gibson and F. C. Bailey, *Macromolecules*, **9**, 10 (1976).
609. H. W. Gibson, *Macromolecules*, **8**, 89 (1975).
610. A. Pelter and K. J. Gould, *Chem. Commun.*, **1974**, 347.
611. B. F. Bowden, K. Picker, E. Ritchie, and W. C. Taylor. *Austral. J. Chem.*, **28**, 2681 (1975).
612. T. Kurihara and T. Hata, *Chem. Lett.*, **1975**, 495.
613. H. Yamanaka, T. Shiraishi, and T. Sakamoto, *Heterocycles*, **3**, 1173 (1975).

614. H. Yamanaka, T. Shiraishi, and T. Sakamoto, *Heterocycles*, **3,** 1175 (1975).
615. H. Yamanaka, T. Sakamoto, and T. Shirashi, *Heterocycles*, **3,** 1065 (1975).
616. H. Yamanaka, T. Shiraishi, and T. Sakamoto, *Heterocycles*, **3,** 1069 (1975).
617. T. Shiraishi and H. Yamanaka, *Heterocycles*, **6,** 535 (1977).
617a. T. Shiraishi, T. Sakamoto, and H. Yamanaka, *Heterocycles*, **6,** 1716 (1977).
618. A. K. Sheinkman, G. V. Samojlenko, N. A. Kluev, Yu. G. Skripnik, and V. I. Zolotarev, *Khim. Geterotsikl. Soedin.*, **1977,** 491.
619. A. K. Sheinkman, G. V. Samojlenko, and N. A. Kiuev, *J. Gen. Chem. USSR.*, **44,** 1472 (1974).
620. A. K. Sheinkman, M. M. Mestechkin, A. P. Kucherenko, N. A. Kluev, V. N. Poltavetz, G. A. Maltzeva, L. A. Palagushkina, and Yu. B. Visotzkij, *Khim. Geterotsikl. Soedin.*, **1974,** 1096.
621. A. K. Sheinkman, M. M. Mestechkin, A. P. Kucherenko, V. V. Artemova, V. N. Poltavetz, and Yu. B. Visotzkij, *Khim. Geterotsikl. Soedin.*, **1974,** 537.
622. A. K. Sheinkman, G. V. Samoilenko, S. N. Baranov, and N. R. Kalnitzky, *Khim. Geterosikl. Soedin.*, **1975,** 1368.
623. H. M. Deubel, D. G. Wolkenstein, and P. Doyle, Brit. pat. 1311336 (1973); *Ref. Zh. Khim.*, **1974,** 22–348.
624. R. P. Rayan, R. A. Hamby, and Y.-H. Wu, *J. Org. Chem.*, **40,** 724 (1975).
625. Y.-H. Wu and R. P. Rayan, US Pat. 3886165 (1975); *Ref. Zh. Khim.*, **1976,** 90128.
626. K. Takayama, K. Harano, and T. Taguchi, *J. Pharm. Soc. Japan*, **94,** 540 (1974).
627. K. Takayama, K. Harano, and T. Taguchi, *J. Pharm. Soc. Japan*, **94,** 548 (1974).
628. J. Bergman and J. E. Bäckvall, *Tetrahedron*, **31,** 2063 (1975).
629. A. P. Kucherenko, S. G. Potashnikova, S. S. Radkova, S. N. Baranov, A. K. Sheinkman, and N. V. Volbushko, *Khim. Geterotsikl. Soedin.*, **1974,** 1257.
630. M. Natsume and M. Wada, *Chem. Pharm. Bull. (Japan)*, **24,** 2651 (1976).
631. R. E. Lyle and D. L. Comins, *J. Org. Chem.*, **41,** 3250 (1976).
632. R. E. Lyle, J. L. Marshall, and D. L. Comins, *Tetrahedron Le.*, **1977,** 1015.

3,3-REARRANGEMENTS OF IMINIUM SALTS

By H. HEIMGARTNER, *Organisch-chemisches Institut der Universität, CH-8001 Zürich, Rämistrasse 76, Switzerland*; H.-J. HANSEN, *Institut de chimie organique de l'université, CH-1705 Fribourg, Pérolles, Switzerland*; and H. SCHMID,* *Organisch-chemisches Institut der Universität, Zürich, Switzerland*

CONTENTS

I. Introduction 655
II. Aza-Cope Rearrangements in Ammonium and Iminium Salts 658
   A. 3-Aza-Cope Ammonium–Iminium Rearrangements 659
      1. Allylation and Propargylation of Enamines 660
      2. Acylation of Enamines 674
      3. Aromatic 3-Aza-Cope Rearrangements in Anilinium Salts 682
   B. 2-Aza-Cope Iminium–Iminium Rearrangements 689
      1. Rearrangements in the Course of Eschweiler–Clarke Methylations of N-Butenylamines 689
      2. Rearrangements of Protonated 1,2-Dihydroisoquinolines 694
      3. Rearrangements Accompanying Heterolytic Fragmentation Reactions . . . 700
   C. Influence of Charge in Ammonium–Iminium and Iminium–Iminium Rearrangements 702
III. 3,4- and 2,5-Diaza-Cope Rearrangements 706
   A. Thermal and Acid-Catalyzed Rearrangements of Enhydrazines 706
   B. 2,5-Diaza-Cope Rearrangements 716
IV. Concluding Remarks 720
   References 724

I. Introduction

The thermal rearrangement of the iminium salts to be discussed in this review corresponds to the basic system of [3,3]-sigmatropic rearrangements (1), namely, the (degenerate) Cope rearrangement of hexa-1,5-diene (**1**) (see

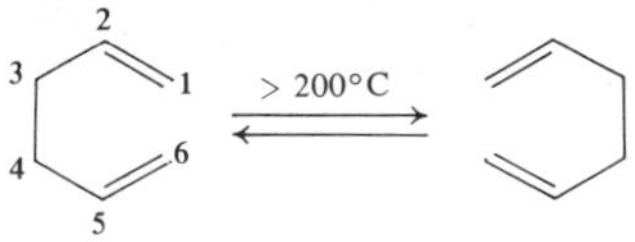

* Deceased (1976).

ref. 2). Substitution of C-3 in **1** by nitrogen leads to *N*-allyl-vinylamines (**2**) (3-azahexa-1,5-dienes). It is known that these compounds rearrange thermally (~250°*) to 4-pentenylimines (**3**) (4–6) ($\Delta\Delta H_f°$ of **2** and **3** amounts to ~7–10 kcal/mole; calculated using enthalpy increments (7)]. When **3** is a strained imine (e.g., *endo*-bicyclo[3.1.0]hex-2-enyl-6-methylimine in which C-2 and C-3 belong to a cyclopropane ring) the reverse rearrangement, that is, **3** → **2**, is also observed (see refs. 8 and 9). Transformations of the type **2** → **3** are henceforth called 3-aza-Cope rearrangements.† It is of particular significance that the ammonium compounds **4** (corresponding to **2** and isoelectronic with **1**) undergo this rearrangement much more easily than **2** and lead to the iminium compounds **5**. $\Delta\Delta H_f°$ of **4** and **5** must be greater than $\Delta\Delta H_f°$ of **2** and **3** since in **4** (cf. **2**) the vinylamine resonance does not exist. On the other hand, the energy level of the bonding π orbital lies lower in **5** than in **3** (cf. Section II-C). Steric factors also play a role because **2** possesses a trivalent nitrogen atom, whereas in **4** the nitrogen atom is tetravalent. Thus there should be a greater release of steric congestion in going from **4** to **5** than in going from **2** to **3**. These ammonium–iminium rearrangements are discussed in more detail in Section II-A.

If C-2 in **1** is substituted by nitrogen, *N*-methylidene-3-butenylamines **6** (2-azahexa-1,5-dienes) are constructed. In the simplest cases these systems should undergo degenerate 2-aza-Cope rearrangements. The position of equilibrium in substituted derivatives of **6** will be determined, as in **1**, by "secondary" structural factors such as steric interactions or conjugative effects in reactant and product. The previously known examples of rearrangements in compounds of type **6** are restricted to the readily occurring transformation of *N*-methylidene-*cis*-vinylcyclopropylamines into dihydroazepines (see ref. 8 and

---

* Reactions of this type occur as low as 80°C if they are accompanied by aromatization (3).

† The notation is derived from IUPAC rule C-0.6. Reactions of this type are also called amino-Claisen rearrangements (cf. ref. 4). Recently "hetero-Cope rearrangements" was chosen as the general name for these rearrangements (6).

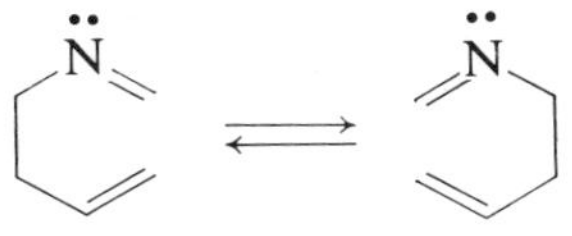

I. INTRODUCTION

(6) ⇌ (6)

literature cited therein). However there exists, corresponding to the *N*-methylideneamine **6**, a whole series of iminium salts **7** which can undergo 2-aza-Cope rearrangements, the simplest cases again being degenerate. As a rule they occur even at room temperature (cf. Section II-B), that is, under much milder conditions than in the isoelectronic Cope system **1** and possibly faster than in the uncharged systems **6**.

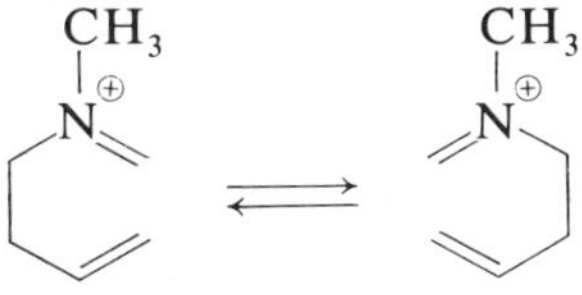

(7) ⇌ (7)

Finally, by substituting C-3 and C-4 or C-2 and C-5 by nitrogen atoms one produces *N,N'*-divinylhydrazines **8** (3,4-diazahexa-1,5-dienes) or *N,N*-dimethylene-1,2-ethylenediamines **9** (2,5-diazahexa-1,5-dienes). These can undergo thermal 3,4-diaza-Cope rearrangements leading to tetramethylenediimines **10** ($\Delta\Delta H_f^\circ$ of **8** and **10** amounts to 15–20 kcal/mole; cf. ref. 7) and thermal 2,5-diaza-Cope rearrangements, of which the simplest example is the degenerate rearrangement of **9**. For thermal rearrangements of the type **8** → **10** see refs. 10 and 11 and literature cited therein; for rearrangements of compounds of type **9**, see refs. 12 and 13. In principle it is possible to construct the mono- and dicationic forms **11–14**, corresponding to both the systems **8** and **9**. Salts **11** and **12** then might exhibit 3,4-diaza-Cope rearrangements, whereas salts **13** and **14** might, if formed, be expected to undergo the 2,5-rearrangements (cf. Sections III-A and III-B).

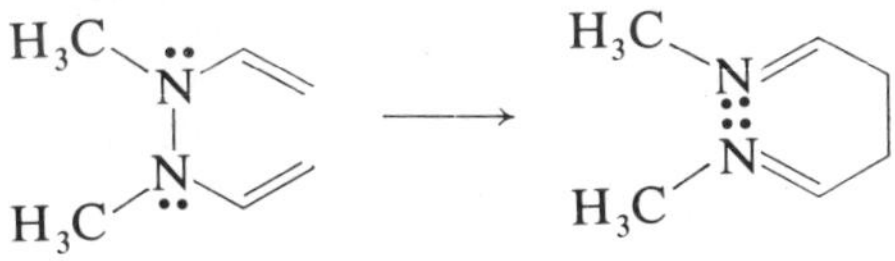

(8) → (10)

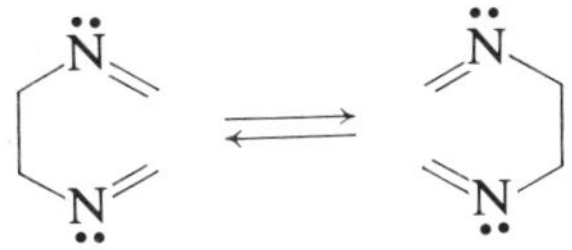

(9) ⇌ (9)

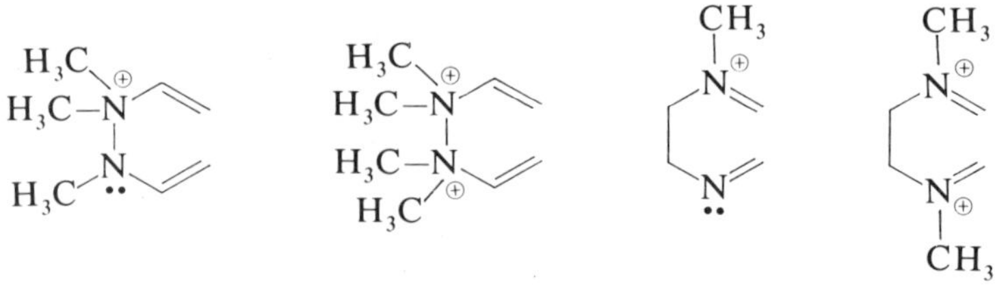

658   3,3-REARRANGEMENTS OF IMINIUM SALTS

(11)   (12)   (13)   (14)

Further diaza-Cope systems can be derived from 1, for example, the N-allyl-hydrazone 15, a 2,3-diaza system, which decomposes at 300°C via 16 to yield the hydrocarbon 17 by loss of nitrogen (14). Rearrangements in charged systems of this type are not known.

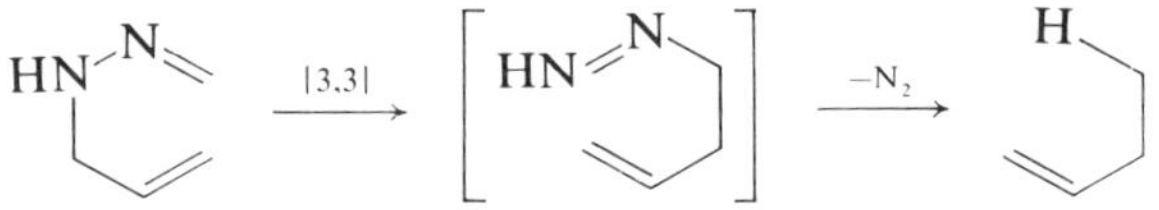

(15)   (16)   (17)

II. Aza-Cope Rearrangements in Ammonium and Iminium Salts

It is known that Cope and Claisen rearrangements occur considerably faster in charged or strongly polarized systems than in uncharged systems. For example, in the purely thermal [3,3]-sigmatropic rearrangement of allylphenyl ether into 2-allylphenol (aromatic 3-oxa-Cope rearrangement) the ether must be heated at 180°C for several hours. However, the conversion is complete within seconds by using boron trichloride at 0°C (15, 16) or in a few hours by using trifluoroacetic acid at 60°C (17, 18). (Note: in the presence of boron trichloride dichloroboric acid 2-allylphenyl ester is formed primarily.) Such reactions which take place in charged systems have recently been called charge-induced (19) or, better, charge-accelerated (18) [3,3]-sigmatropic rearrangements. To the same class of reactions belong also the ammonium–iminium rearrangements, that is, 4 → 5, and the iminium–iminium rearrangements, that is, 7 ⇌ 7, as well as the corresponding rearrangements of the ions 11–14. The essential feature of these charge-accelerated reactions is that the charge does not shift during the course of the rearrangement* but resides, at least formally, at the same atom in reactant and product. As an example the charge-accelerated Cope rearrangement of 4-allylated 4-methyl-cyclohexa-2,5-

---

* The charge-controlled rearrangements (19) [cf., for instance, [3,4]-sigmatropic rearrangements (20)] proceed with a displacement of charge.

II. AZA-COPE REARRANGEMENTS IN AMMONIUM AND IMINIUM SALTS

dien-1-one p-tosylhydrazones **18** into 2-allylated (4-methylphenyl)-p-tosylhydrazines **21** can be mentioned (21). When performed in ethanolic hydrochloric acid at room temperature the rearrangement, which proceeds via **19** and **20** (Scheme 1), is complete within a few hours, the exact time depending on the concentration of the acid. For the purely thermal rearrangement of **18** (R = H) temperatures greater than 100°C and considerably longer periods of time are required. This leads to the formation of several products, which can also be formed by heating the hydrazine **21** (R = H) above 100°C.

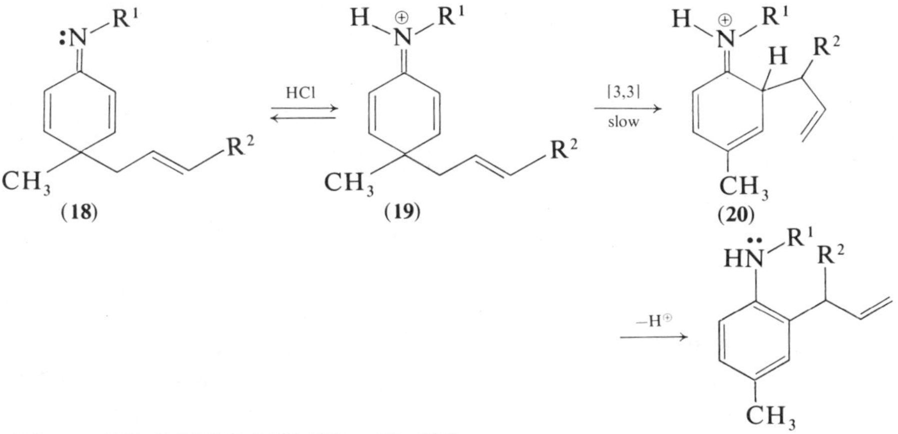

$R^1 = p\text{-}CH_3C_6H_4SO_2NH, R^2 = H, CH_3$

Scheme 1

The accelerating influence of the charge is essentially attributable to three factors: (1) a stronger delocalization of the charge in the activated complex than in the reactant; (2) a decrease of the molecular orbital (MO) energies in the activated complex of the charged aza-Cope rearrangements compared with that of the uncharged species; and (3) an increase of the energy difference between reactant and product by introducing a charge.

A more detailed discussion is given in Section II-C.

A. 3-AZA-COPE AMMONIUM–IMINIUM REARRANGEMENTS

This class of reactions is observed especially in the course of allylation and propargylation as well as acylation of enamines. The reaction of enamines have been discussed recently in a series of reviews (22) and thus only the sigmatropic character of the rearrangements is discussed in the following pages. Aromatic 3-aza-Cope rearrangements have also been observed in the reactions of N-allylated anilinium salts.

1. Allylation and Propargylation of Enamines

The allylation of the pyrrolidine enamines of aldehydes, leading after hydrolysis to α-allylated aldehydes, was described for the first time by Opitz and Mildenberger (23) and by Elkik (24) (see Table I and Scheme 2).

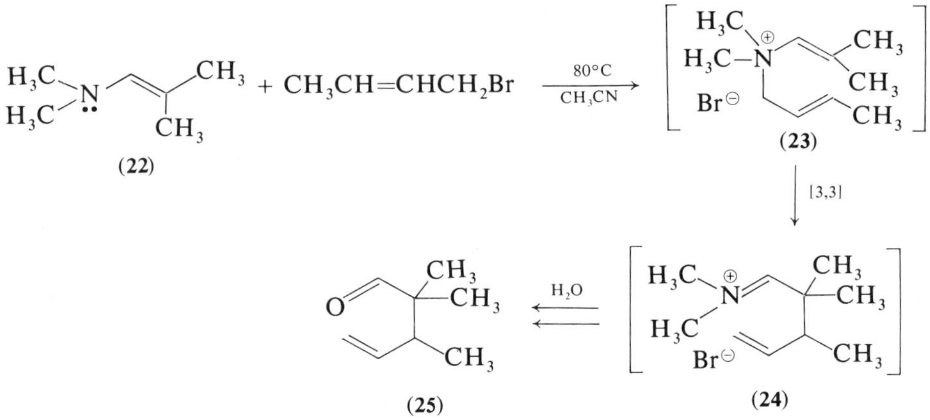

Brannock and Burpitt (30) showed that the reaction of *N,N*-dimethylisobutenylamine (**22**) with 2-butenyl bromide yields, after hydrolysis, 2,2,3-trimethyl-4-pentenal (**25**) (Scheme 2) as the sole product. Thus it was concluded that the enammonium salt **23** was formed by the kinetically controlled *N*-allylation of the enamine **22** by the 2-butenyl bromide. The ammonium compound **23** then, under the reaction conditions, rearranges by a Cope-like process into the thermodynamically favored iminium salt **24**.

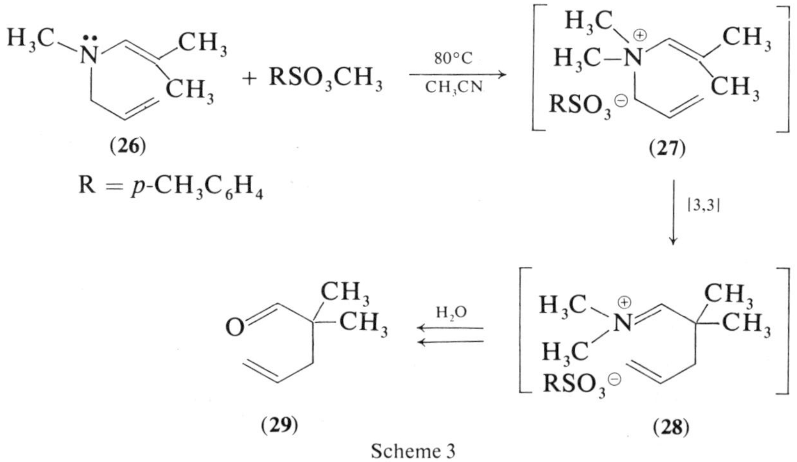

R = *p*-CH$_3$C$_6$H$_4$

Scheme 3

TABLE I

Allylation of Enamines of Aldehydes

Aldehyde	Amine	Alkylating agent	Conditions	α-Allylated aldehyde	Yield, %	Ref.
$CH_3CH_2CH_2CHO$	Pyrrolidine	Allyl bromide	DMF, 20°C	$CH_3CH_2CH-CHO$ $\|$ $CH_2=CH-CH_2$	31	24
	Piperidine	Allyl bromide	Acetonitrile + ethyldicyclo-hexylamine, 50°C	$CH_3CH_2CH-CHO$ $\|$ $CH_2=CH-CH_2$	23	24, 25
	Morpholine	Allyl bromide	DMF, 20°C	$CH_3CH_2CH-CHO$ $\|$ $CH_2=CH-CH_2$	20–25	24
	Morpholine	2-Butenyl bromide[a]	Acetonitrile, 20°C	$CH_3CH_2CH-CHO$ $\quad CH_3CH_2CH-CHO$ $\quad\|\qquad\qquad\qquad\|$ $CH_2=CH-CH-CH_3 \quad CH_3CH=CH-CH_2$ 9 : 1	17	25
$(CH_3)_2CHCHO$	Pyrrolidine	Allyl bromide	Acetonitrile, 80°C	$(CH_3)_2C-CHO$ $\|$ $CH_2=CH-CH_2$	51	26
	Dimethyl-amine	Allyl bromide	Acetonitrile, 80°C[b]	$(CH_3)_2C-CHO$ $\|$ $CH_2=CH-CH_2$	56	27
	Pyrrolidine	β-Methallyl chloride	Acetonitrile, 80°C	$(CH_3)_2C-CHO$ $\|$ $CH_2=C-CH_2$ $\quad\|$ $\quad CH_3$	10	26

TABLE I (cont.)

Aldehyde	Amine	Alkylating agent	Conditions	α-Allylated aldehyde	Yield, %	Ref.
	Pyrrolidine	2-Butenyl bromide[a]	Acetonitrile, 80°C[c]	$(CH_3)_2C-CHO$ $\|$ $CH_2=CH-CH-CH_3$	55	26
	Piperidine	2-Butenyl bromide	Acetonitrile, 80°C[c]	$(CH_3)_2C-CHO$ $\|$ $CH_2=CH-CH-CH_3$	?	28
	Morpholine	2-Butenyl bromide	Acetonitrile, 20°C[d]	$(CH_3)_2C-CHO$ $\|$ $CH_2=CH-CH-CH_3$	?	29
	Dimethylamine	2-Butenyl bromide[a]	Acetonitrile, 80°C	$(CH_3)_2C-CHO$ $\|$ $CH_2=CH-CH-CH_3$	67	30
	Dicyclohexylamine	2-Butenyl bromide	Acetonitrile, 80°C	$(CH_3)_2C-CHO$ $\|$ $CH_3CH=CH-CH_2$	10	28
$CH_3(CH_2)_5CHO$	Pyrrolidine, piperidine, morpholine	Allyl bromide	Acetonitrile or DMF, 20°C	$CH_3(CH_2)_4CH-CHO$ $\|$ $CH_2=CH-CH_2$	43	24
$(CH_3CH_2)_2CHCHO$	Pyrrolidine	Allyl bromide	Acetonitrile, 80°C	$(CH_3CH_2)_2C-CHO$ $\|$ $CH_2=CH-CH_2$	78	26
	Pyrrolidine	2-Butenyl bromide	Acetonitrile, 80°C	$(CH_3CH_2)_2C-CHO$ $\|$ $CH_2=CH-CH-CH_3$	56	26
$\begin{array}{c}CH_3CH_2\\CH_3(CH_2)_3\end{array}\!\!>\!\!CH-CHO$	Pyrrolidine	Allyl bromide	Acetonitrile, 80°C	CH_3CH_2 $\|$ $CH_3(CH_2)_3C-CHO$ $\|$ $CH_2=CH-CH_2$	75	26

	Pyrrolidine	2-Butenyl bromide	Acetonitrile 80°C	CH₃CH₂\\CH₃(CH₂)₃/C—CHO	66	26
	Dicyclo-hexylamine	2-Butenyl bromide	Acetonitrile, 80°C	CH₂=CH—CH—CH₃ | CH₃(CH₂)₃—C—CHO | CH₃—CH=CH—CH₂	17	28
H₅C₆\\CH₃/CH—CHO	Dimethyl-amine	Allyl bromide	Acetonitrile[b]	CH₃ | H₅C₆—C—CHO | CH₂=CH—CH₂	~56	27
	Piperidine	2-Butenyl bromide	Acetonitrile, 80°C	CH₃ | H₅C₆—C—CHO | CH₂=CH—CH—CH₃	85	26
	Dimethyl-amine	Allyl bromide	Acetonitrile[b]	H₅C₆—CH—CHO | CH₂=CH—CH₂	~56	27
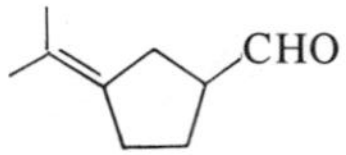 H₅C₆—CH₂—CHO	Pyrrolidine	(E)-2-Butenyl bromide (Z)-2-Butenyl bromide	Acetonitrile, 80°C	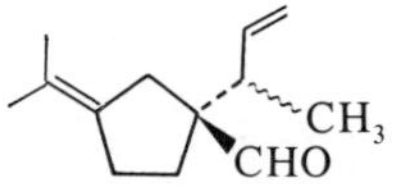 [cyclopentane structure with CH₃, CHO substituents] Stereoisomers 55:45 45:55	75	31

[a] According to the authors the 2-butenyl bromide contained ~15% α-methylallyl bromide.
[b] The N-allylenammonium bromide could be isolated (~90%) at 0°C. It rearranged by heating to yield the corresponding iminium salt in 70% yield.
[c] The N-(2-butenyl)enammonium bromide was isolated.
[d] The enammonium salt was isolated.

Hydrolysis of this salt leads to the aldehyde **25**. In accordance with this mechanism is the observation that *N*-allyl-*N*-methylisobutenylamine (**26**), in the presence of methyl *p*-toluenesulfonate, is transformed via **27** into **28**, which then, after hydrolysis, yields the aldehyde **29** (30) (Scheme 3).

The results of the previously investigated allylations of pyrrolidine, piperidine, morpholine, and dimethylamine enamines of aldehydes are given in Table I. It should be noted that in the reactions with 2-butenyl bromide the products with the inverted allyl chain are formed in all cases. This can be taken as evidence for the occurrence of [3,3]-sigmatropic transformations of the type **23 → 24** (Scheme 2). That the allylation of the enamines with allyl bromide proceeds via the primarily formed enammonium bromides is made evident by the fact that the salts of this type can be isolated by working at low temperature. For example, the ammonium salts **27** and **30–34** (Scheme 4) could be isolated under these conditions. They represent crystalline unstable compounds which slowly rearrange at room temperature into the corresponding iminium salts. This reaction is faster at the melting point temperature or by warming up solutions of the salts in acetonitrile, *N,N*-dimethylformamide, or chloroform. In this way salts **30–32** yield iminium salts with an inverted allyl chain. Hydrolysis of the iminium salts then gives as sole product 2,2,3-trimethyl-4-pentenal (**25**, Scheme 2). The transformations of the salts **27**, **33**, and **34** into the corresponding iminium salts at 60–65°C could also be

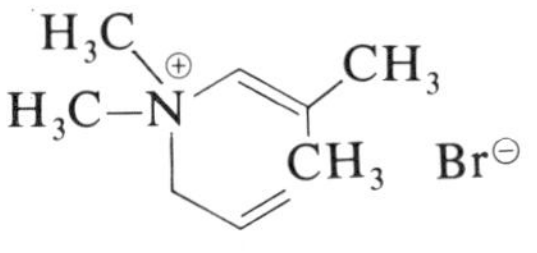

(**27**)[27] 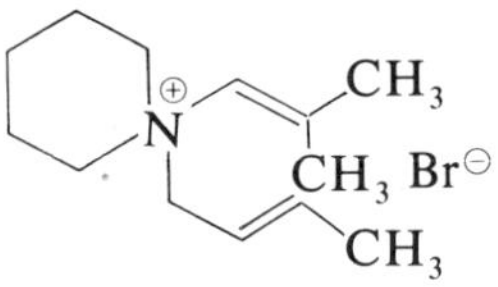   (**30**)[28] 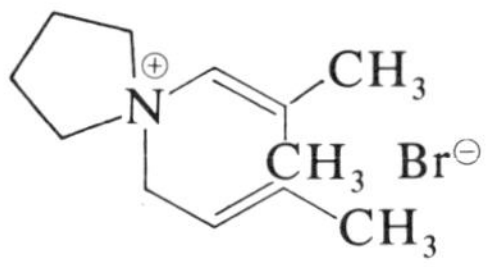   (**31**)[28]

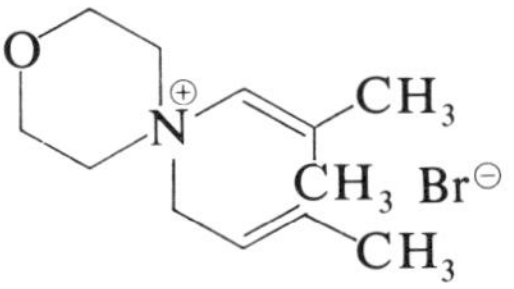

(**32**)[29] 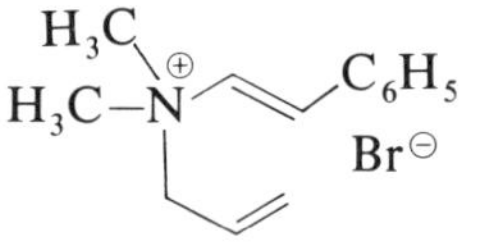   (**33**)[27]

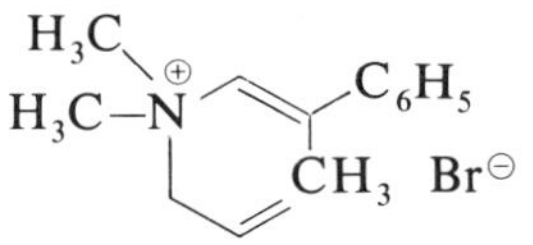

(*E*)-(**34**)[27] 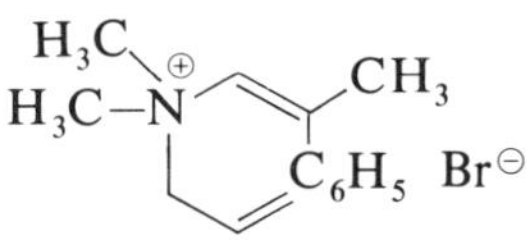   (*Z*)-(**34**)[27]

Scheme 4

II. AZA-COPE REARRANGEMENTS IN AMMONIUM AND IMINIUM SALTS

followed by NMR spectroscopy. The sigmatropic rearrangement **27** bromide to **28** bromide (cf. Scheme 3) seems to be complete within a few hours [exact kinetic data are not given (27b)]. The 3-aza-Cope rearrangement of N-allyl-N-methylisobutenylamine (**26**) (the uncharged species corresponding to the salt **27**) into the imine **35** proceeds only when **26** is heated at 250°C for 1 hr (4). Thus the accelerating influence of the positive charge in the rearrangement **27** bromide → **28** bromide is quite obvious (cf. Section II-C*).

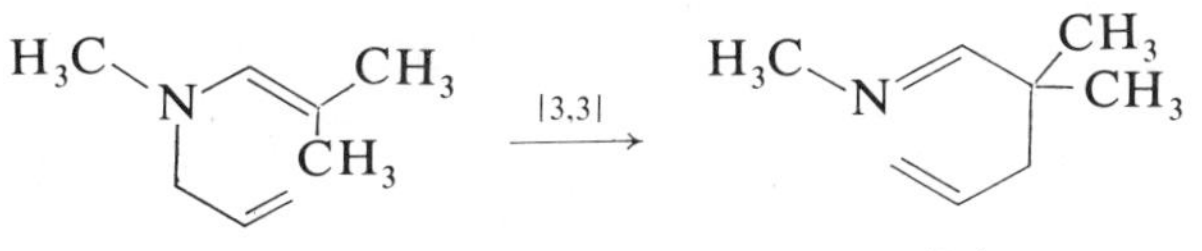

It is apparent from Table I that the N-alkylation of enamines of aldehydes is suppressed when the nitrogen carries bulky alkyl substituents. Here direct C-allylation occurs, giving rise to iminium salts, which are formed in low yields, however. For example, the alkylation of N,N-dicyclohexylisobutenylamine (**36**) with 2-butenyl bromide affords, after hydrolysis, only aldehyde **37** in which the allyl chain is not inverted (Scheme 5) (28).

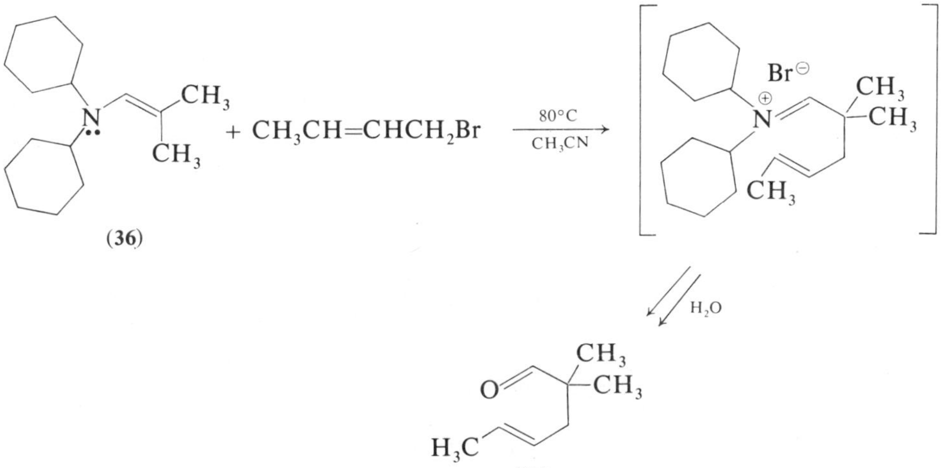

Scheme 5

* The increased steric interactions in the reactant in going from **26** to **27** (geminal methyl groups at the N-atom) cannot be responsible for the strong acceleration of the reaction, since the Cope rearrangement of 3,3-dimethyl-hexa-1,5-diene (**32**) proceeds at a rate which is faster by only a factor of 2 than the rearrangement of 3-methyl-hexa-1,5-diene (gas phase, 249°C) (33).

No uniform results are obtained when the enamines of carbocyclic ketones are allylated (28, 34).* Here, for steric reasons (alkyl chain at C-1), direct *C*-allylation competes effectively with *N*-allylation. For example, the reaction of 1-pyrrolidinocyclopentene (**38**) with 2-butenyl bromide yields, after hydrolysis, a mixture consisting mainly of 2-(2-butenyl)-cyclopentanone (**39**) and minor amounts of 2-(1-methylallyl)cyclopentanone (**40**) (Scheme 6) (34).† Since the direct *C*-allylation of **36** with 2-butenyl bromide yields, after hydrolysis, only aldehyde **37** (Scheme 5) with uninverted allyl chain, it can be concluded that **39** results from direct *C*-allylation. The formation of **40**, however, results from a preceding *N*-allylation with subsequent [3,3]-sigmatropic rearrangement to form the corresponding iminium salt with the inverted allyl chain.‡ This hypothesis is supported by the observation that methylation of **38** (Scheme 6) leads to the formation of 81% iminium salt **41** and 19% enammonium salt **42** (35).

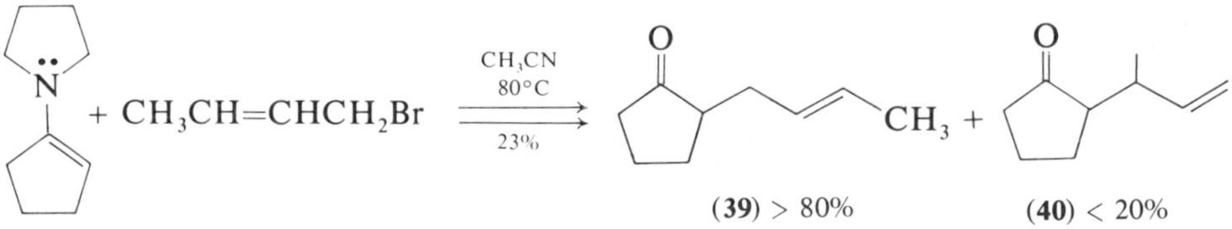

(**38**) (**39**) > 80% (**40**) < 20%

(The same product ratio is observed for 1-morpholinocyclopentene)

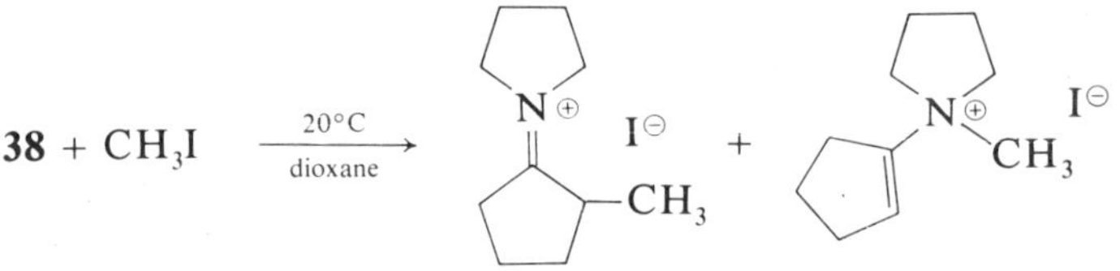

(**41**) 81% (**42**) 19%

Scheme 6

1-Pyrrolidinocyclohexene (**43**) behaves similarly to **38**. Reaction of this enamine with 2-butenyl bromide in boiling chloroform, followed by hydrolysis, results in the formation of a 3 : 2 mixture of 2-(2-butenyl)cyclohexanone (direct *C*-allylation) and 2-(1-methylallyl)cyclohexanone (*N*-allylation followed by

* The allylation of the enamines of open-chain ketones has not yet been closely investigated. Alkylations of those enamines with methyl, ethyl, and sec-butyl iodide as well as benzyl bromide in dioxane lead to *N*- and *C*-alkylated products (35).

† As a rule the allylation of the enamines of cyclic ketones leads also to the formation of diallylated products (34).

‡ It seems unlikely that the formation of **40** is attributable to a direct *C*-allylation with 3-bromo-1-butene which is present in 2-butenyl bromide (cf. footnote a in Table I) since the same bromide reacts with the enamines of aldehydes to give only products with the inverted allyl chain (cf. Scheme 2 and Table I).

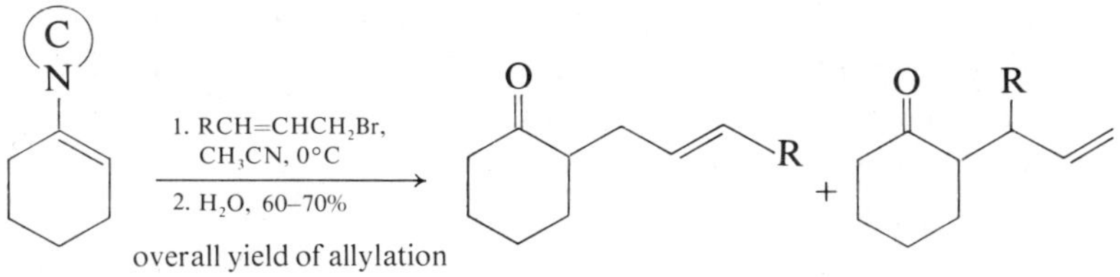

II. AZA-COPE REARRANGEMENTS IN AMMONIUM AND IMINIUM SALTS

(43) C = pyrrolidino
(44) C = piperidyl
(45) C = morpholino

(46) R = CH$_3$
(47) R = C$_6$H$_5$
(direct C-allylation)

(48) R = CH$_3$
(49) R = C$_6$H$_5$
(N-allylation + 3-aza-Cope rearrangement)

Enamine	Relative yield, %			
	46	47	48	49
43	71	87	29	13
44	15	11	85	89
45	10	20	90	80

Scheme 7

rearrangement) (34). Recently this reaction was studied in more detail (36). The results of the reaction of the cyclohexene enamines **43–45** with 2-butenyl bromide and cinnamyl bromide are given in Scheme 7. For synthetic purposes it is important to know that N-allylation followed by the 3-aza-Cope rearrangement is strongly favored when the piperidyl and morpholino enamines **44** and **45** are used. This effect could be attributed to the different charge densities at C-2 of the enamines. The signals in the NMR spectra for H at C-2 and for ^{13}C-2 of **43** appear at higher field than those for H at C-2 and for ^{13}C-2 in **44** and **45** (36) (cf. also the relative thermodynamic stability of methylene cyclopentane versus 1-methylcyclopentene with that of 1-methylcyclohexene versus methylene cyclohexane).

That 3-aza-Cope rearrangements are indeed responsible for the formation of **49** follows from the reaction of 1-(N-cinnamyl-N-ethylamino)cyclohexene (**50**) with methyl iodide. After hydrolysis only **49** and no **47** is obtained (36). The reaction of 1-pyrrolidinocycloheptene and -cyclooctene with 2-butenyl bromide in acetonitrile yielded, after hydrolysis, only the products with the uninverted

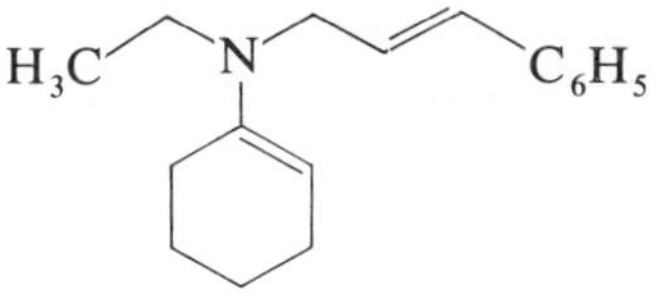

(50)

allyl chain (~50%), that is, the 2-(2-butenyl)alkanones (**34**).* Treatment of 1-pyrrolidinocyclooctene with 2-butenyl bromide at 0°C in ether resulted in the isolation of a salt, to which the structure **51** was assigned (28). Remarkably, this salt could not be rearranged in boiling acetonitrile.*

(**51**)

Interesting products are obtained when diene amines are allylated because in this case 3-aza-Cope rearrangements can be followed by Cope rearrangements. Some results (36) are put forth in Scheme 8. The 6-pyrrolidino diene **52** is alkylated with 2-butenyl bromide as well as with cinnamyl bromide without inversion of the allyl chain, leading, after hydrolysis, to **54** and **55** (cf. the alkylation of **43**, Scheme 7). On the other hand, the 6-morpholino diene **53** reacts with 2-butenyl bromide to give, via N-allylation and 3-aza-Cope rearrangement, the 5-(1-methyl allyl) derivative **57** as the main product and the tautomeric form **56** as a by-product. When **57** is heated at 140°C it rearranges via a Cope reaction to form the 4-(2-butenyl) compound **58**. The diene **53** reacts with cinnamyl bromide to yield, after hydrolysis, the 4-substituted compound **59**, which is the result of N-allylation followed by a 3-aza-Cope rearrangement, and this is in turn followed by a Cope rearrangement. The main product, however, is the bisallylated dione **60**, and a possible pathway by which **60** is formed is given in Scheme 9. The corresponding 6-piperidino and 6-dimethylamino dienes behave similarly to the 6-morpholino diene **53** (36).

The above formulated mechanisms are supported by the observation that the 6-(N-cinnamyl-N-ethylamino) diene **61** with methyl iodide at 0°C yields, via the diene ammonium bromide **62**, finally the enone **59** (Scheme 10) (36).

Reactions of diene amines of type **53** with pentadienyl chlorides would be of great interest because in this case [3,3]-sigmatropic rearrangements (3-aza-Cope and Cope rearrangements) could be accompanied by 5-aza-[5,5] rearrangements [cf. aromatic 5-oxa-[5,5] rearrangements (37)].

Little is known about the stereochemistry of 3-aza-Cope rearrangements. Hill and Gilman (4) showed that the thermal rearrangement of (R)-N-phenyl-N-(1-methylallyl)-2E-phenylpropenylamine (**63**) leads to a 9:1 mixture of the imines E-**64** and Z-**64** with (S)- and (R)-chirality, respectively (Scheme 11). It can be concluded then that the rearrangement—similar to the Claisen

---

*The behavior of the corresponding piperidino and morpholino enamines have not been investigated (cf. Scheme 7).

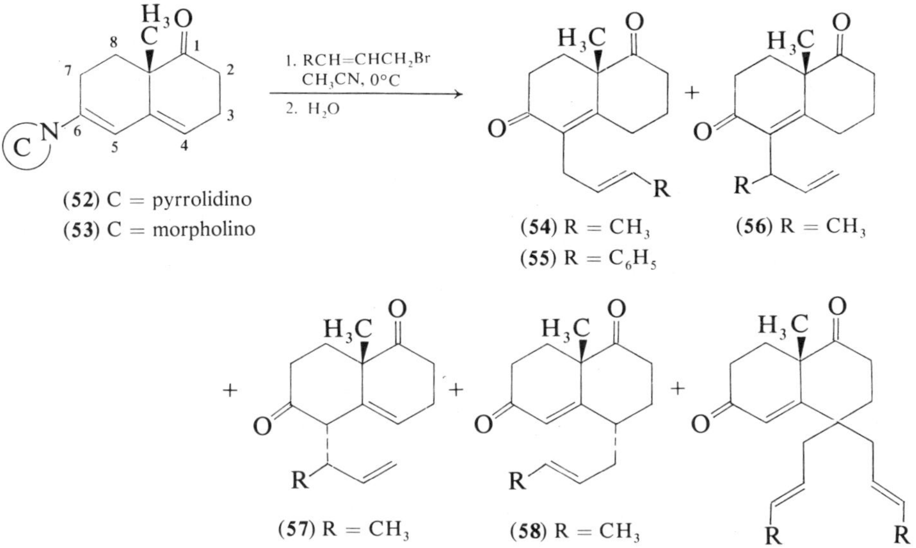

II. AZA-COPE REARRANGEMENTS IN AMMONIUM AND IMINIUM SALTS 669

(52) C = pyrrolidino
(53) C = morpholino

1. RCH=CHCH$_2$Br, CH$_3$CN, 0°C
2. H$_2$O

(54) R = CH$_3$
(55) R = C$_6$H$_5$

(56) R = CH$_3$

(57) R = CH$_3$

(58) R = CH$_3$
(59) R = C$_6$H$_5$

(60) R = C$_6$H$_5$

			Relative yield, %			
Diene amine	54	55	56	57	58	59	60
52	95	100	2	3	—	—	—
53	15	23	19	66	—	21	56

Scheme 8

rearrangement (38)—proceeds mainly via a chairlike but not a boat-like transition state (for further information about the stereochemistry of [3,3]-sigmatropic rearrangements see ref. 39). The stereochemistry of the charge-accelerated 3-aza-Cope rearrangement is not known. It is plausible that a chairlike transition state predominates also in this mode.*

Whereas the reaction of the pyrrolidine and morphiline enamines of cyclic (28, 40) and open-chain ketones (5) with propargyl bromides in acetonitrile leads exclusively to the formation of the corresponding propargyl iminium compounds, the reaction with aldehyde enamines yields, after hydrolysis, the corresponding 2-allenyl aldehydes as the main products (Table II). There is no doubt that these allenyl compounds are the result of 3-aza-Cope rearrangements of the corresponding N-propargyl enammonium compounds, which occur in boiling acetonitrile. The propargylation of isobutyric aldehyde

---

* The experiments discussed in ref. 36 allow no decision (cf. ref. 39).

enamines at 0°C led to the enammonium bromides **65–68** given in Scheme 12. When **65** was heated in acetonitrile at 80°C for 13 hr it rearranged partially to give the corresponding allenyl iminium salt **69** (Scheme 13). Hydrolysis of this salt yielded 15% pure 2,2-dimethyl-3,4-pentadienal (**70**). According to ref. 28,

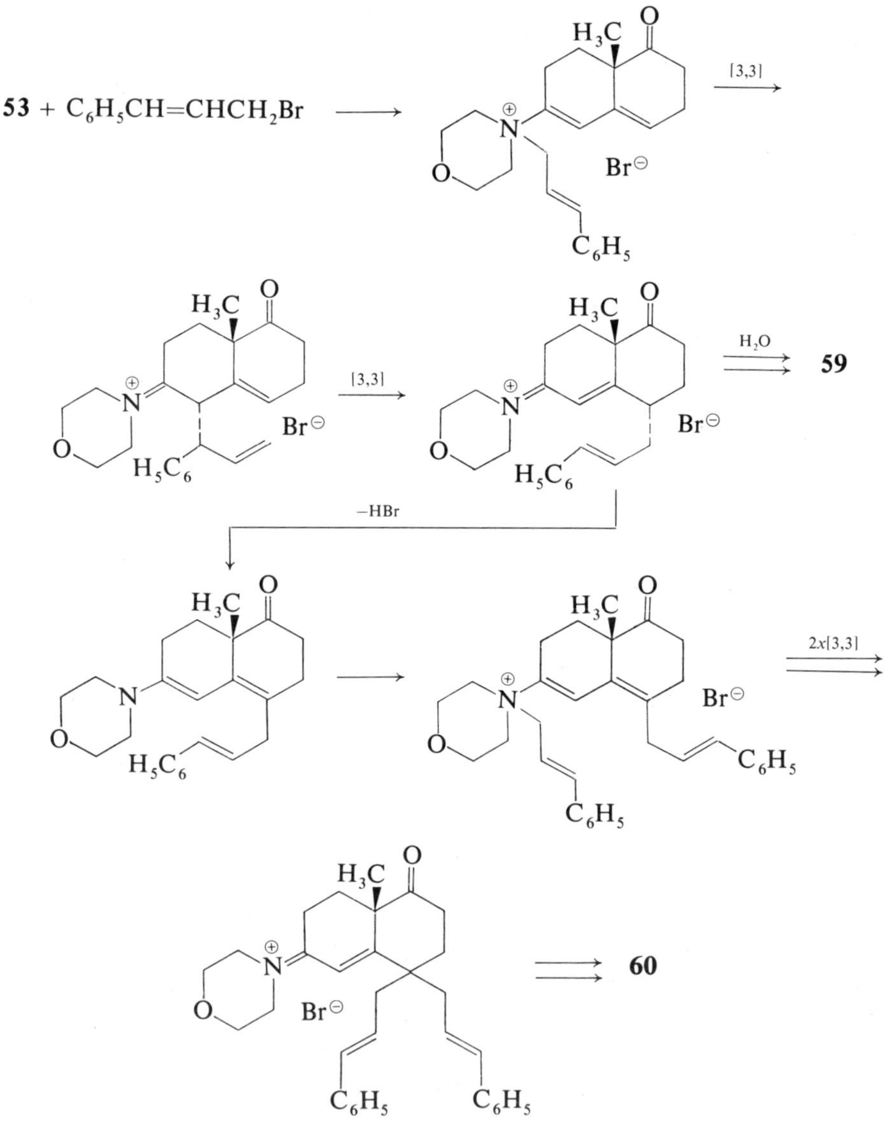

Scheme 9

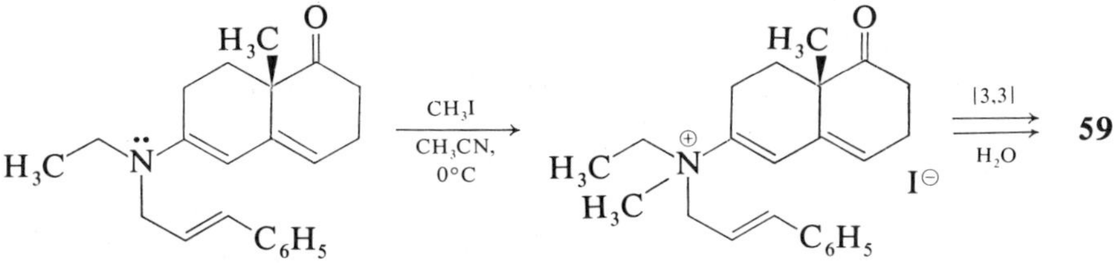

II. AZA-COPE REARRANGEMENTS IN AMMONIUM AND IMINIUM SALTS 671

Scheme 10

the rearrangement of **65** is slower than that of the N-2-butenyl enammonium salts **30** and **31** (Scheme 4). The N-propargyl enammonium salts **66–68** behave as **65** and yield, after heating and hydrolysis, the corresponding allenyl aldehydes (5, 29); that is, **68** forms pure 2,2-dimethyl-3-phenyl-3,4-pentadienal under these conditions. It can be concluded from these experiments that the 2-

Scheme 11

allenyl aldehydes, mentioned in Table II, result from the N-propargyl enammonium salts by a 3-aza-Cope rearrangement, whereas the 2-propargyl aldehydes are obviously formed by a direct C-propargylation (favored by heating) of the corresponding enamines. Thus in order to prepare pure 2-allenyl aldehydes, it is advantageous to make first at 0°C the N-propargyl

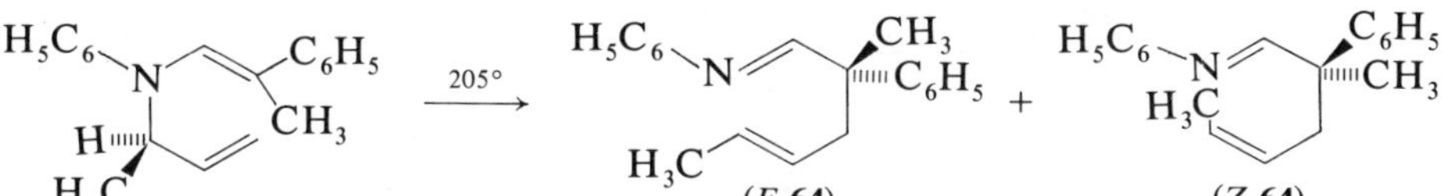

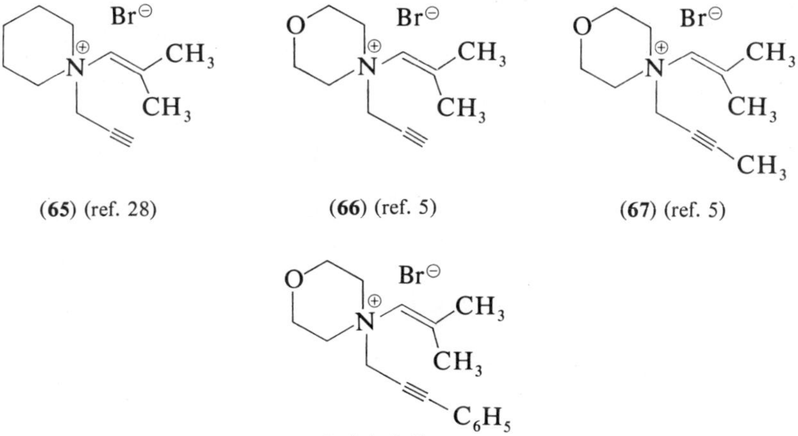

(65) (ref. 28)   (66) (ref. 5)   (67) (ref. 5)

(68) (ref. 5)

Scheme 12

TABLE II

Propargylation of Aldehyde Enamines

Aldehyde	Amine	Propargylating agent	Conditions	α-Allenyl and α-propargyl aldehydes		Yield, %	Ref.
$(CH_3)_2CHCHO$	Pyrrolidine	Propargyl bromide	Acetonitrile, 80°C	$(CH_3)_2C-CHO$ $\|$ $CH_2=C=CH$ 85 : 15	$(CH_3)_2C-CHO$ $\|$ $HC\equiv C-CH_2$	30	23
	Piperidine	Propargyl bromide	Acetonitrile, 80°C[a]		100 : 0	15	28
	Dicyclo-hexylamine	Propargyl bromide	Acetonitrile, 80°C		0 : 100	16	28
	Morpholine[d]	Propargyl bromide	Acetonitrile, 80°C[b]		85 : 15	?	5, 29
	Morpholine	2-Butynyl bromide	Acetonitrile, 80°C[c]	$(CH_3)_2C-CHO$ $\|$ $CH_2=C=CCH_3$ 65 : 35	$(CH_3)_2C-CHO$ $\|$ $CH_3-C\equiv C-CH_2$	?	5, 29

Starting aldehyde	Amine	Bromide	Solvent, Temp	N-Propargylation product	Ratio	C-Propargylation product	Yield (%)	Yield (%)
$CH_3CH_2\!\!>\!\!CH\text{–}CHO$ $CH_3(CH_2)_3$	Morpholine	3-Phenyl-propargyl bromide	Acetonitrile, 80°C [c]	$(CH_3)_2C\text{–}CHO$ $CH_2\!=\!C\!=\!C\text{–}C_6H_5$ CH_3CH_2		$(CH_3)_2C\text{–}CHO$ $H_5C_6\text{–}C\!\equiv\!C\text{–}CH_2$ CH_3CH_2	?	5
	Pyrrolidine	Propargyl bromide	Acetonitrile, 80°C	$CH_3(CH_2)_3\text{–}C\text{–}CHO$ $CH_2\!=\!C\!=\!CH$	53 : 47	$CH_3(CH_2)_3\text{–}C\text{–}CHO$ $HC\!\equiv\!C\text{–}CH_2$	20	26
	Piperidine	Propargyl bromide	Acetonitrile, 80°C		86 : 14		48	28
	Dicyclo-hexylamine	Propargyl bromide	Acetonitrile, 80°C		0 : 100		12	28

[a] The enammonium salt 65 could be isolated when the reaction was carried out at 10°C. The thermal rearrangement of the enammonium salt was not complete after 13 hr.
[b] The enammonium salt was isolated at room temperature. The thermal rearrangement of this salt yielded, after hydrolysis, pure 2,2-dimethyl-3,4-pentadienal.
[c] The enammonium salt was isolated.
[d] Similar results were obtained with piperidine as amine. N,N-Dimethylisobutenylamine (22) was N-propargylated at 0°C (5). At 80°C 22 reacted with 3-phenylpropargyl bromide to give (up to 22%) C-propargylation.

enammonium salts, which then can be rearranged at 80°C. A rearrangement in the reverse direction, that is, of the type **69** → **65** (Scheme 13), has so far not been observed.

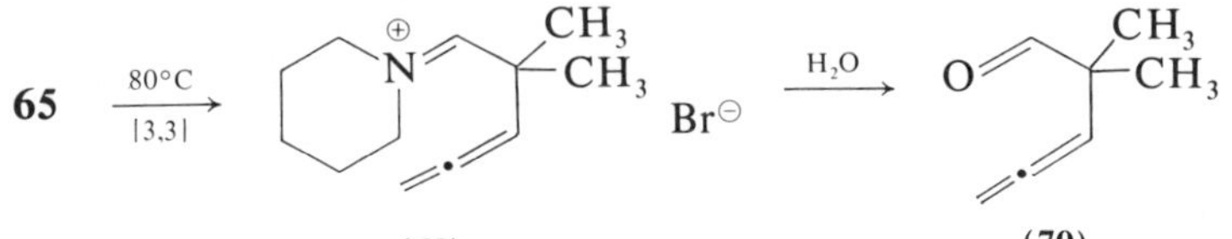

Scheme 13

In conclusion it may be mentioned that Corbier and Cresson (5) succeeded in the preparation of the *N*-methyl-*N*-propargylisobutenylamines **71–73** (Scheme 14). The enamines **71** and **72** could be rearranged, though at high temperature, to the imines **74** and **75**, respectively. No rearrangement has been observed when the enamine **73** was heated at 340°C. Thus in the propargyl–allenyl rearrangements reported a strong acceleration is observed in the charged system.

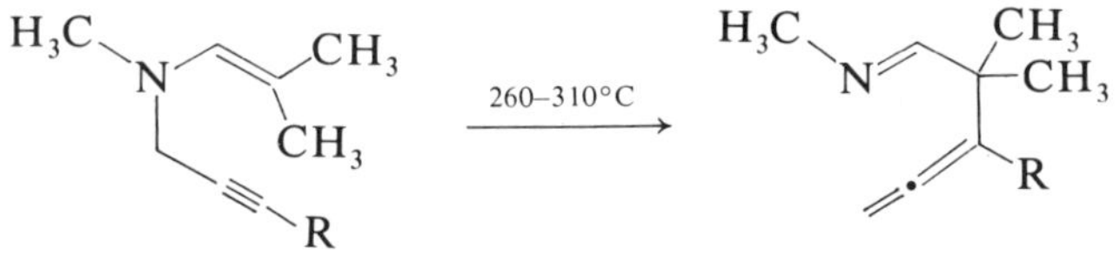

(71) R = H
(72) R = CH_3
(73) R = C_6H_5

(74) R = H
(75) R = CH_3

Scheme 14

2. Acylation of Enamines

Hickmott and Hargreaves (41) were able to isolate, after hydrolysis, bicyclo[3.3.1]nona-2,9-dione (**82**, R = H) in yields of up to 45% from the reaction between 1-morpholinocyclohexene (**45**) and acryloyl chloride (**76**, R = H) in boiling benzene (Scheme 15). Under the same conditions **45** reacted with cinnamoyl chloride to yield the 4-phenyl derivative **82** (R = C_6H_5).*
Further transformations of this type are given in Table III. The *N*-acyl enammonium chlorides **77** (R = H, C_6H_5), given in Scheme 15, could be isolated at 5°C. When they were heated in benzene solution the bicyclic diones **82** were likewise formed. Thus the first reaction step in the formation of **82**

---

* C-acylated products and consecutive products of these were observed as by-products (41–43).

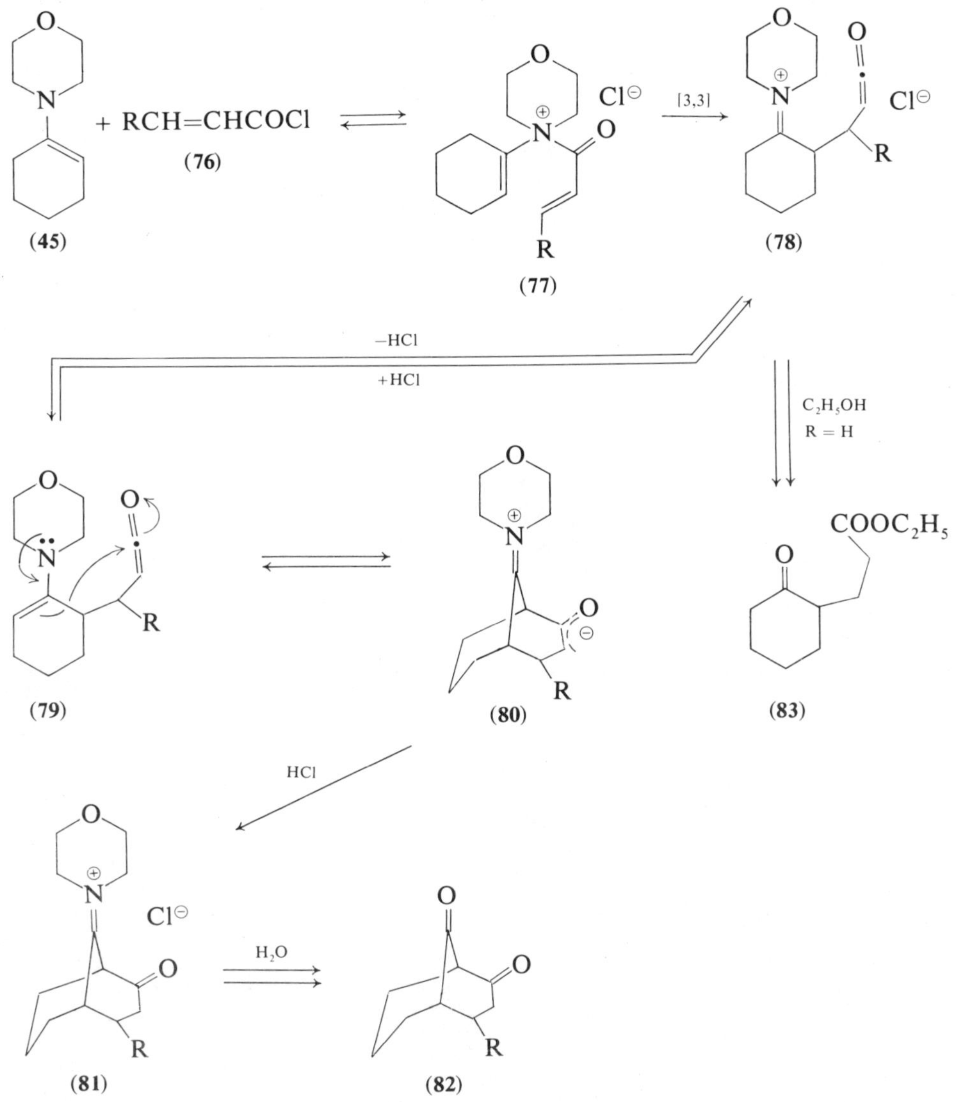

II. AZA-COPE REARRANGEMENTS IN AMMONIUM AND IMINIUM SALTS

Scheme 15

seems to be a 3-aza-Cope rearrangement, that is, **77** → **78**, which leads to the ketene **78** (41). The ketene **78** (R = H) could be trapped with ethanol (i.e., the ester **83** was isolated when **77** was heated in ethanol solution). The further transformations of **78** resulting in the formation of **81** represent well-documented enamine reactions (see refs. 22, 44, and 45). The reaction of the

TABLE III
Acylation of Enamines of Ketones

Ketone	Amine	Acylating agent	Conditions	Product	Yield, %	Ref.
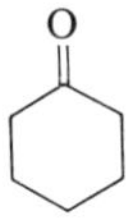 cyclohexanone	Morpholine	Acryloyl chloride	Benzene, refl.	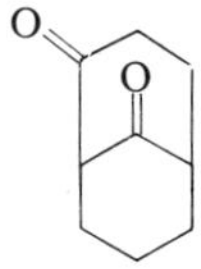 bicyclic diketone	45	41
cyclohexanone	Morpholine	Cinnamoyl chloride	Benzene, refl.	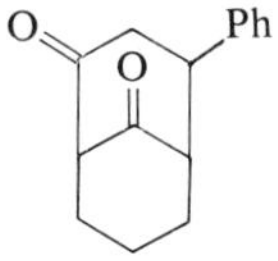 bicyclic diketone with Ph		41
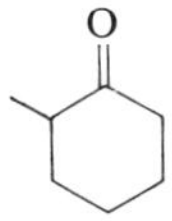 2-methylcyclohexanone	Morpholine	Acryloyl chloride	Benzene, refl.	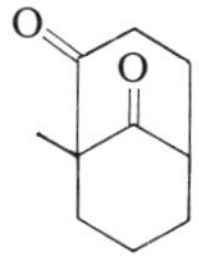 bicyclic diketone		41
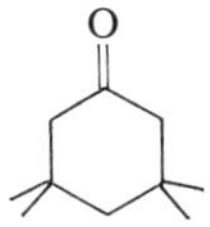 3,3,5,5-tetramethylcyclohexanone [?]	Morpholine	Acryloyl chloride	Benzene, refl.	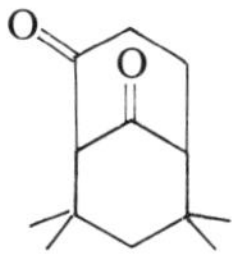 bicyclic diketone		41
2-methyl-3-pentanone	Morpholine	Acryloyl chloride	Benzene, refl.	cyclohexanedione	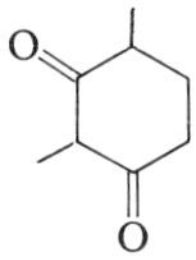 a	42
	Morpholine	Crotonyl chloride	Benzene, refl.	cyclohexanedione	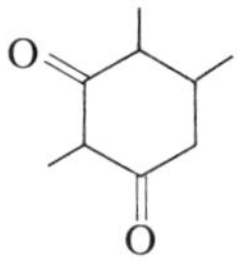 a	42
	Morpholine	Methacryloyl chloride	Benzene, refl.	cyclohexanedione	a	42
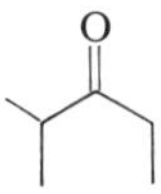 2-methyl-3-pentanone	Morpholine	Acryloyl chloride	Benzene, refl.	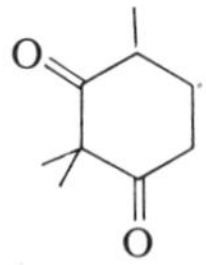 cyclohexanedione		42

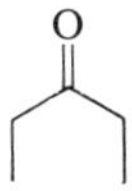

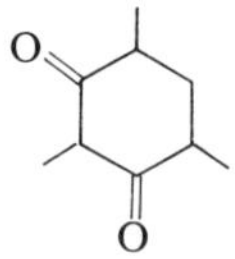

TABLE III (cont.)

Ketone	Amine	Acylating agent	Conditions	Product	Yield, %	Ref.
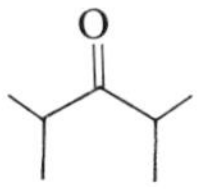 diisopropyl ketone	Dimethyl-amine	Acryloyl chloride	Benzene, refl.	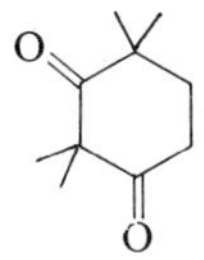 (cyclohexanedione derivative)		42
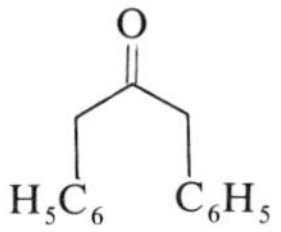 PhCH$_2$COCH$_2$CH$_2$Ph-type ketone	Morpholine	Acryloyl chloride	Benzene, refl.	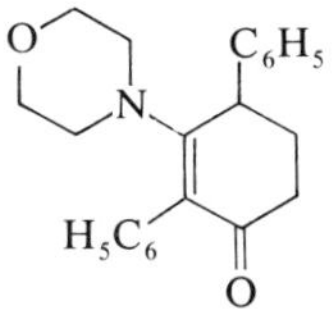 (substituted cyclohexenone)	b	42
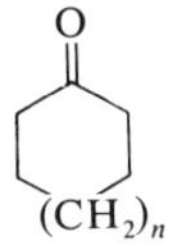 cyclic ketone (CH$_2$)$_n$, $n=2,3,4,5,7$	Morpholine	Acryloyl chloride	Benzene, refl.	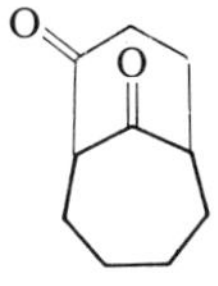 (bicyclic ketone)		42
$n=2$				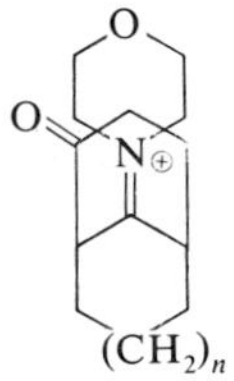	20	
3					52	
4					62	
5					82	
7					63	
7	Pyrrolidine	Acryloyl chloride	Benzene, refl.	(iminium salt, (CH$_2$)$_n$)		
7	Morpholine	Methacryloyl chloride	Benzene, refl.	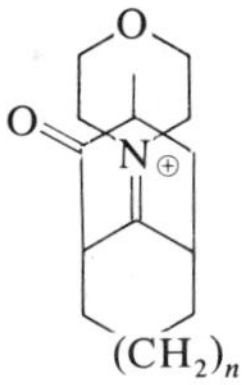 (iminium salt, (CH$_2$)$_n$)	44	43
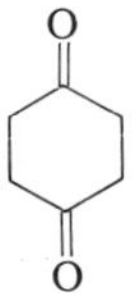 1,4-cyclohexanedione	Morpholine	Acryloyl chloride	Benzene, refl.	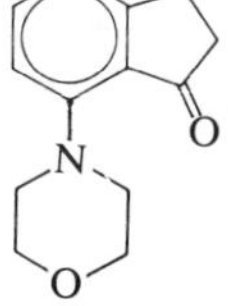 (morpholino indanone)	21	47
	Morpholine	Methacryloyl chloride	Benzene, refl.	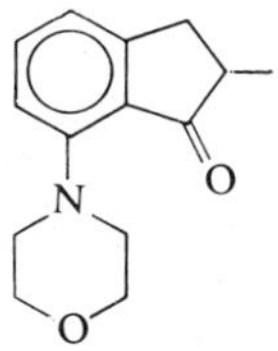 (methyl morpholino indanone)	10	47

TABLE III (cont.)

Ketone	Amine	Acylating agent	Conditions	Product	Yield, %	Ref.
	Morpholine	Crotonoyl chloride	Benzene, refl.	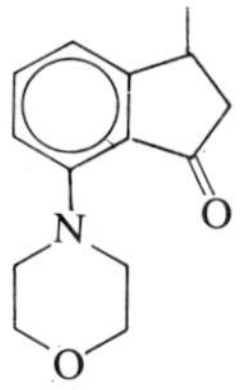	9	47
	Morpholine	Cinnamyloyl chloride	Benzene, refl.	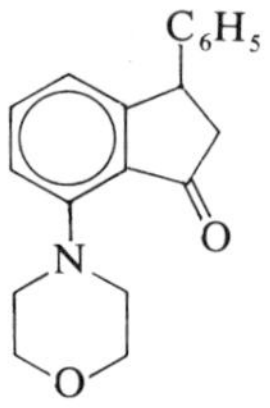	11	47
	Morpholine	β-Chloracryloyl chloride	Benzene, refl.	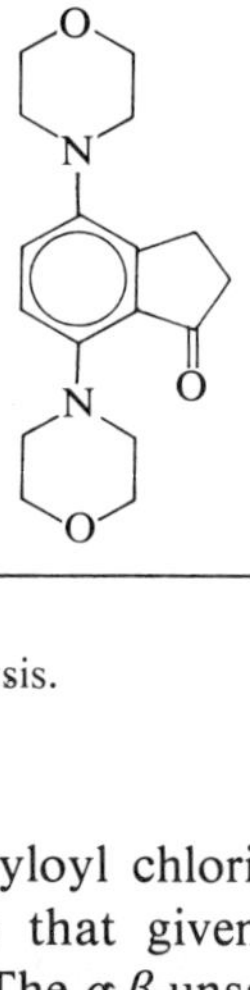	12	47

[a] The intermediate product could be isolated.
[b] The second morpholine moiety could not be removed by hydrolysis.

enamines of open-chain ketones with substituted acryloyl chlorides yield—following a reaction sequence that is analogous to that given in Scheme 15—substituted cyclohexa-1,3-diones (see Table III). The α,β-unsaturated acid chloride can bear an alkyl or phenyl substituent in the α- or β-position. No reaction is observed with β,β-dimethylacrylic acid chloride (senecioic acid chloride), because in this case the rearrangement **77 → 78** is hindered probably for steric reasons, so that the direct C-acylation predominates (46).

Hickmott et al. (47) observed that heating of 1,4-dimorpholino-1,3-cyclohexadiene (**84**) with various acryloyl chlorides yielded 7-morpholino-1-indanones **85** (10–20%), 1,4-dimorpholinobenzene (10–25%), and the corresponding acrylic acid morpholides (28–37%) (Scheme 16). The formation of **85** is thought to follow the mechanism given in Scheme 17. Here again, the first

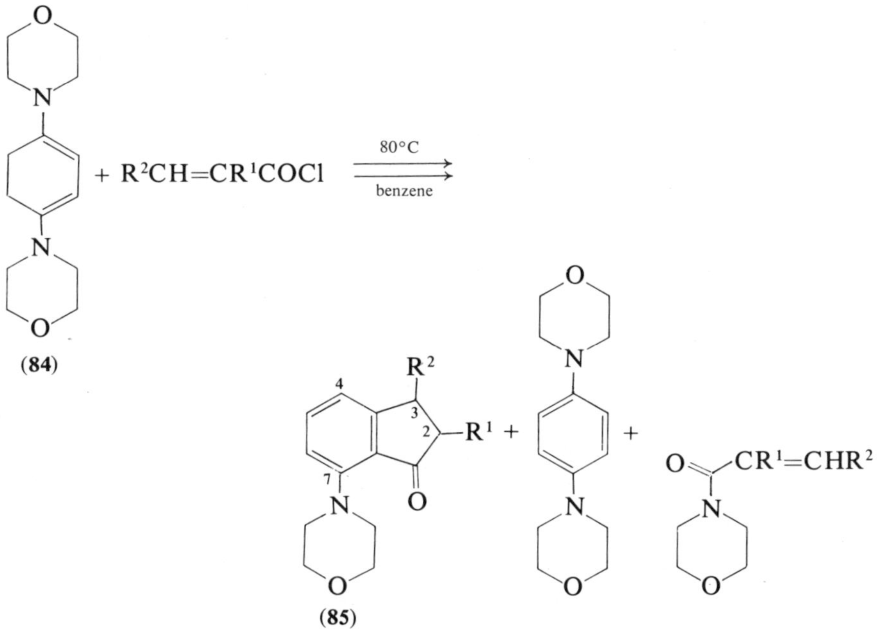

II. AZA-COPE REARRANGEMENTS IN AMMONIUM AND IMINIUM SALTS

[Scheme 16 showing compound (84) + R²CH=CR¹COCl at 80°C in benzene giving compound (85) and related products]

$R^1 = R^2 = H$; $R^1 = CH_3$, $R^2 = H$; $R^1 = H$, $R^2 = CH_3$; $R^1 = H$, $R^2 = C_6H_5$

Scheme 16

step represents a N-acylation followed by a [3,3]-sigmatropic ammonium–iminium rearrangement (**86 → 87**). The iminium compounds cyclize to give the dimorpholinodihydroindanones **88**, which can give **85** by an elimination of the morpholine moiety in position 4. The bisenamine **84** undergoes a similar reaction with β-chloroacryloyl chloride (**76**, $R^1 = H$, $R^2 = Cl$). In this case 4,7-dimorpholino-1-indanone (12%) is obtained by elimination of a second mole of hydrogen chloride from the intermediate **88** ($R^1 = H$, $R^2 = Cl$) (47).

It is assumed by Hickmott and Sheppard (48) that [3,3]-ammonium–iminium rearrangements also play a role in the acylation of imines with acryloyl chlorides to form 2-piperidones. For example, the reaction of acryloyl chloride (**76**, R = H) with N-(2-methylcyclohexylidene)cyclohexylamine (**89**, $R^1 = C_6H_{11}$, $R^2 = CH_3$) in boiling benzene results in the formation of 38% of 2-oxo-$\Delta^{8,8a}$-octahydroquinoline **92** ($R^1 = C_6H_{11}$, $R^2 = CH_3$, $R^3 = H$), 27% of the enamide **90** ($R^1 = C_6H_{11}$, $R^2 = CH_3$, $R^3 = H$), and hydrogen chloride.* The formation of **92** is thought to occur according to the mechanism given in Scheme 18. When $R^2 = H$, octahydroquinolin-2-ones with the double bond

---

* In other examples the yields of **92** are between 10 and 25%.

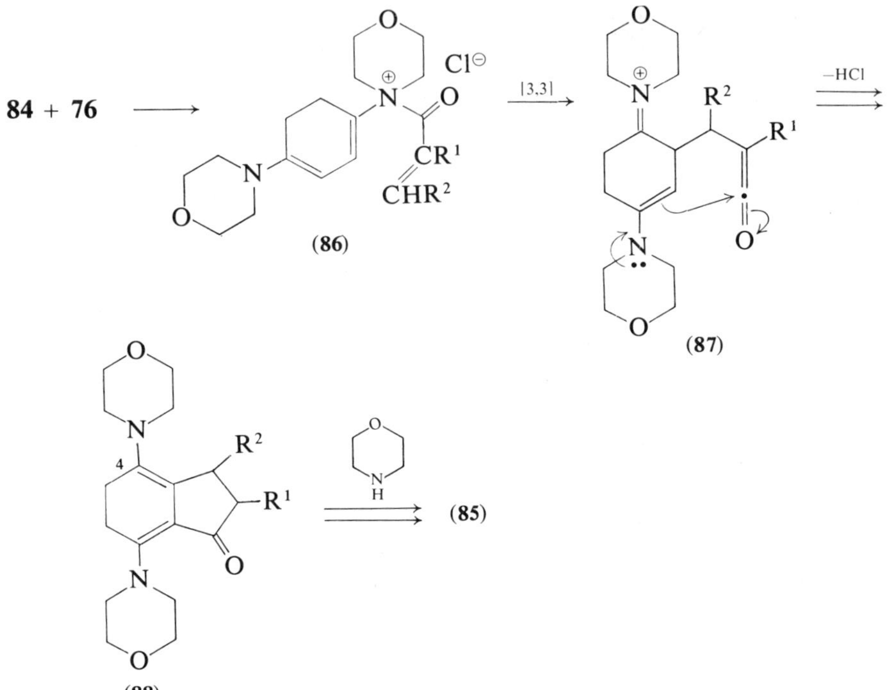

3,3-REARRANGEMENTS OF IMINIUM SALTS

84 + 76 ⟶

Scheme 17

between C-4a and C-8a are obtained. The proposed mechanism implies that loss of the proton in the protonated enamide **90**, induced by any base present in the system, must be slower than the [3,3]-sigmatropic rearrangement leading to **91**.*

---

* In a control experiment it was shown (48) that the enamide **93** (≡ **90**, $R^1 = C_6H_5CH_2$, $R^2 =$ H, $R^3 = CH_3$) in boiling hydrochloric acid does not form the expected 2-oxo-octahydroquinoline **94** but yields the corresponding crotonamide. This result does not invalidate the proposed

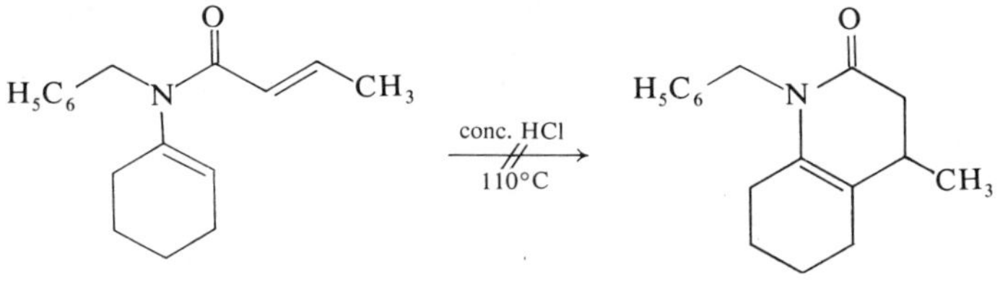

mechanism (Scheme 18) because water is a stronger base than an amide group. Perhaps the rearrangement **93** → **94** could be induced by boiling the enamide with hydrogen chloride in benzene or by treating it with a strong Lewis acid.

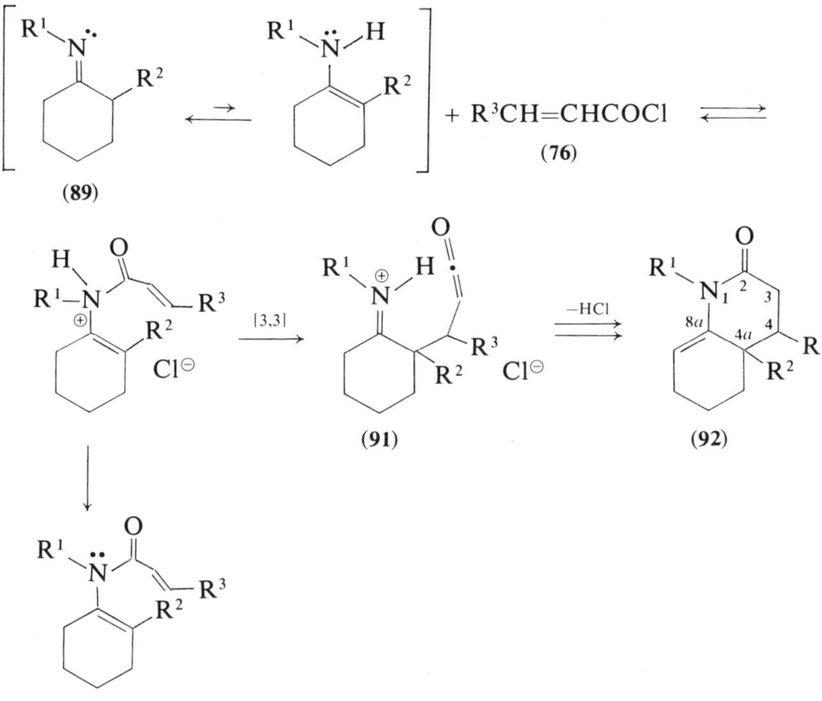

II. AZA-COPE REARRANGEMENTS IN AMMONIUM AND IMINIUM SALTS

Scheme 18

The reaction of cinnamoyl chloride with the imines **89** does not furnish the corresponding octahydroquinolin-2-ones **92** but results in the formation of the enamides **90**. A similar reaction was observed when *N*-(3-pentylidene)-benzylamine was treated with acryloyl chloride, this time leading to the corresponding 2-piperidone (35%) (48a).

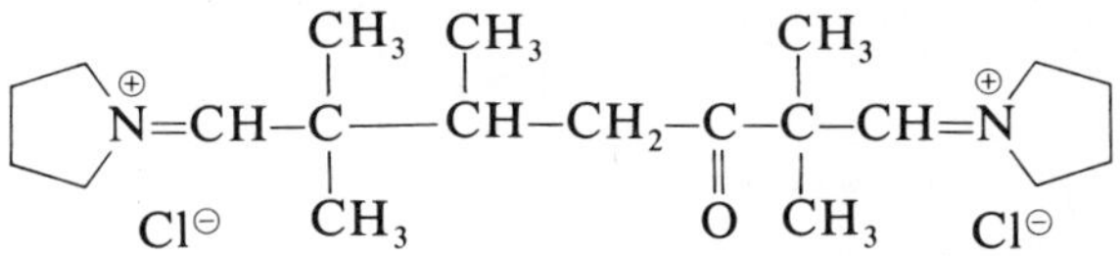

It has not been clarified whether 3-aza-Cope rearrangements of acryloyl enammonium compound are involved in the reactions of the piperidine, morpholine, pyrrolidine, or diisopropyl enamines of isobutyric aldehyde with acryloyl chlorides (49). These form the corresponding acrylic acid amides as main products (30–60%). They could play a role in the complex reaction of acryloyl chloride with 2 molar equivalents of 2-methyl-1-piperidylpropene, which results in the formation of the intermediate compound **95**. Hydrolysis of this salt yields 2,2,6,6-tetramethyl-3-oxoheptanedial in 16% yield (cf. ref. 49).

In conclusion, it follows that 3-aza-Cope rearrangements have been observed only in acryloyl enamides containing a quaternary or protonated nitrogen atom. That such rearrangements do not occur in the acryloyl enamides themselves is plausible if one compares the energy content of reactant and product: these rearrangements should be strongly endothermic. The reverse situation is true for the protonated or quaternized acryloyl enamides, because the amide resonance is lacking in the reactant.*

The reactions of acryloyl chlorides with enamines are of a rather complex nature. For this reason they have only limited synthetic value.

3. *Aromatic 3-Aza-Cope Rearrangements in Anilinium Salts*

Until now 3-aza-Cope rearrangements have been observed only with *N*-allyl-1-naphthylamine and *N*-allyl-2-naphthylamine (50). These compounds rearrange at ~260°C to give 2-allyl-1-naphthylamine and 1-allyl-2-naphthylamine, respectively, in good yields. *N*-Allylpyrroles are transformed into *C*-Allylpyrroles at 400–600°C in part via 3-aza-Cope rearrangements (51). In contrast, no 3-aza-Cope rearrangements have been achieved with *N*-allylanilines when heated at 275°C or when subjected to flash pyrolysis at 700° (52; cf. ref. 53 and literature cited therein).† The isolable products were anilines and compounds derived from the fragmentation of the allyl chain.

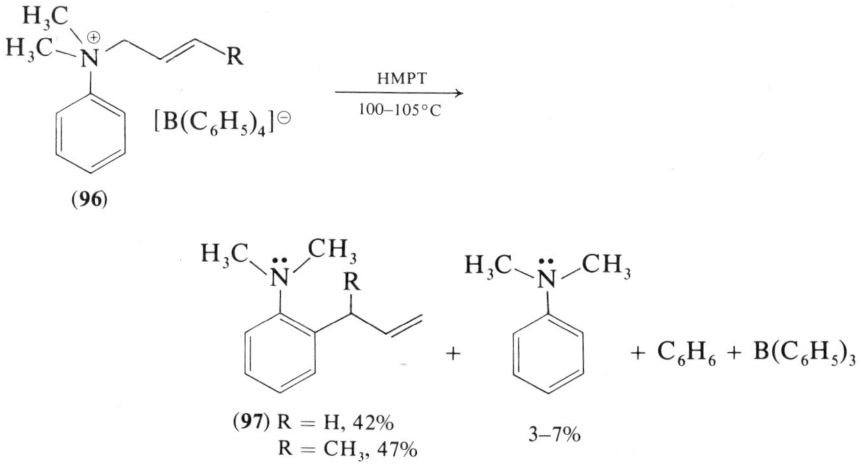

(96) → HMPT, 100–105°C →

(97) R = H, 42%
R = CH$_3$, 47%

+ 3–7% + C$_6$H$_6$ + B(C$_6$H$_5$)$_3$

Scheme 19

---

* The initiation of the rearrangement of acryloyl enamides by Lewis acids has, so far as we know, not yet been investigated.

† *N*-(1,1-Dimethylallyl)anilines rearrange at ~200°C to yield 2-(3,3-dimethylallyl)anilines (54).

II. AZA-COPE REARRANGEMENTS IN AMMONIUM AND IMINIUM SALTS

On the other hand, charge-accelerated 3-aza-Cope rearrangements could be realized in anilinium ions (53). When N-allyl- or N-(2-butenyl)-N,N-dimethylanilinium tetraphenylborate (**96**, R = H, CH$_3$; Scheme 19) is heated in hexamethylphosphoramide (HMPT) at ~100°C a smooth rearrangement occurs to yield the 2-allylated anilines **97** (R = H, CH$_3$). Under these conditions the butenyl salt **96** (R = CH$_3$) gives exclusively the aniline **97** (R = CH$_3$) with the inverted allyl chain. This is an accordance with a [3,3]-sigmatropic process. The rearrangement of the cinnamyl salt **96** (R = C$_6$H$_5$) is less conclusive. Besides the product of a 3-aza-Cope rearrangement [i.e., 2-(1-phenylallyl)-N,N-dimethylaniline] 2- and 4-cinnamylaniline and N,N-dimethylaniline are also formed in comparable amounts. The latter compounds are thought to be products of fragmenting and recombination reactions, which may occur in a complex, built up from N,N-dimethylaniline and the cinnamyl cation (53) (cf. section II-C).

It can be assumed that the transformations of N-allylanilines **98** in the presence of polyphosphoric acid at 200°C (55) or concentrated hydrochloric acid at 180°C (56) include also charge-accelerated 3-aza-Cope rearrangements (Scheme 20). In this case the indolines **100** and the indoles **101** are formed via the 2-allylated anilines **99**.* Similar mechanisms appear to be valid for the rearrangements of N-(2-chloro- and 2-bromoallyl)anilines into indoles, induced by BF$_3$ (57), HF (58), or polyphosphoric acid (59, 60).

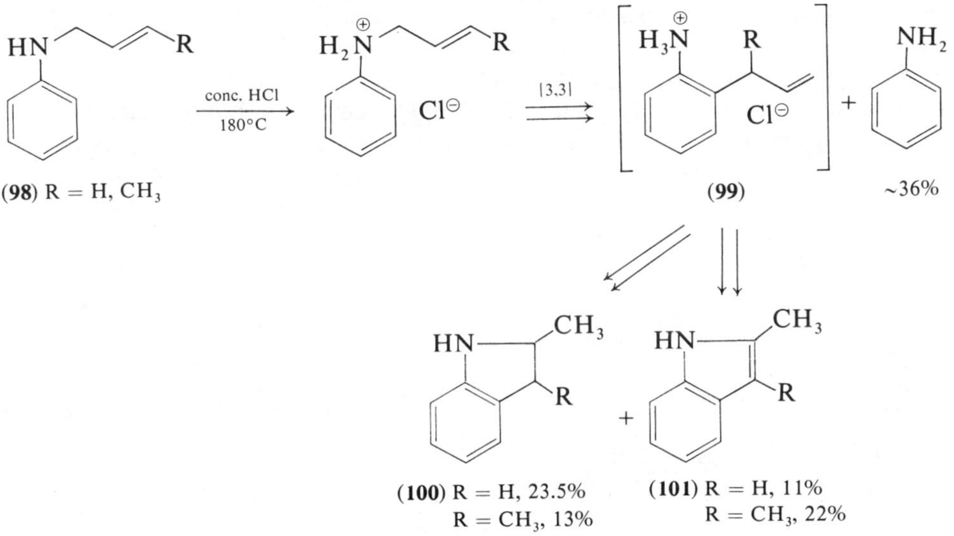

Scheme 20

*The reaction of N-allylaniline (**98**, R = H) produces also 2-methyl-3-propylindole (**100**, R = C$_3$H$_7$) (56).

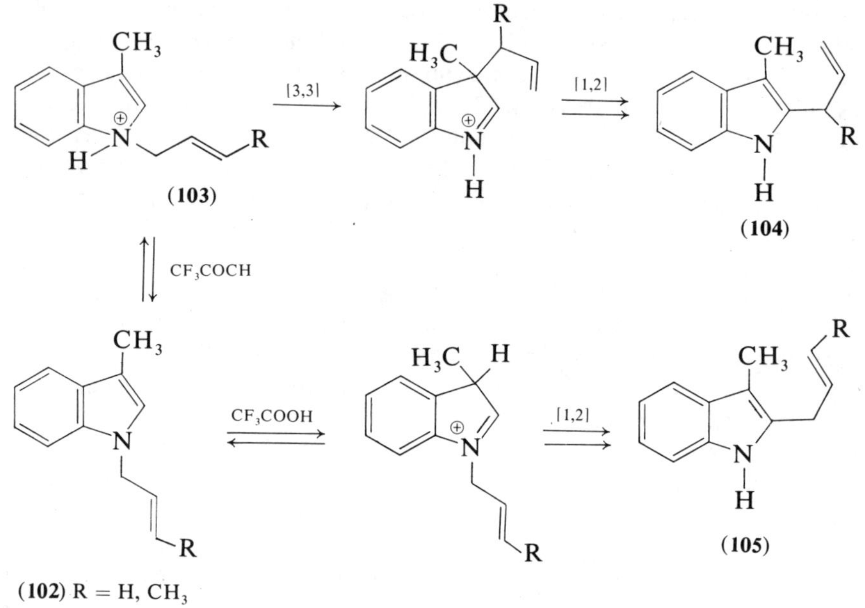

Scheme 21

The intramolecular transformation of *N*-allylated 3-alkylindoles **102** into 3-alkylindoles allylated in the 2 position (**104**) under the influence of Lewis acids, such as BF_3, $SnCl_4$, or $AlCl_3$ in hexane (61), may take place via 3-aza-Cope rearrangements of the indolium ion **103** (Scheme 21). For example, the rearrangement of *N*-(2-butenyl)-3-methylindole (**102**, R = CH_3) in the presence of trifluoroacetic acid leads to a mixture of 56% of 2-(1-methylallyl)-3-methylindole (**104**, R = CH_3) and 22% of 2-(2-butenyl)-3-methylindole (**105**, R = CH_3).

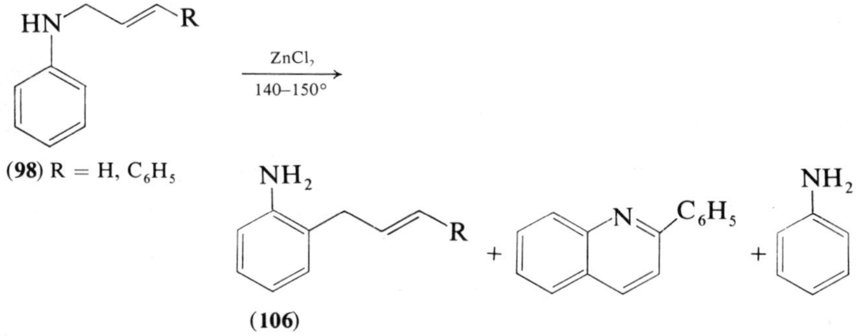

Scheme 22

II. AZA-COPE REARRANGEMENTS IN AMMONIUM AND IMINIUM SALTS

In 1957 Hurd and Jenkins (62) reported that *N*-allylaniline (**98**, R = H) could be rearranged into 2-allylaniline (**106**, R = H; 42%) in the presence of anhydrous $ZnCl_2$ in boiling xylene (Scheme 22). Since the reaction of *N*-cinnamylaniline (**98**, R = C_6H_5) in bromobenzene with $ZnCl_2$ at 150°C led to the formation of 2-cinnamylaniline (**106**, R = C_6H_5; 16%) and 2-phenylquinoline (16%) as well as small amounts of aniline, it was concluded that **106** might be formed by a bimolecular mechanism (62). In the meantime it has been demonstrated that such transformations, as a rule, occur via charge-accelerated 3-aza-Cope rearrangements (53). Thus the rearrangement of *N*-(2-butenyl)-2-methylaniline (**107**, R = CH_3) in the presence of $ZnCl_2$ in xylene results in the formation of 2-(1-methylallyl)-6-methylaniline (**108**, R = CH_3; 55%). Besides this compound 4-(2-butenyl)-2-methylaniline (**109**, R = CH_3; 8.5%), the indoline **110** (R = CH_3) and small amounts of 2-(*E*- and *Z*-1-methylpropenyl)-6-methylaniline (**111**, R = CH_3) are formed as subsequent products from **108** (R = CH_3) (Scheme 23). The amount of **110** was increased when the concentration of $ZnCl_2$ was increased or when the reaction time was prolonged. On the other hand, **108** was transformed into a mixture of **110** and

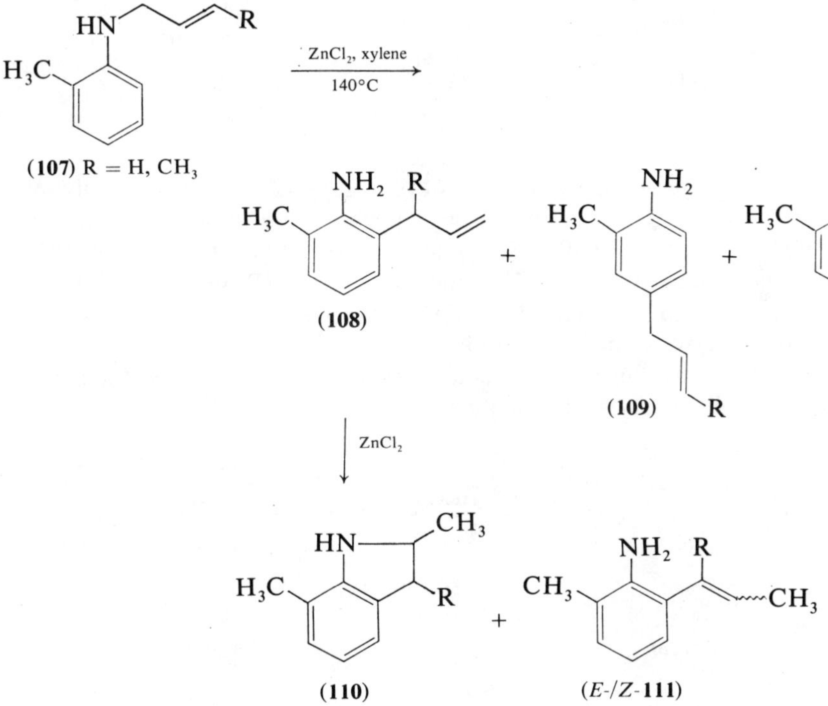

Scheme 23

E-/Z-**111** under the conditions of the reaction. Similar results were obtained when N-allyl-2-methylaniline (**107**, R = H) was treated with $ZnCl_2$ at 140°C. In this case the 2-allylated aniline **108** (R = H; 57%), the 4-allylated aniline **109** (R = H; 7%), and the indoline **110** (R = H; 5.5%) were obtained. The propenylanilines **111** (R = H) could not be detected.

Likewise the tertiary N-allyl-N-methylanilines **112** (R = H, CH_3) undergo rearrangements to give the 2-allyl-N-methylanilines **113** (R = H, CH_3) in the presence of $ZnCl_2$ in xylene at 140°C (Scheme 24) (53).

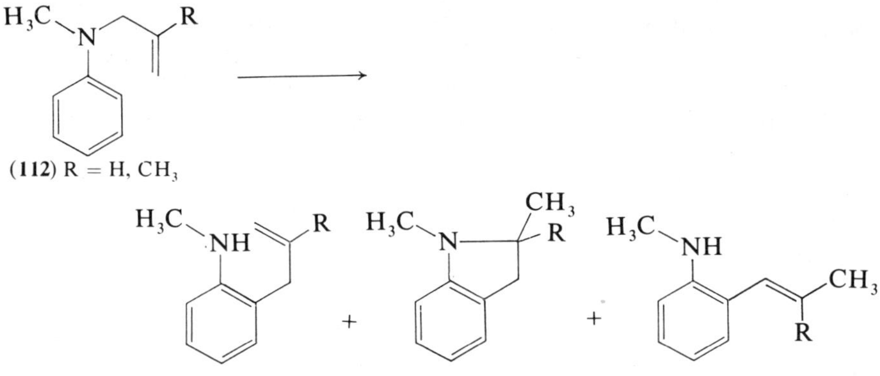

(**112**) R = H, CH_3

(**113**) + (**114**) + (**115**)

Scheme 24

Whereas N-allyl-N-methylaniline (**112**, R = H) yielded 2-allyl-N-methylaniline (**113**, R = H; 70%) in high purity under these conditions, the methylallyl compound **112** (R = CH_3) gave a mixture of 2-(2-methylallyl)-N-methylaniline (**113**, R = CH_3; 23%), the indoline **114** (R = CH_3; 37%), and the isobutenyl compound **115** (R = CH_3; 10.5%). The two last-mentioned compounds are also formed when **113** (R = CH_3) is heated in the presence of $ZnCl_2$ in xylene. The analogous products appear after the rearrangement of **112** (R = H) only in very small quantities.

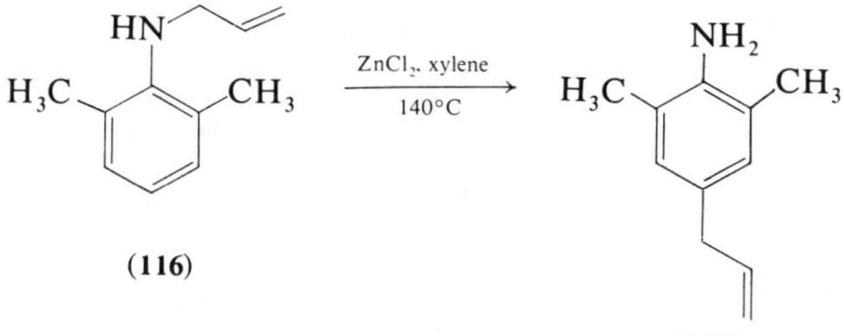

(**116**) →[$ZnCl_2$, xylene, 140°C] (**117**)

Scheme 25

II. AZA-COPE REARRANGEMENTS IN AMMONIUM AND IMINIUM SALTS

N-Allyl-2,6-dimethylaniline (**116**), in the presence of $ZnCl_2$ in xylene at 140°C, afforded 4-allyl-2,6-dimethylaniline (**117**) in 73% yield (Scheme 25) (63). The corresponding *N*-benzyl-2,6-dimethylaniline was recovered unchanged under these conditions (63).

Whereas the rearrangement of *N*-(2-butenyl)-2-methylaniline (**107**, R = CH_3; Scheme 23) in the presence of $ZnCl_2$, as mentioned, yields exclusively the 2-allylated aniline with the inverted allyl chain **108** (R = CH_3), the transformation of *N*-(3-methyl-2-butenyl)aniline (**118**, R = H) with $ZnCl_2$ leads to the formation of both 2- and 4-allylated anilines **119** and **120** (R = H) which possess the uninverted allyl chain (Scheme 26) (64). Similar results were obtained when *N*,*N*-bis-(3-methyl-2-butenyl)aniline (**118**, R = $(CH_3)_2C=CHCH_2$) was treated with $ZnCl_2$.*

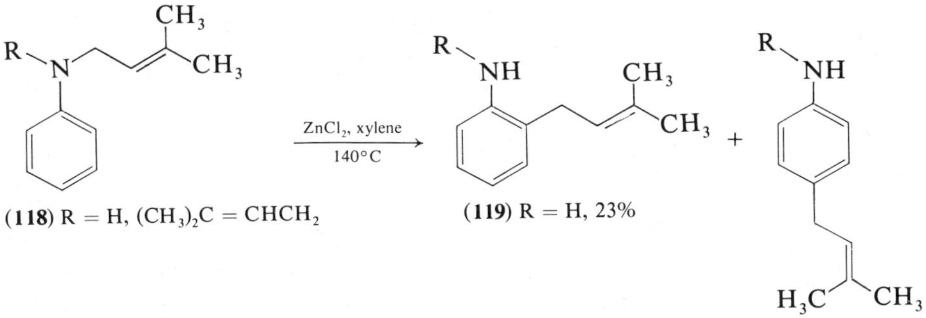

(**118**) R = H, $(CH_3)_2C = CHCH_2$

(**119**) R = H, 23%

(**120**) R = H, 12%

Scheme 26

The formation of the pericyclic transition state of the 3-aza-Cope rearrangement appears to be made more difficult by the two methyl groups at C-3 of the allyl moiety, so that the splitting into a complex consisting of the aniline part associated with $ZnCl_2$ and the relatively stable dimethylallyl cation is favored energetically. Thus the situation is similar to that of the rearrangement of the *N*-cinnamyl-*N*,*N*-dimethylanilinium salt **96** (R = C_6H_5; Scheme 19).

By repeated *N*-allylation followed by $ZnCl_2$-catalyzed rearrangement aniline could be transformed into 2,4,6-triallylaniline in an overall yield of ~3% (64).

It is made clear by these examples that the rearrangements of *N*-allyl and *N*-(2-butenyl)anilines **121**, realized in different ways, represent charge-accelerated aromatic 3-aza-Cope reactions (Scheme 27). 2,4-Dieniminium intermediates **122** are formed which, if R^3 = H, tautomerize rapidly to give the 2-allylated anilines with the inverted allyl chain. When R^3 = alkyl, a charge-accelerated

* Compound **118** (R = $(CH_3)_2C=CHCH_2$) with 1 molar equivalent of $ZnCl_2$ yielded only **120** (R = $(CH_3)_2C=CHCH_2$; 11%), whereas with 2 molar equivalents of $ZnCl_2$ 2,4-bis-(3-methyl-2-butenyl)aniline was obtained in 23% yield (64).

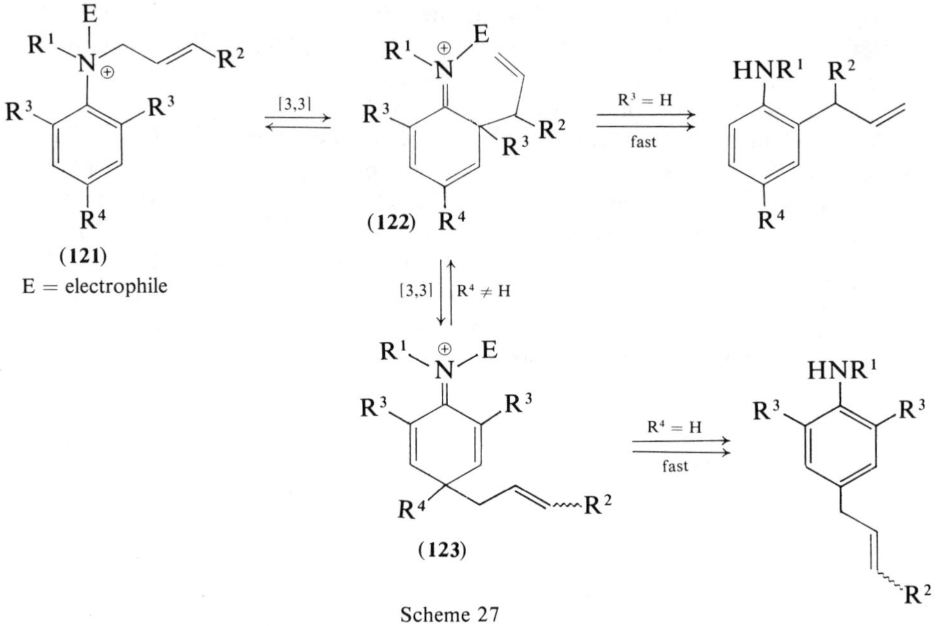

3,3-REARRANGEMENTS OF IMINIUM SALTS

Scheme 27

Cope rearrangement occurs which leads to the 2,5-dieniminium compounds **123**, and when $R^4 = H$ these give rapidly the 4-allylated anilines with a doubly inverted allyl chain, that is, with an uninverted allyl chain. In accordance with this mechanism the cyclohexa-2,4-dienone **124** reacts with p-toluenesulfonic acid hydrazide in methanol to yield the 4-allyl hydrazide **127** via the (not

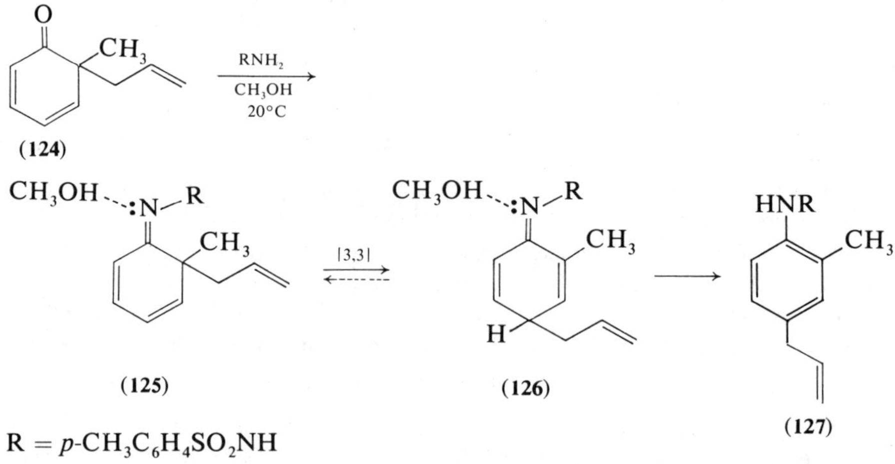

$R = p\text{-}CH_3C_6H_4SO_2NH$

Scheme 28

II. AZA-COPE REARRANGEMENTS IN AMMONIUM AND IMINIUM SALTS

isolable) 2,4-dienimine **125** (Scheme 28) (21). The reaction corresponding to **126** → **125** has already been discussed (Scheme 1).

With respect to the $ZnCl_2$-promoted rearrangements ($E = ZnCl_2$, Scheme 27) it should be added that aniline as well as 2- and 4-methylaniline form complexes with anhydrous $ZnCl_2$ of the type ZnA_2Cl_2 (A = aniline component) (65). The above-mentioned rearrangements probably occur via analogous complexes. These can be the *N*-allylated aniline or *C*-allylated imine complexes. Thus these rearrangements are related to the rearrangements of allyl aryl ethers, 2-allyl-2,4-cyclohexadienones, and also 4-allyl-2,5-cyclohexadienones induced by borone trichloride. All lead to allylphenols (15, 16).

In conclusion it should be stressed that charge-accelerated aromatic 3-aza-Cope rearrangements can be used readily to synthesize 2- and 4-allylated anilines or subsequent products of these.

B. 2-AZA-COPE IMINIUM–IMINIUM REARRANGEMENTS

Iminium compounds of type **7**, which can undergo 2-aza-Cope rearrangements, have until now been postulated only as intermediates (Scheme 29). They may be produced: (1) under Eschweiler–Clarke conditions (cf. ref. 66) during the methylation of *N*-butenylamines **128** (cf. Section II-B-1); (2) by *C*-protonation of *N*-butenylvinylamines **129** (cf. ref. 67 and Section II-B-2); and (3) by heterolytic fragmentation (cf. ref. 68) of 4-piperidine halogenides or sulfonates **130** (cf. Section II-B-3).

1. Rearrangements in the course of Eschweiler–Clarke Methylations of N-Butylenamines

The first example of this type of rearrangement was evidently observed by Horowitz and Geissman (69). When 1-phenyl-3-butenylamine (**131**) was treated with formaldehyde in formic acid at 100°C, benzaldehyde and *N,N*-dimethyl-3-butenylamine (**132**) [but none of the expected *N,N*-dimethyl-1-phenyl-3-butenylamine (**133**)] were obtained after hydrolysis (Scheme 30). The amine **131** with formaldehyde, but without addition of acid, gave benzaldehyde and 3-butenylamine (**137**) (Scheme 31). In accordance with this observation 4-amino-4-phenyl-1,6-heptadiene (**134**) and formaldehyde gave, in the presence of aqueous sulfuric acid, the amine **137** and allyl phenyl ketone. These fragmentations were thought to be the result of Cope-like rearrangements of the intermediate iminium ions **135**, which yield the iminium ions **136**. The shift of the iminium double bond into conjugation with the phenyl ring can be regarded as the driving force of the transformation.

Further reactions of this type were later detected by Winterfeldt and Franzischka (70) in connection with reactions of 1-allyl- and 1-propargyltetrahydronorharmanes. For example, the conversion of the 1-allylated and 1-

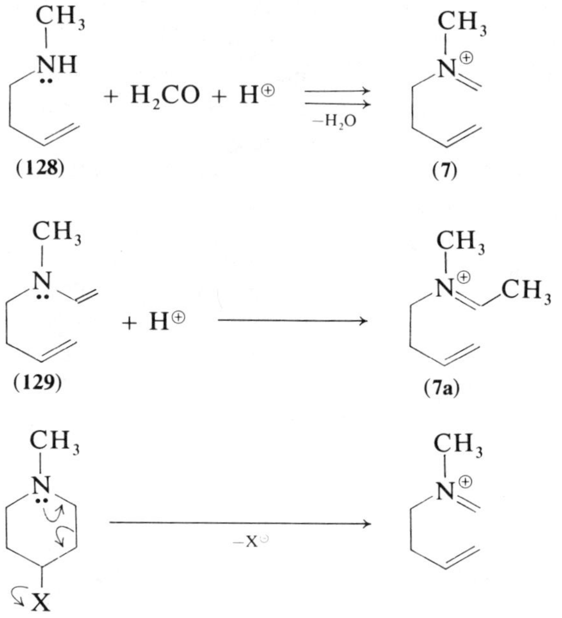

3,3-REARRANGEMENTS OF IMINIUM SALTS

Scheme 29

propargylated tetrahydronorharmane hydrochlorides **138–143** with formaldehyde in glacial acetic acid at 20°C yielded, via the indirectly detectable iminium salts **144** or **146**, the iminium salts **145** or **147** (UV evidence!) (Scheme 32). These salts were reduced *in situ* by $NaBH_4$ to give the 2-substituted tetrahydronorharmanes with an inverted allyl chain (compounds **148–150**) or an allenyl chain (compounds **151–153**). Again, the shift of the iminium double bond into conjugation with the indole system can be regarded as the driving force of these rearrangements. This assumption is supported by

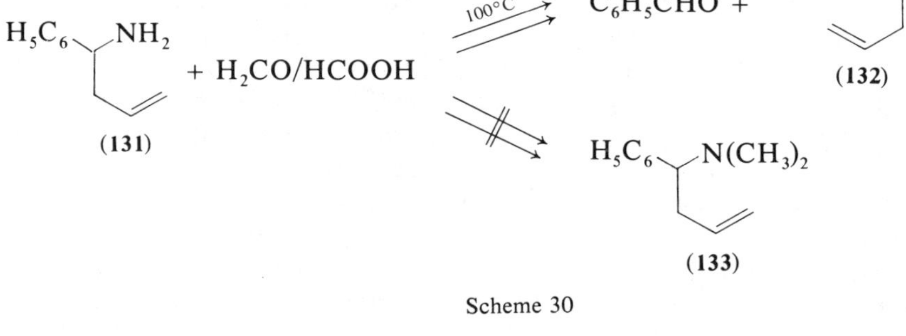

Scheme 30

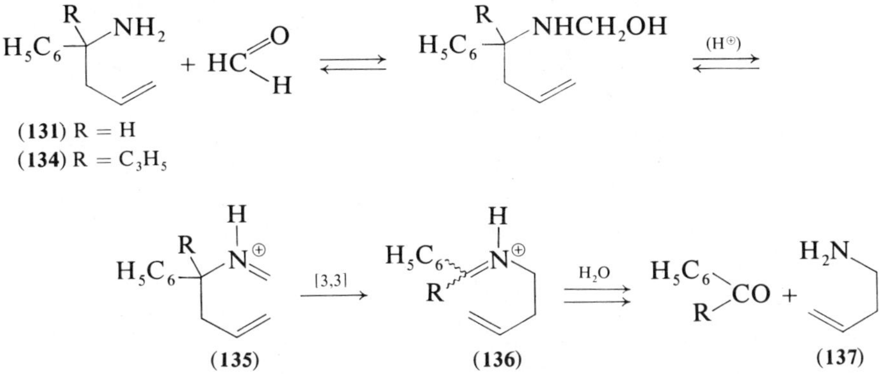

II. AZA-COPE REARRANGEMENTS IN AMMONIUM AND IMINIUM SALTS

Scheme 31

(131) R = H
(134) R = C$_3$H$_5$

the observation (70b) that **138** and **141**, in the presence of butyric aldehyde in glacial acetic acid, undergo the expected rearrangements to yield, via the iminium compounds **154** (R^2 = C$_3$H$_7$), the iminium compounds **155** (R^2 = C$_3$H$_7$). These can then be reduced by NaBH$_4$ to give the tetrahydronorharmanes **156** (Scheme 33). However, the iminium salts **154** (R$_2$ = C$_6$H$_5$), formed from **138** and **141** with benzaldehyde, do not undergo the 2-aza-Cope rearrangement since here the reduction of the reaction mixture leads to the 2-benzyltetrahydronorharmanes **158** in good yields. No reduced products of **157** were obtained.

A smooth 2-aza-Cope rearrangement was observed when 1-(cyclohexen-1-ylmethyl)-1,2,3,4-tetrahydronorharmane hydrochloride [**159**, R^1, R^2 = –(CH$_2$)$_4$–] was treated with formaldehyde in glacial acetic acid (Scheme 34) (71). The iminium salt **160** [R^1, R^2 = –(CH$_2$)$_4$–] was formed in nearly quantitative yield since the reduction of the reaction mixture with NaBH$_4$ furnished the norharmane **161** [R^1, R^2 = –(CH$_2$)$_4$–] in 92% yield. Similarly the methylallyl compound **159** (R^1 = CH$_3$, R^2 = H) was transformed into **161** (R^1 = CH$_3$, R^2 = H).*

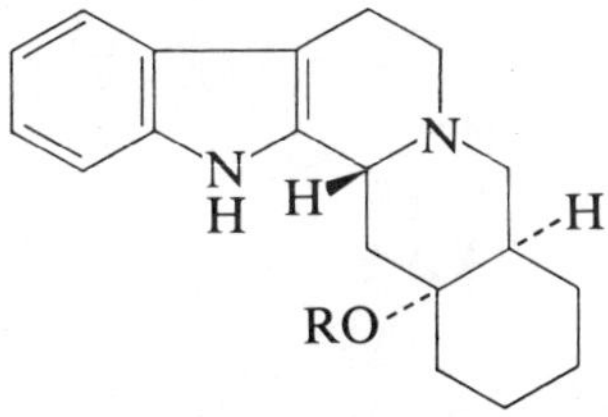

(162) 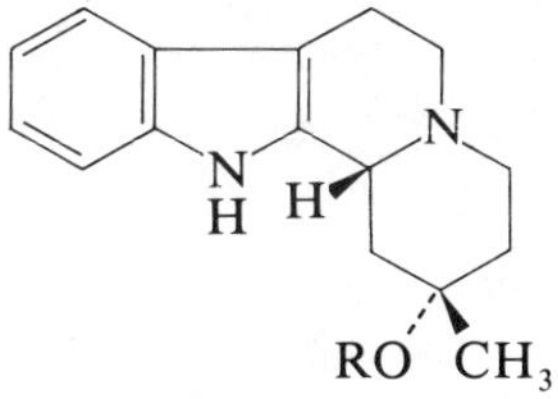   (163)

* When the iminium salts **160** [R^1, R^2 = –(CH$_2$)$_4$– and R^1 = CH$_3$, R^2 = H] are allowed to stand in aqueous or methanolic solution a slowly occurring stereoselective cyclization (retro-process of the corresponding synchronous fragmentation) takes place yielding the epialloyohimbane derivatives **162** (R = H or CH$_3$) and the quinolizine derivatives **163** (R = H or CH$_3$) (71).

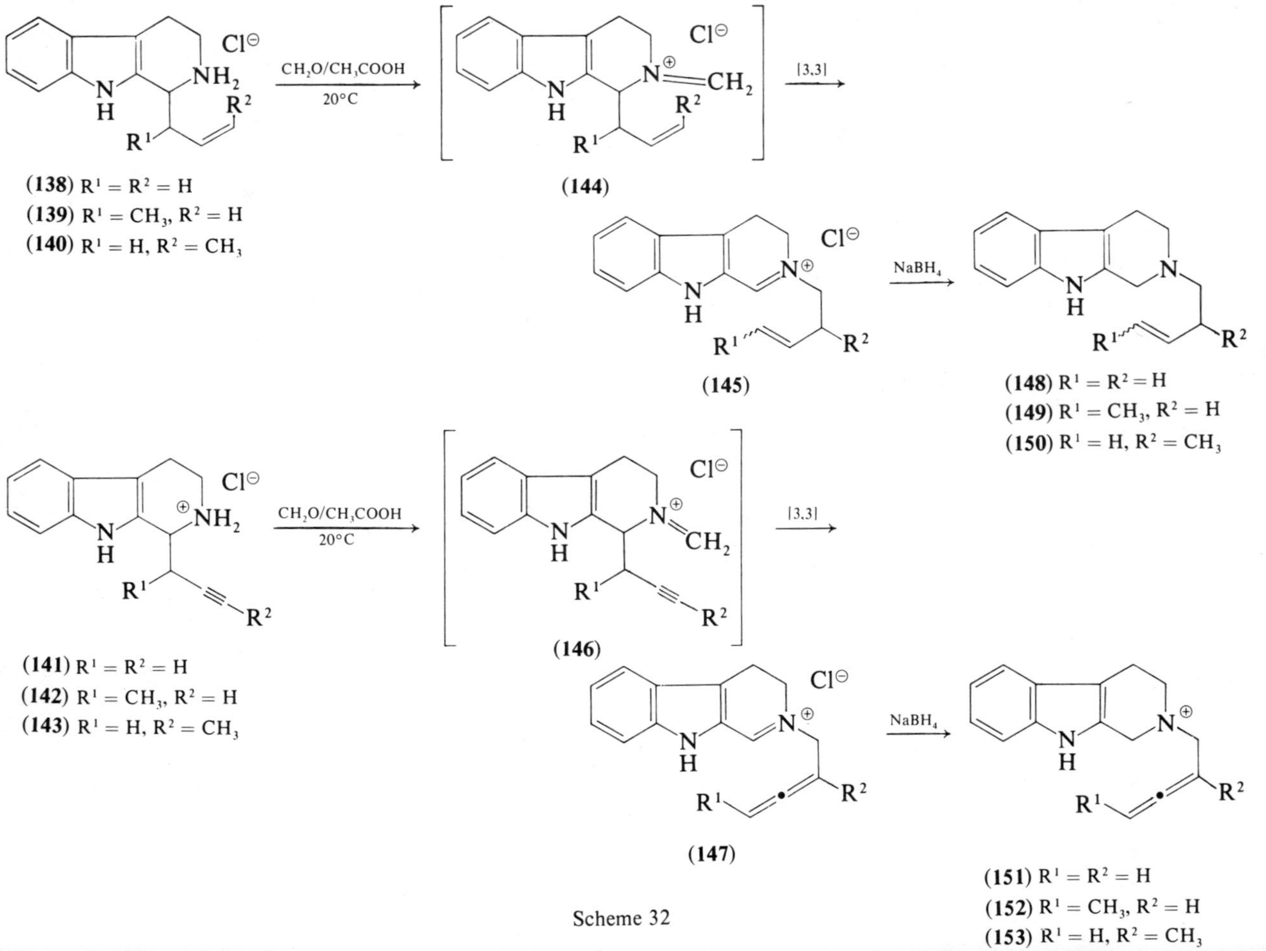
CH$_2$O/CH$_3$COOH
20°C
[3,3]
NaBH$_4$
(138) R^1 = R^2 = H
(139) R^1 = CH$_3$, R^2 = H
(140) R^1 = H, R^2 = CH$_3$
(144)
(145)
(148) R^1 = R^2 = H
(149) R^1 = CH$_3$, R^2 = H
(150) R^1 = H, R^2 = CH$_3$
(141) R^1 = R^2 = H
(142) R^1 = CH$_3$, R^2 = H
(143) R^1 = H, R^2 = CH$_3$
(146)
(147)
(151) R^1 = R^2 = H
(152) R^1 = CH$_3$, R^2 = H
(153) R^1 = H, R^2 = CH$_3$

Scheme 32

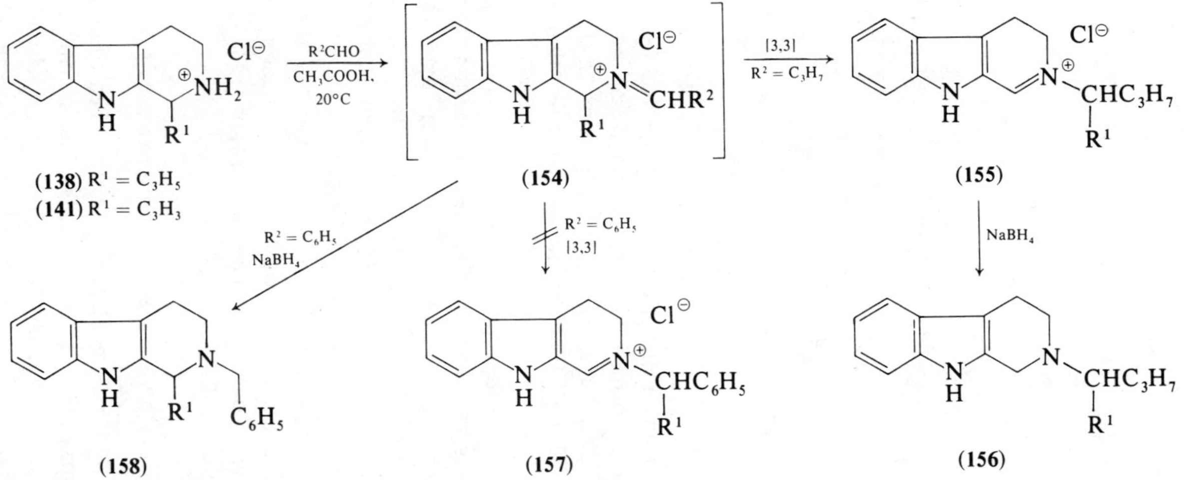
Cl⊖
⊕NH2
N H
R1
(138) R1 = C3H5
(141) R1 = C3H3
R2CHO
CH3COOH, 20°C
(154)
N=CHR2
[3,3]
R2 = C3H7
(155)
CHC3H7
NaBH4
(156)
R2 = C6H5
[3,3]
(157)
CHC6H5
R2 = C6H5
NaBH4
(158)
C6H5

Scheme 33

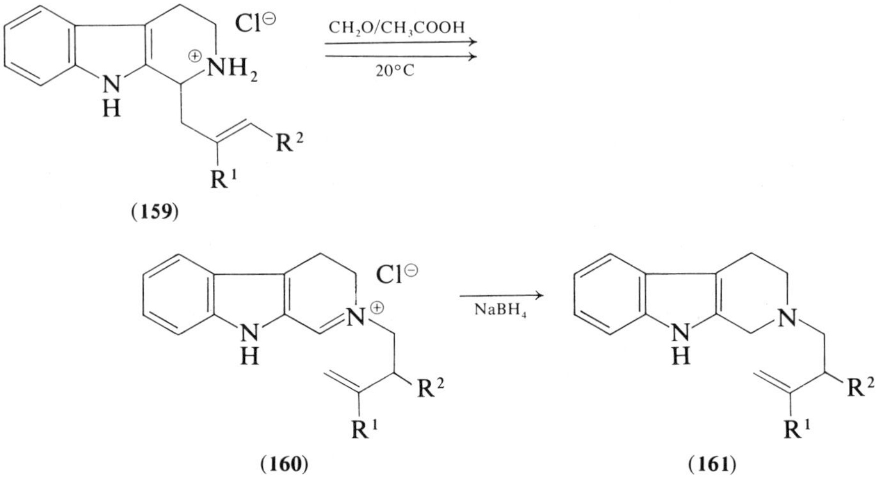

Scheme 34

(159) → (160) → (161)

Reagents: CH₂O/CH₃COOH, 20°C; NaBH₄

The 2-aza-Cope rearrangement of the iminium salt **165** (Scheme 35), which would be expected to be formed from the propiolic acid ester derivative **164** with formaldehyde in glacial acetic acid, seems to be much slower than that of the salt **146** (Scheme 32) (70b). This is shown by the fact that reduction of the reaction mixture afforded, apart from 40% **169**, only 12% of the hydrogenated product of rearrangement **168**. The hydroxymethylation of the nitrogen of the indole moiety represents an independent secondary reaction. The authors claim that the isomerization **165** → **167** occurs faster than the 2-aza-Cope rearrangement **165** → **166**. Hence the main product of the reduction comes from **167**. The retardation of the 2-aza-Cope rearrangement was attributed to the electronic effects of the methoxycarbonyl group (70b).

Finally, it has been reported (72) that treatment of the tetrahydroisoquinoline hydrochloride **170** with formaldehyde in glacial acetic acid yielded 82% of the rearranged dihydroisoquinolinium ion **171** (isolated as the "pseudocyanide") (Scheme 36).

2. Rearrangements of Protonated 1,2-Dihydroisoquinolines

It has been shown by Knabe et al. (73) that, when heated in diluted acid, *N*-methyl-1,2-dihydropapavarine (**172**) is transformed intermolecularly (74,75) by a shift of the 3,4-dimethoxybenzyl residue into the 3-(3,4-dimethoxybenzyl)-6,7-dimethoxy-2-methyl-3,4-dihydroisoquinolinium compound **173** (cf. Section IV as well as refs. 76 and 77) (Scheme 37).

Later it was found by Sainsbury et al. (78) that the comparable 1-allyl-2-methyl-1,2-dihydroisoquinoline (**174**) rearranges via the undetectable dihydro-

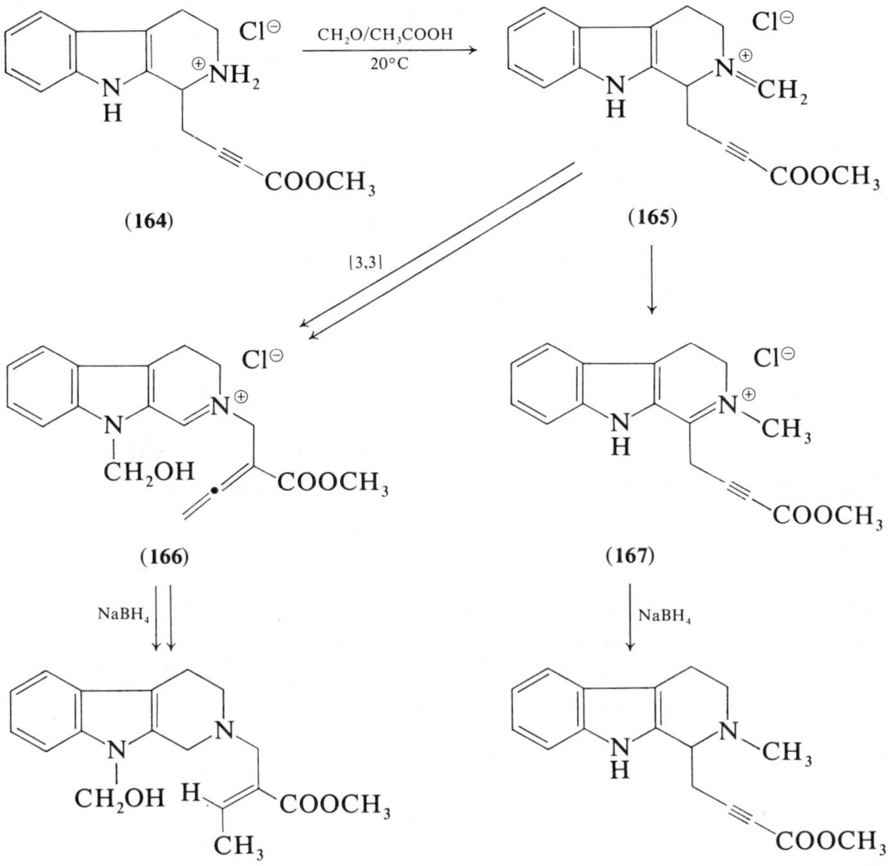

II. AZA-COPE REARRANGEMENTS IN AMMONIUM AND IMINIUM SALTS

Scheme 35

isoquinolinium salt **176** (R = H) by a charge-accelerated 2-aza-Cope reaction to give the 3-allyl-2-methyl-3,4-dihydroisoquinolinium compound **177** (R = H) (Scheme 38). The 6,7-dimethoxy-dihydroisoquinoline **175** showed similar

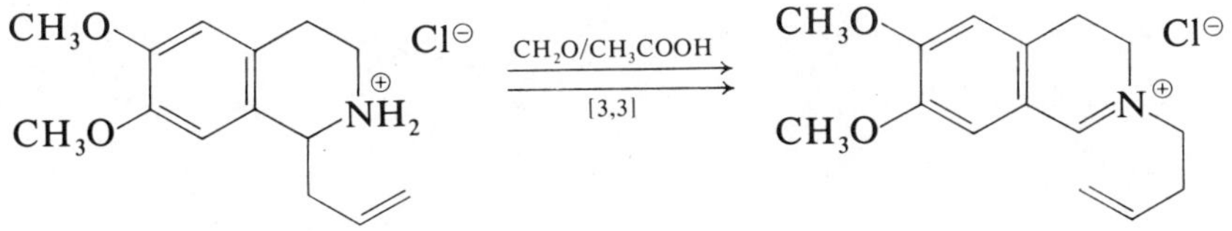

Scheme 36

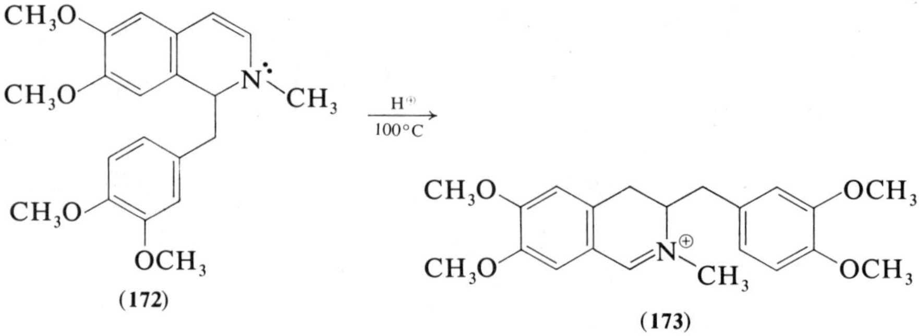

Scheme 37

behavior (79). The iminium salts **177** (R = H, CH$_3$O) could be isolated as the so-called "pseudocyanides" (**178** and **179**). The iminium ions **176** (R = H, CH$_3$O) are obviously formed from **174** and **175** by a C-4-protonation.

That the allyl moiety rearranges with inversion of the carbon chain in these reactions is made obvious by the transformations of the aminoacetals **180–183** with ethanolic hydrochloric acid. These give directly the 3,4-dihydroisoquinolinium salts **185** according to Scheme 39 (72,80). Neither the 1,4-dihydroisoquinolinium salts **184** nor the intermediates leading to them were isolated (cf. ref. 81). From the aminoacetals **182** and **183** only the products with the inverted allyl chain (**187** and **189**, respectively) were obtained.* No

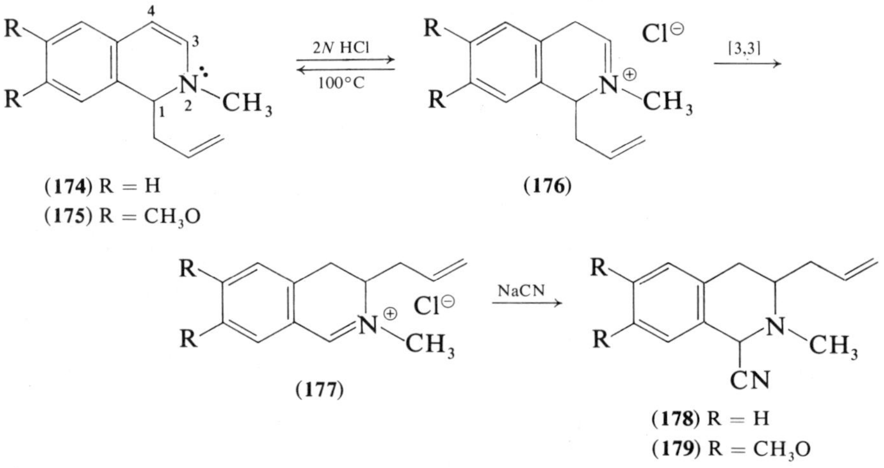

Scheme 38

* It appears that the products of rearrangement, that is, **187** and **189**, possess predominantly a *trans*-configurated butenyl side chain (72).

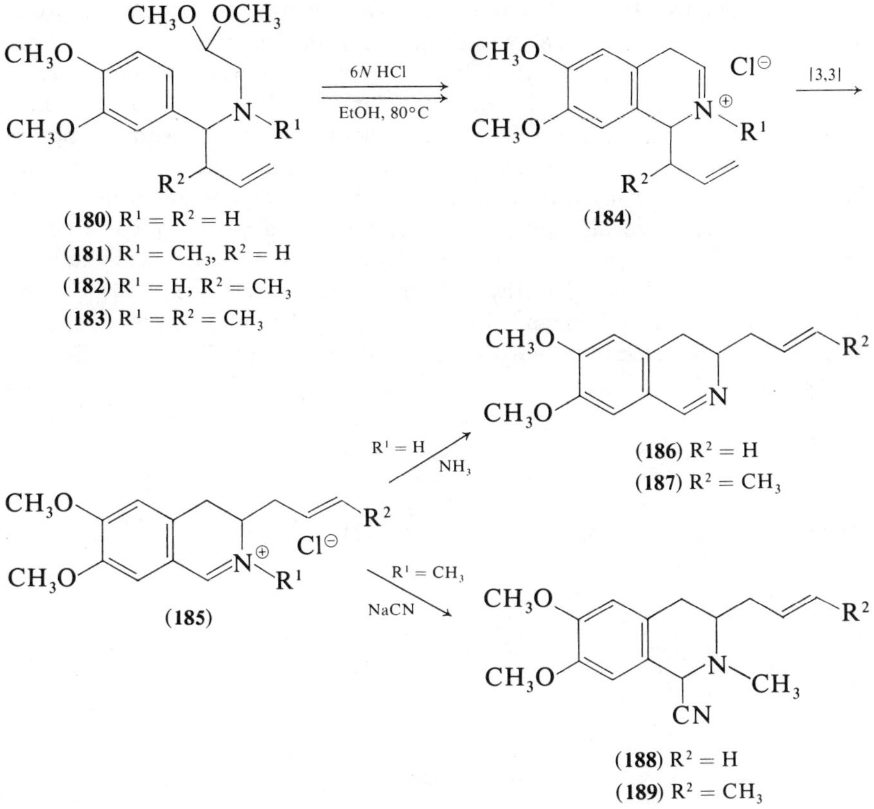

II. AZA-COPE REARRANGEMENTS IN AMMONIUM AND IMINIUM SALTS

(180) $R^1 = R^2 = H$
(181) $R^1 = CH_3, R^2 = H$
(182) $R^1 = H, R^2 = CH_3$
(183) $R^1 = R^2 = CH_3$

(186) $R^2 = H$
(187) $R^2 = CH_3$

(188) $R^2 = H$
(189) $R^2 = CH_3$

Scheme 39

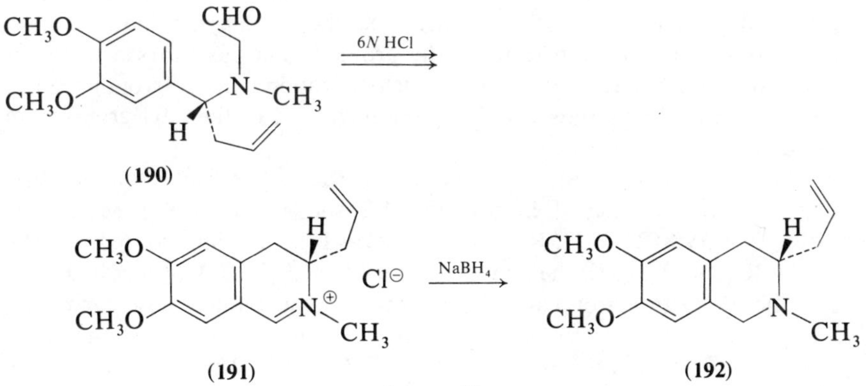

Scheme 40

cross-products **186** and/or **189** were observed when an equimolar mixture of the amino acetals **181** and **182** was subjected to the same reaction conditions. Furthermore, the acid-catalyzed rearrangement of the (R)-configurated aldehyde **190** yielded a dihydroisoquinolinium salt **191**, which on reduction with NaBH$_4$ furnished the (S)-tetrahydroisoquinoline **192** with an optical purity of 97% (Scheme 40) (72). This demonstrates that the allyl group must have shifted suprafacially with respect to the dihydroisoquinolinium ring.

If C-1 in the dihydroisoquinolines bears both an allyl group and a benzyl group, then only the allyl group migrates. This follows from the acid-catalyzed rearrangement of 1-allyl-2-methyl-1,2-dihydropapaverine (**193**), which produces exclusively the compound with the allyl group at C-3 **194**. Reduction of **194** gives the expected tetrahydroisoquinoline derivative in 90% yield (Scheme 41) (82).

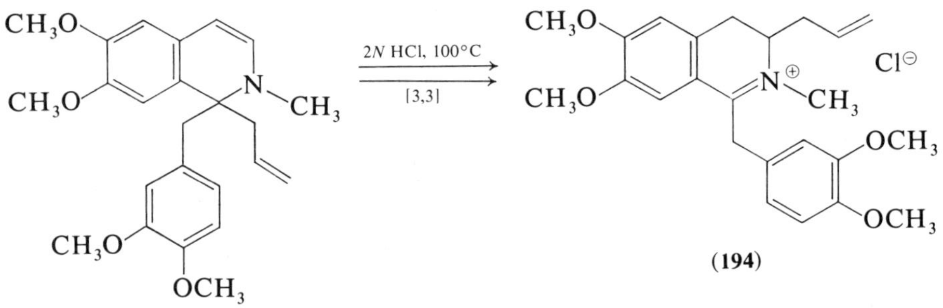

Scheme 41

In the examples discussed so far, the equilibrium between the two iminium salts (cf. **184** and **185**; Scheme 39), accomplished via 2-aza-Cope rearrangements, lies to the side of that compound that bears the iminium bond in conjugation with the aromatic nucleus (i.e., the 3,4-dihydroisoquinolinium salts). This is particularly true when the aromatic nucleus bears a methoxy group in position 6, because the charge delocalization is more pronounced in this case. Steric effects may also favor the migration of the allyl group from C-1 to C-3.

A change of steric and electronic effects can alter this situation. For example, whereas 1-allyl-1,2,3-trimethyl-1,2-isoquinoline (**195**) rearranges smoothly via **198** (R^1 = H, R^2 = CH$_3$) to yield the iminium compound **199** (R^1 = H, R^2 = CH$_3$) (83); the iminium salt **198** (R^1 = R^2 = H, generated from **196**) without a methyl group at C-1 remains unchanged under these conditions (Scheme 42); that is, the equilibrium position lies to the side of the sterically more favorable iminium salt **198** (R^1 = R^2 = H). The position of this equilibrium could be influenced by introduction of a methoxy group in position

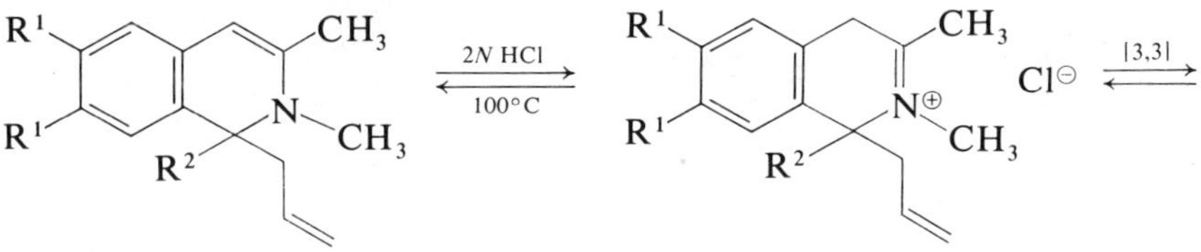

II. AZA-COPE REARRANGEMENTS IN AMMONIUM AND IMINIUM SALTS

Scheme 42

(195) $R^1 = H, R^2 = CH_3$
(196) $R^1 = R^2 = H$
(197) $R^1 = CH_3O, R^2 = H$

(198)

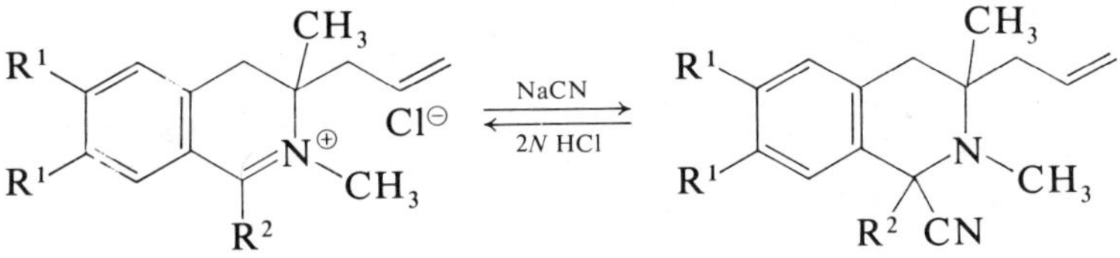

(199)

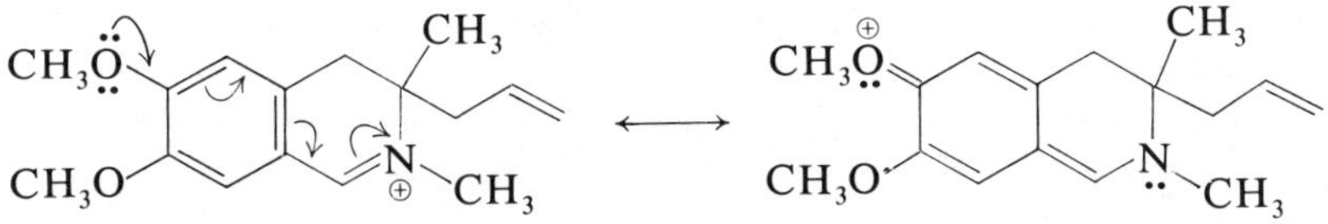

(200) $R^1 = CH_3O, R^2 = H$

(201) $R^1 = CH_3O, R^2 = H$

6 of the dihydroisoquinoline as mentioned earlier (cf. also mesomeric forms of **200**). Indeed, the acid-catalyzed rearrangement of **197** yields a 2:1 mixture of the dihydroisoquinolinium compounds **198** and **199** ($R^1 = CH_3O, R^2 = H$) (83). From this mixture **199** could be separated as the "pseudocyanide" **201**. When this "pseudocyanide" **201** was treated with 2N HCl the 2:1 mixture of **198** and **199** ($R^1 = CH_3O, R^2 = H$) was obtained once more, that is, in this case a migration of the allyl group from C-3 to C-1 was observed.

Presumably a nonintramolecular mode of rearrangement (cf. Section IV) is observed when 1-cinnamyl-6,7-dimethoxy-2-methyl-1,2-dihydroisoquinoline is treated with acid. The 3-cinnamyl-6,7-dimethoxy-2-methyl-3,4-dihydroisoquinolinium ion **185** ($R^1 = CH_3, R^2 = C_6H_5$; Scheme 39) is formed in high yield (79b, 84) since the corresponding "pseudocyanide" could be isolated in 80% yield.

Allyl migrations of the type **184** → **185** (Scheme 39) were also realized recently with β-carbolines (85). When the 1-allyl-β-carbolines **202** were treated with 2N HCl followed by addition of NaCN, the "pseudocyanides" **205** could be isolated in a yield of 35%; that is, the iminium ions **203** and **204** must represent intermediates (Scheme 43).

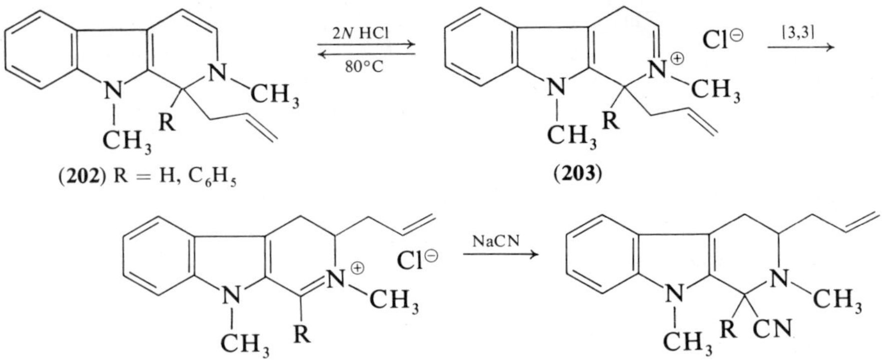

3,3-REARRANGEMENTS OF IMINIUM SALTS

Scheme 43

In conclusion, it may be mentioned that a propargyl → allenyl rearrangement was also observed with dihydroisoquinolines (72) (80b). The reaction of the amino aldehyde **206** with 6N HCl at 100°C yielded, via the corresponding 1-propargyl-1,4-dihydroisoquinolinium compound, the 3-allenyl iminium salt **207**, which was transformed by reduction and methylation into the quaternary ammonium salt **208** in an overall yield of 46% (Scheme 44).

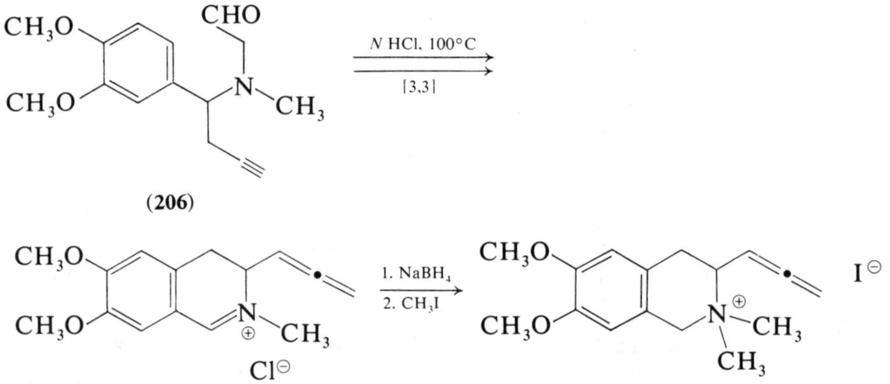

Scheme 44

3. Rearrangements Accompanying Heterolytic Fragmentation Reactions

Reactions of this type, in which the iminium species capable of rearrangement is generated by a fragmentation reaction (cf. Scheme 29), have until now been established in only one system.

II. AZA-COPE REARRANGEMENTS IN AMMONIUM AND IMINIUM SALTS

Treatment of 1-benzyl-4α-mesyloxy-*trans*-decahydroquinoline (**209**, R = C$_6$H$_5$CH$_2$) with NaOH at 20°C in the presence of an excess of NaBH$_4$ affords, in an overall yield of 93%, the products **212**, **213**, and **214** (R = C$_6$H$_5$CH$_2$) in the ratio 21:30:49 (Scheme 45) (86) (cf. also ref. 87). When the solvolytic treatment of **209** (R = C$_6$H$_5$CH$_2$) is conducted at 40°C and NaBH$_4$ is added afterward the products **212**, **213**, and **214** are obtained in a yield of 41% in the ratio 35:65:0. These results may be explained as follows: as first step the mesylate **209** (R = C$_6$H$_5$CH$_2$) suffers a fragmentation to yield by participation of the C-2, C-3 bond the iminium ion **210** (R = C$_6$H$_5$CH$_2$) or by participation of the C-8a–C-4a bond the iminium ion **211** (R = C$_6$H$_5$CH$_2$). In the presence of an excess of NaBH$_4$ these ions are trapped at once to give the reduced compounds **213** and **214** (R = C$_6$H$_5$CH$_2$). Apparently, the bicyclic amine **212** (R = C$_6$H$_5$CH$_2$) is formed by a 1,2 elimination of methanesulfonic acid. In the absence of NaBH$_4$ the iminium ion **211** (R = C$_6$H$_5$CH$_2$) undergoes a charge-accelerated 2-aza-Cope rearrangement to yield the thermodynamically more

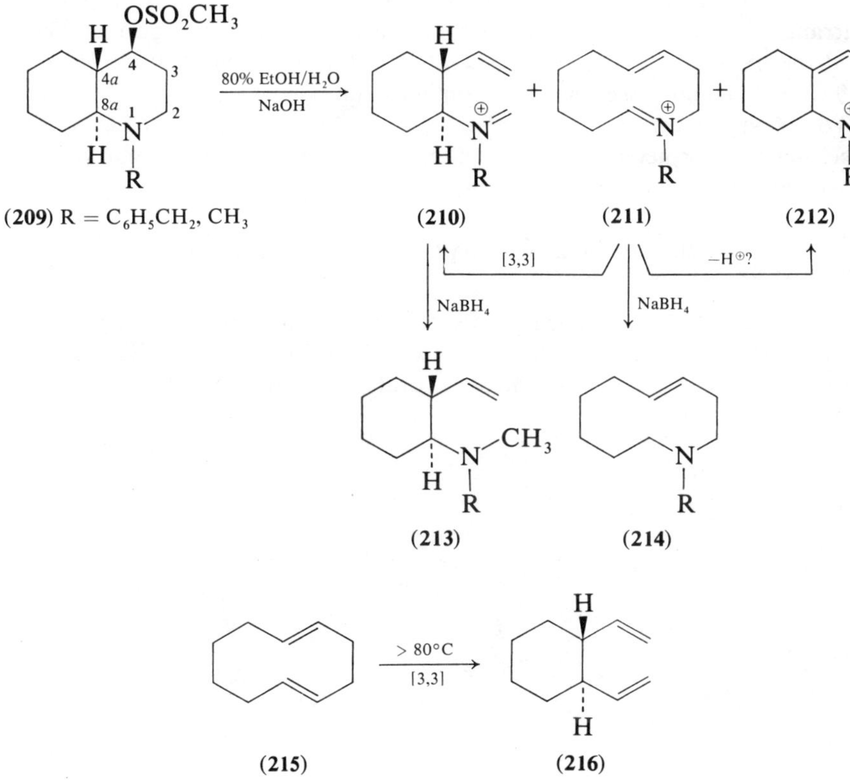

Scheme 45

favorable iminium ion **210** (R = C$_6$H$_5$CH$_2$). The compound **210** lacks the strain associated with the medium-sized ring system of **211**. The iminium salt **211** (R = C$_6$H$_5$CH$_2$) could lead, via loss of a proton, to the bicyclic amine **212** (R = C$_6$H$_5$CH$_2$). The reduction product of **210**, that is, **213** (R = C$_6$H$_5$CH$_2$) possesses a *trans* configuration, hence the charge-accelerated 2-aza-Cope rearrangement (**211** → **210**) must occur via a chairlike transition state. The stereochemistry of **211** is determined by the stereochemistry of the starting material **209** (cf. ref. 68). As analogy, the Cope rearrangement of *E,E*-cyclodeca-1,5-diene (**215**; Scheme 45) (corresponding to the iminium ion **211**) into *trans*-1,2-divinylcyclohexane (**216**) (corresponding to the iminium ion **210**) also proceeds via a chairlike activated complex (88) (cf. ref. 39).

Analogous results were obtained with the *N*-methyl compound **209** (R = CH$_3$) (86). Solvolytic treatment of **209** (R = CH$_3$) at 25°C in the presence of an excess of NaBH$_4$, gives in a yield of 60–70% the compounds **212**, **213**, and **214** (R = CH$_3$) in the ratio of 3.3:1.0:10.5. When NaBH$_4$ is added later **212**, **213**, and **214** (R = CH$_3$) are obtained in 47% yield, now in the ratio 1.4:1.0:1.0. Thus the iminium ion **211** (R = CH$_3$) must also in this case rearrange to a large extent via a charge-accelerated 2-aza-Cope reaction to yield the more stable iminium ion **210** (R = CH$_3$). Again, **213** derived from **210** (R = CH$_3$) possesses the *trans* configuration.

To what extent charge-accelerated 2-aza-Cope rearrangements may accompany other heterolytic fragmentation reactions has so far not yet been investigated (cf. ref. 68).

C. INFLUENCE OF CHARGE ON AMMONIUM–IMINIUM AND IMINIUM–IMINIUM REARRANGEMENTS

In general, Cope rearrangements such as **1** ⇌ **1** are thought to occur via a cyclic transition state which contains six electrons (1, 89–91). This transition state is characterized by the most favorable orbital overlap and degree of

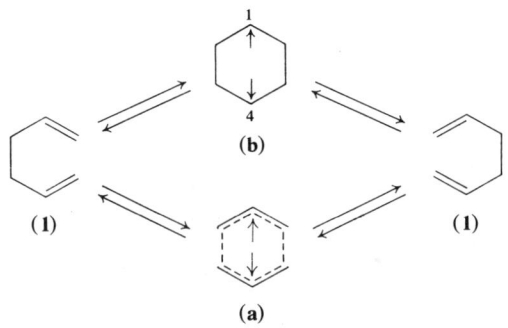

Scheme 46

II. AZA-COPE REARRANGEMENTS IN AMMONIUM AND IMINIUM SALTS

delocalization of the six electrons. It represents the highest state of energy on the reaction coordinate between reactant and product. Calculations based on this model have been performed by several groups (see refs. 1 and 89–91 and literature cited therein). Recently, the possibility was discussed that Cope rearrangements, for example, **1** ⇌ **1**, do not occur via a pericyclic transition state (**a**) but via cyclohexa-1,4-diyles **b** (Scheme 46) (92, 93).

Though in principle a similar possibility exists regarding aza-Cope rearrangements, the following discussion is based on the concerted transition state model which appears, at least until now, better established. In a rough approach this model could be understood as a complex of two allyl radicals. For comparison—keeping in mind that this simplified model can serve only for trend estimations—we can assume that the energy of the transition state is connected with the Hückel π energy (E_π) of both allyl radicals. The π energies of the 1-aza- and 2-aza-allyl system can be obtained by taking into account a first-order perturbation (94) (for nitrogen atom parameters see ref. 95). Scheme 47 contains the MO energy levels (in units of β) for the 1-aza- and 2-aza-allyl system in comparison with the allyl system. It can be seen that the energy of ψ_1 of the 1-aza-allyl system is lower than that of the allyl system and that ψ_2 also becomes bonding, as a result of the asymmetric perturbation. Both these effects are enlarged by the introduction of a positive charge on the nitrogen atom. In the 2-aza-allyl system the energy of the bonding ψ_1 is also lowered in comparison with the allyl system. This effect is strengthened in the positively charged system. On the other hand, ψ_2 remains unchanged, that is, nonbonding. Thus the E_π of the radical partners in the transition states of

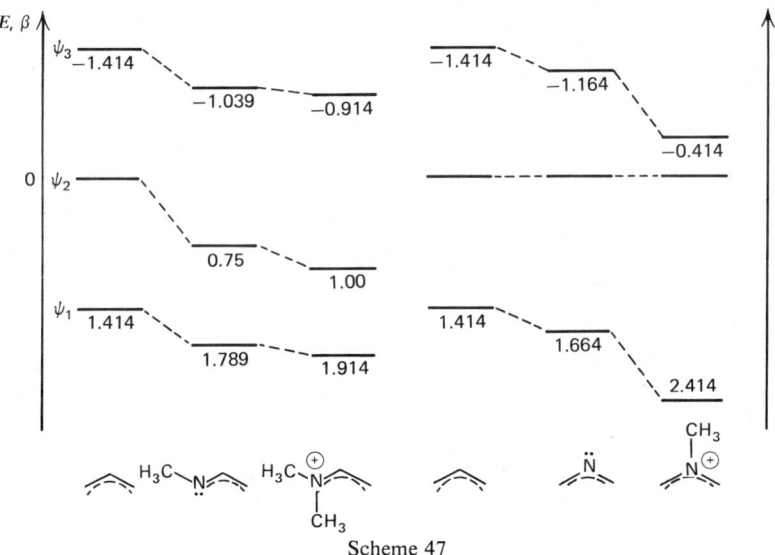

Scheme 47

TABLE IV

Sum of the E_π's (in β Units) of Cope and Aza-Cope Rearrangements

Type of rearrangement	E_π(reactant), β		E_π(transition state), β^a		E_π(product), β	
Cope	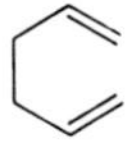 (1)	4.00	5.66	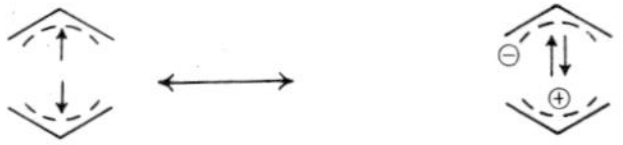 5.66	 (1)	4.00
3-Aza-Cope	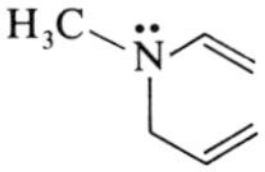 (2)	7.08	7.16	 7.91	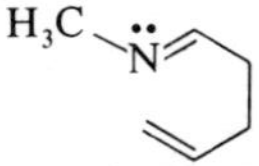 (3)	5.50
	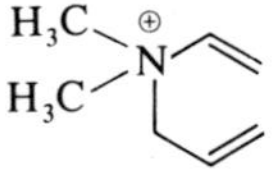 (4)	4.00	7.66	 8.66	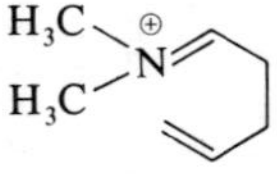 (5)	6.00
2-Aza-Cope	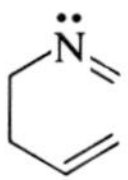 (6)	5.50	6.16	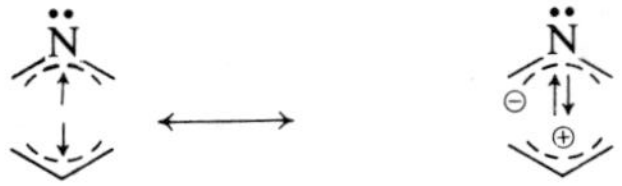 6.16	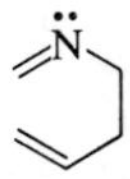 (6)	5.50
	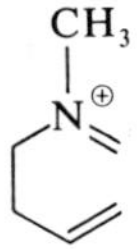 (7)	6.00	7.66	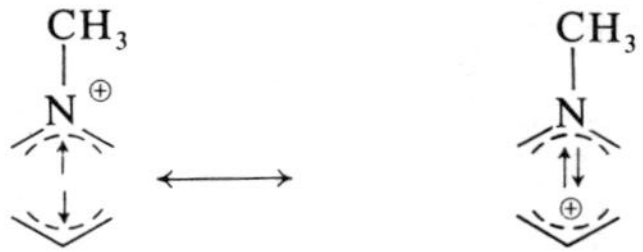 7.66	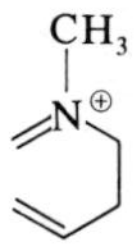 (7)	6.00

[a] Only the electrons that occupy ψ_2 are shown.

Cope and 3- and 2-aza-Cope rearrangements can be estimated on the basis of Scheme 47. E_π of reactants and products are obtained in a similar manner by taking the MO energies of the C–C and C–N double bond (94).

All three factors discussed in Section II play a role in the 3-aza-Cope system. If, as mentioned above, E_π of the radical partners is taken as a measure of the energy of the transition state, it can be seen that in the uncharged 3-aza-Cope system **2** the dipolar resonance structure should not contribute to the energy of the transition state. This is not true for the cationic 3-aza-Cope system **4**. In this case the difference of the E_π (ΔE_π) between reactant and product accounts for -4.66 to -3.66β according to our "rough-trend" estimation. Thus the clearly increased E_π of the transition state is caused by the delocalization of the six electrons involved in the transition state. This is equivalent to a "delocalization" of the positive charge in the transition state (cf. ref. 19) in comparison with the localized charge in the reactant.

In the uncharged 3-aza-Cope system **2** ΔE_π amounts to about -0.1β. The vinylamine resonance which no longer exists in the transition state is mainly responsible for this approximation. The large ΔE_π in the case of system **4** leads to an appreciably stronger reduction of the energy difference between reactant and transition state as compared with system **2**.*

Furthermore it can be seen by inspection of Table IV that the stabilization of the transition state by resonance is more pronounced in the rearrangement **4 → 5** than in the rearrangement **2 → 3**. Moreover, the removal of the vinylamine resonance in **4** increases the ΔE_π value of reactant and product in comparison with **2**. In the uncharged 3-aza-Cope rearrangement (**2 → 3**) ΔE_π amounts to $+1.58\beta$, whereas in the charged Cope system (**4 → 5**) it amounts to -2.0β; that is, $\Delta\Delta E_\pi$ is equal to $+3.58\ \beta$.

The last-mentioned factor is not present in the degenerate 2-aza-Cope rearrangements **6 ⇌ 6** and **7 ⇌ 7**. However, ΔE_π between **6** and its corresponding transition state amounts to $-0.66\ \beta$ and to $-1.66\ \beta$ in the charged system; that is, the last rearrangement should occur appreciably faster than the uncharged 2-aza-Cope rearrangement.

A two-step mechanism can also be postulated for the iminium–iminium rearrangements of type **7 ⇌ 7** (Scheme 48). This mechanism could be of importance in cases where C-5 bears an electron-donating group and where the intermediate **218** is more unstable than the product **217**. The ring opening of the intermediate **218** can be compared with a heterolytic fragmentation reaction. The cyclization of compounds **160–162** (cf. Scheme 34 and the

---

* Based on the data given in Table IV the reverse 3-aza-Cope rearrangement (**3 → 2**) would possess a larger ΔE_π ($\sim-1.7\beta$) than the forward reaction **2 → 3**; that is, the equilibrium position should lie to the side of **2**. Owing to the different characters of bonding (see also ref. 96), however, **3** has, as mentioned in Section I, a lower energy content than **2** ($\Delta\Delta H_f^\circ \sim 7$–$10$ kcal/mole); that is, in reality the equilibrium position lies to the side of **3**.

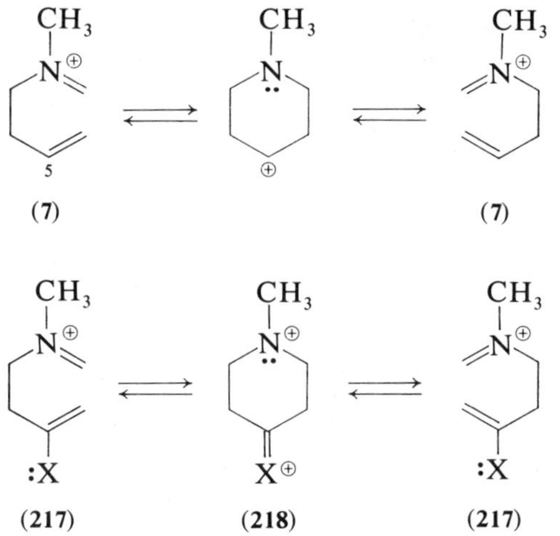

Scheme 48

corresponding footnote) in the presence of nucleophiles could be based on a ring closure reaction of the type **217** → **218**. It has not been investigated whether treatment of **162** with acid causes the reverse reaction to give **160** (71).

III. 3,4- and 2,5-Diaza-Cope Rearrangements

The 3,4 rearrangement type is represented by the C,C-combination step in the Fischer indole synthesis (for recent reviews see ref. 97) and that in the Piloty–Robinson pyrrole synthesis (98, 99). In Section III-A only such enhydrazines and dienhydrazines are discussed which were shown to undergo thermal as well as acid-catalyzed rearrangements to yield indoles or pyrroles. 2,5-Diaza-Cope rearrangements are presented in Section III-B.

A. THERMAL AND ACID-CATALYZED REARRANGEMENTS OF ENHYDRAZINES

Enhydrazines and their N-protonated species have already been postulated as intermediates in the Fischer indole synthesis by Robinson and Robinson (99, 100) (Scheme 49; the rearrangement is exemplified for acetone–phenylhydrazone **219**). The 3,4-diaza-Cope rearrangement **220** → **221** has been compared with the o-benzidine rearrangement (99, 100). Carlin and Fisher (101) pointed to the great similarity between this step and the thermal o-Claisen rearrangement.

Neber et al. (102) were probably the first to obtain an enhydrazine, for example, **223** (Scheme 50), when dibenzyl ketone was reacted with N,N'-

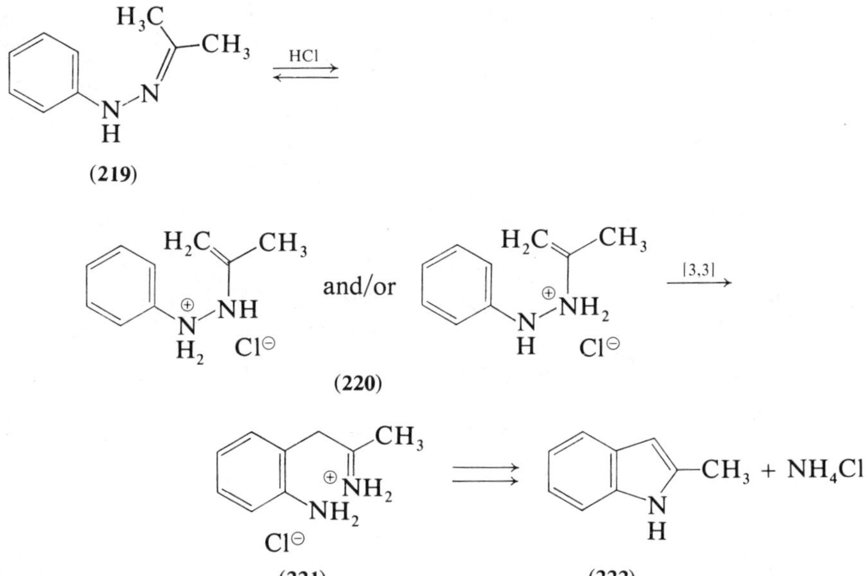

III. 3,4- AND 2,5-DIAZA-COPE REARRANGEMENTS

Scheme 49

dimethyl-(*p*-xylyl)hydrazine in glacial acetic acid.* When compound **223** was treated with ethanolic HCl it rearranged to the indole **224**.

Diels and Reese (104) prepared a series of enhydrazines by the addition of phenyl- and benzyl-substituted hydrazines to acetylene dicarboxylic esters. The

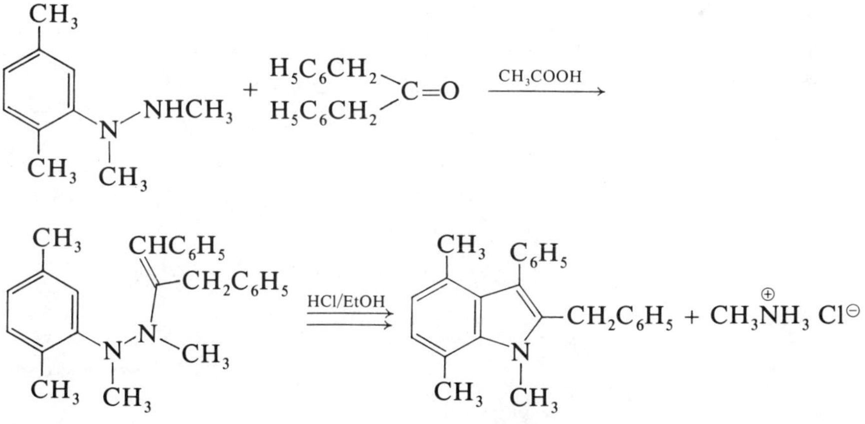

Scheme 50

* On repeating this experiment Elgersma (103) obtained **224** but no **223**.

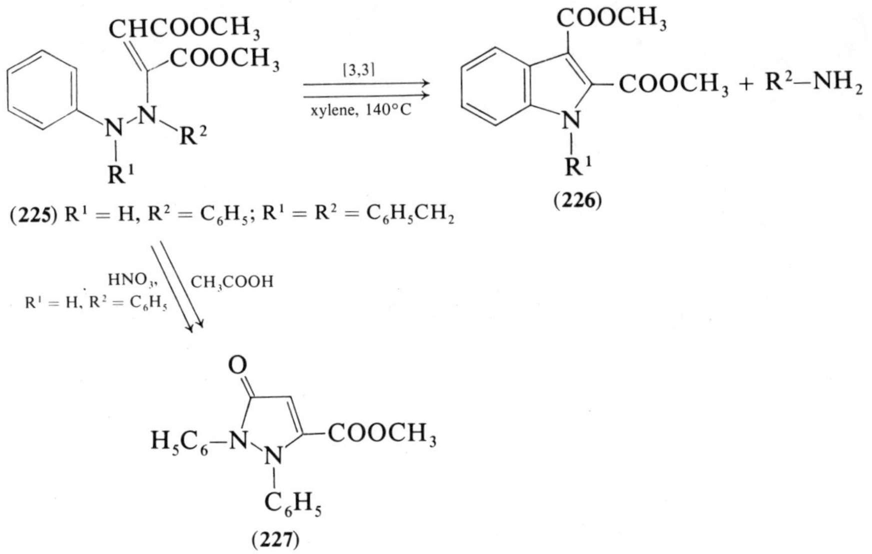

3,3-REARRANGEMENTS OF IMINIUM SALTS

(225) $R^1 = H, R^2 = C_6H_5; R^1 = R^2 = C_6H_5CH_2$

(226)

$R^1 = H, R^2 = C_6H_5$ | HNO$_3$, CH$_3$COOH

(227)

Scheme 51

enhydrazines **225** (Scheme 51), formed in the reaction of hydrazobenzene or N,N'-dibenzyl-N-phenylhydrazine and acetylene dimethylcarboxylate, rearranged in boiling xylene to give the indole derivatives **226** and aniline or

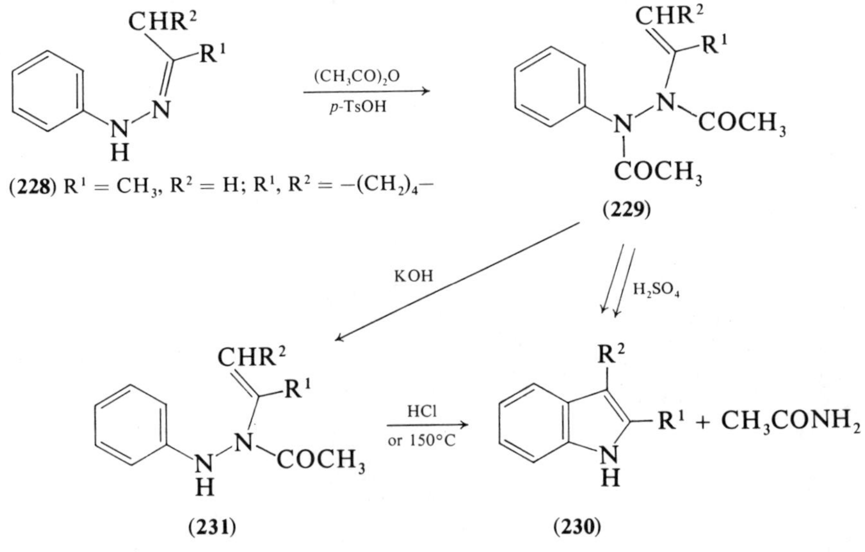

(228) $R^1 = CH_3, R^2 = H; R^1, R^2 = -(CH_2)_4-$

(229)

(231)

(230)

Scheme 52

benzylamine. Under acidic conditions, however, the enhydrazine **225** ($R^1 = H$, $R^2 = C_6H_5$) yielded the pyrazolone **227**, whereas the benzylated enhydrazine **225** ($R^1 = R^2 = C_6H_5CH_2$) rearranged in boiling acetic anhydride to the corresponding indole **226** ($R^1 = C_6H_5CH_2$).

Suvorov et al. (105) obtained the N,N'-diacetylated enhydrazines **229** (Scheme 52) from the reaction of the phenylhydrazones **228** with acetic anhydride in the presence of p-toluenesulfonic acid. These were transformed in sulfuric acid into the indoles **230** (105). When treated with 5% alkali solution the enhydrazines **229** yielded the monoacetylated compounds **231** (105), the structure of which was assigned by Elgersma and Havinga (10, 103). These monoacetylated enhydrazines were also converted into the indoles **230** with loss of acetamide when heated without solvent at 150°C or in the presence of

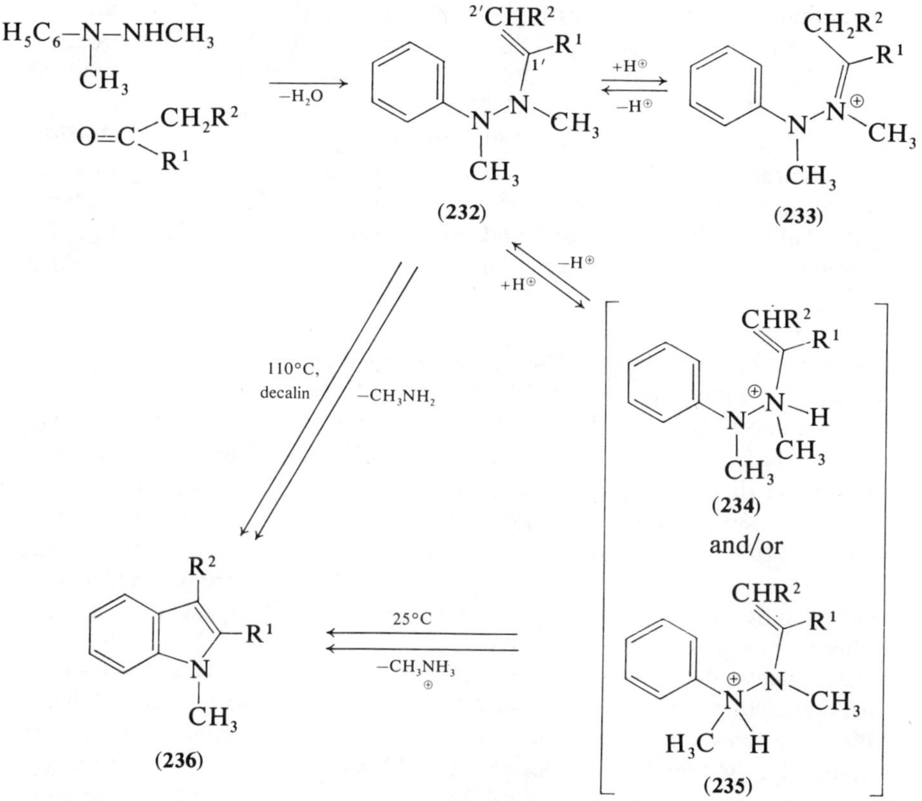

a: $R^1, R^2 = -(CH_2)_4-$
b: $R^1, R^2 = -(CH_2)_3-$
c: $R^1, R^2 = -(CH_2)_2-C_6H_4-$
d: $R^1 = C_2H_5, R^2 = CH_3$
e: $R^1 = H, R^2 = C_2H_5$
f: $R^1 = H, R^2 = C_6H_5$

Scheme 53

mineral acid at room temperature (10, 103, 105). Thus the 3,4-diaza-Cope rearrangement occurs also in this case much faster in the protonated enhydrazines **231** than in the nonprotonated forms (cf. ref. 97).

Recently, reactions of this type were investigated thoroughly by Schiess and Grieder (11, 106) (cf. ref. 107). The reaction of N,N'-dimethyl-N-phenylhydrazine with ketones and aldehydes gave enhydrazines of type **232** (Scheme 53). These enhydrazines are protonated exclusively at C-2' with trifluoroacetic acid in liquid SO_2 at $-60°C$ and form the hydrazonium salts **233** as a rule, enamines are protonated preferentially at the nitrogen atom in ether at $-70°C$ with hydrochloric acid to yield enammonium salts, which rearrange slowly at $-70°C$ and faster on warming to give the corresponding C-protonated iminium salts (67)]. The hydrazonium compounds **233a** and **b** could be isolated as crystalline perchlorates that were stable at room temperature and did not rearrange to the corresponding indoles **236a** and **b**. After warming the solutions of the hydrazonium salts **233** in trifluoroacetic acid in SO_2 the corresponding indoles **236** were observed. They were obviously formed via the hydrazinium salts* **234** and/or **235**, the occurrence of which could not be proved directly (NMR!). The formation of the indoles **236** from the enhydrazines **232** could be followed kinetically in 0.5 N dichloroacetic acid in acetonitrile solution. The relative rates of formation of **236a–f** $[\tau_{1/2}(\mathbf{236a}) = 25$ min] were found to be 1.0, 0.05, 25, 0.5, 8, and 25, respectively. Again, these observations are in accordance with the assumption that the hydrazonium ions **233** are not direct intermediates in the acid-catalyzed indole formation. If salts of type **233** were intermediates the accelerating effect of the phenyl group in **232c** and **f** could hardly be understood because protonation at C-2' should be hindered (the conjugation between the phenyl ring—present in **232c** and **f**—and the double bond is removed).† The comparably fast rearrangement of **232e** can be attributed to the facilitated N-protonation of this compound because there is no alkyl substituent at C-1'. Both the discussed effects work together in the enhydrazine **237a** (E/Z mixture of isomers; Scheme 54). This only slightly stable enhydrazine gives the indoleninium ion **240** instantaneously when dissolved in trifluoroacetic acid solution in SO_2 at $-70°C$. Thus the 3,4-diaza-Cope rearrangement must occur in this case even at $-70°C$ (11, 106).

The phenyl derivative **237a**, which is only slightly stable, undergoes a rearrangement even at room temperature within 55 hr to give the aldimine **238a**, the corresponding p-isomer **241**, and an unknown compound (11, 106). Apparently, the thermal rearrangement of **237a** occurs by a radical pathway. When chromatographed on silica gel **238a** is converted into a *cis/trans* mixture

---

* For nomenclature see ref. 108.

† The occurrence of N,N'-diprotonated species seems very unlikely (11, 106) when one considers the low concentration of acid which induces the rearrangement.

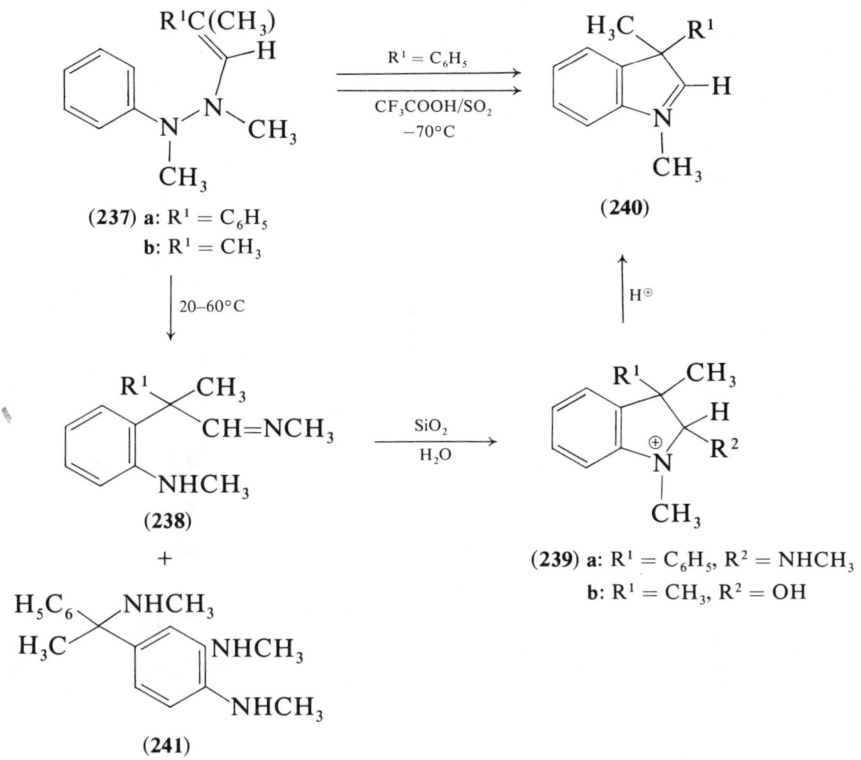

III. 3,4- AND 2,5-DIAZA-COPE REARRANGEMENTS

Scheme 54

(237) a: $R^1 = C_6H_5$
b: $R^1 = CH_3$

(238)

(241)

(239) a: $R^1 = C_6H_5$, $R^2 = NHCH_3$
b: $R^1 = CH_3$, $R^2 = OH$

(240)

of the 2-methylaminoindoline **239a** which in the presence of mineral acid yields the ion **240**. The attempt to synthesize the enhydrazine **237b** from N,N'-dimethyl-N-phenylhydrazine and isobutyric aldehyde at 60°C gave only **238b**. On silica gel this compound cyclized and hydrolyzed to give the 2-hydroxyindoline **239b** (11, 106).

In contrast to **237a** the enhydrazines **232** (having only one alkyl substituent at C-2') appear to undergo a 3,4-diaza-Cope rearrangement when heated in decalin at 110°C. The corresponding indoles **236** are formed by loss of methylamine. In this case the differences in the reactivity of the enhydrazines **232a–f** (relative k-values 1.0–6.0) are small; **232f** shows the fastest and **232a** the slowest rearrangement (11, 106).

The reactions of 1-phenylpyrazolidine (**242**) and carbonyl compounds (Scheme 55) (109) are comparable with the reactions illustrated in Schemes 53 and 54. The indole **243** is formed when **242** is treated with cyclohexanone (a corresponding reaction is observed with cyclopentanone) in methanolic HCl, whereas with isobutyric aldehyde indoline **244** is formed. β-Dicarbonyl compounds react with **242** in acetic acid to give the enhydrazines **245** which

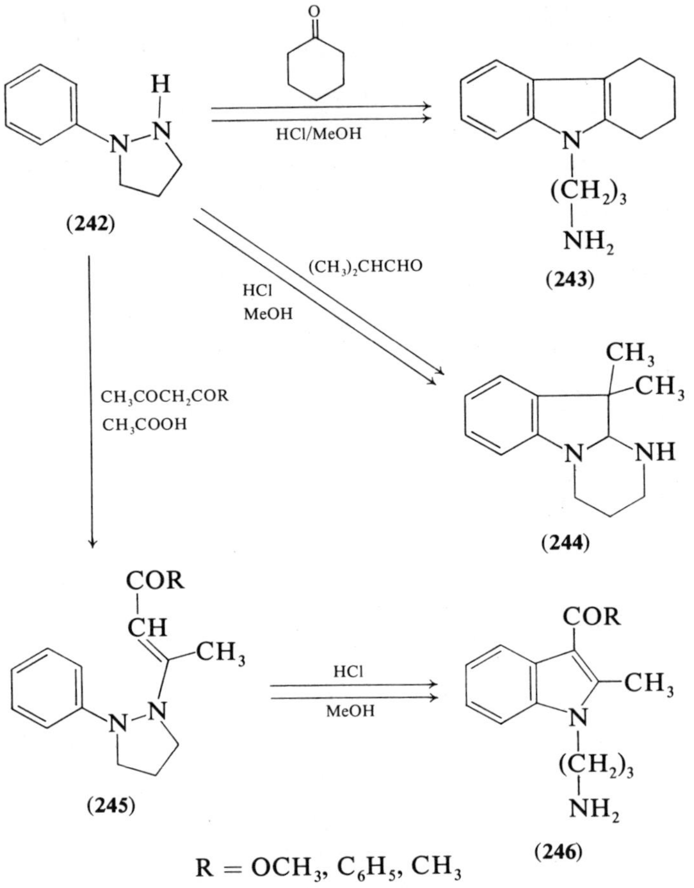

Scheme 55

R = OCH$_3$, C$_6$H$_5$, CH$_3$

are stabilized by the group R–CO. With HCl in methanol these enhydrazines are smoothly converted into the 3-acyl indoles **246** (109).

Recently, Grandberg et al. (110) found an efficient indole synthesis. The alkylation of phenylhydrazones with reactive alkylating reagents such as benzyl chloride, allyl bromide, or dimethyl sulfate in alcohol led to the corresponding indoles. For example, when methylphenylhydrazone **247** (Scheme 56) was boiled with benzyl chloride in alcohol, *N*-methyl-1,2,3,4-tetrahydrocarbazole (**251**) was obtained in 90% yield. The reactive intermediates are probably the benzylated hydrazinium ions **249** and/or **250**, which are formed from the hydrazonium ion **248**.

The rearrangements of enhydrazines, so far cited in this section, have always been considered to be concerted reactions, that is, 3,4-diaza-Cope rearrange-

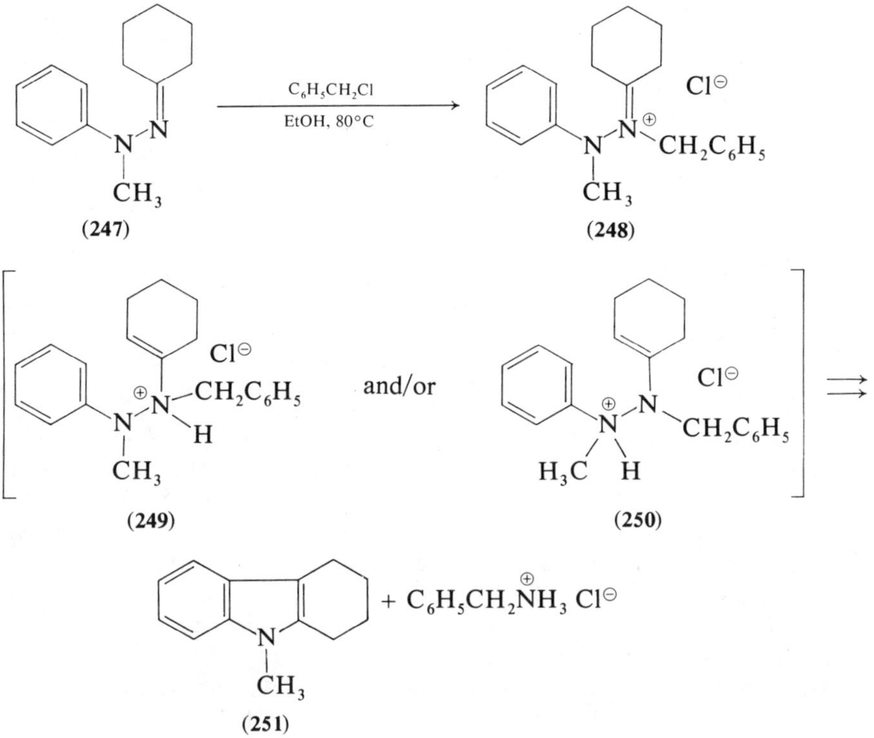

III. 3,4- AND 2,5-DIAZA-COPE REARRANGEMENTS

Scheme 56

ments. The results given in Scheme 54 are in agreement with a radical process. This possibly seems to be open for all enhydrazines that bear two substituents at C-2'.* It is still uncertain to what extent dissociative processes in general play a role in the thermal and acid-catalyzed transformation of enhydrazines into indoles or their precursors.

The synthesis of benzofurans from O-aryl oximes can be compared directly with the previously discussed Fischer indole synthesis from phenylhydrazones. This reaction is catalyzed by acids or boron trifluoride in ether (112–114). The O-aryl oximes can be prepared from O-aryl hydroxyl amines and ketones or by the reaction of the salts of ketone oximes with suitably activated aryl halides. As an example, the synthesis of 2-phenylbenzofuran (253) from acetophenone O-phenyloxime (252) is given. In the course of this transformation a 3,4-oxaza-Cope rearrangement must occur (Scheme 57).

---

* γ,γ-Dimethylallyl phenyl ether which is comparable with the enhydrazine 237 gives, on heating at 214°C in N,N-diethylaniline, exclusively products that follow from sigmatropic rearrangements (111).

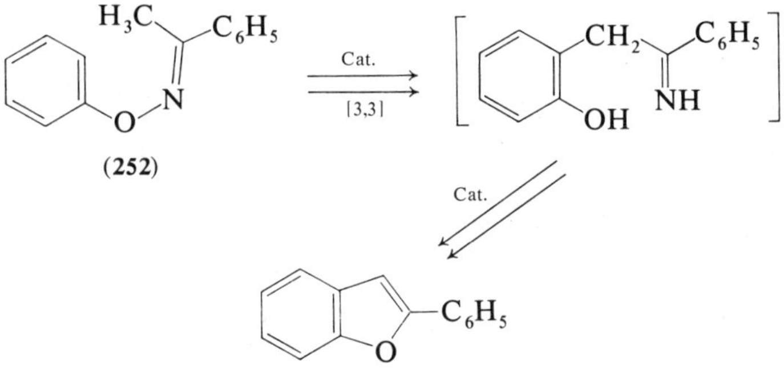

Scheme 57

(252) → [intermediate] → (253)

The aliphatic analogue of the Fischer indole synthesis is the Piloty–Robinson synthesis of pyrroles (98, 99). The azines, for example, **254**, of enolizable ketones ($R^1CH_2COR^2$) are converted into pyrroles in the presence of acidic catalysts such as zinc chloride or hydrochloric acid. The mechanism of this transformation can be formulated as given in Scheme 58. Intermediates are doubtlessly the *N,N'*-dienhydrazines **255** which are protonated or complexed with the Lewis acid present. The compounds **255** can then rearrange via a 3,4-diaza-Cope reaction to give **256** and **257**.

Attempts to synthesize dienhydrazines with alkyl substituents at both nitrogen atoms failed (115–117). The reaction of *N,N'*-dialkylhydrazines* with cyclohexanone [or cycloheptanone, *trans*-2-decalone, or *N*-methyl-4-

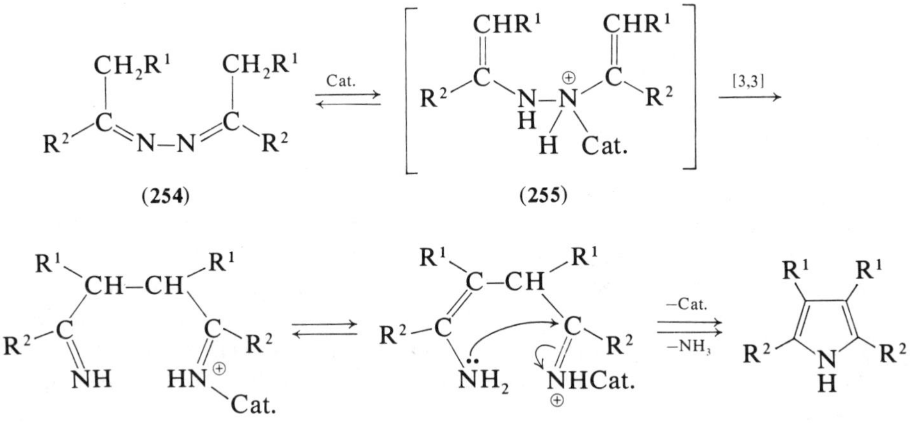

Scheme 58

* Alkyl = CH_3 (115–117), C_2H_5 (116), n-C_4H_9 (115), and $C_6H_5CH_2$ (115).

III. 3,4- AND 2,5-DIAZA-COPE REARRANGEMENTS

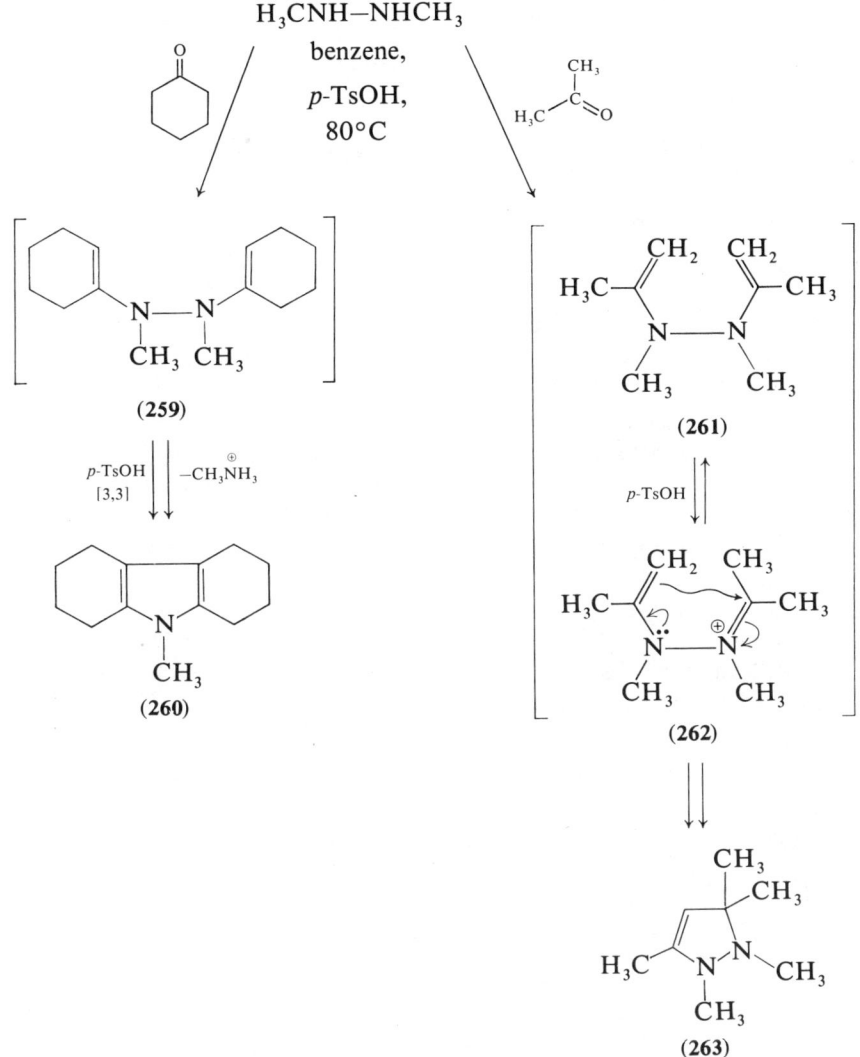

Scheme 59

piperidone (118)] in boiling benzene in the presence of p-TsOH led only to the N-alkylated pyrrole derivatives (e.g., **260** starting with N,N'-dimethylhydrazine; Scheme 59).

A dienhydrazine, for example, **264** (Scheme 60) could be obtained from the reaction of 2-tetralone with N,N'-dimethylhydrazine (evident from NMR). Under mild acid catalysis as well as on heating it rearranges to the corresponding pyrrole **265** (118). The dienhydrazine from phenylacetaldehyde

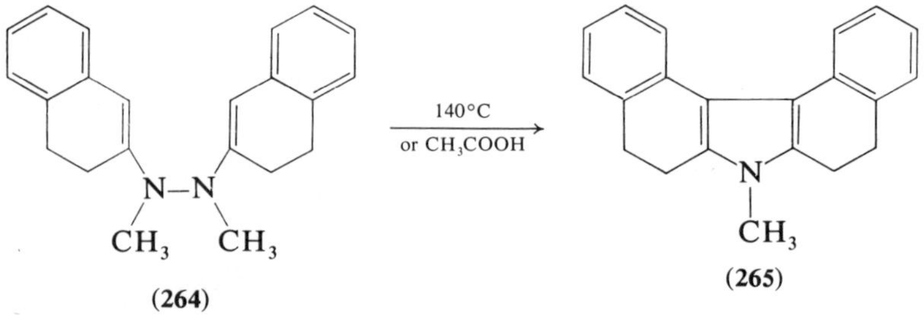

Scheme 60

and N,N'-dimethylhydrazine exhibits a similar behavior (118) (cf. ref. 119). The reaction of aldehydes and ketones with N,N-dimethylhydrazine (benzene, 80°C, p-TsOH) was studied in greater detail (117, 120). It was found that six-membered ring ketones and alkyl benzyl ketones undergo the transformation into the corresponding pyrroles (via protonated dienhydrazines of type **259**; Scheme 59) whereas aryl methyl and alkyl methyl ketones give pyrazolines of type **263** (Scheme 59) probably via C-protonated dienhydrazines of type **262**. With aliphatic aldehydes both N-methylpyrrole derivatives and pyrazolines were obtained. The different behavior of the carbonyl reactants was attributed to steric and electronic effects in the intermediates.

Dimedone reacts with N,N'-dimethylhydrazine to give the "carbonyl-stabilized," thermally stable enhydrazine **266**, which undergoes a further reaction with cyclohexanone (or other carbonyl compounds) in the presence of p-TsOH in benzene to yield the 3-keto pyrrole derivative **267** (Scheme 61) (115, 121).

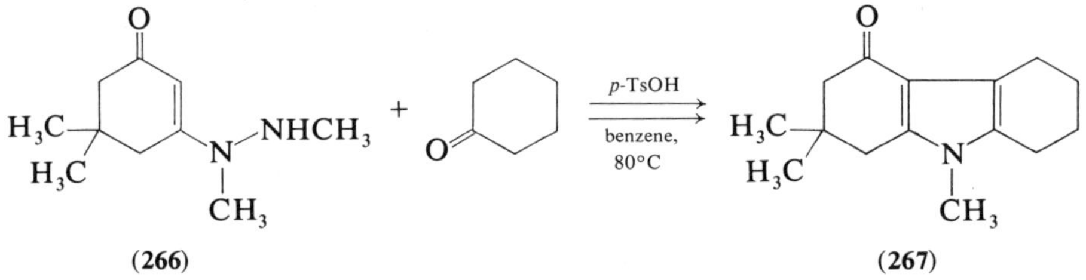

Scheme 61

In conclusion it can be said that the synthesis of pyrroles also occurs via charge-accelerated 3,4-diaza-Cope rearrangement of protonated dienhydrazines.

B. 2,5-DIAZA-COPE REARRANGEMENTS

Double Schiff bases of type **268** (Scheme 62) with *meso* configuration do not yield products of 2,5-diaza-Cope rearrangements (122,123). Neither N,N'-diisopropylidene-1,2-diphenylethylenediamine (**268**; $R^1 = R^2 = CH_3$) nor

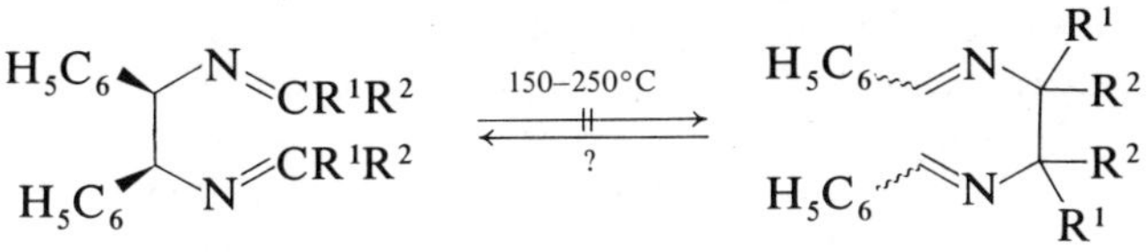

III. 3,4- AND 2,5-DIAZA-COPE REARRANGEMENTS

Scheme 62

$R^1 = R^2 = CH_3$; $R^1 = H$, $R^2 = i\text{-}C_3H_7$

N,N'-diisobutylidene-1,2-diphenylethylenediamine (**268**; $R^1 = H$, $R^2 = i\text{-}C_3H_7$) on heating at 150–250°C gave the corresponding isomeric Schiff bases **269**. The steric interactions of the alkyl substituents in the products could of course influence the equilibrium position despite the fact that two benzaldimine structures are present in the products. On the other hand, N,N'-dialkylidene-

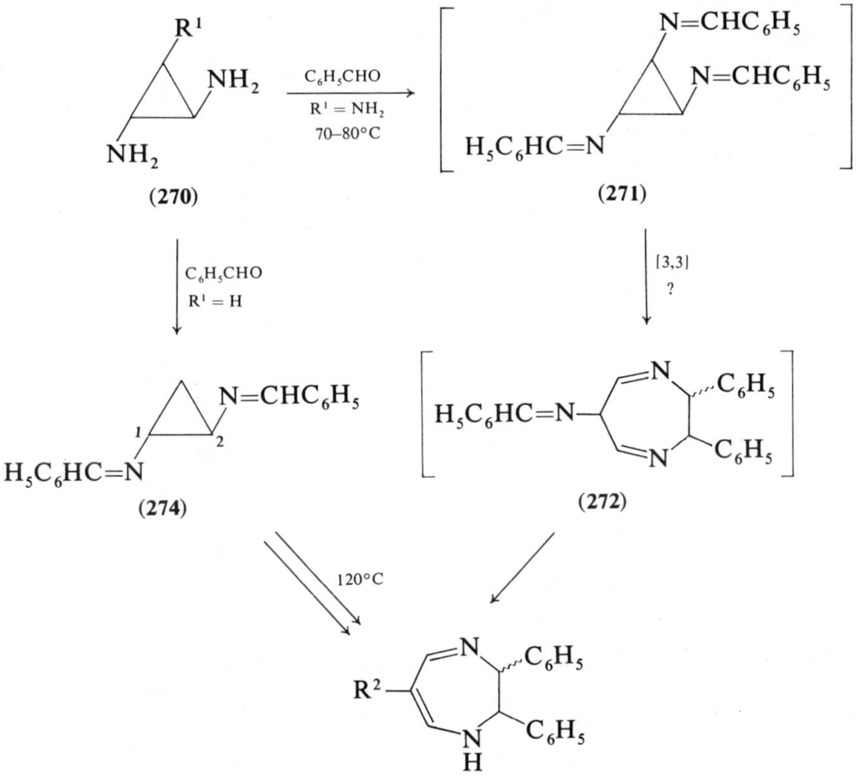

(**273**) $R^2 = C_6H_5CH=N$
(**275**) $R^2 = H$

Scheme 63

cis-1,2-cyclopropanediamines of type **271** do undergo (at least formally) 2,5-diaza-Cope rearrangements (Scheme 63). The reaction of the unstable 1,2,3-cyclopropanetriamine **270** (R = NH$_2$) with benzaldehyde at temperatures up to 80°C affords exclusively the rearrangement product **273** in low yield (123). The precursor of this product is surely the Schiff base **271**. Since it is known that bisbenzylidene derivatives (e.g., **274**) of trans-1,2-cyclopropanediamine (**270**; R^1 = H) rearrange even at 120°C to give dihydroazepins of type **275** (123), it is questionable whether the transformation **271** → **272** is pericyclic. The transformation **274** → **275** is apparently induced by a homolytic cleavage of the C-1–C-2 bond in **274**.

Although the double Schiff bases **268** (Scheme 62) do not undergo 2,5-diaza-Cope rearrangements the analogous forms **277** (Scheme 64), which contain 2-hydroxyphenyl or 2-pyridyl moieties, are easily transformed via 2,5-diaza-Cope rearrangements (124). For example, the reaction of (possibly meso-*) 1,2-di-(2-hydroxyphenyl)ethylenediamine (**276**; R^2 = 2-HO-C$_6$H$_4$) with benzaldehyde in boiling benzene affords directly the double Schiff base **278b** which after hydrolysis yields 2-hydroxybenzaldehyde and meso-1,2-diphenylethylenediamine. The double Schiff base **277b** is without doubt an intermediate in this transformation. It must rearrange to **278b** even at ~80°C. An analogous reaction is observed when the diamine **276** (R^2 = 2-HO-C$_6$H$_4$) is treated with p-methoxybenzaldehyde. The reaction mixture yielded meso-1,2-di-(p-methoxyphenyl)ethylenediamine after hydrolysis. The double Schiff base **277a** from 2-pyridinecarboxaldehyde and 1,2-diphenylethylenediamine (**276**; R^2 = C$_6$H$_5$) is thermally more stable: the rearrangement into the corresponding Schiff base **278a** occurs only slowly in boiling benzene.

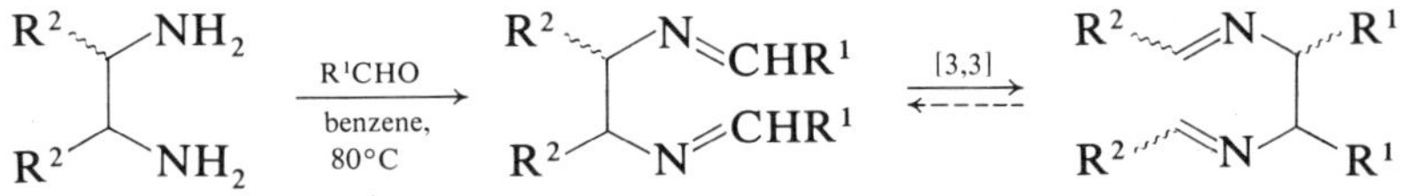

a: R^1 = 2-C$_5$H$_4$N, R^2 = C$_6$H$_5$
b: R^1 = C$_6$H$_5$, R^2 = 2-HO-C$_6$H$_4$
c: R^1 = 4-CH$_3$OC$_6$H$_4$, R^2 = 2-HO–C$_6$H$_4$
d: R^1 = 2-C$_5$H$_4$N, R^2 = 2-HO–C$_6$H$_4$

Scheme 64

The Schiff bases **277** most likely possess the E,E-configuration (cf. ref. 126).

In our opinion, the easily occurring 2,5-diaza-Cope rearrangements **277** → **278** of the double Schiff bases containing a 2-hydroxy group in the phenyl ring might be considered as intramolecularly acid-catalyzed iminium–iminium rearrangements as given in Scheme 65 (e.g., **279** → **280**). On the basis of the

---

* Presumably, the diamines used were prepared according to Tripett's method (125; cf. ref. 122), which leads to meso compounds.

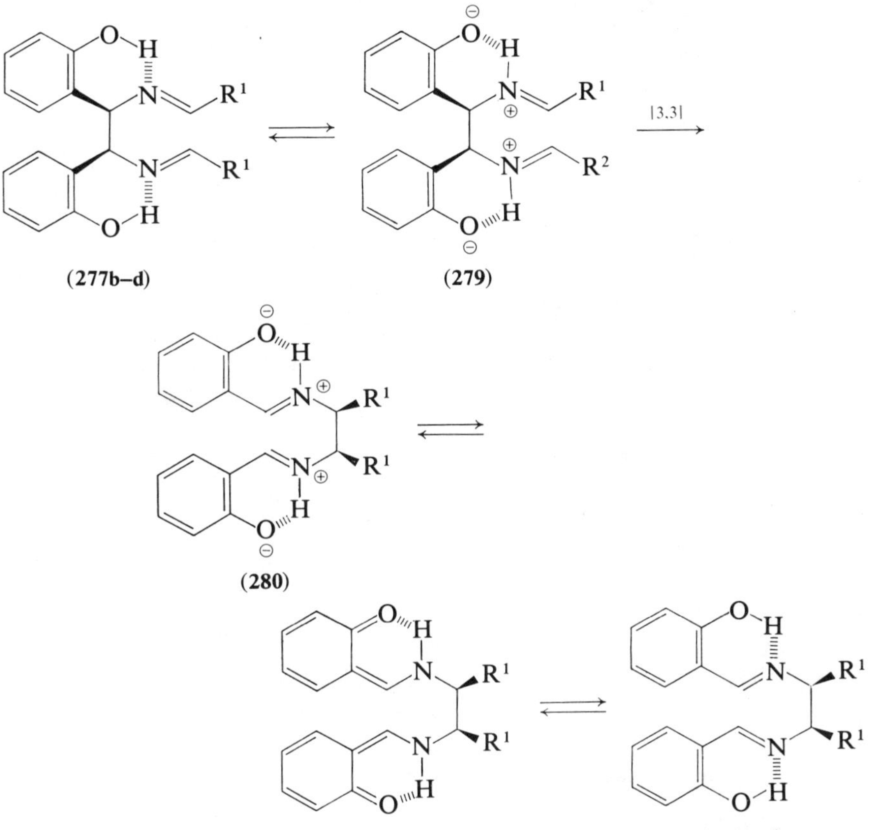

III. 3,4- AND 2,5-DIAZA-COPE REARRANGEMENTS

Scheme 65

simple MO treatment of hetero-Cope rearrangements (see Section II-C and Scheme 47) it is to be expected that 2,5-diaza-Cope systems show a strong rate acceleration in going from the neutral to the charged system. If it is accepted that those species rearrange faster owing to the influence of intramolecular hydrogen bridges (or intramolecular proton transfers) in which the nitrogen atoms bear a positive charge, then it can be concluded that the *meso* Schiff bases (derived from the *meso*-diamines) should be transformed into *meso-E,E* Schiff bases **278** via a boat-like activated complex. Thus hydrolysis of **278** must lead to *meso*-diamines.

On the other hand, *rac-E,E* Schiff bases of type **277** (R^2 = 2-HO-C_6H_4) should rearrange via a chairlike activated complex to yield again the isomeric *rac-E,E* Schiff bases of type **278**. Only when these conditions are met can two intramolecular hydrogen bridges be formed and maintained in the course of the reaction. The rearrangement of a *meso-E,E* Schiff base via a chairlike activated complex would, for steric reasons, allow the formation of only one hydrogen bridge.* It is known that even *meso*-3,4-diphenylhexa-1,5-diene rearranges thermally (120°C) to an extent of 37% via a boat-like transition state leading to *E,E*-1,6-diphenylhexa-1,5-diene (127). In general, the difference in the free energy of activation between the chairlike and boatlike activated complexes amounts to 4–6 kcal/mole (cf. ref. 39) for the Cope rearrangement. Thus it seems reasonable to assume that *meso-E,E* Schiff bases of type **277** (R^2 = 2-HO-C_6H_4) rearrange via a boat-like transition state, which is favored by intramolecular hydrogen bridges.

The equilibrium in the systems **277b–d** ⇌ **278b–d** lies to the side of **278b–d**. The reason for this is that 2-hydroxyphenylimines of type **278b–d** prefer the dienone–amine (vinyl analogue) structures of type **281** (cf. ref. 128).

(282)   (283)

In connection with this it should be noted that calculations performed by Dewar et al. (13) for the degenerate 2,5-diaza-Cope rearrangement of diazasemibullvalene **282** indicated that the "transition state" **283** (i.e., bishomopyrazine) has a lower energy than the reactant **282**.†

IV. Concluding Remarks

Mechanistic investigations dealing with the nature of the 3,3 rearrangements of iminium salts are scant. Generally kinetic data are missing. A complete stereochemical reactant–product correlation has never been performed.

Where it appeared reasonable to us we have treated the 3,3 rearrangements of iminium salts as sigmatropic processes. In some cases there are without doubt other mechanistic possibilities. For example, *N*-benzyl enammonium bromides rearrange—formally via a 1,3-benzyl shift—under the same conditions as the analogous *N*-allyl enammonium bromides to yield the corresponding iminium salts. The sigmatropic mechanism of a charge-accelerated 3-aza-Cope rearrangement can be formulated only for allyl compounds (see Scheme 2). The reaction of *N,N*-dimethylisobutenylamine (**22**)

---

* *rac-E,E* Schiff bases also can accommodate only one hydrogen bridge in a boat-like activated complex.

† The energy of activation for the degenerate Cope rearrangement in semibullvalene itself was calculated to amount to 2.3 kcal/mole (129).

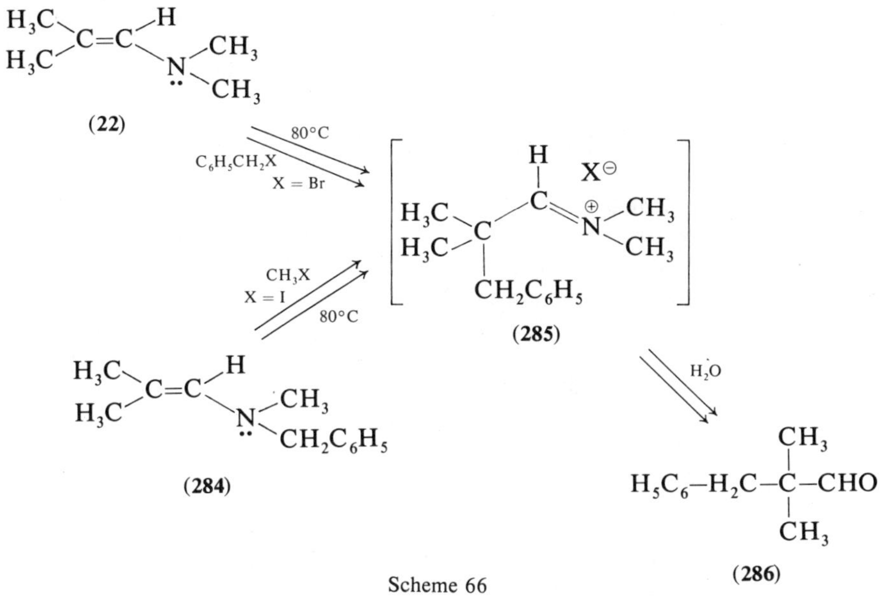

IV. CONCLUDING REMARKS

Scheme 66

with benzyl bromide in boiling acetonitrile yields the same iminium compound **285** as the reaction of *N*-benzyl-*N*-methylisobutenylamine (**284**) with methyl iodide under the same conditions. Both reaction mixtures afforded, after hydrolysis, α,α-dimethylhydrocinnamaldehyde (**286**) in yields of 37 and 47%, respectively (30) (see also refs. 23 and 26) (Scheme 66). Enammonium bromides of type **287** could be obtained directly by the action of benzyl bromide on enamines at low temperature (0°C) (27) (Scheme 67). They represent crystalline, stable compounds. On heating at 80–100°C in acetonitrile or above the melting point they rearrange to the corresponding iminium compounds **288**, whereas at 200°C in vacuum they fragment into benzyl bromide and the corresponding enamine (27). It can be concluded from these experiments that the alkylation of the enamines of aldehydes leads in the first step under kinetic control to *N*-alkyl enammonium salts whereas

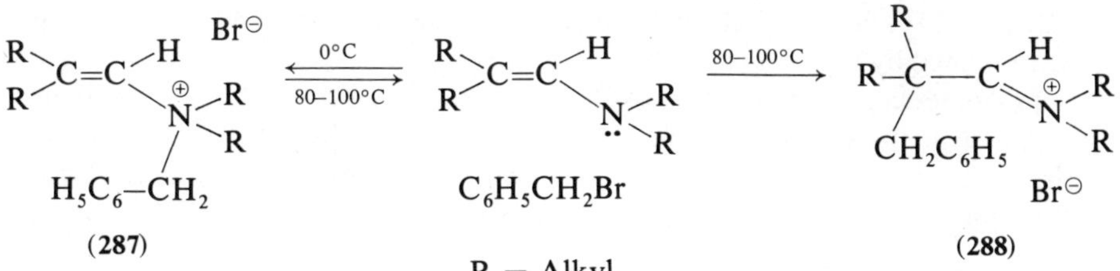

R = Alkyl

Scheme 67

thermodynamic control results in the formation of the *C*-alkylated iminium salts. Thus the $N \rightarrow C$ rearrangement of benzyl groups can be best understood as a combination of a S_N2 displacement by the bromide ion followed by a *C*-alkylation of the enamine (Scheme 67). If it is assumed that a similar two-step mechanism comes into play in the rearrangement of the allyl moiety from nitrogen to carbon (with inversion of the allyl chain), then a reaction sequence such as that presented in Scheme 68 is required. In this case one has to postulate that, for steric reasons (i.e., the bulkiness of the enammonium group), the bromide ion attacks the allyl group of **289** in an S_N2' fashion, followed by an S_N2 reaction to yield the iminium salt **290**.

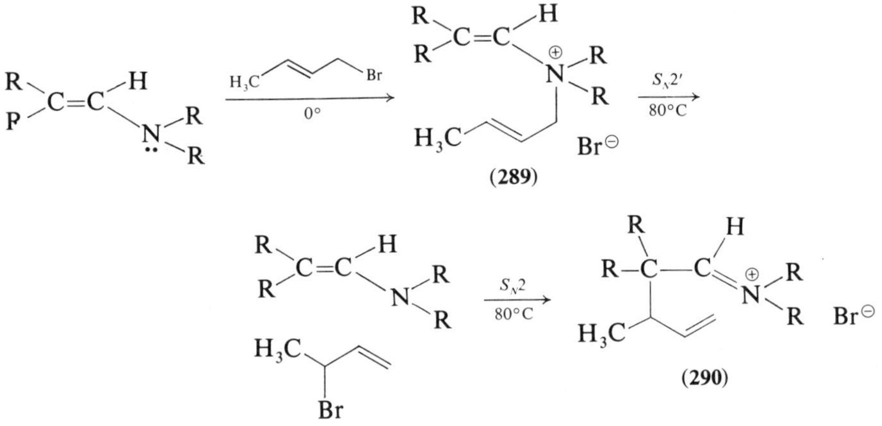

Scheme 68

However, this mechanistic sequence cannot be a general substitute for the concerted one-step 3-aza-Cope rearrangement. Some of the more significant points are as follows: (1) *N*-allyl enammonium bromides as well as salts containing the weakly nucleophilic *p*-toluenesulfonate anion rearrange evidently with comparable rates (see Scheme 3). (2) The bromides of the *N*-allylated *N,N*-dimethylanilinium ions **96** (Scheme 19) react even at 40°C in an S_N2 fashion to yield *N,N*-dimethylaniline and the corresponding allyl bromides (130). (3) The rearrangement of the *N*-propargyl enammonium bromide **65** (Scheme 13) into the *C*-allenyl iminium bromide **66** even at 80°C can hardly be explained by this mechanism.

In some cases there may exist a competition between the sigmatropic rearrangement and the discussed two-step reaction of Scheme 68. Under such circumstances the two-step mechanism should be able to be suppressed by choosing a nonnucleophilic anion (e.g., $B(C_6H_5)_4^{\ominus}$).* Other possible ways of

---

* An investigation of the influence of the anion on the rate of 3,3-rearrangements of enammonium and iminium salts evidently has not yet been performed.

IV. CONCLUDING REMARKS

formulating the activated complex of concerted 3,3-rearrangements have already been discussed in Section II-C (see Schemes 46 and 48).

A dissociative mechanism of rearrangement (analogous to Scheme 68) for the charge-accelerated 2-aza-Cope rearrangements seems unlikely because, in this case, strained aziridines would be expected as intermediates. The mechanism of the intermolecular benzyl shift in protonated 1,2-dihydroisoquinolines of type **172** (Scheme 37) to yield 3-benzylated 3,4-dihydroisoquinolium salts of type **173** is largely unknown. The rearrangement of optically active 1-benzyl-1,2-dihydroisoquinolines with chiral centers at C-1 and C-α is interesting in that it leads to optically active 3-benzylated 3,4-dihydroisoquinolines (75). 1-Cinnamyl-6,7-dimethoxy-2-methyl-1,2-dihydroisoquinoline undergoes, in the presence of acid, a formal 1,3 shift of the cinnamyl moiety (see Section II-B-2) but no [3,3]-sigmatropic rearrangement.

The charge-accelerated 3,4-diaza-Cope rearrangements raise similar mechanistic problems to those of the benzidine rearrangement (see ref. 97c).

In recapitulation it may be said that the charge-accelerated rearrangements of iminium salts open a series of interesting synthetic possibilities. In particular the following reactions may be emphasized: (1) the allylation and propargylation of enamines which occur with inversion and take place under much milder conditions than other methods leading also to α-allylated or α-allenyl aldehydes (cf. the acid-catalyzed decomposition of diallyl acetals; see ref. 131 and literature cited therein); (2) the acylation of enamines which allows the formation of bicyclic diketones; and (3) the Lewis acid-catalyzed rearrangement of N-allylated to C-allylated anilines. This is particularly valuable because, in contrast to the aromatic Claisen rearrangement of allyl aryl ethers, the thermal, uncatalyzed rearrangement of N-allylated anilines does not occur.

The 2-aza-Cope rearrangements are also of synthetic significance, for example, (1) the N-butenylation of tetrahydronorharmanes, tetrahydroisoquinolines, and related systems which can be used to build condensed ring systems (see Scheme 34); and (2) the synthesis of 3-allylated dihydro- or tetrahydroisoquinolines. This reaction should also be applicable to other nitrogen-containing ring systems (see Scheme 43).

It is not necessary to stress the importance of the Fischer indole synthesis and the Piloty–Robinson pyrrole synthesis.

ACKNOWLEDGMENT

We would like to thank Dr. D. G. Leppard (Institut de chimie organique de l'université, CH-1705 Fribourg) for his useful comments on the English text.

Our own work cited here was supported by the Swiss National Research Foundation.

REFERENCES

1. R. B. Woodward and R. Hoffmann, *Angew. Chem.*, **81**, 797 (1969); *Die Erhaltung der Orbitalsymmetrie*, Verlag Chemie GmbH, Weinheim, 1970.
2. W. von E. Doering, V. G. Toscano, and G. H. Beasley, *Tetrahedron*, **27**, 5299 (1971).
3. R. K. Hill and G. R. Newkome, *Tetrahedron Letters*, **1968**, 5059.
4. R. K. Hill and N. W. Gilman, *Tetrahedron Letters*, **1967**, 1421.
5. J. Corbier and P. Cresson, *Compt. Rend.*, **270**, 2077 (1970).
6. E. Winterfeldt, *Fortschr. Chem. Forsch.*, **16**, 75 (1970).
7. S. W. Benson, *Thermochemical Kinetics*, John Wiley and Sons, New York, 1968; S. W. Benson, F. R. Cruickshank, D. M. Golden, G. R. Haugen, N. E. O'Neal, A. S. Rodgers, R. Shaw, and R. Walsh, *Chem. Rev.*, **69**, 279 (1969).
8. G. Maier, *Valenzisomerisierungen*, Verlag Chemie GmbH, Weinheim, 1972, pp. 197ff.
9. L. Hoesch, Thesis, University of Zurich, 1974; J. E. Franz and C. Osuch, *Chem. Ind. London*, **1964**, 2058; A. C. Oehlschlager and L. H. Zalkow, *J. Org. Chem.*, **30**, 4205 (1965); A. G. Anastassiou, *J. Org. Chem.*, **31**, 1131 (1966).
10. R. H. C. Elgersma and E. Havinga, *Tetrahedron Letters*, **1969**, 1735.
11. P. Schiess and A. Grieder, *Tetrahedron Letters*, **1969**, 2097.
12. F. Vögtle and E. Goldschmitt, *Angew. Chem.* **85**, 824 (1973).
13. M. J. S. Dewar, Z. Nahlovska, and B. D. Nahlovsky, *Chem. Commun.*, **1971**, 1377; M. J. S. Dewar and D. H. Lo, *J. Am. Chem. Soc.*, **93**, 7201 (1971).
14. R. V. Stevens, E. E. McEntire, W. E. Barnett, and E. Wenkert, *Chem. Commun.*, **1973**, 662.
15. W. Gerrard, M. F. Lappert, and H. B. Silver, *Proc. Chem. Soc.*, **1957**, 19.
16. R. Barner, J. Borgulya, R. Madeja, P. Fahrni, H.-J. Hansen, and H. Schmid, *Helv. Chim. Acta*, **56**, 14 (1973).
17. U. Svanholm and V. D. Parker, *Chem. Commun.*, **1972**, 645.
18. U. Widmer, H.-J. Hansen, and H. Schmid, *Helv. Chim. Acta*, **56**, 2644 (1973).
19. U. Widmer, J. Zsindely, H.-J. Hansen, and H. Schmid, *Helv. Chim. Acta*, **56**, 75 (1973); U. Widmer, H.-J. Hansen, and H. Schmid, *Helv. Chim. Acta*, **56**, 1895 (1973).
20. H.-J. Hansen, B. Sutter, and H. Schmid, *Helv. Chim. Acta*, **51**, 828 (1968); H. Heimgartner, J. Zsindely, H.-J. Hansen, and H. Schmid, *Helv. Chim. Acta*, **55**, 1113 (1972).
21. M. Schmid, H.-J. Hansen, and H. Schmid, *Helv. Chim. Acta*, **54**, 937 (1971).
22. A. G. Cook, Ed., *Enamines*, Marcel Dekker, New York, 1969; S. Hünig and H. Hoch, *Fortschr. Chem. Forsch.*, **14**, 235 (1970); H. O. House, *Modern Synthetic Reactions*, 2nd ed., W. A. Benjamin, Menlo Park, Calif., 1972, pp. 570ff; S. F. Dyke, *The Chemistry of Enamines*, Cambridge University Press, Cambridge, 1973.
23. G. Opitz and H. Mildenberger, *Angew. Chem.*, **72**, 169 (1960).
24. E. Elkik, *Bull. Soc. Chim. France*, **1960**, 972.
25. G. Opitz and H. Mildenberger, *Ann. Chem.*, **649**, 26 (1961).
26. G. Opitz, H. Hellmann, H. Mildenberger, and H. Suhr, *Ann. Chem.*, **649**, 36 (1961).
27. (a) E. Elkik and A. Kirrmann, *Compt. Rend.*, **267**, 623 (1968); (b) E. Elkik and C. Francesch, *Bull. Soc. Chim. France*, **1969**, 903.
28. G. Opitz, *Ann. Chem.*, **650**, 122 (1961).
29. P. Cresson and J. Corbier, *Compt. Rend.*, **268**, 1614 (1969).
30. K. C. Brannock and R. D. Burpitt, *J. Org. Chem.*, **26**, 3576 (1961).
31. P. M. McCurry, Jr., and R. K. Singh, *Tetrahedron Letters*, **1973**, 3325.
32. H. M. Frey and R. K. Solly, *Trans. Faraday Soc.*, **65**, 1372 (1969).
33. H. M. Frey and R. K. Solly, *Trans. Faraday Soc.*, **64**, 1858 (1968).

34. G. Opitz, H. Mildenberger, and H. Suhr, *Ann. Chem.*, **649**, 47 (1961).
35. M. E. Kuehne and T. Garbacik, *J. Org. Chem.*, **35**, 1555 (1970).
36. P. Houdewind, Thesis, University of Amsterdam, 1973.
37. G. Fráter and H. Schmid, *Helv. Chim. Acta*, **53**, 269 (1970).
38. P. Vittorelli, T. Winkler, H.-J. Hansen, and H. Schmid, *Helv. Chim. Acta,* **51**, 1457 (1968); P. Vittorelli, H.-J. Hansen, and H. Schmid, *Helv. Chim. Acta*, **58**, 1293 (1975).
39. H.-J. Hansen and H. Schmid, *Tetrahedron*, **30**, 1959 (1974).
40. G. Stork, A. Brizzolara, H. Landesman, J. Szmuszkovicz, and R. Terrell, *J. Am. Chem. Soc.*, **85**, 207 (1969).
41. P. W. Hickmott and J. R. Hargreaves, *Tetrahedron*, **23**, 3151 (1967).
42. J. R. Hargreaves, P. W. Hickmott, and B. J. Hopkins, *J. Chem. Soc. C*, **1968**, 2599.
43. J. R. Hargreaves, P. W. Hickmott, and B. J. Hopkins, *J. Chem. Soc. C*, **1969**, 592.
44. N. F. Firrell and P. W. Hickmott, *J. Chem. Soc. C*, **1968**, 2320.
45. A. C. Cope, D. L. Nealy, P. Scheiner, and G. Wood, *J. Am. Chem. Soc.*, **87**, 3130 (1965).
46. P. W. Hickmott, G. J. Miles, G. Sheppard, R. Urbani, and C. T. Yoxall, *J. Chem. Soc. Perkin I*, **1973**, 1514.
47. P. W. Hickmott, B. J. Hopkins, G. Sheppard, and D. J. Barraclough, *J. Chem. Soc. Perkin I*, **1972**, 1639.
48. (a) P. W. Hickmott and G. Sheppard, *J. Chem. Soc. C*, **1971**, 1358; (b) **1971**, 2112.
49. P. W. Hickmott and B. J. Hopkins, *J. Chem. Soc. C*, **1968**, 2918.
50. S. Marcinkiewicz, J. Green, and P. Mamalis, *Tetrahedron*, **14**, 208 (1961); S. Marcinkiewicz, *Bull. Acad. Polon. Sci. Ser. Sci. Chim.*, **19**, 603, 609 (1971).
51. J. M. Patterson, J. W. de Haan, M. R. Boyd, and J. D. Ferry, *J. Am. Chem. Soc.*, **94**, 2487 (1972).
52. F. L. Carnahan and C. D. Hurd, *J. Am. Chem. Soc.*, **52**, 4586 (1930).
53. M. Schmid, H.-J. Hansen, and H. Schmid, *Helv. Chim. Acta*, **56**, 105 (1973).
54. S. Jolidon and H.-J. Hansen, *Chimia*, **30**, 21 (1976).
55. J. E. Hyre and A. R. Bader, *J. Am. Chem. Soc.*, **80**, 437 (1958).
56. A. R. Bader, R. J. Bridgewater, and P. R. Freeman, *J. Am. Chem. Soc.*, **83**, 3319 (1961).
57. C. George, E. W. Gill, and J. A. Hudson, *J. Chem. Soc. C*, **1970**, 74.
58. E. B. Towne and H. M. Hill, U.S. Pat. 2,607,779 (1952); *C.A.*, **47**, 5452 (1953).
59. J. A. Degutis and V. P. Barkanskas, *Khim. Geterotsikl. Soedin.*, **1969**, 1003; *C.A.*, **73**, 35147 (1970); *Russ. J. Heterocycl. Chem.*, **1972**, 752.
60. B. McDonald, A. McLean, and G. R. Proctor, *Chem. Commun.*, **1973**, 208.
61. G. Casnati and A. Pocchini, *Chem. Commun.*, **1973**, 1328.
62. C. D. Hurd and W. W. Jenkins, *J. Org. Chem.*, **22**, 1418 (1957).
63. M. Elliott and N. F. Janes, *J. Chem. Soc. C*, **1967**, 1780.
64. N. Takamatsu, S. Inoue, and Y. Kishi, *Tetrahedron Letters*, **1971**, 4661.
65. J. V. Dubsky and A. Rabas, *Collect. Trav. Chim. Tchecosl.*, **1**, 528 (1926).
66. K. Krauch and W. Kunz, *Reaktionen in der Organischen Chemie*, A. Hüthig Verlag, Heidelberg, 1966, p. 208; J. E. Gowan and T. S. Wheeler, *Name Index of Organic Reactions*, Longmans, London, 1960, p. 81.
67. G. Opitz and A. Griesinger, *Ann. Chem.*, **665**, 101 (1963).
68. C. A. Grob and P. W. Schiess, *Angew. Chem.*, **79**, 1 (1967); C. A. Grob, *Angew. Chem.*, **81**, 543 (1969).
69. R. M. Horowitz and T. A. Geissman, *J. Am. Chem. Soc.*, **72**, 1518 (1950).
70. (a) E. Winterfeldt and W. Franzischka, *Chem. Ber.*, **100**, 3801 (1967); (b) **101**, 2938 (1968).
71. H. Rischke, J. D. Wilcock, and E. Winterfeldt, *Chem. Ber.*, **106**, 3106 (1973).
72. M. Sainsbury, S. F. Dyke, D. W. Brown, and R. G. Kinsman, *Tetrahedron*, **26**, 5265 (1970).

73. J. Knabe, J. Kubitz, and N. Ruppenthal, *Angew. Chem.*, **75**, 981 (1963); J. Knabe and J. Kubitz, *Arch. Pharm.*, **297**, 129 (1964).
74. J. Knabe and K. Detering, *Chem. Ber.*, **99**, 2873 (1966).
75. J. Knabe and R. Dörr, *Arch. Pharm.*, **306**, 784 (1973); see also J. Knabe, R. Dörr, S. F. Dyke, and R. G. Kinsman, *Tetrahedron Letters*, **1972**, 5373.
76. S. F. Dyke, *Advances in Heterocyclic Chemistry*, Vol. 14, 1972, p. 279.
77. J. Knabe, "Iminium Salts in Nature," this book.
78. M. Sainsbury, D. W. Brown, S. F. Dyke, R. G. Kinsman, and B. J. Moon, *Tetrahedron*, **24**, 6695 (1968).
79. (a) J. Knabe and H.-D. Höltje, *Tetrahedron Letters*, **1969**, 433; (b) *Arch. Pharm.*, **303**, 404 (1970).
80. (a) D. W. Brown, S. F. Dyke, and M. Sainsbury, *Tetrahedron*, **25**, 101 (1969); (b) D. W. Brown, S. F. Dyke, R. G. Kinsman, and M. Sainsbury, *Tetrahedron Letters*, **1969**, 1731.
81. J. M. Bobbitt and J. C. Sih, *J. Org. Chem.*, **33**, 856 (1968).
82. S. F. Dyke, R. G. Kinsman, J. Knabe, and H.-D. Höltje, *Tetrahedron*, **27**, 6181 (1971).
83. R. G. Kinsman, S. F. Dyke, and J. Mead, *Tetrahedron*, **29**, 4303 (1973).
84. J. Knabe and H.-D. Höltje, *Tetrahedron Letters*, **1969**, 2107.
85. J. Knabe and R. Saggau, *Arch. Pharm.*, **306**, 500 (1973).
86. J. A. Marshall and J. H. Babler, *J. Org. Chem.*, **34**, 4186 (1969).
87. C. A. Grob, H. R. Kiefer, H. J. Rutz, and H. J. Wilkens, *Helv. Chim. Acta*, **50**, 416 (1967); M. Geisel, C. A. Grob, and R. A. Wohl, *Helv. Chim. Acta*, **52**, 2206 (1969).
88. C. A. Grob, H. Link, and P. W. Schiess, *Helv. Chim. Acta*, **46**, 483 (1963).
89. M. J. S. Dewar, *Angew. Chem.*, **83**, 859 (1971).
90. H. E. Zimmerman, *Accounts Chem. Res.*, **4**, 272 (1971).
91. K. Fukui, *Accounts Chem. Res.*, **4**, 57 (1971); *Fortschr. Chem. Forsch.*, **15**, 1 (1970).
92. W. v. E. Doering, V. G. Toscano, and G. H. Beasley, *Tetrahedron*, **27**, 5299 (1971).
93. M. J. S. Dewar and L. E. Wade, *J. Am. Chem. Soc.*, **95**, 290 (1973).
94. E. Heilbronner and H. Bock, *Das HMO-Modell und seine Anwendung*, Vols. 1–3, Verlag Chemie GmbH, Weinheim, 1968–1970.
95. Ref. 94, Vol. 3, p. 218; see also A. Streitwieser, Jr., *Molecular Orbital Theory for Organic Chemists*, John Wiley & Sons, New York, 1961.
96. P. A. Kollman, "The Electronic Structure of Iminium Ions," *Iminium Salts in Organic Chemistry*, Part 1, H. Böhme and H. G. Viehe, Eds., John Wiley & Sons, New York, 1976, p. 1.
97. (a) B. Robinson, *Chem. Rev.*, **63**, 373 (1963); (b) **69**, 227 (1969); (c) H. J. Shine, *Aromatic Rearrangements*, Elsevier Publishing Company, Amsterdam, 1967, 190.
98. O. Piloty, *Ber. Deut. Chem. Ges.*, **43**, 489 (1910).
99. G. M. Robinson and R. Robinson, *J. Chem. Soc.*, **113**, 639 (1918).
100. G. M. Robinson and R. Robinson, *J. Chem. Soc.*, **125**, 827 (1924).
101. R. B. Carlin and E. E. Fisher, *J. Am. Chem. Soc.*, **70**, 3421 (1948).
102. P. W. Neber, G. Knöller, K. Herbst, and A. Trissler, *Ann. Chem.*, **471**, 113 (1929).
103. R. H. C. Elgersma, Thesis, University of Leiden, 1969.
104. O. Diels and J. Reese, *Ann. Chem.*, **511**, 168 (1934); **519**, 147 (1935).
105. N. N. Suvorov, N. P. Sorokina, and Y. P. Sheinker, *J. Gen. Chem. USSR*, **28**, 1090 (1958); *C.A.*, **52**, 18373 (1958); N. N. Suvorov and N. P. Sorokina, *Dokl. Akad. Nauk SSSR*, **136**, 840 (1961); *C.A.*, **55**, 17621 (1961).
106. A. Grieder and P. Schiess, *Chimia*, **24**, 25 (1970); A. Grieder, Thesis, University of Basel, 1970.
107. I. I. Grandberg and N. M. Przheval'skii, *Khim. Geterotsikl. Soedin.*, **1969**, 943.
108. J. Elguero and C. Marzin, "*N*-Heteroiminium Salts," *Iminium Salts in Organic Chemistry*, Part 1, H. Böhme and H. G. Viehe, Eds., John Wiley & Sons, New York, 1976, p. 533.

109. P. D. Croce, *Ann. Chim. Rome*, **63**, 29 (1973).
110. I. I. Grandberg and D. V. Sibiryakova, USSR Pat. 199,892; *C.A.*, **68**, 95677 (1968); I. I. Grandberg, D. V. Sibiryakova, and D. V. Brovkin, *Khim. Geterotsikl. Soedin.*, **1969**, 94; *C.A.* **70**, 114,931 (1969); I. I. Grandberg, *Izv. Timiryazev. Sel'skokhoz. Akad.*, **1972**, 188.
111. F. Scheinmann, R. Barner, and H. Schmid, *Helv. Chim. Acta*, **51**, 1603 (1968).
112. T. Sheradsky, *Tetrahedron Letters*, **1966**, 5225: *J. Heterocycl. Chem.*, **4**, 413 (1967).
113. A. Mooradian, *Tetrahedron Letters*, **1967**, 407; A. Mooradian and P. E. Dupont, *J. Heterocycl. Chem.*, **4**, 441 (1967).
114. D. Kaminsky, J. Shavel, Jr., and R. I. Meltzer, *Tetrahedron Letters*, **1967**, 859.
115. W. Sucrow, *Chimia*, **23**, 36 (1969).
116. E. Schmitz and H. Fechner, *Org. Prep. Proced.*, **1**, 253 (1969).
117. R. Jacquier, J.-P. Chapelle, J. Elguero, and G. Tarrago, *Chem. Commun.*, **1969**, 752; J.-P. Chapelle, J. Elguero, R. Jacquier, and G. Tarrago, *Bull. Soc. Chim. France*, **1970**, 240.
118. W. Sucrow and G. Chondromatidis, *Chem. Ber.*, **103**, 1759 (1970).
119. W. Sucrow, H. Bethe, and G. Chondromatidis, *Tetrahedron Letters*, **1971**, 1481.
120. J.-P. Chapelle, J. Elguero, R. Jacquier, and G. Tarrago, *Bull. Soc. Chim. France*, **1970**, 3147; see also **1969**, 4464; **1971**, 280.
121. W. Sucrow and E. Wiese, *Chem. Ber.*, **103**, 1767 (1970).
122. H. A. Staab and F. Vögtle, *Chem. Ber.*, **98**, 2681, 2691 (1965).
123. H. A. Staab and F. Vögtle, *Chem. Ber.*, **98**, 2701 (1965).
124. F. Vögtle and E. Goldschmitt, *Chem. Ind. London*, **1973**, 1072.
125. S. Tripett, *J. Chem. Soc.*, **1957**, 4407.
126. J. Bernstein and G. M. J. Schmidt, *J. Chem. Soc. Perkin II*, **1972**, 951.
127. R. P. Lutz, S. Bernal, R. J. Boggio, R. O. Harris, and M. W. McNicholas, *J. Am. Chem. Soc.*, **93**, 3985 (1971).
128. G. O. Dudek and E. P. Dudek, *J. Am. Chem. Soc.*, **86**, 4283 (1964); **88**, 2407 (1966).
129. M. J. S. Dewar and W. W. Schoeller, *J. Am. Chem. Soc.*, **93**, 1481 (1971).
130. E. C. F. Ko and K. T. Leffek, *Can. J. Chem.*, **50**, 1297 (1972).
131. D. J. Faulkner, *Synthesis*, **1971**, 175.

ADDENDUM

To Section I

A classification for sigmatropic rearrangements of "hetero-Cope" systems, including charged species, was suggested (132).

To Section II-A

The thermal and acid-catalyzed rearrangement of a series of 2,3-diacyl-2,3-diaza-bicyclo[2.2.1]hept-5-enes and -hept-5-en-7-ones as well as of 2,3-diacyl-2,3-diaza-bicyclo[2.2.2]oct-5-enes (1,3-oxaza-Cope systems) was studied (133).

Allyl-propynyl ammonium ions were isomerized in boiling water, containing a trace of sodium hydroxide, into the corresponding allyl-allenyl cations which partially underwent a [3,3]-sigmatropic rearrangement (134).

To Section II-A-1

Protonation and deprotonation of enamines was investigated using ^{1}H and ^{13}C NMR spectroscopy (135). The results of the allylation of cyclic enamines and diene amines, already briefly discussed on pp. 617 and 619, respectively, have been published (136, 137). Spiro[4.5]decan-1-one-4-carboxylates were prepared by allylation of pyrrolidine enamines of acetyl cyclopentanes with methyl α-(1-bromoethyl)acrylate (138). Enamines derived from sterically hindered amines undergo mainly C-alkylation (139). Tricyclic hetero-aromatic compounds were prepared by 2-aza-Cope rearrangements (140).

To Section II-A-2

A review on the acylation of enamines has appeared (141). The acylation of enamines of cyclohexanones was further investigated (142).

To Section II-A-3

The thermal and acid-catalyzed amino-Claisen rearrangement of N-allyl, N-(α-methylallyl), and N-(α,α-dimethylallyl)anilines were thoroughly investigated (143). Treatment of N-(2'-deuterioallyl)-N-methylaniline with ZnCl$_2$ in boiling xylene gave 2-(2'-deuterioallyl)aniline (144). Rearrangement of N-allyl-3,5-dimethylaniline into the 2-allyl isomer was achieved by using ZnCl$_2$ in boiling xylene (145). Rearrangement of N-allylanilines into 2-allylanilines was performed in boiling water–alcohol mixtures in the presence of hydrochloric acid or in phosphoric acid (146). 4-(N-β-Chloroallylamino)quinolines rearrange in polyphosphoric acid to yield pyrrolo[2,3-c]quinolines (147).

A new synthesis of 2,3-disubstituted indoles was achieved by treating N-propargylated anilines with m-chloroperbenzoic acid in methylene chloride at room temperature (148). A full paper concerning the acid-catalyzed rearrangements of N-allylated indoles (see scheme 21) has appeared (149). Pure thermal rearrangements are discussed in (150). The rearrangement of N-allyl indoles into 3-allyl indoles catalyzed by ZnCl$_2$, AlCl$_3$ or TiCl$_4$ was investigated (151).

To Section II-B-1

Yohimbane derivatives were synthesized by a reaction sequence including 2-aza-Cope rearrangements in iminium ions (152).

To Section II-B-2

Trapping of an intermediate ion by chloride ions present in the acid-catalyzed rearrangement of 1-allyl-1,2-dihydroisoquinolines was reported (153). For further acid-catalyzed rearrangements of 1,2-dihydroisoquinolines see ref. 154.

5-Allyl-3,3,5-trimethyl-1-pyrroline-1-oxide undergoes thermal charge-accelerated Cope rearrangement to 5-allyl-2,4,4-trimethyl-1-pyrroline-1-oxide and further cyclization to 1,6,6-trimethyl-9-oxa-8-azatricyclo[3.2.1.1^{3,8}]-nonane (155).

To Section II-B-3

1-Methylidene-2,3,6,7-tetrahydroazepinium ions obtained by solvolysis of endo-4-tosyloxy-1-azabicyclo[3.2.1]octane rearrange partially at 20°C presumably via a chairlike transition state to give 1-methylidene-3-vinyl-pyrrolidinium ions (156). The carbon analogue 1-methylidene-cyclohept-4-ene is in equilibrium with 1-methylidene-3-vinyl-cyclopentane at 350°C (157). 1-Methylidene-1,2,3,4,7,8-hexahydroazocinium ions are quantitatively transformed into 1-methylidene-3-vinyl-piperidinium ions at 20°C (156).

To Section III-A

A full paper concerning the reactions discussed in schemes 53 and 54 has appeared (158). The thermal and acid-catalyzed rearrangement of N-methyl-N-phenyl-N'-acetyl-enhydrazines was investigated (159). A study on the thermal and acid-catalyzed Fischer indole synthesis with phenylhydrazones derived from 4-substituted cyclohexanones and with comparable enhydrazines was published (160). The thermal and acid-catalyzed 3,4-diaza-Cope rearrangement of N-phenyl-N'-(9-phenanthryl)hydrazine was studied (161). Fischer indole syntheses with 1-amino-1,2,3,4-tetrahydroquinoline derivatives as hydrazine components were studied. See ref. 162 and literature cited there.

A review concerning Fischer synthesis of indoles and some related reactions has appeared (163). For further examples see ref. 164. Fischer indole syntheses with 1-phenylpyrazolidine as discussed in scheme 55 were already performed by A. N. Kost et al. in 1970 (165). For further examples see ref. 166. Fischer indole syntheses, using 1-phenylpyrazolidines and 1-phenylhexahydropyridazines were reported (167). 1-Phenylpyrazolidin-5-ones and cyclic ketones, in the presence of hydrochloric acid, undergo Fischer indole synthesis (1968).

Benzofurane formation from O-phenyl oxime ethers (see scheme 57) are dependent on the substituents in the oxime part. Hydrogen as substituent favors the formation of o-amino phenol (169). The Piloty–Robinson syntheses of pyrroles were investigated (170). Carbazole derivatives were prepared starting with enhydrazines of 2-cyclohexenones and cyclohexanone (cf. scheme 61) (171).

To Section III-B

Double Schiff bases of a cis-diamino cyclopropane rearrange rapidly already at 0°C to yield the corresponding 2,3-cis-disubstituted 2,3-dihydro-1,4-diazepines. However, the Schiff base of pivalaldehyde and cis-diamino cyclopropane is slowly transformed into the 1,4-diazepine at 100°C (172). 2,5-Diaza-Cope rearrangements of double Schiff bases (see schemes 64 and 65) were further investigated (132, 173).

Diprotonated 2,5-diaza-3,4-homotropilidene which is isoelectronic with 3,4-homotropilidene undergoes already at 30°C a degenerate Cope rearrangement ($\mathit{\Gamma G}_{100°C}^{\neq} \approx 17.5$ kcal/mol). A similar behavior is shown in the case of the 1,6-dimethyl derivative (174).

REFERENCES

132. F. Vögtle and E. Goldschmitt, *Chem. Ber.*, **109**, 1 (1976).
133. C. Y.-J. Chung, D. Mackay, and T. D. Sauer, *Can. J. Chem.*, **50**, 1568 (1972); C. P. R. Jennison and D. Mackay, *Tetrahedron*, **29**, 1255 (1973); D. Mackay and C. W. Pilger, *Can. J. Chem.*, **52**, 1114 (1974); J. A. Campbell, I. Harris, D. Mackay, and T. D. Sauer, *Can. J. Chem.*, **53**, 535 (1975); W. G. Barrett and D. Mackay, *J. Chem. Soc. Perkin I*, **1975**, 1046.
134. T. Laird and W. D. Ollis, *Chem. Commun.*, **1972**, 557.
135. L. Nilsson, R. Carlson, and C. Rappe, *Acta Chem. Scand.*, **B30**, 271 (1976).
136. P. Houdewind, U. K. Pandit, A. K. Bose, R. J. Brambilla, and G. L. Traynor, *Heterocycles*, **1**, 53 (1973).
137. P. Houdewind and U. K. Pandit, *Tetrahedron Lett.*, **1974**, 2359.
138. D. J. Dunham and R. G. Lawton, *J. Am. Chem. Soc.*, **93**, 2074 (1971).
139. T. J. Curphey, J. Ch. Hung, and Ch. Ch. Ch. Chu, *J. Org. Chem.*, **40**, 607 (1975).
140. S. Blechert, R. Gericke, and E. Winterfeldt, *Chem. Ber.*, **106**, 355 (1973).
141. P. W. Hickmott, *Chem. Ind.*, **1974**, 731.
142. P. W. Hickmott, P. J. Cox, and G. A. Sim, *J. Chem. Soc. Perkin I*, **1974**, 2544; P. W. Hickmott, K. N. Woodward, and R. Urbani, *J. Chem. Soc. Perkin I*, **1975**, 1885.
143. S. Jolidon and H.-J. Hansen, *Chimia*, **30**, 23 (1976); **31**, 46 (1977); *Helv. Chim. Acta*, **60**, 978 (1977).
144. U. Koch-Pomeranz, H. Schmid, and H.-J. Hansen, *Helv. Chim. Acta*, **60**, 768 (1977).
145. R. Wehrli, H. Heimgartner, H. Schmid, and H.-J. Hansen, *Helv. Chim. Acta*, **60**, 2034 (1977).
146. G. de Saqui-Sannes, M. Rivière, and A. Lattes, *Tetrahedron Lett.*, **1974**, 2073; K. Krowicki, N. Paillous, M. Rivière, and A. Lattes, *J. Heterocycl. Chem.*, **13**, 555 (1976); A Lattes, in R. N. Castle and M. Tisler, Eds., *Lectures in Heterocyclic Chemistry*, Vol. 3, HeteroCorporation Orem (Utah), 1976, p. S-93.
147. B. G. McDonald and G. R. Proctor, *J. Chem. Soc. Perkin I*, **1975**, 2073.
148. B. S. Thyagarajan, J. B. Hillard, K. V. Reddy, and K. C. Majumdar, *Tetrahedron Lett.* **1974**, 1999; J. B. Hillard, K. V. Reddy, K. C. Majumdar, and B. S. Thyagarajan, *J. Heterocycl. Chem.*, **11**, 369 (1964); B. S. Thyagarajan and K. C. Majumdar, *J. Heterocycl. Chem.*, **12**, 43 (1975).
149. G. Casnati, R. Marchelli, and A. Pocchini, *J. Chem. Soc. Perkin I*, **1974**, 754.

150. J. M. Patterson, A. Wu, Ch. S. Kook, and W. T. Smith, Jr., *J. Org. Chem.*, **39,** 486 (1974); J. M. Patterson, Ch. F. Mayer, and W. T. Smith, Jr., *J. Org. Chem.*, **39,** 1511 (1975).
151. S. Inada, K. Nagai, Y. Takayanagi, and M. Okazaki, *Bull. Chem. Soc. Japan*, **49,** 833 (1976).
152. J. D. Wilcock and E. Winterfeldt, *Chem. Ber.*, **107,** 975 (1974).
153. R. G. Kinsman and S. F. Dyke, *Tetrahedron Lett.*, **1975,** 2231.
154. R. G. Kinsman, A. W. C. White, and S. F. Dyke, *Tetrahedron*, **31,** 449 (1975); J. Knabe and A. Frie, *Arch. Pharm.*, **306,** 592 (1973); J. Knabe and A. Ecker, *Arch. Pharm.*, **307,** 727 (1974).
155. J. B. Bapat, D. St. C. Black, R. F. C. Brown, and C. Ichlov, *Aust. J. Chem.*, **25,** 2445 (1972).
156. W. Kunz, thesis, University of Basel, 1973; P. R. Marbet, thesis, University of Basel, 1975; K. B. Becker and C. A. Grob, in S. Patai, Ed., *The Chemistry of Double-bonded Functional Groups*, Supplement A, Part 2, John Wiley and Sons, New York, 1977, p. 653.
157. R. W. Thies and L. E. Schick, *J. Am. Chem. Soc.*, **96,** 456 (1974).
158. P. Schiess and A. Grieder, *Helv. Chim. Acta*, **57,** 2643 (1974).
159. E. B. Sendi, thesis, University of Basel, 1973.
160. H. Posvic, R. Dombro, H. Ito, and T. Telinski, *J. Org. Chem.*, **39,** 2575 (1974).
161. B. J. Anret, R. G. R. Bacon, R. Bankhead, D. C. H. Bigg, and J. S. Ramsey, J. Chem. Soc. Perkin I, **1974,** 2153.
162. R. Fusco and F. Sannicolo, *Gazz. Chim. Ital.*, **106,** 85 (1976).
163. I. I. Grandberg, *Khim. Geterotsikl. Soedin.*, **1974,** 579; *C. A.*, **81,** 63397 (1974).
164. I. I. Grandberg and T. P. Moskvina, *Khim. Geterotsikl. Soedin.*, **1974,** 90; *C. A.*, **80,** 108292 (1974); N. M. Przheval'skii and I. I. Grandberg, *Khim. Geterotsikl. Soedin.*, **1974,** 1581; *C. A.*, **82,** 86261 (1975).
165. A. N. Kost, L. A. Sviridova, G. A. Golubeva, and Yu. N. Portnov, *Khim. Geterotsikl. Soedin.*, **1974,** 371; *C. A.*, **73,** 25205 (1970).
166. L. G. Yudin, A. N. Kost, and N. B. Chernyskova, *Khim. Geterotsikl. Soedin.*, **1970,** 484; *C. A.*, **73,** 45380 (1970); A. N. Kost, G. A. Golubeva, and Yu. N. Portnov, *Dokl. Akad. Nauk SSSR*, **200,** 342 (1971); *C. A.*, **76,** 34047 (1972); A. N. Kost, N. Yu. Lebedenko, and L. A. Sviridova, *J. Org. Chem. USSR*, **12,** 2374 (1976).
167. M. K. Eberle, G. G. Kahle, and S. M. Talti, *Tetrahedron*, **29,** 4045 (1973); M. K. Eberle and G. G. Kahle, *Tetrahedron*, **29,** 4049 (1973).
168. M. K. Eberle and L. Brzechffa, *J. Org. Chem.*, **41,** 3775 (1976).
169. M. A. Carter and B. Robinson, *Chem. Ind.*, **1974,** 304.
170. R. Baumes, R. Jaquier, and G. Tarrago, *Bull. Soc. Chim. France*, **1973,** 317; **1974,** 1147.
171. W. Sucrow, M. Slopianka, and C. Mentzel, *Chem. Ber.*, **106,** 745 (1973).
172. H. Quast and J. Stawitz, *Tetrahedron Lett.*, **1977,** 2709.
173. F. Vögtle and E. Goldschmitt, *Angew. Chem.*, **86,** 520 (1974).
174. H. Quast and J. Stawitz, *Angew. Chem.*, **89,** 668 (1977).

IMINIUM SALTS IN NATURE

By J. KNABE, *Institut für Pharmazeutische Chemie der Universität, D-66-Saarbrücken 15, Germany*

CONTENTS

I.	Pyridine Derivatives	733
II.	Isoquinoline Derivatives	735
III.	Protoberberine Derivatives	738
IV.	Phenanthridine Derivatives	742
V.	Benzophenanthridine Derivatives	743
VI.	Aporphine Derivatives	745
VII.	Quinoline Derivatives	745
VIII.	Indole Derivatives	747
IX.	Imidazole Derivatives	749
X.	Pyrrole Derivatives	750
XI.	Quinazoline Derivatives	750
XII.	Purine Derivatives	750
XIII.	Betalain Derivatives	751
XIV.	Sterol Derivatives	753
	General References	753
	References	753
	Addendum	757

Iminium salts occur naturally in both animals and plants. Derivatives of isoquinoline and of protoberberines and benzophenanthridines, closely related to these biogenetically, are found relatively often. Other iminium salts encountered are some indole alkaloids and representatives of a group of nitrogen-containing plant coloring materials, grouped together under the name of betalains. Iminium salts occur only sporadically in all the other alkaloid groups, and not at all in some. This chapter deals only with those naturally occurring iminium salts that are stable and isolable. The vast number of iminium salts, postulated as intermediate products in biosyntheses of nitrogen compounds in nature and which function largely as aminoalkylating agents in a sort of Mannich reaction, are not touched upon (1).

Table I shows the classes of iminium salts found in nature up to now.

I. Pyridine Derivatives

Ackermann discovered the *N*-methylpyridinium salt (**1**) in various lower animals as long ago as 1922, and it has been recently isolated from *Vandopsis*

TABLE I
Classes of Naturally Occurring Iminium Salts

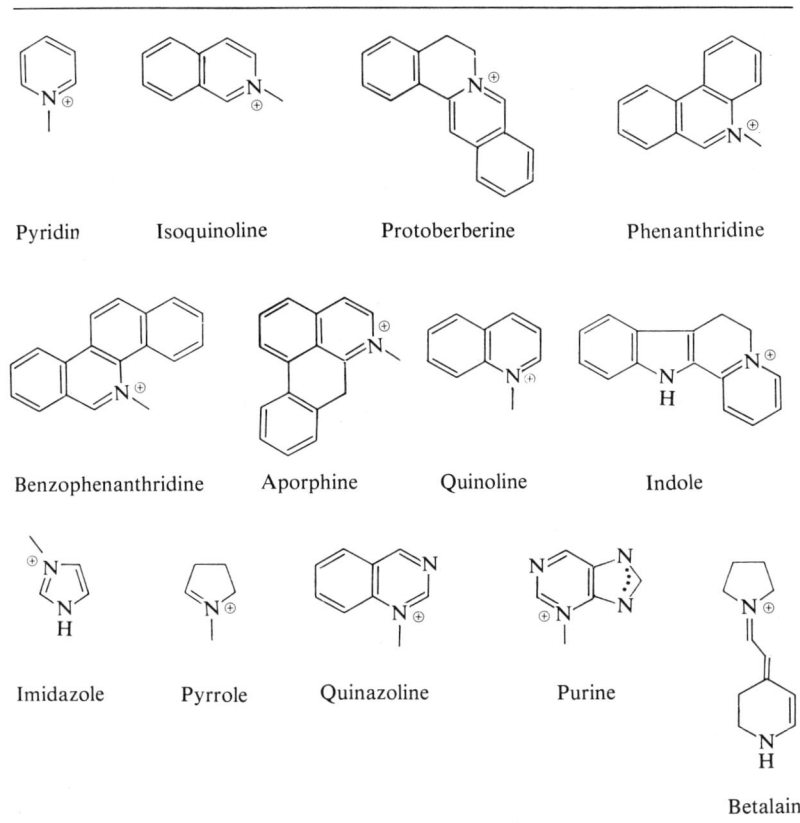

longicaulis (2). Trigonelline (**2**), the *N*-methylbetaine of nicotinic acid, has been known for a longer time. Jahns first discovered it in *Trigonella foenum graecum*, and it was found subsequently in numerous other plants. **2** has been detected also in *Anemonia sulcata* and other animals (3). The isomeric homarine (**3**) has been found in *Homarus vulgaris*. Renz isolated the optically active, levorotatory mimosine (**4**) from *Mimosaceae*; polarographic measurements show that it occurs in aqueous solution as the double zwitterion form **4b**. Leucanol and leucaenine are identical with mimosine.

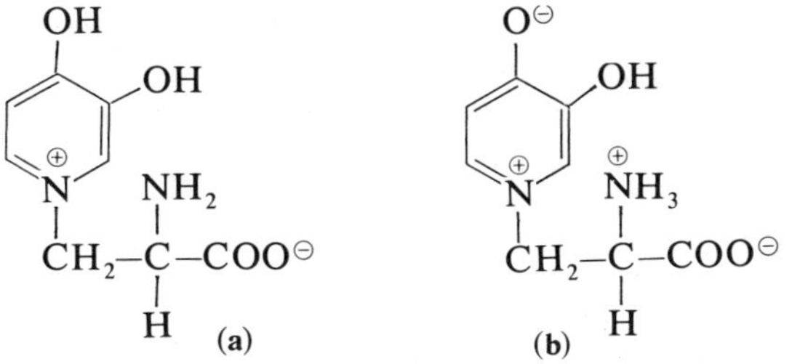

(**4**)

Another pyridinium salt is *N*-methylhalfordine (**5**), isolated from *Halfordia scleroxyla* (4). Its constitution (5, 6) was confirmed by a one-stage synthesis of the basic skeleton, halfordinol (**6**) (7). *N*-Methylhalfordine is the second alkaloid known so far that contains an oxazole ring.

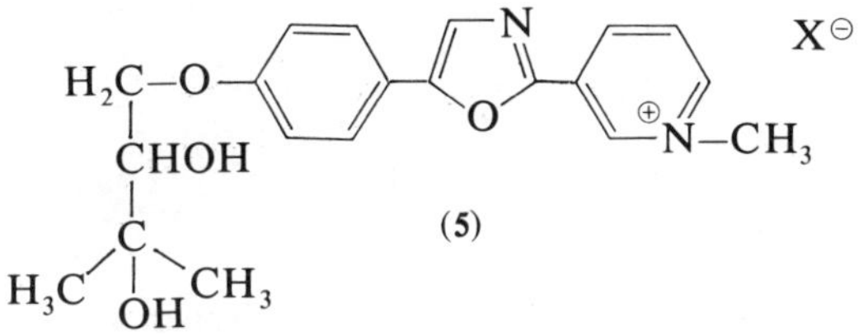

(**5**) 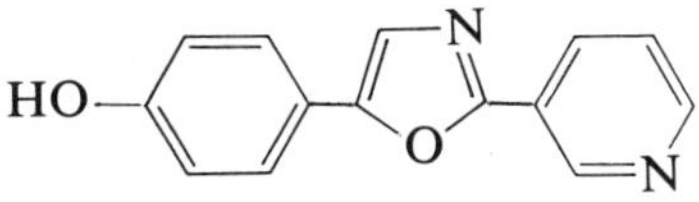   (**6**)

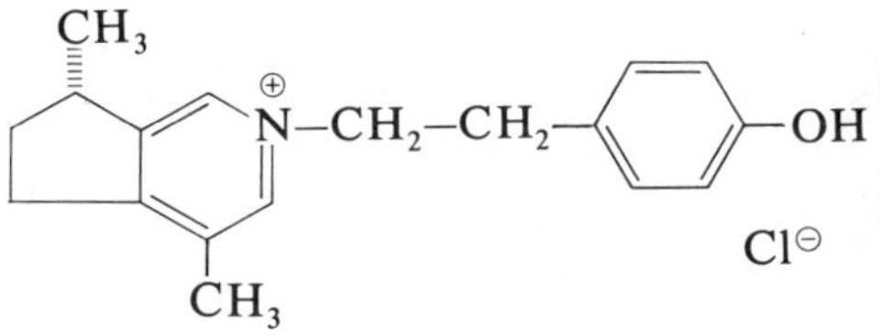

(**7**)

N-β-(*p*-hydroxyphenyl)ethylactinidinium chloride (**7**) has been isolated from *Valeriana officinalis*; it yields on pyrolysis the tertiary alkaloid actinidine and 4-hydroxystyrene (8).

II. Isoquinoline Derivatives

Oxidative degradation of the alkaloids narcotine and hydrastine yields two simple isoquinoline derivatives, cotarnine (**8**) and hydrastinine (**9**), which formerly played a part in gynecology.

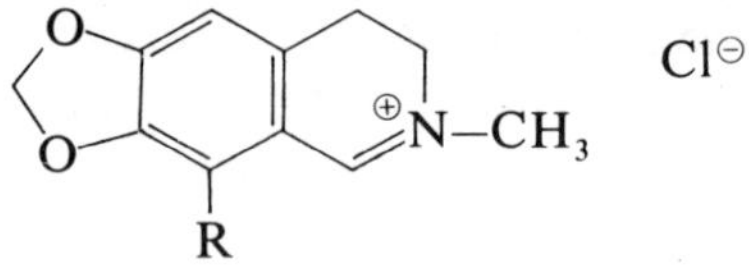

(8) R = OCH₃
(9) R = H

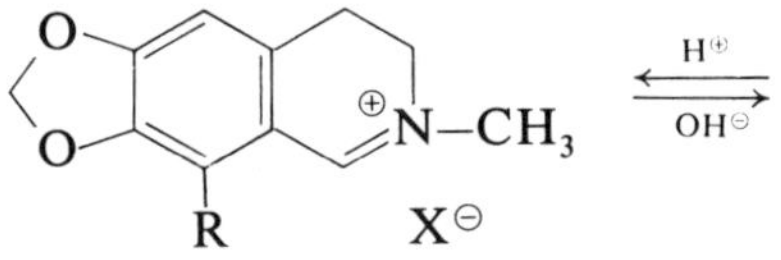

(A)

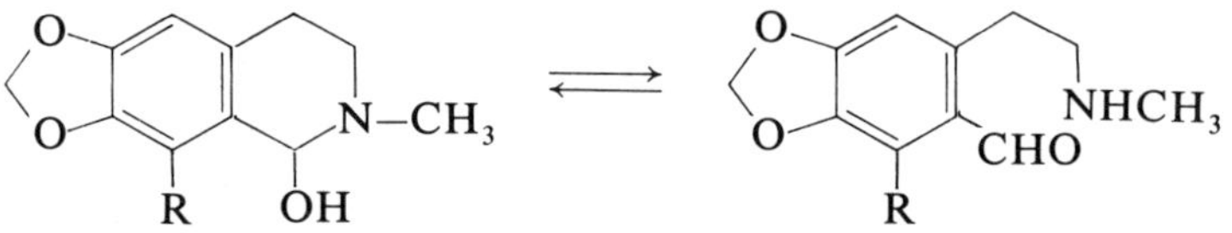

(B) (C)

In acid solution the iminium salt *A* is formed, which yields the carbinolamine *B* with alkali; *B* is not in equilibrium with the tautomeric, open-ring aldehyde form *C*, as was formerly believed (9, 10). Thus the acetone derivative, for example, which was taken to support the open-ring aldehyde form, has a cyclic structure. Cotarnine disproportionates to dihydrocotarnine and the corresponding 1-oxo compound with strong alkalies. The reaction of **8** with aliphatic diazo compounds has been studied exhaustively (11).

1-Benzylisoquinolinium compounds occurring naturally are escholamine (**10**), *N*-methylpalaudinium chloride (**11**), and takatonine (**12**). The structure of **10** has been established by synthesis (14). That of **11** is derived from the fact

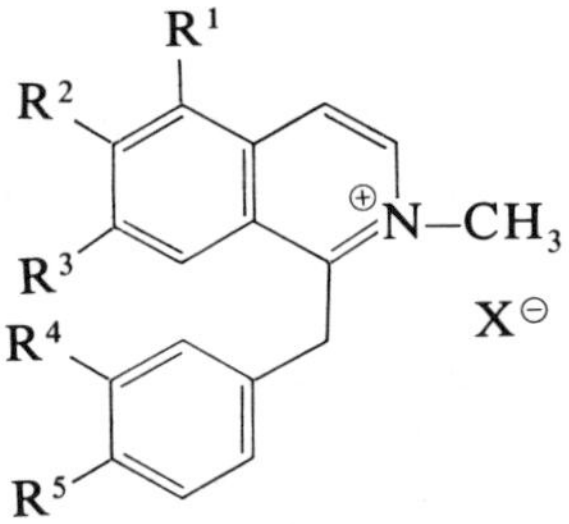

(10–12)

II. ISOQUINOLINE DERIVATIVES

Substance	Origin	Ref.
10: Escholamine $R^1 = H; R^2 + R^3 = R^4 + R^5 = OCH_2O$	*Escholtzia* species	12–14
11: *N*-Methylpalaudinium chloride $R^1 = H; R^2 = R^3 = R^5 = OCH_3; R^4 = OH$	*Thalictrum polyganum*	15
12: Takatonine $R^1 = R^2 = R^3 = R^5 = OCH_3; R^4 = H$	*Thalictrum thunbergii*	14, 16, 17

that it yields the tetrahydroisoquinoline alkaloid laudanine on reduction with NaBH₄ (15). The constitution of **12** was confirmed in 1972 through synthesis (16).

N-Methylxanthaline (**13**) is a 1-benzoylisoquinolinium salt, isolated from *Stephania sasakii* and identified as *N*-methylpapaveraldinium chloride (18). A

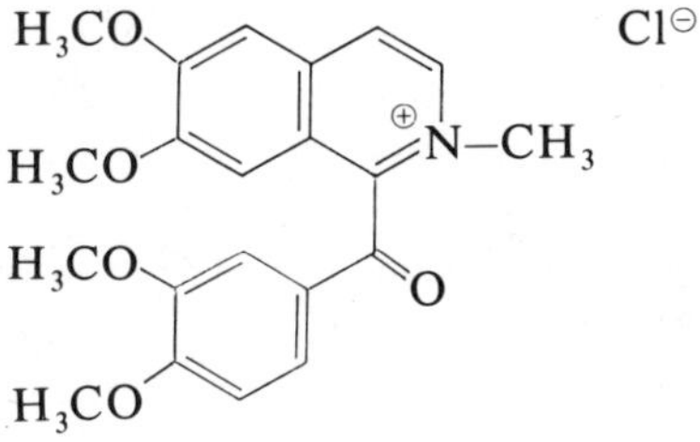

(**13**)

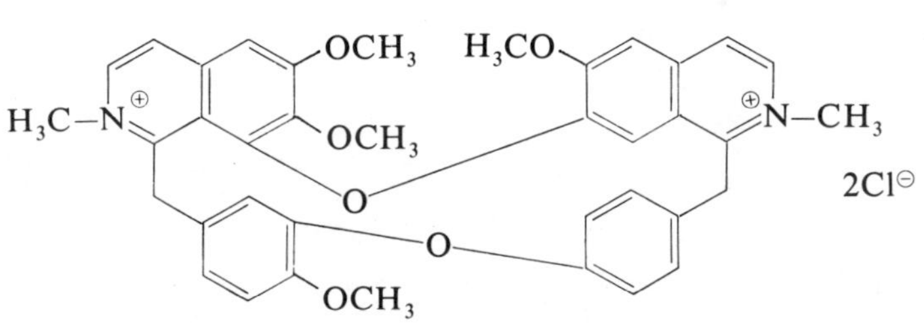

(**14**)

bis(benzylisoquinoline) alkaloid was found in the bark of the tree *Phaeanthus ebracteolatus*, isolated as the chloride salt and given the name of phaeantharine (**14**) (19). Structural elucidation showed that it was a diphenyl ether alkaloid, in which two 1-benzylisoquinolinium units are linked via two ether bridges. The positions of the ether bridges and methoxyl groups correspond to those in the ditertiary isotetrandine (= berbamine methyl ether), which occurs in the same plant (20, 21). Phaeantharine is the sole bis(benzylisoquinoline) alkaloid known up to now with two aromatic isoquinolinium systems.

III. Protoberberine Derivatives

Protoberberines are widely distributed in the vegetable kingdom. They occur in *Anonaceae*, *Aristolochiaceae*, *Berberidaceae*, *Convolvulaceae*, *Menispermaceae*, *Papaveraceae*, *Ranunculaceae*, and *Rutaceae*. Protoberberines contain a dibenzoquinolizidine skeleton with varying substituents. Depending on the substitution, the iminium compounds are yellow to red salts; they can be formulated also as carbinol bases (cf. cotarnine).

On the other hand, controversy rages about the open-ring aldehyde form, formerly assumed to be tautomeric with the carbinolamine and its participation in certain reactions. Even with modern spectroscopic methods it has not been possible to detect it (22). The inability to carry out *N*-acylation of protoberberines also testifies against the aldehyde form (23).

The constitutions of protoberberines have often been elucidated by conversion into known representatives of this group, for example, methylation or reduction to the tertiary tetrahydro derivatives. Spectroscopic methods have also been used successfully in recent times. Berberine is the most extensively distributed alkaloid of this group, isolated in 1837 by Büchner from various plants. The close relationship of protoberberines and isoquinolines is derived from biosynthesis. It has been assumed for a long time that protoberberines are

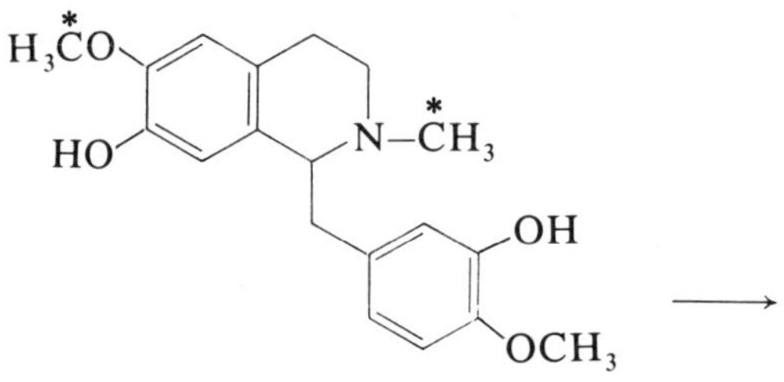

(15)

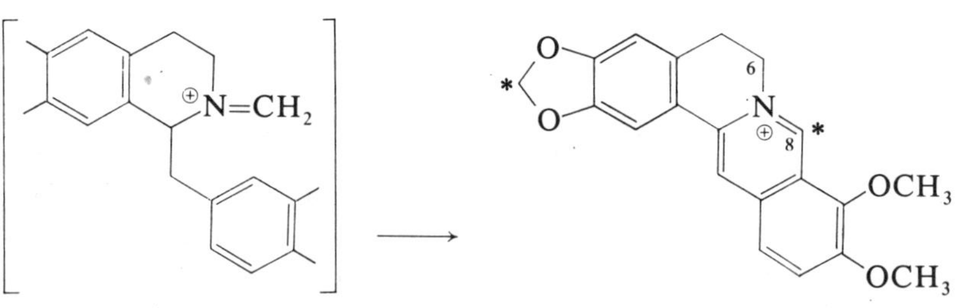

(16)   (17)

III. PROTOBERBERINE DERIVATIVES

formed from prior isoquinoline stages through ring closure with a "formaldehyde equivalent". The opinion that the berberine bridge might be formed by oxidative cyclization of an *N*-methyl group has been confirmed experimentally (24, 25). On feeding reticuline (**15**), labeled at both the 6-methoxyl and *N*-methyl groups, to *Hydrastis canadensis*, radioactive berberine (**17**), labeled at the same positions, was obtained (26). Laudanosoline was labeled at the *N*-methyl group and at the 3 position and fed to *Berberis japonica*. The radioactive berberine isolated was labeled at the 6 and 8 positions (27). Ring closure could occur via the iminium compound **16** (28).

Similar biogenetic routes can be assumed for the other protoberberines.

The research groups of W. H. Perkin, Jr., and J. Gadamer were those principally involved in the structural elucidation of berberine (**17**). Its structure has been confirmed through several syntheses. It is identical with umbellatine. Epiberberine (**18**), in which the methylenedioxy and methoxy groups of berberine are interchanged, was synthesized in 1918 by Perkin but not found in nature until 1951 (in *Berberis floribunda*) (29).

Coptisine (**19**) was first isolated in 1908 by K. Makoshi from *Corydalis ambigua*. E. Späth and R. Posega carried out the structure elucidation in 1929. K. Feist found palmatine (**20**) in *Jatrorrhiza palmata* in 1907. Its constitution was established in 1918. It is encountered in many *Berberis* species and in other plants.

(**17–26**)

	R^1	R^2	R^3	R^4
17: Berberine	$-O-CH_2-O-$		OCH_3	OCH_3
18: Epiberberine	OCH_3	OCH_3	$-O-CH_2-O-$	
19: Coptisine	$-O-CH_2-O-$		$-O-CH_2-O-$	
20: Palmatine	OCH_3	OCH_3	OCH_3	OCH_3
21: Berberrubine	$-O-CH_2-O-$		OH	OCH_3
22: Thalifendine	$-O-CH_2-O-$		OCH_3	OH
23: Columbamine	OH	OCH_3	OCH_3	OCH_3
24: Jatrorrhizine	OCH_3	OH	OCH_3	OCH_3
25: Dehydrocorydalmine	OCH_3	OCH_3	OCH_3	OH
26: Stepharanine	OH	OCH_3	OCH_3	OH

Berberrubine (**21**), obtained earlier by thermolysis of berberine chloride, was discovered by E. Späth and N. Polgar in *Berberis vulgaris* in 1929. G. Frerichs and P. Stoepel proposed the correct formula as long ago as 1916; it was confirmed in 1926 by E. Späth and G. Burger. The constitution is based on its yielding DL-canadine (= tetrahydroberberine) on *O*-methylation and reduction. The position of the phenolic groups was established by ethylation and subsequent oxidative degradation. A protoberberine, isomeric with berberrubine, was found in *Thalictrum fendleri* and named thalifendine (**22**) (29). It also gives tetrahydroberberine on *O*-methylation and reduction, so that the OH and OCH$_3$ groups in positions 9 and 10 in berberrubine must be reversed in thalifendine (30). In columbamine (**23**), which was discovered in *Jatrorrhiza palmata* in 1905 by E. Günzel, the phenolic group is in the 2 position according to the results of an oxidative breakdown by E. Späth and G. Burger. It yields palmatine (**20**) on methylation, yielded equally by methylation of jatrorrhizine (**24**), dehydrocorydalmine (**25**), and stepharanine (**26**). Jatrorrhizine (**24**), an isomer of columbamine (**23**), has the phenolic group in the 3 position. It was discovered in *Jatrorrhiza palmata* by K. Feist in 1907 and has been subsequently found in many plants. Another isomer, dehydrocorydalmine (**25**), occurring in *Corydalis* species, has the phenolic group in the 10 position (31). Stepharanine (**26**), isolated in 1967 from *Stephania glabra*, is the first protoberberine alkaloid with two phenolic OH groups (32).

Lambertine (**27/28**) was isolated in 1953 from *Berberis lamberti* (33) and is identical to 7,8-dihydroberberine (34), obtainable from berberine by reduction with LiAlH$_4$ (35). It has the enamine structure **27** and occurs as salt in the iminium form **28** (36).

The first bisprotoberberine alkaloid known is bisjatrorrhizine (**29**), recently found in *Jatrorrhiza palmata* (37). The link in the 4,4' positions was deduced from spectral and partial synthetic data (37).

Berberastine (**30**) was the first protoberberine alkaloid found with an alcoholic OH group at C-5; it was isolated from *Hydrastis* species (38) and from *Thalictrum fendleri* (38). Its constitution (39, 40) has been confirmed in the meantime by synthesis (41). Biosynthetic experiments have shown that **30** is not formed in the plant through oxidation of berberine (42). The alcoholic

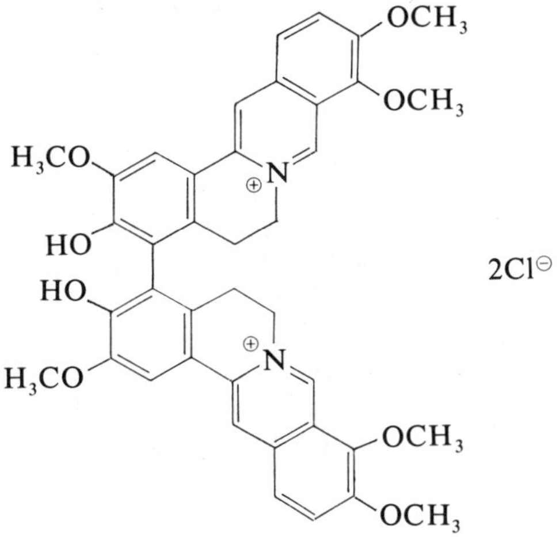

III. PROTOBERBERINE DERIVATIVES

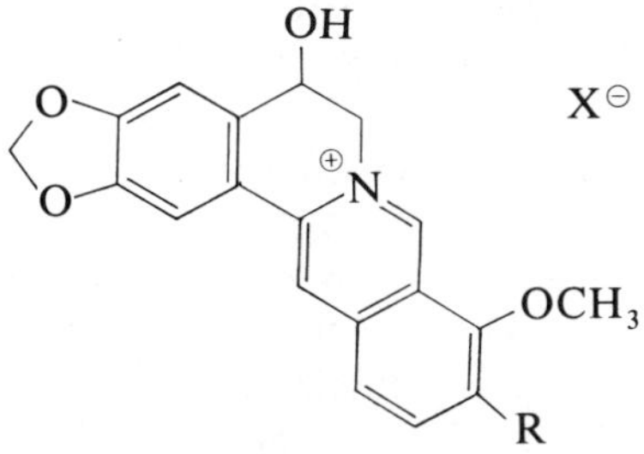

OH group is evidently introduced in a prior biosynthetical step before formation of the benzylisoquinoline skeleton. A further 5-hydroxyprotoberberine is thalidastine (**31**), isolated as the principal alkaloid from *Thalictrum fendleri* (43). It was identified as 5-hydroxythalifendine (43). On standing for some days at room temperature **31** undergoes dehydration to yield a fully aromatic compound (43), which has also been prepared synthetically (44). Such a fully aromatic protoberberine had not been hitherto found in nature. Coralyn chloride (**32**), claimed to be effective against leukemia, is a product of synthesis, available in good yield from papaverine (45).

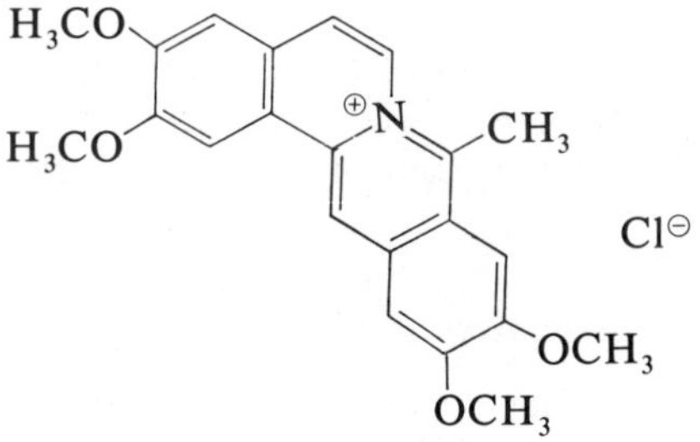

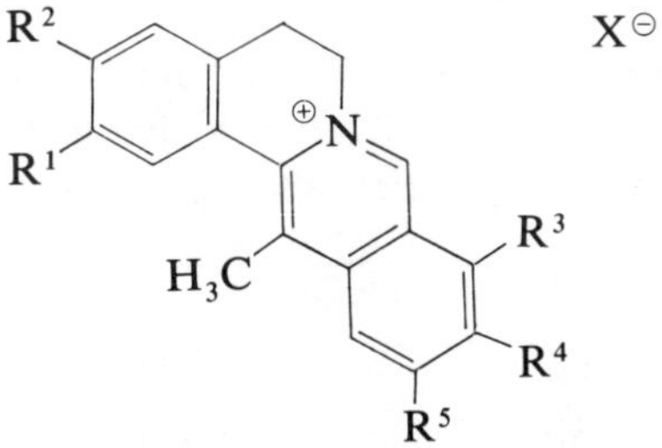

(**33**) $R^1 = R^2 = R^3 = R^4 = OCH_3$, $R^5 = H$
(**34**) $R^1 = R^2 = OCH_3$; $R^3 + R^4 = O-CH_2-O$, $R^5 = H$
(**35**) $R^1 + R^2 = R^3 + R^4 = O-CH_2-O$, $R^5 = H$
(**36**) $R^1 + R^2 = R^4 + R^5 = O-CH_2-O-$, $R^3 = H$

Dehydrocorydaline (33), dehydrothalictrifoline (34), corysamine (35), and worenine (36) are protoberberines possessing a C-13 methyl group. 33 was isolated from *Berberis floribunda* and various *Corydalis* species by K. Makoshi in 1908. Its constitution is revealed by its reduction to corydaline. 34, discovered in *Corydalis thalictrifolia*, can be reduced to thalictrifoline (46), which occurs in the same plant. The structure of 34 has been confirmed by partial synthesis (47). Corysamine (35) was obtained from *Corydalis incisa* (48). The structure elucidation (49) depends on alkylation with methyl iodide of acetonylcoptisine at C-13 to yield corysamine which can be reduced to tetrahydrocorysamine. However, the formula derived from this, 35, had been already attributed to worenine (36), an iminium salt isolated by Z. Kitasato in 1927 from *Coptis japonica*. Yet 35 and worenine and their derivatives are clearly different. The formula 36 was therefore proposed for worenine (48), and this was recently confirmed by synthesis (50).

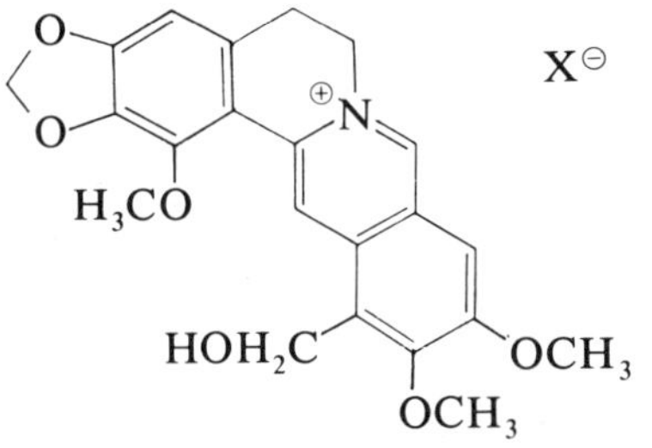

(37) 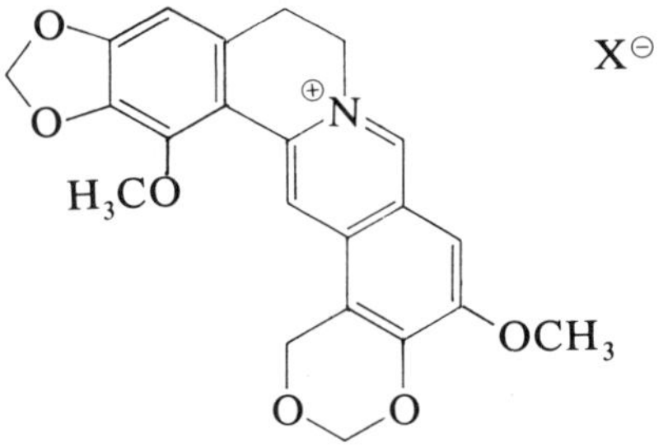 (38)

A protoberberine with six substituents and a primary alcohol group in the 12 position has been isolated from *Papaver alboroseum* and other *Papaver* species (51, 52). It gives oreophiline on reduction, a compound found in these same plants. The constitution of the alkaloid alborine (37) was established in 1970 (53–55).

Another iminium salt, the "alkaloid PO4" was isolated along with alborine (37). Structure elucidation yielded formula 38, containing the unusual cyclic acetal grouping which shows the close biogenetic connection with alborine (37).

IV. Phenanthridine Derivatives

Ungeremine (39) was isolated as an inner iminium salt from the *Amarylidaceae* species *Ungernia minor*; its structure was established by synthesis (56).

Haemanthidine (40) (= 6-hydroxyhaemanthamine) and the epimeric hydroxycrinamine (41) are *Amarylidaceae* alkaloids of the crinine type which are unique among the ethanophenanthridine group as a result of the C-6

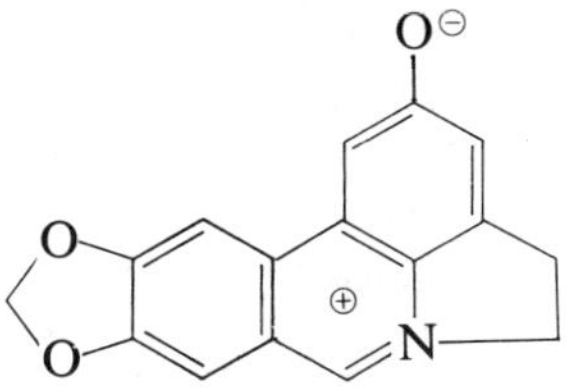

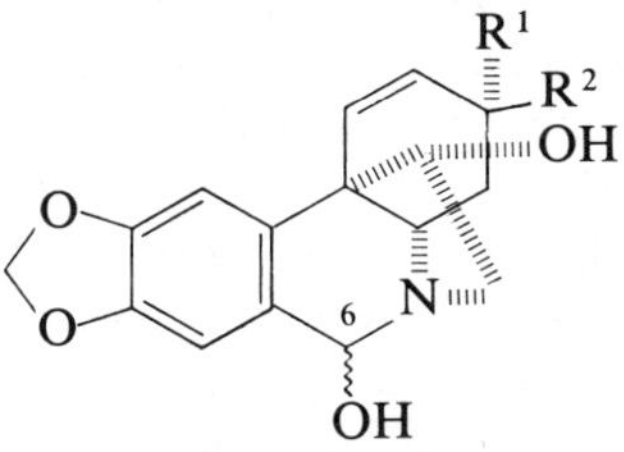

(40) $R^1 = H$, $R^2 = OCH_3$
(41) $R^1 = OCH_3$, $R^2 = H$

hydroxyl substitution. They are only formally carbinolamines since they do not form iminium salts; the nitrogen atom is a bridgehead atom in a 1-azabicyclo[3,2,1]octane system (57). The corresponding 1-oxo compounds, easily obtained by oxidation, possess ketone rather than lactam character.

V. Benzophenanthridine Derivatives

The benzophenanthridine alkaloids are formed biogenetically from 1-benzylisoquinolines via protoberberine intermediates. The fully aromatic

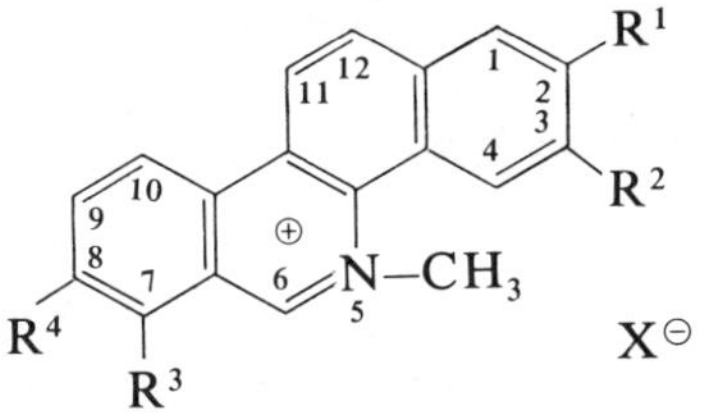

(42) $R^1 + R^2 = O-CH_2-O$, $R^3 = R^4 = OCH_3$
(43) $R^1 + R^2 = R^3 + R^4 = O-CH_2-O$

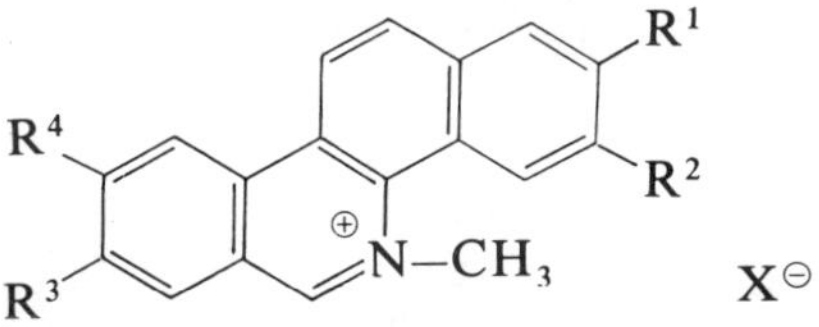

(44–46)

Substance	Origin	Ref.
44: Nitidine $R^1 + R^2 = O-CH_2-O$, $R^3 = R^4 = OCH_3$	*Zanthoxylum nitidum*	62–65
45: Avicine $R^1 + R^2 = R^3 + R^4 = O-CH_2-O$	*Zanthoxylum nitidum*	62, 63
46: Fagaronine $R^1 = OH$; $R^2 = R^3 = R^4 = OCH_3$	*Fagara xanthoxyloides*	66

benzophenanthridine alkaloids chelerythrine (**42**) and sanguinarine (**43**) have been known for a long time. Both occur in numerous *Papaveraceae* and also occasionally in *Rutaceae*.

Chelerythrine (**42**) was found in *Chelidonium majus* by J. M. Probst in 1839, and sanguinarine (**43**) in *Sanguinaria canadensis* by E. Schmidt in 1893. Their constitutions were elucidated in 1930–1931 by F. v. Bruchhausen and H.-W. Bersch and by E. Späth and F. Kuffner. Confirmation was obtained in 1950 through synthesis of a degradation product (58, 59). **42** was obtained in 1956 by a total synthesis (60), and **43** in 1968 in an overall yield of 4% (61). Nitidine (**44**) and avicine (**45**), isomers of chelerythrine and sanguinarine, respectively, were isolated in 1958–1959. The structure of **44** has been established by two independent syntheses (63, 64). Fagaronine (**46**) has been shown to be highly active against leukemia in mice (67).

A benzophenanthridine alkaloid with a methylenedioxy group in the 9,10 positions is bocconine, isolated from *Bocconia cordata* (68). It has been

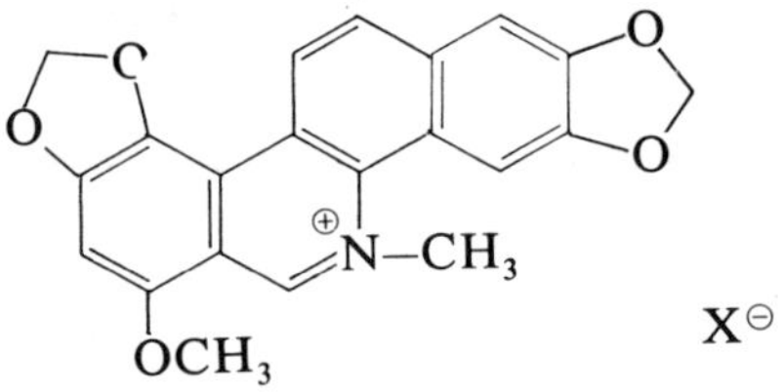

(**47**)

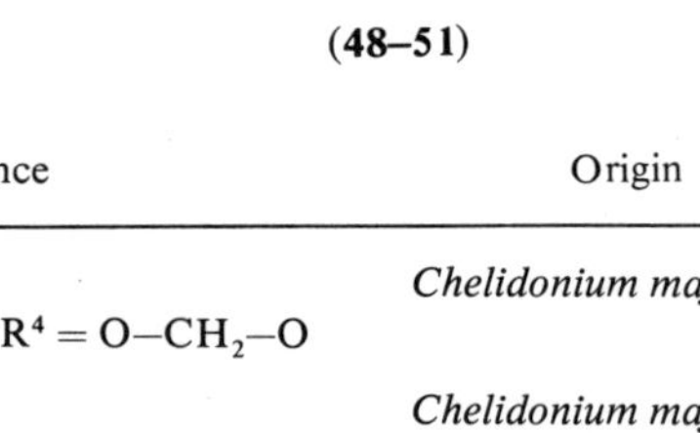

(**48–51**)

Substance	Origin	Ref.
48: Chelilutine $R^1 = R^2 = OCH_3, R^3 + R^4 = O-CH_2-O$	*Chelidonium majus*	70, 72
49: Chelirubine $R^1 + R^2 = R^3 + R^4 = O-CH_2-O$	*Chelidonium majus*	70, 72
50: Sanguirubine $R^1 + R^2 = O-CH_2-O, R^3 = R^4 = OCH_3$	*Sanguinaria canadensis*	71, 72
51: Sanguilutine $R^1 = R^2 = R^3 = R^4 = OCH_3$	*Sanguinaria canadensis*	71, 72

accorded the formula **47**, based on NMR data and nuclear Overhauser effect measurements (69).

Chelilutine (**48**), chelirubine (**49**), sanguirubine (**50**), and sanguilutine (**51**) are closely related benzophenanthridine alkaloids with an additional OCH$_3$ group in the 11 position.

Corynoloxine (**52**), from *Corydalis incisa* (56), occurs as the aminocarbinol ether base **52a**; salt formation yields the iminium salt **52b**. The constitution of

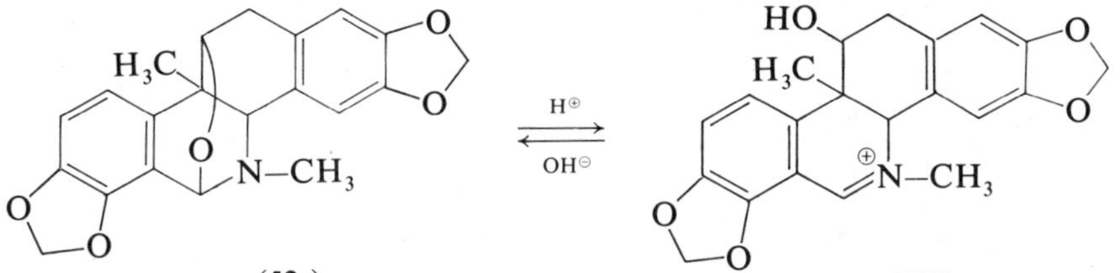

(**52a**)   (**52b**)

52 is deduced from spectroscopic data and from the fact that reduction gives the tertiary alkaloid corynoline, occurring in the same plant (73, 74).

VI. Aporphine Derivatives

Corrunine (**53**), isolated recently from *Glaucium flavum*, is the only example so far; it is a resonance-stabilized inner iminium salt, the structure of which has been established by synthesis (75, 76).

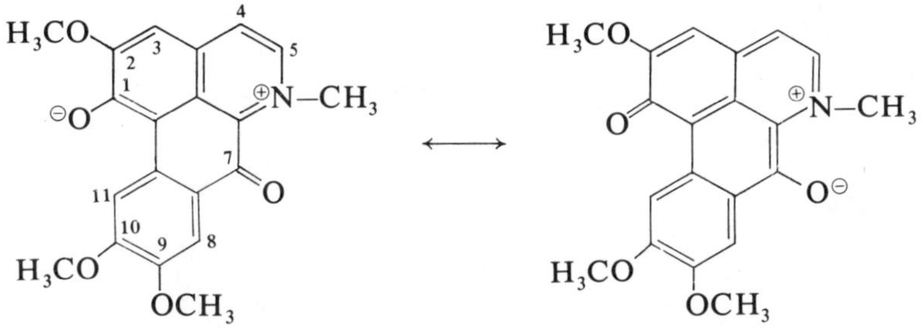

(**53**)

VII. Quinoline Derivatives

Echinorine (**54**) from *Echinops* species, has an especially simple structure (77). Lunasine (**55**) contains an additional fused furan ring (78, 79); it was isolated from *Lunaria* species by W. G. Boorsma at the turn of the century. Its derivatives include the *Rutaceae* alkaloids ribalinium (**56**), the *O*-methylbalfourodinium salt (**57**), and the platydesminium salt (**58**).

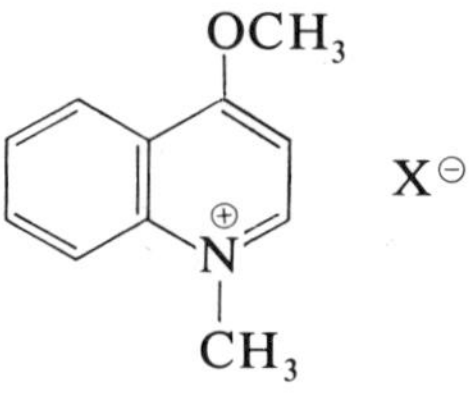

(**54**)

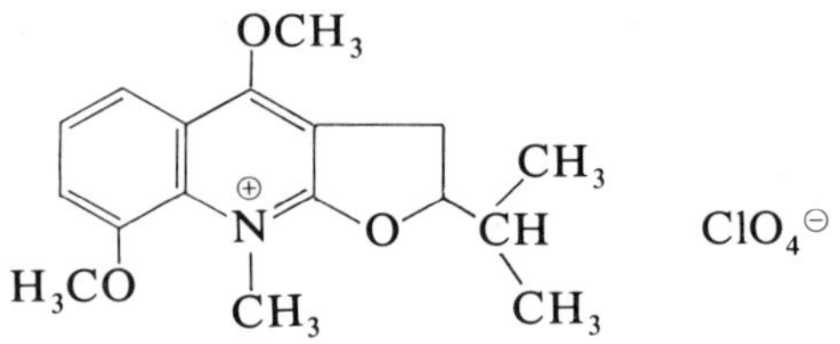

(**55**)

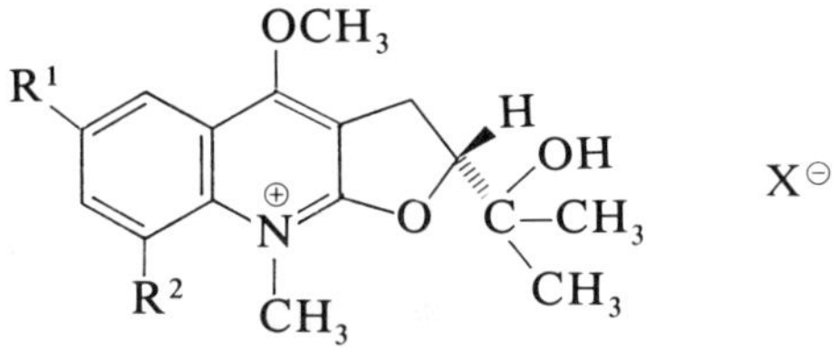

(**56–58**)

Substance	Origin	Ref.
56: Ribalinium $R^1 = OH, R^2 = H$	*Balfourodendron riedelianum*; *Ruta graveolens*	80, 81
57: O-Methylbalfourodinium salt $R^1 = H, R^2 = OCH_3$	*Balfourodendron riedelianum*	82, 83
58: Platydesminium salt $R^1 = R^2 = H$	*Skimmia japonica*; *Ruta graveolens*	84–86

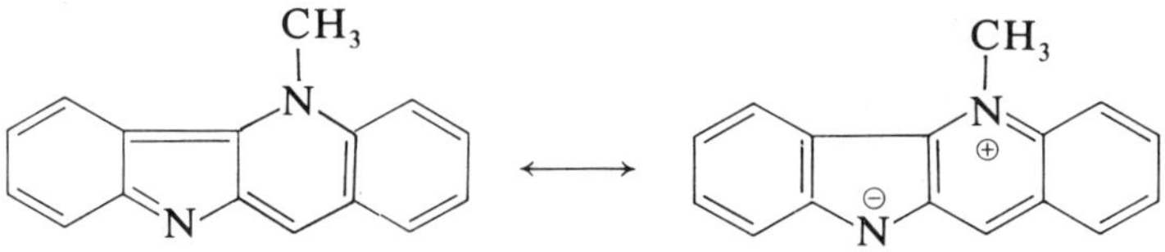

(a)   (b)

(**59**)

The only naturally occurring compound with an indoloquinoline skeleton is cryptolepine (**59**), discovered in *Cryptolepis triangularis* by Clinquart in 1929. It can be formulated as an anhydronium base. The free base is dark violet. Its mesoionic structure has been determined through unambiguous syntheses (87, 88).

VIII. Indole Derivatives

The first harman derivative found to exist as an iminium salt was isolated as its chloride from *Strychnos melioniana* (89). It received the name of melionine F and proved to be identical to *N*-methylharmane (**60**).

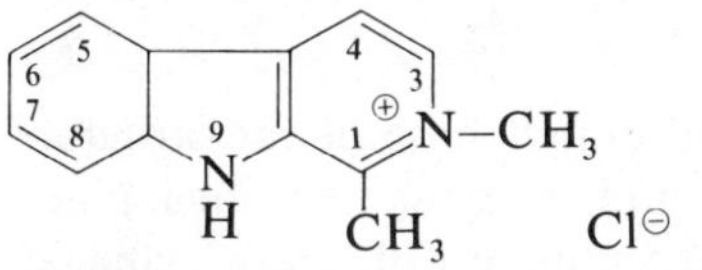

(**60**) 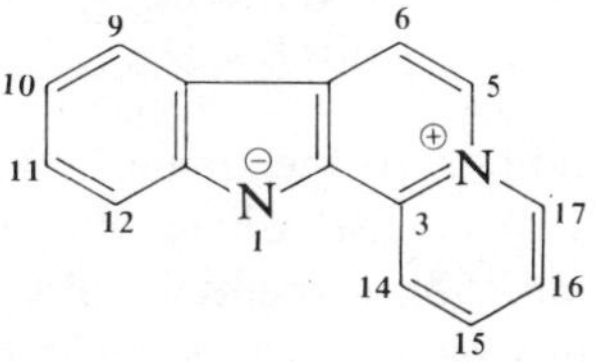   (**61**)

The anhydronium base indolopyridocoline (**61**) was isolated from *Gonomia kamassi* (90). **61** had been prepared synthetically before its discovery in nature (91). The corresponding 16,17-dihydro derivative of **61** was also found as an accompanying alkaloid (90).

Flavopereirine (**62**), also termed melionine G, was obtained from *Geissospermum laeve* (92, 93). Its structure has been established with the help

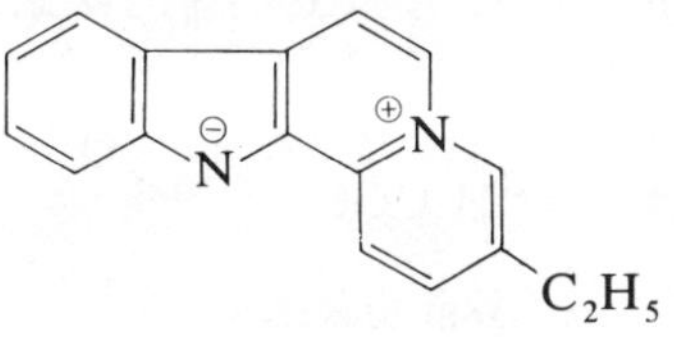

(**62**) 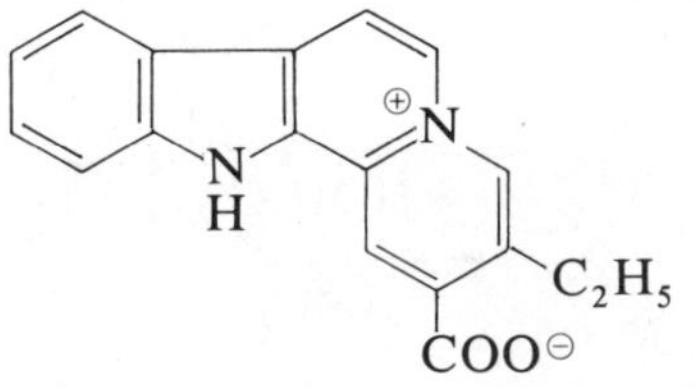   (**63**)

of several independent syntheses (91, 94). A betaine, the UV spectrum of which is identical with that of **62**, was isolated from *Pleiocarpa* species and named flavocarpine (95). The structure **63** was determined by synthesis.

The alkaloids alstoniline (**64**) and ourouparine (**65**) and the anhydronium bases sempervirine (**66**) and melionine E (**67**) are yohimbine types. Alstoniline (**64**) was isolated from *Alstonia constricta* as the red chloride and its structure elucidated (96). An easy synthesis is due to Beisler (97). Ourouparine was discovered in *Uncara gambier* and other plants and identified as demethoxy-alstoniline (**65**) (98). Sempervirine (**66**) was found by Stevenson and Sayre in *Gelsemium sempervirens* in 1915. The correct structure of anhydronium base was suggested in 1949 by R. Bentley and R. B. Woodward independently, the latter carrying out the first synthesis. A three-stage synthesis was performed by Ban and Seo (99).

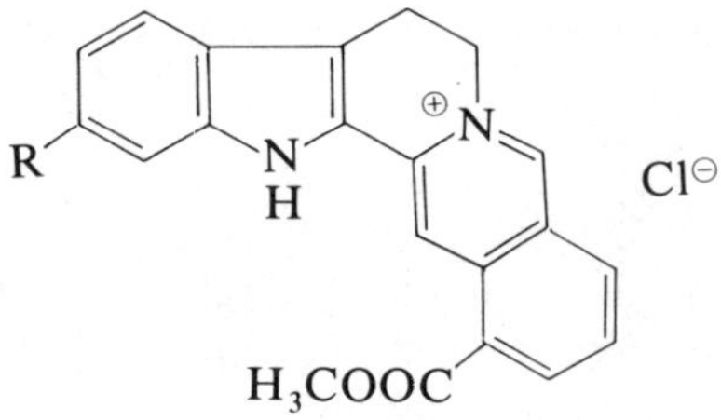

(64) R = OCH$_3$
(65) R = H

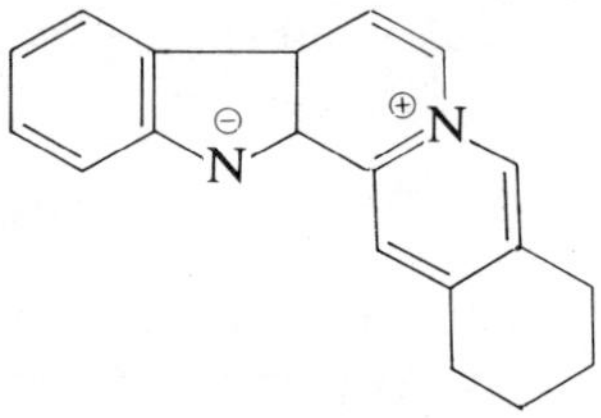

(66)

In addition to melionine F, an anhydronium base of the structure **67** was isolated from *Strychnos melioniana* and was named melionine E (89). Alstonine (**68**), serpentine (**69**), and serpentinine (**70**) are ajmalicine types, a group in which the carbocyclic ring E is replaced by a cyclic ether. Alstonine has been known for a long time, having been isolated in 1865 by Hesse from *Alstonia constricta*. It has been subsequently found in *Rauwolfia* and *Vinca* species.

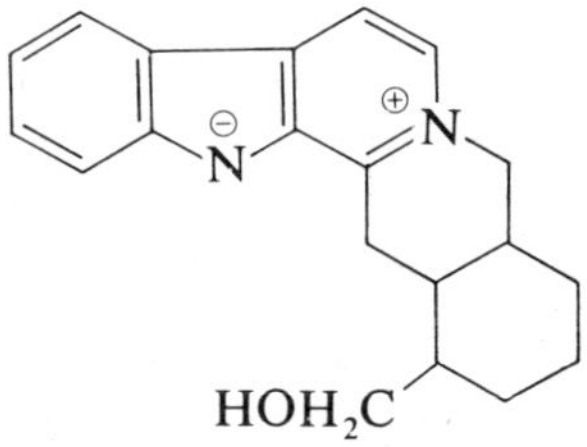

(67)

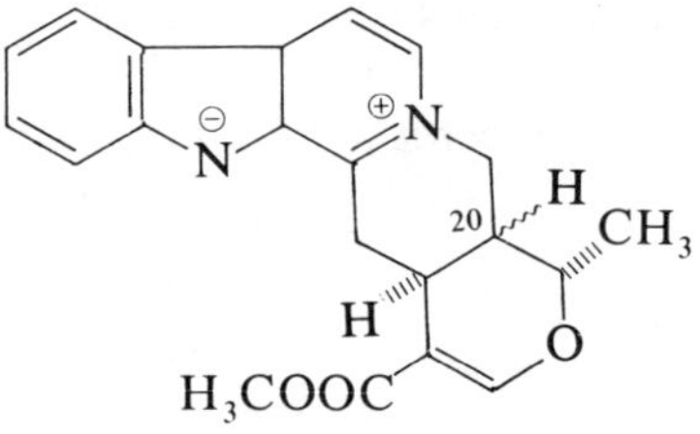

(68) 20-α-H
(69) 20-β-H

68 and serpentine (**69**), isolated by Siddiqui in 1931 from *Rauwolfia serpentina*, are diastereomers. They differ only in the position of the C-20 hydrogen. The rings D/E are linked in the *cis* position in alstonine (**68**) and *trans* in serpentine (**69**) (100–106).

Serpentinine (**70**) is a dimeric alkaloid of the ajmalicine type and was one of the first alkaloids isolated in 1931 from *Rauwolfia serpentina* by Siddiqui. It was not until 1960 that it was recognized as a coupling product of the two *Rauwolfia* alkaloids serpentine and ajmalicine (107).

Ajmaline (**71**), discovered in 1931 by Siddiqui as one of the chief alkaloids in *Rauwolfia serpentina*, is a particular type of indole alkaloid. It is a carbinolamine with a bridgehead nitrogen atom and therefore undergoes the reactions of an amine and an aldehyde, yet is unable to form iminium salts and occurs in acid solution as a bishydroxyamine. A total synthesis of biogenetic type was published in 1970 (108).

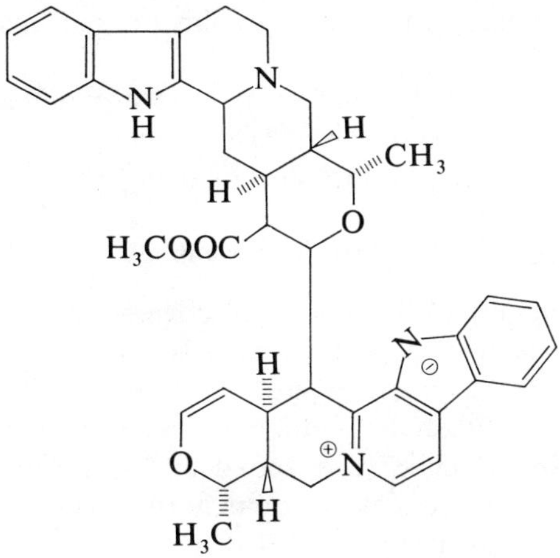

(70)

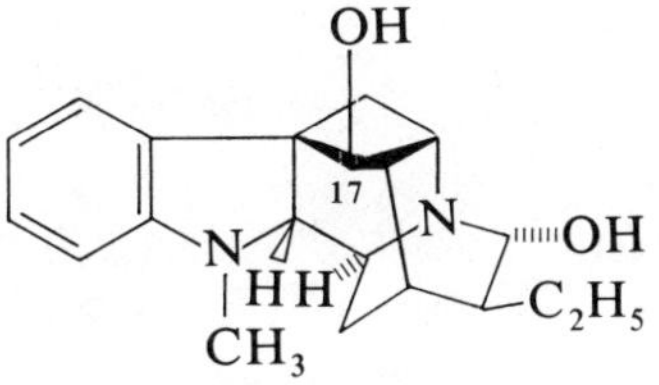

(71)

Ajmalidine, isoajmaline, sandwicine, vomalidine, and vomilenine are alkaloids structurally related to ajmaline. They all formally possess a carbinolamine structure but likewise do not form iminium salts as a result of the nitrogen atom being the bridgehead in a tricyclic system.

IX. Imidazole Derivatives

Up to now imidazole alkaloids have been found only occasionally in higher plants. Zooanemomine, the N,N-dimethylbetaine of imidazoleacetic acid (72)

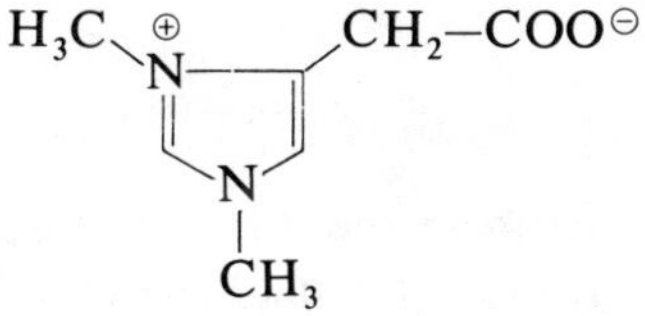

(72) 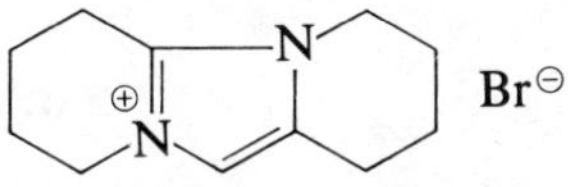   (73)

was isolated by D. Ackermann from the animal organisms *Anemonia sulcata* and *Arca noae*.

The first iminium salt of the imidazole group of plant origin to be isolated was **73**, as the bromide from the orchid *Dendrobium parishii*. Synthesis (109) and X-ray analysis (110) have established its structure.

X. Pyrrole Derivatives

The alkaloid shihunine (**74**) was also discovered in a *Dendrobium* species and its constitution elucidated (111). Synthesis confirmed the structure (112). It occurs as lactone **74a** in the crystalline state or in solution in nonpolar liquids; in aqueous or methanolic solution it changes rapidly and quantitatively into the betaine **74b**, which is reconverted to **74a** on drying. It thus seems probable that shihunine occurs as a betaine in the plant and that the lactone is formed during the isolation procedure (113).

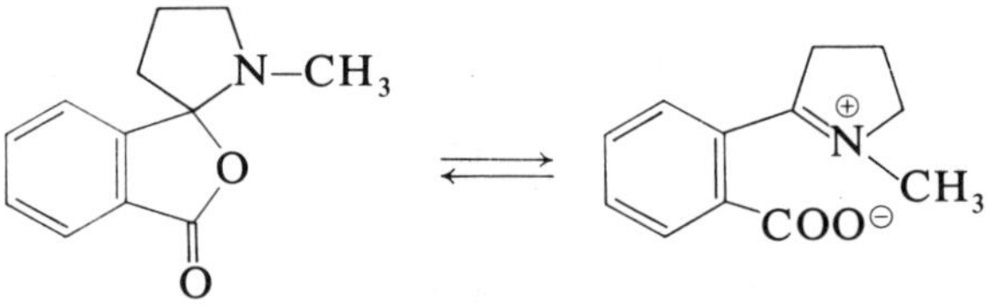

(74a)   (74b)

XI. Quinazoline Derivatives

Glycorine is a quinazoline alkaloid, isolated from *Glycosmis arborea*. Its chloride has the structure **75**, established through NMR spectroscopy (114, 115).

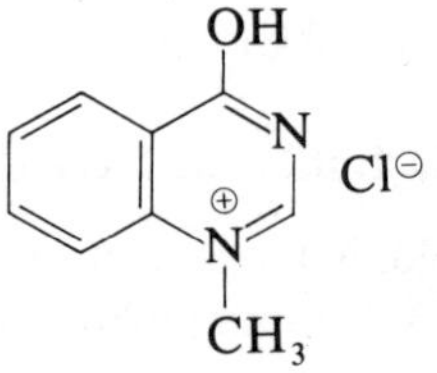

(75)

XII. Purine Derivatives

Two naturally occurring iminium salts from the purine group are known. One was found in the vegetable, the other in the animal kingdom. Triacanthine was isolated from the *Leguminaceae* species, *Gleditsia triacanthus* (116) and

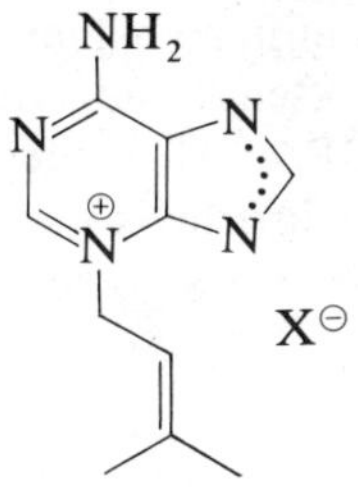

(76) 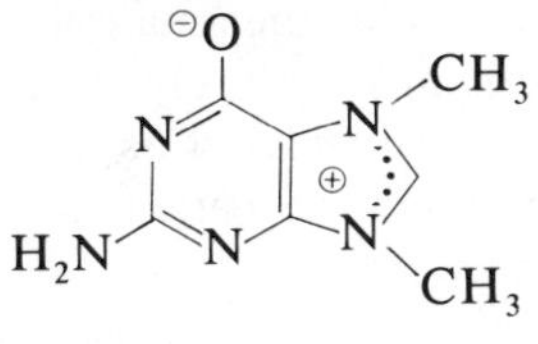   (77)

its structure, **76**, was elucidated later (117). It is stable towards alkali but is easily cleaved to adenine by strong acids. Herbipoline was isolated from *Geodia gigas* and found to be 7,9-dimethylguaninium betaine (**77**) (118).

XIII. Betalain Derivatives

The term betalain is given to a number of structurally closely related alkaloid-like pigments, found in the organs of plants of the *Centrospermae* order (119). They are divided into the red-violet betacyanins and the yellow betaxanthins and have hitherto been discovered in only 10 plant families, in which anthocyanins have never been found to exist as far as is known. The betalains are water-soluble and possess inner iminium salt structures. They are susceptible to the influence of light, oxygen, and alkali. Especially significant in the structure elucidations was a pigment called betanin, isolated from *Beta vulgaris* and finally obtained in 1957 in crystalline form. All betalains contain a dihydropyridinedicarboxylic acid structural element which, depending on configuration at the chirality center, is termed betalamic (**78**) or isobetalamic acid (**79**).

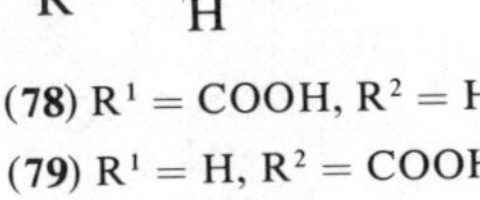

(**78**) R^1 = COOH, R^2 = H
(**79**) R^1 = H, R^2 = COOH

The pigments are formed by condensation of the aldehyde group of these acids [which also exist in uncombined form in various plants (120)] with various amino acids and sometimes also with amines. A dihydroxydihydro-αindolecarboxylic acid is the amino acid component in all betacyanins known; it is variously combined as a glycoside and can often be acylated in the sugar moiety. It can be considered as a cyclized DOPA. All betacyanins yield on hydrolysis the aglucone betanidin (**80**), the epimeric iso compound, or a mixture of both.

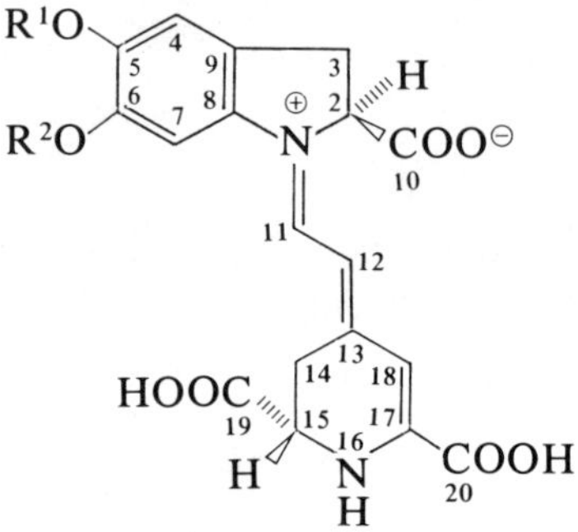

(**80**) $R^1 = R^2 = H$
(**81**) $R^1 = \beta\text{-D-glucosyl}, R^2 = H$

The glucoside betanin (**81**) is the 5-O-β-D-glucopyranoside of betanidin (121–126). In amarantin, prebetanin, phyllocactin, celosianin and iresinin, R^1 is a β-D-glucosyl group, acylated in various ways; and R^2 is hydrogen (127–129). R^1 is hydrogen and R^2 a variously acylated β-D-glucosyl group in the compounds named as gomphrenins (130). The corresponding epimeric iso compounds contain isobetalamic (**79**) instead of betalamic acid (**78**). Indicaxanthin (**82**) is of importance in the structure elucidation of the betaxanthins. It was isolated from the fruit of *Opuntia ficusindica* and identified as a condensation product of betalamic acid and L-proline (131).

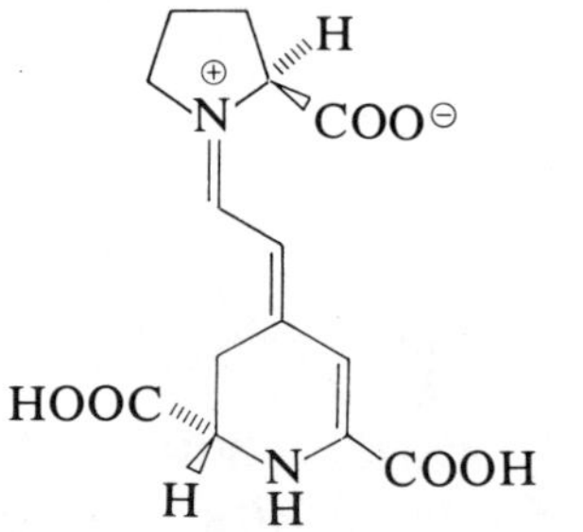

(**82**)

Other betaxanthins are formed by condensation of betalamic acid (**78**) with glutamic acid or glutamine (vulgaxanthins), aspartic acid, and methionine sulfoxide (miraxanthins) (132). Other miraxanthins contain the amines tyramine or dopamine instead of an amino acid component. Very recently intensively colored iminium salts of the betalain type have been isolated from the red skin of the cap of the fly agaric mushroom, *Amanita muscaria* (133); these had formerly been found only in higher plants. In the presence of a weak base, the betalains undergo an exchange reaction in which the betalamic acid is transferred from one amino acid to another amino acid. Betanin (**81**) can be converted into indicaxanthin (**82**) in this way. Similar *in vivo* conversions may take place in plants, so that other naturally occurring betalains may be discovered.

XIV. Sterol Derivatives

Two alkaloids possessing a carbinolamine structure have been found in the poison from the skin glands of salamander species. These are cycloneosamandaridine (**83**) and cycloneosamandione (**84**), the structures having been established through total synthesis and X-ray analysis (134–136). Neither forms an iminium salt since the nitrogen is a bridgehead atom in a bicyclic system. They can, however, react in the open-ring tautomeric aldehyde form; for example, **84** may be *N*-acetylated yielding a neutral product (137).

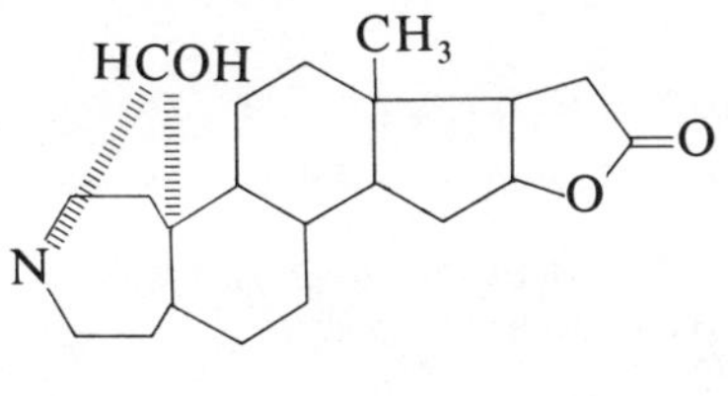

(83) 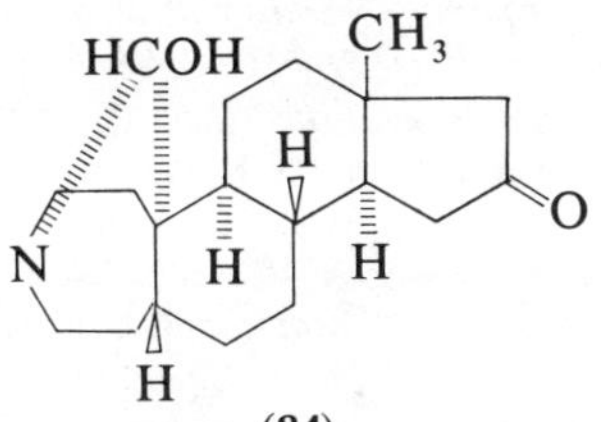   (84)

GENERAL REFERENCES

H.-G. Boit, *Ergebnisse der Alkaloidchemie bis 1960*, Akademie Verlag, Berlin, 1961.
M. Hesse, *Indolalkaloide in Tabellen*, Springer Verlag, Berlin, 1974; Ergänzungswerk 1968.
T. Kametani, *The Isoquinoline Alkaloids*, Elsevier Publishing Company, Amsterdam, 1969.
R. H. F. Manske and H. L. Holmes (Vols. I–V), *The Alkaloids, Chemistry and Physiology*, Vols. I–XIII, Academic Press, New York, 1950–1971.
S. W. Pelletier, *Chemistry of the Alkaloids*, Van Nostrand–Reinhold Co., New York, 1970.

REFERENCES

1. E. Leete, in *Biogenesis of Natural Compounds*, P. Bernfeld, Ed., Pergamon Press, London, 1967.
2. S. Brandänge and B. Lüning, *Acta Chem. Scand.*, **24**, 353 (1970).
3. D. Ackermann, *Hoppe-Seylers Z. Physiol. Chem.*, **295**, 1 (1953).

4. W. D. Crow and J. H. Hodgkin, *Tetrahedron Letters*, **2**, 85 (1963).
5. A. Brossi and E. Wenis, *J. Heterocycl. Chem.*, **2**, 310 (1965).
6. W. D. Crow, J. H. Hodgkin, and J. S. Shannon, *Australian J. Chem.*, **18**, 1433 (1965).
7. T. Onaka, *Tetrahedron Letters*, **1971**, 4393.
8. K. Torssell and K. Wahlberg, *Tetrahedron Letters*, **1966**, 445.
9. D. Beke and K. Harsányi, *Acta Chim. Acad. Sci. Hung.*, **11**, 303, 349 (1957).
10. D. Beke, K. Harsányi, and J. Körösi, *Acta Chim. Acad. Sci. Hung.*, **11**, 310 (1957).
11. B. Göber, S. Pfeifer, G. Dube, G. Engelhardt, and H. Jancke, *Pharmazie*, **28**, 221 (1973).
12. H. Furukawa, T.-H. Yang, and T. J. Lin, *J. Pharm. Soc. Japan*, **85**, 472 (1965).
13. L. Slavikowa and J. Slavik, *Collection Czech. Chem. Commun.*, **31**, 3362 (1966).
14. A. J. Birch, A. H. Jackson, P. V. R. Shannon, and P. S. P. Vakma, *Tetrahedron Letters*, **1972**, 4789.
15. M. Shamma and J. L. Moniot, *J. Pharm. Sci.*, **61**, 295 (1972).
16. E. Fujita and T. Tomimatsu, *J. Pharm. Soc. Japan*, **79**, 1082 (1959).
17. S. Kubota, T. Matsui, E. Fujita, and S. M. Kupchan, *Tetrahedron Letters*, **1965**, 3599; *J. Org. Chem.*, **31**, 516 (1966).
18. J. Kunimoto, E. Yuge, and Y. Nagai, *J. Pharm. Soc. Japan*, **86**, 456 (1966).
19. A. C. Santos, *Arch. Pharm.*, **284**, 360 (1951).
20. F. v. Bruchhausen, A. C. Santos, J. Knabe, and G. A. Santos, *Arch. Pharm.*, **290**, 232 (1957).
21. J. Knabe, *Chem. Ber.*, **91**, 1612 (1958).
22. B. Skinner, *J. Chem. Soc.*, **1950**, 823.
23. D. Beke, *Acta Chim. Acad. Sci. Hung.*, **17**, 463 (1958); for a review see D. Beke, *Advances in Heterocyclic Chemistry*, Vol. 1, 1963, p. 167.
24. A. R. Battersby, *Proc. Chem. Soc.*, **1963**, 189.
25. D. H. R. Barton, *Proc. Chem. Soc.*, **1963**, 293.
26. D. H. R. Barton, R. H. Hesse, and G. W. Kirby, *Proc. Chem. Soc.*, **1963**, 267.
27. A. R. Battersby, R. J. Francis, M. Hirst, and J. Staunton, *Proc. Chem. Soc.*, **1963**, 268.
28. For a review see: A. R. Battersby, in *Festschrift Kurt Mothes, Beiträge zur Biochemie und Physiologie von Naturstoffen*, Gustav Fischer Verlag, Jena 1965, p. 81.
29. R. Chatterjee, *J. Indian Chem. Soc.*, **28**, 225 (1951).
30. M. Shamma, M. A. Greenberg, and B. S. Dudock, *Tetrahedron Letters*, **1965**, 3595.
31. I. Imaseki and H. Taguchi, *J. Pharm. Soc. Japan*, **82**, 1214 (1962).
32. R. W. Doskotch, M. Y. Malik, and J. L. Beal, *J. Org. Chem.* **32**, 3253 (1967).
33. R. Chatterjee and A. Banerjee, *J. Indian Chem. Soc.*, **30**, 705 (1953).
34. R. Chatterjee and P. C. Maiti, *J. Indian Chem. Soc.*, **32**, 609 (1955).
35. H. Schmid and P. Karrer, *Helv. Chim. Acta*, **32**, 960 (1949).
36. B. Witkop, *J. Am. Chem. Soc.*, **78**, 2873 (1956).
37. M. L. Carvalhas, *J. Chem. Soc. Perkin I*, **1972**, 327.
38. M. M. Nijland, *Pharm. Weekblad*, **96**, 640 (1961).
39. M. M. Nijland, *Pharm. Weekblad*, **98**, 301 (1963).
40. M. Shamma and B. S. Dudock, *Tetrahedron Letters*, **1965**, 3825.
41. S. F. Dyke and E. P. Tiley, *Tetrahedron Letters*, **1972**, 5175.
42. I. Moncovic and I. D. Spencer, *J. Am. Chem. Soc.*, **87**, 1137 (1965).
43. M. Shamma and B. S. Dudock, *Tetrahedron Letters*, **1965**, 3825.
44. H. F. Andrew and C. K. Bradsher, *Tetrahedron Letters*, **1966**, 3069.
45. K. Y. Zee-Cheng and C. C. Cheng, *J. Pharm. Sci.*, **61**, 969 (1972).
46. R. F. H. Manske, *Can. J. Res. B*, **21**, 111 (1943).
47. H. Taguchi and I. Imazeki, *J. Pharm. Soc. Japan*, **84**, 955 (1964).
48. C. Tani and N. Takao, *J. Pharm. Soc. Japan*, **82**, 594, 598 (1962).
49. C. Tani, N. Takao, and S. Takao, *J. Pharm. Soc. Japan*, **82**, 748 (1962).

50. T. R. Govindachari, K. Nagarajan, S. Natarajan, and B. R. Bai, *Indian J. Chem.*, **9**, 1313 (1972).
51. S. Pfeifer and D. Thomas, *Pharmazie*, **21**, 701 (1966).
52. M. Maturová, D. Pavlásková, and F. Šantavý, *Plant. Med.*, **14**, 22 (1966).
53. S. Pfeifer, I. Mann, L. Dolejš, V. Hanuš, and A. D. Cross, *Tetrahedron Letters*, **1967**, 83.
54. V. Preininger, L. Hruban, V. Šimanek, and F. Šantavý, *Collection Czech. Chem. Commun.*, **35**, 124 (1970).
55. V. Šimanek, V. Preininger, P. Sedmera, and F. Šantavý, *Collection Czech. Chem. Commun.*, **35**, 1440 (1970).
56. C. Tani and N. Takao, *J. Pharm. Soc. Japan*, **82**, 596 (1962).
57. S. Uyeo, H. M. Fales, R. J. Highet, and W. C. Wildman, *J. Am. Chem. Soc.*, **80**, 2590 (1958).
58. A. S. Bailey and R. Robinson, *J. Chem. Soc.*, **1950**, 1375.
59. A. S. Bailey, R. Robinson, and R. S. Staunton, *J. Chem. Soc.*, **1950**, 2732.
60. A. S. Bailey and R. C. Worthing, *J. Chem. Soc.*, **1956**, 4535.
61. S. F. Dyke, B. J. Moon, and M. Sainsbury, *Tetrahedron Letters*, **1968**, 3933.
62. H. R. Arthur, W. H. Hui, and Y. L. Ng, *Chem. Ind.*, **1958**, 1514; *J. Chem. Soc.*, **1959**, 1840.
63. H. R. Arthur, W. H. Hui, and Y. L. Ng, *J. Chem. Soc.*, **1959**, 4007, 4010.
64. T. Kametani, K. Kigasawa, M. Hiiragi, and O. Kusoma, *J. Heterocycl. Chem.*, **10**, 31 (1973).
65. K. Y. Zee-Cheng and C. C. Cheng, *J. Heterocycl. Chem.*, **10**, 85 (1973).
66. W. M. Messmer, M. Tin-Wa, H. H. S. Fong, C. Bevelle, N. R. Farnsworth, D. J. Abraham, and J. Trojanek, *J. Pharm. Sci.*, **61**, 1858 (1972).
67. F. Fish and P. G. Waterman, *J. Pharm. Pharmacol.*, **23**, 67 (1971).
68. M. Onda, K. Tagiguchi, M. Hirakura, H. Fukushima, M. Akagawa, and F. Naoi, *J. Agr. Chem. Soc. Japan*, **39**, 168 (1965).
69. M. Onda, K. Abe, K. Yonezawa, N. Esumi, and T. Suzuki, *Chem. Pharm. Bull. (Tokyo)*, **18**, 1435 (1970).
70. J. Slavik and L. Slavikowa, *Collection Czech. Chem. Commun.*, **20**, 21 (1954).
71. J. Slavik and L. Slavikowa, *Collection Czech. Chem. Commun.*, **25**, 1667 (1960).
72. J. Slavik, L. Dolejš, V. Hanuš, and D. Cross, *Collection Czech. Chem. Commun.*, **33**, 1619 (1968).
73. N. Takao, *Chem. Pharm. Bull. Tokyo*, **19**, 247 (1971).
74. N. Takao, H.-W. Bersch, and S. Takao, *Chem. Pharm. Bull. Tokyo*, **19**, 259 (1971).
75. I. Ribas, J. Sueiras, and L. Castedo, *Tetrahedron Letters*, **1971**, 3093.
76. I. Ribas, J. Sáa, and L. Castedo, *Tetrahedron Letters*, **1973**, 3617.
77. P. Schröder and M. Luckner, *Arch. Pharm.* **301**, 39 (1968).
78. J. C. Price, *Australian J. Chem.*, **12**, 458 (1959).
79. S. Goodwin and E. C. Horning, *J. Am. Chem. Soc.*, **81**, 1908 (1959).
80. R. A. Corral and O. O. Orazi, *Tetrahedron*, **21**, 909 (1965).
81. K. Szendrei, E. Minker, M. Koltai, J. Reisch, I. Novák, and G. Buzás, *Pharmazie*, **23**, 519 (1968).
82. H. Rapoport and H. G. Holden, *J. Am. Chem. Soc.*, **81**, 3738 (1959).
83. E. A. Clarke and M. F. Grundon, *J. Chem. Soc.*, **1964**, 4196.
84. D. R. Boyd and M. F. Grundon, *Tetrahedron Letters*, **1967**, 2637; *J. Chem. Soc. C*, **1970**, 556.
85. J. Reisch, K. Szendrei, E. Minker, and I. Novák, *Pharmazie*, **24**, 699 (1969).
86. R. M. Bowman and M. F. Grundon, *J. Chem. Soc. C*, **1966**, 1504.
87. E. Gellért, Raymond-Hamet, and E. Schlittler, *Helv. Chim. Acta*, **34**, 642 (1951).
88. C. Schöpf, *Anales Real Soc. Espan. Fis. Quim. B*, **51**, 173 (1955).

89. E. Bächli, C. Vamvacas, H. Schmid, and P. Karrer, *Helv. Chim. Acta*, **40**, 1167 (1957).
90. R. Kaschnitz and G. Spiteller, *Monatsh. Chem.*, **96**, 909 (1965).
91. K. B. Prasad and G. A. Swan, *J. Chem. Soc.*, **1958**, 2024.
92. L. O. Bejar, R. Goutarel, M. M. Janot, and A. Le Hir, *Ann. Pharm. Franç.*, **14**, 334 (1956); *Compt. Rend.*, **244**, 2066 (1957).
93. M. M. Janot, R. Goutarel, A. Le Hir, and L. O. Bejar, *Ann. Pharm. Franç.*, **16**, 38 (1958).
94. A. Le Hir, M. M. Janot, and D. van Stolk, *Bull. Soc. Chim. France*, **1958**, 551, 580.
95. G. Büchi, R. E. Manning, and F. A. Hochstein, *J. Am. Chem. Soc.*, **84**, 3393 (1962).
96. R. C. Elderfield and O. L. McCurdy, *J. Org. Chem.*, **21**, 295 (1956).
97. J. A. Beisler, *Chem. Ber.*, **103**, 3360 (1970).
98. W. I. Taylor and Raymond-Hamet, *Compt. Rend.*, **262 D**, 1141 (1966).
99. Y. Ban and M. Seo, *Tetrahedron*, **16**, 11 (1961).
100. R. C. Elderfield and R. C. Gray, *J. Org. Chem.*, **16**, 506 (1951).
101. F. E. Bader, *Helv. Chim. Acta*, **36**, 215 (1953).
102. E. Wenkert and D. K. Roychaudhuri, *J. Am. Chem. Soc.*, **79**, 1519 (1957); **80**, 1613 (1958).
103. M. Shamma and J. B. Moss, *J. Am. Chem. Soc.*, **83**, 5038 (1961); **84**, 1739 (1962).
104. M. W. Klohs, M. D. Draper, F. Keller, W. Malesh, and F. J. Petracek, *J. Am. Chem. Soc.*, **76**, 1332 (1954).
105. E. Wenkert, B. Wickberg, and C. L. Leicht, *J. Am. Chem. Soc.*, **83**, 5037 (1961).
106. H. Fritz, *Ann. Chem.*, **655**, 148 (1962).
107. H. Kaneko, *J. Pharm. Soc. Japan*, **80**, 1357, 1374, 1493 (1960).
108. E. E. van Tamelen and L. K. Oliver, *J. Am. Chem. Soc.*, **92**, 2136 (1970).
109. K. Leander and B. Lüning, *Tetrahedron Letters*, **1968**, 905.
110. E. Soderberg and P. Kierkegaard, *Acta Chem. Scand.*, **24**, 397 (1970).
111. Y. Inubushi, Y. Ysuda, T. Konida, and S. Matsumoto, *Chem. Pharm. Bull. Tokyo*, **12**, 749 (1964); **16**, 1014 (1968).
112. T. Onaka, *J. Pharm. Soc. Japan*, **85**, 839 (1965).
113. M. Elander, L. Gawell, and K. Leander, *Acta Chem. Scand.*, **25**, 721 (1971).
114. S. C. Pakrashi and J. Bhattacharyya, *J. Sci. Ind. Res. India B*, **21**, 49 (1962); *Ann. Biochem. Exp. Med. Calcutta*, **23**, 123 (1963); see D. Gröger, *Pharmazie*, **25**, 22 (1970).
115. S. C. Pakrashi, J. Bhattacharyya, L. F. Johnson, and H. Budzikiewicz, *Tetrahedron*, **19**, 1011 (1963).
116. A. S. Belikov, A. I. Ban'Kovskii, and M. V. Tsarev, *J. Obsch. Khim.*, **24**, 919 (1954); *C.A.*, **48**, 11727 (1954).
117. N. J. Leonard and J. A. Deyrup, *J. Am. Chem. Soc.*, **82**, 6202 (1960); **84**, 2148 (1962).
118. D. Ackermann and H. P. List, *Hoppe-Seylers Z. Physiol. Chem.*, **308**, 270; **309**, 286 (1957).
119. For a review see T. J. Mabry in S. W. Pelletier, *Chemistry of the Alkaloids*, Van Nostrand-Reinhold Co., New York, 1970, p. 367; K.-H. Köhler, *Pharmazie*, **28**, 18 (1973).
120. T. J. Mabry, L. Kimler, and R. A. Larson, *Hoppe-Seylers Z. Physiol. Chem.*, **353**, 127 (1972).
121. T. J. Mabry, H. Wyler, G. Sassu, M. Mercier, I. Parikh, and A. S. Dreiding, *Helv. Chim. Acta*, **4 5**, 640 (1962).
122. T. J. Mabry, H. Wyler, I. Parikh, and A.S. Dreiding, *Tetrahedron*, **23**, 3111 (1967).
123. H. Wyler, T. J. Mabry, and A. S. Dreiding, *Helv. Chim. Acta*, **46**, 1747 (1963).
124. M. E. Wilcox, H. Wyler, and A. S. Dreiding, *Helv. Chim. Acta*, **48**, 1134 (1965).
125. M. Piattelli, L. Minale, and G. Prota, *Ann. Chim.*, **54**, 955 (1964).
126. M. E. Wilcox, H. Wyler, T. J. Mabry, and A. S. Dreiding, *Helv. Chim. Acta*, **48**, 252 (1965).
127. M. Piattelli, L. Minale, and G. Prota, *Ann. Chim.*, **54**, 963 (1964).

128. H. Wyler, H. Rösler, M. Mercier, and A. S. Dreiding, *Helv. Chim. Acta*, **50**, 545 (1967).
129. L. Minale, M. Piattelli, S. de Stefano, and R. A. Nicolaus, *Phytochemistry*, **5**, 1037 (1966).
130. L. Minale, M. Piattelli, and S. de Stefano, *Phytochemistry*, **6**, 703 (1967).
131. M. Piattelli, L. Minale, and G. Prota, *Tetrahedron*, **20**, 2325 (1964).
132. M. Piatelli, L. Minale, and G. Prota, *Phytochemistry*, **4**, 121 (1965).
133. H. Döpp and H. Musso, *Chem. Ber.*, **106**, 3473 (1973).
134. G. Habermehl and G. Haaf, *Chem. Ber.*, **98**, 3001 (1965).
135. G. Habermehl and S. Göttlicher, *Chem. Ber.*, **98**, 1 (1965).
136. G. Habermehl and A. Haaf, *Z. Naturforsch.*, **23b**, 1551 (1968).
137. C. Schöpf and O. W. Müller, *Ann. Chem.*, **633**, 127 (1960).

ADDENDUM

To Section II

From Eschscholtzia species a new iminium alkaloid related to escholamine (**10**) has been isolated, called escholamidine, formula **10**: $R^4 = OCH_3$, $R^5 = OH$ (138).

To Section V

The structure of fagaronine (**46**) has been confirmed by spectral studies on its N-demethyl derivative (139). The structure of chelirubine (**49**) should be revised to **85** with the methoxy group in 10 position, which has been confirmed by the synthesis of oxychelirubine (140).

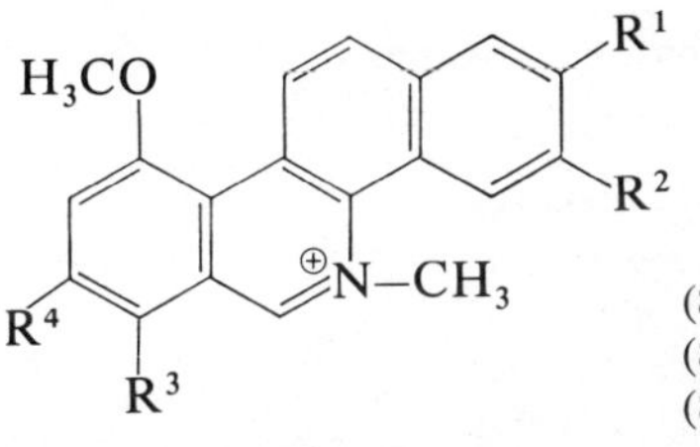

(**85**) Chelirubine: $R^1 + R^2 = R^3 + R^4 = O-CH_2-O$
(**86**) Chelilutine: $R^1 + R^2 = O-CH_2-O$; $R^3 = R^4 = OCH_3$
(**87**) Sanguilutine: $R^1 = R^2 = R^3 = R^4 = OCH_3$

A synthesis of chelilutine and sanguilutine confirms that in the methoxy series also a similar oxygenation pattern is present (141). Chelilutine has the structure **86**, sanguilutine the structure **87**.

To Section VIII

From Vinca elegantissima two new zwitterionic alkaloids, related to flavocarpine (**63**) have been isolated and their structure elucidated by chemical and spectroscopic methods: Vincarpine (**88**) and dihydrovincarpine (**89**) (142).

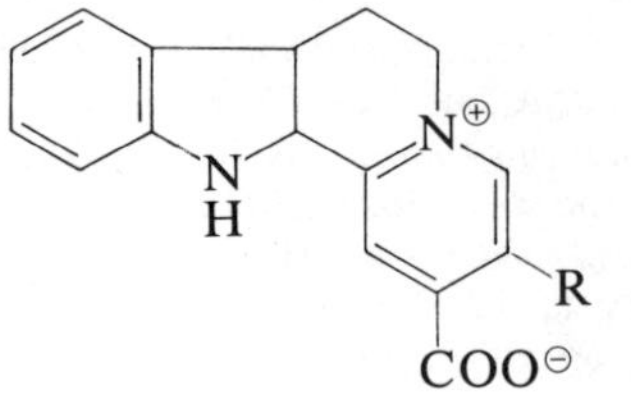

(88) R = CH=CH$_2$
(89) R = C$_2$H$_5$

To Section IX

From Pseudopterogoria americana a new betain has been isolated, which may be regarded as the nor derivative of zooanemonine without the methylene group in the side chain, called norzooanemonine (143). The structure has been confirmed by synthesis.

To Section XIII

A total synthesis of betalains has been achieved by Dreiding and Hermann (144, 145). The key intermediate is dimethyl betalamate semicarbazone. A synthesis of betalamic acid (78) has been performed by Büchi and coworkers (146).

RFERENCES

138. J. Slavík, L. Slavíková, and L. Dolejs, *Collection Czech. Chem. Commun.*, **40,** 1095 (1975).
139. M. Tin-Wa, C. L. Bell, C. Bevelle, H. H. S. Fong, and N. R. Farnsworth, *J. Pharm. Sci.*, **63,** 1476 (1974).
140. H. Ishii, K. Harada, T. Ishida, E. Ueda, and K. Nakajima, *Tetrahedron Lett.*, **1975,** 319.
141. S. V. Kessar, Y. P. Gupta, K. Dhingra, G. S. Sharma, and S. Narula, *Tetrahedron Lett.*, **1977,** 1459.
142. E. Ali, V. S. Giri, and S. C. Pakrashi, *Tetrahedron Lett.*, **1976,** 4887.
143. A. J. Weinheimer, E. K. Metzner, and M. L. Mole, Jr., *tetrahedron*, **29,** 3135 (1973).
144. K. Hermann and A. S. Dreiding, *Helv. Chim. Acta*, **58,** 1805 (1975).
145. K. Hermann and A. S. Dreiding, *Helv. Chim. Acta*, **60,** 673 (1977).
146. G. Büchi, H. Fliri, and R. Shapiro, *J. Org. Chem.*, **42,** 2192 (1977).

EPILOGUE: COMPLEMENTS AND PERSPECTIVES FOR IMINIUM SALT CHEMISTRY

By H. BÖHME, *Universität Marburg, Germany*, and H. G. VIEHE, *Université de Louvain, Louvain-la-Neuve, Belgium*

CONTENTS

I. Introduction 759
II. Activation of Multiple Bonds by Iminium Substituents 760
III. Biosynthesis and Biomimetic Synthesis via Iminium Salts 760
IV. Pyrrolidine and Piperidine Alkaloids 760
V. Isoquinoline and Indole Alkaloids 769
VI. Perspectives for Iminium Salt Chemistry 772
References 772

I. INTRODUCTION

At the close of these two volumes dealing with the chemistry of iminium salts, only comments on perspectives in this field would be in order if our twenty-four-author work were indeed complete. However, all human enterprise is a compromise, and no task of this magnitude can be truly complete. For this reason, we are including in this epilog some additional comments about the role of iminium salt in biosynthesis and in biomimetic synthesis, since this recent development is providing much stimulus to the entire field. Although only leading references are given in this brief discussion of the relevance of iminium salts to nature, we include some of the most important and intriguing structures and reactions dealing with the preparation of iminium salts by the oxidation of amines, enamines, and amides through the intermediacy of inorganic salts, peroxides, or (singlet) oxygen. New developments in stereochemical control of intramolecular additions to double and to triple bonds provide additional insight into biosynthetic pathways, and are stimulating new applications of iminium salt chemistry in the field of synthesis.

We focus attention on the ubiquitous importance of Michael-type addition reactions to multiple bonds activated by the iminium functionality. Cycloadditions to such groupings are also known, but they will be treated independently in a separate volume.

II. Activation of Multiple Bonds by Iminium Substituents

Multiple bonds are activated more strongly towards cycloaddition reactions by iminium substituents than by either carbonyl or nitrile groups. In particular, a comparison between acetylenic esters, amides, and amidium salts (as their alkylation products) showed that the latter were the most reactive (1). It is noteworthy that these iminium substituents need not be tertiary. Both Diels–Alder and 1,3-dipolar cycloaddition reactions have been reported with such activated multiple bonds. The following section on biomimetic synthesis gives many examples of Michael-type addition reactions to such iminium-activated double bonds.

III. Biosynthesis and Biomimetic Synthesis via Iminium Salts

Imines and iminium salts play a central role in the chemistry of two of the most important classes of nitrogen compounds in nature, amino acids and alkaloids. As early as 1912 it was believed that alkaloids were derived from amino acids (2), and structural correlations among the thousands of known alkaloids led to the formulation of possible biosynthetic pathways which, in many instances, have now been verified both *in vivo* and through laboratory syntheses. From the many contributions covering this field some leading references (2–17) have been selected between the years 1917 and 1977, spanning the period from the classical paper by Robinson up to the very recent work of Husson and Potier.

In summary, it has become very clear through the use of carbon, nitrogen, and hydrogen labels in plant feeding experiments that, particularly in short-term biosynthesis, alkaloids are indeed derived either directly from amino acids or from their decarboxylation products. Iminium salts and imines play a central role as intermediates in these biosynthetic pathways.

IV. Pyrrolidine and Piperidine Alkaloids

Some possible biosynthetic pathways to piperidine and pyrrolidine alkaloids from the amino acids lysine or ornithine are given in the following scheme which includes the synthesis of coniine, anabasine, the tropa alkaloids, pelletierine, sparteine, and lobeline. Other biogenetic pathways for pyrrolidine and piperidine alkaloids via imine intermediates are also known (see Scheme 1), such as the facile *in vivo* transformation of octanoic acid to coniine.

In order to account for stereochemical restrictions in these natural products a closely related biogenetic hypothesis starting from Δ^1-piperideine was advanced (see Scheme 2) (18).

The classic biomimetic syntheses of tropinone by Robinson (2b,c) and Schöpf (3a) provide a vivid illustration of the stimulus provided to organic

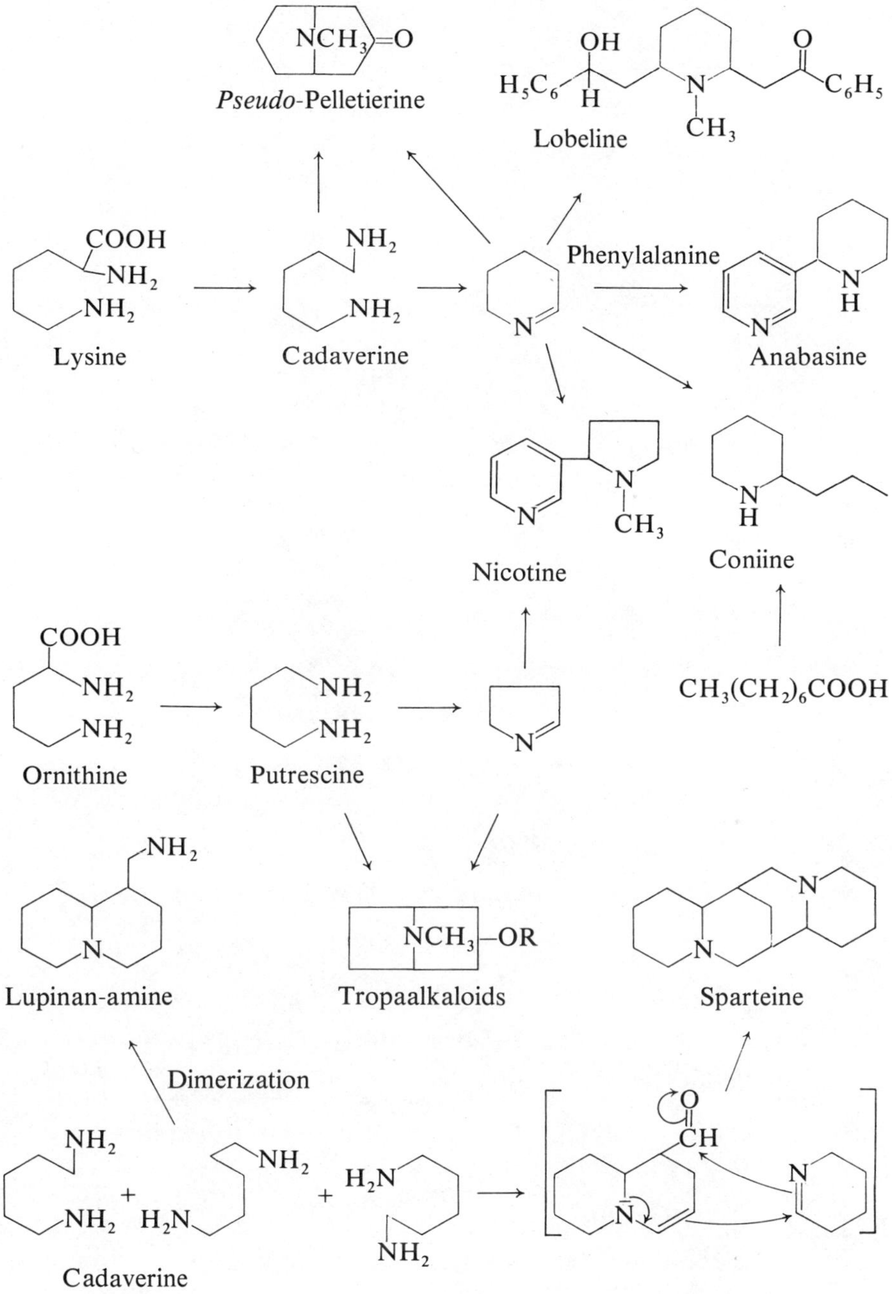

IV. PYRROLIDINE AND PIPERIDINE ALKALOIDS

Scheme 1

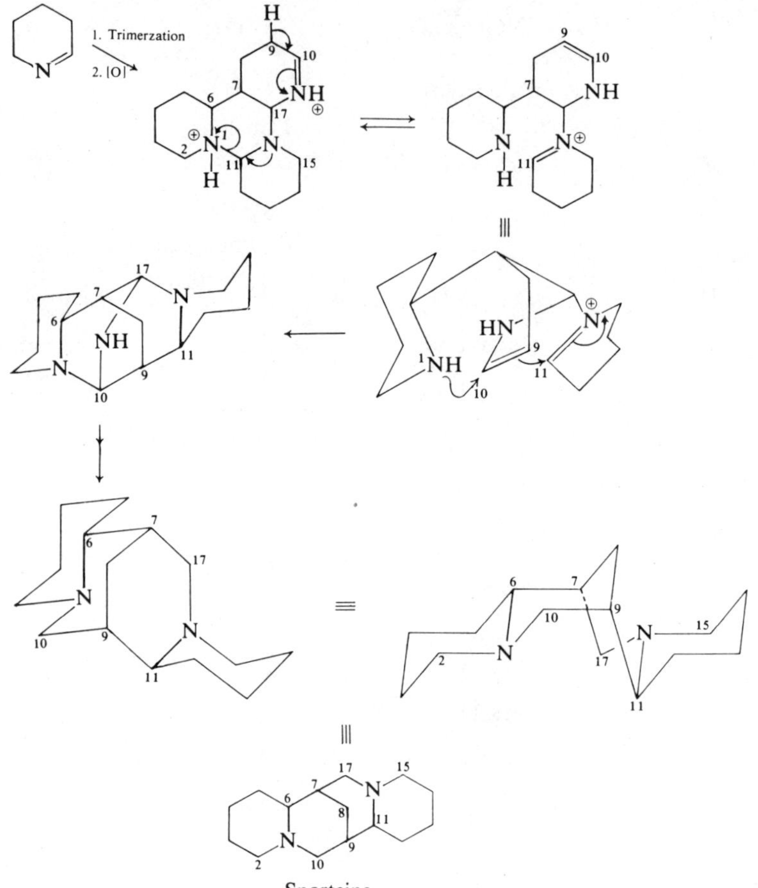
1. Trimerzation
2. [O]
Sparteine

Scheme 2

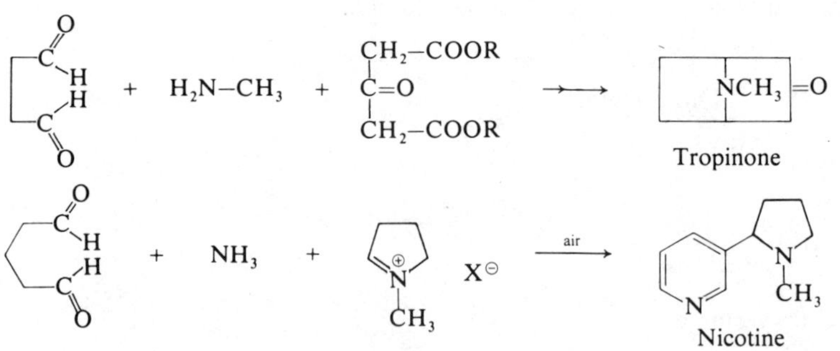
Tropinone
air
Nicotine

Scheme 3

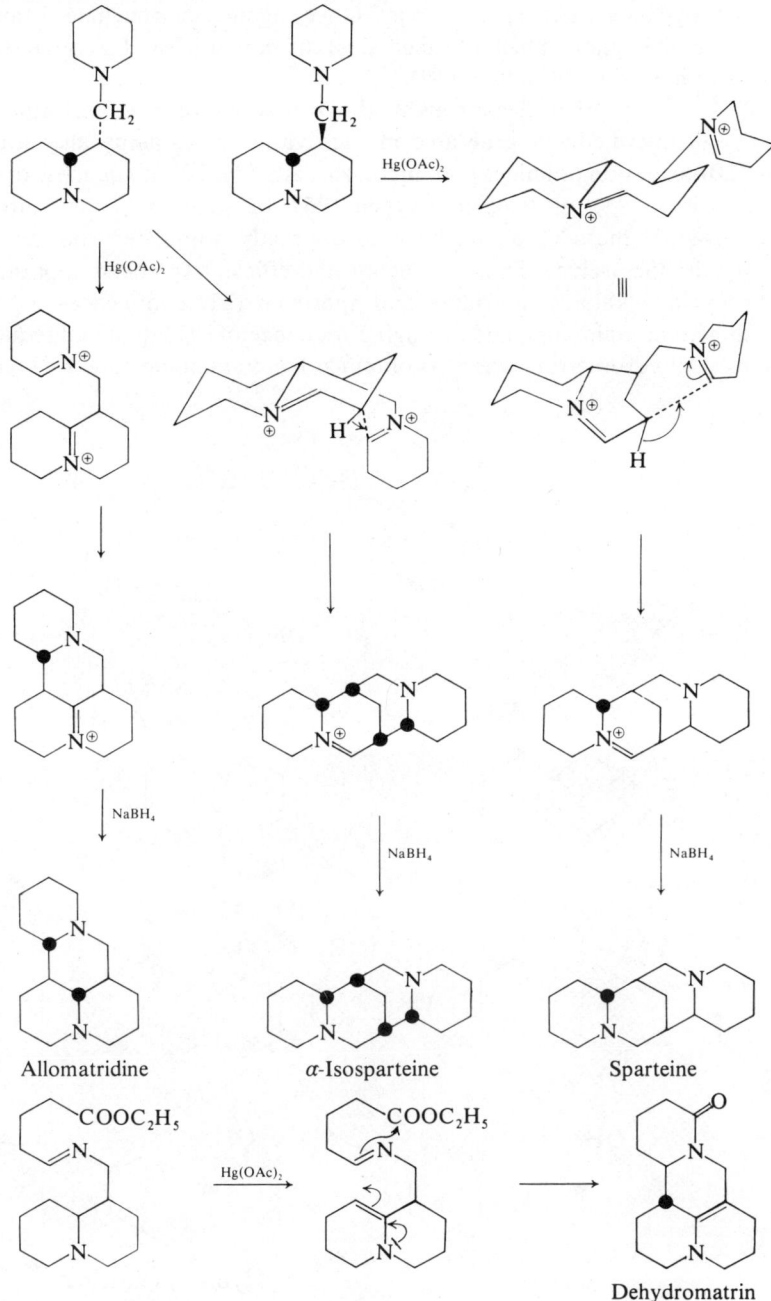

Scheme 4

chemistry by biogenetic speculations. Once again, intermediate Mannich reactions involve iminium salt intermediates; the preparation of nicotine should also be mentioned in this context (19).

Similarly, many other elegant alkaloid syntheses have been performed via iminium salt intermediates generated in such varied ways as by alkylation of imines (20) oxidation of imines with mercuric salts (21, 22), manganese dioxide (23), peroxides (24), or (singlet) oxygen (25). In addition to their intrinsic synthetic value, these reactions have also greatly stimulated the study of iminium salts themselves. Thus an elegant and efficient synthetic approach to the tetracyclic alkaloids matridine and sparteine (21) commences with the stereoisomers of aminolupinane through a sequence of oxidation and reduction reactions. Dehydromatrine was also obtained from aminolupinane by conden-

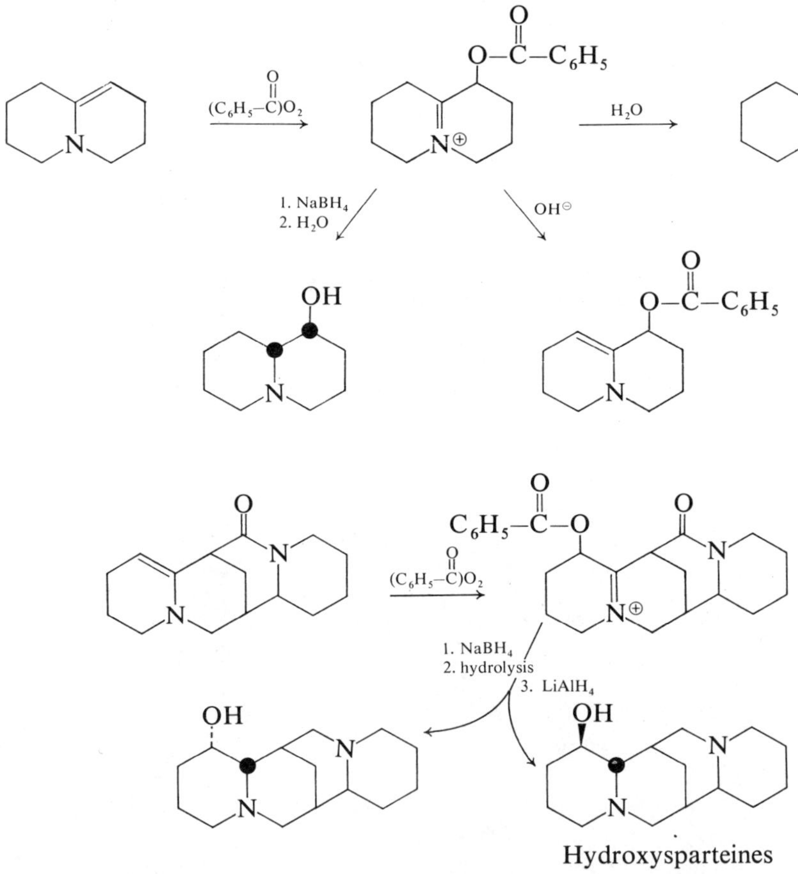

Scheme 5

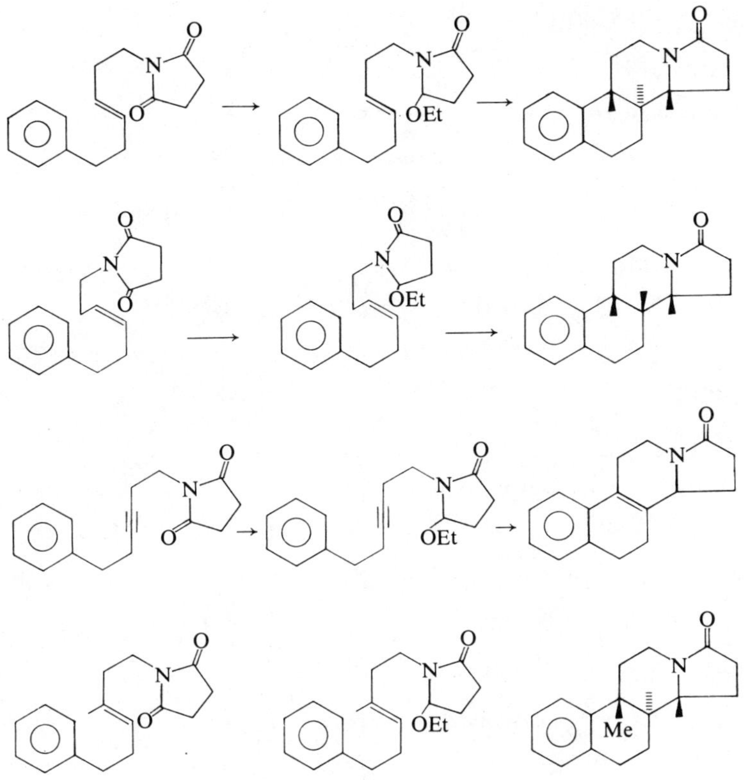

Scheme 6

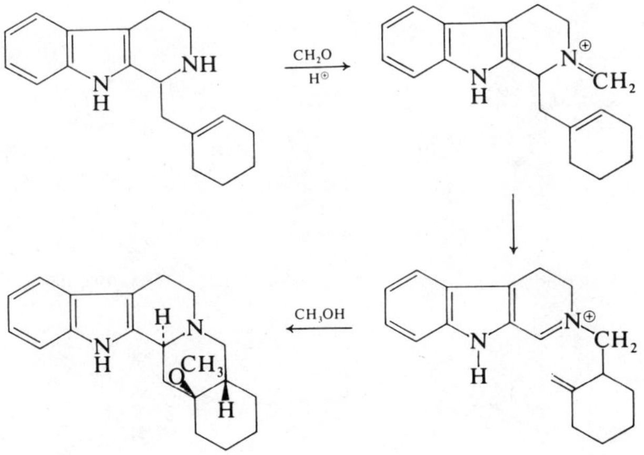

Scheme 7a

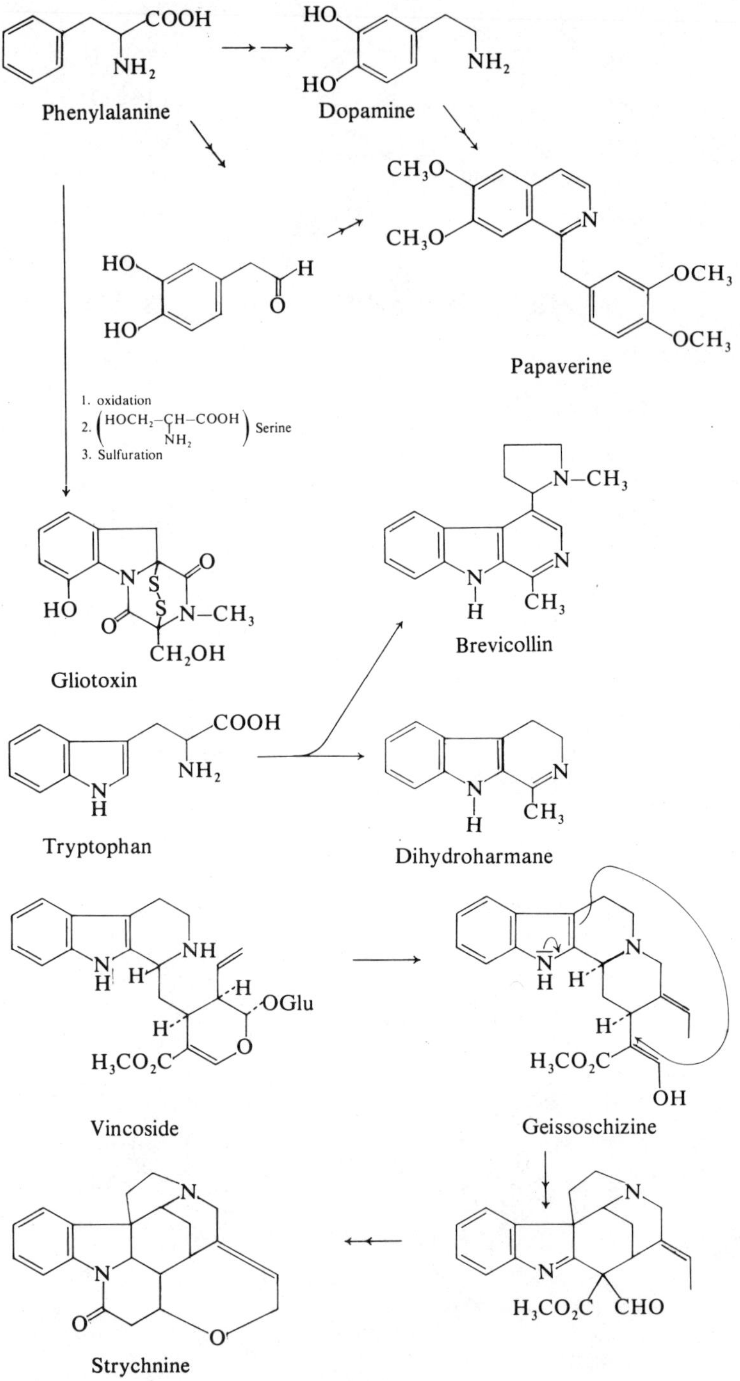
COOH
NH2
Phenylalanine
HO
HO
NH2
Dopamine
CH3O
CH3O
N
OCH3
OCH3
Papaverine
HO
HO
H
O
1. oxidation
2. (HOCH2–CH–COOH NH2) Serine
3. Sulfuration
N–CH3
N
N
H
CH3
Brevicollin
N
S
S
N–CH3
O
O
HO
CH2OH
Gliotoxin
COOH
NH2
N
H
Tryptophan
N
N
H
CH3
Dihydroharmane
NH
N
H
H
H
OGlu
H
H3CO2C
O
Vincoside
N
N
H
H
H
H3CO2C
OH
Geissoschizine
N
N
O
O
Strychnine
N
N
H3CO2C
CHO

Scheme 7b

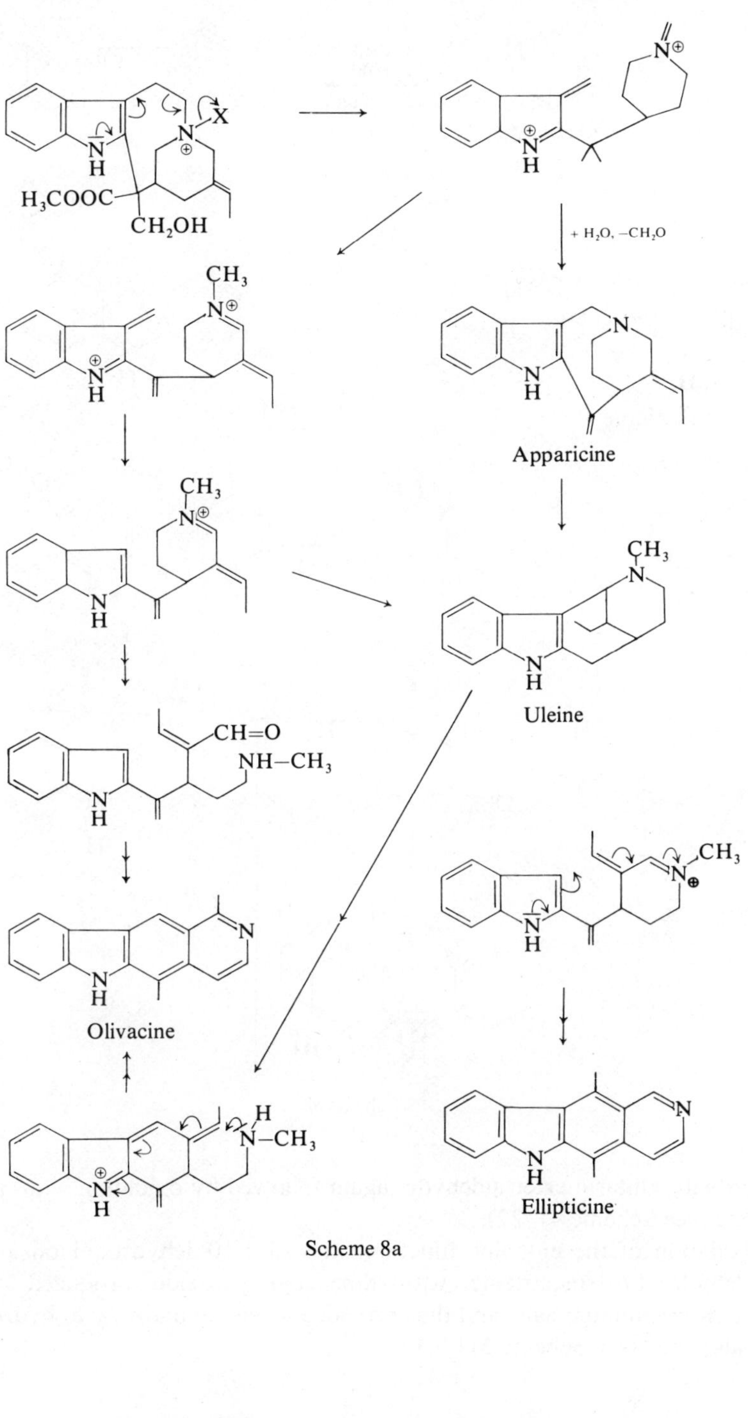
H$_3$COOC
CH$_2$OH
X
+ H$_2$O, −CH$_2$O
CH$_3$
CH=O
NH−CH$_3$
Apparicine
Uleine
Olivacine
Ellipticine
N−CH$_3$

Scheme 8a

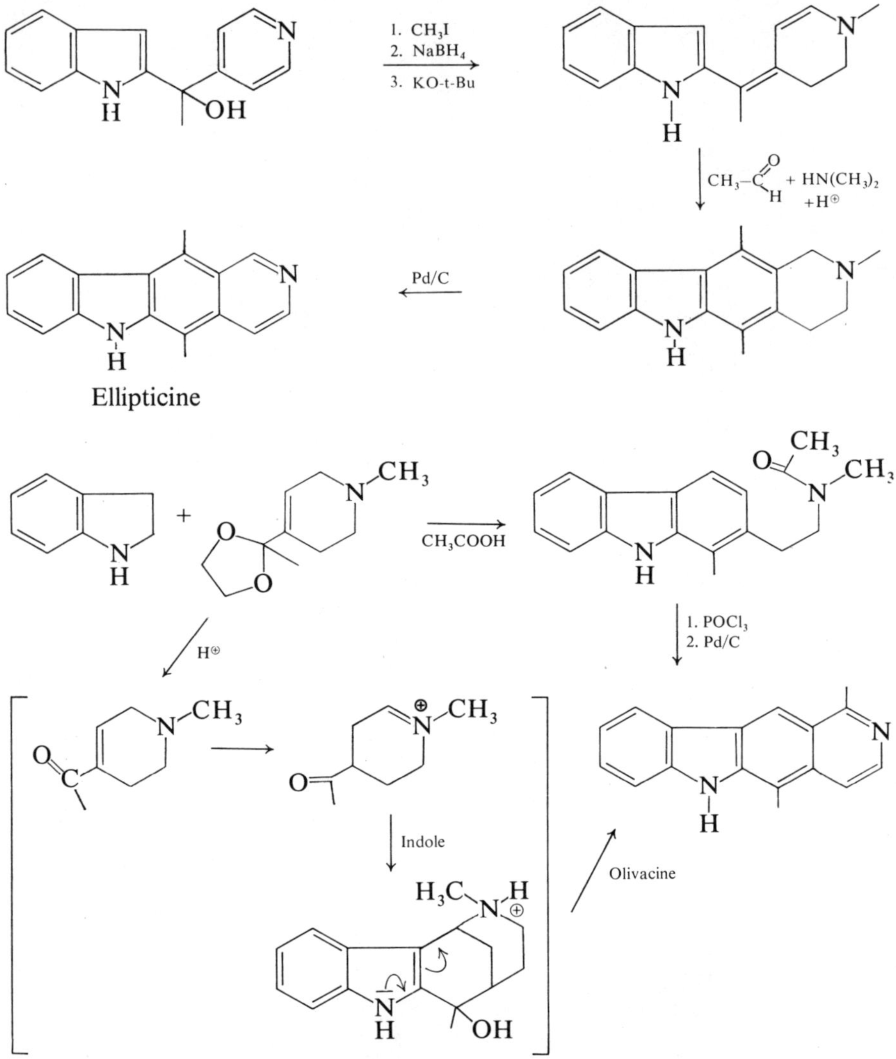

Scheme 8b

Ellipticine

Olivacine

sation with glutaric ester-aldehyde, again followed by oxidation with mercuric acetate (see Scheme 4) (22).

Oxidation of the enamine functionalities of 1,10-dehydroquinolizidine and 5,6-dehydro-17-oxosparteine with dibenzoyl peroxide produced versatile benzoyloxy-iminium salts and thus provided a general pathway to hydroxy and oxo alkaloids (see Scheme 5) (24).

Biomimetic cyclizations involving iminium salts (13) and especially α-acyl iminium salts (26), have been shown to be of great synthetic potential. Many stereoselective hetero- and homocyclizations occur in high yield under very mild conditions (see Scheme 6 for acid catalyzed acyliminium salt reactions).

In the tetrahydrocarboline series, polycyclic indene derivatives of, for example, the yohimbane type have been readily prepared by stereospecific cyclizations involving iminium salt intermediates (see Scheme 7a) (27a).

V. Isoquinoline and Indole Alkaloids

The aromatic amino acids phenylalanine and tryptophan can be transformed into a fascinating variety of alkaloids (10, 12). Scheme 7b depicts possible biosynthetic pathways to papaverine, gliotoxin, dihydroharmane, and brevicollin from phenylalanine or from tryptophan (9, 27b, c), as well as an extension of these concepts to the biosynthesis of strychnine from vincoside via geissoschizine (28).

Biosynthetic studies on the cancerostatic alkaloids ellipticine and olivacine (29) have been accompanied by biomimetic syntheses of these alkaloids (see Schemes 8a and 8b) (30, 31).

The preparation of iminium salts by oxidative pathways has been found to be extremely useful for the preparation of indole alkaloids. The Polonovski reaction (32), in its trifluroacetic acid modification as designed by Potier (33), proceeds as indicated in Scheme 9 (34).

The importance of oxidative methods in the synthesis of bisindole alkaloids via the intermediacy of iminium salts is further illustrated with the coupling of

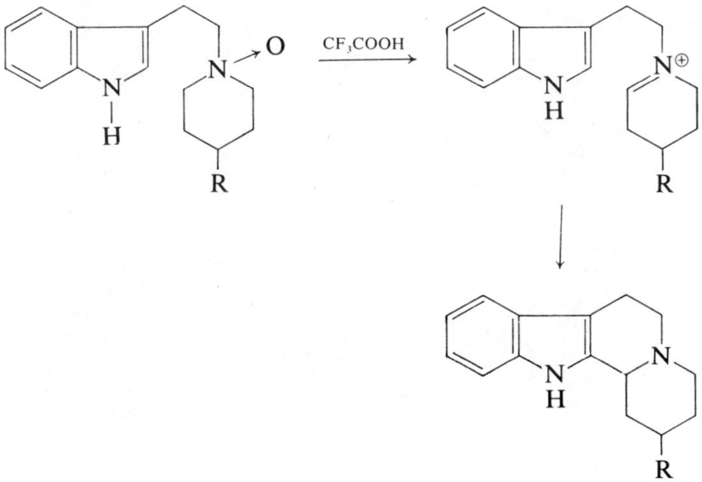

Scheme 9

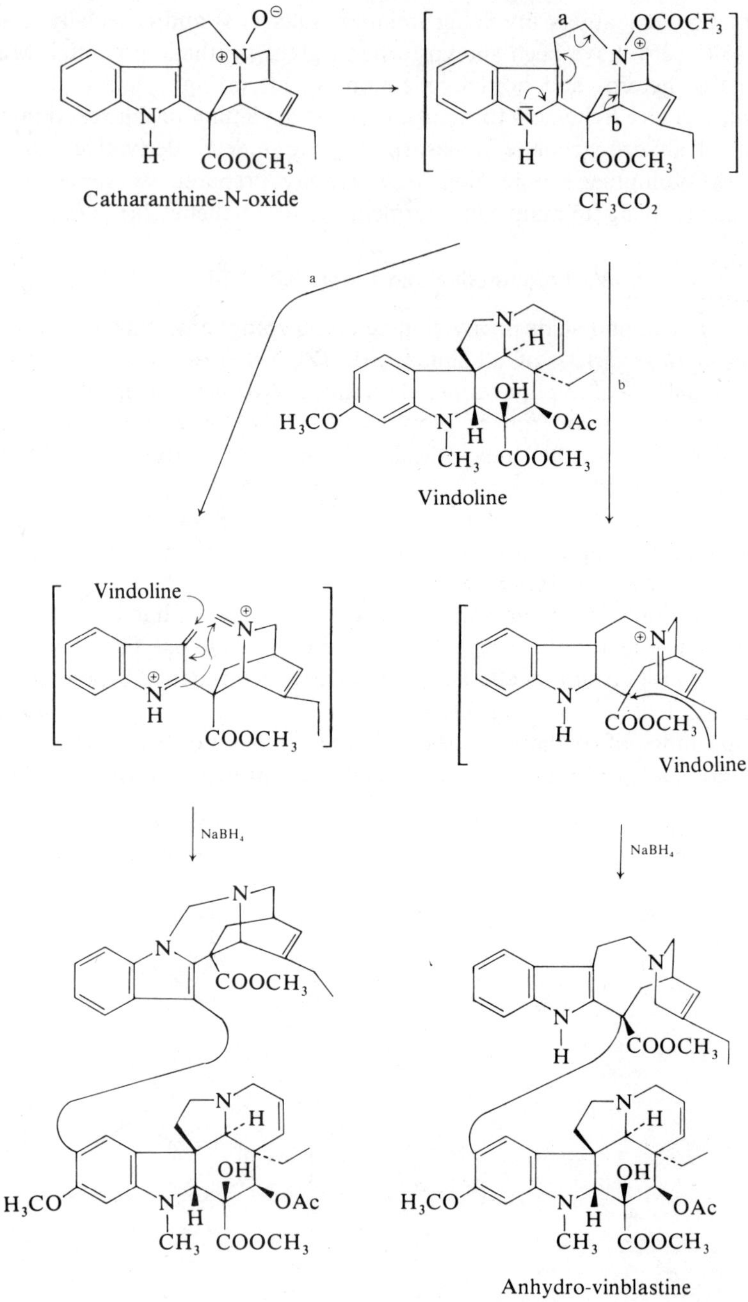

O⊖
⊕N
N
H
COOCH$_3$
Catharanthine-N-oxide
a
OCOCF$_3$
N⊕
N
H
b
COOCH$_3$
CF_3CO_2
a
b
N
H
OH
H_3CO
N
H
OAc
CH_3
COOCH$_3$
Vindoline
Vindoline
⊕N
N⊕
H
COOCH$_3$
⊕N
N
H
COOCH$_3$
Vindoline
NaBH$_4$
NaBH$_4$
Anhydro-vinblastine

Scheme 10

catharanthine-N-oxide with vindoline to give anhydrovinblastine (35). Vinblastine itself is now widely employed in cancer chemotherapy. Similar approaches (36–38) illustrate both the current intense interest in this class of complex alkaloids and the potential of this simple method for the preparation of iminium salts (see Scheme 10).

The methylene blue-sensitized photooxidation of vincamone and of Δ^{14}-vincine provide our final examples of the selective synthesis of alkaloids via iminium salts (see Scheme 11) (25).

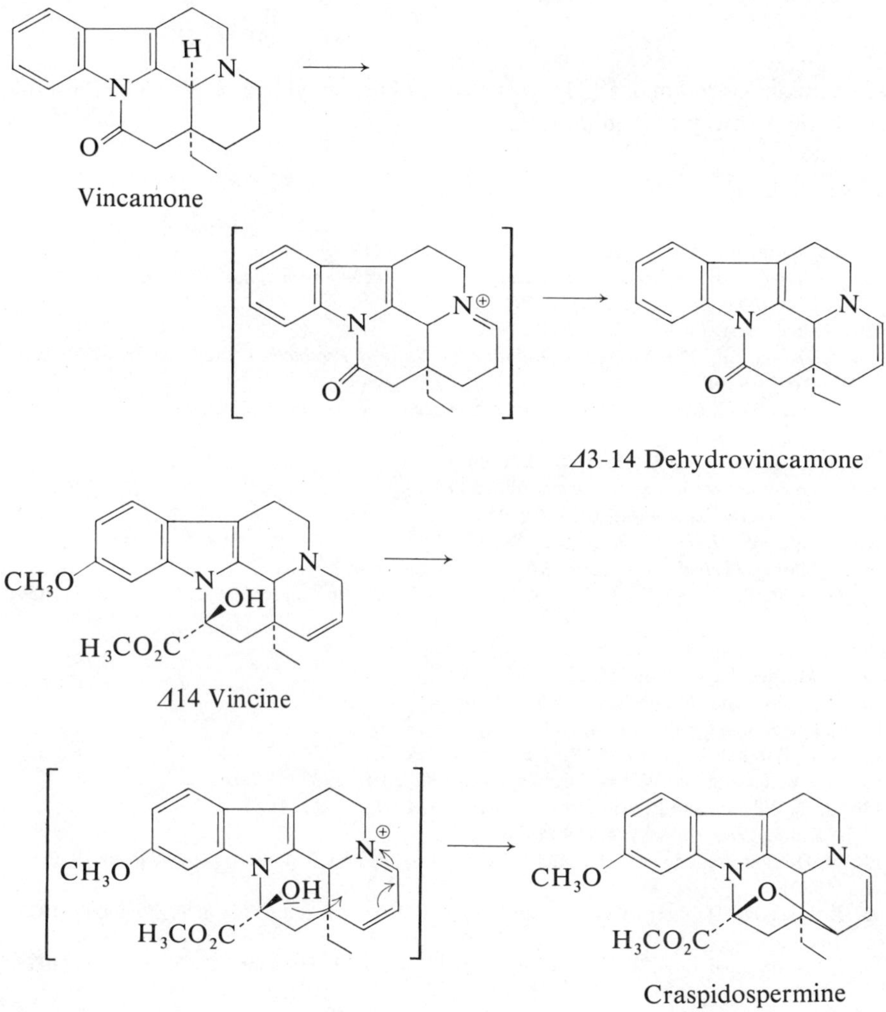

Vincamone

Δ3-14 Dehydrovincamone

Δ14 Vincine

Craspidospermine

Scheme 11

VI. Perspectives for Iminium Salt Chemistry

We are convinced that iminium salts will be increasingly recognized as interesting and versatile reagents in organic chemistry. Novel methods continue to be developed for their synthesis, such as the decarbonylation of α-tertiary amino acids (39), and the electrooxidation of tertiary amines (40). The reaction

of enamides or ynamides (41) with electrophiles may also prove to be a useful synthetic pathway to iminium salts.

REFERENCES

1. J. S. Baum and H. G. Viehe, *J. Org. Chem.* **41**, 183 (1976).
2a. G. Trier, *Über einfache Pfanzenbasen und ihre Beziehungen zum Aufbau der Eiweisstoffe und Lecithine*, Bornträger, Berlin, 1912, p. 117.
2b. R. Robinson, *J. Chem. Soc.*, **1917**, 762, 876.
2c. R. Robinson, *The Structural Relation of Natural Products*, Clarendon Press, Oxford, England, 1955.
3a. C. Schöpf, G. Lehmann, and W. Arnold, *Angew. Chem.*, **50**, 783, 787, 797 (1937).
3b. F. Faltis, L. Holzinger, P. Ita, and R. Schwarz, *Ber. Deut. Chem. Ges.*, **74**, 79 (1941).
4a. R. B. Woodward, *Nature*, **162**, 155 (1948).
4b. R. B. Woodward, *Angew. Chem.*, **68**, 13 (1956).
5a. E. Wenkert, *Experientia*, **10**, 346 (1954).
5b. E. Wenkert, *J. Amer. Chem. Soc.*, **84**, 98 (1962).
6. R. Thomas, *Tetrahedron Lett.*, **1961**, 544.
7. D. H. R. Barton and A. Cohen, *Festschrift A. Stoll.*, Birkhauser, Basel, Switzerland, 1957, p. 117.
8. A. R. Battersby, *Proceedings* (London), **1963**, 189.
9. K. Mothes, *Experientia*, **25**, 225 (1969).
10. A. I. Scott, *Acc. Chem. Res.*, **3**, 151 (1970).
11. E. Leete, *Acc. Chem. Res.*, **4**, 100 (1971).
12. A. R. Battersby, *Acc. Chem. Res.*, **5**, 148 (1972).
13a. E. Winterfeldt and W. Franzischka, *Chem. Ber.*, **101**, 2938 (1968).
13b. J. D. Wilcock and E. Winterfeldt, *Chem. Ber.* **107**, 975 (1974).
14. J. Kutney, *Heterocycles*, **4**, 169 (1976).
15. A. Husson, Y. Langlois, C. Riche, H.-P. Husson, and P. Potier, *Tetrahedron*, **29**, 3095 (1973).
16. J. Stoeckigt, H. P. Husson, C. Kan-Fan, and M. H. Zenk, *J.C.S. Chem. Commun.*, **1977**, 164.
17. M. Andriantsiferana, R. Besselièvre, C. Riche, and H. P. Husson, *Tetrahedron Lett.*, **1977**, 2587.
18. W. M. Golebiewski and I. D. Spenser, *J. Amer. Chem. Soc.*, **98**, 6726 (1976).

19. E. Leete, *J.C.S. Chem. Commun.*, **1972,** 1091.
20. W. Oppolzer, H. Hauth, P. Pfäffli, and R. Wenger, *Helv. Chim. Acta*, **60,** 1801 (1977).
21. F. Bohlmann, E. Winterfeldt, and U. Friese, *Chem. Ber.*, **96,** 2251 (1963).
22. F. Bohlmann, D. Schumann, U. Friese, and E. Poetsch, *Chem. Ber.*, **99,** 3358 (1966).
23. B. Franck and G. Blaschke, *Ann. Chem. Liebigs.*, **668,** 145 (1963).
24. F. Bohlmann and H. Peter, *Chem. Ber.*, **99,** 3362 (1966).
25. R. Beugelmans, D. Herlem, H. P. Husson, F. Khuong-Huu, and M. T. Le Goff, *Tetrahedron Lett.*, **1976,** 435.
26a. J. Dijking and W. Speckamp, *Tetrahedron Lett.*, **1975,** 4039, 4047.
26b. J. Dijking, H. F. Schoemaker, and W. N. Speckamp, *Tetrahedron Lett.*, **1975,** 4043.
26c. J. B. P. A. Wijnberg and W. N. Speckamp, *Tetrahedron Lett.*, **1975,** 3963, 4035.
26d. H. E. Schoemaker, J. Dijking, and W. N. Speckamp, unpublished results, 1977.
26e. J. Dijking and W. N. Speckamp, *Tetrahedron Lett.*, **1977,** 935 and unpublished results, 1977.
27a. V. U. Ahmad, K.-H. Feuerherd, and E. Winterfeldt, *Chem. Ber.*, **110,** 3624 (1977).
27b. R. B. Herbert, in *The Alkaloids*, J. E. Saxton, Ed., 5th ed. (Specialist Periodical Reports), The Chemical Society, London, 1975.
27c. W. H. Müller, R. Preuss, and E. Winterfeldt, *Chem. Ber.*, **110,** 2424 (1977).
28. S. I. Heimberger and A. I. Scott, *J.C.S. Chem. Commun.*, **1973,** 217.
29a. G. Kunesch, C. Poupat, Nguen Van Bac, G. Henry, T. Sévenet, and P. Potier, *Compt. Rend. Acad. Sci. Paris*, **285,** 89 (1977).
29b. H. P. Husson and R. Besselièvre, private communication.
30a. R. Besselièvre, C. Thal, H. P. Husson, and P. Potier, *J.C.S. Chem. Commun.*, **1975,** 90.
30b. R. Besselièvre and H. P. Husson, *Tetrahedron Lett.*, **1976,** 1873.
31. M. Sainsbury, *Synthesis*, **1977,** 437.
32. M. Polonovski and M. Polonovski, *Bull. Soc. Chim. France*, **1927,** 1190.
33. A. Ahond, A. Cavé, C. Kan-Fan, H. P. Husson, J. P. de Rostolan, and P. Potier, *J. Am. Chem. Soc.*, **90,** 5622 (1968).
34. L. Chevolot, A. Husson, C. Kan-Fan, H. P. Husson, and P. Potier, *Bull. Soc. Chim. France*, **1976,** 1222.
35. N. Langlois, F. Guéritte, Y. Langlois, and P. Potier, *J. Am. Chem. Soc.*, **98,** 7017 (1976).
36a. Y. Honma and Y. Ban, *Heterocycles*, **6,** 285 (1977).
36b. Y. Honma and Y. Ban, *Heterocycles*, **6,** 291(1977).
36c. J. P. Kutney, A. V. Joshua, and P. H. Liao, *Heterocycles*, **6,** 297 (1977).
37a. J. P. Kutney, G. H. Bokelman, M. Ichikawa, E. Jahngen, A. V. Joshua, P. H. Liao, and B. R. Worth, *Can. J. Chem.*, **55,** 3227 (1977).
37b. J. P. Kutney, A. V. Joshua, P. H. Liao, and B. R. Worth, *Can. J. Chem.*, **55,** 3235 (1977).
38. P. Mangeney, R. Costa, Y. Langlois, and P. Potier, *Compt. Rend. Acad. Sci. Paris*, **284,** 701 (1977).
39a. V. I. Maksimov, Izvest. Akad. Nauk. S.S.S.R., **112,** (1962); N. A. Poddubnaya and V. I. Maksimov, *J. Gen. Chem. U.S.S.R.* (Eng. Transl.), **29,** 3448 (1968).
39b. R. T. Dean, H. C. Padgett, and H. Rapoport, J. Am. Chem. Soc., **98,** 7448 (1976).
40a. K. Nyberg and R. Severin, *Acta Chem. Scand. Ser B*, **30,** 640 (1976).
40b. G. Minet and A. Rambout, Ph.D. research, unpublished results, 1977.
41. E. Goffin, Y. Legrand, and H. G. Viehe, *J. Chem. Res. (S)*, **1977,** 105.

Author Index

Numbers in parentheses are reference numbers and show that an author's work is referred to although his name is not mentioned in the text. Numbers in *italics* indicate the pages on which the full references appear.

Abe, K., 745(69), *755*
Abgott, R. A., 342(113), *380*
Abraham, D. J., 350(121), *380,* 743(66), *755*
Abramovitch, R. A., 574(3), 595(228), *634, 639*
Abusharab, E., 647(530, 531), *651*
Ackermann, D., 734(3), 751(118), *753, 756*
Ackermann, H., 241(177), *266*
Adam, W., 592(189), *639*
Adams, R., 138(284), *141*
Adduci, J. M., 592(201), *639*
Adkins, H., 575(30), 576(30), 592(30), 593(30), *635*
Adrianova, G. M., 22(74), *52*
Agai, B., 169(63), *172*
Agawa, T., 629(474), *646*
Ah-Kow, G., 57(191), *63*
Ahlbrecht, H., 125(226), *140,* 190(39), 241(178), 243(178), *263, 266*
Ahmad, I., 539(55), *569*
Ahmad, V. U., 769(27a), *773*
Ahmed, M. G., 205(100), 207(106), *265*
Ahmed, S., 59(198), *64*
Ahoud, A., 769(33), *773*
Ainsworth, C. A., 18(56), *51*
Aizawa, S., 298(98), 301(98), *312*
Akagawa, M., 744(68), *755*
Akalaev, A. N., 409(62, 63), 425(62, 63), 426(63), 439(63), 485(63), 499(62), *523*
Albert, A., 11(42), *51*
Albright, J. D., 38(130), 40(130), 41(130), *53*
Alder, R. W., 205(100), 207(106), *265*

Alekperova, Yu. A., 580(98), *636*
Alexander, S. M., 232(169), *266*
Alferova, O. J., 647(516), *651*
Alhaique, F., 125(220), *139,* 269(219), 274(219), *277*
Ali, E., 757(142), *758*
Ali, W., 604(293), *641*
Alimov, P. I., 30(99), *53*
Al-Janabi, M. Y., 556(88), *570*
Allan, J. A. V., 50(166), *54*
Allen, C. F. H., 592(198), *639*
Allenstein, E., 73(56, 57, 58, 59, 60, 61, 62, 63, 64, 65), 74(61, 75), 78(56, 57, 58, 59, 60), 80(56, 57, 58, 59, 60, 61, 62, 63, 64, 65), 83(75, 109), 84(75), 99(75), *121, 122,* 126(233), *140,* 157(46), 158(46), *165,* 174(4), *180,* 182(10), *262,* 214(127, 128, 129), 222(129), *265,* 360(166), *381,* 532(28), 533(28), 535(38), 538(48), 541(62), 554(86), *569, 570*
Allinger, N. L., 648(571), *652*
Altreuter, R., 583(129), *637*
Altreuther, P., 182(9), *262*
Al'tshuler, R. A., 56(178), *63,* 270(222), 272(222), 273(222), *277*
Alumi, S., 6(18), 7(18), *51*
Ambrosini, A., 58(195), *63*
Amimoto, Y., 647(547), *652*
Amin, S. G., 288(70), *312*
Amoros, M., 88(123), 90(123), *123*
Amschler, H., 8(31), 11(31), *51,* 325(42), 326(42), 354(42), 355(42), 356(42), 359(42), 366(42), 375(42), *379*
Anaud, N., 300(100), *312*

775

AUTHOR INDEX

Anders, B., 332(67), *379*
Anderson, A. G., Jr., 61(211), *64*
Anderson, L. B., 191(56), 245(56), 246(56), *264*
Ando, T., 126(236), 139(236), *140*
Andreev, L. N., 316(129), 319(129), *320*
Andrew, H. F., 741(44), *754*
Andrianov, V. G., 268(211), 276(211), *276*
Andriantsiferana, M., 760(17), *772*
Andrieu, L., 279(55a), 284(55a), 286(55a), *311*
Angelici, R. A., 315(127), *320*
Anisimov, O. S., 60(201), *64*
Anret, B. J., 729(161), *731*
Anthony, S. A., 633(506), *647*
Antokhina, L. N., 30(99), *53*
Ántus-Ercsényi, A., 149(31), 153(36), 157(44), 158(31, 36, 44), 159(48), *164*
Appel, R., 24(86, 87, 88, 89, 90, 91, 92), 25(90, 91, 92), 26(86, 87, 88, 89), *52*, 60(202), *64*
Arbuzov, A. E., 212(119), 265, 287(68), 295(68), 297(68), *312*, 633(501), *646*
Archer, W. L., 6(12), *50*
Arens, J. F., 579(79), 609(79), 633(487, 500, 510), *636, 646, 647*
Argersinger, W. J., 597(272), *641*
Arient, J., 96(160), *123*, 323(25), 377(25), *378*
Arndt, G., 313(113, 114), *320*
Arnett, E. M., 568(120), *571*
Arnold, W., 760(3a), *772*
Arnold, Z., 7(21), 16(52), 27(21), 30(21), 44(154), 48(21), *51, 54*, 66(11, 12, 13), 77(13, 98, 99), 78(13), 79(13, 98), 81(13, 99), 83(13, 110), 85(110), 88(131, 132), 89(131, 132), 90(13, 131, 132), 92(13, 98), 93(11, 12), 103(179), 113(13), 114(13), *120, 122, 124*, 137(278, 279), *141*, 170(64), *172*, 214(126, 130), 252(199), *265, 267*, 324(29), 331(63), 367(175), 368(175), *378, 379, 381*, 400(16), 405(27), 406(27, 36), 407(27), 411(27, 36), 445(16), 447(16), 455(27), 457(21), 480(27), 485(16), *522*
Aronova, N. I., 596(260), 626(260, 457), *640, 645*
Artamonov, A. A., 633(492), *646*
Artemova, V. V., 650(621), *654*

Arthur, H. R., 743(62, 63), 744(62, 63), *755*
Arya, V. P., 382(196), *391*
Ashby, C. E., 596(248), *640*
Atavin, A. S., 387(215), *392*
Atlani, P. M., 633(494, 502, 503, 505, 507), *646, 647*
Atta-Ur-Rahman, 59(200), *64*
Aufderhaar, E., 207(107), 212(107), *265, 579(71), 636*
Augstein, L. L., 126(234), 129(234), 133(234), *140*
Aumann, G., 590(167), *638*
Autenrieth, W., 290(85), *312*
Auterhoff, G., 351(136), *381*
Auwers, K., 589(159), *638*
Avison, A. W. D., 593(217), *639*
Ayling, E. E., 74(74), *121*

Baardman, F., 593(211), *639, 648(558), 652*
Babad, H., 579(84), *636*
Babadjamian, A., 317(144), 319(143, 144), *320*
Babichev, F. S., 648(573), *652*
Babin, E. P., 633(492), *646*
Babler, J. H., 701(86), 702(86), *726*
Bachli, E., 747(89), 748(89), *756*
Bachmann, G., 91(142), 100(168), *123*
Bachmann, G. B., 32(107, 108), *53*
Backvall, J.-E., 575(41), 595(41), 596(41), 620(41), 623(41), *635*, 651(628), *654*
Bacon, R. G. R., 729(161), *731*
Bader, A. R., 683(55, 56), *725*
Bader, F. E., 748(101), *756*
Bader, H., 290(86), *312*, 574(8), *634*
Baeza, J., 592(189), *639*
Bafna, S., 603(283), *641*
Bahner, C. T., 386(212), *392*
Bai, B. R., 742(50), *755*
Bailey, A. S., 623(439, 440), *645*, 744(58, 59, 60), *755*
Bailey, F. C., 649(608), *653*
Bailey, P. S., Jr., 590(179), *638*
Balbi, A., 58(195), *63*
Balch, A. L., 336(83), *379*
Bald, R. W., 452(86e), 465(86e), *523*
Ball, A. A., 578(59), *636*
Balli, H., 21(71), *52*, 167(61), 171(61), *172*
Balsiger, R. W., 354(145), *381*

AUTHOR INDEX

Bamberger, E., 31(100), *53,* 322(12), 327(12), *378*
Ban, Y., 184(19, 22), 190(19, 45, 49), 194(19), 244(49), 245(49), 246(181), 254(45), 257(45, 198), 268(213), 269(221), 276(213), *263, 267, 277,* 500(152, 153, 154), 501(154), 502(152, 153), *525,* 747(99), *756,* 771(36a, 36b), *773*
Banerjee, A., 740(33), *754*
Bangert, K. F., 186(31), *263*
Banister, A. J., 581(112), *637*
Bank, J., 110(200), *124*
Bankhead, R., 729(161), *731*
Ban'Kovskii, A. I., 750(116), *756*
Bapat, J. B., 729(155), *731*
Baranov, S. N., 589(156), 617(403, 404), 618(403, 404, 405, 406), 619(403, 404, 406), 620(431, 432, 433), 621(431, 432, 433), 624(444), 625(444, 446, 450, 451, 452), 626(461), 628(471, 472), 630(406), 631(403), 634(513), *638, 644, 645, 646, 647,* 650(622), 651(629), *654*
Barbiroli, G., 609(347), *643*
Bard, R. R., 383(200, 201), *391*
Barkanskas, V. P., 683(59), *725*
Barkhagh, V. A., 597(263), *640*
Barnela, S. B., 55(173), 56(173), *63*
Barner, R., 658(16), 689(16), 713(111), *724, 727*
Barnett, W. E., 658(14), *724*
Barrett, W. G., 727(133), *730*
Barr, J. T., 386(212), *392*
Barraclough, D. J., 677(47), 678(47), 679(47), *725*
Barry, J. A., 42(146), *54*
Bartel, K., 106(187), *124*
Bartlett, P. A., 245(180), *267*
Barton, D. D. R., 760(7), *772*
Barton, D. H. R., 739(25, 26), *754*
Barton, J. M., 232(169), *266*
Barton, T. J., 247(188), *267*
Bartos, J., 609(350), *643*
Basaka, J. K., 595(226), *639*
Basha, A., 59(198, 199, 200), *64*
Basch, H., 182(1), *262*
Batcho, A. D., 406(61), 417(61), 478(61), *522*
Bates, A. C., 342(112), *380*
Batey, J. F., 648(561), *652*

Battenberg, E., 182(12), 192(12), *263*
Battersby, A. R., 739(24, 27, 28), *754,* 760(8, 12), 769(12), *772*
Baudet, P., 594(219), *639*
Bauer, A., 184(18), 226(18), *263*
Bauer, F., 279(15), 283(15), 284(15), 287(15), 290(15), 294(15), 295(15), 299(15), *310*
Bauer, H., 287(617), 307(67), *312,* 620(421), *644*
Bauer, K., 614(385), *643*
Bauer, L., 232(169), *266*
Bauer, W., 444(81), 445(81), 474(81), 475(81), *523,* 621(437), 623(437), *645*
Baum, J., 165(53), 170(53), *171*
Baum, J. S., 126(241), 139(241), *140,* 268(208, 209), 275(208, 209), *276,* 760(1), *772*
Baumann, G., 609(349), *643*
Baumann, M., 287(67), 307(67), *312*
Baumes, R., 729(170), *731*
Baumgarten, H. E., 576(48), 578(48), 590(48), 592(48), *635*
Baumgarten, P., 608(340), *642*
Baur, R., 397(7), 406(7), 407(7), 448(7), 458(7), 464(7), 491(7), 495(7), 496(7), 507(7), 508(7), 509(7), *521*
Bayer, O., 6(4), *50, 54*(155), 138(285), *141,* 331(64), 361(64), 363(64), *379*
Bayer, R., 608(332), *642*
Bayerlin, H. P., 365(173), 369(173), *381*
Beak, P., 271(225), *277,* 279(55), 284(55), 287(55), *311,* 313(115), *320,* 332(68), *379*
Beal, J. L., 740(32), *754*
Beasley, G. H., 659(2), 689(2), 703(92), 722(2), 723(2), *724, 726*
Beck, D., 621(438), *645*
Beck, F., 584(131), 604(131), 605(131), *637*
Beck, G., 232(171), 235(171), *266,* 369(182), 370(182), *382,* 406(23), 407(23), 415(23), 418(23), 419(23), 420(23), 450(23), 494(145), 498(145), 508(145), 509(145), 513(145), 520(145), *522, 525*
Becker, G. A., 608(333), *642*
Becker, K. B., 729(156), *731*
Beckman, E., 96(155), *123,* 322(19), 342(19), *378*

Behr, D., 596(258), 619(258), *640*
Behr, L. C., *320*(142)
Behringer, H., 200(76), 245(76), *264*
Beisler, J. A., 747(97), *756*
Bejar, L. O., 747(92, 93), *756*
Bekasova, N. I., 595(240, 241), *640*
Beke, D., 574(5), *634,* 736(9, 10), 738(23), *754*
Bekhli, A. F., 325(41), 326(41), *379*
Belenkij, L. I., 595(229), *639*
Belg. Pat. 618,751, 91(142), *123*
Belg. Pat. 628,832 (1963), 147(22), *164*
Belg. Pat. 660,941 (1965), 174(2), 175(2), *180*
Belikov, A. S., 750(116), *756*
Bell, A. J., 344(114), 345(114), 347(114), 348(114), 350(122), 354(114), 355(114), 357(114), 358(114), 359(114), 366(114), 375(114), *380*
Bell, B., 581(112), *637*
Bell, C. L., 757(139), *758*
Bellasio, E., 88(135), 90(135), *123,* 147(24), 152(24), 156(24), 157(24), 163(24), *164*
Bello, B., 633(496), *646*
Benary, E., 597(268), *640*
Bender, M., 591(180), 594(180), 599(180), 602(180), 604(180), *638*
Bender, M. L., 604(299), *641*
Bender, P., 191(52), 237(52), 238(52), *264*
Benkovic, S., 599(276), 603(276), *641,* 647(276)
Benkovic, S. J., 364(170), *381*
Benoiton, N. L., 268(214), 274(214), *277*
Benson, R. E., 200(73, 74), 245(73), *264*
Benson, S. W., 657(7), *724*
Benz, E., 38(130), 40(130), 41(130), *53*
Beresin, I. V., 647(543), *652*
Berestetskaja, V. D., 606(304), *641*
Bergman, J., 574(11, 12), 576(42), 595(42), 596(42), 620(42), 623(42), 630(42), *635,* 647(11, 12), 651(628), *654*
Bergmann, J., 575(41), 595(41), 596(41), 620(41), 623(41), *635*
Berk, H., 191(56), 245(56), 246(56), *264*
Bernal, S., 720(127), *727*
Bernstein, J., 718(126), *727*
Bernthsen, A., 94(152), *123,* 279(3, 4), 280(3), 281(4), 282(3, 4), 284(3, 4), 290(4), 294(3, 4), 295(3, 4), 299(3, 4), *310*

Berry, K. H., 632(486), *646*
Bersch, H.-W., 745(74), *755*
Bertele, M., 247(182), *267*
Bertucat, M., 575(33), *635*
Bespalova, G. v., 319(141), *320*
Besselieve, R., 760(17), 769(29h, 30a, 30h), *772, 773*
Besthorn, E., 627(465), *646*
Bestmann, H. J., 23(85), 25(85), 26(85), *52*
Bethe, H., 715(119), *727*
Betrabet, M. A., 23(83), *52*
Betz, W., 186(28, 29, 30), 207(28, 29, 30), *263,* 268(212), 271(212), 275(212), 276(212), *276*
Beugelmans, R., 764(25), 771(25), *773*
Bevelle, C., 743(66), 757(139), *755, 758*
Beyer, K. H., 104(182), 105(182), *124,* 324(30), *378*
Beyerlin, H. P., 194(68), 195(68), 197(68), 198(68), 199(68), 200(68), 201(68), 202(68, 92), 203(92), 220(68), 221(68, 92), 222(68), 232(163), 236(163), 249(163), 250(163), 255(92), 256(163), 257(92), *264, 266,* 455(95), 460(95), 501(95), 519(95), *523*
Beyl, V., 360(166), *381,* 532(28), 533(28), *569*
Beynon, J. H., 74(74), *121*
Bhatt, M. V., 648(567), *652*
Bhattacharyya, J., 750(114, 115), *756*
Bianco, E. J., 39(136), *53,* 596(249), *640*
Biellmann, J., 633(507), *647*
Biellmann, J. F., 633(494, 502, 503, 505), *646*
Biernacki, Z., 633(489), *646*
Bigg, D. C. H., 729(161), *731*
Bilte, H., 72(46), *121,* 538(50), *569*
Binder, H., 582(120), *637*
Bindra, A., 647(531), *651*
Birch, A. J., 736(14), 737(14), *754*
Birke, G., 166(58), *171*
Birum, G. H., 151(35), 153(35), 161(35), 162(35), *165*
Bischoff, G., 279(11), 284(11), *310*
Bitoun, J., 279(55a), 284(55a), 286(55a), *311*
Bitter, I., 149(31), 153(36), 157(44), 158(31, 36, 44), 159(48), *164,* 323(26, 27, 28), 324(31), 361(26, 27, 169), 362(26, 27, 28, 169), 376(31), *378, 381*

Bitter, J., 97(163, 164, 165), 103(181), *123, 124*
Bitzer, D., 203(94), *264,* 285(65), *312*
Black, D. St. C., 729(155), *731*
Blackwood, R. K., 204(97), *265*
Blaschke, G., 764(23), *773*
Blechert, S., 728(140), *730*
Bleibtreu, H., 279(2), 280(2), 281(2), 284(2), 294(2), 295(2), 299(2), *310*
Blohm, T. R., 11(43), *51*
Blount, H., 617(401), *644*
Blount, W., 612(365, 366, 367, 368), 613(367), *643*
Blout, E. R., 633(512), *647*
Bobbitt, J. M., 696(81), *726*
Bock, H., 703(94), 705(94), *726*
Bocz, A. K., 200(79), *264*
Bode, K., 174(3), *180*
Bode, K.-D., 360(167), *381*
Bodenbrenner, K., 190(43), 194(43), *263*
Bodnarchuk, R. D., 619(412), *644*
Boeckmann, R. K., Jr., 60(203), *64*
Bogatkov, S. V., 574(17), 583(128), 600(128), 607(329), *635, 637, 642,* 647(521, 557), *651, 652*
Bogdanov, G. N., 589(152), *638*
Boggio, R. J., 720(127), *727*
Bogoslovski, B. M., 576(50), 578(50), 604(50), *635*
Bohlmann, F., 764(21, 22, 24), 768(22, 24), *773*
Böhme, H., 96(161), 111(161), *123,* 323(22), 344(115), 345(115), 346(115), 347(115), 348(115), 349(115), 351(136), 354(115), 358(22), 361(22), 365(22), 375(115), *378, 380, 381,* 593(212), *639*
Bokelman, G. H., 771(37a), *773*
Bolotyn, B. M., 593(210), *639*
Bolton, I. J., 503(158), 504(158), *525*
Bond, P. M., 648(587, 588), 649(587), *653*
Bongert, A., 597(269), *640*
Bonner, T. G., 587(143), 606(143, 321, 322, 323), 607(143, 321, 322, 323), *637, 642*
Bonnet, R., 322(9), *378*
Bonnett, R., 6(10), *50*
Bonnetti, R., 528(2), 554(2), *568*
Boos, H., 247(182), *267*
Borch, R. F., 184(23), 190(23, 38), 193(38), 218(138), 232(23), 257(23), 260(38, 138), *263, 265,* 406(41), 411(41), 448(41), 478(41), *522,* 658(16), 689(16), *724*
Borlev, I. V., 391(231), *392*
Bormann, D., 190(50), 201(50), 245(50), 247(185), *263, 267*
Borner, P., 182(13), 188(13), 190(13, 43), 192(13), 194(43), 215(13), 217(13), 220(13), 248(13), 252(13), 254(13), *263*
Bornowski, H., 66(22), *120*
Borovnik, V. P., 269(217, 218), 273(217, 218), *277*
Borron, J. A., 332(68), *379*
Borror, A. L., 614(387), *643*
Bos, H. J. T., 279(38), 301(38), *311*
Bose, A. K., 279(54), 284(54), 288(54, 69, 70), 300(54), *311, 312,* 728(136), *730*
Boshart, G. L., 62(214), *64*
Bosshard, H. H., 6(17), 7(17), 8(17), *51,* 66(6, 19), 78(6, 19, 100), 79(100), 80(6), 81(6), 88(19, 124), 89(124), 107(100), 108(100, 197), 113(6, 100), 114(6), *120, 122, 124*
Botsch, H., 8(30), 10(53), 11(30), *51,* 325(39, 40), 326(39), *379*
Botsch, H. J., 405(25), 406(25), 407(25), 410(25), 411(25), 412(25), 480(134, 135), 481(135), *522, 524*
Bottcher, P., 279(15), 283(15), 284(15), 287(15), 290(15), 294(15), 295(15), 299(15), *310*
Botvinik, M. M., 590(169), *638*
Boudet, R., 43(148), *54,* 279(19, 23), *311*
Bouveault, L., 597(269), *640*
Bovey, F. A., 182(1), *262*
Bowden, B. F., 650(611), *653*
Bowman, R. M., 746(86), *755*
Boyd, D. R., 746(84), *755*
Boyd, M. R., 682(51), *725*
Boyle, F. T., 649(603), *653*
Bracke, W., 595(231), 623(231), *640*
Brade, H., 182(2, 3), *262*
Brader, J. J., 284(62), 290(62), *312*
Bradsher, C. K., 741(44), *754*
Bram, G., 647(538), *652*
Brambilla, R. J., 728(136), *730*
Bramley, R., 610(358), 611(358), 612(358), 614(358), *643*
Brandange, S., 596(258), 619(258), *640,* 734(2), *753*

Brandsma, L., 279(38, 39), 301(38, 39), 302(39), *311*
Brannock, K. C., 107(190), *124,* 660(30), 662(30), 664(30), 721(30), *724*
Bratachanskii, P. E., 88(128), *122*
Braun, J. V., 66(4, 20, 52), 68(25), 69(4), 82(5), 96(20), 114(210), 115(211), 118(4, 210, 211, 212, 215), 119(4), *120, 124,* 322(11, 20, 21), 342(20), *378*
Braun, J. W., 96(156, 157, 158), *123*
Braun, R., 134(260), *140*
Braz, G. L., 595(237), *640*
Brechbühler, H., 252(200), *267,* 398(10), 399(10), 406(10), 410(10), 455(10), 465(10, 109), 466(10), 467(10, 109), 469(10, 109), *521, 524*
Bredereck, H., 1(1), *4,* 6(11), 7(11), 8(11, 28, 29, 30), 9(29), 10(11, 29, 53), 11(11, 29, 30), 12(29), 14(29), 15(29), 16(29), 17(11, 29), 18(11, 29), 19(61, 62, 63, 64), 22(29, 75), 31(106), 38(106), 42(11), 43(106, 147, 149), 46(106), 47(11), 48(106), *50, 51, 52, 53,* 107(192, 195), 108(192, 195), 112(206, 207), *124,* 144(3), 149(30), 157(30), *164,* 183(15), 186(15, 32), 188(15), 189(15), 193(15), 194(61, 62, 63, 64, 65, 66, 67, 68), 195(68), 196(61, 62, 63), 197(68), 198(65, 67, 68, 71), 199(65, 68), 200(68), 201(68, 86), 202(68, 92, 93, 94), 203(61, 62, 92, 93, 94, 96), 204(62, 63), 210(62, 65), 211(62), 219(65), 220(15, 64, 65, 68, 71), 221(65, 68, 92), 222(65, 68), 224(65, 144), 231(161), 232(66, 144, 163, 165, 166, 167, 170, 171), 235(67, 165, 166, 167, 171), 236(163, 166), 238(61, 67), 239(63, 67), 241(67, 86), 242(67, 86), 243(86), 244(86), 245(86), 248(15, 61, 64, 71), 249(15, 163), 250(163), 252(15, 64, 71), 253(15), 255(92), 256(15, 163), 257(92), 258(71), 259(86), *263, 264, 265, 266,* 279(28, 51), 281(28, 51), 284(28, 51), 285(65), 295(28), 296(28), 299(28, 51), 304(28), *311, 312,* 325(37, 38, 39, 40), 326(37, 38, 39, 46), 333(37, 69), 338(90), 339(96, 97, 98, 102), 360(69, 168), 361(102), 363(102), 364(171, 172), 365(171, 172, 173), 369(90, 171, 173, 178, 179, 180, 181, 182), 370(90, 179, 182, 184, 185), 371(187), 372(90, 188), 373(90), 375(90), 376(98), *379, 380, 381, 382,* 396(3, 5), 400(15), 401(20, 21, 22), 402(5, 5a, 24), 403(3), 405(5, 25, 26, 28a), 406(23, 24, 25, 26, 28a, 29, 32, 34, 39, 52, 57, 60), 407(23, 24, 25, 26, 32, 34, 39, 52), 408(21, 22, 26, 54, 55), 409(29), 410(25, 33), 411(25, 26, 34), 412(25, 57), 413(32, 57), 414(32), 415(5, 23, 32), 416(32, 39), 417(26), 418(23, 57), 419(23), 420(23, 60), 421(24), 422(54, 55), 423(29), 435(52), 436(52), 437(52), 440(32, 69, 70, 71, 72), 441(60, 69, 70, 71), 442(29, 32, 69, 70, 71), 443(32, 55, 60, 70, 72), 444(32, 70), 445(5, 15, 82), 446(3, 69, 82, 85), 447(82), 448(20, 69), 449(69), 450(23), 453(89), 455(94, 95, 96), 457(5), 458(69), 459(69), 460(95, 101), 461(3, 101), 470(3, 89), 471(24), 472(24), 473(21), 474(22, 122, 123, 124, 125), 475(20, 21, 22, 70, 71, 122, 126), 476(22, 70, 71, 123, 127), 477(54, 55, 129, 130, 131, 132, 133), 478(26, 28a), 480(57, 69, 135), 481(57, 135), 482(57), 485(70, 71, 72a), 491(85), 492(101), 493(101), 494(85, 145), 495(85), 496(22, 101, 146, 147), 497(101), 498(101, 145), 499(69), 501(95), 507(147, 165, 166, 167), 508(145, 147, 165, 166, 167), 509(101, 145, 147, 167), 510(101, 147, 168, 169), 511(169), 512(89, 169), 513(145), 514(170), 515(170), 516(26, 170), 517(5a, 170, 171), 519(29, 95, 98), 520(145, 167), *521, 522, 523, 524, 525*
Bredereck, H. J., 112(205), *124,* 400(15), 445(15), 497(150), *521, 525*
Bredereck, K., 8(29), 9(29), 10(29), 11(29), 12(29), 14(29), 15(29), 16(29), 17(29), 18(29), 22(29, 73), *51, 52,* 107(192), 108(192), *124,* 144(3), 145(12), 153(12), 158(12), 162(12), *164,* 186(32), *263,* 325(38), 326(38), *379,* 557(93), 558(93), 559(93), *570*
Breest, F., 597(267), *640*
Brehm, L., 88(134), 89(134), *123*
Brendle, T., 112(206, 207), *124,* 232(167), 235(167), *266,* 338(90), 369(90), 370(90), 372(90, 188), 373(90), 375(90), *380, 382,* 396(3), 403(3), 406(46, 60),

420(46, 60), 440(76), 441(46, 60, 76), 443(46, 60), 446(3), 461(3), 470(3, 46), 485(46), 499(46), 507(46), *521, 522*
Brenner, M., 137(280), *141*
Brenner, S., 130(253), *140*
Bretschneider, H., 305(108), *313*
Breuer, S. W., 200(80), *264*
Breuning, W., 579(69), 608(69), *636*
Brewster, J. H., 591(187), 592(187), 596(187), *638*
Bridgewater, R. J., 683(56), *725*
Briere, R., 633(494, 505), *646*
Brinkmann, H., 648(564), *652*
Brion, J.-D., 389(226), *392*
Brit. Pat. 448,469 (1936), 290(88), *312*
Brizzolara, A., 669(40), *725*
Brodka, S., 620(422), 621(422), 625(422), *644*
Brooks, J., 595(234), *640*
Brooks, R. E., 137(282), *141*
Brossi, A., 735(5), *754*
Brouwer, D. M., 182(5), *262*
Brovkin, D. V., 712(110), *727*
Brown, D., 647(522), *651*
Brown, D. J., 201(91), 228(91), *264*
Brown, D. M., 593(214), *639*
Brown, D. W., 694(72, 78), 696(72, 80), 698(72), 700(72, 80), *725, 726*
Brown, H. C., 627(464), *646*
Brown, M., 400(18, 19), 485(18), *521, 522*
Brown, M. H., 66(15, 16), 68(15, 16), 108(15), 111(15, 16), *120,* 144(4), *164*
Brown, R. D., 589(154), *638*
Brown, R. F. C., 729(155), *731*
Bruchhausen, F. v., 738(20), *754*
Brück, B., 406(35), 408(35), 410(35), *522*
Bruckenstein, S., 567(115), *571*
Bruice, T., 599(276), 603(276), *641,* 647(276)
Bruice, T. C., 594(223), *639*
Bruice, Th. C., 647(556), 648(556), *652*
Bruma, M., 595(230), *640*
Bruning, A., 290(85), *312*
Brunner, O., 613(376), *643*
Bryant, W. M. D., 591(188), *639*
Brzechffa, L., *731*(168)
Buchanan, G. L., 615(389), *643*

Buchanan, J. B., 308(110), *313*
Buchel, K. H., 167(60), 169(60), 170(60), *172,* 200(79), *264*
Büchi, H., 252(200), *267,* 398(10), 399(10), 406(10), 410(10), 455(10), 465(10, 109), 466(10), 467(10, 109), 469(10, 109), *521, 524,* 747(95), *756,* 758(146), *758*
Buchler, J. W., 351(124, 125, 126), *380*
Buchner, W., 279(10), 284(10), *310*
Buckley, A. J., 623(440), *645*
Budanova, L. I., 56(178), *63,* 270(222), 272(222), 273(222), *277*
Buder, W., 391(230), *392*
Budy, K., 578(64), *636*
Budzikiewicz, H., 750(115), *756*
Buehler, C. A., 57(184), *63,* 138(285), *141*
Buhler, W., 620(421), *644*
Buhner, A., 194(60), 198(60), 199(60), 241(60), 242(60), 243(60), *264*
Bukovsky, M. I., 609(351), *643*
Bulakh, E. L., 88(125), *122*
Bullard, W. P., 364(170), *381*
Bullock, C., 648(561), *652*
Buncel, E., 581(104), 588(104), 590(104), 591(104), *637*
Bunting, J. W., 610(355), *643*
Bunton, C. A., 602(281), 604(281, 297, 298), *641*
Buono, G., 26(94), *52*
Bureš, E. E., 9(34), *51,* 325(34), *378*
Burmeister, J. L., 556(88), *570*
Burpitt, R. D., 107(190), *124,* 660(30), 662(30), 664(30), 721(30), *724*
Buryak, V. S., 88(125), *122*
Buryanov, Y. B., 562(103), *571*
Butin, W. E., 18(56), *51*
Butler, A. R., 603(284), 604(284), *641,* 649(589), *653*
Buyle, R., 70(36), *120*
Buza, D., 318(137), *320*
Buzas, G., 746(81), *755*
Buzykin, B. I., 20(68), *52*
Bycroft, B. W., 279(45), 284(45), 299(45), 300(45), 302(45), *311*

Caillaux, B., 126(243), 128(243), 136(243), *140*
Cairns, T. L., 200(73, 74), 245(73), *264*
Calder, I., 596(259), *640*

Callender, M. E., 18(56), *51*
Campbell, J. A., 727(133), *730*
Cannon, J. G., 205(99), *265*
Caple, G., 608(338), *642*
Carayon-Gentil, A., 279(32), 284(32), 286(32), 294(32), *311*
Carboni, R. A., 554(87), *570*
Carlin, R. B., 706(101), *726*
Carlsen, P., 106(188), *124*
Carlson, R., 728(135), *730*
Carlson, R. D., 119(218), *124*
Carnaham, E., 317(135), *320*
Carnahan, F. L., 682(52), *725*
Carnahan, J. E., 400(18), 485(18), *521*
Carney, A. L., 201(90), 241(90), *264*
Carpenter, F. H., 592(200), *639*
Carpenter, W., 351(129, 130), *380*
Carr, R. L. K., 44(153), *54,* 210(115), 226(115), *265*
Carre, P., 579(80), *636*
Carriuolo, J., 600(278), *641*
Carter, M. A., 729(169), *731*
Carvalhas, M. L., 740(37), *754*
Casanova, J., 39(134), *53*
Casassas, E., 609(348), *643*
Casnati, G., 684(61), *725,* 728(149), *730*
Castedo, L., 745(75, 76), *755*
Castillo, R., 125(223), 136(223), *139*
Castle, R. N., Ed., 728(146), *730*
Castro, E. A., 604(288), 605(288), *641,* 648(587), 649(587), *653*
Catala, A., 615(391, 392), *644*
Cava, M. P., 614(382), *643*
Cave, A., 769(33), *773*
Centner, K., 608(334), *642*
Chabra, S. R., 649(602), *653*
Chabrier, P., 279(12, 13, 14, 16, 32, 56), 283(16), 284(12, 14, 32), 286(32), 294(32), 295(13, 14), 300(99), 304(105), 306(105), *310, 311, 312, 313,* 581(108), *637*
Chadha, S. L., 581(110), 587(142), *637*
Chadra, S., 584(135), 585(135), *637*
Chakrabarti, J. K., 596(253), *640*
Chakrabarti, S., 575(21), *635*
Chander, K., 582(121), *637*
Chang, Y.-W., 91(140), *123*
Chanon, M., 317(144), 319(143, 144), *320*
Chantooni, M. K., 567(115), *571*

Chapelle, J.-P., 714(117), 716(117, 120), *727*
Chapman, N. B., 613(380), *643*
Charalambous, J., 581(105), *637*
Chatterjee, R., 739(29), 740(29, 33, 34), *754*
Chaturvedi, R. K., 279(43), 295(43, 91), 306(91), *311, 312*
Chawla, H. P. S., 288(69), *312*
Cheema, A. S., 319(145), *320*
Chen, F. M. F., 268(214), 274(214), *277*
Cheng, C. C., 741(45), 743(65), *754, 755*
Cheplanova, I. V., 30(99), *53*
Cherkasova, E. M., 583(128), 600(128), *637,* 647(521, 557), *651, 652*
Chernov, V. A., 190(46), 191(46), 245(46), *263*
Chernyskova, N. B., 729(166), *731*
Chervinsky, A. Yu., 647(545), *652*
Chevolot, L., 769(34), *773*
Chew, C., 342(110), *380*
Chib, J. S., 288(69), *312*
Chittum, J. W., 648(578), *653*
Chizhova, E. A., 606(319), *642*
Chladek, S., 440(74), 441(74), 451(86a), 456(74, 75e, 97a), 461(74), 499(74), *523, 524*
Chondromatidis, G., 715(118), 716(118, 119), *727*
Christensen, G., 279(55c), *311*
Christie, C. F., 86(117), *122*
Christmann, O., 203(95), 205(95), *264*
Christophersen, S., 106(188), *124*
Chu, Ch. Ch. Ch., 728(139), *730*
Chum, M. C., 623(439), *645*
Chung, B. C., 26(95), *52*
Chung, C. Y.-J., 727(133), *730*
Chupakhin, O. N., 647(520), *651*
Ciba, Ltd., Brit. Pat. 911,475 (1962), 108(196), *124*
Ciotti, C. J., Jr., 591(187), 592(187), 596(187), *638*
Ciuffarin, E., 604(291), *641*
Ciusa, W., 609(347), *643*
Claisen, L., 531(27), 533(31), *569,* 74(68, 69), *121,* 597(262), *640*
Clamot, B., 133(257), *140*
Clarke, E. A., 746(83), *755*
Clarke, K., 613(380), *643*
Claxton, G. P., 11(43), *51*

Clemens, D. H., 338(92, 93, 94), 342(92), 343(92), 344(114), 345(114), 347(114), 348(114), 350(122), 354(92, 93, 114), 355(114), 357(92, 114), 358(114), 359(92, 114), 365(92), 366(114), 368(92), 369(92), 370(92), 371(92), 372(92), 375(92, 114), 377(92), *380, 470*(113), 487(113), 495(113), *524*
Clinch, P. W., 579(81), 588(81), *636*
Cloke, J. B., 290(89), *312*
Coates, V., 648(578), *653*
Cobb, R. L., 612(361), 613(361), 614(361), 616(361), 628(361), *643*
Cochran, E., 595(233), 598(274), *640, 641*
Cockerill, A. F., 579(85), *636*
Coetzee, J. F., 568(119, 121), *571*
Coffey, G. P., 184(24), *263*
Coffman, D. D., 68(26), 73(53), *120, 121,* 144(5), *164,* 350(123), *380,* 400(18), 485(18), *521,* 554(87), *570*
Cohen, A., 760(7), *772*
Cojocuriu, I., 56(180), *63*
Cojokaru, Z., 56(180), *63*
Colchester, J. E., 579(82), 632(82), *636*
Collins, L. R., 126(230), *140*
Colonna, M., 613(378), *643*
Colson, J. G., 210(115), 226(115), *265*
Comins, D. L., 651(631, 632), *654*
Compaigne, E., 6(12), *50*
Condo, F. E., 279(9), 280(9), 290(9), *310*
Conway, T. T., 232(169), *266*
Cook, A. G., 659(22), 675(22), *724*
Cook, D., 576(44), 583(44), 584(44), 585(44), 586(44), 598(44), *635*
Cook, J. W., 615(389), *643*
Cooksey, C. J., 580(94), 589(94), 611(94), *636*
Cooper, J. W., 579(83), 597(83), 628(83, 473), *636, 646*
Cope, A. C., 675(45), *725*
Corbier, J., 662(29), 669(5), 671(5, 29), 672(5, 29), 673(5), 674(5), *724*
Corey, E. J., 39(132), *53,* 614(387), *643*
Corley, R. S., 633(512), *647*
Corral, R. A., 746(80), *755*
Correira, Y., 200(81), 244(81), 245(81), *264*
Costa, R., 771(38), *773*
Costin, R. C., 273(236), *277*
Costisella, B., 262(206), *267,* 374(190),

382, 450(87), 453(87), 454(87), 460(87), 461(87), 462(87), 464(87), 465(87), 466(87), 468(87), 470(87, 114, 115, 116, 117), 493(144), *523, 524, 525*
Cotter, R. J., 63(215), *64,* 588(144), 590(144), *637*
Cox, M. V., 31(101), *53,* 322(18), 327(18), *378*
Cox, P. J., 728(142), *730*
Cozzani, R. G. E., 125(220), *139,* 269(219), 274(219), *277*
Crabbe, P., 596(251), *640*
Cram, D. J., 566(114), *571*
Cramer, F., 27(96, 97), 28(97), 29(96, 97), 30(97), *52,* 92(143), *123*
Creasey, S. E., 591(185), *638,* 648(574), *652*
Cresson, P., 254(203), *267,* 669(5, 29), 671(5, 29), 672(5, 29), 673(5), 674(5), *724*
Crivetz, L., 595(230), *640*
Croce, P. D., 579(72), *636,* 711(109), 712(109), *727*
Cross, A. D., 742(53), *755*
Cross, D., 744(72), *755*
Crow, W. D., 735(4, 6), *754*
Crowe, D. F., 648(560), *652*
Cruickshank, F. R., 657(7), *724*
Cruickshank, P. A., 62(214), *64*
Csuros, Z., 97(163, 164, 165), 103(181), *123, 124,* 149(31), 153(36), 157(44), 158(31, 36, 44), 159(48), *164,* 323(26, 27, 28), 324(31), 361(26, 27, 169), 362(26, 27, 28, 169), 376(31), *378, 381*
Culbertson, T. P., 504(161), *525*
Curphey, T. J., 728(138), *730*
Czernecki, S., 60(204), *64*

Dadaly, Ch., 647(527), *651*
Dadaly, V. A., 647(525, 539), 649(525), *651, 652*
Dahne, S., 357(157), *381*
Dains, F. B., 74(77), 86(77), *121*
Dallacker, F., 39(135), *53*
D'Andrea, F., 647(544), *652*
Danyluk, S. S., 73(54), *121,* 54(61), *570*
Darre, F., 315(123), *320*
Daub, J., 186(28, 29, 30), 207(28, 29, 30),

263, 268(212), 271(212), 275(212), 276(212), *276*
Davenport, J., 18(56), *51*
Davidson, D., 49(165), *54*
Davies, G. L. O., 579(85), *636*
Davies, M., 583(123), *637*
Davis, C. B., 596(247), *640*
Davis, C. S., 16(51), *51*
Davis, P., *643*(363)
Davis, T. L., 9(35), 31(35, 102), *51, 53,* 325(35), 327(35), 376(35), *379*
Dayal, B., 288(69), *312*
Dean, R. T., 772(39b), *773*
DeBernardo, S. L., 406(47), 448(47), *522*
Declercq, J. P., 165(55), 168(55), *171*
Degutis, J. A., 683(59), *725*
de Haan, J. W., 682(51), *725*
Dehn, W. M., 575(32), 577(32), 578(59), 580(95), *635, 636*
Deikalo, A. A., 620(426, 427, 433, 434, 435), 621(426, 433, 434), 623(435), 624(444), 625(444, 450, 451, 452), 631(435), *644, 645*
De La Mater, G., 279(57), 284(57), 285(57, 66), 286(66), 289(57), 295(57), *312*
de Meheas, M. R., 6(6), *50*
Denivelle, L., 590(161), *638*
Dennstedt, M., 575(22), *635*
Deno, N. C., 567(118), *571*
de Radzitsky, P., 205(102), *265,* 284(62), 290(62), *312*
De Roocker, A., 205(102), *265*
de Rostolan, J. P., 769(33), *773*
Desai, P., 27(96), 29(96), *52,* 92(143), *123*
de Saqui-Sannes, G., 728(146), *730*
Deslongchamps, P., 224(148), 225(148), *266,* 272(233), *277*
de Stefano, S., 752(129, 130), *757*
Detering, K., 694(74), *726*
Deubel, H., 620(422, 423, 428), 621(422, 428), 623(423, 428), 625(422), *644, 645*
Deubel, H. M., 650(623), *654*
Deubel, H. M. D., 620(424), *644*
Devdhar, P. B., 232(169), *266*
de Voghel, G. J., 126(242), 133(257), *140*
Devynck, J., 582(122), *637*
Dewar, M. J. S., 657(13), 702(89), 703(89, 93), 720(13, 129), *724, 726, 727*
Dewhurst, B. B., 648(571), *652*

DeWolfe, R.H., 226(152, 153), 228(152, 153), *266,* 338(89), *380,* 395(2), 461(2), *521*
Deyrup, J. A., 75(84, 85), *121,* 126(231), *140,* 751(117), *756*
Dhingra, K., 757(141), *758*
Diana, G. D., 269(220), *277*
Dickinson, C. L., 40(143), *54*
Dickore, K., 49(162), *54,* 574(9), *634*
Didich, M. N., 581(111), *637*
Diels, O., 707(104), *726*
Dienel, B., 191(52), 237(52), 238(52), *264*
Diepers, W., 207(107), 212(107), *265*
Dierbach, K., 11(39), *51,* 325(43), *379*
Dietl, A., 619(417), 620(417), *644*
Dietrich, M. A., 386(213), 387(213), *392*
Digenis, G. A., 44(152), 46(152), *54,* 210(114), *265,* 590(168), *638*
Dijking, J., 769(26a, 26b, 26d, 26e), *773*
Dimroth, O., 6(1), *50,* 632(484, 485), *646*
Dimroth, P., 32(109), *53,* 87(119), *122*
Dingwall, J. G., 9(32), *51*
Dinkeldein, U., 184(21), 185(21), 189(21), 253(21), *263,* 399(14), 450(14), 485(14), *521*
Dixon, K. W., 647(518), 649(518), *651*
Dobeneck, H. v., 185(26), *263,* 620(422, 423, 428, 429), 621(422, 428, 429), 623(423, 428, 429), 624(429), 625(422), 630(429), *644, 645*
Dobner, O., 387(219, 220), *392*
Dobrzynski, E. D., 315(127), *320*
Doering, W. E., 583(130), 597(130), 624(130), *637*
Doering, W. v. E., 703(92), *726*
Dolejs, L., 742(53), 744(72), *755,* 757(138), *758*
Dolgushina, I. Yu., 75(92), 83(92), *122*
Dolphin, D., 245(180), *267*
Dombro, R., 729(160), *731*
Donleavy, J. J., 279(7), 284(7), *310*
Dopp, H., 227(156), *266,* 753(133), *757*
Dorofeenko, G. N., 6(5), *50*
Dorr, R., 694(75), 723(75), *726*
Doskotch, R. W., 740(32), *754*
Dougherty, T. J., 357(153), *381*
Douglas, K. T., 647(519, 542), *651, 652*
Dounzon, P., 609(346), *643*
Dowalo, F., 191(52), 237(52), 238(52), *264*

Dowens, R. W., 32(107), *53*
Downer, J. D., 290(86), *312*
Doyle, D., 620(424), *644*
Doyle, K. N., 279(53), 284(53), 306(53), *311*
Doyle, P., 650(623), *654*
Drach, B. S., 75(92), 83(92), *122*, 387(216), *392*
Draper, M. D., 748(104), *756*
Dreiding, A. S., 752(121, 122, 123, 124, 126, 128), *756*, 758(144, 145), *758*
Driver, P., 290(86), *312*
Drozdov, N. S., 325(41), 326(41), *379*
Dryden, H., 590(164), *638*
Dube, G., 736(11), *754*
Dube, S., 272(233), *277*
Dubov, S. S., 144(6), *164*
Dubs, P., 279(42b), 299(42b), 300(42b), *311*
Dubsky, J. V., 689(65), *725*
Dudek, E. P., 720(128), *727*
Dudek, G. O., 720(128), *727*
Dudock, B. S., 740(30, 40), 741(43), *754*
Dunham, D. J., 728(138), *730*
Dunitz, J. D., 247(182), *267*
Dunn, R. T., 533(30), *569*
Dupont, P. E., 713(113), *727*
Duquette, L. G., 553(84), 560(101), *570*
Durmashkina, V. V., 608(337), *642*
Dury, K., 32(109), *53*, 87(119), *122*
Dyke, S. F., 694(72, 75, 76, 78), 696(72, 80), 698(72, 82), 699(83), 700(72, 80), 723(75), *725, 726,* 728(153, 154), *731,* 740(41), 744(61), *754, 755*
Dzhafarov, Zh. I., 580(98), *636*
Dziembowska, T., 613(377), *643*

Easley, T., 594(220), 595(220), *639*
Eastwood, F. W., 521(178), *525*
Eberle, M. K., 729(167), *731*(168)
Eberson, L., 617(399), *644*
Ebert, G., 608(334), *642*
Ecker, A., 728(154), *731*
Edelmann, N. K., 406(30), 420(30), 432(30), 439(30), *522*
Edgell, W. F., 591(183), *638*
Edwards, J. O., 75(88), *121,* 137(282), *141*
Effenberger, F., 8(30), 10(53), 11(30), 43(149), *51, 54,* 107(195), 108(195), 112(206, 207), *124,* 194(64, 65, 66, 67, 68), 195(68), 197(68), 198(65, 67, 68), 199(65, 68), 200(68), 201(68), 202(68), 210(65), 219(65), 220(64, 65, 68), 221(65, 68), 222(65, 68), 224(65), 232(66, 163, 165, 166, 167), 235(67, 165, 166, 167), 236(163, 166), 238(67), 239(67), 241(67), 242(67), 248(64), 249(163), 250(163), 252(64), 256(163), *264, 266,* 325(39, 40), 326(39), 333(69), 338(90), 360(69), 364(171, 172), 365(171, 172, 173), 369(90, 171, 173, 178), 370(90, 185), 372(90), 373(90), 375(90), *379, 380, 381, 382,* 396(3), 400(15), 401(21), 403(3), 405(25), 406(25, 60), 407(25), 408(21, 54), 410(25), 411(25), 412(25), 421(60), 422(54), 440(70, 71, 72), 441(60, 70, 71), 442(70, 71), 443(60, 70, 71, 72), 444(70), 445(15), 446(3), 455(94, 95), 460(95), 461(3), 470(3), 473(21), 474(123, 124, 125), 475(21, 70, 71, 126), 476(70, 71, 123, 124, 125), 477(54, 128, 129, 130, 131, 132, 133), 480(135), 481(135), 485(70, 71, 72a), 496(146), 501(95), 519(95), *521, 522, 523, 524, 525*
Ege, G., 87(120), *122,* 215(132), *265,* 461(102), *524*
Eger, K., 132(255), *140*
Eggerichs, T. L., 133(257), *140*
Eholzer, U., 19(59), 26(59), 36(59), *52,* 105(186), 106(186), *124,* 596(254), *640*
Ehrhart, W. A., 337(87, 88), *380*
Eiden, F., 50(168), *54,* 91(141), *123*
Eilingsfeld, H., 18(54, 55), 22(54, 76), *51, 52,* 66(7, 14), 68(7), 76(7, 97), 77(97), 78(7, 97), 79(7), 80(7, 14, 97), 81(7, 14, 97), 82(7, 14), 84(7, 14), 85(7, 14), 87(7), 89(7), 91(7), 94(7, 14), 95(14), 96(7), 98(97), 99(7, 14), 100(7), 103(7, 180), 104(182), 105(7, 182), 107(7, 14, 193), 108(7, 14, 193), 109(7, 14), 115(7, 97), 116(7, 97), 117(97), *120, 122, 124,* 144(1, 2), 145(1, 2), 146(1, 2), 151(1, 2), 152(1, 2), 153(1, 2), 154(1,2), 155(1, 2, 40), 156(2), 157(1, 2), 158(2), 162(1, 2), *164,* 174(1, 2), 175(1, 2), 176(1), 177(1), 178(1), *180,* 213(124), 214(124, 125), 224(125), 230(125), 240(125), *265,* 290(83), 294(906),

295(83), 297(83), *312,* 323(23), 324(23, 30), 326(23, 47), 328(47), 359(23), 361(23), 363(23), *378, 379,* 436(68), 437(68), 460(100), 465(110), 485(68), *523*(88), *524*
Einhorn, A., 589(157), 592(203), *638, 639*
Eisner, U., 632(482), *646*
Eitner, P., 529(12), 530(12, 15), *568, 569*
Ekstrin, F. A., 40(140), *53,* 331(62), *379*
El-Agamey, A.-G., 56(181), *63*
Elam, E. U., 55(173), 56(173), *63*
Elander, M., 750(113), *756*
El-Berins, Ramadan, 220(141), 259(141), *266*
Elderfield, R. C., 747(96), 748(100), *756*
Elgavi, A., 126(246), *140,* 165(54), 169(54), *171*
Elgersma, R. H. C., 657(10), 707(103), 709(10, 103), 710(10, 103), *724, 726*
Elguero, J., 710(108), 714(117), 716(117, 120), *726, 727*
El-Kerdawy, M., 56(181), *63*
Elkik, E., 660(24), 661(24, 27), 662(24), 663(27), 721(27), *724*
Elliot, D. F., 305(109), *313*
Elliot, I. W., 610(357), 617(397), 618(397), 624(397), *643, 644*
Elliott, I. W., Jr., 613(374), *643*
Elliott, M., 687(63), *725*
Eloy, F., 88(123), 90(123), *123*
Elser, W., 279(42, 42a), 281(42), 282(42), 284(42), 291(90), 292(90), 293(90), 294(42), 301(42, 103), *311, 312, 313*
Elsinger, F., 247(182), *267*
Emmons, W. D., 338(92, 93), 342(92), 343(92), 354(92, 93), 357(92), 359(92), 365(92), 368(92), 369(92), 370(92), 371(92), 372(92), 375(92), 377(92), *380,* 470(113), 487(113), 495(113), *524*
Enders, E., 11(37), *51,* 609(352), *643*
Engelfried, O., 596(255), *640*
Engelhardt, G., 736(11), *754*
Engelhardt, V. A., 40(143), *54,* 68(27), 73(52, 53), *120, 121*
England, D. C., *320*(142), 386(213), 387(213), *392*
Engler, C., 72(44), *121*
Englin, M. A., 144(6), *164*
Entelis, S. G., 575(18), *635*
Entenmann, G., 82(105), 101(105), *122,* 174(5), 179(5), *180,* 528(8), *568,* 648(565), *652*
Erdmann, H., 167(60), 169(60), 170(60), *172*
Erickson, D., 61(211), *64*
Erickson, J. G., 339(100), 376(100), *380*
Erlenmeyer, H., 279(10, 11), 284(10, 11), *310*
Ermilii, A., 58(194, 195), *63*
Ernst, P., 619(415), *644*
Eschenmoser, A., 247(182, 185, 186), 252(200), *267,* 279(42b), 299(42b), 300(42b), *311,* 398(10), 399(10), 406(10), 410(10), 455(10), 465(10, 109), 466(10), 467(10, 109), 469(10, 109), 503(156), 504(156), 505(156), *521, 524, 525*
Esumi, N., 745(69), *755*
Etienne, A., 200(81), 244(81), 245(81), *264*
Etling, H., 19(66), 23(66), *52*
Euler, H. v., 88(133), 89(133), *123*
Evans, R. F., 627(464), *646*

Fabian, J., 279(27), 294(27), *311,* 355(148), *381*
Fahey, J. D., 279(54), 284(54), 288(54), 300(54), *311*
Fahrenholtz, K. E., 468(111), *524*
Fahrni, P., 658(16), 689(16), *724*
Fales, H. M., 743(57), *755*
Faltis, F., 760(3b), *772*
Fanshawe, W., 621(437), 623(437), *645*
Farina, L. J., 364(170), *381*
Farina, P. R., 364(170), *381*
Farkharov, A. A., 580(98), *636*
Farnsworth, N. R., 743(66), *755,* 757(139), *758*
Fassero, A., 279(9), 280(9), 290(9), *310*
Fatome, M., 279(55a), 284(55a), 286(55a), *311*
Fatutta, S., 613(378), *643*
Faulkner, D. J., 723(131), *727*
Fawcett, F. S., 68(26), *120,* 144(5), *164*
Fechner, H., 714(116), *727*
Fedder, J. E., 45(159), *54,* 72(39, 42), *121*
Federov, B. P., 22(74), *52,* 57(190), *63,* 231(160), *266,* 406(48), 407(48), 421(48), 443(48), 452(48), 457(48), 471(48, 118), *522, 524*

AUTHOR INDEX

Fedin, E. I., 607(325), *642*
Fehlhammer, W. P., 106(187), *124*, 135(271), *141*
Feinauer, R., 207(107), 212(107), *265*, 395(1), 461(1, 105), *521, 524*
Feldman, H. B., 595(232), *640*
Fel'dman, L. Kh., 85(114), *122*
Felix, D., 503(156), 504(156), 505(156), *525*
Fellrath, E., 96(155), *123*, 322(19), 342(19), *378*
Felner, J., 247(182, 185), *267*
Fendler, J. H., 602(281), 604(281), *641*
Fengeas, C., 471(120a), *524*
Ferles, M., 631(478), *646*
Ferre, G., 68(29), *120*
Ferry, J. D., 682(51), *725*
Fersht, A. R., 576(46), 586(46, 138), 587(46), 600(46, 138, 279), 601(46, 138), 602(279), 604(46), 605(279), *635, 641*, 647(279)
Fetzer, U., 19(59), 26(59), 39(59), *52*, 105(186), 106(186), *124*, 596(254), *640*
Feuer, H., 110(200), *124*
Feuerherd, K.-H., 769(27a), *773*
Feugeas, C., 252(201), *267*
Fialkov, Y. A., 562(103), *571*
Field, L., 580(102, 103), 591(102, 103), *637*
Fields, E. K., 75(83, 86), 83(83), *121*
Fierce, W. L., 539(53), *569*
Filatova, L. S., 269(217), 273(217), *277*
Filipenko, T. Ya., 60(201), *64*
Filleux, M. L., 7(22), 8(22), *51*, 68(33), *120*
Fink, C. M., 628(473), *646*
Finkbeiner, H., 47(171), *54*
Firell, N. F., 675(44), *725*
Firestone, R. A., 279(55c), *311*
Fischer, F., 289(81), *312*
Fischer, G. W., 574(2), 607(2), *634*, 647(551), *652*
Fischer, H., 619(415), *644*
Fischer, H. O. L., 613(372a), *643*
Fish, F., 744(67), *755*
Fisher, A. M., 592(204), *639*
Fisher, E. E., 706(101), *726*
Flagg, E. M., 289(79), *312*
Fleming, I., 626(460), *645*
Fligge, M., 619(418), 620(418), *644*

Fliri, H., 758(146), *758*
Florian, W., 107(194), 108(194), *124*, 182(14), 188(14), 190(14), 191(14), 192(14), 193(14), 215(14), 220(14), 248(14), 249(14), 250(14), 252(14), 254(14), 255(14), 256(14), 257(14), 258(14), *263*, 396(4), 406(4), 409(4), 410(4) 411(4), 423(4), 424(4), 425(4), 427(4), 428(4), 429(4), 430(4), 431(4), 440(4), 441(4), 422(4), 443(4), 445(4), 453(4), 455(4), 460(4), 461(4), 469(4), 487(4), 503(4), 519(4), *521*
Fluck, E., 582(120), *637*
Fodor, G., 66(23), 68(23), 78(23), 79(23), 80(23), 81(23), 97(23), 114(23), 118(23, 216), *120, 124, 378*(33)
Foglia, T., 614(387), *643*
Föhlisch, B., 19(61, 62, 63, 64), *52*, 134(260), *140*, 406(38), 411(38), 458(38), 479(38), *522*
Fong, H. H. S., 743(66), *755*, 757(139), *758*
Foster, C. H., 55(173), 56(173), *63*
Foster, H. E., 595(227), *639*
Foucaud, A., 61(210), *64*
Fowler, F. W., 633(495), *646*
Fraenkel, G., 182(4), *262*, 579(83), 597(83), 628(83, 473), *636, 646*
Francesch, C., 661(27), 663(27), 721(27), *724*
Francis, R. J., 739(27), *754*
Franck, B., 764(23), *773*
Franke, W., 6(20), 20(20), *51*, 92(146), *123*
Frankel, M. B., 110(200), *124*
Franz, J. E., 656(9), *724*
Franzischka, W., 689(70), 691(70), 694(70), *725*, 760(13a), 769(13a), *772*
Frater, G., 668(37), *725*
Frazor, M. J., 581(105), *637*
Freeman, F., 556(89), *570*
Freeman, P. R., 683(56), *725*
Freeman, R. C., 74(79), 115(79), *121*, 386(214), *392*, 563(105), *571*
Freenor, F. J., 61(211), *64*
Frehel, D., 88(123), 90(123), *123*
Freidlin, E. G., 88(125), *122*
Freudenberg, K., 575(34), 576(34), 578(34), *635*
Frey, H. M., 665(32, 33), *724*
Frey, H. O., 87(120), *122*, 215(132), *265*, 461(102), *524*

Frey, W., 612(364), *643*
Freytag, H., 608(344), *642*
Frie, A., 614(384), 616(384), *643,* 728(154), *731*
Friese, U., 764(21, 22), 768(22), *773*
Frister, F., 632(485), *646*
Fritz, H., 8(26), *51,* 103(178), *124,* 748(106), *756*
Fritz, J. S., 592(191), 595(191), *639*
Fröhlich, I., 111(204), *124*
Frunze, T. M., 590(172), 595(235), *638, 640*
Fry, D. J., 279(20), *311*
Fry, H. S., 579(76), *636*
Fuchs, O., 182(13), 188(13), 190(13), 192(13), 215(13), 217(13), 220(13), 248(13), 252(13), 254(13), *263*
Fuentes, O., 55(174), *63*
Fuentes-Duchemin, J., 609(348), *643*
Fuertes, M., 319(140), *320*
Fujii, T., 270(224), *277*
Fujimori, K., 353(140), *381*
Fujita, E., 737(16, 17), *754*
Fujita, T., 101(172), *123,* 542(65), 548(65), 550(79), *570*
Fuks, R., 333(76, 77, 78, 79), 334(79), 335(78), 336(78), 348(119), 349(119), 364(77), 365(77), *379, 380,* 384(203), 385(204), 390(203), *391,* 547(75), 560(100), *570*
Fukuda, M., 62(212), *64*
Fukui, K., 702(91), 703(91), *726*
Fukui, M., 269(221), *277*
Fukui, Y., 43(150), *54,* 210(112), *265*
Fukumoto, K., 595(244), *640*
Fukumura, M., 595(226), *639*
Fukushima, H., 744(68), *755*
Fuller, N., 604(298), *641*
Funke, B., 229(159), *266,* 406(32, 59), 407(32), 413(32), 414(32), 415(32), 416(32), 440(32), 442(32), 443(32), 444(32), *522*
Funke, W., 110(203), *124*
Furukawa, H., 737(12), *754*
Furukawa, M., 126(237), 129(237), *140*
Furukawa, Y., 132(254), *140*
Furumoto, S., 36(124), *53*
Fusco, R., 579(72), *636,* 729(162), *731*

Gabb, E. G., 587(143), 606(143), 607(143), *637*
Gabriel, S., 295(95), *312*
Gadalla, K. Z., 19(57), *52*
Gadamer, J., 579(77), *636*
Gadreau, C., 61(210), *64*
Gal, H., 74(67), *121*
Gal, J., 66(23), 68(23), 78(23), 79(23), 80(23, 101), 81(23), 97(23), 114(23), 118(23), *120, 122,* 378(33)
Galibey, V. I., 647(555), *652*
Gallo, R., 317(144), 319(144), *320*
Galushko, L. Ya., 606(308, 311, 312, 314), *641, 642*
Gambill, C. R., 580(100), *636*
Ganem, B., 60(203), *64*
Garanzha, V. G., 319(141), *320*
Garbacik, T., 666(35), *725*
Garcia-Lopez, T., 319(140), *320*
Garcia-Munoz, G., 319(140), *320*
Garson, L. R., 355(150), *381*
Gassmann, A., 613(375), *643*
Gaston, D. W., 628(470), *646*
Gattermann, L., 74(76), *121*
Gattow, G., 317(136), *320*
Gaugler, R. W., 567(118), *571*
Gautier, A., 72(43), 74(67), *121,* 542(66), *570*
Gavin, D. F., 159(49), *165*
Gavin, G., 154(37), *165*
Gawell, L., 750(113), *756*
Gebauer-Fulnegg, E., 580(99), 584(99), 609(99), 632(99), *636*
Geiger, B., 476(127), *524*
Geisel, M., 701(87), *726*
Geiser, T., 227(156), *266*
Geissman, T. A., 689(69), *725*
Gellert, E., 747(87), *755*
George, C., 683(57), *725*
George, P., 126(243), 128(243), 136(243), *140*
Georgoulis, C., 60(204), *64*
Gèrard, G. J., 54(170), *63,* 128(248), *140*
Gerard, W., 581(105, 106, 107), *637*
Gerhardt, G., 322(10), *378*
Gerhart, F., 187(35), *263,* 336(82), *379*
Gericke, R., 728(140), *730*
Germain, G., 165(55), 168(55), *171*
Gerngross, O., 647(536), *651*
Ger. Pat. 109,933 (1900), 575(24), 579(24), *635*
Ger. Pat. 116,386 (1901), 575(25), 579(25), *635*

AUTHOR INDEX

Ger. Pat. 117,267 (1901), 592(195), *639*
Ger. Pat. 117,625, 118,566, 579(75), 590(75), *636*
Ger. Pat. 118,537, 579(74), 590(74), *636*
Ger. Pat. 201,325 (1908), 592(196), *639*
Ger. Pat. 372,842 (1924), 329(53), *379*
Ger. Pat. F. R. 2,219,002 (1972), 595(231), 623(231), *640*
Gerrard, W., 658(15), 689(15), *724*
Geske, D. H., 351(127), *380*
Gewald, K., 319(139), *320*
Ghigi, E., 624(442, 443), *645*
Ghosez, L., 74(80), 76(80), 106(80), 116(80), *121*, 125(221), 136(221), *139*
Gibron, J. D., 386(212), *392*
Gibson, H. W., 649(608, 609), *653*
Gierschner, K., 609(349), *643*
Gil, P. S., 583(127), 584(127), *637*
Gilak, A., 24(92), 25(92), *52*
Gilbert, H. G., 530(20), *569*
Gilchrist, M., 602(282), 605(282), 606(282), *641*
Gilkerson, W. R., 597(272), *641*
Gill, E. W., 683(57), *725*
Gilman, N. W., 665(4), 668(4), *724*
Gingrich, H. L., 126(231), *140*
Ginsburg, D., 200(80), *264*
Ginzburg, A. O., 620(432), 621(432), *645*
Giori, C., 595(236), *640*
Giri, V. S., 757(142), *758*
Giron, H. M., 578(67), 599(67), 609(67), *636*
Gitis, S. S., 88(125), *122*
Glinskaja, O. V., 647(523), *651*
Gloede, J., 66(24), 118(24), *120*, 406(37), 450(87), 453(87), 454(87), 460(87), 461(87), 462(87), 464(87), 465(87), 466(87), 468(87), 470(37, 87), 471(37), 473(37), *522, 523*
Glogger, I., 182(3), *262*
Glos, M., 614(386), *643*
Gluchoedov, N. P., 575(20), *635*
Glushenkova, V. R., 336(80), *379*
Glushkov, R. G., 56(178), *63,* 69(35), 115(35), *120,* 190(46, 47, 48), 191(46, 48, 51), 201(83, 84, 85), 213(47, 122, 123), 220(84), 244(190), 245(46, 47, 48, 123), 246(51), 250(47, 122), 254(47, 83, 85), 255(122), *263, 264, 265, 267,*
270(222), 272(222), 273(222), *277,* 397(8), 409(62, 63, 64), 425(62, 63), 426(63), 439(63), 485(63), 495(8), 499(62), 502(155), *521, 523, 525*
Gnehm, R., 66(2a), *120*
Gober, B., 736(11), *754*
Godefroy, E. F., 88(129), 89(129), *122*
Goerdeler, J., 61(205), *64,* 289(71, 72, 73), *312,* 333(73), *379*
Goffin, E., 126(244), 128(244), 138(244), *140,* 772(41), *773*
Göknel, E., 496(147, 149), 497(149), 507(147, 165), 508(147, 165), 509(147, 149), 510(147), *525*
Gold, H., 33(118), 34(118), 35(118), 36(118), 37(118), *53, 54*(155, 156), 182(12), 192(12), *263,* 331(64, 65), 361(64), 363(64), 367(65, 176), 368(65, 176), *379, 381,* 443(80), *523*
Gold, V., 587(140), 603(140), 283, 284), 604(140, 284, 287), *637, 641*
Goldberg, A. A., 295(93), *312*
Golden, D. M., 657(7), *724*
Goldfarb, Ya. L., 595(229), *639*
Golding, B., 247(186), *267*
Golding, B. T., 272(229), 273(229), *277*
Goldman, L., 38(130), 40(130), 41(130), *53*
Goldschmidt, B. M., 647(552), *652*
Goldschmitt, E., 657(12), 718(124), 727(132), *724, 727,* 730(132, 173), *730, 731*
Goldstein, J. A., 575(37, 38, 39, 40), 576(37, 39), 577(37, 39), 587(37, 38, 39, 40), 589(37, 38, 39, 40), *635*
Goldstein, R. O., 647(552), *652*
Goldstein, S. B., 575(19), *635*
Golebiewski, W. M., 760(18), *772*
Golfier, M., 126(229), *140*
Golovina, Z. P., 647(521, 557), *651, 652*
Goltzsche, W., 620(429), 621(429), 623(429), 624(429), 630(429), *645*
Golubeva, G. A., 13(47), *51,* 729(165, 166), *731*
Gompper, R., 1(1), *4,* 6(11), 7(11), 8(11, 28), 10(11), 11(11), 17(11), 18(11), 22(75), 31(28, 106), 38(106), 42(11), 43(106, 147), 46(106), 47(11), 48(106), *50, 51, 52, 53, 54,* 182(9), 194(61, 62, 63), 196(61, 62, 63), 202(93, 94),

203(93, 94, 95, 96), 204(62, 63),
205(95), 210(65), 211(62), 238(61),
239(63), 248(61), *262, 264, 265,*
279(28, 33, 36, 37, 42, 42a), 281(28,
42), 282(42), 283(33), 284(28, 33,
36, 37, 42), 285(65), 287(33, 37),
291(90), 292(90), 293(90), 295(28),
296(28), 299(28, 33), 301(42, 42a, 103),
302(33, 36, 37, 104), 304(28), 315(124),
319(124), *311, 312, 313, 320,* 325(37),
326(37), 333(37), 360(168), *379, 381,*
401(20, 22), 408(22), 448(20), 473(22),
474(22, 122), 475(20, 22, 122),
476(22, 127), 496(22, 146), *522, 524,*
583(129), *637*
Goodman, L., 357(156), *381,* 647(530,
531), *651*
Goodwin, S., 745(79), *755*
Gorbatenko, V. I., 133(256), *140*
Gorbunov, G. V., 648(563), *652*
Gordon, J. E., 218(139), 219(139), *266,*
333(75), *379*
Gordon, M., 589(153), *638*
Gospodinov, N., 633(493), *646*
Gota, D. J., 528(11), 548(11), *568*
Gotschi, E., 279(42b), 299(42b),
300(42b), *311*
Gottfried, R., 289(81), *312*
Gottlicher, S., 753(135), *757*
Gould, K. J., *653*(610)
Goulden, J. D. S., 279(22), 294(22), *311*
Gousson, T., 304(107), 305(107), *313*
Goutard, R., 747(92, 93), *756*
Govindachari, T. R., 742(50), *755*
Grabowski, G., 322(4), 354(4), *378*
Graefe, J., 111(204), 126(238), *124, 140*
Graf, R., 608(342), *642*
Graham, J. C., 33(117), *53*
Grandberg, I. I., 710(107), 712(110),
726, 727, 729(163, 164), *731*
Granger, R., 279(55a), 284(55a), 286(55a),
311
Granik, E. M., 502(155), *525*
Granik, V., 409(62), 425(62), 444(62),
499(62), *523*
Granik, V. G., 56(178), *63,* 190(46, 47, 48),
191(46, 48, 51), 201(83, 84), 213(47),
220(84), 244(190), 245(46, 47, 48),
246(51), 250(47), 254(47, 83), *263, 264,*
267, 270(222), 272(222), 273(222), *277,*

397(8), 409(63, 64), 425(63), 426(63),
439(63), 485(63), 495(8), 502(155),
523, 525
Grashey, R., 287(67), 307(67), *312*
Gray, R. C., 748(100), *756*
Grdinic, M., 66(8), 74(8), 78(8), 79(8),
81(8), 82(8), 97(8), *120*
Green, J., 682(50), *725*
Greenberg, M. A., 740(30), *754*
Greenwald, R. B., 75(84, 85), *121*
Gregory, W. A., 88(123), 89(123), *122*
Greiner, H., 627(462), *645*
Grekov, A. P., 606(320), *642*
Grenan, M. M., 232(169), *266*
Greth, W. E., 44(151), *54,* 208(110),
209(110), 224(110), *265*
Gribi, H. P., 247(182), *267*
Griebenow, W., 406(39, 40, 57), 407(39),
412(57), 413(57), 416(39), 418(40, 57),
439(40), 480(40, 57), 481(40, 57),
482(40, 57), *522*
Grieder, A., 657(11), 710(11, 106), 711(11,
106), *724, 726,* 729(158), *731*
Grieshaber, P., 198(71), 220(71), 248(71),
252(71), 258(71), *264,* 369(180),
370(180), *382,* 455(96), 457(96), *523*
Griesinger, A., 689(67), 710(67), *725*
Grigat, E., 530(19), 537(42), 557(19),
560(94), 563(106), 568(19), *569, 570, 571*
Grill, W., 72(50), 78(50), *121,* 538(51),
541(51), *569*
Grinev, A. N., 55(175), *63,* 126(232), *140*
Grisar, J. M., 11(43), *51*
Grishina, E. V., 648(572), *652*
Grivas, J. C., 355(149), *381*
Grob, C. A., 689(68), 701(87), 702(68,
88), *725, 726,* 729(156), *731*
Gröbner, P., 145(14), *164*
Grosheintz, J. M., 613(372a), *643*
Gross, H., 66(22, 24), 118(24), *120,*
148(28), 159(28), *164,* 262(206), *267,*
374(190), *382,* 406(37), 470(37, 114,
115, 116, 117), 471(37), 473(37),
493(144), *522, 524, 525*
Gross, P., 183(16), 184(16), 188(16),
189(16), 195(16), 196(16), 197(16),
198(16), 201(16), 220(16), 222(16),
229(158), *263, 266*
Grossman, R., 322(3), 325(3), 342(109),
354(3), *378, 380*

Grudzinskas, C. V., 184(23), 190(23), 232(23), 257(23), *263,* 406(41), 411(41), 448(41), 478(41), *522*
Grundmann, C., 74(73), *121,* 531(26), 533(29), 540(26), 542(26), 544(26), *569*
Grundnes, J., 528(5), 557(5), *568*
Grundon, M. F., 746(83, 84, 86), *755*
Gschwend, H., 247(182), *267*
Gubanova, L. A., 269(218), 273(218), *277*
Guepet, P., 279(32), 284(32), 286(32), 294(32), *311*
Gueritte, F., 771(35), *773*
Guibe-Jampel, E., 579(86), 580(87, 88), 581(51, 114), 583(86), 585(86), 586(51, 86, 87), 587(51, 86), 590(88), 594(87, 88), 595(86, 87, 88), 604(51, 86, 87), *635, 636, 637,* 647(538, 549), *652*
Guillerez, M. G., 126(229), *140*
Günther, E., 9(33), *51*
Gupta, Y. P., 757(141), *758*
Gurne, D., 633(489), *646*
Gutbrod, H. D., 184(20), 188(20), 195(20), 198(20, 72), 226(20, 72), 227(20), 228(20), 229(20, 158), 230(20), 231(20), *263, 264, 266*
Guthrie, R. D., 591(185), *638*
Gutrie, R. R., 648(574), *652*
Gutsche, K., 94(149), *123*
Guyot, M., 593(209), *639*
Guzman, A., 596(251), *640*
Gynane, M. J. S., 167(59), *171,* 385(205), *391*

Haack, A., 6(2), *50*
Haaf, A., 753(136), *757*
Haaf, G., 753(134), *757*
Haake, P., 250(193), *267*
Haase, E., 597(262), *640*
Haase, L., 406(37), 470(37), 471(37), 473(37), *522*
Habeck, D., 125(222), *139*
Habermehl, G., 753(134, 135, 136), *757*
Hafner, K., 186(31), 191(55), 246(55), *263, 264*
Hagedorn, J., 39(133), *53*
Hahn, V., 66(8), 74(8), 78(8), 79(8), 81(8), 82(8), 97(8), *120*
Hajek, M., 440(72, 72a), 443(72), 485(72a), *523*
Hall, A. J., 317(134), *320*

Hall, D. R., 272(229), 273(229), *277*
Hall, H. K., 38(129), 42(129), *53*
Hallmann, F., 66(9), 82(9), 88(9), *120*
Halski, L., 633(689), *646*
Hamada, Y., 614(388), *643*
Hamana, M., 574(7), *634*
Hamby, R. A., 55(177), *63,* 650(624), *654*
Hammond, G. S., 357(152, 153, 161), *381, 566(114), 571*
Hanessian, S., 225(149), *266,* 456(97), 461(97), *524*
Hanewald, K., 317(136), *320*
Hanford, W. E., 574(14), 588(14), *635*
Hanna, J. G., 592(192, 193), *639*
Hansen, H.-J., 658(16, 18, 19, 20), 659(21), 669(38, 39), 682(54), 702(39), 705(19), 720(39), *724, 725,* 728(143, 144, 145), *730*
Hansen, J. F., 191(56), 245(56), 246(56), *264*
Hansen, P. E., 227(156), *266*
Hantke, K., 57(187), *63*
Hantzsch, A., 72(49), *121,* 538(45, 49), 540(49), 541(49), *569*
Hanus, V., 742(53), 744(72), *755*
Hara, Sh., 648(586), *653*
Harada, K., 757(140), *758*
Harano, K., 623(441), *645,* 650(626, 627), *654*
Harbuck, J. W., 595(224), 596(224), *639*
Harcourt, R. D., 589(154), *638*
Hardy, E. M., 148(25), *164*
Hargrove, R. J., 648(581), *653*
Harrington, K. J., 521(178), *525*
Harris, D. L., 337(86), 357(158), *380, 381*
Hatris, I., 727(133), *730*
Harris, R. L. N., 178(6), *180,* 293(90a), 295(90a), 297(90a), 304(90a), 307(90a), *312*
Harris, R. O., 720(127), *727*
Harrison, C. R., 61(209), *64*
Harrison, R. G., 58(193), *63,* 503(158), 504(158), *525*
Harsanyl, K., 38(127), *53,* 736(9, 10), *754*
Hartel, H., 33(116), *53*
Hartemink, M., 547(75), *570*
Hartigan, R. T., 290(89), *312*
Hartmann, H., 59(196), *64,* 101(173), *123,* 134(261, 262, 263), 135(261, 262,

263, 264, 265, 266, 267, 268, 269, 270),
  140, 232(162), *266,* 279(46), 280(46),
  281(46), 282(46), 283(46), 290(82),
  297(46), *311, 312,* 359(165), 364(165),
  381, 388(222), 389(222), *392*
Hartner, H., 597(265), *640*
Hasedorn, I., 19(65, 66), 23(66), *52*
Hasek, W. R., 68(27), *120*
Hasselquist, H., 88(133), 89(133), *123*
Hassner, A., 39(138), *53*
Haszczyc, B., 648(562), *652*
Hata, T., 650(612), *653*
Hatz, E., 252(200), *267,* 398(10), 399(10),
  406(10), 410(10), 455(10), 465(10, 109),
  466(10), 467(10, 109), 469(10, 109),
  521, 524
Haug, E., 382(195), *391*
Haugen, G. R., 657(7), *724*
Haughton, E., 279(55b), 284(55b),
  288(55b), *311*
Hausberg, H. H., 57(187), *63*
Hauser, Ch., 647(527), *651*
Hauser, W., 72(51), *121,* 539(52), 541(52),
  561(52), 564(52), *569*
Haussermann, J., 574(16), 575(16), *635*
Hauth, H., 764(20), *773*
Hauthal, G., 490(141a), *525*
Haveaux, B., 74(80), 76(80), 106(80),
  116(80), *121*
Havel, J. J., 386(207), *391*
Havinga, E., 657(10), 709(10), 710(10),
  724
Haworth, R. D., 614(381), *643*
Hayama, H., 101(171), *123,* 558(95), *570*
Hayashi, Y., 531(24), *569*
Haymaker, A., 351(130), *380*
Hayre, J. E., 683(55), *725*
Hechelhammer, W., 208(109), *265*
Hechenblikner, I., 535(34, 35, 36, 37), *569*
Hecht, S. S., 182(11), *262*
Hedayatullah, M., 590(161), *638*
Hederich, V., 215(133), *265,* 487(140),
  525
Heene, R., 632(484), *646*
Heffe, W., 354(145), *381*
Heideman, Ch., 588(147), 608(147), *638*
Heilbronner, E., 703(94), 705(94), *726*
Heilmann, H. H., 74(66), *121*
Heimberger, S. I., 769(28), *773*
Heimgartner, H., 728(145), *730*

Heine, H. W., 83(107), *122,* 137(281), *141*
Heinsleit, E., 194(66, 67), 198(67), 232(66,
  166), 235(67, 166), 236(166), 238(67),
  239(67), 241(67), 242(67), *264, 266*
Heinze, B., 341(108), *380*
Heise, H., 496(146), *525*
Held, P., 54(171), *63*
Heldt, W. Z., 200(75), 245(75), *264*
Hell, P. M., 289(80), *312*
Hellmann, H., 207(107), 212(107), *265,*
  661(26), 662(26), 663(26), 673(26),
  721(26), *724*
Henckel, E., 461(105), *524*
Hendrickson, J. B., 566(114), *571*
Henecka, H., 322(6), 329(6), 333(6),
  364(6), *378*
Henklein, P., 442(78), *523*
Hennige, H., 184(18), 191(53, 54), 226(18),
  245(53, 54), *263, 264*
Henry, G., 769(29a), *773*
Henry, L., 72(45), *121*
Hensleleit, E., 333(69), 360(69), 364(172),
  365(172), *379, 381*
Hepburne, D. R., 84(111), 85(111),
  88(111), *122,* 129(250), *140,* 215(131),
  265
Herbert, R. B., 769(27b), *773*
Herbertz, M., 325(44), *379*
Herberz, M., 11(40), *51*
Herbst, K., 706(102), *726*
Herlem, D., 764(25), 771(25), *773*
Herlinger, H., 544(70), 550(70), *570*
Hermann, K., 758(144, 145), *758*
Herrendorfer, E., 580(90), 612(90), *636*
Herrmann, F., 576(49), 590(49), *635*
Herschkowitz, R. L., 26(95), *52*
Hertler, W., 39(132), *53*
Herzog, J., 578(63, 64), 592(63, 202),
  595(63), 608(63), *636, 639*
Hesse, R. H., 739(26), *754*
Hettohe, O., 627(466), *646*
Heuer, W., 336(81), 337(84), *379*
Heuser, G., 608(343), *642*
Heymann, P., 295(95), *312*
Heymons, A., 66(4), 69(4), 96(158),
  118(4), 119(4), *120, 123,* 322(21),
  378
Hickmott, P. W., 674(41, 42, 43), 675(41,
  44), 676(41, 42), 677(42, 43, 47),
  678(46, 47), 679(47, 48), 680(48),

681(49), 682(49), 725, 728(141, 142), 730
Higashi, F., 581(115, 116, 117, 118, 119), 582(115, 116), 586(116, 117, 119), 594(115, 118, 119), 637, 648(582, 584, 585), 653
Higashimura, T., 647(547), 652
Highet, R. J., 743(57), 755
Higuchi, M., 13(48), 32(48), 51, 484(137), 524
Hildebrand, R., 442(79), 523
Hill, A. J., 31(101, 104), 53, 322(18), 327(18, 48), 378, 379
Hill, H. M., 683(58), 725
Hill, R. K., 503(159), 504(159), 505(159), 525, 656(3), 659(3), 665(4), 668(4), 689(3), 722(3), 723(3), 724
Hillard, J. B., 728(148), 730
Hillier, K., 606(323), 607(323), 642
Hinkel, E. T., 279(9), 280(9), 290(9), 310
Hinkel, L. E., 73(55), 74(55, 71, 72, 74), 121, 533(30), 542(69), 569, 570
Hinshaw, W. B., 269(220), 277
Hiragi, M., 743(64), 744(64), 755
Hirakura, M., 744(68), 755
Hiratani, K., 454(93), 487(93), 523
Hirst, M., 739(27), 754
Hjeds, H., 88(134), 89(134), 123
Ho, T.-L., 319(138), 320, 554(77), 570, 593(208), 639
Hobbs, C. F., 160(51), 161(51), 165, 327(50), 328(50), 338(91), 344(116), 345(116, 117), 346(117, 118), 347(116), 354(50, 116), 355(91, 116), 356(50, 116), 357(50, 116), 359(50, 91, 116, 117, 118), 366(174), 367(50), 375(50, 116), 379, 380, 381, 461(103), 524
Hoch, H., 659(22), 675(22), 724
Hochstein, F. A., 747(95), 756
Höcker, J., 250(197), 267, 301(102), 313, 461(104), 524
Hodge, P., 61(209), 64
Hodgkin, J. H., 735(4, 6), 754
Hoechst, 88(122), 89(122), 122
Hoerger, E., 201(91), 228(91), 264
Hoesch, L., 656(9), 724
Hofer, R. M., 504(161), 525
Hoffe, G., 625(454, 455, 456), 645
Hoffmann, H., 23(84), 24(84), 52, 198(71), 220(71), 248(71), 252(71), 258(71), 264, 279(51), 281(51), 284(51), 299(51), 311, 369(179), 370(179, 184), 371(187), 382, 396(5), 402(5, 24), 405(5), 406(24), 407(24), 415(5), 421(24), 445(5), 455(96), 457(5, 96), 471(24), 472(24), 498(151), 499(151), 507(151), 508(151), 521, 522, 523, 525
Hoffmann, H. M. R., 520(175, 176, 177), 525
Hoffmann, M., 118(213), 124, 231(161), 266
Hoffmann, R., 655(1), 659(1), 689(1), 702(1), 703(1), 722(1), 723(1), 724
Hoffmann, R. W., 351(134), 380
Hofle, G., 590(162, 163, 165), 596(162, 261), 625(162, 261), 626(162, 261), 638, 640
Höfle, J., 648(570), 652
Hofmann, A., 401(21), 408(21), 440(70, 71), 441(70, 71), 442(70, 71), 443(70, 71), 444(70, 71), 473(21), 475(21, 70, 71, 126), 476(70, 71), 485(70, 71), 522, 523, 524
Hofmann, D., 100(168), 101(170), 123
Höhle, A., 61(208), 64
Hohle, M., 191(57), 246(57), 264
Holden, H. G., 746(82), 755
Holecek, J., 575(43), 583(125, 134), 584(125, 134), 585(134), 586(134), 600(125), 637
Hollandt, F., 589(157), 638
Holmberg, B., 284(61), 312
Holtje, H.-D., 696(79), 698(82), 699(79, 84), 726
Holy, A., 7(21), 27(21), 30(21), 48(21), 51, 66(13), 77(13, 99), 78(13), 79(13), 81(13, 99), 83(13), 90(13), 92(13), 113(13), 114(13), 120, 122, 214(126), 265, 406(36), 411(36), 440(74, 75), 441(74, 75), 448(86), 450(86), 451(86, 86a, 86c), 452(86e), 456(74, 75, 75e, 97a), 461(74, 75), 465(86e), 461(74, 75), 472(121a), 499(74), 522, 523, 524
Holzinger, L., 760(3b), 772
Honda, M., 269(221), 277
Hong, Ng. D., 452(86e), 465(86e), 523
Hong Thu, Nguyen, 304(106), 313
Honk, K. N., 250(195), 267
Honma, Y., 771(36a, 36b), 773
Hopkins, B. J., 674(42, 43), 676(42),

677(42, 43, 47), 678(47), 679(47), 681(49), 682(49), *725*
Hopkins, T., 579(78), *636*
Hoppe, B., 62(213), *64*
Hoppe, I., 129(252), *140*
Horie, M., 385(206), *391*
Horn, P., 198(71), 220(71), 248(71), 252(71), 258(71), *264,* 339(96, 97), 369(179), 370(179), *380, 382,* 397(6), 402(5a), 406(52), 407(52), 435(52), 436(52), 437(52), 455(96), 457(96), 458(6), 491(6), 517(5a), *521, 523*
Horner, L., 23(84), 24(84), *52*
Horning, D. E., 46(160), 47(160), 48(160), *54,* 109(198), *124*
Horning, E. C., 745(79), *755*
Hornung, K. H., 20(67), *52,* 92(147, 148), *123*
Hornyak, Gy., 170(64), *172*
Horowitz, R. M., 689(69), *725*
Horstmann, H., 289(71, 72), *312*
Hoshino, T., 627(466), *646*
Hotten, T. M., 596(253), *640*
Houben-Weyl, 218(136), 226(136), 228(136), 229(136), *265,* 589(158), 592(158), 594(158), 595(158), 596(158), *638*
Houdewind, P., 667(36), 668(36), 669(36), *725,* 728(136, 137), *730*
Houlihan, W. J., 125(222), *139*
Hover, W., 344(115), 345(115), 346(115), 347(115), 348(115), 349(115), 351(136), 354(115), 375(115), *380, 381*
Howard, E. G., 554(87), *570*
Howells, R. D., 648(578), *653*
Hruban, L., 742(54), *755*
Huckings, B. J., 584(137), 585(137), 588(137), 611(137, 360), *637, 643*
Hudson, H. R., 84(111), 85(111), 88(111), *122,* 129(250), *140,* 215(131), *265,* 579(81), 588(81), *636*
Hudson, J. A., 683(57), *725*
Hughes, W. L., 201(90), 241(90), *264*
Hui, W. H., 743(62, 63), 744(62, 63), *755*
Huisgen, R., 38(131), *53,* 182(2, 3), *262,* 579(71), *636*
Hull, R., 611(359), 615(359), *643,* 649(603, 604, 605), *653*
Hullin, R. P., 74(71), *121*
Hung, J. Ch., 728(139), *730*

Hunig, S., 222(143), 223(143), 248(143), 258(143), *266,* 279(30a), 284(30a), 299(30a), 307(30a), *311,* 659(22), 675(22), *724*
Hurd, C. D., 682(52), 685(62), *725*
Hurd, R. N., 279(57), 284(57), 285(57, 66), 286(66), 289(57), 295(57), *312*
Hurst, J., 595(227), *639*
Husson, A., 760(15), 769(34), *772, 773*
Husson, H.-P., 760(15, 16, 17), 764(25), 769(29b, 30a, 30b, 33, 34), 771(25), *772, 773*
Husted, D. R., 144(7), *164*
Huwyler, S., 592(194), *639*
Huys-Francotte, M., 126(245), 128(245), 130(245), 131(245), *140*
Hyde, C. L., 342(112), *380*

Ibele, J., 627(465), *646*
Ichiba, M., 56(179), *63*
Ichikawa, E., 168(62), *172*
Ichikawa, K., 597(264), 598(264, 273), *640, 641*
Ichikawa, M., 771(37a), *773*
Ichimura, K., 75(87), 83(87), 86(87), 92(87), 99(87), 137(87, 283), *121, 141*
Ichino, M., 68(30), *120*
Ichlov, C., 729(155), *731*
Ignasiak, B., 135(272), *141*
Ignasiak, T., 135(272), *141*
Ikawa, K., 43(150), *54,* 210(112), *265*
Ikeda, F., 13(50), *51*
Ikehara, M., 92(145), *123*
Iljina, Ju. N., 634(514), *647*
Imanishi, Y., 647(547), *652*
Imaseki, I., 740(31), *754*
Imazeki, I., 742(47), *754*
Inada, S., 728(151), *731*
Inomata, K., 308(111), 309(111), 310(111), *313*
Inoue, S., 687(64), *725*
Inubushi, Y., 750(111), *756*
Irisbaev, A., 55(176), *63*
Irngartner, H., 250(192), *267*
Isaka, H., 212(120), *265*
Isegawa, K., 13(50), *51*
Ishida, T., 757(140), *758*
Ishii, H., 757(140), *758*
Isineev, A. A., 595(240, 241), *640*
Isobe, M., 623(441), *645*

Ispirjan, R. M., 595(229), *639*
Issleib, K., 113(208), 124, 261(205), 262(205), *267,* 373(189), 374(189), *382*
Istratov, E. N., 647(535), *651*
Ita, P., 760(3b), *772*
Ito, H., 729(160), *731*
Ito, K., 595(244), *640*
Ito, Y., 383(197), *391*
Itoh, M., 210(113), 229(113), *265*
Ivanov, V. A., 634(513, 515), *647*
Ivanova, A. I., 231(160), *266*
Ivanova, I. A., 471(118, 119), *524*
Ivanova, J. A., 406(48), 407(48), 421(48), 443(48), 452(48), 457(48), 471(48), *522*
Ivanova, L. V., 608(337), *642*
Iwakura, C., 568(122), *571*

Jackson, A. H., 614(383), *643,* 736(14), 737(14), *754*
Jacob, K., 619(419), *644*
Jacob, T. M., 92(144), *123*
Jacobsen, P., 57(188), *63,* 383(198), *391*
Jacquier, R., 714(117), 716(117, 120), *727*
Jahngen, E., 771(37a), *773*
Jahnke, D., 279(34), 284(34), *311*
Jakovleva, M. Ja., 625(447), *645*
Jakubovich, A. Ja., 595(237), *640*
James, G. H., 205(100), *265*
Jameson, G. W., 581(113), 586(113), 587(113), 603(113), 604(113), *637*
Jamieson, W. B., 268(210), 274(210), *276*
Jancke, H., 736(11), *754*
Jander, J., 33(116), *53*
Janes, N. F., 687(63), *725*
Jankov, L., 633(491), *646*
Janot, M. M., 747(92, 93, 94), *756*
Janousek, Z., 76(96), *122,* 126(243, 245), 128(243, 245), 130(245), 131(245), 136(243), *140,* 150(33), 158(47), *164, 165,* 563(107), *571*
Janus, J. W., 201(89), *264*
Janz, G. J., 73(54), *121,* 539(55), 541(61), *569, 570*
Jaquier, R., 729(170), *731*
Jarovenko, N. N., 75(82), *121*
Jastorff, B., 590(170), *638*
Jaus, H., 195(69), 198(69), 199(69), *264*
Jaus, J., 399(11), 406(11), 427(11), 428(11), 430(11), 432(11), 433(11), 439(11), 440(11), 460(11), *521*

Jefferson, E. G., 587(140), 603(140), 604(140, 287), *637, 641*
Jen, T., 191(52), 237(52), 238(52), *264*
Jencks, W. P., 576(46), 586(46, 138), 587(46), 599(275), 600(46, 138, 278, 279), 601(46, 138), 602(279, 282), 603(275), 604(46), 605(279, 282), 606(282), *635, 637, 641, 647*(275, 279), *647*
Jenkins, G. L., 16(51), *51*
Jenkins, W. W., 685(62), *725*
Jennison, C. P. R., 727(133), *730*
Jenny, E. J., 78(100), 79(100), 107(100), 108(100), 113(100), *122*
Jensen, K. A., 284(63), 297(63), *312*
Jentsch, W., 23(82), *52,* 71(37), 83(108), 84(108), 93(108), 98(108), 102(37), 103(180), 107(108), 109(108), *120, 122, 124,* 324(32), 330(58), 331(59), 341(103), *378, 379, 380*
Jerchel, D., 627(463), 633(463), *645*
Jesaitis, M. A., 614(386), *643*
Johnson, D. S., 342(112, 113), *380*
Johnson, F., 528(7), 552(7, 82, 83, 84), 554(7), 560(101), 563(7), *568, 570*
Johnson, L. F., 750(115), *756*
Johnson, M., 580(94), 589(94), 611(94), *636*
Johnson, M. D., 580(63), 584(137), 585(137), 588(137), 589(155), 610(358), 611(63, 137, 155, 358, 360), 612(358), 614(358), *636, 637, 638, 643*
Johnson, R. M., 556(88), *570*
Johnson, S., 591(181), 594(181), 599(181), 604(181), *638*
Johnson, S. L., 578(66, 67, 68), 583(66), 584(66, 132), 585(66, 68), 586(66), 599(67, 277), 600(68, 132), 602(280), 603(280), 604(280), 605(280), 609(67, 68, 353, 354), 610(68), 633(506), *636, 637, 641, 643, 647*
Johnson, T. B., 212(117, 118), *265*
Jokisch, H., 620(422), 621(422), 625(422), *644*
Jolidon, S., 682(54), *725,* 728(143), *730*
Jolley, M. R. J., 58(193), *63*
Jonas, Y., 357(155), *381*
Jones, J. I., 530(17), *569*
Jones, V., 357(159), *381*
Josan, J. S., 521(178), *525*

Joshi, K., 300(100), *312*
Joshua, A. V., 771(36c, 37a, 37b), *773*
Jostes, F., 96(158), 114(210), 118(210), *123, 124,* 322(21), *378*
Jugie, G., 54(169), *63,* 128(248), *140*
Junov, A., 54(171), *63*
Jutz, C., 6(15), 8(15, 31), 11(31), *50, 51,* 325(42), 326(42), 350(120), 354(42), 355(42), 356(42), 359(42), 366(42), 375(42), *379, 380*

Kabuss, S., 194(59), *264,* 406(31), 434(31), 435(31), 453(90, 91), 454(92), 485(31), 487(31), *522, 523*
Kacprowizc, A., 126(239), 129(239), *140*
Kadaba, P. K., 75(88), *121*
Kadunce, R. E., 11(38), 40(38), 47(38), *51*
Kadyrov, Ch. Sh., 55(176), *63*
Kahle, G. G., 729(167), *731*
Kaiser, L., 314(120), *320*
Kaiser, P., 626(459), *645*
Kakihana, T., 247(187), *267*
Kalenda, N. W., 38(128), *53*
Kalikhuman, I. D., 387(215), *392*
Kalinowski, H. O., 156(41, 43), 160(41), *165*
Kalish, R., 188(41), 189(41), 220(41), 232(41), 234(41), 235(41), 236(41), *263*
Kallianos, A., 592(206), *639*
Kalnitzky, N. R., 650(622), *654*
Kalugina, Z. G., 268(207), *276*
Kamensky, O., 66(1), 102(1), *120*
Kametani, T., 595(244), *640,* 743(64), 744(64), *755*
Kaminsky, D., 713(114), *727*
Kamo, Y., 12(45), *51*
Kane, J., 317(135), *320*
Kaneko, H., 748(107), *756*
Kan-Fan, C., 760(16), 769(33, 34), *772, 773*
Kantlehner, W., 80(104), 102(104), *122,* 183(15), 186(15), 188(15), 189(15), 193(15), 198(71), 201(86), 220(15, 71, 140), 228(157), 229(158, 159), 234(157), 241(86), 242(86), 243(86), 244(86), 245(86), 248(15, 71), 249(15), 250(191), 252(15, 71, 140, 202), 253(15), 256(15), 257(140), 258(71), 259(86, 140), *263, 264, 266, 267, 277,* 330(57), 339(98,

102), 361(102), 363(102), 368(177), 369(177, 180), 370(177, 180), 376(98), *379, 380, 381, 382*(195), *382, 391,* 398(9), 406(29), 409(29), 423(29), 440(69), 441(69), 442(29, 69), 446(69, 85), 448(69), 449(69), 455(96), 457(96), 458(69, 99), 459(69, 99), 480(69), 491(85), 494(85), 495(85), 499(69), 507(9), 508(9), 518(172, 173), 519(29, 173), *521, 522, 523, 524, 525*
Kapasakalidis, J., 406(51), 407(51), 448(51), 480(51), 482(51), *522*
Kapaun, G., 514(170), 515(170), 516(170), 517(170, 171), *525*
Kapkan, L. M., 647(545), *652*
Kapur, J. C., 288(70), *312*
Karasuski, W., 318(137), *320*
Kardash, I. E., 575(20), *635*
Karelora, A. V., 558(97), *570*
Karo, W., 226(151), 228(151), 241(179), *266, 267,* 322(7), 329(7), 333(7), 338(7), 364(7), *378*
Kărpăti, E. A., 103(181), *124,* 159(48), *165,* 324(31), 376(31), *378*
Karrer, P., 740(35), 747(89), 748(89), *754, 756*
Kaschnitz, R., 747(90), *756*
Kasheva, T. N., 134(258), *140*
Kashiwagi, T., 212(120), *265*
Kashtanova, T. N., 589(156), *638*
Kast, H., 279(36), 284(36), 302(36), *311*
Katagiri, I., 535(39), 536(39), 539(39, 54), 540(39), 542(65), 548(65), 558(95, 96), 562(54), 563(39), *569, 570*
Katagiri, J., 101(172), *123*
Kato, T., 187(36, 37), 190(36), 237(36), *263,* 269(216), *277,* 315(128), *320*
Katritzky, A. R., 574(6), *634*
Katt, R., 594(220), 595(220), *639*
Katz, L. E., 615(390), *643*
Kaufman, J. E., 227(156), *266*
Kawabata, J., 648(584), *653*
Kawahara, S., 212(120), *265*
Kawashima, T., 68(30, 31), *120*
Kawashiro, K., 57(185), *63*
Kay, I. T., 125(228), *140*
Kazarinova, N. F., 587(139), 617(139), 618(139), 619(139, 411), 629(139), 630(139), 633(492), *637, 644, 646*
Keck, H., 31(106), 38(106), 43(106),

46(106), 48(106), *53,* 194(61, 62),
196(61, 62), 204(62), 211(62), 238(61),
248(61), *264, 265,* 360(168), *361,*
401(20), 448(20), 475(20), 496(146),
522, 525
Keese, R., 247(186), *267*
Kefeli, T. Ja., 608(335), *642*
Keller, F., 748(104), *756*
Keller, H., 145(13), *164*
Kelly, W., 295(93), *312*
Kendall, J. D., 279(20), *311*
Kerfanto, M., 339(101), *380*
Kern, R., 100(168), 101(170), *123*
Kershaw, J. R., 616(395), *644,* 649(602), *653*
Kersting, F., 21(71), *52*
Kerzhner, B. K., 125(224), *140*
Kessar, S. V., 757(141), *758*
Kessler, H., 156(41, 43), 160(41), *165*
Khachaturjan, A. A., 592(205), *639*
Khalilova, F. I., 580(98), *636*
Kharizomenova, I. A., 55(175), *63,* 126(232), *140*
Khedija, H., 50(167), *54*
Khidojatov, A. A., 590(171), *638*
Khlebnikov, A. F., 75(93), 83(93), 119(93), *122,* 126(235), 137(235), *140*
Khmelnitzky, R. A., 620(427), *644*
Khorana, H. G., 92(144), *123*
Khorlin, A. Ya., 548(78), *570*
Khullar, K. K., 232(169), *266*
Khuong-Huu, F., 764(25), 771(25), *773*
Khurana, H., 581(110), *637*
Kibler, C. J., 592(198), *639*
Kieckmann, W., 597(266, 267), *640*
Kiefer, H., 316(131), *320*
Kiefer, H. R., 701(87), *726*
Kienitz, L., 192(58), *264,* 399(12), 406(12), 427(12), 432(12), 433(12), 444(12), 448(12), *521*
Kierkegaard, P., 750(110), *756*
Kierstead, R. W., 468(111), *524*
Kigasawa, K., 743(64), 744(64), *755*
Kikugawa, K., 68(30), *120*
Kikui, J., 126(236), 139(236), *140*
Kikukawa, K., 68(31), *120*
Kilbourn, E. E., 406(45), 410(45), 448(45), 478(45), *522*
Kilminster, K. N., 633(508), *647*
Kim, M., 205(101), *265*

Kim, M.-G., 647(548), *652*
Kimler, L., 751(120), *756*
Kim-Su, M., 132(255), *140*
Kimura, K., 568(122), *571*
Kimura, M., 184(22), *263*
King, G. S. D., 348(119), 349(119), *380*
King, J. A., 279(25), *311*
King, R. B., 351(132), *380*
Kinsman, R. G., 694(72, 75, 78), 696(72, 80), 698(72, 80), 699(83), 700(72, 80), 723(75), 728(153, 154), *725, 726, 731*
Kiovsky, T. E., 564(108), *571*
Kira, M. A., 19(57), *52*
Kirby, G. W., 739(26), *754*
Kirichenko, A. I., 606(300, 301, 302, 303, 304, 305, 307, 308, 311, 312, 314, 315), *641, 642*
Kirkien-Konasiewicz, A. M., 604(292, 293, 294, 295), *641*
Kiro, Z. B., 33(115), *53,* 355(151), 357(151), *381*
Kirrmann, A., 661(27), 663(27), 721(27), *724*
Kirsanov, V. A., 387(216), *392*
Kishi, Y., 687(64), *725*
Kiso, Y., 628(469), *646*
Kiss, P., 38(127), *53*
Klaboe, P., 528(5), 557(5), *568*
Klages, F., 72(50, 51), 78(50), *121,* 538(51), 539(52), 541(51, 52), 561(52), 564(52), *569,* 583(124), 586(124), 608(341), *637, 642*
Klamann, D., 75(89), *121,* 591(186), *638*
Klauke, E., 332(67), *379*
Klebnikov, A. F., 129(249), 133(249), 136(249), 138(249), *140*
Klein, F., 208(111), *265,* 290(84), *312,* 341(105), *380*
Kleinstück, R., 24(86, 87, 88), 26(86, 87, 88), *52*
Klemm, K., 6(11), 7(11), 8(11, 28), 10(11), 11(11), 17(11), 18(11), 22(75), 31(28, 106), 38(106), 42(11), 43(106), 46(106), 47(11), 48(106), *50, 51, 52, 53,* 194(61, 62, 63), 196(61, 62, 63), 204(62, 63), 211(62), 238(61), 239(63), 248(61), *264, 265,* 326(46), 360(168), *379, 381,* 401(20), 448(20), 475(20), *522*
Klilorka, J., 575(43), 584(134), 585(134), 586(134), *635, 637*

Klinedinst, P. E., Jr., 631(476), *646*
Klinger, H., 66(26), *120*
Klinot, J., 596(252), *640*
Klinowski, C. W., 615(390), *643*
Klocker, W., 11(40), *51,* 325(44), *379*
Klohs, M. W., 748(104), *756*
Kluev, N. A., 577(52), 619(52, 408), 620(427, 434), 621(434), 626(52), 629(52), 634(515), *635, 644, 645,* 647(529, 534, 535), *647,* 650(618, 619, 620), 651(529), *651, 654*
Kluttz, R. Q., 386(207), *391*
Knabe, J., 614(384), 616(384), *643,* 694(73, 74, 75, 77), 696(79), 698(82), 699(79, 84, 85), 723(75), *726,* 728(154), *731,* 738(20, 21), *754*
Kneval, A. D., 16(51), *51*
Knight, J. A., 531(25), *569*
Knoch, F., 579(77), *636*
Knoller, G., 706(102), *726*
Knollpfeiffer, F., 627(467), *646*
Knunjanz, I. L., 608(335), *642*
Knupfer, H., 19(59), 26(59), 39(59), *52,* 105(186), 106(186), *124,* 596(254), *640*
Ko, E. C. F., 722(130), *727*
Kobayashi, S., 22(79, 80, 81), 31(80, 81), 39(79), *52,* 58(192), *63*
Kobbert, M., 322(4), 354(4), *378*
Kober, E., 159(49), *165*
Kober, W., 271(226), *277*
Kochendorfer, G., 66(10), *120*
Kocher, R. C., 6(14), *50*
Kochetkov, N. K., 548(78), *570,* 590(169), *638*
Kochman, J., 648(580), *653*
Koch-Pomeranz, U., 728(144), *730*
Koda, A., 212(120), *265*
Kodama, Y., 547(74), *570*
Koenigs, E., 617(398), 627(462), *644, 645*
Koganty, R. R., 44(152), 46(152), *54,* 210(114), *265*
Kohler, K.-H., 751(119), *756*
Koizumi, M., 595(244), *640*
Kojtscheff, T., 37(126), 38(126), *53*
Kollman, P. A., 705(96), *726*
Koltai, M., 746(81), *755*
Kolthoff, I. M., 567(115), *571*
Komarova, L. I., 607(324), *642*
Kommissarova, G., 88(128), *122*
Komori, S., 82(106), 101(106, 171, 172), *122, 123,* 528(9), 535(38, 39), 536(39), 539(39, 54), 540(39), 541(60), 542(65), 548(65), 550(9, 79), 558(95, 96), 562(54), 563(39), 565(112), *568, 569, 570, 571*
Konida, T., 750(111), *756*
König, K. H., 44(157), *54,* 331(66), *379*
König, W., 608(331, 332, 333, 334), *642*
Kononenko, V. E., 589(156), *638*
Kook, Ch. S., 728(150), *731*
Korbonits, D., 38(127), *53*
Korchachenko, N. F., 649(601), 650(601), *653*
Kornblum, N., 184(24), 204(97), *263, 265*
Kornilov, M., 252(199), *267,* 367(175), 368(175), *381,* 405(27), 406(27), 407(27), 411(27), 455(27), 457(27), 480(27), *522*
Koroleva, V. N., 391(231), *392*
Korosi, J., 200(78), 224(78), 245(78), *264,* 736(10), *754*
Korshak, V. V., 583(128), 584(136), 586(136), 590(136, 172, 173, 174, 175), 595(240, 241), 600(128), 607(136, 173, 175, 324, 325, 326, 327, 328, 329), *637, 638, 640, 642*
Korte, F., 200(79), *264*
Koshar, R. J., 144(7), *164*
Koshland, D. E., Jr., 576(45), 587(45), *635*
Koshmina, N. V., 631(478), *646*
Koskiallio, J., 603(285, 286), 604(286), 605(285, 286), *641*
Kosler, W., 71(38), *120*
Kosloski, C. L., 148(25), *164*
Kosower, E. M., 631(476, 477), 632(480), *646*
Kost, A. N., 13(47), *51,* 547(1), 575(36), 576(36), 577(36, 52), 587(36, 139), 589(1, 36, 149, 150, 151), 603(1), 607(1), 616(1), 617(1, 139, 402, 407), 618(139, 402), 619(52, 139, 402, 407, 412, 414), 620(432), 621(432), 626(52), 629(1, 52, 139), 630(139, 407), *634, 635, 637, 638, 644, 645,* 648(566, 569), *652,* 729(165, 166), *731*
Kostikov, R. R., 75(93), 83(93), 119(93), *122,* 126(235), 129(249), 133(249), 136(249), 137(235), 138(249), *140*
Kostyuclenko, N. P., 190(48), 191(48), 245(48), *263*

Kotelenez, M.I., 633(492), *646*
Kothe, N., 61(208), *64*
Koutecky, J., 631(479), *646*
Kovac, J., 576(49), 590(49), *635*
Kovalev, V. A., 387(216), *392*
Kozlov, L. B., 590(172), *638*
Kozlov, N. S., 75(95), 83(95), *122*
Kramer, W., 565(113), *571*
Krantz, A., 62(213), *64*
Kranz, J., 240(174, 175, 176), *266,* 530(16), 531(22), *569*
Kränzlein, G., 145(13), *164*
Krapcho, A. P., 333(72), *379*
Krauch, K., 689(66), *725*
Krebs, T., 590(170), *638*
Kreher, R., 184(18), 191(53, 54), 226(18), 245(53, 54), *263, 264*
Krehuak, V., 16(52), *51*
Kresze, G., 597(265), *640*
Kretschmar, H. J., 487(138), 494(138), 512(138), *524*
Kretzschmar, G., 279(34), 284(34), *311*
Kreutzberger, A., 74(73), *121,* 533(29), *569*
Krimen, L. I., 528(11), 548(11), *568*
Krisanov, A. V., 165(57), 168(57), *171*
Krivun, S. V., 647(516), *651*
Krkocsca, P., 576(49), 590(49), *635*
Krohn, J., 279(44, 49), 280(44, 49), 281(44), 282(44, 49), 283(44), 284(44, 49), 285(44), 288(49), 289(44, 79), 290(49), 291(44), 294(44), 295(44), 296(44), 299(44, 49), *311, 312,* 333(74), *379*
Kröhnke, F., 49(162, 163, 164), *54,* 184(17), 222(17), 224(17), 239(17), *263,* 574(9), 588(148), 591(182), 608(148), *634, 638*
Krommer, H., 218(136a), 238(136a), *265*
Kroogsgaard, P., 88(134), 89(134), *123*
Krowicki, K., 728(146), *730*
Krueger, R. A., 91(139), *123*
Kruglikova, R. I., 574(17), *635*
Kubitz, J., 694(73), *726*
Kubota, S., 737(17), *754*
Kucherenko, A. P., 577(54), 617(404), 618(404), 619(404), 620(433), 621(54, 433), 624(444), 625(444), *636, 644, 645,* 647(529), 650(620, 621), 651(529, 620), *651, 654*

Kuehne, M. E., 666(35), *725*
Kugelman, M., 648(560), *652*
Kühle, E., 37(125), 38(125), *53,* 102(175), 117(175), 118(175), *123,* 154(38), *165,* 330(56), 332(56), 377(191), *379, 382*
Kuhn, M., 534(32), *569*
Kuhn, St. J., 6(7), *50*
Kukhar', V. P., 134(258), *140,* 150(34, 34a), 163(34a), *164,* 165(56, 57), 168(57), *171*
Kukurina, I. V., 647(529), 651(529), *651*
Kulikova, A. E., 40(140), *53,* 331(62), *379*
Kulikova, L. K., 319(141), *320*
Kumagai, R., 550(79), *570*
Kumura, M., 268(213), 276(213), *277*
Kunckel, F., 322(2), *378*
Kundera, M., 9(34), *51,* 325(34), *378*
Kunert, F., 190(43), 194(43), *263*
Kunesch, G., 769(29a), *773*
Kunimoto, J., 737(18), *754*
Kunitake, T., 647(546), 649(591), *652, 653*
Kunz, W., 689(66), *725,* 729(156), *731*
Kupchan, S. M., 737(17), *754*
Kurabayashi, M., 559(99), *570*
Kurashev, V. V., 595(235), *640*
Kurihara, T., 650(612), *653*
Kurita, A., 21(72), *52*
Kurita, K., 648(577), *653*
Kurjakova, L. G., 649(597), *653*
Kurkjy, R. P., 588(144), 590(144), *637*
Kurkutova, E. N., 647(529), 651(529), *651*
Kurtz, P., 57(183), *63,* 136(275), *141,* 322(6), 329(6), 333(6), 364(6), *378*
Kuryatov, N. S., 190(47), 213(47), 245(47), 250(47), 254(47), *263,* 409(64), 502(155), *523, 525*
Kurze, J., 619(420), 620(420, 421), *644*
Kurzer, F., 279(53), 284(53), 306(53), *311*
Kusoma, O., 743(64), 744(64), *755*
Kuthan, J., 87(118), 97(118), *122,* 631(478), 632(482), *646*
Kutney, J., 760(14), *772*
Kutney, J. P., 771(36c, 37a, 37b), *773*
Kutter, E., 279(36), 284(36), 302(36), *311*
Kuwata, K., 351(127), *380*
Kuzovkin, V. A., 56(178), *63,* 270(222), 272(222), 273(222), *277*

Kvitko, Ya., 11(44), *51*
Kwon, S., 13(50), *51*

Laasch, P., 113(209), *124,* 218(137), *265,* 551(91), 560(102), 564(102), *570*
Labes, M. M., 204(98), *265*
Lafferty, R. H., Jr., 386(212), *392*
Lagowski, J. M., 574(6), *634*
Laird, T., 727(134), *730*
Lakhanpal, M., 583(126), 584(126), *637*
LaLonde, R. T., 596(247), *640*
Lamazouere, A. M., 315(123), *320*
Land, E. J., 631(477), *646*
Lander, G. D., 107(191), *124*
Landesman, H., 669(40), *725*
Landon, W., 279(45), 284(45), 299(45), 300(45), 302(45), *311*
Landsberg, J. M., 250(194, 195), *267*
Lang, A. H., 284(64), 285(64), *312*
Lang, G., 484(136), 514(136), 515(136), *524*
Lang, S. A., Jr., 191(56), 245(56), 246(56), *264*
Langerman, N. R., 208(110), 209(110), 224(110), *265*
Langermann, N. R., 44(151), *54*
Langlois, N., 771(35), *773*
Langlois, Y., 760(15), 771(35, 38), *772, 773*
Lanzilotti, A. E., 38(130), 40(130), 41(130), *53*
Lape, H. E., 269(220), *277*
Lappert, M. F., 167(59), *171,* 385(205), *391,* 658(15), 689(15), *724*
Lapshin, S. A., 647(525, 528, 539), 649(525), *651, 652*
Larson, R. A., 751(120), *756*
Lasne, M. C., 205(104), 206(104), *265*
Latscha, H. P., 145(16), *164,* 166(58), *171*
Lattes, A., 728(146), *730*
Laurie, V. W., 547(76), *570*
Lavrukhina, A. D., 574(17), *635*
Lawaczeck, P., 279(8), 284(8), 299(8), *310*
Lawlor, J. M., 581(113), 586(113), 587(113), 603(113), 604(113), *637*
Lawton, E. A., 100(167), *123*
Lawton, R. G., 728(138), *730*
Lazaris, A. Y., 541(59), 542(59), 552(80), *569, 570*
Leander, K., 750(109, 113), *756*

Lebedenko, N. Yu., 729(166), *731*
Lebreux, C., 224(148), 225(148), *266, 272(233), 277*
Lecher, H. Z., 148(25), *164*
Le Clef, B., 126(246), *140*
Le Clerc, A., 609(346), *643*
Leclerq, D., 648(579), *653*
Ledus, P., 581(108), *637*
Lee, D.-Y., 647(530), *651*
Lee, J., 279(55), 284(55), 287(55), *311*
Lee, J. K., 271(225), *277,* 313(115), *320*
Leete, E., 733(1), *753,* 760(11), 764(19), *772, 773*
Leffek, K. T., 722(130), *727*
Lefort, D., 647(554), *652*
Le Goff, M. T., 764(25), 771(25), *773*
Legrand, M., 279(27), 294(27), *311,* 355(148), *381*
Legrand, Y., 126(242, 244), 128(244), 138(244), *140,* 772(41), *773*
Le Guilly, L., 648(583), *653*
LeHir, A., 747(92, 93, 94), *756*
Lehmann, G., 760(3a), *772*
Lehmann, H. D., 279(37), 284(37), 287(37), 302(37), *311*
Lehr, H., 279(10), 284(10), *310*
Leicht, C. L., 748(105), *756*
Leimgruber, W., 406(47, 61), 417(61), 448(47), 478(61), *522*
Lemaire, H., 633(494), *646*
Lemke, G., 96(157), *123*
Lengfeld, F., 148(26), *164*
Leo, H., 94(153), *123*
Leon, N. H., 315(125), 317(125), *320*
Leonard, N. J., 119(217), *124,* 751(117), *756*
Lepschy, J., 648(570), *652*
Lerch, U., 619(420), 620(420, 421), *644*
Lesinka, J., 75(91), *122*
Letourneau, F., 66(23), 68(23), 78(23), 79(23), 80(23), 81(23), 97(23), 114(23), 118(23, 216), *120, 124,* 378(33)
Letters, R., 593(216), *639*
Levchenko, L. V., 558(97), *570*
Leveque, J. C., 590(161), *638*
Levey, G., 137(282), *141*
Levi, Sh., 633(493), *646*
Levkoev, I. I., 608(337), *642*
Levkovskaya, G. G., 387(215), *392*
Lewis, A. D., 279(25), *311*

Leznoff, C. C., 590(166), *638*
Liao, P. H., 771(36c, 37a, 37b), *773*
Lichtel, K. E., 19(66), 23(66), *52*
Liebscher, J., 59(196), *64,* 101(173), *123,* 134(261, 262, 263), 135(261, 262, 263, 264, 265, 266, 267, 268, 269, 270), *140,* 359(165), 364(165), *381,* 388(222), 389(222), *392*
Lienert, J., 23(85), 25(85), 26(85), *52*
Lienhard, G. E., 279(40), 281(40), 294(40), 295(40), 302(40), *311*
Limuell, I., 88(133), 89(133), *123*
Limure, T., 385(206), *391*
Lin, T. J., 737(12), *754*
Linch, S. M., 589(153), *638*
Linda, P., 6(18), 7(18, 19), *51*
Lindgren, B. O., 591(184), *638*
Lindner, C., 61(205), *64*
Lindsay, R. V., Jr., 386(213), 387(213), *392*
Lindstrom, B., 595(225), 596(225, 258), 619(258), *639, 640*
Lindstrom, J.-O., 575(41), 595(41), 596(41), 620(41), 623(41), *635*
Lindwurm, F., 633(504), *646*
Link, H., 702(88), *726*
Lipp, M., 39(135), *53*
Lira, E. P., 595(245), 608(245), *640*
Lischewski, M., 113(208), *124,* 261(205), 262(205), *267,* 373(189), 374(189), *382*
List, H. P., 751(118), *756*
Little, E. L., 73(53), *121*
Litvinenko, L. M., 606(300, 301, 302, 303, 304, 305, 306, 307, 308, 309, 310, 311, 312, 313, 314, 315, 316, 317, 318), *641, 642,* 647(525, 539), 649(525, 590, 594, 597, 598, 600, 601), 650(601), *651, 652, 653*
Liu, J.-Ch., 592(189), *639*
Lloyd, D., 382(194), *391*
Lo, D. H., 657(13), 720(13), *724*
Loer, B., 191(52), 237(52), 238(52), *264*
Loginova, T. M., 11(44), *51*
Lohaus, G., 39(137), *53*
Lohmar, K., 126(233), *140*
Loiseau, P., 314(118), 316(118), *320*
Loliger, P., 247(186), *267*
Lonshehakova, T. I., 20(68), *52*
Look, A. G., 75(83), 83(83), *121*

Loop, W., 580(101), 583(101), 591(101), *637*
Lopatina, K. I., 548(78), *570*
Lora-Tamayo, M., 531(21), *569*
Lorenz, A., 6(16), *51*
Lorenzen, J., 31(100), *53,* 322(12), 327(12), *378*
Loseva, M. V., 593(210), *639*
Lossen, W., 237(172), *266,* 322(4), 354(4), *378*
Loudon, J., 615(389), *643*
Luccarelli, A., 7(19), *51*
Luckner, M., 745(77), *755*
Lugosi, P., 169(63), *172*
Luhrs, E., 580(101), 583(101), 591(101), *637*
Lukasiewicz, A., 75(90, 91), *122*
Luknitzkij, F. I., 574(15), 588(15), *635*
Luning, B., 734(2), 750(109), *753, 756*
Lurie, M., 468(111), *524*
Lurje, C. I., 576(47), 592(47), 595(47), *635*
Lussi, H., 200(82), 244(82), 245(82), *264*
Lutz, R. P., 720(127), *727*
Lwowski, W., 237(173), *266*
Lyle, R. E., 651(631, 632), *654*
Lythgoe, B., 503(158), 504(158), *525*

Mabry, T. J., 751(119, 120), 752(121, 122, 123, 126), *756*
McCall, J. M., 269(215), *277*
McClelland, R. A., 182(7), 224(147), *262, 266,* 272(232), *277*
McCowen, M. C., 18(56), *51*
McCurdy, O. L., 747(96), *756*
McCurry, P. M., Jr., 663(31), *724*
McDermott, J. P., 285(66), 286(66), *312*
McDonald, B., 683(60), *725*
McDonald, B. G., 728(147), *730*
McDonald, R. N., 91(139), *123*
McElvain, S. M., 596(257), 597(257), *640*
McEntire, E. E., 658(14), *724*
McEwen, W. E., 583(130), 597(130, 270, 271, 272), 612(361), 613(361, 371), 614(361), 616(361), 617(361, 397), 618(397), 624(130, 397, 445), 625(271), 628(361), *637, 640, 641,* 643(363), *643, 644, 645*
McGuire, D. K., 568(121), *571*

Maccoll, A., 604(294, 295), *641*
Machinskaja, I. V., 597(263), *640*
Maciejevicz, N. S., 279(55c), *311*
Mackay, D., 727(133), *730*
McKay, G. A., Jr., 596(257), 597(257), *640*
McKennis, J. S., 342(111), 355(111), *380*
McKenzie, R. D., 11(43), *51*
McLachlin, D. M., 592(198), *639*
McLean, A., 683(60), *725*
McMahon, A. E., 295(91), 306(91), *312*
McManaman, J. L., 126(234), 129(234), 133(234), *140*
McNab, H., 382(194), *391*
McNamara, P. M., 587(143), 606(143, 321, 322), 607(143, 321, 322), *637, 642*
McNicholas, M. W., 720(127), *727*
McRitchie, D. D., 100(167), *123*
Madden, M. J., 342(113), *380*
Madeja, R., 658(16), 689(16), *724*
Madronero, R., 528(7), 552(7), 554(7), 563(7), *568*
Maerten, G., 95(154), *123*
Maffraud, J. P., 88(123), 90(123), *123*
Magouchi, S., 41(145), *54*
Mah, Heduck, 220(141), 259(141), *266*
Mai, L., 538(45), *569*
Maier, G., 656(8), *724*
Maiti, P. C., 740(34), *754*
Majda-Grabowska, H., 580(96, 97), 591(96, 97), *636*
Majumdar, K. C., 728(148), *730*
Makarov, S. P., 144(6), *164*
Makaruk, M. S., 581(111), *637*
Make, S., 592(199), *639*
Makhova, N. N., 596(260), 626(260, 457), *640, 645*
Maki, Y., 594(222), *639*
Makitra, R. G., 134(259), *140,* 581(111), *637*
Makosza, M., 126(239), 129(239), *140,* 616(394), *644*
Maksimov, V. I., 772(39a), *773*
Maleeva, N. T., 649(597), *653*
Malesh, W., 748(104), *756*
Malhotra, S. S., 337(85), *380*
Malik, M. Y., 740(32), *754*
Malisheva, L. A., 606(319), *642*
Mallams, A. K., 648(560), *652*
Maltzeva, A., 650(620), *654*

Maltzeva, G. A., 620(427, 434), 621(434), 634(515), *644, 645, 647*
Mamaev, V. P., 269(217, 218), 273(217, 218), *277*
Mamalis, P., 682(50), *725*
Mamedova, P. S., 580(98), *636*
Mangeney, P., 771(38), *773*
Manhas, B. S., 575(35), 578(35), 582(35), *635*
Manhas, M. S., 288(69, 70), *312*
Mann, I., 742(53), *755*
Mann, K., 66(17), 81(17), 82(17), 94(17), 113(17), *120*
Manning, R. E., 747(95), *756*
Mannschreck, A., 250(192), *267*
Manske, R. F. H., 742(46), *754*
Marbet, P. R., 729(156), *731*
Marchelli, R., 728(149), *730*
Marchenko, N. B., 56(178), *63,* 270(222), 272(222), 273(222), *277*
Marcinkiewicz, S., 682(50), *725*
Marecek, J. F., 649(526), *651*
Mariam, L., 88(136), 90(136), *123*
Marino, G., 6(18), 7(18, 19), *51,* 620(436), *645*
Markgraf, J., 594(220), 595(220), *639*
Markowskii, L. N., 133(256), *140,* 148(29), *164*
Marquarding, D., 135(273), *141*
Marsh, J. P., Jr., 357(156), *381*
Marshall, J. A., 701(86), 702(86), *726*
Marshall, J. L., 315(126), 319(126), *320,* 651(632), *654*
Martin, D., 493(143), 530(18), *525,* 537(43, 44), 560(18, 43, 44), *569*
Martin, G., 7(23), 8(23), *51,* 54(170), *63,* 125(225), 128(225), *140*
Martin, G. J., 7(22, 24), 8(22, 24, 27), 31(27), *51,* 68(33), *120,* 316(133), *320*
Martin, M., 7(23), 8(23), *51,* 54(170), *63*
Martin, M. L., 125(225), 128(225), *140*
Martin, M. M., 100(169), *123*
Martin, R. B., 295(92), *312*
Martinek, K., 647(543), *652*
Martinov, V. P., 648(572), *652*
Marumoto, R., 22(79), 39(79), *52*
Marvel, C. S., 100(169), *123,* 284(62), 290(62), *312*
Marvell, E. N., 608(338), *642*
Marzin, C., 710(108), *726*

Mashevskii, V. V., 75(95), 83(95), *122*
Mashkovskii, M. D., 69(35), 115(35), *120,* 213(123), 245(123), *265*
Mason, J. B., 626(460), *645*
Masson, S., 205(103, 104), 206(103, 104), *265*
Masronero, R., 531(21), *569*
Masugi, T., 594(222), *639*
Masui, M., 617(400), *644*
Mathew, S. I., 310(112), *313*
Matkovic, B., 80(102), *122,* 541(64), *570*
Matsuda, Y., 568(122), *571*
Matsui, T., 737(17), *754*
Matsumoto, I., 86(116), *122*
Matsumoto, S., 750(111), *756*
Matsumura, T., 13(48), 32(48), *51*
Matsuoka, H., 614(388), *643*
Matthews, F., 74(68, 69), *121*
Matthews, W., 531(27), 533(31), *569*
Maturova, M., 742(52), *755*
Matyushecheva, G. I., 24(93), *52*
Matzner, M., 588(144), 590(144), *637*
Maul, R., 167(61), 171(61), *172*
Maun, K., 279(52), 284(52), 294(52), 296(52), 297(52), *311*
May, P., 279(6), 284(6), 296(6), *310*
Mayer, Ch. F., 728(150), *731*
Mayr, A., 135(271), *141*
Mazzei, M., 58(194), *63*
Mead, J., 699(83), *726*
Meathrel, W. G., 610(355), *643*
Mecke, R., 534(32), *569*
Medvedev, S. S., 575(19, 20), 595(238, 239), *635, 640*
Mee, J. D., 107(189), *124*
Meerwein, H., 107(194), 108(194), 113(209), *124,* 182(12, 13, 14), 188(13, 14), 190(13, 14, 43), 191(14), 192(12, 13, 14), 193(14), 194(43), 215(13, 14, 133), 217(13), 218(136, 137), 220(13, 14), 226(136), 228(136), 229(136), 248(13, 14), 249(14), 250(14), 252(13, 14), 254(13, 14), 255(14), 256(14), 257(14), 258(14), *263, 265,* 396(4), 406(4), 409(4), 410(4), 411(4), 423(4), 424(4), 425(4), 427(4), 428(4), 429(4), 430(4), 431(4), 440(4), 441(4), 442(4), 443(4), 445(4), 453(4), 455(4), 460(4), 461(4), 469(4), 487(4, 140), 503(4), 519(4), *521, 525,* 551(91), 560(102), 564(102), *570*
Meese, C. O., 279(50), 280(50), 283(50), 289(50), *311,* 314(122), 316(132), *320*
Mehdi Nafissi, M., 190(44), 220(44), 221(44), 244(44), *263*
Meier, H., 200(76), 245(76), *264*
Meier zu Kocker, I., 39(135), *53*
Meiklejohn, R. A., 144(7), *164*
Meilahn, M. K., 126(234), 129(234), 133(234), *140*
Meindl, H., 241(177), *266*
Melby, L. R., 386(213), 387(213), *392*
Meltzer, R. I., 279(25), *311,* 713(114), *727*
Menard, G., 389(226), *392*
Mentzel, C., 729(171), *731*
Mercier, M., 752(121, 128), *756, 757*
Merenyi, R., 384(203), 390(203), *391*
Mersch, R., 113(209), *124,* 218(137), *265,* 551(91), *570*
Mersh, R., 560(102), 564(102), *570*
Merten, R., 250(197), *267,* 301(102), *313,* 461(104), *524*
Merz, M. M., 232(169), *266*
Mesnard, P., 575(33), *635*
Messerschmitt, T., 620(422), 621(422), 625(422), *644*
Messmer, W. M., 743(66), *755*
Mestechkin, M. M., 650(620, 621), *654*
Metzger, J., 317(144), 319(143, 144), *320*
Metzner, E. K., 758(143), *758*
Meyer, E. F., 247(182), *267*
Meyer, E. V., 578(65), *636*
Meyer, H., 138(287), *141*
Meyer, H.-W., 289(76, 78), *312,* 314(116, 117), *320*
Meyers, A. I., 528(4), 531(4), 552(4), *568*
Meyr, R., 19(58, 60), 39(59), *52,* 105(184, 185), *124*
Michael, A., 542(67), 551(67), *570*
Michelmann, J. S., 250(195, 196), *267*
Michelson, A. M., 593(216), *639*
Middleton, W. J., 40(143), *54,* 73(52, 53), *121*
Mierau, F., 322(4), 354(4), *378*
Migachev, G. I., 33(112, 113, 114), *53,* 619(410), *644*
Mikawa, H., 591(184), *638*
Mikhailov, V. S., 24(93), *52*
Mildenberger, H., 660(23), 661(25, 26),

662(26), 663(26), 666(34), 667(34), 668(34), 672(23), 673(26), 672(23), 673(26), 721(23, 26), *724, 725*
Miles, G. J., 678(46), *725*
Miljkovic, D., 247(186), *267*
Miller, M., 574(16), 575(16), *635*
Miller, S. I., 629(474), *646*
Miller, W. B., 250(193), *267*
Minakova, S. M., 190(46), 191(46), 245(46), *263*
Minale, L., 752(125, 127, 129, 130, 131, 132), *756, 757*
Minet, G., 772(406), *773*
Minker, E., 746(81, 85), *755*
Minkin, V. I., 6(5), *50*
Minunni, G., 575(23), 582(23), 592(23), *635*
Mioduszewska, M., 633(489), *646*
Mirskova, A. N., 387(215), *392*
Mitajshvili, T. I., 583(128), 584(136), 586(136), 590(136), 600(128), 607(136, 324, 326), *637, 642*
Mitchell, J., Jr., 596(248), *640*
Mitin, Y. V., 336(80), *379,* 647(523), *651*
Miyake, M., 101(172), *123*
Miyashita, O., 132(254), *140*
Miyataka, K., 279(24), *311*
Miyoshi, M., 580(89), 594(89), *636*
Mobius, L., 174(1, 2), 175(1, 2), 176(1), 177(1), 178(1), *180,* 294(90b), *312*
Moderhack, D., 336(81), 337(84), *379*
Mohr, P., 74(66), *121*
Mojzis, C., 587(141), 592(141), *637*
Moldi, J., 592(206), *639*
Mole, M. L., Jr., 758(143), *758*
Molenaar, E., 126(240), 139(240), *140*
Molho, D., 593(209), *639*
Moller, W., 608(343), *642*
Moncovic, I., 740(42), *754*
Moniot, J. L., 737(15), *754*
Monteiro, H. J., 260(204), *267*
Mooberry, J. B., 503(157), 504(157), *525*
Moodie, R. B., 604(288), 605(288), 648(587, 588), 649(587), *641, 653*
Moon, B. J., 694(78), *726,* 744(61), *755*
Mooney, R. B., 538(47), *569*
Mooradian, A., 713(113), *727*
Moore, D. W., 351(130), *380,* 632(486), *646*
Moore, J. A., 575(37, 38, 39, 40), 576(37, 39), 577(37, 39), 587(37, 38, 39, 40), 589(37, 38, 39, 40), *635*
Moore, L. J., 581(112), *637*
Moore, W., 595(234), *640*
Moraliogeu, E., 456(97), 461(97), *524*
Morawetz, H., 647(517), 649(517), *651*
Morck, H., 66(17, 18), 80(18), 81(17), 82(17, 18), 94(17), 113(17, 18), *120,* 186(34), 233(34), *263*
Moreau, R. C., 279(30), 280(30), 282(30), 283(30), 284(30), 304(106, 107), 305(107), *311, 313,* 314(118), 316(118), *320*
Mori, M., 184(19), 190(19, 45), 194(19), 254(45), 257(45), *263,* 500(153), 502(153), *525*
Moriarty, R. M., 190(40), 245(40), *263*
Morimoto, S., 57(185), *63*
Morita, K., 22(78, 79, 80), 31(78, 80), 39(79), 41(154), *52, 54,* 129(251), *140*
Moriyama, R., 190(49), 244(49), 245(49), *263*
Moron, J., 633(507), *647*
Morozova, L. N., 584(133), 632(133), *637*
Morr, D. H., 33(117), *53*
Morris, L. R., 126(230), *140*
Morrison, D. L., 578(68), 585(68), 600(68), 609(68), 610(68), *636*
Morrison, R. W., Jr., 11(41), *51,* 325(45), *379*
Morrow, C. J., 273(236), *277*
Morrow, D. F., 504(161), *525*
Morschel, H., 215(133), *265,* 487(140), *525*
Morszak, I., 386(211), *392*
Mort, M., 648(575), *652*
Morthey, E., 595(242), *640*
Mory, R., 66(19), 78(19), 88(19, 121), *120, 122*
Mosettig, E., 138(284), *141*
Moskvina, T. P., 729(164), *731*
Moss, J. B., 748(103), *756*
Mothes, K., 760(9), 769(9), *772*
Mott, L., 23(85), 25(85), 26(85), *52*
Moulines, J., 648(579), *653*
Muchowski, J. M., 31(105), 46(160), 47(160), 48(160), *53, 54,* 109(198), *124*
Mueller-Westerhoff, U., 185(27), *263,* 353(139), *381*
Muenter, J. S., 547(76), *570*
Muffler, H., 232(167), 235(167), *266,*

338(90), 369(90), 370(90), 372(90), 373(90), 375(90), *380,* 396(3), 403(3), 446(3), 461(3), 470(3), *521*
Muha, G. M., 632(486), *646*
Muhlstadt, M., 111(204), *124*
Mukaiyama, T., 298(96, 97, 98), 301(98), 304(96), 305(96), 308(111), 309(96, 111), 310(111), *312, 313,* 426(49, 50), 444(49), 480(49), *522*
Mukerjee, R., 279(47), 284(47), 296(47), *311*
Muller, C., 68(25), *120*
Muller, D. S., 279(55), 284(55), 287(55), *311*
Muller, E., 350(120), *380*
Muller, F., 279(30a), 284(30a), 299(30a), 307(30a), *311*
Muller, K., 247(186), *267*
Muller, O. W., 753(137), *757*
Muller, S., 295(94), *312*
Muller, W. H., 769(27c), *773*
Mulliken, R. S., 567(117), *571*
Mumm, O., 580(90), 612(90), *636*
Munch, W., 114(210), 118(210), *124*
Murakami, M., 212(120), *265*
Murakami, S., 500(154), 501(154), *525*
Murakami, Y., 21(72), *52*
Murata, Y., 598(273), *641*
Muren, J. F., 628(468), *646*
Murray, F. E., 567(116), *571*
Murray, R. W., 110(201), *124*
Musso, H., 753(133), *757*
Mutafceva, E., 633(493), *646*
Muxfeldt, H., 225(150), 227(156), *266,* 503(157), 504(157), *525*

Nagai, K., 728(151), *731*
Nagai, M., 190(49), 244(49), 245(49), *263,* 500(152), 502(152), *525*
Nagai, N., 246(181), *267*
Nagai, Y., 737(18), *754*
Nagamatsu, T., 484(137), *524*
Nagarajan, K., 742(50), *755*
Nahlovska, Z., 657(13), 720(13), *724*
Nahlovsky, B. D., 657(13), 720(13), *724*
Nakajima, K., 757(140), *758*
Nakai, T., 300(101), *313,* 454(93), 487(93), *523,* 565(109), 567(109), *571*
Nakakimura, H., 190(45), 254(45), 257(45), *263,* 500(153), 502(153), *525*

Nakao, M., 594(221), *639*
Nakatani, T., 595(226), *639*
Nakav, Y., 597(264), 598(264), *640*
Nakayama, T., 257(198), *267*
Naoi, F., 744(68), *755*
Narula, S., 757(141), *758*
Nasutavicus, W. A., 553(83), 560(101), *570*
Natarajan, S., 742(50), *755*
Natsume, M., 651(630), *654*
Nealy, D. L., 675(45), *725*
Neber, P. W., 579(70), *636,* 706(102), *726*
Nedelec, J. Y., 647(554), *652*
Nedeleva, L., 633(491), *646*
Nef, J. U., 74(70), *121*
Nehring, R., 207(107), 212(107), *265*
Neidhardt, G., 88(130), 90(130), *122*
Neidlein, R., 191(57), 246(57), *264*
Neilson, D. G., 212(121), 218(121), 226(121), 228(121), 230(121), 232(121), 240(121), *265,* 272(230), 273(230), *277,* 316(130), 318(130), 319(130), *320,* 329(51), *379*
Nelken, A., 632(483), *646*
Nelson, M. F., Jr., 633(499), *646*
Nentwig, J., 560(102), 564(102), *570*
Nerdel, F., 75(89), *121*
Nesmeyanov, A. N., 268(211), 276(211), *276*
Nesterov, O. V., 575(18), *635*
Neubauer, G., 144(2), 145(2), 146(2), 151(2), 152(2), 153(2), 154(2), 155(2), 156(2), 157(2), 158(2), 162(2), *164,* 460(100), *524*
Neumann, F. W., 96(159), *123,* 322(5), 329(5), 333(5), 344(5), 354(5), 357(5), 364(5), 375(5), *378*
Neumann, H., 519(174), *525*
Neumann, R. C., 357(153, 155), *381*
Neumann, R. C., Jr., 357(152, 159, 160, 161), *381*
Neumeyer, J. L., 205(99), *265*
Neveu, M. C., 604(299), *641*
Newkome, G. R., 656(3), 659(3), 689(3), 722(3), 723(3), *724*
Newton, L. S., 232(169), *266*
Ng, Y. L., 743(62, 63), 744(62, 63), *755*
Nichikawa, M., 41(145), *54*
Nicolaus, A., 578(65), *636*
Nicolaus, B. J. R., 88(135, 136), 90(135, 136), *123*

Nicolaus, R. A., 752(129), *757*
Niczato, H., 206(105), *265*
Nielsen, A. T., 632(486), 633(511), *646, 647*
Niemann, C., 182(4), *262*
Nijland, M. M., 740(38, 39), *754*
Nikishina, N. I., 649(600), *653*
Nikolenko, L. N., 355(151), 357(151), *381*
Nilsson, L., 728(135), *730*
Nishigaki, S., 56(179), *63*
Nishikawa, M., 129(251), *140*
Nistor, C., 56(180), *63*
Nivard, R. J., 110(202), *124*
Nivard, R. J. F., 383(199), *391,* 400(16), 445(16), 447(16), 485(16), *521*
Niwano, M., 648(584), *653*
Nizik, G. E., 102(176), *123*
Nofal, Z. N., 19(57), *52*
Noguchi, S., 129(251), *140*
Nohira, H., 298(96), 304(96), 305(96), 309(96), *312*
Nommensen, E. W., 119(217), *124*
Nordmann, H. G., 49(163, 164), *54,* 184(17), 222(17), 224(17), 239(17), *263*
Novak, I., 746(81, 85), *755*
Novitskaya, N. A., 69(35), 115(35), *120,* 213(123), 245(123), *265*
Nuesslein, J., 290(87), *312*
Nyberg, K., 772(40a), *773*

Oakenfull, D. G., 647(540), *652*
O'Brian, J. L., 350(122), *380*
O'Brien, E., 648(561), *652*
O'Brien, J. L., 344(114), 345(114), 347(114), 348(114), 354(114), 355(114), 357(114), 358(114), 359(114), 366(114), 375(114), *380*
O'Brieu, D. H., 182(6), *262*
Ochiai, M., 6(8), 22(79, 80), 31(80), 39(79), *50, 52,* 190(49), 244(49), 245(49), 246(181), 257(198), *263, 267,* 500(152), 502(152), *525*
Oda, N., 613(372), *643*
Oda, R., 34(121), *53,* 531(24), *569*
Odo, K., 168(62), *172*
Oediger, H., 23(84), 24(84), *52*
Oehl, R., 8(26), *51,* 103(178), *124*
Oehlmann, L., 57(186), *63*
Oehlschlager, A. C., 656(9), *724*

Oeser, E., 357(163), *381*
Offermann, K., 19(59), 26(59), 39(59), *52,* 105(186), 106(186), *124,* 596(254), *640*
Ogata, N., 647(524), 648(524), *651*
Ogloblin, A., 75(93), 83(93), 119(93), *122*
Ogloblin, K. A., 126(235), 129(249), 133(249), 136(249), 137(235), 138(249), *140*
Ohme, R., 41(144), 46(144), *54,* 109(199), *124,* 211(116), *265*
Ohmori, H., 617(400), *644*
Ohoka, M., 101(72), *123,* 535(38, 39), 536(39), 539(39, 54), 540(39), 541(60), 542(65), 548(65), 550(79), 562(54), 563(39), 565(112), *569, 570, 571*
Ohorodnik, A., 619(416), *644*
Ohta, M., 75(87), 83(87), 86(87), 92(87), 99(87), 137(87), *121*
Oishi, T., 184(19, 22), 190(19, 45, 49), 194(19), 244(49), 245(49), 246(181), 254(45), 257(45, 198), *263, 267,* 268(213), 269(221), 276(213), *277,* 500(152, 153, 154), 501(154), 502(152, 153), *525*
Oka, K., 648(586), *653*
Okahara, M., 541(60), 565(112), *569, 571*
Okahata, Y., 649(591), *653*
Okamoto, Y., 59(197), 62(212), *64*
Okano, S., 595(226), *639*
Okawara, M., 300(101), *313,* 455(94), *523*
Okawara, T., 126(237), 129(237), *140*
Okazaki, H., 649(526), *651*
Okazaki, M., 728(151), *731*
Okon, K., 580(96, 97), 591(96, 97), *636*
Oku, M., 529(14), *569*
Okui, K., 595(244), *640*
Okuyama, T., 224(145), *266*
Okyama, T., 224(146), *266*
Olah, G. A., 6(7), *50,* 182(6), *262,* 536(40), 545(73), 564(108), 565(40), *569, 570, 571,* 578(61, 62), 586(61), 587(61, 62), 590(62), 609(62), 631(61, 62), *636*
Olah, J. A., 578(62), 587(62), 590(62), 609(62), 631(62), *636*
Olejnik, N. M., 606(313), *642,* 649(590), *653*
Oles, S. R., 188(41), 189(41), 220(41), 232(41), 234(41), 235(41), 236(41), *263*
Oliver, L. K., 748(108), *756*
Ollis, W. D., 727(134), *730*

Olofson, R. A., 250(194, 195, 196), *267*
Olschwang, D., 252(201), *267,* 471(120, 120a), *524*
Olson, V. R., 595(232), *640*
Onaka, T., 735(7), 750(112), *754, 756*
Onda, M., 744(68), 745(69), *755*
O'Neal, N. E., 657(7), *724*
Ongley, P. A., 201(88), *264*
Onishi, T., 580(89), 594(89), *636*
Onuma, T., 190(49), 244(49), 245(49), *263*
Opitz, G., 660(23), 661(25, 26), 662(26, 28), 663(26, 28), 665(28), 666(28, 34), 667(34), 668(28, 34), 669(28), 670(28), 671(28), 672(23, 28), 673(26, 28), 689(67), 710(67), 721(23, 26), *724, 725*
Oppolzer, W., 764(20), *773*
Orazi, O. O., 746(80), *755*
Orfanos, V., 186(31), *263*
Oripov, E., 55(176), *63*
Ortiz de Montellano, P., 596(251), *640*
Ortiz de Montellano, P. R., 125(223), 136(223), *139*
Orudgeva, I. A., 580(98), *636*
Osipov, A. P., 647(543), *652*
Osuch, C., 656(9), *724*
Otah, M., 137(283), *141*
Otohiko, Tsuge, 351(135), *381*
Otroshko, G. V., 606(320), *642*
Otten, C., 594(219), *639*
Overberger, C. G., 647(518), 649(518), *651*
Overchuk, N. A., 578(62), 587(62), 590(62), 609(62), 631(62), *636*
Owen, N. E. T., 61(211), *64*
Oxley, P., 333(70, 71), *379*

Padgett, H. C., 772(39b), *773*
Pagani, G., 88(136), 90(136), *123*
Paige, J. N., 207(108), *265,* 284(60), *312*
Paillous, N., 728(146), *730*
Paisley, H. M., 530(17), *569*
Pak, V. D., 75(95), 83(95), *122*
Pakhomov, V. P., 190(47), 213(47), 245(47), 250(47), 254(47), *263,* 409(64), 502(155), *523, 525*
Pakrashi, S. C., 750(114), 757(142), *756, 758*
Palagushkina, L. A., 650(620), *654*
Palaitis, W., 647(541), *652*
Palinkas, J., 97(163, 164, 165), *123,*
323(26, 27), 361(26, 27, 169), 362(26, 27, 169), *378, 381*
Palomo, A. L., 68(29), *120*
Pandit, R. S., 54(172), 55(172), *63*
Pandit, U. K., 728(136, 137), *730*
Panov, E. V., 649(596), *653*
Paquette, L. A., 191(56), 245(56), 246(56), 247(184, 187, 188), *267,* 279(31), 284(31), 300(31), *311*
Parcell, A., 295(92), *312*
Parikh, I., 752(121, 122), *756*
Parker, S. H., 88(126), *122*
Parker, V. D., 617(399), *644,* 658(17), *724*
Parker, W. L., 560(101), *570*
Parks, J. E., 336(83), *379*
Parshin, V. A., 270(222), 272(222), 273(222), *277*
Parts, L., 591(183), *638*
Passerini, N., 58(195), *63*
Pasternak, V. I., 150(34), *164,* 165(56, 57), 168(57), *171*
Pastour, P., 57(191), *63*
Patai, S., 382(192), *391,* 729(156), *731*
Patchett, A. A., 387(217), *392*
Patchornik, A., 648(576), *653*
Paterson, T. R., 592(197), *639*
Pat. Hung. 2,387 (1971), 633(504), *646*
Patison, V. A., 647(553), *652*
Patterson, D. R., 272(233), *277*
Patterson, J. M., 682(51), *725,* 728(150), *731*
Pattison, V. A., 44(153), *54,* 210(115), 226(115), *265*
Paudler, W. W., 55(174), *63*
Paukstelis, J. V., 205(101), *265,* 647(548), *652*
Paul, R. C., 575(31, 35, 55, 56), 576(31), 577(31), 578(35, 56), 581(110), 582(31, 35, 55, 56, 121), 583(31, 126, 127), 584(56, 126, 127, 135), 585(135), 587(142), *635, 636, 637*
Paulmier, C., 57(191), *63*
Pavlaskova, D., 742(52), *755*
Pavlenkov, G. N., 150(34), *164*
Pavlik, J., 584(134), 585(134), 586(134), *637*
Paytin, B. M., 190(48), 191(48, 51), 213(122), 245(48), 246(51), 250(122), 255(122), *263, 264, 265*
Peak, D. A., 279(21), 281(21), 282(21),

283(21), 284(21), 294(21), 297(21), 301(21), 304(21), 306(21), *311*
Pearson, D. E., 57(184), *63,* 138(286), *141*
Pearson, R. C., 554(85), *570*
Pearson, R. G., 632(481), *646*
Pechmann, H. V., 118(214), *124,* 322(13, 14, 15), 341(108), *378, 380*
Pedersen, C., 284(63), 297(63), *312*
Pelletier, S. W., 751(119), *756*
Pelter, A., *653*(610)
Pennanen, S., 20(69), 21(69), *52,* 86(115), *122*
Peregudov, G. V., 575(19), *635*
Peresleni, E. M., 60(201), *64,* 190(48), 191(48), 245(48), *263*
Perfilov, Ju. I., 590(175), 607(175, 327), *638, 642*
Perkin, W. H., Jr., 614(381), *643*
Perret, A., 538(46), *569*
Perret, R., 538(46), *569*
Perrin, C. L., 357(162), *381*
Perry, S. G., 604(298), *641*
Pershin, G. N., 69(35), 115(35), *120,* 213(123), 245(123), *265*
Persianova, I. V., 502(155), *525*
Persianova, J. V., 190(48), 191(48), 245(48), *263*
Pesaro, M., 247(182, 185), *267*
Peter, H., 764(24), 768(24), *773*
Petermann, K. E., 125(219), 137(219), *139*
Peters, D., 575(34), 576(34), 578(34), *635*
Petersen, S., 200(77), 247(183), *264, 267,* 279(29), 284(29), 300(29), *311*
Peterson, D. A., 184(23), 190(23), 232(23), 257(23), *263,* 406(41), 411(41), 448(41), 478(41), *522*
Peterson, S. W., 80(102, 103), *122,* 541(63, 64), *570*
Petracek, F. J., 748(104), *756*
Petrov, K. A., 316(129), 319(129), *320*
Petrovskaja, L. Yu., 647(534), *651*
Pettit, G. R., 11(38), 40(38), 47(38), *51,* 355(150), *381*
Petz, W., 171(66), *172*
Pfaffli, P., 764(20), *773*
Pfeifer, S., 736(11), 742(51, 53), *754, 755*
Pfeiffer, P., 609(352), *643*
Pfeil, E., 182(12), 192(12), *263*
Philbin, E. M., 565(110), *571*

Philipps, B. A., 66(23), 68(23), 78(23), 79(23), 80(23, 101), 81(23), 97(23), 114(23), 118(23), *120, 122, 378*(33)
Piattelli, M., 752(125, 127, 129, 130, 131, 132), *756, 757*
Picard, P., 648(579), *653*
Piccirilli, R., 616(395a), *644*
Picker, K., 650(611), *653*
Pickorski, G., 305(108), *313*
Piers, E., 649(607), *653*
Pigamol, P., 279(32), 284(32), 286(32), 294(32), *311*
Pilger, C. W., 727(133), *730*
Pilgrim, F. J., 39(136), *53,* 596(249), *640*
Pilott, A., 188(42), 190(42), 192(42), 197(42), 199(42), 218(42), 219(42), 220(42), 221(42), 241(42), 242(42), 243(42), 244(42), 245(42), *263*
Piloty, O., 706(98), 714(98), *726*
Pinkernelle, W., 66(20), 96(20), *120*
Pinner, A., 84(112), *122,* 208(111), 218(135), *265,* 290(84), 322(1), 341(105), *312, 378, 380*
Pirath, P., 94(151), *123*
Pittman, C. U., Jr., 545(73), *570*
Pletcher, T. C., 224(145), *266*
Plieninger, H., 220(141), 259(141), *266,* 619(420), 620(420, 421), *644*
Plöger, W., 32(110), *53*
Pocchini, A., 684(61), *725,* 728(149), *730*
Pocker, Y., 545(72), *570*
Poddubnaya, N. A., 772(39a), *773*
Podstata, J., 96(160), *123,* 323(25), 377(25), *378*
Poetsch, E., 764(22), 768(22), *773*
Pogrebnyak, A. A., 268(211), 276(211), *276*
Poignant, S., 7(22, 24), 8(22, 24, 27), 31(27), *51,* 54(170), *63,* 68(33), *120,* 125(225), 128(225), *140*
Poirier, P., 279(27), 294(27), *311,* 355(148), *381*
Polanskij, N. B., 648(563), *652*
Polievktov, M. K., 201(83, 84), 220(84), 254(83), *264,* 397(8), 495(8), *521,* 584(133), 632(133), *637*
Pollock, E. J., 604(299), *641*
Polonovski, M., 769(32), *773*
Pommer, H., 44(157), *54,* 331(66), *379*
Popp, F. D., 612(362, 365, 366, 367, 368,

369, 370), 613(367, 371), 614(362, 370), 615(362, 365, 390, 391, 392, 393), 616(362, 370, 395a, 396), *643, 644*
Porcel, E., 647(532), *651*
Porkert, H., 447(83), 453(89), 470(89), 510(168, 169), 511(169), 512(89, 169), *523, 525*
Porrmann, H., 333(73), *379*
Portnov, Yu. N., 13(47), *51,* 729(165, 166), *731*
Portnova, S. A., 575(36), 576(36), 577(36), 587(36), 589(36), *635*
Postovskij, I. Ya., 647(520), *651*
Posvic, H., 729(160), *731*
Potashnikova, S. G., 618(405, 406), 619(406), 620(431, 432), 621(431, 432), 624(444), 625(444), 630(406), *644, 645,* 651(629), *654*
Potier, P., 760(15), 769(29a, 30a, 33, 34), 771(35, 38), *772, 773*
Potts, K. I., 317(135), *320*
Potts, K. T., 279(55b), 284(55b), 288(55b), *311,* 315(126), 319(126), *320*
Pougny, J. R., 271(228), *277*
Poupat, C., 769(29a), *773*
Povolotskii, M. I., 148(29), 150(34), *164*
Powers, J. C., 620(430), 621(430), 623(430), *645*
Pozharskij, A. F., 391(231), *392,* 647(535, 537), *651, 652*
Pracejus, G., 590(178), *638*
Prasad, K. B., 747(91), *756*
Prashar, R., 583(126), 584(126), *637*
Pravednikov, A. N., 575(20), *635*
Preckel, M., 406(43), 413(43), 485(43), *522*
Preininger, V., 742(54, 55), *755*
Preuss, R., 769(27c), *773*
Prey, V., 575(29), 576(29), 582(29), *635*
Price, J. C., 745(78), *755*
Priewe, H., 94(149), *123*
Prilepskaja, A. N., 577(52), 617(402, 407), 618(402), 619(52, 402, 407, 409, 414), 625(449), 626(52), 629(52, 449, 475), 630(407), *635, 644, 645, 646*
Pritzkow, H., 33(116), *53*
Proctor, G. R., 683(60), *725,* 728(147), *730*
Prota, G., 752(125, 127, 131, 132), *756, 757*

Protopovova, V. T., 339(95), *380*
Pruett, R. L., 386(212), *392*
Przheval'skii, N. M., 710(107), *726,* 729(164), *731*
Psaltyra, S. A., 609(351), *643*
Pura, J. L., 521(178), *525*
Putter, R., 537(42), 560(94), 563(106), *569, 570, 571*
Pyman, F. L., 341(106, 107), 342(107), *380*
Pyman, L., 342(110), *380*

Quast, H., 730(172, 174), *731*
Quemeneuer, M. T., 7(22), 8(22), *51,* 68(33), *120*
Quis, P., 73(63, 64), 80(63, 64), *121,* 157(46), 158(46), *165,* 174(4), *180,* 214(129), 222(129), 265, 535(38), ' 554(86), *569, 570*

Raap, R., 279(41), 301(41), *311*
Rabas, A., 689(65), *725*
Rabiller, C., 316(133), *320*
Rabinowitz, I., 31(104), *53,* 327(48), *379*
Rackham, D. M., 579(85), *636*
Radkova, S. S., 651(629), *654*
Raetz, R., 28(98), 29(98), *52*
Rahman, A., 59(198, 199), *64*
Rainer, G., 477(130, 131, 132, 133), *524*
Raison, C. G., 11(36), 22(36), 31(36), *51,* 325(36), 326(36), 327(36), 342(36), 343(36), 354(36), 375(36), *379*
Raksa, M. A., 75(82), *121*
Rambacher, P., 592(199), *639*
Rambout, A., 772(40h), *773*
Ramey, K. C., 190(40), 245(40), *263*
Ramirez, F., 649(526), *651*
Ramirez, R. S., 592(201), *639*
Ramsey, J. S., 729(161), *731*
Rao, C. G., 648(568), *652*
Rao, V. A., 300(100), *312*
Rapoport, H., 273(276), *277,* 595(225), 596(225), *639,* 746(82), *755,* 772(39b), *773*
Rapp, K. E., 386(212), *392*
Rapp, K. M., 268(212), 271(212), 275(212), 276(212), *276*
Rappe, C., 728(135), *730*
Rasmussen, J. K., 39(138), *53*
Rassat, A., 633(494, 505), *646*

Rauft, J., 357(157), *381*
Ray, J. N., 31(103), *53*
Ray, S., 56(182), *63*
Rayan, R. P., 650(624, 625), *654*
Raymond-Hamet, 747(87, 98), *755, 756*
Razmak, S. L., 289(79), *312*
Rebek, J., 647(522), *651*
Rebsdat, S., 198(71), 220(71), 248(71), 252(71), 258(71), *264,* 455(96), 457(96), 460(101), 461(101), 492(101), 493(101), 496(101, 148), 497(101), 498(101), 507(166, 167), 508(166, 167), 509(101, 167), 510(101), 518(172), 520(167), *523, 524, 525*
Reddy, K. V., 728(148), *730*
Reese, J., 707(104), *726*
Rehn, H., 8(30), 11(30), *51,* 480(135), 481(135), *524*
Reich, L., 633(490), *646*
Reicheneder, F., 32(109), *53,* 87(119), *122*
Reid, D. H., 9(32), *51*
Reid, E. E., 279(58), *312*
Reid, H. G., 538(47), *569*
Rein, B. M., 26(95), *52*
Reinhard, H., 620(423), 623(423), *644*
Reinhard, W., 27(96), 29(96), *52,* 92(143), *123*
Reinheckel, H., 279(34), 284(34), *311*
Reisch, J., 746(81, 85), *755*
Reissert, A., 578(60), 610(60, 356), 612(60, 356), 614(60, 356), *636, 643*
Reiter, F., 597(268), *640*
Reitzenstein, F., 579(69), 608(69), *636*
Relles, H. M., 588(145, 146), *638*
Rempfer, H., 6(11), 7(11), 8(11), 10(11), 11(11), 17(11), 18(11), 31(106), 38(106), 42(11), 43(106), 46(106), 47(11), 48(106), *50, 53,* 194(61, 62, 63), 196(61, 62, 63), 204(62, 63), 211(62), 238(61), 239(63), 248(61), *264, 265,* 325(37), 326(37), 333(37), 360(168), *379, 381,* 401(20), 448(20), 475(20), *522*
Renard, S. H., 279(12, 13, 14, 16, 56), 283(16), 284(12, 14), 295(13, 14), 300(99), 304(105), 306(105), *310, 311, 313*
Renault, J., 279(26), *311*
Renault, J. H., 279(18), 306(18), *311*
Rengaraju, S., 648(568), *652*

Renou, Y. P., 316(133), *320*
Renson, M., 125(227), *140*
Renstrom, L., 574(12), 647(12), *635*
Resemann, W., 474(125), 476(125), *524*
Retuert, P. J., 582(120), *637*
Reuss, R. H., 649(606), *653*
Reuterhall, A., 188(42), 190(42), 192(42), 197(42), 199(42), 218(42), 219(42), 220(42), 221(42), 241(42), 242(42), 243(42), 244(42), 245(42), *263*
Reynaud, P., 389(226), *392*
Reynaud, R., 279(30), 280(30), 282(30), 283(30), 284(30), 304(106, 107), 305(107), *311, 313*
Reynen, W. A., 648(559), *652*
Reynolds, C. A., 595(233), 597(271), 598(274), 625(271), *640, 641*
Reynolds, G. A., 50(166), *54*
Reynolds, W. F., 182(7), *262*
Ribachenko, V. I., 647(545), *652*
Ribachenko, V. T., 647(525, 528), 649(525), *651*
Ribas, I., 745(75, 76), *755*
Ribenko, L. A., 647(533), *651*
Ricci, A., 156(42), *165*
Riccieri, F. M., 125(220), *139,* 269(219), 274(219), *277*
Riche, C., 760(15, 17), *772*
Richenbacher, H. R., 137(280), *141*
Richter, R., 145(12), 149(30), 153(12), 157(30), 158(12), 162(12), *164,* 441(77), 486(77), *523,* 557(93), 558(93), 559(93), *570*
Richter, W., 197(70), 199(70), 234(70), 236(70), 237(70), *264*
Ricolleau, G., 54(170), *63,* 125(225), 128(225), *140*
Ridd, J. H., 589(155), 611(155), *638*
Ried, W., 61(208), *64,* 88(130), 90(130), *122,* 227(155), 247(155), *266,* 271(227), *277,* 279(35), 284(35), 289(35), 299(35), *311,* 314(120, 121), *320*
Riese, H., 436(67), *523*
Riesenfeld, F., 580(99), 584(99), 609(99), 632(99), *636*
Rischke, H., 691(71), 706(71), *725*
Ritchie, C. D., 649(592), *653*
Ritchie, E., 650(611), *653*
Rittner, S., 27(96), 29(96), *52,* 92(143), *123*

Riviere, M., 728(146), *730*
Robbe, Y., 279(55a), 284(55a), 286(55a), *311*
Roberts, E. M., 11(43), *51*
Roberts, T. D., 580(100), *636*
Robertson, T. H., 649(589), *653*
Robin, M. B., 182(1), *262*
Robins, R. K., 13(49), 32(49), *51*
Robinson, B., 706(97), 710(97), 723(97), 729(169), *731*
Robinson, G. M., 706(99, 100), 714(99), *726*
Robinson, M. A., 159(49), *165*
Robinson, R., 706(99, 100), 714(99), *726*, 744(58, 59), *755*, 760(2b, 2c), *772*
Rodgers, A. S., 657(7), *724*
Roedig, A., 45(158), *54*, 66(21), *120*
Rogalski, W., 225(150), *266*
Roger, R., 212(121), 218(121), 226(121), 228(121), 230(121), 232(121), 240(121), *265*, 329(51), *379*
Rogers, G. A., 594(223), *639*, 647(556), 648(556), *652*
Rogers, M. A., 279(17), 284(17), 295(17), 304(17), 305(17), *311*
Rogers, R. B., 595(228), *639*
Rogers, W. J., 61(209), *64*
Rogne, O., 604(289, 290, 296), 605(289, 290, 296), *641*, 649(595), *653*
Roh, N., 66(10), *120*
Rohloff, C., 171(65), *172*, 386(208, 209), *391*
Roma, G., 58(194, 195), *63*
Ronwin, E., 565(111), *571*
Roos, R., 323(26, 27, 28), 324(31), 361(26, 27), 362(26, 27, 28), 376(31), *378*
Rosenberg, A. N., 589(150, 151), *638*
Rosenberg, A. Ya., 592(205), *639*
Rosenthal, T. C., 342(112), *380*
Rosler, H., 752(128), *757*
Ross, A., 649(593), *653*
Ross, S. D., 204(98), *265*
Ross, W. J., 268(210), 274(210), *276*
Roth, A., 632(483), *646*
Roth, H., 132(255), *140*
Roth, M., 279(42b), 299(42b), 300(42b), *311*
Rothman, E. S., 182(11), *262*
Roychaudhuri, D. K., 748(102), *756*

Rubinstein, M., 648(576), *653*
Rüchardt, Ch., 648(564), *652*
Ruckman, M., 594(220), 595(220), *639*
Rudakov, E. S., 606(300), *641*
Rudolph, W., 145(14), *164*
Rudy, D. D., 88(126), *122*
Ruess, K. P., 353(141, 142, 143, 144), 354(141, 142, 143, 144), *381*
Ruggli, P., 386(211), *392*
Ruhnau, R., 72(51), *121*, 539(52), 541(52), 561(52), 564(52), *569*
Rule, H. G., 592(197), *639*
Rumon, K. A., 578(66), 583(66), 584(66, 132), 585(66), 586(66), 599(132, 277), 600(132), 609(353, 354), *636, 637, 641, 643*
Rupe, H., 612(364), 613(375), *643*
Ruppelt, E., 617(398), *644*
Ruppenthal, N., 694(73), *726*
Rusche, J., 66(22), *120*
Ruske, W., 528(6), 563(6), *568*
Rutz, H. J., 701(87), *726*
Rütz, W., 461(108), 462(108), 464(108), *524*
Ryan, J. J., 66(23), 68(23), 78(23), 79(23), 80(23), 81(23), 97(23), 114(23), 118(23, 216), *120, 124, 378*(33)
Ryan, R. P., 55(177), *63*
Rybin, L. V., 268(211), 276(211), *276*
Rybinskaya, M. I., 268(211), 276(211), *276*
Ryskiewicz, E. E., 6(14), *50*

Saa, J., 745(76), *755*
Sadovskaja, T. N., 606(309), *641*
Saegusa, T., 383(197), *391*
Safir, S., 621(437), 623(437), *645*
Saggau, R., 699(85), *726*
Sagitullin, R. S., 648(566), *652*
Saha, J. G., 574(3), *634*
Sahn, D. J., 224(145, 146), *266*
Sainsbury, M., 633(508), *647*, 694(72, 78), 696(72, 80), 698(72), 700(72, 80), *725, 726*, 744(61), *755*, 769(31), *773*
Sajapina, S. V., 647(516), *651*
Sakamoto, T., 650(613, 614, 616, 617a), *653, 654*(615)
Sakhs, E. R., 648(572), *652*
Sakurai, H., 62(212), *64*
Sal'nikova, T. N., 268(211), 276(211), *276*

Salov, B. V., 270(223), *277*
Salvi, A., 579(72), *636*
Sammy, G. M., 604(295), *641*
Samoilenko, G. V., 626(461), 628(471, 472), 650(618, 619, 622), *645, 646, 654*
Samoilenko, T. V., 619(408), *644*
Sandhu, S. S., 575(31, 35), 576(31), 577(31), 578(35), 582(31, 35), 583(31), *635*
Sandler, S. R., 226(151), 228(151), 241(179), *266, 267,* 322(7), 329(7), 333(7), 338(7), 364(7), *378*
Sandner, W. J., 539(53), *569*
Sandri, J. M., 75(86), *121*
Sandstrom, J., 357(154), *381*
Sannicolo, F., 729(162), *731*
Santavy, F., 742(52, 54, 55), *755*
Santini, S., 6(18), 7(18), *51*
Santos, A. C., 737(19), 738(20), *754*
Santos, G. A., 738(20), *754*
Sardessai, M. S., 23(83), *52*
Sarfert, O., 322(2), *378*
Saroff, H. A., 201(90), 241(90), *264*
Sasaki, A., 227(156), *266*
Sasse, H. J., 182(13), 188(13), 190(13), 192(13), 215(13), 217(13), 220(13), 248(13), 252(13), 254(13), *263*
Sasse, W., 596(259), *640*
Sassu, G., 752(121), *756*
Sat, R., 387(218), *392*
Satchell, D. P. N., 317(134), *320*
Sauer, J. C., 574(14), 588(14), *635*
Sauer, T. D., 727(133), *730*
Sauers, C. K., 63(215), *64*
Sauliova, J., 16(52), *51*
Saunders, J. C., 58(193), *63*
Saunders, M., 110(201), *124*
Saur, H., 406(56), 410(56), 412(56), 481(56), 482(56), 485(56), *522*
Sauter, R., 474(123, 124), 476(123, 124), *524*
Savchenko, A. S., 606(305, 308, 311, 312, 314, 315), *641, 642*
Savelli, G., 6(18), 7(18, 19), *51*
Savelova, V. A., 606(306, 309, 310, 316, 317, 318), *641, 642,* 649(597, 598, 599, 600, 601), 650(601), *653*
Sav'jalov, S. I., 596(260), 626(260), *640*
Sawada, S., 503(159), 504(159), 505(159), *525*

Sawhney, R. N., 587(142), *637*
Saxton, J. E., Ed., 769(27b), *773*
Sayigh, A. A., 105(183), *124*
Sayigh, A. A. R., 88(137), 90(137), 102(177), *123,* 144(8, 9, 10, 11), 145(8, 9, 10, 11), 146(8, 9, 10, 11, 17, 18, 19), 147(11, 18, 19, 21, 23), 148(9, 11, 18, 21, 23), 151(9), 152(9), 154(18), 157(9), 162(9, 11), 163(17), *164,* 329(54), *379*
Schafer, F. C., 72(48), 82(48), *121,* 333(72), *379,* 528(3), 529(3), *568*
Schafer, H., 319(139), *320*
Schaumann, E., 216(134), *265,* 384(202), *391*
Scheeren, J. W., 110(202), *124,* 383(199), *391,* 400(16), 445(16), 447(16), 485(16), *521*
Scheffold, R., 247(182), *267*
Scheider, W. G., 567(116), *571*
Scheiner, P., 675(45), *725*
Scheinmann, F., 713(111), *727*
Schenck, H. U., 232(170), *266,* 369(181), 370(181), *382,* 445(82), 446(82), 447(82), *523*
Schenk, G. H., 587(141), 592(141, 191), 595(191), *637, 639*
Schenker, K., 621(438), *645*
Scherowsky, G., 341(104), *380*
Schick, L. E., 729(157), *731*
Schied, D., 490(141a), *525*
Schiess, P., 657(11), 710(11, 106), 711(11, 106), 729(158), *724, 726, 731*
Schiess, P. W., 689(68), 702(68, 88), *725, 726*
Schiff, H., 145(13), *164*
Schilling, C. L., Jr., 226(154), *266*
Schindbauer, H., 40(141), *53,* 331(61), *379*
Schindler, N., 32(110, 111), *53*
Schlack, P., 197(70), 199(70), 234(70), 236(70), 237(70), *264*
Schlittler, E., 747(87), *755*
Schmid, H., 658(16, 18, 19, 20), 659(21), 668(37), 669(38, 39), 682(53), 683(53), 685(53), 686(53), 702(39), 705(19), 713(111), 720(39), *724, 725, 727,* 728(144, 145), *730,* 740(35), 747(89), 748(89), *754, 756*
Schmid, M., 66(19), 78(19), 88(19), *120 121), 120, 122,* 659(21), 682(53),

683(53), 685(53), 686(53), 689(21), *724, 725*
Schmidt, A., 73(61, 62, 65), 74(61, 75), 80(61, 62, 65), 83(75, 109), 85(75), 99(75), *121, 122,* 166(58), *171,* 182(10), 214(127, 128), *262, 265,* 360(166), *381,* 391(230), *392,* 532(28), 533(28), 538(48), 541(62), *569*
Schmidt, E., 227(155), 247(155), *266,* 271(227), *277,* 279(35), 284(35), 289(35), 299(35), *311,* 314(121), *320*
Schmidt, E. A., 520(175, 176, 177), *525*
Schmidt, F., 191(55), 246(55), *264*
Schmidt, G. M. J., 718(126). *727*
Schmidt, R. R., 34(119, 120), *53,* 279(33), 283(33), 284(33), 287(33), 299(33), 302(33, 104), *311, 313,* 557(92), 559(98), *570*
Schmiedel, R., 279(52), 284(52), 294(52), 296(52), 297(52), *311*
Schmir, G. L., 224(145, 146), *266,* 272(231), *277,* 295(91), 306(91), *312*
Schmir, G. O., 279(43), 295(43), *311*
Schmitz, E., 41(144), 46(144), *54,* 109(199), *124,* 211(116), *265,* 714(116), *727*
Schnabel, W. J., 159(49), *165*
Schneider, J., 227(156), *266*
Schneider, K., 627(467), *646*
Schneider, R. S., 503(157), 504(157), *525*
Schnekenburger, J., 626(458, 459), *645*
Schnitzspahn, K., 74(76), *121*
Schoeller, W. W., 720(129), *727*
Schöllkopf, U., 57(187, 189), *63,* 130(252), *140,* 187(35), *263,* 336(82), *379*
Scholz, D., 165(53), 170(53), *171*
Schön, N., 107(194), 108(194), *124,* 182(14), 188(14), 190(14), 191(14), 192(14), 193(14), 215(14), 220(14), 248(14), 249(14), 250(14), 252(14), 254(14), 255(14), 256(14), 257(14), 258(14), *263,* 396(4), 406(4), 409(4), 410(4), 411(4), 423(4, 65), 424(4, 65), 425(4), 427(4), 428(4), 429(4), 430(4, 65), 431(4), 440(4), 441(4), 442(4), 443(4), 445(4), 453(4), 455(4), 460(4), 461(4), 469(4), 473(65), 487(4), 503(4), 519(4), *521, 523*
Schopf, C., 747(88), 753(137), *755, 757,* 760(3a), *772*

Schosser, H. P., 477(131), *524*
Schrader, G., 593(215), *639*
Schran, H., 593(212), *639*
Schreiber, J., 252(200), *267,* 398(10), 399(10), 406(10), 410(10), 455(10), 465(10, 109), 466(10), 467(10, 109), 469(10, 109), *521, 524*
Schriver, L., 536(41), *569*
Schroder, P., 745(77), *755*
Schrodt, H., 182(13), 188(13), 190(13), 192(13), 215(13), 217(13), 220(13), 248(13), 252(13), 254(13), *263*
Schuh, H. G. v., 1(1), *4,* 203(96), *265,* 401(22), 408(22), 473(22), 474(22), 475(22), 476(22), 496(22), *522*
Schuijl, P. J. W., 279(38, 39), 301(38, 39), 302(39), *311*
Schultze, K. W., 134(260), *140*
Schulze, W., 54(171), *63*
Schumann, D., 764(22), 768(22), *773*
Schunack, W., 233(171a), *266*
Schuster, R. E., 39(134), *53*
Schwartz, G. N., 580(95), *636*
Schwarz, R., 760(3b), *772*
Schwarzenbach, G., 608(339), *642*
Schweitzer, R., 61(208), *64*
Schweizer, D., 183(15), 186(15), 188(15), 189(15), 193(15), 220(15), 248(15), 249(15), 252(15), 253(15), 256(15), *263,* 406(29), 409(29), 423(29), 442(29), 508(164), 519(29), *522, 525*
Schweizer, E. H., 477(129), *524*
Scott, A. I., 760(10), 769(10, 28), *772, 773*
Scott, M. D., 35(122), 36(122), *53,* 97(162), *123*
Scott, P. L., 42(146), *54*
Seckinger, K., 232(168), 234(168), *266,* 323(24), 364(24), 365(24), *378*
Seconi, G., 156(42), *165*
Sedavkina, V. A., 319(141), *320*
Sedmera, P., 742(55), *755*
Seefelder, M., 18(54, 55), 22(54, 76, 77), 31(77), *51, 52,* 66(7, 14), 68(7), 76(7, 97), 77(97), 78(7, 97), 79(7), 80(7, 14, 97), 81(7, 14, 97), 82(7, 14), 84(7, 14), 85(7, 14), 87(7), 89(7), 91(7), 94(7, 14), 95(14), 96(7), 98(97), 99(7, 14), 100(7), 103(7, 180), 105(7), 107(7, 14, 193), 108(7, 14, 193), 109(7, 14), 115(7, 97), 116(7, 97), 117(97), *120, 122, 124,*

144(1, 2), 145(1, 2), 146(1, 2), 151(1, 2), 152(1, 2), 153(1, 2), 154(1, 2), 155(1, 2, 39), 156(2), 157(1, 2), 158(2), 162(1, 2), *164, 165,* 213(124), 214(124, 125), 224(125), 230(125), 240(125), *265,* 290(83), 295(83), 297(83), *312,* 323(23), 324(23, 32), 326(23, 47), 328(47), 339(99), 340(99), 341(103), 359(23), 360(99), 361(23), 363(23), 364(99), 376(99), *378, 379, 380,* 386(210), *391,* 436(68), 437(68), 460(100), 465(110), 473(121), 485(68), *523, 524*
Seeliger, W., 207(107), 212(107), *265*
Seide, S., 531(26), 540(26), 542(26), 544(26), *569*
Seidel, F., 596(255), *640*
Seidel, M. C., 406(44, 45), 410(44, 45), 448(44, 45), 478(45), *522*
Seitz, G., 66(17, 18), 80(18), 81(17), 82(17, 18), 94(17), 113(17, 18), *120,* 186(34), 233(34), *263,* 279(52), 284(52), 294(52), 296(52), 297(52), *311*
Seitz, G. H. R., 313(113, 114), *320*
Seitz, H., 43(147), *54,* 279(28), 281(28), 284(28), 295(28), 296(28), 299(28), 304(28), *311*
Seliger, H., 590(167), *638*
Semenova, R. G., 647(528, 545), *651, 652*
Semicheva, G. S., 85(114), *122*
Sen, A. K., 56(182), *63*
Sen, M., 31(103), *53*
Senatore, L., 604(291), *641*
Sendega, R. V., 134(259), *140*
Sendi, E. B., 729(159), *731*
Senga, Keitaro, 13(48), 32(48), *51,* 56(179), *63*
Seo, M., 747(99), *756*
Serafinowa, B., 633(489), *646*
Serdjuk, A. I., 647(532), *651*
Seredrnitskaya, M. N., 134(259), *140*
Serova, K. L., 595(235), *640*
Seshadri, S., 23(83), *52,* 54(172), 55(172), *63*
Settlage, P. H., 580(103), 591(103), *637*
Seuffert, R., 592(203), *639*
Sevenet, T., 769(29a), *773*
Severin, R., 772(40a), *773*
Severin, Th., 406(35), 408(35), 410(35), *522*
Seyferth, D., 75(94), 119(94), *122*

Shahidi, I., 608(338), *642*
Shakhidoyatov, Kh. M., 55(176), *63*
Shakhrovich, A. I., 270(223), *277*
Shambhu, M. B., 44(152), 46(152), *54,* 210(114), *265,* 590(168), *638*
Shamma, M., 737(15), 740(30, 40), 741(43), 748(103), *754, 756*
Shand, E. W., 542(68), *570*
Shannon, J. S., 735(6), *754*
Shannon, P. V. R., 736(14), 737(14), *754*
Shapiro, R., 758(146), *758*
Sharma, G. S., 757(141), *758*
Sharma, K. S., 613(380), *643*
Shatskaja, V. A., 606(306, 309, 310), *641,* 649(600, 601), 650(601), *653*
Shavel, J., Jr., 713(114), *727*
Shaw, E., 574(4), *634*
Shaw, R., 657(7), *724*
Shechter, H., 580(100), *636*
Sheehan, J. C., 62(214), *64,* 190(44), 220(44), 221(44), 244(44), *263*
Sheeto, J., 351(128), *380*
Sheinker, Y. N., 190(48), 191(48), 245(48), *263*
Sheinker, Yu. N., 60(201), *64,* 575(36), 576(36), 577(36), 587(36), 589(36), *635*
Sheinker, Y. P., 709(105), 710(105), *726*
Sheinkman, A. K., 574(1, 13), 575(36), 576(36), 577(52, 53), 578(36, 53), 584(36, 53, 133), 585(36, 53), 586(36), 587(139), 589(1, 53, 149, 150, 151, 152, 156), 603(1), 607(1), 616(1, 13), 617(1, 13, 53, 139, 402, 403, 404, 407), 618(139, 402, 403, 404, 405, 406), 619(52, 53, 139, 402, 403, 404, 406, 407, 408, 409, 411, 412, 413, 414), 620(426, 427, 431, 432, 433, 434, 435), 621(426, 431, 432, 433, 434), 623(435), 624(444), 625(444, 446, 447, 448, 449, 450, 451, 452), 626(52, 53, 461), 628(13, 471, 472), 629(1, 13, 52, 53, 139, 449, 475), 630(13, 139, 406, 407), 631(13, 403, 435), 632(133), 633(492), 634(513, 514, 515), 647(529, 532, 533, 534, 535, 537, 550), 650(618, 619, 620, 621, 622), 651(529, 629), *634, 635, 636, 637, 644, 645, 646, 647, 651, 652, 654*
Sheitshenko, V. I., 589(151), *638*
Sheldon, J. C., 581(109), *637*
Shenoy, S., 382(196), *391*

AUTHOR INDEX

Sheppard, G., 677(47), 678(46, 47), 679(47, 48), 680(48), *725*
Sheradsky, T., 713(112), *727*
Shermolovich, Ju. G., 133(256), *140,* 148(29), *164*
Shevchenko, M., 150(34a), 163(34a), *164*
Shevchenko, M. V., 165(57), *171*
Shevchenko, V. I., 133(256), *140,* 148(29), *164*
Shevedov, V. I., 126(232), *140*
Shibuya, I., 559(99), *570*
Shie, M. D., 648(578), *653*
Shigorin, D. N., 355(147), *381*
Shih, H., 75(94), 119(94), *122*
Shijan, Zh. B., 617(402), 618(402), 619(402), *644*
Shima, S., 132(254), *140*
Shimadzu, H., 22(79, 80), 31(80), 39(79), *52*
Shimidzu, T., 580(91, 92), *636*
Shimizu, Y., 279(48), 281(48), 308(48), *311*
Shine, H. J., 706(97), 710(97), 723(97), *726*
Shiner, V. J., Jr., 604(297, 298), *641*
Shinkai, S., 647(546), *652*
Shirai, H., 613(372), *643*
Shiraishi, T., 650(613, 614, 616, 617a), *653, 654*(615)
Shishkin, V. E., 212(119), *265*
Shkaev, V. S., 268(207), *276*
Shoemaker, H. F., 769(26b), *773*
Sholtissek, C., 590(176), 592(176), 595(176), *638*
Short, W. F., 333(70, 71), *379*
Shpanko, I. V., 606(304), *641*
Shreeve, J. M., 125(219), 137(219), *139*
Shriner, R. L., 96(159), *123,* 279(9), 280(9), 290(9), *310,* 322(5), 329(5), 333(5), 344(5), 354(5), 357(5), 364(5), 375(5), *378*
Shropshire, E. Y., 338(92), 342(92), 343(92), 354(92), 357(92), 359(92), 365(92), 368(92), 369(92), 370(92), 371(92), 372(92), 375(92), 377(92), *380,* 470(113), 487(113), 495(113), *524*
Shtepanek, A. S., 165(57), 168(57), *171*
Shvedov, V. I., 55(175), *63*
Sibiryakova, D. V., 712(110), *727*

Siebenthall, F., 289(79), *312*
Siedel, P., 135(273), *141*
Sieveking, S., 216(134), *265,* 384(202), *391*
Siggia, S., 592(190, 192, 193), *639*
Sih, J. C., 696(81), *726*
Sikkema, D. J., 126(240), 139(240), *140*
Silver, H. B., 658(15), 689(15), *724*
Silver, J. L., 556(88), *570*
Silverstein, R. N., 6(14), *50*
Sim, G. A., 728(142), *730*
Simaga, K., 595(226), *639*
Simalty, M., 50(167), *54*
Simanek, V., 742(54, 55), *755*
Simanenko, Yu. S., 647(525, 528, 539), 649(525), *651, 652*
Simchen, G., 43(149), *54,* 82(105), 101(105), 107(195), 108(195), *122, 124,* 174(5), 179(5), *180,* 194(64, 65), 198(65), 199(65), 201(86), 210(65), 219(65), 220(64, 65, 71), 221(65), 222(65), 224(65, 144), 228(157), 232(144, 165, 170, 171), 234(157), 235(165, 171), 241(86), 242(86), 243(86), 244(86), 245(86), 248(64, 71), 250(191), 252(64, 71), 258(71), 259(86), *264, 266, 267,* 271(226), *277,* 279(51), 281(51), 284(51), 299(51), *311,* 339(96, 97, 98), 364(171), 365(171), 369(171, 178, 179, 180, 181, 182), 370(179, 180, 181, 182), 371(187), 376(98), *380, 381, 382,* 396(5), 399(13), 402(5, 5a, 24), 405(5, 26), 406(23, 24, 26, 32, 33, 34, 39, 52, 53, 57), 407(23, 24, 26, 32, 33, 34, 39, 52), 408(26, 53, 55), 410(33), 411(26, 34), 412(57), 413(32, 57), 414(32), 415(5, 23, 32), 416(32, 39), 417(26), 418(23, 57), 419(23), 420(23), 421(24), 422(55), 435(52), 436(52), 437(52), 440(32, 69), 441(69), 442(32, 69), 443(32, 55), 444(32), 445(5, 82), 446(69, 82, 85), 447(82), 448(69), 449(69), 450(23), 453(89), 455(94, 96), 457(5, 96, 98), 458(69), 459(69), 460(101), 461(101), 470(89), 471(24), 472(24), 477(55), 478(26), 480(57, 69), 481(57), 482(57), 483(33), 485(33, 53), 487(139), 491(85), 492(101), 493(101), 494(85, 145), 495(85), 496(101, 147), 497(101),

498(101, 145), 499(69), 507(147, 165, 166, 167), 508(145, 147, 165, 166, 167), 509(101, 145, 147, 167), 510(101, 147, 168, 169), 511(169), 512(89, 169), 513(145), 514(170), 515(170), 516(26, 170), 517(5a, 170, 171), 518(172, 173), 519(98, 173), 520(145, 167), *521, 522, 523, 524, 525,* 528(8), 565(113), *568, 571*
Simpson, M., 592(206), *639*
Sinay, P., 271(228), *277*
Singh, D., 575(31), 576(31), 577(31), 582(31), 583(31), *635*
Singh, G., 575(56), 478(56), 582(56, 121), 584(56), *636, 637*
Singh, H., 319(145), *320*
Singh, J., 583(126, 127), 584(126, 127), *637*
Singh, R. K., 663(31), *724*
Singh, S., 319(145), *320*
Singh, U. P., 279(55b), 284(55b), 288(55b), *311,* 317(135), *320*
Sinitsa, A. D., 75(92), 83(92), *122*
Sinnott, M. L., 205(100), *265*
Sipos, F., 628(470), *646*
Sircar, J. C., 528(4), 531(4), 552(4), *568*
Siskin, V. E., 287(68), 295(68), 297(68), *312*
Skinner, B., 738(22), *754*
Skoldinov, A. P., 339(95), *380*
Skorna, G., 135(274), *141*
Skovronek, H., 49(165), *54*
Skripka, A. V., 649(598), *653*
Skripnik, Yu. G., 649(596), 650(618), *653, 654*
Slates, H. L., 625(453), *645*
Slavik, J., 737(13), 744(70, 71, 72), *754, 755,* 757(138), *758*
Slavikowa, L., 737(13), 744(70, 71), *754, 755,* 757(138), *758*
Slavinskoya, R. A., 558(97), *570*
Slopianka, M., 729(171), *731*
Smalley, R. K., 593(207, 213), *639*
Smarzewska, K., 279(14, 16), 283(16), 284(14), 295(14), *310*
Smerz, O., 474(122), 475(122), *524*
Smethrust, B., 606(321), 607(321), *642*
Smets, F., 126(242), *140*
Smirnova, V. G., 69(35), 115(35), *120,* 213(123), 245(123), *265*

Smirt, J., 440(74, 75), 441(74, 75), 456(74), 75), 461(74, 75), 499(74), *523*
Smith, A. B., 83(107), *122,* 137(281), *141*
Smith, C. V. Z., 13(49), 32(49), *51*
Smith, D. M., 591(188), *639*
Smith, F., 137(282), *141*
Smith, G. F., 6(13), *50*
Smith, J.A.S., 54(169), *63,* 128(248), *140*
Smith, L. R., 45(159), *54,* 69(34), 71(34), 72(34, 39, 40, 41, 42), 74(34, 78), 115(34), *120, 121*
Smith, N. G., 649(606), *653*
Smith, P. A. S., 38(128), *53,* 342(111), 355(111), *380*
Smith, R., 80(101), *122*
Smith, R. F., 342(112, 113), *380*
Smith, T. D., 8(25), 30(25), 43(25), *51,* 68(32), *120,* 222(142), *266*
Smith, Th. W., 647(518), 649(518), *651*
Smith, V. F., Jr., 272(231), *277*
Smith, W. C., 68(27, 28), *120*
Smith, W. T., Jr., 728(150), *731*
Soderberg, E., 750(110), *756*
Soenderop, H., 597(268), *640*
Sokolov, L. B., 595(238, 239), *640*
Sokolov, V. I., 647(537), *652*
Sokolova, D. F., 595(239), *640*
Soldan, F., 96(161), 111(161), *123,* 323(22), 358(22), 361(22), 365(22), *378*
Solly, R. K., 665(32, 33), *724*
Solomko, K. A., 633(492), *646*
Solomojchenko, T. N., 606(316, 317, 318), *642,* 649(599), *653*
Solomon, W., 633(497), *646*
Solson, J. G., 44(153), *54*
Soman, R., 503(159), 504(159), 505(159), *525*
Son, P., 350(121), *380*
Songstad, J., 554(85), *570,* 632(481), *646*
Soos, R., 97(163, 164, 165), 103(181), *123, 124,* 149(31), 153(36), 157(44), 158(31, 36, 44), 159(48), *164,* 361(169), 362(169), *381*
Sorba, J., 647(554), *652*
Sorm, F., 66(13), 93(13), *120,* 440(75), 441(75), 456(75), 461(75), *523*
Sorokin, M. N., 606(313), *642*
Sorokina, N. P., 709(105), 710(105), *726*
Sotiropoulos, J., 315(123), *320*

Soto, A., 612(369), 615(369), *643*
Soucy, M., 649(607), *653*
Southwick, P. L., 185(25), 241(25), *263*
Spach, G., 648(583), *653*
Spath, E., 613(376), *643*
Speckamp, W., 769(26a, 26c), *773*
Speckamp, W. N., 769(26b, 26c, 26d, 26e), *773*
Spedding, H., 35(122), 36(122), *53*, 97(162), *123*
Speh, P., 80(104), 102(104), *122*, 220(140), 252(140), 257(140), 259(140), *266*, 368(177), 369(177), 370(177), *381*, 458(99), 459(99), *524*
Spenser, I. D., 740(42), *754*, 760(18), *772*
Speziale, A. J., 45(159), *54*, 69(34), 71(34), 72(34, 39, 40, 41, 42), 74(34, 78, 79), 94(78), 115(34, 79), *120, 121*, 386(214), *392*, 563(105), *571*
Spille, J., 113(209), *124*, 182(13), 188(13), 190(13), 192(13), 215(13), 217(13), 218(137), 220(13), 248(13), 252(13), 254(13), *263, 265*, 551(91), *570*
Spiteller, G., 747(90), *756*
Spitzner, E. B., 406(42), 417(42), 479(42), *522*
Spraque, J., 596(256), *640*
Sprague, J. M., 284(64), 285(64), *312*
Sprenger, H. E., 186(33), *263*
Srinivasan, M., 614(382), *643*
Staab, H. A., 250(192), *267*, 574(10), *634*, 716(122, 123), 718(122, 123), *727*
Stacey, B. E., 648(575), *652*
Stachel, H. D., 148(27), 159(27), *164*
Stadelbauer, K., 289(73), *312*
Stang, P. J., 648(581), *653*
Stangeland, L. J., 604(291), *641*
Stankevičius, A. P., 648(569), *652*
Stansfield, F., 279(21), 281(21), 282(21), 283(21), 284(21), 294(21), 297(21), 301(21), 304(21), 306(21), 310(112), *311, 313*
Starodubzeva, M. P., 649(596), *653*
Staunton, J., 739(27), *754*
Staunton, R. S., 744(59), *755*
Stavrova, S. D., 575(19), *635*
Stavrovskaya, A. V., 339(95), *380*
Stawitz, J., 730(172, 174), *731*

Steen, K., 503(156), 504(156), 505(156), *525*
Steglich, W., 590(162, 163, 165), 596(162, 261), 625(162, 261, 454, 455, 456), 626(162, 261), *638, 640, 645*, 648(570), *652*
Steiger, N., 40(142), *53*
Stein, R., 597(266), *640*
Steindorff, A., 145(15), *164*
Steinkopf, W., 295(94), *312*
Stelander, B., 126(247), 138(247), *140*, 165(55), 168(55), *171*
Stepanov, B. I., 33(112, 113, 114, 115), *53*, 355(151), 357(151), *381*, 619(410), *644*
Stephens, C. R., 39(136), *53*, 596(249), *640*
Stevens, R. V., 658(14), *724*
Stevens, W., 589(160), *638*
Stewart, G. W., 614(383), *643*
Stieglitz, J., 148(26), *164*
Stilz, W., 136(277), *141*
Stochlin, E., 88(121), *122*
Stoeckigt, J., 760(16), *772*
Stojanovic, F. W., 231(160), *266*, 406(48), 407(48), 421(48), 443(48), 452(48), 457(48), 471(48), *522*
Stopp, G., 107(194), 108(194), *124*, 182(14), 188(14), 190(14), 191(14), 192(14), 193(14), 215(14), 220(14), 248(14), 249(14), 250(14), 252(14), 254(14), 255(14), 256(14), 257(14), 258(14), *263*, 396(4), 406(4), 409(4), 410(4), 411(4), 423(4), 424(4, 66), 425(4, 66), 427(4, 66), 428(4, 66), 429(4, 66), 430(4), 431(4, 66), 440(4), 441(4), 422(4), 443(4), 445(4), 453(4), 455(4), 460(4), 461(4), 469(4), 485(66), 487(4), 503(4), 519(4), *521, 523*
Stork, G., 669(40), *725*
Stoyanovich, F. M., 22(74), *52*, 57(190), *63*, 471(118, 119), *524*
Strauss, M. J., 383(200, 201), *391*
Streeting, I. T., 125(228), *140*
Streitwieser, A., Jr., 703(95), *726*
Strizhakov, O. D., 541(59), 542(59), 562(104), *569, 571*
Struchkov, Yu T., 268(211), 276(211), *276*
Strzelecka, H., 50(167), *54*
Stud, M., 319(140), *320*

Stupnikova, T. N., 620(434, 435), 621(434), 623(435), 625(452), 631(435), *645, 647(532, 533, 534, 535, 537, 550), 651, 652*
Sturn, H. J., 531(23), *569*
Stutz, R. L., 597(271), 625(271), *641*
Suama, M., 597(264), 598(264, 273), *640*
Sucrow, W., 504(162, 163), *525,* 714(115), 715(118), 716(115, 118, 119, 121), *727,* 729(171), *731*
Sueiras, J., 745(75), *755*
Sugihara, T., 647(547), *652*
Sugiyama, Y., 353(140), *381*
Suhr, H., 661(26), 662(26), 663(26), 664(34), 667(34), 668(34), 673(26), 721(26), *724, 725*
Sukhorukova, S. A., 606(320), *642*
Sukhoruskin, A. G., 409(64), *523*
Sumarokova, T. N., 558(97), *570*
Suminov, S. I., 547(1), 589(1), 603(1), 607(1), 616(1), 617(1), 629(1), *634*
Summers, G. H., 74(72), *121*
Suranyi, L., 85(113), *122*
Suschitzky, H., 593(207), *639*
Sutter, B., 658(20), *724*
Suvorov, N. N., 709(105), 710(105), *726*
Suydam, F. H., 44(151), *54,* 208(110), 209(110), 224(110), *265*
Suzuki, T., 279(48), 281(48), 308(48), *311,* 745(69), *755*
Suzuki, Y., 540(56), 544(71), 550(71), *569, 570*
Suzuko, J., 135(272), *141*
Svanholm, U., 658(17), *724*
Sviridova, L. A., 729(165, 166), *731*
Svoboda, M., 137(279), *141,* 170(64), *172*
Swain, M. R., 649(604, 605), *653*
Swallow, A. J., 631(477), *646*
Swan, G. A., 747(91), *756*
Sweeny, J. G., 504(160), *525*
Sweeting, O. J., 28(98), 29(98), *52*
Szafran, M., 613(377), *643*
Szendrei, K., 746(81, 85), *755*
Szilagyi, P. J., 578(61), 586(61), 587(61), *636*
Szmuszkovica, J., 596(250), *640,* 669(40), *725*

Tafel, J., 279(8), 284(8), 299(8), *310*

Tagiguchi, K., 744(68), *755*
Taguchi, H., 11(42), *51,* 740(31), 742(47), *754*
Taguchi, T., 623(441), *645,* 650(626, 627), *654*
Taillefer, R., 224(148), 225(148), *266,* 272(235), *277*
Taillefer, R. J., 272(233), *277*
Takada, A., 187(36, 37), 190(36), 237(36), *263,* 269(216), *277,* 315(128), *320*
Takahashi, Ch., 387(218), *392*
Takahashi, K., 212(120), *265*
Takamatsu, N., 687(64), *725*
Takami, F., 43(150), *54,* 210(112), *265*
Takanobu, K., 212(120), *265*
Takao, N., 742(48, 49, 56), 745(56, 73, 74), *754, 755*
Takao, S., 742(49), 745(74), *754, 755*
Takayama, K., 623(441), *645,* 650(626, 627), *654*
Takayanagi, Y., 728(151), *731*
Takemura, S., 206(105), *265*
Takeuchi, J., 614(388), *643*
Takikawa, Y., 387(218), *392*
Takizawa, S., 387(218), *392*
Takizawa, T., 168(62), *172*
Talaty, E. R., 88(127), 89(127), *122*
Talti, S. M., 729(167), *731*
Tamás, J., 149(31), 157(44), 158(31, 44), *164*
Tamura, H., 568(122), *571*
Tanabe, M., 648(560), *652*
Tanaka, H., 647(524), 648(524), *651*
Tani, C., 742(48, 49, 56), 745(56), *754, 755*
Tanli, R., 552(81), *570*
Tao, E. V. P., 86(117), *122*
Tarasov, A. I., 590(175), 607(175, 328, 329), *638, 642*
Tarnow, H., 332(67), *379*
Tarrago, G., 714(117), 716(117, 120), *727,* 729(170), *731*
Tataruch, F., 126(243), 128(243), 136(243), *140,* 165(53), 170(53), *171*
Taub, D., 387(217), *392*
Taurin, A., 552(81), *570*
Taurins, A., 355(149), *381*
Taylor, E. C., 11(41), *51,* 325(45), 337(87, 88), *379, 380*
Taylor, W. C., 650(611), *653*
Taylor, W. I., 747(98), *756*

Tcherkasova, E. M., 574(17), *635*
Teichmann, H., 61(206), *64*
Telinski, T., 729(160), *731*
Ten Brink, R. E., 269(215), *277*
Tenenbaum, L. E., 633(490), *646*
Tengi, J. P., 406(47), 448(47), *522*
Teramura, K., 126(236), 139(236), *140*
Terawaki, Y., 126(237), 129(237), *140*
Terehova, S. E., 649(590), *653*
Terol, A., 279(55a), 284(55a), 286(55a), *311*
Terrell, R., 669(40), *725*
Terss, R. H., 617(397), 618(397), 624(397, 445), *644, 645*
Testa, E., 88(135, 136), 90(135, 136), *123*
Teterin, Y. A., 355(151), 357(151), *381*
Thai, A., 61(206), *64*
Thal, C., 769(30a), *773*
Theilig, G., 1(1), *4,* 203(93, 96), *264, 265,* 401(22), 408(22), 473(22), 474(22), 475(22), 476(22), 496(22), *522*
Theiling, L. F., 647(527), *651*
Thelen, P. J., 633(499), *646*
Thier, W., 207(107), 212(107), *265*
Thies, R. W., 729(157), *731*
Thimm, K., 314(119), 316(119), *320*
Thoma, U., 354(145), *381*
Thomas, D., 742(51), *755*
Thomas, K., 627(463), 633(463), *645*
Thomas, R., 760(6), *772*
Thompson, Q. E., 573(30), 576(30), 592(30), 593(30), 596(246), *635, 640*
Thompson, W. R., 250(196), *267*
Thornton, E. R., 647(541), *652*
Thu, N. H., 279(30), 280(30), 282(30), 283(30), 284(30), *311*
Thullier, A., 205(104), 206(104), *265*
Thun, W. E., 351(131), *380*
Thurman, J. C., 100(166), *123*
Thyagarajan, B. S., 728(148), *730*
Tierney, B., 648(575), *652*
Tietze, E., 200(77), 247(183), *264, 267,* 279(29), 284(29), 300(29), *311*
Tiley, E. P., 740(41), *754*
Tille, H., 373(189), 374(189), *382*
Tilley, J. N., 144(9), 145(9), 146(9, 17), 148(9), 151(9), 152(9), 157(9), 162(9), 163(17), *164*
Tindall, Ch. G., 648(578), *653*
Tin-Wa, M., 743(66), 757(139), *755, 758*

Tischli, A., 247(185), *267*
Tisler, M., 728(146), *730*
Titov, E. B., 647(528), *651*
Titov, E. V., 647(545), *652*
Tjulenev, S. S., 287(68), 295(68), 297(68), *312*
Tobey, S. W., 553(83), *570*
Tokarev, A. K., 617(403), 618(403), 619(403, 413), 624(444), 625(444, 446, 447, 448, 449), 629(449), 631(403), *645*
Tokuyama, K., 43(150), *54,* 210(112), *265*
Tolba, M. N., 56(181), *63*
Tolman, R. L., 13(49), 32(49), *51*
Tolpigina, T. A., 647(555), *652*
Tomalia, D. A., 207(108), *265*
Tomasz, J., 633(504), *646*
Tomimatsu, T., 737(16), *754*
Tonellato, U., 647(544), *652*
Tönjes, H., 39(133), *53*
Torgov, I. V., 590(169), *638*
Torsell, K., 188(42), 190(42), 192(42), 197(42), 199(42), 218(42), 219(42), 220(42), 221(42), 241(42), 242(42), 243(42), 244(42), 245(42), *263,* 735(8), *754*
Toscano, V. G., 659(2), 689(2), 703(92), 722(2), 723(2), *724, 726*
Towne, E. B., 683(58), *725*
Toye, J., 125(221), 136(221), *139*
Trager, H., 608(341), *642*
Trani, A., 147(24), 152(24), 156(24), 157(24), 163(24), *164*
Traut, H., 408(55), 422(55), 443(55), 477(55), *522*
Traynor, G. L., 728(136), *730*
Treharne, G. J., 542(69), *570*
Treharne, G. N., 73(55), 74(55), *121*
Treiber, H. J., 408(54), 422(54), 477(54), *522*
Treibs, A., 574(8), 619(416, 417, 418, 419), 620(417, 418, 425), *634, 644*
Tribolett, R., 619(419), *644*
Trier, G., 760(2a), *772*
Tripett, S., 718(125), *727*
Trissler, A., 706(102), *726*
Tritschler, W., 453(90, 91), 454(92), *523*
Troeger, J., 540(57), 541(58), 542(58), *569*
Trojanek, J., 743(66), *755*
Tronich, W., 75(94), 119(94), *122*

Trost, B. M., 406(43), 413(43), 485(43), 522
Truce, W. E., 350(121), *380*, 590(179), *638*
Tsarev, M. V., 750(116), *756*
Tsataronis, G., 477(128), *524*
Tshitshibabin, A. E., 575(28), *635*
Tsivunin, V. S., 20(68), *52*
Tsuge, O., 385(206), *391*
Tsuji, T., 12(45), 13(46), *51*
Tsunoda, M., 540(56), 544(71), 550(71), *569, 570*
Tsutae, K., 190(49), 244(49), 245(49), *263*
Tucker, B., 88(137), 90(137), 105(183), *123, 124,* 146(18), 147(18), 148(18), 154(18), *164*
Tullier, A., 205(103), 206(103), *265*
Tullock, C. W., 68(26), *120,* 144(5), *164*
Turchin, K. F., 60(201), *64*
Turell, G. C., 218(139), 219(139), *266,* 333(75), *379*
Turicina, N. F., 608(336, 337), *642*
Tyree, S. Y., 581(109), *637*

Ueda, E., 757(140), *758*
Ueda, K., 613(379), 630(379), *643*
Ueda, S., 531(24), *569*
Ueda, T., 59(197), *64,* 187(36, 37), 190(36), 237(36), *263,* 269(216), *277,* 315(128), *320*
Ueno, Y., 206(105), *265*
Uff, B. C., 616(395), 649(602), *644, 653*
Ugi, I., 19(58, 59, 60), 26(59), 39(59), *52,* 135(273, 274), *141,* 584(131), 596(254), 604(131), 605(131), *637, 640*
Ugi, J., 105(184, 185, 186), 106(186), *124*
Uhl, A., 185(26), *263*
Ulery, H. E., 40(139), *53,* 330(60), *379*
Ulm, K., 75(89), *121*
Ulrich, H., 6(9), *50,* 88(137), 90(137), 102(177), 105(183), *123, 124,* 144(8, 9, 10, 11), 145(8, 9, 10, 11), 146(8, 9, 10, 11, 17, 18, 19), 147(11, 18, 19, 21, 23), 148(9, 11, 18, 21, 23), 151(9), 152(9), 154(18), 157(9), 162(9, 11), 163(17), *164,* 290(87), *312,* 322(8), 329(54), *378, 379,* 441(77), 486(77), *523,* 528(10), 565(10), *568*
Ulrich, W.-R., 315(124), 319(124), *320*
Umo, H., 92(145), *123*

Unags, P. C., 185(25), 241(25), *263*
Urban, R., 135(273), *141*
Urbani, R., 678(46), *725,* 728(142), *730*
Urbanski, J., 633(489), *646*
Urbanski, T., 633(489), *646*
Uritskaya, M. Ya., 60(201), *64*
Urry, W. H., 351(128), *380*
Ursprung, J. J., 269(215), *277*
Uyeo, S., 743(57), *755*

Vagin, V. V., 595(240, 241), *640*
Vakma, P. S. P., 736(14), 737(14), *754*
Vamvacas, C., 747(89), 748(89), *756*
Van Bac, N., 769(29a), *773*
van den Brock, P. J., 649(604, 605), *653*
van der Eycken, C. A. M., 88(129), 89(129), *122*
van der Plas, H. C., 627(464), *646*
van der Vennen, D., 633(498), *646*
van de Westeringh, C., 88(129), 89(129), *122*
van Doorn, J. A., 182(5), *262*
Van Dorp, D. A., 579(79), 609(79), 633(500, 510), *646, 647*
van Duuren, B. L., 647(552), *652*
Van Es, A., 589(160), *638*
van Guldener, D. B., 126(240), 139(240), *140*
Van Heyningen, E. M., 469(112), *524*
Van Hoeven, H., 191(52), 237(52), 238(52), *264*
Van Meersche, M., 165(55), 168(55), *171*
van Reijendam, J. W., 593(211), *639,* 648(558), *652*
van Stolk, D., 747(94), *756*
van Tamelen, E. E., 748(108), *756*
Vasileff, R. T., 469(112), *524*
Vasil'eva, V. K., 126(232), *140*
Vasiliev, A. E., 592(205), *639*
Vasiljev, A. V., 607(326), *642*
Vasisht, S. K., 581(110), *637*
Vasnev, E. I., 607(325), *642*
Vasnev, V. A., 583(128), 584(136), 586(136), 590(136, 173, 174, 175), 600(128), 607(136, 173, 175, 324, 325, 326, 327, 328, 329), *637, 638, 642*
Vaughan, W. R., 119(218), *124*
Vernay, H. F., 648(560), *652*
Viehe, H. G., 70(36), 74(80, 81), 76(80, 81, 96), 106(80), 112(81), 113(81),

AUTHOR INDEX

116(80, 81), *120, 121, 122,* 126(241, 242, 243, 244, 245, 246, 247), 128(241, 243, 244, 245), 130(245, 253), 131(245), 133(257), 136(243), 138(244, 247), *140,* 149(32), 150(33), 158(47), *164, 165,* 165(53, 54, 55), 168(55), 169(54), 170(53), *171,* 268(208, 209), 275(208, 209), *276,* 348(119), 349(119), *380,* 384(203), 385(204), 390(203), *391,* 563(107), *571,* 705(96), 710(108), *726,* 760(1), *772,* 772(41), *773*
Vilkas, M., 581(114), *637,* 647(538), *652*
Villemin, D., 205(103), 206(103), *265*
Vilsmeier, A., 6(2, 3), *50*
Vinkler, P., 314(119), 316(119), *320*
Vinogradova, S. V., 583(128), 584(136), 586(136), 590(136, 173, 174, 175), 600(128), 607(136, 173, 175, 324, 325, 326, 327, 328, 329), *637, 638, 642*
Visgert, R. V., 649(596), *653*
Visotzkij, Yu. B., 650(620, 621), *654*
Vittorelli, P., 156(42), *165,* 669(38), *725*
Vlasov, G. P., 336(80), *379*
Vo, B. J., 314(119), 316(119), *320*
Vogtle, F., 657(12), 716(122, 123), 718(122, 123, 124), *724,* 727(132), *727,* 730(132, 172), *730*
Volbushko, N. V., 651(629), *654*
Volke, J., 631(478), *646*
Volkov, V. S., 13(47), *51*
Voloshanovskij, I. S., 647(555), *652*
Voloveiskij, L. N., 625(446, 447), *645*
Volovenko, Yu. M., 648(573), *652*
Volskova, V. A., 190(46), 191(46), 245(46), *263*
Vompe, A. F., 608(336, 337), *642*
von Chamier, C., 156(41), 160(41), *165*
Vonderheit, C., 125(226), *140,* 190(39), 241(178), 243(178), *263, 266*
von E. Doering, W., 659(2), 689(2), 722(2), 723(2), *724*
Vorbrüggen, H., 461(106, 107), 462(106, 107), 464(107), 465(106, 107), 466(107), 467(107), 468(106, 107), 489(107), 490(107, 142), *524, 525*
Vorobiev, N. K., 606(319), *642*
Voss, J., 312(59)
Voznesenskaja, N. N., 595(237), *640*
Vromen, S., 633(509), *647*
Vystrcil, A., 596(252), *640*

Vyve, T. v., 149(32), *164*

Wache, H., 75(89), *121*
Wada, M., 651(630), *654*
Wade, K., 9(32), *51*
Wade, L. E., 703(93), *726*
Wagner, F., 406(58), 412(58), 414(58), 415(58), 416(58), 452(58), 489(141), 490(141), 514(141), 517(141), 518(141), *522, 525*
Wagner, H. U., 357(164), *381*
Wagner, J. S., 351(137, 138), 354(138), 375(138), *381*
Wagner, K., 57(186), *63*
Wagner, Kh.-Yu., 649(590), *653*
Waheed, N., 59(198), *64*
Wahl, R., 198(71), 220(71), 248(71), 252(71), 258(71), *264,* 369(179, 180), 370(179, 180), *382,* 402(5a), 405(26), 406(26), 407(26), 408(26), 411(26), 417(26), 455(96), 457(96), 478(26, 26a), 516(26), 517(5a, 171), *522, 523, 525*
Wahlberg, K., 735(8), *754*
Wajcht, J., 648(580), *653*
Wakabayashi, K., 540(56), 544(71), 550(71), *569, 570*
Wakselman, M., 578(51, 57, 58), 579(86), 580(88), 581(51, 114), 583(58, 86), 584(57), 585(57, 58, 86), 586(51, 57, 58, 86), 587(51, 58, 86), 590(88), 594(88), 595(86, 88), 601(58), 604(51, 57, 58, 86), *635, 636, 637,* 647(549), *652*
Walborsky, H. M., 102(176), *123*
Walczak, A., 648(580), *653*
Walczyk, W., 648(562), *652*
Walker, J., 354(146), *381*
Walker, S. F. H., 595(233), 598(274), *640, 641*
Wallach, O., 66(1, 2, 3), 94(150, 151), 96(150), 102(1), 118(213), *120, 123, 124,* 279(1, 2), 280(1, 2), 281(2), 284(1, 2), 294(1, 2), 295(1, 2), 299(1, 2), *310,* 322(16, 17), 329(52), *378, 379*
Wallbilich, G., 579(71), *636*
Wallenfels, K., 100(168), 101(170), *123*
Walsh, R., 657(7), *724*
Walter, R., 342(109), *380*

Walter, W., 95(154), *123,* 171(65), 174(3), *172, 180,* 216(134), *265,* 279(44, 49, 50, 59), 280(44, 49, 50), 281(44, 49), 282(44, 49), 283(44, 50), 284(44, 49, 59), 285(44, 59), 288(49), 289(44, 50, 59, 74, 75, 76, 77, 78, 80), 290(49), 291(44), 294(44), 295(44), 296(44), 299(44, 49, 75), *311, 312,* 314(116, 117), *320,* 333(74), 353(141, 142, 143, 144), 354(141, 142, 143, 144), 360(167), *379, 381,* 386(208, 209), *391*

Walther, R., 322(3), 325(3), 354(3), *378*

Walton, R. A., 557(90), 560(90), *570*

Walz, H., 182(3), *262*

Walz, K., 19(61, 62, 63, 64), *52*

Wang, G. L., 648(571), *652*

Wang, T., 232(171b), 233(171b), *266*

Wang, T. C., 279(40), 281(40), 294(40), 295(40), 302(40), *311*

Wanzlick, H. W., 400(17), *521*

Warning, K., 24(89, 90, 91, 92), 25(90, 91, 92), 26(89), *52*

Warr, W. A., 623(440), *645*

Warren, W. H., 102(174), *123,* 330(55), *379*

Waterman, P. G., 744(67), *755*

Wawzonek, S., 633(499), *646*

Webb, R. L., 406(42), 417(42), 479(42), *522*

Webber, J. A., 469(112), *524*

Weber, H., 405(28), 406(28), 417(28), 448(28), 478(28), *522*

Weber, L. D., 184(23), 190(23), 232(23), 257(23), *263,* 406(41), 411(41), 448(41), 478(41), *522*

Weber, O., 279(11), 284(11), *310*

Weber, R., 125(227), *140,* 608(339), *642*

Weber, S., 625(453), *645*

Wedekind, E., 574(16), 575(16, 26, 27), 592(27), 594(218), 619(218), *635, 639*

Wedgwood, J. J., 623(439, 440), *645*

Wefer, J., 612(370), 614(370), 616(370), *643*

Wefer, J. M., 615(390, 391, 392), 616(396), *643, 644*

Wegler, R., 154(38), *165,* 590(177), *638*

Wegner, K., 233(171a), *266*

Wegner, P., 247(189), *267*

Wehrli, P., 247(186), *267*

Wehrli, R., 728(145), *730*

Weidinger, H., 18(54, 55), 21(70), 22(54, 76), *51, 52,* 66(7, 14), 68(7), 76(7, 97), 77(97), 78(7, 97), 79(7), 80(7, 14, 97), 81(7, 14, 97), 82(7, 14), 84(7, 14), 85(7, 14), 87(7), 89(7), 91(7), 94(7, 14), 95(14), 96(7), 98(97), 99(7, 14), 100(7), 103(7, 180), 104(182), 105(7, 182), 107(7, 14, 193), 108(7, 14, 193), 109(7, 14), 115(7, 97), 116(7, 97), 117(97), *120, 122, 124,* 144(1, 2), 145(1, 2), 146(1, 2), 151(1, 2), 152(1, 2), 153(1, 2), 154(1, 2), 155(1, 2, 39, 40), 156(2), 158(2), 162(1, 2), *164, 165,* 213(124), 214(124, 125), 224(125), 230(125), 240(125, 174, 175, 176), *265, 266,* 290(83), 295(83), 297(83), *312,* 323(23), 324(23, 30), 326(23, 47), 328(47), 359(23), 361(23), 363(23), *378, 379,* 436(68), 437(68), 460(100), 465(110), 485(68), *523*(88), *524,* 530(16), 531(22, 23), *569*

Weigele, M., 406(47), 448(47), *522*

Weinelt, A., 135(273), *141*

Weingarten, H., 160(50, 51), 161(51), *165,* 327(50), 328(50), 338(91), 344(116), 345(116, 117), 346(117, 118), 347(116), 351(137, 138), 354(50, 116, 138), 355(91, 116), 356(50, 116), 357(50, 116), 359(50, 91, 116, 117, 118), 366(174), 367(50), 370(186), 375(50, 116, 138), *379, 380, 381, 382,* 401(22a), 406(30), 420(30), 432(30), 439(30), 461(103), *522, 524*

Weinheimer, A. J., 758(143), *758*

Weintraub, L., 188(41), 189(41), 220(41), 232(41), 234(41), 235(41), 236(41), *263*

Weise, A., 493(143), *525,* 530(18), 537(43, 44), 560(18, 43, 44), *569*

Weiss, S., 218(136a), 238(136a), *265*

Weissbach, K., 96(156), *123,* 322(11), *378*

Weisse, G., 531(26), 540(26), 542(26), 544(26), *569*

Weissman, A., 628(468), *646*

Weisswange, W., 574(16), 575(16), *635*

Weitz, E., 632(483), *646*

Wellenreuther, G., 21(70), *52*

Wellman, K. M., 357(158), 337(86), *380, 381*

Wendler, N. L., 625(453), *645*

Wenger, R., 764(20), *773*

Wenis, E., 735(5), *754*
Wenkert, E., 406(42), 417(42), 479(42), *522,* 658(14), *724,* 748(102, 105), *756,* 760(5a, 5b), *772*
Wenzel, W., 45(158), *54,* 66(21), *120*
Werner, E. A., 201(87), 202(87), *264*
Werner, N. D., 39(134), *53*
Westland, R. D., 232(169), *266*
Westphal, G., 442(78), *523*
Westwood, D., 593(213), *639*
Wheeler, H. L., 212(117, 118), *265*
Wheeler, T. S., 565(110), *571*
White, A. N., 182(6), *262*
White, A. W. C., 728(154), *731*
White, E. H., 182(8), *262*
White, W. A., 160(50), *165,* 370(186), *382*
Whitehurst, P. W., 190(40), 245(40), *263*
Whiting, M. C., 205(100), *265,* 337(85), *380*
Wibaut, J. P., 633(487, 488, 498, 509), *646, 647*
Wiberg, N., 351(124, 125, 126, 133), *380, 381*
Wicherink, S. C., 383(199), *391*
Wick, A., 247(185), *267*
Wick, A. E., 245(180), *267,* 503(156), 504(156), 505(156), *525*
Wickberg, B., 748(105), *756*
Widmer, U., 658(18, 19), 705(19), *724*
Wiechert, K., 74(66), *121*
Wieland, H., 627(466), *646*
Wieland, T., 386(210), *391*
Wiese, E., 716(121), *727*
Wijnberg, J. B. P. A., 769(26c), *773*
Wilbert, G., 633(490), *646*
Wilcock, J. D., 691(71), 706(71), *725,* 728(152), *731,* 760(13b), *772*
Wilcox, M. E., 752(124, 126), *756*
Wildman, W. C., 743(57), *755*
Wiley, G. A., 26(95), *52*
Wiley, R. H., *320*(142)
Wilk, H., 85(113), *122*
Wilkens, H. J., 701(87), *726*
Will, W., 147(20), *164*
Willert, C., 6(14), *50*
Willfang, G., 182(12), 192(12), *263*
Williams, A., 647(519, 542), *651, 652*
Williams, J. M., 80(102, 103), *122,* 541(63, 64), *570,* 648(561), *652*

Willstatter, R., 279(5), 281(5), 284(5), 295(5), *310*
Wilschwitz, L., 648(570), *652*
Wilson, C. V., 592(198), *639*
Wilson, F. E., 102(174), *123,* 330(55), *379*
Wilson, J. D., 151(35), 153(35), 161(35), 162(35), *165,* 327(50), 328(50), 354(50), 356(50), 357(50), 359(50), 367(50), 375(50), *379*
Winberg, H., 232(164), 254(164), *266,* 370(183), *382*
Winberg, H. E., 350(123), *380,* 400(18), 440(73), 441(73), 442(73), 443(73), 444(73), 445(84), 449(84), 485(18), *521, 523*
Wines, P., 587(141), 592(141), *637*
Wing, J. F., 542(67), 551(67), *570*
Winkler, T., 669(38), *725*
Winnacker, E. L., 247(185), *267*
Winter, M., 27(97), 28(97), 29(97), 30(97), *52*
Winterfeldt, E., 656(6), 689(70), 691(70, 71), 694(70), 706(71), *724, 725,* 728(140, 152), *730, 731,* 760(13a, 13b), 764(21), 769(13a, 27a, 27c), *772, 773*
Winters, L. J., 649(606), *653*
Winthrop, S. O., 154(37), *165*
Winzinger, R., 6(16), *51*
Wirth, T., 279(5), 281(5), 284(5), 295(5), *310*
Wisotsky, M. J., 567(118), *571*
Witkop, B., 740(36), *754*
Witte, J., 38(131), *53*
Wittig, G., 614(386), *643*
Wittmann, R., 36(123), *53*
Wohl, R. A., 701(87), *726*
Wolf, F., 9(33), 37(126), 38(126), *51, 53*
Wolkenstein, D., 620(422, 423, 424), 621(422), 623(423), 625(422), *644, 645*
Wolkenstein, D. G., 650(623), *654*
Wolkoff, P., 61(207), *64*
Wolter, G., 9(33), 37(126), 38(126), *51, 53,* 71(38), *120*
Wong, C. M., 319(138), *320,* 593(208), *639*
Wong, J. Y., 590(166), *638*
Wood, G., 675(45), *725*
Woodward, K. N., 728(142), *730*
Woodward, R. B., 613(373), *643,* 655(1),

659(1), 689(1), 702(1), 703(1), 722(1), 723(1), *724,* 760(4a, 4b), *772*
Worner, H., 579(70), *636*
Worth, B. R., 771(37a, 37b), *773*
Worthing, R. C., 744(60), *755*
Wovsi, B. A., 574(15), 588(15), *635*
Wright, P. E., 597(270), *640*
Wu, A., 728(150), *731*
Wu, M. T., 387(217), *392*
Wu, Mon-Tai, 250(192), *267*
Wu, Y.-H., 55(177), *63,* 650(624, 625), *654*
Wunderlich, K., 190(43), 194(43), 215(133), *263, 265,* 487(140), *525*
Wyler, H., 752(121, 122, 123, 124, 126, 128), *756, 757*

Yagupol'skii, L. M., 24(93), *52,* 125(224), *140*
Yagupol'skii, Yu., 125(224), *140*
Yajima, H., 628(469), *646*
Yakhontov, L. N., 60(201), *64*
Yakubovich, A. Ya., 144(6), *164*
Yamada, K., 270(224), *277*
Yamada, Y., 247(186), *267*
Yamaguchi, T., 279(48), 281(48), 298(96, 97, 98), 301(98), 304(96), 305(96), 308(48, 111), 309(96, 111), 310(111), *311, 312, 313,* 426(49, 50), 444(49), 480(49), *522*
Yamamoto, H., 594(221), *639*
Yamamoto, K., 34(121), *53*
Yamanaka, H., 126(236), 139(236), *140,* 650(613, 614, 616, 617, 617a), *653, 654*(615)
Yamazaki, N., 581(115, 116, 117, 118, 119), 582(115, 116), 586(116, 117, 119), 594(115, 118, 119), *637,* 648(582, 584, 585), *653*
Yanagida, S., 82(106), 101(106, 171, 172), *122, 123,* 528(9), 535(38, 39), 536(39), 539(39, 54), 540(39), 541(60), 542(65), 548(65), 550(9, 79), 558(95, 96), 562(54), 563(39), 565(112), *568, 569, 570, 571*
Yandovskij, V. N., 648(567), *652*
Yang, T.-H., 737(12), *754*
Yatsimiski, A. K., 647(543), *652*
Yeh, C. L., 190(40), 245(40), *263*
Yelland, W. E., 9(35), 31(35, 102), *51, 53,* 325(35), 327(35), 376(35), *379*
Yokoe, M., 535(39), 536(39), 539(39, 54), 540(39), 558(95, 96), 562(54), 563(39), *569, 570, 571*
Yoneda, F., 13(48), 32(48), *51,* 56(179), *63,* 484(137), *524*
Yoneya, O., 21(72), *52*
Yonezawa, K., 745(69), *755*
Yoschikawa, T., 279(24), *311*
Yoshida, H., 57(185), *63*
Yoshifuji, Sh., 270(224), *277*
Yoshimura, J., 353(140), *381*
Young, L. B., 357(160), *381*
Yoxall, C. T., 678(46), *725*
Ysuda, Y., 750(111), *756*
Yudin, L. G., 729(166), *731*
Yuge, E., 737(18), *754*
Yun, L. M., 55(176), *63*

Zahradnik, H., 631(479), *646*
Zaitseva, K. A., 69(35), 115(35), *120,* 213(123), 245(123), *265*
Zalkow, L. H., 656(9), *724*
Zange, E., 583(124), 586(124), *637*
Zaoral, M., 88(131, 132), 89(131, 132), 90(131, 132), *122*
Zasosov, V. A., 595(243), *640*
Zav'jalov, S. I., 626(457), *645*
Zawadzki, S., 609(345), *643*
Zborucki, Z., 648(580), *653*
Zeawin, J. M., 592(204), *639*
Zee-Cheng, K. Y., 741(45), 743(65), *754, 755*
Zeeh, B., 316(131), *320*
Zeiler, A. G., 579(84), *636*
Zelazko, D., 633(489), *646*
Zemlicka, J., 103(179), *124,* 324(29), *378,* 440(74, 75), 441(74, 75), 448(86), 450(86), 451(86, 86a, 86b, 86c, 86d), 456(74, 75, 75e, 97a), 461(74, 75), 485(86), 499(74), *523, 524*
Zenk, M. H., 760(16), *772*
Zherebchenko, V. I., 647(533), *651*
Zhidkova, A. M., 190(47), 213(47), 245(47), 250(47), 254(47), *263*
Zhilnikov, V. G., 647(534), *651*
Zhokhovets, N. G., 85(114), *122*
Ziegenbein, W., 6(20), 20(20, 67), *51, 52,* 92(146, 147, 148), *123,* 186(33), *263*
Ziegler, C., 596(256), *640*

Ziegler, F. E., 504(160), *525*
Ziehn, K. D., 24(86, 87, 88, 89, 90, 91, 92), 25(90, 91, 92), 26(86, 87, 88, 89), *52*
Zil'berman, E. N., 72(47), 82(47), *121, 528*(1), 529(13), 538(1), 541(1, 59), 542(59), 552(80), 562(1), 567(1), *568, 569*
Ziman, S. D., 165(52), 170(52), *171*
Zimmerman, H. E., 702(90), 703(90), *726*
Zimmerman, J., 575(22), *635*
Zimmerman, S., 647(522), *651*
Zincke, T., 608(330, 343), *642*
Zinner, G., 148(28), 159(28), *164,* 336(81), 337(84), *379*

Zobacova, A., 136(276), *141*
Zoeppritz, E., 6(1), *50*
Zollinger, H., 6(17), 7(17), 8(17), *51,* 66(6, 19), 78(6, 19, 100), 79(100), 80(6), 81(6), 88(19, 124), 89(124), 107(100), 108(100, 197), 113(6, 100), 114(6), *120, 122, 124*
Zolotarev, V. I., 650(618), *654*
Zook, H. D., 531(25), *569*
Zsindely, J., 658(19), 705(19), *724*
Zukauskaite, L. N., 648(569), *652*
Zumach, G., 332(67), *379*
Zvaric, S., 649(590), *653*
Zwanenburg, D. J., 648(559), *652*
Zymalkowski, F., 382(193), *391*

Subject Index

Acetals, O,N and N,N analogs, 436-437, 490
  preparation of, 435, 490
Acetamide, α,α-dichloro-N,N-dimethyl, 328
  triethoxy, 184, 185
Acetamidinium acetates, 337
Acetamidinium salts, N,N,N',N'-tetraethyl, 365
  N,N,N',N'-tetramethyl, 160, 367
Acetoacetates, O-acylation, 598
Acetonitrile, codimerization of, 545
  reaction with, acetyl chloride, 558
    hydrogen chloride, 542, 547
    sulfuric acid, 529, 567
    sulfur trioxide, 530
Acetonitrile [bis(dialkylamino)], conductivity, 397
  preparation of, 339, 375-376, 446, 448, 449, 495
  reaction with, alcoholates, 369, 458, 459
    aldehydes, 490
    H_2S, 360
    isothiocyanates, 496
    phenols, 464
  reactivity, 407
  transamination, 363
Acid chlorides, preparation of, 88-90
  reaction with pyridines, 573
  use in ketene synthesis, 590
  vinylogues, 619
Acridine, 620, 624, 628
Acridinium salts, N-acyl, 575, 576
Acryloyl chloride, 674
Actinidine, 735
Actinidinium chloride N-β(p-hydroxyphenyl-ether), 735
Acyl azides, 48, 494
Acylcyanides, 339, 493, 593
  O,N-acetals, 220, 254

Acyl transfer reactions, 602
Adenine, 23, 39
Adipic acid bisamidinium salts, 326
Adipic acid bis dimethylamide, adduct with $POCl_3$, 9
  alkylation of, 197
Ajmalicine, 749
Ajmalidine, 749
Aldehydes, allenyl, 661-663, 671
Alkaloid, 733-755, 760-771
Alkaloid PO_4, 742
Alkoxybis (dialkylamino)methanes, 395
  condensation with CH-active compounds, 412
  preparation, 369, 370, 458
  reaction with acyl cyanides, 495
    alcohols, 457
    N-alkylformamides, 452
    arynes, 516
    CO_2 and CS_2, 499
    isothiocyanates, 497
    Schiff Bases, 514
    secondary amines, 445-447, 452
Alkoxydialkylamino carboxylic acid nitriles, 395
  reaction from acyl cyanides, 493
  reaction with alcohols, 458, 459
    secondary amines, 448, 449
Alkoxymethyleniminium salts, 181
  N-acyl, 212, 228
  alkylsulfates, 197
  from amide acetals and BF_3, 215, 220
  conductivity measurements in, 220
  from cyanates and HX, 537
  fluoroborates, 188
  heterocyclizations, 239
  hexachloroantimonates, preparation of, 214

828 SUBJECT INDEX

hydrolysis of, 224
IR spectra, 221
methylsulfates, 231
reaction with, alkoxides, 248-258
  amines, 232, 328
  cyanides, 258
  reduction, 260
structure, 219
synthesis via alkylation, 182, 502
synthesis by chloroformate method, 208, 209
thiolysis, 230, 471
use in synthesis of, anhydroanreomycin, 227
  tetracyclines, 225
Alkoxytris (dialkylamino)methanes, 395
  reaction with CH-active compounds, 426
Alkylating agents, 183, 284
Alkylmercaptomethyleniminium salts, 279
  dimeric, 536
  hydrolysis and alcoholysis, 294
  N-primary from isothiocyanates and HX, 536
  properties and structure, 294
  proton abstraction with bases, 299
  reaction with, amine derivatives, 304
    Grignard reagents, 308
  synthesis from, mercaptoformamide chlorides, 293
    thioamides, 279-289
    thioiminoesters, 287
Alkylthiobis (dialkylamino)methanes, 395
Allophanyl chlorides, 145
N-Allylanilines, rearrangement into 2-allyl-anilines, 685
  transformation in polyphosphoric acid, 683
Allylation of enamines and aldehydes, 660-663
N-Allyl-enammonium bromides, thermal rearrangement of, 664
Allylic rearrangement, 503
N-Allyl-N-methylisobutenylamine, 665
Allyl systems aza, π-energy of, 703-705
Alstoniline, 747
Aluminium trichloride, 532, 557
Amide acetals, activation by N-quaternization, 434, 487
  N-alkylation, 434, 487
  in carbanion alkylation, 436

condensation with active methylene compounds, 410
exchange of alkoxy groups, 455
preparation from, alkoxyiminium salts, 252-257
  amide chloride, 108-111
  formamidinium salts, 368
reaction with, activated alkenes and alkynes, 500
  allylic alcohols, 503
  amines, 440
  N-haloamides, 510
  heterocumulenes, 496, 506
  see also Dimethylformamide acetals
Amide acylations, 208
  adducts with acid anhydrides, 49
    acid halides, 42
    chlorophosphoric ester, 27-30
    chlorosulfonamide, 42
    cyanuric chloride and ClCN, 33
    phosphorus nitrile chloride, 33
    $POCl_3$, 6
      reactions with, amines, 9, 325
        hydrazines and hydroxylamines, 17
  adducts with PX_3, 30-33, 327
    sulfonamides, 11
    sulfonylchloride, 38
    sulfurylchloride, 37
    triphenylphosphine dihalide, 23
Amide alkylations, 183
  O-versus N-alkylation, 207
  reaction with acylpyridinium salts, 595
  salts, 183
  tertiary, self-condensation of, 22
Amide azines, 99
Amide bromides, 68, 77
  aminolysis, 98, 99
  anion exchange, 114
  mass spectra, 80
  physical properties, 80
  thermal degradation, 118
Amide chlorides, anion exchange, 113
  chlorination of, 70, 76, 115
  N-chloromethylene, 538
  coupling with diazonium salts, 77, 116
  α,α-dichloro, 76
  physical properties, 80
  N-primary from nitriles and HX, 538, 541
  reaction with, alcohols, 213
    alkoxides, 107

SUBJECT INDEX

arylcyanates, 563
carboxylic and sulfonic acids, 88
cyclic ethers and acetals, 92
NH_2 groups, 96-105, 322-324
thiols, 94, 95, 290
self-condensation of, 78, 326
see also Chloromethyleniminium salts
Amide oximes, 99, 176
Amide thioacetals, 395
condensation with active CH_2-compounds, 421
reaction with alcohols, 457
synthesis of, 471
Amidines, acylation of, 341
α-chloro, 99
β-ketoacids, 354
reaction with amide acetals, 485
salt formation, 322
synthesis of, 11-15, 31, 47, 96-99, 232, 234, 304, 439-444
Amidinium salts, 321
aminoethinyl, 107
anion exchange, 375
bisamidinium, 345, 351
chlorides, 96, 98
N-(α-chloroalkenyl) from chlorinated nitriles and HX, 542-552
from codimerization of nitriles, 544-548
from dimerization of nitriles, 528, 531-544
hydrolysis, thiolysis, 357
IR spectra, 355
methylsulfates, 234
NMR data, 355, 358
reaction of N-(α-chloroalkenyl)alkyl-, 548-552
structure and properties, 364
synthesis, 322-354
transamination, 361
treatment with bases, 365
α,β-unsaturated, 348
UV spectra, 359
vinylogous, 239, 350
see also Formamidinium salts
Amidrazones, 307, 342, 364, 441-443
Aminal, insertion to isonitriles, 337
Aminal esters, *see* Alkoxybis (dialkylamino)methanes
Amino acid, N-acylation, 594
esterification, 469

formylation of, 16
Amino acid esters, reaction with nitrilium salts, 335
Aminolupinane, 764
Aminopyrimidines, formylation of, 11
Ammonium-iminium and iminium-iminium rearrangements, influence of charge, 702
Anabasine, 760
Anisole, 37
Antimony pentachloride, 532, 537, 538, 557, 562
Arbuzov's rearrangement, 628
Avicine, 743
Azatrimethinium chlorides, 34, 37, 159
Aziridine, alcoholysis, 86
aminolysis, 99
2,2-dihalo, 75
hydrolysis of, 82-83
reduction, 119
2,2,3,3-tetrachloro, 75, 119

Balfourodinium alkaloids, 746
Benzenesulfonyl chloride, 38
Benzonitrile, codimerization of, 545
reaction with, hydrogen chloride, 538-539
hydrogen chloride and phosgene, 558
N-phenyltrichloromethylimidoyl chloride, 564
sulfuric acid, 529, 567
sulfur trioxide, 530
Benzophenanthridine alkaloids, 743
Benzoxazole, synthesis of, 16, 177
Benzoyl cyanide, dimer, 593
Benzthiazoles, 177
Berberine, 738
epi-isomer, 739
radioactive, 739
thermolysis of hydrochloride, 740
Berberrubine, 739
Betacyanins, 751, 752
Betalains, 733, 753
Betalamic acid, 751, 752
Betanidin, 752
Betanin, 751, 752
Betaxanthins, 751, 752
Biguanide derivatives, 535
Biomimetic synthesis, 759
Biosynthesis, 759
Bisamidines, from bisamides, 12

SUBJECT INDEX

Bis(dialkylamino) carboxylic acid nitriles,
    see Acetonitrile [bis(dialkylamino)]
Bis(dimethylamino) difluoromethane, 144
    reaction with alcohol, 155
Bocconine, 744
Butadiene-1,3, 1,1,4,4-tetrakis(dimethylamino), 346
2-Butenylbromide, reaction with enamines, 665-667
t-Butoxycarbonylchloride, 579

Cadmium halides, 556
Canadine-DL, 740
Cancertherapy, 771
Carbamic acid acetals, 395
    reaction with alcohols, 460
    thermolysis of, 399
Carbamoyl halides, preparation of, 33, 37
    reaction with, formamides, 331
        pyridines, 578, 615
Carbazole N-methyl-1,2,3,4-tetrahydro, 712
Carbene, bis(dimethylamino), 161, 251, 301, 515
Carbodiimides, addition of HCl, 148
    preparation of, 26, 36, 152
    reaction with oxalyl chloride, 148
    N-sulfonyl, 147
Carbolines, 1-allyl, acid catalysed rearrangement of, 699
Carbonyl bromide, 68
Carbonyl fluoride, 68, 144
Carboxonium ions, as alkylating agents, 194
    aminolysis of, 217
Catalysis nucleophilic, 607
Catalytic effect of pyridine bases, 603
Catharantine N-oxide, 771
Charge transfer complexes, 631
Chelethyrine, 744
Chelirubine, 745
Chelitutine, 745
Chloroacetonitrile, reaction with, hydrogen halides, 541-543, 554
    sulfuric acid, 567
β-Chloroalkylformates, synthesis of, 205
Chloro(alkylmercapto)methyleniminium salts, generation in situ, 179
    heterocyclizations, 176, 177
    reaction with phenols and amines, 175

S_E or heterocycles, 178
    synthesis of, 173
N-(α-Chlorobenzylidene), 558
α-Chloroenamines, 69, 106, 566
    addition of HX, 74
    β-chlorocarbonyl, 70
Chloroformamidines, 25
Chloroformamidinium salts, 143
    alcoholysis, 153
    N-chlorocarbonyl, 147
    formation from dialkylcyanamides and HX, 534, 535
    N-guanyl, 535, 564
    hydrolysis of, 152
    reaction with, NH groups, 157
        organometallics, 160
        organophosphorus compounds, 161
        SH groups, 155
    structure and properties, 150
    synthesis, 144-150
Chloroformates, reaction with, amides, 44, 208
    ketene S,N-acetals, 292
    pyridines, 579
    tetramethylurea, 332
Chloromethyleniminium salts, 66
    fluoroborate, 77
    generation in situ from $CHCl_3$, 95
    IR spectra, 78
    NMR spectra, 78
    structure of, 78
    synthesis of, 66
    thermal degradation of, 118
    see also N,N-Dimethylchloromethyleniminium chloride
Cinnamoyl chloride, 676, 678
Claisen rearrangement, 107, 668
    amino, 656
    thio, 302
    see also Cope rearrangement
Claisen-type condensation, 554
Cofarnine, 735
    disproportination of, 736
Columbamine, 739, 740
Coniine, 760
Cope rearrangement, 655
    aza, 656, 658, 681, 688, 689
    3-aza and 2-aza cope rearrangements, comparison with, 704
    boron trichloride catalysis in, 658

SUBJECT INDEX

as C-C forming step in Fisher indole and Robinson-Piloty synthesis, 706
diaza, 657, 710, 716
of E,E-cyclodeca-1,5-diene, 702
hetero, 656
iminium ions in, 702, 720
3-oxa aromatic, 658
stereochemistry of, 668
transition states, 702
two-step mechanism of, 705
Coptisine, 739
Coralyn chloride, 741
Corrunine, 745
Corydaline, 742
  cycloneosamandarine, 753
  cycloneosamandione, 753
Corynoloxine, 745
Corysamine, 742
Crotonylchloride, 674, 676
Cyanamides, addition of Hx, 218, 533-535, 564
  codimerization of, 548, 564
  formation and reaction of diazapyrylium salts from, 560
  reaction with, $COCl_2$ and SCl_2, 149
    Lewis acids and benzoylchloride, 557
    sulfur trioxide, 530
Cyanates, reaction with, acids, 537, 567
  Lewis acids and acylchloride, 557, 558
  sulfur trioxide, 530
Cyanogen bromide, 580, 593, 608, 619
Cyanogen chloride, 33, 608
  codimerization of, 544, 546, 547
  reaction with, hydrogen chloride, 538, 568
  sulfur trioxide, 530
Cyanuric chloride, 33
Cyclobutenylium salt, 97
Cyclohexadiene, 1,4-dimorpholino, 679
Cyclopropane diamines, N,N'-dialkyldine, rearrangement of, 718

Dakin-West reaction, 548, 577
Dehydrochinolizidine, 768
Dehydrocorydaline, 742
Dehydrocorydalmine, 739, 740
Dehydroharmane, 769
Dehydromatridine, 764
(Dialkoxymethyl)ammonium salts, 434

reaction with NH compounds, 453
Diazapyrylium salts, 557-560
Diazasemibullvalene, 720
Dichlorofuranones, 72
α,α-Dichlorolactames, 69
Dichloromalonitrile, 556
2,2-Dichloropropionitrile, 540, 544
Dicyano compounds, cyclization of, 552-557
Dieneamines, allylation of, 668
α,α-Difluoroamines, 68, 79
  N,N-dimethyl, 77, 92
  reaction with, alkoholates, 111
    boron trifluoride, 79
    carboxylic acids, 92
Dihalocarbene additions, 75
Dimethylacetamide, 38
  diethylacetal, 249
  dimethyl sulfate adduct, 195
  methylation with fluorosulfonate, 205
Dimethylaniline, 42
N,N-Dimethylchloromethyleniminium chloride, as condensing reagent, 92
  as dehydrating reagent, 99
  exchange with DMF, 103
  hexachloroantimonate, 214
  reaction with, alcohols, 83-85
    ammonium chloride, 323
    carboxylic acids, 88-91
    dimethylamine, 323
    metalated amides, 111
    nitriles, 101
    phenol, 88, 215
    pyrocatechols, 93, 215
Dimethylformamide, acylation with chloroformate, 210
  catalysis, 85, 87, 99
  dimethyl sulfate adduct, 195
  methylation with fluorosulfonate, 205
  reaction with acetic anhydride, 50
    acylbromides, 42
    benzenesulfonylchloride, 38
    chloroformates, 43
    cyanuric chloride, 34
    perchlorobutyryl chloride, 45
  self-condensation of, 9
  thionylchloride adduct, NMR, 69
Dimethylformamide acetals, 367, 368, 398
  activation by N-quaternization, 434

condensation with active CH-compounds, 410
cyclic, elimination of DMF, 519-521
dineopentylacetal, 465
di-*t*-butyl, procedure, 457
H-D exchange in, 518
reaction with, Grignard reagents, 473
   N-haloamides, 510
   heterocumulenes, 498, 506
   isocyanates, 507
   ketones, 489
   NH-compounds, 439
   phenols, 461
   thioacetals, 421, 452
use in esterification, 467
Dimethylformamide/POCl$_3$ adduct, 6
kinetic studies, 8
NMR spectra of, 7, 8
reaction with, acyloins, 20
   alkoxides and mercaptans, 22
   amide oximes, 19
   azomethines, 19
   bases, 19
   epoxides and cyclic ethers, 20
   hydroxyl groups, 21
   NH$_2$OH, 18
structure of, 6
Dimethylresorcinol, 619
Dimethylthioformamide/POCl$_3$ adduct, 9
Dimroth reduction, 589, 634
Dissociation constant, 567-568
Di-*t*-butylphenol, 619
Dithiocarbamates, 173
   cyclic, 174

Echinorine, 745
Einhorn acylation, 589
Ellipticine, 769
Enamides, 772
Enamines, 308
N-acyl, 422
acylation with acryolylchloride, 674
N-allyl, Cope rearrangement, 656
allylation and propargylation of, 660, 664
α-alkoxy, 500, 502
α-chloro, 69, 70, 74, 106, 566
α-cyano, 309
α-halo, 566
β-nitro, 483, 484

synthesis from, alkoxytris (dialkylamino) methanes, 432
amide acetals, 352, 406, 408, 410-422, 423, 430, 490
1,2,2-trichloro, 563
Enammonium salts, 664, 669, 670
N-allyl and propargyl rearrangement of, 722
N-benzyl, rearrangement of, 720
N-2-butenyl, 671
Enhydrazines, *N,N*-diacyl, rearrangement of, 707-708
rearrangement into indoles, 706, 710
Enol ethers, 435
EPR, 629
Escholamine, 736
Eschweiler-Clarke methylations, rearrangements in, 689
Ethylene diamines, 1, 2, 517

Fagaronine, 743
Ferrocenes, carbamoyl, 186
hetorylation of, 626
Fischer indole synthesis, 706
Flavocarpine, 747
Flavopereine, 747
Formamide, acylation, 210
alkylation of, 196
Formamide benzoylchloride adduct, 211
Formamidines, 340, 353, 364, 365
N,N'-diaryl, 453
N-formyl, 473
N,N,N'-trialkyl, 440
Formamidinium salts, bis, 332
conversion to formamide acetals, 368
N-diacyl, 339
N-dichloromethyl, 71, 330
N,N'-disubstituted, 331
N-formyl, 71, 533
perchlorates, 376
preparation of, 44, 239, 323, 337
reaction with, alcaliamides, 372
   mercaptides, 371
   phosphides and phosphonates, 373
tetraalkyl, 96, 102, 470
N,N,N',N'-tetramethyl, 330, 331, 397
transamination, 363
Formamidrazone, 18
N-formyl, 17
Formic acid orthoamides, 405, 406

SUBJECT INDEX 833

reaction with, carboxylic acids, 465
    NH-compounds, 439
Formylation, 407
Fulvene bis(dimethylamino), 353
Fumanic acid amide acetal, 442
Furane, 617
Furoine, 86

Gliotoxine, 769
Glutaconic dialdehyde, 608
Glycine, 233
Glycorine, 750
Guanidines, N-acyl, 444
    from chloroformamidinium salts, 157
    reaction with amide acetals, 485
    from ureas/$POCl_3$, 16
Guanidinium salts, preparation, 238

Haemanthidine, 742
Halfordine N-methyl, 735
Halfordinol, 735
Halide *cis* addition to olefins, 206
Hammet equation, 606, 611
Harmane N-methyl, 747
H-D exchange reaction, 565, 566, 571
Herbipoline, 751
Heteroarylation, 616
    EPR signals in, 629
    mechanism of, 629
Homarine, 630
Homophtalic anhydride, 626
Houben-Hoesch reaction, 528, 531, 560, 568
Hydantoins, 484
Hydrastinione, 735
Hydrazine, 1,2-dialkyl, reaction with cyclic ketones, 714
    reaction with, aryl methyl ketones, 716
    dimedone, 716
    2-tetralone, 714
Hydrazonium salts, 710
Hydrocinnamaldehyde α,α-dimethyl, 721
Hydrogen cyanide sesquihalides, 360, 531
Hydroxyl groups, exchange with, bromine, 32, 85
    chlorine, 21, 83-85

Imidazoles, 233, 335, 480
    N-acetyl, 601
    4-ethoxy, 2-aryl, 187

Imidazoline-2-aryloxy, 154
    2-arylthio, 156
    2-chloro, N-silyl, 163
Imidazoline-2,4-diarylimino, 508
Imidazoline-4,5-diones-2,2-dichloro, 148
    alcoholysis of, 154
    aminolysis of, 159
Imidazoline-2,4-dioxo, 507
Imidazolium salts, 328
Imidazolones, 193
Imidoyl halides, 547, 565-567
    preparation of, 26, 70, 72
    reaction with amines, 322, 579
Iminium group, activating effect in cycloadditions, 759
Iminium ion, 528, 562-563, 571
Iminium salts, aza Cope rearrangement, 658
    from cyano compounds, 531-541, 547, 563
    dimeric, *see* Amidinium salts, from dimerization of nitriles
    as natural products, 733
    structure of, 541
Imino esters, N-acyl, 212
    alkylation and acylation of, 211
    N-cyano, 237
    synthesis from alkoxyiminium salts, 241
    thio, acylation and alkylation of, 287
Indole, 478, 596, 617
    alkaloids, 747
    N-allyl, rearrangement of, 684
    from enehydrazines, 710
    hetarylation, 621
    from phenylhydrazones, 712
    pyridylation, 623
Indolopyridocoline, 747
Iodomethyleniminium iodide, 77
Isobutenylamine, N,N-dimethyl, N-benzyl, N-methyl, reaction with methyl iodide, 721
    N-propargyl, rearrangement of, 674
    reaction with benzylbromide, 720-721
    reaction with 2-butenyl bromide, 665
Isocyamide dichlorides, 75, 117
Isocyanamide, 19
Isocyanates, acyl, 72
    acyl O,N-acetals, 511
    preparation, 25, 38, 48, 105
Isonitriles, preparation of, 19, 26, 30, 32, 36, 39, 102, 105

reaction with amines, 336
Isoquinolines, 1-allyl-1,2,3-trimethyl, acid catalyzed rearrangement of, 698
  dihydro, propargyl-allenyl rearrangement of, 700
  hetarylation, 617, 620
  nature occurring, 735
  reaction with thiophosgene, 610, 615
Isoquinolinium salts, 349, 580
  1-allyl-1,2-dihydro, aza Cope rearrangement, 694
  N-cyano, 611
  3,4-dihydro, pseudo cyanides of, 696
  IR-spectrum, 585
  reaction with, bases, 610
    with cyanide, 612
Isotetrandine, 737
Isothioureas, 175

Jatrorrhizine, 739, 740

Ketene N,N-acetals, alkylation and acylation of, 344, 349
  2-arylamino, 514-516
  coupling with diazonium salts, 346
  oxidation of, 351
  preparation, 305, 328, 366, 434
  reaction with sulphur, 350
Ketene O,N-acetals, formation of, 248, 249, 369, 500, 519
  reaction with electrophilic reagents, 217, 502
Ketene O,O-acetals, benzoylation, 596
Ketene S,N-acetals, formation, 301
  reaction with electrophiles, 291
Ketenimine, 566, 571
  preparation, 26
β-Ketoacids, amidines, 354
Ketoamide chloride arylhydrazones, 77, 98, 116

Lactam acetals, 395
  condensation with active CH_2-groups, 409
  reactions with alkenes, 500
Lactam alkylation, 190, 200
Lactam imides from lactames, 14
Lactam thioacetals, 395
Lactim esters, survey of syntheses, 244
Lambertine, 740

Landanine, 737
Landanosoline, 739
Lepidinium salts, N-acyl, 626
Leucaenine, 735
Leucanol, 735
Lewis acids, 528, 538, 539, 557
Lobeline, 760
Lunasine, 745
Lysine, 760

Malonic acid bis(dimethylamide), adduct with $POCl_3$, 8
  alkylation of, 186
Malonitrile, aminated, 259, 310, 495
  cyclization, 554-557
  reaction with pyrylium salts, 559
Matridine, 764
Melionine, 747, 748
Mercaptoformamidinium bromides, 174
Mesoionic compounds, 288
N-Methylcaprolactam, 201
Methylene blue, 771
Methyleniminium salts, from N-tertiary aminoacids, 772
  reaction with, ketene aminals, 346
    ynamines, 349
Methyllithium, 160
N-Methylpyrrolidone, 14, 38
  alkylation, 190, 200
  diethylacetal, 250
Michael addition, 759
Mimosine, 735
Miraxanthin, 752
Morpholine enamines, 664, 666
  dienamine, 679
  reaction with acryoylchloride, 674

Naphtylamines, N-allyl, rearrangement of, 682
  formylation of, 11
Nicotinamide, 578
Nicotinamide N,N-dimethyl, 184
  N-acylation, 578
  N-carbamoyl, 585
Nicotine, 764
Nitidine, 743
Nitriles, α-amino, α-aryl, 491
  via dehydration of amides, 101
  reaction with, ammonium salts, 333
    HX, 72

SUBJECT INDEX 835

Nitrilium N-acyl, 557
 complex with Lewis acids, 557
 dipolar, 568
 intermediate, 528, 548, 562-564, 567, 571
 structure of, 564
 N-substituted, 560
Nitrilium salts, addition of, alcohols, 218
 amines, 333
 reaction with amino acid esters, 335
Norharmanes, tetrahydro, 1-allyl and propargyl, 689-690
 1-(cyclohexen-1-ylmethyl), 691

Olefines, stereospecific synthesis of, 592
Olivacine, 769
Oreophiline, 742
Ornithine, 760
Orthoamides, 393
 acylations, 492
 basicity of, 398
 dimerisation, 400
 dismutation, 401
 dissociation of, 396
 1,3-eliminations, 519
 heterocyclizations, 473
 hydrolysis of, 460
 reactions with, activated CH-groups, 406
  active methine groups, 435
  alcohols, 454
  electrophiles, 487
  NH-compounds, 439-452
  nucleophiles, 403
  PH-functions, 470
  secondary amides, 450
  strong acids, 466
 reactivity of, 406
 stability of, 398
Orthoesters, formation of, 451
 reactions with amines, 337
Ortoleva-King reaction, 617, 623
Oxadiazoles, 533, 551, 559
Oxadiazoles-1,2,4, 18, 240, 480
Oxadiazoles-1,3,4, 176, 240
Oxalyl chloride, 66, 71, 145
1,2,3,5-Oxathiadiazine-2,2-dioxides, reaction of, 530
Oxazoles, 480
Δ^2-Oxazolines, formation from N-(2-halo-ethyl)amides, 207
 pyridylation, 625
Oximes, O-aryl, rearrangement into benzofurans, 713
Oxonium salts in alkylations, 182

Palandinium chloride N-methyl, 736
Palmatine, 739, 740
Papaveraldinium chloride N-methyl, 737
Papaverine, 769
Papaverine 1,2-dihydro, 1-allyl, 698
 rearrangement of, 694
Pearson's concept HSAB, 599, 614, 616, 629, 632
Pelletierine, 760
Peptide esterification, 469
 N-acylation, 594
 synthesis of, 628
Perchloric acid, 568
Perchlorobutyryl chloride, 45, 66
Phaeantharine, 737
Phenantridine alkaloids, 742
Phenols, acylation of, 589, 598
 alkylation by orthoamides, 462
 arylsulfonation, 591
 formylation of, 464
Phenylacetamide N,N-diethyl, reaction with CCl_3COCl, 45
Phenylacetonitriles, 2-amino-2-alkylthio, 309
Phenylalanine, 769
2-Phenylbenzofuran from acetophenone O-phenyloxime, 713
1-Phenyl-3-butenylamine, rearrangement under Eschweiler-Clark conditions, 689
Phenylhydrazine, N,N'-dimethyl, 711
 reaction with ketones and aldehydes, 710
Phenyllithium, 160, 366
Phenylmercury trihalomethanes, 75
Phenyltetrachlorophosphorane, 24
Phosgene, 66, 70, 85, 88, 107, 144, 146, 173, 329, 550, 558, 596, 608
Phosgeniminium salts, 76, 149, 150, 563
Phosphite triethyl, 161
Phosphorus nitrile chloride, 33
Phosphorus pentabromide, 66, 68, 69, 145, 329, 608
Phosphorus pentachloride, 66, 562
Phyllocactin, 752

Picolinium salts, N-acyl, 575, 576
  carbamoyl, 599
  IR spectrum, 585
Picrylchloride, 342
Piloty-Robinson pyrrole synthesis, 706, 714
Pinner's imino ester synthesis, 208, 228, 528
Δ^1-Piperideine, 760
Piperidine enamines, 664
Piperidinium halides, 553
Pivalic acid dimethylamide, 183, 327
Platydesminium alkaloids, 746
Polonovski reaction, 769
Polymethine dyes, 609
Prebetanin, 752
Propargylation of enamines and aldehydes, 660, 671-673
N-Propargylenammonium salts, rearrangement of, 670
Protoberberine derivatives, 738
Pseudobases, 610, 611, 612, 614
Purines, 477, 483, 484
Pyridine, acylations, 574, 575
  derivatives of, as natural products, 734
  dihydro derivatives, 579
  4-dimethylamino, 578, 579, 581, 585, 590, 626
  hetarylation, 617
  reaction with RCOCl/phosphite, 628
  synthesis, 478, 479
Pyridine derivatives, 556
Pyridinium salts, N-acyl, 573
  N-arylsulphonyl, 580
  N-benzoyl, 578, 593
  N-carbamoyl, 578, 583, 585, 599
  conductivity measurements in, 583
  IR spectra of, 584, 585
  NMR spectra, 587
  occurrence of radicals in, 630
  one electron transfer in, 632
  N-phosphonium, 581
  N-phosphoryl, 603
  polarographic reduction of, 584
  reaction with, acids, 592
    CH-acidic compounds, 624
    cyanides, 612
    Grignard reagents, 628
  reactivity, 587
  ring-openings, 588, 608

  structure of, 582
  synthesis via acylation, 574-582
  use in acylations, 589
  UV spectra, 586
"Pyridylation," 617, 619
Pyrimidine, 2,4-dioxotetrahydro, 451
  preparation of, 23, 101, 361, 471, 472, 476-478
Pyrimidine derivatives, 531, 550, 556, 558
Pyrocatechol, cyclization with $DMF/COCl_2$ reagent, 87
Pyrocatechol dichloromethylenether, 66
Pyrocatechol phosphorus trichloride, 66
Pyrrole, C-acylation, 596
  N-allyl, 682
  hetarylation, 617, 619
  hydroxy, 620
Pyrrolidine enamines, 664
  reaction with 2-butenylbromide, 666, 667
Pyrrolidinium halides, 553
Pyrylium, see Diazapyrylium salts

Quinaldinium salts, N-acyl, 575, 576, 626
Quinazoline derivatives, 560, 564
Quinoline, alkaloids, 745
  1-benzyl 4a-mesyloxy-trans decahydro, fragmentation of, 701
  hetarylation, 617
  octahydro, 2-oxo, 679
  reaction with thiophosgene, 615
Quinolinium salts, N-acyl, 575, 576
  aza Cope rearrangement, 694
  N-cyano, 580, 589, 611
  reaction with, bases, 610
    cyanide, 612
Quinomethides, 302, 303

Reissert compounds, 612
  application in synthesis, 616
  deprotonation, 616
Reticuline, 739
Ribalinium alkaloids, 746
Ribonucleosides, 440
Ritter reaction, 518, 528, 529, 548
Rosemund reduction, 616

Sandwicine, 749
Sanguilutine, 745
Sanguirubine, 745

Schiff bases double, rearrangement of, 716, 718
Schotten-Baumann procedure, 578
Selenoamides, alkylation of, 280
　preparation of, 297
Selenodimethylformamide, 232
Selenophene, 620, 623
Sempervirine, 747
Senecioic acid chloride, 678
Serpentine, 748
Serpentinine, 748
Sesquichlorides of HCN, 360, 531
Shihunine, 750
Silver acetylenide, 629
Singlet oxygen, 759, 764
Sparteine, 760, 764
Squaric acid derivative, amides, alkylation, 186
　aminolysis, 233
　reaction with, phosgene, 80
　thioamides, 94, 280, 297
Stanic tetrachloride, 534, 560
Stepharanine, 739, 740
Stephen reduction, 531
Steroids, 3-keto, 625
Succinic acid, bis(dimethylamide), adduct with $POCl_3$, 8
　alkylation of, 186
Sugars, acetalization of, 456
Sulfenyl chlorides, 118
　reaction with thioamides, 289
Sulfide contraction, 354
Sulfonyl chlorides, 580
Sulfur dichloride, 148
Sulfuric acid, 529, 567
　fuming, 529
Sulfur trioxide, 529, 530
Symbiotic effect, 554

Takatonine, 736
Terephtalic acid, bis(dimethylamide), adduct with $POCl_3$, 9
　alkylation of, 186, 197
Testosterone, 41
Tetraaminoethylenes, 350, 370, 400, 401
Tetrafluoroethylene, 75
Tetrakis(dialkylamino)methanes, 160, 395
Tetrakis(dimethylamino)methane, 160
Tetrazine, dihydro, 306
Thalictrifoline, 742

Thalidastine, 741
Thalifendine, 739, 740
　5-hydroxy, 741
Thiadiazole-1,3,4, 306
Thiadiazole-1,2,4, 377
1,3-Thiazetine-2-one, 146
Thioamides, N-acyl, 333
　acylation with chloroformates, 208
　alkylation of, 279, 353
　preparation of, 230, 360
　reaction with β-chlorovinylmethyleniminium salts, 290
Thiocyanates, addition of HX, 174
　aryl, 106
　codimerization of, 544
　reaction with hydrogen halides, 535, 536
Thioesters, 95, 230
Thioformamidines, 95
Thiohydroxamic esters, 308
Thionyl chloride, 66, 71, 85, 88, 145, 330, 581
Thioparabanic acids, 147
Thiophosgene, 608, 610
Thioureas, reaction with $COCl_2$, 146
　synthesis of, 155
　trioxides, 354
Titanium tetrachloride, 327, 367, 575
Triacanthine, 750
S-Triazines, 475, 476, 485
　from diazapyrylium, 560
　oxathiadiazine dioxide, 530, 532
　hydrochloride, 533
　preparation of, 241
　6-substituted-bis (trichloromethyl)-, 545, 549, 550
　2,4,6-triaryl-, 539, 549, 562
　2,4,6-triaryloxy-, 537
　2,4,6-trihalo-, 538
　2,4,6-tris(alkylthio)-, 536
　2,4,6-tris(chloromethyl)-, 542
　2,4,6-tris(dichloromethyl)-, 549
　2,4,6-tris(trichloromethyl)-, 540
　2,4,6-tris(trifluoromethyl)-, 540, 545
Triazoles, 533, 551, 559
Trichloroacetonitrile, codimerization of, 544, 545, 548
　reaction with, hydrogen halides, 540
　sulfuric acid, 529
　trimerization of, 550
Trichloroacetylchloride, 45

Trichlorovinylamines, 74
Triethoxyacetamides, 184, 185
Trifluoroacetic anhydride, 606
Trifluoroacetonitrile, 540, 545, 567
Triformamidomethane, 31, 32, 38, 48, 196, 204, 210, 360, 401
  condensation with active CH_2-compounds, 422
  use in heterocyclizations, 473
Trigonelline, 734
Triphenylphosphine/CCl_4 reagent, 24
Tris(dialkylamino)methanes, 111, 112, 160, 338, 370, 372, 373, 395
  condensation with active CH_2-compounds, 420
  mixed, 403
  reaction with, alcohols, 457
    secondary amines, 446
Tropinone, 760
Tryptophane, 769

Ungeremine, 742
Uracil derivatives, 241, 474
Urea, N,N-bis(dialkylcarbamoyl), 153
  N-guanyl, 535
Urea acetals, 257, 258, 395
  reaction with, alcohols, 460
    CH-acidic compounds, 427
Urea alkylations of, 192, 201

Urea/$POCl_3$ adducts, 8, 9
Urethanes, alkylation of, 193, 202

Vilsmeier formylation, 71
Vilsmeier reagents, vinylogous, 49. *See also* N,N-Dimethylchloromethyleniminium chloride
Vinylpyrrolidone, 14
Vomalidine, 749
Vomilenine, 749
Vinblastine, 771
Vincamone, 771
Vincine, 771
Vindoline, 771

Wibaut reaction, 633
Worenine, 742

Xanthaline N-methyl, 737

Ynamides, 5
Ynamines, addition of HX, 74
  preparation of, 112
  reaction with iminium salts, 349
Yohimbine, 41

Zinc chloride, 556, 557
Zincke-Koenig reactions, 588
Zooanemomine, 749

QD
251
A36
v.9
pt.2
1979

MAR 14 1979